JACARANDA MATHS QUEST

MATHEMATICAL METHODS 12

VCE UNITS 3 AND 4 | THIRD EDITION

MARGARET SWALE

ELIZABETH KEMPTON

SUE MICHELL

RAYMOND ROZEN

jacaranda
A Wiley Brand

Third edition published 2023 by
John Wiley & Sons Australia, Ltd
155 Cremorne Street, Cremorne, Vic 3121

First edition published 2016
Second edition published 2019

Typeset in 10.5/13 pt TimesLTStd

ISBN: 978-1-119-87656-4

The covers of the *Jacaranda Maths Quest VCE Mathematics* series are the work of Victorian artist Lydia Bachimova.

Lydia is an experienced, innovative and creative artist with over 10 years of professional experience, including 5 years of animation work with Walt Disney Studio in Sydney. She has a passion for hand drawing, painting and graphic design.

Illustrated by diacriTech and Wiley Composition Services

Typeset in India by diacriTech

NATIONAL LIBRARY OF AUSTRALIA

A catalogue record for this book is available from the National Library of Australia

Printed in Singapore
M121039_110822

Contents

About this resource

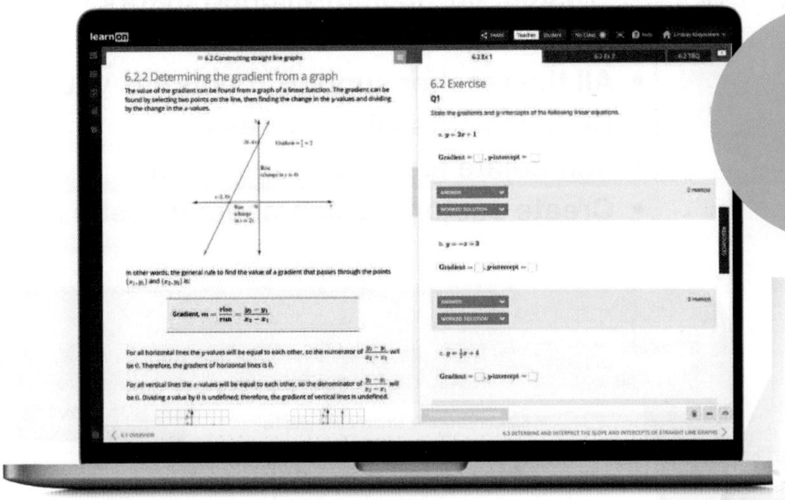

Everything you need for your students to succeed

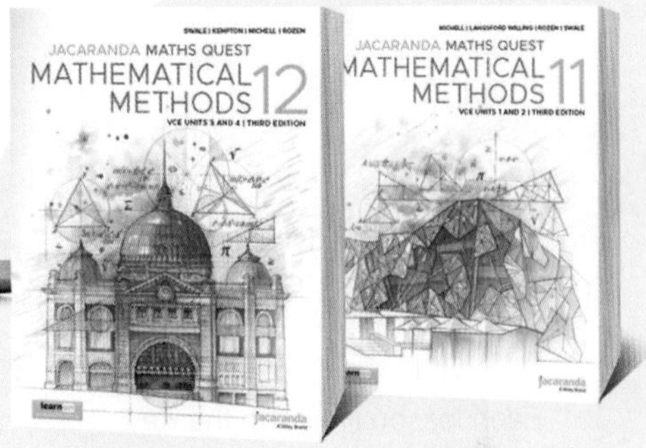

JACARANDA MATHS QUEST
MATHEMATICAL METHODS 12
VCE UNITS 3 AND 4 | THIRD EDITION

Developed by expert Victorian teachers for VCE students

Tried, tested and trusted. The NEW Jacaranda VCE Mathematics series continues to deliver curriculum-aligned material that caters to students of all abilities.

Completely aligned to the VCE Mathematics Study Design

Our expert author team of practising teachers and assessors ensures 100 per cent coverage of the new VCE Mathematics Study Design (2023–2027).

Everything you need for your students to succeed, including:

- NEW! Access targeted question sets including exam-style questions and all relevant past VCAA exam questions since 2013. Ensure assessment preparedness with practice school-assessed coursework.
- NEW! Be confident your students can get unstuck and progress, in class or at home. For every question online they receive immediate feedback and fully worked solutions.
- NEW! Teacher-led videos to unpack concepts, plus VCAA exam questions, and exam-style questions to fill learning gaps after COVID-19 disruptions.

Learn online with Australia's most

Everything you need for each of your lessons in one simple view

- Trusted, curriculum-aligned theory
- Engaging, rich multimedia
- All the teacher support resources you need
- Deep insights into progress
- Immediate feedback for students
- Create custom assignments in just a few clicks.

Practical teaching advice and ideas for each lesson provided in teachON.

Each lesson linked to the Key Knowledge (and Key Skills) from the VCE Mathematics Study Design.

Reading content and rich media including embedded videos and interactivities.

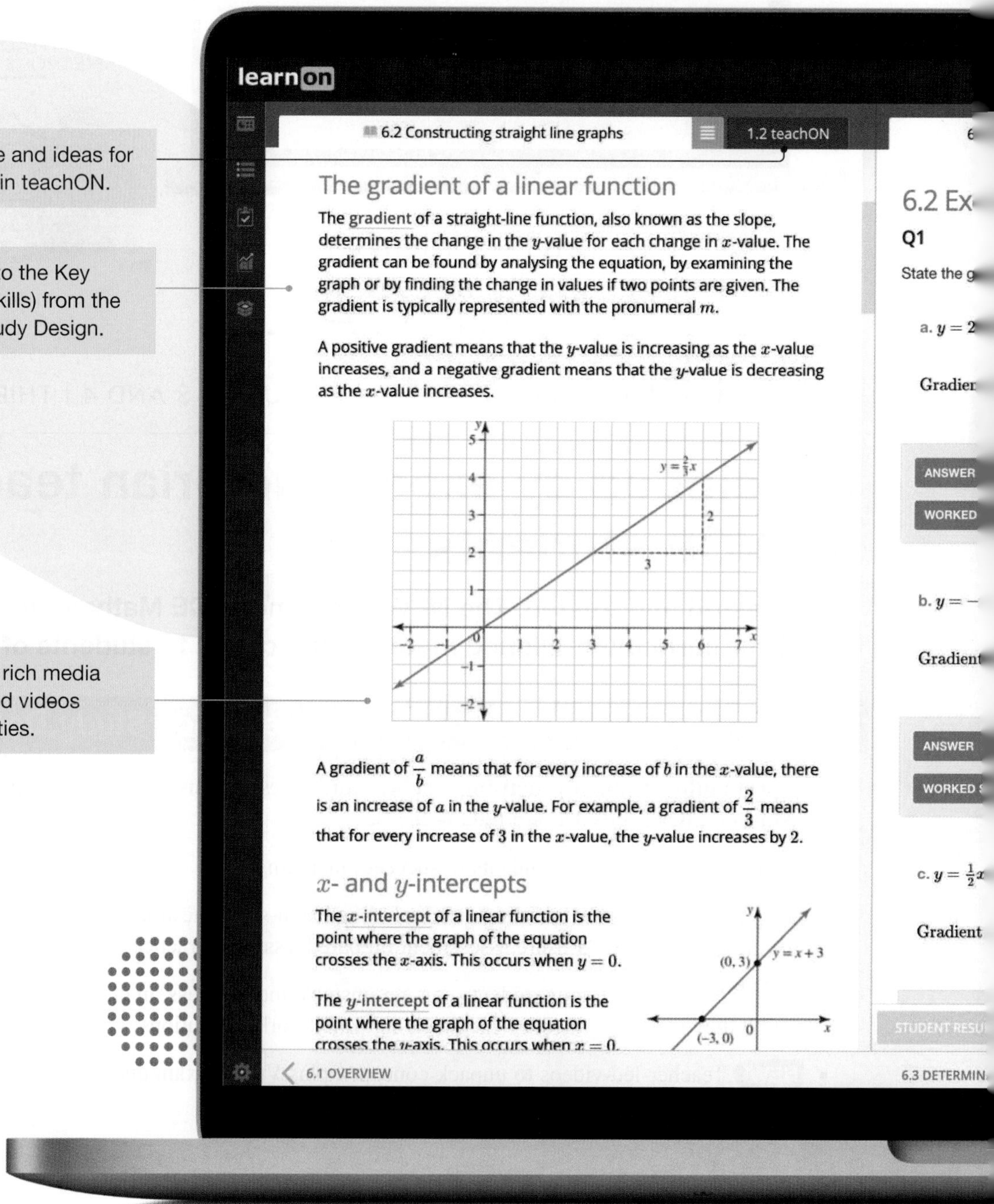

powerful learning tool, learnON

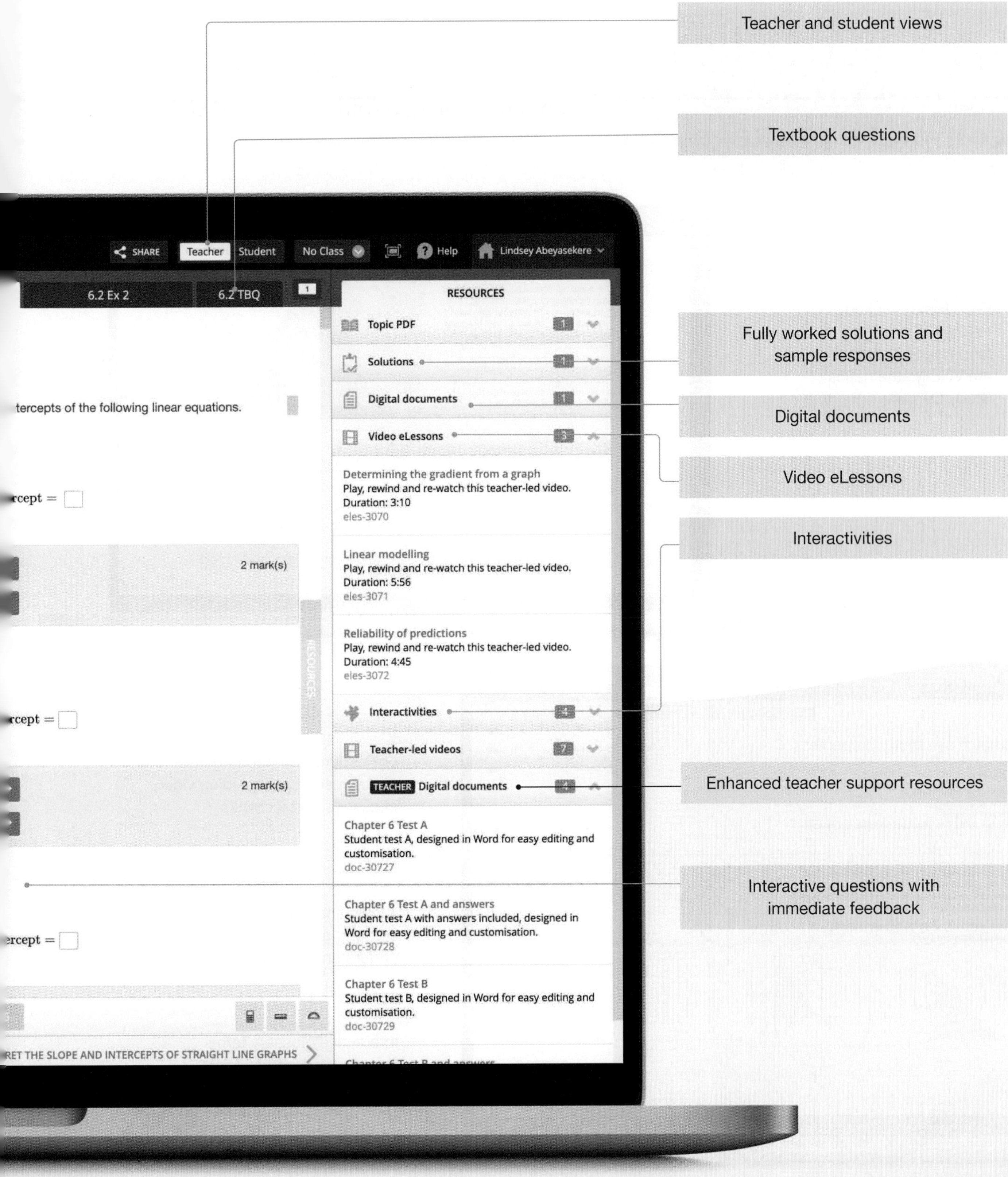

Get the most from your online resources

Online, these new editions are the **complete package**

Trusted Jacaranda theory, plus tools to support teaching and make learning more engaging, personalised and visible.

Each topic is linked to Key Knowledge (and Key Skills) from the VCE Mathematics Study Design.

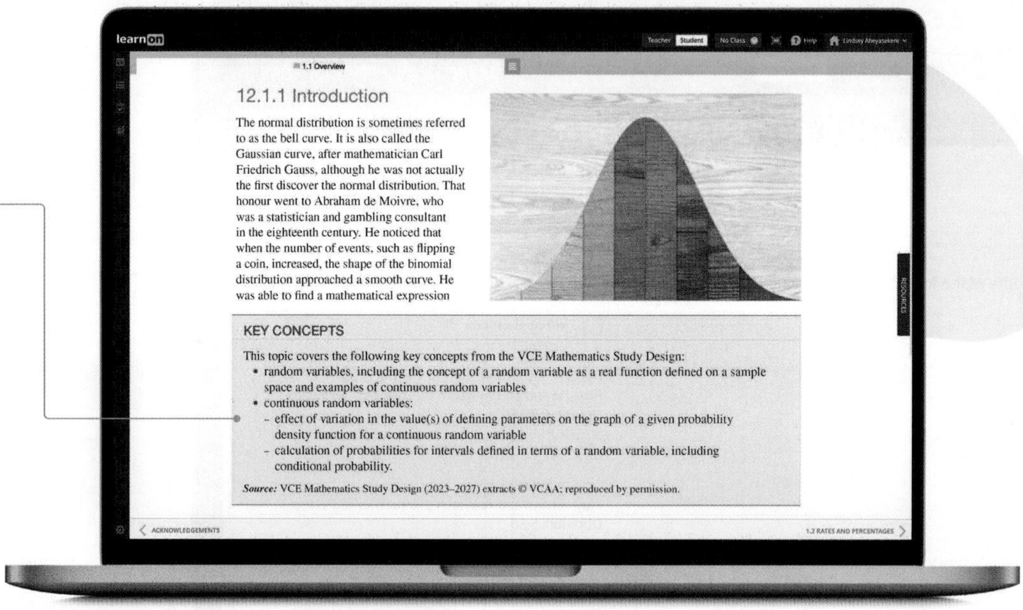

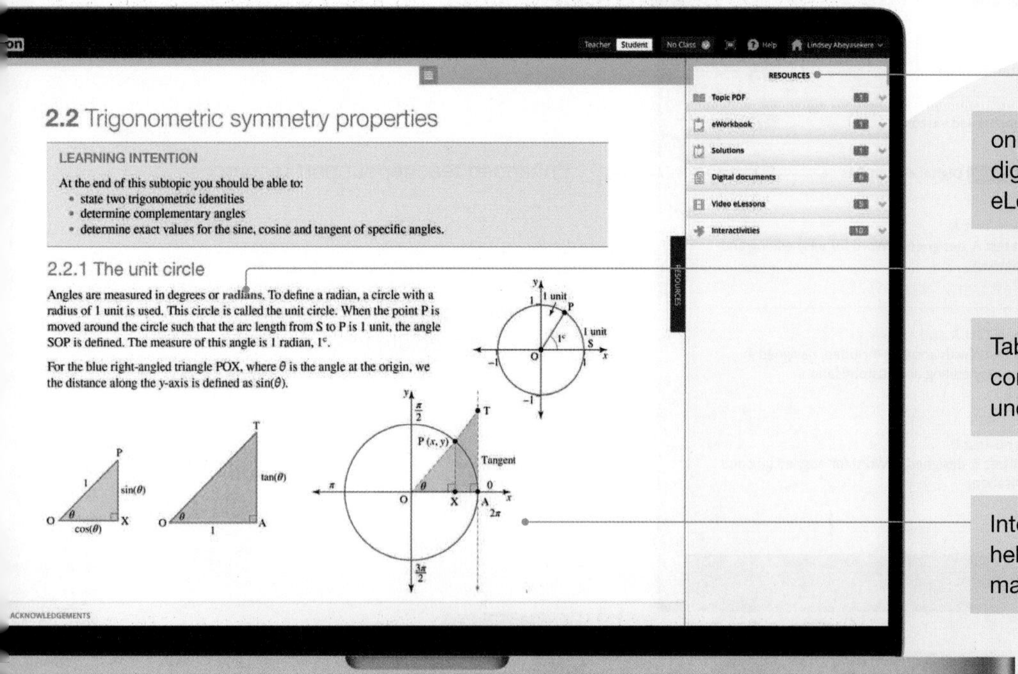

onResources link to targeted digital resources including video eLessons and weblinks.

Tables and images break down content, allowing students to understand complex concepts.

Interactive glossary terms help develop and support mathematical literacy.

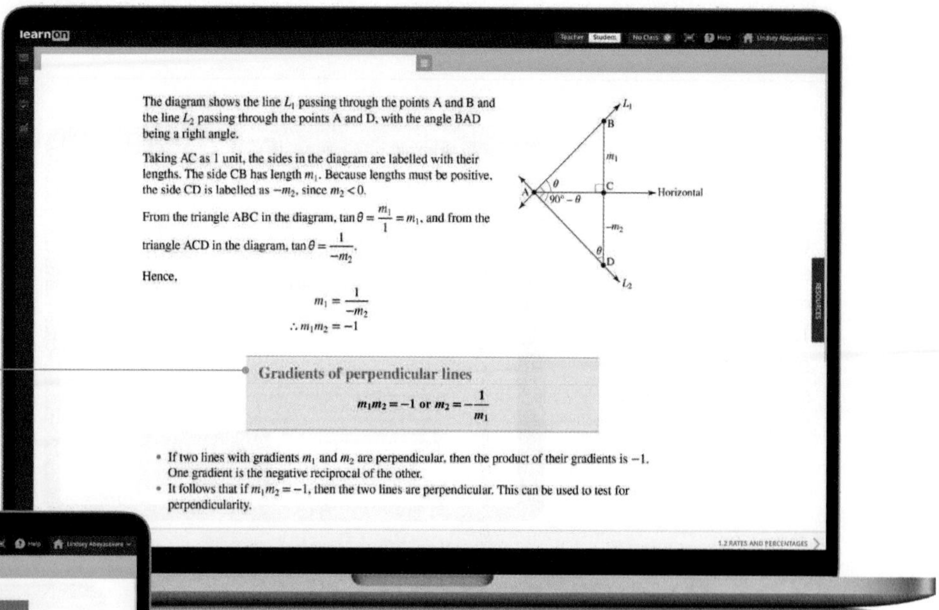

Pink highlight boxes summarise key information and provide tips for VCE Mathematics success.

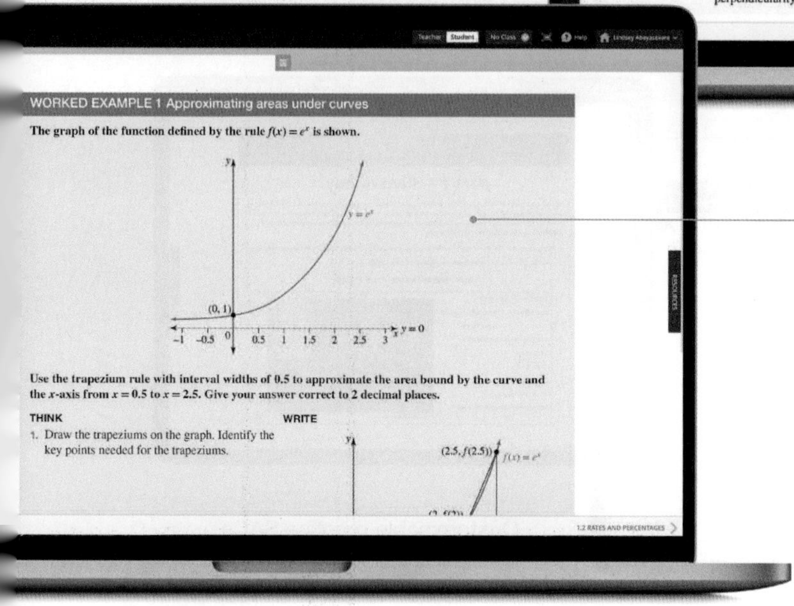

WORKED EXAMPLE 1 Approximating areas under curves
The graph of the function defined by the rule $f(x) = e^x$ is shown.
$y = e^x$
(0, 1)
$y = 0$
Use the trapezium rule with interval widths of 0.5 to approximate the area bound by the curve and the x-axis from $x = 0.5$ to $x = 2.5$. Give your answer correct to 2 decimal places.
THINK
WRITE
1. Draw the trapeziums on the graph. Identify the key points needed for the trapeziums.
(2.5, f(2.5))
$f(x) = e^x$

Worked examples break down the process of answering questions using a think/write format.

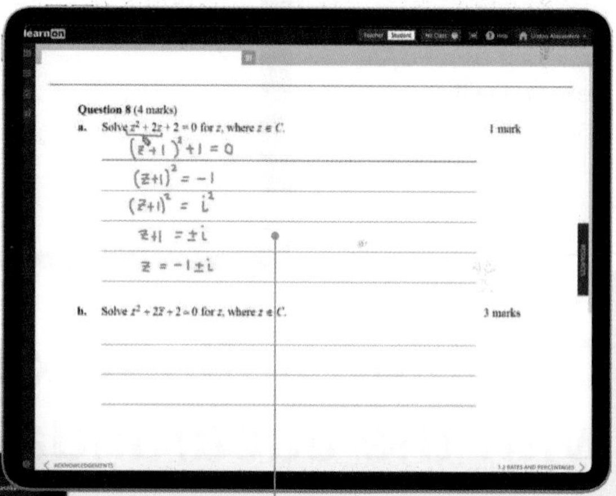

Teacher-led videos that explain how to approach exam questions, including VCAA exam questions.

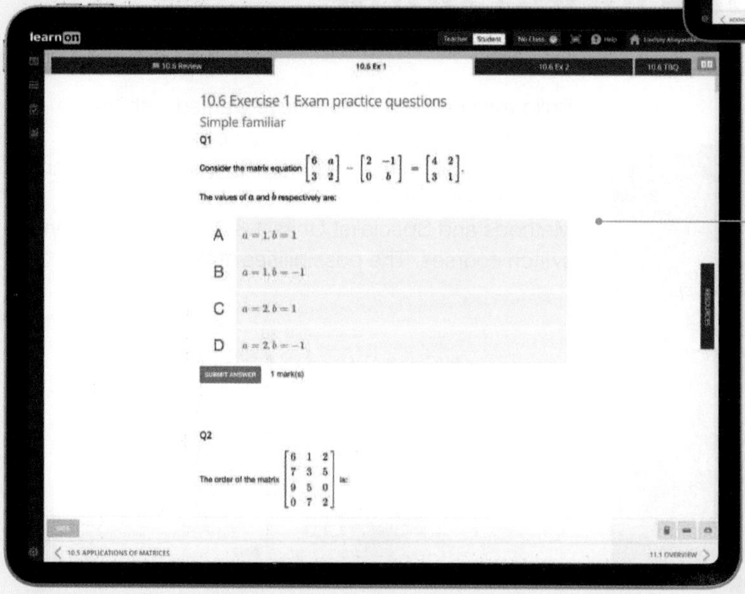

- Online and offline question sets contain practice questions, exam-style questions and past VCAA exam questions with exemplary responses and marking guides.
- Every question has immediate, corrective feedback to help students to overcome misconceptions as they occur and to study independently — in class and at home.

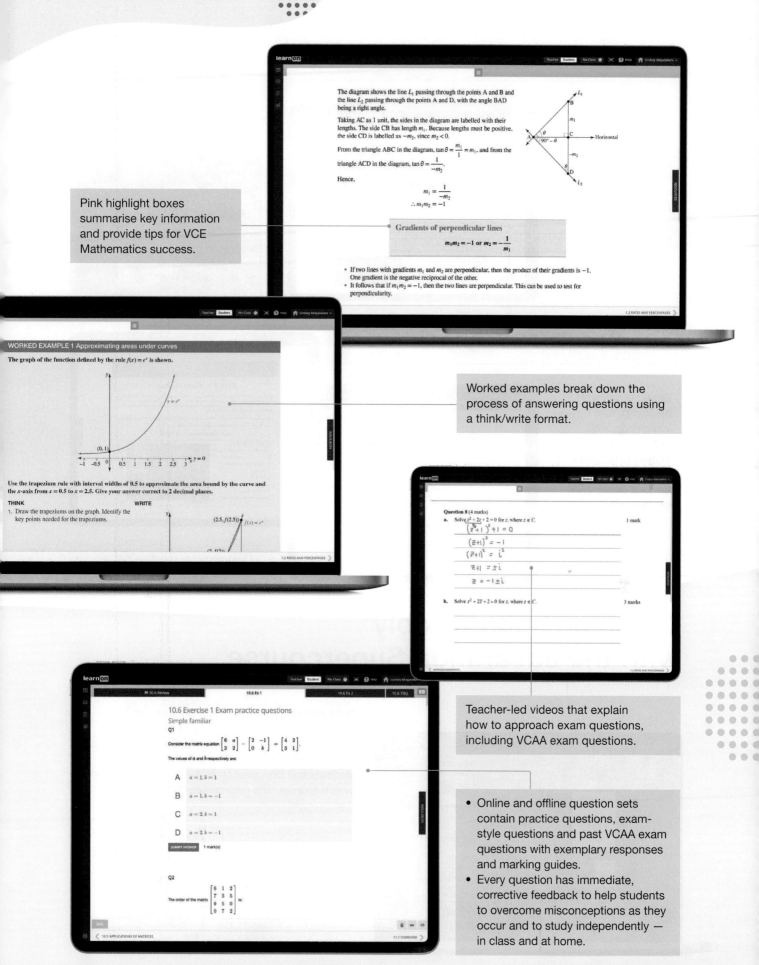

Topic reviews

Topic reviews include online summaries and topic-level review exercises that cover multiple concepts. Topic-level exam questions are structured just like the exams.

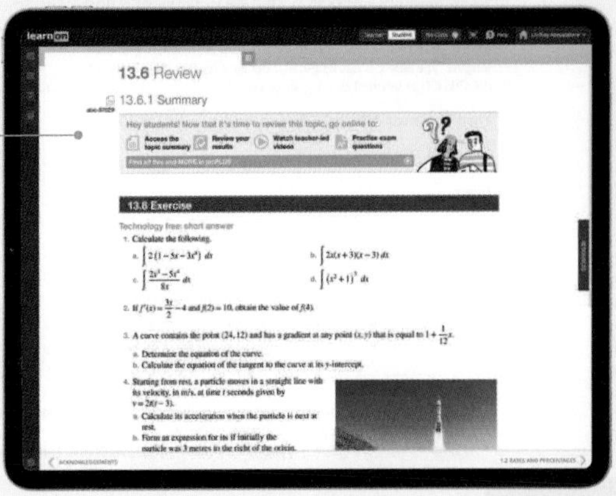

End-of-topic exam questions include relevant past VCE exam questions and are supported by teacher-led videos.

Get exam-ready!

Students can start preparing from lesson one, with exam questions embedded in every lesson — with relevant past VCAA exam questions since 2013.

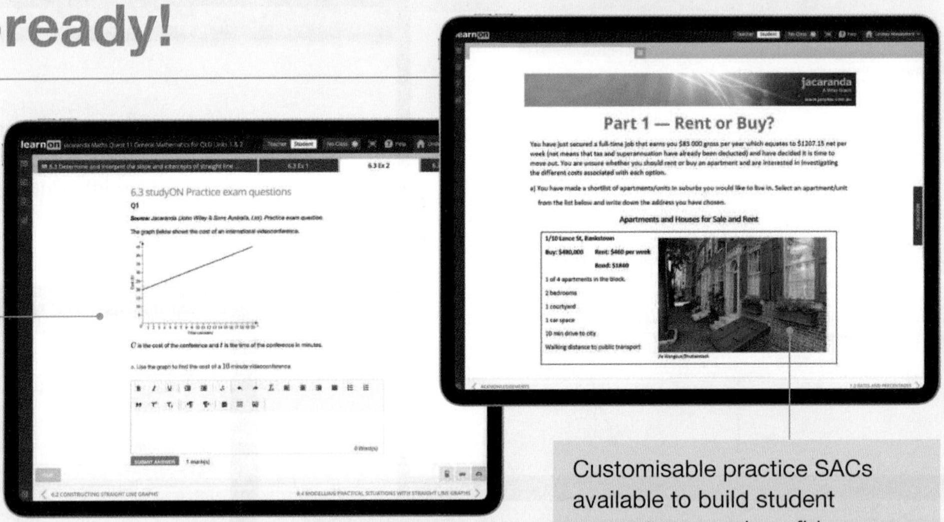

Customisable practice SACs available to build student competence and confidence.

Combine units flexibly with the Jacaranda Supercourse

Build the course you've always wanted with the Jacaranda Supercourse. You can combine all Mathematical Methods Units 1 to 4, so students can move backwards and forwards freely. Or combine Methods and Specialist Units 1 & 2 for when students switch courses. The possibilities are endless!

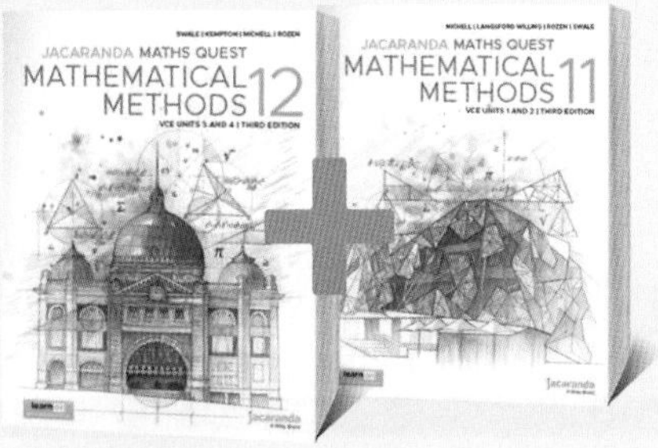

A wealth of teacher resources

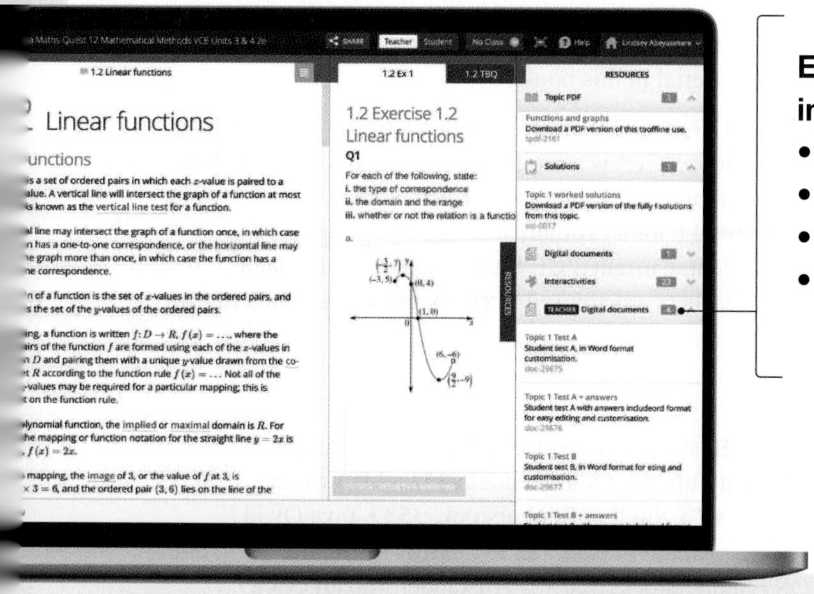

Enhanced teacher support resources, including:

- work programs and curriculum grids
- teaching advice and additional activities
- quarantined topic tests (with solutions)
- quarantined SACs (with worked solutions and marking rubrics)

Customise and assign

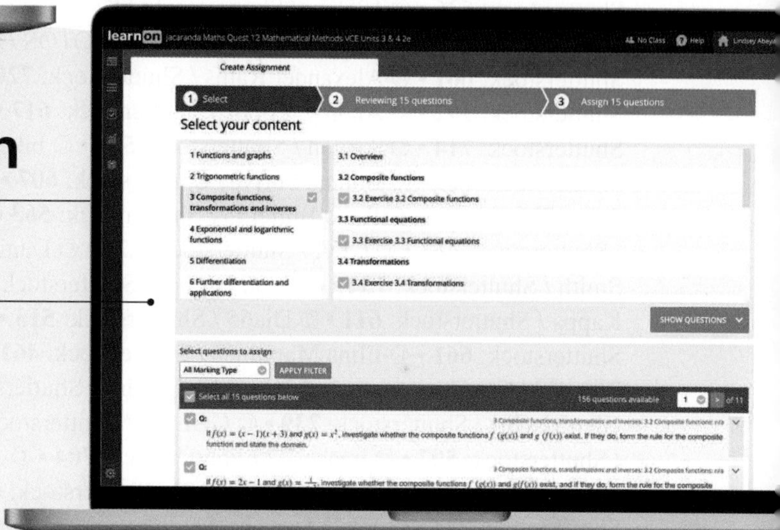

A testmaker enables you to create custom tests from the complete bank of thousands of questions (including past VCAA exam questions).

Reports and results

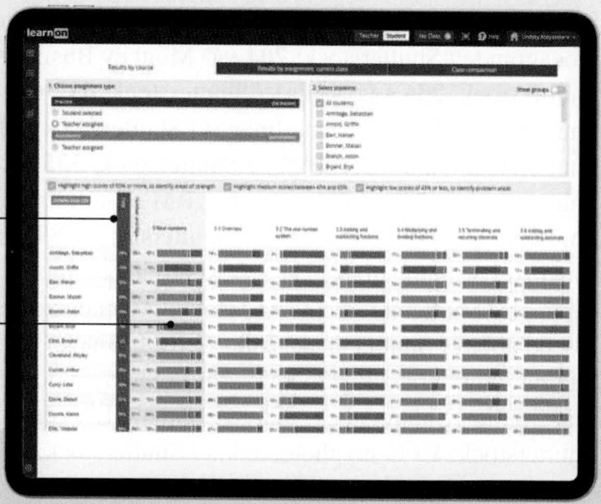

Data analytics and instant reports provide data-driven insights into performance across the entire course.

Show students (and their parents or carers) their own assessment data in fine detail. You can filter their results to identify areas of strength and weakness.

Acknowledgements

The authors and publisher would like to thank the following copyright holders, organisations and individuals for their assistance and for permission to reproduce copyright material in this book.

Selected extracts from the VCE Mathematics Study Design (2023–2027) are copyright Victorian Curriculum and Assessment Authority (VCAA), reproduced by permission. VCE* is a registered trademark of the VCAA. The VCAA does not endorse this product and makes no warranties regarding the correctness and accuracy of its content. To the extent permitted by law, the VCAA excludes all liability for any loss or damage suffered or incurred as a result of accessing, using or relying on the content. Current VCE Study Designs and related content can be accessed directly at www.vcaa.vic.edu.au. Teachers are advised to check the VCAA Bulletin for updates.

Images

• Billion Photos: **604** • aslysun / Shutterstock: **606** • Franck Boston / Shutterstock: **103** • gorillaimages / Shutterstock: **150** • Mauro Carli / Shutterstock: **672** • Maxim Tupikov / Shutterstock: **151** • MIA Studio / Shutterstock: **354** • TORWAISTUDIO / Shutterstock: **702** • VTT Studio / Shutterstock: **353** • Taya Ovod / Shutterstock: **536** • © 185buy / Shutterstock: **587** • © 54613 / Shutterstock: **552** • © aarrows / Shutterstock: **606** • © Africa Studio / Shutterstock: **732** • © AG-PHOTOS / Shutterstock: **526** • © Alexander Mazurkevich / Shutterstock: **661** • © Alexander Raths / Shutterstock: **720** • © AlexAnton / Shutterstock: **2** • © Always Joy / Shutterstock: **392** • © Andrey_Popov / Shutterstock: **617** • © Arthimedes / Shutterstock: **312** • © Arthimedes / Shutterstock: **714** • © Artifan / Shutterstock: **514** • © bikeriderlondon / Shutterstock: **610** • © bonchan / Shutterstock: **719** • © Bonma Suriya / Shutterstock: **607** • © Brenda Carson / Shutterstock: **705** • © Bulls-Eye Arts / Shutterstock: **619** • © Butter45 / Shutterstock: **563** • © carlos castilla / Shutterstock: **584** • © Cbenjasuwan / Shutterstock: **461** • © CKA / Shutterstock: **224** • © Daniel Korzeniewski / Shutterstock: **730** • © David P. Smith / Shutterstock: **734** • © Dean Drobot / Shutterstock: **600** • © deb talan / Shutterstock: **80** • © Denise Kappa / Shutterstock: **611** • © Dja65 / Shutterstock: **513** • © eakasarn / Shutterstock: **618** • © Elena Elisseeva / Shutterstock: **661** • © Elina Manninen / Shutterstock: **461** • © ESB Professional / Shutterstock: **734** • © FiledIMAGE / Shutterstock: **736** • © FXQuadro / Shutterstock: **727** • © George Rudy / Shutterstock: **608** • © GlenIrisGirl / Shutterstock: **739** • © GOLFX / Shutterstock: **545** • © guteksk7 / Shutterstock: **572** • © huyangshu / Shutterstock: **507** • © images72 / Shutterstock: **704** • © Jag_cz / Shutterstock: **236** • © Jalin / Shutterstock: **586** • © jax10289 / Shutterstock: **456** • © JNP / Shutterstock: **605** • © Joanne Harris and Daniel Bubnich / Shutterstock: **274** • © Johan Larson / Shutterstock: **314** • © Judith Andrews / Shutterstock: **425** • © Juice Dash / Shutterstock: **326** • © Kim D. Lyman / Shutterstock: **513** • © l i g h t p o e t / Shutterstock: **78** • © Leena Robinson / Shutterstock: **400** • © LightField Studios / Shutterstock: **609** • © M. Unal Ozmen / Shutterstock: **655** • © MarcelClemens / Shutterstock: **273** • © Maria Sbytova / Shutterstock: **514** • © Martynova Anna / Shutterstock: **268** • © mavo / Shutterstock: **608** • © Maxim Khytra / Shutterstock: **705** • © Menno Schaefer / Shutterstock: **263** • © Miks Mihails Ignats / Shutterstock: **457** • © mkmakingphotos / Shutterstock: **695** • © mocagrande / Shutterstock: **704** • © Monkey Business Images / Shutterstock: **516, 606** • © monticello / Shutterstock: **236** • © Nadja1 / Shutterstock: **412** • © Nattika / Shutterstock: **400** • © Neale Cousland / Shutterstock: **735** • © Odua Images / Shutterstock: **687** • © Olesya Kuznetsova / Shutterstock: **609** • © Orla / Shutterstock: **456** • © Pablo77 / Shutterstock: **535** • © paulrommer / Shutterstock: **699** • © Petr Malyshev / Shutterstock: **458** • © Photographer RM / Shutterstock: **471** • © photoyh / Shutterstock: **588** • © Phuketian.S / Shutterstock: **679** • © Pincasso / Shutterstock: **589** • © Pot of Grass Productions / Shutterstock: **611** • © Rawpixel / Shutterstock: **715** • © Remigiusz / Shutterstock: **583** • © Robert Kneschke / Shutterstock: **602** • © robin.ph / Shutterstock: **237** • © Robyn Mackenzie / Shutterstock: **544** • © Rokas Tenys / Shutterstock: **515** • © roseed abbas / Shutterstock: **288** • © Ruth Black / Shutterstock: **719** • © safakcakir / Shutterstock: **696** • © Sahroe / Shutterstock: **426** • © serhio / Shutterstock: **545** • © SFIO CRACHO / Shutterstock: **713** • © Shooting Star Studio / Shutterstock: **326** • © Shuang Li / Shutterstock: **575** • © Shutterstock: **681** • © anystock / Shutterstock: **1** • © myphotobank.a / Shutterstock: **686** • © Scott Rothstein / Shutterstock: **562** • © spainter_vfx /

Shutterstock: **168** • © SIAATH / Shutterstock: **606** • © Simon Bratt / Shutterstock: **655** • © sirtravelalot / Shutterstock: **601** • © sportpoint / Shutterstock: **459** • © StevanZZ / Shutterstock: **287** • © Tania Zbrodko / Shutterstock: **517** • © tartaruga1988 / Shutterstock: **588** • © TK Kurikawa / Shutterstock: **602** • © Tortoon Thodsapol / Shutterstock: **727** • © TZIDO SUN / Shutterstock: **671** • © v.s.anandhakrishna / Shutterstock: **561** • © VIAVAL TOURS / Shutterstock: **411** • © Vorobyeva / Shutterstock: **720** • © WAYHOME studio / Shutterstock: **237** • © Willyam Bradberry / Shutterstock: **312** • © wpap / Shutterstock: **288** • © yanik88 / Shutterstock: **327** • © Yulia Davidovich / Shutterstock: **722** • © zlikovec / Shutterstock: **601** • © Zoom Team / Shutterstock: **698** • © alaaddin / Shutterstock: **552** • © Brittny / Shutterstock: **563** • © ChameleonsEye / Shutterstock: **266** • © dien / Shutterstock: **551** • © ER_09 / Shutterstock: **697** • © Gordon Bell / Shutterstock: **325** • © graphit / Shutterstock: **717** • © harikarn / Shutterstock: **237** • © Iakov Filimonov / Shutterstock: **699** • © Jarrod Erbe / Shutterstock: **342** • © Katherine Welles / Shutterstock: **736** • © Kzenon / Shutterstock: **267** • © Leigh Prather / Shutterstock: **223** • © Macrovector / Shutterstock: **703** • © martiapunts / Shutterstock: **686** • © Maximilian Pawlikowsky / Shutterstock: **167** • © mhatzapa / Shutterstock: **695** • © oBebee / Shutterstock: **587** • © Palo_ok / Shutterstock: **313** • © seeyou / Shutterstock: **600** • © Sonulkaster / Shutterstock: **551** • © stockphoto mania / Shutterstock: **79** • © Susan Flashman / Shutterstock: **312** • © Tobik / Shutterstock: **694** • © wavebreakmedia / Shutterstock: **706** • © Wildnerdpix / Shutterstock: **457** • © yuris / Shutterstock: **679** • © Zaharia Bogdan Rares / Shutterstock: **266**

Every effort has been made to trace the ownership of copyright material. Information that will enable the publisher to rectify any error or omission in subsequent reprints will be welcome. In such cases, please contact the Permissions Section of John Wiley & Sons Australia, Ltd.

Shutterstock, 168 • © NPAVEH [illegible] 606 • © Simon Pratt / Alamy [illegible] 655 • © [illegible] / Shutterstock, 681 • © [illegible] / Shutterstock, 159 • © Sarawut [illegible] / Shutterstock, 267 • © [illegible] Shutterstock, 512 • © [illegible] 195 • Shutterstock, 588 • © [illegible] / Shutterstock, 602 • © [illegible] / Shutterstock, 722 • © TYRONE [illegible] / Shutterstock, 671 • [illegible] Shutterstock, 561 • © [illegible] 411 • © [illegible] / Shutterstock, 120 • [illegible] Shutterstock, [illegible] • [illegible] / Shutterstock, 512 • [illegible] 323 • [illegible] Shutterstock, 722 • © [illegible] 694 • © [illegible]

[illegible] 608 • [illegible]

[illegible] 222 • © [illegible] / Shutterstock, 707 • © [illegible] / Shutterstock, 688 • © Maximilian [illegible] / Shutterstock, 167 • © [illegible] / Shutterstock, 695 • © [illegible] / Shutterstock, 657 • © Fabio_dc / Shutterstock, 313 • © [illegible] / Shutterstock, 680 • © [illegible] / Shutterstock, 551 • © stock photo mania / Shutterstock, 78 • © Susan Flashman / Shutterstock, 312 • © Tobik / Shutterstock, 694 • © wavebreakmedia / Shutterstock, 706 • © Wildnerdpix / Shutterstock, 157 • © [illegible] / Shutterstock, 679 • © Zabotnova Bogdan Baress / Shutterstock, 206

Every effort has been made to trace the ownership of copyright material. Information that will enable the publisher to rectify any error or omission in subsequent reprints will be welcome. In such cases, please contact the Permissions Section of John Wiley & Sons Australia, Ltd.

1 Functions and graphs

LEARNING SEQUENCE

Fully worked solutions for this topic are available online.

1.1 Overview

Hey students! Bring these pages to life online

Watch videos

Engage with interactivities

Answer questions and check results

Find all this and MORE in jacPLUS

1.1.1 Introduction

Solving algebraic problems is one of the oldest processes in mathematics; however, it wasn't until the fifteenth century that the practical notation we use today was created. Before this, all equations were written out in words. As far back as ancient Egypt and Babylon, people were solving linear and quadratic equations, and the current solving processes are similar to the original methods. It took until the sixteenth century for a group of Italian mathematicians to solve the general cubic equation.

Polynomials have many applications in a range of industries: in engineering, a polynomial might be used to model the curves of rollercoasters or bridges; in economics, a combination of polynomial functions might be used to do cost analyses; and in physics, polynomials are used to describe energy, inertia and voltage difference, as well as the trajectories of moving objects.

KEY CONCEPTS

This topic covers the following key concepts from the VCE Mathematics Study Design:

- graphs of polynomial functions and their key features
- graphs of the following functions: power functions, $y = x^n, n \in Q$; exponential functions, $y = a^x, a \in R^+$, in particular $y = e^x$; logarithmic functions, $y = \log_e(x)$ and $y = \log_{10}(x)$; and circular functions, $y = \sin(x)$, $y = \cos(x)$ and $y = \tan(x)$ and their key features
- graphs of sum, difference, product and composite functions involving functions of the types specified above (not including composite functions that result in reciprocal or quotient functions)
- modelling of practical situations using polynomial, power, exponential and logarithmic functions, simple transformation and combinations of these functions, including simple piecewise (hybrid) functions
- solution of polynomial equations with real coefficients of degree n having up to n real solutions, including numerical solutions
- solution of equations of the form $f(x) = g(x)$ over a specified interval, where f and g are functions of the type specified in the 'Functions, relations and graphs' area of study, by graphical, numerical and algebraic methods, as applicable
- solution of simple systems of simultaneous linear equations, including consideration of cases where no solution or an infinite number of possible solutions exist (geometric interpretation only required for two equations in two variables).

Note: *Concepts shown in grey are covered in other topics.*

1.2 Linear functions

LEARNING INTENTION

At the end of this subtopic you should be able to:
- define a function
- sketch and determine a linear function
- determine the midpoint and distance between two points.

1.2.1 Functions

A function is a set of ordered pairs in which each x-value is paired to a unique y-value. A vertical line will intersect the graph of a function at most once. This is known as the **vertical line test** for a function.

A horizontal line may intersect the graph of a function once, in which case the function has a one-to-one correspondence, or the horizontal line may intersect the graph more than once, in which case the function has a many-to-one correspondence.

The **domain** of a function is the set of x-values in the ordered pairs, and the **range** is the set of the y-values of the ordered pairs.

As a mapping, a function is written $f: D \rightarrow R, f(x) = \ldots$, where the ordered pairs of the function f are formed using each of the x-values in the domain D and pairing them with a unique y-value drawn from the **co-domain** set R according to the function rule $f(x) = \ldots$ Not all of the available y-values may be required for a particular mapping; this is dependent on the function rule.

For any polynomial function, the **implied** or **maximal domain** is R. For example, the mapping or function notation for the straight line $y = 2x$ is $f: R \rightarrow R, f(x) = 2x$.

Under this mapping, the **image** of 3, or the value of f at 3, is $f(3) = 2 \times 3 = 6$, and the ordered pair $(3, 6)$ lies on the line of the function.

If only that part of the line $y = 2x$ where the x-values are positive was required, then this straight line function would be defined on a **restricted domain**, a subset of the maximal domain, and this would be written as $g: R^+ \rightarrow R, g(x) = 2x$.

WORKED EXAMPLE 1 Recognising functions

Part of the graph of the parabola $y = x^2$ is shown in the diagram.

a. Explain why the graph is a function and state the type of correspondence.

b. State the domain and range.

c. Express the given parabola using function notation.

d. Calculate the value of y when $x = -\sqrt{2}$.

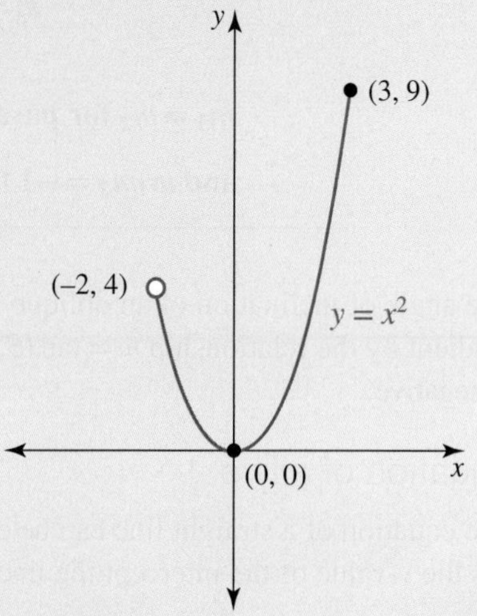

THINK	WRITE
a. 1. Use the vertical line test to explain why the graph is of a function.	a. This is a function because any vertical line that intersects the graph does so in exactly one place.
2. State the type of correspondence.	A horizontal line could cut the graph in up to two places. The correspondence is many-to-one.
b. 1. State the domain.	b. Reading from left to right horizontally in the x-axis direction, the domain is $(-2, 3]$.
2. State the range.	Reading from bottom to top vertically in the y-axis direction, the range is $[0, 9]$.
c. Use the domain and the function rule to form the mapping.	c. Let the function be f. As a mapping, it is $f:(-2,3] \rightarrow R, f(x)=x^2$.
d. Calculate the required value.	d. $f(x)=x^2$ Let $x=-\sqrt{2}$. $f\left(-\sqrt{2}\right)=\left(-\sqrt{2}\right)^2$ $=2$

1.2.2 The linear polynomial function

Two points are needed in order to determine the equation of a line. When sketching an oblique line by hand, usually the two points used are the x- and y-intercepts. If the line passes through the origin, then one other point needs to be determined from its equation.

Gradient

The **gradient**, or slope, of a line may be calculated from $m=\dfrac{y_2-y_1}{x_2-x_1}$.

This remains constant between any pair of points (x_1, y_1) and (x_2, y_2) on the line. The linear function either increases or decreases steadily.

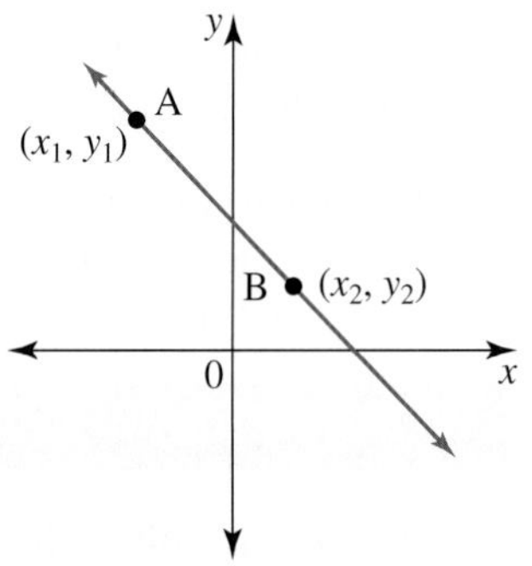

Parallel lines have the same gradient, and the product of the gradients of perpendicular lines is equal to -1. That is:

Gradient

$$m=\frac{y_2-y_1}{x_2-x_1}$$

$m_1=m_2$ for parallel lines

and $m_1m_2=-1$ for perpendicular lines.

The angle of inclination of an oblique line with the positive direction of the x-axis can be calculated from the gradient by the relationship $m=\tan(\theta)$. The angle θ is acute if the gradient is positive and obtuse if the gradient is negative.

Equation of a line

The equation of a straight line can be expressed in the form $y=mx+c$, where m is the gradient of the line and c is the y-value of the intercept the line makes with the y-axis.

If a point (x_1, y_1) and the gradient m are known, the equation of a line can be calculated from the point–gradient form, $y - y_1 = m(x - x_1)$.

Oblique lines are one-to-one functions.

Horizontal lines run parallel to the x-axis and have the equation $y = c$. These are many-to-one functions.

Vertical lines rise parallel to the y-axis and have the equation $x = k$. These lines are not functions.

WORKED EXAMPLE 2 Sketching a linear graph

Consider the line L where $L = \{(x, y) : 2x + 3y = 12\}$.

a. Sketch the line.

b. Calculate the gradient of the line.

THINK	WRITE
a. 1. Calculate the x- and y-intercepts.	a. $2x + 3y = 12$ y-intercept: let $x = 0$. $3y = 12$ $y = 4$ The y-intercept is $(0, 4)$. x-intercept: Let $y = 0$. $2x = 12$ $x = 6$ The x-intercept is $(6, 0)$.
2. Sketch the graph.	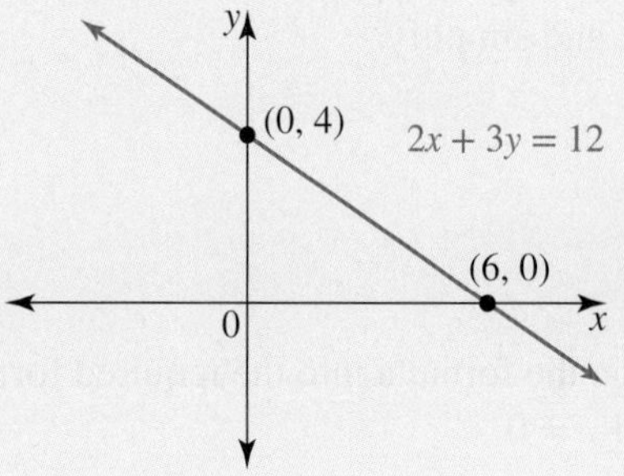
b. Rearrange the equation in the form $y = mx + c$ and state the gradient. *Note:* The gradient could also be calculated using $m = \frac{\text{rise}}{\text{run}}$ from the diagram.	b. $2x + 3y = 12$ $3y = -2x + 12$ $y = \frac{-2x}{3} + 4$ The gradient is $m = -\frac{2}{3}$.

WORKED EXAMPLE 3 Determining linear equations

Determine the equation, in the form $ax + by + c = 0$, for the line:

a. passing through the point (2, 3) and parallel to the line with equation $y - 3x + 5 = 0$

b. passing through the point (−1, 6) and perpendicular to the line with equation $2y + 4x = -10$.

THINK	WRITE
a. 1. Convert the equation into the form $y = mx + c$ and state the gradient of the parallel line. This is also the gradient of the desired line.	a. $y - 3x + 5 = 0$ $y = 3x - 5$ $\therefore m = 3$

2. State the gradient–point form of a straight line equation. $y - y_1 = m(x - x_1)$

3. Substitute the given point (2, 3) and the gradient, and simplify.

$$y - 3 = 3(x - 2)$$
$$y - 3 = 3x - 6$$
$$y = 3x - 3$$

4. Rearrange the formula into the required form, $ax + by + c = 0$.

$$y = 3x - 3$$
$$y - 3x + 3 = 0$$

b. 1. Convert the equation into the form $y = mx + c$ and state the gradient of the perpendicular line.

b. $2y + 4x = -10$
$$2y = -4x - 10$$
$$y = -2x - 5$$
$$\therefore m = -2$$

2. Calculate the gradient of the desired line.

$$m_1 \times m_2 = -1$$
$$-2 \times m_2 = -1$$
$$\therefore m_2 = \frac{1}{2}$$

3. State the gradient–point form of a straight line equation. $y - y_1 = m(x - x_1)$

4. Substitute the given point $(-1, 6)$ and the gradient, and simplify.

$$y - 6 = \frac{1}{2}(x + 1)$$
$$y - 6 = \frac{1}{2}x + \frac{1}{2}$$
$$y = \frac{1}{2}x + \frac{13}{2}$$

5. Rearrange the formula into the required form, $ax + by + c = 0$.

$$y = \frac{1}{2}x + \frac{13}{2}$$
$$2y = x + 13$$
$$2y - x - 13 = 0$$

WORKED EXAMPLE 4 Gradients and linear equations

a. Calculate, correct to 1 decimal place, the angle made with the positive direction of the x-axis by the line that passes through the points $(-3, -2)$ and $(4, 1)$.

b. Determine the equation of the line that passes through the point $(5, 2)$ at an angle of 45°.

THINK / **WRITE**

a. 1. Determine the gradient of the line passing between the given points.

a. $m = \frac{y_2 - y_1}{x_2 - x_1}$
$$= \frac{1 + 2}{4 + 3}$$
$$= \frac{3}{7}$$

2. State the relationship between the angle and the gradient.

$\tan(\theta) = m$

$\tan(\theta) = \frac{3}{7}$

3. Calculate θ, correct to 1 decimal place.

$\theta = \tan^{-1}\left(\frac{3}{7}\right)$

$= 23.2°$

b. 1. Determine the gradient from the given angle.

b. $\tan(\theta) = m$

$m = \tan(45°)$

$= 1$

2. State the gradient–point form of a straight line equation.

$y - y_1 = m(x - x_1)$

3. Substitute the given point (5, 2) and the gradient, and simplify.

$y - 2 = 1(x - 5)$

$y - 2 = x - 5$

$y = x - 3$

1.2.3 Midpoint and distance between two points

Midpoint between two points

The midpoint, M, of a line cuts the line exactly in half, so M is equidistant from points A and B.

The coordinates of M are found by averaging the x- and y-coordinates of points A and B.

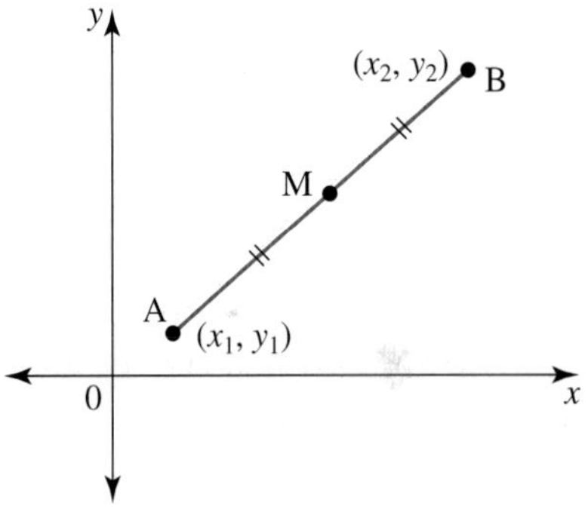

Midpoint between two points

$$M = \left(\frac{x_1 + x_2}{2}, \frac{y_1 + y_2}{2}\right)$$

Distance between two points

The distance between two points is the length of that line segment.

Essentially, a right-angled triangle can be constructed with vertical height $y_2 - y_1$ and horizontal length $x_2 - x_1$. Applying Pythagoras' theorem, the formula for the distance between two points is as follows.

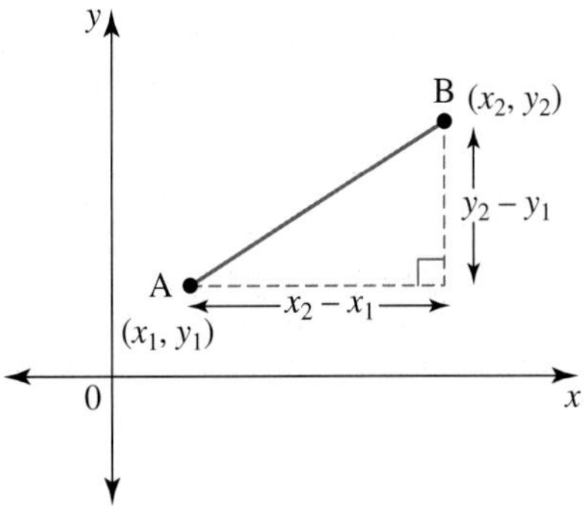

Distance between two points

$$d = \sqrt{(x_2 - x_1)^2 + (y_2 - y_1)^2}$$

WORKED EXAMPLE 5 Calculating the midpoint and distance between two points

Given the points (−3, 5) and (4, −6), calculate:

a. the midpoint, M, between the two points

b. the distance between the two points.

THINK	WRITE
a. 1. Write the formula for finding the midpoint between two points. Substitute the x- and y-coordinates and simplify. *Note:* It doesn't matter which is point 1 and which is point 2.	**a.** $M = \left(\dfrac{x_1 + x_2}{2}, \dfrac{y_1 + y_2}{2}\right)$ $M = \left(\dfrac{-3+4}{2}, \dfrac{5-6}{2}\right)$ $= \left(\dfrac{1}{2}, -\dfrac{1}{2}\right)$
b. 1. Write the distance formula.	**b.** $d = \sqrt{(x_2 - x_1)^2 + (y_2 - y_1)^2}$
2. Substitute the two points into the equation and simplify. *Note:* It doesn't matter which is point 1 and which is point 2.	$d = \sqrt{(4+3)^2 + (-6-5)^2}$ $= \sqrt{7^2 + (-11)^2}$ $= \sqrt{49 + 121}$ $= \sqrt{170}$
3. State the final answer.	The distance is $\sqrt{170}$ units.

on Resources

Interactivities Equations from point–gradient and gradient–y-intercept form (int-2551)
Midpoint of a line segment and the perpendicular bisector (int-2553)

1.2 Exercise

Technology free

1. **WE1** Part of the graph of the parabola $y = x^2$ is shown in the diagram.
   a. Explain why the graph shows a function and state the type of correspondence.
   b. State the domain and range.
   c. Express the given parabola using function notation.
   d. Calculate the value of y when $x = -2\sqrt{3}$.

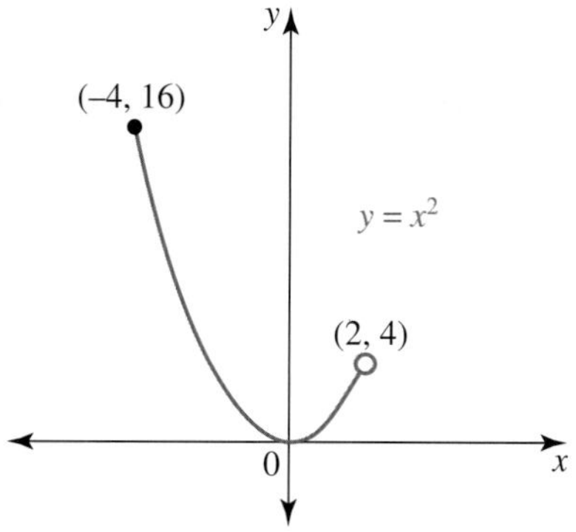

2. For each of the following, state:
   i. the type of correspondence
   ii. the domain and the range
   iii. whether or not the relation is a function.

a.

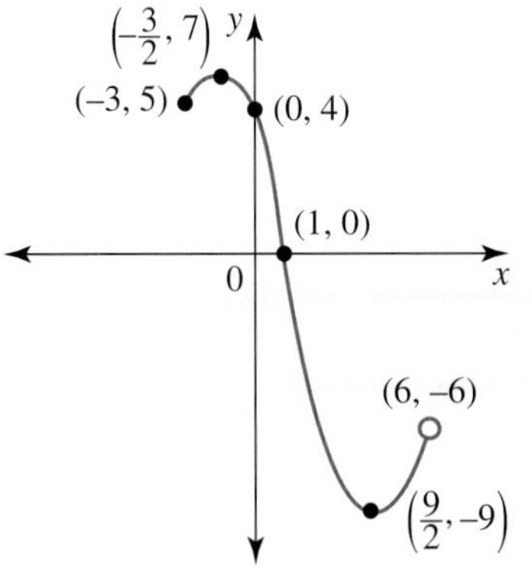

b.

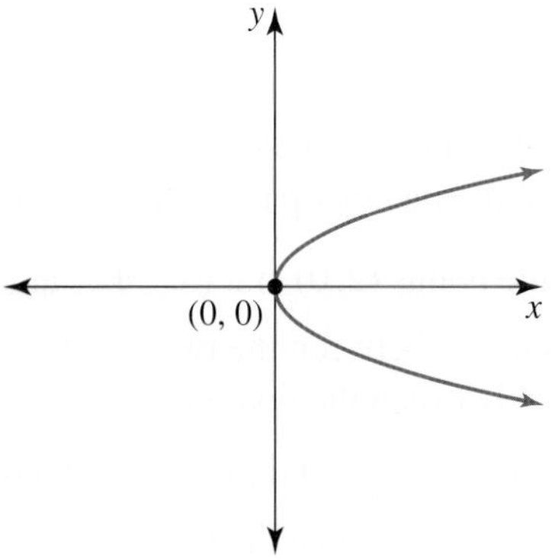

c.

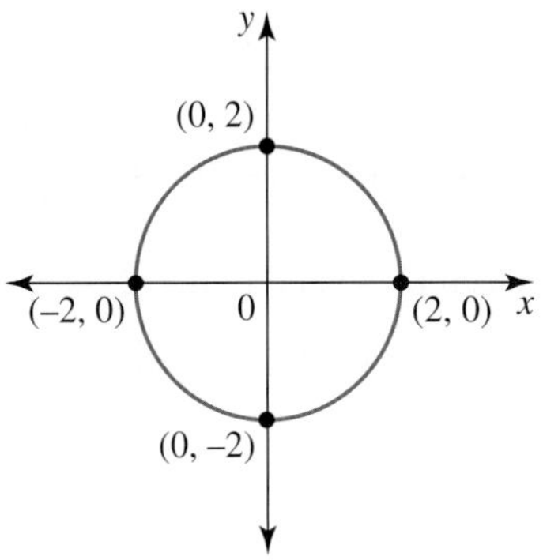

d.

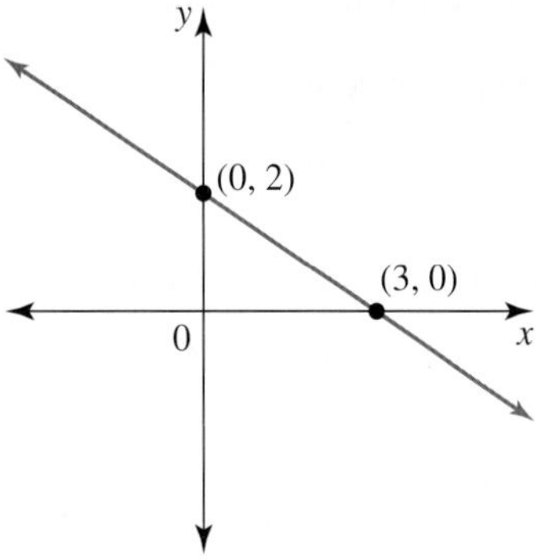

e.

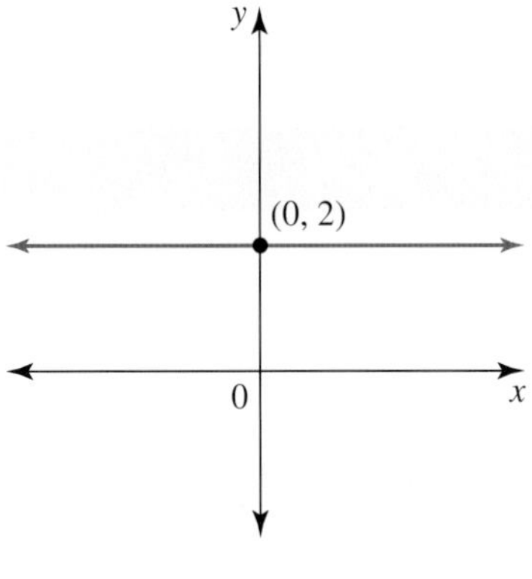

f.

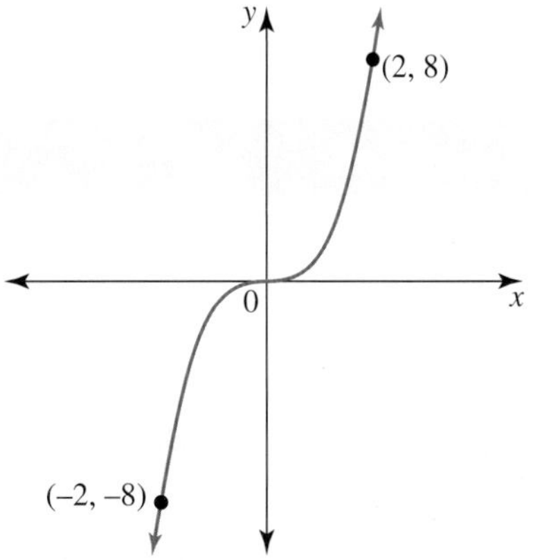

3. **WE2** Consider the line L where $L = \{(x, y) : 3x - 4y = 12\}$.
   a. Sketch the line.
   b. Calculate the gradient of the line.

4. Sketch the following linear functions and state the range of each.
   a. $f : R \to R, f(x) = 9 - 4x$
   b. $g : (-3, 5] \to R, g(x) = \dfrac{3x}{5}$
   c. $2x + y = 4, \;\; x \in [-2, 4)$
   d. $y = \dfrac{2x}{3} + 5, \;\; x \in [-1, 5]$

5. Determine the equation of the line:
   a. with a gradient of -3 and passing through the point $(2, 6)$
   b. passing through the points $(-2, -4)$ and $(1, 5)$.

6. WE3 Determine the equation, in the form $ax + by + c = 0$, for the line:

a. passing through the point $(-1, 4)$ and parallel to the line with equation $y + 2x - 3 = 0$

b. passing through the point $(2, 3)$ and perpendicular to the line with equation $3y - 6x = 4$.

Technology active

7. WE4 Determine:

a. the angle made with the positive direction fo the x-axis by the line that passes through the points (2,1) and (8,–4) correct to 1 decimal place

b. the equation of the line that passes through the point $(-2, 4)$ at an angle of 45°.

8. a. WE5 Given the points $(2, 0)$ and $(6, -4)$, calculate:

i. the midpoint, M, between the two points

ii. the distance between the two points.

b. Given the points $(-3, -2)$ and $(4, 3)$, calculate:

i. the midpoint, M, between the two points

ii. the distance between the two points.

9. Determine the value of a when:

a. the midpoint of $(a, 4)$ and $(10, -2)$ is $(8, 1)$

b. the midpoint of $(6, a)$ and $(-3, -2)$ is $\left(\frac{3}{2}, \frac{5}{2}\right)$

c. the distance between $(1, a)$ and $(4, 6)$ is $\sqrt{73}$

d. the distance between $(a, 4)$ and $(-2, -2)$ is $\sqrt{45}$.

10. Consider the three points A $(5, -3)$, B $(7, 8)$ and C $(-2, p)$. The line through A and C is parallel to $9x + 7y = 24$.

a. Calculate the value of p.

b. Determine the equation of the line through B that is perpendicular to AC.

c. Calculate the shortest distance from B to AC, expressing the value to 1 decimal place.

1.2 Exam questions

Question 1 (1 mark) TECH-ACTIVE

Source: VCE 2013, Mathematical Methods (CAS) Exam 2, Section 1, Q2; © VCAA.

MC The midpoint of the line segment that joins $(1, -5)$ to $(d, 2)$ is

A. $\left(\frac{d+1}{2}, -\frac{3}{2}\right)$

B. $\left(\frac{1-d}{2}, -\frac{7}{2}\right)$

C. $\left(\frac{d-4}{2}, 0\right)$

D. $\left(0, \frac{1-d}{3}\right)$

E. $\left(\frac{5+d}{2}, 2\right)$

Question 2 (1 mark) TECH-ACTIVE

Source: VCE 2014, Mathematical Methods (CAS) Exam 2, Section 1, Q2; © VCAA.

MC The linear function $f: D \to R, f(x) = 4 - x$ has range $[-2, 6)$.

The domain D of the function is

A. $[-2, 6)$

B. $[-2, 2)$

C. R

D. $(-2, 6]$

E. $[-6, 2)$

Question 3 (1 mark) TECH-ACTIVE

MC The gradient of a line **perpendicular** to the line that passes through $(3, 1)$ and $(0, -5)$ is

A. $\frac{1}{2}$

B. 2

C. $-\frac{1}{2}$

D. -6

E. -2

More exam questions are available online.

1.3 Solving systems of equations

LEARNING INTENTION

At the end of this subtopic you should be able to:
- solve simultaneous equations with two and three variables.

1.3.1 Solving simultaneous equations with two variables

Three possible scenarios exist when we are dealing with two linear **simultaneous equations**. There may be one solution only, there may be no solutions, or there may be infinitely many solutions.

One solution	If the two straight lines intersect each other at only one place, we have one solution. This indicates that the gradients of the two equations are different.	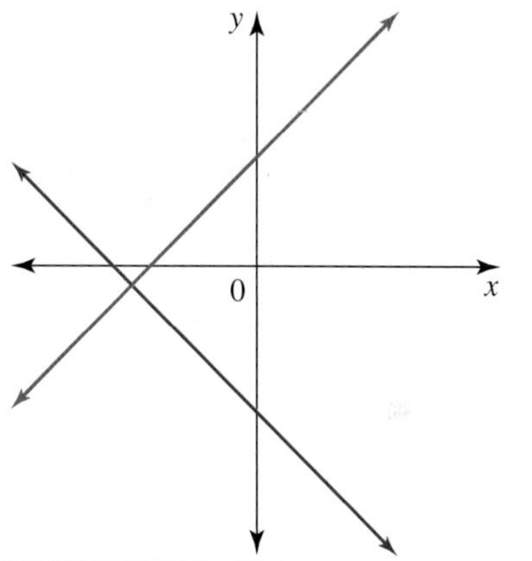
No solution	If the two straight lines have the same gradient, they are parallel lines, so they never meet. Therefore, there are no solutions to the simultaneous equations. Although the gradients of the lines are the same, the y-intercepts are different.	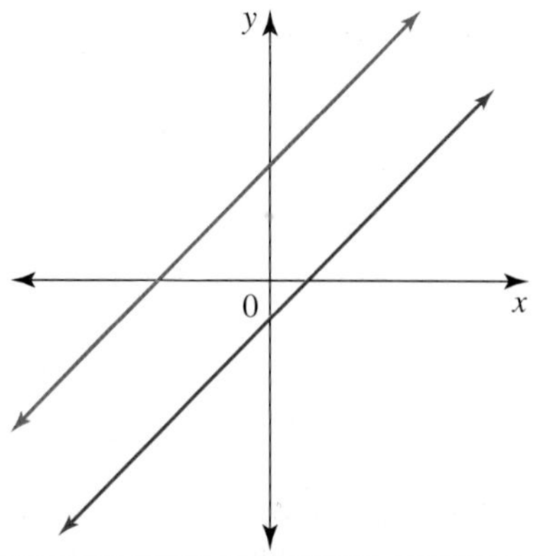
Infinitely many solutions	If the two straight lines have the same equation, one line lies on top of the other and there are infinitely many solutions. Both the gradients and the y-intercepts are identical.	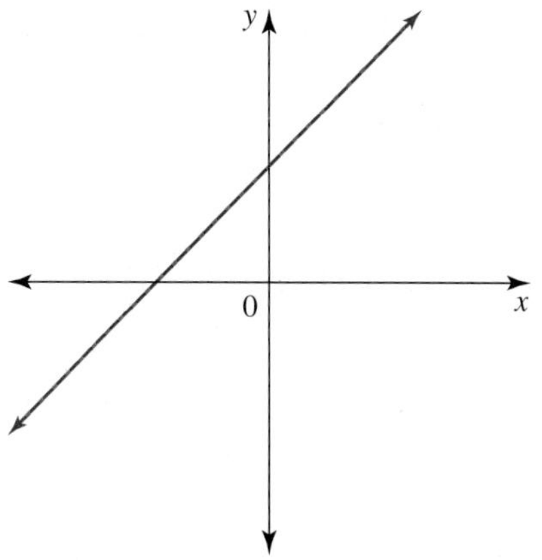

on Resources

Interactivities Solving systems of equations (int-2549)
Intersecting, parallel and identical lines (int-2552)

WORKED EXAMPLE 6 Determining unique solutions

Determine the value of k for which the following simultaneous equations have a unique solution.

$$kx + 3y = 1$$
$$4x + 3ky = 0$$

THINK	WRITE
1. Label the equations.	$kx + 3y = 1 \quad [1]$ $4x + 3ky = 0 \quad [2]$
2. There will be a unique solution for all values of k, except when the gradients of the two lines are the same. To find the gradient, write the equations in the general form, $y = mx + c$.	$[1] \Rightarrow kx + 3y = 1$ $3y = 1 - kx$ $y = \frac{1 - kx}{3}$ $\therefore m = -\frac{k}{3}$ $[2] \Rightarrow 4x + 3ky = 0$ $3ky = -4x$ $y = \frac{-4x}{3k}$ $\therefore m = -\frac{4}{3k}$
3. Equate the gradients and solve for k.	$-\frac{k}{3} = -\frac{4}{3k}$ $3k^2 = 12$ $k^2 = 4$ $k = \pm 2$
4. Write the solution. This solution tells us that if $k = \pm 2$, the equations will have the same gradient, so for any other value of k, there will be a unique solution.	$k \in R \setminus \{-2, 2\}$

1.3.2 Simultaneous equations with three variables

An equation with two variables defines a line. An equation with three variables defines a **plane**. If an equation has three variables, there needs to be three different equations for us to be able to solve for the point at which the three planes intersect (if in fact they do intersect at a single point).

There are a number of different possible outcomes when planes intersect.

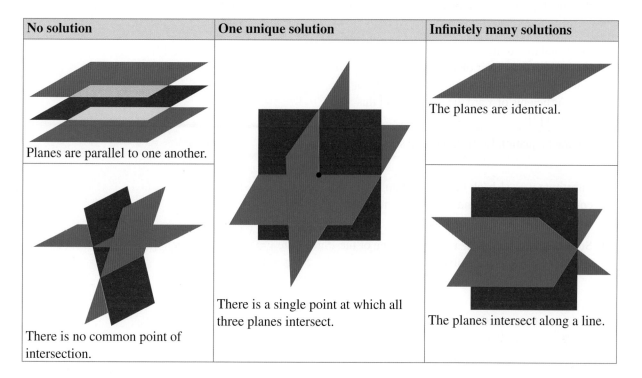

<table>
<tr><th>No solution</th><th>One unique solution</th><th>Infinitely many solutions</th></tr>
<tr><td>
Planes are parallel to one another.</td><td rowspan="2">
There is a single point at which all three planes intersect.</td><td>
The planes are identical.</td></tr>
<tr><td>
There is no common point of intersection.</td><td>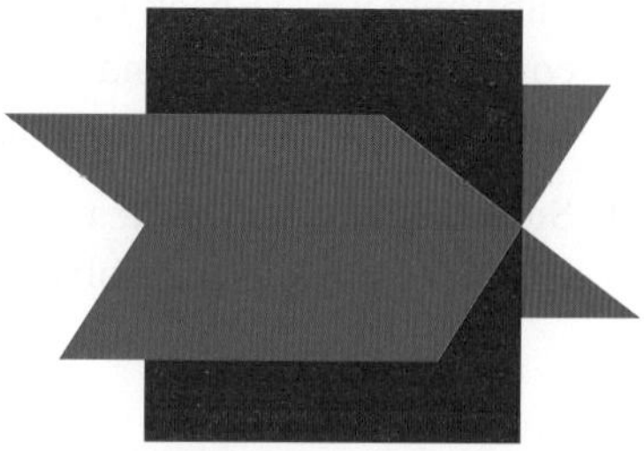
The planes intersect along a line.</td></tr>
</table>

When solving three simultaneous equations without technology, the strategy is to eliminate one of the variables and reduce the three equations with three unknowns to two equations with two unknowns.

Solving simultaneous equations with technology becomes a straightforward problem in CAS by using the inbuilt functions.

WORKED EXAMPLE 7 Solving simultaneous equations with three unknowns

Solve the following system of simultaneous equations.

$$\begin{aligned} 2x - 3y + 2z &= -5 \\ x - 5y + z &= 1 \\ 2x + 3y + z &= -2 \end{aligned}$$

THINK	WRITE
1. Label the equations and determine which of the three pronumerals you are going to eliminate. Either x or z would be appropriate choices, as the coefficients in all three equations are either the same or a multiple of the other. Let us eliminate z.	$2x - 3y + 2z = -5 \quad [1]$ $x - 5y + z = 1 \quad [2]$ $2x + 3y + z = -2 \quad [3]$
2. Subtract equation [2] from [3] to eliminate z and label this equation [4].	$[3] - [2] \Rightarrow \quad x + 8y = -3 \quad [4]$
3. We need another equation without z. In order to subtract equation [1] from [3], multiply equation [3] by 2. Label this equation [5].	$[3] \times 2 \Rightarrow \quad 4x + 6y + 2z = -4 \quad [5]$
4. Subtract equation [1] from the newly formed [5] to eliminate z and label this equation [6].	$[5] - [1] \Rightarrow \quad 2x + 9y = 1 \quad [6]$

5. We now have two equations with only x and y.

$$x+8y=-3 \quad [4]$$
$$2x+9y=1 \quad [6]$$

6. The standard elimination method will be used to solve this pair of simultaneous equations. Multiply equation [4] by 2 so that the coefficients of x are the same. Label this equation [7].

$$[4]\times 2 \Rightarrow \quad 2x+16y=-6 \quad [7]$$

7. Subtract equation [6] from [7] and solve for y.

$$[7]-[6] \Rightarrow \quad 7y=-7$$
$$y=-1$$

8. Substitute $y=-1$ back into the system of equations in order to find x. Choose one of the equations containing only x and y.

Substitute $y=-1$ into [4]:
$$x-8=-3$$
$$x=5$$

9. Substitute the values for x and y into one of the original equations and solve for z.

Substitute $y=-1$ and $x=5$ into [2]:
$$5+5+z=1$$
$$10+z=1$$
$$z=-9$$

10. Write the final solution. Alternatively, CAS can be used to solve the three simultaneous equations if the question is technology active.

$x=5,\ y=-1,\ z=-9$

TI | THINK

1. On a Calculator page, press MENU, then select:
3: Algebra
7: Solve System of Equations
1: Solve System of Equations …
Complete the entry line as:
$$\text{solve}\left(\begin{cases}2x-3y+2z=-5\\ x-5y+z=1\\ 2x+3y+z=-2\end{cases},\{x,y,z\}\right)$$
then press ENTER.

DISPLAY/WRITE

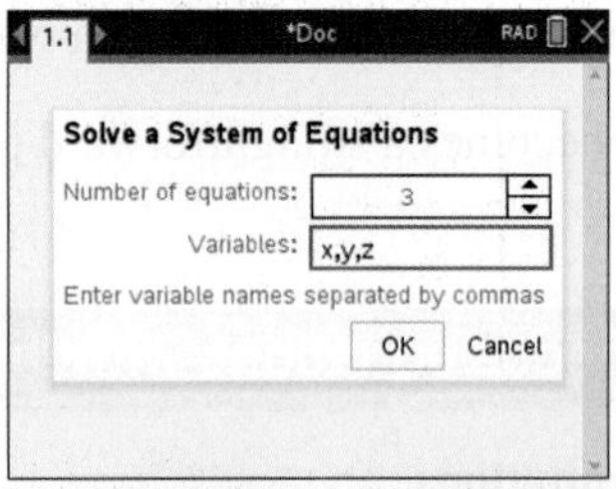

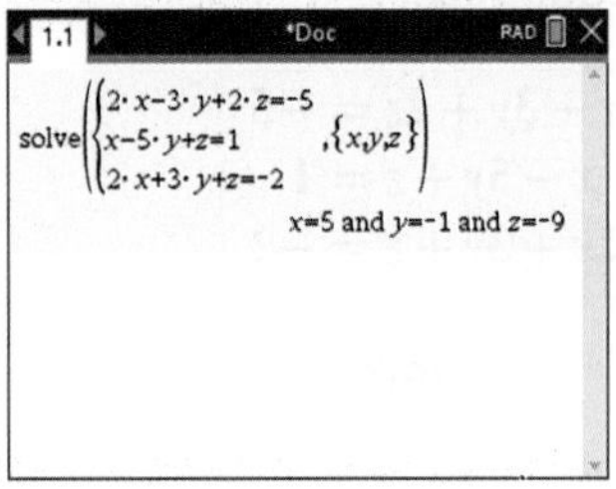

2. The answer appears on the screen.

$x=5, y=-1, z=-9$

CASIO | THINK

1. On a Main screen, complete the entry line as:
Solve $(\{2x-3y+2z=-5,$
$x-5y+z=1,$
$2x+3y+z=-2\},$
$\{x,y,z\}$
then press EXE.

DISPLAY/WRITE

2. The answer appears on the screen.

$x=5, y=-1, z=-9$

on Resources

Interactivity Equations in 3 variables (int-2550)

1.3.3 Simultaneous equations involving parameters

When there are infinitely many solutions to a system of equations, such as when planes intersect along a line, we can describe the set of solutions through the use of a **parameter**. Conventionally, the parameter chosen is λ.

WORKED EXAMPLE 8 Solving simultaneous equations using a parameter

The simultaneous equations shown have infinitely many solutions.

$$\begin{aligned} 2x + y - 4z &= 2 \\ x + y + 3z &= -1 \end{aligned}$$

a. Eliminate y by subtracting the second equation from the first equation.
b. Let $z = \lambda$ and solve the equations in terms of λ.
c. Explain what this solution represents.

THINK	WRITE
a. 1. Label the equations.	**a.** $2x + y - 4z = 2$ [1] $x + y + 3z = -1$ [2]
2. Subtract equation [2] from equation [1].	$[1] - [2] \Rightarrow x - 7z = 3$
b. 1. Substitute $z = \lambda$ and solve for x.	**b.** $z = \lambda$ $x - 7\lambda = 3$ $x = 3 + 7\lambda$
2. Substitute $z = \lambda$ and $x = 3 + 7\lambda$ into equation [2] and solve for y. *Note:* Equation [1] could have also been chosen.	Substitute $z = \lambda$ and $x = 3 + 7\lambda$ into [2]: $3 + 7\lambda + y + 3\lambda = -1$ $y + 10\lambda + 3 = -1$ $y = -4 - 10\lambda$
3. Write the solution.	$x = 3 + 7\lambda,\ y = -4 - 10\lambda,\ z = \lambda$
c. Interpret the solution.	**c.** This solution describes the line along which the two planes intersect.

TI | THINK

b. 1. On a Calculator page, press MENU, then select:
3: Algebra
7: Solve System of Equations
1: Solve System of Equations . . .
Complete the entry line as:

$$\text{solve}\left(\begin{cases} 2x + y - 4z = 2 \\ x + y + 3z = -1, \{x, y, z\} \\ z = \lambda \end{cases}\right)$$

then press ENTER.
Note: The λ symbol can be found by pressing ctrl+ [book key]

DISPLAY/WRITE

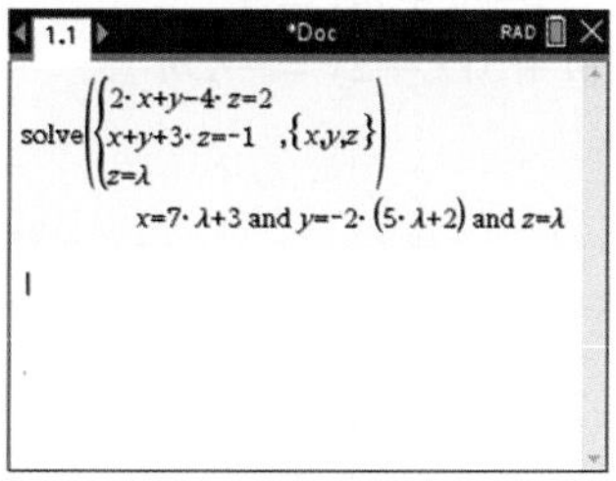

2. The answer appears on the screen.

$x = 7\lambda + 3,$
$y = -2(5\lambda + 2),$
$z = \lambda$

CASIO | THINK

b. 1. On a Main screen, complete the entry line as:
solve $(\{2x + y - 4z = 2,$
$x + y + 3z = -11,$
$z = \lambda\}, \{x, y, z\}$
then press EXE.
Note: The λ symbol can be found in the Keyboard menu by selecting abc then selecting $\alpha\beta\gamma$.

DISPLAY/WRITE

2. The answer appears on the screen.

$x = 7\lambda + 3,$
$y = -10\lambda - 4,$
$z = \lambda$

1.3 Exercise

Students, these questions are even better in jacPLUS

Receive immediate feedback and access sample responses

Access additional questions

Track your results and progress

Find all this and MORE in jacPLUS

Technology free

1. WE6 Determine the value of k for which the following simultaneous equations have a unique solution.

$$\begin{aligned} 2x + ky &= 4 \\ (k-3)x + 2y &= 0 \end{aligned}$$

2. Determine the value of m for which the following simultaneous equations have infinitely many solutions.

$$\begin{aligned} mx - 2y &= 4 \\ x + (m-3)y &= m \end{aligned}$$

3. Determine the value of m for which the following simultaneous equations have no solution.

$$\begin{aligned} x + my &= 3 \\ 4mx + y &= 0 \end{aligned}$$

4. Determine the value of k for which the following simultaneous equations have a unique solution.

$$\begin{aligned} x + 3ky &= 2 \\ (k-1)x - 1 &= -6y \end{aligned}$$

5. Calculate the value of m for which the following simultaneous equations have:
   a. a unique solution
   b. no solution
   c. an infinite number of solutions.

$$\begin{aligned} -2x + my &= 1 \\ (m+3)x - 2y &= -2m \end{aligned}$$

Technology active

6. WE7 Solve the following system of simultaneous equations.

$$\begin{aligned} 2m - 4n - p &= 1 \\ 4m + n + p &= 5 \\ 3m + 3n - 2p &= 22 \end{aligned}$$

7. Solve the following system of simultaneous equations.

$$\begin{aligned} 2d - e - f &= -2 \\ 3d + 2e - f &= 5 \\ d + 3e + 2f &= 11 \end{aligned}$$

8. Solve the following systems of simultaneous equations.

a. $$\begin{aligned} 2x + y - z &= 12 \\ -x - 3y + z &= -13 \\ -4x + 3y - z &= -2 \end{aligned}$$

b. $$\begin{aligned} m + n - p &= 6 \\ 3m + 5n - 2p &= 13 \\ 5m + 4n - 7p &= 34 \end{aligned}$$

c. $$\begin{aligned} u + 2v - 4w &= 23 \\ 3u + 4v - 2w &= 37 \\ 3u + v - 2w &= 19 \end{aligned}$$

d. $$\begin{aligned} a + b + c &= 4 \\ 2a - b + 2c &= 17 \\ -a - 3b + c &= 3 \end{aligned}$$

9. The measure of the largest angle of a triangle is 20° more than the smallest angle, and the sum of the largest and smallest angles is 60° more than the third angle. Calculate the angle sizes of the triangle using simultaneous equations.

10. Solve the following system of simultaneous equations.

$$\begin{aligned} w - 2x + 3y - z &= 10 \\ 2w + x + y + z &= 4 \\ -w + x + 2y - z &= -3 \\ 3w - 2x + y &= 11 \end{aligned}$$

11. Solve the following system of simultaneous equations in terms of a.

$$\begin{aligned} 2x - y + az &= 4 \\ (a + 2)x + y - z &= 2 \\ 6x + (a + 1)y - 2z &= 4 \end{aligned}$$

12. **WE8** The simultaneous equations shown have infinitely many solutions.

$$\begin{aligned} x + 2y + 2z &= 1 \\ 2x - 2y + z &= 2 \end{aligned}$$

a. Eliminate y by adding the second equation to the first equation.
b. Let $z = \lambda$ and solve the equations in terms of λ.
c. Explain what this solution represents.

13. Solve the following pair of simultaneous equations through the use of the parameter λ.

$$\begin{aligned} x + 2y + 4z &= 2 \\ x - y - 3z &= 4 \end{aligned}$$

14. Solve the following pair of simultaneous equations through the use of the parameter λ.

$$\begin{aligned} x + y - 2z &= 5 \\ x - 2y + 4z &= 1 \end{aligned}$$

15. Solve the following pair of simultaneous equations through the use of the parameter λ.

$$\begin{aligned} -2x + y + z &= -2 \\ x - 3z &= 0 \end{aligned}$$

16. Determine the values of m and n for which the equations below have:

a. a unique solution
b. an infinite number of solutions
c. no solution.

$$\begin{aligned} 3x + 2y &= -1 \\ mx + 4y &= n \end{aligned}$$

1.3 Exam questions

Question 1 (1 mark) TECH-ACTIVE

Source: *VCE 2014, Mathematical Methods (CAS) Exam 2, Section 1, Q17; © VCAA.*

MC The simultaneous linear equations $ax - 3y = 5$ and $3x - ay = 8 - a$ have **no solution** for

A. $a = 3$
B. $a = -3$
C. both $a = 3$ and $a = -3$
D. $a \in R \setminus \{3\}$
E. $a \in R \setminus [-3, 3]$

Question 2 (1 mark) TECH-ACTIVE

MC The simultaneous linear equations

$$-2x - my = -4 \text{ and}$$
$$(m-1)x + 6y = 2(m-1),$$

where m is a real constant, have a unique solution for

A. $m = 4$ or $m = -3$
B. $m = 4$ only
C. $m \in R \setminus (4, -3)$
D. $m \neq 4$
E. $m = -3$ only

Question 3 (1 mark) TECH-ACTIVE

MC A unique solution for solving a system of three simultaneous equations in three variables represents

A. three planes intersecting along a line.
B. three planes intersecting along a plane.
C. three planes intersecting at a point.
D. three planes having no common intersection.
E. three planes intersecting at a point, a line or a plane.

More exam questions are available online.

1.4 Quadratic functions

LEARNING INTENTION

At the end of this subtopic you should be able to:
- factorise and solve quadratic functions
- sketch quadratic functions in different forms.

1.4.1 Factorisation

Review of quadratic expressions

The following techniques are used to **factorise** quadratic expressions.
- Perfect squares: $a^2 \pm 2ab + b^2 = (a \pm b)^2$
- Difference of perfect squares: $a^2 - b^2 = (a - b)(a + b)$
- Product and sum (trinomials): To factorise $ax^2 + bx + c$, we look for two numbers that multiply to give ac and add to form the middle term, b.

For example, in the trinomial $x^2 - x - 6$, $ac = -6$ and $b = -1$. The two numbers are -3 and 2.

Therefore, we split the middle term: $x^2 - 3x + 2x - 6$

Then pair and factorise:

$$x^2 - 3x + 2x - 6$$
$$x(x-3) + 2(x-3)$$
$$(x-3)(x+2)$$

- Completing the square: The method of completing the square will work for any **quadratic** that can be factorised.

WORKED EXAMPLE 9 Factorising quadratic expressions

Use an appropriate technique to factorise each of the following quadratic expressions.

a. $9a^2 - 24ab + 16b^2$ **b.** $6x^2 - 17x + 7$ **c.** $2t^2 + 8t - 14$

THINK	WRITE
a. The first and last terms are perfect squares, so check if the expression fits the perfect square formula.	**a.** $9a^2 - 24ab + 16b^2 = (3a)^2 - 2(3a)(4b) + (4b)^2$ $= (3a - 4b)^2$
b. 1. Factorise by product and sum before applying the method of completing the square.	**b.** $6x^2 - 17x + 7$ $ac = 42, b = -17$
2. Find two numbers that have a product of ac and a sum of b.	The two numbers are -14 and -3.
3. Split the middle term.	$6x^2 - 3x - 14x + 7$
4. Pair and factorise.	$6x^2 - 3x - 14x + 7$ $3x(2x-1) - 7(2x-1)$ $(3x-7)(2x-1)$
5. Write the answer.	$6x^2 - 17x + 7 = (3x-7)(2x-1)$
c. 1. Take out the common factor of 2.	**c.** $2t^2 + 8t - 14 = 2\left(t^2 + 4t - 7\right)$
2. Product and sum is not an appropriate method here, as the only factors of 7 are 1 and 7, and these cannot be combined to give a middle coefficient of 4. Thus, the method of completing the square is required.	$= 2\left(t^2 + 4t + (2)^2 - (2)^2 - 7\right)$ $= 2\left((t+2)^2 - 4 - 7\right)$ $= 2\left((t+2)^2 - 11\right)$ $= 2\left(t+2-\sqrt{11}\right)\left(t+2+\sqrt{11}\right)$

on Resources

Interactivities Perfect square form of a quadratic (int-2558)
Completing the square (int-2559)

1.4.2 Solving quadratic equations

Polynomial equations, whether they be quadratics, **cubic** polynomials, **quartic** polynomials or polynomials of a higher degree, can be solved using the **Null Factor Law** (that is, if the product of two numbers is zero, one of them must equal zero).

For example,

$$0 = x^2 - 5x + 6$$

Factorising gives

$$0 = (x - 3)(x - 2)$$

Applying the Null Factor Law,

$$x = 3 \text{ or } x = 2$$

The quadratic formula

Quadratic equations of the form $0 = ax^2 + bx + c$ can also be solved by using the quadratic formula.

Solving quadratic equations

$$x = \frac{-b \pm \sqrt{b^2 - 4ac}}{2a}$$

The discriminant $= \Delta$

$$= b^2 - 4ac$$

If $\Delta > 0$, there are two real solutions to the equation.
If $\Delta = 0$, there is one real solution to the equation.
If $\Delta < 0$, there are no real solutions to the equation.

Generally, we check if the discriminant is a perfect square; if it is not, the quadratic formula is required.

WORKED EXAMPLE 10 Solving quadratic equations

Solve the following equations for x.

a. $6x^2 - 7x - 3 = 0$

b. $x^2 + 8x + 1 = 0$

THINK	WRITE
a. 1. First try to factorise by two brackets.	**a.** $6x^2 - 7x - 3 = 0$
2. Apply the Null Factor Law to solve for x.	$(3x + 1)(2x - 3) = 0$ $x = -\frac{1}{3}, \frac{3}{2}$
b. 1. First try to factorise by two brackets. If this doesn't work, check the discriminant. As the discriminant is not a perfect square, the quadratic formula must be used to solve the equation.	**b.** $\Delta = b^2 - 4ac$ $= 8^2 - 4 \times 1 \times 1$ $= 60$ $\Delta > 0, \therefore 2$ solutions

2. Apply the quadratic formula.

$$x = \frac{-b \pm \sqrt{b^2 - 4ac}}{2a}$$
$$= \frac{-8 \pm \sqrt{60}}{2}$$
$$= \frac{-8 \pm 2\sqrt{15}}{2}$$

3. Write the answer.

$$x = -4 \pm \sqrt{15}$$

TI | THINK	DISPLAY/WRITE	CASIO | THINK	DISPLAY/WRITE
b. 1. On a Calculator page, press MENU, then select: 3: Algebra 1: Solve Complete the entry line as: solve $(x^2 + 8x + 1 = 0, x)$ then press ENTER.	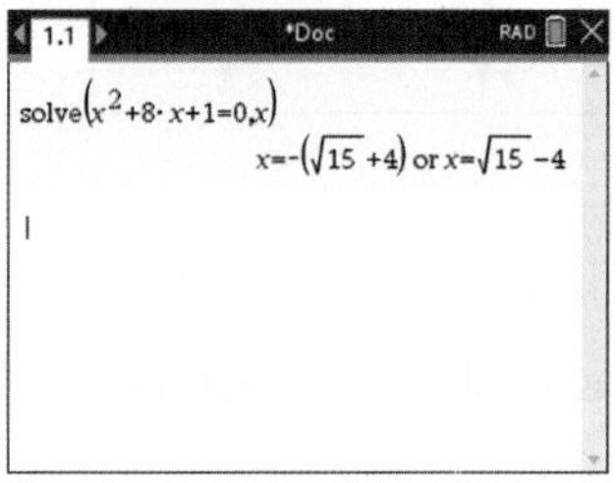	**b. 1.** On a Main screen, complete the entry line as: solve $(x^2 + 8x + 1 = 0, x)$ then press EXE.	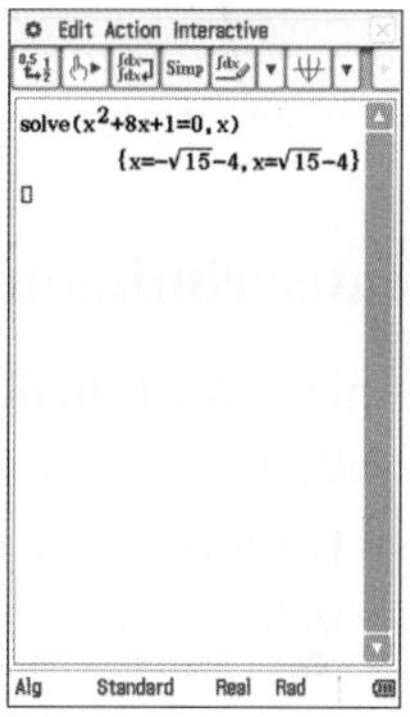
2. The answer appears on the screen.	$x = -\sqrt{15} - 4, x = \sqrt{15} - 4$	**2.** The answer appears on the screen.	$x = -\sqrt{15} - 4, x = \sqrt{15} - 4$

on Resources

Interactivities The discriminant (int-2560)
The quadratic formula (int-2561)

1.4.3 Graphing quadratic functions

The function $f: R \to R, f(x) = ax^2 + bx + c$, where $a, b, c \in R$ and $a \neq 0$, is the quadratic polynomial function. If $a > 0$, the graph of the function is a concave-up parabola with a minimum turning point; if $a < 0$, the graph of the function is a concave-down parabola with a maximum turning point.

General form, $y = ax^2 + bx + c$

As the x-intercepts of the graph of $y = ax^2 + bx + c$ are the roots of the quadratic equation $ax^2 + bx + c = 0$, there may be zero, one or two x-intercepts as determined by the **discriminant** $\Delta = b^2 - 4ac$.

	$\Delta < 0$	$\Delta = 0$	$\Delta > 0$
$a > 0$	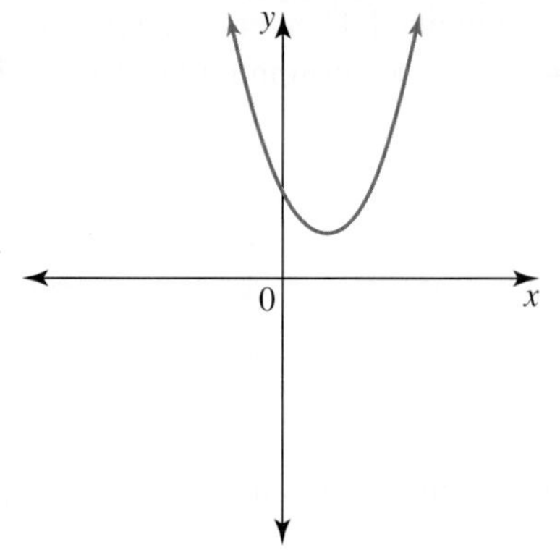	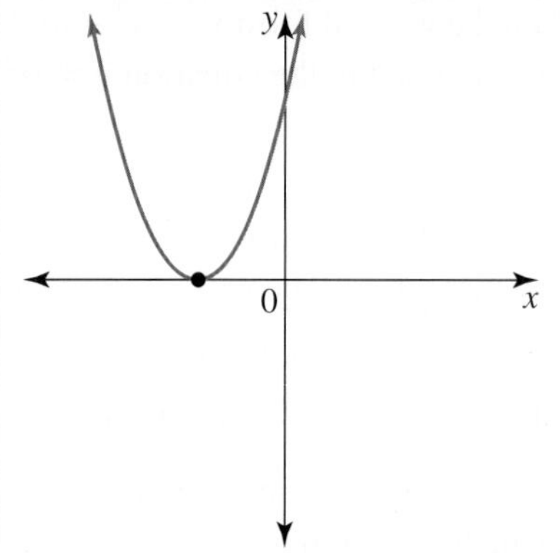	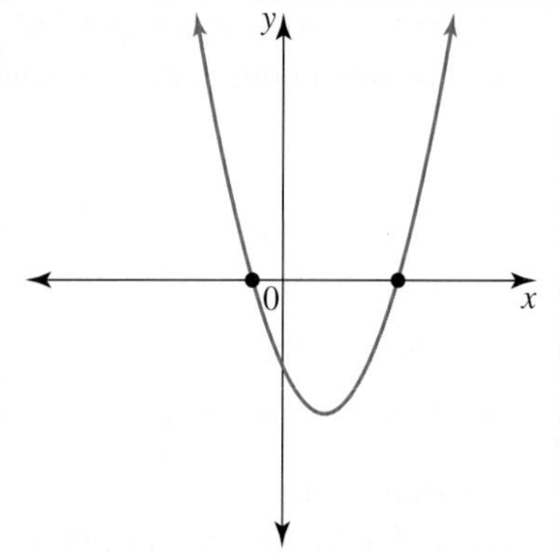

(continued)

(continued)

	$\Delta < 0$	$\Delta = 0$	$\Delta > 0$
$a < 0$	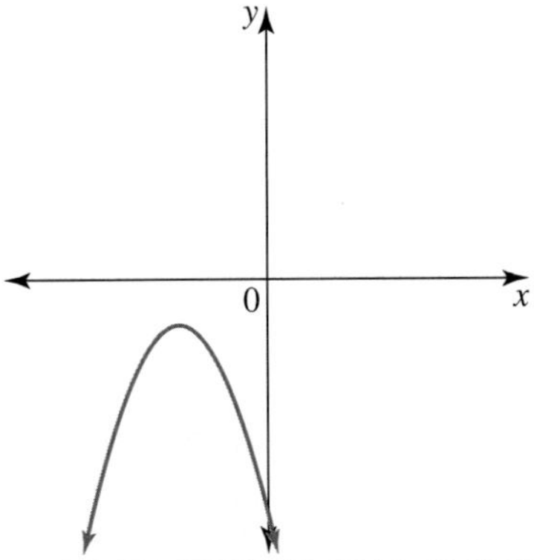	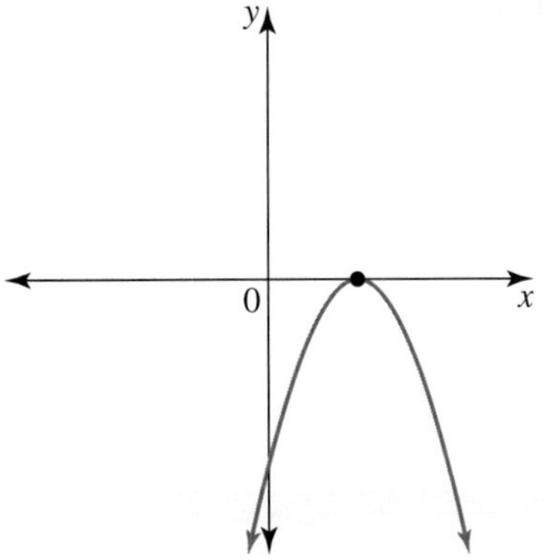	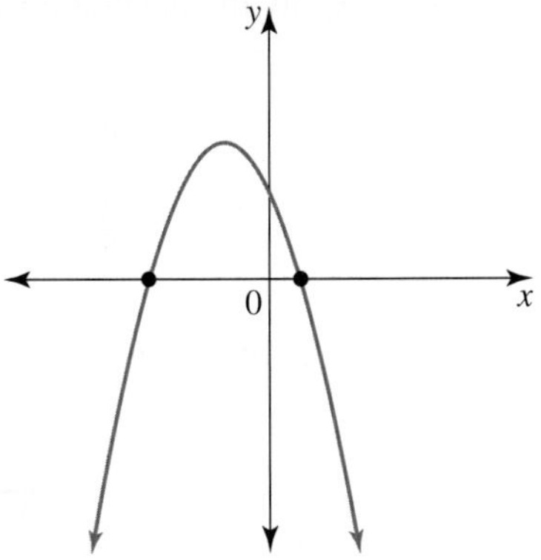

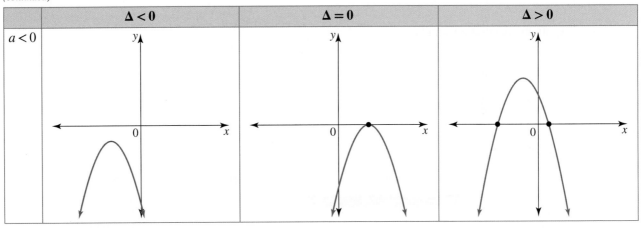

The discriminant

The quadratic function is either positive or negative, depending whether $a > 0$ or $a < 0$ respectively.
If $\Delta < 0$, there are no x-intercepts.
If $\Delta = 0$, there is one x-intercept, a turning point where the graph touches the x-axis.
If $\Delta > 0$, there are two distinct x-intercepts and the graph crosses the x-axis at these places.

Axis of symmetry

As the roots of the quadratic equation are given by $x = \dfrac{-b \pm \sqrt{b^2 - 4ac}}{2a}$, the axis of symmetry of the parabola has the equation $x = \dfrac{-b}{2a}$.

This is also the x-coordinate of the turning point, so by substituting this value into the parabola's equation, the y-coordinate of the turning point can be calculated.

Turning point form, $y = a(x - h)^2 + k$

The simplest parabola has the equation $y = x^2$. Its turning point is the origin, $(0, 0)$, which is unaltered by a **dilation** from the x-axis in the y-direction. However, if the graph of this parabola undergoes a horizontal translation of h units and a vertical translation of k units, the turning point moves to the point (h, k).

Thus, $y = a(x - h)^2 + k$ is the equation of a parabola with turning point (h, k) and axis of symmetry $x = h$.

If $y = a(x - h)^2 + k$ is expanded, then the general form $y = ax^2 + bx + c$ is obtained. Conversely, when the technique of completing the square is applied to the equation $y = ax^2 + bx + c$, the turning point form is obtained.

x-intercept form, $y = a(x - x_1)(x - x_2)$

When the equation of a quadratic function is expressed as the product of its two linear factors, the x-intercepts at $x = x_1$ and $x = x_2$ can be obtained by inspection. The axis of symmetry lies midway between the intercepts, so the equation for this axis must be $x = \dfrac{x_1 + x_2}{2}$, and this gives the x-coordinate of the turning point. The y-coordinate of the turning point can be calculated from the equation once the x-coordinate is known.

Expanding the equation $y = a(x - x_1)(x - x_2)$ will return it to general form, and factorising the general equation $y = ax^2 + bx + c$ will convert it to x-intercept form.

Key features of the graph of a quadratic function

When sketching the graph of a parabola by hand, identify:

- **the y-intercept**
- **any x-intercepts**
- **the turning point**
- **the axis of symmetry, if it is helpful to the sketch**
- **any end-point coordinates if the function is given on a restricted domain.**

The methods used to identify the key features of a graph will depend on the form in which the equation of the graph is expressed.

Similarly, when determining the equation of a parabola given a key feature, you should select the form of the equation that emphasises that key feature.

- If the turning point is given, use the $y = a(x-h)^2 + k$ form.
- If the x-intercepts are given, use the $y = a(x-x_1)(x-x_2)$ form.
- Otherwise, use the $y = ax^2 + bx + c$ form.

Three pieces of information are always required to determine the equation, as each form involves 3 constants or parameters.

WORKED EXAMPLE 11 Sketching a quadratic graph

Sketch the graph of $y = 9 - (2x+1)^2$ and state its domain and range.

THINK	WRITE
1. Rewrite the equation so it is in a standard form (turning point form).	$y = 9 - (2x+1)^2$ $y = -(2x+1)^2 + 9$ or $y = -\left(2\left(x+\frac{1}{2}\right)\right)^2 + 9$ $y = -4\left(x+\frac{1}{2}\right)^2 + 9$
2. State the coordinates and type of turning point.	The graph has a maximum turning point at $\left(-\frac{1}{2}, 9\right)$.
3. Calculate the y-intercept.	y-intercept: let $x = 0$. $y = 9 - (1)^2$ $y = 8$ The y-intercept is $(0, 8)$.
4. Calculate any x-intercepts.	As the graph has a maximum turning point with a positive y-value, there will be x-intercepts. Let $y = 0$. $9 - (2x+1)^2 = 0$ $(2x+1)^2 = 9$ $2x + 1 = \pm 3$ $2x = -4$ or 2 $x = -2$ or 1 The x-intercepts are $(-2, 0)$ and $(1, 0)$.

5. Sketch the graph.

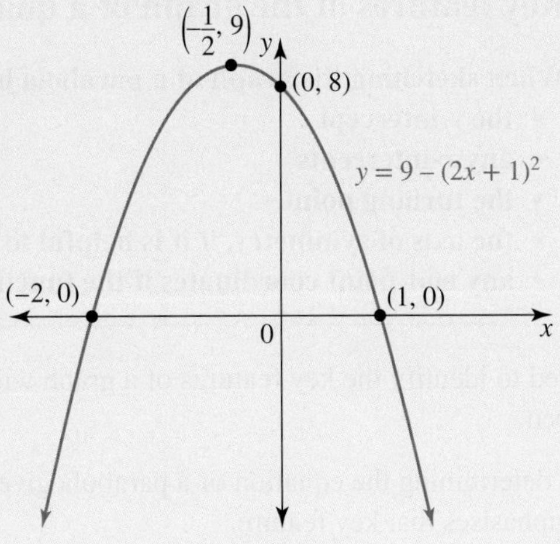

6. State the domain and range. The domain is R and the range is $(-\infty, 9]$.

TI | THINK — **DISPLAY/WRITE**

1. On a Graphs page, complete the entry line for function 1 as: $f1(x) = 9 - (2x + 1)^2$ then press ENTER.

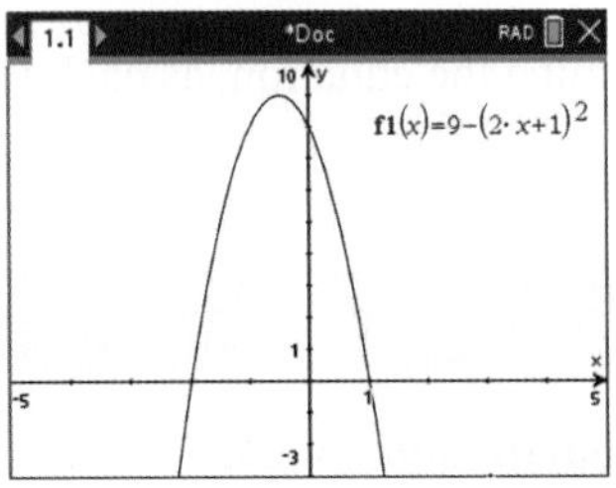

2. To find the x-intercepts, press MENU, then select:
6: Analyze Graph
1: Zero
Move the cursor to the left of the x-intercept when prompted for the lower bound, then press ENTER. Move the cursor to the right of the x-intercept when prompted for the upper bound, then press ENTER.
Repeat this step to find the other x-intercept.

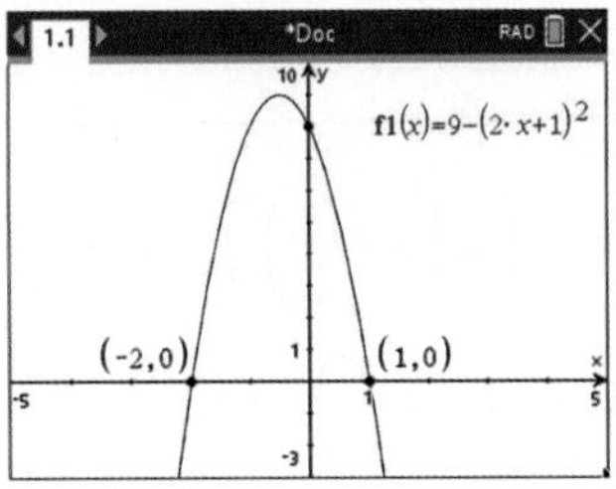

3. To find the y-intercept, press MENU, then select:
5: Trace
1: Graph Trace
Type 0, then press ENTER twice.

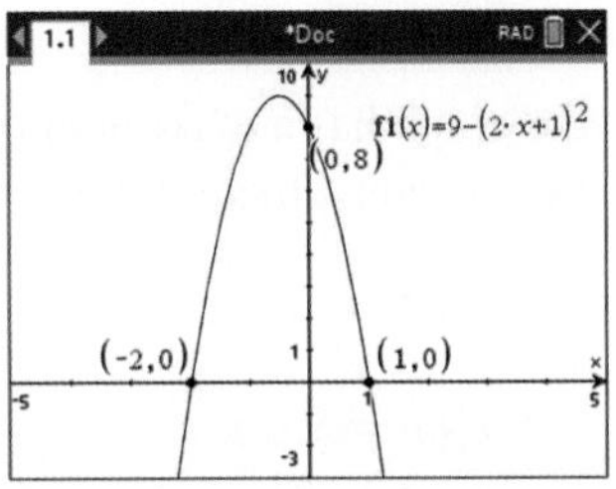

CASIO | THINK — **DISPLAY/WRITE**

1. On a Graph & Table screen, complete the entry line for y1 as:
$y1 = 9 - (2x + 1)^2$
then press EXE.
Select the Graph icon to draw the graph.

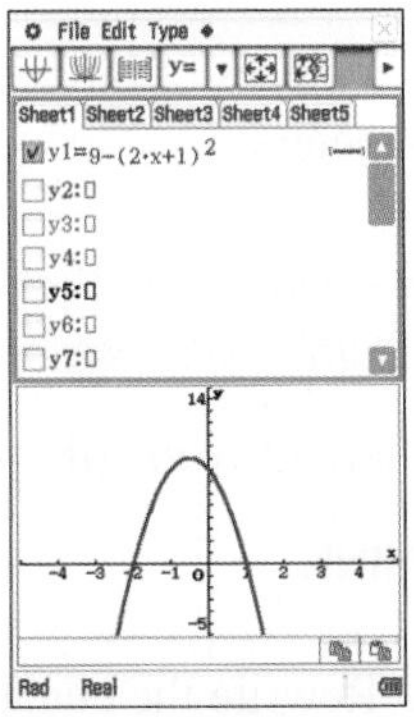

2. To find the x-intercepts, select:
- Analysis
- G-Solve
- Root

With the cursor on the first x-intercept, press EXE. Use the left/right arrows to move to the other x-intercept, then press EXE.

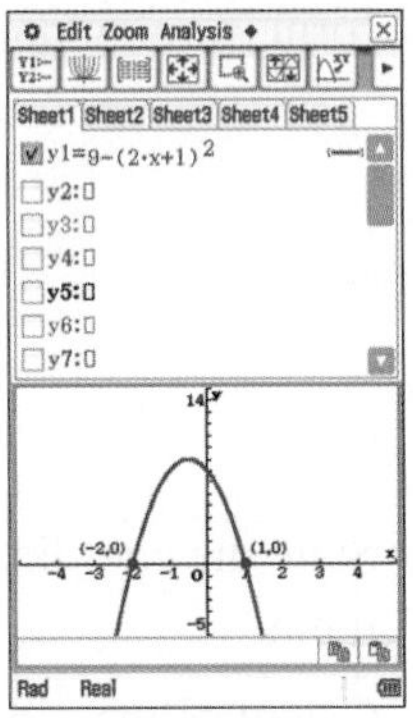

3. To find the y-intercept, select:
- Analysis
- G-Solve
- Y-Intercept

then press EXE.

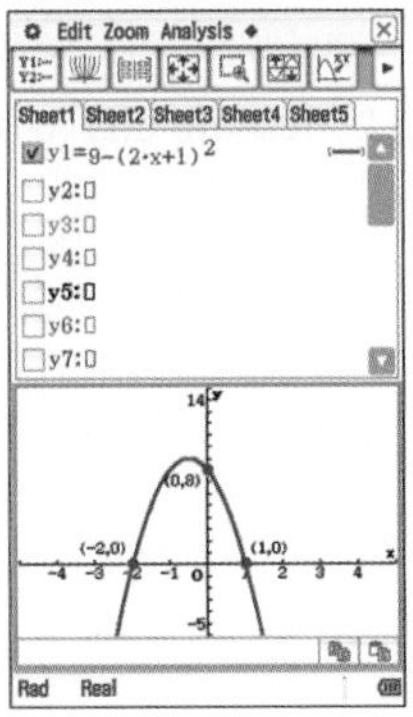

4. To find the maximum, press MENU, then select:
6: Analyze Graph
3: Maximum
Move the cursor to the left of the maximum when prompted for the lower bound, then press ENTER. Move the cursor to the right of the maximum when prompted for the upper bound, then press ENTER.

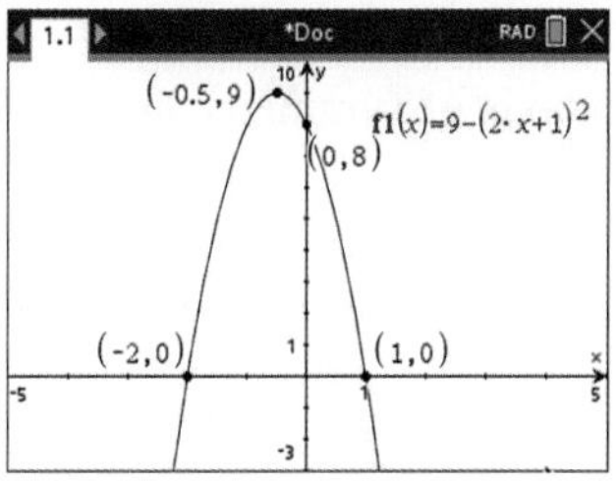

5. The domain and range can be read from the graph.

The domain is R and the range is $(-\infty, 9]$.

4. To find the maximum, select:
- Analysis
- G-Solve
- Max

then press EXE.

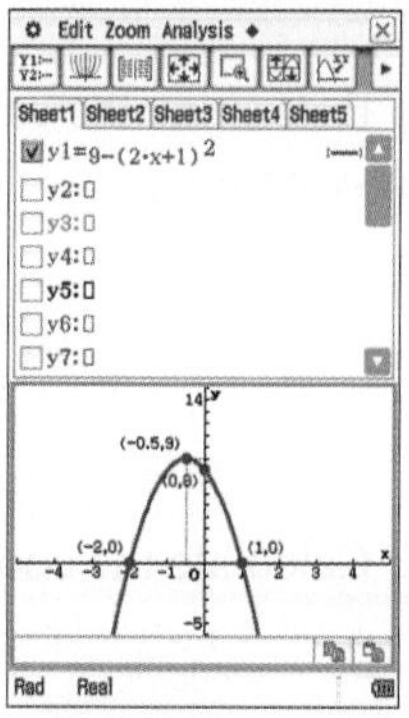

5. The domain and range can be read from the graph.

The domain is R and the range is $(-\infty, 9]$.

WORKED EXAMPLE 12 Determining the equation of a quadratic graph

Determine the equation of the given graph and hence obtain the coordinates of the turning point.

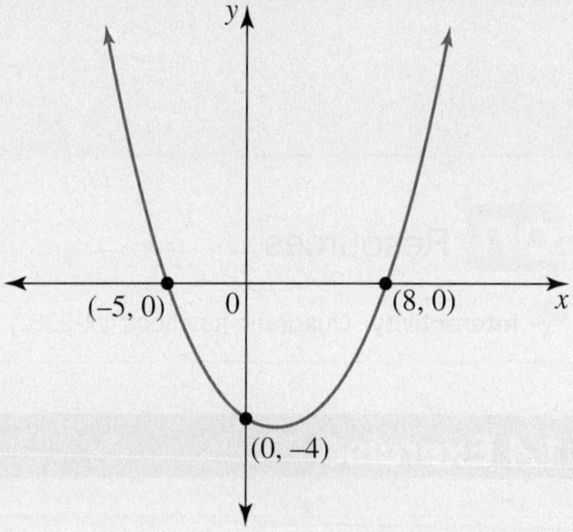

THINK	WRITE
1. Select a form of the equation.	As the two x-intercepts are known, the x-intercept form of the equation will be used. $y = a(x - x_1)(x - x_2)$
2. Use the key features to partially determine the equation.	There is an x-intercept at $x = -5$. $\Rightarrow (x + 5)$ is a factor. There is an x-intercept at $x = 8$. $\Rightarrow (x - 8)$ is a factor. The equation is $y = a(x + 5)(x - 8)$.
3. Use the third piece of information to fully determine the equation.	The point $(0, -4)$ lies on the graph. Substitute this point in $y = a(x + 5)(x - 8)$. $-4 = a(5)(-8)$ $-4 = -40a$ $a = \frac{1}{10}$ The equation is $y = \frac{1}{10}(x + 5)(x - 8)$.

4. Determine the equation of the axis of symmetry.	The axis of symmetry lies midway between the x-intercepts. $\therefore x = \dfrac{-5+8}{2}$ $= \dfrac{3}{2}$
5. Calculate the coordinates of the turning point.	The x-coordinate of the turning point is $\dfrac{3}{2}$. Substitute $x = \dfrac{3}{2}$ in the equation of the graph. $y = \dfrac{1}{10}\left(\dfrac{3}{2}+5\right)\left(\dfrac{3}{2}-8\right)$ $y = \dfrac{1}{10} \times \dfrac{13}{2} \times \dfrac{-13}{2}$ $y = -\dfrac{169}{40}$ The turning point is $\left(\dfrac{3}{2}, -\dfrac{169}{40}\right)$.

on Resources

Interactivity Quadratic functions (int-2562)

1.4 Exercise

Students, these questions are even better in jacPLUS

Receive immediate feedback and access sample responses

Access additional questions

Track your results and progress

Find all this and MORE in jacPLUS

Technology free

1. **WE9** Use an appropriate technique to factorise each of the following quadratic expressions.
   a. $15u^2 - u - 2$ b. $6d^2 - 28d + 16$ c. $3j^2 + 12j - 6$ d. $b^2 - 1$
2. Use an appropriate method to factorise each of the following quadratic expressions.
   a. $f^2 - 12f - 28$ b. $g^2 + 3g - 4$
3. **WE10** Solve the following equations for x.
   a. $8x^2 + 2x - 3 = 0$ b. $2x^2 - 4x + 1 = 0$
4. Fully factorise and solve the following quadratic equations over R.
   a. $81y^2 = 1$ b. $4z^2 + 28z + 49 = 0$ c. $5m^2 + 3 = 10m$ d. $x^2 - 4x = -3$
5. Fully factorise and solve the following quadratic equations over R.
   a. $48p = 24p^2 + 18$ b. $39k = 4k^2 + 77$ c. $m^2 + 3m = 4$ d. $4n^2 = 8 - 5n$
6. **WE11** Sketch the graph of $y = 2(3x-2)^2 - 8$ and state its domain and range.

7. a. Given the function $f: (-2, 2] \to R, f(x) = 3(1-x)^2 + 2$, state the range of the graph.
   b. Sketch the graphs of the following quadratic functions. State the range for each one.
      i. $y = (x-2)(2x+3),\ x \in [-2, 3]$
      ii. $y = -x^2 + 4x + 2,\ x \in R$
      iii. $y = -2(x+1)^2 - 3,\ x \in (-3, 0]$
      iv. $y = \frac{1}{2}(2x-3)^2 - 1,\ x \in R$

Technology active

8. Consider the quadratic function $f: R^+ \cup \{0\} \to R, f(x) = 4x^2 - 8x + 7$.
   a. Determine the number of intercepts the graph of $y = f(x)$ makes with the x-axis.
   b. Express the equation of the function in the form $f(x) = a(x+b)^2 + c$.
   c. Sketch the graph of $y = f(x)$ and state its domain and range.

9. WE12 Determine the equation of the given graph and hence obtain the coordinates of the turning point.

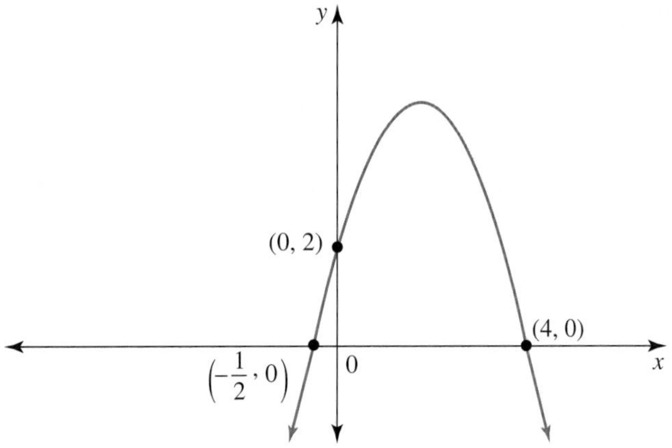

10. Determine the equations of the following quadratic functions.
   a. The turning point has coordinates $(-6, 12)$ and the graph of the function passes through the point $(4, -3)$.
   b. The points $(-7, 0)$, $(0, -20)$ and $\left(-2\frac{1}{2}, 0\right)$ lie on the graph.
   c. The minimum value of the function is -5 and it contains the points $(-8, 11)$ and $(8, 11)$

11. a. Express $-x^2 + 2x - 5$ in the form $a(x+b)^2 + c$.
   b. Hence, state the coordinates of the turning point of the graph of $y = -x^2 + 2x - 5$.
   c. Sketch the graph of $y = -x^2 + 2x - 5$ and state its range.
   d. Use a graphical method to show that the graphs of $y = x + 3$ and $y = -x^2 + 2x - 5$ never intersect.
   e. Determine the value of k so that the graphs of $y = x + k$ and $y = -x^2 + 2x - 5$ will touch exactly once.

12. Determine the possible values of k such that the graphs of $y = 2x^2$ and $y = kx - 2$ do not intersect.

13. Determine the values of m for which the line $y = -3 - 2mx$ intersects the parabola $y = x^2 - 1$ twice.

14. Show that $y = 2x - 3$ is a tangent to the graph $y = x^2 - 2$.

15. a. Find the values of k for which the equation $kx^2 - 3x + k = 0$ has no solutions.
   b. If $kx^2 + 4x - k + 2 = 0$, show that the equation has a solution for all values of k.

16. A quadratic equation has the rule $(m-1)x^2 + \left(\frac{5-2m}{2}\right)x + 2m = 0$. Find the value(s) of m for which the quadratic equation has two solutions.

1.4 Exam questions

Question 1 (1 mark) TECH-ACTIVE

Source: *VCE 2019, Mathematical Methods Exam 2, Section A, Q2; © VCAA.*

MC The set of values of k for which $x^2 + 2x - k = 0$ has two real solutions is

A. $\{-1, 1\}$
B. $(-1, \infty)$
C. $(-\infty, -1)$
D. $\{-1\}$
E. $[-1, \infty)$

Question 2 (1 mark) TECH-ACTIVE

Source: *VCE 2018, Mathematical Methods Exam 2, Section A, Q17; © VCAA.*

MC The turning point of the parabola $y = x^2 - 2bx + 1$ is closest to the origin when

A. $b = 0$
B. $b = -1$ or $b = 1$
C. $b = -\frac{1}{\sqrt{2}}$ or $b = \frac{1}{\sqrt{2}}$
D. $b = \frac{1}{2}$ or $b = -\frac{1}{2}$
E. $b = \frac{1}{4}$ or $b = -\frac{1}{4}$

Question 3 (1 mark) TECH-ACTIVE

Source: *VCE 2015, Mathematical Methods (CAS) Exam 2, Section 1, Q21; © VCAA.*

MC The graphs of $y = mx + c$ and $y = ax^2$ will have no points of intersection for all values of m, c and a such that

A. $a > 0$ and $c > 0$
B. $a > 0$ and $c < 0$
C. $a > 0$ and $c > -\frac{m^2}{4a}$
D. $a < 0$ and $c > -\frac{m^2}{4a}$
E. $m > 0$ and $c > 0$

More exam questions are available online.

1.5 Cubic functions

LEARNING INTENTION

At the end of this subtopic you should be able to:

- factorise and solve cubic equations
- sketch cubic functions in different forms.

1.5.1 Factorisation

For polynomials of degree 3, it is necessary to remember the perfect cube patterns as well as the sum and difference of two cubes.

Cubic factorisations

Perfect cubes:

$$a^3 + 3a^2b + 3ab^2 + b^3 = (a + b)^3$$
$$a^3 - 3a^2b + 3ab^2 - b^3 = (a - b)^3$$

Sum and difference of two cubes:

$$a^3 + b^3 = (a + b)(a^2 - ab + b^2)$$
$$a^3 - b^3 = (a - b)(a^2 + ab + b^2)$$

WORKED EXAMPLE 13 Factorising cubic expressions

Use an appropriate method to factorise each of the following cubic expressions.

a. $27y^3 - 27y^2 + 9y - 1$
b. $x^3 + 8$
c. $3y^3 - 81$
d. $8m^3 + 60m^2 + 150m + 125$

THINK	WRITE
a. This is a perfect cube pattern. Check to see that it has the pattern of $a^3 - 3a^2b + 3ab^2 - b^3 = (a - b)^3$.	a. $27y^3 - 27y^2 + 9y - 1$ $= (3y)^3 - 3(3y)^2(1) + 3(3y)(1)^2 - (1)^3$ $= (3y - 1)^3$
b. This is a sum of two cubes pattern.	b. $x^3 + 8 = x^3 + 2^3$ $= (x + 2)\left(x^2 - 2x + 2^2\right)$ $= (x + 2)\left(x^2 - 2x + 4\right)$
c. 1. Take out a common factor first.	c. $3y^3 - 81 = 3\left(y^3 - 27\right)$
2. Now factorise using the difference of two cubes pattern.	$= 3\left(y^3 - 3^3\right)$ $= 3(y - 3)\left(y^2 + 3y + 3^2\right)$ $= 3(y - 3)\left(y^2 + 3y + 9\right)$
d. This is a perfect cube pattern. Check to see that it has the pattern of $a^3 + 3a^2b + 3ab^2 + b^3 = (a + b)^3$.	d. $8m^3 + 60m^2 + 150m + 125$ $= (2m)^3 + 3(2m)^2(5) + 3(2m)(5)^2 + (5)^3$ $= (2m + 5)^3$

Factorising cubics using the technique of grouping

To factorise $x^3 - 3x^2 + 4x - 12$, start by grouping two groups of two terms in the following manner.

$$\underbrace{x^3 - 3x^2}_{\text{TWO}} + \underbrace{4x - 12}_{\text{TWO}}$$
$$= x^2(x - 3) + 4(x - 3)$$
$$= (x - 3)(x^2 + 4)$$

The other possible method is to group three and one. For example, to factorise $x^2 - z^2 + 4x + 4$, rearrange the expression as shown.

$$x^2 + 4x + 4 - z^2$$
$$\underbrace{x^2 + 4x + 4}_{\text{THREE}} \underbrace{- z^2}_{\text{ONE}}$$
$$= (x + 2)^2 - z^2$$
$$= (x + 2 - z)(x + 2 + z)$$

WORKED EXAMPLE 14 Factorising cubic expressions

Fully factorise the following polynomials.

a. $m^2 - n^2 - 36 - 12n$

b. $p^3 + 2p^2 - 4p - 8$

THINK	WRITE
a. 1. Group the polynomial one and three and rearrange, taking out −1 as a common factor.	a. $m^2 - n^2 - 36 - 12n$ $= m^2 - n^2 - 12n - 36$ $= \underbrace{m^2}_{\text{ONE}} - \underbrace{(n^2 + 12n + 36)}_{\text{THREE}}$
2. Factorise the group of three terms as a perfect square.	$= m^2 - (n+6)^2$
3. Apply the difference of perfect squares method.	$= (m - (n+6))(m + (n+6))$ $= (m - n - 6)(m + n + 6)$
b. 1. Group the polynomial two and two.	b. $\underbrace{p^3 + 2p^2}_{\text{TWO}} \underbrace{- 4p - 8}_{\text{TWO}}$
2. Factorise each pair.	$p^3 + 2p^2 - 4p - 8$ $= p^2(p+2) - 4(p+2)$ $= (p+2)(p^2 - 4)$
3. Finish the factorisation by applying the difference of perfect squares method.	$= (p+2)(p-2)(p+2)$ $= (p+2)^2(p-2)$

Factor theorem

When the previous methods are not appropriate for a third degree polynomial or a higher degree polynomial, then knowledge of the **factor theorem** is essential. The factor theorem is an algebraic theorem that links the zeros of a polynomial. It states the following:

> **Factor theorem**
>
> **A polynomial, $P(x)$, has a factor $(x - a)$ if and only if $P(a) = 0$; that is, if a is a root of the polynomial.**

Consider the factorisation of $x^3 + 3x^2 - 13x - 15$.

Let $P(x) = x^3 + 3x^2 - 13x - 15$.

By substituting integer values of x that are factors of the constant term, we aim to achieve a zero remainder, that is, to achieve $P(x) = 0$. If this is so, we have found one linear factor of the polynomial.

$$\begin{aligned} P(1) &= 1^3 + 3(1)^2 - 13(1) - 15 \\ &= 1 + 3 - 13 - 15 \\ &\neq 0 \end{aligned}$$

$$\begin{aligned} P(-1) &= (-1)^3 + 3(-1)^2 - 13(-1) - 15 \\ &= -1 + 3 + 13 - 15 \\ &= 0 \end{aligned}$$

Thus, $(x+1)$ is a factor. The quadratic factor can then be found by long division or by inspection.

$$
\begin{array}{r}
x^2 + 2x - 15 \\
x+1 \overline{\smash{\big)}\, x^3 + 3x^2 - 13x - 15} \\
-\underline{(x^3 + x^2)} \\
2x^2 - 13x - 15 \\
-\underline{(2x^2 + 2x)} \\
-15x - 15 \\
-\underline{(-15x - 15)} \\
\underline{0}
\end{array}
$$

$$-15x + 2x$$

$$\text{or } x^3 + 3x^2 - 13x - 15 = (x+1)(x^2 + 2x - 15)$$

$$2x^2 + x^2$$

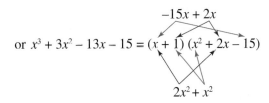

Completing the factorisation gives:

$$
\begin{aligned}
x^3 + 3x^2 - 13x - 15 &= (x+1)\left(x^2 + 2x - 15\right) \\
&= (x+1)(x+5)(x-3)
\end{aligned}
$$

WORKED EXAMPLE 15 Factorising a cubic expression

Fully factorise $2x^3 - 3x^2 - 8x - 3$.

THINK	WRITE
1. Let $P(x)$ equal the cubic polynomial.	$P(x) = 2x^3 - 3x^2 - 8x - 3$
2. Try $P(1)$, $P(-1)$, $P(2)$ and so on to get a zero remainder.	$P(1) = 2 - 3 - 8 - 3 \neq 0$ $P(-1) = -2 - 3 + 8 - 3 = 0$ Therefore, $x + 1$ is a factor.
3. Use long division to obtain a quadratic factor.	$\begin{array}{r} 2x^2 - 5x - 3 \\ x+1 \overline{\smash{\big)}\, 2x^3 - 3x^2 - 8x - 3} \\ -\underline{(2x^3 + 2x^2)} \\ -5x^2 - 8x \\ -\underline{(-5x^2 - 5x)} \\ -3x - 3 \\ -\underline{(-3x - 3)} \\ \underline{0} \end{array}$
4. Complete the cubic factorisation by factorising the quadratic factor into its two linear factors.	$P(x) = 2x^3 - 3x^2 - 8x - 3$ $= (x+1)\left(2x^2 - 5x - 3\right)$ $= (x+1)(2x+1)(x-3)$

TI | THINK	DISPLAY/WRITE
1. On a Calculator page, press MENU, then select: 3: Algebra 2: Factor Complete the entry line as: factor $\left(2x^3-3x^2-8x-3\right)$ then press ENTER.	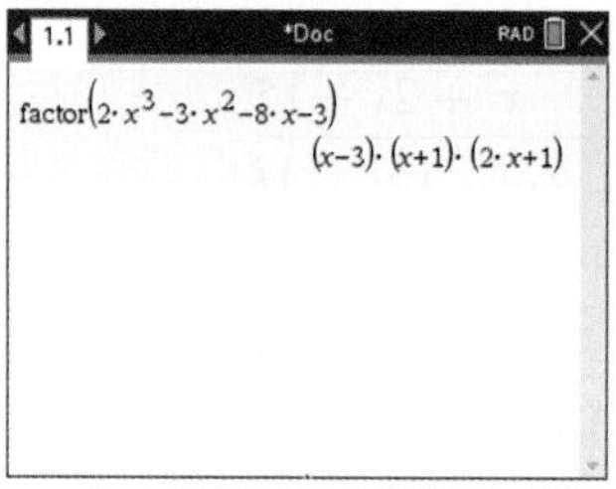
2. The answer appears on the screen.	$2x^3-3x^2-8x-3=$ $(x-3)(x+1)(2x+1)$

CASIO | THINK	DISPLAY/WRITE
1. On a Main screen, select: • Action • Transformation • factor • factor Complete the entry line as: factor $\left(2x^3-3x^2-8x-3\right)$ then press EXE.	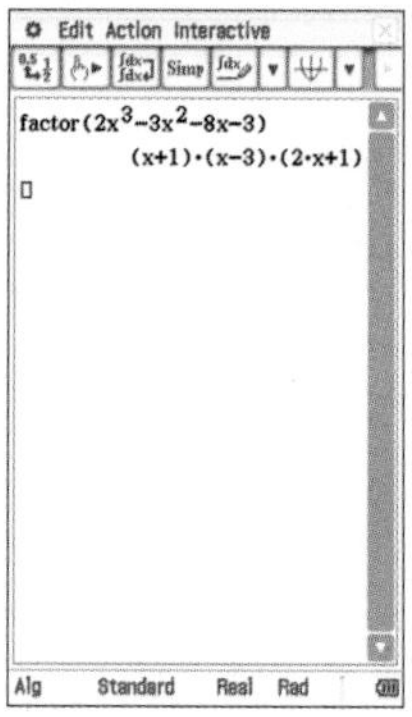
2. The answer appears on the screen.	$2x^3-3x^2-8x-3=$ $(x+1)(x-3)(2x+1)$

on Resources

Interactivities Long division of polynomials (int-2564)
The remainder and factor theorems (int-2565)

1.5.2 Solving polynomial equations

Consider again the cubic polynomial $x^3+3x^2-13x-15$.

$$\begin{aligned} x^3+3x^2-13x-15 &= (x+1)\left(x^2+2x-15\right) \\ &= (x+1)(x+5)(x-3) \end{aligned}$$

We will equate it to zero so that we have a cubic equation to solve.

$$\begin{aligned} x^3+3x^2-13x-15 &= 0 \\ (x+1)\left(x^2+2x-15\right) &= 0 \\ (x+1)(x+5)(x-3) &= 0 \end{aligned}$$

Applying the Null Factor Law, $x=-1$, $x=-5$ or $x=3$.

WORKED EXAMPLE 16 Solving a cubic equation

Solve $2x^3-3x^2-8x-3=0$ for x.

THINK	WRITE
1. The cubic expression was factorised in Worked example 15.	$0=2x^3-3x^2-8x-3$ $=(x+1)\left(2x^2-5x-3\right)$ $=(x+1)(2x+1)(x-3)$
2. Use the Null Factor Law to solve the cubic equation for x.	$x=-1,-\frac{1}{2},3$

Equality of polynomials

Two polynomials, $P(x)$ and $Q(x)$, are such that:

$P(x)=a_nx^n+a_{n-1}x^{n-1}+a_{n-2}x^{n-2}+\ldots+a_2x^2+a_1x+a_0$ and
$Q(x)=b_nx^n+b_{n-1}x^{n-1}+b_{n-2}x^{n-2}+\ldots+b_2x^2+b_1x+b_0$.

$P(x)$ is identically equal to $Q(x)$ for all values of x (that is, $P(x) \equiv Q(x)$) if and only if:

$$a_n = b_n, a_{n-1} = b_{n-1}, a_{n-2} = b_{n-2} ..., a_2 = b_2, a_1 = b_1 \text{ and } a_0 = b_0.$$

For instance, $3x^3 + (m-2)x^2 + (m+n)x \equiv kx^3 + x^2$. You are required to find the values of m, n and k. As the polynomials are equal to each other, we can equate coefficients to give:

$$\begin{array}{lll} k = 3 & m - 2 = 1 & m + n = 0 \\ & m = 3 & 3 + n = 0 \\ & & n = -3 \end{array}$$

WORKED EXAMPLE 17 Equating coefficients

If $(n-3)x^3 + (2n+p)x^2 + (p+q)x \equiv -2x^3 + x^2$, find the values of n, p and q.

THINK	WRITE
1. Equate the coefficients of the x^3 terms and solve for n.	$n - 3 = -2$ $n = 1$
2. Equate the coefficients of the x^2 terms and solve for p.	$2n + p = 1$ $2 + p = 1$ $p = -1$
3. Equate the coefficients of the x terms and solve for q. *Note:* As there is no x term on the right-hand side, the coefficient is zero.	$p + q = 0$ $-1 + q = 0$ $q = 1$
4. Write the answer.	$n = 1, p = -1, q = 1$

1.5.3 Graphing cubic functions

The function $f: R \to R, f(x) = ax^3 + bx^2 + cx + d, a, b, c, d \in R, a \neq 0$ is the cubic polynomial function. Although the shape of its graph may take several forms, for its maximal domain the function has a range of R. Its long-term behaviour is dependent on the sign of the coefficient of the x^3 term.

If $a > 0$, then as $x \to \infty, y \to \infty$ and as $x \to -\infty, y \to -\infty$.

If $a < 0$, then as $x \to \infty, y \to -\infty$ and as $x \to -\infty, y \to \infty$.

This behaviour is illustrated in the graph of $y = x^3$, the simplest cubic function, and that of $y = -x^3$.

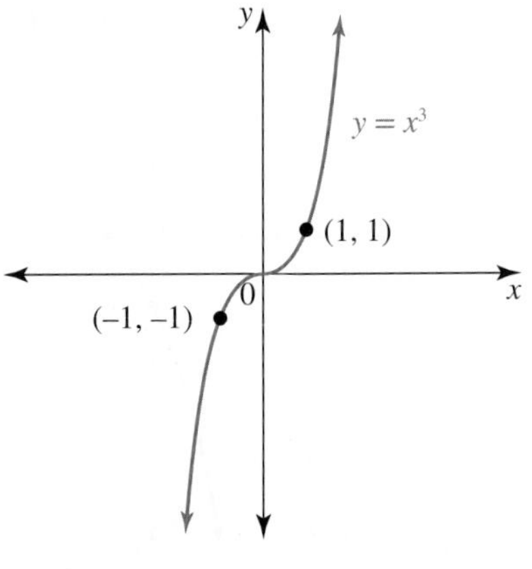

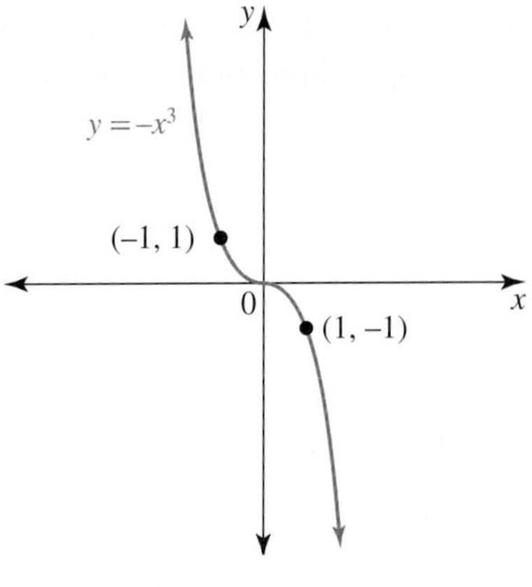

Cubic functions of the form $y = a(x - h)^3 + k$

A significant feature of the graphs of $y = x^3$ and $y = -x^3$ is the stationary point of inflection at the origin. This point is constant under a dilation but becomes the point (h, k) following a horizontal and vertical translation of h and k units respectively.

> **Cubic functions of the form $y = a(x - h)^3 + k$**
>
> **These cubic functions have:**
> - **a stationary point of inflection at (h, k)**
> - **one x-intercept**
> - **long-term behaviour dependent on the sign of a.**

The coordinates of the stationary point of inflection are read from the equation in exactly the same way the turning points of a parabola are read from its equation in turning point form.

Cubic functions expressed in factorised form

A cubic function may have one, two or three x-intercepts, and hence its equation may have up to three linear factors. When the equation can be expressed as the product of linear factors, we can readily deduce the behaviour of the function and sketch its graph without finding the positions of any turning points. Unlike a quadratic function, the turning points in a cubic function are not symmetrically placed between pairs of x-intercepts.

> **Cubic functions — factorised form**
>
> - **If there are three linear factors, that is $y = (x - m)(x - n)(x - p)$, the graph cuts the x-axis at $x = m$, $x = n$ and $x = p$.**
> - **If there is one factor of multiplicity 2 and one other linear factor, that is $y = (x - m)^2(x - n)$, the graph touches the x-axis at a turning point at $x = m$ and cuts the x-axis at $x = n$.**

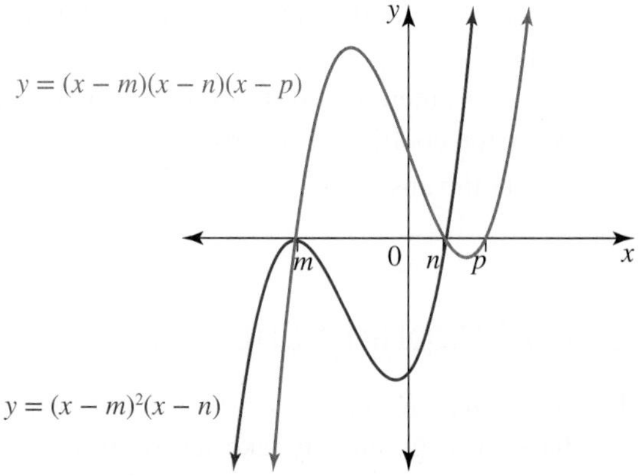

If the equation of the cubic function has one linear factor and one irreducible quadratic factor, it is difficult to deduce its behaviour without either technology or calculus. For example, the following diagram shows the graphs of $y = (x + 3)\left(x^2 + 1\right)$ and $y = \left(x^2 + 3\right)(x - 1)$.

The intercepts made with the coordinate axes can be located and the long-term behaviour is known. However, at this stage we could not predict that $y = \left(x^2 + 3\right)(x - 1)$ has no turning points or stationary point of inflection (it has a non-stationary point of inflection). Nor could we predict, without numerical calculations, that there is a maximum and a minimum turning point on the graph of $y = (x + 3)\left(x^2 + 1\right)$.

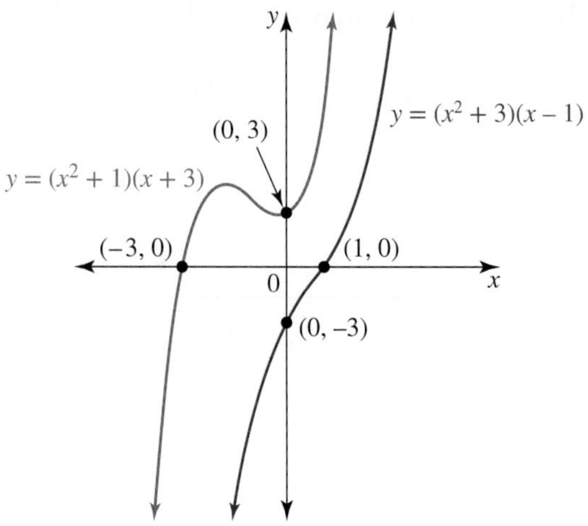

Determining the equation of a cubic function from its graph

Depending on the information given, one form of the cubic equation may be preferable over another.

- If there is a stationary point of inflection given, use the $y = a(x-h)^3 + k$ form.
- If the x-intercepts are given, use the $y = a(x-m)(x-n)(x-p)$ form, or the repeated factor form $y = a(x-m)^2(x-n)$ if there is a turning point at one of the x-intercepts.
- If an x-intercept occurs at $x = \frac{b}{c}$, then $\left(x - \frac{b}{c}\right)$ is a factor. Alternatively, the rational root theorem allows this factor to be expressed as $(cx - b)$.
- Use the general form $y = ax^3 + bx^2 + cx + d$ if, for example, neither x-intercepts nor a stationary point of inflection are given.

WORKED EXAMPLE 18 Sketching a cubic graph

Sketch the graph of $y = 2(x-1)^3 + 8$, labelling the intercepts with the coordinate axes with their exact coordinates.

THINK	WRITE
1. State the key feature that can be deduced from the equation.	$y = 2(x-1)^3 + 8$ This equation shows there is a stationary point of inflection at $(1, 8)$.
2. Calculate the y-intercept.	y-intercept: let $x = 0$. $y = 2(-1)^3 + 8$ $y = 6$ The y-intercept is $(0, 6)$.
3. Calculate the x-intercept in exact form.	x-intercept: let $y = 0$. $2(x-1)^3 + 8 = 0$ $(x-1)^3 = -4$ $x - 1 = \sqrt[3]{-4}$ $x = 1 + \sqrt[3]{-4}$ $x = 1 - \sqrt[3]{4}$ The x-intercept is $\left(1 - \sqrt[3]{4}, 0\right)$.
4. Sketch the graph and label the intercepts with the coordinate axes.	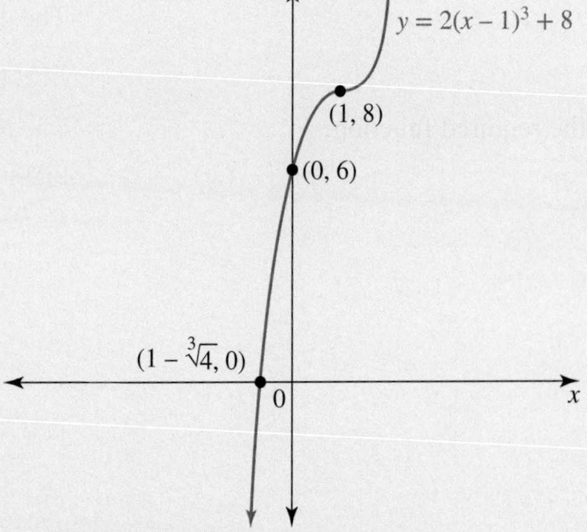

WORKED EXAMPLE 19 Determining the equation of a cubic graph

Determine the function *f* whose graph is shown in the diagram, expressing its rule as the product of linear factors with integer coefficients.

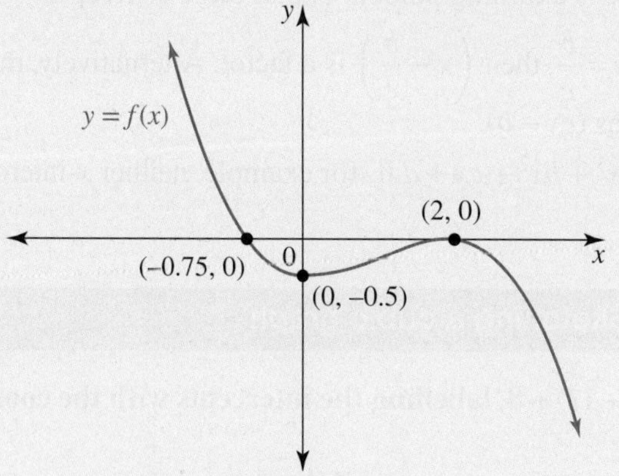

THINK

1. Obtain a linear factor of the equation of the graph that has integer coefficients.

WRITE

The graph has an x-intercept at $x = -0.75$.

In fraction form, this is $x = -\frac{3}{4}$.

Therefore, $(4x + 3)$ is a factor.

2. State a second factor.

The graph has a turning point on the x-axis at $x = 2$.

This means $(x - 2)^2$ is a factor.

3. State the form of the equation.

The equation is of the form $y = a(4x + 3)(x - 2)^2$.

4. Determine the equation fully.

The point $(0, -0.5)$ or $\left(0, -\frac{1}{2}\right)$ lies on the graph.

Substitute one point into

$y = a(4x + 3)(x - 2)^2$.

$$-\frac{1}{2} = a(3)(-2)^2$$

$$-\frac{1}{2} = 12a$$

$$a = -\frac{1}{24}$$

The graph has the equation

$y = -\frac{1}{24}(4x + 3)(x - 2)^2$.

5. State the required function.

The domain of the graph is R. Hence, the function f is given by

$f: R \rightarrow R, f(x) = -\frac{1}{24}(4x + 3)(x - 2)^2$.

Resources

Interactivities Cubic polynomials (int-2566)
x-intercepts of cubic graphs (int-2567)

1.5 Exercise

Technology free

1. **WE13** Use an appropriate method to factorise each of the following cubic expressions.
   a. $125a^3 - 27b^3$
   b. $2c^3 + 6c^2d + 6cd^2 + 2d^3$
   c. $40p^3 - 5$
   d. $8x^3 - 12x^2 + 6x - 1$

2. Use an appropriate method to factorise each of the following cubic expressions.
   a. $27z^3 - 54z^2 + 36z - 8$
   b. $m^3n^3 + 64$

3. **WE14** Fully factorise the following polynomials.
   a. $3x^2 - xy - 3x + y$
   b. $3y^3 + 3y^2z^2 - 2zy - 2z^3$

4. Fully factorise the following polynomials.
   a. $9a^2 - 16b^2 - 12a + 4$
   b. $n^2p^2 - 4m^2 - 4m - 1$

5. **WE15** Fully factorise $x^3 - 2x^2 - 21x - 18$.

6. Factorise the following expressions.
   a. $7r^3 - 49r^2 + r - 7$
   b. $36v^3 + 6v^2 + 30v + 5$
   c. $2m^3 + 3m^2 - 98m - 147$
   d. $2z^3 - z^2 + 2z - 1$
   e. $4x^2 - 28x + 49 - 25y^2$
   f. $16a^2 - 4b^2 - 12b - 9$
   g. $v^2 - 4 - w^2 + 4w$
   h. $4p^2 - 1 + 4pq + q^2$

7. **WE16** Solve $2x^3 - x^2 - 10x + 5 = 0$ for x.

Technology active

8. Solve each of the following equations over R.
   a. $b^3 + 5b^2 + 2b - 8 = 0$
   b. $-2m^3 + 9m^2 - m - 12 = 0$
   c. $2x^3 - x^2 - 6x + 3 = 0$
   d. $2x^3 + 7x^2 + 2x - 3 = 0$

9. a. Show that $3t^3 + 22t^2 + 37t + 10$ is divisible by $(t + 5)$ and hence solve the equation $3t^3 + 22t^2 + 37t + 10 = 0$.
   b. Show that $3d^3 - 16d^2 + 12d + 16$ is divisible by $(d - 2)$ and hence solve the equation $3d^3 - 16d^2 + 12d + 16 = 0$.

10. **WE17** If $Ax^3 + (B - 1)x^2 + (B + C)x + D \equiv 3x^3 - x^2 + 2x - 7$, determine the values of A, B, C and D.

11. If $x^3 + 9x^2 - 2x + 1 \equiv x^3 + (dx + e)^2 + \frac{8}{9}$, determine the values of d and e.

12. a. Given that $P(z) = 5z^3 - 3z^2 + 4z - 1$ and $Q(z) = az^3 + bz^2 + cz + d$, calculate the values of a, b, c and d if $P(z) \equiv Q(z)$.
   b. Given that $P(x) = x^3 - 6x^2 + 9x - 1$ and $Q(x) = x(x + a)^2 - b$, calculate the values of a and b if $P(x) \equiv Q(x)$.

13. If $2x^3 - 5x^2 + 5x - 5 \equiv a(x-1)^3 + b(x-1)^2 + c(x-1) + d$, determine the values of a, b, c and d and hence express $2x^3 - 5x^2 + 5x - 5$ in the form $a(x-1)^3 + b(x-1)^2 + c(x-1) + d$.

14. Given $(x + 3)$ and $(x - 1)$ are factors of $ax^3 + bx^2 - 4x - 3$, calculate the values of a and b.

15. **WE18** Sketch the graph of $y = -4(x + 2)^3 + 16$, labelling the intercepts with the coordinate axes with their exact coordinates.

16. Consider the function $f: [-2, 4] \to R, f(x) = 4x^3 - 8x^2 - 16x + 32$.
   a. Factorise $4x^3 - 8x^2 - 16x + 32$.
   b. Sketch the graph of $y = f(x)$.
   c. State the maximum and minimum values of the function f.

17. Sketch the graphs of the following cubic functions without attempting to locate any turning points that do not lie on the coordinate axes.
   a. $y = x^3 - x^2 - 6x$
   b. $y = 1 - \frac{1}{8}(x + 1)^3, x \in [-3, 2)$
   c. $y = 12(x + 1)^2 - 3(x + 1)^3$

18. **WE19** Determine the function f whose graph is shown in the diagram, expressing its rule as the product of linear factors with integer coefficients.

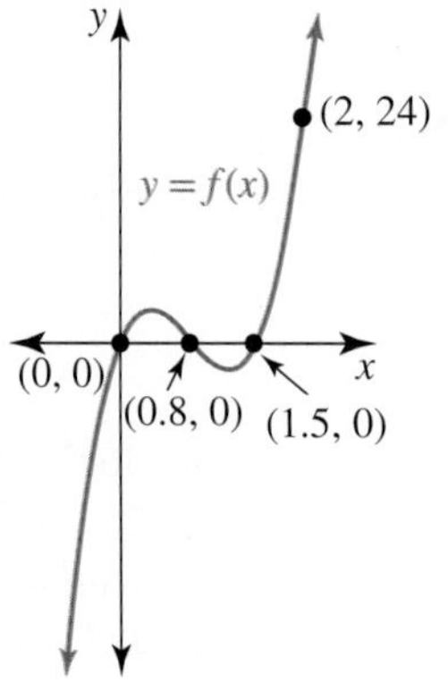

19. Form a possible equation for the cubic graph shown.

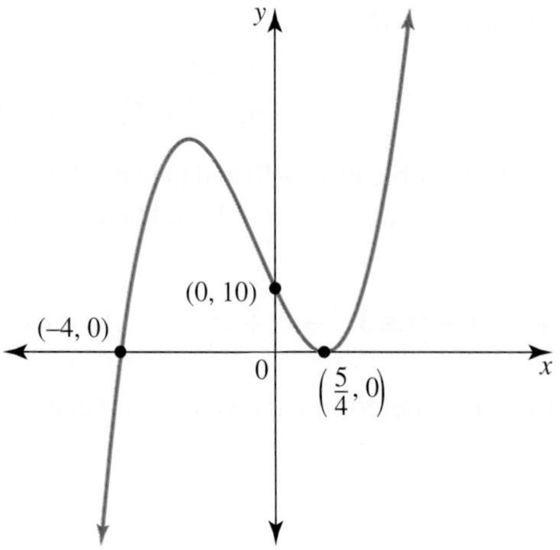

20. a. Show that the graph of $y=f(x)$ where $f(x)=-2x^3+9x^2-24x+17$ has exactly one x-intercept.
b. Show that there is no stationary point of inflection on the graph.
c. State the long-term behaviour of the function.
d. Given the function has a one-to-one correspondence, draw a sketch of the graph.

1.5 Exam questions

Question 1 (1 mark) TECH-ACTIVE

Source: *VCE 2020, Mathematical Methods Exam 2, Section A, Q2; © VCAA.*

MC Let $p(x)=x^3-2ax^2+x-1$ where $a \in R$. When p is divided by $x+2$, the remainder is 5.

The value of a is

A. 2 **B.** $-\frac{7}{4}$ **C.** $\frac{1}{2}$ **D.** $-\frac{3}{2}$ **E.** -2

Question 2 (4 marks) TECH-FREE

Source: *VCE 2017, Mathematical Methods Exam 1, Q3; © VCAA.*

Let $f:[-3,0] \to R, f(x)=(x+2)^2(x-1)$.

a. Show that $(x+2)^2(x-1)=x^3+3x^2-4$. **(1 mark)**

b. Sketch the graph of f on the axes below. Label the axis intercepts and any stationary points with their coordinates. **(3 marks)**

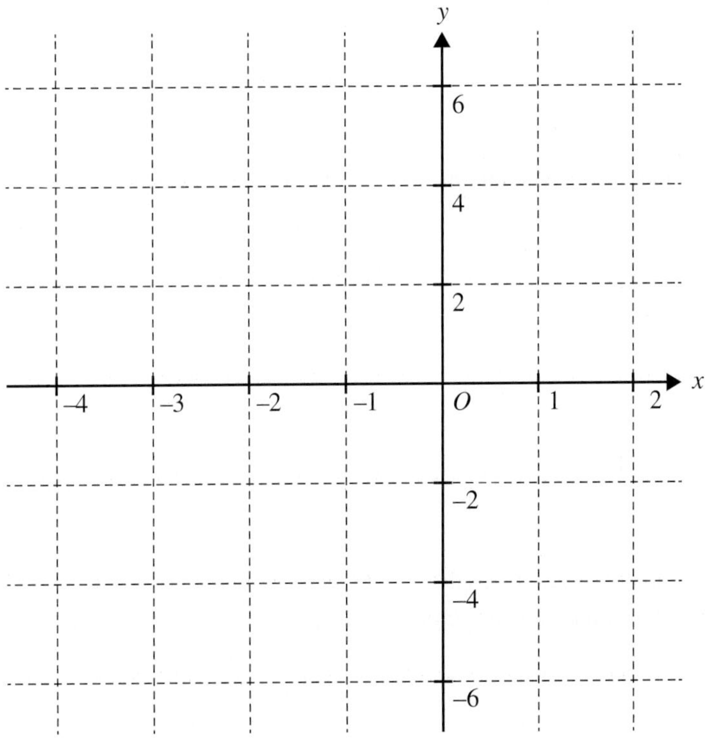

Question 3 (1 mark) TECH-ACTIVE

Source: *VCE 2015, Mathematical Methods (CAS) Exam 2 Section 1, Q17; © VCAA.*

MC A graph with rule $f(x)=x^3-3x^2+c$, where c is a real number, has three distinct x-intercepts.

The set of all possible values of c is

A. R **B.** R^+ **C.** $\{0,4\}$ **D.** $(0,4)$ **E.** $(-\infty,4)$

More exam questions are available online.

1.6 Higher degree polynomials

LEARNING INTENTION

At the end of this subtopic you should be able to:
- factorise and solve quartic functions
- graph quartic functions
- sketch polynomial functions of higher degrees.

1.6.1 Factorisation and solving

Many of the factorising resolving techniques covered in subtopics 1.4 and 1.5 also apply to polynomials of degree 4 or higher.

WORKED EXAMPLE 20 Factorising a quartic expression

Fully factorise $x^4 - 4x^3 - x^2 + 16x - 12$.

THINK	WRITE
1. Let $P(x)$ equal the quartic polynomial.	$P(x) = x^4 - 4x^3 - x^2 + 16x - 12$
2. Try $P(1)$, $P(-1)$, $P(2)$, $P(-2)$ etc. to get a zero remainder.	$P(1) = 1^4 - 4(1)^3 - (1)^2 + 16(1) - 12$ $= 17 - 17$ $= 0$ Thus, $(x - 1)$ is a factor.
3. Use long division to obtain the cubic factor.	$$\begin{array}{r} x^3 - 3x^2 - 4x + 12 \\ x-1\overline{)x^4 - 4x^3 - x^2 + 16x - 12} \\ -(x^4 - x^3) \qquad\qquad\qquad \\ \hline -3x^3 - x^2 + 16x - 12 \\ -(-3x^3 + 3x^2) \qquad\qquad \\ \hline -4x^2 + 16x - 12 \\ -(-4x^2 + 4x) \qquad \\ \hline 12x - 12 \\ -(12x - 12) \\ \hline 0 \end{array}$$
4. Let $H(x)$ equal the cubic polynomial. Apply the factor theorem again to find a linear factor of the cubic.	$H(x) = x^3 - 3x^2 - 4x + 12$ $H(1) = 1^3 - 3(1)^2 - 4(1) + 12$ $= 13 - 7$ $\neq 0$ $H(2) = 2^3 - 3(2)^2 - 4(2) + 12$ $= 20 - 20$ $= 0$ Thus, $(x - 2)$ is a factor.

5. Use long division to obtain the quadratic factor.	$$\begin{array}{r} x^2 - x - 6 \\ x-2\overline{)x^3 - 3x^2 - 4x + 12} \\ -\underline{(x^3 - 2x^2)} \\ -x^2 - 4x + 12 \\ -\underline{(-x^2 + 2x)} \\ -6x + 12 \\ -\underline{(-6x + 12)} \\ 0 \end{array}$$
6. Complete the quartic factorisation by factorising the quadratic factor into its two linear factors.	$$\begin{aligned} P(x) &= x^4 - 4x^3 - x^2 + 16x - 12 \\ &= (x-1)(x-2)\left(x^2 - x - 6\right) \\ &= (x-1)(x-2)(x-3)(x+2) \end{aligned}$$

WORKED EXAMPLE 21 Solving quartic equations

a. **Solve $x^4 - 4x^3 - x^2 + 16x - 12 = 0$.**

b. **Solve $2a^4 - 5a^2 - 3 = 0$.**

THINK	WRITE
a. 1. The quartic expression was factorised in Worked example 20.	**a.** $x^4 - 4x^3 - x^2 + 16x - 12 = 0$ $(x-1)(x-2)(x-3)(x+2) = 0$
2. Use the Null Factor Law to solve the quartic polynomial for x.	$x = 1, 2, 3, -2$
b. 1. The left-hand side is in quadratic form. Let $m = a^2$ to help with the factorisation.	**b.** $2a^4 - 5a^2 - 3 = 0$ Let $m = a^2$.
2. Factorise the quadratic.	$2m^2 - 5m - 3 = 0$ $(2m+1)(m-3) = 0$
3. Substitute $m = a^2$ and factorise further where possible. *Note:* There is no factorisation technique for the addition of perfect squares.	$\left(2a^2+1\right)\left(a^2-3\right) = 0$ $\left(2a^2+1\right)\left(a-\sqrt{3}\right)\left(a+\sqrt{3}\right) = 0$
4. Solve the equation.	$2a^2 + 1 = 0$ has no real solution. $\therefore a = \pm\sqrt{3}$

1.6.2 Graphing quartic and higher degree polynomial functions

The function $f: R \to R, f(x) = ax^4 + bx^3 + cx^2 + dx + e$, where $a, b, c, d, e \in R$, $a \neq 0$, is the general form of a quartic polynomial function. Its graph can take various shapes, but all of them exhibit the same long-term behaviour. If the x^4 term has a positive coefficient, $y \to \infty$ as $x \to \pm\infty$; if the x^4 term has a negative coefficient, $y \to -\infty$ as $x \to \pm\infty$. Particular forms of the quartic equation enable some shapes of the graphs to be predicted.

Quartic functions of the form $y = a(x-h)^4 + k$

The simplest quartic function is $y = x^4$. It has a graph that has much the same shape as $y = x^2$, as shown in the diagram.

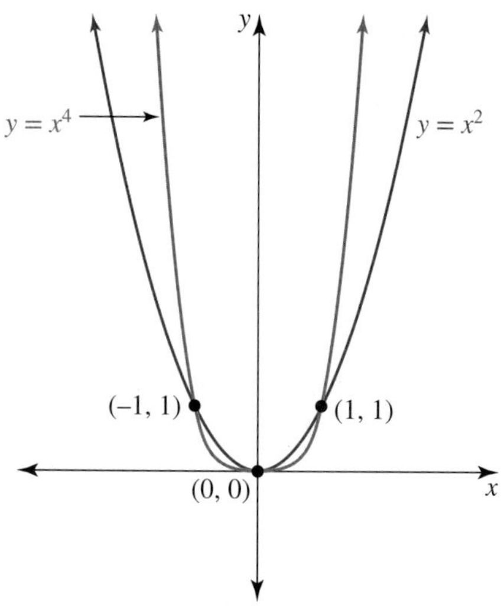

This leads to the conclusion that the graph of $y = a(x-h)^4 + k$ will be much the same shape as that of $y = a(x-h)^2 + k$ and will have the following characteristics.

Quartic functions in turning point form

For $y = a(x-h)^4 + k$:
- **if $a > 0$, the graph will be concave up with a minimum turning point (h, k)**
- **if $a < 0$, the graph will be concave down with a maximum turning point (h, k)**
- **the axis of symmetry has the equation $x = h$**
- **there may be zero, one or two x-intercepts.**

Quartic functions with linear factors

Not all quartic functions can be factorised. However, if it is possible to express the equation as the product of linear factors, then the multiplicity of each factor will determine the behaviour of its graph.

Quartic functions with linear factors

A quartic polynomial may have up to 4 linear factors as it is of fourth degree. The possible combinations of these linear factors are:
- **four distinct linear factors: $y = (x-a)(x-b)(x-c)(x-d)$**
- **one repeated linear factor: $y = (x-a)^2(x-b)(x-c)$, where the graph has a turning point that touches the x-axis at $x = a$**
- **two repeated linear factors: $y = (x-a)^2(x-b)^2$, where the graph has turning points that touch the x-axis at $x = a$ and $x = b$**
- **one factor of multiplicity three: $y = (x-a)^3(x-b)$, where the graph has a stationary point of inflection that cuts the x-axis at $x = a$.**

The factorised forms may be derived from the general equation using standard algebraic techniques. Technology or calculus is required to accurately identify the position of turning points that do not lie on the x-axis.

WORKED EXAMPLE 22 Sketching a quartic graph

Sketch the graph of $y = -x^4 + 8x^2 - 7$ and hence determine graphically the number of solutions to the equation $x^4 - 8x^2 + 3 = 0$.

THINK	WRITE
1. Express the equation in factorised form.	$y = -x^4 + 8x^2 - 7$ This is a quadratic in x^2. $y = -\left(x^4 - 8x^2 + 7\right)$ Let $a = x^2$. $y = -\left(a^2 - 8a + 7\right)$ $= -(a-7)(a-1)$ Substitute back for a: $y = -\left(x^2 - 7\right)\left(x^2 - 1\right)$ $= -\left(x + \sqrt{7}\right)\left(x - \sqrt{7}\right)(x+1)(x-1)$

2. State the x- and y-values of the intercepts with the axes.

x-intercepts: let $y = 0$.

$$-\left(x+\sqrt{7}\right)\left(x-\sqrt{7}\right)(x+1)(x-1)=0$$

$\therefore x = \pm\sqrt{7}, x = \pm 1$

y-intercept:

$y = -x^4 + 8x^2 - 7$

Let $x = 0$.

$\therefore y = -7$.

3. State the long-term behaviour.

As the coefficient of x^4 is negative, $y \to -\infty$ as $x \to \pm\infty$.

4. Sketch the graph.

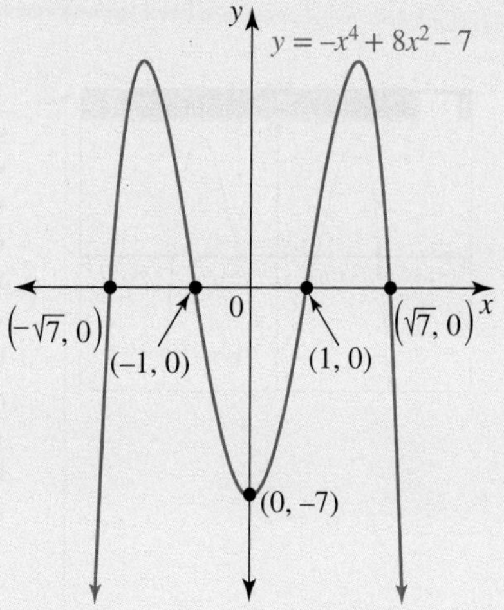

5. Rearrange the given equation so that the graph's equation appears on one of its sides.

The given equation is $x^4 - 8x^2 + 3 = 0$.

This rearranges to

$$3 = -x^4 + 8x^2$$

$$3 - 7 = -x^4 + 8x^2 - 7$$

$$-x^4 + 8x^2 - 7 = -4$$

6. Explain how the number of solutions to the equation could be solved graphically.

The number of intersections of the graph of $y = -x^4 + 8x^2 - 7$ with the horizontal line $y = -4$ will determine the number of solutions to the equation $x^4 - 8x^2 + 3 = 0$.

7. Specify the number of solutions.

The line $y = -4$ lies parallel to the x-axis between the origin and the y-intercept of the graph $y = -x^4 + 8x^2 - 7$.

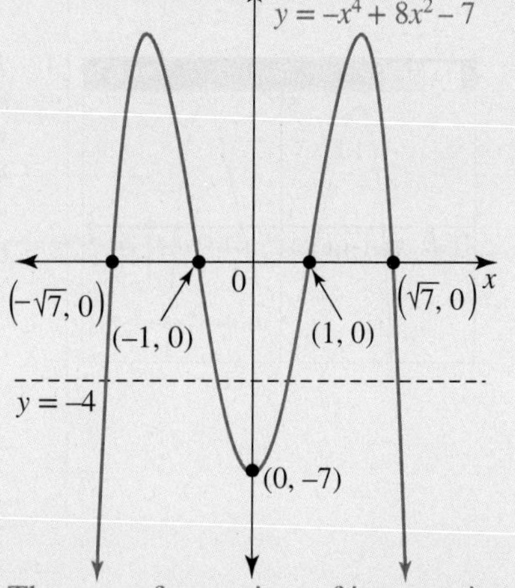

There are four points of intersection, so there are four solutions to the equation $x^4 - 8x^2 + 3 = 0$.

TI | THINK	DISPLAY/WRITE
1. On a Graphs page, complete the entry line for function 1 as: $f1\,(x) = -x^4 + 8x^2 - 7$ then press ENTER.	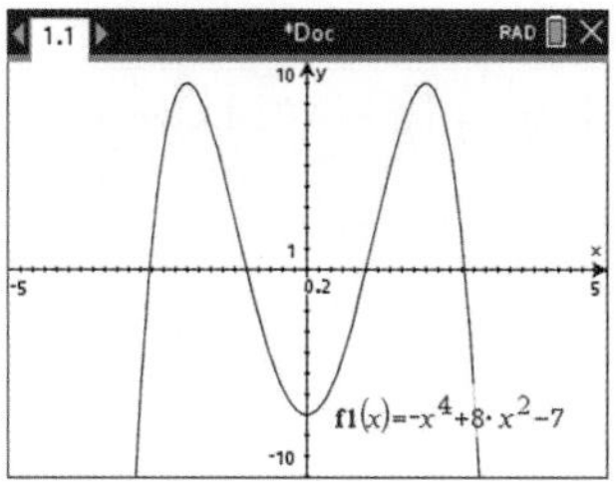
2. To find the x-intercepts, press MENU, then select: 6: Analyze Graph 1: Zero Move the cursor to the left of the x-intercept when prompted for the lower bound, then press ENTER. Move the cursor to the right of the x-intercept when prompted for the upper bound, then press ENTER. Repeat for all x-intercepts. *Note:* The calculator will show approximate values for the x-intercepts, not exact values.	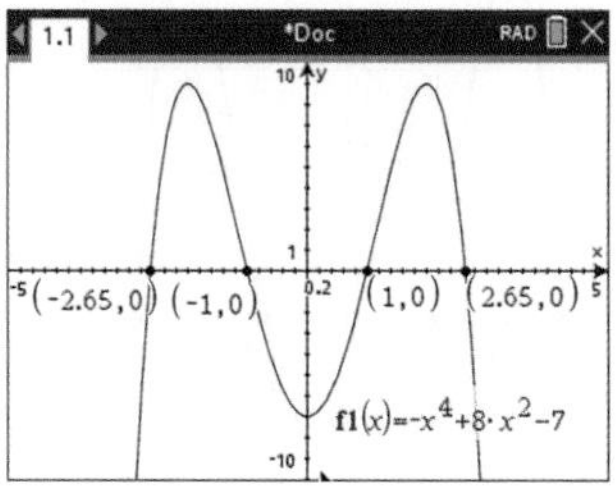
3. To calculate exact values for the x-intercepts, open a Calculator page and complete the entry line as: solve $\left(-x^4 + 8x^2 - 7 = 0, x\right)$ then press ENTER.	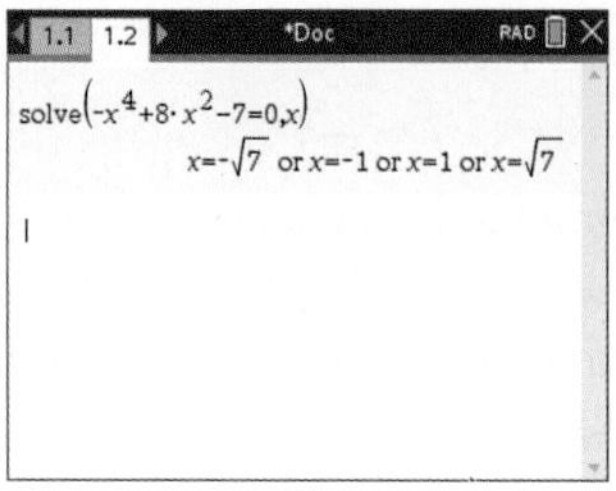
4. Return to the Graphs page, double click on the value $-2.64\ldots$, change the value to $-\sqrt{7}$, then press ENTER. Double click on the value $2.64\ldots$, change the value to $\sqrt{7}$, then press ENTER.	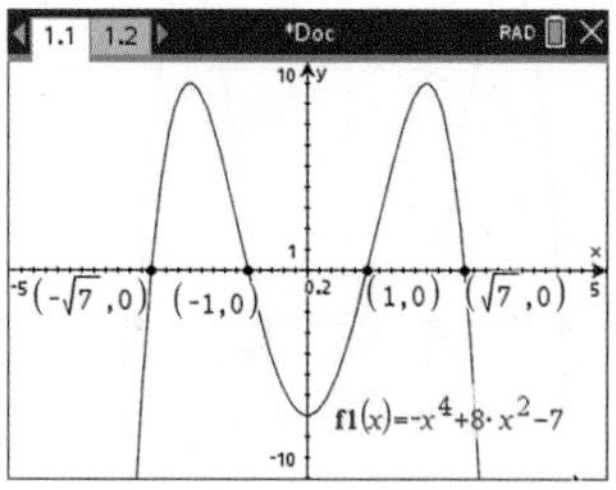

CASIO | THINK	DISPLAY/WRITE
1. On a Graph & Table screen, complete the entry line for $y1$ as: $y1 = -x^4 + 8x^2 - 7$ then press EXE. Select the Graph icon to draw the graph.	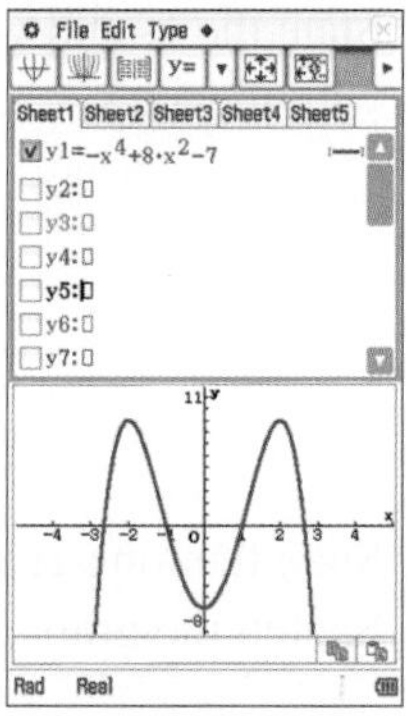
2. To find the x-intercepts, select: • Analysis • G-Solve • Root With the cursor on the first x-intercept, press EXE. Use the left/right arrows to move to the next x-intercept, then press EXE. Repeat for all x-intercepts. *Note:* The calculator will show approximate values for the x-intercepts, not exact values.	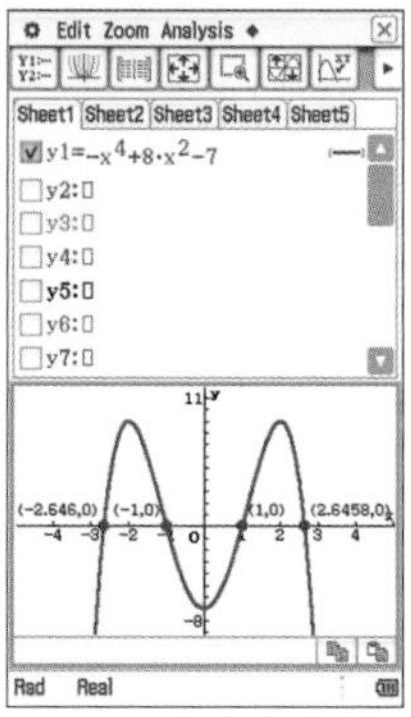
3. To calculate exact values for the x-intercepts, go to the Main screen and complete the entry line as: solve $\left(-x^4 + 8x^2 - 7 = 0, x\right)$ then press ENTER.	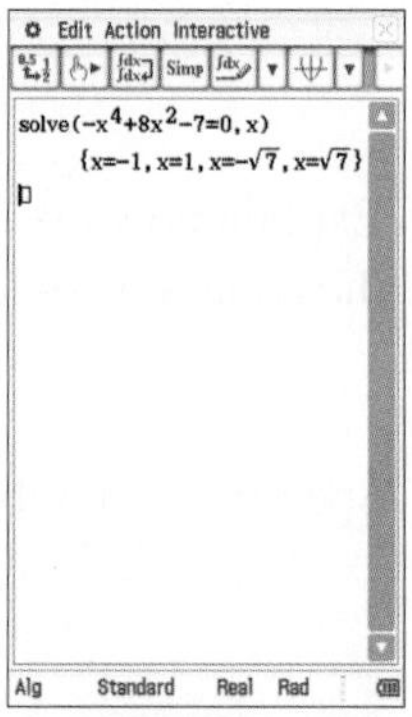
4. To find the y-intercept, select: • Analysis • G-Solve • Y-Intercept then press EXE.	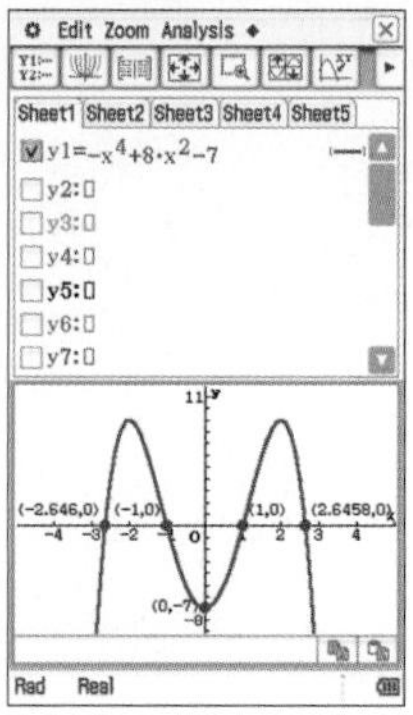

5. To find the y-intercept, press MENU, then select:
5: Trace
1: Graph Trace
Type '0', then press ENTER twice.

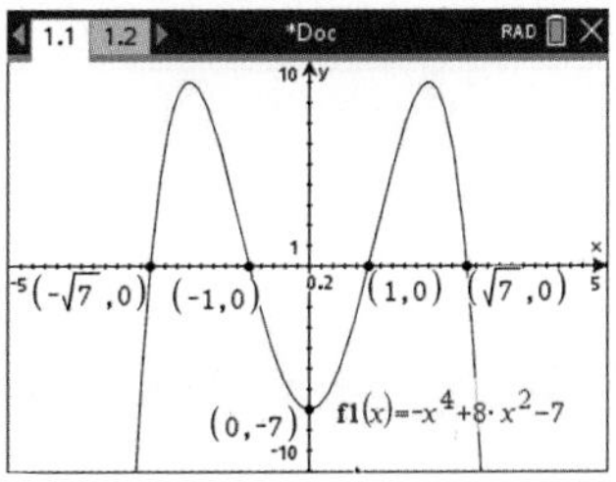

6. Answer the question.

The graph shows four x-intercepts. Hence, there are four solutions to the equation $-x^4 + 8x^2 - 7 = 0$.

5. Answer the question.

The graph shows four x-intercepts. Hence, there are four solutions to the equation $-x^4 + 8x^2 - 7 = 0$.

WORKED EXAMPLE 23 Determining the equation of a quartic function

A quartic function has the equation $y = a(x+b)^4 + c$. The points $(0, 5)$, $(-2, 9)$ and $(4, 9)$ lie on the graph of the function. Calculate the values of a, b and c, and state the coordinates of the turning point.

THINK	WRITE
1. Deduce the equation of the axis of symmetry.	$y = a(x+b)^4 + c$ As the points $(-2, 9)$ and $(4, 9)$ have the same y-value, the axis of symmetry must pass midway between them. The axis of symmetry is the line $x = \frac{-2+4}{2}$ $x = 1$ $\therefore b = -1$
2. Use the given points given to form a pair of simultaneous equations.	The equation is $y = a(x-1)^4 + c$. Substitute the point $(4, 9)$: $a(3)^4 + c = 9$ [1] $81a + c = 9$ Substitute the point $(0, 5)$: $a(-1)^4 + c = 5$ [2] $a + c = 5$
3. Solve the equations.	Subtract equation [2] from equation [1]: $80a = 4$ $a = \frac{1}{20}$ $\therefore c = 5 - \frac{1}{20}$ $c = \frac{99}{20}$
4. Give the coordinates of the turning point.	The equation is $y = \frac{1}{20}(x-1)^4 + \frac{99}{20}$. The minimum turning point is $\left(1, \frac{99}{20}\right)$.

on Resources

Interactivity Polynomials of higher degrees (int-2569)

1.6.3 The family of polynomial functions $y = x^n$ where $n \in N$

One classification of the polynomial functions is to group them according to whether their degree is even or odd.

The graph of $y = x^n$, where n is an even positive integer

The similarities shown between the graphs of $y = x^2$, and $y = x^4$ continue to hold for all polynomial functions of even degree. A comparison of the graphs of $y = x^2$, $y = x^4$ and $y = x^6$ is shown in the diagram.

The graphs each have a minimum turning point at $(0, 0)$ and each contains the points $(-1, 1)$ and $(1, 1)$. They exhibit the same long-term behaviour that as $x \to \pm\infty$, $y \to \infty$.

The graph of the function with the highest degree, $y = x^6$, rises more steeply than the other two graphs for $x < -1$ and $x > 1$. However, for $-1 < x < 0$ and $0 < x < 1$, the function with the highest degree lies below the other graphs.

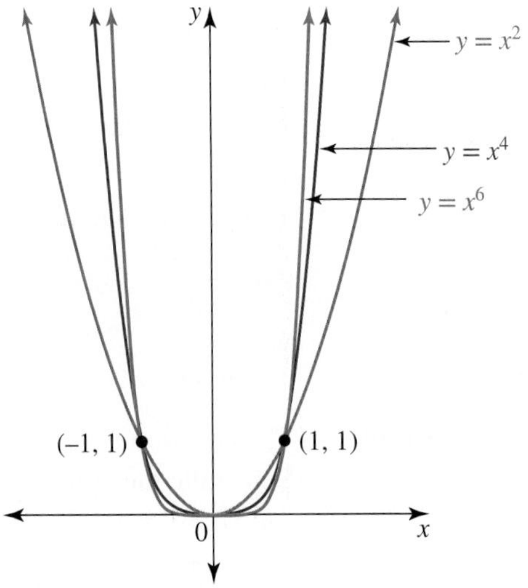

The graph of $y = x^n$, where n is an even positive integer

For $y = a(x - h)^n + k$, where n is an even positive integer:

- **if $a > 0$, the graph will be concave up with a minimum turning point (h, k)**
- **if $a < 0$, the graph will be concave down with a maximum turning point (h, k)**
- **the axis of symmetry has the equation $x = h$**
- **there may be zero, one or two x-intercepts**
- **the shape of the graph will be similar to that of $y = a(x - h)^2 + k$**
- **if $a > 0$, the range is $[k, \infty)$**
- **if $a < 0$, the range is $(-\infty, k]$.**

The graph of $y = x^n$, where n is an odd positive integer, $n > 1$

Polynomials of odd degree also share similarities, as the graphs of $y = x^3$ and $y = x^5$ illustrate. Both $y = x^3$ and $y = x^5$ have a stationary point of inflection at $(0, 0)$, and both pass through the points $(-1, -1)$ and $(1, 1)$, as does the linear function $y = x$. The three graphs display the same long-term behaviour that as $x \to \pm\infty$, $y \to \pm\infty$.

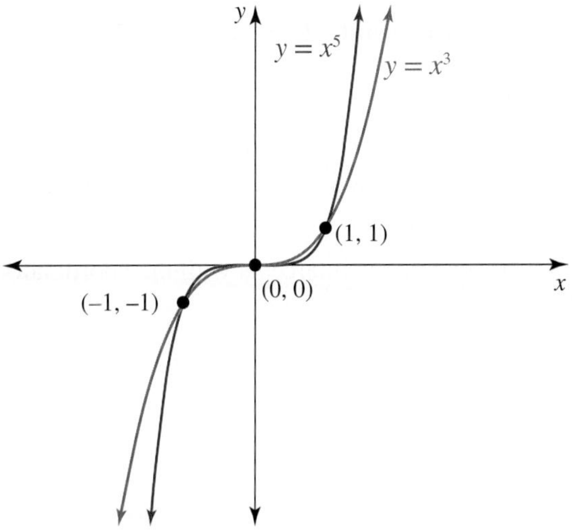

As observed for even degree polynomials, the graph of the function with the highest degree, $y = x^5$, rises more steeply than the other two graphs for $x < -1$ and $x > 1$. However, for $-1 < x < 0$ and $0 < x < 1$, the function with the highest degree lies closer to the x-axis.

The graphs of $y = a(x - h)^n + k$, where n is an odd positive integer, $n \neq 1$, have the following characteristics.

> **The graph of $y = x^n$, where n is an odd positive integer, $n > 1$**
>
> **For $y = a(x - h)^n + k$, where n is an odd positive integer and $n > 1$:**
> - **there is a stationary point of inflection at (h, k)**
> - **if $a > 0$, the long-term behaviour is as $x \to \pm\infty$, $y \to \pm\infty$**
> - **if $a < 0$, the long-term behaviour is as $x \to \pm\infty$, $y \to \mp\infty$**
> - **there will be one x-intercept**
> - **the shape of the graph is similar to that of the cubic function $y = a(x - h)^3 + k$.**

Polynomial functions that can be expressed as the product of linear factors

A degree n polynomial function may have up to n linear factors and therefore up to n intercepts with the x-axis. Where the polynomial can be specified completely as the product of linear factors, its graph can be drawn by interpreting the multiplicity of each linear factor together with the long-term behaviour determined by the sign of the coefficient of x^n.

For example, consider $y = (x + 2)^3(1 - x)(x - 3)^2$. The equation indicates there are x-intercepts at -2, 1 and 3. The x-intercept $(-2, 0)$ has a multiplicity of 3, meaning that there is a stationary point of inflection at this point. The x-intercept $(3, 0)$ has a multiplicity of 2, so this point is a turning point. The point $(1, 0)$ is a standard x-intercept. The polynomial is of degree 6 and the coefficient of x^6 is negative; therefore, as $x \to \pm\infty$, $y \to -\infty$.

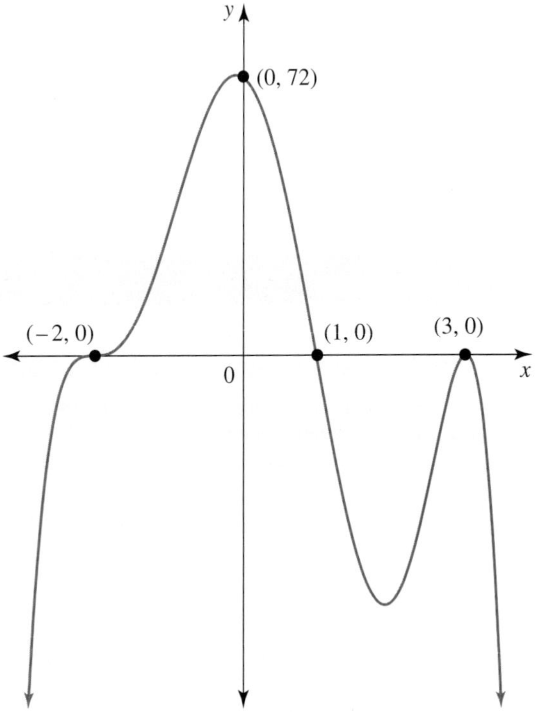

WORKED EXAMPLE 24 Sketching graphs of higher degree polynomials

Sketch the graph of $y = (x - 1)^5 - 32$.

THINK | **WRITE**

1. State whether the graph has a turning point or a point of inflection, and give the coordinates of the key point.

$y = (x-1)^5 - 32$

As the degree is odd, the graph will have a stationary point of inflection at $(1, -32)$.

2. Calculate the intercepts with the coordinate axes.

y-intercept: let $x = 0$.

$y = (-1)^5 - 32$

$= -33$

The y-intercept is $(0, -33)$.

x-intercepts: let $y = 0$.

$0 = (x-1)^5 - 32$

$(x-1)^5 = 32$

$x - 1 = 2$

$x = 3$

The x-intercept is $(3, 0)$.

3. Sketch the graph.

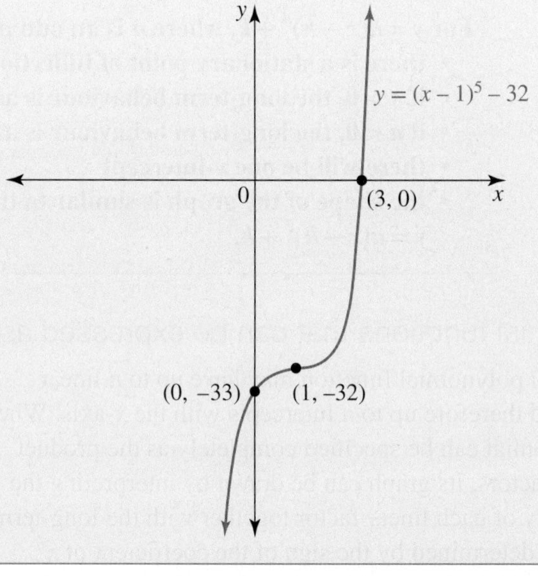

1.6 Exercise

Students, these questions are even better in jacPLUS

Receive immediate feedback and access sample responses

Access additional questions

Track your results and progress

Find all this and MORE in jacPLUS

Technology free

1. **WE20** Fully factorise $x^4 - 5x^3 - 32x^2 + 180x - 144$.

2. **WE21** Solve the following.

a. $x^4 - 8x^3 + 17x^2 + 2x - 24 = 0$

b. $a^4 + 2a^2 - 8 = 0$

3. Solve each of the following equations over R.
   a. $l^4 - 17l^2 + 16 = 0$
   b. $c^4 + c^3 - 10c^2 - 4c + 24 = 0$
4. Solve each of the following equations over R.
   a. $a^4 - 10a^2 + 9 = 0$
   b. $4k^4 - 101k^2 + 25 = 0$
   c. $9z^4 - 145z^2 + 16 = 0$
   d. $\left(x^2 - 2x\right)^2 - 47\left(x^2 - 2x\right) - 48 = 0$

Technology active

5. If $(x + 2)$, $(x - 3)$ and $(x + 4)$ are factors of $x^4 + ax^3 + bx^2 + cx + 24$, determine the values of a, b and c.
6. **WE22** Sketch the graph of $y = x^2 - x^4$ and hence determine graphically the number of solutions to the equation $x^4 - x^2 + x - 2 = 0$.
7. Sketch the graph of $y = x^4 - 6x^3$ and hence state the number of intersections the graph of $y = x^4 - 6x^3 + 1$ would make with the x-axis.
8. **WE23** A quartic function has the equation $y = a(x + b)^4 + c$. The graph of the function cuts the x-axis at $x = -9$ and $x = -3$. The range of the graph is $(-\infty, 7]$. Calculate the values of a, b and c, and state the coordinates of the turning point.
9. Use CAS technology to sketch the graphs of $y = x^4 - 2$ and $y = 2 - x^3$, and hence state to 2 decimal places the values of the roots of the equation $x^4 + x^3 - 4 = 0$.
10. Use CAS technology to obtain the coordinates of any turning points or stationary points of inflection on the graphs of the following. Express answers to 2 decimal places where appropriate.
    a. $y = \left(x^2 + x + 1\right)\left(x^2 - 4\right)$
    b. $y = 1 - 4x - x^2 - x^3$
    c. $y = \dfrac{1}{4}((x - 2)^5(x + 3) + 80)$
11. **WE24** Sketch the graph of $y = (x + 1)^6 + 10$.
12. Sketch the graph of $y = (x + 4)\ (x + 2)^2(x - 2)^3\ (x - 5)$.
13. a. A quartic function has exactly one turning point at $(-5, 12)$ and also contains the point $(-3, -36)$. Form its equation.
    b. Sketch the graph of $y = (2 + x)\ (1 - x)^3$.
    c. i. Factorise $-x^4 + x^3 + 10x^2 - 4x - 24$.
       ii. Hence, sketch $y = -x^4 + x^3 + 10x^2 - 4x - 24$.
14. a. i. Sketch the graphs of $y = x^6$ and $y = x^7$ on the same set of axes, labelling any points of intersection with their coordinates.
       ii. Hence, state the solutions to $\{x : x^6 - x^7 \geq 0\}$.
    b. Sketch the graphs of $y = 16 - (x + 2)^4$ and $y = 16 - (x + 2)^5$ on the same set of axes, identifying the key features of each graph and any points of intersection.
    c. Consider the graph of the polynomial function shown.
       i. Assuming the graph is a monic polynomial that maintains the long-term behaviour suggested in the diagram, give a possible equation for the graph and state its degree.
       ii. In fact, the graph cuts straight through the x-axis once more at $x = 10$. This is not shown on the diagram. Given this additional information, state the degree and a possible equation for the function.

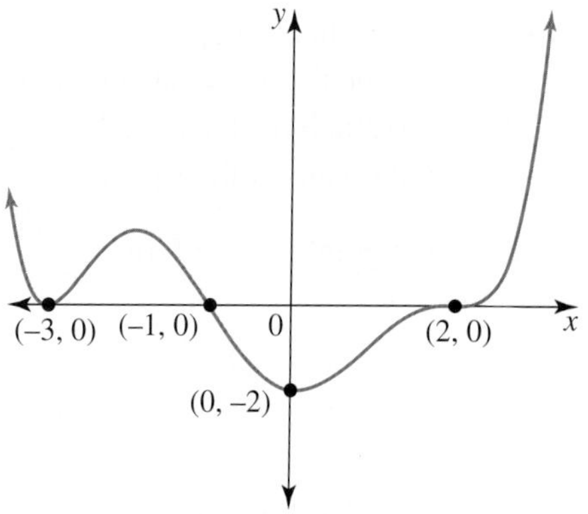

1.6 Exam questions

Question 1 (1 mark) TECH-ACTIVE

Source: *VCE 2015, Mathematical Methods (CAS) Exam 2 Section 1, Q3; © VCAA.*

MC

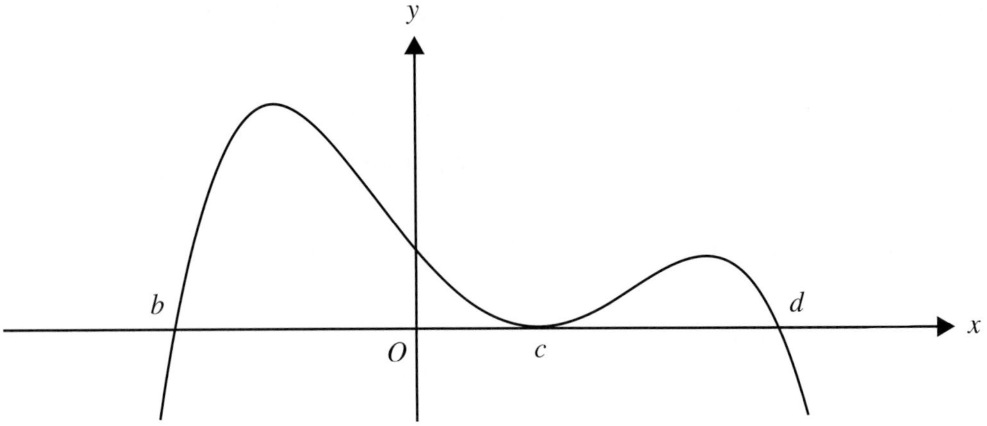

The rule for a function with the graph above could be

A. $y = -2(x+b)(x-c)^2(x-d)$

B. $y = 2(x+b)(x-c)^2(x-d)$

C. $y = -2(x-b)(x-c)^2(x-d)$

D. $y = 2(x-b)(x-c)(x-d)$

E. $y = -2(x-b)(x+c)^2(x+d)$

Question 2 (2 marks) TECH-FREE

Solve $-x^4 + 7x^3 - 12x^2 \geq 0$.

Question 3 (1 mark) TECH-ACTIVE

MC A quartic graph crosses the x-axis at $x = -3$, $x = -1$, $x = 2$ and $x = 4$, and crosses the y-axis at $y = -24$. The equation of the graph is

A. $y = -(x-3)(x-1)(x+2)(x+4)$

B. $y = (x-3)(x-1)(x+2)(x+4)$

C. $y = -(x+3)(x+1)(x-2)(x-4)$

D. $y = -(x-3)(x+1)(x-2)(x-4)$

E. $y = (x+3)(x+1)(x-2)(x-4)$

More exam questions are available online.

1.7 Other algebraic functions

LEARNING INTENTION

At the end of this subtopic you should be able to:

- determine the maximal domain of a function
- sketch graphs of the hyperbola and truncus functions
- sketch graphs of the square root and cube root functions
- sketch graphs of the form $y = x^{\frac{p}{q}}$.

The powers of the variable in a polynomial function must be natural numbers. In this subtopic we consider functions where the power of the variable may be rational.

1.7.1 Maximal domain

The maximal domain of any function must exclude:

- any value of x for which the denominator would become zero
- any value of x that would create a negative term under an even root sign.

The maximal or implied domain of rational functions of the form $y = \frac{g(x)}{f(x)}$, where both $f(x)$ and $g(x)$ are polynomials, must exclude any values of x for which $f(x) = 0$. The domain would be $R \setminus \{x : f(x) = 0\}$.

Likewise, the maximal domain of square root functions of the form $y = \sqrt{f(x)}$ would be $\{x : f(x) \geq 0\}$.

For a function of the form $y = \frac{g(x)}{\sqrt{f(x)}}$, the maximal domain would be $\{x : f(x) > 0\}$.

WORKED EXAMPLE 25 Determining the maximal domain

Determine the maximal domains of each of the following functions.

a. $y = \sqrt{3 - 4x}$

b. $y = 2 - \frac{5}{x - 4}$

THINK	WRITE
a. 1. Identify what must be excluded from the domain. The restrictive element to the equation is the square root function.	The term under the square root cannot be negative. $\therefore 3 - 4x \geq 0$
2. Solve the inequation to find x.	$3 - 4x \geq 0$ $-4x \geq -3$ $x \leq \frac{3}{4}$
3. State the maximal domain.	Domain = $\left(-\infty, \frac{3}{4}\right]$
b. 1. Identify what must be excluded from the domain. The restrictive element to the equation is the denominator of the fraction.	The denominator cannot be equal to zero. $\therefore x - 4 \neq 0$
2. Solve the equation.	$x \neq 4$
3. State the maximal domain.	Domain = $x \in R \setminus \{4\}$

1.7.2 The rectangular hyperbola

The equation of the simplest hyperbola is $y = \frac{1}{x}$. In power form this is written as $y = x^{-1}$. Its maximal domain is $R \setminus \{0\}$, as the function is undefined if $x = 0$.

The graph of this function has the following characteristics.

- There is a vertical **asymptote** with equation $x = 0$.
- There is a horizontal asymptote with equation $y = 0$.
- As $x \to \infty$, $y \to 0$ from above the horizontal asymptote, and as $x \to -\infty$, $y \to 0$ from below the horizontal asymptote.
- As $x \to 0^+$, $y \to \infty$, and as $x \to 0^-$, $y \to -\infty$.
- The function has one-to-one correspondence.
- The domain is $R \setminus \{0\}$ and the range is $R \setminus \{0\}$.

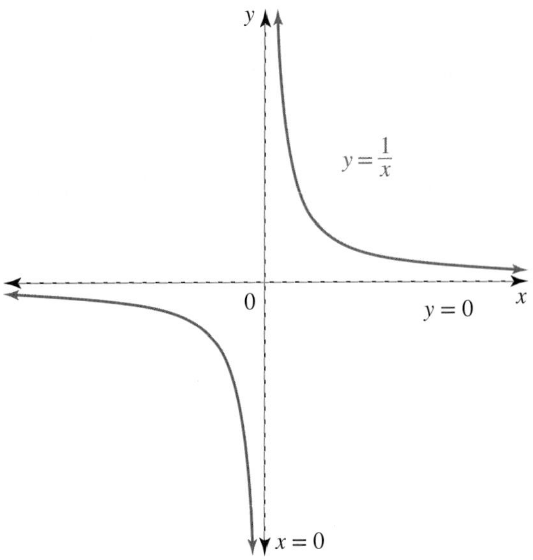

As the asymptotes are perpendicular to each other, the graph is called a rectangular hyperbola. The graph lies in the first and third quadrants formed by its asymptotes. The graph of $y = -\frac{1}{x}$ would lie in the second and fourth quadrants.

Hyperbolas of the form $y = \frac{a}{x-h} + k$

The asymptotes are the key feature of the graph of a hyperbola. Their positions are unaffected by a dilation, but if the graph of $y = \frac{1}{x}$ is horizontally or vertically translated, then the vertical and horizontal asymptotes are moved accordingly.

> **Hyperbolas of the form $y = \frac{a}{x-h} + k$**
>
> **The graph of $y = \frac{a}{x-h} + k$ has:**
> - **a vertical asymptote at $x = h$**
> - **a horizontal asymptote at $y = k$**
> - **a domain of $R \setminus \{h\}$**
> - **a range of $R \setminus \{k\}$.**

If $a > 0$, the graph lies in quadrants 1 and 3 as formed by its asymptotes.

If $a < 0$, the graph lies in quadrants 2 and 4 as formed by its asymptotes.

Identifying the asymptotes

The presence of a vertical asymptote at $x = h$ on the graph of $y = \frac{a}{x-h} + k$ could also be recognised by solving $x - h = 0$. The hyperbola $y = \frac{a}{bx+c}$ has a vertical asymptote when $bx + c = 0$, and its maximal domain is $R \setminus \left\{-\frac{c}{b}\right\}$.

The horizontal asymptote is identified from the equation of a hyperbola expressed in proper rational form, that is, when the numerator is of lower degree than the denominator. The equation $y = \frac{1+2x}{x}$ should be rewritten as $y = \frac{1}{x} + 2$ in order to identify the horizontal asymptote $y = 2$.

WORKED EXAMPLE 26 Sketching and determining the equations of hyperbolas

a. Determine an appropriate equation for the hyperbola shown.

b. i. Obtain the maximal domain of $y = \frac{2x+5}{x+1}$.

ii. Sketch the graph of $y = \frac{2x+5}{x+1}$ and state its range.

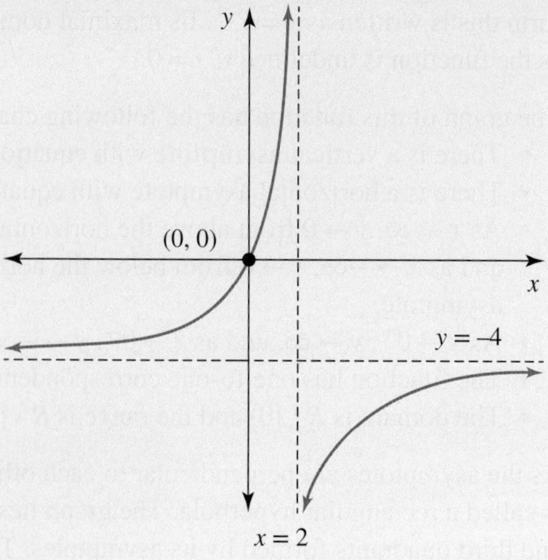

THINK	WRITE
a. 1. Write the general equation of a hyperbola.	**a.** Let the equation be $y = \dfrac{a}{x-h} + k$.
2. Identify the asymptotes and enter them into the equation.	The graph shows there is a vertical asymptote at $x = 2$. $\therefore y = \dfrac{a}{x-2} + k$ There is a horizontal asymptote at $y = -4$. $\therefore y = \dfrac{a}{x-2} - 4$
3. Identify the known point through which the graph passes and use this to fully determine the equation.	The graph passes through the origin. Substitute $(0, 0)$: $0 = \dfrac{a}{-2} - 4$ $4 = -\dfrac{a}{2}$ $a = -8$ The equation is $y = \dfrac{-8}{x-2} - 4$.
b. i. 1. Identify what must be excluded from the domain.	**b. i.** $y = \dfrac{2x+5}{x+1}$ The function is undefined if its denominator is zero. When $x + 1 = 0$, $x = -1$. This value must be excluded from the domain.
2. State the maximal domain.	The maximal domain is $R \setminus \{-1\}$.
ii. 1. Express the equation in proper rational form.	**ii.** $\dfrac{2x+5}{x+1} = \dfrac{2(x+1)+3}{x+1}$ $= \dfrac{2(x+1)}{x+1} + \dfrac{3}{x+1}$ $= 2 + \dfrac{3}{x+1}$ The equation is $y = \dfrac{3}{x+1} + 2$.
2. State the equations of the asymptotes.	The graph has a vertical asymptote at $x = -1$ and a horizontal asymptote at $y = 2$.
3. Calculate any intercepts with the coordinate axes.	x-intercept: let $y = 0$ in $y = \dfrac{2x+5}{x+1}$. $0 = \dfrac{2x+5}{x+1}$ $0 = 2x + 5$ $x = -\dfrac{5}{2}$ The x-intercept is $\left(-\dfrac{5}{2}, 0\right)$

y-intercept: let $x = 0$.

$$y = \frac{5}{1}$$
$$= 5$$

The y-intercept is $(0, 5)$.

4. Sketch the graph.

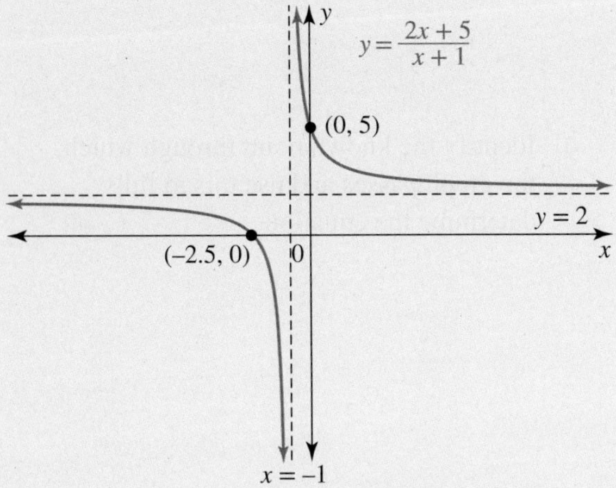

5. State the range.

The range is $R \setminus \{2\}$.

on Resources

Interactivity Hyperbola (int-2573)

1.7.3 The truncus

The graph of the function $y = \frac{1}{x^2}$ is called a truncus. Its rule can be written as a power function, $y = x^{-2}$.

The graph of this function has the following characteristics.

- There is a vertical asymptote with equation $x = 0$.
- There is a horizontal asymptote with equation $y = 0$.
- The domain is $R \setminus \{0\}$.
- The range is R^+.
- The function has many-to-one correspondence.
- The graph is symmetric about its vertical asymptote.

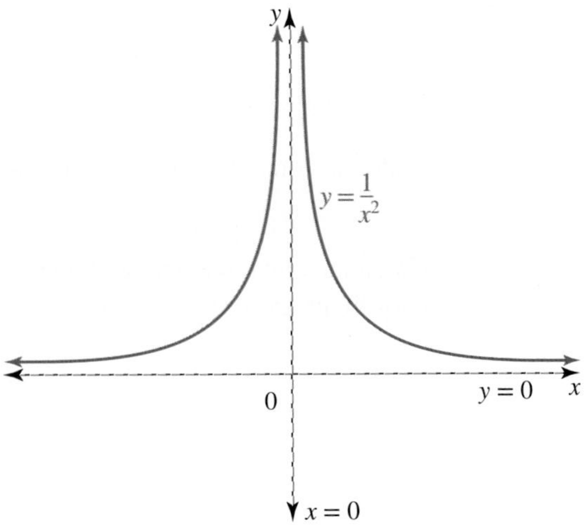

The graph of $y = \frac{1}{x^2}$ lies in the first and second quadrants that are created by its asymptotes. The graph of $y = -\frac{1}{x^2}$ would lie in the third and fourth quadrants.

The truncus is steeper than the hyperbola for $x \in (-1, 0)$ and $x \in (0, 1)$. However, a similar approach is taken to sketching both functions.

The general form of the truncus $y = \dfrac{a}{(x-h)^2} + k$

The graph of $y = \dfrac{a}{(x-h)^2} + k$ has the following characteristics.

- **There is a vertical asymptote at $x = h$.**
- **There is a horizontal asymptote at $y = k$.**
- **The domain is $R \setminus \{h\}$.**
- **If $a > 0$, then the range is (k, ∞).**
- **If $a < 0$, then the range is $(-\infty, k)$.**

WORKED EXAMPLE 27 Sketching the graph of a truncus

Sketch the graph of $y = 8 - \dfrac{2}{(x-3)^2}$ and state its domain and range.

THINK	WRITE
1. State the equations of the asymptotes.	$y = 8 - \dfrac{2}{(x-3)^2}$ The vertical asymptote is $x = 3$. The horizontal asymptote is $y = 8$.
2. Calculate the y-intercept.	y-intercept: let $x = 0$. $y = 8 - \dfrac{2}{(-3)^2}$ $y = 7\dfrac{7}{9}$ The y-intercept is $\left(0, 7\dfrac{7}{9}\right)$.
3. Calculate any x-intercepts.	x-intercepts: let $y = 0$. $0 = 8 - \dfrac{2}{(x-3)^2}$ $\dfrac{2}{(x-3)^2} = 8$ $2 = 8(x-3)^2$ $(x-3)^2 = \dfrac{1}{4}$ $x - 3 = \pm\dfrac{1}{2}$ $x = 2\dfrac{1}{2}$ or $x = 3\dfrac{1}{2}$ The x-intercepts are $\left(2\dfrac{1}{2}, 0\right), \left(3\dfrac{1}{2}, 0\right)$.

4. Sketch the graph.

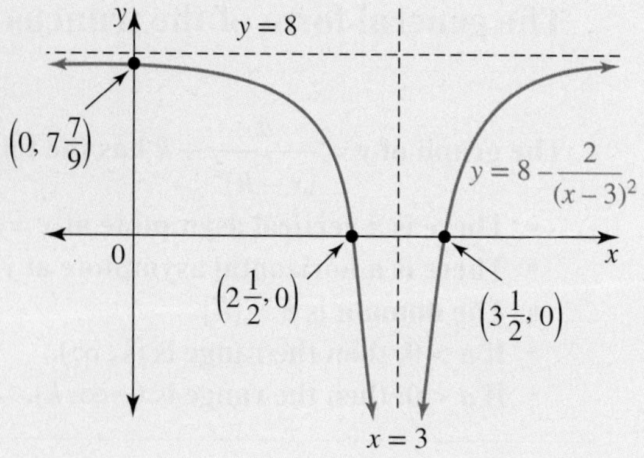

5. State the domain and range.

The domain is $R \setminus \{3\}$ and the range is $(-\infty, 8)$.

1.7.4 The square root and cube root functions

The square root function has the rule $y = \sqrt{x}$, and the rule for the cube root function is $y = \sqrt[3]{x}$. As power functions these rules can be expressed as $y = x^{\frac{1}{2}}$ and $y = x^{\frac{1}{3}}$ respectively.

The maximal domain of $y = \sqrt{x}$ is $[0, \infty)$, because negative values under a square root must be excluded. However, cube roots of negative numbers are real, so the maximal domain of the cube root function $y = \sqrt[3]{x}$ is R.

The graph of the square root function

The function $y = \sqrt{x}$ is the top half of the 'sideways' parabola $y^2 = x$. The bottom half of this parabola is the function $y = -\sqrt{x}$.

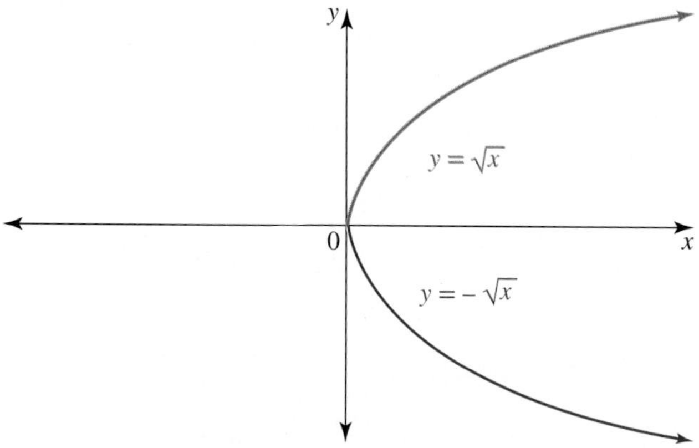

The parabola $y^2 = x$ is not a function, but its two halves are. The equation $y^2 = x$ could also be written as $y = \pm\sqrt{x}$. The turning point or vertex of the parabola is the end point for the square root functions $y = \sqrt{x}$ and $y = -\sqrt{x}$. These functions both have domain $[0, \infty)$, but their ranges are $[0, \infty)$ and $(-\infty, 0]$ respectively.

The parabola $y^2 = -x$ would open to the left of its vertex. Its two branches would be the square root functions $y = \sqrt{-x}$ and $y = -\sqrt{-x}$, with domain $(-\infty, 0]$ and ranges $[0, \infty)$ and $(-\infty, 0]$ respectively.

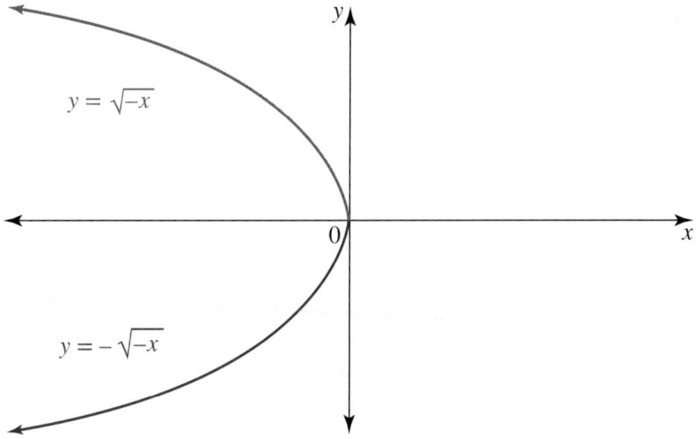

The four square root functions show the different orientations that can be taken. Calculation of the maximal domain and the range will identify which form a particular function takes.

Square root functions

Square root functions of the form $y = a\sqrt{x-h} + k$ have the following characteristics.
- **The end point is (h, k).**
- **The domain is $[h, \infty)$.**
- **If $a > 0$, the range is $[k, \infty)$; if $a < 0$, the range is $(-\infty, k]$.**

Square root functions of the form $y = a\sqrt{-(x-h)} + k$ have the following characteristics.
- **The end point is (h, k).**
- **The domain is $(-\infty, h]$.**
- **If $a > 0$, the range is $[k, \infty)$; if $a < 0$, the range is $(-\infty, k]$.**

The graph of the cube root function

The graph of the cubic function $y = x^3$ has a stationary point of inflection at the origin. The graph of $y^3 = x$ has a 'sideways' orientation but still has a point of inflection at the origin.

The rule $y^3 = x$ can also be expressed as $y = \sqrt[3]{x}$. The graph of $y = \sqrt[3]{x}$ is shown in the diagram.

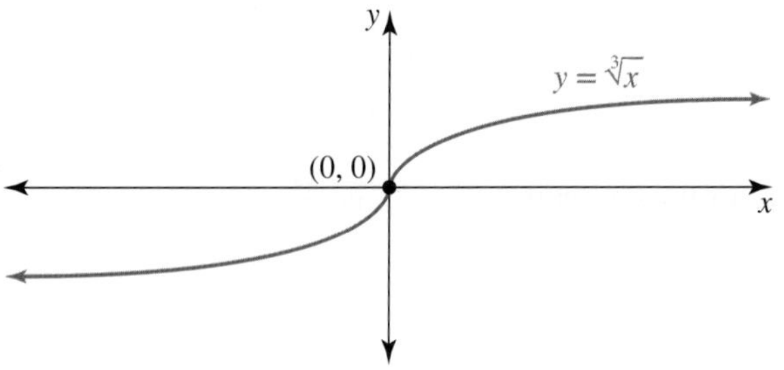

The graph $y = \sqrt[3]{x}$ has the following characteristics.
- There is a point of inflection at $(0, 0)$ where the tangent drawn to the curve would be vertical.
- The domain is R and the range is R.
- The function has one-to-one correspondence.

The graph of $y = -\sqrt[3]{x}$ is the reflection of $y = \sqrt[3]{x}$ in the x axis.

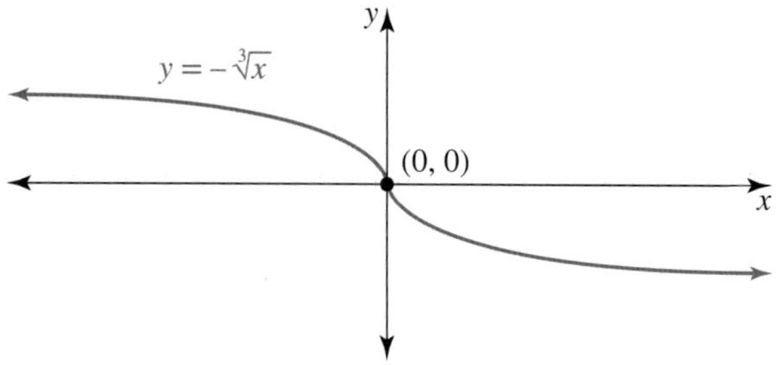

This is also the graph of $y = \sqrt[3]{-x}$, as $\sqrt[3]{-x} = -\sqrt[3]{x}$.

Cube root functions

The general equation $y = a\sqrt[3]{x - h} + k$ shows the graph has the following characteristics.

- **There is a point of inflection at (h, k).**
- **The domain is R and the range is R.**
- **One x-intercept can be located by solving $a\sqrt[3]{x - h} + k = 0$.**
- **If $a > 0$, the long-term behaviour is $x \to \pm\infty, y \to \pm\infty$.**
- **If $a < 0$, the long-term behaviour is $x \to \pm\infty, y \to \mp\infty$.**

The long-term behaviour of the cube root function resembles that of the cubic function.

WORKED EXAMPLE 28 Cube root and square root graphs

a. i. State the maximal domain of $y = \sqrt{4 - x} - 1$.

ii. Sketch the graph of $y = \sqrt{4 - x} - 1$ and state its range.

b. The graph of a cube root function has its point of inflection at $(1, 5)$ and the graph cuts the y-axis at $(0, 2)$. Determine the rule and sketch the graph.

THINK	WRITE
a. i. Form the maximal domain.	a. i. $y = \sqrt{4 - x} - 1$ The term under the square root cannot be negative. $4 - x \geq 0$ $x \leq 4$ The maximal domain is $(-\infty, 4]$.
ii. 1. State the coordinates of the end point.	ii. The end point is $(4, -1)$.
2. Calculate the y-intercept, if there is one.	With the domain $(-\infty, 4]$, the graph opens to the left, so it will cut the y-axis. y-intercept: let $x = 0$. $y = \sqrt{4} - 1$ $y = 1$ The y-intercept is $(0, 1)$.

3. Calculate the x-intercept, if there is one.

The end point lies below the x-axis and the y-intercept lies above the x-axis. There will be an x-intercept.

x-intercept: let $y = 0$.

$$0 = \sqrt{4-x} - 1$$
$$\sqrt{4-x} = 1$$
$$4 - x = 1$$
$$x = 3$$

The x-intercept is $(3, 0)$.

4. Sketch the graph.

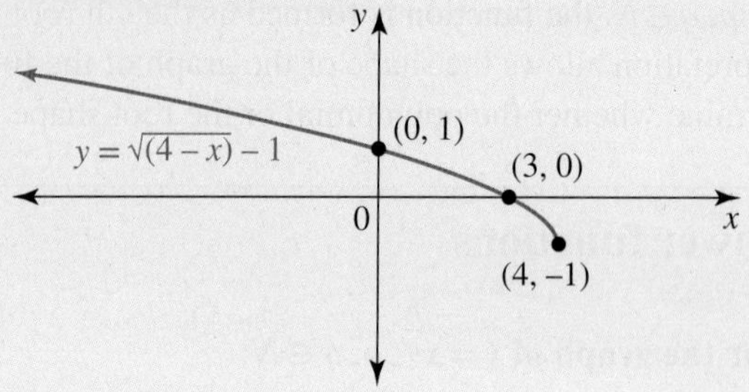

5. State the range.

The range is $[-1, \infty)$.

b. 1. Write the general equation of a cube root function.

b. Let the equation be $y = a\sqrt[3]{x-h} + k$.

2. Insert the information about the point of inflection.

The point of inflection is $(1, 5)$.

$\therefore y = a\sqrt[3]{x-1} + 5$

3. Fully determine the equation using the other piece of information given.

Substitute the point $(0, 2)$:

$$2 = a\sqrt[3]{-1} + 5$$
$$2 = -a + 5$$
$$a = 3$$

The equation is $y = 3\sqrt[3]{x-1} + 5$.

4. Calculate the x-intercept.

x-intercept: let $y = 0$.

$$0 = 3\sqrt[3]{x-1} + 5$$
$$\sqrt[3]{x-1} = \frac{-5}{3}$$
$$x - 1 = \left(-\frac{5}{3}\right)^3$$
$$x = 1 - \frac{125}{27}$$
$$x = -\frac{98}{27}$$

The x-intercept is $\left(-\frac{98}{27}, 0\right)$.

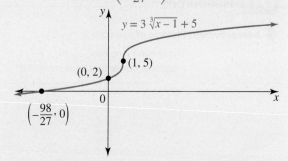

5. Sketch the graph.

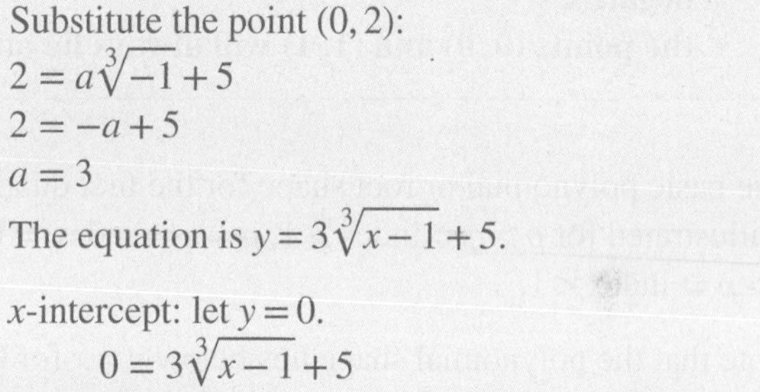

1.7.5 Power functions of the form $y = x^{\frac{p}{q}}$, $p, q \in N$

The square root and cube root functions are examples of power functions of the form $y = x^{\frac{p}{q}}$, $p, q \in N$. For the square root function, $y = \sqrt{x} = x^{\frac{1}{2}}$, so $p = 1$ and $q = 2$; for the cube root function, $y = \sqrt[3]{x} = x^{\frac{1}{3}}$, so $p = 1$ and $q = 3$.

In this section, we consider some other functions that have powers that are positive rational numbers and deduce the shape of their graphs through an analysis based on index laws.

Index laws enable $x^{\frac{p}{q}}$ to be expressed as $\sqrt[q]{x^p}$.

With $p, q \in N$, the function is formed as the qth root of the polynomial x^p. As polynomial shapes are known, this interpretation allows the shape of the graph of the function to be deduced. Whichever is the larger of p and q will determine whether the polynomial or the root shape will be the **dominant function**.

Power functions

For the graph of $y = x^{\frac{p}{q}}, p, q \in N$:

- **if $p > q$, the polynomial shape dominates, because the index $\frac{p}{q} > 1$**
- **if $q > p$, the root shape dominates, because the index must be in the interval $0 < \frac{p}{q} < 1$**
- **if $p = q$, the index is 1 and the graph is $y = x$**
- **even roots of the polynomial x^p cannot be formed in any section where the polynomial graph is negative**
- **the points (0, 0) and (1, 1) will always lie on the graph.**

The basic polynomial or root shape for the first quadrant is illustrated for $p > q \Rightarrow$ index > 1, $p = q \Rightarrow$ index $= 1$ and $q > p \Rightarrow$ index < 1.

Note that the polynomial shape lies below $y = x$ for $0 < x < 1$ and above $y = x$ for $x > 1$, whereas the root shape lies above $y = x$ for $0 < x < 1$ and below $y = x$ for $x > 1$. It is always helpful to include the line $y = x$ when sketching a graph of the form $y = x^{\frac{p}{q}}$.

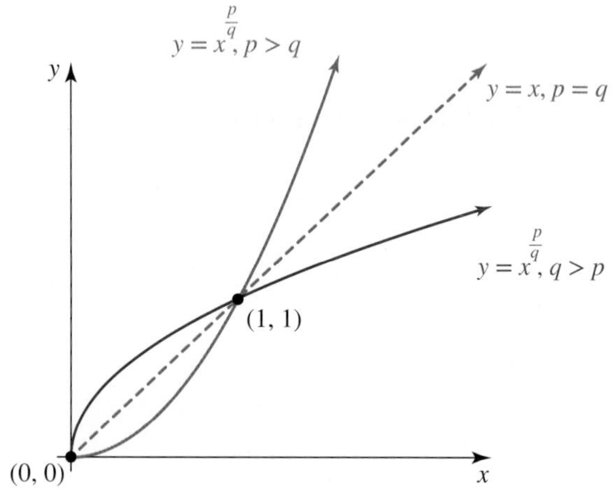

on Resources

Interactivities Patterns of functions (int-6415)

The relation $y^2 = x$ (int-2574)

Fractional power functions (int-6521)

WORKED EXAMPLE 29 Graphs of power functions

Give the domain and deduce the shape of the graph of each of following.

a. $y = x^{\frac{2}{3}}$ **b.** $y = x^{\frac{3}{2}}$

THINK

WRITE

a. 1. Express the function rule in surd form and deduce how the function can be formed.

a. $y = x^{\frac{2}{3}}$

$= \sqrt[3]{x^2}$

The function is formed as the cube root of the quadratic polynomial $y = x^2$.

2. Use the nature of the operation forming the function to determine the domain of the function.

Cube roots of both positive and negative numbers can be calculated. However, the graph of $y = x^2$ lies in quadrants 1 and 2 and is never negative. Therefore, there will be two non-negative branches to the power function, giving it a domain of R.

3. Reason which shape, the root or the polynomial, will dominate.

As $3 > 2$ (or as the index is less than 1), the root shape dominates the graph. This means the graph lies above $y = x$ for $0 < x < 1$ and below it for $x > 1$.

4. Draw the required graph, showing its position relative to the line $y = x$.
Note: There is a sharp point at the origin.

The points (0, 0) and (1, 1) lie on the graph, and by symmetry the graph will also pass through the point (−1, 1).

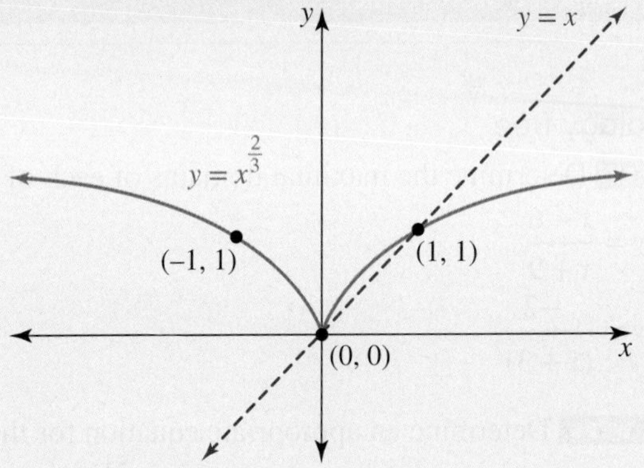

b. 1. Express the function rule in surd form and deduce how the function can be formed.

b. $y = x^{\frac{3}{2}}$

$= \sqrt{x^3}$

The function is formed as the square root of the cubic polynomial $y = x^3$.

2. Use the nature of the operation forming the function to determine the domain of the function.

The graph of $y = x^3$ is positive in quadrant 1 and negative in quadrant 3, so the square root can only be taken of the section in quadrant 1. There will be one branch and its domain will be $R^+ \cup \{0\}$.

3. Reason which shape, the root or the polynomial, will dominate.

As $3 > 2$ (or as the index is greater than 1), the polynomial shape dominates. The graph will lie below $y = x$ for $0 < x < 1$ and above it for $x > 1$.

4. Draw the required graph, showing its position relative to the line $y = x$.

The points $(0, 0)$ and $(1, 1)$ lie on the graph.

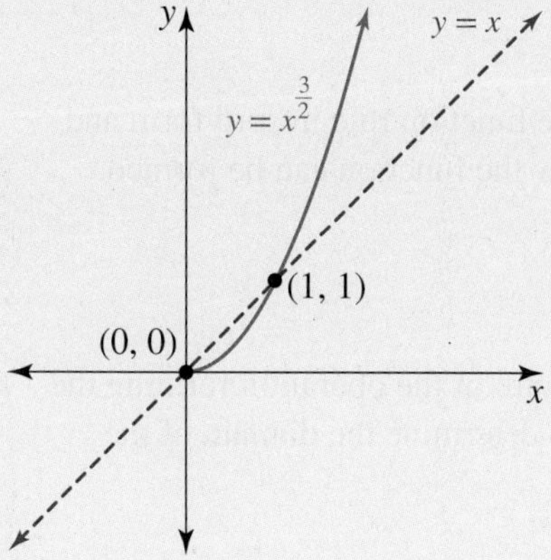

1.7 Exercise

Technology free

1. **WE25** Determine the maximal domains of each of the following functions.

   a. $y = \dfrac{x-6}{x+9}$

   b. $y = \sqrt{1-2x}$

   c. $y = \dfrac{-2}{(x+3)^2}$

   d. $y = \dfrac{1}{x^2+3}$

2. a. **WE26** Determine an appropriate equation for the hyperbola shown.

   b. i. Obtain the maximal domain of $y = \dfrac{5x-2}{x-1}$.

   ii. Sketch the graph of $y = \dfrac{5x-2}{x-1}$ and state its range.

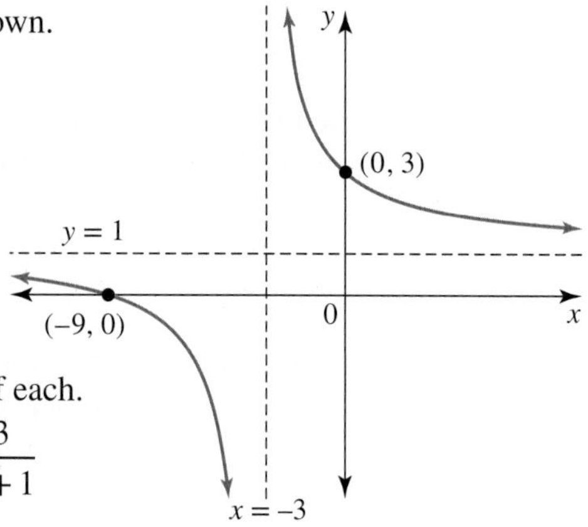

3. Sketch the graph of $y = \dfrac{4}{1-2x}$, stating its domain and range.

4. Sketch the following hyperbolas and state the domain and range of each.

   a. $y = \dfrac{4}{x} + 5$

   b. $y = 2 - \dfrac{3}{x+1}$

   c. $y = \dfrac{4x+3}{2x+1}$

   d. $xy + 2y + 5 = 0$

   e. $y = \dfrac{10}{5-x} - 5$

5. a. The graph of a hyperbola has a vertical asymptote at $x=-3$ and a horizontal asymptote at $y=6$. The point $(-4, 8)$ lies on the graph. Form the equation of this graph.
b. Form a possible equation for the graph shown.

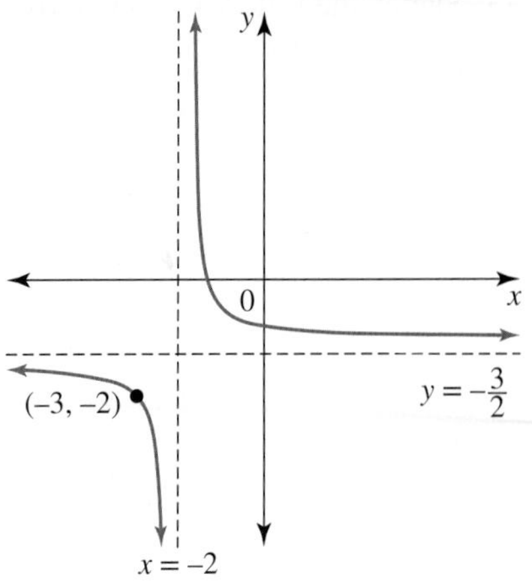

6. WE27 Sketch the graph of $y=\dfrac{8}{(x+2)^2}-2$ and state its domain and range.

7. Sketch each of the following and state the domain and range of each.

a. $y=\dfrac{2}{(3-x)^2}+1$
b. $y=\dfrac{-3}{4(x-1)^2}-2$
c. $y=\dfrac{1}{(2x+3)^2}-1$
d. $y=\dfrac{25x^2-1}{5x^2}$

8. Determine an appropriate equation for the truncus shown.

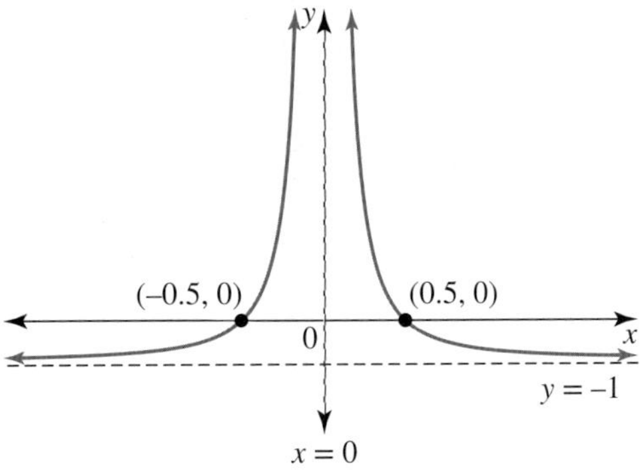

Technology active

9. a. The diagram shows the graph of a truncus. Form its equation.
b. A function f defined on its maximal domain has a graph $y=f(x)$ in the shape of a truncus with range $(-4, \infty)$. Given $f(-1)=8$ and $f(2)=8$, determine the equation of the graph and state the function f using function notation.

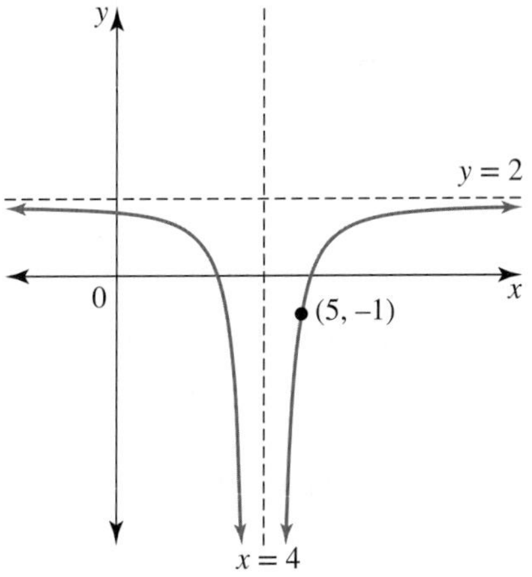

10. WE28
a. i. State the maximal domain of $y=-\sqrt{x+9}+2$.
ii. Sketch the graph of $y=-\sqrt{x+9}+2$ and state its range.
b. The graph of a cube root function has its point of inflection at $(1, 3)$ and the graph cuts the y-axis at $(0, 1)$. Determine its rule and sketch its graph, locating its x-intercept.

11. a. Determine the maximal domain and the range of $y=3\sqrt{4x-9}-6$, and sketch its graph.
b. State the coordinates of the point of inflection of the graph of $y=(10-3x)^{\frac{1}{3}}$ and sketch the graph.

12. Give the equations of the two square root functions that form the branches of each of the following 'sideways' parabolas, and state the domain and range of each function.

a. $(y-2)^2=4(x-3)$
b. $y^2+2y+2x=5$

13. Sketch the following square root functions and state the domain and range of each.

a. $y = 1 - \sqrt{3x}$

b. $y = 2\sqrt{-x} + 4$

c. $y = 2\sqrt{4 + 2x} + 3$

d. $y = -\sqrt{3} - \sqrt{12 - 3x}$

14. a. The graph of the function $f : [5, \infty) \rightarrow R, f(x) = a\sqrt{x + b} + c$ is shown in the diagram. Determine the values of a, b and c.

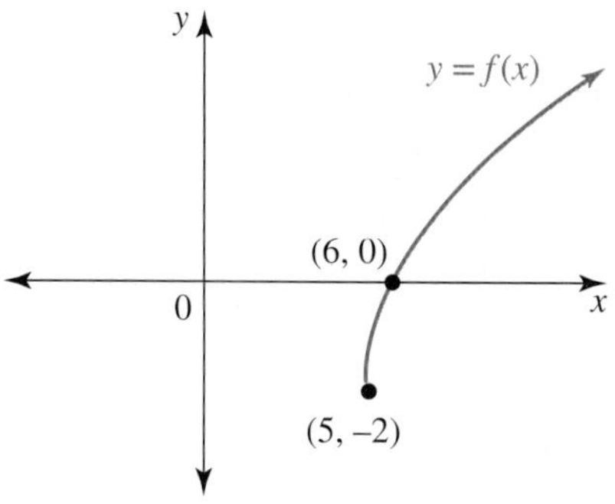

b. The graph of the function $f : (-\infty, 2] \rightarrow R, f(x) = \sqrt{ax + b} + c$ is shown in the diagram.

i. Determine the values of a, b and c.

ii. State the equation of the reflection if the graph of $y = f(x)$ is reflected in the x-axis.

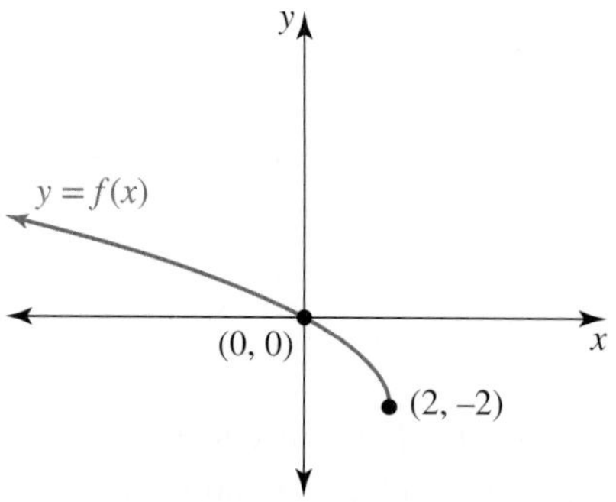

15. a. Sketch the graph of $\left\{(x, y) : y = \sqrt[3]{x + 2} - 1\right\}$, labelling the intercepts with the coordinate axes with their exact coordinates.

b. Sketch the graph of $y = f(x)$ where $f(x) = \dfrac{1 - \sqrt[3]{x + 8}}{2}$, stating its implied domain and range.

c. Sketch the graph of $g : [-3, 6] \rightarrow R,\ g(x) = \sqrt[3]{-x + 5}$ and state its domain and range.

d. Form a possible equation for the cube root function shown in the diagram.

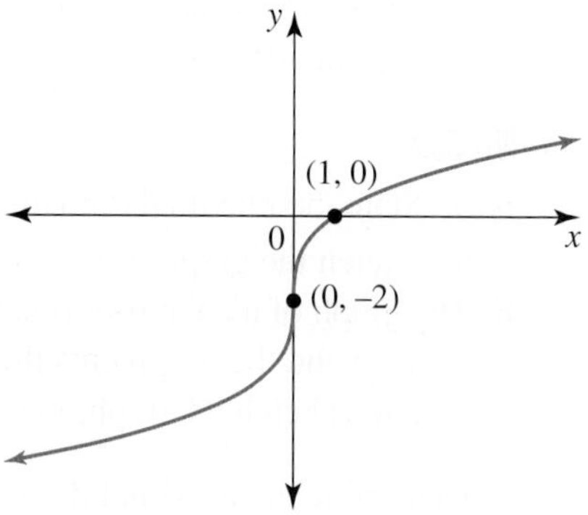

e. The graph of a cube root function passes through the points $(-9, 5)$ and $(-1, -2)$. At the point $(-1, -2)$, the tangent drawn to the curve is vertical. Determine the equation of the graph.

f. Express y as the subject of the equation $(y + 2)^3 = 64x - 128$ and hence state the coordinates of the point of inflection of its graph.

16. State the maximal domain of the function $y = \sqrt{(2 - x)(x + 3)}$. Use CAS technology to investigate the shape of the graph.

17. **WE29** Give the domain and deduce the shape of the graph of each of the following.

a. $y = x^{\frac{3}{4}}$

b. $y = x^{\frac{4}{3}}$

18. Give the domain and deduce the shape of the graph of each of the following.

a. $y = x^{\frac{1}{5}}$

b. $y = x^{\frac{1}{8}}$

19. For each of the following, identify the domain and the quadrants in which the graph lies, and sketch the graph, showing its position relative to the line $y = x$.

a. $y = x^{\frac{5}{2}}$

b. $y = x^{\frac{5}{3}}$

c. $y = x^{\frac{3}{5}}$

d. $y = x^{0.25}$

20. a. Explain how the graph of $y = x^{\frac{1}{3}}$ could be drawn using the graph of $y = x$.

b. On the same set of axes, sketch the graphs of $y = x$ and $y = x^{\frac{1}{3}}$.

c. Hence, obtain $\left\{x : x^{\frac{1}{3}} - x > 0\right\}$.

1.7 Exam questions

Question 1 (1 mark) TECH-ACTIVE

Source: *VCE 2020, Mathematical Methods Exam 2, Section A, Q5; © VCAA.*

MC The graph of function $f : D \to R, f(x) = \dfrac{3x+2}{5-x}$, where D is the maximal domain, has asymptotes

A. $x = -5,\ y = \dfrac{3}{2}$

B. $x = -3,\ y = 5$

C. $x = \dfrac{2}{3},\ y = -3$

D. $x = 5,\ y = 3$

E. $x = 5,\ y = -3$

Question 2 (1 mark) TECH-ACTIVE

Source: *VCE 2020, Mathematical Methods Exam 2, Section A, Q18; © VCAA.*

MC Let $a \in (0, \infty)$ and $b \in R$.

Consider the function $h : [-a, 0) \cup (0, a] \to R,\ h(x) = \dfrac{a}{x} + b$.

The range of h is

A. $[b-1, b+1]$

B. $(b-1, b+1)$

C. $(-\infty, b-1) \cup (b+1, \infty)$

D. $(-\infty, b-1] \cup [b+1, \infty)$

E. $[b-1, \infty)$

Question 3 (1 mark) TECH-ACTIVE

Source: *VCE 2018, Mathematical Methods Exam 2, Section A, Q2; © VCAA.*

MC The maximal domain of the function f is $R \backslash \{1\}$.

A possible rule for f is

A. $f(x) = \dfrac{x^2-5}{x-1}$

B. $f(x) = \dfrac{x+4}{x-5}$

C. $f(x) = \dfrac{x^2+x+4}{x^2+1}$

D. $f(x) = \dfrac{5-x^2}{1+x}$

E. $f(x) = \sqrt{x-1}$

More exam questions are available online.

1.8 Combinations of functions

LEARNING INTENTION

At the end of this subtopic you should be able to:
- sketch and determine hybrid functions
- sketch sums, differences and products of functions.

By combining together pieces of different functions defined over restricted domains, a 'piecewise' function can be created. By combining together different functions using arithmetic operations, other functions can be created. In this section we consider some of these combinations.

1.8.1 Hybrid functions

A **hybrid function**, or piecewise function, is a function whose rule takes a different form over different subsets of its domain. An example of a hybrid function is the one defined by the rule

$$f(x)=\begin{cases}\sqrt[3]{x}, & x\le 0\\ 2, & 0<x<2.\\ x, & x\ge 2\end{cases}$$

To sketch its graph, the three functions that combine to form its branches, $y=\sqrt[3]{x}$, $y=2$ and $y=x$, are drawn on their respective restricted domains on the same set of axes. If the branches do not join, then it is important to indicate which end points are open and which are closed, as each of the x-values of any function must have a unique y-value. The graph of this hybrid function $y=f(x)$ is shown in the diagram.

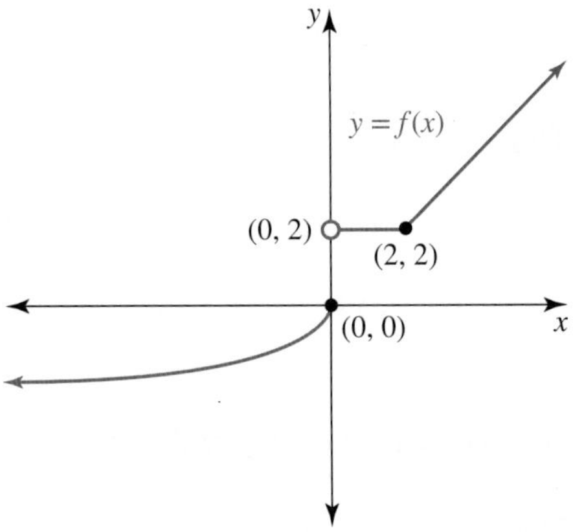

The function is not **continuous** when $x=0$ as the branches do not join for that value of x. It is said to be discontinuous at that point of its domain. As the rule shows, $x=0$ lies in the domain of the cube root section, the point $(0,0)$ is closed and the point $(0,2)$ is open.

The function is continuous at $x=2$ as there is no break or gap in the curve. There is no need for a closed point to be shown at $x=2$, because its two neighbouring branches run 'naturally' into each other at this point.

To calculate the value of the function for a given value of x, choose the function rule of that branch defined for the section of the domain to which the x-value belongs.

WORKED EXAMPLE 30 Sketching hybrid functions

Consider the function for which $f(x)=\begin{cases}\sqrt{-x}, & x\le -1\\ 2-x^2, & -1<x<1.\\ \sqrt{x}+1, & x\ge 1\end{cases}$

a. **Evaluate** $f(-1), f(0)$ **and** $f(4)$.
b. **Sketch the graph of** $y=f(x)$.
c. **State:**
   i. **any value of** x **for which the function is not continuous**
   ii. **the domain and range.**

THINK	WRITE
a. For each x-value, decide which section of the domain it is in and calculate its image using the branch of the hybrid function's rule applicable to that section of the domain.	**a.** $f(x) = \begin{cases} \sqrt{-x}, & x \le -1 \\ 2 - x^2 & -1 < x < 1 \\ \sqrt{x} + 1, & x \ge 1 \end{cases}$ $f(-1)$: Since $x = -1$ lies in the domain section $x \le -1$, use the rule $f(x) = \sqrt{-x}$. $f(-1) = \sqrt{-(-1)}$ $= \sqrt{1}$ $= 1$ $f(0)$: Since $x = 0$ lies in the domain section $-1 < x < 1$, use the rule $f(x) = 2 - x^2$. $f(0) = 2 - 0^2$ $= 2 - 0$ $= 2$ $f(4)$: Since $x = 4$ lies in the domain section $x \ge 1$, use the rule $f(x) = \sqrt{x} + 1$. $f(4) = \sqrt{4} + 1$ $= 2 + 1$ $= 3$
b. 1. Obtain the information needed to sketch each of the functions forming the branches of the hybrid function.	**b.** $y = \sqrt{-x}, x \le -1$ is a square root function. The points $(-1, 1)$ and $(-4, 2)$ lie on its graph. $y = 2 - x^2, -1 < x < 1$ is a parabola with maximum turning point $(0, 2)$. At $x = -1$ or $x = 1$, $y = 1$. The points $(-1, 1)$ and $(1, 1)$ are open for the parabola. $y = \sqrt{x} + 1, x \ge 1$ is a square root function. The points $(1, 2)$ and $(4, 3)$ lie on its graph.
2. Sketch each branch on the same set of axes to form the graph of the hybrid function.	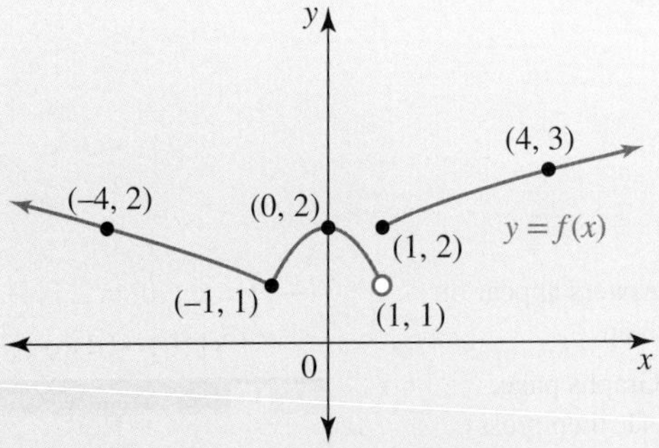
c. i. State any value of x where the branches of the graph do not join.	**c.** The function is not continuous at $x = 1$.
ii. State the domain and range.	The domain is R. The range is $[1, \infty)$.

TI | THINK | **DISPLAY/WRITE**

a. 1. On a Calculator page, press MENU, then select:
1: Actions
1: Define
Complete the entry line as:
Define $f1(x) =$
$\begin{cases} \sqrt{-x}, x \le -1 \\ 2 - x^2, -1 < x < 1 \\ \sqrt{x} + 1, x \ge 1 \end{cases}$
then press ENTER.
Note: The hybrid function template can be found by pressing the Template button.

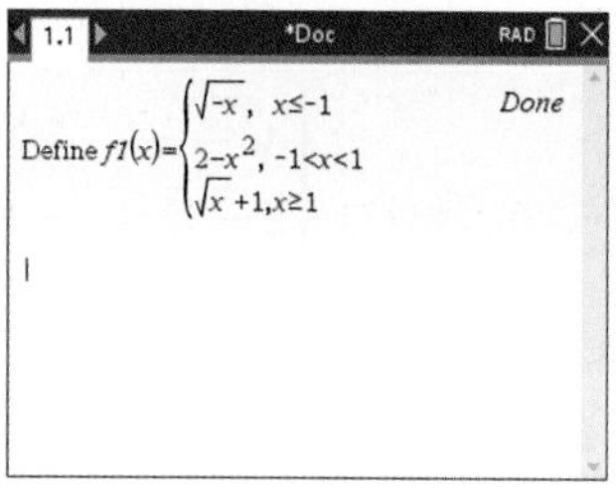

2. Complete the next entry line as:
$f1(-1)$
then press ENTER.
Complete the next entry line as:
$f1(0)$
then press ENTER.
Complete the next entry line as:
$f1(4)$
then press ENTER.

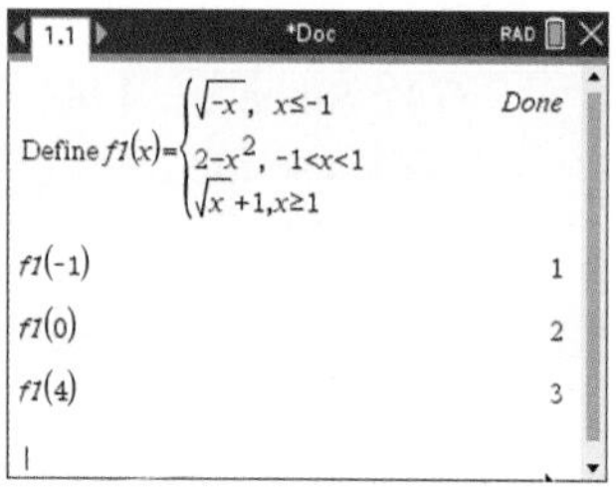

3. The answers appear on the screen.

$f(-1) = 1, f(0) = 2, f(4) = 3$

b. 1. On a Graphs page, select f1, then press ENTER.
Note: The hybrid function automatically appears in the entry line for f1 as it was previously defined on the Calculator page.

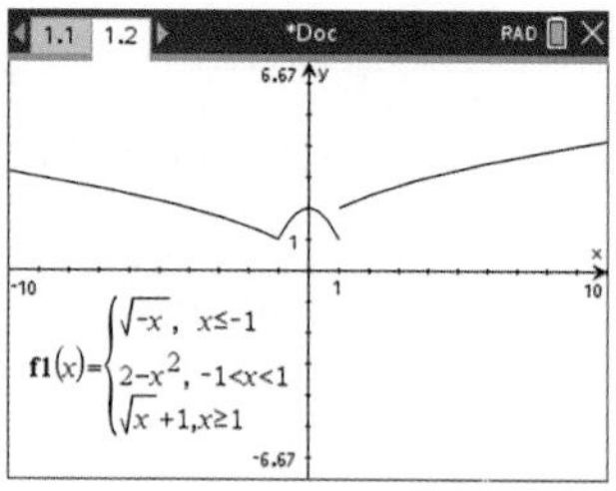

CASIO | THINK | **DISPLAY/WRITE**

a. 1. On a Graph & Table screen, complete the entry line for $y1$ as:
$y1 =$
$\begin{cases} \sqrt{-x}, x \le -1 \\ 2 - x^2, -1 < x < 1 \\ \sqrt{x} + 1, x \ge 1 \end{cases}$
then press EXE.
Note: The hybrid function template can be found in the Math 3 tab of the Keyboard menu. Click it twice for three equations.

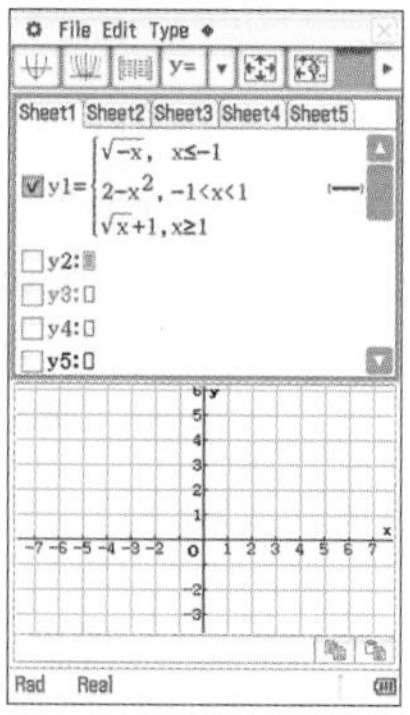

2. Click the Table Input icon and complete the fields as:
Start: -1
End: 4
Step: 1
then select OK.

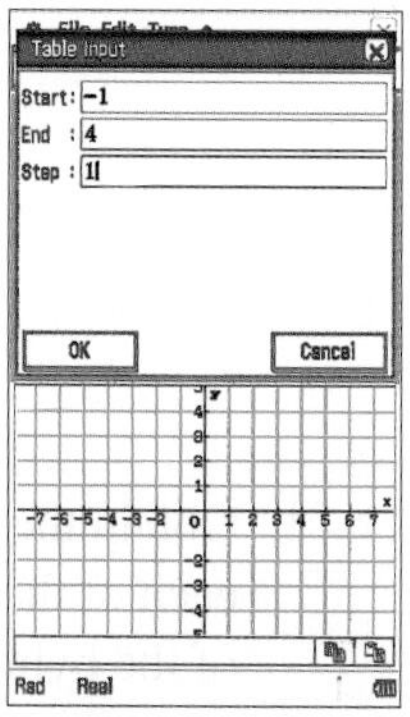

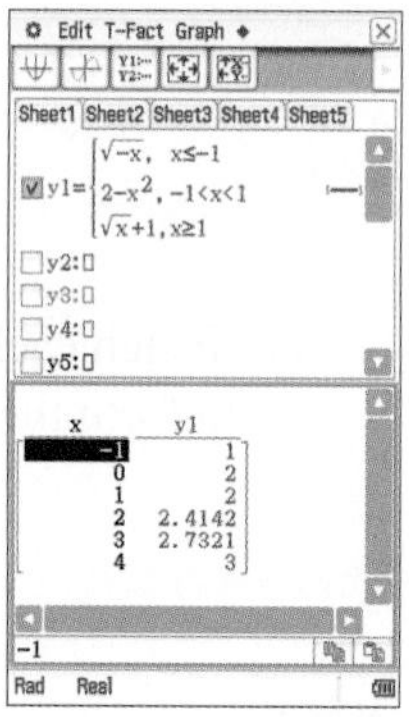

3. The answers appear on the screen.

$f(-1) = 1, f(0) = 2,$
$f(4) = 3$

b. 1. Click the Graph icon.

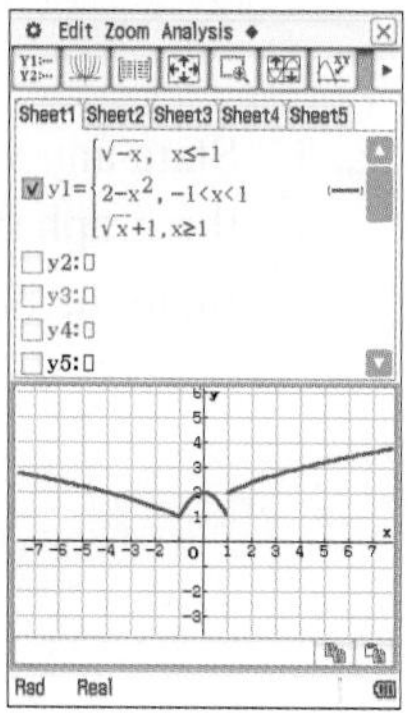

TI | THINK

2. Press MENU, then select:
5: Trace
1: Graph Trace
Type '−1', then press ENTER twice to mark the point $(-1, 1)$.
Type '1', then press ENTER twice to mark the point $(1, 2)$.
Press the down arrow to move to the point $(1, 1)$, then press ENTER.
Note: The point $(1, 1)$ should be drawn as an open circle.

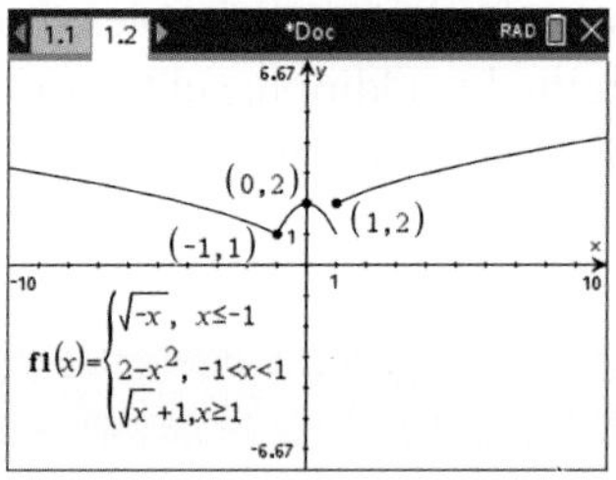

3. To find the maximum, press MENU, then select:
6: Analyze Graph
3: Maximum
Move the cursor to the left of the maximum when prompted for the lower bound, then press ENTER. Move the cursor to the right of the maximum when prompted for the upper bound, then press ENTER.

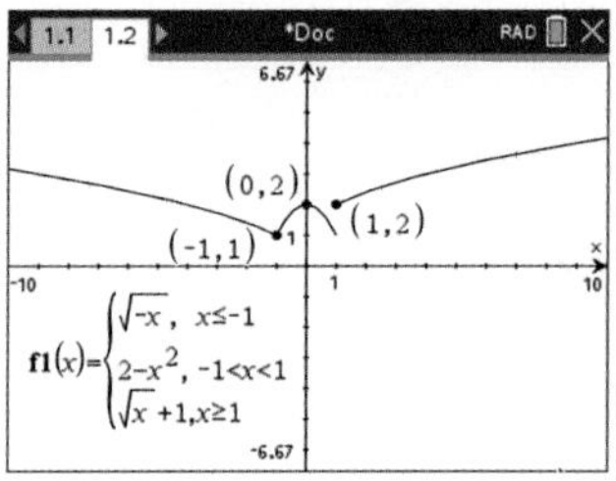

c. i. Look for points where a break in the graph occurs. — The graph is discontinuous at $x = 1$.

ii. State the domain and range. — The domain is R and the range is $[1, \infty]$.

CASIO | THINK

2. Select:
• Analysis
• Trace
Type '−1', select OK, then press EXE to mark the point $(-1, 1)$.
Type '1', select OK, then press EXE to mark the point $(1, 2)$.
Note: The point $(1, 1)$ should be drawn as an open circle.

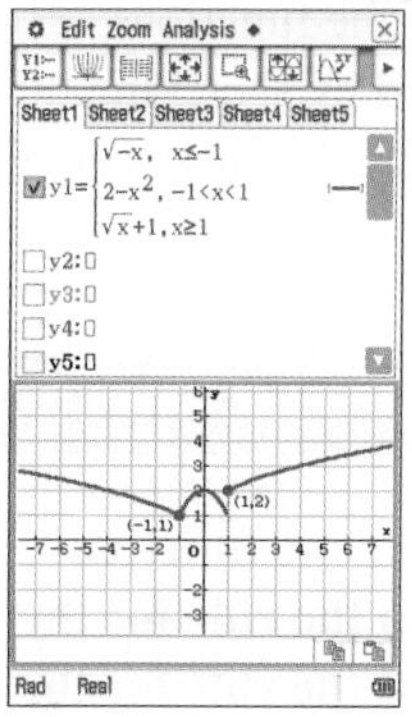

To find the maximum, select:
• Analysis
• G-Solve
• Max
then press EXE.

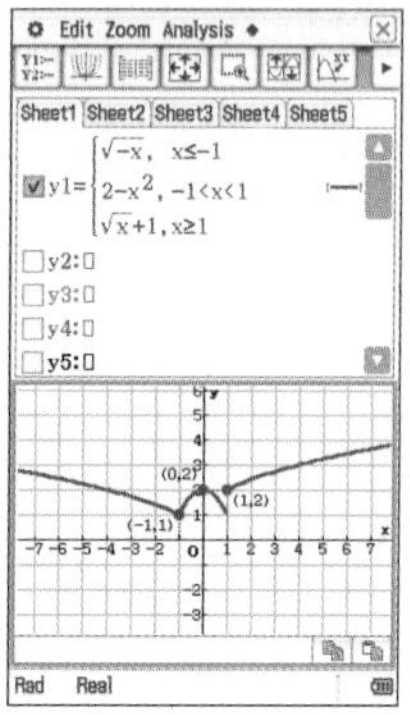

c. i. Look for points where a break in the graph occurs. — The graph is discontinuous at $x = 1$.

ii. State the domain and range. — The domain is R and the range is $[1, \infty]$.

on Resources

Interactivity Hybrid functions (int-6414)

1.8.2 Sums, differences and products of functions

New functions are formed when two given functions are combined together under the operations of addition, subtraction and multiplication. The given functions can only be combined where they both exist, so the domain of the new function formed must be the domain common to both the given functions. For functions f and g with domains d_f and d_g respectively, the common domain is $d_f \cap d_g$.

Sums, differences and products of functions

- **The sum and difference functions $f \pm g$ are defined by $(f \pm g)(x) = f(x) \pm g(x)$ with domain $d_f \cap d_g$.**
- **The product function fg is defined by $(fg)(x) = f(x)g(x)$ with domain $d_f \cap d_g$.**

Graphs of the functions $f \pm g$ and fg may be able to be recognised from their rules. If not, the graphs may be deduced by sketching the graphs of f and g and combining by addition, subtraction or multiplication, as appropriate, the values of $f(x)$ and $g(x)$ for selected x-values in their common domain. The difference function $f - g$ can be considered to be the sum function $f + (-g)$.

WORKED EXAMPLE 31 Determining sums and products of functions

Consider the functions f and g defined by $f(x) = \sqrt{4+x}$ and $g(x) = \sqrt{4-x}$ respectively.

a. Form the rule for the sum function $f+g$, stating its domain, and sketch the graph of $y = (f+g)(x)$.

b. Form the rule for the product function fg and state its domain and range.

THINK / **WRITE**

a. 1. Write the domains of the functions f and g.

a. $f(x) = \sqrt{4+x}$

Domain: $4 + x \geq 0$

$x \geq -4$

$d_f = [-4, \infty)$

$g(x) = \sqrt{4-x}$

Domain: $4 + x \geq 0$

$x \leq 4$

$d_g = (-\infty, 4]$

2. State the common domain.

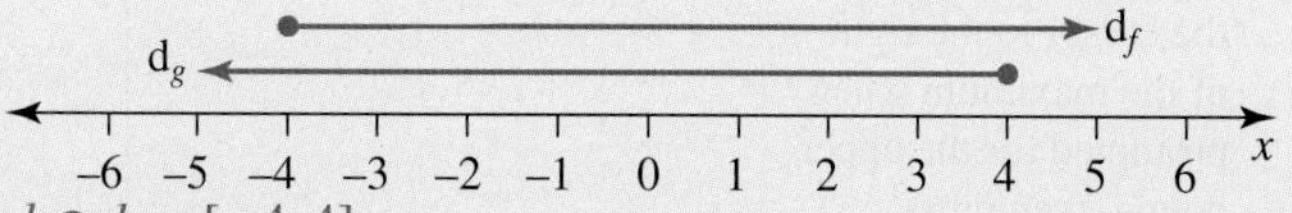

$d_f \cap d_g = [-4, 4]$

3. Form the sum function and state its domain.

$(f+g)(x) = f(x) + g(x)$

$= \sqrt{4+x} + \sqrt{4-x}$

$d_{f+g} = [-4, 4]$

4. Sketch the graphs of $y = f(x)$ and $y = g(x)$ on the same set of axes. Add the y-coordinates of key points together to form the graph of $y = (f+g)(x)$.

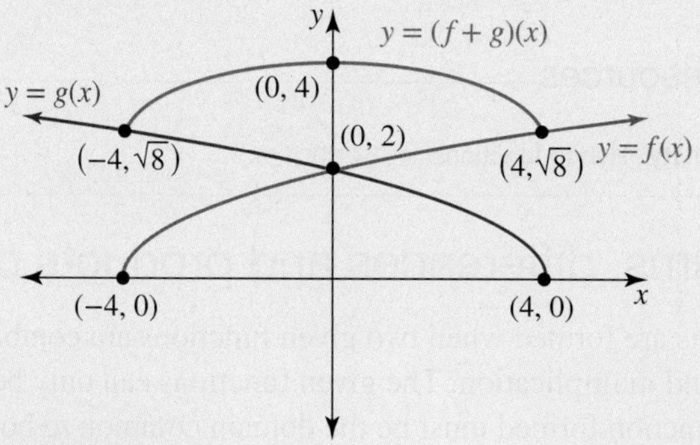

x	-4	0	4
$f(x)$	0	2	$\sqrt{8}$
$g(x)$	$\sqrt{8}$	2	0
$f(x) + g(x)$	$\sqrt{8}$	4	$\sqrt{8}$

b. 1. Form the product function and state its domain.

b. $(fg)(x) = f(x)\,g(x)$

$$= \left(\sqrt{4+x}\right) \times \left(\sqrt{4-x}\right)$$
$$= \sqrt{(4+x)(4-x)}$$
$$= \sqrt{16-x^2}$$
$$d_{fg} = [-4, 4]$$

2. State the range of the function.

The rule $(fg)(x) = \sqrt{16-x^2}$ is that of the top half of a semicircle with centre $(0, 0)$ and radius 4. Therefore, the range is $[0, 4]$.

on Resources

Interactivity Sums, differences and products of functions (int-6416)

1.8.3 Graphical techniques

Given the graphs of functions whose rules are not necessarily known, it may be possible to deduce the shape of the graph of the function that is the sum or other combination of the functions whose graphs are given.

Addition of ordinates

Given the graphs of $y_1 = f(x)$ and $y_2 = g(x)$, the graphing technique known as addition of ordinates adds together the y-values, or ordinates, of the two given graphs over the common domain to form the graph of the sum function $y = y_1 + y_2 = f(x) + g(x)$.

Graphical techniques

Note the following points when applying the addition of ordinates technique over the common domain $d_f \cap d_g$:

- **If the graphs of f and g intersect at (a, b), then the point $(a, 2b)$ lies on the graph of $f + g$.**
- **Where $f(x) = -g(x)$, the graph of $f + g$ cuts the x-axis.**
- **If one of $f(x)$ or $g(x)$ is positive and the other is negative, the graph of $f + g$ lies between the graphs of f and g.**
- **If one of $f(x)$ or $g(x)$ is zero, then the graph of $f + g$ cuts the other graph.**
- **If $f(x) \to 0^+$, then the graph of $f + g$ approaches the graph of f from above.**
- **If $f(x) \to 0^-$, then the graph of $f + g$ approaches the graph of g from below.**
- **Any vertical asymptote of f or g will be a vertical asymptote on the graph of $f + g$.**

The subtraction of ordinates is usually simpler to achieve as the addition of the ordinates of $y_1 = f(x)$ and $y_2 = -g(x)$.

Squaring ordinates

Given the graph of $y = f(x)$, the graph of $y = (f(x))^2$ can be deduced by squaring the y-values, or ordinates, noting in particular that $0^2 = 0$, $1^2 = 1$ and $(-1)^2 = 1$.

- The graph of f and its square will intersect at any point on f where $y = 0$ or $y = 1$.
- If the point $(a, -1)$ is on the graph of f, then $(a, 1)$ lies on the graph of the squared function.
- The squared function's graph can never lie below the x-axis.
- Where $0 < f(x) < 1$, $(f(x))^2 < f(x)$, and where $f(x) > 1$ or $f(x) < -1$, $(f(x))^2 > f(x)$.

Similar reasoning about the ordinates and their square roots and the domain will allow the graph of $y = \sqrt{f(x)}$ to be deduced.

These graphing techniques can be applied to combinations of known functions where the first step would be to draw their graphs.

WORKED EXAMPLE 32 Addition of ordinates

The graphs of the functions f and g are shown.

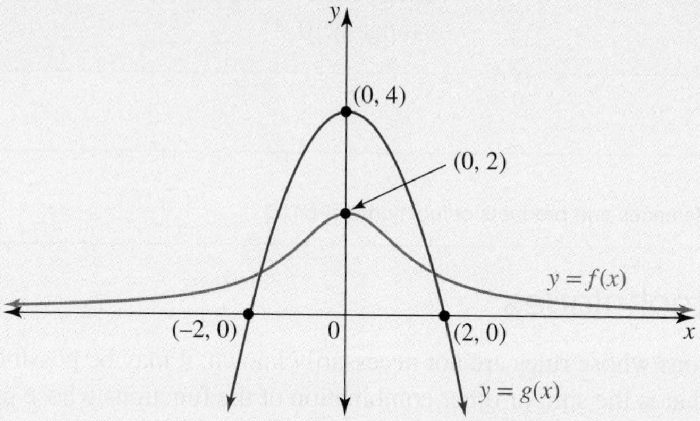

Draw the graph of $y = (f + g)(x)$.

THINK	WRITE
1. State the domain common to both functions.	Both of the functions have a domain of R, so $d_f \cap d_g = R$.
2. Determine the coordinates of a key point on the required graph.	At $x = 0, f(x) = 2$ and $g(x) = 4$. Hence, the point $(0, 6)$ lies on the graph of $f + g$.
3. Deduce the behaviour of the required graph where one of the given graphs cuts the x-axis.	At $x = \pm 2$, $g(x) = 0$. Hence, the graph of $f + g$ will cut the graph of f when $x = \pm 2$.
4. Use the long-term behaviour of one of the given graphs to deduce the long-term behaviour of the required graph.	As $x \to \pm\infty, f(x) \to 0^+$. Hence, $(f + g)(x) \to g(x)$ from above as $x \to \pm\infty$.
5. Draw a sketch of the required graph.	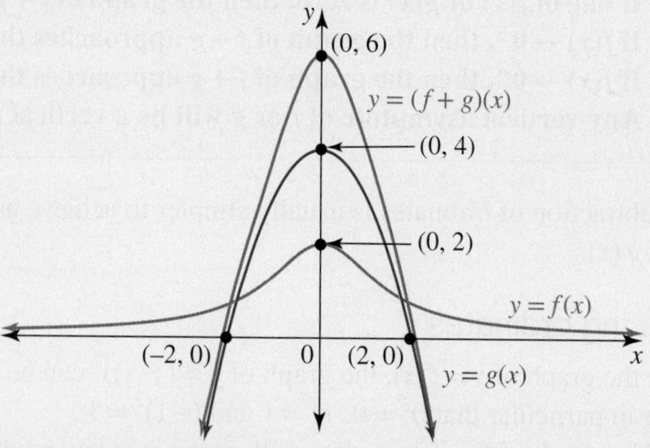

1.8 Exercise

Technology free

1. WE30 Consider the function for which:

$$f(x)=\begin{cases}-\sqrt[3]{x}, & x<-1\\ x^3 & -1\le x\le 1.\\ 2-x, & x>1\end{cases}$$

a. Evaluate $f(-8), f(-1)$ and $f(2)$.
b. Sketch the graph of $y=f(x)$.
c. State:
   i. any value of x for which the function is not continuous
   ii. the domain and range.

2. Sketch the graphs of each of the following hybrid functions and state their domains, ranges and any points of discontinuity.

a. $y=\begin{cases}-2x, & x\le 0\\ 4-x^2, & x>0\end{cases}$

b. $y=\begin{cases}\sqrt[3]{x}, & x<1\\ \dfrac{1}{x}, & x\ge 1\end{cases}$

3. A hybrid function is defined by:

$$f(x)=\begin{cases}\dfrac{1}{(x+1)^2}, & x<-1\\ x^2-x, & -1\le x\le 2\\ 8-2x, & x>2\end{cases}$$

a. Evaluate:
   i. $f(-2)$
   ii. $f(2)$.
b. Sketch the graph of $y=f(x)$.
c. State the domain over which the hybrid function is continuous.

4. Consider the following function.

$$f: R\to R,\ f(x)=\begin{cases}\dfrac{1}{9}x^3+5, & x<-3\\ \sqrt{1-x}, & -3\le x\le 1.\\ x-2, & x>1\end{cases}$$

a. Show the function is not continuous at $x=1$.
b. Sketch the graph of $y=f(x)$ and state the type of correspondence it displays.
c. Determine the value(s) of x for which $f(x)=4$.

5. Form the rule for the hybrid function shown in the diagram.

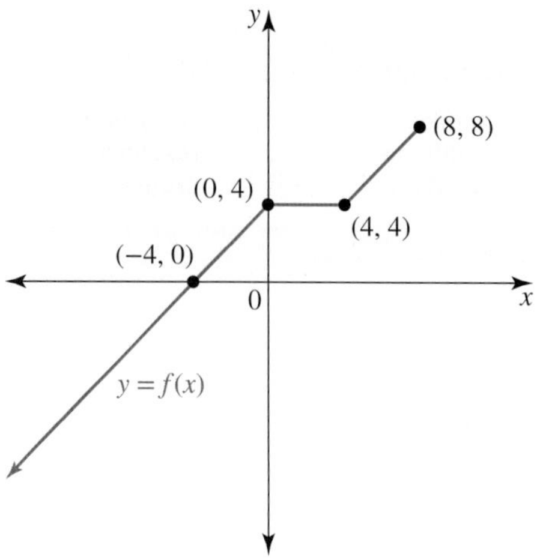

6. Form the rule for the function whose graph is shown in the diagram.

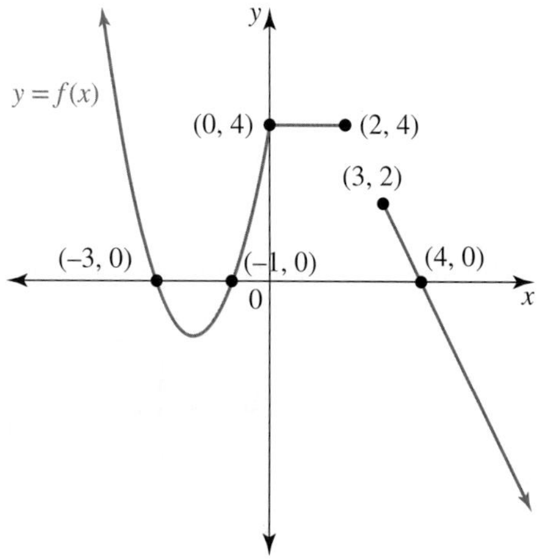

Technology active

7. **WE31** Consider the functions f and g defined by $f(x) = -\sqrt{1+x}$ and $g(x) = -\sqrt{1-x}$ respectively.
   a. Form the rule for the sum function $f + g$, stating its domain, and sketch the graph of $y = (f + g)(x)$.
   b. Form the rule for the product function fg and state its domain and range.

8. Given $f(x) = x^3$ and $g(x) = x^2$, form the rule $(f - g)(x)$ for the difference function and sketch the graphs of $y = f(x)$, $y = -g(x)$ and $y = (f - g)(x)$ on the same set of axes. Comment on the relationship of the graphs at the places where $y = (f - g)(x)$ cuts the axes.

9. **WE32** The graphs of the functions f and g are shown. Draw the graph of $y = (f + g)(x)$.

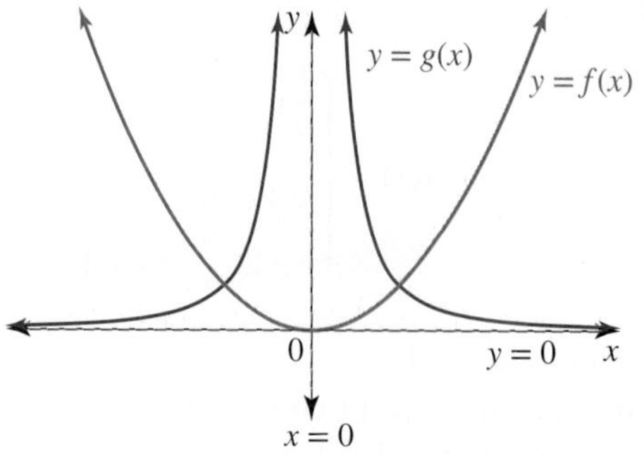

10. The graphs of two functions $y = f(x)$ and $y = g(x)$ are drawn in the following diagrams. Use the addition of ordinates technique to sketch $y = f(x) + g(x)$ for each diagram.

a. 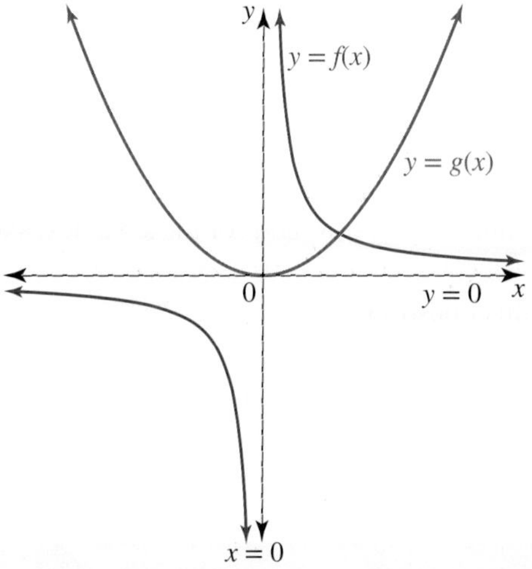

b. 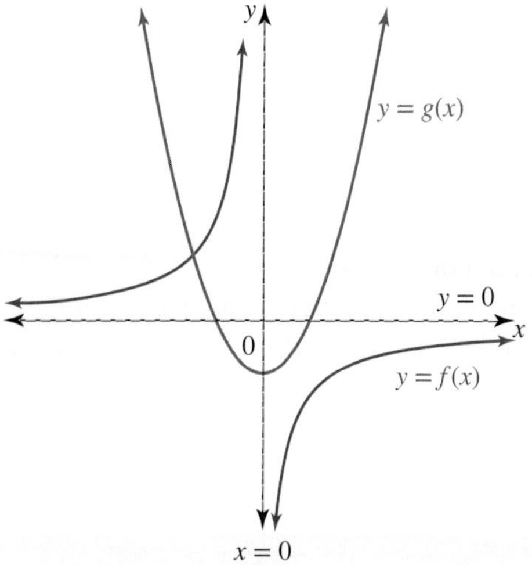

11. Consider the functions f and g defined by $f(x) = 5 - 2x$ and $g(x) = 2x - 2$ respectively. For each of the following, give the rule, state the domain and the range, and sketch the graph.

a. $y = (f + g)(x)$
b. $y = (f - g)(x)$
c. $y = (fg)(x)$

12. Consider the functions $f(x) = x^2 - 1$ and $g(x) = \sqrt{x + 1}$.

a. Evaluate:

i. $(g - f)(3)$
ii. $(gf)(8)$.

b. State the domain of the function $f + g$.
c. Draw a possible graph for each of the following functions.

i. $f + g$
ii. $g - f$
iii. fg

13. Use addition of ordinates to sketch $y = x + \sqrt{-x}$.

14. Sketch the graph of $y = x^2 - 1$ and hence draw the graph of $y = \left(x^2 - 1\right)^2$, stating the domain and range.

15. a. Consider the function defined by $g(x) = (2x - 1)^3$. Sketch the graph of $y = g(x)$ and hence sketch $y = (g(x))^2$.
b. Calculate the coordinates of the points of intersection of the graphs of $y = f(x)$ and $y = (f(x))^2$ if $f(x) = x^3 - 2x$.

16. Use CAS technology to draw on screen the hybrid function defined by the rule:

$$f(x) = \begin{cases} -x, & x < -1 \\ 1, & -1 \le x \le 1 \\ (2x - 1)(x - 3), & x > 1 \end{cases}$$

State the range of the function.

17. A hybrid function is defined by:

$$f(x)=\begin{cases} x+a, & x\in(-\infty,-8) \\ \sqrt[3]{x}+2, & x\in(-8,8) \\ \dfrac{b}{x}, & x\in(8,\infty) \end{cases}$$

a. Determine the values of a and b so that the function is continuous for $x \in R$, and for these values, sketch the graph of $y=f(x)$.
Use the values of a and b from part **a** for the remainder of this question.

b. Determine the values of k for which the equation $f(x)=k$ has:

i. no solutions
ii. one solution
iii. two solutions.

c. Find $\{x : f(x)=1\}$.

1.8 Exam questions

Question 1 (1 mark) TECH-ACTIVE

MC The diagram shows the graphs of three functions: $y=f(x)$, $y=g(x)$ and $y=h(x)$. State which of the following statements is true.

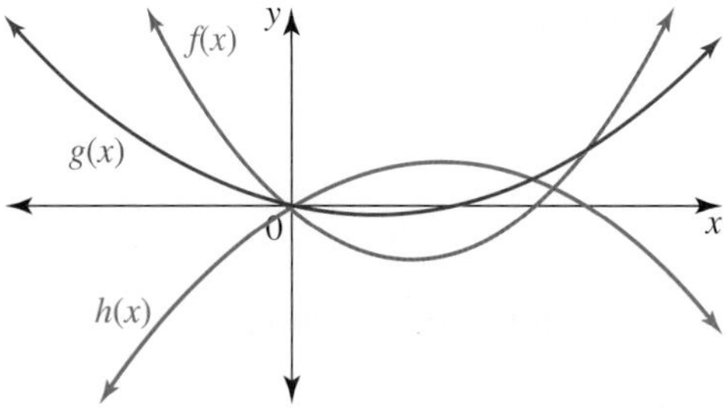

A. $f(x)=g(x)-h(x)$
B. $g(x)=f(x)-h(x)$
C. $g(x)=h(x)-f(x)$
D. $h(x)=f(x)-g(x)$
E. $h(x)=g(x)-f(x)$

Question 2 (1 mark) TECH-ACTIVE

MC Given the functions $f(x)=\sqrt{x+3}$ and $g(x)=\sqrt{1-x}$, the graph of $y=f(x)-g(x)$ has

A. a maximal domain of $[-3,\infty)$ and a range of R.
B. a maximal domain of $[1,\infty)$ and a range of R^+.
C. a maximal domain of $(-\infty,-3]\cup[1,\infty)$ and a range of R.
D. a maximal domain of $[-3,1]$ and a range of R^+.
E. a maximal domain of $[-3,1]$ and a range of $[-2,2]$.

Question 3 (1 mark) TECH-ACTIVE

MC For the function defined by $f(x)=\begin{cases} x-2, & x\geq 1 \\ -1, & -1<x<1 \\ x, & x\leq -2 \end{cases}$ then

A. $f(1)=-1$ and the range is R.
B. $f(1)=-1$ and the range is $[-2,-1]$.
C. $f(-1)=-1$ and the range is $R\backslash(-2,-1)$.
D. $f(-1)$ does not exist and the range is $R\backslash(-2,-1]$.
E. $f(-1)$ does not exist and the range is $R\backslash[-2,-1)$.

More exam questions are available online.

1.9 Modelling and applications

LEARNING INTENTION

At the end of this subtopic you should be able to:
- apply various functions to practical situations
- determine the best function to fit a set of data.

People in research occupations, such as scientists, engineers and economists, analyse data though the use of mathematical models in order to increase our understanding of natural phenomena and to draw inferences about future behaviour. In this subtopic we consider some applications of the functions that are discussed earlier in this topic.

1.9.1 Modelling with data

Consider the set of data shown in the table.

t	2	4	7	9
h	15	23.5	21	8.5

In deciding what type of model this data might best fit, a linear model would be ruled out as the data is not steadily increasing or decreasing. The values increase and then decrease; there are no obvious signs that the data is oscillating or showing asymptotic behaviour. Observations such as these would rule out an exponential model, a trigonometric model, and a hyperbola or truncus model.

The data is likely to be a polynomial model with a many-to-one correspondence. Plotting the points can help us recognise a possible model. If the variables t and h are time and height respectively, then we may suspect the polynomial would be a quadratic one. Three of the data points could be used to form the model in the form $h = at^2 + bt + c$, or the entire set of data could be used to obtain the model through a quadratic regression function on CAS. The quadratic model $h(t) = -0.9t^2 + 9.3t$ shows a good fit with the data.

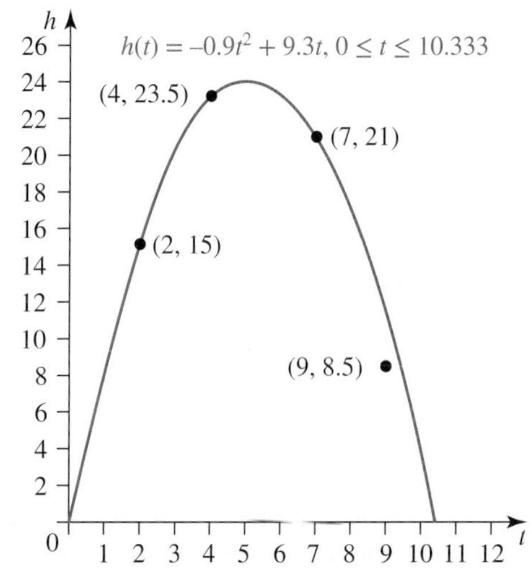

Applications of mathematical models

The variables in a mathematical model are usually treated as continuous, even though they may represent a quantity that is discrete in reality, such as the number of foxes in a region. Values obtained using the model need to be considered in context and rounded to whole numbers where appropriate.

Domain restrictions must also be considered. A variable representing a physical quantity such as length must be positive. Similarly, a variable representing time usually cannot be negative. However, it is important to read carefully how the variables are defined. For example, if t is the time in hours after 10 am, then $t = -2$ would be possible as it refers to the time 8 am.

1.9 Exercise

Students, these questions are even better in jacPLUS

Receive immediate feedback and access sample responses

Access additional questions

Track your results and progress

Find all this and MORE in jacPLUS

Technology active

1. Consider the data points shown.

x	0	1	3	4
y	4	2	10	8

a. Discuss why neither a linear, trigonometric, exponential nor a power function of the form $y = x^n$ is a likely fit for the data.

b. Assuming the data set fits a hyperbola of the form $y = \dfrac{a}{x-2} + k, x \in [0, \infty) \setminus \{2\}$:

i. use the data to determine the equation of the hyperbola

ii. sketch the model, showing the data points.

2. The population, in thousands, of bees in a particular colony increases as shown in the table.

Month (t)	1	2	3	4
Population in thousands (P)	36	38.75	42.5	45

a. Plot the data points p against t and suggest a likely model for the data.

b. Use the values when $t = 2$ and $t = 4$ to form a rule for the model expressing P in terms of t.

c. If the variable t measures the number of months since January, determine how many bees were in the colony in January, according to the model.

d. State the rate of increase in the population of bees according to the model.

3. A parabolic skate ramp has been built at a local park. It is accessed by climbing a ladder to a platform as shown. The platform is 2 metres long. The horizontal distance from the origin is x and the vertical distance from the origin is y. The lowest point on the skate ramp is at $(5, 0)$ and the highest point is at $\left(9, \dfrac{32}{9}\right)$.

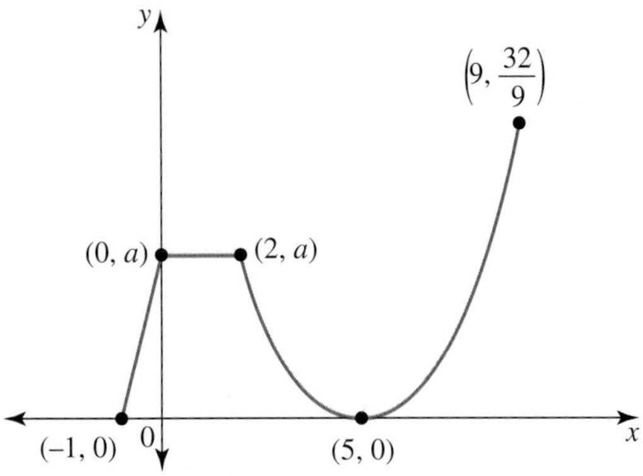

a. Calculate the value of a where $(0, a)$ is the point where the ladder connects with the platform.

b. Find the coordinates of the point where the platform and the skate ramp meet.

c. Determine the equation of the parabolic section of the skate ramp.

d. Write a hybrid function rule to define the complete skate ramp system for $\{x : -1 \le x \le 9\}$.

e. Determine the exact values of x when the skateboarder is 1.5 metres above the ground.

4. ABCD is a square field of side length 40 metres. The points E and F are located on AD and DC respectively so that ED = DF = x m. A gardener wishes to plant an Australian native garden in the region that is shaded in the diagram.

a. Show that the area, A m^2, to be used for the Australian native garden is given by $A = 800 + 20x - \frac{1}{2}x^2$.

b. State the restrictions that must be placed on x.

c. i. Calculate the value of x for which the area of the Australian native garden is greatest.

ii. Calculate the greatest possible area of the native garden.

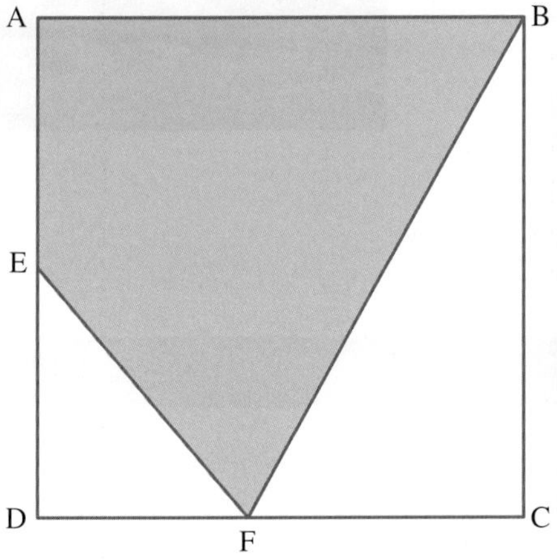

5. The graph of $y = g(x)$ is shown. The graph has a stationary point of inflection at the origin and also crosses the x-axis at the points where $x = -\sqrt{5}$ and $x = \sqrt{5}$. The coordinates of the maximum and minimum turning points $\left(-\sqrt{3}, 12\sqrt{3}\right)$ and $\left(\sqrt{3}, -12\sqrt{3}\right)$ respectively.

a. Use the above information to form the equation of the graph.

b. Hence, show that $g(x) = 2x^5 - 10x^3$.

c. A water slide is planned for a new theme park. Its cross-sectional shape is to be designed using a horizontal and vertical translation of the curve $g(x) = 2x^5 - 10x^3$.

The image of the point A, the maximum turning point of the original curve, now lies on the y-axis. The image of the point B, the minimum turning point of the original curve, now lies 1 unit above the x-axis. The point C is the image of the origin $(0, 0)$ after the original curve is translated.

The water slide is modelled by the section of the curve from A to B with the x-axis as the water level.

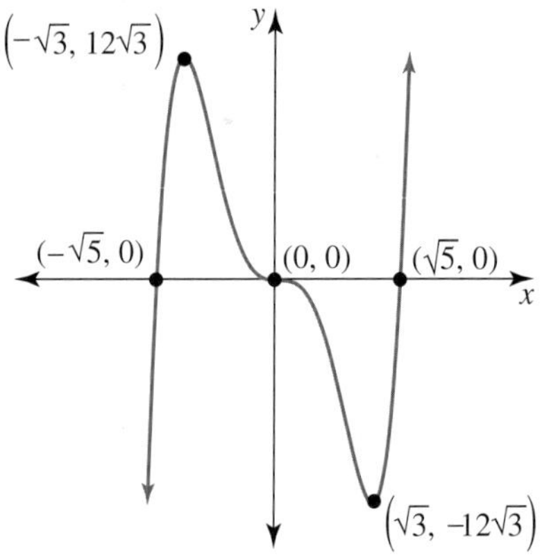

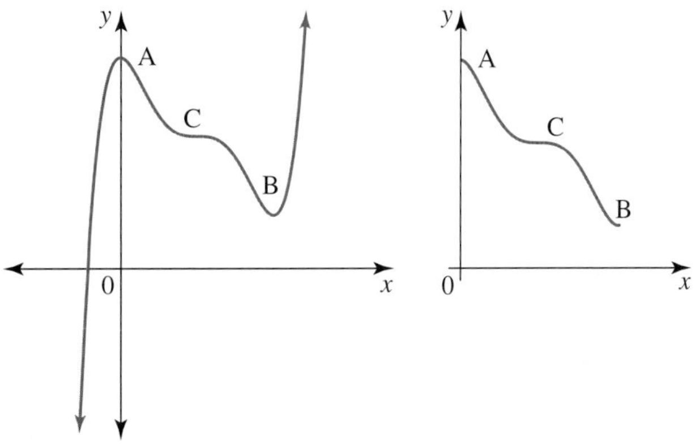

i. State the values of the horizontal and vertical translations required to achieve this model.
ii. Give the height of A above the water level to 1 decimal place.
iii. State the coordinates of the points C and B.

6. In an effort to understand more about the breeding habits of a species of quoll, 10 quolls were captured and relocated to a small reserve where their behaviour could be monitored. After 5 years the population size grew to 30 quolls.
A model for the size of the quoll population, N, after t years on the reserve is thought to be defined by the function $N : R^+ \cup \{0\} \to R$, $N(t) = \dfrac{at+b}{t+2}$.

a. Calculate the values of a and b.
b. Sketch the graph of N against t.
c. Hence or otherwise, determine how large the quoll population can grow.

1.9 Exam questions

Question 1 (1 mark) TECH-ACTIVE

***Source:** VCE 2014, Mathematical Methods (CAS) Exam 2, Section 1, Q15; © VCAA.*

MC Zoe has a rectangular piece of cardboard that is 8 cm long and 6 cm wide. Zoe cuts squares of side length x centimetres from each of the corners of the cardboard, as shown in the diagram below.

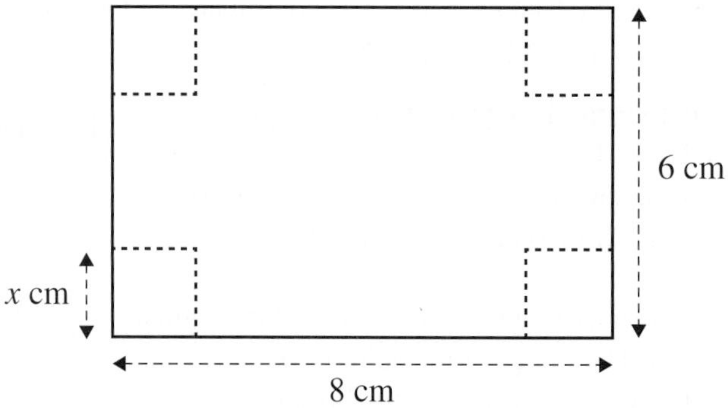

Zoe turns up the sides to form an open box.

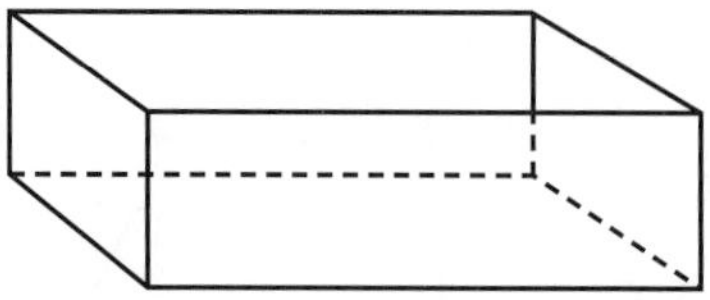

The value of x for which the volume of the box is a maximum is closest to

A. 0.8
B. 1.1
C. 1.6
D. 2.0
E. 3.6

Question 2 (1 mark) TECH-ACTIVE

MC The graph shows the relationship between two variables, x and y.

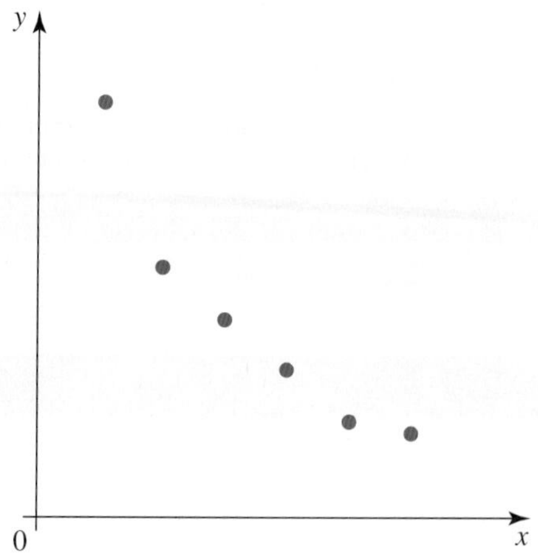

If a is a positive constant, the equation relating x and y is most likely

A. $y = a\sqrt{x}$

B. $y = ax^2$

C. $y = \frac{a}{x}$

D. $y = ae^x$

E. $y = a\log_e(x)$

Question 3 (6 marks) TECH-ACTIVE

A right circular cone is inscribed in a sphere of radius 4 cm, as shown in the cross-section.

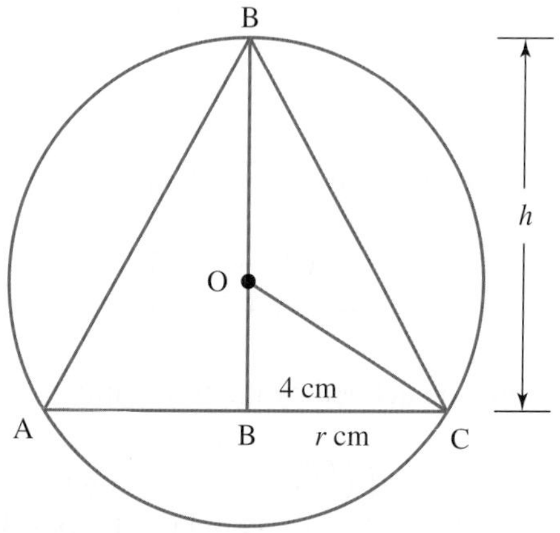

a. Express the radius, r cm, of the cone in terms of h. **(1 mark)**

b. Write an equation expressing the volume of the cone, $V\text{cm}^3$, in terms of h and state any restrictions on h. **(2 marks)**

c. Sketch the graph of V versus h. **(2 marks)**

d. Use the graph to find the maximum volume for the cone to the nearest cm^3. **(1 mark)**

More exam questions are available online.

1.10 Review

1.10.1 Summary

doc-37030

Hey students! Now that it's time to revise this topic, go online to:

Access the topic summary | Review your results | Watch teacher-led videos | Practise VCAA exam questions

Find all this and MORE in jacPLUS

1.10 Exercise

Technology free: short answer

1. Solve the following equations.
   a. $3x^2 - 5x = 4$
   b. $4x^3 - 8x^2 - 3x + 6 = 0$
   c. $2m^4 - 5m^2 - 3 = 0$
   d. $-2x^3 + 12x^2 - 22x + 12 = 0$
2. a. Show that $m - 2$ is a factor of $4m^4 - 11m^3 - 19m^2 + 44m + 12$.
   b. Hence, solve the equation $4m^4 - 11m^3 - 19m^2 + 44m + 12 = 0$.
3. If $(x + 1)$, $(x + 2)$ and $(x + 3)$ are factors of $x^4 + ax^3 + bx^2 + cx - 6$, find the values of a, b and c.
4. Sketch the graphs of each of the following, stating the domain and range.
   a. $y = \frac{1}{81}(x - 2)^4 - 1$
   b. $y = 1 - \frac{4}{(x - 2)^2}$
   c. $y = (2x + 1)^3 + 8$
5. Solve the following system of simultaneous equations.

$$\begin{aligned} m + n - p &= -2 \\ 3m + 5n - 2p &= 1 \\ 5m + 4n + 2p &= 9 \end{aligned}$$

6. Solve the pair of simultaneous equations through the use of a parameter, λ.

$$\begin{aligned} 2x + y - z &= 7 \\ -x - y + 3z &= 1 \end{aligned}$$

7. a. Sketch the graph of the hybrid function defined by the following.

$$f(x) = \begin{cases} \sqrt[3]{x + 1}, & x \le 0 \\ (3 - x)(x + 1), & 0 < x \le 3 \\ x + 3, & x > 3 \end{cases}$$

   b. State where this function is discontinuous.
   c. State the domain and range of the function.
8. Consider the functions f and g with respective rules $f(x) = \sqrt{x + 2}$ and $g(x) = \sqrt{4 - x^2}$.
   a. Give the maximal domain of each function and hence state their common domain, D.
   b. On the same axes, sketch $y = f(x)$ and $y = g(x)$, and hence sketch the graph of the function $y = (f + g)(x)$.

Technology active: multiple choice

9. **MC** When fully factorised, $4x^3 - 2x^2 - 36x + 18$ becomes:

A. $(2x-1)\left(2x^2-18\right)$
B. $2(2x-1)(x-3)^2$
C. $4(x-1)(x-3)(x+3)$
D. $2(2x-1)(x-3)(x+3)$
E. $2(2x-1)\left(x^2-9\right)$

10. **MC** $(x-1)$ is a linear factor of $6x^3 - 5x^2 - 2x + 1$. The solutions to the equation $6x^3 - 5x^2 - 2x + 1 = 0$ are:

A. $x = 1, -\frac{1}{2}$ and $\frac{1}{3}$
B. $x = 1, -\frac{1}{2}$ and $-\frac{1}{3}$
C. $x = 1, \frac{1}{2}$ and $\frac{1}{3}$
D. $x = 1, \frac{1}{2}$ and $-\frac{1}{3}$
E. $x = -1, -\frac{1}{2}$ and $-\frac{1}{3}$

11. **MC** If $ax^3 + bx^2 + cx + d \equiv 3(x-2)^3 + 2(x-2)$, then a, b, c and d are respectively:

A. $3, -18, 36$ and -24
B. $3, 18, -36$ and 24
C. $3, -18, 38$ and 28
D. $3, -18, 38$ and -28
E. $3, 18, 38$ and -28

12. **MC** Find the value of k for which the simultaneous equations

$$2kx + (k+2)y = 1$$
$$4x + (5k+1)y = 2$$

have infinitely many solutions.

A. $k = -\frac{4}{5}$
B. $k \neq -\frac{4}{5}$ and $k \neq -1$
C. $k = -1$
D. $k \neq -1$ and $k = \frac{4}{5}$
E. $k = 1$

13. **MC** A possible equation for the graph shown is:

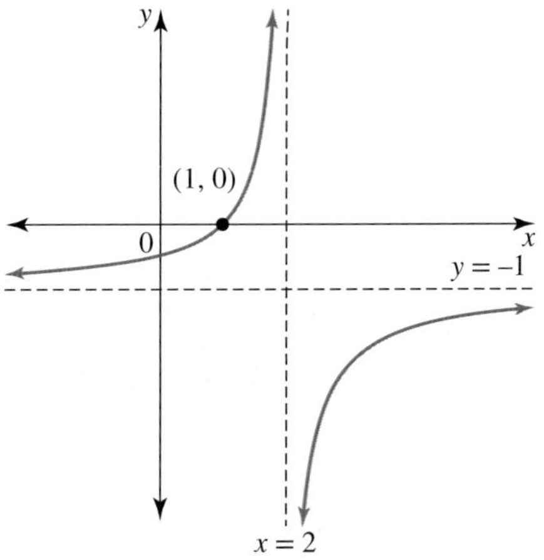

A. $y = \frac{1}{x-2} - 1$
B. $y = -\frac{1}{x-2} - 1$
C. $y = \frac{1}{(x-2)^2} - 1$
D. $y = -\frac{1}{(x-2)^2} - 1$
E. $y = -\frac{1}{x-1} - 2$

14. **MC** A possible equation for the graph shown is:

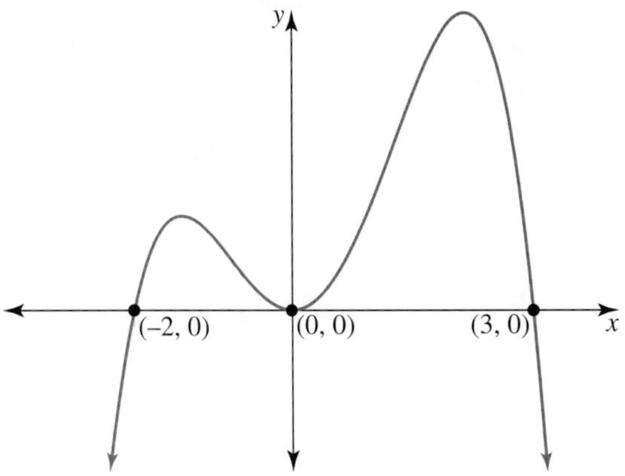

A. $y = x^2(x+2)(x-3)$
B. $y = x^2(x-2)(x+3)$
C. $y = x^2(x+2)(3-x)$
D. $y = -x^2(x-2)(x+3)$
E. $y = -x(x+2)(x-3)$

15. **MC** The maximal domain of the function with the rule $y = \dfrac{4}{\sqrt[3]{x-64}}$ is:

A. $R \setminus \{4\}$
B. $R \setminus \{64\}$
C. $(4, \infty)$
D. $(64, \infty)$
E. R

16. **MC** Data about the relationship between two variables, x and y, are plotted in the diagram.

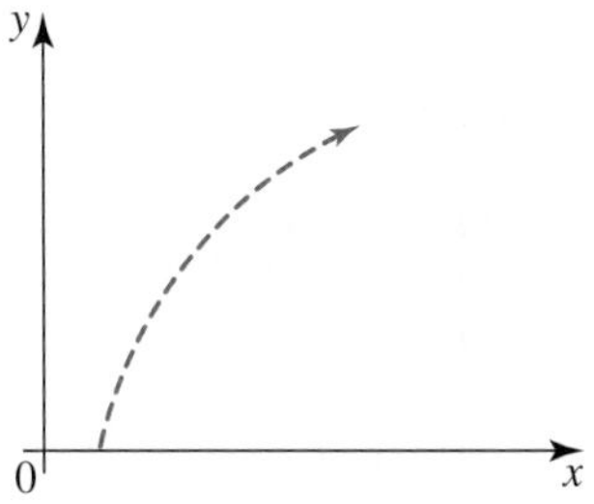

The most likely model for the relationship is:

A. linear of the form $y = x - a$
B. exponential of the form $y = a^x$
C. power function of the form $y = (x-a)^{\frac{1}{2}}$
D. trigonometric of the form $y = \cos(x-a)$
E. polynomial of the form $y = (x-a)^4$

17. **MC** The graphs of $y = f(x)$ and $y = g(x)$ are shown.

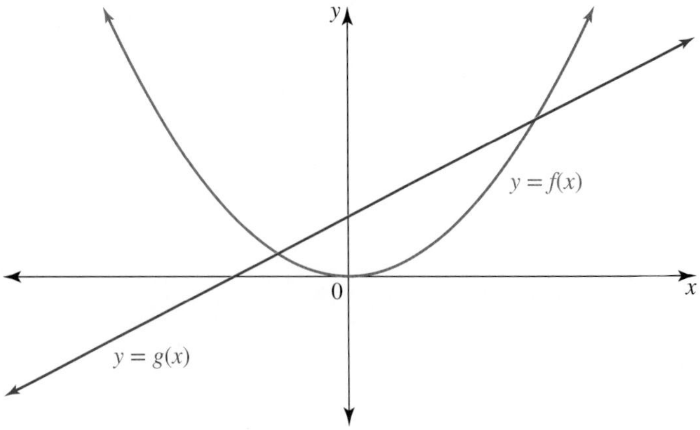

Select the correct statement about the graph of $y = g(x) - f(x)$.

A. The graph of $y = g(x) - f(x)$ has no x-intercepts.
B. The graph of $y = g(x) - f(x)$ has one x-intercept.
C. The graph of $y = g(x) - f(x)$ has two x-intercepts.
D. The graph of $y = g(x) - f(x)$ has three x-intercepts.
E. The graph of $y = g(x) - f(x)$ has four x-intercepts.

Technology active: extended reponse

18. a. If $am^4 + bm^3 + cm^2 + dm + e \equiv \left(m^2 + 2\right)^2 - 13\left(m^2 + 2\right) + 42$, find the values of a, b, c, d and e.
    b. Hence, or otherwise, solve $\left(m^2 + 2\right)^2 - 13\left(m^2 + 2\right) + 42 = 0$ over R.

19. Determine the values of m for which the equation $(m + 1)x^2 + 2x + 3m = 0$ has two solutions.

20. Solve the following system of simultaneous equations.

$$\begin{aligned} x - y + z + w &= 5 \\ 2x + y - z + 2w &= 1 \\ x + 2z - w &= 0 \\ 2y - 3z - 2w &= -11 \end{aligned}$$

21. In the figure shown, the parabola is defined by $y = ax^2 + bx + c$, where a, b and c are constants.

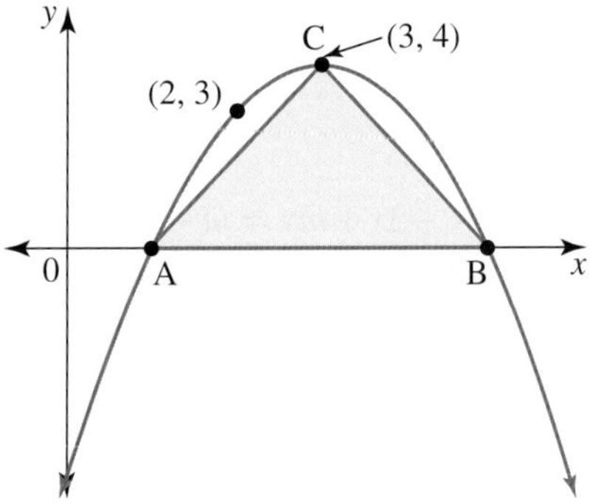

The curve intersects the x-axis at points A and B. Point C is the vertex of the parabola. The curve passes through the point (2, 3) and has a maximum turning point at (3, 4).

a. Calculate the values of a, b and c.
b. Find the coordinates of points A and B.
c. Hence, calculate the area of the triangle ABC.
d. i. Show that every member of the family of quadratic polynomials defined by $y = ax^2 - 6ax + 8a + 3$, $a \in R \setminus \{0\}$ passes through the point (2, 3).
   ii. Determine the value(s) of a for which these parabolas will have two x-intercepts.

22. Consider the polynomial $y = 8 + 4x - 2x^2 - x^3$.
    a. Factorise $8 + 4x - 2x^2 - x^3$.
    b. Sketch the graph of $y = 8 + 4x - 2x^2 - x^3$ and state an interval, using integer values for x, in which its maximum turning point must lie.
    c. The cross-section of a mountain range is modelled by the curve $y = \sqrt{8 + 4x - 2x^2 - x^3}$ for $-2 \le x \le a$, where x km is the horizontal distance and y is the vertical height in hundreds of metres.
       i. State the largest value a can take.
       ii. Express $y = \sqrt{8 + 4x - 2x^2 - x^3}$ as the product of a polynomial f and a square root function g, specifying the rule for each function.
       iii. Using the largest value that a can take, state the ranges of the functions f and g.
       iv. Sketch the functions f and g.
       v. Hence, draw a sketch of the cross-section of the mountain range.

1.10 Exam questions

Question 1 (1 mark) TECH-ACTIVE

Source: *VCE 2018, Mathematical Methods Exam 2, Section A, Q3; © VCAA.*

MC Consider the function $f[a, b] \rightarrow R,\ f(x) = \frac{1}{x}$, where a and b are positive real numbers. The range of f is

A. $\left[\frac{1}{a}, \frac{1}{b}\right)$ **B.** $\left(\frac{1}{a}, \frac{1}{b}\right]$ **C.** $\left[\frac{1}{b}, \frac{1}{a}\right)$
D. $\left(\frac{1}{b}, \frac{1}{a}\right]$ **E.** $[a, b)$

Question 2 (1 mark) TECH-ACTIVE

Source: *VCE 2015, Mathematical Methods (CAS) Exam 2, Section 1, Q7; © VCAA.*

MC The range of the function $f: (-1, 2] \rightarrow R,\ f(x) = -x^2 + 2x - 3$ is

A. R **B.** $(-6, -3]$ **C.** $(-6, -2]$
D. $[-6, -3]$ **E.** $[-6, -2]$

Question 3 (1 mark) TECH-ACTIVE

Source: *VCE 2014, Mathematical Methods (CAS) Exam 2, Section 1, Q18; © VCAA.*

MC The graph of $y = kx - 4$ intersects the graph of $y = x^2 + 2x$ at two distinct points for

A. $k = 6$ **B.** $k > 6$ or $k < -2$ **C.** $-2 \le k \le 6$
D. $6 - 2\sqrt{3} \le k \le 6 + 2\sqrt{3}$ **E.** $k = -2$

Question 4 (1 mark) TECH-ACTIVE

MC The simultaneous linear equations

$$-3x + my = m - 1 \text{ and}$$
$$(m + 1)x - 10y = -8,$$

where m is a real constant, have an infinite number of solutions for

A. $m = 5$ or $m = -6$ **B.** $m = 5$ only **C.** $m \in R \setminus \{-6, 5\}$
D. $m \neq -6$ **E.** $m \neq 5$

Question 5 (2 marks) TECH-FREE

Find the value(s) of a for which the simultaneous equations $3x + ay = 5$ and $(a + 2)x + 5y = a$ have no solution.

More exam questions are available online.

Answers

Topic 1 Functions and graphs

1.2 Linear functions

1.2 Exercise

1. a. Many-to-one correspondence
   b. Domain: $[-4, 2)$; range: $[0, 16]$
   c. $f: [-4, 2) \to R, f(x) = x^2$
   d. $y = 12$
2. a. i. Many-to-one   ii. $[-3, 6), [-9, 7]$
   iii. Yes
   b. i. One-to-many   ii. $[0, \infty), R$ iii. No
   c. i. Many-to-many   ii. $[-2, 2], [-2, 2]$
   iii. No
   d. i. One-to-one   ii. R, R iii. Yes
   e. i. Many-to-one   ii. $R, \{2\}$ iii. Yes
   f. i. One-to-one   ii. R, R iii. Yes
3. a.

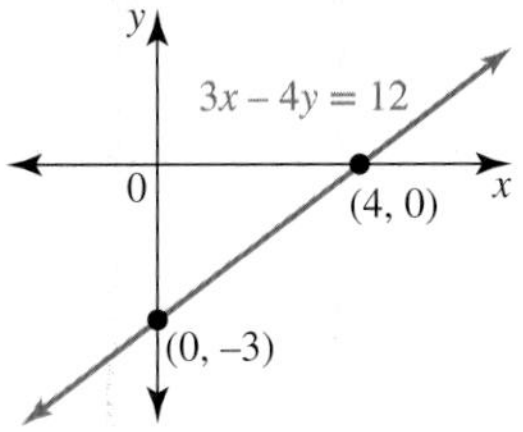

   b. $\frac{3}{4}$
4. a.

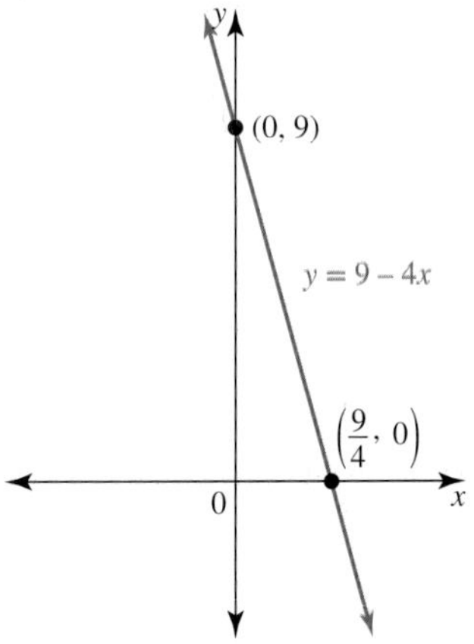

   Range R

   b.

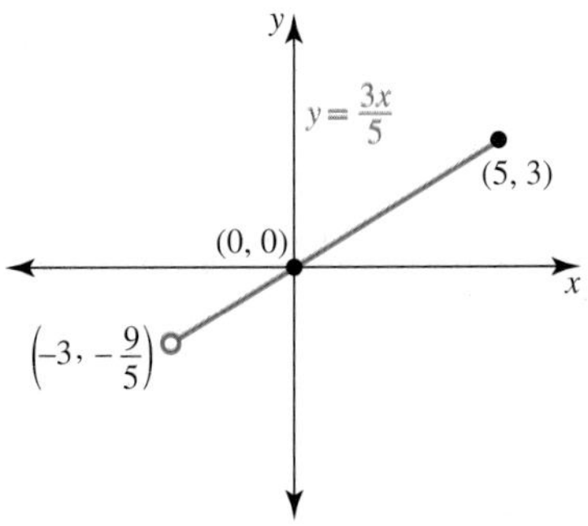

   Range $\left(-\frac{9}{5}, 3\right]$

   c.

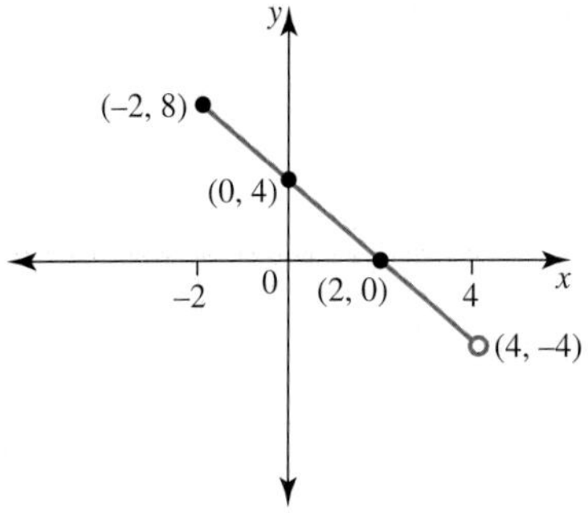

   Range $(-4, 8]$

   d.

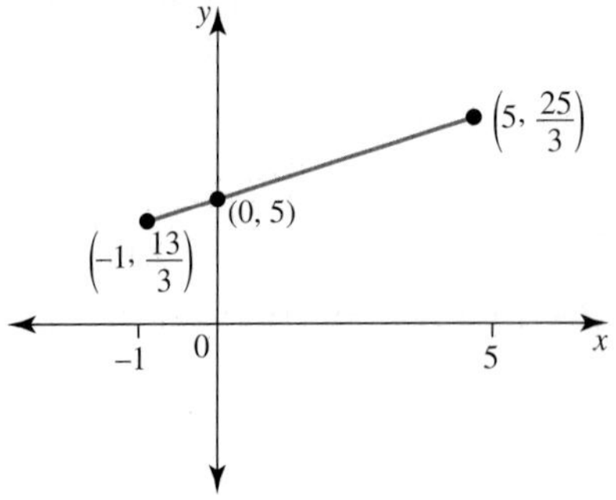

   Range $\left[\frac{13}{3}, \frac{25}{3}\right]$

5. a. $y = -3x + 12$ b. $y = 3x + 2$
6. a. $y + 2x - 2 = 0$ b. $2y + x - 8 = 0$
7. a. $140.2°$ b. $y = x + 6$
8. a. i. $(4, -2)$ ii. $4\sqrt{2}$ units
   b. i. $\left(\frac{1}{2}, \frac{1}{2}\right)$ ii. $\sqrt{74}$ units
9. a. $a = 6$ b. $a = 7$
   c. $a = -2, 14$ d. $a = -5, 1$
10. a. $p = 6$ b. $9y - 7x = 23$
    c. 8.3 units

1.2 Exam questions

Note: Mark allocations are available with the fully worked solutions online.

1. A   2. D   3. C

1.3 Solving systems of equations

1.3 Exercise

1. $k \in R \setminus \{-1, 4\}$
2. $m = 2$
3. $m = \pm\frac{1}{2}$
4. $k \in R \setminus \{-1, 2\}$
5. a. $m \in R \setminus \{-4, 1\}$ b. $m = -4$
   c. $m = 1$
6. $m = 2, n = 2$ and $p = -5$
7. $d = 1, e = 2$ and $f = 2$
8. a. $x = 3, y = 2$ and $z = -4$
   b. $m = 7, n = -2$ and $p = -1$

c. $u = 3, v = 6$ and $w = -2$

d. $a = \dfrac{13}{2}, b = -3$ and $c = \dfrac{1}{2}$

9. The largest angle is 70°, the smallest angle is 50° and the third angle is 60°.

10. $w = 1, x = -3, y = 2$ and $z = 3$

11. $x = \dfrac{2(a+2)}{a(a+4)}, y = \dfrac{4(a+2)}{a(a+4)}, z = \dfrac{4}{a}$

12. a. $3x + 3z = 3$

b. $x = 1 - \lambda, y = -\dfrac{\lambda}{2}, z = \lambda$

c. This solution describes the line along which the two planes are intersecting.

13. $x = \dfrac{2(\lambda + 5)}{3}, y = -\dfrac{7\lambda + 2}{3}, z = \lambda$

14. $x = \dfrac{11}{3}, y = \dfrac{2(3\lambda + 2)}{3}, z = \lambda$

15. $x = 3\lambda, y = 5\lambda - 2, z = \lambda$

16. a. $m \in R \setminus \{6\}, n \in R$ b. $m = 6, n = -2$

c. $m = 6, n \in R \setminus \{-2\}$

1.3 Exam questions

Note: Mark allocations are available with the fully worked solutions online.

1. B   2. C   3. C

1.4 Quadratic functions

1.4 Exercise

1. a. $(5u - 2)(3u + 1)$

b. $2(3d - 2)(d - 4)$

c. $3(j + 2 - \sqrt{6})(j + 2 + \sqrt{6})$

d. $(b - 1)(b + 1)$

2. a. $(f - 14)(f + 2)$ b. $(g + 4)(g - 1)$

3. a. $x = -\dfrac{3}{4}, \dfrac{1}{2}$ b. $x = \dfrac{2 \pm \sqrt{2}}{2}$

4. a. $y = \pm \dfrac{1}{9}$ b. $z = -\dfrac{7}{2}$

c. $m = \dfrac{5 \pm \sqrt{10}}{5}$ d. $x = 1$ or 3

5. a. $p = \dfrac{1}{2}$ or $\dfrac{3}{2}$ b. $k = \dfrac{11}{4}$ or 7

c. $m = -4$ or 1 d. $n = \dfrac{-5 \pm 3\sqrt{17}}{8}$

6.

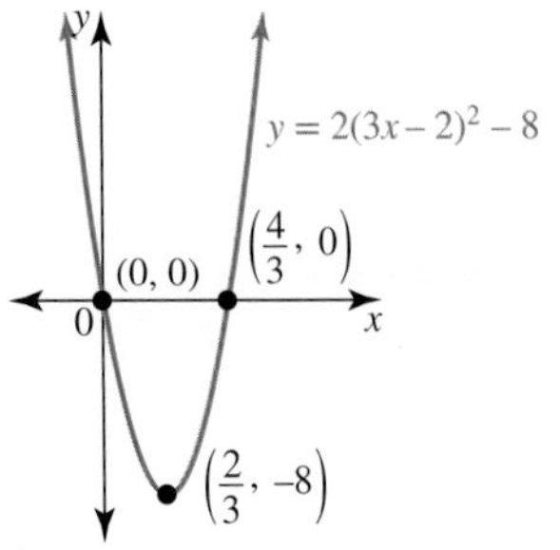

Domain R, range $[-8, \infty)$.

7. a. $[2, 29)$

b. i.

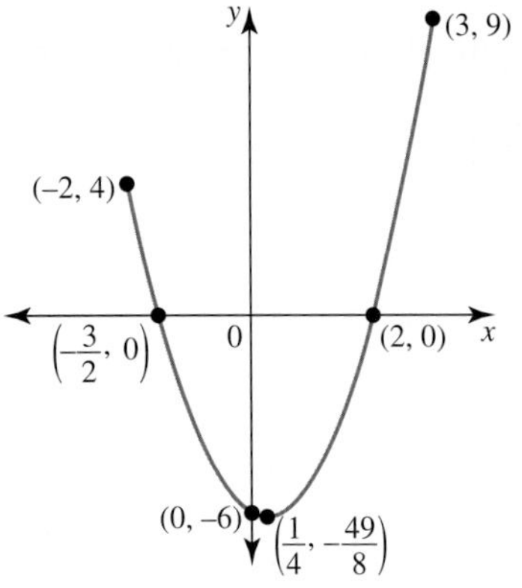

Range = $\left[-\dfrac{49}{8}, 9\right]$

ii.

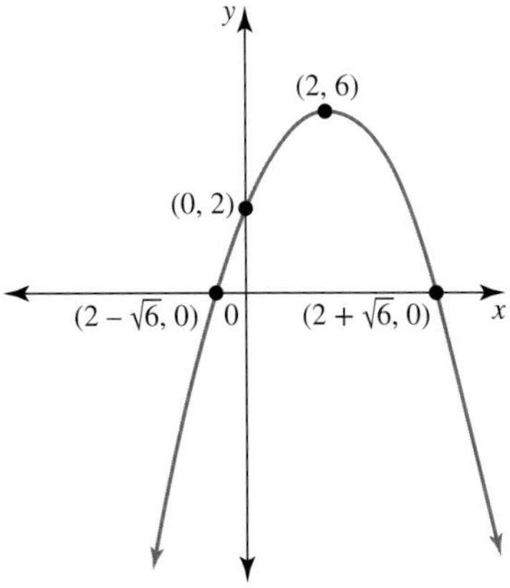

Range = $(-\infty, 6]$

iii.

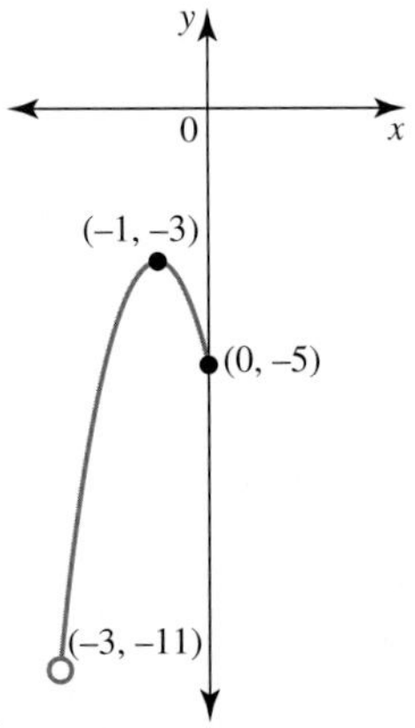

Range = $(-11, -3]$

iv.

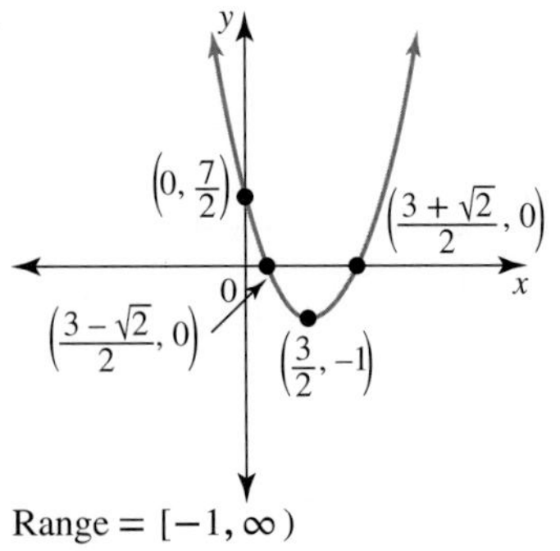

Range = $[-1, \infty)$

8. a. None

b. $f(x) = 4(x-1)^2 + 3$

c.

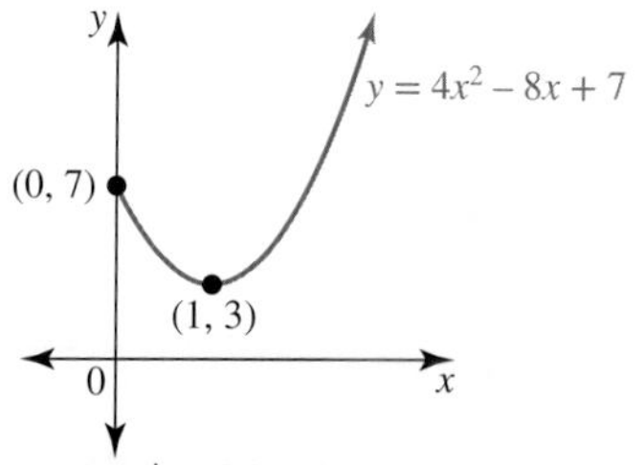

Domain $R^+ \cup \{0\}$, range $[3, \infty]$.

9. $y = -\frac{1}{2}(2x+1)(x-4)$, TP $= \left(\frac{7}{4}, \frac{81}{16}\right)$

10. a. $y = -\frac{3}{20}(x+6)^2 + 12$

b. $y = -\frac{4}{7}(x+7)(2x+5)$

c. $y = \frac{1}{4}x^2 - 5$

11. a. $-(x-1)^2 - 4$

b. $(1, -4)$

c.

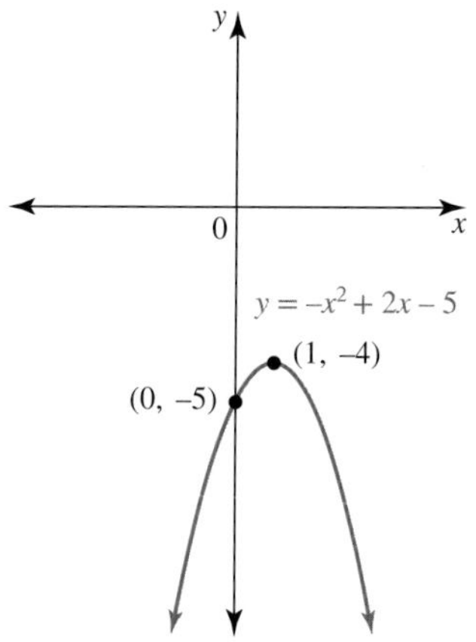

Range $(-\infty, -4]$

d.

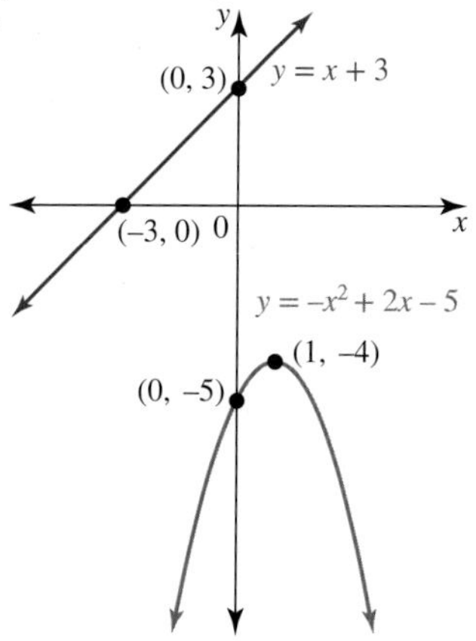

e. $k = -\frac{19}{4}$

12. $k \in (-4, 4)$

13. $m \in \left(-\infty, \sqrt{2}\right) \cup \left(\sqrt{2}, \infty\right)$

14. $x^2 - 2 = 2x - 3$

$$\begin{aligned}
x^2 - 2x + 1 &= 0 \\
\Delta &= b^2 - 4ac \\
&= (-2)^2 - 4 \times 1 \times 1 \\
&= 4 - 4 \\
&= 0
\end{aligned}$$

As $\Delta = 0$, the graph of $y = 2x - 3$ is a tangent to the graph of $y = x^2 - 2$.

15. a. $k \in \left(-\infty, -\frac{3}{2}\right) \cup \left(\frac{3}{2}, \infty\right)$

b. $kx^2 + 4x - k + 2 = 0$

$$\begin{aligned}
\Delta &= 16 - 4 \times k \times (-k+2) \\
&= 16 + 4k^2 - 8k \\
&= 4\left(k^2 - 2k + 4\right) \\
&= 4\left(k^2 - 2k + 1^2 - 1^2 + 4\right) \\
&= 4\left[(k+1)^2 + 3\right] \\
&= 4(k+1)^2 + 12
\end{aligned}$$

As $(k+1)^2 \geq 0$,

$\therefore 4(k+1)^2 \geq 0$

and $4(k+1)^2 + 12 > 0$.

Δ is always greater than zero. Therefore, the equation will always have a solution for all values of k.

16. $m \in \left(\frac{3 - 2\sqrt{46}}{14}, \frac{3 + 2\sqrt{46}}{14}\right) \setminus \{1\}$

1.4 Exam questions

Note: Mark allocations are available with the fully worked solutions online.

1. B
2. C
3. D

1.5 Cubic functions

1.5 Exercise

1. a. $(5a - 3b)\left(25a^2 + 15ab + 9b^2\right)$

b. $2(c+d)^3$

c. $5(2p-1)\left(4p^2 + 2p + 1\right)$

d. $(2x-1)^3$

2. a. $(3z-2)^3$

b. $(mn+4)\left(m^2n^2 - 4mn + 16\right)$

3. a. $(x-1)(3x-y)$

b. $\left(y + z^2\right)\left(3y^2 - 2z\right)$

4. a. $(3a - 2 - 4b)(3a - 2 + 4b)$

b. $(np - 2m - 1)(np + 2m + 1)$

5. $(x+1)(x-6)(x+3)$

6. a. $(r-7)\left(7r^2 + 1\right)$

b. $(6v+1)\left(6v^2 + 5\right)$

c. $(2m+3)(m-7)(m+7)$

d. $(2z-1)\left(z^2 + 1\right)$

e. $(2x - 7 - 5y)(2x - 7 + 5y)$

f. $(4a - 2b - 3)(4a + 2b + 3)$

g. $(v - w + 2)(v + w - 2)$

h. $(2p + q - 1)(2p + q + 1)$

7. $x = \pm\sqrt{5}, \frac{1}{2}$

8. a. $b = -4, -2, 1$

b. $m = -1, \frac{3}{2}, 4$

c. $x = \pm\sqrt{3}, \frac{1}{2}$

d. $x = -3, -1$ and $\frac{1}{2}$

9. a. Let $P(t) = 3t^3 + 22t^2 + 37t + 10$

$$\begin{aligned} P(-5) &= 3(-5)^3 + 22(-5)^2 + 37(-5) + 10 \\ &= 3 \times 125 + 22 \times 25 - 185 + 10 \\ &= -375 + 550 - 175 \\ &= 0 \end{aligned}$$

$\therefore t + 5$ is a factor.

$t = -5, -2$ and $-\frac{1}{3}$

b. Let $P(d) = 3d^3 - 16d^2 + 12d + 16$.

$$\begin{aligned} P(2) &= 3(2)^3 - 16(2)^2 + 12(2) + 16 \\ &= 24 - 64 + 24 + 16 \\ &= 0 \end{aligned}$$

$\therefore d - 2$ is a factor

$d = -\frac{2}{3}, 2$ and 4

10. $A = 3, B = 0, C = 2$ and $D = -7$

11. $d = \pm 3, e = \pm\frac{1}{3}$

12. a. $a = 5, b = -3, c = 4$ and $d = -1$

b. $a = -3$ and $b = 1$

13. $a = 2, b = 1, c = 1$ and $d = -3$;
$2(x-1)^3 + (x-1)^2 + (x-1) - 3$

14. $a = 2, b = 5$

15.

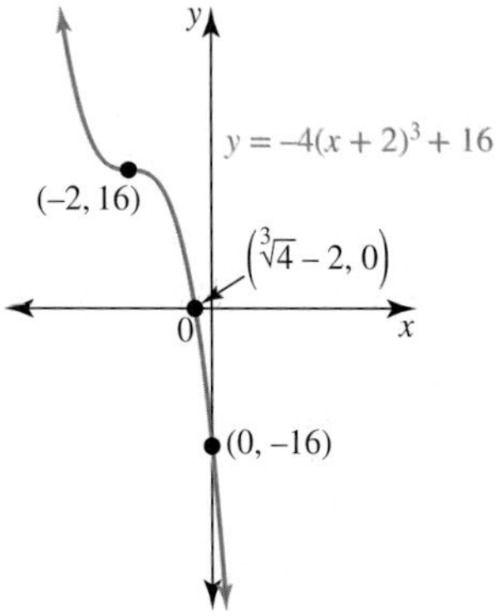

16. a. $4(x-2)^2(x+2)$

b.

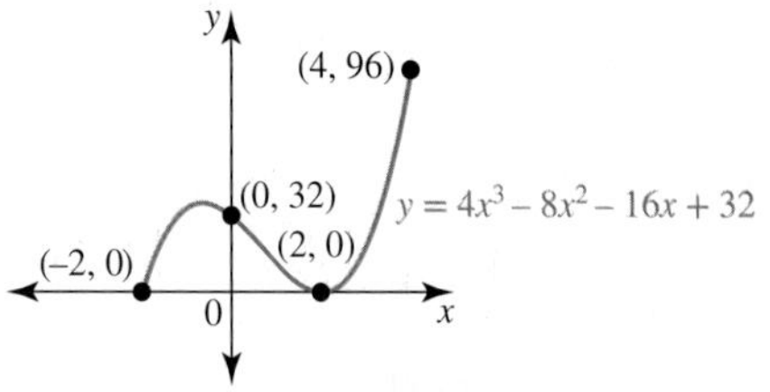

c. Maximum value 96, minimum value 0.

17. a.

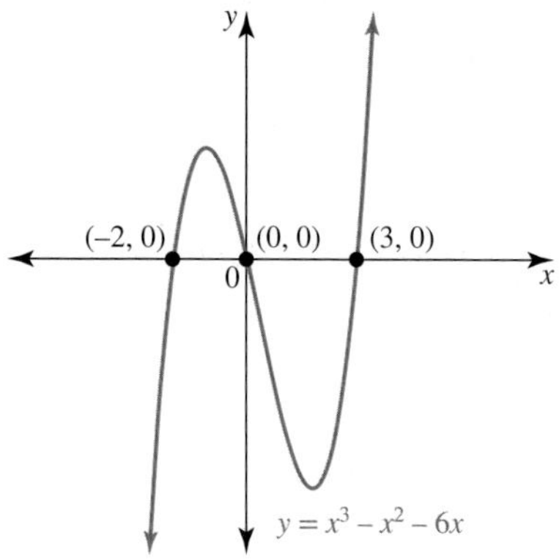

b.

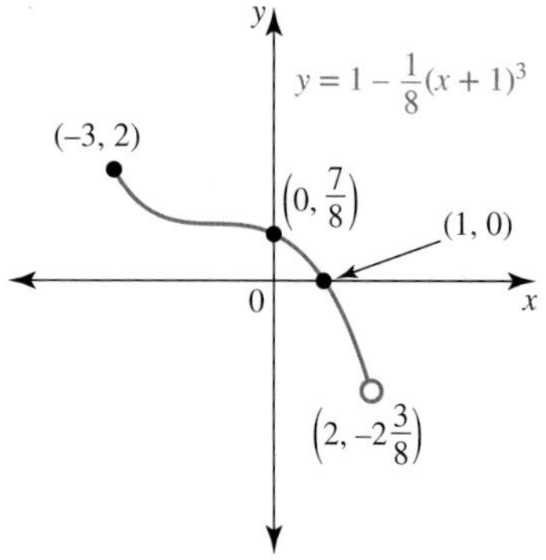

c.

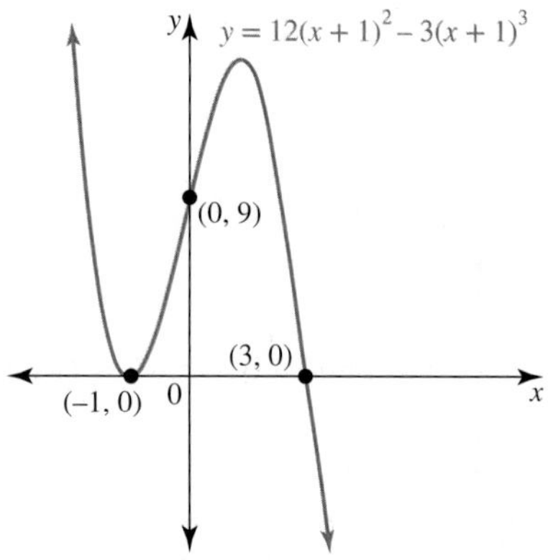

18. $y = 2x(5x-4)(2x-3)$

19. $y = \frac{1}{10}(x+4)(4x-5)^2$

20. a. $f(x) = -2x^3 + 9x^2 - 24x + 17$
$f(1) = -2 + 9 - 24 + 17 = 0$

$\therefore (x-1)$ is a factor.

By inspection,

$-2x^3 + 9x^2 - 24x + 17 = (x-1)\left(-2x^2 + 7x - 17\right)$.

Consider the discriminant of the quadratic factor $-2x^2 + 7x - 17$.

$$\begin{aligned} \Delta &= 49 - 4(-2)(-17) \\ &= 49 - 136 \\ &< 0 \end{aligned}$$

As the discriminant is negative, the quadratic cannot be factorised into real linear factors; therefore, it has no real zeros.

For the cubic, this means there can only be one x-intercept, the one which comes from the only linear factor, $(x-1)$.

b. For there to be a stationary point of inflection, the equation of the cubic function must be able to be written in the form $y = a(x+b)^3 + c$.
Let $-2x^3 + 9x^2 - 24x + 17 = a(x+b)^3 + c$
By inspection, the value of a must be -2.

$$\begin{aligned} &\therefore -2x^3 + 9x^2 - 24x + 17 \\ &= -2\left(x^3 + 3x^2b + 3xb^2 + b^3\right) + c \end{aligned}$$

Equate coefficients of like terms:

$$x^2 : 9 = -6b \Rightarrow b = -\frac{3}{2}$$

$$x : -24 = -6b^2 \Rightarrow b^2 = 4$$

It is not possible for b to have different values.
Therefore, it is not possible to express the equation of the function in the form. $y = a(x+b)^3 + c$.
There is no stationary point of inflection on the graph of the function.

c. $x \to \pm\infty, y \to \mp\infty$

d.

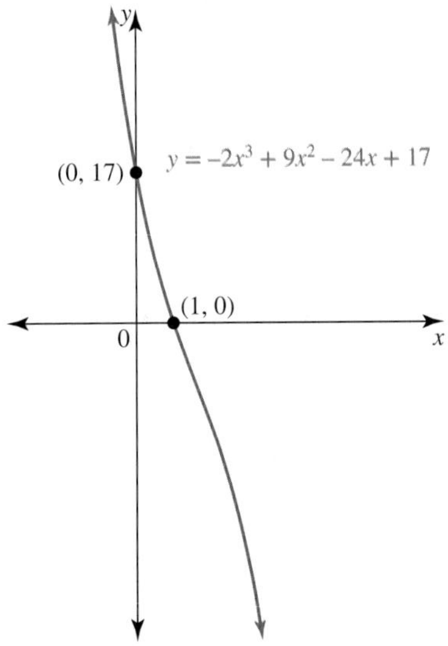

1.5 Exam questions

Note: Mark allocations are available with the fully worked solutions online.

1. E
2. a. $f(x) = (x+2)^2(x-1)$

$$\begin{aligned} &= \left(x^2 + 4x + 4\right)(x-1) \\ &= x^3 - x^2 + 4x^2 - 4x + 4x - 4 \\ &= x^3 + 3x^2 - 4 \end{aligned}$$

b.

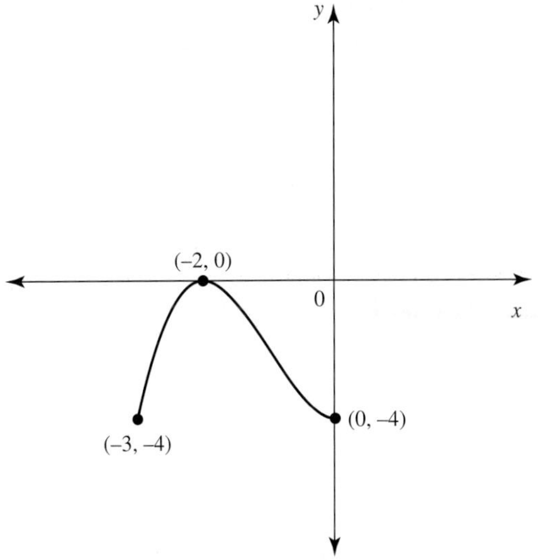

3. D

1.6 Higher degree polynomials

1.6 Exercise

1. $(x-1)(x-4)(x-6)(x+6)$
2. a. $x = -1, 2, 3, 4$
   b. $a = \pm\sqrt{2}$
3. a. $l = \pm 4$ and ± 1
   b. $c = -3$ and ± 2
4. a. $a = \pm 1$ and ± 3
   b. $k = \pm\frac{1}{2}$ and ± 5
   c. $z = \pm\frac{1}{3}$ and ± 4
   d. $x = 8, -6, 1$
5. $a = 2, b = -13, c = -14$
6.

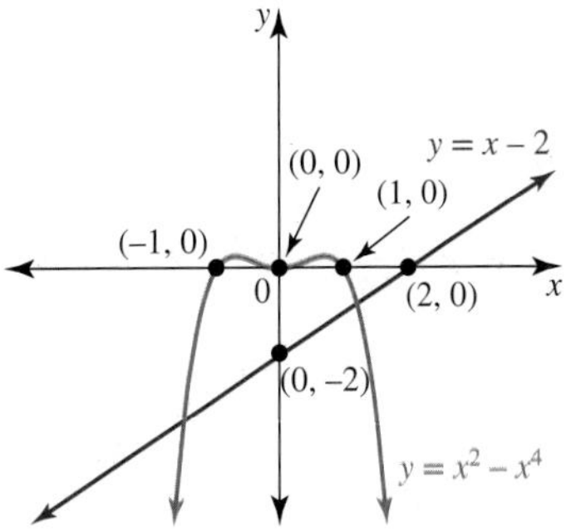

2 solutions

7.

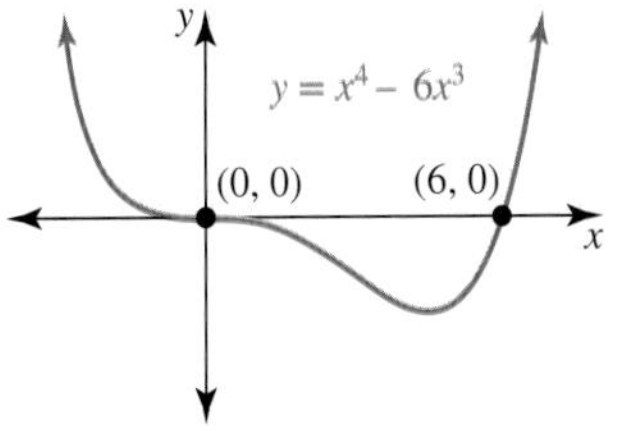

$y = x^4 - 6x^3 + 1$ will make 2 intersections with the x-axis.

8. $a = -\frac{7}{81}, b = 6, c = 7, (-6, 7)$

9. $x = -1.75, x = 1.22$

10. a. Minimum turning points $(-1.31, -3.21)$ and $(1.20, -9.32)$, maximum turning point $(-0.64, -2.76)$

b. None

c. Minimum turning point $(-2.17, -242)$, stationary point of inflection $(2, 20)$

11.

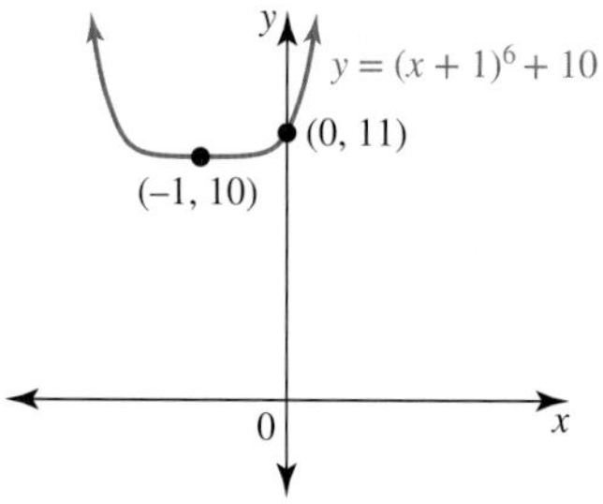

12.

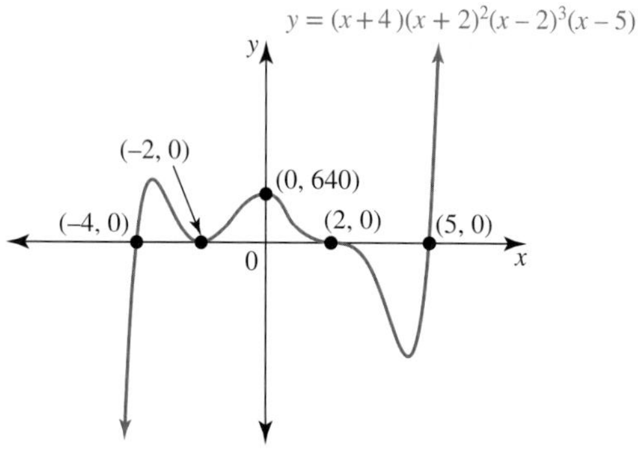

13. a. $y = -3(x+5)^4 + 12$

b.

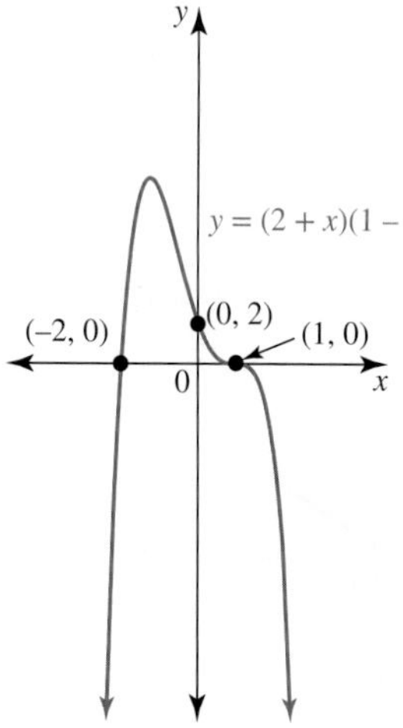

c. i. $-(x+2)^2(x-2)(x-3)$

ii.

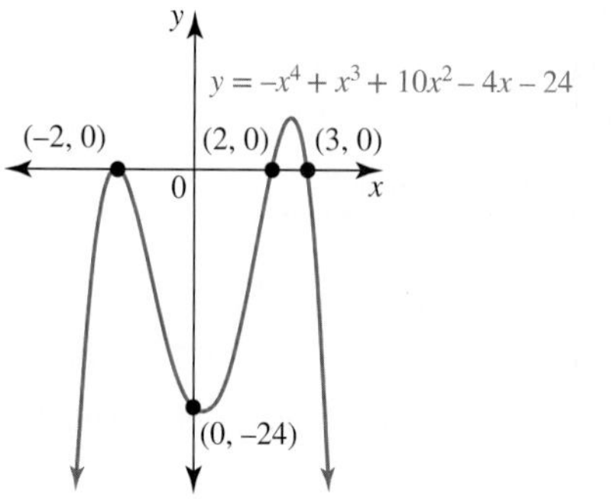

14. a. i.

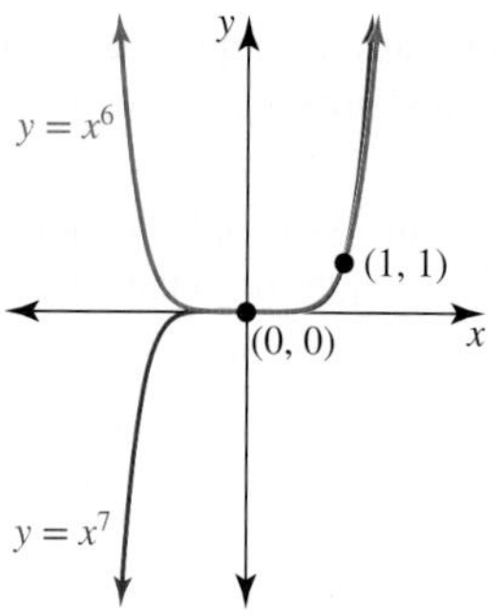

ii. $\{x : x \leq 1\}$

b.

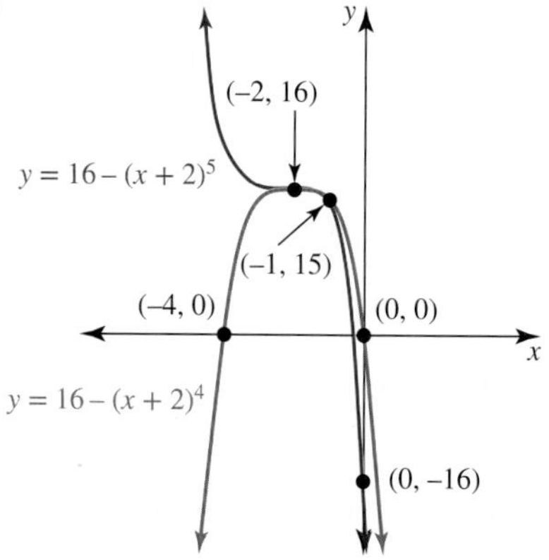

c. i. $y = (x+3)^2(x+1)(x-2)^3$, degree 6

ii. $y = (x+3)^2(x+1)(x-2)^3(10-x)$, degree 7

1.6 Exam questions

Note: Mark allocations are available with the fully worked solutions online.

1. C
2. $x \in [3, 4] \cup \{0\}$
3. C

1.7 Other algebraic functions

1.7 Exercise

1. a. $R \setminus \{-9\}$ b. $\left(-\infty, \dfrac{1}{2}\right]$

c. $R \setminus \{-3\}$ d. R

2. a. $y = \dfrac{6}{x+3} + 1$

b. i. Maximal domain $R \setminus \{1\}$

ii. Range $R \setminus \{5\}$

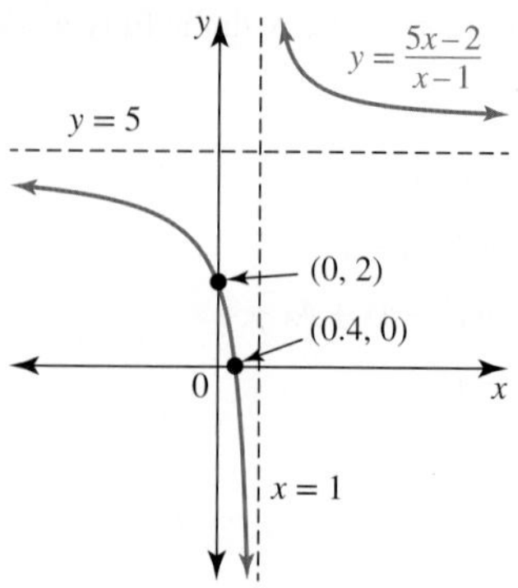

3.

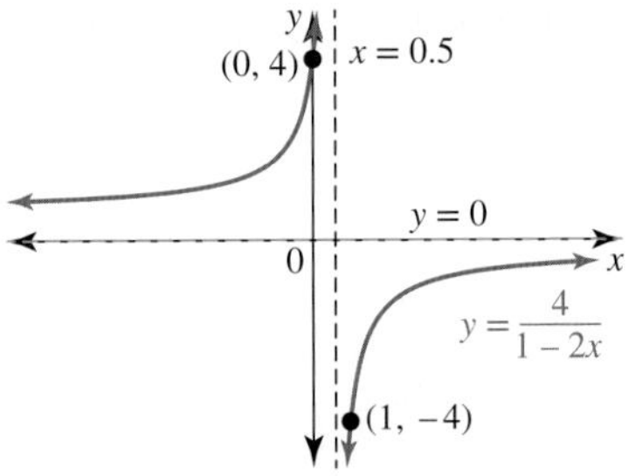

Domain $R \setminus \left\{\frac{1}{2}\right\}$, range $R \setminus \{0\}$.

4. a.

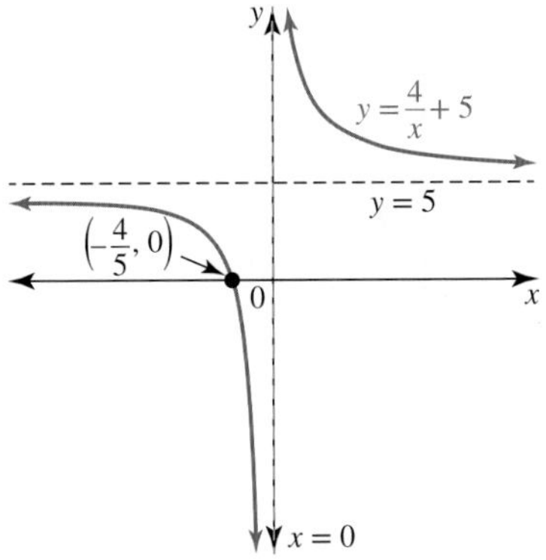

Domain $R \setminus \{0\}$, range $R \setminus \{5\}$

b.

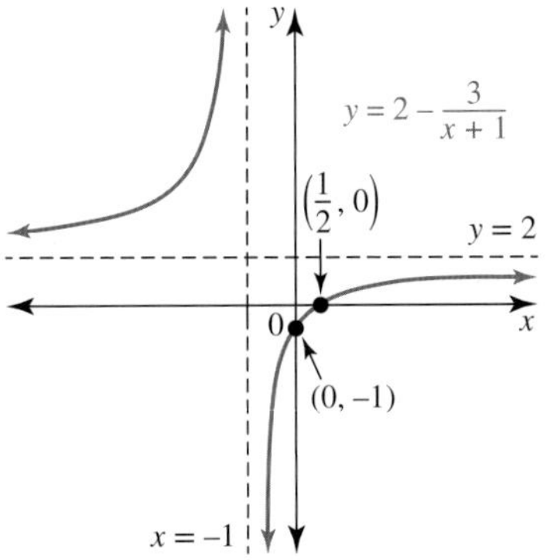

Domain $R \setminus \{-1\}$, range $R \setminus \{2\}$

c.

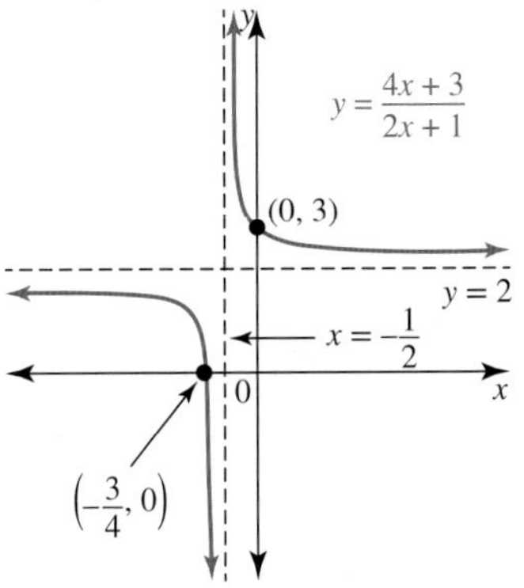

Domain $R \setminus \left\{-\frac{1}{2}\right\}$, range $R \setminus \{2\}$

d.

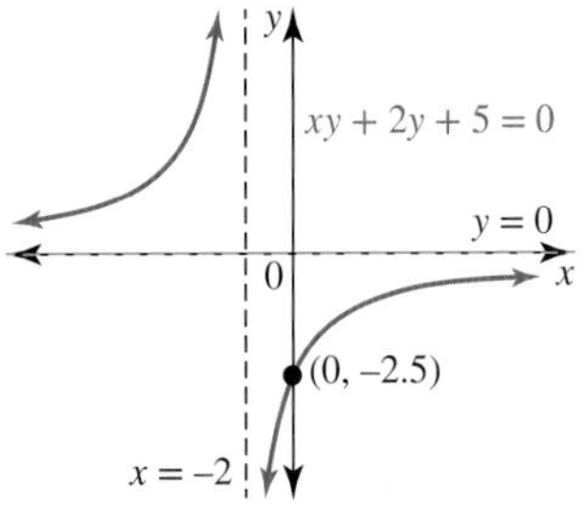

Domain $R \setminus \{-2\}$, range $R \setminus \{0\}$

e.

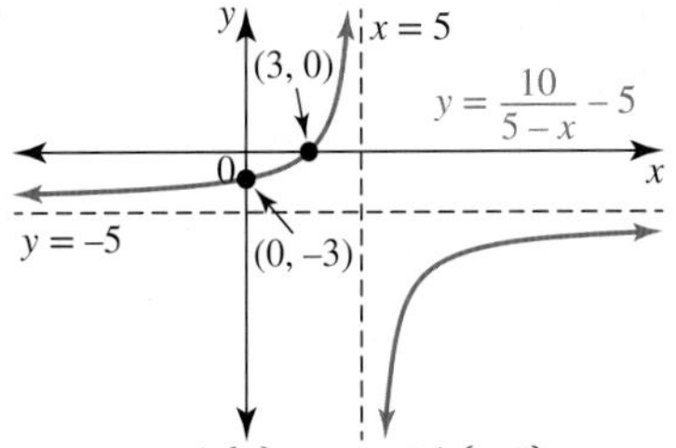

Domain $R \setminus \{5\}$, range $R \setminus \{-5\}$

5. a. $y = \dfrac{-2}{x + 3} + 6$

b. $y = \dfrac{1}{2(x + 2)} - \dfrac{3}{2}$

6.

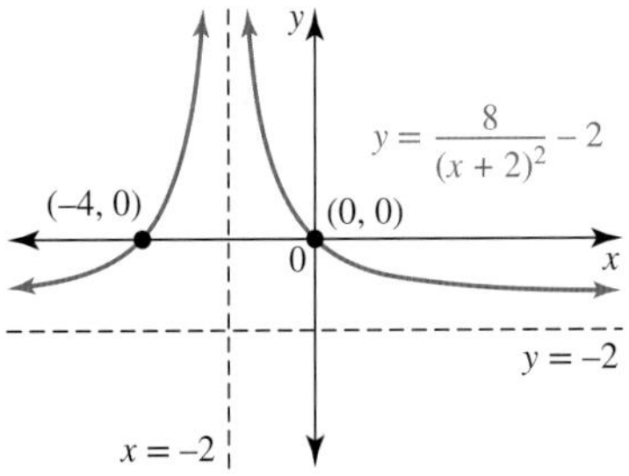

Domain $R \setminus \{-2\}$, range $(-2, \infty)$.

7. a.

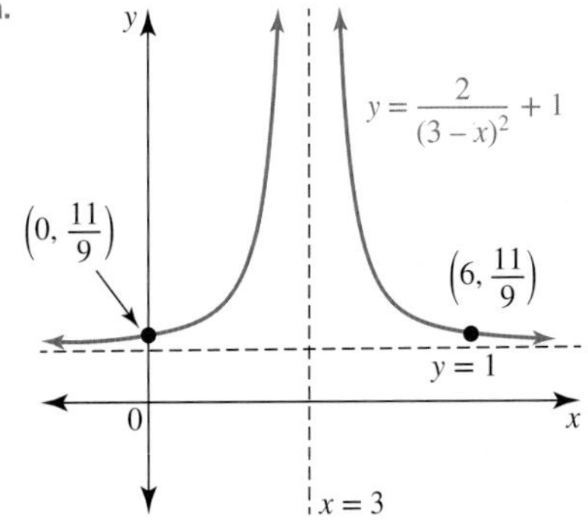

Domain $R \setminus \{3\}$, range $(1, \infty)$

b.

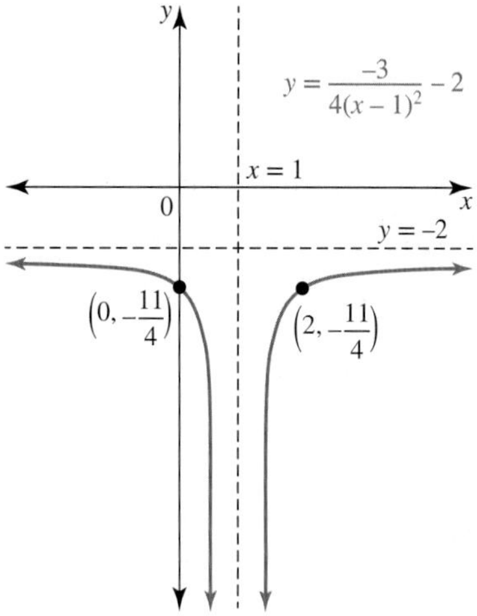

Domain $R \setminus \{1\}$, range $(-\infty, -2)$

c.

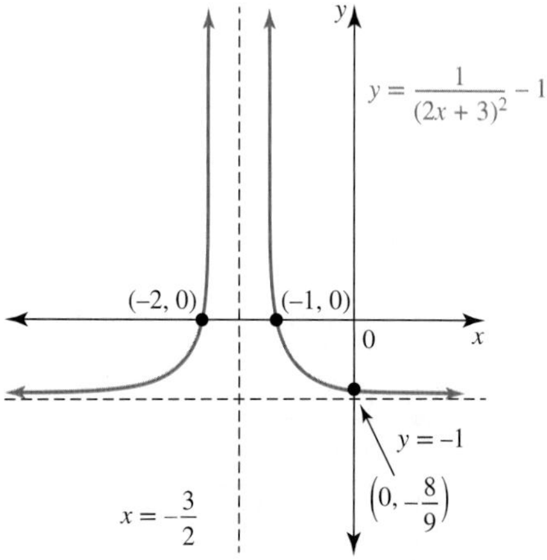

Domain $R \setminus \left\{-\frac{3}{2}\right\}$, range $(-1, \infty)$

d.

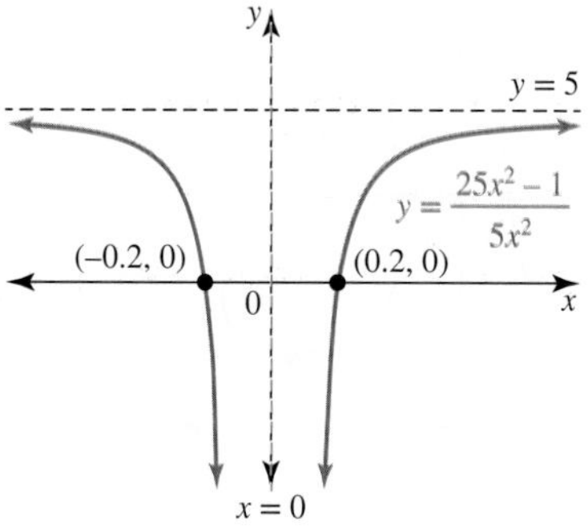

Domain $R \setminus \{0\}$, range $(-\infty, 5)$

8. $y = \frac{1}{4x^2} - 1$

9. a. $y = \frac{-3}{(x-4)^2} + 2$

b. $y = \frac{108}{(2x-1)^2} - 4$;

$f: R \setminus \left\{\frac{1}{2}\right\} \to R, f(x) = \frac{108}{(2x-1)^2} - 4$

10. a. i. Maximal domain $[-9, \infty)$

ii. Range $(-\infty, 2]$

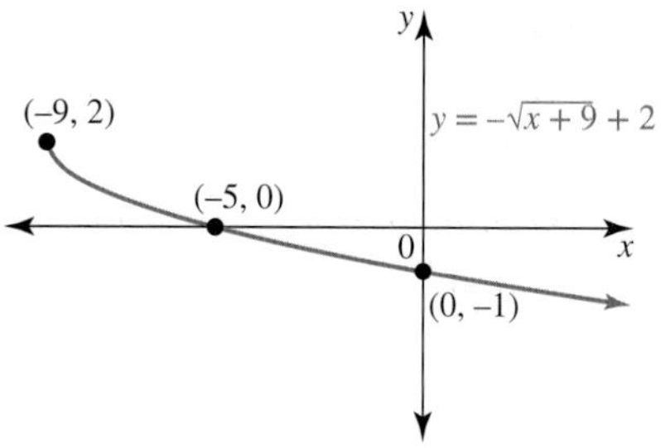

b. $y = 2\sqrt[3]{x-1} + 3$

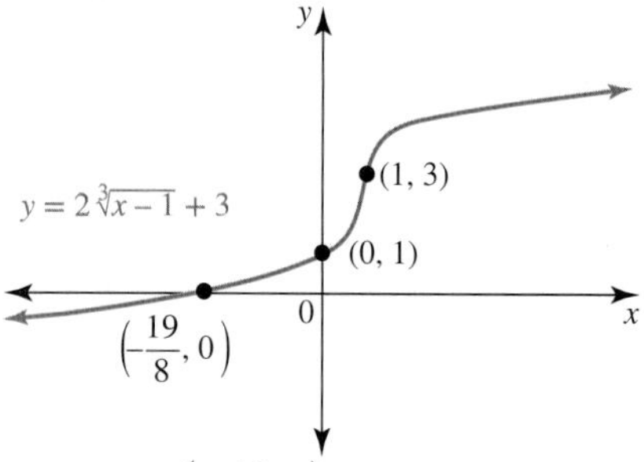

x-intercept $\left(-\frac{19}{8}, 0\right)$.

11. a. Maximal domain $\left[\frac{9}{4}, \infty\right)$, range $[-6, \infty)$

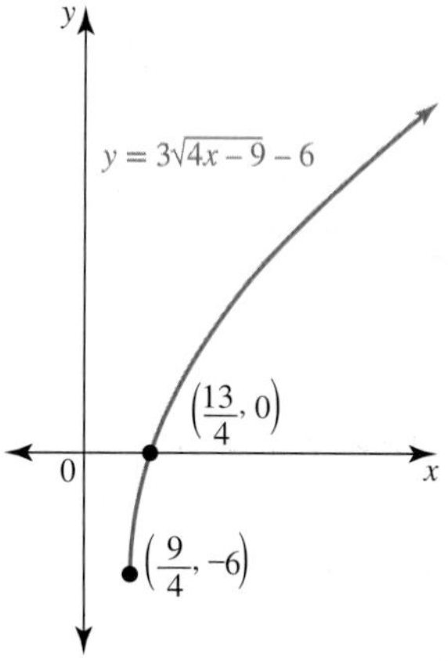

b. $\left(\frac{10}{3}, 0\right)$

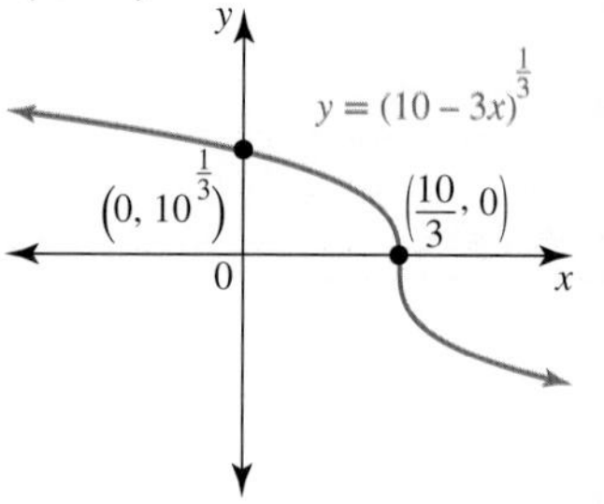

12. a. $y = 2\sqrt{(x-3)} + 2$, domain $[3, \infty)$, range $[2, \infty)$;
$y = -2\sqrt{(x-3)} + 2$, domain $[3, \infty)$, range $(-\infty, 2]$.

b. $y = \sqrt{-2(x-3)} - 1$, domain $(-\infty, 3]$, range $[-1, \infty)$;
$y = -\sqrt{-2(x-3)} - 1$, domain $(-\infty, 3]$, range $(-\infty, -1]$.

13. a.

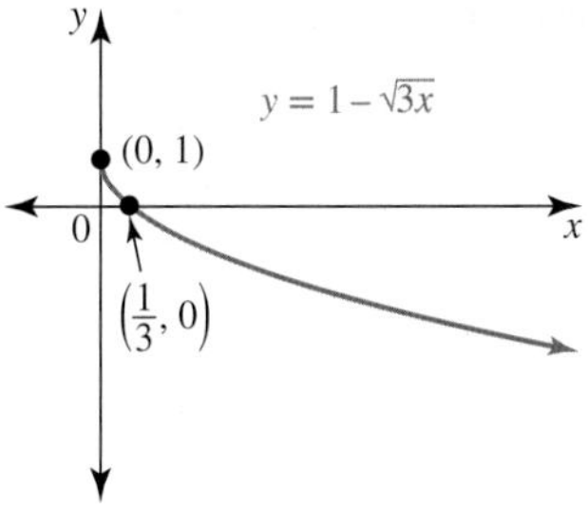

Domain $[0, \infty)$, range $(-\infty, 1]$

b.

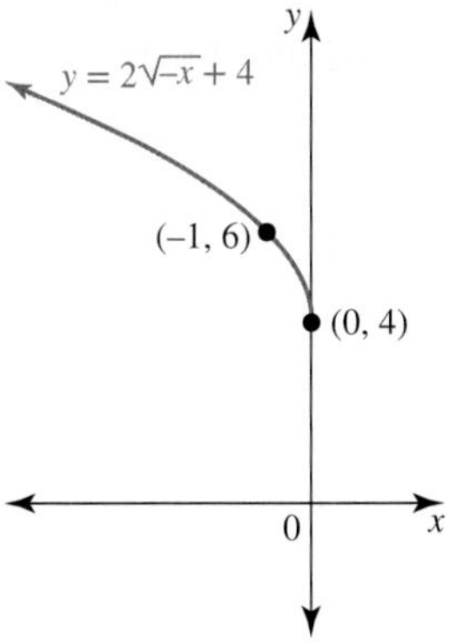

Domain $(-\infty, 0]$, range $[4, \infty)$

c.

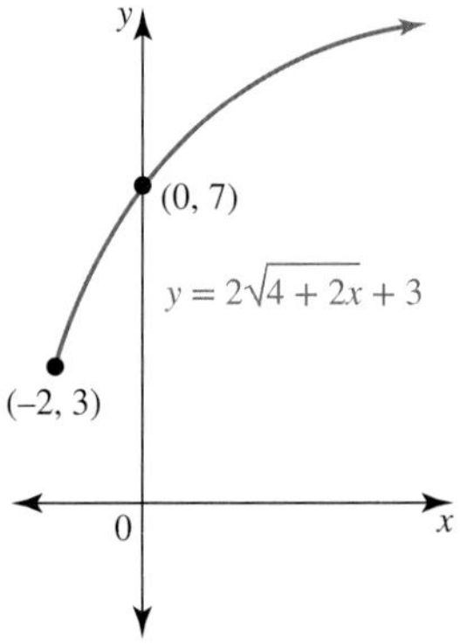

Domain $[-2, \infty)$, range $[3, \infty)$

d.

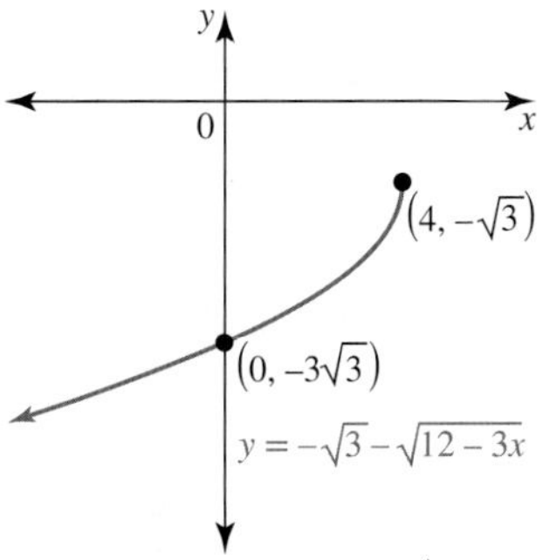

Domain $(-\infty, 4]$, range $\left(-\infty, -\sqrt{3}\right]$

14. a. $a = 2, b = -5, c = -2$

b. i. $a = -2, b = 4, c = -2$

ii. $y = -\sqrt{-2x + 4} + 2$

15. a.

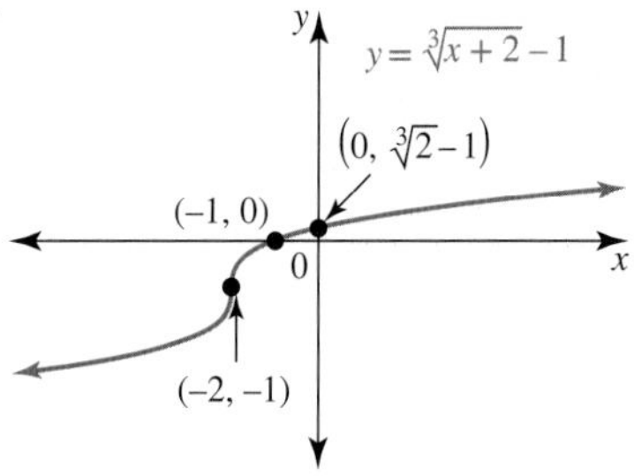

b.

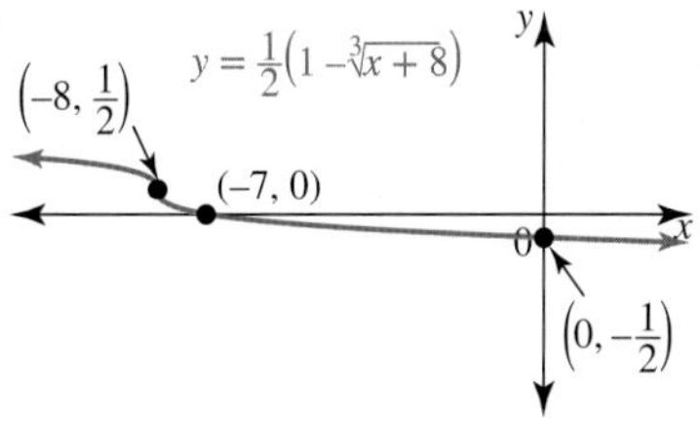

Domain R, range R

c.

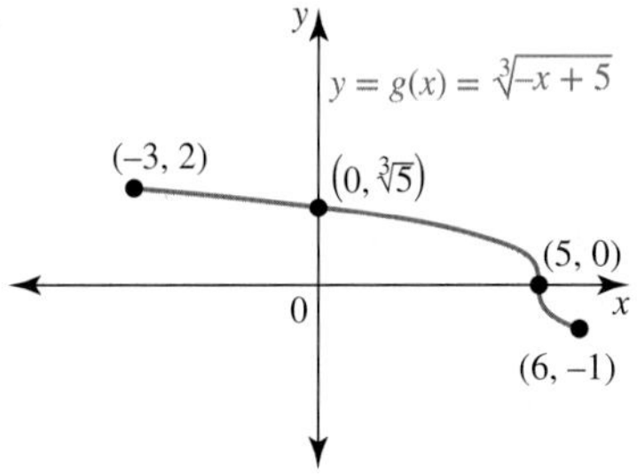

Domain $[-3, 6]$, range $[-1, 2]$

d. $y = 2\sqrt[3]{x} - 2$

e. $y = -\dfrac{7\sqrt[3]{x+1}}{2} - 2$

f. $y = 4\sqrt[3]{(x-2)} - 2, (2, -2)$

16. The maximal domain is $x \in [-3, 2]$.

17. a. Domain $R^+ \cup \{0\}$

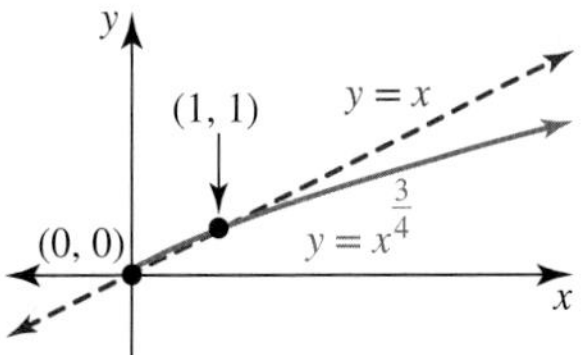

b. Domain R

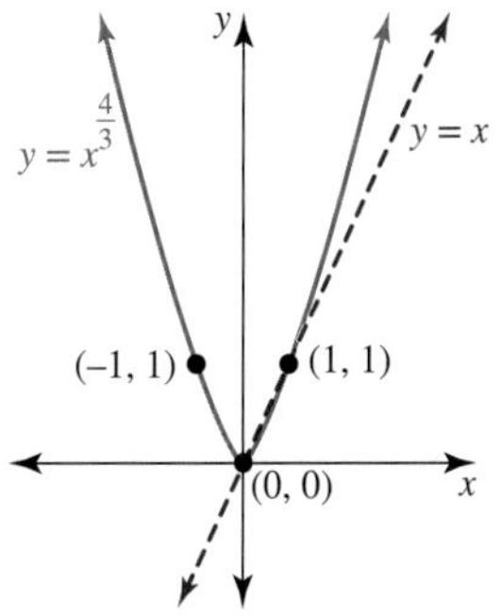

18. a. Domain R

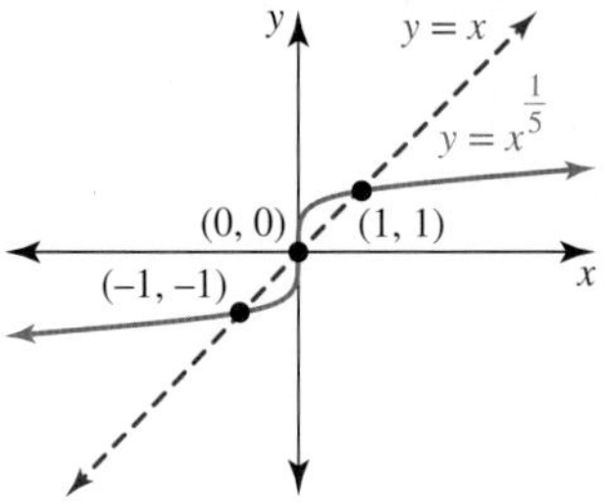

b. Domain $R^+ \cup \{0\}$

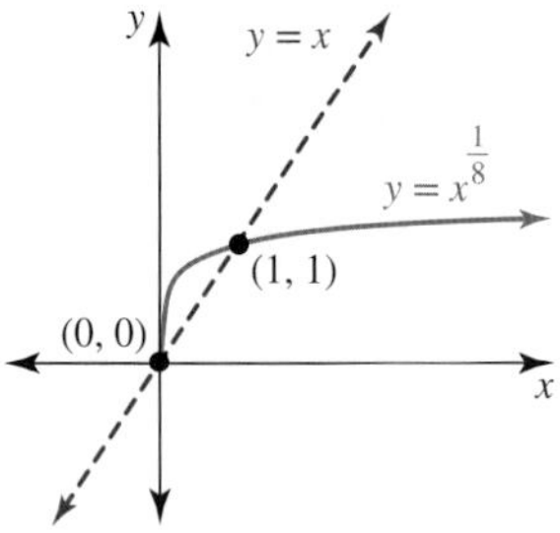

19. a. Domain $[0, \infty)$, quadrant 1

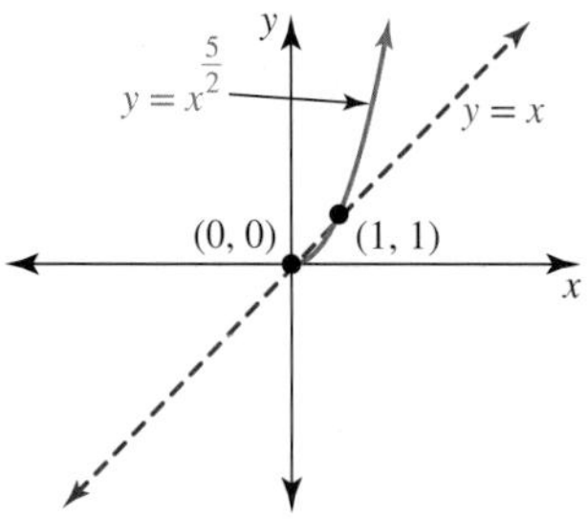

b. Domain R, quadrants 1 and 3

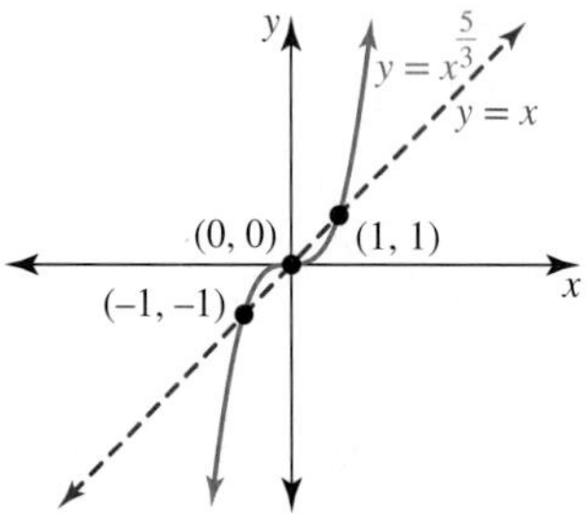

c. Domain R, quadrants 1 and 3

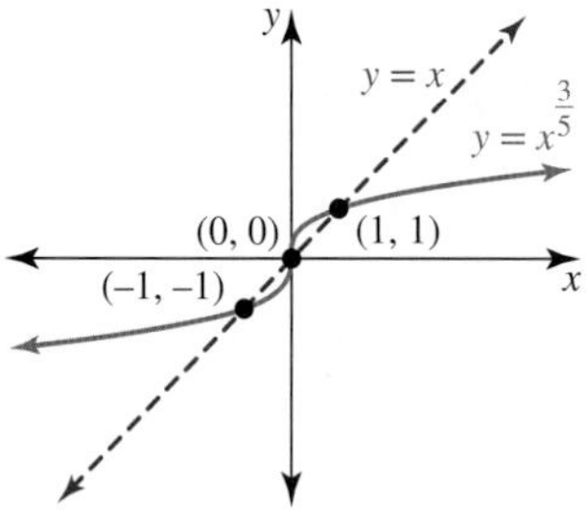

d. Domain $[0, \infty)$, quadrant 1

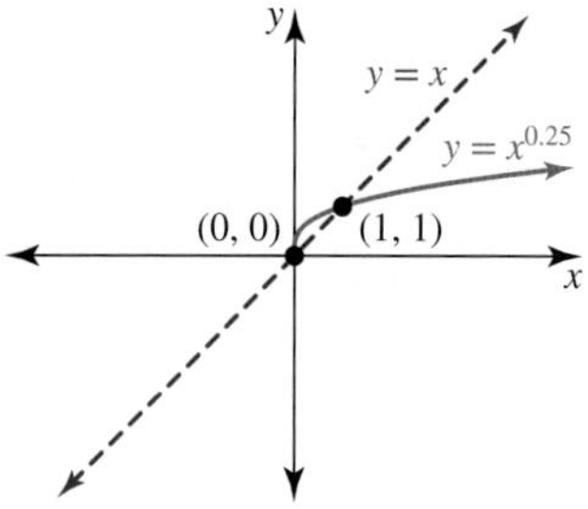

20. a. Draw $y = x$ and determine the cube roots of the appropriate y-values to construct the shape.

b.

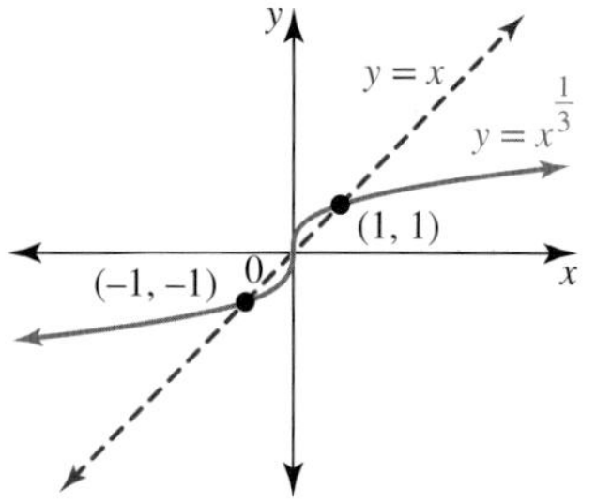

c. $\{x : x < -1\} \cup \{x : 0 < x < 1\}$

1.7 Exam questions

Note: Mark allocations are available with the fully worked solutions online.

1. E
2. D
3. A

1.8 Combinations of functions

1.8 Exercise

1. a. $f(-8) = 2, f(-1) = -1, f(2) = 0$

b.

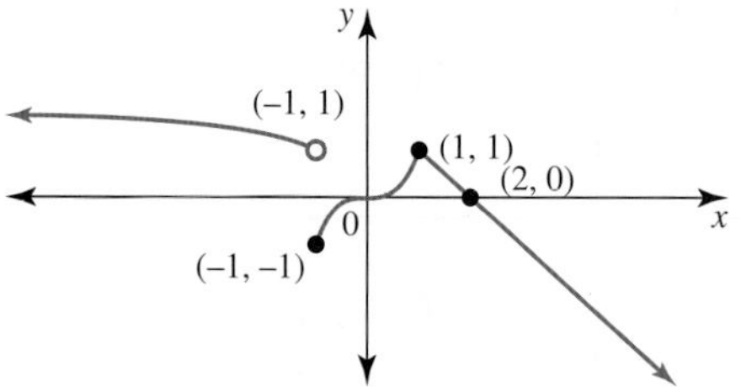

c. i. $x = -1$ ii. Domain R, range R.

2. a.

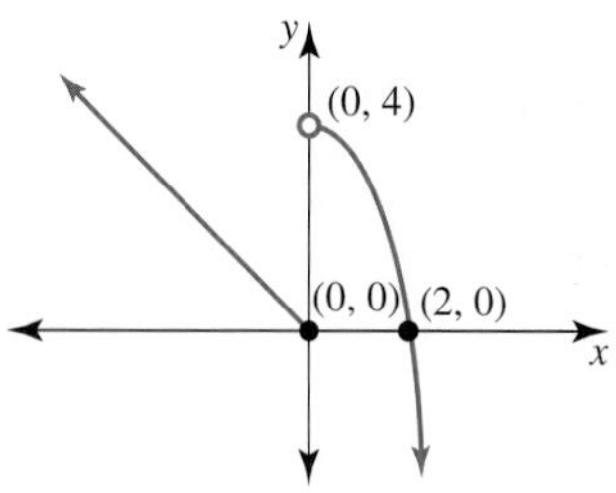

Domain R, range R, $x = 0$

b.

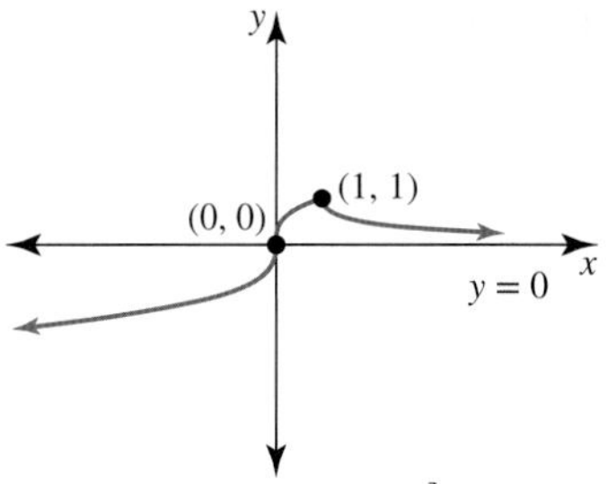

Domain R, range $(-\infty, 1]$, no point of discontinuity

3. a. i. 1

ii. 2

b.

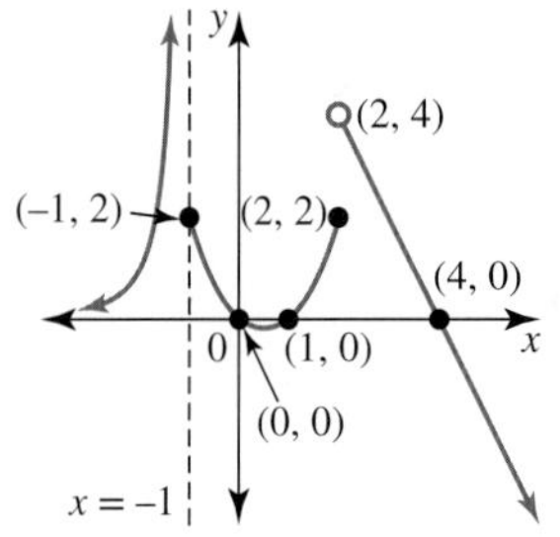

c. $R \setminus \{-1, 2\}$

4. a. The branch to the left of $x = 1$ has the rule $f(x) = \sqrt{1 - x}$, so $f(1) = 0$.
The branch to the right of $x = 1$ has the rule $f(x) = x - 2$, so $f(1) \to -1$ (open circle).
These branches do not join, so the hybrid function is not continuous at $x = 1$.

b.

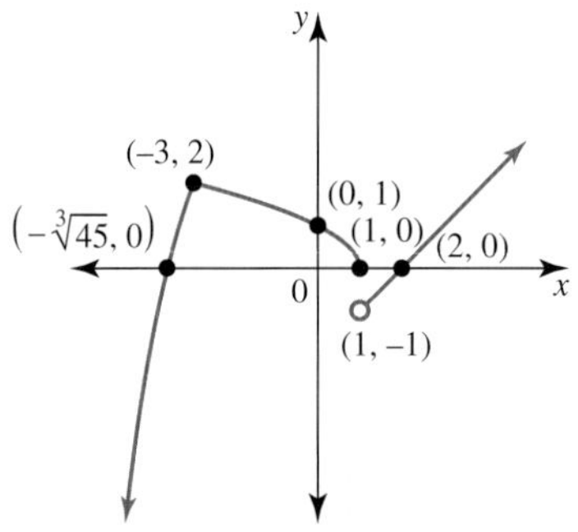

Many-to-one correspondence

c. $x = 6$

5. $y = \begin{cases} x + 4, & x < 0 \\ 4, & 0 \leq x < 4 \\ x, & 4 \leq x \leq 8 \end{cases}$

6. $y = \begin{cases} \frac{4}{3}(x + 3)(x + 1), & x < 0 \\ 4, & 0 \leq x \leq 2 \\ -2x + 8, & x \geq 3 \end{cases}$

7. a. $y = -\sqrt{1 + x} - \sqrt{1 - x}$, domain $[-1, 1]$

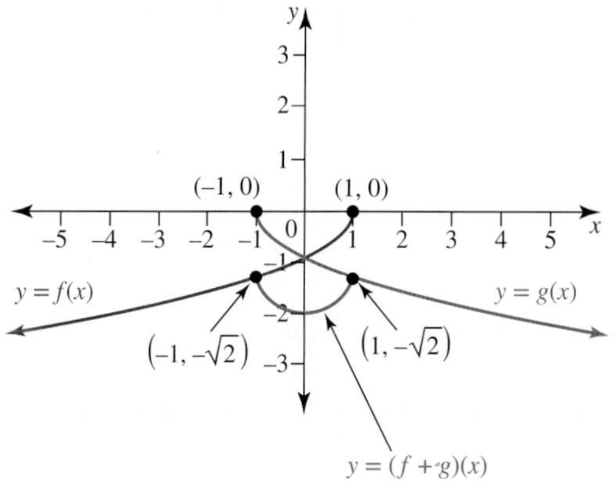

b. $y = \sqrt{1 - x^2}$, domain $[-1, 1]$, range $[0, 1]$.

8. $(f - g)(x) = x^3 - x^2$

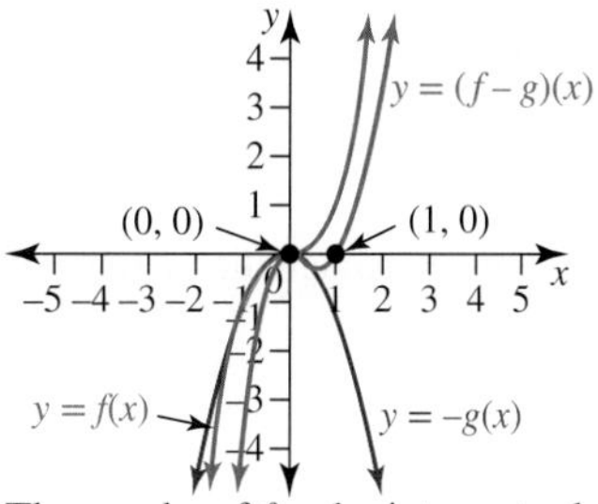

The graphs of f and g intersect when $x = 0$, $x = 1$, which gives the places where the difference function has x-intercepts.

9.

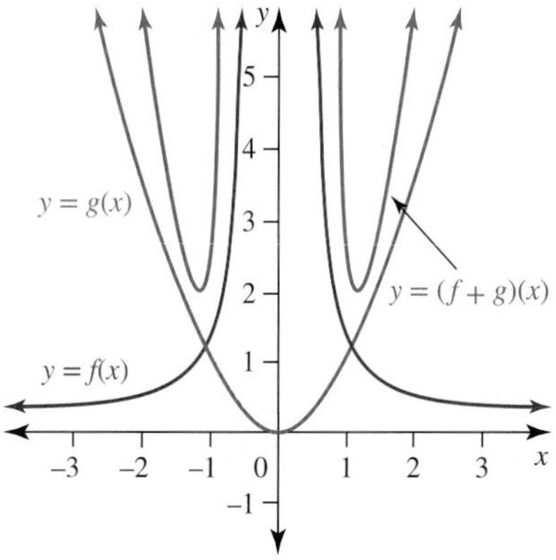

10. a.

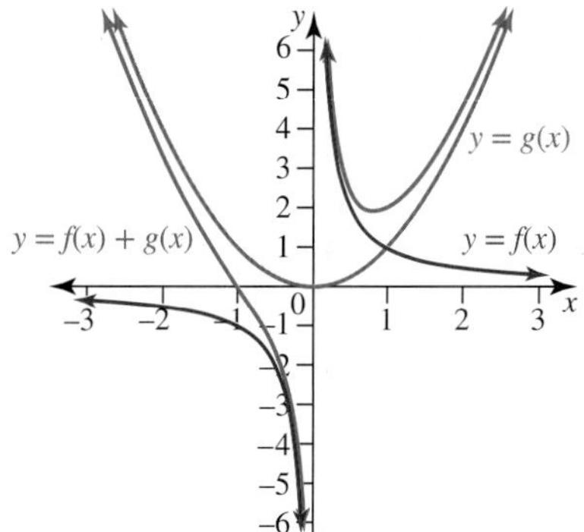

b.

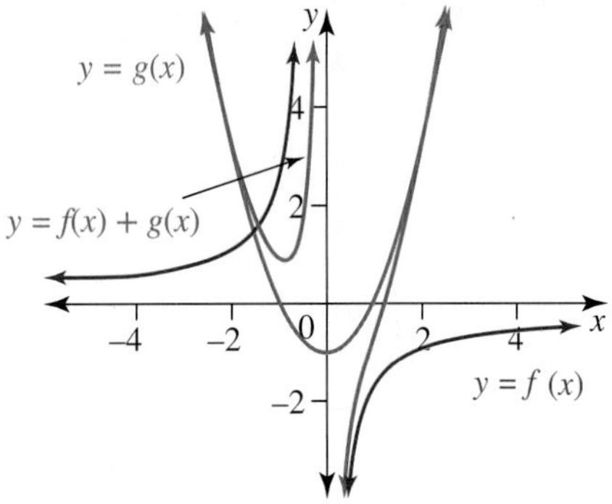

11. a. $y = 3$, domain R, range $\{3\}$, horizontal line through $(0, 3)$

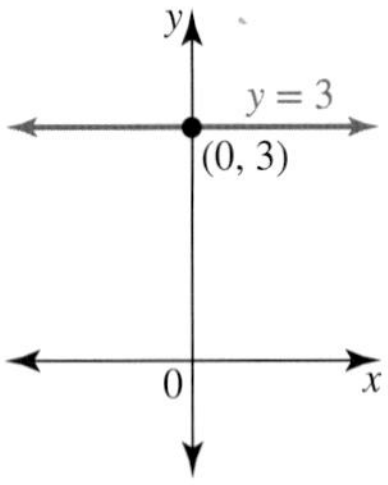

b. $y = 7 - 4x$, domain R, range R, straight line through $(0, 7)$ and $\left(\frac{7}{4}, 0\right)$

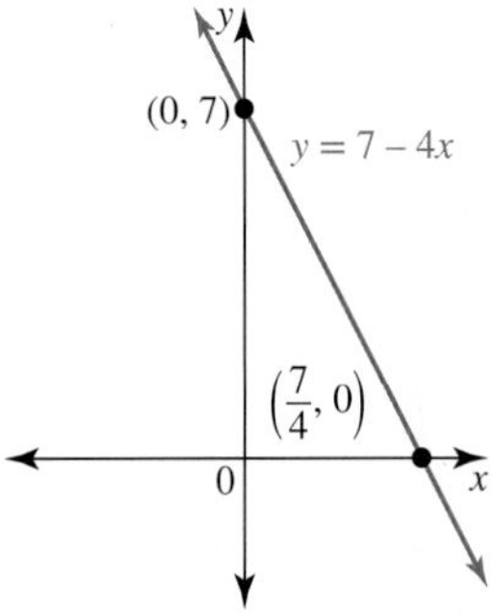

c. $y = 2(5 - 2x)(x - 1)$, domain R, range $\left(-\infty, \frac{9}{4}\right]$, concave down parabola with turning point $\left(\frac{7}{4}, \frac{9}{4}\right)$ and passing through $(0, -10)$, $(1, 0)$ and $(2.5, 0)$

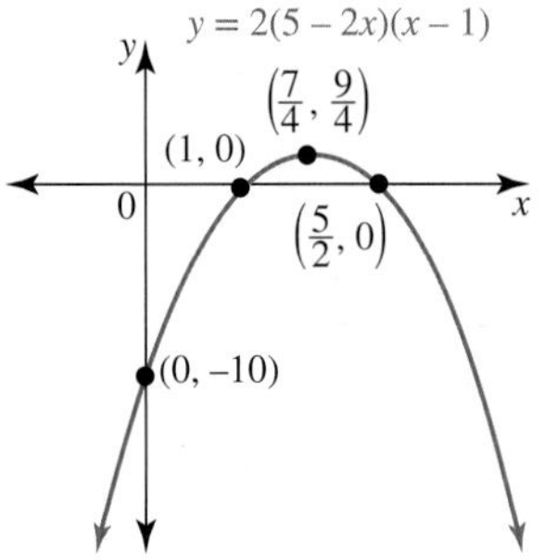

12. a. i. -6 ii. 189

b. $[-1, \infty)$

c. i.

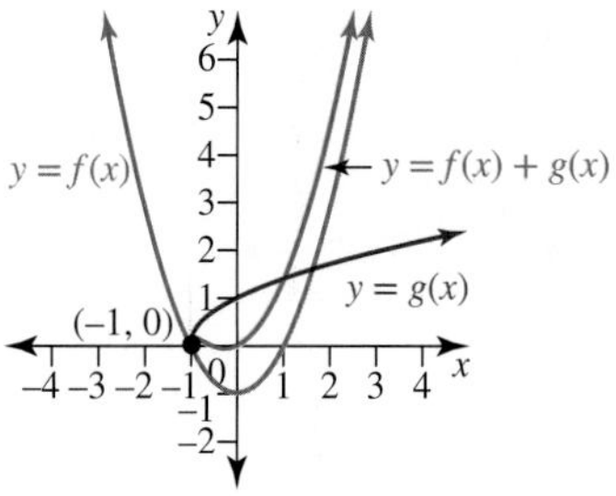

ii.

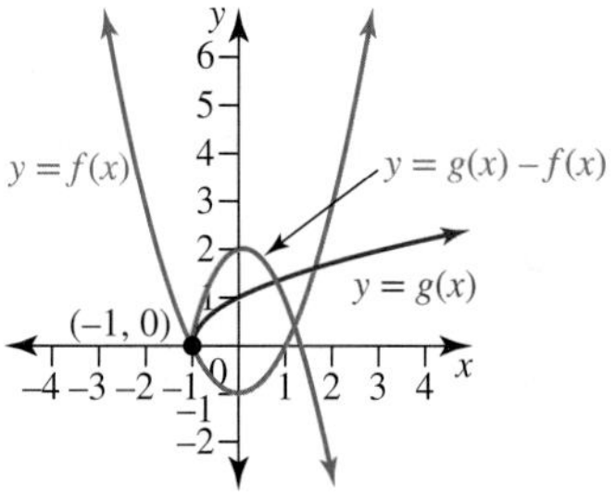

iii.

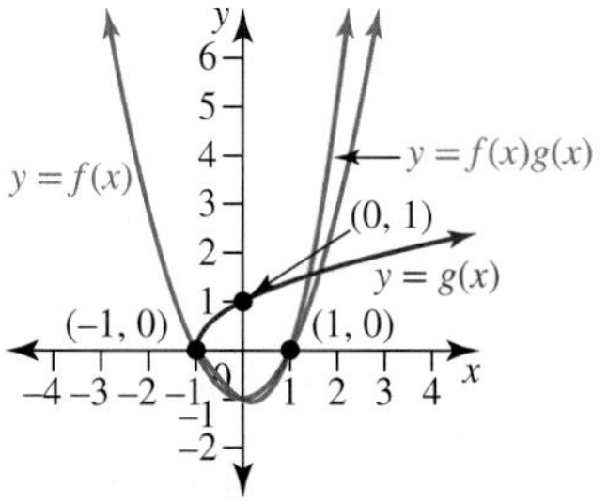

13.

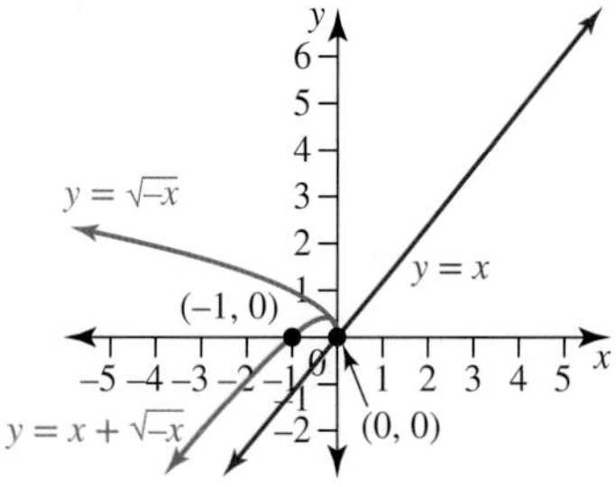

14.

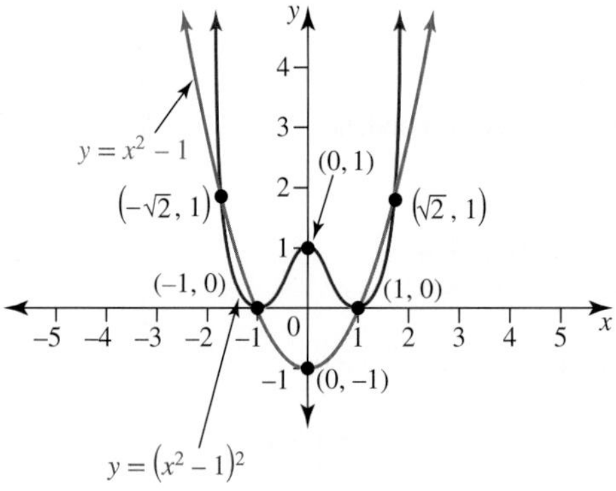

Domain R, range $[0, \infty)$

15. a.

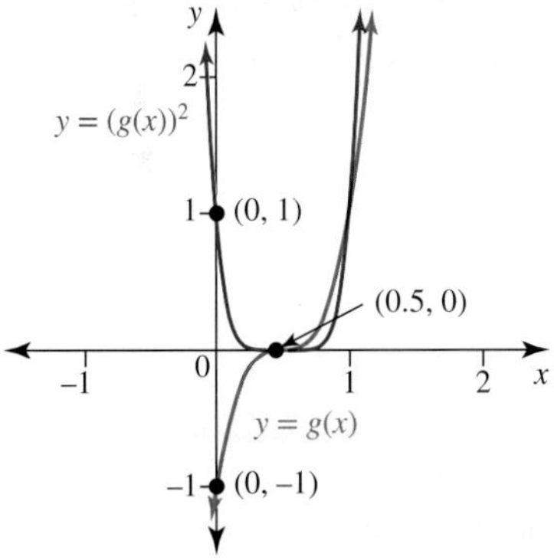

b. $(0,0), \left(\pm\sqrt{2},0\right), (-1,1), \left(\frac{1\pm\sqrt{5}}{2},1\right)$

16.

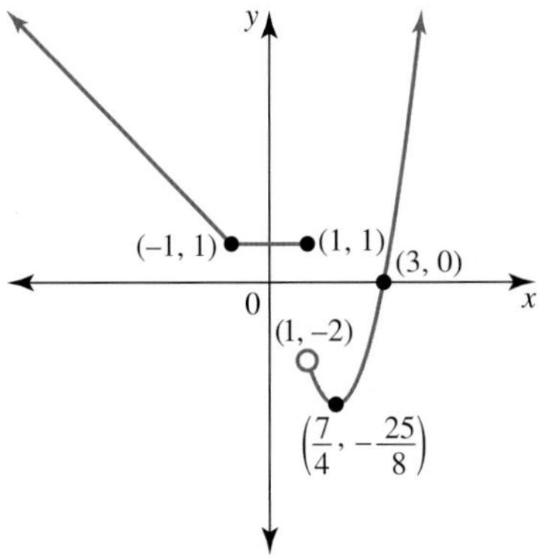

Range $= \left[-\frac{25}{8}, \infty\right)$

17. a. $a = 8, b = 32$

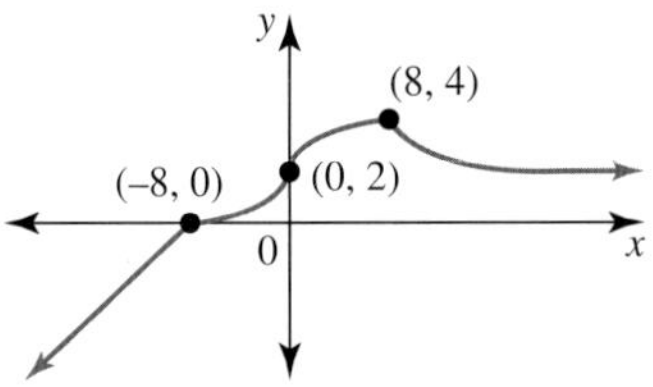

b. i. $k > 4$ ii. $k = 4$ or $k \leq 0$

iii. $0 < k < 4$

c. $\{-1, 32\}$

1.8 Exam questions

Note: Mark allocations are available with the fully worked solutions online.

1. E 2. E 3. D

1.9 Modelling and applications

1.9 Exercise

1. a. The data points increase and decrease, so they cannot be modelled by a one-to-one function. Neither a linear model nor an exponential model is possible.
The data is not oscillating, so it is unlikely to be trigonometric. The jump between $x = 1$ and $x = 3$ is a concern, but the data could be modelled by a polynomial such as a cubic with a turning point between $x = 1$ and $x = 3$. However, $y = x^n$ requires the point $(0, 0)$ to be on it and that is not true for the data given.

b. i. $y = \frac{4}{x-2} + 6$

ii.

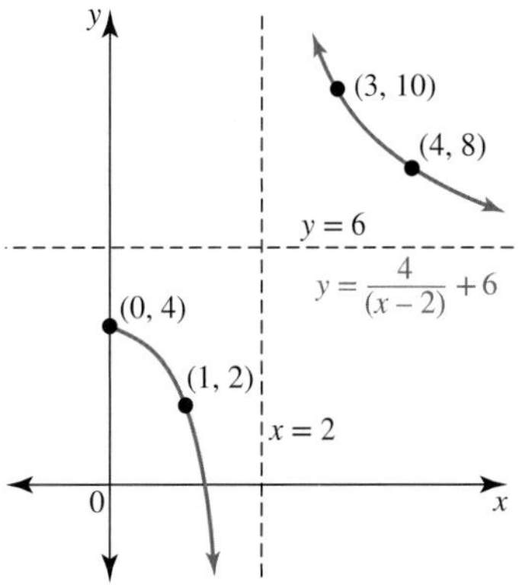

2. a.

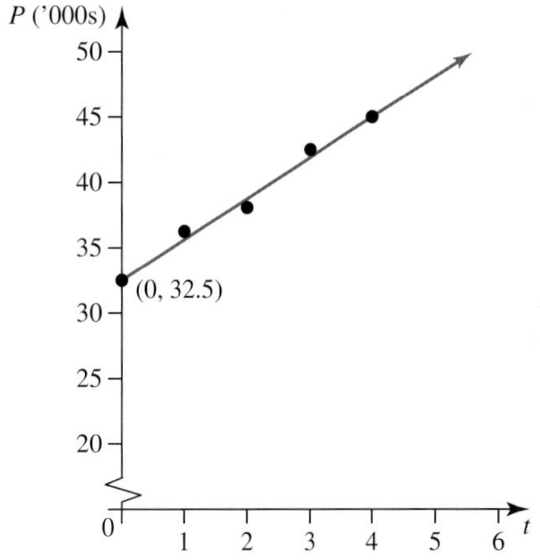

The data appears to be linear.

b. $P = 3.125t + 32.5$

c. 32 500 bees

d. 3.125 thousand per month

3. a. $a = 2$

b. $(2, 2)$

c. $y = \frac{2}{9}(x-5)^2, 2 \leq x \leq 9$

d. $y = \begin{cases} 2x+2, & -1 < x < 0 \\ 2, & 0 \leq x \leq 2 \\ \frac{2}{9}(x-5)^2, & 2 < x \leq 9 \end{cases}$

e. $x = 5 \pm \frac{3\sqrt{3}}{2}$ or $x = -\frac{1}{4}$

4. a. The garden area is the area of the entire square minus the area of the two right-angled triangles.

$$A = 40\times40 - \frac{1}{2}\times x\times x - \frac{1}{2}\times(40-x)\times40$$
$$= 1600 - \frac{1}{2}x^2 - 20(40-x)$$
$$= 1600 - \frac{1}{2}x^2 - 800 + 20x$$
$$= -\frac{1}{2}x^2 + 20x + 800$$

b. $0 < x < 40$

c. i. 20

ii. 1000 m^2

5. a. $y = 2x^3\left(x - \sqrt{5}\right)\left(x + \sqrt{5}\right)$

b. $y = 2x^3\left(x^2 - 5\right)$
$= 2x^5 - 10x^3$

c. i. Horizontal translation of $\sqrt{3}$ units to the right and vertical translation of $12\sqrt{3} + 1$ units upward

ii. $\left(24\sqrt{3} + 1\right) \approx 42.6$ metres

iii. $B\left(2\sqrt{3}, 1\right)$, $C\left(\sqrt{3}, 12\sqrt{3} + 1\right)$

6. a. $a = 38$, $b = 20$

b.

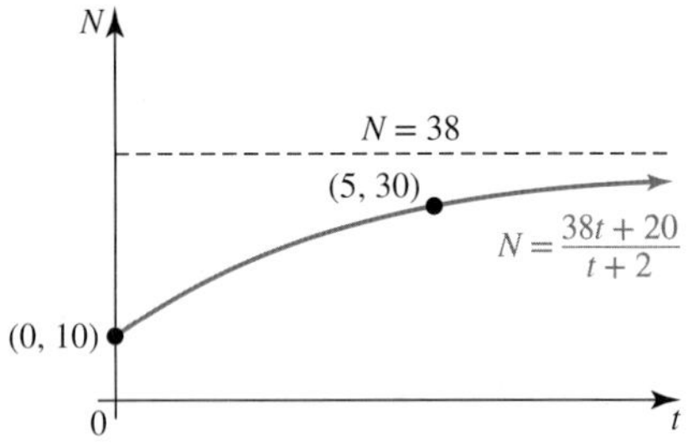

c. The population will never exceed 38.

1.9 Exam questions

Note: Mark allocations are available with the fully worked solutions online.

1. B
2. C
3. a. $r = \sqrt{8h - h^2}$

b. $V = \dfrac{1}{3}\pi h^2(8 - h)$, $0 < h < 8$

c.

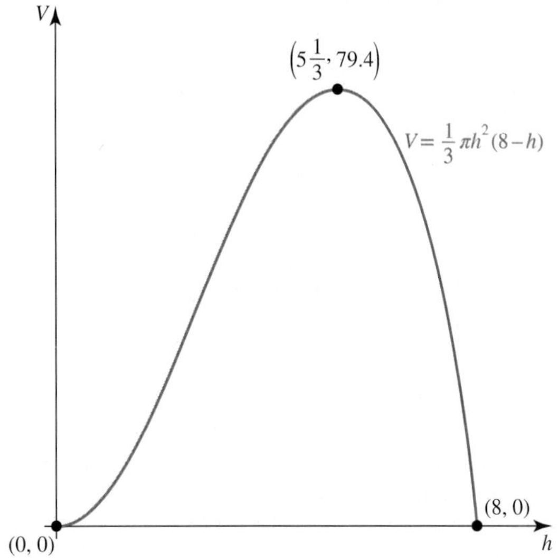

d. 79 cm^3

1.10 Review

1.10 Exercise

Technology free: short answer

1. a. $x = \dfrac{5 \pm \sqrt{73}}{6}$ b. $x = 2, \pm\dfrac{\sqrt{3}}{2}$

c. $m = \pm\sqrt{3}$ d. $x = 1, 2, 3$

2. a. Let $P(m) = 4m^4 - 11m^3 - 19m^2 + 44m + 12$.

If $m - 2$ is a factor, then $P(2) = 0$.

$$\begin{aligned}P(2) &= 4(2)^4 - 11(2)^3 - 19(2)^2 + 44(2) + 12\\ &= 4 \times 16 - 11 \times 8 - 19 \times 4 + 88 + 12\\ &= 64 - 88 - 76 + 88 + 12\\ &= 76 - 88 - 76 + 88\\ &= 0\end{aligned}$$

$\therefore m - 2$ is a factor.

b. $m = -\dfrac{1}{4}$, ± 2 and 3

3. $a = 5$, $b = 5$ and $c = -5$

4. a. Domain R, range $[-1, \infty)$

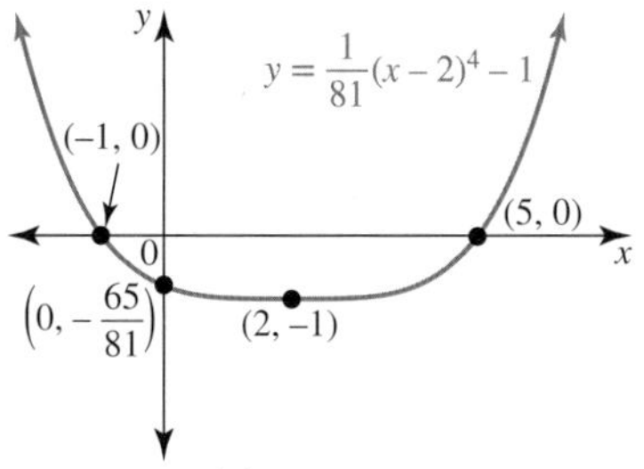

b. Domain $R \setminus \{2\}$, range $(-\infty, 1)$

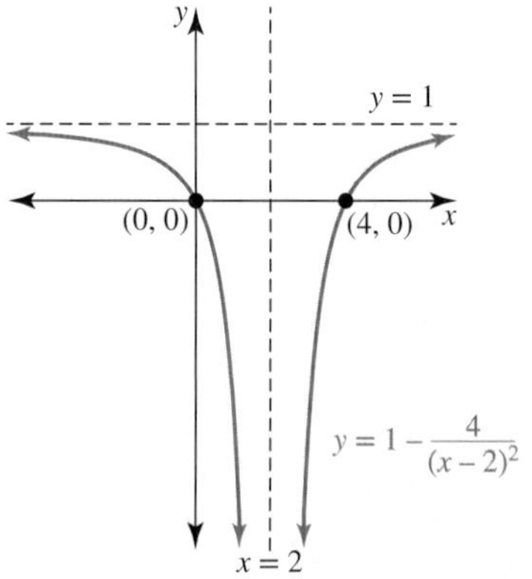

c. Domain R, range R

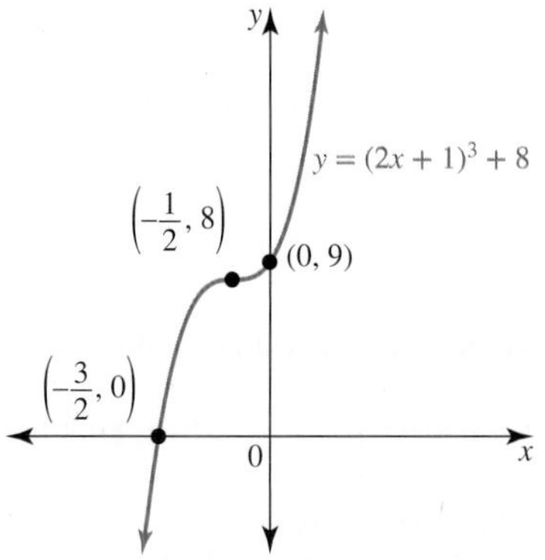

5. $m = -1$, $n = 2$ and $p = 3$

6. $x = 8 - 2\lambda$, $y = 5\lambda - 9$ and $z = \lambda$

7. a.

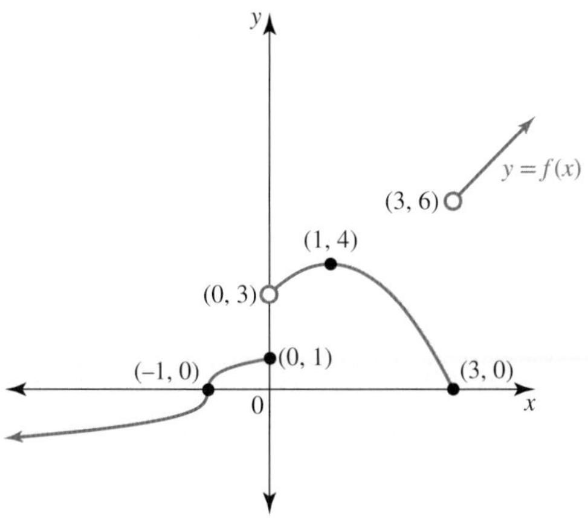

b. $x = 0$ and $x = 3$

c. Domain R, range $R \setminus (4, 6]$

8. a. $d_f = [-2, \infty)$, $d_g = [-2, 2]$, $d_f \cap d_g = D = [-2, 2]$

b.

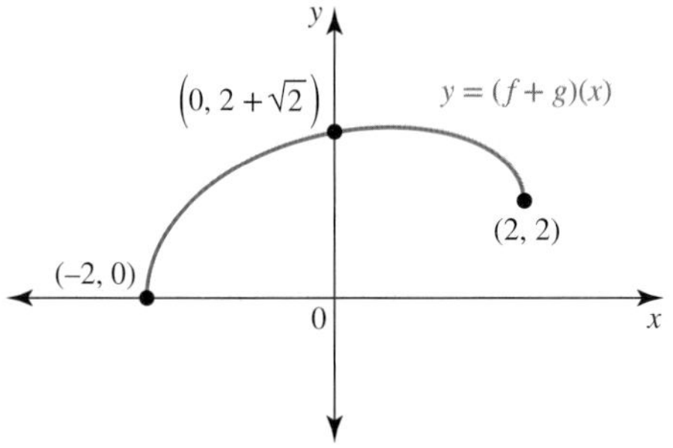

Technology active: multiple choice

9. D
10. A
11. D
12. E
13. B
14. C
15. B
16. C
17. C

Technology active: extended response

18. a. $a = 1, b = 0, c = -9, d = 0$ and $e = 20$

b. $m = \pm\sqrt{5}$ and ± 2

19. $m \in \left(\dfrac{3-\sqrt{21}}{-6}, \dfrac{3+\sqrt{21}}{-6}\right)$

20. $x = 0, y = -2, z = 1, w = 2$

21. a. $a = -1, b = 6, c = -5$

b. A $(1, 0)$ and B $(5, 0)$

c. 8 units^2

d. i. $y = ax^2 - 6ax + 8a + 3, a \in R \setminus \{0\}$

Let $x = 2$:

$$y = 4a - 12a + 8a + 3$$
$$= 3$$

Every parabola in this family passes through the point $(2, 3)$.

ii. $a < 0$ or $a > 3$

22. a. $(2 + x)^2 (2 - x)$

b.

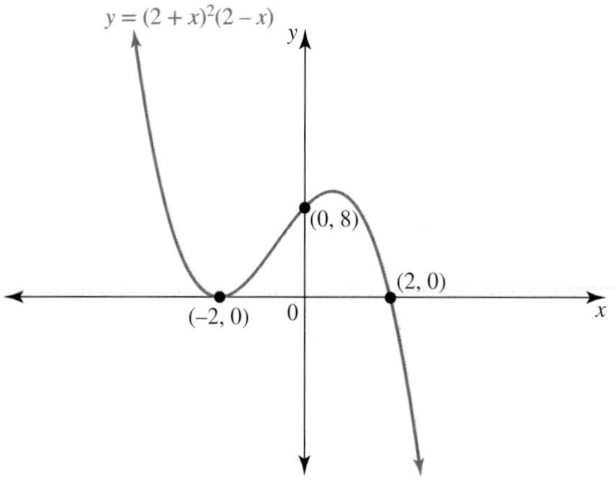

The maximum turning point lies in the interval $-2 \le x \le 2$.

c. i. $a = 2$

ii. $f(x) = 2 + x, x \in [-2, 2]$; $g(x) = \sqrt{2 - x}$, $x \in [-2, 2]$

iii. $\text{ran}_f = [0, 4]$, $\text{ran}_g = [0, 2]$

iv.

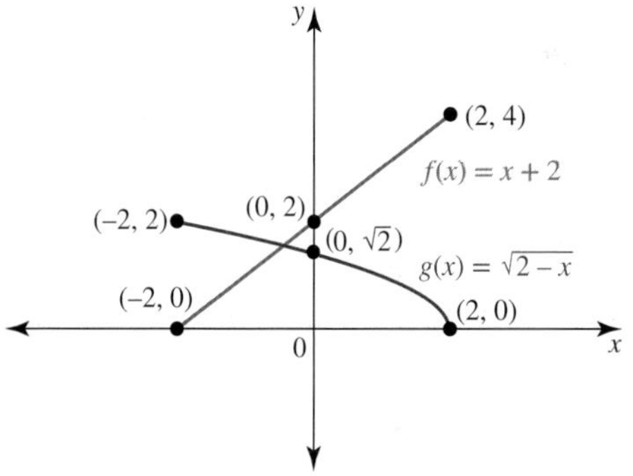

v.

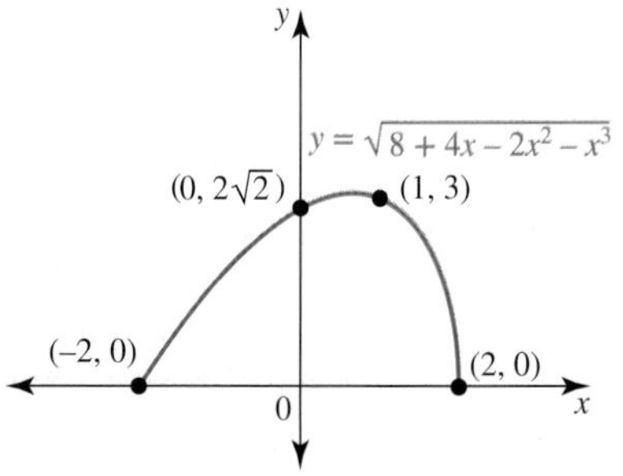

1.10 Exam questions

Note: Mark allocations are available with the fully worked solutions online.

1. D
2. C
3. B
4. B
5. $a = -5, a = 3$

No solutions, $a = 3$

2 Trigonometric (circular) functions

LEARNING SEQUENCE

Fully worked solutions for this topic are available online.

2.1 Overview

Hey students! Bring these pages to life online

Watch videos | Engage with interactivities | Answer questions and check results

Find all this and MORE in jacPLUS

2.1.1 Introduction

Trigonometry comes from two Greek words meaning 'triangle' and 'measure'.

Trigonometry has its origins in astronomy and was used to give coordinates for stars using spherical triangles — the stars were thought to be fixed on a crystal sphere of a large size. It has been hypothesised that the beginnings of trigonometry began with Hipparchus of Nicaea around 150 BCE, as he was the first to create a table of chords from a circle — he tabulated corresponding arc and chord lengths for different angles. He is known as the 'father of trigonometry'. In the early ninth century a Persian astronomer, al-Khwarizmi, produced accurate sine and cosine tables; he also created the first table of tangents.

Trigonometry has been used for centuries by surveyors and engineers. It is also used in developing digital music. Sound travels in the form of a wave, and this wave pattern can be described by a sine or cosine function. Other applications include calculating the heights of buildings; in microbiology, investigating how the depth of sunlight affects algae in terms of photosynthesis; and in oceanography, calculating the heights of ocean tides and waves.

KEY CONCEPTS

This topic covers the following key concepts from VCE Mathematics Study Design:

- graphs of the following functions: power functions $y = x^n$, $n \in Q$; exponential functions, $y = a^x$, $a \in R^+$, in particular $y = e^x$; logarithmic functions, $y = \log_e(x)$ and $y = \log_{10}(x)$; and circular functions, $y = \sin(x)$, $y = \cos(x)$ and $y = \tan(x)$ and their key features
- transformation from $y = f(x)$ to $y = Af(n(x+b)) + c$, where A, n, b and $c \in R$, $A, n \neq 0$, and f is one of the functions specified above, and the inverse transformation
- modelling of practical situations using polynomial, power, circular, exponential and logarithmic functions, simple transformation and combinations of these functions, including simple piecewise (hybrid) functions
- solution of equations of the form $f(x) = g(x)$ over a specified interval, where f and g are functions of the type specified in the 'Functions, relations and graphs' area of study, by graphical, numerical and algebraic methods, as applicable
- solution of literal equations and general solution of equations involving a single parameter.

Note: *Concepts shown in grey are covered in other topics.*

Source: VCE Mathematics Study Design (2023–2027) extracts © VCAA; reproduced by permission.

2.2 Trigonometric symmetry properties

LEARNING INTENTION

At the end of this subtopic you should be able to:
- state two trigonometric identities
- understand the symmetry properties of the unit circle
- determine complementary angles
- determine exact values for the sine, cosine and tangent of specific angles.

2.2.1 The unit circle

Angles are measured in degrees or **radians**. To define a radian, a circle with a radius of 1 unit is used. This circle is called the unit circle. When the point P is moved around the circle such that the arc length from S to P is 1 unit, the angle SOP is defined. The measure of this angle is 1 radian, 1^c.

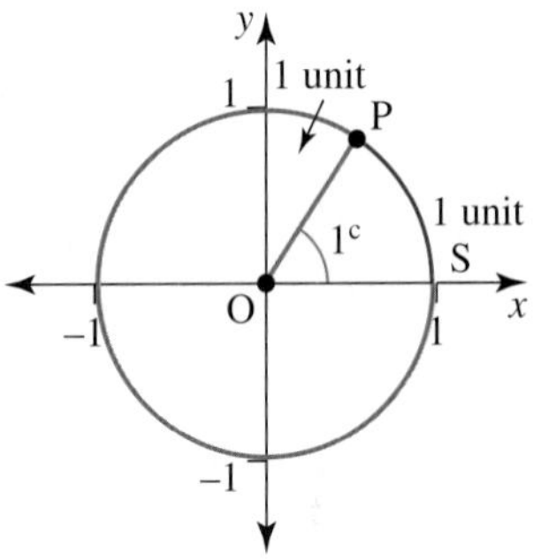

For the blue right-angled triangle POX, where θ is the angle at the origin, we know that, by definition, the distance along the x-axis is defined as $\cos(\theta)$ and the distance along the y-axis is defined as $\sin(\theta)$. In addition, if we consider the similar triangles POX (blue) and TOA (pink), the following important facts can be observed.

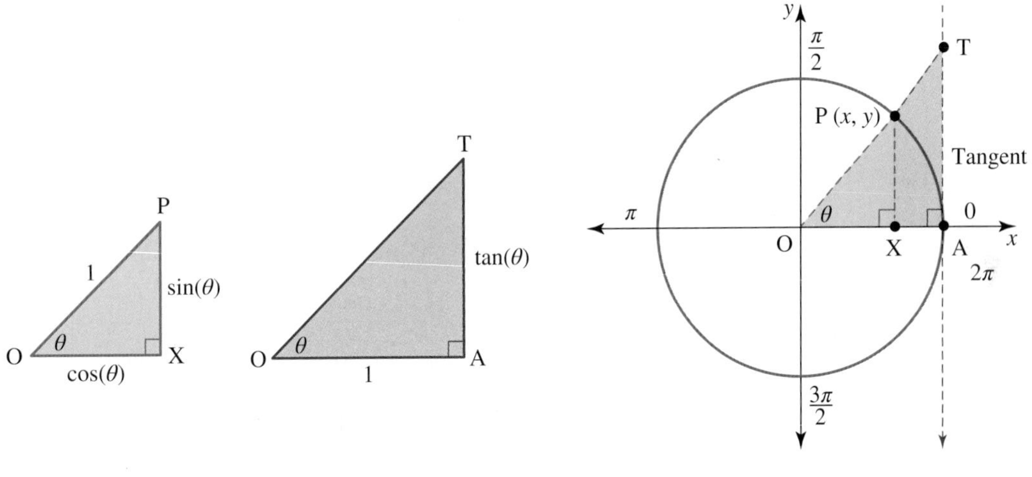

For triangle TOA, by definition, $\tan(\theta) = \dfrac{TA}{OA} = \dfrac{TA}{1}$; hence, $TA = \tan(\theta)$.

Using similar triangles, we can say that

$$\frac{\tan(\theta)}{\sin(\theta)} = \frac{1}{\cos(\theta)}$$

So

$$\tan(\theta) = \frac{\sin(\theta)}{\cos(\theta)}$$

This result is known as one of the trigonometric identities. You should learn and remember it, as it will be used frequently in later sections.

Also, if we consider the triangle POX, then by **Pythagoras' theorem**,

$$(\sin(\theta))^2 + (\cos(\theta))^2 = 1$$

or

$$\sin^2(\theta) + \cos^2(\theta) = 1$$

This is known as the **Pythagorean identity** and should also be learned and remembered.

Trigonometric identities

$$\tan(\theta) = \frac{\sin(\theta)}{\cos(\theta)}$$

$$\sin^2(\theta) + \cos^2(\theta) = 1$$

Special values for sine, cosine and tangent

Using the unit circle and rotating anticlockwise, we can determine the values of sine and cosine for the angles $0, \frac{\pi}{2}, \pi, \frac{3\pi}{2}$ and 2π by reading off the x-axis or the y-axis. The value for tangent is determined by the identity $\tan(\theta) = \frac{\sin(\theta)}{\cos(\theta)}$.

Angle (θ)	$\sin(\theta)$	$\cos(\theta)$	$\tan(\theta)$
0	0	1	0
$\frac{\pi}{2}$	1	0	Undefined
π	0	-1	0
$\frac{3\pi}{2}$	-1	0	Undefined
2π	0	1	0

The first quadrant: $0° < \theta < 90°$ or $0 < \theta < \frac{\pi}{2}$

For $0° < \theta < 90°$ or $0 < \theta < \frac{\pi}{2}$,

$$\cos(\theta) = \frac{x}{1} = x$$

$$\sin(\theta) = \frac{y}{1} = y$$

$$\tan(\theta) = \frac{\sin(\theta)}{\cos(\theta)} = \frac{y}{x}$$

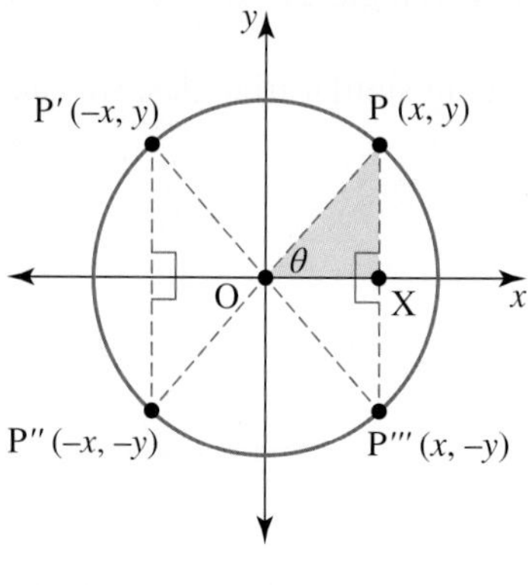

The second quadrant: $90° < \theta < 180°$ or $\frac{\pi}{2} < \theta < \pi$

Consider the point P′ in the second quadrant. When θ is the angle in the blue triangle at the origin, angles in the second quadrant are usually expressed as $180° - \theta$ or $\pi - \theta$. The angle refers to the angle made with respect to the positive direction of the x-axis and in an anticlockwise direction.

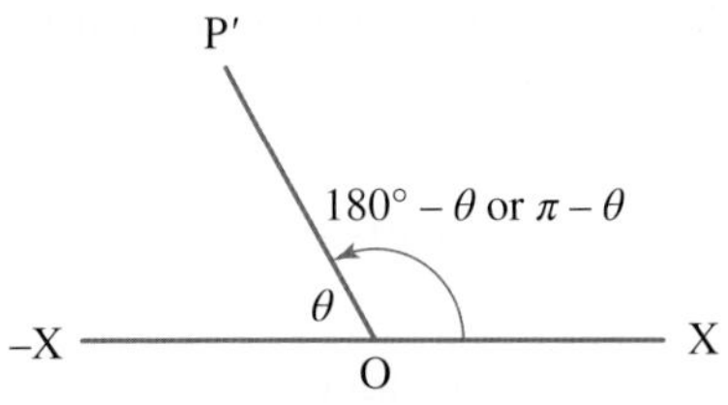

All angles in the second, third and fourth quadrants can be related back to the first quadrant.

Remember that in the first quadrant, $x = \cos(\theta)$, $y = \sin(\theta)$ and $\tan(\theta) = \frac{y}{x}$.

So in the second quadrant, using symmetry, the angles are:

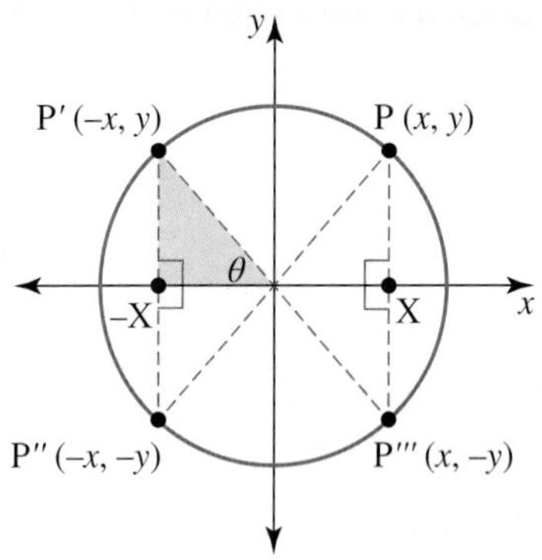

$$
\begin{aligned}
\cos(\pi - \theta) &= -x = -\cos(\theta) \\
\sin(\pi - \theta) &= y = \sin(\theta) \\
\tan(\pi - \theta) &= \frac{\sin(\pi - \theta)}{\cos(\pi - \theta)} \\
&= \frac{y}{-x} \\
&= -\tan(\theta)
\end{aligned}
$$

The third quadrant: $180° < \theta < 270°$ or $\pi < \theta < \frac{3\pi}{2}$

When the point P″ is in the third quadrant and θ is the angle in the blue triangle at the origin, angles in the third quadrant are usually expressed as $180° + \theta$ or $\pi + \theta$. The angle refers to the angle made with respect to the positive direction of the x-axis and in an anticlockwise direction.

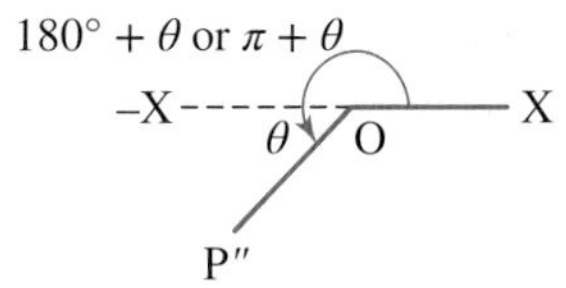

In the first quadrant, $x = \cos(\theta)$, $y = \sin(\theta)$ and $\tan(\theta) = \frac{y}{x}$.

So in the third quadrant, using symmetry, the angles are:

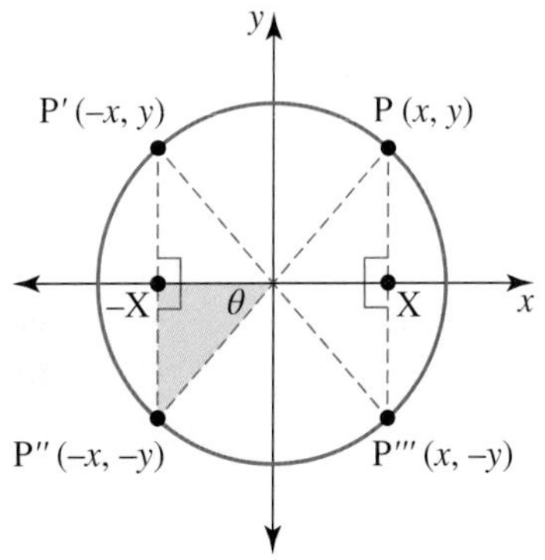

$$
\begin{aligned}
\cos(\pi + \theta) &= -x = -\cos(\theta) \\
\sin(\pi + \theta) &= -y = -\sin(\theta) \\
\tan(\pi + \theta) &= \frac{\sin(\pi + \theta)}{\cos(\pi + \theta)} \\
&= \frac{-y}{-x} \\
&= \tan(\theta)
\end{aligned}
$$

The fourth quadrant: $270° < \theta < 360°$ or $\frac{3\pi}{2} < \theta < 2\pi$

When P‴ is a point in the fourth quadrant and θ is the angle in the blue triangle at the origin, angles in the fourth quadrant are usually expressed as $360° - \theta$ or $2\pi - \theta$. The angle refers to the angle made with respect to the positive direction of the x-axis and in an anticlockwise direction.

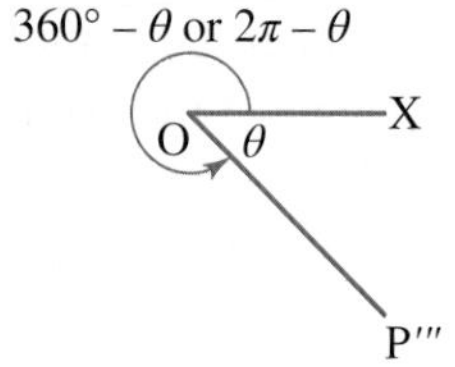

In the first quadrant, $x = \cos(\theta)$, $y = \sin(\theta)$ and $\tan(\theta) = \frac{y}{x}$.

So in the fourth quadrant, using symmetry, the angles are:

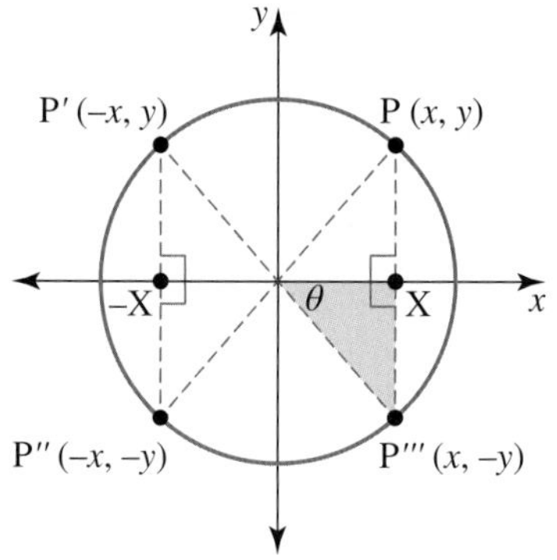

$$\begin{aligned}
\cos(2\pi - \theta) &= x = \cos(\theta) \\
\sin(2\pi - \theta) &= -y = -\sin(\theta) \\
\tan(2\pi - \theta) &= \frac{\sin(2\pi - \theta)}{\cos(2\pi - \theta)} \\
&= \frac{-y}{x} \\
&= -\tan(\theta)
\end{aligned}$$

The summary of the results from all four quadrants is as follows.

Symmetry properties of the trigonometric functions

2nd quadrant $\sin(\pi - \theta) = \sin(\theta)$ $\cos(\pi - \theta) = -\cos(\theta)$ $\tan(\pi - \theta) = -\tan(\theta)$ **S** **Sin positive**	**1st quadrant** $\sin(\theta)$ $\cos(\theta)$ $\tan(\theta)$ **A** **All positive**
T **Tan positive** $\sin(\pi + \theta) = -\sin(\theta)$ $\cos(\pi + \theta) = -\cos(\theta)$ $\tan(\pi + \theta) = \tan(\theta)$ **3rd quadrant**	**C** **Cos positive** $\sin(2\pi - \theta) = -\sin(\theta)$ $\cos(2\pi - \theta) = \cos(\theta)$ $\tan(2\pi - \theta) = -\tan(\theta)$ **4th quadrant**

Negative angles

Angles measured in a clockwise direction rather than in an anticlockwise direction are called negative angles.

$$\begin{aligned}
\cos(-\theta) &= x = \cos(\theta) \\
\sin(-\theta) &= -y = -\sin(\theta) \\
\tan(-\theta) &= \frac{\sin(-\theta)}{\cos(-\theta)} \\
&= \frac{-y}{x} \\
&= -\tan(\theta)
\end{aligned}$$

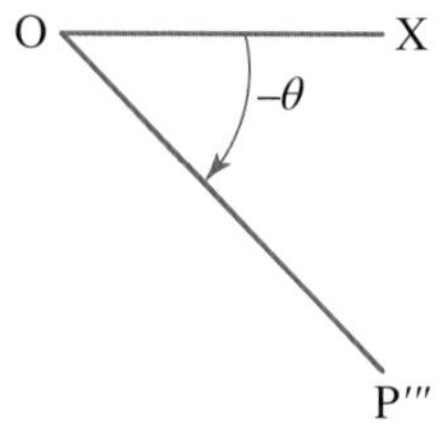

The following relationships apply for negative angles.

Negative angles of trigonometric functions

$$\sin(-\theta) = -\sin(\theta)$$
$$\cos(-\theta) = \cos(\theta)$$
$$\tan(-\theta) = -\tan(\theta)$$

***Note:* These relationships are true no matter what quadrant the negative angle is in.**

WORKED EXAMPLE 1 Symmetry properties

If $\sin(\theta) = \frac{\sqrt{3}}{2}$ and $\cos(\alpha) = \frac{4}{5}$, and θ and α are in the 1st quadrant, determine the exact values of the following trigonometric expressions.

a. $\sin(\pi + \theta)$
b. $\cos(-\alpha)$
c. $\tan(\theta)$
d. $\cos(\pi - \theta)$
e. $\sin(\pi + \alpha)$
f. $\tan(2\pi - \alpha)$

THINK	WRITE
a. 1. $(\pi + \theta)$ means the 3rd quadrant, where sine is negative.	a. $\sin(\pi + \theta) = -\sin(\theta)$
2. Substitute the appropriate value.	$= -\frac{\sqrt{3}}{2}$
b. 1. $(-\alpha)$ means the 4th quadrant, where cosine is positive.	b. $\cos(-\alpha) = \cos(\alpha)$
2. Substitute the appropriate value.	$= \frac{4}{5}$
c. 1. Use the Pythagorean identity to find the value of $\cos(\theta)$.	c. $\sin^2(\theta) + \cos^2(\theta) = 1$ $\cos^2(\theta) = 1 - \sin^2(\theta)$ $\cos^2(\theta) = 1 - \left(\frac{\sqrt{3}}{2}\right)^2$ $\cos^2(\theta) = 1 - \frac{3}{4}$ $\cos^2(\theta) = \frac{1}{4}$ $\cos^2(\theta) = \pm\frac{1}{2}$ θ is in the 1st quadrant, so cosine is positive. Hence, $\cos(\theta) = \frac{1}{2}$.

2. Use the identity $\tan(\theta)=\dfrac{\sin(\theta)}{\cos(\theta)}$ to find $\tan(\theta)$.

$$\tan(\theta) = \frac{\sin(\theta)}{\cos(\theta)}$$
$$= \frac{\frac{\sqrt{3}}{2}}{\frac{1}{2}}$$
$$= \frac{\sqrt{3}}{2} \times \frac{2}{1}$$
$$= \sqrt{3}$$

d. 1. $(\pi - \theta)$ means the 2nd quadrant, where cosine is negative.

d. $\cos(\pi - \theta) = -\cos(\theta)$

2. In part **c** we determined $\cos(\theta)=\dfrac{1}{2}$, so we can substitute this value.

$$= -\frac{1}{2}$$

e. 1. In order to find the value of $\sin(\alpha)$, apply the Pythagorean identity.

e. $\sin^2(\alpha) + \cos^2(\alpha) = 1$
$$\sin^2(\alpha) = 1 - \cos^2(\alpha)$$
$$\sin^2(\alpha) = 1 - \left(\frac{4}{5}\right)^2$$
$$\sin^2(\alpha) = 1 - \frac{16}{25}$$
$$\sin^2(\alpha) = \frac{9}{25}$$
$$\sin(\alpha) = \pm\frac{3}{5}$$

α is in the 1st quadrant, so sine is positive.
Hence, $\sin(\alpha)=\dfrac{3}{5}$.

2. $(\pi + \alpha)$ means the 3rd quadrant, where sine is negative.

$\sin(\pi + \alpha) = -\sin(\alpha)$

3. Substitute the appropriate value.

$$= -\frac{3}{5}$$

f. Use the identity $\tan(\theta)=\dfrac{\sin(\theta)}{\cos(\theta)}$ and simplify.

f. $\tan(2\pi - \alpha) = \dfrac{\sin(2\pi - \alpha)}{\cos(2\pi - \alpha)}$
$$= \frac{-\sin(\alpha)}{\cos(\alpha)}$$
$$= \frac{-\frac{3}{5}}{\frac{4}{5}}$$
$$= -\frac{3}{4}$$

Resources

Interactivities The unit circle (int-2582)
Symmetry points and quadrants (int-2584)
All sin cos tan (int-2583)

2.2.2 Complementary angles

Sometimes angles are named relative to the y-axis rather than the x-axis, for example $\frac{\pi}{2} \pm \theta$ or $\frac{3\pi}{2} \pm \theta$. These are special cases, and great care should be taken with these types of examples.

The first quadrant: reference angle $90° - \theta$ or $\frac{\pi}{2} - \theta$

Remember that in the first quadrant the distance along the x-axis is defined as $\cos(\theta)$, while the distance along the y-axis is defined as $\sin(\theta)$.

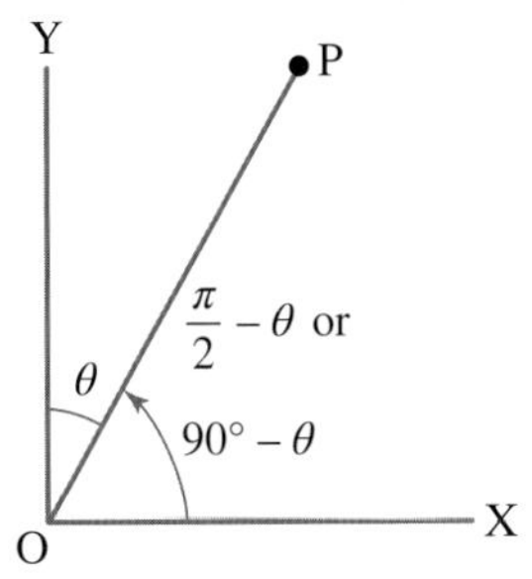

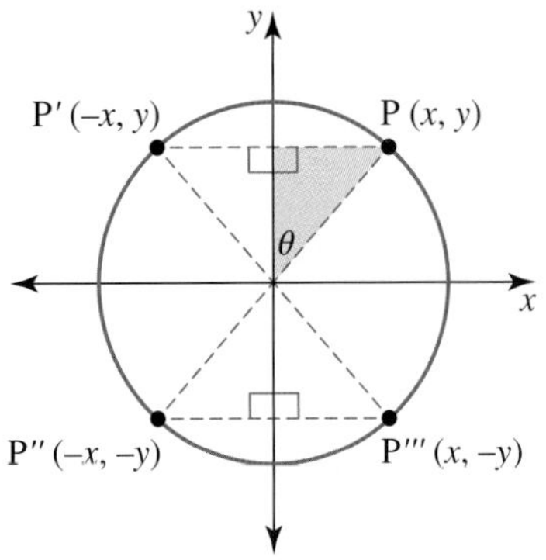

$$\cos\left(\frac{\pi}{2} - \theta\right) = \sin(\theta)$$

$$\sin\left(\frac{\pi}{2} - \theta\right) = \cos(\theta)$$

$$\tan\left(\frac{\pi}{2} - \theta\right) = \frac{\sin\left(\frac{\pi}{2} - \theta\right)}{\cos\left(\frac{\pi}{2} - \theta\right)} = \frac{\cos(\theta)}{\sin(\theta)} = \frac{1}{\tan(\theta)}$$

Note: $\sin\left(\frac{\pi}{2} - \theta\right) = \cos(\theta)$ is a **complementary relationship** because the sum of the angles adds to $\frac{\pi}{2}$.

The second quadrant: reference angle $90° + \theta$ or $\frac{\pi}{2} + \theta$

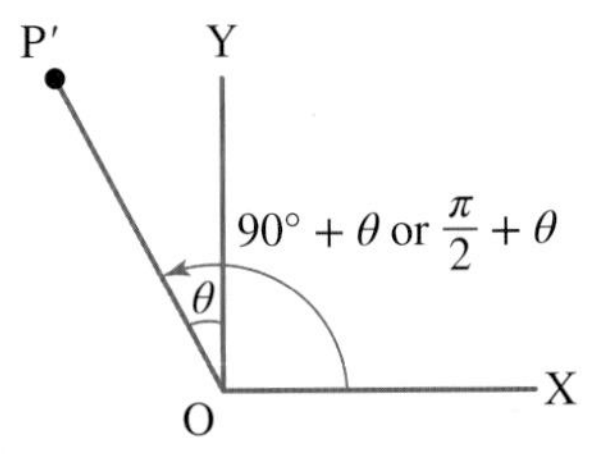

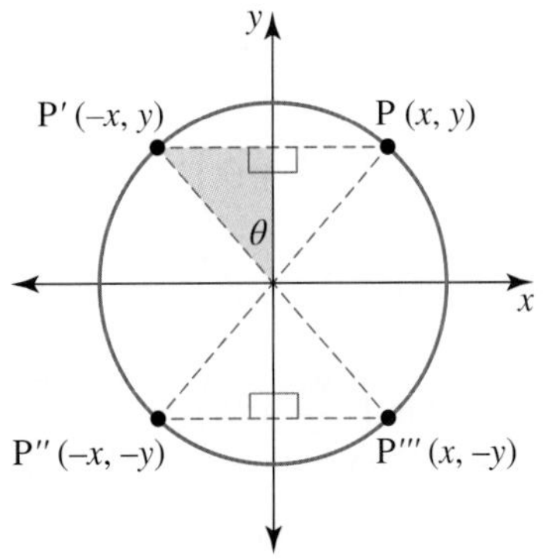

$$\cos\left(\frac{\pi}{2}+\theta\right) = -\sin(\theta)$$

$$\sin\left(\frac{\pi}{2}+\theta\right) = \cos(\theta)$$

$$\tan\left(\frac{\pi}{2}+\theta\right) = \frac{\sin\left(\frac{\pi}{2}+\theta\right)}{\cos\left(\frac{\pi}{2}+\theta\right)} = \frac{\cos(\theta)}{-\sin(\theta)} = -\frac{1}{\tan\theta}$$

The third quadrant: reference angle $270° - \theta$ or $\dfrac{3\pi}{2} - \theta$

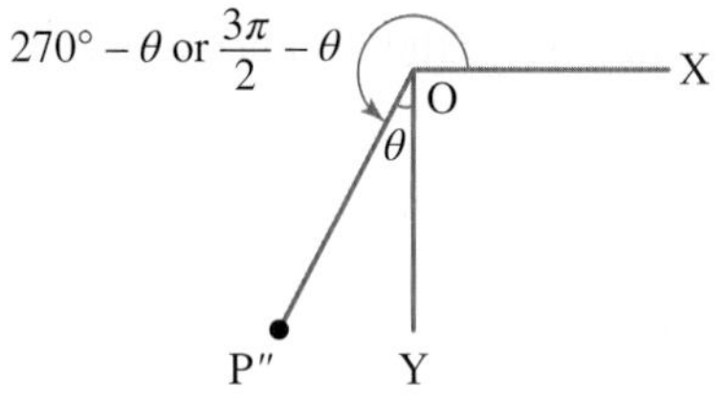

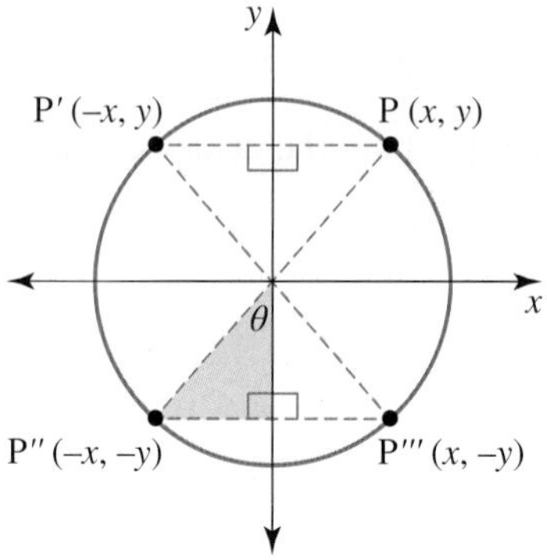

$$\cos\left(\frac{3\pi}{2}-\theta\right) = -\sin(\theta)$$

$$\sin\left(\frac{3\pi}{2}-\theta\right) = -\cos(\theta)$$

$$\tan\left(\frac{3\pi}{2}-\theta\right) = \frac{\sin\left(\frac{3\pi}{2}-\theta\right)}{\cos\left(\frac{3\pi}{2}-\theta\right)} = \frac{-\cos(\theta)}{-\sin(\theta)} = \frac{1}{\tan\theta}$$

The fourth quadrant: reference angle $270° + \theta$ or $\dfrac{3\pi}{2} + \theta$

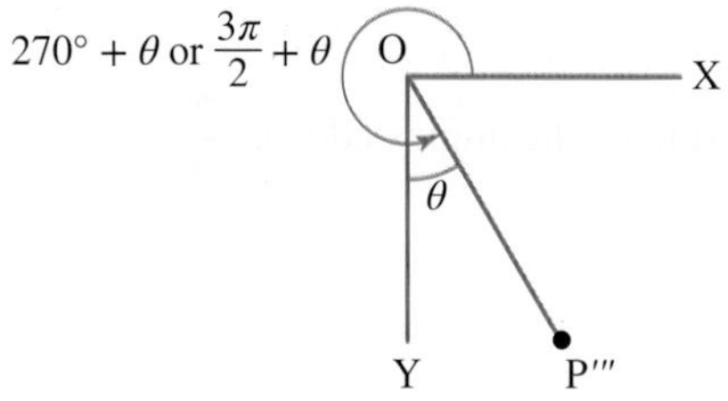

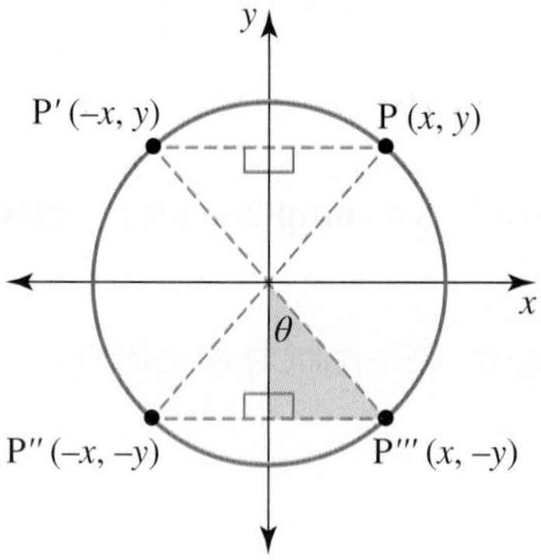

$$\cos\left(\frac{3\pi}{2}+\theta\right) = \sin(\theta)$$

$$\sin\left(\frac{3\pi}{2}+\theta\right) = -\cos(\theta)$$

$$\tan\left(\frac{3\pi}{2}+\theta\right) = \frac{\sin\left(\frac{3\pi}{2}+\theta\right)}{\cos\left(\frac{3\pi}{2}+\theta\right)} = \frac{-\cos(\theta)}{\sin(\theta)} = -\frac{1}{\tan(\theta)}$$

WORKED EXAMPLE 2 Complementary angles

If $\cos(\theta) = 0.5300$ and θ is in the 1st quadrant, determine the values of the following, correct to 4 decimal places.

a. $\sin\left(\frac{\pi}{2} - \theta\right)$ **b.** $\cos\left(\frac{3\pi}{2} - \theta\right)$ **c.** $\tan\left(\frac{3\pi}{2} + \theta\right)$

THINK	WRITE
a. 1. $\left(\frac{\pi}{2} - \theta\right)$ is in the 1st quadrant, so all trigonometric ratios are positive.	**a.** $\sin\left(\frac{\pi}{2} - \theta\right) = \cos(\theta)$
2. Substitute the appropriate value.	$= 0.5300$
b. 1. $\left(\frac{3\pi}{2} - \theta\right)$ is in the 3rd quadrant, so sine is negative.	**b.** $\cos\left(\frac{3\pi}{2} - \theta\right) = -\sin(\theta)$
2. Use the Pythagorean identity to find $\sin(\theta)$.	$\sin^2\theta + \cos^2(\theta) = 1$ $\sin^2\theta = 1 - \cos^2(\theta)$ $\sin^2\theta = 1 - (0.5300)^2$ $\sin(\theta) = \pm\sqrt{1 - (0.5300)^2}$ $\sin(\theta) = 0.8480$ as θ is in the 1st quadrant.
3. Substitute the appropriate values to determine $\cos\left(\frac{3\pi}{2} - \theta\right)$.	$\cos\left(\frac{3\pi}{2} - \theta\right) = -\sin(\theta)$ $= -0.8480$
c. 1. $\left(\frac{3\pi}{2} + \theta\right)$ is in the 4th quadrant, so tangent is negative.	$\tan\left(\frac{3\pi}{2} + \theta\right) = -\frac{1}{\tan\theta}$
2. Use the identity $\tan\theta = \frac{\sin(\theta)}{\cos(\theta)}$ to find the reciprocal.	$\frac{1}{\tan(\theta)} = \frac{\cos(\theta)}{\sin(\theta)}$ $= \frac{0.5300}{0.8480}$ $= 0.6250$ $\therefore \tan\left(\frac{3\pi}{2} + \theta\right) = -0.6250$

on Resources

Interactivity Complementary properties of sin and cos (int-2979)

2.2.3 Exact values

In Mathematical Methods Unit 2, you studied the exact trigonometric ratios for the angles $\frac{\pi}{6}$, $\frac{\pi}{4}$ and $\frac{\pi}{3}$ (30°, 45° and 60°). These values come from an isosceles triangle and an equilateral triangle.

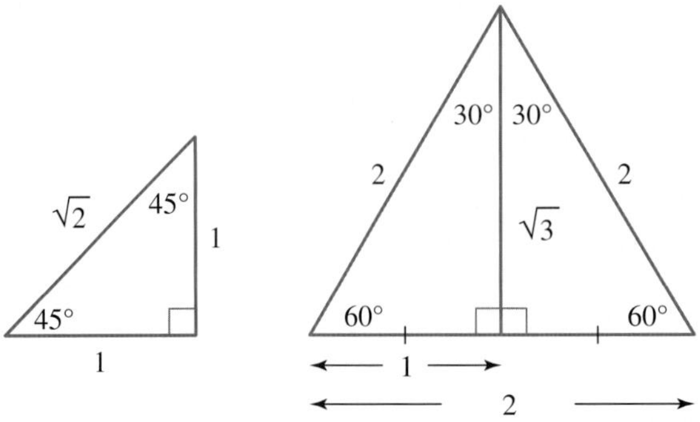

The following table provides a summary of these angles and their ratios.

Angle (θ)	sin(θ)	cos(θ)	tan(θ)
30° or $\frac{\pi}{6}$	$\frac{1}{2}$	$\frac{\sqrt{3}}{2}$	$\frac{1}{\sqrt{3}}$
45° or $\frac{\pi}{4}$	$\frac{1}{\sqrt{2}}$	$\frac{1}{\sqrt{2}}$	1
60° or $\frac{\pi}{3}$	$\frac{\sqrt{3}}{2}$	$\frac{1}{2}$	$\sqrt{3}$

WORKED EXAMPLE 3 Exact values

Give exact values for the following trigonometric expressions.

a. $\cos\left(\frac{2\pi}{3}\right)$ b. $\tan\left(\frac{7\pi}{4}\right)$ c. $\cos\left(-\frac{\pi}{6}\right)$ d. $\sin\left(\frac{11\pi}{3}\right)$

THINK	WRITE
a. 1. Rewrite the angle in terms of π and find the corresponding angle in the 1st quadrant.	a. $\cos\left(\frac{2\pi}{3}\right) = \cos\left(\pi - \frac{\pi}{3}\right)$
2. The angle is in the 2nd quadrant, so cosine is negative.	$= -\cos\left(\frac{\pi}{3}\right)$
3. Write the answer.	$= -\frac{1}{2}$
b. 1. Rewrite the angle in terms of 2π and find the corresponding angle in the 1st quadrant.	b. $\tan\left(\frac{7\pi}{4}\right) = \tan\left(2\pi - \frac{\pi}{4}\right)$

2. The angle is in the 4th quadrant, so tangent is negative. $= -\tan\left(\frac{\pi}{4}\right)$

3. Write the answer. $= -1$

c. 1. Rewrite the negative angle as $\cos(-\theta) = \cos(\theta)$.

c. $\cos\left(-\frac{\pi}{6}\right) = \cos\left(\frac{\pi}{6}\right)$

2. Write the answer. $= \frac{\sqrt{3}}{2}$

d. 1. Rewrite the angle in terms of a multiple of 2π.

d. $\sin\left(\frac{11\pi}{3}\right) = \sin\left(4\pi - \frac{\pi}{3}\right)$

2. Subtract the extra multiple of 2π so the angle is within one revolution of the unit circle. $= \sin\left(2\pi - \frac{\pi}{3}\right)$

3. The angle is in the 4th quadrant, so sine is negative. $= -\sin\left(\frac{\pi}{3}\right)$

4. Write the answer. $= -\frac{\sqrt{3}}{2}$

2.2 Exercise

Technology free

1. WE3 Give exact values for the following trigonometric expressions.

a. $\tan\left(\frac{3\pi}{4}\right)$
b. $\cos\left(\frac{5\pi}{6}\right)$
c. $\sin\left(-\frac{\pi}{4}\right)$
d. $\cos\left(\frac{7\pi}{3}\right)$
e. $\tan\left(-\frac{\pi}{3}\right)$
f. $\sin\left(\frac{11\pi}{6}\right)$

2. Give exact values for the following trigonometric expressions.

a. $\tan\left(\frac{5\pi}{6}\right)$
b. $\cos\left(\frac{14\pi}{3}\right)$
c. $\tan\left(-\frac{5\pi}{4}\right)$
d. $\cos\left(-\frac{3\pi}{4}\right)$
e. $\sin\left(-\frac{2\pi}{3}\right)$
f. $\sin\left(\frac{17\pi}{6}\right)$

3. Determine exact values for the following trigonometric expressions.

a. $\sin\left(\frac{7\pi}{3}\right)$
b. $\cos\left(\frac{7\pi}{3}\right)$
c. $\tan\left(\frac{5\pi}{6}\right)$
d. $\sin(150°)$
e. $\cos\left(\frac{7\pi}{6}\right)$
f. $\tan\left(-\frac{7\pi}{6}\right)$

4. Determine exact values for the following trigonometric expressions.

a. $\cos\left(\frac{\pi}{2}\right)$
b. $\tan(270°)$
c. $\sin(-4\pi)$
d. $\tan(\pi)$
e. $\cos(-6\pi)$
f. $\sin\left(\frac{3\pi}{2}\right)$

5. Simplify the following.

a. $\sin(\pi - \theta)$
b. $\cos(6\pi - \theta)$
c. $\tan(\pi + \theta)$
d. $\cos(-\theta)$
e. $\sin(180° + \theta)$
f. $\tan(720° - \theta)$

6. Calculate the exact values of the following.

a. $\cos\left(\frac{7\pi}{6}\right) + \cos\left(\frac{2\pi}{3}\right)$
b. $2\sin\left(\frac{7\pi}{4}\right) + 4\sin\left(\frac{5\pi}{6}\right)$
c. $\sqrt{3}\tan\left(\frac{5\pi}{4}\right) - \tan\left(\frac{5\pi}{3}\right)$
d. $\sin^2\left(\frac{8\pi}{3}\right) + \sin\left(\frac{9\pi}{4}\right)$
e. $2\cos^2\left(-\frac{5\pi}{4}\right) - 1$
f. $\dfrac{\tan\left(\frac{17\pi}{4}\right)\cos(-7\pi)}{\sin\left(-\frac{11\pi}{6}\right)}$

Technology active

7. **WE1** Evaluate the following expressions correct to 4 decimal places, given that $\sin(\theta) = 0.4695$, $\cos(\alpha) = 0.5592$ and $\tan(\beta) = 0.2680$, where θ, α and β are in the first quadrant.

a. $\sin(2\pi - \theta)$
b. $\cos(\pi - \alpha)$
c. $\tan(-\beta)$
d. $\sin(\pi + \theta)$
e. $\cos(2\pi - \alpha)$
f. $\tan(\pi + \beta)$

8. Evaluate the following expressions correct to 4 decimal places, given that $\sin(\theta) = 0.4695$ and $\cos(\alpha) = 0.5592$, where θ and α are in the first quadrant.

a. $\cos(-\theta)$
b. $\tan(180° - \theta)$
c. $\sin(360° + \alpha)$
d. $\tan(360° - \alpha)$

9. Simplify the following.

a. $\cos\left(\frac{\pi}{2} - \alpha\right)$
b. $\tan(90° + \alpha)$
c. $\sin(270° - \alpha)$
d. $\tan\left(\frac{11\pi}{2} - \alpha\right)$
e. $\cos\left(\frac{3\pi}{2} + \alpha\right)$
f. $\sin(90° - \alpha)$

10. **WE2** Evaluate the following expressions correct to 4 decimal places, given that $\cos(\theta) = 0.8829$ and $\sin(\alpha) = 0.1736$, where θ and α are in the first quadrant.

a. $\sin\left(\frac{\pi}{2} + \theta\right)$
b. $\cos\left(\frac{3\pi}{2} - \theta\right)$
c. $\tan\left(\frac{\pi}{2} - \theta\right)$
d. $\sin\left(\frac{3\pi}{2} + \alpha\right)$
e. $\sin\left(\frac{\pi}{2} - \alpha\right)$
f. $\tan\left(\frac{3\pi}{2} + \alpha\right)$

11. Evaluate the following expressions correct to 4 decimal places, given that $\sin(\theta) = 0.8290$ and $\cos(\beta) = 0.7547$, where θ and β are in the first quadrant.

a. $\sin(90° - \theta)$
b. $\cos(270° + \theta)$
c. $\tan(90° + \theta)$
d. $\sin(270° - \beta)$
e. $\tan(90° - \beta)$
f. $\cos(270° - \beta)$

12. Given that $\sin(\theta) = 0.9511$ and θ is in the first quadrant, evaluate the following correct to 4 decimal places.

a. $\sin(2\pi - \theta)$
b. $\sin(\pi - \theta)$
c. $\cos\left(\frac{\pi}{2} - \theta\right)$
d. $\tan(\theta)$
e. $\cos(3\pi + \theta)$
f. $\tan(2\pi - \theta)$

13. Given that $\cos(\alpha) = 0.8572$ and α is in the first quadrant, evaluate the following correct to 4 decimal places.

a. $\cos(180° + \alpha)$ b. $\cos(-\alpha)$ c. $\sin\left(\frac{3\pi}{2} + \alpha\right)$

d. $\tan(180° - \alpha)$ e. $\cos(360° - \alpha)$ f. $\tan\left(\frac{\pi}{2} + \alpha\right)$

14. If $\sin(\beta) = \frac{4}{5}$ and $\frac{\pi}{2} < \beta < \pi$, determine the exact values of:

a. $\cos(\beta)$ b. $\tan(\beta)$ c. $\cos^2(\beta) + \sin^2(\beta)$ d. $\cos^2(\beta) - \sin^2(\beta)$

15. For the given triangle, determine the values of:

a. $\sin(\theta)$ b. $\tan(\theta)$ c. $\cos(\theta)$
d. $\sin(90° - \theta)$ e. $\cos(90° - \theta)$ f. $\tan(90° - \theta)$

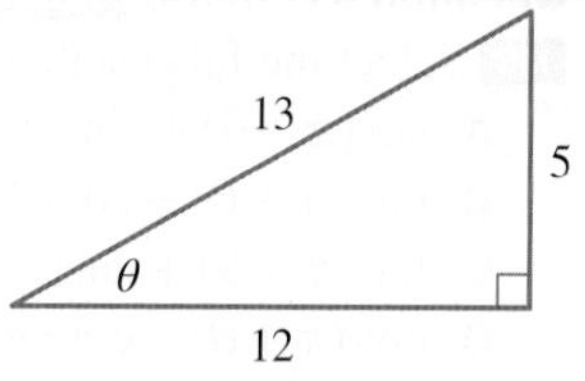

16. a. Use the Pythagorean identity $\sin^2(x) + \cos^2(x) = 1$ to show that:

$$\tan^2(x) + 1 = \frac{1}{\cos^2(x)}$$

b. Hence, determine the value of $\tan(x)$ correct to 4 decimal places, given that $\sin(x) = 0.6157$ and $0 \le x \le \frac{\pi}{2}$.

17. The weight on a spring moves in such a way that its speed, v cm/s, is given by the rule $v = 12 + 3\sin\left(\frac{\pi t}{3}\right)$.

Determine:

a. the initial speed of the weight
b. the exact value of the speed of the weight after 5 seconds
c. the exact value of the speed of the weight after 12 seconds.

18. The height, $h(t)$ metres, that water reaches up the side of the bank of the Yarra river is determined by the rule

$$h(t) = 0.5\cos\left(\frac{\pi t}{12}\right) + 1.0$$

where t is the number of hours after 6 am. Calculate the height of the water up the side of the bank at:

a. 6 am b. 2 pm c. 10 pm.

Give your answers in exact form.

2.2 Exam questions

Question 1 (1 mark) TECH-ACTIVE

MC Select the expression from the following that is equal to $\sin\left(-\frac{4\pi}{3}\right)$.

A. $\frac{\sqrt{3}}{2}$ B. $-\frac{\sqrt{3}}{2}$ C. $-\frac{1}{2}$

D. $\frac{1}{2}$ E. $\sin\left(\frac{4\pi}{3}\right)$

Question 2 (1 mark) TECH-ACTIVE

MC Select the false statement from the following.

A. $\sin\left(\frac{\pi}{6}\right)+\cos\left(\frac{\pi}{6}\right)=\sin\left(\frac{\pi}{3}\right)+\cos\left(\frac{\pi}{3}\right)$

B. $\cos^3(\pi)+\sin^3(\pi)=1$

C. $\cos^2(\pi)+\sin^2(\pi)=1$

D. $\cos(\pi)+\sin(\pi)=-1$

E. $\cos\left(\frac{\pi}{2}\right)+\sin\left(\frac{\pi}{2}\right)=1$

Question 3 (1 mark) TECH-ACTIVE

MC Select the false statement from the following.

A. $\sin(\pi+\theta)+\sin(\pi-\theta)=0$

B. $\cos(\pi+\theta)+\cos(2\pi-\theta)=0$

C. $\tan(\pi+\theta)+\tan(2\pi-\theta)=0$

D. $\cos(\pi+\theta)-\cos(\pi-\theta)=0$

E. $\sin(\pi+\theta)+\sin(2\pi-\theta)=0$

More exam questions are available online.

2.3 Trigonometric equations

LEARNING INTENTION

At the end of this subtopic you should be able to:
- solve trigonometric equations over specific domains
- solve trigonometric equations by first changing the domain
- solve trigonometric equations that can be written as polynomials by substitution.

2.3.1 Solving trigonometric equations

Trigonometric equations frequently involve working with the special angles that have exact values previously discussed, but may also require the use of CAS.

To solve the basic equation $\sin(x)=a, 0\le x\le 2\pi$, remember the following:
- Identify the quadrants in which solutions lie from the sign of a:
  - if $a>0$, x must lie in quadrants 1 and 2 where sine is positive
  - if $a<0$, x must be in quadrants 3 and 4 where sine is negative.
- Obtain the base value or first quadrant value by solving $\sin(x)=a$ if $a>0$ or ignoring the negative sign if $a<0$ (to ensure the first quadrant value is obtained).
- Use the base value to generate the values for the quadrants required from their symmetric forms.

The basic equations $\cos(x)=a$ or $\tan(x)=a$, $0\le x\le 2\pi$, are solved in a similar manner, with the sign of a determining the quadrants in which solutions lie.

For $\cos(x)=a$: if $a>0$, x must lie in quadrants 1 and 4 where cosine is positive; if $a<0$, x must be in quadrants 2 and 3 where cosine is negative.

For $\tan(x)=a$: if $a>0$, x must lie in quadrants 1 and 3 where tangent is positive; if $a<0$, x must be in quadrants 2 and 4 where tangent is negative.

In technology active questions, by defining the domain of the equation, the CAS technology will solve the problem without having to determine a base value or first quadrant value.

WORKED EXAMPLE 4 Trigonometric equations

Solve the following equations.

a. $\sqrt{2}\cos(x)+1=0,\ 0\le x\le 2\pi$

b. $2\sin(x)=-1.5,\ 0°\le x\le 720°$, **correct to 2 decimal places**

c. $\tan(\theta)-1=0,\ -\pi\le\theta\le\pi$

THINK	WRITE
a. 1. Express the equation with the trigonometric function as the subject.	a. $\sqrt{2}\cos(x)+1=0$ $\sqrt{2}\cos(x)=-1$ $\cos(x)=-\dfrac{1}{\sqrt{2}}$
2. Identify the quadrants in which the solutions lie.	Cosine is negative in quadrants 2 and 3. 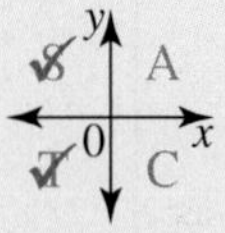
3. Use knowledge of exact values to state the first quadrant base.	The base is $\dfrac{\pi}{4}$, since $\cos\left(\dfrac{\pi}{4}\right)=\dfrac{1}{\sqrt{2}}$.
4. Generate the solutions using the appropriate quadrant forms.	$x=\pi-\dfrac{\pi}{4},\ \pi+\dfrac{\pi}{4}$
5. Calculate the solutions from their quadrant forms.	$x=\dfrac{3\pi}{4},\ \dfrac{5\pi}{4}$
b. 1. Express the equation with the trigonometric function as the subject.	b. $2\sin(x)=-1.5$ $\sin(x)=-0.75$
2. Identify the quadrants in which the solutions lie.	Sine is negative in quadrants 3 and 4. 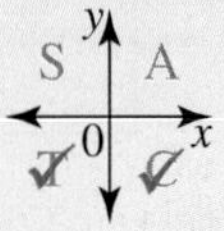
3. Calculate the base using CAS, as an exact value is not possible.	The base is $\sin^{-1}(0.75)=48.59°$.
4. Generate the solutions using the appropriate quadrant forms. As $x\in[0°, 720°]$, there will be four positive solutions from two anticlockwise rotations.	$x=180°+48.59°,\ 360°-48.59°,$ $540°+48.59°,\ 720°-48.59°$
5. Calculate the solutions from their quadrant forms. Alternatively, the solve function on CAS can be used to find the solutions (but remember to define the domain).	$x=228.59°,\ 311.41°,\ 588.59°,\ 671.41°$
c. 1. Express the equation with the trigonometric function as the subject.	c. $\tan(\theta)-1=0$ $\tan(\theta)=1$
2. Identify the quadrants in which the solutions lie.	Tangent is positive in quadrants 1 and 3. 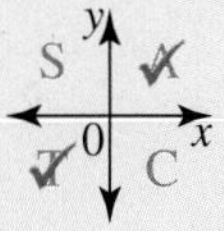

3. Use knowledge of exact values to state the first quadrant base.

The base is $\frac{\pi}{4}$, since $\tan\left(\frac{\pi}{4}\right) = 1$.

4. Generate the solutions using the appropriate quadrant forms. As the domain is $x \in [-\pi, \pi]$, there will be one positive solution and one negative solution.

$x = \frac{\pi}{4}, -\pi + \frac{\pi}{4}$

5. Calculate the solutions from their quadrant forms.

$x = \frac{\pi}{4}, \frac{-3\pi}{4}$

TI | THINK / **DISPLAY/WRITE**

a. 1. Put the Calculator into Radian mode.
On a Calculator page, press MENU, then select:
3: Algebra
1: Solve
Complete the entry line as:
solve $\left(\sqrt{2}\cos(x) + 1 = 0, x\right)$
$| 0 \leq x \leq 2\pi$
then press ENTER.

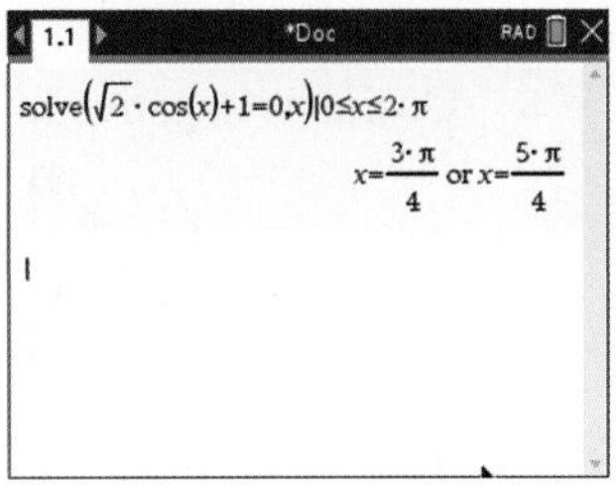

2. The answers appear on the screen.

$x = \frac{3\pi}{4}, x = \frac{5\pi}{4}$

b. 1. Put the Calculator into Degree mode by toggling between RAD and DEG in the top right corner.
On a Calculator page, press MENU, then select:
3: Algebra
1: Solve
Complete the entry line as:
solve $(2\sin(x) = -1.5, x)$
$|0 \leq x \leq 720$
then press ENTER.

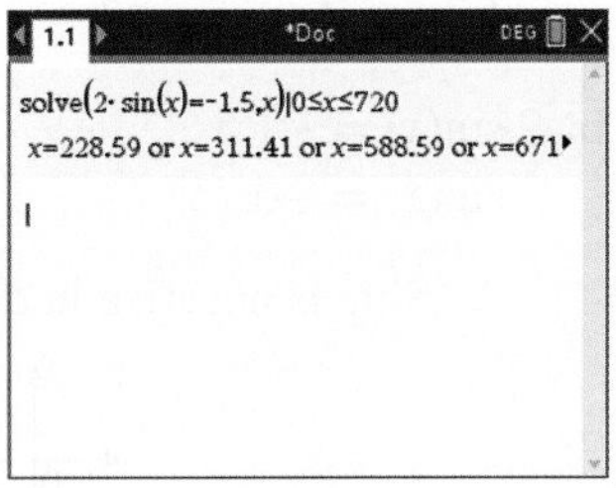

2. The answers appear on the screen.

$x = 228.59°, 311.41°, 588.59°, 671.41°$

CASIO | THINK / **DISPLAY/WRITE**

a. 1. Put the Calculator into Radian mode.
On a Main screen, complete the entry line as:
solve $(\sqrt{2}\cos(x) + 1 = 0, x)$
$| 0 \leq x \leq 2\pi$
then press EXE.

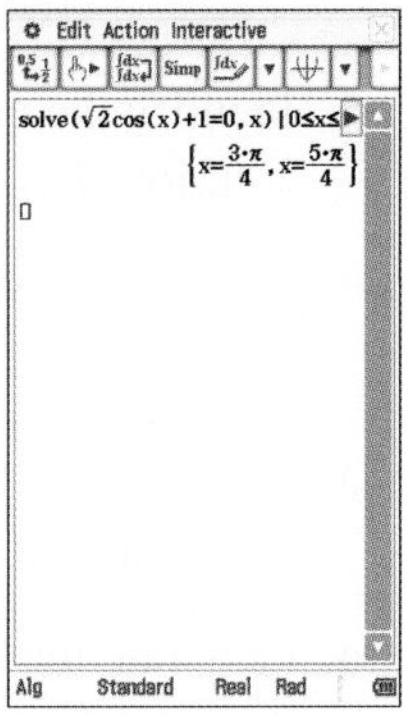

2. The answers appear on the screen.

$x = \frac{3\pi}{4}, x = \frac{5\pi}{4}$

b. 1. Put the Calculator into Degree mode.
On a Main screen, complete the entry line as:
solve $(2\sin(x) = -1.5, x) \mid 0 \leq x \leq 720$
then press EXE.

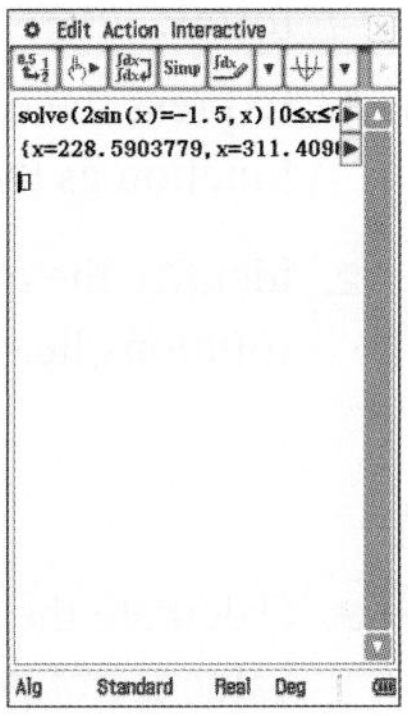

2. The answers appear on the screen.

$x = 228.59°, 311.41°, 588.59°, 671.41°$

2.3.2 Changing the domain

Equations such as $\sin(2x) = 1, 0 \leq x \leq 2\pi$ can be expressed in the basic form by the substitution $\theta = 2x$. However, the accompanying domain must be changed to be the domain for θ. This requires the end points of the domain for x to be multiplied by 2. Hence, $0 \leq x \leq 2\pi \Rightarrow 2 \times 0 \leq 2x \leq 2 \times 2\pi$ gives the domain requirement for θ as $0 \leq \theta \leq 4\pi$.

This allows the equation to be written as $\sin(\theta) = 1, 0 \leq \theta \leq 4\pi$.

WORKED EXAMPLE 5 Solving by changing the domain

Solve the following for *x*.

a. $2\sin(2x) - 1 = 0,\ 0 \le x \le 2\pi$

b. $2\cos(2x - \pi) - 1 = 0,\ -\pi \le x \le \pi.$

THINK	WRITE
a. 1. Change the domain to be that for the given multiple of the variable.	**a.** $2\sin(2x) - 1 = 0, 0 \le x \le 2\pi$ Multiply each value by 2: $2\sin(2x) - 1 = 0, 0 \le 2x \le 4\pi$
2. Express the equation with the trigonometric function as the subject.	$2\sin(2x) - 1 = 0$ $2\sin(2x) = 1$ $\sin(2x) = \frac{1}{2}$
3. Solve the equation for $2x$. As $2x \in [0, 4\pi]$, each of the 2 revolutions will generate 2 solutions, giving a total of 4 values for $2x$.	Sine is positive in quadrants 1 and 2. 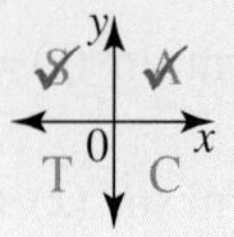 The base is $\frac{\pi}{6}$. $2x = \frac{\pi}{6}, \pi - \frac{\pi}{6}, 2\pi + \frac{\pi}{6}, 3\pi - \frac{\pi}{6}$ $2x = \frac{\pi}{6}, \frac{5\pi}{6}, \frac{13\pi}{6}, \frac{17\pi}{6}$
4. Calculate the solutions for x. *Note:* By dividing by 2 at the very end, the solutions lie back within the domain originally specified, namely $0 \le x \le 2\pi$.	$x = \frac{\pi}{12}, \frac{5\pi}{12}, \frac{13\pi}{12}, \frac{17\pi}{12}$
b. 1. Change the domain to that for the given multiple of the variable.	**b.** $2\cos(2x - \pi) - 1 = 0, -\pi \le x \le \pi$ Multiply each value by 2: $2\cos(2x - \pi) - 1 = 0, -2\pi \le 2x \le 2\pi$ Subtract π from each value: $2\cos(2x - \pi) - 1 = 0, -3\pi \le 2x - \pi \le \pi$
2. Express the equation with the trigonometric function as the subject.	$2\cos(2x - \pi) - 1 = 0$ $2\cos(2x - \pi) = 1$ $\cos(2x - \pi) = \frac{1}{2}$
3. Solve the equation for $(2x - \pi)$. The domain of $[-3\pi, \pi]$ involves 2 complete rotations of the unit circle, so there will be 4 solutions, 3 of which will be negative and 1 of which will be positive.	Cosine is positive in quadrants 1 and 4. 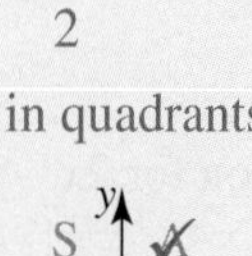 The base is $\frac{\pi}{3}$. $2x - \pi = \frac{\pi}{3}, -\frac{\pi}{3}, -2\pi + \frac{\pi}{3}, -2\pi - \frac{\pi}{3}$ $2x - \pi = \frac{\pi}{3}, -\frac{\pi}{3}, -\frac{5\pi}{3}, -\frac{7\pi}{3}$

4. Calculate the solutions for x.

$$2x = \frac{\pi}{3} + \pi, -\frac{\pi}{3} + \pi, -\frac{5\pi}{3} + \pi, -\frac{7\pi}{3} + \pi$$
$$= \frac{4\pi}{3}, \frac{2\pi}{3}, -\frac{2\pi}{3}, -\frac{4\pi}{3}$$
$$x = \frac{2\pi}{3}, \frac{\pi}{3}, -\frac{\pi}{3}, -\frac{2\pi}{3}$$

2.3.3 Further types of trigonometric equations

Trigonometric equations may require algebraic techniques or the use of relationships between the functions before they can be reduced to the basic form $f(x) = a$, where f is either sin, cos or tan.

- Equations of the form $\sin(x) = a\cos(x)$ can be converted to $\tan(x) = a$ by dividing both sides of the equation by $\cos(x)$.
- Equations of the form $\sin^2(x) = a$ can be converted to $\sin(x) = \pm\sqrt{a}$ by taking the square roots of both sides of the equation.
- Equations of the form $\sin^2(x) + b\sin(x) + c = 0$ can be converted to standard quadratic equations by using the substitution $A = \sin(x)$.

Because $-1 \le \sin(x) \le 1$ and $-1 \le \cos(x) \le 1$, neither $\sin(x)$ nor $\cos(x)$ can have values greater than 1 or less than -1. This may have implications requiring the rejection of some steps when working with sine or cosine trigonometric equations. As $\tan(x) \in R$, there is no restriction on the values the tangent function can take.

WORKED EXAMPLE 6 Further trigonometric equations

Solve the following equations.

a. $\sin(2x) = \cos(2x), 0 \le x \le 2\pi$

b. $2\sin^2(\theta) + 3\sin(\theta) - 2 = 0, 0 \le x \le 2\pi$

c. $\cos^2(2\alpha) - 1 = 0, -\pi \le \alpha \le \pi$

THINK

a. 1. Change the domain to that for the given multiple of the variable.

2. Reduce the equation to one trigonometric function by dividing through by $\cos(2x)$.

3. Solve the equation for $2x$.

WRITE

a. $0 \le x \le 2\pi$

Multiply through by 2:

$0 \le 2x \le 4\pi$

$\sin(2x) = \cos(2x)$

$$\frac{\sin(2x)}{\cos(2x)} = \frac{\cos(2x)}{\cos(2x)} \text{ providing } \cos(2x) \ne 0$$

$\tan(2x) = 1$

Tangent is positive in quadrants 1 and 3.

The base is $\frac{\pi}{4}$.

$$2x = \frac{\pi}{4}, \pi + \frac{\pi}{4}, 2\pi + \frac{\pi}{4}, 3\pi + \frac{\pi}{4}$$
$$= \frac{\pi}{4}, \frac{5\pi}{4}, \frac{9\pi}{4}, \frac{13\pi}{4}$$

4. Calculate the solutions for x. Note that the answers are within the prescribed domain of $0 \le x \le 2\pi$.

$$x = \frac{\pi}{8}, \frac{5\pi}{8}, \frac{9\pi}{8}, \frac{13\pi}{8}$$

b. 1. Use substitution to form a quadratic equation.

b. $2\sin^2(\theta) + 3\sin(\theta) - 2 = 0$

Let $A = \sin(\theta)$.

$2A^2 + 3A - 2 = 0$

2. Solve the quadratic equation.

$(2A - 1)(A + 2) = 0$

$A = \frac{1}{2}$ or $A = -2$

But $A = \sin(\theta)$.

$\sin(\theta) = \frac{1}{2}$ or $\sin(\theta) = -2$

3. Solve each trigonometric equation separately.

$\sin(\theta) = \frac{1}{2}$

Sine is positive in quadrants 1 and 2.

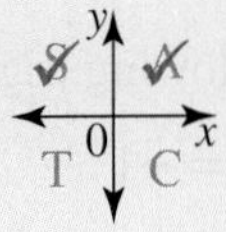

The base is $\frac{\pi}{6}$.

$$\theta = \frac{\pi}{6}, \pi - \frac{\pi}{6}$$

$$\theta = \frac{\pi}{6}, \frac{5\pi}{6}$$

$\sin(\theta) = -2$

There is no solution as $-1 \le \sin(\theta) \le 1$.

4. Write the answer.

$$\theta = \frac{\pi}{6}, \frac{5\pi}{6}$$

c. 1. Change the domain to that for the given multiple of the variable.

c. $-\pi \le \alpha \le \pi$

Multiply through by 2:

$-2\pi \le 2\alpha \le 2\pi$

2. Use substitution to form a quadratic equation and factorise by applying the difference of perfect squares method.

$\cos^2(2\alpha) - 1 = 0$

Let $A = \cos(2\alpha)$.

$A^2 - 1 = 0$

3. Solve the quadratic equation.

$(A - 1)(A + 1) = 0$

$A = 1, -1$

But $A = \cos(2\alpha)$.

$\therefore \cos(2\alpha) = 1$ or $\cos(2\alpha) = -1$

4. Solve each trigonometric equation separately.

$\cos(2\alpha) = 1$

$$2\alpha = -2\pi, 0, 2\pi$$

$$\alpha = -\pi, 0, \pi$$

$\cos(2\alpha) = -1$

$$2\alpha = -\pi, \pi$$

$$\alpha = -\frac{\pi}{2}, \frac{\pi}{2}$$

5. Write the answers in numerical order.

$$\therefore \alpha = -\pi, -\frac{\pi}{2}, 0, \frac{\pi}{2}, \pi$$

2.3 Exercise

Technology free

1. **WE4** Solve the following equations.
   a. $2\cos(\theta)+\sqrt{3}=0$ for $0\le\theta\le 2\pi$
   b. $\tan(x)+\sqrt{3}=0$ for $0°\le x\le 720°$
2. **WE5** Solve the following for x.
   a. $2\cos(3\theta)-\sqrt{2}=0$ for $0\le\theta\le 2\pi$.
   b. $2\sin(2x+\pi)+\sqrt{3}=0$ for $-\pi\le x\le\pi$.
3. Solve the following trigonometric equations for $0\le\theta\le 2\pi$.
   a. $\sqrt{2}\sin(\theta)=-1$
   b. $2\cos(\theta)=1$
   c. $\tan(3\theta)-\sqrt{3}=0$
   d. $\tan\left(\theta-\frac{\pi}{2}\right)+1=0$
4. Solve the following trigonometric equations for $0°\le x\le 360°$.
   a. $2\cos(x)+1=0$
   b. $2\sin(2x)+\sqrt{2}=0$
5. Solve the following for θ given that $-\pi\le\theta\le\pi$.
   a. $2\sin(2\theta)+\sqrt{3}=0$
   b. $\sqrt{2}\cos(3\theta)=1$
   c. $\tan(2\theta)+1=0$
   d. $2\cos(\theta)=1$ for $-\pi\le\theta\le\pi$
6. Solve the following for x.
   a. $2\sin\left(2x+\frac{\pi}{4}\right)=\sqrt{2},\ x\in[-\pi,\pi]$
   b. $2\cos(x+\pi)=\sqrt{3},\ x\in[-\pi,\pi]$
   c. $\tan(x-\pi)=-1,\ x\in[-\pi,\pi]$
   d. $2\cos\left(3x-\frac{\pi}{2}\right)+\sqrt{3}=0,\ x\in[0,2\pi]$

Technology active

7. Solve the following, correct to 2 decimal places.
   a. $3\sin(\theta)-2=0$ given that $0\le\theta\le 2\pi$.
   b. $7\cos(x)-2=0$ given that $0°\le x\le 360°$.
8. a. Solve the equation $\sin(\theta)+0.5768=0$, $0°\le\theta\le 360°$, correct to 2 decimal places.
   b. Solve $\sin(x)=1$, $-2\pi\le x\le 2\pi$.
9. Solve $\sin(3\theta)=\cos(2\theta)$ for $0\le\theta\le 2\pi$, correct to 3 decimal places.
10. Solve $2\sin(2x)-1=-\frac{1}{2}x+1$ for $0\le x\le 2$, correct to 3 decimal places.
11. **WE6** Solve the equation $\cos^2(\theta)-\sin(\theta)\cos(\theta)=0$ for $0\le\theta\le 2\pi$.
12. **MC** Solve $\{\theta : 2\cos^2(\theta)+3\cos(\theta)=-1, 0\le\theta\le 2\pi\}$. The sum of the solutions is:
    A. 2π
    B. $\frac{7\pi}{3}$
    C. 3π
    D. $\frac{13\pi}{6}$
    E. $\frac{2\pi}{3}$

13. Solve the following for θ given that $0 \le \theta \le 2\pi$.

a. $\tan^2(\theta) - 1 = 0$

b. $4\sin^2(\theta) - (2 + 2\sqrt{3})\sin(\theta) + \sqrt{3} = 0$

14. Solve the following for α where $-\pi \le \alpha \le \pi$.

a. $\sin(\alpha) - \cos^2(\alpha)\sin(\alpha) = 0$

b. $\sin(2\alpha) = \sqrt{3}\cos(2\alpha)$

c. $\sin^2(\alpha) = \cos^2(\alpha)$

d. $4\cos^2(\alpha) - 1 = 0$

2.3 Exam questions

Question 1 (1 mark) TECH-ACTIVE

Source: *VCE 2019, Mathematical Methods Exam 2, Section A, Q19; © VCAA.*

MC Given that $\tan(\alpha) = d$, where $d > 0$ and $0 < \alpha < \frac{\pi}{2}$, the sum of the solutions to $\tan(2x) = d$, where $0 < x < \frac{5\pi}{4}$ in terms of α, is

A. 0 **B.** 2α **C.** $\pi + 2\alpha$ **D.** $\frac{\pi}{2} + \alpha$ **E.** $\frac{3(\pi + \alpha)}{2}$

Question 2 (1 mark) TECH-ACTIVE

Source: *VCE 2017, Mathematical Methods Exam 2, Section A, Q12; © VCAA.*

MC The sum of the solution of $\sin(2x) = \frac{\sqrt{3}}{2}$ over the interval $[-\pi, d]$ is $-\pi$.

The value of d could be

A. 0 **B.** $\frac{\pi}{6}$ **C.** $\frac{3\pi}{4}$ **D.** $\frac{7\pi}{6}$ **E.** $\frac{3\pi}{2}$

Question 3 (2 marks) TECH-FREE

Source: *VCE 2014, Mathematical Methods (CAS) Exam 1, Q3; © VCAA.*

Solve $2\cos(2x) = -\sqrt{3}$ for x, where $0 \le x \le \pi$.

More exam questions are available online.

2.4 General solutions of trigonometric equations

LEARNING INTENTION

At the end of this subtopic you should be able to:

- determine general solutions for sine, cosine and tangent equations in terms of a parameter.

2.4.1 General solutions

All of the trigonometric equations solved so far have been solved over a specific domain and therefore have defined numbers of solutions. However, if no domain is given, there will be an infinite number of solutions to the general equation. This is because multiples of 2π can be added and subtracted to any solutions within a specific domain. In cases such as this, a general solution is given in terms of the parameter n, where n is an integer.

The general solution for the sine function

Consider $\sin(\theta) = a$, where a is a positive value. The solutions are found in quadrants 1 and 2, and the basic angle is in quadrant 1 and determined by $\theta = \sin^{-1}(a)$. The angle in quadrant 2 is found by $(\pi - \theta)$. If we keep cycling around the unit circle in either direction, the two solutions can be summarised as even numbers of π adding on θ and odd numbers of π subtracting θ.

The general solution is as follows.

The general solution for the sine function

$$\theta = 2n\pi + \sin^{-1}(a) \text{ or } \theta = (2n+1)\pi - \sin^{-1}(a)$$

where $n \in Z$ and $a \in [-1, 1]$.

The solutions if a is positive are represented in the diagram.

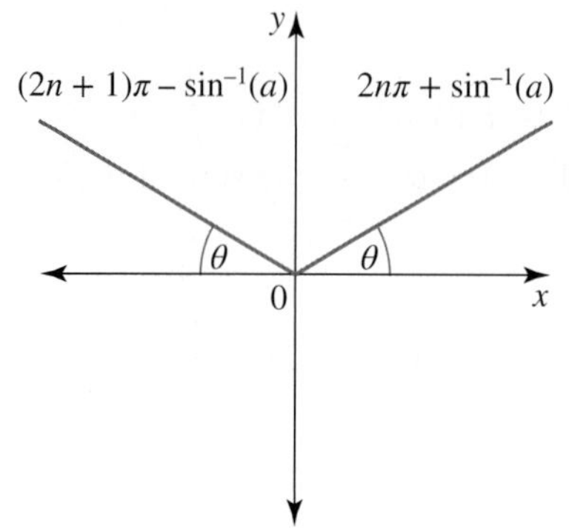

Note: If a is negative, choose the basic angle to be in quadrant 4 (therefore a negative angle).

The general solution for the cosine function

Consider $\cos(\theta) = a$, where a is a positive value. The solutions are found in quadrants 1 and 4, and the basic angle is in quadrant 1 and determined by $\theta = \cos^{-1}(a)$. The angle in quadrant 4 is found by $(2\pi - \theta)$. If we keep cycling around the unit circle in either direction, the two solutions can be summarised as even numbers of π adding on θ or subtracting θ.

The general solution is as follows.

The general solution for the cosine function

$$\theta = 2n\pi \pm \cos^{-1}(a)$$

where $n \in Z$ and $a \in [-1, 1]$.

The solutions if a is positive are represented in the diagram.

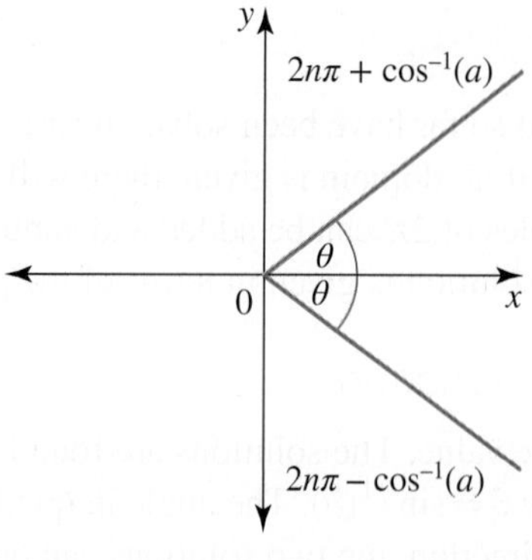

Note: If a is negative, choose the basic angle to be in quadrant 2 (therefore a positive angle).

The general solution for the tangent function

Consider $\tan(\theta) = a$, where a is a positive value. The solutions are found in quadrants 1 and 3, and the basic angle is in quadrant 1 and determined by $\theta = \tan^{-1}(a)$. The angle in quadrant 3 is found by $(\pi + \theta)$. If we keep cycling around the unit circle in either direction, the two solutions can be summarised as multiples of π adding on θ.

The general solution isas follows.

> **The general solution for the tangent function**
>
> $$\theta = n\pi + \tan^{-1}(a)$$
>
> **where $n \in Z$ and $a \in R$.**

The solutions if a is positive are represented in the diagram.

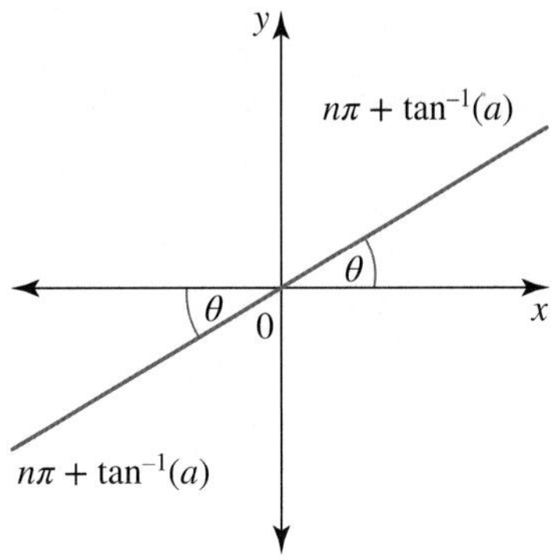

Note: If a is negative, choose the basic angle to be in quadrant 4 (therefore a negative angle).

We can summarise the general solutions for sine, cosine and tangent as follows.

> **Summary**
>
> - **If $\sin(\theta) = a$, then $\theta = 2n\pi + \sin^{-1}(a)$ or $\theta = (2n+1)\pi - \sin^{-1}(a)$, where $a \in [-1, 1]$ and $n \in Z$.**
> - **If $\cos(\theta) = a$, then $\theta = 2n\pi \pm \cos^{-1}(a)$, where $a \in [-1, 1]$ and $n \in Z$.**
> - **If $\tan(\theta) = a$, then $\theta = n\pi + \tan^{-1}(a)$, where $a \in R$ and $n \in Z$.**

WORKED EXAMPLE 7 General trigonometric solutions

a. Determine the general solution of the equation $\tan(x) - \sqrt{3} = 0$.

b. Determine the general solution of the equation $2\cos(2\theta) - \sqrt{2} = 0$ and hence find all the solutions for $\theta \in [-\pi, \pi]$.

THINK	WRITE
a. 1. Express the equation with the trigonometric function as the subject.	**a.** $\tan(x) - \sqrt{3} = 0$ $\tan(x) = \sqrt{3}$

2. Recognise the exact value and determine the quadrants in which the tangent function is positive.

Tangent is positive in quadrants 1 and 3.

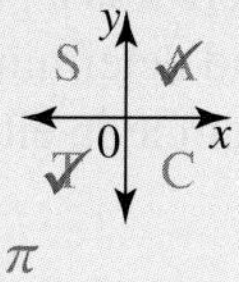

The base angle is $\frac{\pi}{3}$.

3. Write the general solution for $\tan(x) = a$.

$$x = n\pi + \tan^{-1}(a)$$
$$= n\pi + \tan^{-1}(\sqrt{3})$$

4. Substitute the basic angle for $\tan^{-1}(\sqrt{3})$ and simplify.
Note that $n \in Z$ must always be included as part of the solution.

$$= n\pi + \frac{\pi}{3}, n \in Z$$
$$= \frac{3n\pi + \pi}{3}$$
$$= \frac{(3n+1)\pi}{3}, n \in Z$$

b. 1. Express the equation with the trigonometric function as the subject.

b. $$2\cos(2\theta) - \sqrt{2} = 0$$
$$2\cos(2\theta) = \sqrt{2}$$
$$\cos(2\theta) = \frac{\sqrt{2}}{2}$$

2. Recognise the exact value and determine the quadrants in which the cosine function is positive.

Cosine is positive in quadrants 1 and 4.

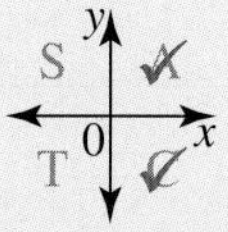

The base is $\frac{\pi}{4}$.

3. Write the general solution for $\cos(2\theta) = a$.

$$2\theta = 2n\pi \pm \cos^{-1}(a)$$
$$= 2n\pi \pm \cos^{-1}\left(\frac{\sqrt{2}}{2}\right)$$

Substitute the basic angle for $\cos^{-1}\left(\frac{\sqrt{2}}{2}\right)$ and simplify.

$$= 2n\pi \pm \frac{\pi}{4}, n \in Z$$
$$= \frac{8n\pi}{4} \pm \frac{\pi}{4}$$
$$= \frac{8n\pi + \pi}{4}, \frac{8n\pi - \pi}{4}$$
$$= \frac{(8n+1)\pi}{4}, \frac{(8n-1)\pi}{4}, n \in Z$$

4. Divide through by 2 to find the solution for θ. This is always best done once the solutions are written with common denominators.

$$\theta = \frac{(8n+1)\pi}{8}, \frac{(8n-1)\pi}{8}, n \in Z$$

5. Substitute appropriate values of n to achieve solutions for $\theta \in [-\pi, \pi]$.	If $n = -1$, $\theta = -\frac{9\pi}{8}$ or $\theta = -\frac{7\pi}{8}$. $\theta = -\frac{9\pi}{8}$ is outside the domain. If $n = 0$, $\theta = \frac{\pi}{8}$ or $\theta = -\frac{\pi}{8}$. Both are within the domain. If $n = 1$, $\theta = \frac{7\pi}{8}$ or $\theta = \frac{9\pi}{8}$. $\theta = \frac{9\pi}{8}$ lies outside the domain. Therefore, the solutions for $\theta \in [-\pi, \pi]$ are $\theta = -\frac{7\pi}{8}, -\frac{\pi}{8}, \frac{\pi}{8}, \frac{7\pi}{8}$.

TI | THINK	DISPLAY/WRITE	CASIO | THINK	DISPLAY/WRITE
a. 1. Put the Calculator into Radian mode. On a Calculator page, press MENU, then select: 3: Algebra 1: Solve Complete the entry line as: solve $(\tan(x) - \sqrt{3} = 0, x)$ then press ENTER.	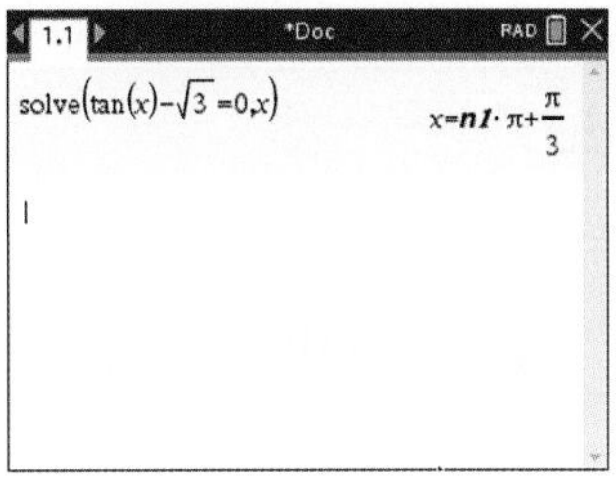	a. 1. Put the Calculator into Radian mode. On a Main screen, complete the entry line as: solve $(\tan(x) - \sqrt{3} = 0, x)$ then press EXE.	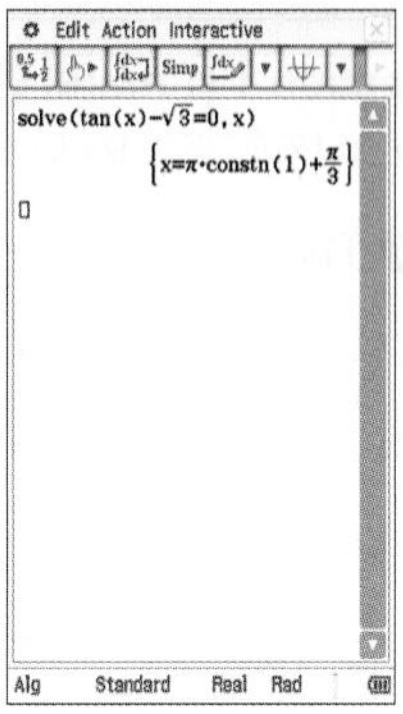
2. The answers appear on the screen.	$x = n\pi + \frac{\pi}{3}, n \in Z$	2. The answers appear on the screen.	$x = n\pi + \frac{\pi}{3}, n \in Z$

on Resources

Interactivity Trigonometric equations and general solutions (int-6413)

2.4 Exercise

Students, these questions are even better in jacPLUS

Receive immediate feedback and access sample responses

Access additional questions

Track your results and progress

Find all this and MORE in jacPLUS

Technology free

1. a. **WE7** Determine the general solution of the equation $2\sin(\theta) - \sqrt{3} = 0$.
   b. Determine the general solution of the equation $\sqrt{3}\tan(2\theta) + 1 = 0$ and hence find all the solutions for $\theta \in [-\pi, \pi]$.

2. Calculate the general solutions for the following.

a. $2\cos(x)+1=0$

b. $2\sin(x)-\sqrt{2}=0$

3. Calculate the general solution of $2\sin(2x)+1=0$ and hence find all solutions for $0\leq x\leq 2\pi$.

Technology active

4. Determine the general solution of $\sqrt{3}\sin\left(x+\frac{\pi}{2}\right)=\cos\left(x+\frac{\pi}{2}\right)$ and hence find all solutions for $-\pi\leq x\leq\pi$.

5. Determine the general solution of $2\cos(2\theta)-2=0$.

6. Determine the general solution of $3\tan\left(2\theta+\frac{\pi}{6}\right)=0$.

2.4 Exam questions

Question 1 (1 mark) TECH-ACTIVE

Source: *VCE 2020, Mathematical Methods Exam 2, Section A, Q4; © VCAA.*

MC The solutions of the equation $2\cos\left(2x-\frac{\pi}{3}\right)+1=0$ are

A. $x=\frac{\pi(6k-2)}{6}$ or $x=\frac{\pi(6k-2)}{6}$, for $k\in z$

B. $x=\frac{\pi(6k-2)}{6}$ or $x=\frac{\pi(6k+5)}{6}$, for $k\in z$

C. $x=\frac{\pi(6k-1)}{6}$ or $x=\frac{\pi(6k+2)}{6}$, for $k\in z$

D. $x=\frac{\pi(6k-1)}{6}$ or $x=\frac{\pi(6k+3)}{6}$, for $k\in z$

E. $x=\pi$ or $x=\frac{\pi(6k+2)}{6}$, for $k\in z$

Question 2 (2 marks) TECH-ACTIVE

Find the general solution of $\sin\left(x-\frac{\pi}{3}\right)=\frac{\sqrt{3}}{2}$.

Question 3 (1 mark) TECH-ACTIVE

MC $x=n\pi-\frac{\pi}{4}$, $n\in z$ is the general solution to the equation

A. $\sin(2x)=1$

B. $\sin(x)=1$

C. $\sin(2x)=-1$

D. $\sin(x)=-1$

E. $\cos(x)=1$

More exam questions are available online.

2.5 The sine and cosine functions

LEARNING INTENTION

At the end of this subtopic you should be able to:

- state and understand the key features of sine and cosine graphs
- sketch sine and cosine graphs, including those that have undergone transformations.

2.5.1 Features of the sine and cosine graphs

Circular functions, or trigonometric functions, are periodic functions such as $y=\sin(x)$, $y=\cos(x)$ and $y=\tan(x)$.

The graph of the sine function has a wave shape that repeats itself every 2π units. Its **period** is 2π as shown in its graph.

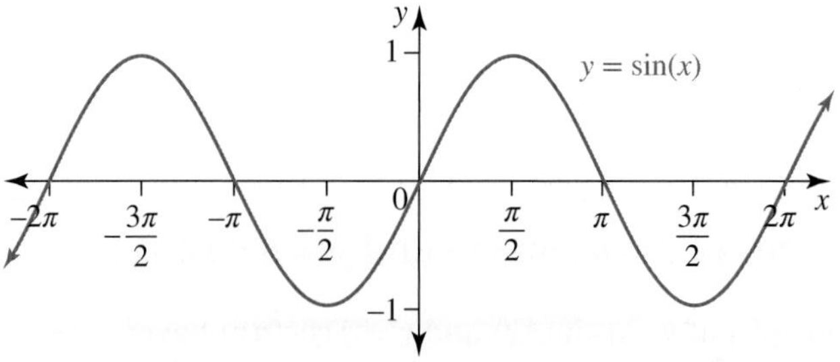

The graph oscillates about the line $y = 0$ (the x-axis), rising and falling by up to 1 unit. This gives the graph its range of $[-1, 1]$ with a mean, or equilibrium, position of $y = 0$ and an **amplitude** of 1.

The graph of the cosine function has the same wave shape with period 2π.

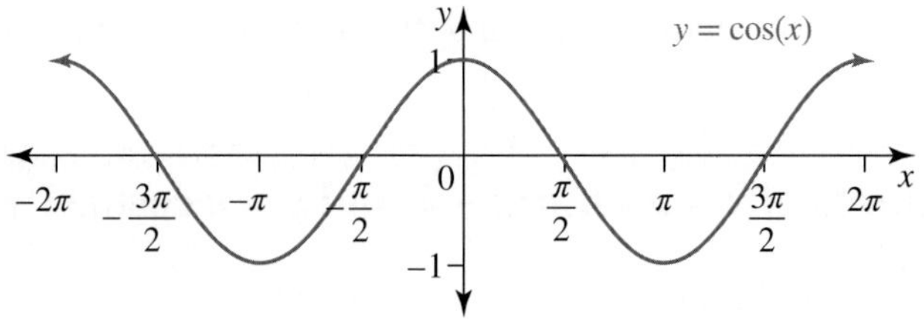

Key features of the sine and cosine functions

The graphs of $y = \sin(x)$ and $y = \cos(x)$ have:

- **period 2π**
- **amplitude 1**
- **mean position $y = 0$**
- **domain R**
- **range $[-1, 1]$**
- **many-to-one correspondence.**

Although the domain of both the sine and cosine functions is R, they are usually sketched on a given restricted domain.

The two graphs of $y = \sin(x)$ and $y = \cos(x)$ are 'out of phase' by $\frac{\pi}{2}$; that is, $\cos\left(x - \frac{\pi}{2}\right) = \sin(x)$. In other words, a horizontal shift of the cosine graph by $\frac{\pi}{2}$ units to the right gives the sine graph. Likewise, a horizontal shift of the sine graph by $\frac{\pi}{2}$ units to the left gives the cosine graph; $\sin\left(x + \frac{\pi}{2}\right) = \cos(x)$.

The periodicity of the functions is expressed by $f(x) = f(x + n2\pi)$, $n \in Z$, where f is sin or cos.

Graphs of $y = a\sin(nx)$ and $y = a\cos(nx)$

The value of a affects the amplitude of the sine and cosine functions.

Because $-1 \leq \sin(nx) \leq 1$, $-a \leq a\sin(nx) \leq a$. This means the graphs of $y = a\sin(nx)$ and $y = a\cos(nx)$ have amplitude $|a|$.

As the amplitude measures a distance — the rise or fall from the mean position — it is always positive. If $a<0$, the graphs will be inverted, or reflected in the x-axis.

The value of n affects the period of the sine and cosine functions.

Since one cycle of $y=\sin(nx)$ is completed for $0\le x\le 2\pi$, one cycle of $y=\sin(nx)$ is completed for $0\le nx\le 2\pi$. This means one cycle is covered over the interval $0\le x\le \frac{2\pi}{n}$, assuming $n>0$.

Features of $y=a\sin(nx)$ and $y=a\cos(nx)$

The graphs of $y=a\sin(nx)$ and $y=a\cos(nx)$ have:
- **period $\frac{2\pi}{n}$**
- **amplitude $|a|$**
- **range $[-a,a]$.**

The graphs of $y=a\sin(nx)+k$ and $y=a\cos(nx)+k$

Any vertical translation affects the equilibrium or mean position about which the sine and cosine graphs oscillate.

Features of $y=a\sin(nx)+k$ and $y=a\cos(nx)+k$

The graphs of $y=a\sin(nx)+k$ and $y=a\cos(nx)+k$ have:
- **mean position $y=k$**
- **range $[k-a,k+a]$.**

Where the graph of $y=f(x)$ crosses the x-axis, the intercepts are found by solving the trigonometric equation $f(x)=0$.

WORKED EXAMPLE 8 Sketching trigonometric graphs

a. Sketch the graph of $y=3\sin(2x)+4,\ 0\le x\le 2\pi$.

b. The diagram shows the graph of a cosine function. State its mean position, amplitude and period, and give a possible equation for the function.

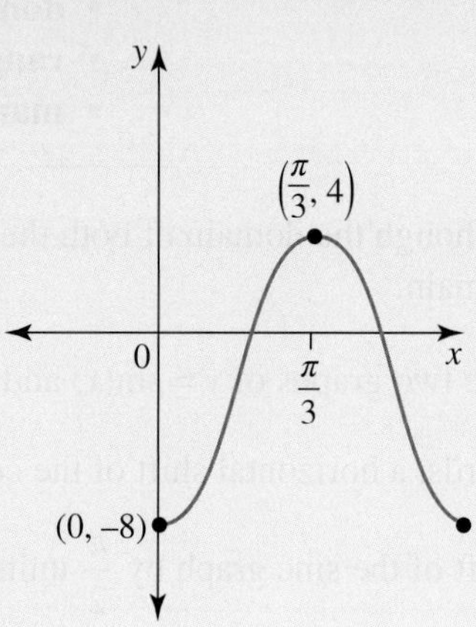

THINK	WRITE
a. 1. State the period and amplitude of the graph.	a. $y=3\sin(2x)+4,\ 0\le x\le 2\pi$ The period is $\frac{2\pi}{2}=\pi$. The amplitude is 3.

2. State the mean position and the range.

The mean position is $y = 4$.
The range of the graph is $[4 - 3, 4 + 3] = [1, 7]$.

3. Construct appropriate scales on the axes and sketch the graph.

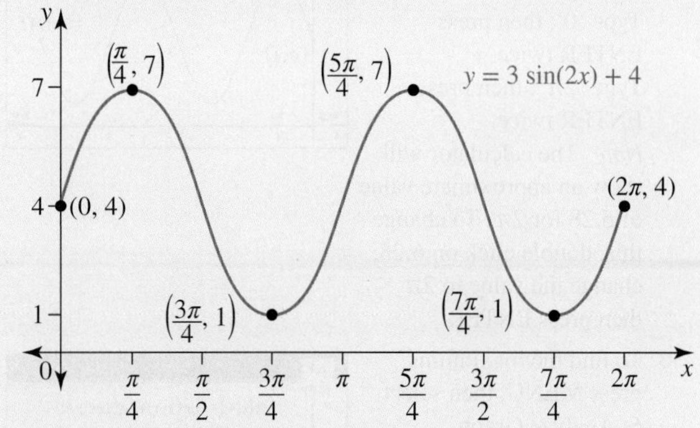

b. 1. Deduce the mean position.

b. The minimum value is -8 and the maximum value is 4, so the mean position is

$$y = \frac{-8 + 4}{2} = -2.$$

2. State the amplitude.

The amplitude is the distance from the mean position to either its maximum or minimum.
The amplitude is 6.

3. State the period.

At $x = \frac{\pi}{3}$, the graph is halfway through its cycle, so its period is $\frac{2\pi}{3}$.

4. Determine a possible equation for the given graph.

Let the equation be $y = a\cos(nx) + k$.

The graph is an inverted cosine shape, so $a = -6$.

The period is $\frac{2\pi}{n}$.

$$\frac{2\pi}{n} = \frac{2\pi}{3}$$
$$n = 3$$

The mean position is $y = -2$, so $k = -2$.
The equation is $y = -6\cos(3x) - 2$.

TI | THINK

a. 1. Put the Calculator into Radian mode.
On a Graphs page, complete the entry line for function 1 as:
$f1(x) = 3\sin(2x) + 4$,
$0 \le x \le 2\pi$
then press ENTER.
Note: You may need to adjust the Window/Zoom to see the full domain.

DISPLAY/WRITE

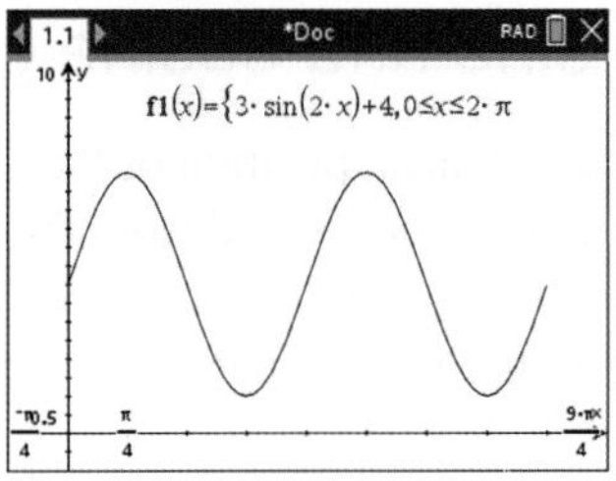

CASIO | THINK

a. 1. Put the Calculator into Radian mode.
On a Graphs screen, complete the entry line for y1 as:
$f1(x) = 3\sin(2x) + 4 = 0|$
$0 \le x \le 2\pi$
then press EXE.
Select the Graph icon to draw the graph.

DISPLAY/WRITE

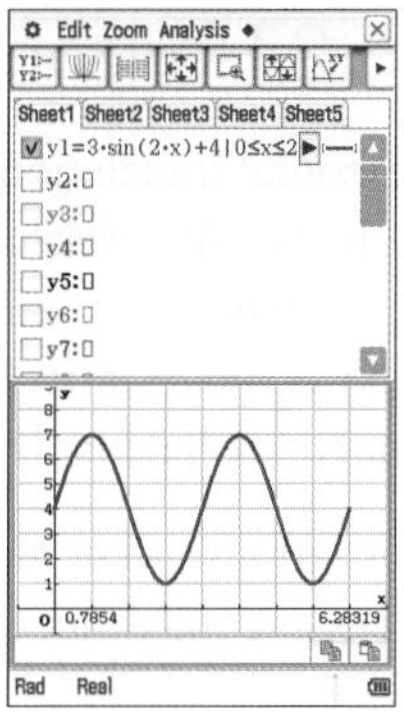

2. To find the end points, press MENU, then select:
5: Trace
1: Graph Trace
Type '0', then press ENTER twice.
Type '2π', then press ENTER twice.
Note: The calculator will show an approximate value of 6.28 for 2π. To change this, double click on 6.28, change the value to 2π, then press ENTER.

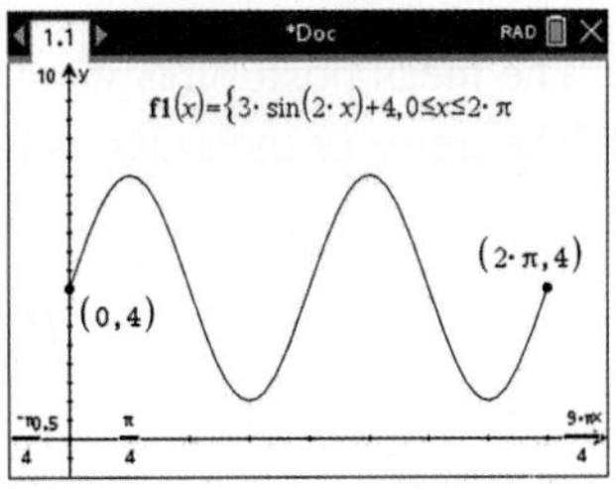

3. To find the maximums, press MENU, then select:
6: Analyze Graph
3: Maximum
Move the cursor to the left of the maximum when prompted for the lower bound, then press ENTER. Move the cursor to the right of the maximum when prompted for the upper bound, then press ENTER. Repeat this process to find the other maximum.

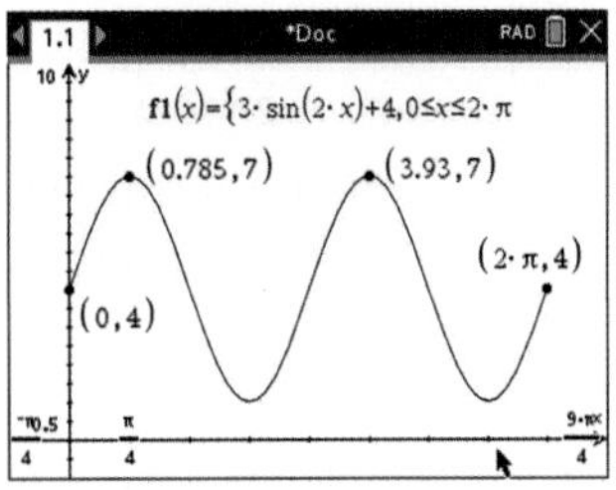

4. To find the minimums, press MENU, then select:
6: Analyze Graph
2: Minimum
Move the cursor to the left of the minimum when prompted for the lower bound, then press ENTER. Move the cursor to the right of the minimum when prompted for the upper bound, then press ENTER. Repeat this process to find the other minimum.

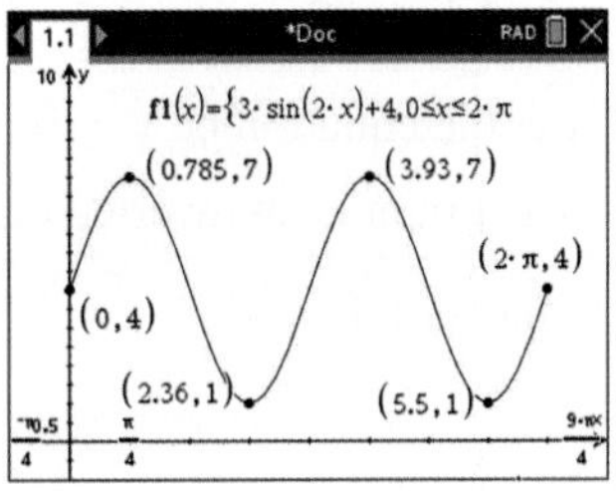

2. To find the end points, select:
- Analysis
- Trace

Type '0', then select OK and press EXE. Type '2π', then select OK and press EXE.

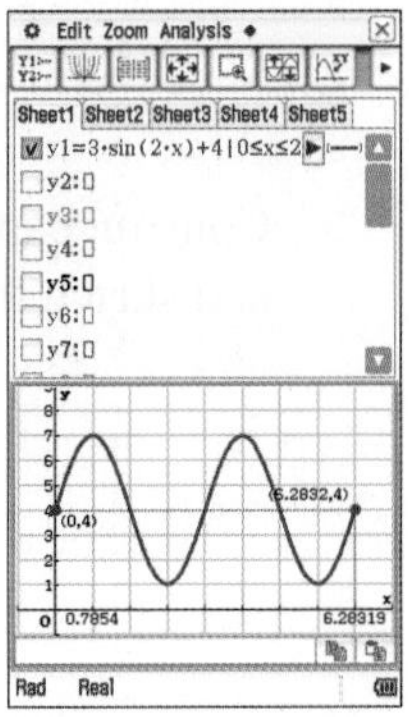

3. To find the maximums, select:
- Analysis
- G-Solve
- Max

then press EXE. Use the left/right arrows to move to the next maximum, then press EXE.

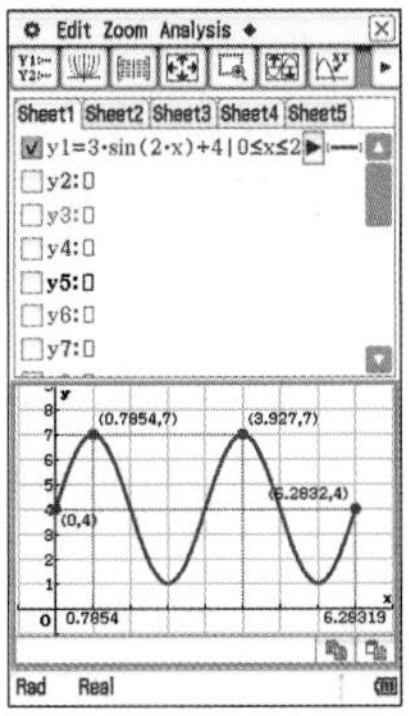

4. To find the minimums, select:
- Analysis
- G-Solve
- Min

then press EXE. Use the left/right arrows to move to the next minimum, then press EXE.

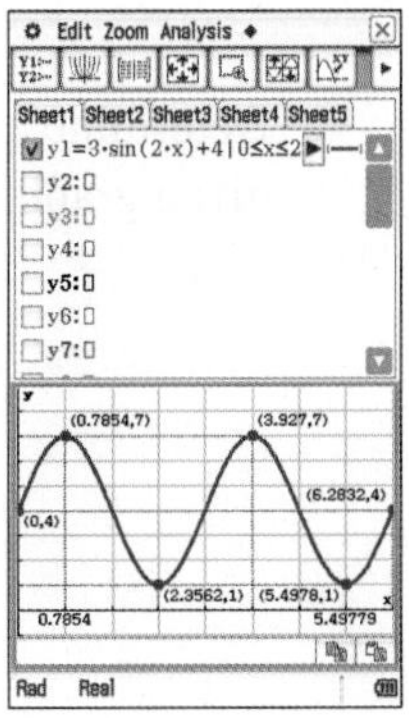

2.5.2 Horizontal translations of the sine and cosine graphs

Horizontal translations do not affect the period, amplitude or mean position of the graphs of sine or cosine functions. The presence of a horizontal translation of h units is recognised from the equation in the form given as $y = a\sin(n(x-h)) + k$ in exactly the same way it is for any other type of function. The graph will have the same shape as $y = a\sin(nx) + k$, but it will be translated to the right or to the left, depending on whether h is positive or negative, respectively.

Translations affect the position of the maximum and minimum points and any x- and y-intercepts. However, successive maximum points would remain one period apart, as would successive minimum points.

The equation in the form $y = a\sin(nx - b) + k$ must be rearranged into the form $y = a\sin\left(n\left(x - \frac{b}{n}\right)\right) + k$ to identify the horizontal translation $h = \frac{b}{n}$.

WORKED EXAMPLE 9 Sketching trigonometric graphs with a horizontal translation

Sketch the graph of the function $f: \left[0, \frac{3\pi}{2}\right] \to R, f(x) = 4\cos\left(2x + \frac{\pi}{3}\right)$.

THINK	WRITE
1. State the period, amplitude, mean position and horizontal translation.	$f: \left[0, \frac{3\pi}{2}\right] \to R, f(x) = 4\cos\left(2x + \frac{\pi}{3}\right)$ $f(x) = 4\cos\left(2\left(x + \frac{\pi}{6}\right)\right)$ The period is $\frac{2\pi}{2} = \pi$. The amplitude is 4. The mean position is $y = 0$. The horizontal translation is $\frac{\pi}{6}$ to the left.
2. Sketch the graph without the horizontal translation, $y = 4\cos(2x)$.	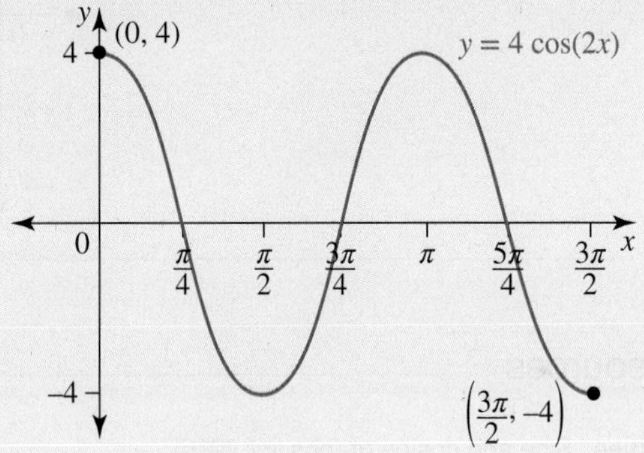
3. Calculate the coordinates of the end points of the domain of the given function.	$f(0) = 4\cos\left(\frac{\pi}{3}\right)$ $= 4 \times \frac{1}{2}$ $= 2$ $f\left(\frac{3\pi}{2}\right) = 4\cos\left(3\pi + \frac{\pi}{3}\right)$ $= 4 \times \frac{-1}{2}$ $= -2$ The end points of the graph are $(0, 2)$ and $\left(\frac{3\pi}{2}, -2\right)$.

4. Calculate or deduce the positions of the x-intercepts.

Each x-intercept on $y = 4\cos(2x)$ is translated $\frac{\pi}{6}$ units to the left.

Alternatively, let $y = 0$.

$$4\cos\left(2x + \frac{\pi}{3}\right) = 0$$

$$\cos\left(2x + \frac{\pi}{3}\right) = 0, \frac{\pi}{3} \le 2x + \frac{\pi}{3} \le 3\pi + \frac{\pi}{3}$$

$$2x + \frac{\pi}{3} = \frac{\pi}{2}, \frac{3\pi}{2}, \frac{5\pi}{2}$$

$$2x = \frac{\pi}{6}, \frac{7\pi}{6}, \frac{13\pi}{6}$$

$$x = \frac{\pi}{12}, \frac{7\pi}{12}, \frac{13\pi}{12}$$

5. Apply the horizontal translation to key points on the graph already sketched and hence sketch the function over its given domain.

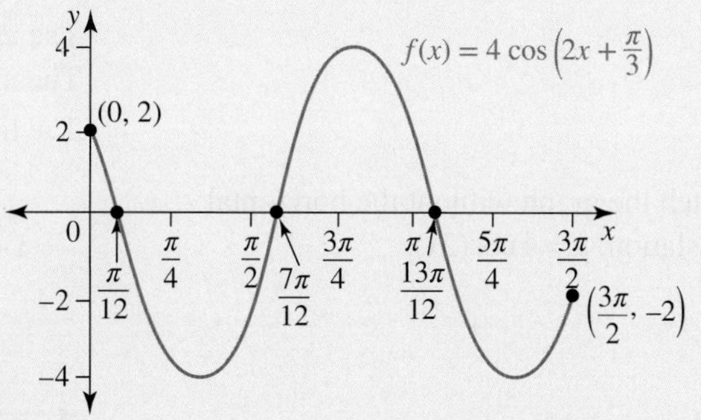

on Resources

Interactivities Sine and cosine graphs (int-2976)
The unit circle, sine and cosine graphs (int-6551)
Oscillation (int-2977)

2.5.3 Combinations of the sine and cosine functions

Trigonometric functions such as $y = \sin(x) + \cos(x)$ can be sketched using addition of ordinates. In this example, both of the functions being combined under addition have the same period. If the functions have different periods, then to observe the periodic nature of the sum function the graphs should be sketched over a domain that allows both parts to complete at least one full cycle.

For example, the function $y = \sin(2x) + \cos(x)$ would be drawn over $[0, 2\pi]$, with two cycles of the sine function and one cycle of the cosine function being added together.

WORKED EXAMPLE 10 Combinations of trigonometric functions

Sketch the graph of $y = \cos(x) + \frac{1}{2}\sin(2x)$ for $x \in [0, 2\pi]$.

THINK	WRITE
1. Identify the two functions forming the sum function.	$y = \cos(x) + \frac{1}{2}\sin(2x)$ $y = y_1 + y_2$ where $y_1 = \cos(x)$ and $y_2 = \frac{1}{2}\sin(2x)$.
2. State the key features of the two functions.	$y_1 = \cos(x)$ has period 2π and amplitude 1. $y_2 = \frac{1}{2}\sin(2x)$ has period π and amplitude $\frac{1}{2}$.
3. Sketch the two functions on the same set of axes and add together the y-values of known points.	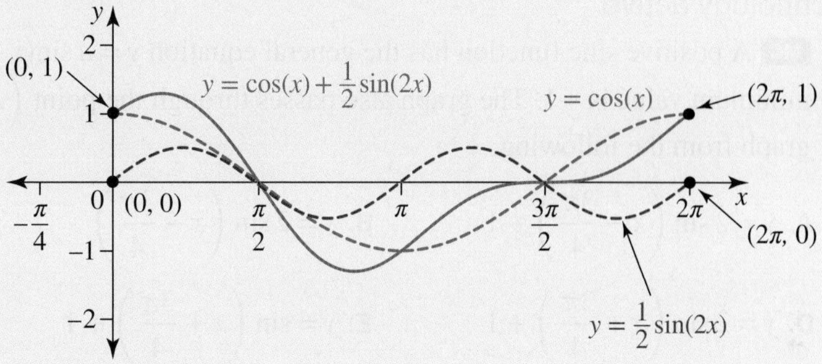

2.5 Exercise

Technology free

1. State the period, amplitude and range of each of the following.

   a. $y = 6\sin(8x)$ b. $y = 2 - 3\cos\left(\frac{x}{4}\right)$ c. $y = -\sin(3x - 6)$ d. $y = 3(5 + 2\cos(6\pi x))$

2. a. WE8 Sketch the graph of $y = 2\cos(4x) - 3$, $0 \le x \le 2\pi$.

   b. The diagram shows the graph of a sine function.

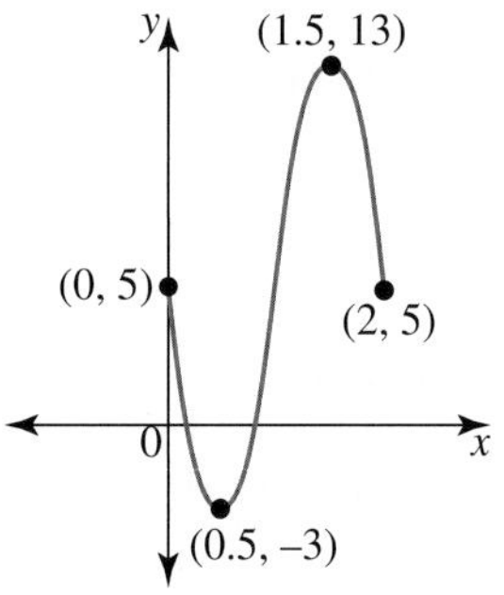

State its mean position, amplitude, and period, and give a possible equation for the function.

3. Sketch the graph of $f: [0, 2\pi] \to R, f(x) = 1 - 2\sin\left(\frac{3x}{2}\right)$, locating any intercepts with the coordinate axes.

4. WE9 Sketch the graph of the function $f: \left[0, \frac{3\pi}{2}\right] \to R, f(x) = -6\sin\left(3x - \frac{3\pi}{4}\right)$.

5. Sketch the following over the intervals specified.

a. $y = -7\cos(4x), 0 \le x \le \pi$

b. $y = 5 - \sin(x), 0 \le x \le 2\pi$

c. $y = \frac{1}{2}\cos(2x) + 3, -\pi \le x \le 2\pi$

d. $y = 2 - 4\sin(3x), 0 \le x \le 2\pi$

e. $y = 2\sin\left(x + \frac{\pi}{4}\right), 0 \le x \le 2\pi$

f. $y = -4\cos\left(3x - \frac{\pi}{2}\right) + 4, -\frac{\pi}{2} \le x \le \frac{3\pi}{2}$

Technology active

6. MC A positive sine function has the general equation $y = a\sin(x - h) + k$. The maximum value is 3 and the minimum value is -1. The graph also passes through the point $\left(\pi, \sqrt{2} + 1\right)$. Select the equation of the graph from the following.

A. $y = 2\sin\left(x - \frac{3\pi}{4}\right) + 1$

B. $y = 2\sin\left(x - \frac{3\pi}{4}\right)$

C. $y = 2\sin\left(x - \frac{\pi}{4}\right) + \sqrt{2} + 1$

D. $y = 2\sin\left(x + \frac{3\pi}{4}\right) + 1$

E. $y = \sin\left(x + \frac{3\pi}{4}\right) + 1$

7. a. i. Solve the equation $2\sin(2x) + \sqrt{3} = 0$ for $x \in [0, 2\pi]$.

ii. Sketch the graph of $y = \sin(2x)$ for $x \in [0, 2\pi]$.

iii. Hence, find $\{x: \sin(2x) < -\frac{\sqrt{3}}{2}, 0 \le x \le 2\pi\}$.

b. State the maximum value of the function $f(x) = 2 - 3\cos\left(x + \frac{\pi}{12}\right)$ and give the first positive value of x for when this maximum occurs.

8. WE10 Sketch the graph of $y = \cos(2x) - 3\cos(x)$ for $x \in [0, 2\pi]$.

9. Sketch the graphs of $y = (\sin(x))^2$ and $y = \sin(x)$ for $x \in [-\pi, \pi]$.

10. Use addition of ordinates to sketch the graphs of the following.

a. $y = \sin(2x) - 4\sin(x), 0 \le x \le 2\pi$

b. $y = x + \sin(x), 0 \le x \le 2\pi$

2.5 Exam questions

Question 1 (1 mark) TECH-ACTIVE

Source: *VCE 2019, Mathematical Methods Exam 2, Section A, Q1; © VCAA.*

MC Let $f: R \to R, f(x) = 3\sin\left(\frac{2x}{5}\right) - 2$.

The period and range of f are respectively

A. 5π and $[-3, 3]$

B. 5π and $[-5, 1]$

C. 5π and $[-1, 5]$

D. $\frac{5\pi}{2}$ and $[-5, 1]$

E. $\frac{5\pi}{2}$ and $[-3, 3]$

Question 2 (5 marks) TECH-FREE

Source: *VCE 2018, Mathematical Methods Exam 1, Q3; © VCAA.*

Let $f: [0, 2\pi] \to R,\ f(x) = 2\cos(x) + 1$.

a. Solve the equation $2\cos(x) + 1 = 0$ for $0 \le x \le 2\pi$. **(2 marks)**

b. Sketch the graph of the function f on the axes below. Label the end points and local minimum point with their coordinates. **(3 marks)**

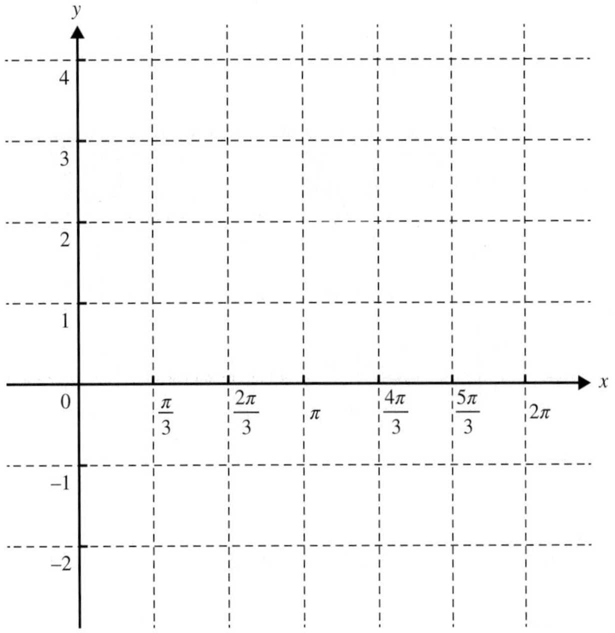

Question 3 (1 mark) TECH-ACTIVE

Source: *VCE 2016, Mathematical Methods Exam 2, Section A, Q2; © VCAA.*

MC Let $f: R \to R,\ f(\pi) = 1 - 2\cos\left(\dfrac{\pi x}{2}\right)$.

The period and range of this function are respectively

A. 4 and $[-2, 2]$
B. 4 and $[-1, 3]$
C. 1 and $[-1, 3]$
D. 4π and $[-1, 3]$
E. 4π and $[-2, 2]$

More exam questions are available online.

2.6 The tangent function

LEARNING INTENTION

At the end of this subtopic you should able to:

- state and understand the key features of the graph of the tangent function
- sketch graphs of the tangent function, including those that have undergone transformations.

2.6.1 The graph of $y = \tan(x)$

The domain of the tangent function, $y = \tan(x)$, can be deduced from the relationship $\tan(x) = \dfrac{\sin(x)}{\cos(x)}$. Whenever $\cos(x) = 0$, the tangent function will be undefined and its graph will have vertical asymptotes. Because $\cos(x) = 0$ when x is an odd multiple of $\dfrac{\pi}{2}$, the domain is $R\backslash\left\{x : x = (2n+1)\dfrac{\pi}{2}, n \in Z\right\}$.

If $\sin(x) = 0$, then $\tan(x) = 0$; therefore, its graph will have x-intercepts when $x = n\pi$, $n \in Z$.

The graph of $y = \tan(x)$ is shown.

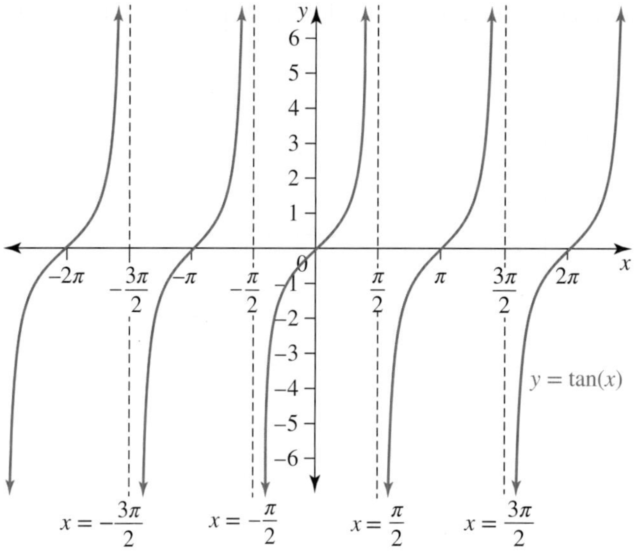

Key features of the tangent function

The graph of $y = \tan(x)$ has:

- **period π**
- **range R, which implies it is not meaningful to refer to an amplitude**
- **vertical asymptotes at $x = \pm\frac{\pi}{2}, \pm\frac{3\pi}{2}, \ldots$, i.e. $x = (2k+1)\frac{\pi}{2}, k \in Z$**
- **asymptotes spaced one period apart**
- **x-intercepts at $x = 0, \pm\pi, \pm 2\pi, \ldots$, i.e. $x = k\pi, k \in Z$**
- **mean position $y = 0$**
- **domain $R\backslash\left\{x : x = (2k+1)\frac{\pi}{2}, k \in Z\right\}$**
- **many-to-one correspondence.**

The graphs of $y = \tan(nx)$ and $y = a\tan(x)$

The period of $y = \tan(x)$ is π, so the period of $y = \tan(nx)$ will be $\frac{\pi}{n}$.

Altering the period alters the position of the vertical asymptotes, as these will now be $\frac{\pi}{n}$ units apart. An asymptote occurs when $nx = \frac{\pi}{2}$. Once one asymptote is found, others can be generated by adding or subtracting multiples of the period.

The mean position remains at $y = 0$, so the x-intercepts will remain midway between successive pairs of asymptotes.

The dilation factor a affects the steepness of the tangent graph $y = a\tan(x)$. Its effect is illustrated by comparing the values of the functions $f(x) = \tan(x)$ and $g(x) = 2\tan(x)$ at the point where $x = \frac{\pi}{4}$.

Because $f\left(\frac{\pi}{4}\right) = \tan\left(\frac{\pi}{4}\right) = 1$ and $g\left(\frac{\pi}{4}\right) = 2\tan\left(\frac{\pi}{4}\right) = 2$, the point $\left(\frac{\pi}{4}, 1\right)$ lies on the graph of $y = f(x)$ but the point $\left(\frac{\pi}{4}, 2\right)$ lies on the dilated graph $y = g(x)$.

Features of $y = a\tan(nx)$

The graph of $y = a\tan(nx)$ has:

- **period $\frac{\pi}{n}$**
- **vertical asymptotes $\frac{\pi}{n}$ units apart**
- **mean position $y = 0$ with x-intercepts on this line midway between pairs of successive asymptotes.**

Also note that the graph has an inverted shape if $a < 0$.

The x-intercepts can be located using their symmetry with the asymptotes. Alternatively, they can be calculated by solving the equation $a\tan(nx) = 0$.

WORKED EXAMPLE 11 Sketching $y = a\tan(nx)$

Sketch the graph of $y = 2\tan(3x)$ for $x \in [0, \pi]$.

THINK	WRITE
1. State the period.	$y = 2\tan(3x)$ The period is $\frac{\pi}{3}$.
2. Calculate the positions of the asymptotes.	An asymptote occurs when $3x = \frac{\pi}{2} \Rightarrow x = \frac{\pi}{6}$. Others are formed by adding multiples of the period. Asymptotes occur at $x = \frac{\pi}{6} + \frac{\pi}{3} = \frac{\pi}{2}$ and $x = \frac{\pi}{2} + \frac{\pi}{3} = \frac{5\pi}{6}$ within the domain constraint $x \in [0, \pi]$.
3. Calculate the positions of the x-intercepts. *Note:* An alternative method is to let $y = 0$ and solve the trigonometric equation for x.	The mean position is $y = 0$, and the x-intercepts occur midway between the asymptotes. One occurs at $x = \frac{1}{2}\left(\frac{\pi}{6} + \frac{\pi}{2}\right) = \frac{\pi}{3}$. The next is a period apart at $x = \frac{\pi}{3} + \frac{\pi}{3} = \frac{2\pi}{3}$, and the one after that is at $x = \frac{2\pi}{3} + \frac{\pi}{3} = \pi$.
4. Sketch the graph.	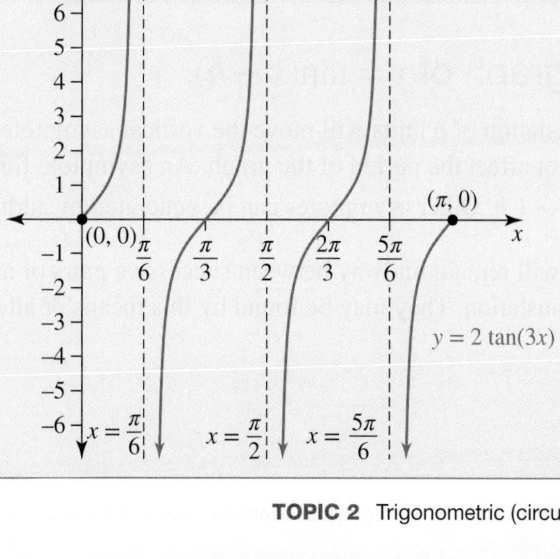

TI | THINK — **DISPLAY/WRITE**

a. 1. Put the Calculator into Radian mode.
On a Graphs page, complete the entry line for function 1 as:
$f1(x) = 2\tan(3x) \mid$
$0 \leq x \leq \pi$
then press ENTER.

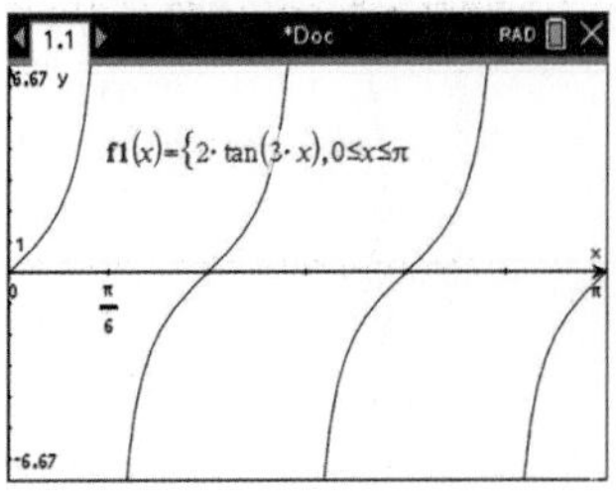

2. To find the x-intercepts, press MENU, then select:
6: Analyze Graph
1: Zero
Move the cursor to the left of the x-intercept when prompted for the lower bound, then press ENTER. Move the cursor to the right of the x-intercept when prompted for the upper bound, then press ENTER. Repeat this process to find the other x-intercepts.

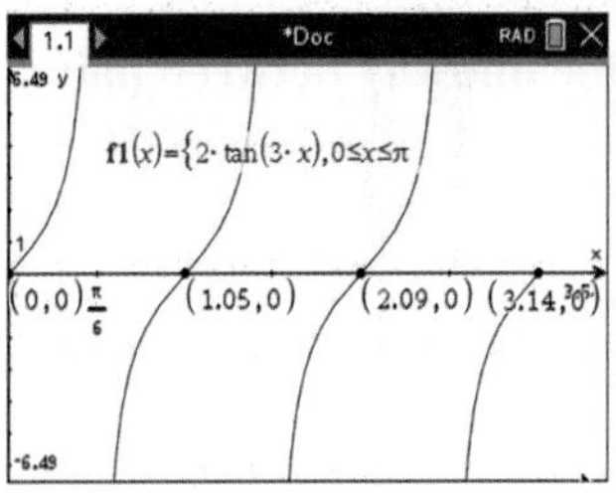

3. To draw the asymptotes, press MENU, then select:
8: Geometry
4: Construction
1: Perpendicular
Click on the x-axis, then click on the point on the x-axis where $x = \dfrac{\pi}{6}$.
Press MENU, then select:
1: Actions
4: Attributes
Click on the asymptote, press the down arrow, then press the right arrow twice to select dashed line.
Repeat this process to draw the other asymptotes.

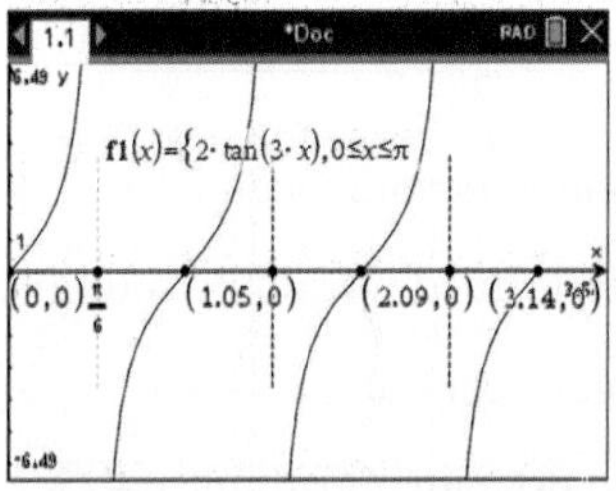

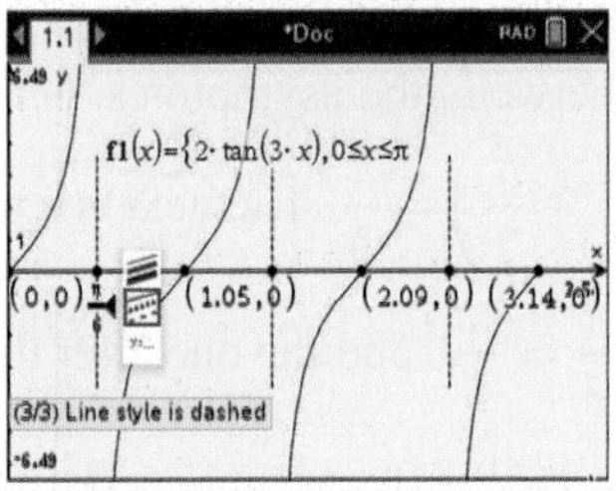

CASIO | THINK — **DISPLAY/WRITE**

a. 1. Put the Calculator into Radian mode.
On a Graphs screen, complete the entry line for y1 as:
$f1(x) = 2\tan(3x) = 0 \mid 0 \leq x \leq \pi$
then press EXE.
Select the Graph icon to draw the graph.

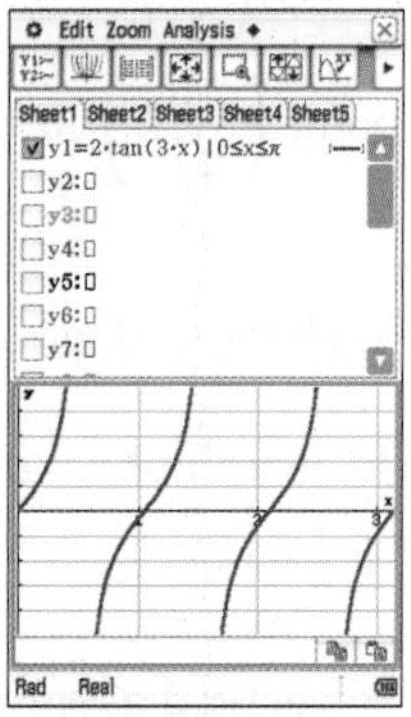

2. To find the x-intercepts, select:
- Analysis
- G-Solve
- Root

Press EXE. Use the left/right arrows to move to the next x-intercept, then press EXE. Repeat this process to find the other x-intercepts.

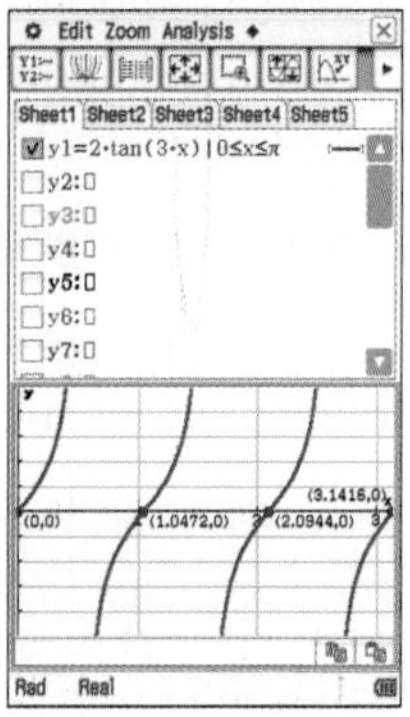

3. To draw the asymptotes, select:
- Analysis
- Sketch
- Vertical

then click on the point on the x-axis where $x = \dfrac{\pi}{6}$.
Repeat this process to draw the other asymptotes.

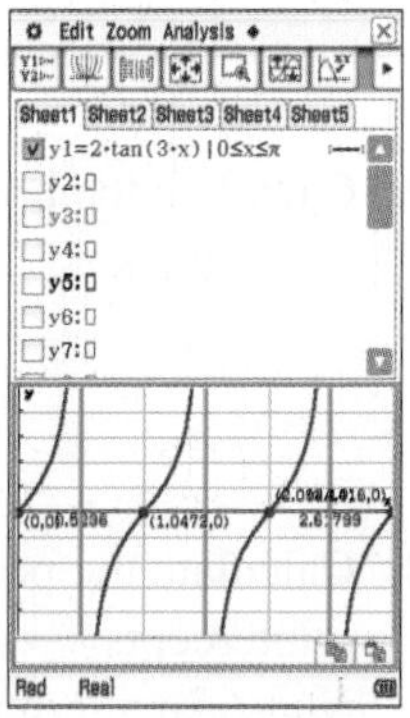

2.6.2 The graph of $y = \tan(x - h)$

A horizontal translation of h units will move the vertical asymptotes h units in the same direction, but the translation will not affect the period of the graph. An asymptote for the graph of $y = \tan(x - h)$ occurs when $x - h = \dfrac{\pi}{2} \Rightarrow x = \dfrac{\pi}{2} + h$. Other asymptotes can be generated by adding or subtracting multiples of the period π.

The x-intercepts will remain midway between successive pairs of asymptotes, as the mean position is unaffected at a horizontal translation. They may be found by this means or alternatively found by solving the equation $\tan(x - h) = 0$.

Features of $y = a\tan(nx - b)$

The graph of $y = a\tan(nx - b)$ has:

- **period $\dfrac{\pi}{n}$**
- **horizontal translation of $h = \dfrac{b}{n}$, as the equation is $y = a\tan\left(n\left(x - \dfrac{b}{n}\right)\right)$**
- **mean position $y = 0$.**

WORKED EXAMPLE 12 Sketching $y = a\tan(nx - b)$

Sketch the graph of $y = -\tan\left(2x + \dfrac{\pi}{2}\right)$ for $x \in (0, 2\pi)$.

THINK	WRITE
1. State the period.	$y = -\tan\left(2x + \dfrac{\pi}{2}\right)$ $\therefore y = -\tan\left(2\left(x + \dfrac{\pi}{4}\right)\right)$ The period is $\dfrac{\pi}{2}$.
2. Calculate the positions of the asymptotes.	An asymptote occurs when $2x + \dfrac{\pi}{2} = \dfrac{\pi}{2} \Rightarrow x = 0$. Adding multiples of the period, another occurs at $x = 0 + \dfrac{\pi}{2} = \dfrac{\pi}{2}$, another at $x = \dfrac{\pi}{2} + \dfrac{\pi}{2} = \pi$, another at $x = \pi + \dfrac{\pi}{2} = \dfrac{3\pi}{2}$, and another at $x = \dfrac{3\pi}{2} + \dfrac{\pi}{2} = 2\pi$. The asymptotes are $x = 0$, $x = \dfrac{\pi}{2}$, $x = \pi$, $x = \dfrac{3\pi}{2}$, $x = 2\pi$.
3. Calculate the positions of the x-intercepts.	The mean position is $y = 0$, and the x-intercepts are midway between the asymptotes. x-intercepts occur at $x = \dfrac{\pi}{4}, x = \dfrac{3\pi}{4}, x = \dfrac{5\pi}{4}, x = \dfrac{7\pi}{4}$.

4. Sketch the graph, noting its orientation.

The graph is inverted due to the presence of the negative coefficient in its equation, $y = -\tan\left(2x + \frac{\pi}{2}\right)$.

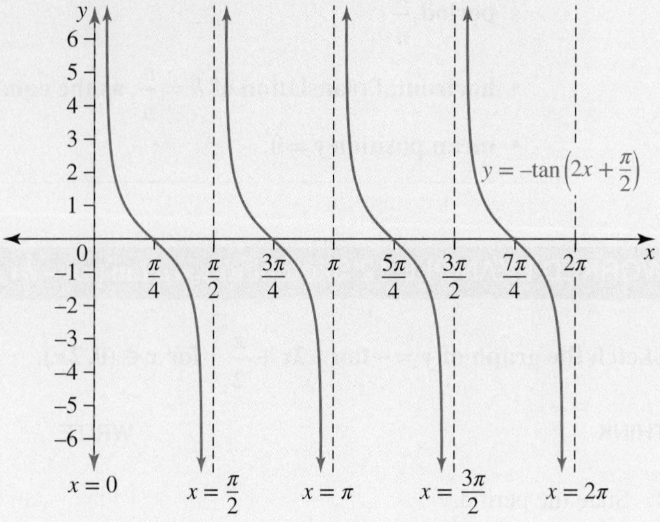

The graph of $y = a\tan(n(x - h)) + k$

Under a vertical translation of k units, the mean position becomes $y = k$. The points that are midway between the asymptotes will now lie on this line $y = k$, not on the x-axis, $y = 0$. The x-intercepts must be calculated by letting $y = 0$ and solving the ensuing trigonometric equation this creates.

The vertical translation does not affect either the asymptotes or the period.

Features of $y = a\tan(n(x - h) + k$

The graph of $y = a\tan(n(x - h)) + k$ has:

- **period $\frac{\pi}{n}$**
- **vertical asymptotes when $n(x - h) = (2k + 1)\frac{\pi}{2}, k \in Z$**
- **mean position $y = k$**
- **x-intercepts where $a\tan(n(x - h) + k = 0$.**

WORKED EXAMPLE 13 Sketching $y = a\tan(n(x - h)) + k$

Sketch the graph of $y = 3\tan(2\pi x) + \sqrt{3}$ over the interval $-\frac{7}{8} \leq x \leq \frac{7}{8}$.

THINK	WRITE
1. State the period and mean position.	$y = 3\tan(2\pi x) + \sqrt{3}$ The period is $\frac{\pi}{2\pi} = \frac{1}{2}$. The mean position is $y = \sqrt{3}$.

2. Calculate the positions of the asymptotes.

An asymptote occurs when $2\pi x = \frac{\pi}{2} \Rightarrow x = \frac{1}{4}$.

Others are formed by adding and subtracting a period.

For the interval $-\frac{7}{8} \leq x \leq \frac{7}{8}$, the asymptotes occur at

$x = \frac{1}{4} + \frac{1}{2} = \frac{3}{4}$, $x = \frac{1}{4} - \frac{1}{2} = -\frac{1}{4}$ and $x = -\frac{1}{4} - \frac{1}{2} = -\frac{3}{4}$.

The asymptotes are $x = -\frac{3}{4}$, $x = -\frac{1}{4}$, $x = \frac{1}{4}$, $x = \frac{3}{4}$.

3. Calculate the positions of the x-intercepts.

x-intercepts: let $y = 0$.

$3\tan(2\pi x) + \sqrt{3} = 0, -\frac{7}{8} \leq x \leq \frac{7}{8}$

$$\tan(2\pi x) = -\frac{\sqrt{3}}{3}, -\frac{7\pi}{4} \leq 2\pi x \leq \frac{7\pi}{4}$$

$$2\pi x = -\frac{\pi}{6}, -\pi - \frac{\pi}{6}, \pi - \frac{\pi}{6}, \cancel{2\pi - \frac{\pi}{6}}$$

$$2\pi x = -\frac{\pi}{6}, -\frac{7\pi}{6}, \frac{5\pi}{6}$$

$$x = -\frac{1}{12}, -\frac{7}{12}, \frac{5}{12}$$

4. Obtain the y-intercept.

When $x = 0$, $y = 3\tan(0) + \sqrt{3} = \sqrt{3}$.

The point $(0, \sqrt{3})$ is on the mean position.

5. Calculate the coordinates of the end points.

End points: let $x = -\frac{7}{8}$.

$$\begin{aligned} y &= 3\tan\left(2\pi \times -\frac{7}{8}\right) + \sqrt{3} \\ &= 3\tan\left(-\frac{7\pi}{4}\right) + \sqrt{3} \\ &= 3 \times 1 + \sqrt{3} \\ &= 3 + \sqrt{3} \end{aligned}$$

One end point is $\left(-\frac{7}{8}, 3 + \sqrt{3}\right)$.

Let $x = \frac{7}{8}$.

$$\begin{aligned} y &= 3\tan\left(\frac{7\pi}{4}\right) + \sqrt{3} \\ &= 3 \times -1 + \sqrt{3} \\ &= -3 + \sqrt{3} \end{aligned}$$

The other end point is $\left(\frac{7}{8}, -3 + \sqrt{3}\right)$.

6. Sketch the graph.

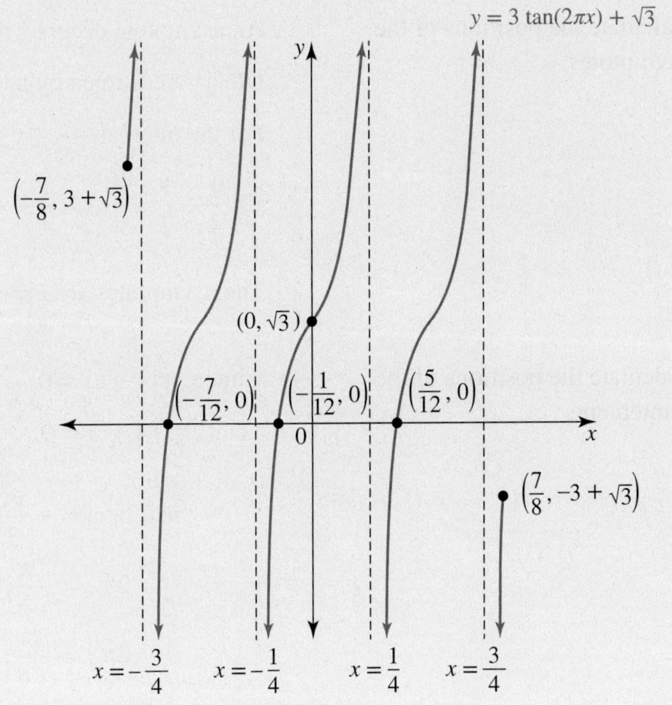

on Resources

Interactivity The tangent function (int-2978)

2.6 Exercise

Students, these questions are even better in jacPLUS

Receive immediate feedback and access sample responses

Access additional questions

Track your results and progress

Find all this and MORE in jacPLUS

Technology free

1. State the period and calculate the equation of the first positive asymptote for each of the following.

   a. $y = \tan(4x)$

   b. $y = 9 + 8\tan\left(\frac{x}{7}\right)$

   c. $y = -\frac{3}{2}\tan\left(\frac{4x}{5}\right)$

   d. $y = 2\tan(6\pi x + 3\pi)$

2. **WE11** Sketch the graph of $y = 3\tan\left(\frac{x}{2}\right)$ for $x \in [-\pi, \pi]$.

3. **WE12** Sketch the graph of $y = -\tan(2x - \pi)$ for $x \in [-\pi, \pi]$.

4. Sketch the following graphs over the intervals specified.

a. $y = -\tan(2x), x \in [0, \pi]$

b. $y = 3\tan\left(x + \frac{\pi}{4}\right), x \in [0, 2\pi]$

c. $y = \tan\left(\frac{x}{3}\right) + \sqrt{3}, x \in [0, 6\pi]$

d. $y = 5\sqrt{3}\tan\left(\pi x - \frac{\pi}{2}\right) - 5, x \in (-2, 3)$

5. **WE13** Sketch the graph of $y = 3\tan(2\pi x) - \sqrt{3}$ over the interval $-\frac{7}{8} \le x \le \frac{7}{8}$.

6. Sketch the graph of $y = 1 - \tan\left(x + \frac{\pi}{6}\right)$ over the interval $0 \le x \le 2\pi$.

Technology active

7. The graph of $y = a\tan(nx)$ has the domain $\left(-\frac{\pi}{3}, \frac{\pi}{3}\right)$ with vertical asymptotes at $x = -\frac{\pi}{3}$ and $x = \frac{\pi}{3}$ only. The graph passes through the origin and the point $\left(-\frac{\pi}{6}, -\frac{1}{2}\right)$. Determine its equation.

8. The graph of $y = \tan(x)$ undergoes a set of transformations to form that of the graph shown.

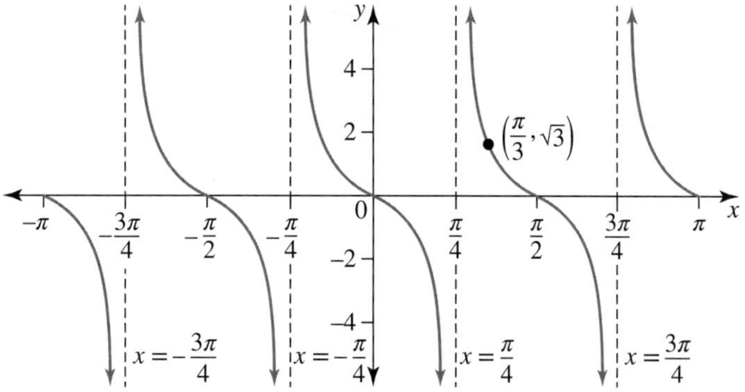

a. Explain why there was no vertical translation among the set of transformations applied to $y = \tan(x)$ to obtain this graph.
b. State the period of the graph shown.
c. Form a possible equation for the graph.

9. a. Use CAS technology to find the coordinates of the points of intersection of the graphs of $y = \sin(2x)$ and $y = \tan(x)$ for $-2\pi \le x \le 2\pi$.
b. Hence or otherwise, give the general solution to the equation $\sin(2x) = \tan(x), x \in R$.

10. A hybrid function is defined by the rule

$$f(x) = \begin{cases} -\sin(x), & -2\pi \le x \le -\frac{\pi}{2} \\ \tan(x), & -\frac{\pi}{2} < x < \frac{\pi}{2} \\ \cos(x), & \frac{\pi}{2} \le x \le 2\pi \end{cases}.$$

a. Evaluate:

i. $f\left(\frac{\pi}{3}\right)$ ii. $f(\pi)$ iii. $f\left(-\frac{\pi}{2}\right)$.

b. Sketch the graph of $y = f(x)$.
c. Identify any points of the domain where the function is not continuous.
d. State the domain and range of the function.

2.6 Exam questions

Question 1 (1 mark) TECH-ACTIVE

Source: *VCE 2021, Mathematical Methods Exam 2, Section A, Q1: © VCAA.*

MC The period of the function with rule $y = \tan\left(\frac{\pi x}{2}\right)$ is

A. 1 **B.** 2 **C.** 4 **D.** 2π **E.** 4π

Question 2 (3 marks) TECH-FREE

Source: *VCE 2020, Mathematical Methods Exam 1, Q3; © VCAA.*

Shown below is part of the graph of a period of the function of the form $y = \tan(ax + b)$.

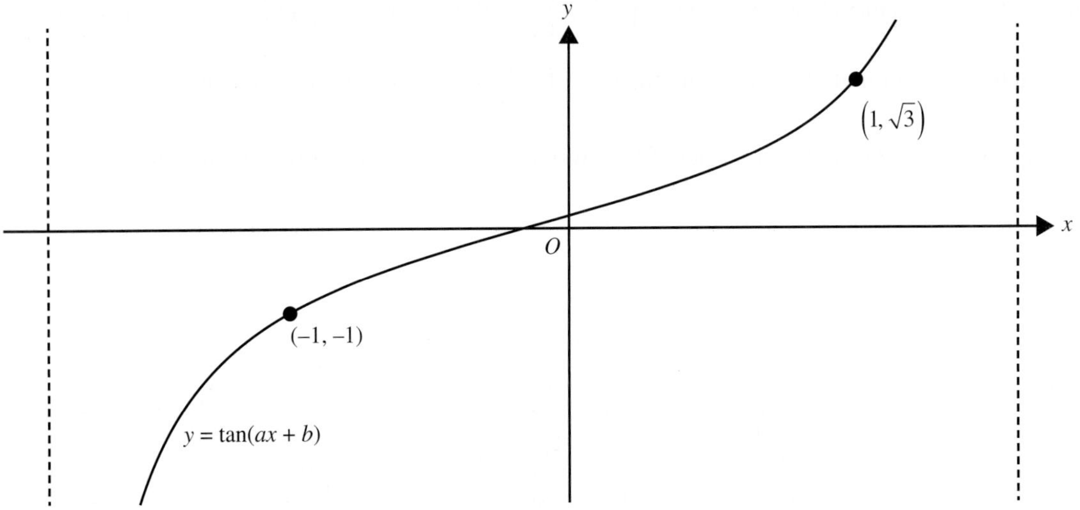

The graph is continuous for $x \in [-1, 1]$.

Find the value of a and the value of b, where $a > 0$ and $0 < b < 1$.

Question 3 (1 mark) TECH-ACTIVE

Source: *VCE 2018, Mathematical Methods Exam 2, Section A, Q11; © VCAA.*

MC The graph of $y = \tan(ax)$, where $a \in R^+$, has a vertical asymptote $x = 3\pi$ and has exactly one x-intercept in the region $(0, 3\pi)$. The value of a is

A. $\frac{1}{6}$ **B.** $\frac{1}{3}$ **C.** $\frac{1}{2}$ **D.** 1 **E.** 2

More exam questions are available online.

2.7 Modelling and applications

LEARNING INTENTION

At the end of this subtopic you should be able to:
- apply trigonometric functions to practical situations.

2.7.1 Real-life scenarios

There are many real-life scenarios where a trigonometric function can be used to model the situation.

WORKED EXAMPLE 14 Trigonometric modelling

The depth of water, *d* metres, at the end of a pier at time *t* hours after 6 am is modelled by

$$d(t) = \sin\left(\frac{\pi t}{6}\right) + 2.5$$

a. **State the period of the function.**
b. **State the maximum and minimum depths of the water.**
c. **Calculate is the depth at 7 am.**
d. **Sketch one cycle of the graph of the function.**
e. **The local council is interested in when the depth of water is at least 3.1 m. Determine for how long, correct to the nearest minute, during a 12-hour period the water is at least 3.1 m deep.**

THINK	WRITE
a. 1. The period of a sine function is given by $\frac{2\pi}{n}$.	$\text{Period} = \frac{2\pi}{n} = \frac{2\pi}{\frac{\pi}{6}}$
2. Write the answer, including units.	$= 12$ hours
b. The maximum and minimum depths take into account the amplitude of the function and the upward translation of the graph.	Maximum depth $= 2.5 + 1 = 3.5$ m Minimum depth $= 2.5 - 1 = 1.5$ m
c. $t = 0$ is at 6 am; therefore, 7 am is equal to $t = 1$.	$d(1) = \sin\left(\frac{1\pi}{6}\right) + 2.5 = \sin\left(\frac{\pi}{6}\right) + 2.5 = 0.5 + 2.5 = 3$ m
d. Use the information already gathered to sketch the graph for one cycle, or 12 hours. Maximum = 3.5 m Minimum = 1.5 m	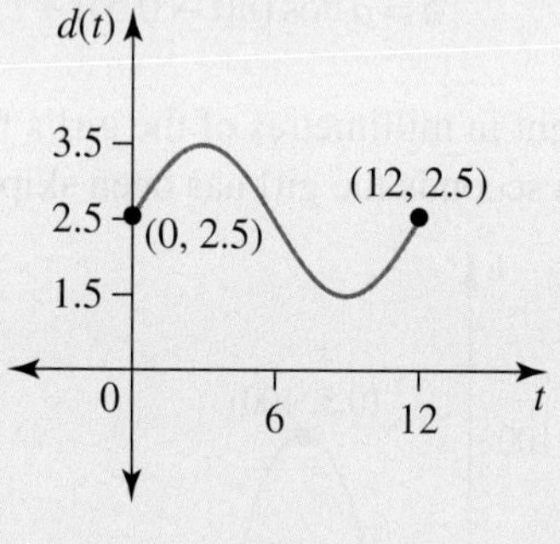
e. 1. Solve $3.1 = d(t)$. Make sure you include the domain in your CAS entry, otherwise you will obtain a general solution.	$3.1 = \sin\left(\frac{\pi t}{6}\right) + 2.5, 0 \leq t \leq 12$ $t = 1.23, 4.77$ hours
2. Find the difference between the times to find the total time the water is above 3.1 m.	$\text{Time} = 4.77 - 1.23 = 3.54$
3. Convert the time to minutes. Multiply the hour fraction by 60 to convert to minutes.	$0.54 \times 60 = 33$ minutes
4. Write the answer.	The depth will be at least 3.1 m for 3 hours 33 minutes.

2.7 Exercise

Technology active

1. WE14 James is in a boat out at sea fishing. The weather makes a change for the worse and the water becomes very choppy. The depth of water above the sea bed can be modelled by the function with equation $d = 1.5\sin\left(\frac{\pi t}{12}\right) + 12.5$, where d is the depth of water in metres and t is the time in hours since the change of weather began.
   a. State the period of the function.
   b. State the maximum and minimum heights of the boat above the sea bed.
   c. Calculate how far from the sea bed the boat was when the change of weather began.
   d. Sketch one cycle of the graph of the function.
   e. If the boat is h metres above the seabed for a continuous interval of 4 hours, calculate h correct to 1 decimal place.
   f. James has heard on the radio that the cycle of weather should have passed within 12 hours, and when the height of water above the sea bed is at a minimum after that, it will be safe to return to shore. If the weather change occurred at 9:30 am, determine when James will be able to return to shore.

2. A very young girl is learning to skip. The graph showing this skipping for one cycle is given.

   The general equation for this graph is given by

$$h = a\cos(n(t - 0.5)) + c$$

   where h is the height in millimetres of the girl's feet above the ground and t is the time in seconds the girl has been skipping.

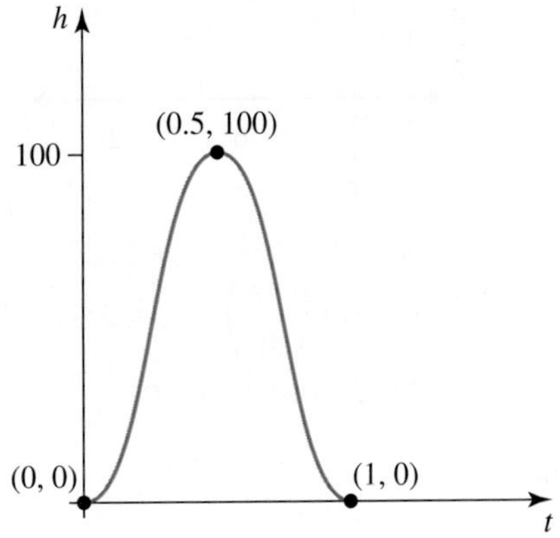

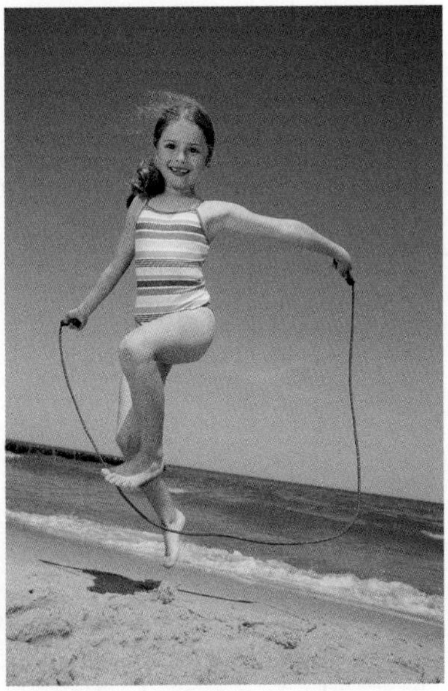

   a. Determine the values of the constants a, n and c, and hence restate the equation for one cycle of the skipping.
   b. Determine how soon after the beginning of the turn the girl's feet are 40 mm off the ground. Give your answer correct to 1 decimal place.

3. The height, h metres, above ground level of a chair on a rotating Ferris wheel is modelled by the function

$$h = 5 - 3.5\cos\left(\frac{\pi t}{30}\right)$$

where t is measured in seconds.

a. People can enter a chair when it is at its lowest position, at the bottom of the rotation. They enter the chair from a platform. Calculate how high the platform is above ground level.
b. Find the highest point reached by the chair.
c. State how long 1 rotation of the wheel takes.
d. Determine the length of time during a rotation that a chair is higher than 7 m off the ground. Give your answer to 1 decimal place.

4. A young girl and boy are lifted onto a seesaw in a playground. At this time the seesaw is horizontal with respect to the ground.

Initially the girl's end of the seesaw rises. Her height above the ground, h metres, t seconds after the seesaw starts to move is modelled by

$$h(t) = a\sin(nt) + k.$$

The greatest height above the ground that the girl reaches is 1.7 metres, and the least distance above the ground that she reaches is 0.7 metres. It takes 2 seconds for her to seesaw between these heights.

a. Find the values of a, n and k.
b. Draw the graph showing the height of the girl above the ground for $0 \le t \le 6$.
c. Determine the length of time during the first 6 seconds of the motion of the seesaw that the girl's height above the ground is 1.45 metres or higher.
d. Sketch the graph showing the height of the boy above the ground during the first 6 seconds and state its equation.

5. The water level in a harbour, h metres below a level jetty, at time t hours after 7 am is given by

$$h = 3 - 2.5\sin\left(\frac{1}{2}(t-1)\right).$$

a. Calculate how far below the jetty the water level in the harbour is at 7:30 am. Give your answer correct to 3 decimal places.
b. Determine the greatest and least distances below the jetty.
c. Sketch the graph of h versus t and hence determine the values of t at which the low and high tides first occur. Give your answers correct to 2 decimal places.
d. A boat ties up to the jetty at high tide. State how much extra rope will have to be left so that the boat is still afloat at low tide.

6. At a suburban shopping centre, one of the stores sells electronic goods such as digital cameras, laptop computers and printers. The store had a one-day sale towards the end of the financial year. The doors opened at 7:55 am and the cash registers opened at 8:00 am. The store closed its doors at 11:00 pm. The total number of people queuing at the six cash registers at any time during the day once the cash registers opened could be modelled by the equation

$$N(t) = 45\sin\left(\frac{\pi t}{5}\right) - 35\cos\left(\frac{\pi t}{3}\right) + 68, 0 \le t \le 15$$

where $N(t)$ is the total number of people queuing t hours after the cash registers opened at 8:00 am.

a. Many people ran into the store and quickly grabbed bargain items. Calculate how many people were queuing when the cash registers opened.
b. Determine the quietest time of the day and how many people were in the queue at this time.
c. Determine how many people were in the queue at midday.
d. Determine the maximum number of people in the queue between 3:00 pm and 7:00 pm.

2.7 Exam questions

Question 1 (1 mark) TECH-ACTIVE

Source: *VCE 2020, Mathematical Methods Exam 2, Section A, Q12; © VCAA.*

MC A clock has a minute hand that is 10 cm long and a clock face with a radius of 15 cm, as shown below.

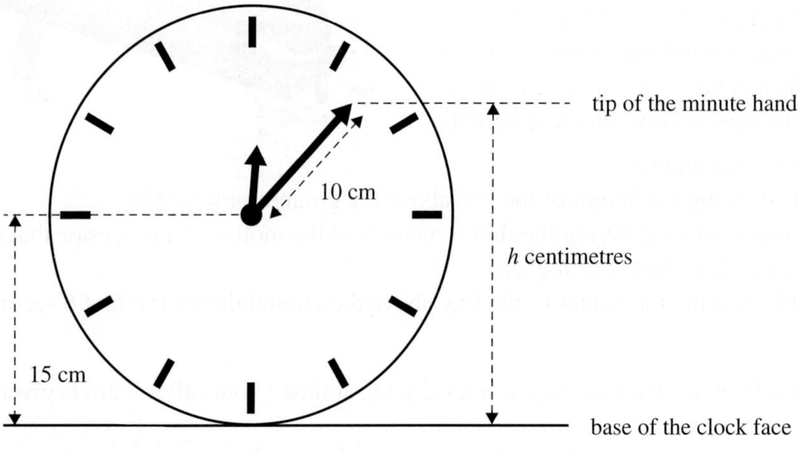

At 12.00 noon, both hands of the clock point vertically upwards and the tip of the minute hand is at its maximum distance above the base of the clock face.

The height, h centimetres, of the tip of the minute hand above the base of the clock face t minutes after 12.00 noon is given by

A. $h(t) = 15 + 10\sin\left(\frac{\pi t}{30}\right)$

B. $h(t) = 15 - 10\sin\left(\frac{\pi t}{30}\right)$

C. $h(t) = 15 + 10\sin\left(\frac{\pi t}{60}\right)$

D. $h(t) = 15 + 10\cos\left(\frac{\pi t}{60}\right)$

E. $h(t) = 15 + 10\cos\left(\frac{\pi t}{30}\right)$

Question 2 (1 mark) TECH-ACTIVE

Source: *VCE 2016, Mathematical Methods Exam 2, Section A, Q8; @ VCAA.*

MC The UV index, y, for a summer day in Melbourne is illustrated in the graph below, where t is the number of hours after 6 am.

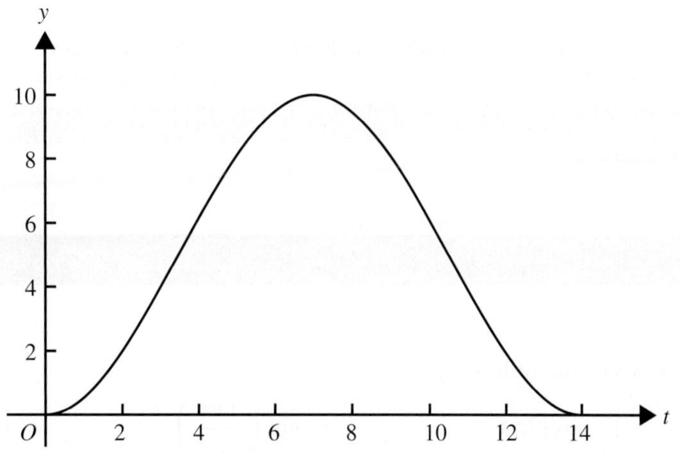

The graph is most likely to be the graph of

A. $y = 5 + 5\cos\left(\frac{\pi t}{7}\right)$

B. $y = 5 - 5\cos\left(\frac{\pi t}{7}\right)$

C. $y = 5 + 5\cos\left(\frac{\pi t}{14}\right)$

D. $y = 5 - 5\cos\left(\frac{\pi t}{14}\right)$

E. $y = 5 + 5\sin\left(\frac{\pi t}{14}\right)$

Question 3 (3 marks) TECH-FREE

Source: *VCE 2015, Mathematical Methods (CAS) Exam 1 Q5; @ VCAA.*

On any given day, the depth of water in a river is modelled by the function

$$h(t) = 14 + 8\sin\left(\frac{\pi t}{12}\right),\ 0 \le t \le 24$$

where h is the depth of water, in metres, and t is the time, in hours, after 6 am.

a. Find the minimum depth of the water in the river. **(1 mark)**

b. Find the values of t for which $h(t) = 10$. **(2 marks)**

More exam questions are available online.

2.8 Review

2.8.1 Summary

doc-37031

Hey students! Now that it's time to revise this topic, go online to:

Access the topic summary | Review your results | Watch teacher-led videos | Practise VCAA exam questions

Find all this and MORE in jacPLUS

2.8 Exercise

Technology free: short answer

1. Determine the exact values of the following.

   a. $\sin\left(-\dfrac{2\pi}{3}\right)$ b. $\cos(7\pi)$ c. $\sin\left(\dfrac{11\pi}{6}\right)$ d. $\tan\left(-\dfrac{\pi}{6}\right)$

2. Evaluate the following expressions, given that $\sin(\theta) = 0.25$ and θ is in the first quadrant.

   a. $\sin(-\theta)$ b. $\cos(3\pi + \theta)$ c. $\tan(\pi + \theta)$ d. $\sin\left(\dfrac{\pi}{2} - \theta\right)$

3. a. If $\cos(\alpha) = \dfrac{2}{3}$ and $\dfrac{3\pi}{2} \le \alpha \le 2\pi$, evaluate the exact values of:

   i. $\sin(\alpha)$ ii. $\tan(\alpha)$.

   b. Hence, show that $\sin^2(\alpha) + \cos^2(\alpha) = 1$.

4. Solve each of the following equations over the given domain.

   a. $\cos(2x) = -1,\ -2\pi \le x \le 2\pi$

   b. $4\cos^2(\theta) - \left(2\sqrt{2} - 2\right)\cos(\theta) - \sqrt{2} = 0,\ 0 \le \theta \le 2\pi$

   c. $\sqrt{3}\sin(2x) = -\cos(2x),\ -\pi \le x \le \pi$

   d. $\sqrt{3}\sin(\theta) - 2\sin(\theta)\cos(\theta) = 0,\ 0 \le \theta \le 2\pi$

   e. $2\sin\left(2\left(\theta + \dfrac{\pi}{12}\right)\right) + 1 = 0,\ -\pi \le \theta \le \pi$

   f. $2\cos\left(x + \dfrac{\pi}{6}\right) = \sqrt{2}$ over R

5. a. State the period and amplitude of the graph of $y = 3\sin(2 - 5x) + 6$.

   b. State the period and the equation of the first positive asymptote of the graph of $y = 2 - 3\tan(3x)$.

6. a. Sketch the graph of $f: [-\pi, \pi] \to R, f(x) = 2\sin(2x) + 1$.

   b. Sketch the graph of $f: [0, 4\pi] \to R, f(x) = 5 + 5\cos\left(\dfrac{x}{2} - \dfrac{\pi}{4}\right)$.

Technology active: multiple choice

7. MC Given that $\cos(\theta) = 0.362$ and θ is in the first quadrant, determine which of the following is **not** true.

   A. $\cos(5\pi + \theta) = -0.362$
   B. $\cos(4\pi - \theta) = 0.362$
   C. $\cos(\pi + \theta) = -0.362$
   D. $\cos(-\theta) = 0.362$
   E. $\cos(3\pi - \theta) = 0.362$

8. **MC** The value of $\cos\left(-\frac{7\pi}{3}\right)$ is:

A. $\frac{1}{2}$ B. $-\frac{\sqrt{3}}{2}$ C. $-\frac{1}{2}$ D. $\frac{\sqrt{2}}{2}$ E. $\frac{\sqrt{3}}{2}$

9. **MC** The exact value of $\cos\left(\frac{13\pi}{4}\right) - 2\sin\left(\frac{5\pi}{6}\right) + \sqrt{3}\tan\left(\frac{4\pi}{3}\right)$ is:

A. $-\frac{1}{\sqrt{2}}$ B. $2 - \frac{1}{\sqrt{2}}$ C. $2 + \frac{1}{\sqrt{2}}$ D. $4 - \frac{1}{\sqrt{2}}$ E. $3 - \frac{\sqrt{3}}{2} + \frac{1}{\sqrt{2}}$

10. **MC** When simplified, $\frac{2\sin(\pi - \theta)\sin\left(\frac{\pi}{2} - \theta\right)}{2 - 2\cos^2(\theta)}$ becomes:

A. $\frac{\sin(\theta)\cos(\theta)}{1 - \cos^2(\theta)}$ B. $\tan(\theta)$ C. $\frac{1}{\tan(\theta)}$ D. $\cos(\theta)$ E. $\frac{1}{\sin(\theta)}$

11. **MC** Given that $\cos(\theta) = 0.6402$ and $0 \le \theta \le \frac{\pi}{2}$, find the value of $\sin\left(\frac{3\pi}{2} - \theta\right)$ correct to 4 decimal places.

A. 0.6402 B. −0.6402 C. −0.3598
D. 1.6402 E. None of the above

12. **MC** If $2\cos\left(\theta - \frac{\pi}{2}\right) - 1 = 0$ and $0 \le \theta \le 2\pi$, then θ is equal to:

A. $\frac{\pi}{6}$ and $\frac{5\pi}{6}$ B. $\frac{5\pi}{6}$ C. $\frac{2\pi}{3}$ D. $-\frac{\pi}{3}$ and $\frac{\pi}{3}$ E. $-\frac{\pi}{6}$ and $-\frac{5\pi}{6}$

13. **MC** The general solution for $2\sin(x) - \sqrt{3} = 0$ is:

A. $x = \frac{12n\pi + \pi}{6}$ and $\frac{12n\pi + 5\pi}{6}$, where $n \in Z$
B. $x = \frac{\pi}{6}$ and $\frac{5\pi}{6}$, where $n \in Z$
C. $x = \frac{6n\pi + \pi}{3}$ and $\frac{6n\pi + 2\pi}{3}$, where $n \in Z$
D. $x = \frac{\pi}{3}$ and $\frac{2\pi}{3}$, where $n \in Z$
E. $x = \frac{6n\pi - \pi}{3}$ and $\frac{6n\pi + 4\pi}{3}$, where $n \in Z$

14. **MC** A possible equation for the graph shown is:

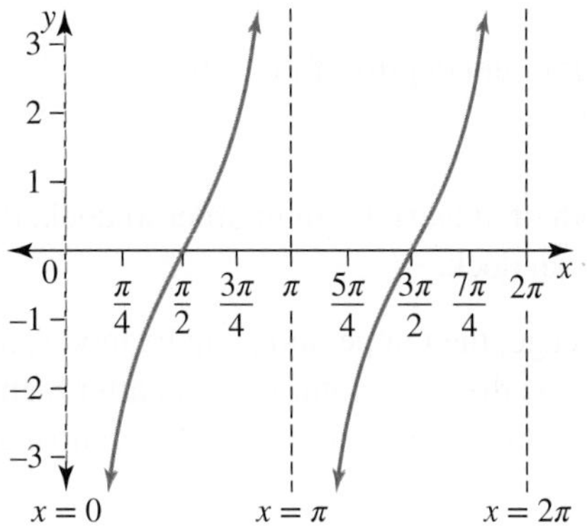

A. $y = -\tan(x)$ B. $y = 2\tan(x)$ C. $y = \tan(2x)$
D. $y = 2\tan\left(x + \frac{\pi}{2}\right)$ E. $y = \tan\left(2\left(x - \frac{\pi}{2}\right)\right)$

15. **MC** If a tangent function has an equation $y = 2\tan\left(\frac{\pi x}{4}\right) - 1$, the equation of the first positive asymptote is:

A. $x = 8$ B. $x = 4\pi$ C. $x = 4$ D. $x = 2\pi$ E. $x = 2$

16. **MC** The range of the function with the rule $f(x) = 3 - 8\cos\left(4x - \frac{\pi}{3}\right),\ 0 \le x \le \pi$ is:

A. $\left[\frac{\pi}{12}, \frac{13\pi}{2}\right]$ B. $\left[\frac{\pi}{3}, \frac{4\pi}{3}\right]$ C. $[-1, 11]$ D. $[-5, 11]$ E. $[-11, -5]$

Technology active: extended response

17. The temperature in degrees Celsius at Thredbo on a day in the middle of winter can be modelled by the equation

$$T = 2 - 6\cos\left(\frac{\pi t}{12}\right)$$

where t is the number of hours after 4:00 am.

a. Calculate the minimum and maximum temperatures.
b. Calculate the time(s) of the day at which the temperature is 0 °C. Give your answer(s) to the nearest minute.
c. Determine when is the temperature at its maximum.
d. Calculate the temperature at 8:00 am.

18. a. On the same set of axes, sketch the graphs of $y = -\cos(2x)$ and $y = \sqrt{3}\sin(2x)$ for $x \in \left[-\frac{\pi}{2}, \frac{\pi}{2}\right]$.

b. Calculate the exact coordinates of the points of intersection of the two graphs.

c. Hence, state $\left\{x : \sqrt{3}\sin(2x) + \cos(2x) \ge 0, -\frac{\pi}{2} \le x \le \frac{\pi}{2}\right\}$.

d. i. For the function defined by $f(x) = 2\sin\left(2x - \frac{\pi}{6}\right)$, calculate $f(0)$ and $f\left(\frac{\pi}{2}\right)$.

ii. Sketch the graph of $y = f(x)$ for $x \in \left[-\frac{\pi}{2}, \frac{\pi}{2}\right]$ and hence state $\left\{x : 2\sin\left(2x - \frac{\pi}{6}\right) \ge 0, -\frac{\pi}{2} \le x \le \frac{\pi}{2}\right\}$.

19. The depth of water, d metres, at the end of a pier at time t hours after 10 am is modelled by

$$d(t) = 2\sin\left(\frac{\pi t}{6} - \frac{\pi}{3}\right) + 2.5$$

a. Calculate the maximum and minimum depths of the water.
b. State the period of the function.
c. Determine the depth at 1 pm.
d. If a particular boat needs a depth of at least 4.2 m in order to dock, determine the times, to the nearest minute, during which the boat can dock.

20. For a particular incubated animal cage, the temperature can be modelled by a positive cosine curve, $T = a\cos(n(x - h)) + k$, where T°C is the temperature t hours after 8 am. The maximum daily temperature was 28 °C and occurred at 2 pm and the minimum was 22 °C, occurring at 8 pm.

a. State the equation of the curve.
b. Calculate the temperature at 10 am.
c. The creature that lives in the incubated cage sleeps when the temperature falls below 24 °C. Determine for how many hours a day, correct to 1 decimal place, the creature sleeps.

2.8 Exam questions

Question 1 (5 marks) TECH-FREE

Source: *VCE 2021, Mathematical Methods Exam 1, Q3; © VCAA.*

Consider the function $g : R \to R, g(x) = 2\sin(2x)$.

a. State the range of g. **(1 mark)**

b. State the period of g. **(1 mark)**

c. Solve $2\sin(2x) = \sqrt{3}$ for $x \in R$. **(3 marks)**

Question 2 (4 marks) TECH-FREE

Source: *VCE 2019, Mathematical Methods Exam 1, Q4; © VCAA.*

a. Solve $1 - \cos\left(\frac{x}{2}\right) = \cos\left(\frac{x}{2}\right)$ for $x \in [-2\pi, \pi]$. **(2 marks)**

b. The function $f : [-2\pi, \pi] \to R, f(x) = \cos\left(\frac{x}{2}\right)$ is shown on the axes below.

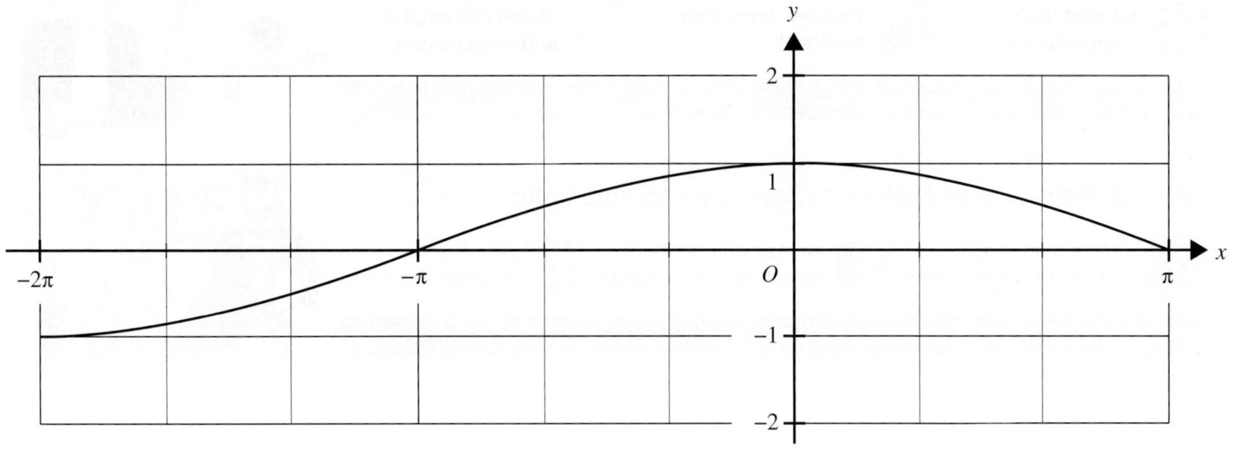

Let $g : [-2\pi, \pi] \to R, g(x) = 1 - f(x)$

Sketch the graph of g on the axes above. Label all points of intersection of the graphs of f and g, and the endpoints of g, with their coordinates. **(2 marks)**

Question 3 (1 mark) TECH-ACTIVE

Source: *VCE 2018, Mathematical Methods Exam 2, Section A, Q1; © VCAA.*

MC Let $f : R \to R, f(x) = 4\cos\left(\frac{2\pi x}{3}\right) + 1$.

The period of this function is

A. 1
B. 2
C. 3
D. 4
E. 5

Question 4 (3 marks) TECH-FREE

Source: *VCE 2017, Mathematical Methods Exam 1, Q6; © VCAA.*

Let $(\tan(\theta) - 1)\left(\sin(\theta) - \sqrt{3}\cos(\theta)\right)\left(\sin(\theta) + \sqrt{3}\cos(\theta)\right) = 0$.

a. State all possible values of $\tan(\theta)$. **(1 mark)**

b. Hence, find all possible solutions for $(\tan(\theta) - 1)\left(\sin^2(\theta) - 3\cos^2(\theta)\right) = 0$, where $0 \le \theta \le \pi$. **(2 marks)**

Question 5 (1 mark) TECH-ACTIVE

Source: *VCE 2015, Mathematical Methods (CAS) Exam 2, Section 1, Q1; © VCAA.*

MC Let $f: R \rightarrow R, f(x) = 2\sin(3x) - 3$.

The period and range of this function are respectively

A. period $= \dfrac{2\pi}{3}$ and range $= [-5, -1]$

B. period $= \dfrac{2\pi}{3}$ and range $= [-2, 2]$

C. period $= \dfrac{\pi}{3}$ and range $= [-1, 5]$

D. period $= 3\pi$ and range $= [-1, 5]$

E. period $= 3\pi$ and range $= [-2, 2]$

More exam questions are available online.

Answers

Topic 2 Trigonometric (circular) functions

2.2 Trigonometric symmetry properties

2.2 Exercise

1. a. -1 b. $-\frac{\sqrt{3}}{2}$ c. $-\frac{1}{\sqrt{2}}$
d. $\frac{1}{2}$ e. $-\sqrt{3}$ f. $-\frac{1}{2}$

2. a. $-\frac{1}{\sqrt{3}}$ b. $-\frac{1}{2}$ c. -1
d. $-\frac{1}{\sqrt{2}}$ e. $-\frac{\sqrt{3}}{2}$ f. $\frac{1}{2}$

3. a. $\frac{\sqrt{3}}{2}$ b. $\frac{1}{2}$ c. $-\frac{1}{\sqrt{3}}=-\frac{\sqrt{3}}{3}$
d. $\frac{1}{2}$ e. $-\frac{\sqrt{3}}{2}$ f. $-\frac{1}{\sqrt{3}}=-\frac{\sqrt{3}}{3}$

4. a. 0 b. Undefined c. 0
d. 0 e. 1 f. -1

5. a. $\sin(\theta)$ b. $\cos(\theta)$ c. $\tan(\theta)$
d. $\cos(\theta)$ e. $-\sin(\theta)$ f. $-\tan(\theta)$

6. a. $-\frac{(\sqrt{3}+1)}{2}$ b. $-\sqrt{2}+2$ c. $2\sqrt{3}$
d. $\frac{3+2\sqrt{2}}{4}$ e. 0 f. -2

7. a. -0.4695 b. -0.5592 c. -0.2680
d. -0.4695 e. 0.5592 f. 0.2680

8. a. 0.8829 b. -0.5318 c. 0.8290
d. -1.4825

9. a. $\sin(\alpha)$ b. $-\frac{1}{\tan(\alpha)}$ c. $-\cos(\alpha)$
d. $\frac{1}{\tan(\alpha)}$ e. $\sin(\alpha)$ f. $\cos(\alpha)$

10. a. 0.8829 b. -0.4696 c. 1.8803
d. -0.9848 e. 0.9848 f. -5.6729

11. a. 0.5592 b. 0.8290 c. -0.6746
d. -0.7547 e. 1.1503 f. -0.6561

12. a. -0.9511 b. 0.9511 c. 0.9511
d. 3.0792 e. -0.3089 f. -3.0792

13. a. -0.8572 b. 0.8572 c. -0.8572
d. -0.6008 e. 0.8572 f. -1.6645

14. a. $-\frac{3}{5}$ b. $-\frac{4}{3}$ c. 1 d. $-\frac{7}{25}$

15. a. $\frac{5}{13}$ b. $\frac{5}{12}$ c. $\frac{12}{13}$
d. $\frac{12}{13}$ e. $\frac{5}{13}$ f. $\frac{12}{5}$

16. a. $\tan^2(x)+1=\frac{\sin^2(x)}{\cos^2(x)}+\frac{\cos^2(x)}{\cos^2(x)}$
$=\frac{\sin^2(x)+\cos^2(x)}{\cos^2(x)}$
$=\frac{1}{\cos^2(x)}$
b. 0.7814

17. a. 12 cm/s b. $12-\frac{3\sqrt{3}}{2}$ cm/s
c. 12 cm/s

18. a. 1.5 m b. 0.75 m c. 0.75 m

2.2 Exam questions

Note: Mark allocations are available with the fully worked solutions online.

1. A 2. B 3. E

2.3 Trigonometric equations

2.3 Exercise

1. a. $\theta=\frac{5\pi}{6}$ and $\frac{7\pi}{6}$
b. $x=120°, 300°, 480°, 660°$

2. a. $\theta=\frac{\pi}{12},\frac{7\pi}{12},\frac{3\pi}{4},\frac{5\pi}{4},\frac{17\pi}{12},\frac{23\pi}{12}$
b. $x=-\frac{5\pi}{6},-\frac{2\pi}{3},\frac{\pi}{6},\frac{\pi}{3}$

3. a. $\theta=\frac{5\pi}{4},\frac{7\pi}{4}$
b. $\theta=\frac{\pi}{3},\frac{5\pi}{3}$
c. $\theta=\frac{\pi}{9},\frac{4\pi}{9},\frac{7\pi}{9},\frac{10\pi}{9},\frac{13\pi}{9},\frac{16\pi}{9}$
d. $\theta=\frac{\pi}{4},\frac{5\pi}{4}$

4. a. $x=120°$ and $240°$
b. $x=112.5°, 157.5°, 292.5°, 337.5°$

5. a. $\theta=-\frac{\pi}{3},-\frac{\pi}{6},\frac{2\pi}{3},\frac{5\pi}{6}$
b. $\theta=-\frac{3\pi}{4},-\frac{7\pi}{12},-\frac{\pi}{12},\frac{\pi}{12},\frac{7\pi}{12},\frac{3\pi}{4}$
c. $\theta=-\frac{5\pi}{8},-\frac{\pi}{8},\frac{3\pi}{8},\frac{7\pi}{8}$
d. $\theta=-\frac{\pi}{3},\frac{\pi}{3}$

6. a. $x=-\pi,-\frac{3\pi}{4},0,\frac{\pi}{4},\pi$
b. $x=-\frac{5\pi}{6},\frac{5\pi}{6}$
c. $x=-\frac{\pi}{4},\frac{3\pi}{4}$
d. $x=\frac{4\pi}{9},\frac{5\pi}{9},\frac{10\pi}{9},\frac{11\pi}{9},\frac{16\pi}{9},\frac{17\pi}{9}$

7. a. $\theta=0.73$ and 2.41
b. $x=73.40°$ and $286.60°$

8. a. $\theta = 215.23°, 324.77°$

b. $x = -\frac{3\pi}{2}, \frac{\pi}{2}$

9. $\theta = 0.314, 1.571, 2.827, 4.084, 5.341$

10. $x = 0.526, 1.179$

11. $\theta = \frac{\pi}{4}, \frac{\pi}{2}, \frac{5\pi}{4}$, and $\frac{3\pi}{2}$

12. C

13. a. $\theta = \frac{\pi}{4}, \frac{3\pi}{4}, \frac{5\pi}{4}, \frac{7\pi}{4}$

b. $\theta = \frac{\pi}{6}, \frac{\pi}{3}, \frac{2\pi}{3}, \frac{5\pi}{6}$

14. a. $\alpha = -\pi, 0, \pi$

b. $\alpha = -\frac{5\pi}{6}, -\frac{\pi}{3}, \frac{\pi}{6}, \frac{2\pi}{3}$

c. $\alpha = -\frac{3\pi}{4}, -\frac{\pi}{4}, \frac{\pi}{4}, \frac{3\pi}{4}$

d. $\alpha = -\frac{2\pi}{3}, -\frac{\pi}{3}, \frac{\pi}{3}, \frac{2\pi}{3}$

2.3 Exam questions

Note: Mark allocations are available with the fully worked solutions online.

1. E

2. C

3. $x = \frac{5\pi}{12}$ and $\frac{7\pi}{12}$

2.4 General solutions of trigonometric equations

2.4 Exercise

1. a. $\theta = \frac{(6n+1)\pi}{3}$ and $\frac{(3n+1)2\pi}{3}, n \in Z$

b. $\theta = \frac{(6n-1)\pi}{12}, n \in Z$. Solutions within $[-\pi, \pi]$ are

$\theta = -\frac{7\pi}{12}, -\frac{\pi}{12}, \frac{5\pi}{12}, \frac{11\pi}{12}$.

2. a. $x = \frac{6n\pi \pm 2\pi}{3}, n \in Z$

b. $x = \frac{8n\pi + \pi}{4}, \frac{8n\pi + 3\pi}{4}, n \in Z$

3. $x = \frac{12n\pi - \pi}{12}, x = \frac{12n\pi + 7\pi}{12}, n \in Z;$

$x = \frac{7\pi}{12}, \frac{11\pi}{12}, \frac{19\pi}{12}, \frac{23\pi}{12}$

4. $x = \frac{3n\pi - \pi}{3}, n \in Z; x = -\frac{\pi}{3}, \frac{2\pi}{3}$

5. $\theta = n\pi, n \in Z$

6. $\theta = \frac{6n\pi - \pi}{12}, n \in Z$

2.4 Exam questions

Note: Mark allocations are available with the fully worked solutions online.

1. D

2. $x = \frac{2\pi}{3} + 2\pi n$ and $x = (2n+1)\pi$, where $n \in Z$

3. C

2.5 The sine and cosine functions

2.5 Exercise

1. a. Period $\frac{\pi}{4}$, amplitude 6, range $[-6, 6]$

b. Period 8π, amplitude 3, range $[-1, 5]$

c. Period $\frac{2\pi}{3}$, amplitude 1, range $[-1, 1]$

d. Period $\frac{1}{3}$, amplitude 6, range $[9, 21]$

2. a.

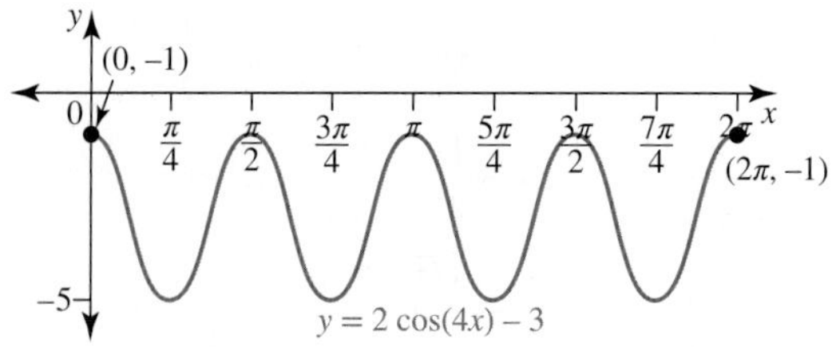

b. Mean position = 5, amplitude = 8, period = 2, $y = -8\sin(\pi x) + 5$

3.

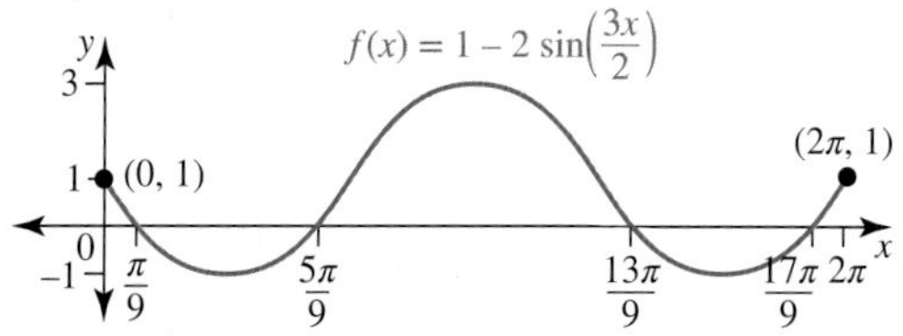

x-intercepts at $x = \frac{\pi}{9}, \frac{5\pi}{9}, \frac{13\pi}{9}, \frac{17\pi}{9}$; y-intercept = (0, 1)

4.

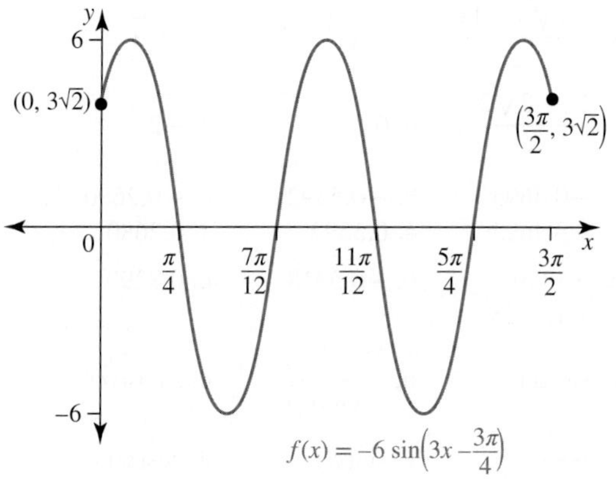

5. a.

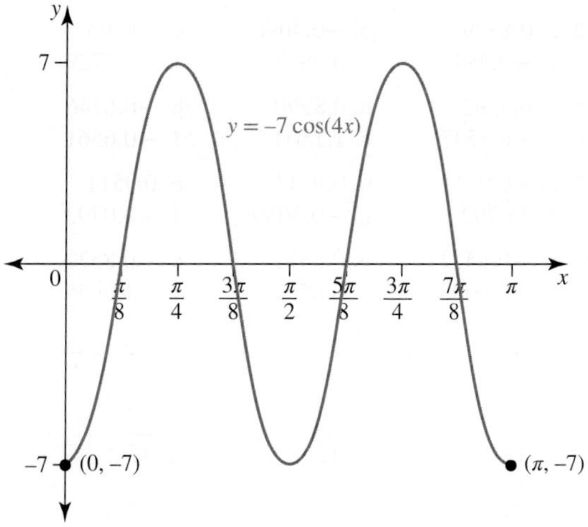

b.

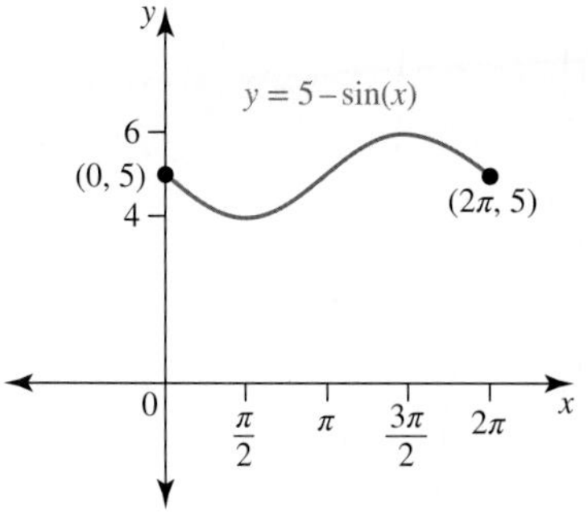

c.

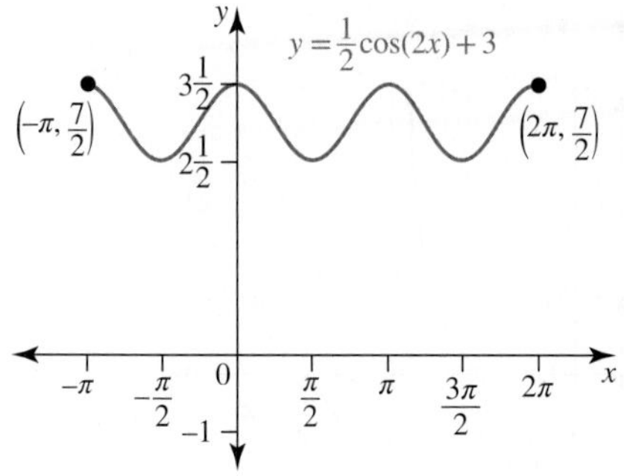

d.

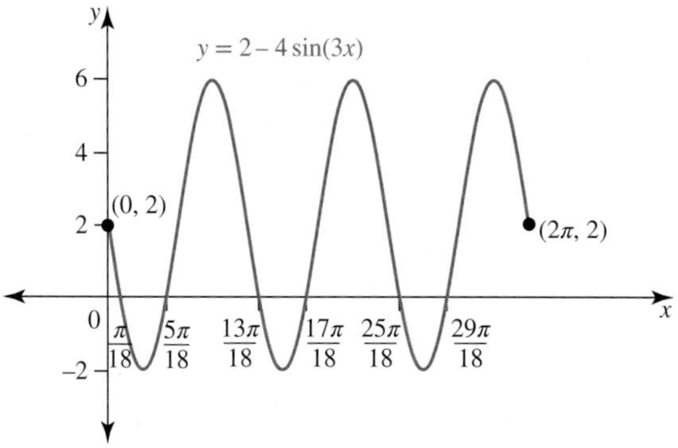

e.

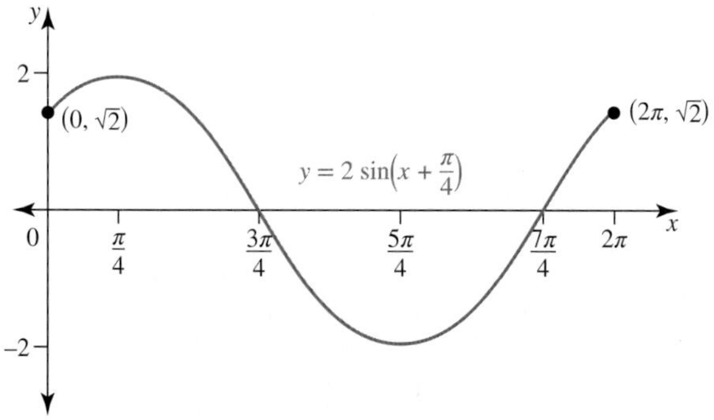

f.

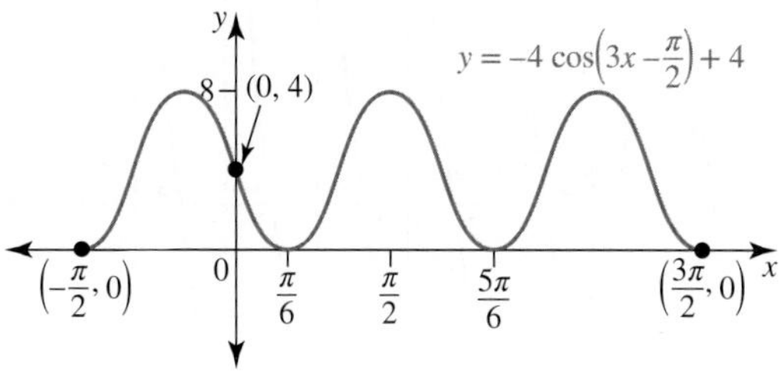

6. A

7. a. i. $x = \frac{2\pi}{3}, \frac{5\pi}{6}, \frac{5\pi}{3}, \frac{11\pi}{6}$

ii.

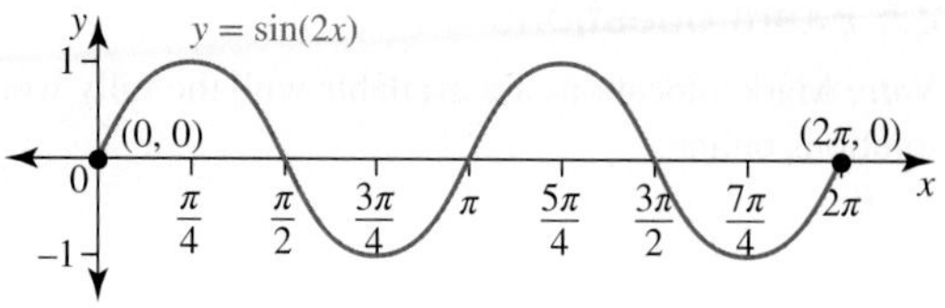

iii. $\left\{x : \frac{2\pi}{3} < x < \frac{5\pi}{6}\right\} \cup \left\{x : \frac{5\pi}{3} < x < \frac{11\pi}{6}\right\}$

b. $\max = 5, x = \frac{11\pi}{12}$

8.

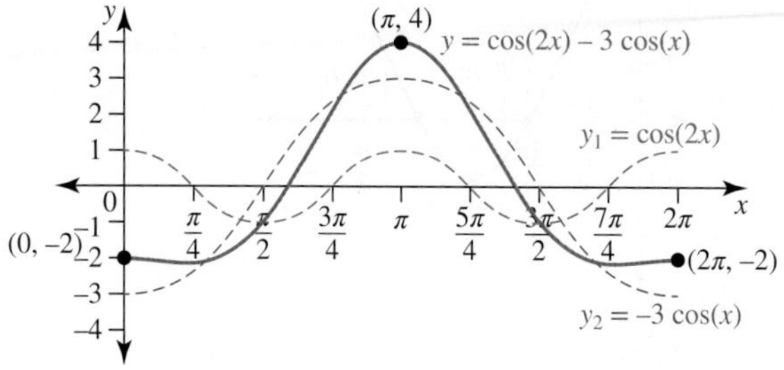

9.

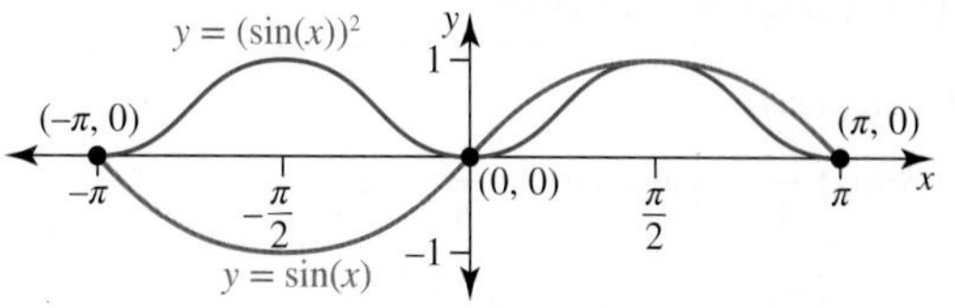

10. a.

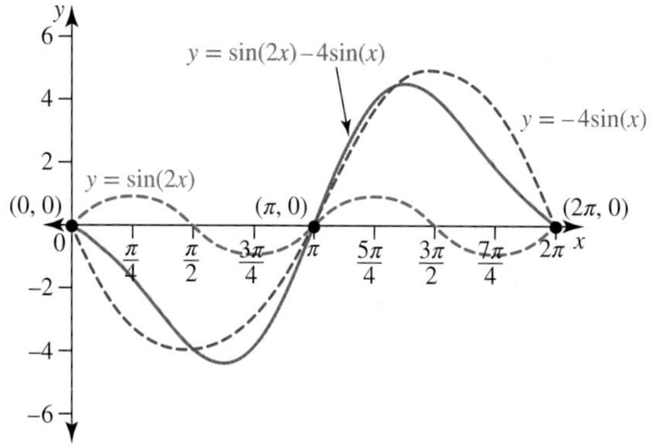

b.

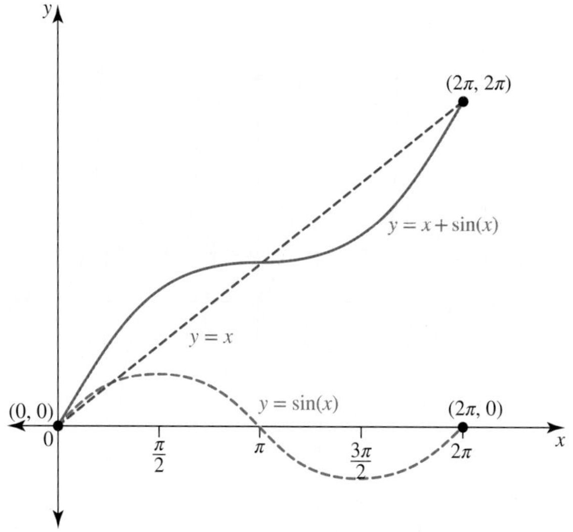

2.5 Exam questions

Note: Mark allocations are available with the fully worked solutions online.

1. B
2. a. $x = \frac{2\pi}{3}, \frac{4\pi}{3}$

   b.

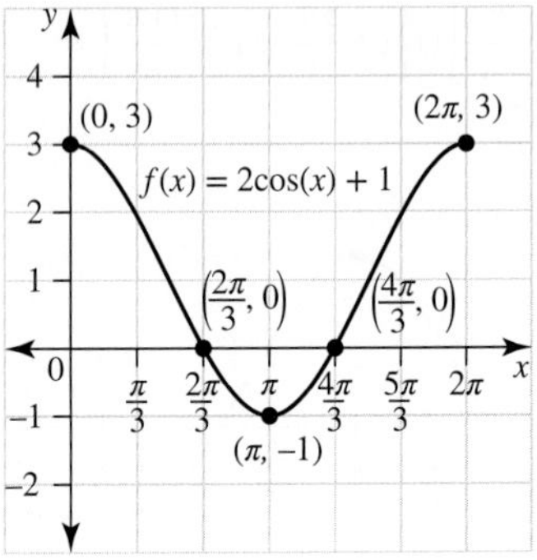

3. B

2.6 The tangent function

2.6 Exercise

1. a. Period $\frac{\pi}{4}$, asymptote $x = \frac{\pi}{8}$

   b. Period 7π, asymptote $x = \frac{7\pi}{2}$

   c. Period $\frac{5\pi}{4}$, asymptote $x = \frac{5\pi}{8}$

   d. Period $\frac{1}{6}$, asymptote $x = \frac{1}{12}$

2.

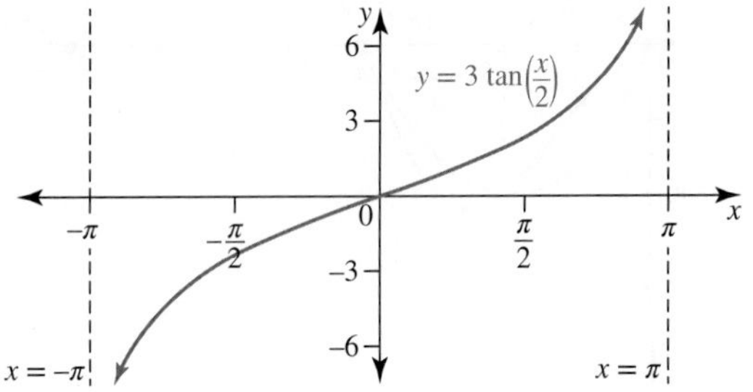

3.

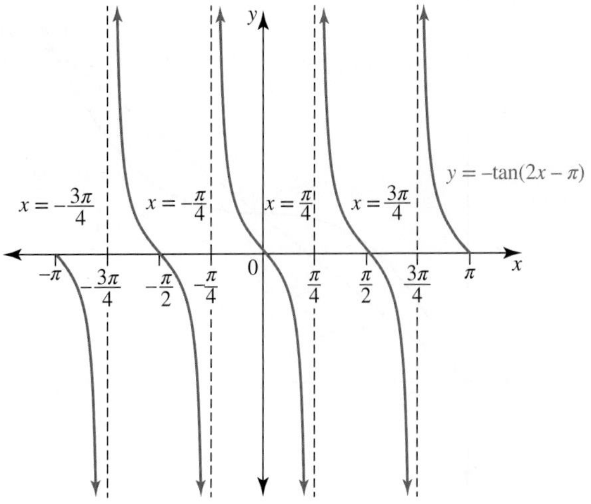

4. a.

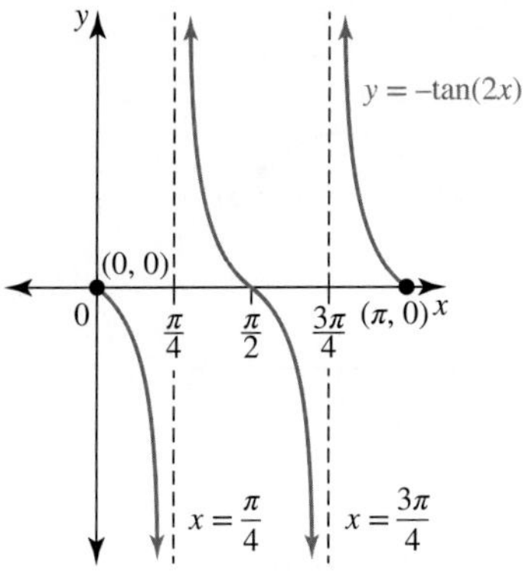

   b.

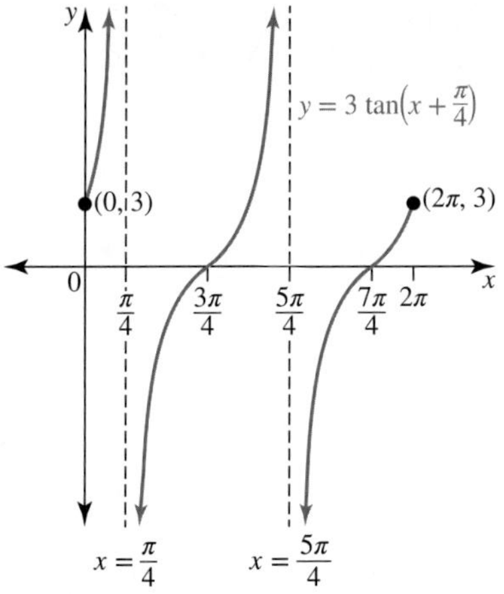

   c.

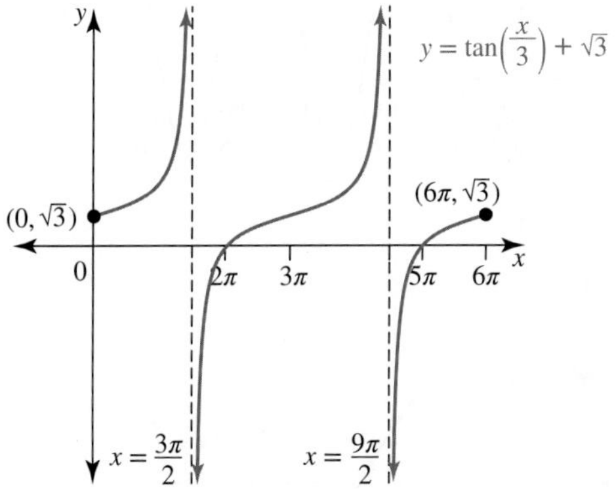

   d.

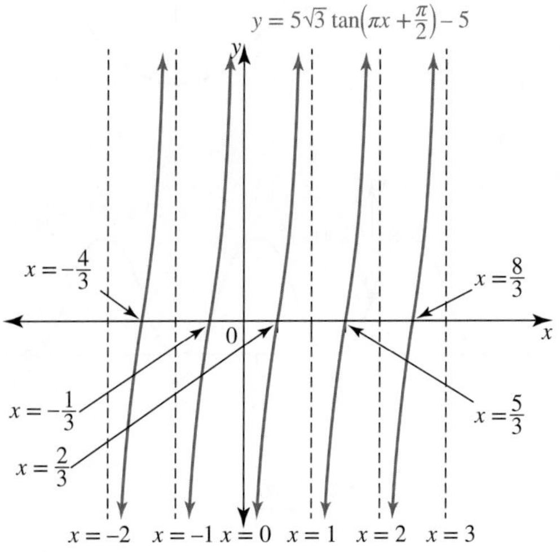

5.

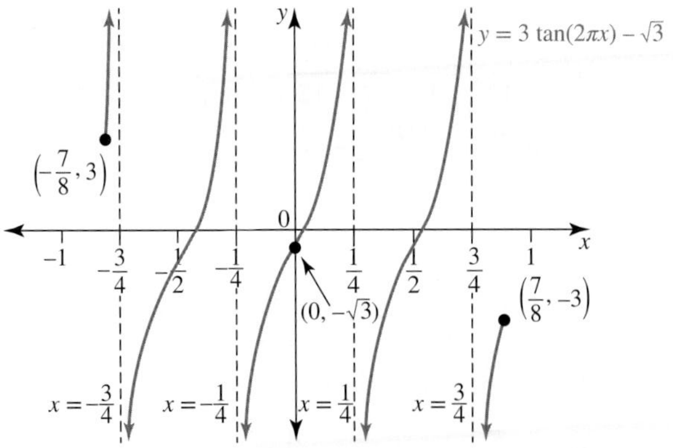

6.

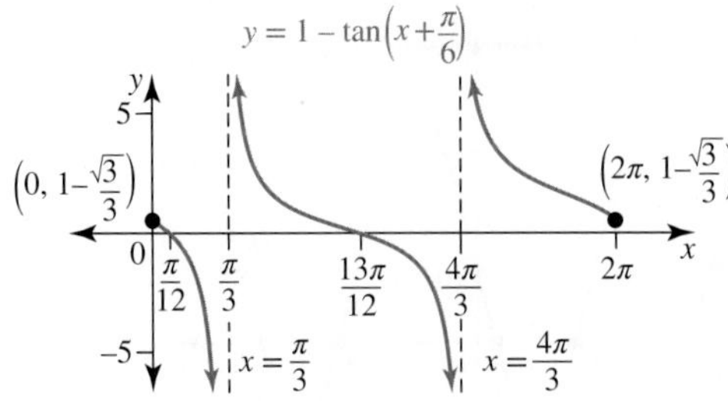

7. $y = \frac{1}{2}\tan\left(\frac{3x}{2}\right)$

8. a. Mean position unaltered

b. $\frac{\pi}{2}$

c. $y = -\tan(2x)$ (other answers possible)

9. a. $(\pm 2\pi, 0)$, $(\pm\pi, 0)$, $(0, 0)$, $\left(-\frac{5\pi}{4}, -1\right)$, $\left(-\frac{\pi}{4}, -1\right)$, $\left(\frac{3\pi}{4}, -1\right)$, $\left(\frac{7\pi}{4}, -1\right)$, $\left(-\frac{7\pi}{4}, 1\right)$, $\left(-\frac{3\pi}{4}, 1\right)$, $\left(\frac{\pi}{4}, 1\right)$, $\left(\frac{5\pi}{4}, 1\right)$.

b. $x = n\pi, n \in Z$ or $x = (2n+1)\frac{\pi}{4}, n \in Z$

10. a. i. $\sqrt{3}$ ii. -1 iii. 1

b.

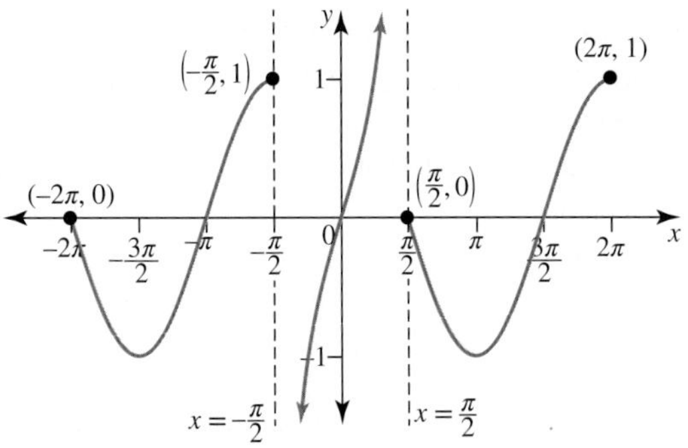

c. Not continuous at $x = \pm\frac{\pi}{2}$

d. Domain $[-2\pi, 2\pi]$, range R

2.6 Exam questions

Note: Mark allocations are available with the fully worked solutions online.

1. B

2. $b = \frac{\pi}{24}$, $a = \frac{7\pi}{24}$

3. C

2.7 Modelling and applications

2.7 Exercise

1. a. 24 hours

b. Maximum 14 metres, minimum 11 metres

c. 12.5 metres

d.

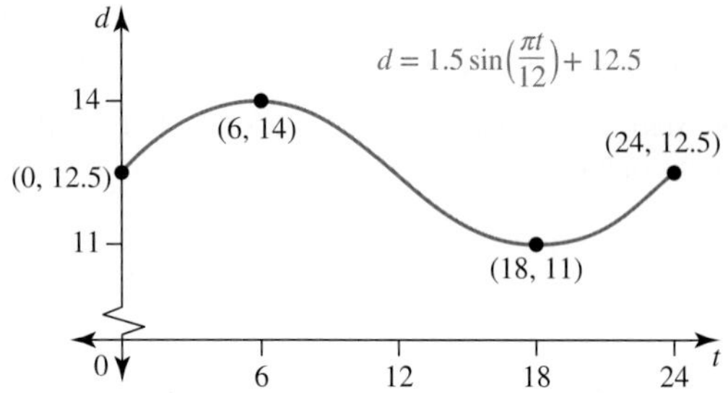

e. $h = \frac{3\sqrt{3} + 50}{4} \approx 13.8$ m

f. 3:30 am the following day

2. a. $h = 50\cos(2\pi(t - 0.5)) + 50$

b. 0.3 seconds

3. a. 1.5 m b. 8.5 m c. 60 s d. 18.4 s

4. a. $a = 0.5, n = \frac{\pi}{2}, k = 1.2$

b.

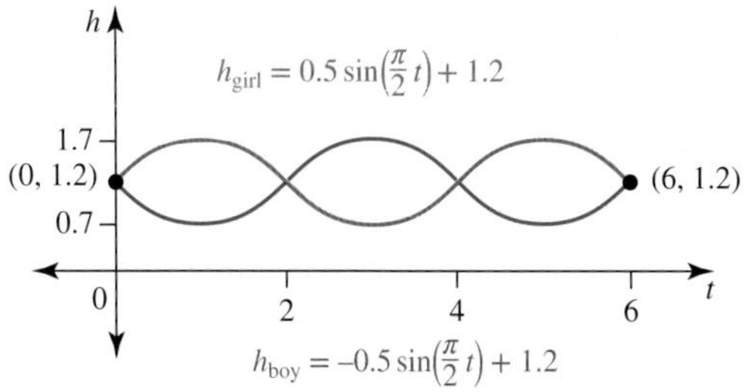

c. $\frac{8}{3}$ seconds

d. $h_{boy} = -0.5\sin\left(\frac{\pi}{2}t\right) + 1.2$. The graph is sketched on the same axes in part **b**.

5. a. 3.619 m below the jetty

b. 5.5 m and 0.5 m

c.

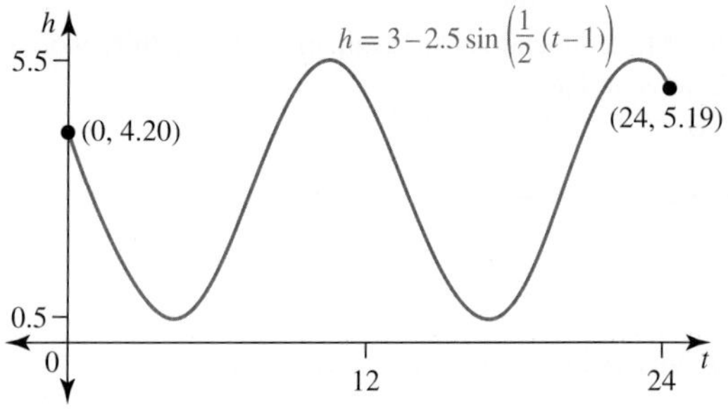

First maximum (high tide) at $t \approx 10.42$, first minimum (low tide) at $t \approx 4.14$

d. 5 m of extra rope

6. a. 33 b. 1 person at 2.28 pm ($t = 6, 46$)
c. 112 d. 86

2.7 Exam questions

Note: Mark allocations are available with the fully worked solutions online.

1. E
2. B
3. a. 6 m b. $t = 14, 22$

2.8 Review

2.8 Exercise

Technology free: short answer

1. a. $-\frac{\sqrt{3}}{2}$ b. -1 c. $-\frac{1}{2}$ d. $-\frac{1}{\sqrt{3}}$

2. a. -0.25 b. $-\frac{\sqrt{15}}{4}$ c. $\frac{1}{\sqrt{15}}$ d. $\frac{\sqrt{15}}{4}$

3. a. i. $\sin(\alpha) = -\frac{\sqrt{5}}{3}$ ii. $\tan(\alpha) = -\frac{\sqrt{5}}{2}$

b. $\sin^2(\alpha) + \cos^2(\alpha) = \left(-\frac{\sqrt{5}}{3}\right)^2 + \left(\frac{2}{3}\right)^2$
$= \frac{5}{9} + \frac{4}{9}$
$= 1$

4. a. $x = -\frac{3\pi}{2}, -\frac{\pi}{2}, \frac{\pi}{2}, \frac{3\pi}{2}$

b. $\theta = \frac{\pi}{4}, \frac{2\pi}{3}, \frac{4\pi}{3}, \frac{7\pi}{4}$

c. $x = -\frac{7\pi}{12}, -\frac{\pi}{12}, \frac{5\pi}{12}, \frac{11\pi}{12}$

d. $\theta = 0, \frac{\pi}{6}, \pi, \frac{11\pi}{6}, 2\pi$

e. $\theta = -\frac{\pi}{2}, -\frac{\pi}{6}, \frac{\pi}{2}, \frac{5\pi}{6}$

f. $x = 2n\pi - \frac{5\pi}{12}, x = 2n\pi + \frac{\pi}{12}$ where $n \in Z$

5. a. Period $\frac{2\pi}{5}$, amplitude 3

b. Period $\frac{\pi}{3}$, asymptote $x = \frac{\pi}{6}$

6. a.

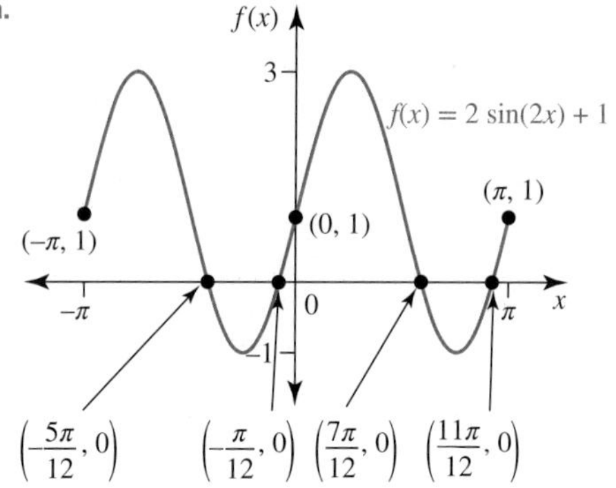

b. See the figure at the bottom of the page.*

Technology active: multiple choice

7. E
8. A
9. B
10. C
11. B
12. A
13. C
14. D
15. E
16. D

Technology active: extended response

17. a. Maximum temperature 8 °C, minimum temperature −4 °C
b. 8:42 am and 11:18 pm
c. 4:00 pm
d. −1 °C

*6. b.

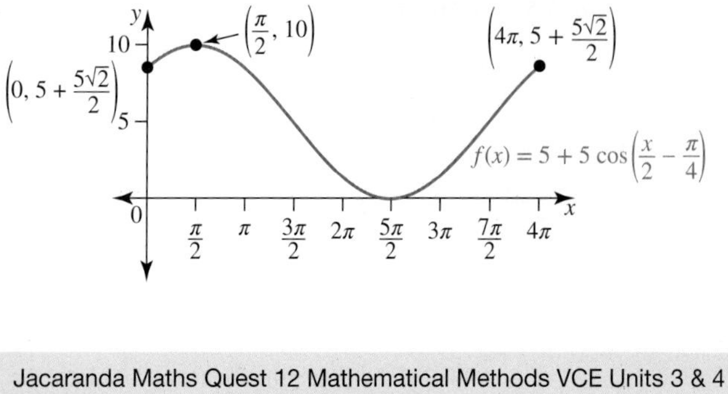

18. a.

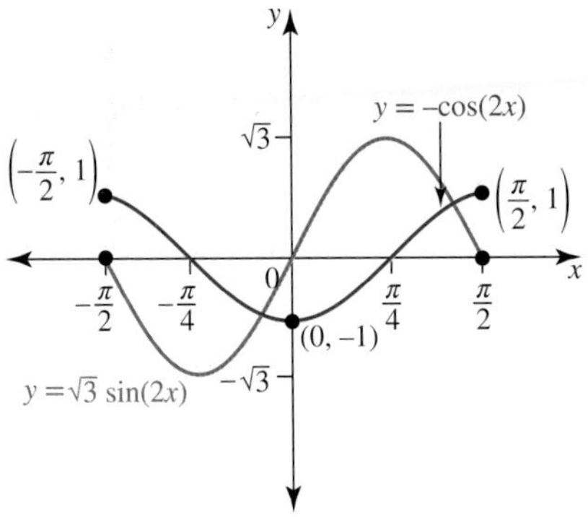

b. $\left(-\frac{\pi}{12}, -\frac{\sqrt{3}}{2}\right)$ and $\left(\frac{5\pi}{12}, \frac{\sqrt{3}}{2}\right)$

c. $\left\{x : -\frac{\pi}{12} \leq x \leq \frac{5\pi}{12}\right\}$

d. i. $f(0) = -1, f\left(\frac{\pi}{2}\right) = 1$

ii.

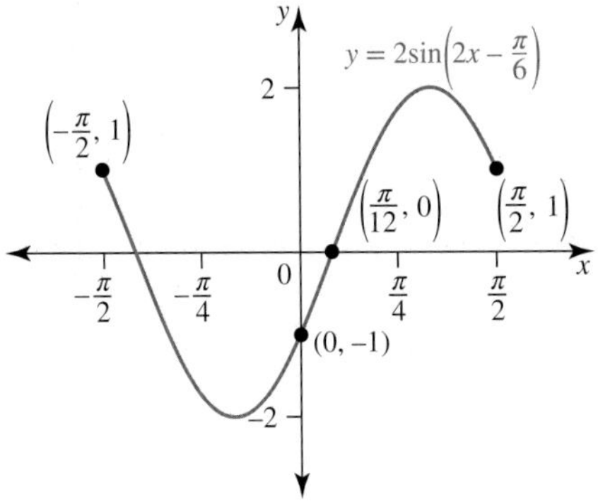

$\left\{x : -\frac{\pi}{2} \leq x \leq -\frac{5\pi}{12}\right\} \cup \left\{x : \frac{\pi}{12} \leq x \leq \frac{\pi}{2}\right\}$

19. a. Maximum depth = 4.5
Minimum depth = 0.5

b. 12 hours

c. 3.5 m

d. From 1:56 pm to 4:04 pm and 1:56 am to 4:04 am

20. a. $T = 3\cos\left(\frac{\pi}{6}(x-6)\right) + 25$

b. 23.5 °C

c. 9.4 hours

2.8 Exam questions

Note: Mark allocations are available with the fully worked solutions online.

1. a. $[-2, 2]$

b. π

c. $x = \frac{\pi}{6} + k\pi, \frac{\pi}{3} + k\pi, k \in z$

2. a. $x = \frac{2\pi}{3}, -\frac{2\pi}{3}$

b.

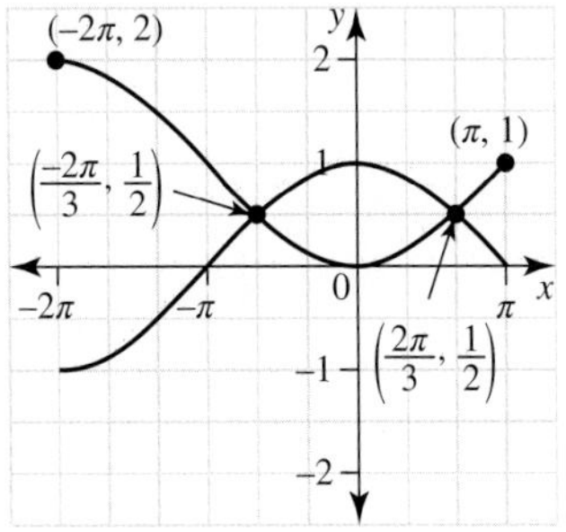

3. C

4. a. $\tan(\theta) = 1, \pm\sqrt{3}$

b. $\theta = \frac{\pi}{4}, \frac{\pi}{3}, \frac{2\pi}{3}$

5. A

3 Composite functions, transformations and inverses

LEARNING SEQUENCE

Fully worked solutions for this topic are available online.

3.1 Overview

Hey students! Bring these pages to life online

Watch videos

Engage with interactivities

Answer questions and check results

Find all this and MORE in jacPLUS

3.1.1 Introduction

Although functions themselves have been around for hundreds of years, it was mathematician Gottfried Wilhelm Leibniz, in 1673, who was the first to use the term 'function' to describe these different mathematical relationships. Around 1734, Leonhard Euler extended this further and introduced the notation for a function, $y = f(x)$. The modern definition of a function — a relation where every x element has a unique y element — was provided in the 1830s by Nikolai Lobachevsky and Peter Dirichlet.

The notation for an inverse function ($y = f^{-1}(x)$) was first used by John Herschel in 1813. Inverse functions are important because they 'undo' the effects of a given function. For example, if you type something into a computer, such as a number, the computer converts this number to binary for internal storage, then converts it back to the number again where it appears on the screen. Converting money from Australian dollars to US dollars and vice versa also uses inverse functions.

KEY CONCEPTS

This topic covers the following key concepts from the VCE Mathematics Study Design:

- graphs of the following functions: power functions, $y = x^n$, $n \in Q$; exponential functions, $y = a^x$, $a \in R^+$ in particular $y = e^x$; logarithmic functions, $y = \log_e(x)$ and $y = \log_{10}(x)$; and circular functions, $y = \sin(x)$, $y = \cos(x)$ and $y = \tan(x)$ and their key features
- transformation from $y = f(x)$ to $y = Af(n(x + b)) + c$, where A, n, b and $c \in R$, $A, n \neq 0$, and the inverse transformation
- the relation between the graph of an original function and the graph of a corresponding transformed function (including families of transformed functions for a single transformation parameter)
- graphs of sum, difference, product and composite functions involving functions of the types specified above (not including composite functions that result in reciprocal or quotient functions)
- functions and their inverses, including conditions for the existence of an inverse function, and use of inverse functions to solve equations involving exponential, logarithmic, circular and power functions
- composition of functions, where f composite g, $f \circ g$, is defined by $(f \circ g)(x) = f(g(x))$ given $r_g \subseteq d_f$
- solution of literal equations and general solution of equations involving a single parameter.

Note: *Concepts shown in grey are covered in other topics.*

3.2 Composite functions

LEARNING INTENTION

At the end of this subtopic you should be able to:
- determine the equations of composite functions
- determine if a composite function exists
- restrict the domain of the inner function, if needed, in order for a composite function to exist.

3.2.1 The composite function

Composition of two functions occurs when the output of one function becomes the input for a second function.

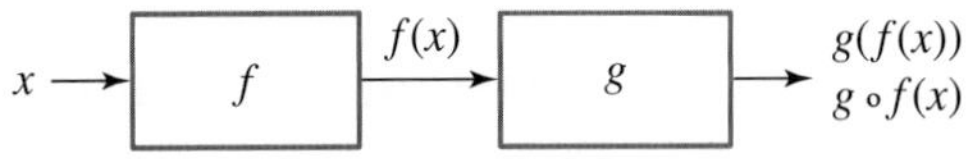

Suppose $f(x) = x^2$ and $g(x) = 3x - 1$.

The **composite function** $f \circ g(x) = f(g(x))$ (pronounced 'f of g of x') involves expressing $f(x)$ in terms of $g(x)$.

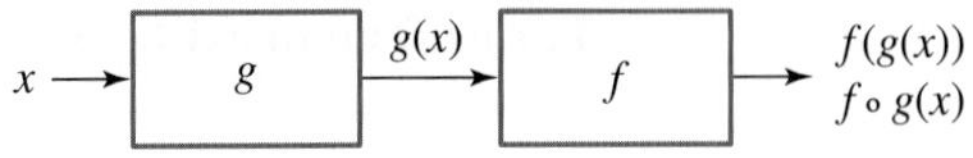

$$\begin{aligned} f \circ g(x) = f(g(x)) &= f(3x-1) \\ &= (3x-1)^2 \\ &= 9x^2 - 6x + 1 \end{aligned}$$

The domain of $f(g(x)) = \text{dom } g(x) = R$.

On the other hand, the composite function $g \circ f(x) = g(f(x))$ involves expressing $g(x)$ in terms of $f(x)$.

$$\begin{aligned} g \circ f(x) = g(f(x)) &= g(x^2) \\ &= 3(x^2) - 1 \\ &= 3x^2 - 1 \end{aligned}$$

The domain of $g(f(x)) = \text{dom } f(x) = R$.

Note that the order of the composition will affect the result of the composition. This means that $f(g(x))$ will generally not equal $g(f(x))$.

WORKED EXAMPLE 1 Composite functions

If $f(x) = \sqrt{3-x}$, $g(x) = 2x^3 - 5$ and $h(x) = x + 4$, determine:

a. $g(h(x))$

b. $h \circ f(x)$

THINK	WRITE
a. Determine $g(h(x))$ by substituting $h(x)$ into $g(x)$.	$g(h(x)) = g(x+4)$ $= 2(x+4)^3 - 5$
b. 1. Rewrite $h \circ f(x)$ as $h(f(x))$.	$h \circ f(x) = h(f(x))$
2. Determine $h(f(x))$ by substituting $f(x)$ into $h(x)$.	$= h\left(\sqrt{3-x}\right)$ $= \sqrt{3-x} + 4$

on Resources

Interactivity Composite functions (int-6417)

3.2.2 Existence of composite functions

When two functions are composed, the output of the first function (the inner function) becomes the input for the second function (the outer function). This means that if the composition is possible, the range of the inner function must be a subset of or equal to the domain of the outer function. It may be necessary to restrict the domain of the inner function to ensure that its range lies completely within the domain of the outer function. The domain of the inner function is always the domain of the composite function.

Existence of composite functions

For $f(g(x))$ to exist, ran $g \subseteq$ dom f. The domain of $g(x) =$ dom $f(g(x))$.

For $g(f(x))$ to exist, ran $f \subseteq$ dom g. The domain of $f(x) =$ dom $g(f(x))$.

Looking again at the functions $f(x) = x^2$ and $g(x) = 3x - 1$, we can investigate why they exist by listing the domains and ranges.

Function	Domain	Range
$f(x)$	R	$[0, \infty)$
$g(x)$	R	R

For $f(g(x))$, the range of g is R, which is equal to the domain of f, R.

$$R \subseteq R$$
$$\text{ran } g \subseteq \text{dom } f$$

Therefore, $f(g(x))$ exists.

For $g(f(x))$, the range of f is $[0, \infty)$, which is a subset of the domain of g, R.

$$[0, \infty) \subseteq R$$
$$\text{ran } f \subseteq \text{dom } g$$

Therefore, $g(f(x))$ exists.

WORKED EXAMPLE 2 Existence of composite functions

If $f(x) = e^x$ and $g(x) = \sqrt{x - 2}$, investigate whether the composite functions $f(g(x))$ and $g(f(x))$ exist. If they do, form the rule for the composite function and state the domain.

THINK

1. Construct a table to investigate the domains and ranges of the two functions.

WRITE

$f(x) = e^x$ and $g(x) = \sqrt{x - 2}$

Function	Domain	Range
$f(x)$	R	$(0, \infty)$
$g(x)$	$[2, \infty)$	$[0, \infty)$

2. Investigate whether $f(g(x))$ exists by comparing the range of g to the domain of f.

$[0, \infty) \subseteq R$
ran $g \subseteq$ dom f
Therefore, $f(g(x))$ exists.

3. Form the rule for $f(g(x))$ and state the domain.

$$f(g(x)) = f\left(\sqrt{x-2}\right)$$
$$= e^{\sqrt{x-2}}$$
Domain = dom g
$= [2, \infty)$

4. Investigate whether $g(f(x))$ exists.

$(0, \infty) \nsubseteq [2, \infty)$
ran $f \nsubseteq$ dom g
Therefore, $g(f(x))$ does not exist.

TI | THINK / **DISPLAY/WRITE**

1. On a Calculator page, press MENU, then select:
1: Actions
1: Define
Complete the entry line as:
Define $f(x) = e^x$
then press ENTER.
Complete the next entry line as:
Define $g(x) = \sqrt{x-2}$
then press ENTER.

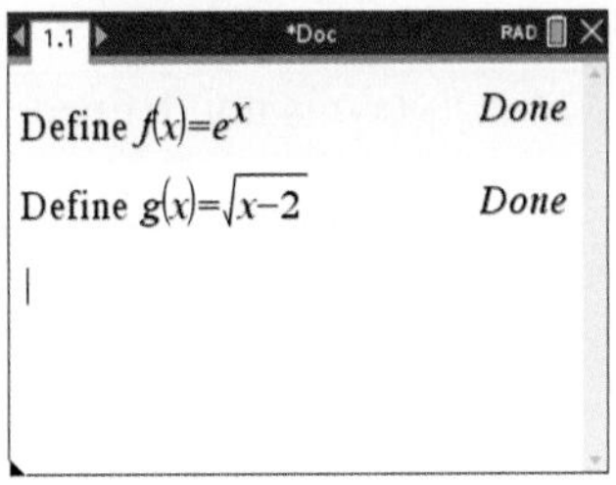

2. Investigate whether $f(g(x))$ and $g(f(x))$ exists.

$[0, \infty) \subseteq R$
ran $g \subseteq$ dom f.
Therefore, $f(g(x))$ exists.
$(0, \infty) \nsubseteq [2, \infty)$
ran $f \nsubseteq$ dom g
Therefore, $g(f(x))$ does not exist.

3. To find $f(g(x))$, complete the next entry line as:
$f(g(x))$
then press ENTER.

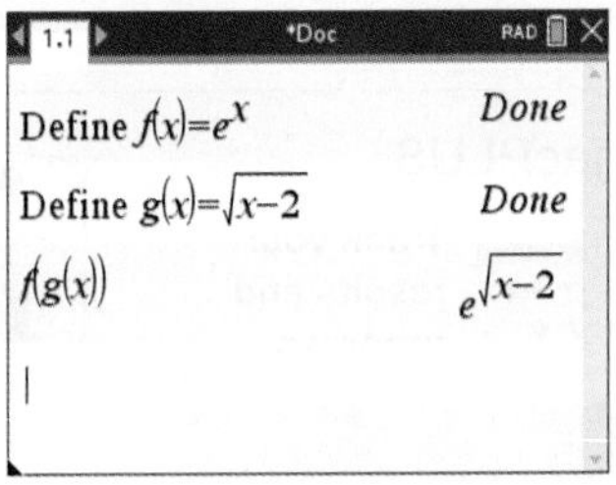

4. The answer appears on the screen.

$f(g(x)) = e^{\sqrt{x-2}}$

CASIO | THINK / **DISPLAY/WRITE**

1. On a Main screen, select:
- Action
- Command
- Define

Complete the entry line as:
Define $f(x) = e^x$
then press EXE.
Complete the next entry line as:
Define $g(x) = \sqrt{x-2}$
then press EXE.

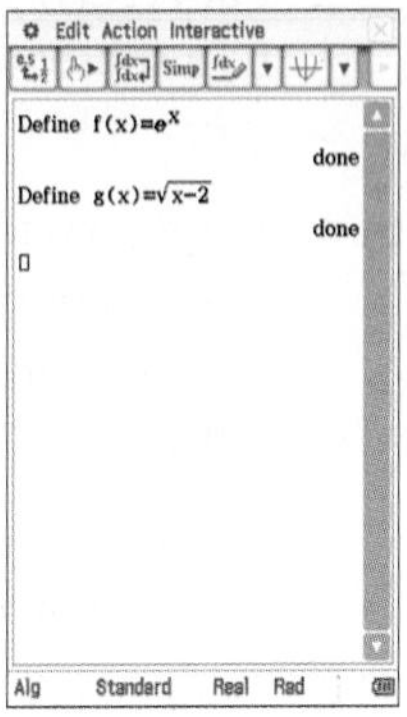

2. Investigate whether $f(g(x))$ and $g(f(x))$ exists.

$[0, \infty) \subseteq R$
ran dom$f$
Therefore, $f(g(x))$ exists.
$(0, \infty) \nsubseteq [2, \infty)$
ran $f \nsubseteq$ dom g
Therefore, $g(f(x))$ does not exist.

3. To find $f(g(x))$, complete the next entry line as:
$f(g(x))$
then press EXE.

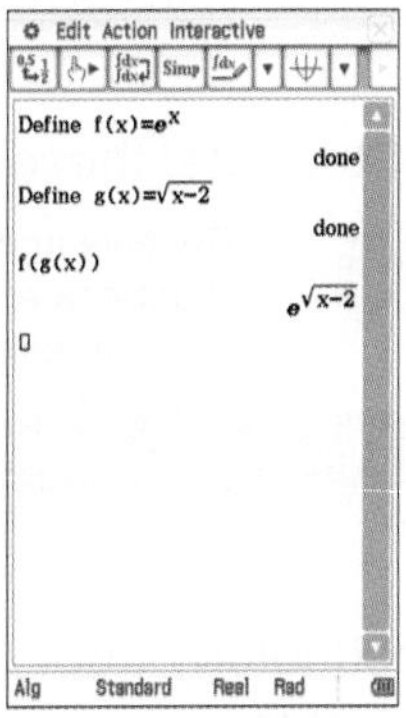

4. The answer appears on the screen.

$f(g(x)) = e^{\sqrt{x-2}}$

WORKED EXAMPLE 3 Restricting the domain

For the functions $f(x) = \sqrt{4 - x}$ and $g(x) = x - 1$:

a. state why $f(g(x))$ is not defined

b. restrict the domain of $g(x)$ to form a new function, $h(x)$, such that $f(h(x))$ is defined

c. determine $f(h(x))$.

THINK	WRITE
a. 1. Construct a table to investigate the domains and ranges of the two functions.	a. $f(x) = \sqrt{4 - x}$ and $g(x) = x - 1$

Function	Domain	Range
$f(x)$	$(-\infty, 4]$	$[0, \infty)$
$g(x)$	R	R

THINK	WRITE
2. For $f(g(x))$ to be defined, the range of g must be a subset of the domain of f.	$R \not\subseteq (-\infty, 4]$ ran $g \not\subseteq$ dom f $\therefore f(g(x))$ is not defined.
b. 1. For $f(g(x))$ to be defined, the range g must be a subset of or equal to the domain of y. The maximal range of g will be when ran $g =$ dom f.	b. We want ran $g =$ dom $f = (-\infty, 4]$.
2. Use the restriction of the range of g to solve for the new domain of g.	$x - 1 \le 4$ $x \le 5$
3. Define $h(x)$.	$h(x) = x - 1, x \in (-\infty, 5]$
c. Find $f(h(x))$ by substituting $h(x)$ into $f(x)$. Make sure the domain is stated.	c. $f(h(x)) = \sqrt{4 - (x - 1)}$ $= \sqrt{5 - x},\ x \in (-\infty, 5]$

3.2 Exercise

Technology free

1. WE1 If $f(x) = -x^2 + 1$, $g(x) = 2x - 3$ and $h(x) = \sqrt{x}$, determine:

   a. $g(h(x))$

   b. $f \circ g(x)$

2. If $f(x) = \dfrac{1}{x + 2}$, $g(x) = (x - 3)^2$ and $h(x) = 1 - x^4$, determine:

   a. $g(h(x))$

   b. $h \circ f(x)$

3. WE2 If $f(x) = (x - 1)(x + 3)$ and $g(x) = x^2$, investigate whether the composite functions $f(g(x))$ and $g(f(x))$ exist. If they do, form the rule for the composite function and state the domain.

4. If $f(x)=2x-1$ and $g(x)=\dfrac{1}{x-2}$, investigate whether the composite functions $f(g(x))$ and $g(f(x))$ exist, and if they do, form the rule for the composite function.

5. For the functions $f(x)=x^2+1$, $g(x)=\sqrt{x}$, and $h(x)=\dfrac{1}{x}$, determine whether the following compositions are defined or undefined. If the composite function exists, identify its domain.

a. $f \circ g(x)$
b. $g(f(x))$
c. $h(g(x))$
d. $h \circ f(x)$

6. For the functions $f(x)=x^2$, $g(x)=\sqrt{x}$, and $h(x)=-\dfrac{1}{x}$, determine whether the following compositions are defined or undefined. If the composite function exists, state the rule and its domain.

a. $f \circ g(x)$
b. $g(f(x))$
c. $h(f(x))$
d. $g(h(x))$

7. The functions f and g are defined by $f: R \to R, f(x)=x^2+1$ and $g: [-2, \infty) \to R, g(x)=\sqrt{x+2}$. Show that $f(g(x))$ exists and find the rule for $f(g(x))$, stating its domain and range.

8. If $f: (0, \infty) \to R, f(x)=\dfrac{1}{x}$ and $g: R \to R, g(x)=\dfrac{1}{x^2}$:

a. prove that $g(f(x))$ exists
b. determine $g(f(x))$ and state its domain and range
c. sketch the graph of $y=g(f(x))$.

9. WE3 For the functions $f(x)=\sqrt{x+3}$ and $g(x)=2x-5$:

a. state why $f(g(x))$ is not defined
b. restrict the domain of g to form a new function, $h(x)$, such that $f(h(x))$ is defined
c. determine $f(h(x))$.

10. For the functions $f(x)=x^2$ and $g(x)=\dfrac{1}{x-4}$:

a. state why $g(f(x))$ is not defined
b. restrict the domain of f to form a new function, $h(x)$, such that $g(h(x))$ is defined
c. determine $g(h(x))$.

Technology active

11. If $g(x)=\dfrac{1}{(x-3)^2}-2$ and $f(x)=\sqrt{x}$:

a. prove that $f(g(x))$ is not defined
b. restrict the domain of g to obtain a function $g_1(x)$ such that $f(g_1(x))$ exists.

12. For the equations $f: (-\infty, 2] \to R,\ f(x)=\sqrt{2-x}$ and $g: R\backslash[1] \to R, g(x)=-\dfrac{1}{x-1}+2$:

a. prove that $g(f(x))$ is not defined
b. restrict the domain of f to obtain a function $f_1(x)$ such that $g(f_1(x))$ exists
c. determine $g(f_1(x))$.

13. For the equations $f: [4, \infty) \to R,\ f(x)=\sqrt{x-4}$ and $g: R \to R, g(x)=x^2-2$:

a. prove that $g(f(x))$ is defined
b. determine the rule for $g(f(x))$ and state the domain
c. sketch the graph of $y=g(f(x))$
d. prove that $f(g(x))$ is not defined
e. restrict the domain of g to obtain a function $g_1(x)$ such that $f(g_1(x))$ exists
f. determine $f(g_1(x))$.

14. If $f: [1, \infty) \to R,\ f(x)=-\sqrt{x}+k$ and $g: (-\infty, 2] \to R, g(x)=x^2+k$, where k is a positive constant, calculate the value(s) for k such that both $f(g(x))$ and $g(f(x))$ are defined.

3.2 Exam questions

Question 1 (1 mark) TECH-ACTIVE

Source: *VCE 2020 Mathematical Methods Exam 2, Section A, Q1; © VCAA.*

MC Let f and g be functions such that $f(-1) = 4$, $f(2) = 5$, $g(-1) = 2$, $g(2) = 7$ and $g(4) = 6$.

The value of $g(f(-1))$ is

A. 2 **B.** 4 **C.** 5 **D.** 6 **E.** 7

Question 2 (1 mark) TECH-ACTIVE

Source: *VCE 2018, Mathematical Methods Exam 2, Section A, Q6; © VCAA.*

MC Let f and g be two functions such that $f(x) = 2x$ and $g(x + 2) = 3x + 1$.

The function $f(g(x))$ is

A. $6x - 5$ **B.** $6x + 1$ **C.** $6x^2 + 1$ **D.** $6x - 10$ **E.** $6x + 2$

Question 3 (1 mark) TECH-ACTIVE

MC If $f(x) = 4x^2$ and $g(x) = 3x + 1$, then $f(g(a))$ is equal to

A. $4(3a + 1)^2$ **B.** $36a^2 + 4$ **C.** $(3a + 1)^2$ **D.** $3a + 1$ **E.** $12a^2 + 1$

More exam questions are available online.

3.3 Transformations

LEARNING INTENTION

At the end of this subtopic you should be able to:
- sketch the graph resulting from a set of transformations
- determine the equation of a transformed function
- describe the transformations that have been applied to a function.

3.3.1 Dilations, reflections and translations

There are three commonly used transformations.

- **Dilations:**
  The point $(1, 2)$ when dilated by factor 2 parallel to the y-axis, or from the x-axis, becomes the point $(1, 4)$.
  The point $(2, 7)$ when dilated by factor $\frac{1}{2}$ parallel to the x-axis, or from the y-axis, becomes the point $(1, 7)$.
- **Reflections:**
  When the point $(1, 2)$ is reflected in the x-axis, it becomes the point $(1, -2)$.
  When the point $(1, 2)$ is reflected in the y-axis, it becomes the point $(-1, 2)$.
- **Translations:**
  The point $(2, 7)$ when translated 2 units in the positive x-direction becomes $(4, 7)$.
  The point $(2, 7)$ when translated 4 units in the negative y-direction becomes $(2, 3)$.

on Resources

Interactivity Transformations of functions (int-2576)

The general rule for transformations

When a function of the form $y = f(x)$ has a number of different transformations applied to it, the general equation becomes

$$y = Af(n\,(x + b)) + c.$$

The following transformations have been applied to $y = f(x)$:

- It has been dilated by a factor of $|A|$ parallel to the y-axis or from the x-axis.
  Each y-value has been multiplied by $|A|$, so each point is now $\left(x,\ |A|y\right)$.
- It has been dilated by a factor of $\frac{1}{|n|}$ parallel to the x-axis or from the y-axis.
  Each x-value has been multiplied by $\frac{1}{|n|}$, so each point is now $\left(\frac{x}{|n|}, y\right)$.
- If A is negative, the graph has been reflected in the x-axis.
  Each y-value has changed sign, so each point is now $(x, -y)$.
- If n is negative, the graph has been reflected in the y-axis.
  Each x-value has changed sign, so each point is now $(-x, y)$.
- It has been translated b units parallel to the x-axis.
  Each x-value has increased by b, so each point is now $(x - b, y)$.
- It has been translated c units parallel to the y-axis.
  Each y-value has increased by c, so each point is now $(x, y + c)$.

As a general rule, when functions are written in the form $y = Af(n(x + b)) + c$, transformations should be read from left to right, as the order is important. Sometimes there may be more than one way to describe the order of transformations, but reading from left to right is a consistent and safe approach.

WORKED EXAMPLE 4 Key transformations

i. **Describe the transformation that has been applied to the graph of $y = x^2$ in each of the following examples. Sketch both graphs on the one set of axes.**
ii. **Determine the image of the point (2, 4) after it has undergone each of the transformations.**

a. $y = 2x^2$ b. $y = (2x)^2$ c. $y = -x^2$ d. $y = (x + 1)^2$ e. $y = (x - 2)^2$ f. $y = x^2 - 2$

THINK	WRITE
a. i. Specify the transformation that has been applied to $y = x^2$, then sketch both graphs on the one set of axes.	a. $y = x^2$ has been dilated by factor 2 parallel to the y-axis or from the x-axis. 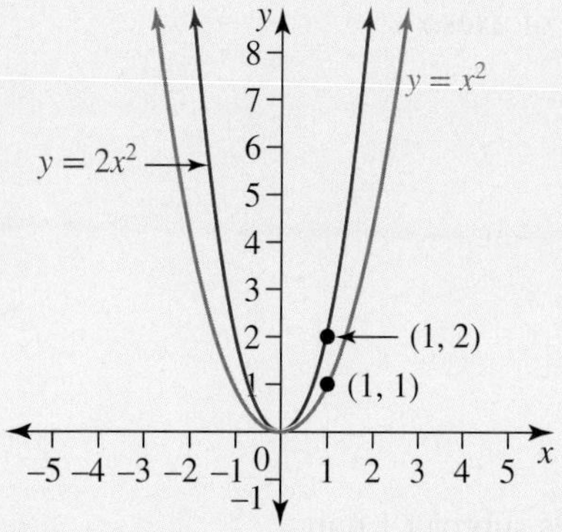
ii. Each y-value is doubled for its corresponding x-value.	$(2, 4) \rightarrow (2, 8)$

b. i. Specify the transformation that has been applied to $y = x^2$, then sketch both graphs on the one set of axes.

b. $y = x^2$ has been dilated by factor $\frac{1}{2}$ parallel to the x-axis or from the y-axis.

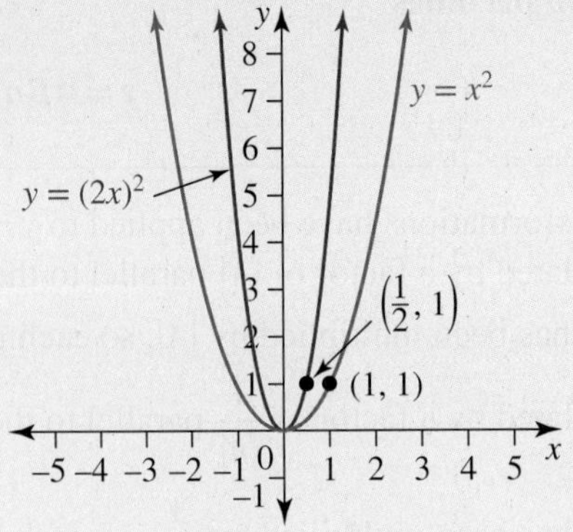

ii. Each x-value is halved for its corresponding y-value.

$(2, 4) \rightarrow (1, 4)$

c. i. Specify the transformation that has been applied to $y = x^2$, then sketch both graphs on the one set of axes.

c. $y = x^2$ has been reflected in the x-axis.

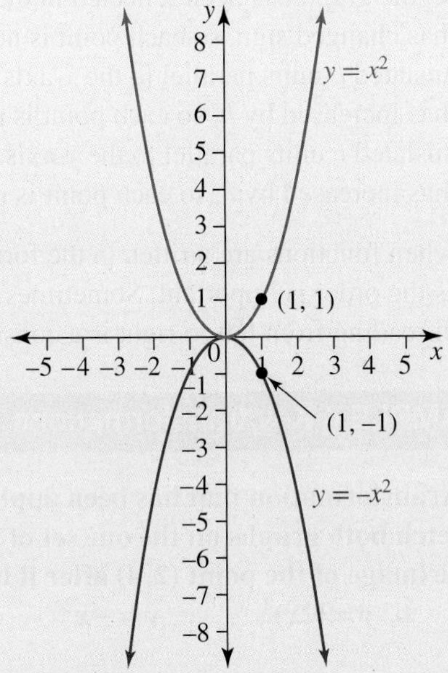

ii. All y-values change sign.

$(2, 4) \rightarrow (2, -4)$

d. i. Specify the transformation that has been applied to $y = x^2$, then sketch both graphs on the one set of axes.

d. $y = x^2$ has been translated 1 unit to the left.

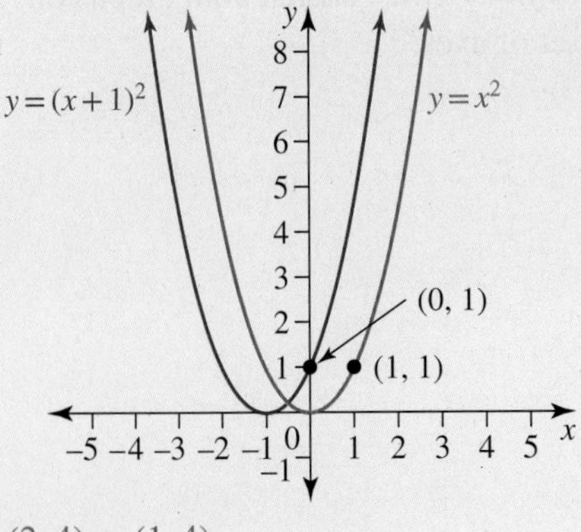

ii. All x-values subtract 1 unit.

$(2, 4) \rightarrow (1, 4)$

e. i. Specify the transformation that has been applied to $y = x^2$, then sketch both graphs on the one set of axes.

e. $y = x^2$ has been translated 2 units to the right.

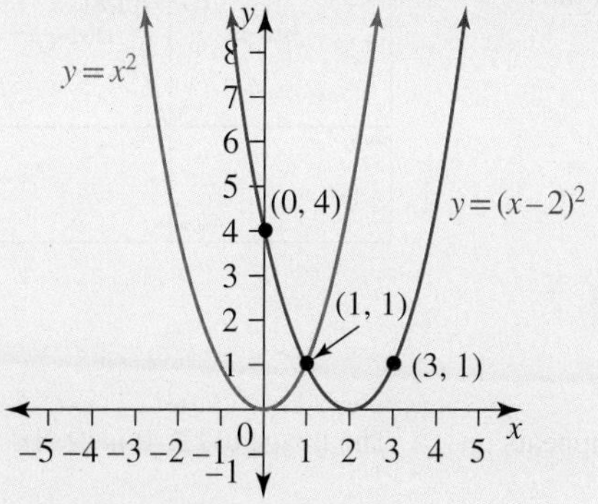

ii. All x-values gain 2 units.

$(2,4) \to (4,4)$

f. i. Specify the transformation that has been applied to $y = x^2$, then sketch both graphs on the one set of axes.

f. $y = x^2$ has been translated down 2 units.

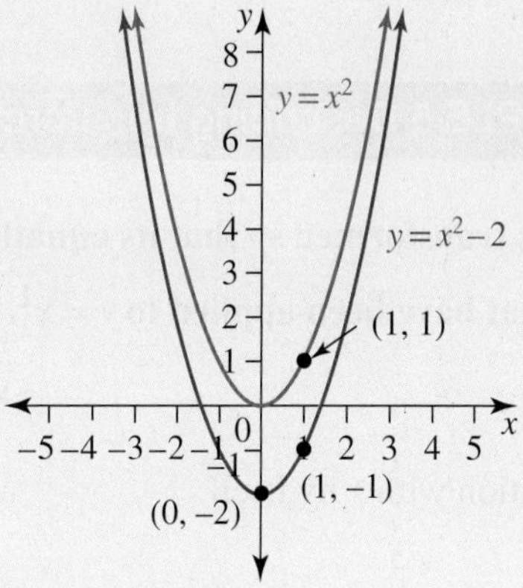

ii. All y-values subtract 2 units.

$(2,4) \to (2,2)$

TI | THINK

a. 1. On a Graphs page, complete the entry line for function 1 as:
$f1(x) = x^2$
then press ENTER.
Complete the entry line for function 2 as:
$f2(x) = x^2$
then press ENTER.

DISPLAY/WRITE

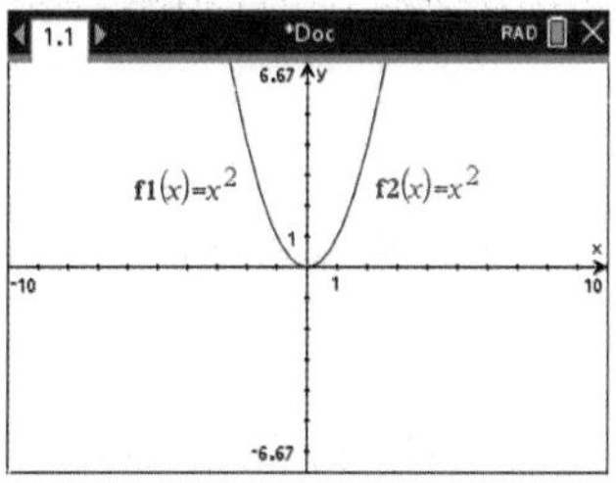

2. Press MENU, then select:
5: Trace
1: Graph Trace
Select the graph of function 1, then type '2' and press ENTER twice. Use the up/down arrows to move to the graph of function 2, then press ENTER.

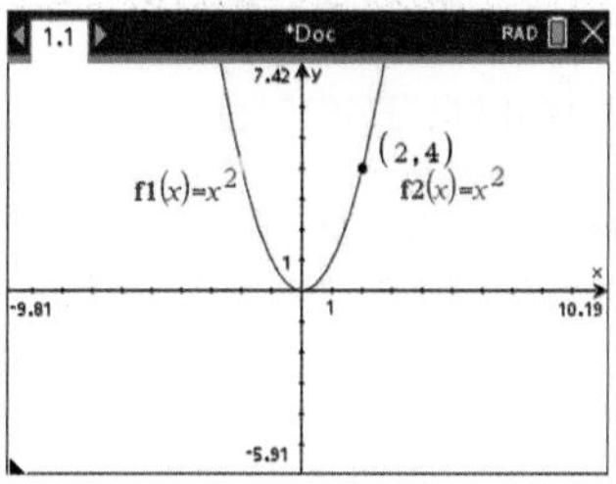

CASIO | THINK

a. 1. On a Graph & Table screen, complete the entry line for y1 as:
$y1 = x^2$
then press EXE.
complete the entry line for y2 as:
$y2 = x^2$
then press EXE.

DISPLAY/WRITE

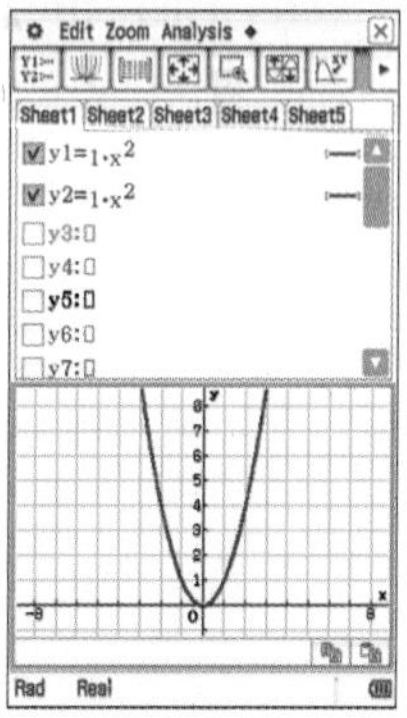

2. Select:
- Analysis
- Trace

then type '2', select OK and press EXE.

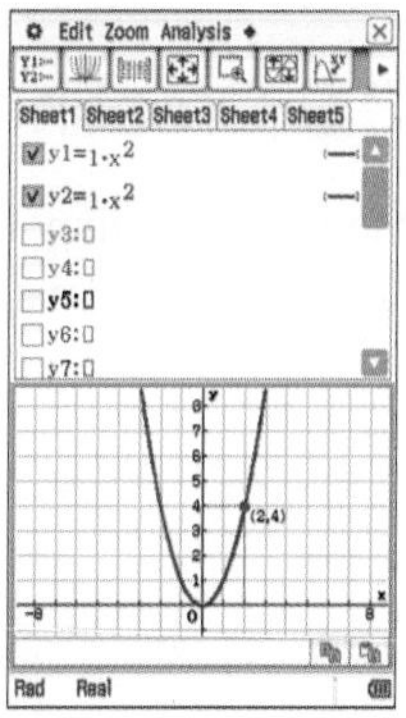

3. Click and drag the graph of function 2 until its equation becomes $f2(x) = 2x^2$.	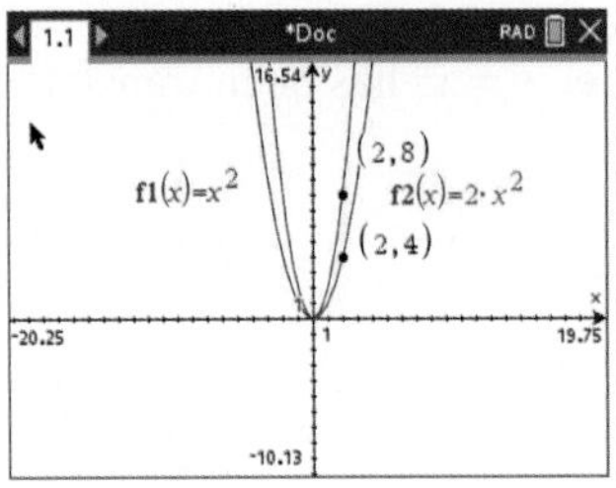	3. Select: • Analysis • Modify then set the Step value to 1 and select OK. Highlight the '1' in the function entry line at the bottom of the screen and press EXE, then use the left/right arrows modify the graph until its equation becomes $2x^2$.	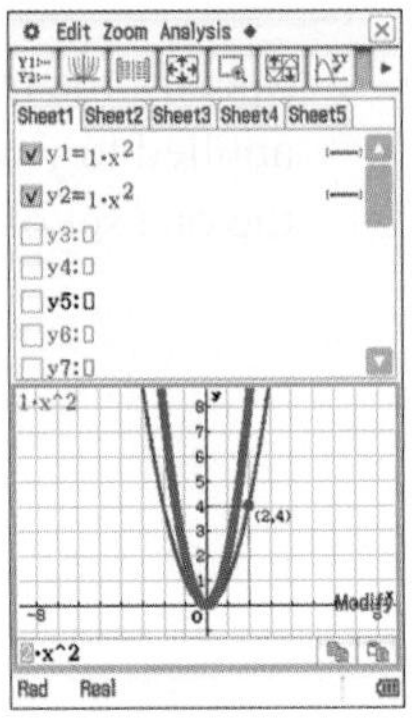
4. The answer appears on the screen.	The image of $(2, 4)$ is $(2, 8)$.	4. The image of $(2, 4)$ can be read from the screen.	The modified graph goes through the point $(2, 8)$; hence, the image of $(2, 4)$ is $(2, 8)$.

WORKED EXAMPLE 5 Describing transformations

The graph $y = x^2$ is transformed so that its equation becomes $y = \frac{1}{2}(2x + 3)^2 - 4$. Define the transformations that have been applied to $y = x^2$.

THINK	WRITE
1. Rewrite the equation with x by itself.	$y = \frac{1}{2}\left(2\left(x + \frac{3}{2}\right)\right)^2 - 4$
2. Define the transformations, reading from left to right. *Note*: There are other ways of writing the order of the transformations. However, the best method is to read the transformations from left to right.	$y = x^2$ has been: 1. dilated by factor $\frac{1}{2}$ parallel to the y-axis or from the x-axis 2. dilated by factor $\frac{1}{2}$ parallel to the x-axis or from the y-axis 3. translated $\frac{3}{2}$ units to the left or $\frac{3}{2}$ units in the negative x-direction 4. translated 4 units down or 4 units in the negative y-direction.

3.3.2 Creating equations from transformations

Sometimes you may be asked to build up an equation from a series of transformations. In these cases, you must apply the transformation in the order that it is mentioned.

WORKED EXAMPLE 6 Equations resulting from transformations

The graph of $g(x) = \sqrt{x}$ undergoes the following transformations:

- **translation 2 units right**
- **dilation of factor 3 from the x-axis**
- **dilation of factor $\frac{1}{2}$ from the y-axis**
- **translation 1 unit down**
- **reflection in the x-axis.**

Determine the rule of the image of the graph.

THINK	WRITE
1. Translation 2 units right means to replace x with $x-2$ in the equation.	$\sqrt{x} \to \sqrt{x-2}$
2. Dilation of factor 3 from the x-axis means multiply the equation by 3.	$\sqrt{x-2} \to 3\sqrt{x-2}$
3. Dilation of factor $\frac{1}{2}$ from the y-axis means replace x with $2x$ in the equation.	$3\sqrt{x-2} \to 3\sqrt{2x-2}$
4. Translation 1 unit down means to subtract 1 unit from the equation.	$3\sqrt{2x-2} \to 3\sqrt{2x-2} - 1$
5. Reflection in the x-axis means to multiply the equation through by –1.	$3\sqrt{2x-2} - 1 \to -3\sqrt{2x-2} + 1$
6. Write the final answer. *Note:* $g(x)$ has not been used to denote the transformed equation because $g(x)$ has already been defined as $g(x) = \sqrt{x}$.	$g(x) = \sqrt{x} \to h(x) = -3\sqrt{2x-2} + 1$

3.3 Exercise

Technology free

1. **WE4** i. Describe the transformation that has been applied to the graph of $y = x^3$ in each of the following examples. Sketch both graphs on the one set of axes.

   ii. Determine the image of the point $(-2, -8)$ after it has undergone each of the following transformations.

   a. $y = 3x^3$ **b.** $y = (x+2)^3$ **c.** $y = -x^3$ **d.** $y = x^3 + 1$

2. Describe the transformation that has been applied to the graph of $y = \sin(x)$ for $x \in [0, 2\pi]$ in each of the following cases. In each case, sketch both graphs on the one set of axes.

   a. $y = 4\sin(x)$ **b.** $y = \sin(2x)$ **c.** $y = \sin\left(x + \frac{\pi}{2}\right)$ **d.** $y = \sin(x) + 2$

3. Describe and sketch the transformation that has been applied to the graph of $y = \cos(x)$ for $x \in [0, 2\pi]$ in each of the following cases.

a. $y = \frac{1}{2}\cos(x)$ b. $y = \cos(2x)$ c. $y = -\cos(x)$ d. $y = \cos(x) - 1$

4. Describe and sketch the transformation that has been applied to the graph of $f(x) = \frac{1}{x}$ in each of the following cases. Give the equations of any asymptotes in each case.

a. $y = f(x - 2)$ b. $y = -f(x)$ c. $y = 3f(x)$ d. $y = f(2x)$

5. **WE5** The graph of $y = \sin(x)$ is transformed so that its equation becomes $y = -2\sin\left[2x - \frac{\pi}{2}\right] + 1$. Define the transformations that have been applied to $y = \sin(x)$.

6. The graph of $y = e^x$ is transformed so that its equation becomes $y = \frac{1}{3}e^{\left(\frac{x+1}{2}\right)} - 2$. Define the transformations that have been applied to $y = e^x$.

7. State the sequence of transformations that has been applied to the first function in order to achieve the transformed function.

a. $y = x^2 \rightarrow y = \frac{1}{3}(x+3)^2 - \frac{2}{3}$ b. $y = x^3 \rightarrow y = -2(1-x)^3 + 1$ c. $y = \frac{1}{x} \rightarrow y = \frac{3}{(2x+6)} - 1$

8. For the corresponding sequence of transformations in question 7, determine the image of the point:

a. $(-2, 4)$ b. $(1, 1)$ c. $\left(2, \frac{1}{2}\right)$.

9. State the sequence of transformations that has been applied to the first function in order to obtain the second function.

a. $y = \cos(x) \rightarrow y = 2\cos\left[2\left(x - \frac{\pi}{2}\right)\right] + 3$ b. $y = \tan(x) \rightarrow y = -\tan(-2x) + 1$

c. $y = \sin(x) \rightarrow y = \sin(3x - \pi) - 1$

Technology active

10. **WE6** **MC** The graph of $g(x) = x^2$ undergoes the following transformations:
- reflection in the y-axis
- translation 4 units right
- dilation of factor 2 from the y-axis
- translation 3 units down
- dilation of factor $\frac{1}{3}$ from the x-axis.

Determine the rule of the image of the graph.

A. $f(x) = \frac{1}{3}\left(\frac{x-8}{2}\right)^2 - 1$ B. $f(x) = -\frac{1}{3}(2x-4)^2 - 3$ C. $f(x) = -\frac{1}{3}\left(\frac{x-8}{2}\right)^2 - 1$

D. $f(x) = 3\left(\frac{x-8}{2}\right)^2 - 3$ E. $f(x) = \frac{1}{3}\left(\frac{x-4}{2}\right)^2 - 3$

11. **MC** The graph of $h(x) = \frac{1}{x}$ undergoes the following transformations:

- dilation of factor 3 parallel to the x-axis
- translation 2 units up
- reflection in the y-axis
- translation 1 unit left
- reflection in the x-axis.

Determine the rule of the image of the graph.

A. $f(x) = \frac{3}{x+1} + 2$ B. $f(x) = -\frac{3}{1-x} - 2$ C. $f(x) = \frac{1}{3(x+1)} - 2$

D. $f(x) = \frac{3}{x+1} - 2$ E. $f(x) = \frac{1}{3x+1} + 2$

12. The graph of $h(x) = \sqrt[3]{x}$ undergoes the following transformations:

reflection in the y-axis, then a translation of 3 units in the positive x-direction, followed by a dilation of factor 2 parallel to the x-axis.

Determine the rule of the image of the graph.

13. The graph of $h(x) = \frac{1}{x^2}$ undergoes the following transformations:

translation of 2 units left and 3 units down, then a reflection in the x-axis, followed by a dilation of factor 3 from the x-axis, and a reflection in the y-axis.

Determine the rule of the image of the graph.

14. The graph of $h(x) = 2x^2 - 3$ undergoes the following transformations:

reflection in the x-axis, then a dilation of factor $\frac{1}{3}$ from the y-axis, followed by a translation of 1 unit in the positive x-direction and 2 units in the negative y-direction.

Determine the rule of the image of the graph.

15. The graph of $h(x) = \frac{1}{x+2}$ undergoes the following transformations:

dilation of factor $\frac{1}{2}$ parallel to the x-axis, then a translation of 3 units down and 3 units left, then a reflection in the y-axis, followed by a dilation of factor 2 from the x-axis.

Determine the rule of the image of the graph.

16. Show that $\frac{2x-5}{x-1} = 2 - \frac{3}{x-1}$ and hence describe the transformations that have been applied to $y = \frac{1}{x}$. Sketch the graph of $y = \frac{2x-5}{x-1}$. State the domain and range and give the equations of any asymptotes.

17. State the transformations that have been applied to the first function in order to obtain the second function. *Hint:* Remember dilations can affect translations.

$$y = 3 - \sqrt{\frac{5-x}{2}} \rightarrow y = \sqrt{x}$$

18. State the transformations that have been applied to the first function in order to obtain the second function.

$$y = -2(3x-1)^2 + 5 \rightarrow y = (x+2)^2 - 1$$

3.3 Exam questions

Question 1 (1 mark) TECH-ACTIVE

Source: *VCE 2019, Mathematical Methods Exam 2, Section A, Q13; © VCAA.*

MC The graph of the function f passes through the point $(-2, 7)$.

If $h(x) = f\left(\frac{x}{2}\right) + 5$, then the graph of the function h must pass through the point

A. $(-1, -12)$ **B.** $(-1, 19)$ **C.** $(-4, 12)$ **D.** $(-4, -14)$ **E.** $(3, 3.5)$

Question 2 (1 mark) TECH-ACTIVE

Source: *VCE 2016, Mathematical Methods Exam 2, Section A, Q12; © VCAA.*

MC The graph of a function f is obtained from the graph of the function g with rule $g(x) = \sqrt{2x-5}$ by a reflection in the x-axis followed by a dilation from the y-axis by a factor of $\frac{1}{2}$.

Which one of the following is the rule for the function f?

A. $f(x) = \sqrt{5-4x}$ **B.** $f(x) = \sqrt{x-5}$ **C.** $f(x) = \sqrt{x+5}$
D. $f(x) = -\sqrt{4x-5}$ **E.** $f(x) = -\sqrt{4x-10}$

Question 3 (1 mark) TECH-ACTIVE

Source: *VCE 2015, Mathematical Methods (CAS) Exam 2, Section 1, Q11; © VCAA.*

MC The transformation that maps the graph of $y = \sqrt{8x^3+1}$ onto the graph of $y = \sqrt{x^3+1}$ is a

A. dilation by a factor of 2 from the y-axis.

B. dilation by a factor of 2 from the x-axis.

C. dilation by a factor of $\frac{1}{2}$ from the x-axis.

D. dilation by a factor of 8 from the y-axis.

E. dilation by a factor of $\frac{1}{2}$ from the y-axis.

More exam questions are available online.

3.4 Inverse graphs

LEARNING INTENTION

At the end of this subtopic you should be able to:

- define an inverse
- state the conditions required for an inverse function
- sketch the graph of an inverse function.

Note: Relations are not specifically covered in the VCE Mathematics Study Design. However, the content has been included in this topic to help with understanding functions.

3.4.1 Inverses

The **relation** $A = \{(-1, 4), (0, 3), (1, 5)\}$ is formed by the mapping

$$-1 \to 4$$
$$0 \to 3$$
$$1 \to 5$$

The **inverse relation** is formed by the 'undoing' mapping:

$$4 \rightarrow -1$$
$$3 \rightarrow 0$$
$$5 \rightarrow 1$$

The inverse of A is the relation $\{(4, -1), (3, 0), (5, 1)\}$.

The x- and y-coordinates of the points in relation A have been interchanged in its inverse. This causes the domains and ranges to be interchanged also.

The domain of $A = \{-1, 0, 1\}$ = the range of its inverse, and the range of $A = \{3, 4, 5\}$ = the domain of its inverse.

Inverses

- **For any relation, the inverse is obtained by interchanging the x- and y-coordinates of the ordered pairs.**
- **Domains and ranges are interchanged between a pair of inverse relations.**

Graphs of inverses

When finding the inverse of a relation graphically, we reflect the relation in the line $y = x$. Consider the equation $y = 2x + 1$.

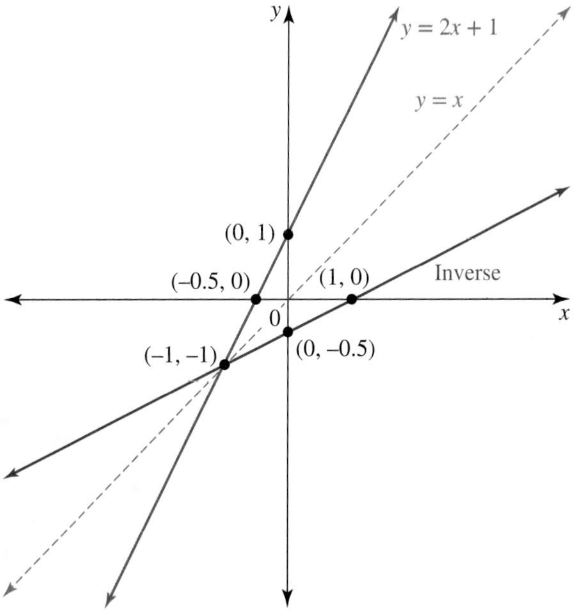

The line $y = x$ acts as a mirror. The inverse is the same distance from the line as the original graph. The coordinates of known points, such as the axial intercepts, are interchanged by this reflection through the mirror. We can see that the line $y = 2x + 1$ cuts the y-axis at $(0, 1)$ and the x-axis at $(-0.5, 0)$, whereas the inverse graph cuts the x-axis at $(1, 0)$. When sketching a graph and its inverse, the line $y = x$ should always be sketched. If the graphs intersect, they will do so on the line $y = x$, since interchanging the coordinates of any point on $y = x$ would not cause any alteration to the coordinates.

WORKED EXAMPLE 7 Sketching inverses

a. Sketch the graph of $y = x^2 - 2x - 3$.
b. On the same set of axes, sketch the graph of the inverse.
c. State the type of mapping for the parabola and its inverse, and whether the relations are functions.
d. Give the domain and range for each of the graphs.

THINK

a. 1. Determine where the given function cuts the x- and y-axes.

WRITE

a. y-intercept, $x = 0$:
$y = -3$
x-intercept, $y = 0$:
$x^2 - 2x - 3 = 0$
$(x - 3)(x + 1) = 0$
$x = 3$ or $x = -1$
A turning point occurs when $x = 1$:
$y = (1)^2 - 2(1) - 3$
$= -4$
$\therefore$ TP $= (1, -4)$

2. Sketch the graph of the parabola.
Note: When sketching graphs and their inverses, the scales on both axes need to be relatively accurate so that distortions do not occur.

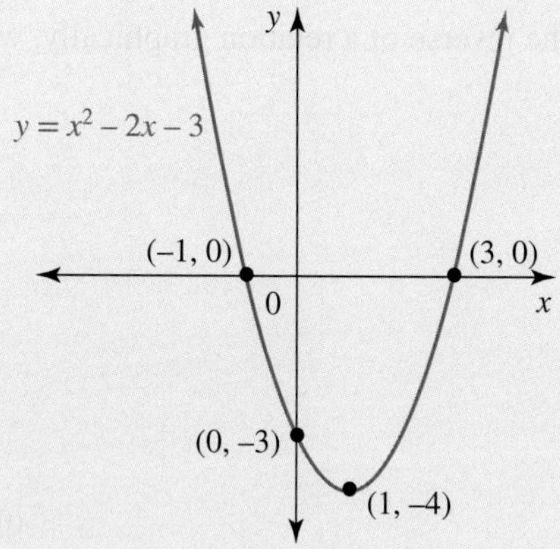

b. On the same set of axes, sketch the inverse by interchanging the coordinates of all important points such as axial intercepts and the turning point.

b.

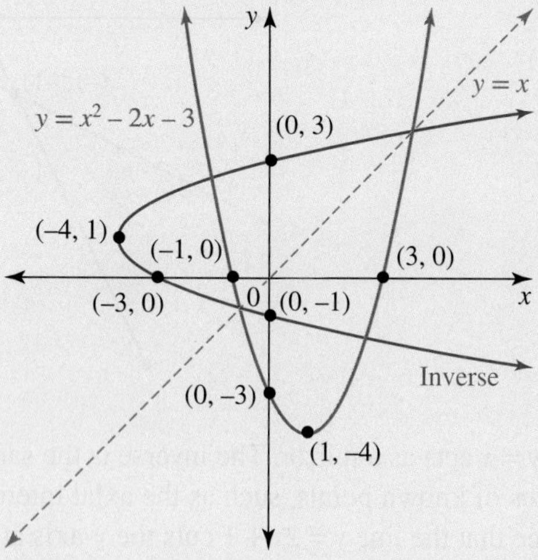

c. Comment on the types of mapping for the two graphs.

c. The parabola $y = x^2 - 2x - 3$ is a many-to-one mapping, so it is a function. However, the inverse is a one-to-many mapping and as such is not a function.

d. State the domains and ranges for both graphs. Remember that the domain of the original graph becomes the range of the inverse and vice versa.

d. For $y = x^2 - 2x - 3$:
Domain $= R$
Range $= [-4, \infty)$
For the inverse:
Domain $= [-4, \infty)$
Range $= R$

TI | THINK	DISPLAY/WRITE
a. 1. On a Graphs page, complete the entry line for function 1 as: $f1(x) = x^2 - 2x - 3$ then press ENTER.	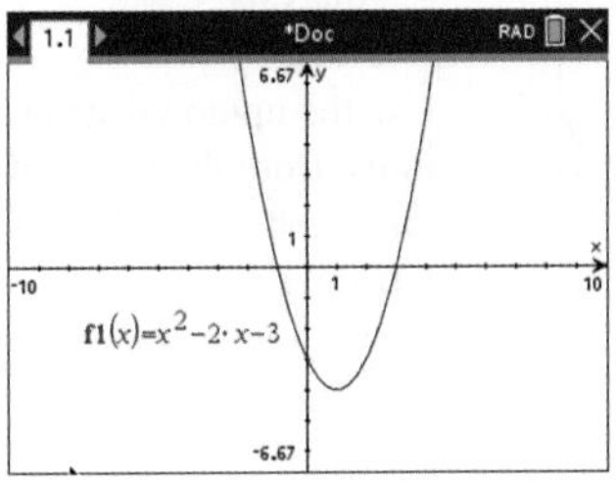
2. To find the x-intercepts, press MENU, then select: 6: Analyze Graph 1: Zero Move the cursor to the left of the x-intercept when prompted for the lower bound, then press ENTER. Move the cursor to the right of the x-intercept when prompted for the upper bound, then press ENTER. Repeat this step to find the other x-intercept.	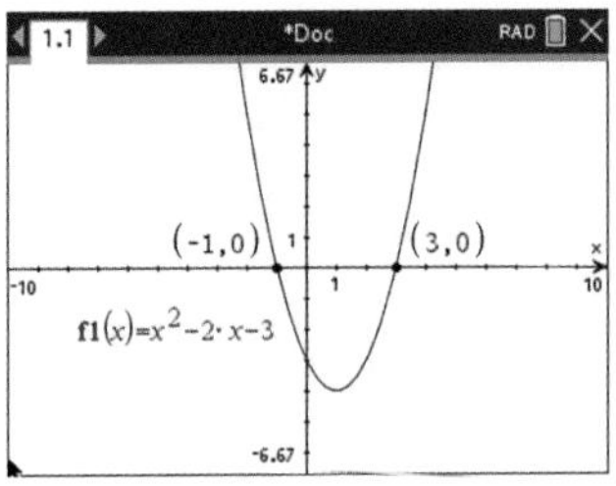
3. To find the y-intercept, press MENU, then select: 5: Trace 1: Graph Trace Type '0', then press ENTER twice.	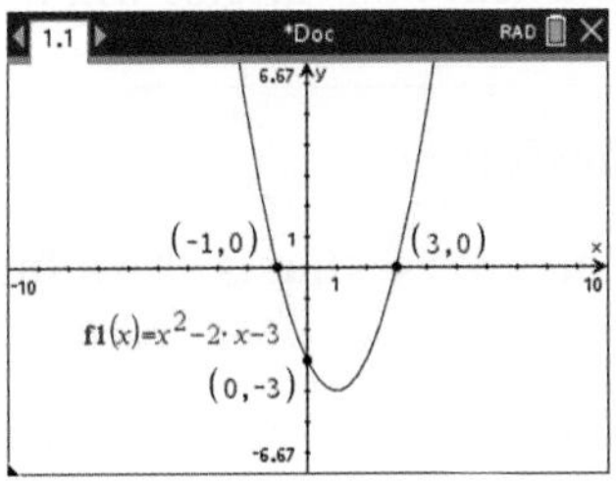

CASIO | THINK	DISPLAY/WRITE
a. 1. On a Graph & Table screen, complete the entry line for y1 as: $y1 = x^2 - 2x - 3$ then press EXE. Select the $y =$ icon and change the input format to $x =$. Complete the entry line for $x2$ as: $x2 = y1(y)$ then press EXE. Select the graph icon.	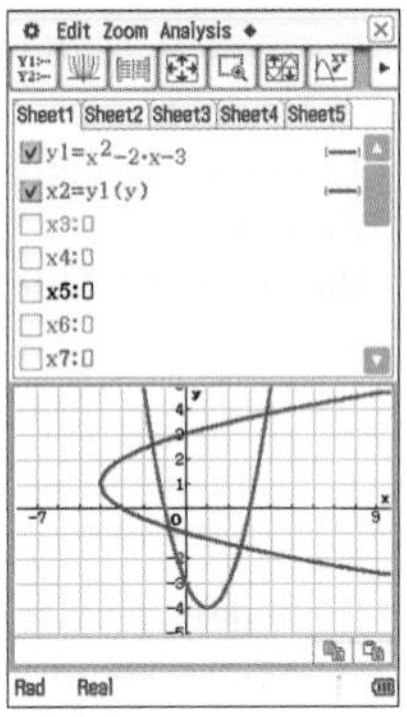
2. To find the x-intercepts of y1, select: • Analysis • G-Solve • Root Press EXE to mark the coordinates of the first x-intercept on the graph, use the left/right arrows to move to the next x-intercept, then press EXE.	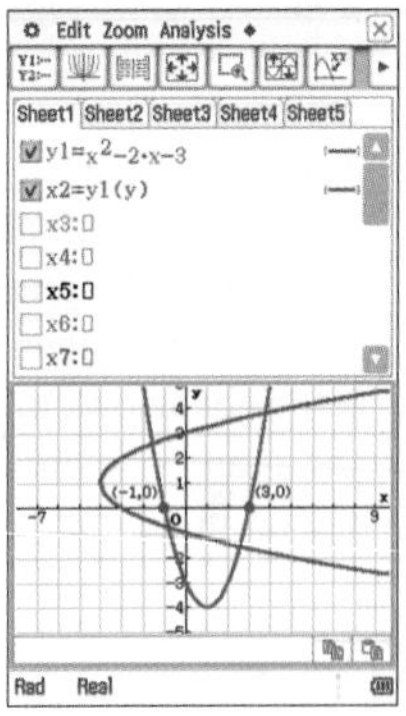
3. To find the y-intercept of y1, select: • Analysis • G-Solve • y-Intercept then press EXE.	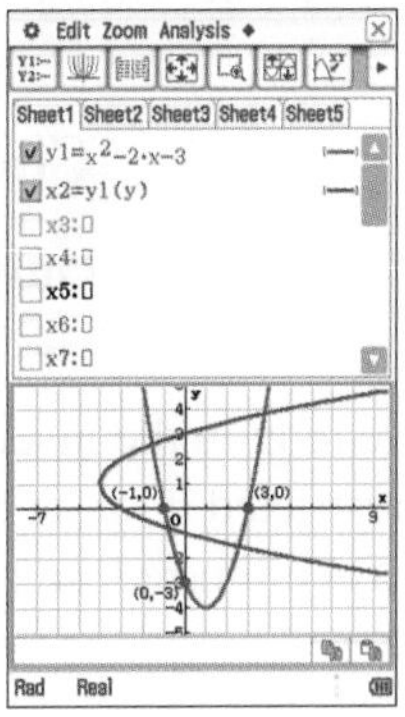

4. To find the minimum, press MENU, then select:
6: Analyze Graph
2: Minimum
Move the cursor to the left of the minimum when prompted for the lower bound, then press ENTER. Move the cursor to the right of the minimum when prompted for the upper bound, then press ENTER.

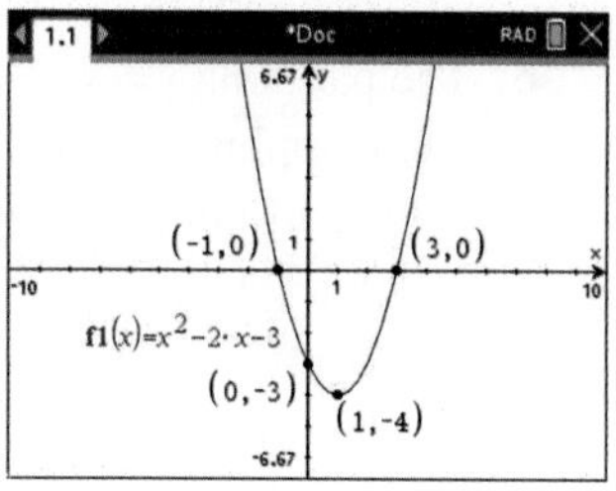

b. 1. To sketch the graph of the inverse, press MENU, then select:
3: Graph Entry/Edit
2: Relation
Complete the entry line as:
$f1(y) = x$
then press ENTER.

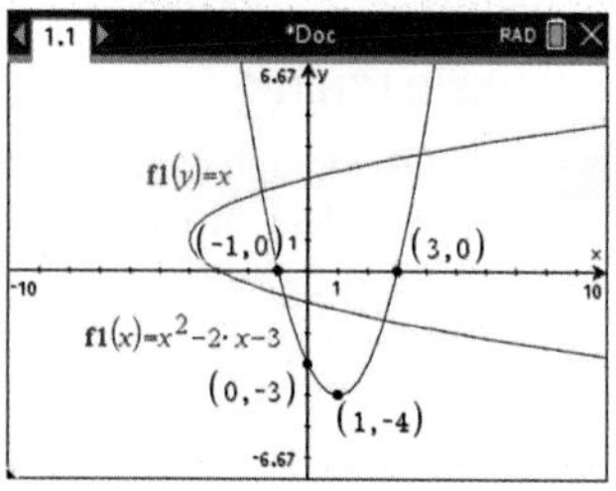

2. To find the points of interest on the inverse graph, press MENU, then select:
5: Trace
1: Graph Trace
Use the up/down arrows to move from the graph to the inverse graph, then use the left/right arrows to locate the x- and y-intercepts and the turning point.

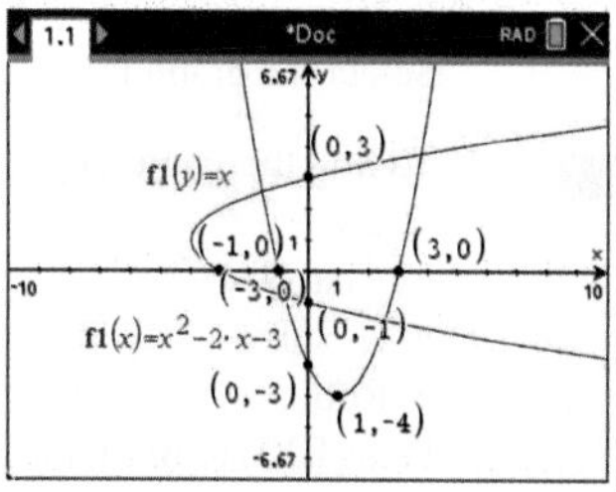

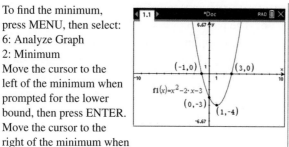

4. To find the minimum, select:
- Analysis
- G-Solve
- Min

then press EXE.

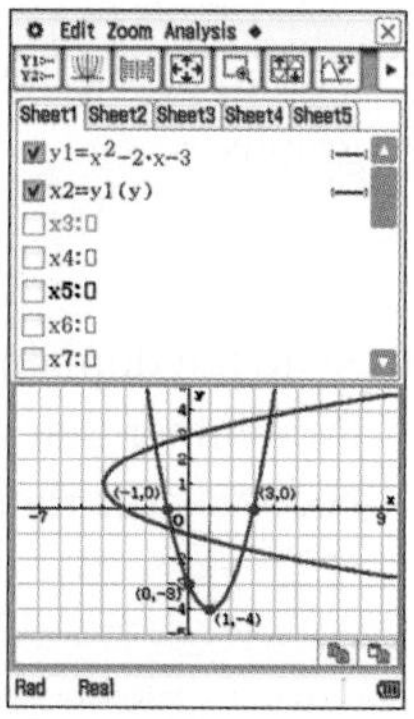

b. 1. To find the points of interest on the inverse graph, select:
- Analysis
- Trace.

Use the up/down arrows to move from the graph of y1 to the graph of x2, then use the left/right arrows to locate the x- and y-intercepts and the turning point.

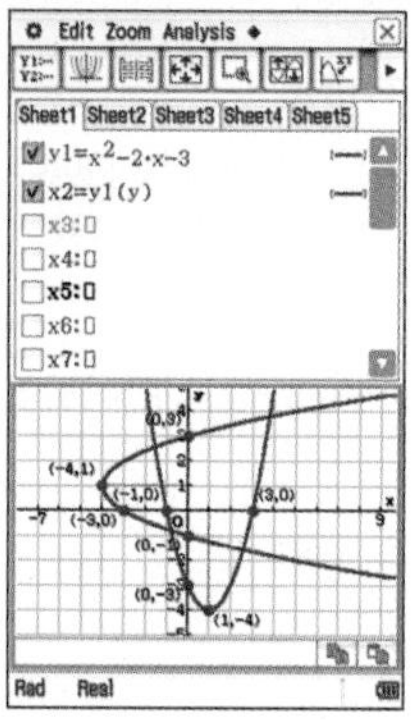

3.4.2 Inverse functions

As we have seen in the previous examples, the inverses produced are not always functions. Any function that is many-to-one will have an inverse that is one-to-many, and hence this inverse will not be a function.

> **Conditions for an inverse function**
>
> **Only one-to-one functions will have an inverse that is also a function.**

If we require the inverse of a many-to-one function to also be a function, the domain of the original graph must be restricted in order to ensure its correspondence is one-to-one. Achieving the maximum possible domain is always preferred, so many-to-one graphs are often restricted about the turning point or an asymptote.

WORKED EXAMPLE 8 Sketching inverse functions

a. **Consider the graph of $y = x^2 - 2x - 3$ from Worked example 7. The domain is restricted to $x \in (-\infty, a]$, where a is the largest possible value such that the inverse function exists. Determine the value of a.**

b. **Sketch the restricted graph of y and its inverse on the same set of axes.**

c. **Give the domain and range for both graphs.**

THINK	WRITE
a. The turning point is $(1, -4)$, so to maximise the domain, we restrict y about this point.	**a.** The x-value of the turning point is 1, so $a = 1$.
b. Sketch the graph of $y = x^2 - 2x - 3$ for $x \in (-\infty, 1]$. Due to the restriction, there is only one x-intercept. Interchange the coordinates of the x-intercept and turning point, and sketch the graph of the inverse by reflecting the graph in the line $y = x$.	**b.** For $y = x^2 - 2x - 3$, $x \in (-\infty, 1]$: x-intercept $= (-1, 0)$, y-intercept $= (0, -3)$ and TP $= (1, -4)$. For the inverse, x-intercept $= (-3, 0)$, y-intercept $= (0, -1)$ and sideways TP $= (-4, 1)$.

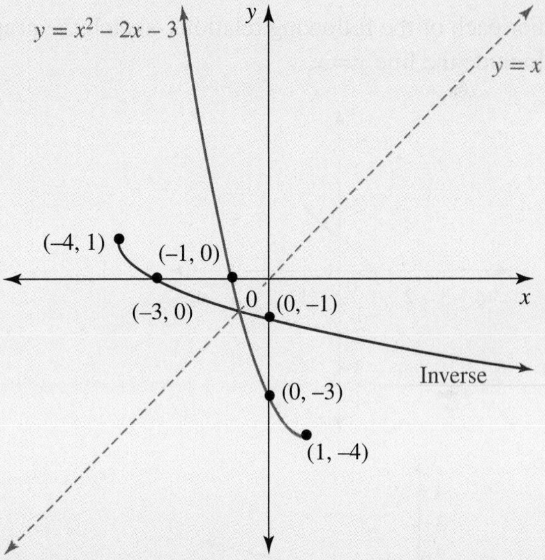

THINK	WRITE
c. State the domain and range for this function and its inverse.	**c.** For $y = x^2 - 2x - 3$: Domain: $x \in (-\infty, 1]$ Range: $y \in (-4, \infty]$ Inverse: Domain: $x \in (-4, \infty]$ Range: $y \in (-\infty, 1]$

3.4 Exercise

Students, these questions are even better in jacPLUS

Receive immediate feedback and access sample responses

Access additional questions

Track your results and progress

Find all this and MORE in jacPLUS

Technology free

1. **WE7** a. Sketch the graph of $y = (1 - x)(x + 5)$ and its inverse on the one set of axes. Show all axis intercepts and turning point coordinates.
   b. State the mapping for each graph and whether it is a function or a relation.
   c. Give the domain and range for the function and its inverse.

2. a. Sketch the graph of $y = \sqrt{x}$.
   b. By reflecting this function in the line $y = x$, sketch the graph of the inverse relation.
   c. State the type of mapping for y and its inverse and state whether the inverse is a relation or a function.

3. For each of the following relations, sketch the graph and its inverse on the same set of axes. Include the line $y = x$.

a.

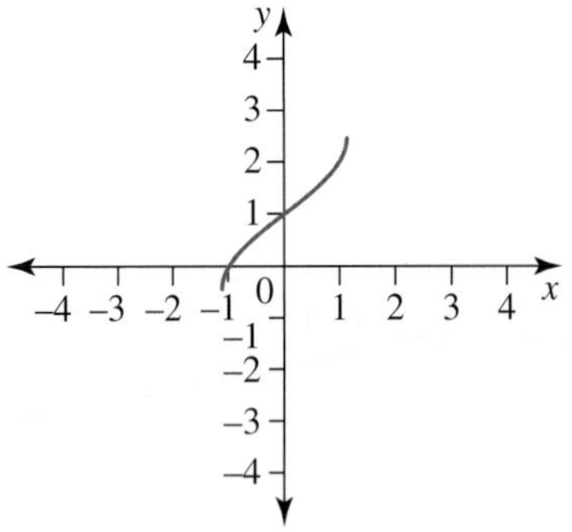

b.

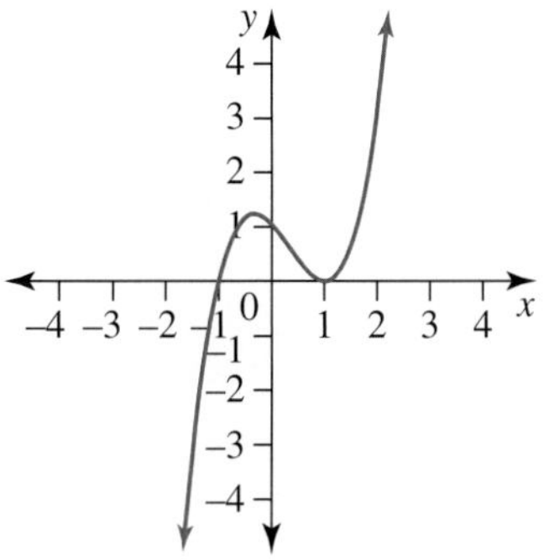

c.

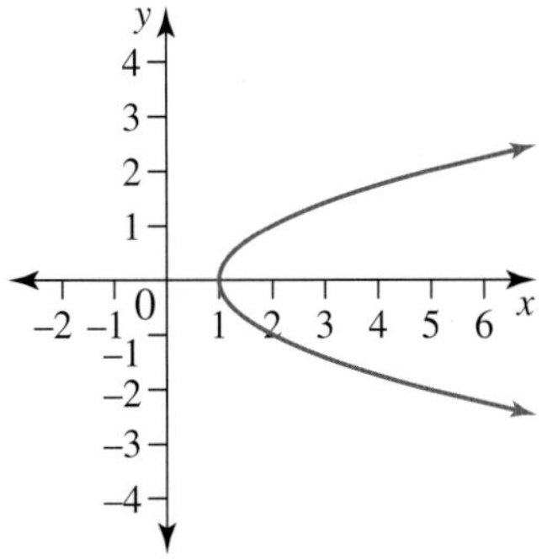

d.

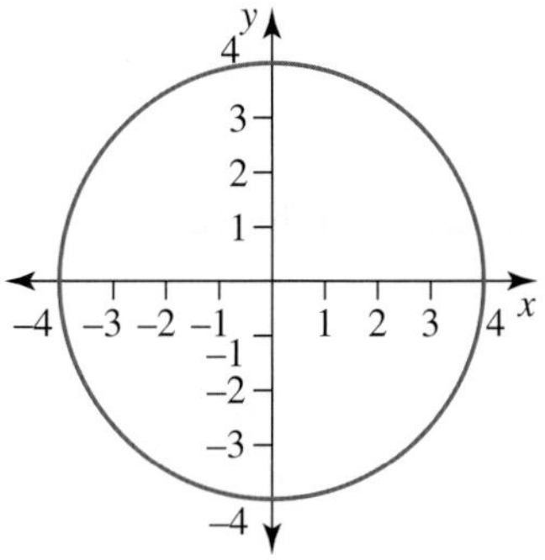

4. a. Sketch the graph of $y = -\frac{1}{3}x - 1$, showing all important features.
   b. On the same set of axes, sketch the inverse function, again showing axis intercepts.

5. Identify which of the following functions are one-to-one functions.
   a. $f(x) = \cos(x)$
   b. $g(x) = 1 - x^3$
   c. $h(x) = 4 - x^2$
   d. $k(x) = 2 + \frac{1}{x - 3}, x \neq 3$

Technology active

6. **MC** Identify which of the following functions has an inverse that is a function.

A. $y = x^2 - 1$

B. $y = \dfrac{1}{(x+2)^2}$

C. $y = \dfrac{1}{x-1}$

D. $y = x^3 - x^2$

E. $y = 10$

7. **MC** Identify the function and inverse function pair.

A.

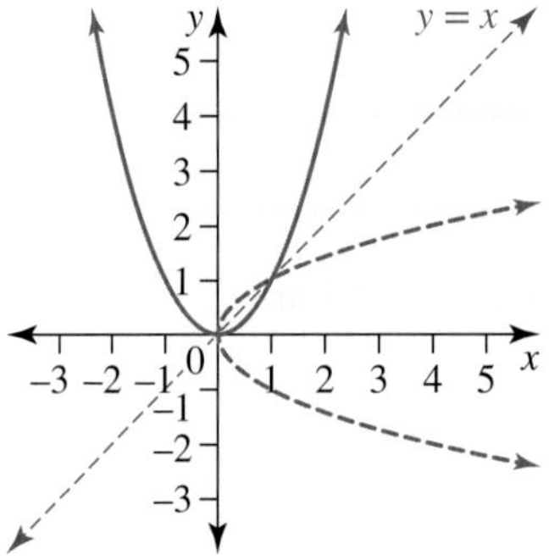

B.

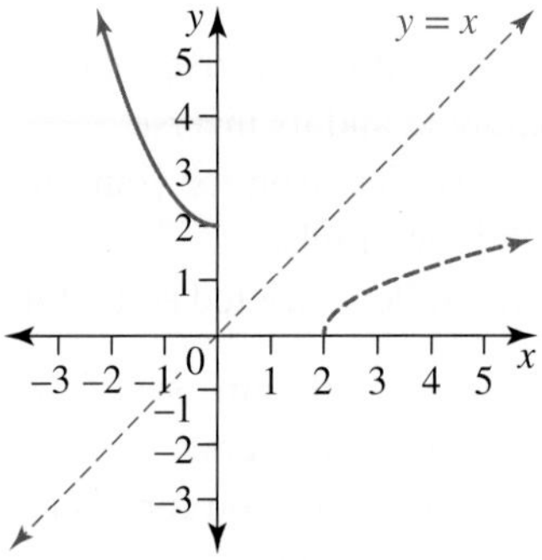

C.

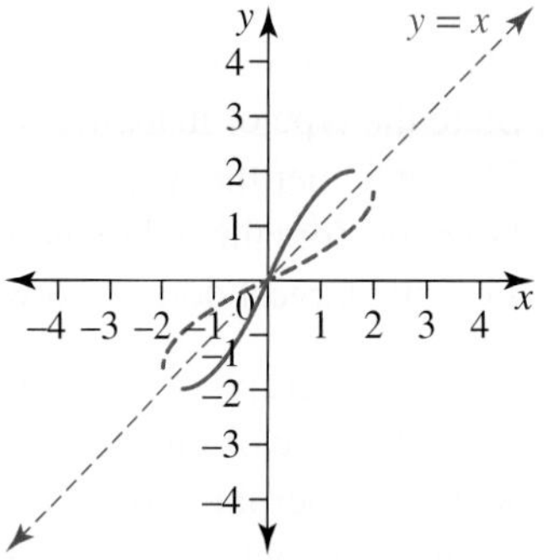

D.

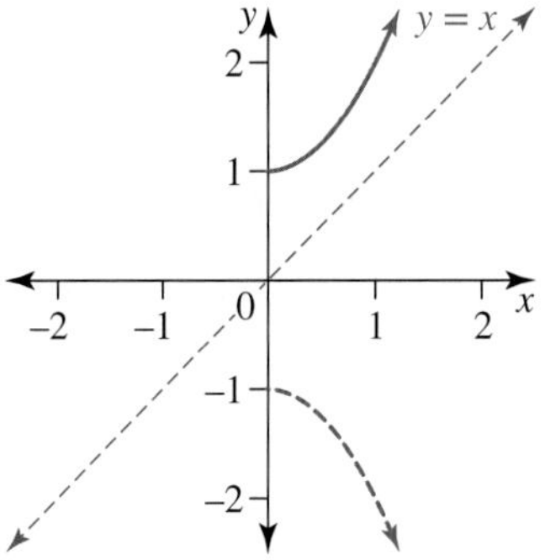

E.

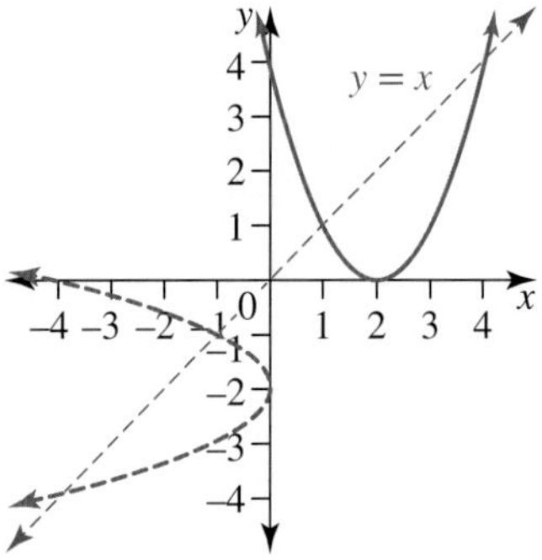

8. The graph of $x = (y-2)^2$ is shown in blue. The inverse relation is one of the other two graphs shown. Choose whether option **A** or option **B** is the inverse, giving clear reasons for your decision.

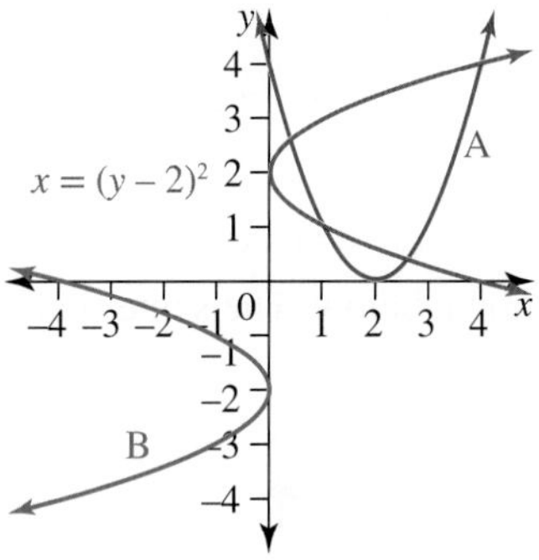

9. a. Use CAS to sketch the graph of $y = 9x - x^3$ and its inverse on the one set of axes.
   b. Determine the points of intersection of y and its inverse that occur along the line $y = x$, correct to 3 decimal places.

10. a. Use CAS to sketch the graph of $y = 3 + \dfrac{1}{\sqrt{x}}, x \in R^+$ and its inverse on one set of axes.
    b. Determine the points of intersection of y and its inverse, correct to 3 decimal places.

11. WE8 a. Consider the graph of $y = (1 - x)(x + 5)$. The domain is restricted to $x \in (-\infty, a]$, where a is the largest possible value such that the inverse function exists. Determine the value of a.
b. Sketch the restricted graph of y and its inverse on the same set of axes.
c. Give the domain and range for both graphs.

12. Consider the graph of $y = -(x - 3)^2$. State the largest positive domain for the given function so that its inverse is a function. Sketch the restricted function with its inverse on the one set of axes.

13. a. Sketch the graph of $y = (x + 4)(x - 2)$.
b. On the same set of axes, sketch the graph of the inverse relation.
c. State the type of mapping for the parabola and its inverse.
d. State whether the inverse a function. Give a reason for your answer.
e. Give the domain and range for each of the graphs.
f. Find the largest domain to which y could be restricted so that its inverse is a function.

14. a. Sketch the graph of $y = x^3$ and its inverse on the same set of axes.
b. State the type of mapping for the graph and its inverse.
c. State whether the inverse is a function. Give a reason for your answer.
d. Give the domain and range for each of the graphs.

15. a. Sketch the graph of $y = \frac{1}{x^2}$.
b. On the same set of axes, sketch the inverse relation.
c. State the mapping for each graph and indicate whether the rule describes a function or a relation.
d. Restrict the domain of $y = \frac{1}{x^2}$, where x consists of negative values only, so that its inverse is a function. State this domain.
e. Using this restricted domain for y, sketch the graph of y and its inverse on a new set of axes. State the domain and range of each function.

16. Given $y = 2x^2 - 12x + 13$ with a domain of $(-\infty, a]$, find the largest value of a so that the inverse of y is a function.

3.4 Exam questions

Question 1 (1 mark) TECH-ACTIVE

Source: *VCE 2017 Mathematical Methods Exam 2, Section A, Q6; © VCAA.*

MC Part of the graph of the function f is shown below. The same scale has been used on both axes.

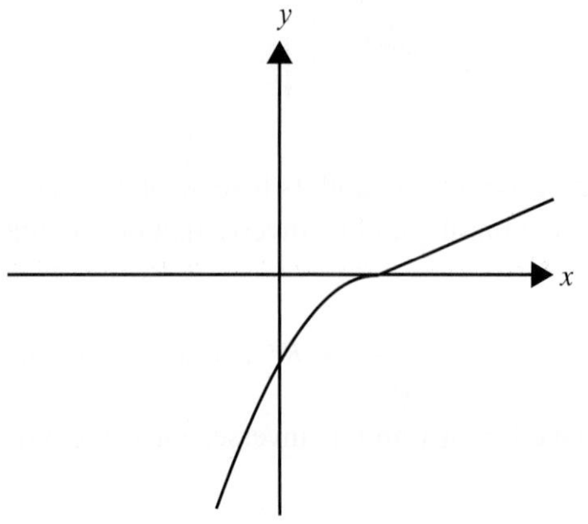

The corresponding part of the graph of the inverse function f^{-1} is best represented by

A.

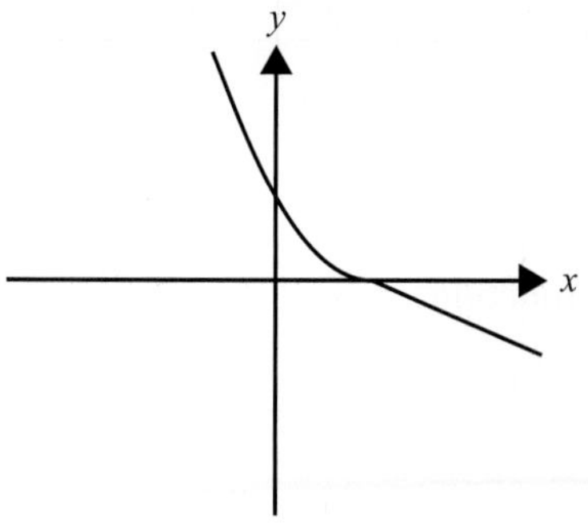

B.

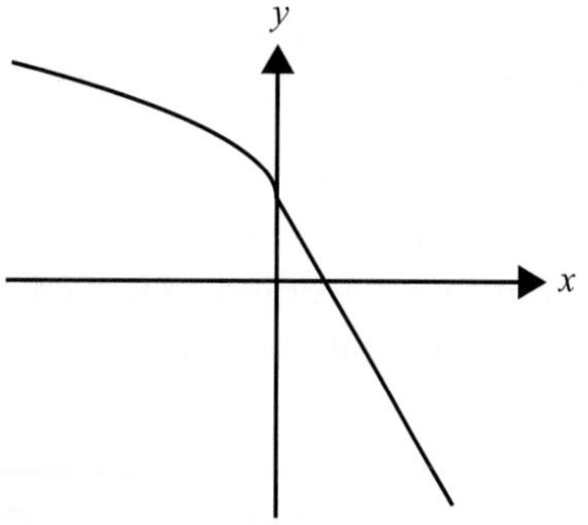

C.

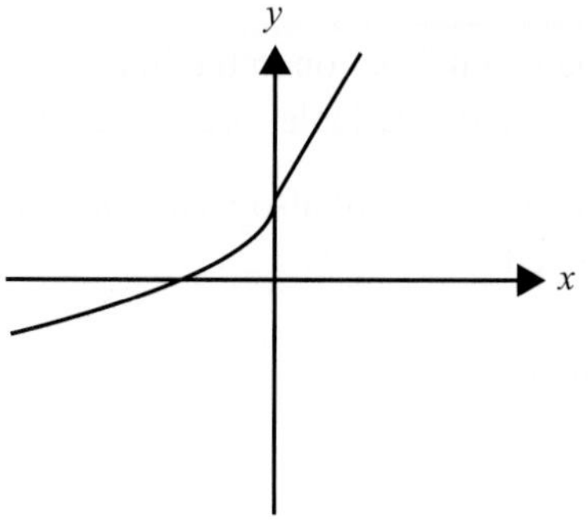

D.

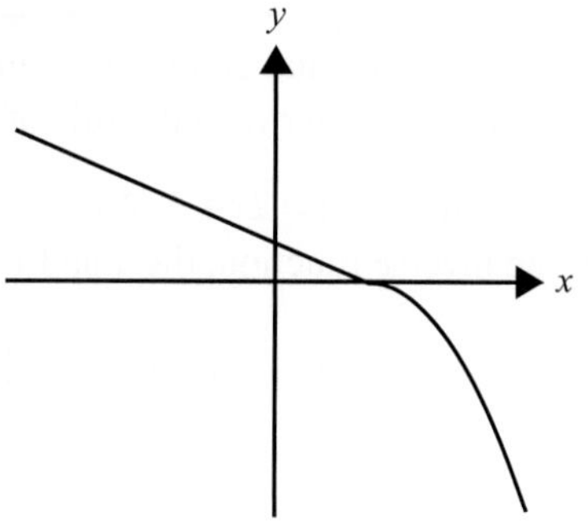

E.

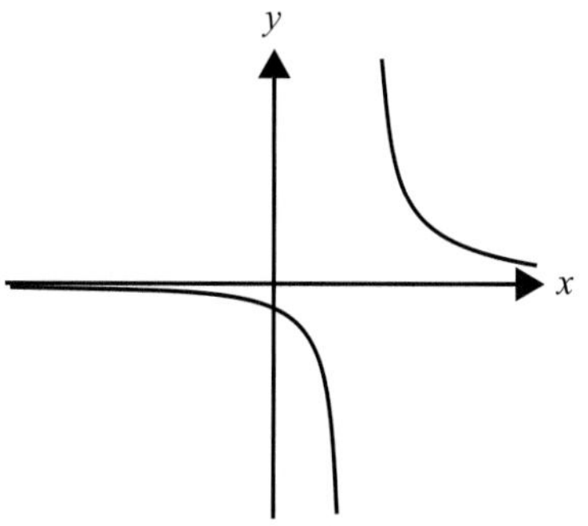

Question 2 (1 mark) TECH-ACTIVE

Source: *VCE 2013, Mathematical Methods (CAS) Exam 2, Section 1, Q7; © VCAA.*

MC The function $g: [-a, a] \to R, g(x) = \sin\left(2\left(x - \frac{\pi}{6}\right)\right)$ has an inverse function.

The maximum possible value of a is

A. $\frac{\pi}{12}$ **B.** 1 **C.** $\frac{\pi}{6}$ **D.** $\frac{\pi}{4}$ **E.** $\frac{\pi}{2}$

Question 3 (1 mark) TECH-ACTIVE

MC Select the rule that does not describe a function.

A. $y = x^2 - 5$

B. $y = -5$

C. $x = -5$

D. $y = \sqrt{5 - x^2}$

E. $y = \frac{1}{x - 5}$

More exam questions are available online.

3.5 Inverse functions

LEARNING INTENTION

At the end of this subtopic you should be able to:
- determine the equation of an inverse function
- restrict the domain of the original function, so that the inverse is a function
- determine the point of intersection between a function and its inverse.

3.5.1 Finding the equation of an inverse

In the previous section, we saw that an inverse is graphed by reflecting the given function in the line $y = x$. We follow the same procedure to determine the rule of an inverse. That is, the x and y variables are interchanged.

Consider the linear function $f(x) = 2x + 1$. As f is a one-to-one function, its inverse will also be a function. To obtain the rule for the inverse function, the x and y variables are interchanged.

$$\text{Inverse:} \quad \text{Let } y = f(x), \text{ swap } x \text{ and } y.$$
$$x = 2y + 1$$

Rearrange the rule to make y the subject of the equation.

$$2y = x - 1$$
$$y = \frac{1}{2}(x - 1)$$

The function $f(x) = 2x + 1$ has a domain of R and range of R.

The inverse function $y = \frac{1}{2}(x - 1)$ has a domain of R and range of R.

Notation for inverse functions

If the inverse of a function f is itself a function, then the inverse function is denoted by f^{-1}.

For example, the equation of the inverse of the square root function, $f(x) = \sqrt{x}$, can be written as $f^{-1}(x) = x^2,\ x \geq 0$.

In mapping notation, if $f: [0, \infty) \rightarrow R,\ f(x) = \sqrt{x}$, then the inverse function is $f^{-1}: [0, \infty) \rightarrow R,\ f^{-1}(x) = x^2$.

The domain of f^{-1} equals the range of f, and the range of f^{-1} equals the domain of f; that is, $d_{f^{-1}} = r_f$ and $r_{f^{-1}} = d_f$.

Note that f^{-1} is a function notation and thus cannot be used for relations that are not functions.

Note also that the inverse function f^{-1} and the reciprocal function $\frac{1}{f}$ represent different functions: $f^{-1} \neq \frac{1}{f}$.

WORKED EXAMPLE 9 Determining equations of inverses

Consider the function $y = (x + 2)^2$. Determine the rule for the inverse and indicate whether this inverse is a function or a relation. Give the domain and range for both.

THINK	WRITE
1. To obtain the inverse, interchange the x and y variables.	$y = (x + 2)^2$ Inverse: swap x and y. $x = (y + 2)^2$

2. Rearrange to make y the subject of the equation.

$$(y+2)^2 = x$$
$$y+2 = \pm\sqrt{x}$$
$$y = \pm\sqrt{x} - 2$$

3. Comment on whether the inverse is a function or a relation.

As $y=(x+2)^2$ is a many-to-one function, the inverse will be a one-to-many relation. Therefore, it is not a function.

4. State the domain and range for both rules.

$y=(x+2)^2$ has a domain of R and a range of $[0, \infty)$.
$y=\pm\sqrt{x}-2$ has a domain of $[0, \infty)$ and a range of R.

TI | THINK

1. On a Calculator page, press MENU, then select:
1: Actions
1: Define
Complete the entry line as:
Define $f(x) = (x + 2)^2$
then press ENTER.

DISPLAY/WRITE

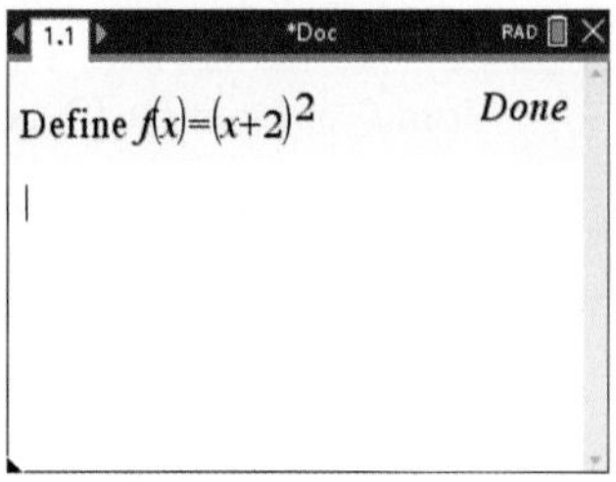

2. Press MENU, then select:
3: Algebra
1: Solve
Complete the entry line as:
solve($f(y) = x, y$)
then press ENTER.

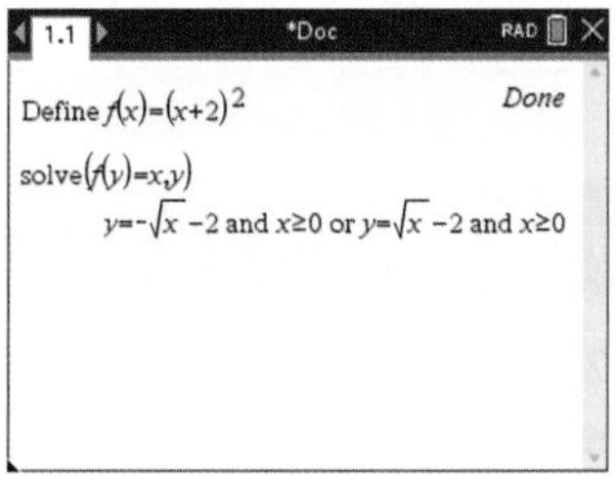

3. The answer appears on the screen.

The inverse function is $y = \pm\sqrt{x} - 2$ with domain $[0, \infty)$ and range R. It is not a function.

CASIO | THINK

1. On a Main screen, select:
- Action
- Command
- Define

Complete the entry line as:
Define $f(x) = (x + 2)^2$
then press EXE.

DISPLAY/WRITE

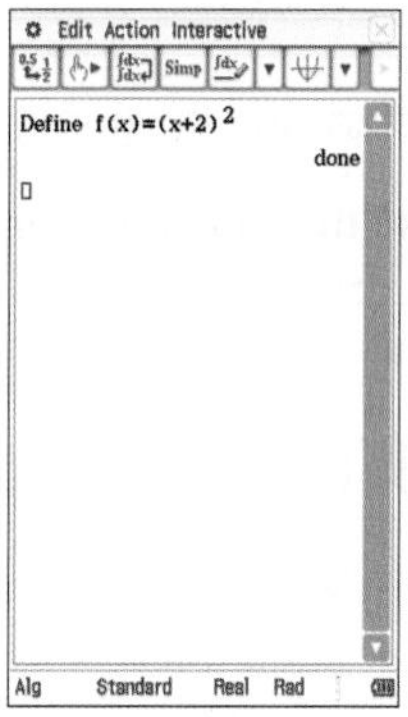

2. Complete the next entry line as:
solve ($f(y) = x, y$)
then press EXE.

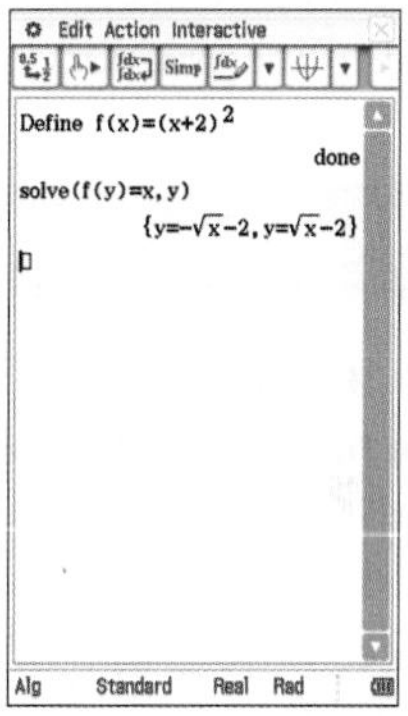

3. The answer appears on the screen.

The inverse function is $y = \pm\sqrt{x} - 2$ with domain $[0, \infty)$ and range R. It is not a function.

on Resources

Interactivity Inverse functions (int-2575)

3.5.2 Restricting domains

In some cases, the domain will need to be included when we state the equation of the inverse. For example, to find the equation of the inverse of the function $y=\sqrt{x}$, interchanging coordinates gives $x=\sqrt{y}$. Expressing $x=\sqrt{y}$ with y as the subject gives $y=x^2$. This rule is not unexpected since 'square root' and 'squaring' are inverse operations. However, as the range of the function $y=\sqrt{x}$ is $[0, \infty)$, this must be the domain of its inverse. Hence, the equation of the inverse of $y=\sqrt{x}$ is $y=x^2$ with the restriction that $x \geq 0$.

Other examples involve restricting the domain of f so that the inverse is a function.

WORKED EXAMPLE 10 Equations of inverse functions

Consider the function $f: [0, \infty) \to R, f(x) = x^2 + 2$. Fully define the inverse, f^{-1}.

THINK	WRITE
1. Let $y = f(x)$, then interchange the x and y variables.	Let $y = f(x)$. Swap x and y. Inverse: $x = y^2 + 2$
2. Rearrange to make y the subject of the equation.	$y^2 = x - 2$ $y = \pm\sqrt{x-2}$
3. Use the domain of $f(x)$ to determine the inverse.	$\text{dom} f = \text{ran} f^{-1}$ $\therefore y = \sqrt{x-2}$
4. Determine the domain of f^{-1}.	$\text{dom} f^{-1} = \text{ran} f = [2, \infty]$
5. Use the full function notation to define the inverse.	$f^{-1}: [2, \infty) \to R, f^{-1}(x) = \sqrt{x-2}$

3.5.3 The point of intersection of $f(x)$ and $f^{-1}(x)$

The point where $f(x)$ intersects with its inverse can be found by solving $f(x) = f^{-1}(x)$. However, this can often be a difficult equation to solve. As $y = f(x)$ intersects with $y = f^{-1}(x)$ along the line $y = x$, there is actually a three-way point of intersection: $f(x) = f^{-1}(x) = x$. Therefore, it is preferable to solve either $f(x) = x$ or $f^{-1}(x) = x$ to find the point of intersection.

WORKED EXAMPLE 11 Intersection of $f(x)$ and $f^{-1}(x)$ (1)

Consider the quadratic function defined by $f(x) = 2 - x^2$.

a. Form the rule for its inverse and explain why the inverse is not a function.

b. If the domain of f is restricted to $(-\infty, a)$, determine the maximum value of a so that the inverse exists.

c. Sketch the graph of $f(x) = 2 - x^2$ over this restricted domain and use this to sketch its inverse on the same diagram.

d. Form the equation of the inverse, $y = f^{-1}(x)$.

e. Determine the point at which the two graphs intersect.

THINK	WRITE
a. 1. Interchange x and y coordinates to form the rule for the inverse.	a. Let $y = f(x)$. Inverse: swap x and y. $x = 2 - y^2$ $y^2 = 2 - x$ $y = \pm\sqrt{2-x}$
2. Explain why the inverse is not a function.	The quadratic function is many-to-one, so its inverse has a one-to-many correspondence. Therefore, the inverse is not a function.

b. To maximise the domain, restrict the graph about the turning point.

b. TP = (0, 2)
Therefore, $a = 0$.

c. 1. Sketch the graph of the function for the restricted domain.

c. $f(x) = 2 - x^2$
y-intercept: (0, 2)
x-intercept: let $y = 0$.

$$2 - x^2 = 0$$
$$x^2 = 2$$
$$x = \pm\sqrt{2}$$
$$\Rightarrow x = -\sqrt{2} \text{ since } x \in (-\infty, 0).$$

x-intercept: $(-\sqrt{2}, 0)$
Turning point: (0, 2)

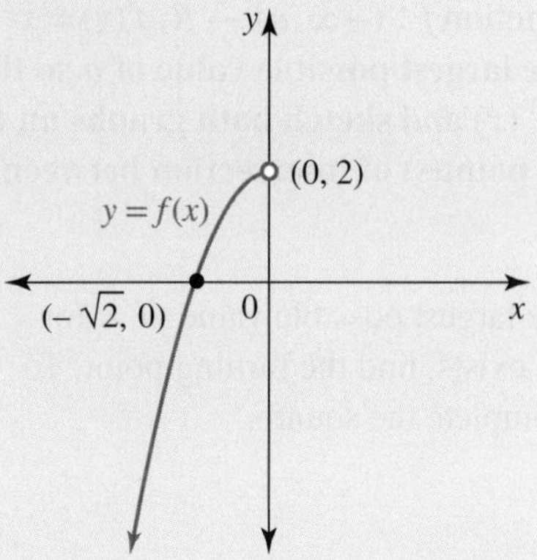

2. Deduce the key features of the inverse. Sketch its graph and the line $y = x$ on the same diagram as the graph of the function.

For the inverse, (2, 0) is an open point on the x-axis and $(0, -\sqrt{2})$ is the y-intercept.
Its graph is the reflection of the graph of $f(x) = 2 - x^2, x \in (-\infty, 0)$ in the line $y = x$.

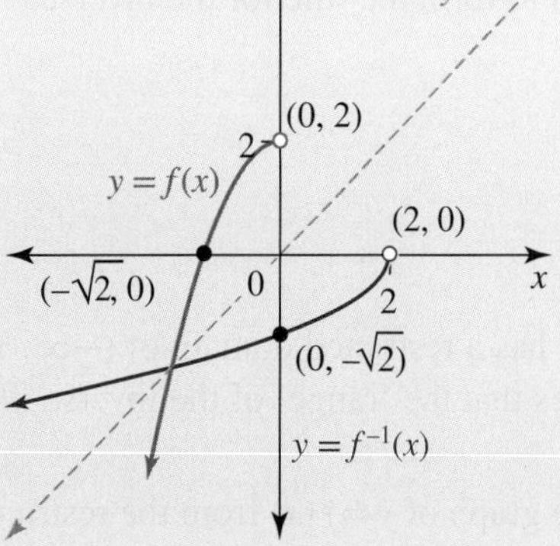

d. Use the range of the inverse to help deduce its equation. *Note*: When you write the answer, the domain must also be included.

d. From part **a**, the inverse of $f(x) = 2 - x^2$ is:

$$y^2 = 2 - x$$
$$\therefore\ y = \pm\sqrt{2 - x}$$

The range of the inverse must be $(-\infty, 0)$ (the domain of the original graph), so the branch with the negative square root is required.
Therefore, the equation of the inverse is
$y = -\sqrt{2 - x}$.
$f^{-1}(x) = -\sqrt{2 - x}$, domain $= (-\infty, 2)$ ▶

e. Choose two of the three equations that contain the required point and solve this system of simultaneous equations. *Note*: As the graph and its inverse intersect along the line $y = x$, the y-value of the coordinate will be the same as the x-value.

e. The point of intersection lies on $y = x$.
Solving $x = f(x)$:

$$x = 2 - x^2, x \in (-\infty, 0)$$
$$x^2 + x - 2 = 0$$
$$(x+2)(x-1) = 0$$
$$x = -2, 1$$

Reject $x = 1$ since $x \in (-\infty,\ 0)$; therefore, $x = -2$.
Therefore, the point of intersection is $(-2, -2)$.

WORKED EXAMPLE 12 Intersection of $f(x)$ and $f^{-1}(x)$ (2)

Consider the function $f: (-\infty, a] \to R, f(x) = x^2 - 6x + 4$.

a. Determine the largest possible value of a so that f^{-1} exists.

b. Determine $f^{-1}(x)$ and sketch both graphs on the same set of axes.

c. Calculate the point(s) of intersection between $y = f(x)$ and $y = f^{-1}(x)$.

THINK	WRITE
a. 1. To find the largest possible value of a for which f^{-1} exists, find the turning point. To do this, complete the square.	**a.** Let $y = f(x)$ $\Rightarrow y = x^2 - 6x + 4$ $y = (x-3)^2 - 9 + 4$ $y = (x-3)^2 - 5$
2. State the turning point.	TP $= (3, -5)$
3. The largest possible value of a for which f^{-1} exists is the x-value of the TP.	$\therefore a = 3$
4. Rewrite the full function.	$f: (-\infty, 3] \to R,\quad f(x) = x^2 - 6x + 4$
b. 1. Interchange x- and y-coordinates in turning point form to form the rule for the inverse.	$y = (x-3)^2 - 5$ Inverse: swap x and y. $x = (y-3)^2 - 5$ $x + 5 = (y-3)^2$ $\pm\sqrt{x+5} = y - 3$ $3 \pm \sqrt{x+5} = y$
2. Since $f(x)$ has a restricted domain of $(-\infty, 3]$, this implies that the 'range' of the inverse is $(-\infty, 3]$.	$f^{-1}(x) = 3 - \sqrt{x+5}$ $f^{-1}: [-5,\ \infty) \to R, f^{-1}(x) = 3 - \sqrt{x+5}$
3. Sketch the graph of $y = f(x)$ from the restricted domain.	$f(x) = (x-3)^2 - 5$ y-intercept: $(0,\ 4)$ TP: $(3,\ -5)$ x-intercept: $0 = (x-3)^2 - 5$ $5 = (x-3)^2$ $\pm\sqrt{5} = x - 3$ $3 \pm \sqrt{5} = x$ Since $x \le 3$, the x-intercept is $\left(3 - \sqrt{5},\ 0\right)$.

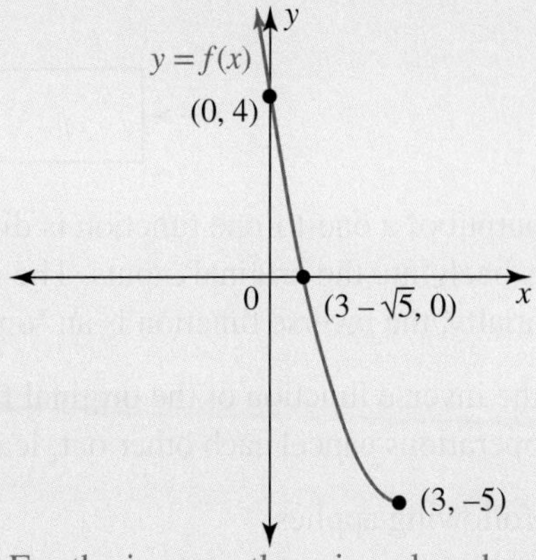

4. Deduce the features in the inverse function. Sketch its graph on the same axes as the graph of $y = f(x)$ and the line $y = x$.

For the inverse, there is a closed point at $(-5,\ 3)$. There is a point $(4,\ 0)$ on the x-axis and a point $\left(0,\ 3-\sqrt{5}\right)$ on the y-axis. $x \in [-5,\ \infty]$.

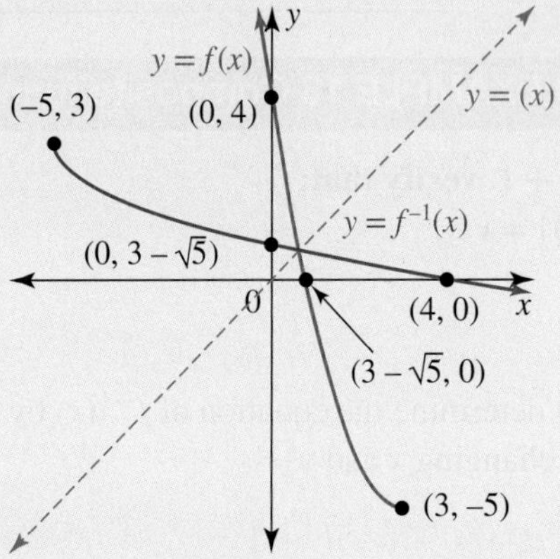

c. To determine the point(s) of intersection between $y = f(x)$ and $y = f^{-1}(x)$, let $f(x) = x$ as any point of intersection will lie on the line $y = x$. Rewrite the equation into general form and solve using the quadratic formula.

$$f(x) = (x-3)^2 - 5$$

Let $f(x) = x$.

$$x^2 - 6x + 4 = x$$

$$x^2 - 7x + 4 = 0$$

$$x = \frac{-(-7) \pm \sqrt{(-7)^2 - 4(1)(4)}}{2(1)}$$

$$x = \frac{7 \pm \sqrt{33}}{2}$$

Disregard $x = \frac{7+\sqrt{33}}{2}$ as $\frac{7+\sqrt{33}}{2} > 3$.

When $x = \frac{7-\sqrt{33}}{2}$,

$$y = \frac{7-\sqrt{33}}{2}$$

Therefore, the point of intersection is $\left(\frac{7-\sqrt{33}}{2}, \frac{7-\sqrt{33}}{2}\right)$.

3.5.4 Composite functions with inverse functions

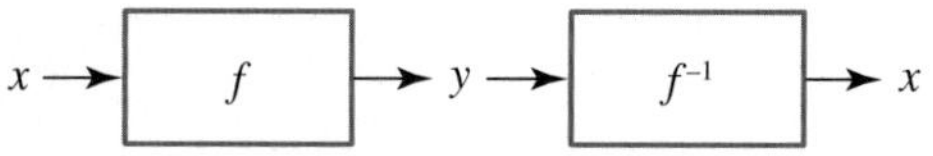

Because each output of a one-to-one function is different for each input, it is possible to reverse the process and turn the outputs back into the original inputs. The inverse is the function that results from reversing a one-to-one function. Essentially, the inverse function is an 'undoing' function.

So, if we take the inverse function of the original function or evaluate the function of the inverse function, in effect the two operations cancel each other out, leaving only x.

Therefore, the following applies.

> **Composite function properties**
>
> $$f(f^{-1}(x)) = x \text{ and } f^{-1}(f(x)) = x.$$

WORKED EXAMPLE 13 Composite functions with inverses

If $f(x) = 3x + 1$, verify that:

a. $f\left(f^{-1}(x)\right) = x$

b. $f^{-1}(f(x)) = x$

THINK

WRITE

a. 1. First determine the equation of $f^{-1}(x)$ by interchanging x and y.

Let $y = 3x + 1$.
Inverse: swap x and y.

$$\begin{aligned} x &= 3y + 1 \\ x - 1 &= 3y \\ y &= \frac{x-1}{3} \\ \therefore f^{-1}(x) &= \frac{x-1}{3} \end{aligned}$$

2. Substitute $f^{-1}(x)$ into $f(x)$ and evaluate.

$$\begin{aligned} f\left(f^{-1}(x)\right) &= f\left(\frac{x-1}{3}\right) \\ &= 3\left(\frac{x-1}{3}\right) + 1 \\ &= x - 1 + 1 \\ &= x \end{aligned}$$

3. State the answer.

Therefore, $f\left(f^{-1}(x)\right) = x$.

b. 1. Substitute $f(x)$ into $f^{-1}(x)$ and evaluate.

$$\begin{aligned} f^{-1}(f(x)) &= f^{-1}(3x + 1) \\ &= \frac{3x + 1 - 1}{3} \\ &= \frac{3x}{3} \\ &= x \end{aligned}$$

2. State the answer.

Therefore, $f^{-1}(f(x)) = x$.

3.5 Exercise

Students, these questions are even better in jacPLUS

Receive immediate feedback and access sample responses

Access additional questions

Track your results and progress

Find all this and MORE in jacPLUS

Technology free

1. WE9 Consider the function $y = x^3$. Determine the rule for the inverse and indicate whether this inverse is a function or a relation. Give the domain and range for both.

2. Consider the function $y = \frac{1}{x^2}$. Determine the rule for the inverse and indicate whether this inverse is a function or a relation. Give the domain and range for both.

3. For each of the following functions, determine the rule for the inverse and indicate whether this inverse is a function or a relation. Give the domain and range for the inverse.

   a. $y = \frac{1}{3}(x - 3)$
   b. $y = (x - 5)^2$
   c. $y = \sqrt[3]{x + 1} - 2$
   d. $y = (x - 1)^3$
   e. $y = \sqrt{x}$
   f. $y = \frac{1}{(x - 1)^2} + 2$

4. WE10 Consider the function $f: (-\infty, 2) \to R, f(x) = -\frac{1}{(x - 2)^2}$. Fully define the inverse, f^{-1}.

5. Consider the function $f: [3, \infty) \to R, f(x) = \sqrt{x - 3}$. Fully define the inverse, f^{-1}.

6. WE11 Consider the quadratic function $f(x) = (x + 1)^2$ defined on its maximal domain.

   a. Form the rule for its inverse and explain why the inverse is not a function.
   b. If the domain of f is restricted to $[b, \infty)$, find the minimum value of b so that the inverse exists.
   c. Sketch the graph of $f(x) = (x + 1)^2$ over this restricted domain and use this to sketch its inverse on the same diagram.
   d. Form the equation of the inverse, $y = f^{-1}(x)$.
   e. Determine the point at which the two graphs intersect.

7. Calculate the point of intersection between $f(x) = 2\sqrt{x + 2}$ and its inverse.

8. WE12 Consider the function $f: (-\infty, a] \to R, f(x) = x^2 - 2x - 1$.

   a. Determine the largest possible value of a so that f^{-1} exists.
   b. Determine $f^{-1}(x)$ and sketch both graphs on the same set of axes.
   c. Calculate the point(s) of intersection between $y = f(x)$ and $y = f^{-1}(x)$.

9. WE13 If $f(x) = \frac{1}{x + 2}, x \neq -2$, verify that:

   a. $f(f^{-1}(x)) = x$
   b. $f^{-1}(f(x)) = x$.

10. If $k(x) = x^3 - 1$, verify that:

    a. $k(k^{-1}(x)) = x$
    b. $k^{-1}(k(x)) = x$.

11. Indicate whether each of the following functions has an inverse function. In each case, give a reason for your decision. If the inverse is a function, write the rule for the inverse in function notation and sketch $y = f(x)$ and $y = f^{-1}(x)$ on the one set of axes, including the point of intersection if it exists.

a. $f: R \rightarrow R, f(x) = x^4$

b. $f: R \rightarrow R, f(x) = 2x^2 - 7x + 3$

c. $f: R\backslash\{3\} \rightarrow R, f(x) = \dfrac{1}{(x-3)^2} + 2$

d. $f: [-2, \infty) \rightarrow R, f(x) = \sqrt{x+2}$

12. Given $f(x) = \dfrac{4x-7}{x-2}$, determine the rule for f^{-1}, then sketch $y = f(x)$ and $y = f^{-1}(x)$ on the same set of axes. Include the point(s) of intersection on your graph.

Technology active

13. Given $f(x) = (x+2)^2$, restrict the maximal domain of f to only negative x-values so that its inverse is also a function. Write the inverse in function notation.

14. Given that $f(x) = \sqrt{1 - \dfrac{x^2}{4}}$, use CAS to view the graph and hence define two inverse functions, f^{-1}, using function notation with maximal domains. Sketch each pair of functions on separate axes.

15. Given $f(x) = x^2 - 10x + 25$ with a domain of $[a, \infty)$, determine:

a. the smallest value of a so that f^{-1} exists

b. $f^{-1}(x)$.

16. Consider $f: [-2, 4) \rightarrow R, f(x) = 1 - \dfrac{x}{3}$.

a. State the domain and determine the range of f.
b. Obtain the rule for f^{-1} and state its domain and range.
c. Sketch $y = f(x)$ and $y = f^{-1}(x)$ on the same diagram.
d. Calculate the coordinates of any point of intersection of the two graphs.

17. Consider $f: D \rightarrow R, f(x) = \sqrt{1-3x}$.

a. Determine D, the maximal domain of f.
b. Obtain the rule for $f^{-1}(x)$ and state its domain and range.
c. Evaluate the point(s) of intersection between $y = f(x)$ and $y = f^{-1}(x)$.
d. Sketch $y = f(x)$ and $y = f^{-1}(x)$ on the same set of axes.

18. Consider the function $f: [1, \infty] \rightarrow R, f(x) = \sqrt{x-1}$.

a. Determine $f^{-1}(x)$.
b. Sketch the graph of $y = f^{-1}(f(x))$ over its maximal domain.
c. Evaluate $f^{-1}\left(f\left(\dfrac{x+2}{3}\right)\right)$.

3.5 Exam questions

Question 1 (3 marks) TECH-FREE

***Source:** VCE 2018, Mathematical Methods Exam 1, Q5; © VCAA.*

Let $f: (2, \infty) \rightarrow R$, where $f(x) = \dfrac{1}{(x-2)^2}$.

State the rule and domain of f^{-1}.

Question 2 (1 mark) TECH-ACTIVE

***Source:** VCE 2016, Mathematical Methods Exam 2, Section A, Q5; © VCAA.*

MC Which one of the following is the inverse function of $[3, \infty) \rightarrow R, g(x) = \sqrt{2x-6}$?

A. $g^{-1}: [3, \infty) \rightarrow R,\ g^{-1}(x) = \dfrac{x^2+6}{2}$

B. $g^{-1}: [0, \infty) \rightarrow R,\ g^{-1}(x) = (2x-6)^2$

C. $g^{-1}: [0, \infty) \rightarrow R,\ g^{-1}(x) = \sqrt{\dfrac{x}{2}+6}$

D. $g^{-1}: [0, \infty) \rightarrow R,\ g^{-1}(x) = \dfrac{x^2+6}{2}$

E. $g^{-1}: R \rightarrow R,\ g^{-1}(x) = \dfrac{x^2+6}{2}$

Question 3 (1 mark) TECH-ACTIVE

***Source:** VCE 2015, Mathematical Methods (CAS) Exam 2, Section 1, Q2; © VCAA.*

MC The inverse function of $f: (-2, \infty) \rightarrow R,\ f(x) = \dfrac{1}{\sqrt{x+2}}$ is

A. $f^{-1}: R^+ \rightarrow R \qquad f^{-1}(x) = \dfrac{1}{x^2} - 2$

B. $f^{-1}: R \backslash \{0\} \rightarrow R \qquad f^{-1}(x) = \dfrac{1}{x^2} - 2$

C. $f^{-1}: R^+ \rightarrow R \qquad f^{-1}(x) = \dfrac{1}{x^2} + 2$

D. $f^{-1}: (-2, \infty) \rightarrow R \qquad f^{-1}(x) = x^2 + 2$

E. $f^{-1}: (2, \infty) \rightarrow R \qquad f^{-1}(x) = \dfrac{1}{x^2 - 2}$

More exam questions are available online.

3.6 Literal equations

LEARNING INTENTION

At the end of this subtopic you should be able to:
- solve literal equations
- solve simultaneous literal equations.

3.6.1 Solutions of literal equations

Equations with several pronumerals are called **literal equations**. Rather than the solution having a numerical answer, the solution will be expressed in terms of pronumerals, also called parameters.

WORKED EXAMPLE 14 Solving literal equations

Solve the following equations for x.

a. $mx + ny = kx - z$

b. $\dfrac{p}{x} - \dfrac{2m}{m+x} = \dfrac{3y}{x}$

THINK / **WRITE**

a. 1. Collect the x terms on the left-hand side.

a. $mx + ny = kx - z$

$mx - kx = -z - ny$

2. Take out the common factor of x to leave only one instance of x on the left-hand side.

$x(m - k) = -z - ny$

3. Divide both sides by $m - k$.

$$x = \frac{-z - ny}{m - k}$$
$$= -\frac{z + ny}{m - k}$$
$$= \frac{z + ny}{k - m}$$

b. 1. Multiply both sides by the common denominator of $x(m + x)$.

b. $\dfrac{p}{x} - \dfrac{2m}{m+x} = \dfrac{3y}{x}$

$p(m + x) - 2mx = 3y(m + x)$

2. Expand the brackets.

$pm + px - 2mx = 3my + 3xy$

3. Collect the x terms on the left-hand side.

$px - 2mx - 3xy = 3my - pm$

4. Take out the common factor of x to leave only one instance of x on the left-hand side.

$x(p - 2m - 3y) = 3my - pm$

5. Divide both sides by $p - 2m - 3y$.

$$x = \frac{3my - pm}{p - 2m - 3y}$$

TI | THINK / **DISPLAY/WRITE**

a. 1. On a Calculator page, press MENU, then select:
3: Algebra
1: Solve
Complete the entry line as:
solve $(m \times x + n \times y = k \times x - z, x)$
then press ENTER.
Note: Be sure to include the multiplication operator between the variables in the terms mx, ny and kx.

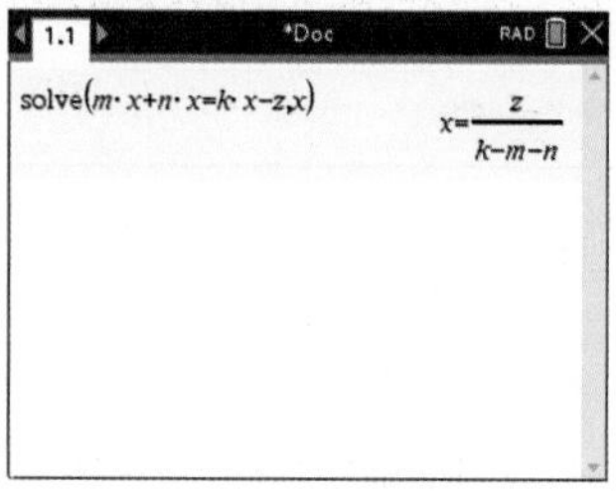

2. The answer appears on the screen.

$x = \dfrac{ny + z}{k - m}$

CASIO | THINK / **DISPLAY/WRITE**

a. 1. On a Main screen, complete the entry line as:
solve $(m \times x + n \times y = k \times x - z, x)$
then press EXE.
Select the simplify icon.
Note: Be sure to include the multiplication operator between the variables in the terms mx, ny and kx.

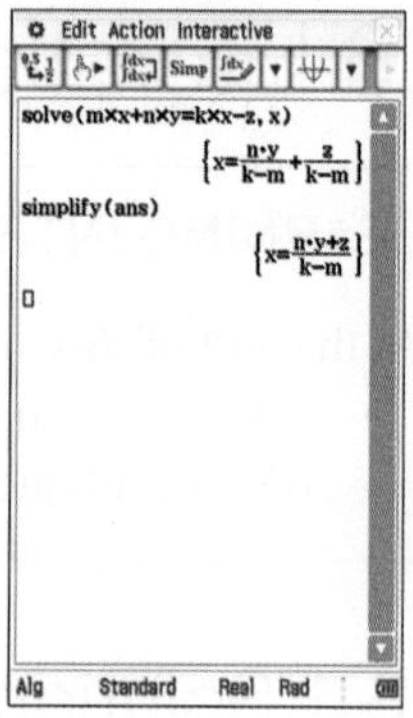

2. The answer appears on the screen.

$x = \dfrac{ny + z}{k - m}$

3.6.2 Simultaneous literal equations

These equations are solved by applying the methods of elimination and substitution. Once again, the solutions will be in terms of parameters. As a rule, if you are solving for n pronumerals, you will need n equations to solve for all the unknowns.

WORKED EXAMPLE 15 Solving simultaneous literal equations

Solve the pair of simultaneous equations for x and y.

$$mx - y = k$$
$$x + ny = 2d$$

THINK	WRITE
1. Label the equations.	$mx - y = k \quad [1]$ $x + ny = 2d \quad [2]$
2. Use the elimination method to solve these equations. Multiply equation [2] by m so that the coefficients of x are the same in both equations, and label this equation [3].	$[2] \times m$: $\Rightarrow mx + mny = 2dm \quad [3]$
3. Subtract [3] from [1] to eliminate the x terms.	$mx - y = k \quad [1]$ $mx + mny = 2dm \quad [3]$ $[1] - [3]$: $-y - mny = k - 2dm$
4. Take out the common factor of y to leave only one instance of y on the left-hand side.	$y(-1 - mn) = k - 2dm$
5. Divide both sides by $-1 - mn$ and simplify.	$y = \dfrac{k - 2dm}{-1 - mn}$ $= -\dfrac{k - 2dm}{1 + mn}$
6. Substitute $y = -\dfrac{k - 2dm}{1 + mn}$ into [1]. *Note*: Equation [2] could have also been chosen.	$mx - \left(-\dfrac{k - 2dm}{1 + mn}\right) = k$ $mx + \dfrac{k - 2dm}{1 + mn} = k$ $mx = k - \dfrac{k - 2dm}{1 + mn}$
7. Simplify the right-hand side.	$mx = k - \dfrac{k - 2dm}{1 + mn}$ $x = \dfrac{k(1 + mn)}{1 + mn} - \dfrac{k - 2dm}{1 + mn}$ $x = \dfrac{k + kmn - (k - 2dm)}{1 + mn}$ $x = \dfrac{kmn + 2dm}{1 + mn}$ $x = \dfrac{m(kn + 2d)}{1 + mn}$
8. Divide both sides by m.	$x = \dfrac{m(kn + 2d)}{m(1 + mn)}$ $x = \dfrac{kn + 2d}{1 + mn}$

TI | THINK	DISPLAY/WRITE
1. On a Calculator page, press MENU, then select: 3: Algebra 1: Solve Complete the entry line as: solve ($m \times x - y = k$ and $x + n \times y = 2d, x, y$) then press ENTER. *Note*: Be sure to include the multiplication operator between the variables in the terms mx and ny.	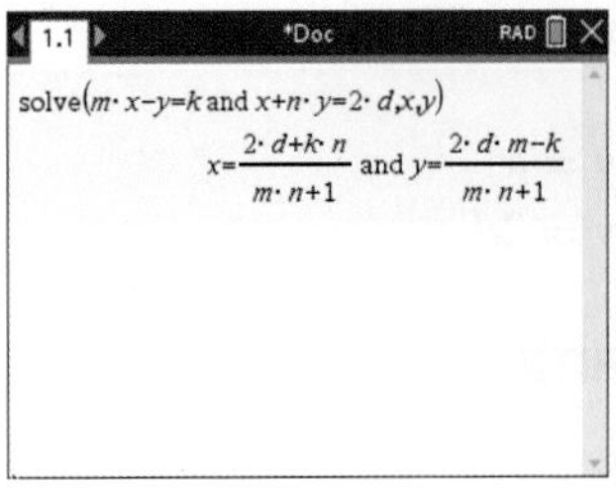
2. The answer appears on the screen.	$x = \dfrac{2d + kn}{mn + 1}$ and $y = \dfrac{2dm - k}{mn + 1}$

CASIO | THINK	DISPLAY/WRITE
1. On a Main screen, complete the entry line as: solve ({m × x - y = k, x + n × y = 2d}, {x, y}) then press EXE. *Note*: Be sure to include the multiplication operator between the variables in the terms mx and ny.	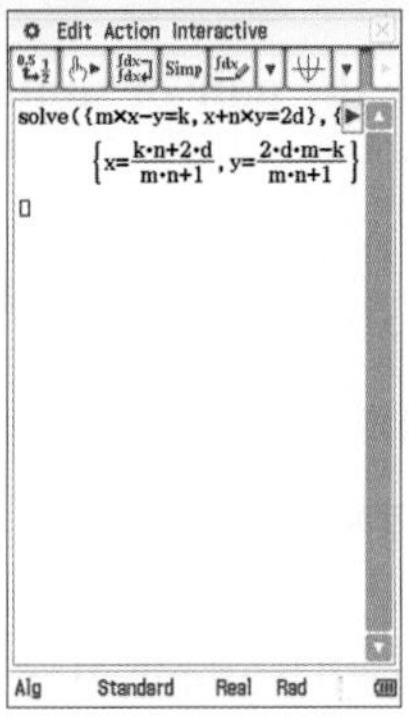
2. The answer appears on the screen.	$x = \dfrac{2d + kn}{mn + 1}$ and $y = \dfrac{2dm - k}{mn + 1}$

3.6 Exercise

Students, these questions are even better in jacPLUS

Receive immediate feedback and access sample responses

Access additional questions

Track your results and progress

Find all this and MORE in jacPLUS

Technology free

1. **WE14** Solve the following equations for x.

   a. $my - nx = 4x + kz$

   b. $\dfrac{2p}{x} - \dfrac{m}{x - c} = \dfrac{3c}{x}$

2. Given that $\dfrac{x - my}{px + y} = 2$, solve the equation for y.

3. Solve the following equations for x.

   a. $\dfrac{kx + dy}{x + 3y} = -2k$

   b. $\dfrac{mx + ny}{p} = x + q$

   c. $\dfrac{m}{x} - k = \dfrac{3k}{x} + m$

   d. $\dfrac{k}{m + x} = \dfrac{2d}{m - x}$

4. Given that $b\sqrt{ax} + cd = 3$, solve the equation for x.

5. **WE15** Solve the pair of simultaneous equations for x and y.

$$x + y = 2k$$
$$mx + ny = d$$

Technology active

6. Solve the following pairs of simultaneous equations for x and y.

   a. $nx - my = k$
      $nx + my = 2d$

   b. $nx + my = m$
      $mx + ny = n$

7. Solve the following pairs of simultaneous equations for x and y.

a. $2mx + ny = 3k$
$mx + ny = -d$

b. $\dfrac{x}{2a} + \dfrac{y}{b} = 2$
$\dfrac{2x}{b} + \dfrac{4y}{a} = 8$

8. Solve the following system of simultaneous equations in terms of a.

$$\begin{aligned} 2x - y + az &= 4 \\ (a+2)x + y - z &= 2 \\ 6x + (a+1)y - 2z &= 4 \end{aligned}$$

3.6 Exam questions

Question 1 (1 mark) TECH-ACTIVE

MC The solution to the equation $\dfrac{1}{x+a} = \dfrac{b}{x}$ in terms of x is

A. $1 + \dfrac{b}{a}$ B. $\dfrac{ab}{1-b}$ C. $\dfrac{ab}{1+b}$ D. $\dfrac{a}{1-b}$ E. $\dfrac{1}{a+b}$

Question 2 (1 mark) TECH-ACTIVE

MC $mx + n = nx + m$ solved for x is

A. $\dfrac{m+n}{m-n}$ B. $\dfrac{m-n}{m+n}$ C. $\dfrac{m}{n}$ D. 0 E. 1

Question 3 (1 mark) TECH-ACTIVE

MC The solutions for the pair of simultaneous equations $ax + by = r$ and $ax - by = s$ in terms of x and y are

A. $x = \dfrac{r+s}{2a}$, $y = \dfrac{-s+r}{2b}$

B. $x = \dfrac{r-by}{a}$, $y = \dfrac{r-ax}{b}$

C. $x = \dfrac{s+by}{a}$, $y = \dfrac{s-ax}{-b}$

D. $x = \dfrac{r+s}{2a}$, $y = \dfrac{s+r}{2b}$

E. $x = \dfrac{r-s}{2a}$, $y = \dfrac{-s+r}{2b}$

More exam questions are available online.

3.7 Review

3.7.1 Summary

doc-37032

Hey students! Now that it's time to revise this topic, go online to:

Access the topic summary | Review your results | Watch teacher-led videos | Practise VCAA exam questions

Find all this and MORE in jacPLUS

3.7 Exercise

Technology free: short answer

1. Consider the equations $f(x) = \sqrt{x+2}$ and $g(x) = 2x^2 - 5$. Determine if $f(g(x))$ or $g(f(x))$ exist. If they do exist, state the rule for the composite function and state the domain and range.

2. A function has the rule $y = \dfrac{x-1}{x-2}, x \neq 2$.

   a. Evaluate the values of m and n if $\dfrac{x-1}{x-2}$ is expressed in the form $\dfrac{m}{x-2} + n$.

   b. Describe the transformations that have been applied to $y = \dfrac{1}{x}$ to obtain $y = \dfrac{x-1}{x-2}, x \neq 2$.

   c. Sketch the graph of $y = \dfrac{x-1}{x-2}, x \neq 2$. State the domain and range, and give the equations of any asymptotes.

   d. Determine the rule for the inverse, and state its domain and range.

   e. Specify whether the inverse is a function or a relation. Give reasons for your answer.

   f. Sketch the graph of the inverse on the same set of axes as the original function. Include the points of intersection on your graph

3. Indicate whether each of the following functions has an inverse function. In each case, give a reason for your decision. If the inverse is a function, write the rule for the inverse function in function notation.

   a. $f: R \to R, f(x) = \dfrac{x^3}{3}$

   b. $f: R \to R, f(x) = 2x^4$

   c. $f: R \to R, f(x) = (3x-1)^2$

   d. $f: [3, \infty] \to R, f(x) = \sqrt{x-3}$

4. a. The graph of $f(x) = \sqrt{x}$ undergoes the following transformations:
   reflection in the x-axis, then a dilation of factor 2 from the y-axis, followed by a translation of 3 units right and 1 unit down, and a dilation of factor $\dfrac{1}{2}$ from the x-axis.
   State the equation of the transformed function, $g(x)$.

   b. The graph of $f(x) = (x-2)^2$ undergoes the following transformations:
   dilation of factor $\dfrac{1}{3}$ parallel to the x-axis, reflection in the x-axis, and a translation of 3 units in the negative x-direction and 2 units in the positive y-direction.
   State the equation of the transformed function, $g(x)$.

5. State the transformations that have been applied to the first function in order to obtain the second function.

   a. $y = x^2 \to y = 3(2x-5)^2 + 1$

   b. $y = -\sqrt[3]{\dfrac{5-x}{2}} + 1 \to \sqrt[3]{x}$

6. Given that $\dfrac{5-cd}{x+2} = -\dfrac{2k}{x}$, solve the equation for x.

Technology active: multiple choice

7. **MC** If $g(x) = 2x - 1$ and $h(x) = (x + 1)^2$, then $g(h(x))$ is equal to:

A. $2x^2 + 4x + 1$ B. $4x^2$ C. $2x^2 + 4x - 1$ D. $(2x - 1)(x + 1)^2$ E. $4(x + 1)^2$

8. **MC** For the functions below, state which of the following compositions is not defined.

$$f(x) = \sqrt{x} + 1$$
$$g(x) = x^2 - 1$$
$$h(x) = 2x + 1$$

A. $g(h(x))$ B. $g(f(x))$ C. $h(f(x))$ D. $f(g(x))$ E. $h(g(x))$

9. **MC** If $g(x) = \sqrt{x - 1}$, then $g(h(x))$ would exist if:

A. $h: R\backslash\{0\} \to R,\ h(x) = \frac{1}{x^2} + 1$ B. $h: R \to R,\ h(x) = (x - 1)^2$

C. $h: [-1, \infty) \to R,\ h(x) = -(x + 1)^2$ D. $h: R\backslash\{-1\} \to R,\ h(x) = \frac{1}{x + 1}$

E. $h: R \to R,\ h(x) = x$

10. **MC** The graph of $y = f(x)$ has been transformed so that its equation becomes $y = -f(3x + 1) - 2$. The image of the point $(3, 5)$ is:

A. $\left(\frac{26}{3}, -7\right)$ B. $(0, -7)$ C. $\left(\frac{2}{3}, -7\right)$ D. $(0, 3)$ E. $\left(-\frac{4}{3}, 3\right)$

11. **MC** If the graph of $y = \sin(x)$ is reflected in the x-axis and dilated by a factor of 4 from the x-axis and by a factor of $\frac{1}{3}$ from the y-axis, then the resulting equation will be:

A. $y = 4\sin\left(-\frac{x}{3}\right)$ B. $y = -4\sin\left(\frac{x}{3}\right)$ C. $y = 4\sin(-3x)$

D. $y = -\frac{1}{4}\sin(3x)$ E. $y = -4\sin(3x)$

12. **MC** The graph of the function $f(x) = x^3$ is transformed so that its new rule is $f(x) = \frac{1}{2}(2(x - 1))^3 + 4$. The transformations that have been applied to $f(x) = x^3$ are:

A. dilation by a factor of $\frac{1}{2}$ parallel to the y-axis, dilation by a factor of 2 parallel to the x-axis, a translation of 1 unit in the negative x-direction and a translation of 4 units up.

B. dilation by a factor of $\frac{1}{2}$ parallel to the y-axis, dilation by a factor of 2 parallel to the x-axis, a translation of 1 unit in the positive x-direction and a translation of 4 units up.

C. dilation by a factor of $\frac{1}{2}$ parallel to the y-axis, dilation by a factor of $\frac{1}{2}$ parallel to the x-axis, a translation of 1 unit in the negative x-direction and a translation of 4 units up.

D. dilation by a factor of $\frac{1}{2}$ parallel to the y-axis, dilation by a factor of $\frac{1}{2}$ parallel to the x-axis, a translation of 1 unit in the positive x-direction and a translation of 4 units up.

E. dilation by a factor of 2 parallel to the y-axis, dilation by a factor of $\frac{1}{2}$ parallel to the x-axis, a translation of 1 unit in the negative x-direction and a translation of 4 units up.

13. **MC** The rule for the inverse of the graph shown would be:

A. $y=\frac{1}{x}+1$

B. $y=\frac{1}{x+1}$

C. $y=\frac{1}{x}-1$

D. $y=\frac{1}{x-1}$

E. $y=\frac{1}{x-1}-1$

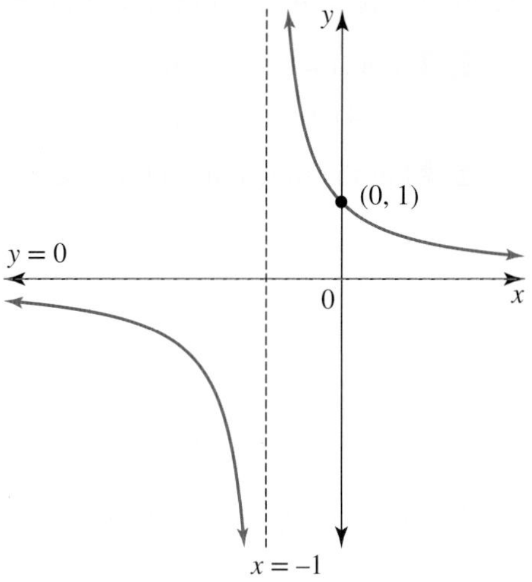

14. **MC** For the function $f(x)=(x+1)(x-3)$ to have an inverse function, its maximal domain:

A. must be restricted to $[0, \infty)$
B. must be restricted to $[1, \infty)$
C. must be restricted to $[-4, \infty)$
D. is R
E. must be restricted to $(-\infty, 0]$

15. **MC** The inverse of the function defined by $f:[-1,\infty]\to R, f(x)=(x+1)^2$ would be:

A. $f^{-1}:[-1,\infty)\to R, f^{-1}(x)=\sqrt{x}-1$

B. $f^{-1}:[-1,\infty)\to R, f^{-1}(x)=-\sqrt{x}-1$

C. $f^{-1}:[-1,\infty)\to R, f^{-1}(x)=(x+1)^2$

D. $f^{-1}:[0,\infty)\to R, f^{-1}(x)=\sqrt{x}-1$

E. $f^{-1}:[0,\infty)\to R, f^{-1}(x)=-\sqrt{x}-1$

16. **MC** Solve $\frac{p}{2x+n}=\frac{n}{2x+p}$ for x.

A. $x=\frac{n^2-p^2}{2(p-n)}$

B. $x=\frac{n+p}{2}$

C. $x=\frac{-(n+p)}{2}$

D. $x=\frac{n-p}{2}$

E. $x=\frac{n+p}{n-p}$

Technology active: extended response

17. Consider the equations $f: R\to R, f(x)=x^2-4$ and $g:(2,\infty)\to R, g(x)=\frac{1}{x-2}$.

a. Prove that $f(g(x))$ is defined.
b. Determine the rule for $f(g(x))$ and state the domain and range.
c. Prove that $g(f(x))$ is not defined.
d. Restrict the domain of $f(x)$ to obtain a function $f_1(x)$ such that $g(f_1(x))$ exists.
e. Determine $g(f_1(x))$ and state the domain.

18. Consider the function defined by the rule $f: D\to R, f(x)=\sqrt{(3x-6)}-1$, where D is the maximal domain for f.

a. Determine D.
b. Describe the transformations that would have been applied to $y=\sqrt{x}$ in order to achieve $y=f(x)$.
c. Define the rule for the inverse function f^{-1} and give its domain and range.
d. Sketch the graphs of $y=f(x)$ and $y=f^{-1}(x)$ on the same set of axes.

19. If $f:[3,\infty)\to R, f(x)=x^2+k$ and $g:[2,\infty)\to R, g(x)=\frac{1}{x}+k$, where k is a positive constant, calculate the value(s) of k such that both $f(g(x))$ and $g(f(x))$ are defined.

20. Consider the function defined by $f(x)=2(x-3)^2$.

a. Sketch this graph, giving the domain and range of the function.
b. Determine the rule for the inverse.
c. Sketch this inverse on the same set of axes that you used for $f(x)=2(x-3)^2$.
d. Restrict the domain of f to the form of $[a,\infty]$ so that the inverse is also a function.
e. State the rules for the restricted f and f^{-1} using function notation.
f. Sketch the graphs of f and f^{-1} on one set of axes.
g. Show that $f\left(f^{-1}(x)\right)=x$.

3.7 Exam questions

Question 1 (1 mark) TECH-ACTIVE

Source: *VCE 2017, Mathematical Methods Exam 2, Section A, Q4; © VCAA.*

MC Let f and g be functions such that $f(2)=5$, $f(3)=4$, $g(2)=5$, $g(3)=2$ and $g(4)=1$.

The value of $f(g(3))$ is

A. 1 **B.** 2 **C.** 3 **D.** 4 **E.** 5

Question 2 (1 mark) TECH-ACTIVE

Source: *VCE 2015, Mathematical Methods (CAS) Exam 2, Section 1, Q5; © VCAA.*

MC Part of the graph of $y=f(x)$ is shown below.

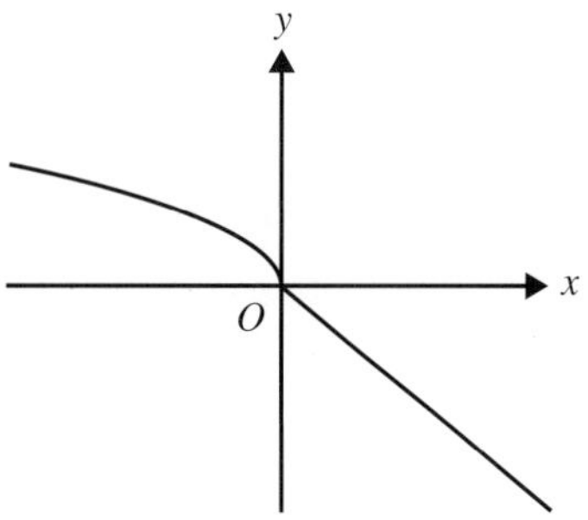

The corresponding part of the graph of the inverse function $y=f^{-1}(x)$ is best represented by

A.

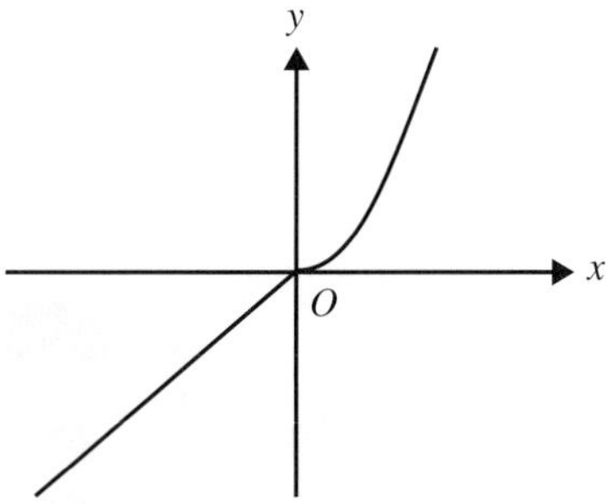

B.

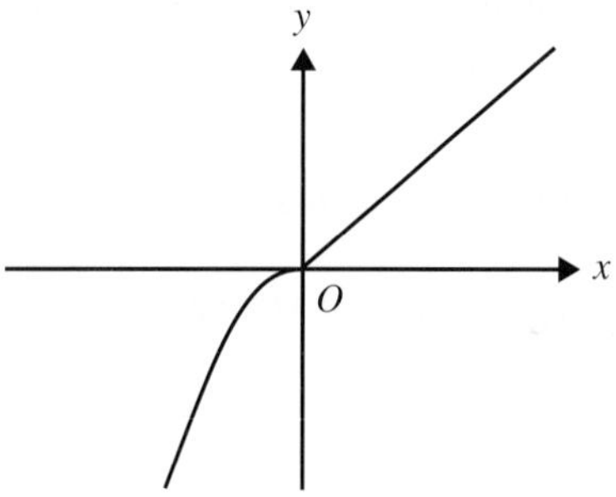

C.

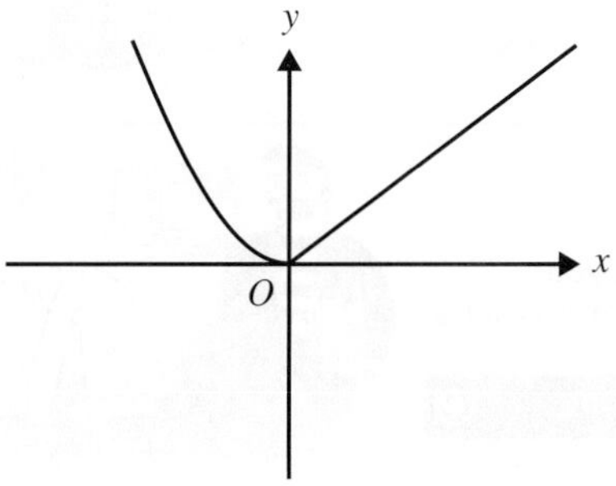

D.

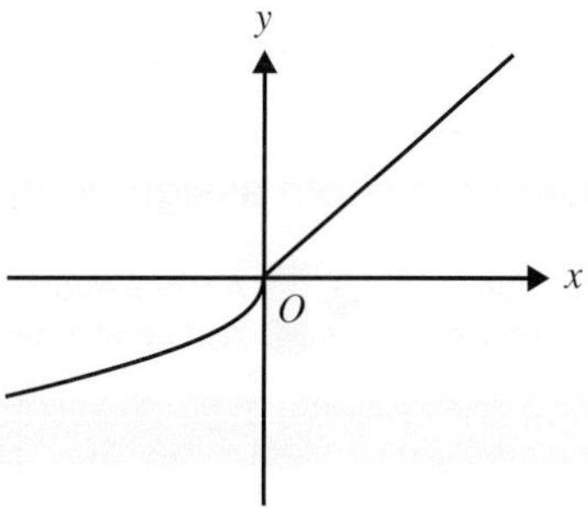

E.

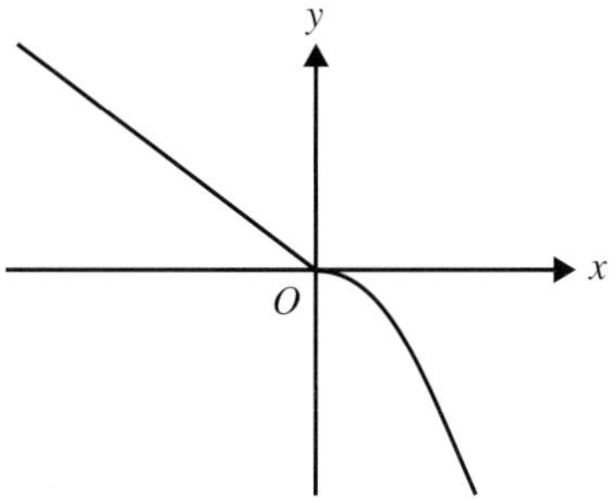

Question 3 (1 mark) TECH-ACTIVE

Source: *VCE 2018, Mathematical Methods Exam 2, Section A, Q4; © VCAA.*

MC The point $A\,(3, 2)$ lies on the graph of the function f. A transformation maps the graph of f to the graph of g, where $g\,(x) = \frac{1}{2}f\,(x - 1)$. The same transformation maps the point A to the point P.

The coordinates of the point P are

A. $(2, 1)$
B. $(2, 4)$
C. $(4, 1)$
D. $(4, 2)$
E. $(4, 4)$

Question 4 (3 marks) TECH-FREE

Source: *VCE 2019, Mathematical Methods Exam 1, Q2; © VCAA.*

Let $f: R\backslash\left\{\frac{1}{3}\right\} \to R,\ f(x) = \dfrac{1}{3x - 1}$.

a. Find the rule of f^{-1}. **(2 marks)**
b. State the domain of f^{-1}. **(1 mark)**

Question 5 (5 marks) TECH-FREE

Source: *VCE 2017, Mathematical Methods Exam 1, Q7; © VCAA.*

Let $f: [0, \infty) \to R,\ f(x) = \sqrt{x + 1}$

a. State the range of f. **(1 mark)**
b. Let $g: (-\infty, c] \to R, g\,(x) = x^2 + 4x + 3$, where $c < 0$.

i. Find the largest possible value of c such that the range of g is a subset of the domain of f. **(2 marks)**
ii. For the value of c found in part **b.i.**, state the range of $f(g\,(x))$. **(1 mark)**

c. Let $h: R \to R, h\,(x) = x^2 + 3$.
State the range of $f(h\,(x))$. **(1 mark)**

More exam questions are available online.

Answers

Topic 3 Composite functions, transformations and inverses

3.2 Composite functions

3.2 Exercise

1. a. $g\,(h(x)) = 2\sqrt{x} - 3$
   b. $f \circ g(x) = -\,(2x-3)^2 + 1$
2. a. $g\,(h(x)) = \left(x^4 + 2\right)^2$
   b. $h \circ f(x) = 1 - \dfrac{1}{(x+2)^4}$
3. $f(g(x)) = (x-1)(x+1)(x^2+3)$, domain $= R$
   $g(f(x)) = (x-1)^2(x+3)^2$, domain $= R$
4. $f(g(x)) = \dfrac{2}{x-2} - 1$, domain $= R\backslash\{2\}$
   $g(f(x))$ does not exist.
5. a. $f \circ g(x)$ is defined, domain $= [0, \infty)$.
   b. $g(f(x))$ is defined, domain $= R$.
   c. $h(g(x))$ is not defined.
   d. $h \circ f(x)$ is defined, domain $= R$.
6. a. $f \circ g(x)$ is defined, $f \circ g(x) = x$, domain $= [0, \infty)$.
   b. $g(f(x))$ is defined, $g(f(x)) = |x|$, domain $= R$.
   c. $h\left(f(x)\right)$ is not defined.
   d. $g(h(x))$ is not defined.
7. ran $g \subseteq$ dom f
   $[0, \infty) \subseteq R$
   Therefore, $f(g(x))$ is defined.
   $f(g(x)) = x + 3$, where domain $= [-2, \infty)$ and range $= [1, \infty)$.
8. a. ran $f \subseteq$ dom g
   $(0, \infty) \subseteq R\backslash\{0\}$
   Therefore, $g(f(x))$ is defined.
   b. $g(f(x)) = x^2$, domain $x \in (0, \infty)$, range $= (0, \infty)$
   c.

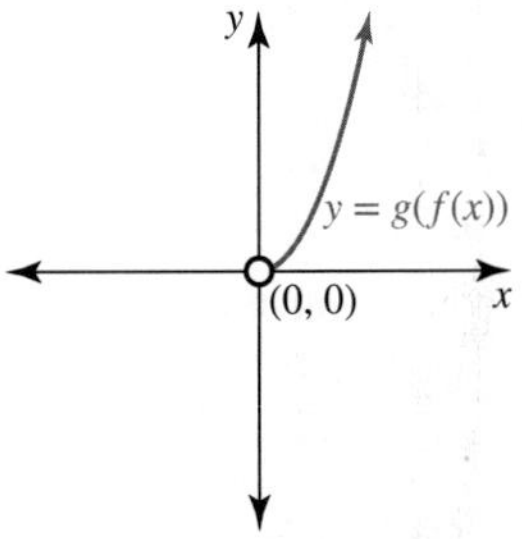

9. a. ran $g \not\subseteq$ dom f
   $R \not\subseteq [-3, \infty)$
   b. $h(x) = 2x - 5, x \in [1, \infty)$
   c. $f(h(x)) = \sqrt{2x-2}, x \in [1, \infty)$
10. a. ran $f \not\subseteq$ dom g
    $[0, \infty) \not\subseteq R\backslash\{4\}$
    b. $h(x) = x^2, x \in R\backslash\{-2, 2\}$
    c. $g(h(x)) = \dfrac{1}{x^2 - 4}, x \in R\backslash\{-2, 2\}$
11. a. ran $g \subseteq$ dom f
    $(2, \infty) \not\subseteq [0, \infty)$
    Therefore, $g(f(x))$ is not defined.
    b. $g_1(x) = \dfrac{1}{(x-3)^2} - 2, x \in \left[3 - \dfrac{1}{\sqrt{2}}, 3\right) \cup \left(3, 3 + \dfrac{1}{\sqrt{2}}\right]$
12. a. ran $f \not\subseteq$ dom g
    $[0, \infty) \not\subseteq R\backslash\{1\}$
    Therefore, $g(f(x))$ is not defined.
    b. $f_1(x) = \sqrt{2-x}, x \in (-\infty, 2] \backslash \{1\}$
    c. $g(f_1(x)) = -\dfrac{1}{\sqrt{2-x} - 1} + 2, x \in (-\infty, 2] \backslash\{1\}$
13. a. ran $f \subseteq$ dom g
    $[0, \infty) \subseteq R$
    Therefore, $g(f(x))$ is defined.
    b. $g(f(x)) = x - 6$, domain $= [4, \infty)$
    c.

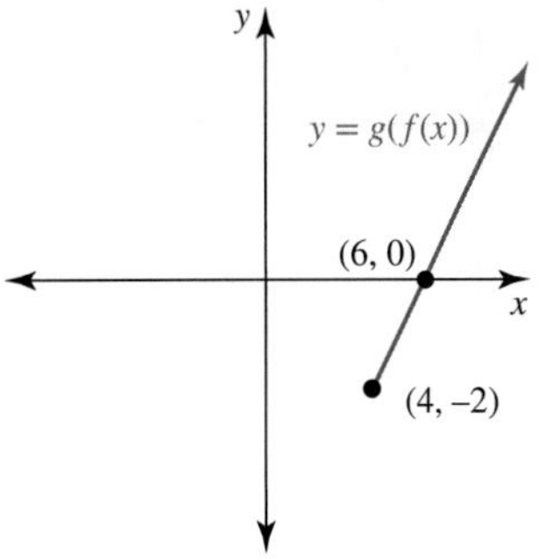

    d. ran $g \not\subseteq$ dom f
    $[-, 2, \infty) \not\subseteq [4, \infty)$
    Therefore, $f(g(x))$ is not defined.
    e. $g_1\,(x) = x^2 - 2, x \in \left(-\infty, -\sqrt{6}\right] \cup \left[\sqrt{6}, \infty\right)$
    $f(g_1(x)) = \sqrt{x^2 - 6}$.
    f. Domain $= \left(-\infty, -\sqrt{6}\right) \cup \left[\sqrt{6}, \infty\right]$
14. $k \in [1, 3]$

3.2 Exam questions

Note: Mark allocations are available with the fully worked solutions online.

1. D
2. D
3. A

3.3 Transformations

3.3 Exercise

1. a. i. Dilated by factor 3 parallel to the y-axis or from the x-axis

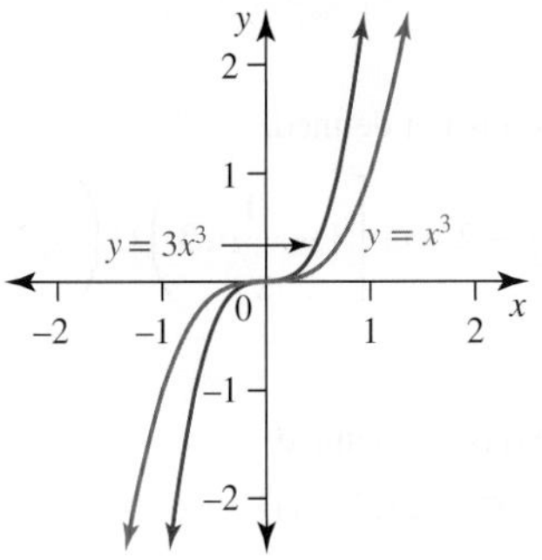

ii. $(-2, -8) \rightarrow (-2, -24)$

b. i. Translated 2 units to the left or in the negative x-direction

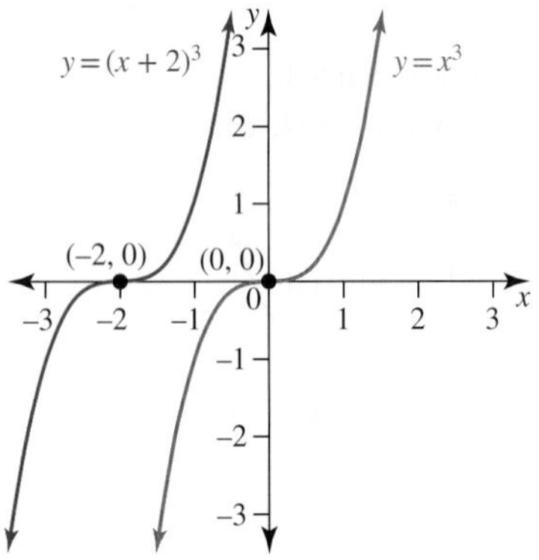

ii. $(-2, -8) \rightarrow (-4, -8)$

c. i. Reflected in the x-axis

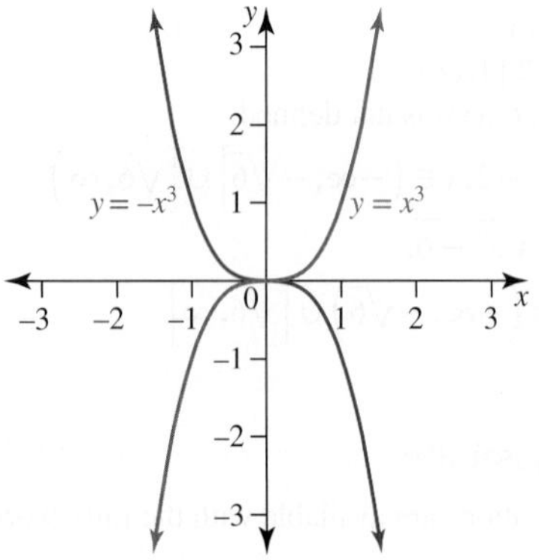

ii. $(-2, -8) \rightarrow (-2, 8)$

d. i. Translated up 1 unit or in the positive y-direction

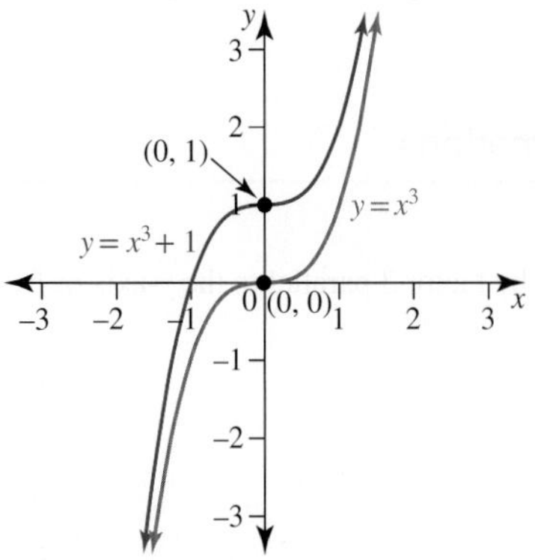

ii. $(-2, -8) \rightarrow (-2, -7)$

2. a. Dilated by factor 4 parallel to the y-axis or from the x-axis

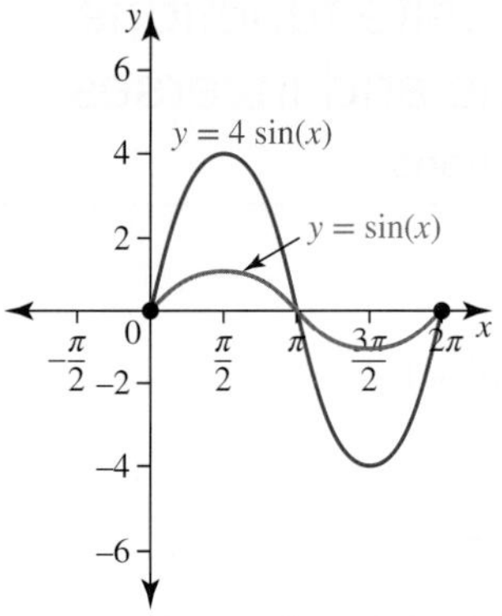

b. Dilated by factor $\frac{1}{2}$ parallel to the x-axis or from the y-axis

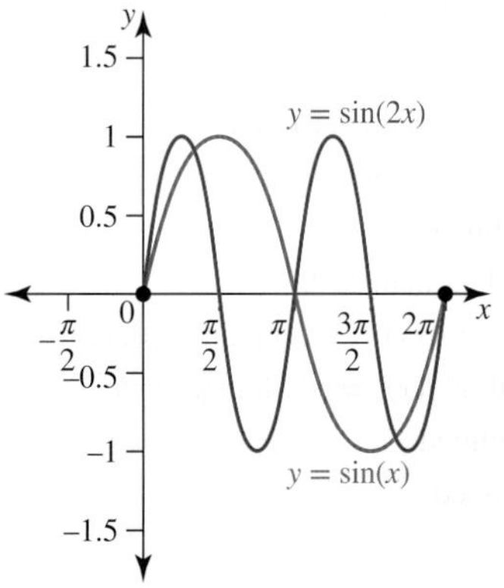

c. Translated $\frac{\pi}{2}$ units to the left or in the negative x-direction

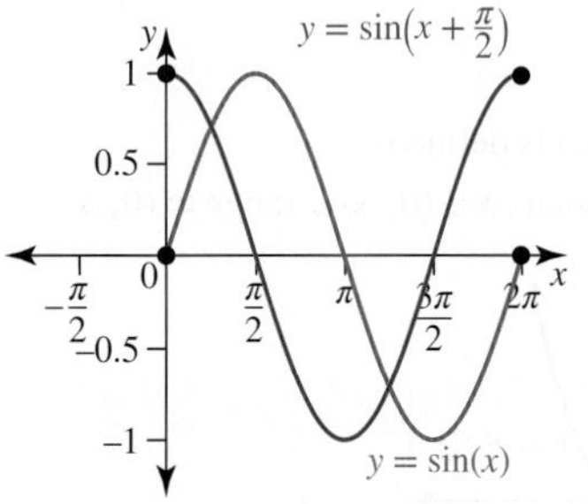

d. Translated up 2 units or in the positive y-direction

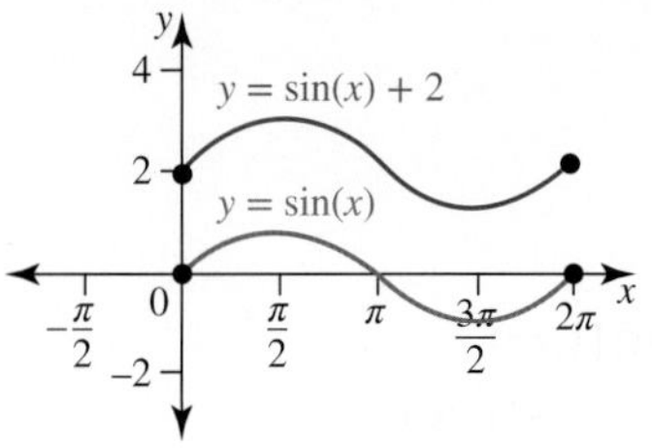

3. a. Dilated by factor $\frac{1}{2}$ parallel to the y-axis or from the x-axis

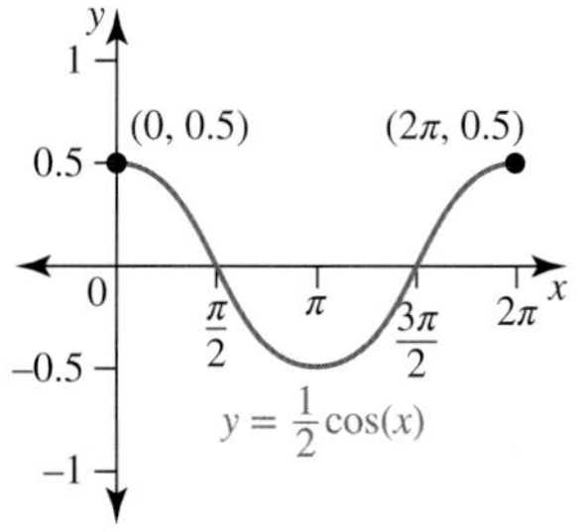

b. Dilated by factor $\frac{1}{2}$ parallel to the x-axis or from the y-axis

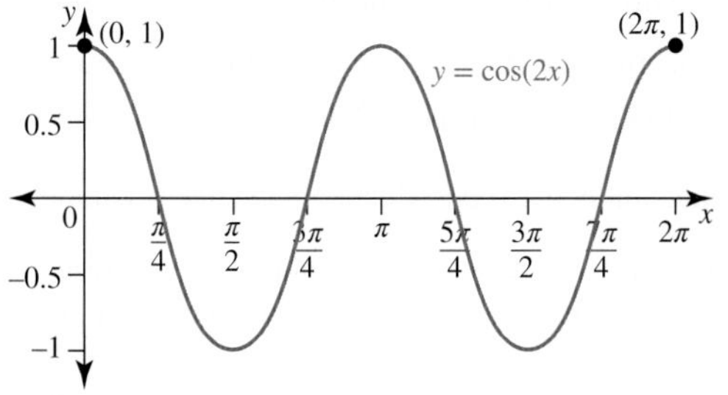

c. Reflected in the x-axis

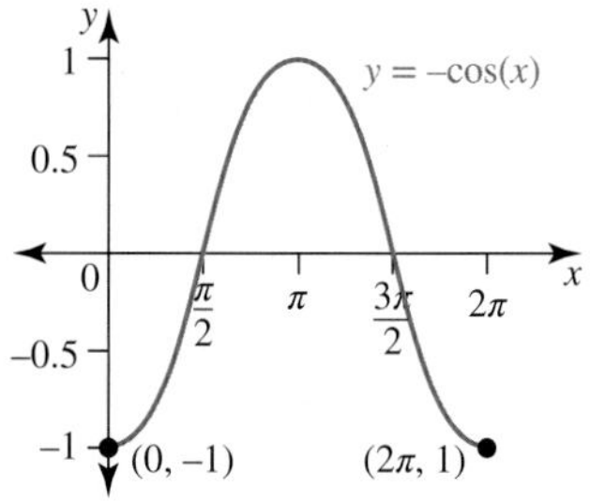

d. Translated down 1 unit or in the negative y-direction

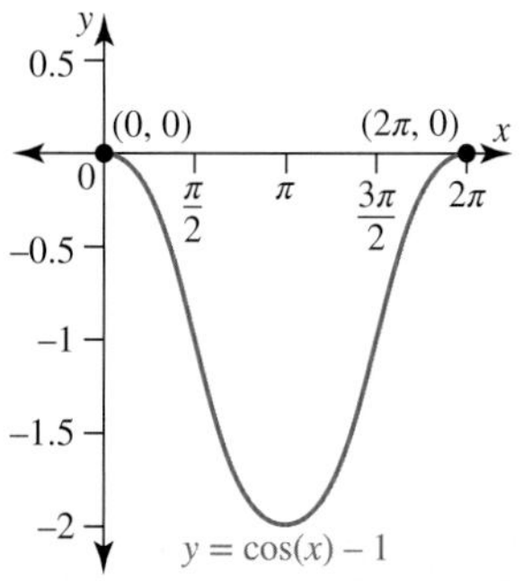

4. a. Translated 2 units to the right or in the positive x-direction, $y = \frac{1}{x-2}$; asymptotes $x = 2$, $y = 0$

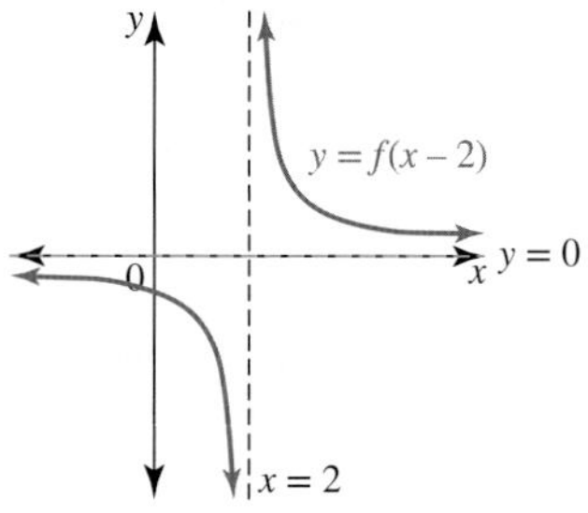

b. Reflected in the x axis, $y = -\frac{1}{x}$; asymptotes $x = 0$, $y = 0$

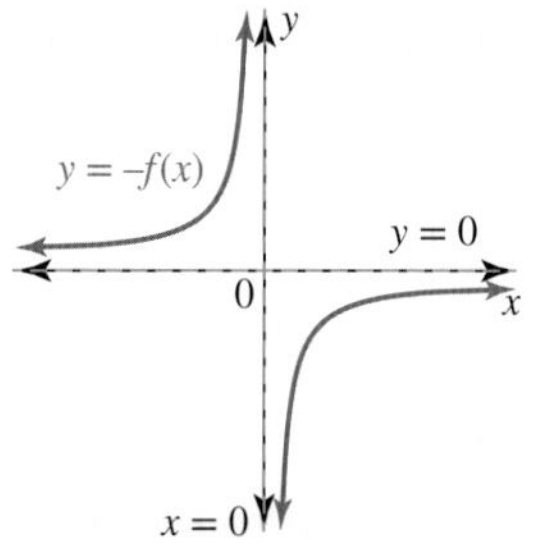

c. Dilated by factor 3 parallel to the y-axis or from the x-axis, $y = \frac{3}{x}$; asymptotes $x = 0$, $y = 0$

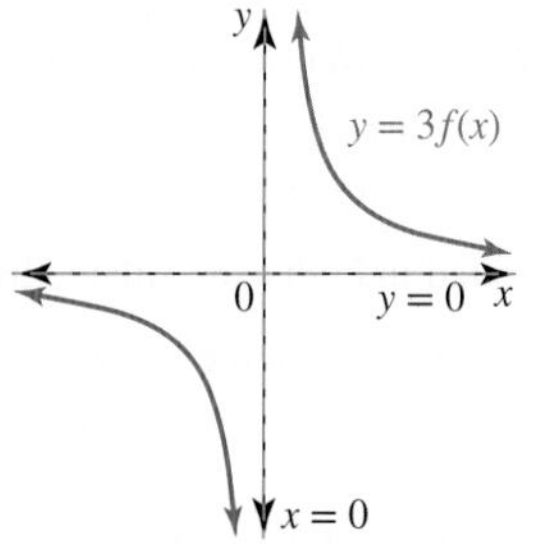

d. Dilated by factor $\frac{1}{2}$ parallel to the x-axis or from the y-axis, $y = \frac{1}{2x}$; asymptotes $x = 0$, $y = 0$

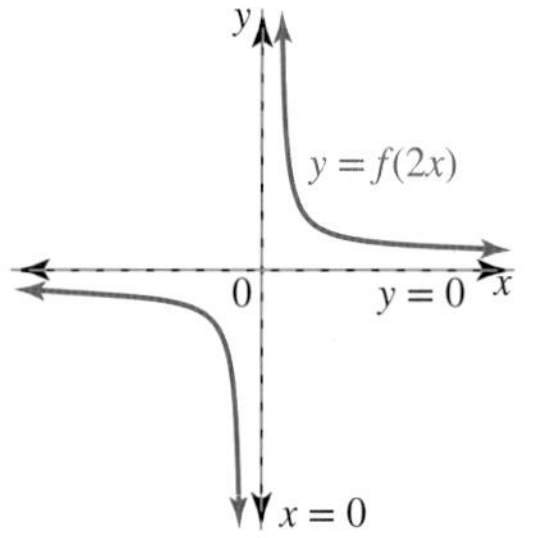

5. Reflected in the x-axis, dilated by factor 2 parallel to the y-axis or from the x-axis, dilated by factor $\frac{1}{2}$ parallel to the x-axis or from the y-axis, translated $\frac{\pi}{4}$ units to the right or in the positive x-direction and translated up 1 unit or in the positive y-direction

6. Dilated by factor $\frac{1}{3}$ parallel to the y-axis or from the x-axis, dilated by factor 2 parallel to the x-axis or from the y-axis, translated 1 unit to the left or in the negative x-direction and translated down 2 units or in the negative y-direction

7. a. $y = x^2$ has been dilated by factor $\frac{1}{3}$ parallel to the y-axis or from the x-axis, translated 3 units to the left or in the negative x-direction, and translated down $\frac{2}{3}$ units or in the negative y-direction.

b. $y = x^3$ has been reflected in the x-axis, dilated by factor 2 parallel to the y-axis or from the x-axis, reflected in the y-axis, translated 1 unit to the right or in the positive x-direction, and translated 1 unit up or in the positive y-direction.

c. $y = \frac{1}{x}$ has been dilated by a factor of 3 parallel to the y-axis or from the x-axis, dilated by factor $\frac{1}{2}$ parallel to the x-axis or from the y-axis, translated 3 units to the left or in the negative x-direction, and translated down 1 unit or in the negative y-direction.

8. a. $(-2, 4) \to \left(-5, \frac{2}{3}\right)$ b. $(1, 1) \to (0, -1)$

c. $\left(2, \frac{1}{2}\right) \to (-2, 0)$

9. a. $y = \cos(x)$ has been dilated by factor 2 parallel to the y-axis or from the x-axis, dilated by factor $\frac{1}{2}$ parallel to the x-axis or from the y-axis, translated $\frac{\pi}{2}$ units to the right or in the positive x-direction, and translated up 3 units up or in the positive y-direction.

b. $y = \tan(x)$ has been reflected in both axes, dilated by factor $\frac{1}{2}$ parallel to the x-axis or from the y-axis, and translated up 1 unit or in the positive y-direction.

c. $y = \sin(x)$ has been dilated by factor $\frac{1}{3}$ parallel to the x-axis or from the y-axis, translated $\frac{\pi}{3}$ units to the right or in the positive x-direction, and translated down 1 unit or in the negative y-direction.

10. A

11. D

12. $f(x) = \sqrt[3]{-\frac{x-6}{2}}$

13. $f(x) = -\frac{3}{(2-x)^2} + 9$

14. $f(x) = -18(x-1)^2 + 1$

15. $f(x) = \frac{1}{4-x} - 6$

16. $$\begin{aligned} y &= \frac{2x-5}{x-1} \\ &= \frac{2(x-1)-3}{x-1} \\ &= \frac{2(x-1)}{x-1} - \frac{3}{x-1} \\ &= 2 - \frac{3}{x-1} \end{aligned}$$

Relative to $y = \frac{1}{x}$, $y = \frac{2x-5}{x-1}$ has been reflected in the y-axis or the x-axis, dilated by factor 3 parallel to the y-axis or from the x-axis, translated 1 unit to the right or in the positive x-direction, and translated 2 units up or in the positive y-direction. Domain $= R\backslash\{1\}$ and range $= R\backslash\{2\}$; asymptotes $x = 1$ and $y = 2$

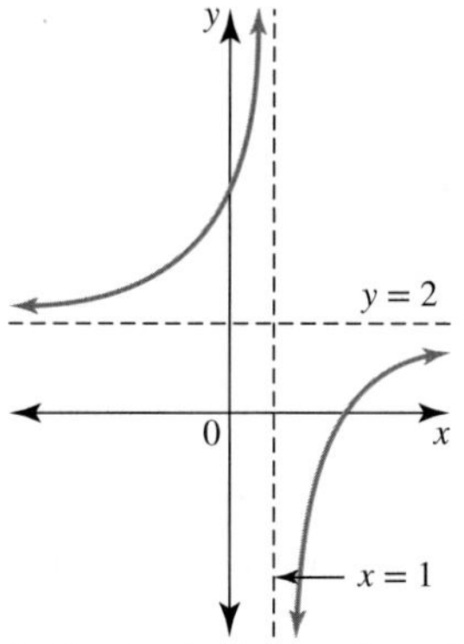

17. Reflection in the x-axis, reflection in the y-axis, translation 5 units right, dilation by factor $\frac{1}{2}$ parallel to the x-axis or from the y-axis, translation 3 units up.

18. Reflection in the x-axis, dilation by factor $\frac{1}{2}$ parallel to the y-axis or from the x-axis, dilation by factor 3 parallel to the x-axis or from the y-axis, translation 3 units left, translation $\frac{3}{2}$ units up.

3.3 Exam questions

Note: Mark allocations are available with the fully worked solutions online.

1. C
2. D
3. A

3.4 Inverse graphs

3.4 Exercise

1. a.

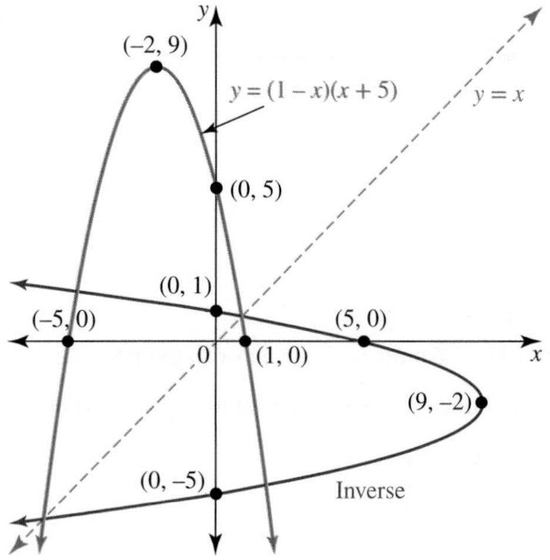

b. $y = (1-x)(x+5)$ is a many-to-one function.
The inverse is a one-to-many relation.

c. $y = (1-x)(x+5)$: domain $= R$, range $= (-\infty, 9]$
Inverse: domain $= (-\infty, 9]$, range $= R$

2. a. and b.

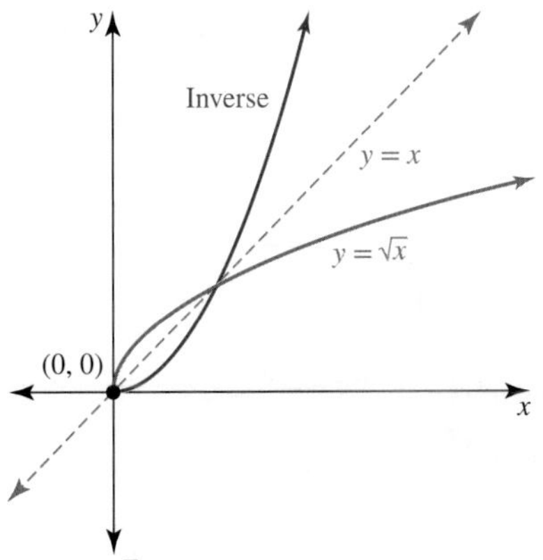

c. $y = \sqrt{x}$ is a one-to-one function. The inverse is a one-to-one function.

3. a.

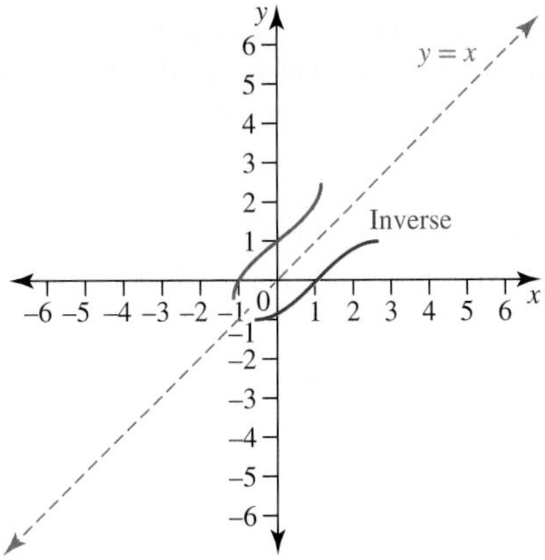

b.

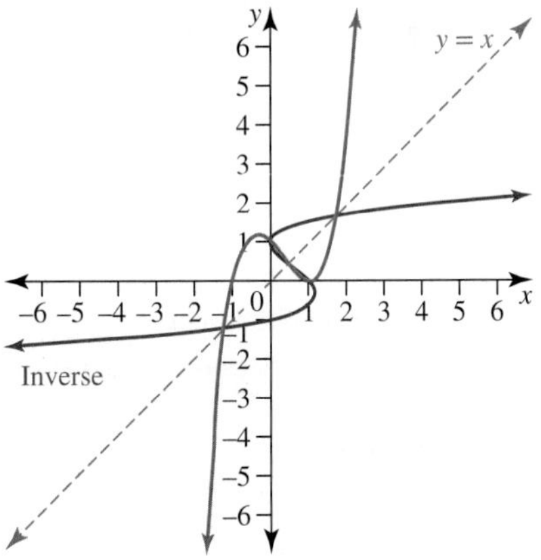

c.

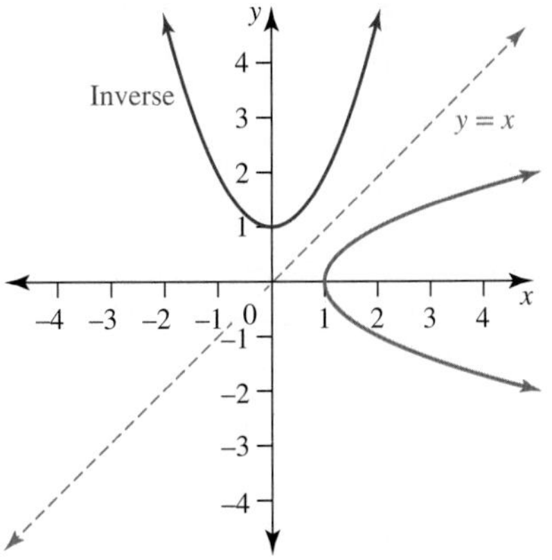

d.

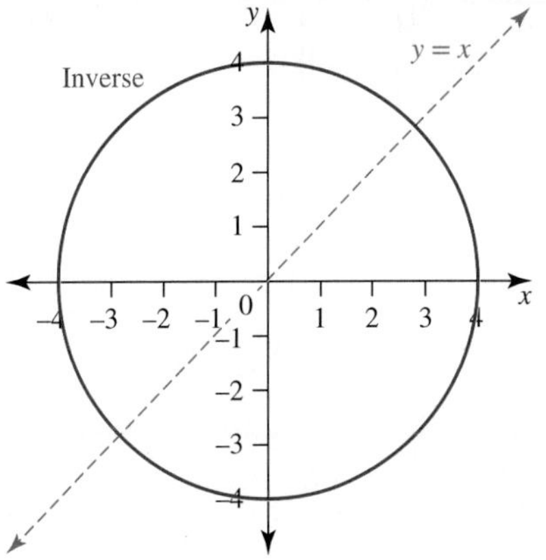

4. a. and b.

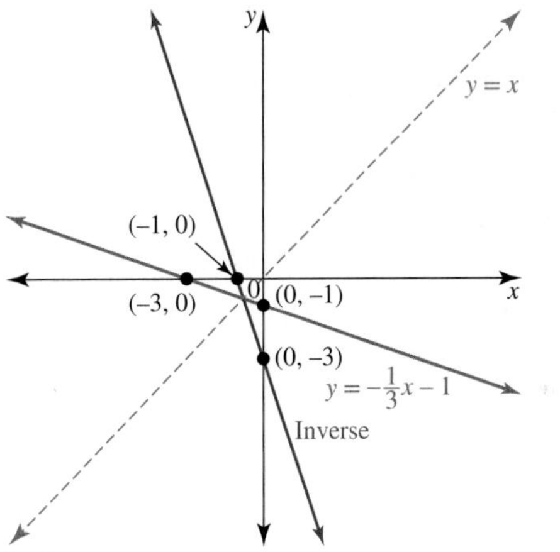

5. b, d

6. C

7. C

8. A is the correct option, as the given function has a turning point at $(0, 2)$ and option A has a turning point at $(2, 0)$.

9. a.

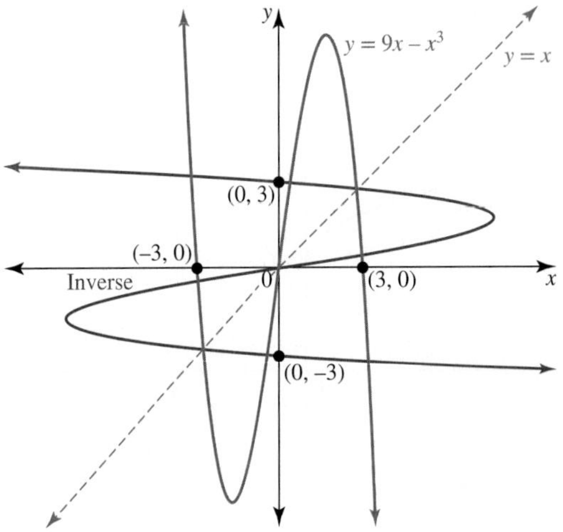

b. $(2.828, 2.828), (0, 0), (-2.828, -2.828)$

10. a.

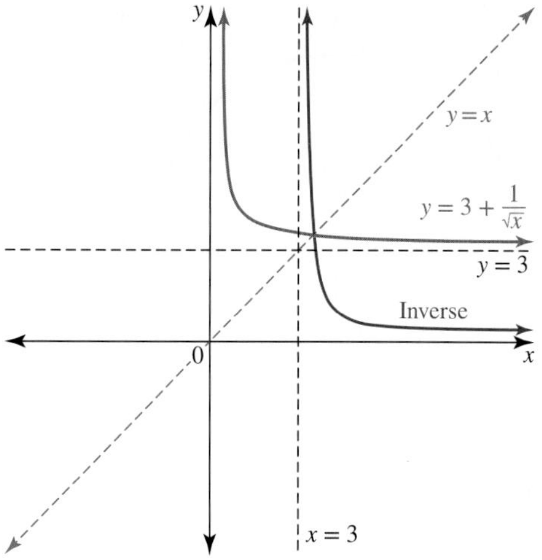

b. Point of intersection = (3.532, 3.532)

11. a. $a = -2$

b.

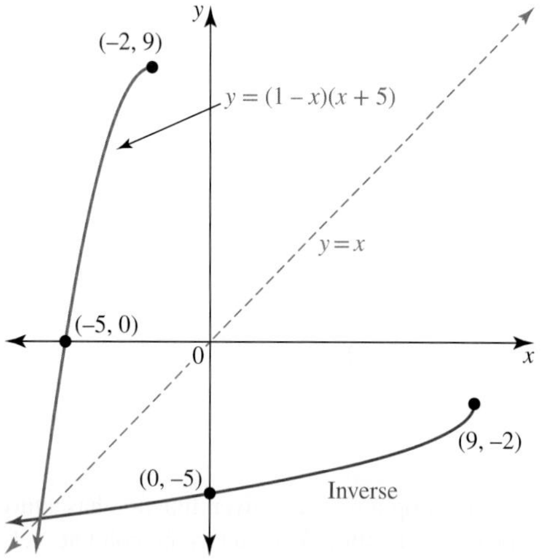

c. y: domain $(-\infty, -2]$, range $(-\infty, 9]$
Inverse: domain $(-\infty, 9]$, range $(-\infty, -2]$

12. Domain = $[3, \infty)$

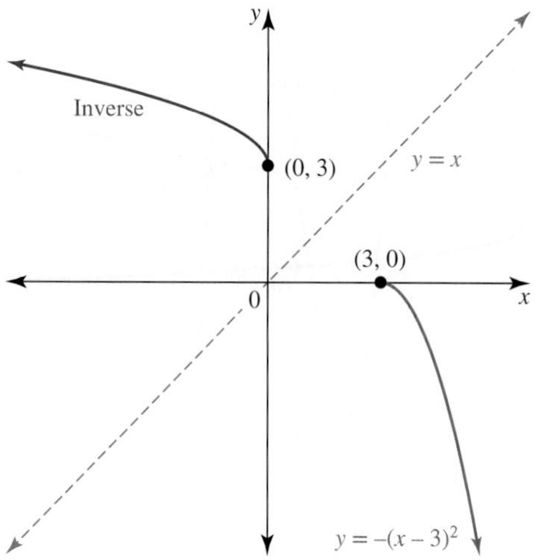

13. a. and b.

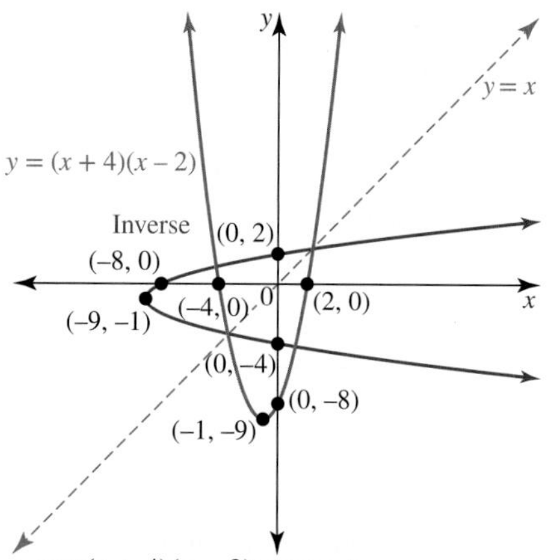

c. $y = (x + 4)(x - 2)$: many-to-one
Inverse: one-to-many

d. The inverse is not a function, as a one-to-many correspondence indicates a relation. Also, the inverse can only be a function if the original graph is a one-to-one function, and this graph is a many-to-one function.

e. y: domain = R, range = $[-9, \infty)$
Inverse: domain = $[-9, \infty)$, range = R

f. $(-\infty, -1]$ or $[-1, \infty)$

14. a.

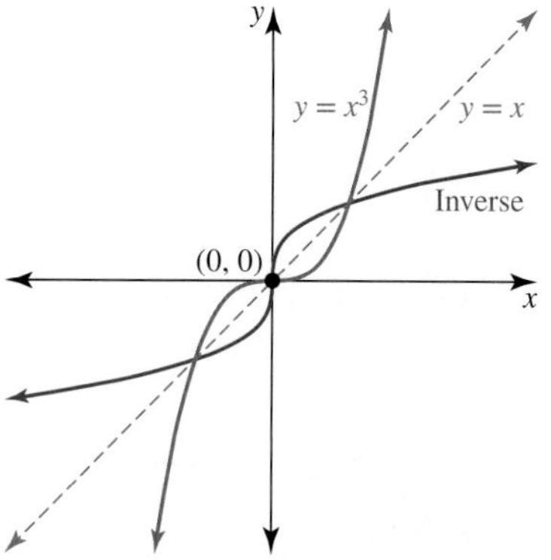

b. $y = x^3$: one-to-one; inverse: one-to-one

c. The inverse of y is a function because y is a one-to-one function.

d. $y = x^3$: domain = R, range = R
Inverse: domain = R, range = R

15. a. and b.

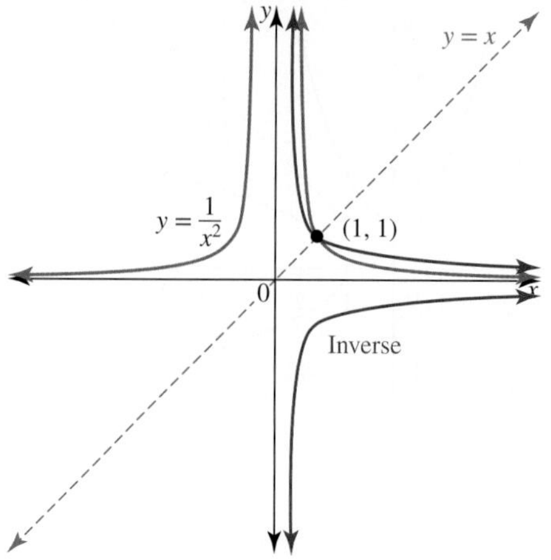

c. $y = \dfrac{1}{x^2}$: many-to-one function
Inverse: one-to-many relation

d. $(-\infty, 0)$

e. $y = \dfrac{1}{x^2}$: domain $= (-\infty, 0)$, range $= (0, \infty)$
Inverse: domain $= (0, \infty)$, range $= (-\infty, 0)$

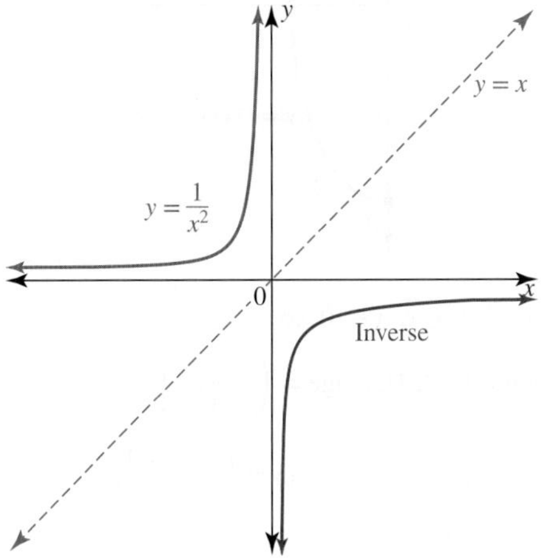

16. $a = 3$

3.4 Exam questions

Note: Mark allocations are available with the fully worked solutions online.

1. C
2. A
3. C

3.5 Inverse functions

3.5 Exercise

1. Inverse: $y = \sqrt[3]{x}$, one-to-one function.
Domain for both $= R$, range for both $= R$

2. Inverse: $y = \pm \dfrac{1}{\sqrt{x}}$, one-to-many relation (therefore not a function)
y: domain $= R \backslash \{0\}$, range $= (0, \infty)$
Inverse: domain $= (0, \infty)$, range $= R \backslash \{0\}$

3. a. $y = 3(x+1)$; one-to-one function with domain $= R$ and range $= R$
b. $y = 5 \pm \sqrt{x}$; one-to-many relation with domain $= [0, \infty)$ and range $= R$
c. $y = (x+2)^3 - 1$; one-to-one function with domain $= R$ and range $= R$
d. $y = 1 + \sqrt[3]{x}$; one-to-one function with domain $= R$ and range $= R$
e. $y = x^2$; one-to-one function with domain $= [0, \infty)$ and range $= [0, \infty)$
f. $y = 1 \pm \dfrac{1}{\sqrt{x-2}}$; one-to-many relation with domain $= (2, \infty)$ and range $= R \backslash \{1\}$

4. $f^{-1}: (-\infty, 0) \to R, f^{-1}(x) = -\sqrt{-\dfrac{1}{x}} + 2$

5. $f^{-1}: [0, \infty) \to R, f^{-1}(x) = x^2 + 3$

6. $f^{-1}(x) = \dfrac{1}{x} - 2$

a. $f(f^{-1}(x)) = \dfrac{1}{\frac{1}{x} - 2 + 2}$
$= \dfrac{1}{\frac{1}{x}}$
$= x$

b. $f^{-1}(f(x)) = \dfrac{1}{\frac{1}{x+2}} - 2$
$= x + 2 - 2$
$= x$

7. $k^{-1}(x) = \sqrt[3]{x+1}$

a. $k(k^{-1}(x)) = \left(\sqrt[3]{x+1}\right)^3 - 1$
$= x + 1 - 1$
$= x$

b. $k^{-1}(k(x)) = \sqrt[3]{x^3 - 1 + 1}$
$= \sqrt[3]{x^3}$
$= x$

8. a. $y = \pm\sqrt{x} - 1$; the inverse is not a function as $f(x)$ is not a one-to-one function.
b. $b = -1$
c.

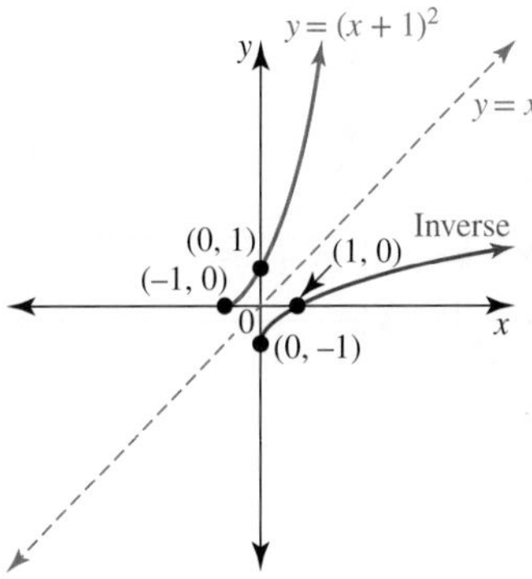

d. $f^{-1}(x) = \sqrt{x} - 1$, domain $= [0, \infty)$
e. No intersection

9. $(2 + 2\sqrt{3}, 2 + 2\sqrt{3})$

10. a. $a = 1$
b. $f^{-1}(x) = -\sqrt{x+2} + 1, x \in [-2, \infty)$

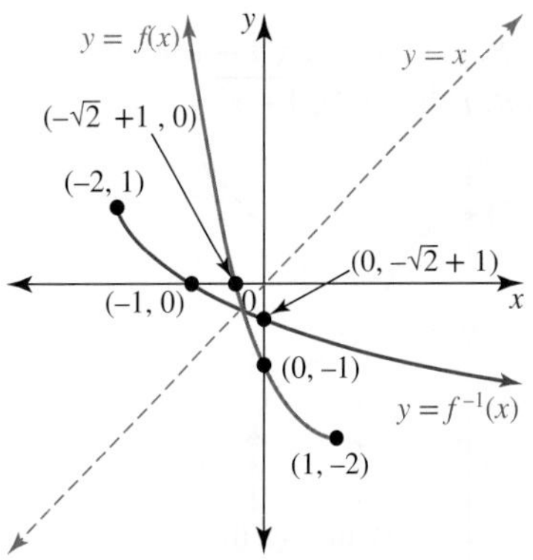

c. The point of intersection is $\left(\dfrac{3-\sqrt{13}}{2}, \dfrac{3-\sqrt{13}}{2}\right)$.

11. a. $f(x) = x^4$: many-to-one function; inverse: one-to-many relation

b. $f(x) = 2x^2 - 7x + 3$: many-to-one function; inverse: one-to-many relation

c. $f(x) = \dfrac{1}{(x-3)^2} + 2$; many-to-one function; inverse: one-to-many relation.

d. $f(x) = \sqrt{x+2}, x \in [-2, \infty)$: one-to-one function; $f^{-1} : [0, \infty) \to R, f^{-1}(x) = x^2 - 2$

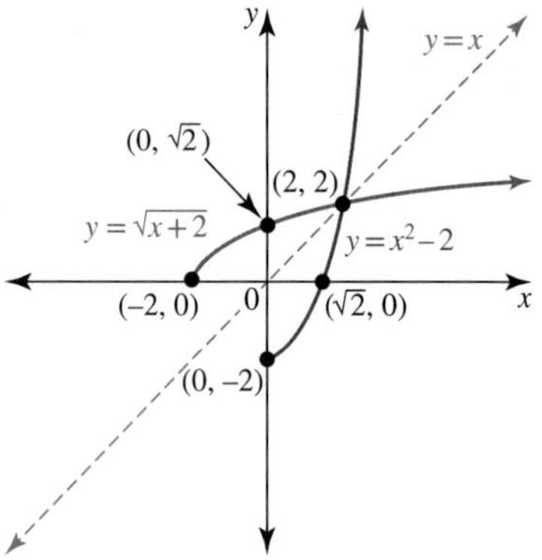

12. $f^{-1}(x) = 2 + \dfrac{1}{x-4}$

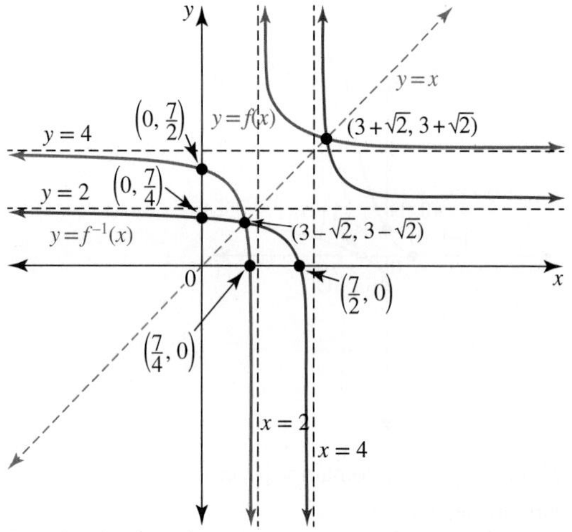

13. Restrict the domain to $(-\infty, -2]$.
$f^{-1} : [0, \infty) \to R, f^{-1}(x) = -\sqrt{x} - 2$

14. Two possible domains are [–2, 0] and [0, 2]. Both have a range of [0, 1].
$f^{-1}(x) : [0, 1] \to R, f^{-1}(x) = -2\sqrt{1-x^2}$ or
$f^{-1}(x) : [0, 1] \to R, f^{-1}(x) = 2\sqrt{1-x^2}$.

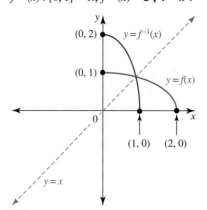

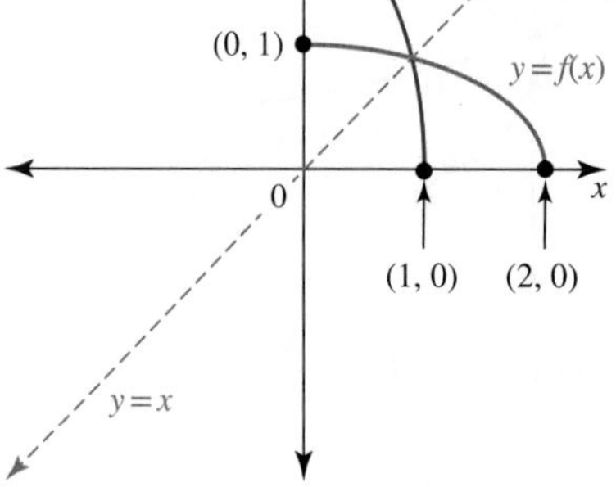

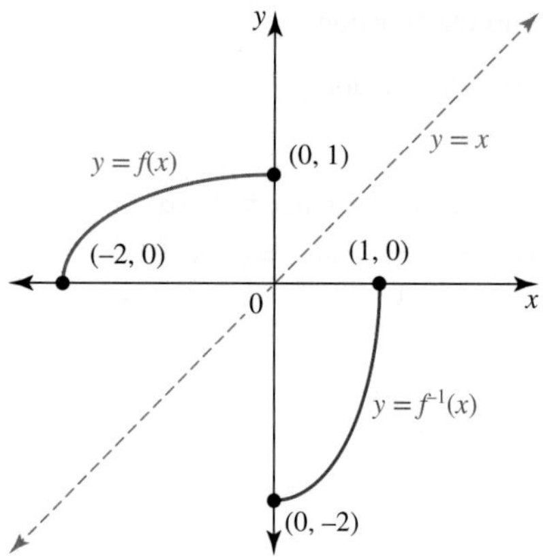

15. a. $a = 5$

b. $f^{-1}(x) = 5 + \sqrt{x}, \ x \in [0, \infty)$

16. a. Domain $= [-2, 4)$, range $= \left(-\dfrac{1}{3}, \dfrac{5}{3}\right]$

b. $f^{-1}(x) = -3(x-1)$; domain $= \left(-\dfrac{1}{3}, \dfrac{5}{3}\right]$, range $= [-2, 4)$

c.

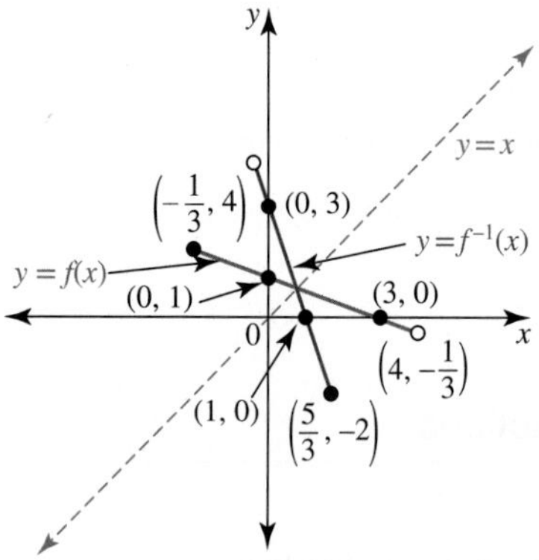

d. The point of intersection is $\left(\dfrac{3}{4}, \dfrac{3}{4}\right)$.

17. a. $D = \left(-\infty, \dfrac{1}{3}\right]$

b. $f^{-1}(x) = \dfrac{1}{3} - \dfrac{x^2}{3}$; domain $= [0, \infty)$, range $= \left(-\infty, \dfrac{1}{3}\right]$

c. The point of intersection is $\left(\dfrac{-3+\sqrt{13}}{2}, \dfrac{-3+\sqrt{13}}{2}\right)$.

d.

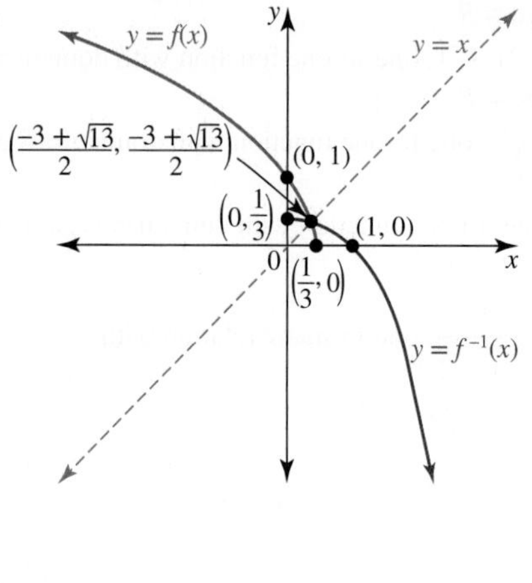

18. a. $f^{-1}(x) = x^2 + 1,\ x \in [0, \infty)$

b. $f^{-1}(f(x)) = x$; domain $= [1, \infty)$

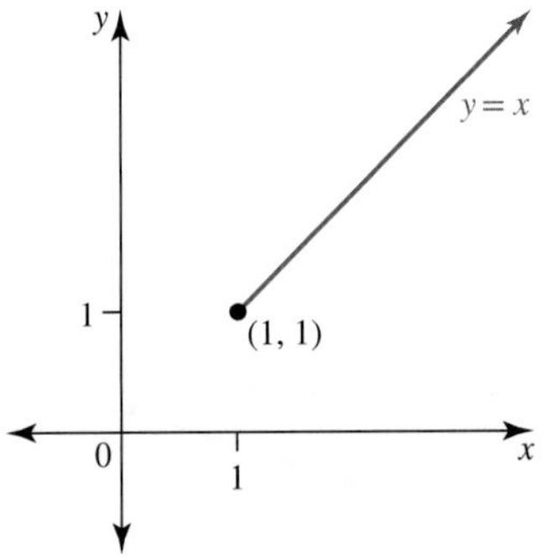

c. $\dfrac{x+2}{3}$

3.5 Exam questions

Note: Mark allocations are available with the fully worked solutions online.

1. $f^{-1}(x) = 2 + \dfrac{1}{\sqrt{x}}$, dom $= (0, \infty)$
2. D
3. A

3.6 Literal equations

3.6 Exercise

1. a. $x = \dfrac{my - kz}{n + 4}$

   b. $x = \dfrac{2pc - 3c^2}{2p - m - 3c}$

2. $y = \dfrac{x(1 - 2p)}{m + 2}$

3. a. $x = -\dfrac{y(6k + d)}{3k}$

   b. $x = \dfrac{pq - ny}{m - p}$

   c. $x = \dfrac{m - 3k}{m + k}$

   d. $x = \dfrac{km - 2dm}{2d + k}$

4. $x = \dfrac{9 - 6cd + c^2d^2}{ab^2}$

5. $x = \dfrac{d - 2kn}{m - n},\ y = \dfrac{2km - d}{m - n}$

6. a. $x = \dfrac{k + 2d}{2n},\ y = \dfrac{2d - k}{2m}$

   b. $x = 0, y = 1$

7. a. $x = \dfrac{3k + d}{m},\ y = -\dfrac{2d + 3k}{n}$

   b. $x = \dfrac{4ab}{a + b},\ y = \dfrac{2ab}{a + b}$

8. $x = \dfrac{2(a + 2)}{a(a + 4)},\ y = \dfrac{4(a + 2)}{a(a + 4)},\ z = \dfrac{4}{a}$

3.6 Exam questions

Note: Mark allocations are available with the fully worked solutions online.

1. B
2. E
3. A

3.7 Review

3.7 Exercise

Technology free: short answer

1. $f(g(x))$ does not exist; $g(f(x))$ exists.
   $g(f(x)) = 2x - 1$. The domain is $[-2, \infty)$ and the range is $[-5, \infty)$.

2. a. $m = 1$ and $n = 1$

   b. Translation of 2 units right and 1 unit up

   c. $y = \dfrac{1}{x - 2} + 1$; domain $= R\backslash\{2\}$ and range $= R\backslash\{1\}$

   Asymptotes: $x = 2$ and $y = 1$

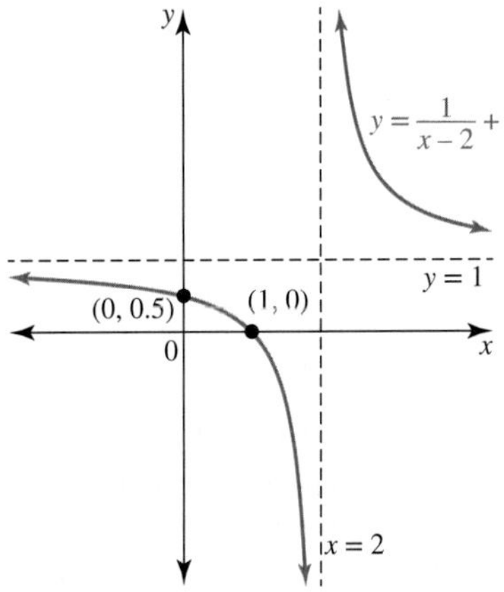

   d. $y = \dfrac{1}{x - 1} + 2$, domain $= R\backslash\{1\}$ and range $= R\backslash\{2\}$

   e. The inverse is a one-to-one function.

   f.

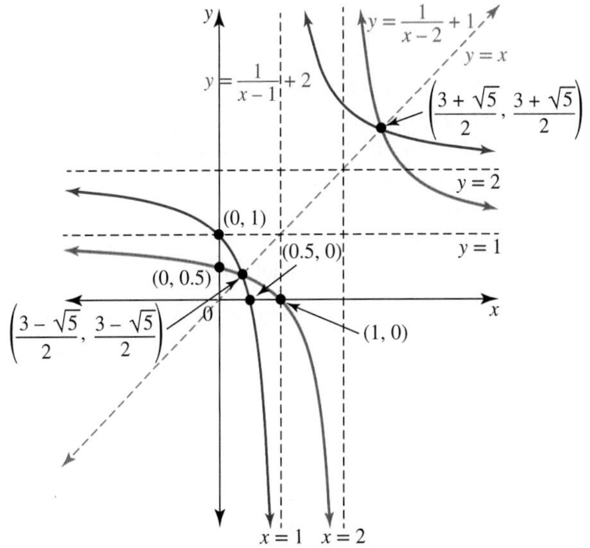

3. a. One-to-one inverse function: $f^{-1}: R \to R, f^{-1}(x) = \sqrt[3]{3x}$

   b. Not a function, as it is a one-to-many mapping

c. Not a function, as it is a one-to-many mapping

d. One-to-one inverse function:
f^{-1}: $[0, \infty) \to R, f^{-1}(x) = x^2 + 3$

4. a. $g(x) = -\frac{1}{2}\sqrt{\frac{x-3}{2}} - \frac{1}{2}$

b. $g(x) = -(3x + 7)^2 + 2$

5. a. Dilation of factor 3 from the x-axis or parallel to the y-axis, dilation of factor $\frac{1}{2}$ from the y-axis or parallel to the x-axis, and translation of $\frac{5}{2}$ units right and 1 unit up

b. Translation down 1 unit, reflection in the x-axis and y-axis, translation 5 units right, and dilation of factor $\frac{1}{2}$ from the y-axis or parallel to the x-axis

6. $x = \frac{4k}{cd - 2k - 5}$

Technology active: multiple choice

7. A
8. D
9. A
10. C
11. E
12. D
13. C
14. B
15. D
16. C

Technology active: extended response

17. a. For $f\left(g(x)\right)$ to exist, the range of the inner function, $g(x)$, must be a subset of or equal to the domain of the outer function, $f(x)$.

$(0, \infty) \subseteq R$
ran $g \subseteq$ dom f

Therefore, $f\left(g(x)\right)$ is defined.

b. $f\left(g(x)\right) = \frac{1}{(x-2)^2} - 4$

Domain $= (2, \infty)$, range $= (-4, \infty)$

c. For $g\left(f(x)\right)$ to exist the range of the inner function, $f(x)$ must be a subset of or equal to the domain of the outer function, $g(x)$.
$[-4, \infty] \not\subseteq (2, \infty)$
ran $f \not\subseteq$ dom g
Therefore, $g\left(f(x)\right)$ is not defined.

d. f_1: $\left(-\infty, -\sqrt{6}\right) \cup \left(\sqrt{6}, \infty\right) \to R, f(x) = x^2 - 4$

e. $g\left(f_1(x)\right) = \frac{1}{x^2 - 6}$

Domain $= \left(-\infty, -\sqrt{6}\right) \cup \left(\sqrt{6}, \infty\right)$

18. a. $D = [2, \infty)$

b. One possible answer is:
Dilated by a factor of $\frac{1}{3}$ parallel to the x-axis or from the y-axis, translated 2 units to the right or in the positive x-direction and translated 1 unit down or in the negative y-direction

c. f^{-1}: $[-1, \infty] \to R, f^{-1}(x) = \frac{1}{3}(x + 1)^2 + 2$ with range $= [2, \infty)$

d.

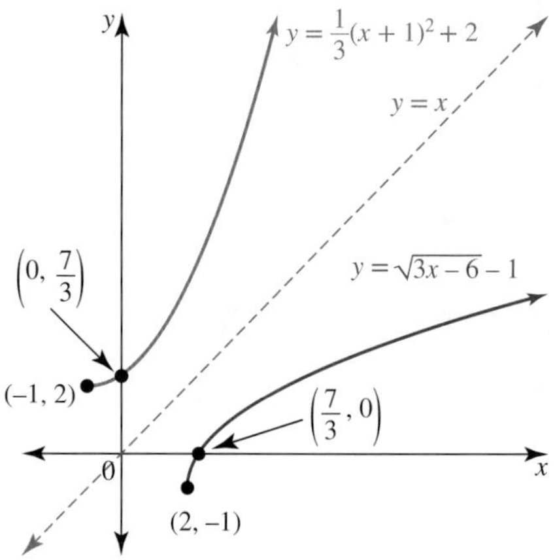

19. $k \geq 3$

20. a. and c. The domain of f is R and the range of f is $[0, \infty)$.

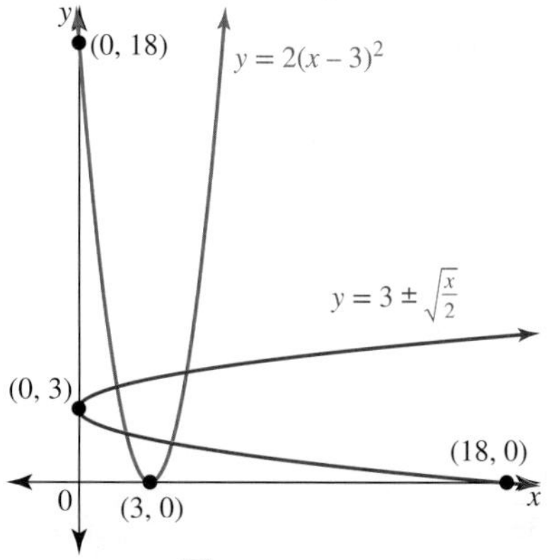

b. $y = 3 \pm \sqrt{\frac{x}{2}}$; domain = $[0, \infty)$ and range = R

d. $[3, \infty)$.

e. f: $[3, \infty) \to R, f(x) = 2(x - 3)^2$
f^{-1}: $[0, \infty) \to R, f(x) = \sqrt{\frac{x}{2}} + 3$

f.

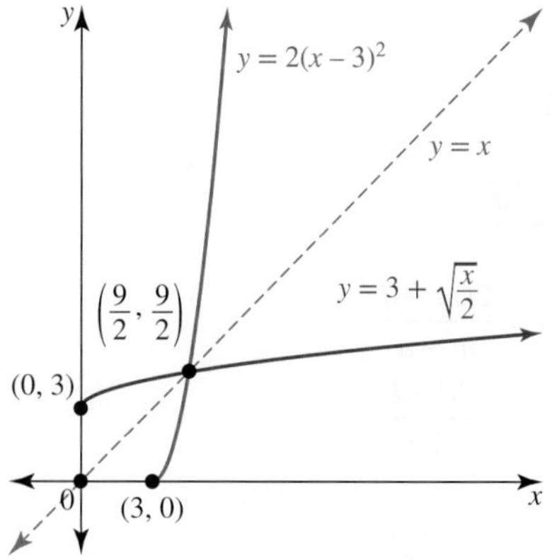

g. $f\left(f^{-1}(x)\right) = 2\left(\sqrt{\frac{x}{2}} + 3 - 3\right)^2$

$= 2\left(\sqrt{\frac{x}{2}}\right)^2$

$= 2 \times \frac{x}{2}$

$= x$

3.7 Exam questions

Note: Mark allocations are available with the fully worked solutions online.

1. E
2. E
3. C
4. a. $\dfrac{x+1}{3x}$

   b. Domain of f^{-1} = range of $f = R\backslash\{0\}$
5. a. Range of $f = [1, \infty)$

   b. i. $c = -3$

   ii. Range of $f(g(x)) = [1, \infty)$

   c. Range of $f(h(x)) = [2, \infty)$

4 Exponential and logarithmic functions

LEARNING SEQUENCE

Fully worked solutions for this topic are available online.

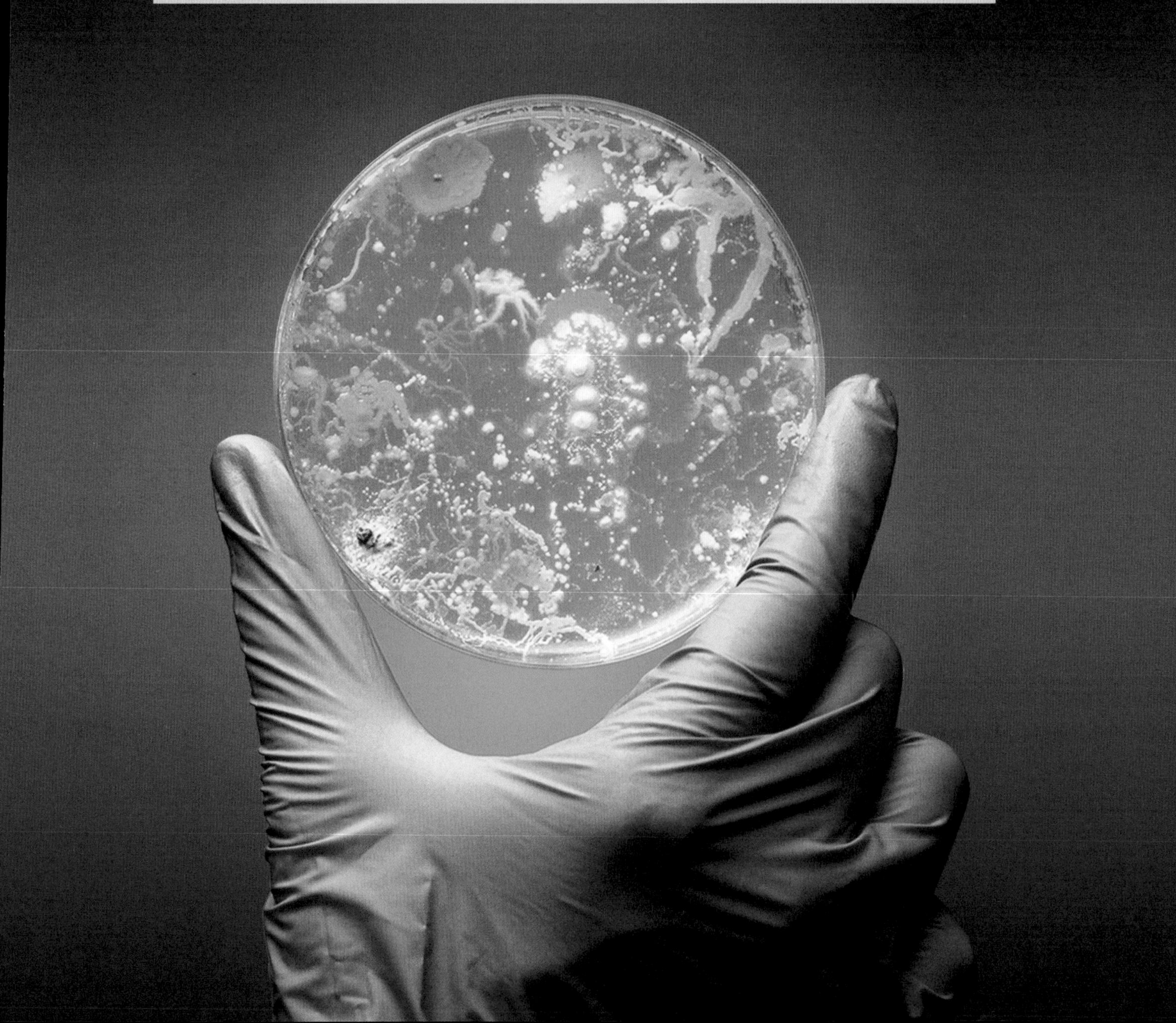

4.1 Overview

Hey students! Bring these pages to life online

Watch videos | Engage with interactivities | Answer questions and check results

Find all this and MORE in jacPLUS

4.1.1 Introduction

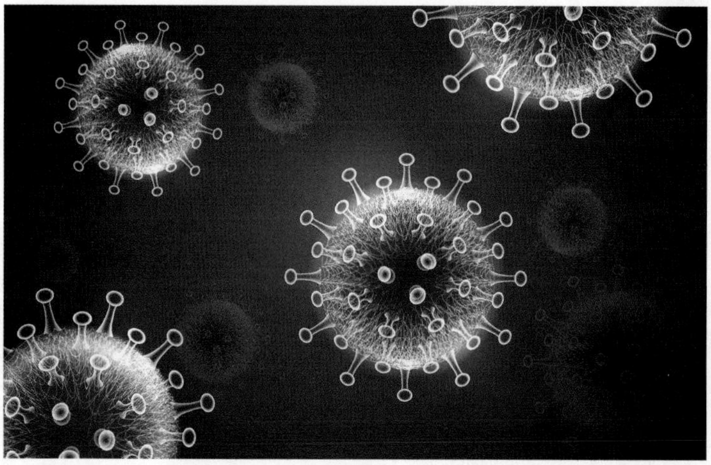

Exponential functions are used to describe situations where a quantity grows or decays at a particular rate that is directly proportional to the value of the function. Examples of this occur around us every day, for example the growth of bacteria on food or the spread of a virus (such as COVID-19).

Exponential functions can have any base; however, the one of most interest is when the base is e. e is a mathematical constant, equal to 2.718 281 828… Swiss mathematician Jacob Bernoulli first discovered the value in 1683 while studying compound interest, but it was another Swiss mathematician, Leonhard Euler, who gave the value the symbol we still use today. Consequently, the constant e is also called Euler's number.

Exponential functions naturally occur in many aspects of life, such as calculating compound interest, and are frequently used in scientific research. In chemistry, the rate at which radioactive substances decay — that is, how long it takes for an element to decay by 50% (its half-life) — is described by exponential functions. This value is constant, no matter how much of the element is present at the beginning. For example, one of the main radioactive materials released in the Chernobyl disaster is caesium-137, which has a half-life of 30 years, meaning that 100 g of caesium-137 will decay to 50 g after 30 years. It will take another 30 years for the 50 g to become 25 g, and so on. Therefore, it is estimated that the land around the reactor that exploded at Chernobyl may not be safe for humans to live on for another 300 years.

KEY CONCEPTS

This topic covers the following key concepts from the VCE Mathematics Study Design:

- graphs of the following functions: power functions, $y = x^n, n \in Q$; exponential functions, $y = a^x, a \in R^+$, in particular $y = e^x$; logarithmic functions, $y = \log_e(x)$ and $y = \log_{10}(x)$; and circular functions, $y = \sin(x)$, $y = \cos(x)$ and $y = \tan(x)$ and their key features
- modelling of practical situations using polynomial, power, circular, exponential and logarithmic functions, simple transformation and combinations of these functions, including simple piecewise (hybrid) functions
- functions and their inverses, including conditions for the existence of an inverse function, and use of inverse functions to solve equations involving exponential, logarithmic, circular and power functions.

Note: *Concepts shown in grey are covered in other topics.*

4.2 Logarithm laws and equations

LEARNING INTENTION

At the end of this subtopic you should be able to:

- recognise and apply index and logarithm laws
- use algebraic and logarithmic properties to simplify and solve equations involving logarithms
- verify the solutions to an equation involving logarithms over its implied domain.

4.2.1 Introduction to logarithms

Logarithm is another name for the **exponent** or **index**. Consider the following indicial equations.

Exponent or index ↓ Base number → $10^2 = 100$ Exponent or index ↓ Base number → $e^x = y$

Written as logarithms, they become:

$\log_{10} 100 = 2$ ← Exponent or index $\log_e(y) = x$ ← Exponent or index
↑ Base number   ↑ Base number

Euler's number, *e*

The number e is known as Euler's number after the eminent Swiss mathematician Leonhard Euler, who first used the symbol. Euler's number is defined as $\lim\limits_{n\to\infty}\left(1+\dfrac{1}{n}\right)^n \approx 2.718\,281\,828\,45\ldots$ Like π, e is an irrational number of great importance in mathematics. Most calculators have keys for both π and e (the latter is not to be confused with the key for the letter e).

The expression $\log_e(x)$ or $\ln(x)$ is called the natural or Napierian logarithm, and can be found on your calculator as 'ln'. The expression $\log_{10}(x)$ is the standard logarithm, which traditionally is written as $\log(x)$ and can be found on your calculator as log.

The logarithms have laws that have been developed from the indicial laws.

4.2.2 Laws of logarithms

Law 1. $a^m \times a^n = a^{m+n} \Leftrightarrow \log_a(m) + \log_a(n) = \log_a(mn)$

To prove this law:

Let $x = \log_a(m)$ and $y = \log_a(n)$.

So $a^x = m$ and $a^y = n$.

Now $a^m \times a^n = a^{m+n}$

or $mn = a^{x+y}$.

By applying the definition of a logarithm to this statement, we get

$\log_a(mn) = x + y$

or $\log_a(mn) = \log_a(m) + \log_a(n)$.

Law 2. $a^m \div a^n = a^{m-n} \Leftrightarrow \log_a(m) - \log_a(n) = \log_a\left(\frac{m}{n}\right)$

To prove this law:

Let $x = \log_a(m)$ and $y = \log_a(n)$.

So $a^x = m$ and $a^y = n$.

Now $\frac{a^x}{a^y} = a^{x-y}$

or $\frac{m}{n} = a^{x-y}$.

By converting the equation into logarithm form, we get

$\log_a\left(\frac{m}{n}\right) = x - y$

or $\log_a\left(\frac{m}{n}\right) = \log_a(m) - \log_a(n)$.

Note: Before the first or second law can be applied, each logarithmic term must have a coefficient of 1.

Law 3. $(a^m)^n = a^{mn} \Leftrightarrow \log_a(m^n) = n\log_a(m)$

To prove this law:

Let $x = \log_a(m)$.

So $a^x = m$.

Now $(a^x)^n = m^n$

or $a^{nx} = m^n$.

By converting the equation into logarithm form, we have

$\log_a(m^n) = nx$

or $\log_a(m^n) = n\log_a(m)$.

Applying these laws, we can also see that:

Law 4. As $a^0 = 1$, then by the definition of a logarithm, $\log_a(1) = 0$.

Law 5. As $a^1 = a$, then by the definition of a logarithm, $\log_a(a) = 1$.

Law 6. $a^x > 0$, therefore, $\log_a(0)$ is undefined, and $\log_a(x)$ is only defined for $x > 0$ and $a \in R^+ \backslash \{1\}$.

Another important fact related to the definition of a logarithm is

$a^{\log_a(m)} = m$.

This can be proved as follows:

Let $y = a^{\log_a(m)}$.

Converting index form to logarithm form, we have

$\log_a(y) = \log_a(m)$.

Therefore, $y = m$.

Consequently, $a^{\log_a(m)} = m$.

Logarithm laws

In summary, the logarithm laws are:

1. $\log_a(m) + \log_a(n) = \log_a(mn)$
2. $\log_a(m) - \log_a(n) = \log_a\left(\dfrac{m}{n}\right)$
3. $\log_a(m^n) = n\log_a(m)$
4. $\log_a(1) = 0$
5. $\log_a(a) = 1$
6. $\log_a(0)$ **is undefined**
7. $\log_a(x)$ **is defined for** $x > 0$ **and** $a \in R^+ \setminus \{1\}$
8. $a^{\log_a(m)} = m$.

WORKED EXAMPLE 1 Simplifying expressions using the index and logarithm laws

Simplify the following.

a. $\log_4(64) + \log_4(16) - \log_4(256)$

b. $2\log_3(7) - 2\log_3(21)$

c. $\dfrac{\log_3(16)}{\log_3(64)}$

d. $\log_5\left(\sqrt[4]{\dfrac{1}{625}}\right)$

THINK	WRITE
a. 1. Express all the numbers in base 4 and, where possible, apply the log law $\log_a(m^n) = n\log_a(m)$.	a. $\log_4(64) + \log_4(16) - \log_4(256)$ $= \log_4(4^3) + \log_4(4^2) - \log_4(4^4)$ $= 3\log_4(4) + 2\log_4(4) - 4\log_4(4)$
2. Apply $\log_a(a) = 1$ and simplify.	$= 3 \times 1 + 2 \times 1 - 4 \times 1$ $= 1$
b. 1. Apply the law $n\log_a(m) = \log_a(m^n)$.	b. $2\log_3(7) - 2\log_3(21)$ $= \log_3(7^2) - \log_3(21^2)$
2. Apply the law $\log_a(m) - \log_a(n) = \log_a\left(\dfrac{m}{n}\right)$ and simplify.	$= \log_3\left(\dfrac{7^2}{21^2}\right)$ $= \log_3\left(\dfrac{7}{21}\right)^2$ $= 2\log_3\left(\dfrac{1}{3}\right)$
3. Convert $\dfrac{1}{3}$ to 3^{-1} and apply $\log_a(a) = 1$.	$= 2\log_3(3^{-1})$ $= -2\log_3(3)$ $= -2$

c. 1. Apply the law $n\log_a(m) = \log_a(m^n)$.
Note: The 16 and 64 cannot be cancelled, as when they are with the log function, they represent single numbers. Therefore, the 16 and 64 cannot be separated from their logarithm components.

c. $\dfrac{\log_3(16)}{\log_3(64)}$

$= \dfrac{\log_3(2^4)}{\log_3(2^6)}$

$= \dfrac{4\log_3(2)}{6\log_3(2)}$

2. Cancel the logs as they are identical.

$= \dfrac{4}{6}$

$= \dfrac{2}{3}$

d. 1. Convert the surd into a fractional power and simplify.

d. $\log_5\left(\sqrt[4]{\dfrac{1}{625}}\right)$

$= \log_5\left(\left(\dfrac{1}{5^4}\right)^{\frac{1}{4}}\right)$

$= \log_5\left((5^{-4})^{\frac{1}{4}}\right)$

$= \log_5(5^{-1})$

2. Apply the laws $n\log_a(m) = \log_a(m^n)$ and $\log_a(a) = 1$.

$= -\log_5(5)$

$= -1$

Solving logarithmic equations involves the use of the logarithm laws as well as converting to index form. As $\log_a(x)$ is only defined for $x > 0$ and $a \in R^+ \setminus \{1\}$, always check the validity of your solution.

Values for which a logarithmic function is defined

- **$\log_a(x)$ is defined for $x > 0$ where $a \in R^+$ {1}.**
- **$\log_a(f(x))$ is only defined where $f(x) > 0$ where $a \in R^+ \setminus \{1\}$.**

WORKED EXAMPLE 2 Solving equations using the index and logarithm laws

Solve the following equations for x.

a. $\log_4(64) = x$

b. $\log_2(3x) + 3 = \log_2(x + 2)$

c. $(\log_2(x))^2 = 3 - 2\log_2(x)$

d. $\log_e(2x) + \log_e(x + 2) = \log_e(6)$

THINK	WRITE
a. 1. Convert the equation into index form.	**a.** $\log_4(64) = x$ $4^x = 64$
2. Convert 64 to base 4 and evaluate.	$4^x = 4^3$ $\therefore x = 3$
b. 1. Rewrite 3 in log form, given $\log_2(2) = 1$.	**b.** $\log_2(3x) + 3 = \log_2(x+2)$ $\log_2(3x) + 3\log_2(2) = \log_2(x+2)$
2. Apply the law $\log_a(m^n) = n\log_a(m)$.	$\log_2(3x) + \log_2(2^3) = \log_2(x+2)$
3. Simplify the left-hand side by applying $\log_a(mn) = \log_a(m) + \log_a(n)$.	$\log_2(3x \times 8) = \log_2(x+2)$
4. Equate the logs and simplify.	$24x = x + 2$ $23x = 2$ $x = \frac{2}{23}$
c. 1. Identify the quadratic form of the log equation. Let $a = \log_2(x)$ and rewrite the equation in terms of a.	**c.** $(\log_2(x))^2 = 3 - 2\log_2(x)$ Let $a = \log_2(x)$. $a^2 = 3 - 2a$
2. Solve the quadratic.	$a^2 + 2a - 3 = 0$ $(a-1)(a+3) = 0$ $a = 1, -3$
3. Substitute in $a = \log_2(x)$ and solve for x.	$\log_2(x) = 1$ $\quad$ $\log_2(x) = -3$ $x = 2^1$ $\quad$ $\therefore x = 2^{-3}$ $\therefore x = 2, \frac{1}{8}$
d. 1. Simplify the left-hand side by applying $\log_a(mn) = \log_a(m) + \log_a(n)$.	**d.** $\log_e(2x) + \log_e(x+2) = \log_e(6)$ $\log_e(2x(x+2)) = \log_e(6)$
2. Equate the logs and solve for x.	$2x(x+2) = 6$ $2x^2 + 4x - 6 = 0$ $x^2 + 2x - 3 = 0$ $(x-1)(x+3) = 0$ $x = 1, -3$
3. Check the validity of both solutions.	$x = -3$ is not valid, as $x > 0$.
4. Write the answer.	$x = 1$

TI | THINK — **DISPLAY/WRITE**

d. 1. On a Calculator page, press MENU, then select:
3: Algebra
1: Solve
Complete the entry line as:
solve $(\ln(2x) + \ln(x + 2) = \ln(6), x)$
then press ENTER.

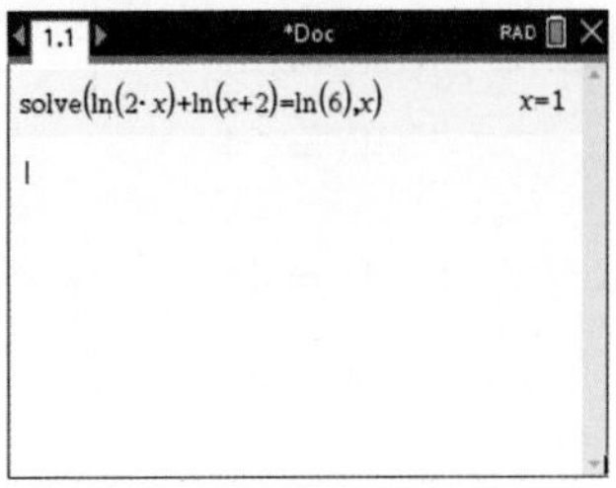

2. The answer appears on the screen.

$x = 1.$

CASIO | THINK — **DISPLAY/WRITE**

d. 1. On a Main screen, complete the entry line as:
solve $(\ln(2x) + \ln(x + 2) = \ln(6), x)$
then press EXE.

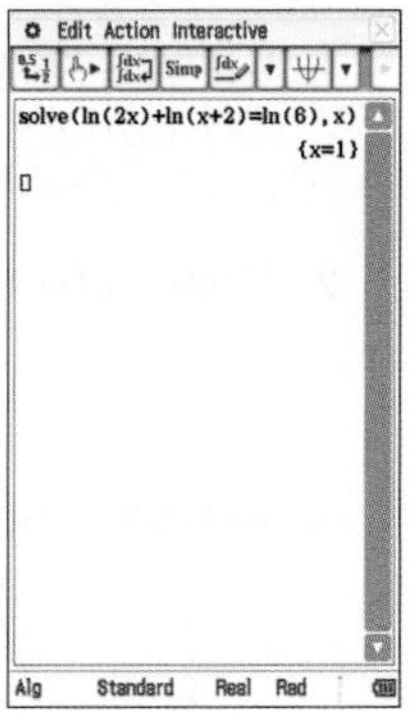

2. The answer appears on the screen.

$x = 1.$

4.2.3 Change of base rule

The definition of a logarithm, together with the logarithmic law $n\log_a(m) = \log_a(m^n)$, is important when looking at the change of base rule.

Suppose $\quad y = \log_a(m).$

By definition, $\quad a^y = m.$

Take the logarithm to the same base of both sides.

$$\log_b(a^y) = \log_b(m)$$

$$y\log_b(a) = \log_b(m)$$

$$y = \frac{\log_b(m)}{\log_b(a)}$$

Therefore, the change of base rule is as follows.

Change of base

$$\log_a(m) = \frac{\log_b(m)}{\log_b(a)}$$

Note: The change of base rule is required to evaluate a logarithm that is not in base 10 or base e on a scientific calculator. Thus, it is an important rule in science subjects.

It also follows that if we start with $m = a^{\log_a(x)}$, taking the logarithm of both sides gives:

$$\log_b(m) = \log_b\left(a^{\log_a(x)}\right)$$

$$\log_b(m) = \log_a(x)\log_b(a)$$

$$\frac{\log_b(m)}{\log_b(a)} = \log_a(x)$$

Applying the change of base rule $\log_a(m) = \log_a(x)$,

$$m = x$$

giving the result $x = a^{\log_a(x)}$.

Applying the change of base rule

$$y = a^{\log_a(y)}$$

$$y = e^{\ln(y)}$$

WORKED EXAMPLE 3 Using the change of base rule to evaluate logarithms

a. Evaluate the following, correct to 4 decimal places.

i. $\log_7(5)$ **ii. $\log_{\frac{1}{3}}(11)$**

b. If $p = \log_5(x)$, find the following in terms of p.

i. x **ii. $\log_x(81)$**

THINK	WRITE
a. i. Input the logarithm into your calculator.	a. i. $\log_7(5) = 0.8271$
ii. Input the logarithm into your calculator.	ii. $\log_{\frac{1}{3}}(11) = -2.1827$
b. i. Rewrite the logarithm in index form to find an expression for x.	b. i. $p = \log_5(x)$ $x = 5^p$
ii. 1. Rewrite $\log_x(81)$ using $\log_a(m^n) = n\log_a(m)$.	ii. $\log_x(81)$ $= \log_x(9^2)$ $= 2\log_x(9)$
2. Apply the change of base rule so that x is no longer a base. *Note:* Although 9 has been chosen as the base in this working, a different value could be applied, giving a different final answer.	$= 2\dfrac{\log_9(9)}{\log_9(x)}$ $= 2\dfrac{1}{\log_9(x)}$
3. Replace x with 5^p and apply the law $\log_a(m^n) = n\log_a(m)$.	$= 2\dfrac{1}{\log_9(5^p)}$ $= \dfrac{2}{p\log_9(5)}$

TI | THINK — **DISPLAY/WRITE**

a. i. 1. On a Calculator page, complete the entry line as:
$\log_{\frac{1}{3}}(11)$
then press ENTER.

2. Press MENU, then select:
2: Number
1: Convert to decimal
then press Enter.

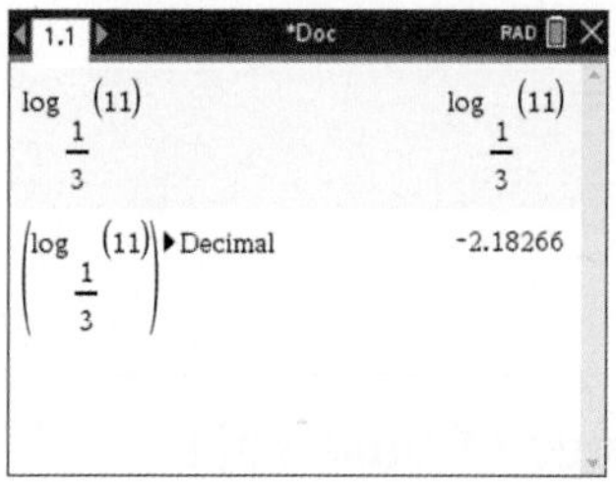

3. The answer appears on the screen. — $\log_{\frac{1}{3}}(11) = -2.1827$

CASIO | THINK — **DISPLAY/WRITE**

a. i. 1. On a Main screen, complete the entry line as:
$\log_{\frac{1}{3}}(11)$
then press EXE.

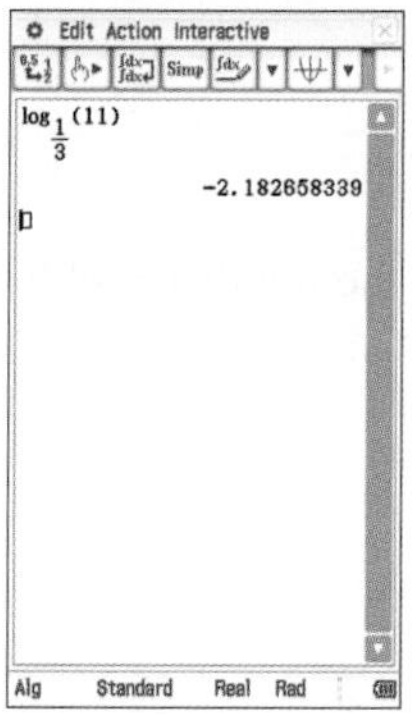

2. The answer appears on the screen. — $\log_{\frac{1}{3}}(11) = -2.1827$

4.2 Exercise

Technology free

1. Express each of the following in logarithmic form.
   a. $6^3 = 216$
   b. $2^8 = 256$
   c. $3^4 = 81$
   d. $10^{-4} = 0.0001$
   e. $5^{-3} = 0.008$
   f. $7^1 = 7$

2. **WE1** Simplify the following.
   a. $\log_7(49) + \log_2(32) - \log_5(125)$
   b. $5\log_{11}(6) - 5\log_{11}(66)$
   c. $\dfrac{\log_4 25}{\log_4 625}$
   d. $\log_2\left(\sqrt[7]{\dfrac{1}{128}}\right)$

3. Simplify the following.
   a. $7\log_4(x) - 9\log_4(x) + 2\log_4(x)$
   b. $\log_7(2x-1) + \log_7(2x-1)^2$
   c. $\log_{10}(x-1)^3 - 2\log_{10}(x-1)$

4. Simplify the following.
   a. $\log_2(256) + \log_2(64) - \log_2(128)$
   b. $5\log_7(49) - 5\log_7(343)$
   c. $\log_4\left(\sqrt[6]{\dfrac{1}{64}}\right)$
   d. $\log_4\left(\dfrac{16}{256}\right)$
   e. $\dfrac{\log_5(32)}{3\log_5(16)}$
   f. $\dfrac{6\log_2(\sqrt[3]{x})}{\log_2(x^5)}$

5. Simplify the following.
   a. $\log_3(x-4) + \log_3(x-4)^2$
   b. $\log_e(2x+3)^3 - 2\log_e(2x+3)$
   c. $\log_5(x^2) + \log_5(x^3) - 5\log_5(x)$
   d. $\log_4(5x+1) + \log_4(5x+1)^3 - \log_4(5x+1)^2$

6. **WE2** Solve the following equations for x.

a. $\log_5(125) = x$

b. $\log_4(x-1) + 2 = \log_4(x+4)$

c. $3(\log_2(x))^2 - 2 = 5\log_2(x)$

d. $\log_e(4x) + \log_e(x-3) = \log_e(7)$

7. Solve the following for x.

a. $\log_3(x) = 5$

b. $\log_3(x-2) - \log_3(5-x) = 2$

8. Calculate the value of x in each of the following equations.

a. $\log_3(81) = x$

b. $\log_6\left(\frac{1}{216}\right) = x$

c. $\log_x(121) = 2$

d. $\log_2(-x) = 7$

9. Solve the following for x.

a. $\log_e(2x-1) = -3$

b. $\log_e\left(\frac{1}{x}\right) = 3$

c. $\log_3(4x-1) = 3$

d. $\log_{10}(x) - \log_{10}(3) = \log_{10}(5)$

e. $3\log_{10}(x) + 2 = 5\log_{10}(x)$

f. $\log_{10}(x^2) - \log_{10}(x+2) = \log_{10}(x+3)$

10. Solve the following for x.

a. $2\log_e(x) - \log_e(2x-3) = \log_e(x-2)$

b. $\log_{10}(2x) - \log_{10}(x-1) = 1$

c. $\log_3(x) + 2\log_3(4) - \log_3(2) = \log_3(10)$

d. $(\log_{10}(x))(\log_{10}(x^2)) - 5\log_{10}(x) + 3 = 0$

e. $(\log_e x)^2 = \log_e(x) + 2$

f. $\log_6(x-3) + \log_6(x+2) = 1$

11. Rewrite the following in terms of base 10.

a. $\log_5(9)$

b. $\log_{\frac{1}{2}}(12)$

Technology active

12. **WE3** a. Evaluate the following, correct to 4 decimal places.

i. $\log_7(12)$

ii. $\log_3\left(\frac{1}{4}\right)$

b. If $z = \log_3(x)$, find the following in terms of z.

i. $2x$

ii. $\log_x(27)$

13. Evaluate the following, correct to 4 decimal places.

a. $\log_3(7)$

b. $\log_2\left(\frac{1}{121}\right)$

14. If $n = \log_5(x)$, find the following in terms of n.

a. $5x$

b. $\log_5(5x^2)$

c. $\log_x(625)$

15. Solve the following for x, correct to 3 decimal places.

a. $e^{2x} - 3 = \log_e(2x+1)$

b. $x^2 - 1 = \log_e(x)$

16. Find x, correct to 4 decimal places, if $(3\log_3(x))(5\log_3(x)) = 11\log_3(x) - 2$.

17. Express y in terms of x for the following equations.

a. $\log_{10}(y) = 2\log_{10}(2) - 3\log_{10}(x)$

b. $\log_4(y) = -2 + 2\log_4(x)$

18. Express y in terms of x for the following equations.

a. $\log_9(3xy) = 1.5$

b. $\log_8\left(\frac{2x}{y}\right) + 2 = \log_8(2)$

19. Solve the equation $8\log_x(4) = \log_2(x)$ for x.

20. a. Find the value of x in terms of m for which $3\log_m(x) = 3 + \log_m 27$, where $m > 0$ and $x > 0$.

b. If $\log_{10}(m) = x$ and $\log_{10}(n) = y$, show that $\log_{10}\left(\dfrac{100n^2}{m^5\sqrt{n}}\right) = 2 + \dfrac{3y}{2} - 5x$.

21. Solve the following for x to 4 decimal places.

a. $13 + \log_9\left(0.2^{x^3}\right) > 7$

b. $\left(2^{\log_4(5x)}\right)^3 = 9$

22. Let $f: R \to R, f(x) = e^{2x} + 2$.

a. Determine the rule and the domain of the inverse function.

b. Sketch the graph of $y = f\left(f^{-1}(x)\right)$ for its maximal domain.

c. Determine $f\left(-f^{-1}(2x)\right)$ in the form $\dfrac{ax}{bx + c}$ where a, b and c are real constants.

4.2 Exam questions

Question 1 (3 marks) TECH-FREE

Source: *VCE 2020, Mathematical Methods Exam 1, Q4; © VCAA.*

Solve the equation $2\log_2(x + 5) - \log_2(x + 9) = 1$.

Question 2 (1 mark) TECH-ACTIVE

Source: *VCE 2020, Mathematical Methods Exam 2, Section A, Q10; © VCAA.*

MC Given that $\log_2(n + 1) = x$, the values of n for which x is a positive integer are

A. $n = 2^k, k \in Z^+$
B. $n = 2^k - 1, k \in Z^+$
C. $n = 2^{k-1}, k \in Z^+$
D. $n = 2k - 1, k \in Z^+$
E. $n = 2k, k \in Z^+$

Question 3 (1 mark) TECH-ACTIVE

Source: *VCE 2019, Mathematical Methods Exam 2, Section A, Q20; © VCAA.*

MC The expression $\log_x(y) + \log_y(z)$, where x, y and z are all real numbers greater than 1, is equal to

A. $-\dfrac{1}{\log_y(x)} - \dfrac{1}{\log_z(y)}$

B. $\dfrac{1}{\log_x(y)} + \dfrac{1}{\log_y(z)}$

C. $-\dfrac{1}{\log_x(y)} - \dfrac{1}{\log_y(z)}$

D. $\dfrac{1}{\log_y(x)} - \dfrac{1}{\log_z(y)}$

E. $\log_y(x) + \log_z(y)$

More exam questions are available online.

4.3 Logarithmic scales

LEARNING INTENTION

At the end of this subtopic you should be able to:

- use index laws and logarithm laws
- apply index laws and logarithm laws in non-routine contexts involving logarithms, requiring problem-solving and modelling to analyse these applications.

4.3.1 Problem solving with logarithmic scales

Logarithmic scales are used in the calculation of many scientific and mathematical quantities, such as the loudness of sound, the strength (magnitude) of an earthquake, octaves in music, pH in chemistry and the intensity of the brightness of stars.

WORKED EXAMPLE 4 Application of a logarithmic scale

Loudness, in decibels (dB), is related to the intensity, I, of the sound by the equation

$$L = 10\log_{10}\left(\frac{I}{I_0}\right)$$

where I_0 is equal to 10^{-12} watts per square metre (W/m^2). (This value is the lowest intensity of sound that can be heard by human ears.)
An ordinary conversation has a loudness of 60 dB. Calculate the intensity in W/m^2.

THINK	WRITE
1. Substitute $L = 60$ and simplify.	$L = 10\log_{10}\left(\frac{I}{I_0}\right)$ $60 = 10\log_{10}\left(\frac{I}{10^{-12}}\right)$ $60 = 10\log_{10}(10^{12}I)$ $6 = \log_{10}(10^{12}I)$
2. Convert the logarithm to index form and solve for I.	$10^6 = 10^{12}I$ $I = 10^{-6}\ \text{W/m}^2$

WORKED EXAMPLE 5 Impact of a change in a logarithmic scale

Using the same formula as in Worked example 4, if the intensity is doubled, determine the change in the loudness, correct to 2 decimal places.

THINK	WRITE
1. Determine an equation for L_1.	$L_1 = 10\log_{10}\left(\frac{I_1}{10^{-12}}\right)$ $= 10\log_{10}(10^{12}I_1)$ $= 10\log_{10}(10^{12}) + 10\log_{10}(I_1)$ $= 120\log_{10}(10) + 10\log_{10}(I_1)$ $= 120 + 10\log_{10}(I_1)$
2. The intensity has doubled; therefore, $I_2 = 2I_1$. Determine an equation for L_2.	$L_2 = 10\log_{10}\left(\frac{2I_1}{10^{-12}}\right)$ $= 10\log_{10}(2\times 10^{12}I_1)$ $= 10\log_{10}(2) + 10\log_{10}(10^{12}) + 10\log_{10}(I_1)$ $= 3.010 + 120\log_{10}(10) + 10\log_{10}(I_1)$ $= 3.01 + 120 + 10\log_{10}(I_1)$
3. Replace $120 + 10\log_{10}(I_1)$ with L_1.	$= 3.01 + L_1$
4. Answer the question.	Doubling the intensity increases the loudness by 3.01 dB.

4.3 Exercise

Technology free

1. WE4 The loudness, L, of a jet taking off about 30 metres away is known to be 130 dB. Using the formula $L = 10\log_{10}\left(\frac{I}{I_0}\right)$, where I is the intensity measured in W/m^2 and I_0 is equal to 10^{-12} W/m^2, calculate the intensity in W/m^2 for this situation.

2. Your eardrum can be ruptured if it is exposed to a noise which has an intensity of 10^4 W/m^2. Using the formula $L = 10\log_{10}\left(\frac{I}{I_0}\right)$, where I is the intensity measured in W/m^2 and I_0 is equal to 10^{-12} W/m^2, calculate the loudness, L, in decibels that would cause your eardrum to be ruptured.

Questions 3–5 relate to the following information.

Chemists define the acidity or alkalinity of a substance according to the formula

$$pH = -\log_{10}[H^+]$$

where $[H^+]$ is the hydrogen ion concentration measured in moles/litre.

Solutions with a pH less than 7 are acidic, whereas solutions with a pH greater than 7 are basic. Solutions with a pH of 7, such as pure water, are neutral.

3. Lemon juice has a hydrogen ion concentration of 0.001 moles/litre. Find the pH and determine whether lemon juice is acidic or basic.

4. Find the hydrogen ion concentration for each of the following.
   a. Battery acid has a pH of zero.
   b. Tomato juice has a pH of 4.
   c. Sea water has a pH of 8.
   d. Soap has a pH of 12.

Technology active

5. Hair conditioner works on hair in the following way. Hair is composed of the protein called keratin, which has a high percentage of amino acids. These acids are negatively charged. Shampoo is also negatively charged. When shampoo removes dirt, it removes natural oils and positive charges from the hair. Positively charged surfactants in hair conditioner are attracted to the negative charges in the hair, so the surfactants can replace the natural oils.
   a. A brand of hair conditioner has a hydrogen ion concentration of 0.000 015 8 moles/litre. Calculate the pH of the hair conditioner.
   b. A brand of shampoo has a hydrogen ion concentration of 0.000 002 75 moles/litre. Calculate the pH of the shampoo.

6. The moment magnitude scale measures the magnitude, M, of an earthquake in terms of energy released, E, in joules, according to the formula

$$M = 0.67 \log_{10}\left(\frac{E}{K}\right)$$

where K is the minimum amount of energy used as a basis of comparison.

a. An earthquake that measures 5.5 on the moment magnitude scale releases 10^{13} joules of energy. Determine the value of K, correct to the nearest integer.
b. An earthquake of magnitude 9.0 occurred in Japan in 2011, releasing about 10^{17} joules of energy. Find the value of K correct to 2 decimal places.
c. Two earthquakes, about 10 kilometres apart, occurred in Iran on 11 August 2012. One measured 6.3 on the moment magnitude scale, and the other one was 6.4 on the same scale. Compare the energy released, in joules, by the two earthquakes.

7. **WE5** Determine how many decibels louder to the human ear a 20 W/m^2 amplifier is compared to a 500 W/m^2 amplifier. Use the formula $L = 10\log_{10}\left(\frac{I}{I_0}\right)$, where L is measured in dB, I is measured in W/m^2 and $I_0 = 10^{-12}$ W/m^2. Give your answer correct to 2 decimal places.

8. The number of atoms of a radioactive substance present after t years is given by

$$N(t) = N_0 e^{-mt}.$$

a. The half-life is the time taken for the number of atoms to be reduced to 50% of the initial number of atoms. Show that the half-life is given by $\frac{\log_e(2)}{m}$.
b. Radioactive carbon-14 has a half-life of 5750 years. The percentage of carbon-14 present in the remains of plants and animals is used to determine how old the remains are. Determine the age of a skeleton that has lost 70% of its carbon-14 atoms. Give your answer correct to the nearest year.

9. A basic observable quantity for a star is its brightness. The apparent magnitudes, m_1 and m_2, for two stars are related to the corresponding brightnesses, b_1 and b_2, by the equation

$$m_2 - m_1 = 2.5\log_{10}\left(\frac{b_1}{b_2}\right).$$

The star Sirius is the brightest star in the night sky. It has an apparent magnitude of −1.5 and a brightness of −30.3. The planet Venus has an apparent magnitude of −4.4. Calculate the brightness of Venus, correct to 2 decimal places.

4.3 Exam questions

Question 1 (2 marks) TECH-ACTIVE

Octaves in music can be measured in cents, n. The frequencies of two notes, f_1 and f_2, are related by the equation

$$n = 1200 \log_{10}\left(\frac{f_2}{f_1}\right).$$

Middle C on the piano has a frequency of 256 hertz; the C an octave higher has a frequency of 512 hertz. Calculate the number of cents between these two Cs.

Question 2 (3 marks) TECH-ACTIVE

Prolonged exposure to sounds above 85 decibels can cause hearing damage or loss. A gunshot from a .22 rifle has an intensity of about $(2.5 \times 10^{13})I_0$.

Calculate the loudness, in decibels, of the gunshot sound and state if ear protection should be worn when a person goes to a rifle range for practice shooting. Use the formula $L = 10 \log_{10}\left(\frac{I}{I_0}\right)$, where I_0 is equal to 10^{-12} W/m^2, and give your answer correct to 2 decimal places.

Question 3 (4 marks) TECH-ACTIVE

Early in the 20th century, San Francisco had an earthquake that measured 8.3 on the magnitude scale. In the same year, another earthquake was recorded in South America that was four times stronger than the one in San Francisco. Using the equation $M = 0.67 \log_{10}\left(\frac{E}{K}\right)$, where M is the magnitude of the earthquake and $\frac{E}{K}$ is the ratio between the largest and smallest waves, calculate the magnitude of the earthquake in South America, correct to 1 decimal place.

More exam questions are available online.

4.4 Indicial equations

LEARNING INTENTION

At the end of this subtopic you should be able to:

- simplify and solve indicial equations.

4.4.1 Solving indicial equations

When we solve an equation such as $3^x = 81$, the technique is to convert both sides of the equation to the same base. For example, $3^x = 3^4$; therefore, $x = 4$.

When we solve an equation such as $x^3 = 27$, we write both sides of the equation with the same index. In this case, $x^3 = 3^3$; therefore, $x = 3$.

If an equation such as $5^{2x} = 2$ is to be solved, then we must use logarithms, as the sides of the equation cannot be converted to the same base or index. To remove x from the power, we take the logarithm of both sides.

$$\begin{aligned}\log_5(5^{2x}) &= \log_5(2)\\ 2x &= \log_5(2)\\ x &= \frac{1}{2}\log_5(2)\end{aligned}$$

Note: If $a^x = b$, a solution for x exists only if $b > 0$.

Index laws

- $a^m \times a^n = a^{m+n}$
- $a^m \div a^n = a^{m-n}$
- $(a^m)^n = a^{mn}$
- $(ab)^m = a^m b^m$
- $\left(\dfrac{a}{b}\right)^m = \dfrac{a^m}{b^m}, b \neq 0$
- $a^0 = 1, a \neq 0$
- $a^{-m} = \dfrac{1}{a^m}, a \neq 0$
- $a^{\frac{1}{m}} = \sqrt[m]{a}$
- $a^{\frac{n}{m}} = \sqrt[m]{a^n}$

Also remember that $a^x > 0$ for all x.

WORKED EXAMPLE 6 Solving indical equations

Solve the following equations for x, giving your answers in exact form.

a. $4^{3x} \times 16^{3-x} = 256$

b. $7^{x-3} - 3 = 0$

c. $(5^x - 25)(5^x + 1) = 0$

d. $3^{2x} - 9(3^x) + 14 = 0$

THINK	WRITE
a. 1. Convert the numbers to the same base.	a. $4^{3x} \times 16^{3-x} = 256$ $4^{3x} \times (4^2)^{3-x} = 4^4$
2. Simplify and add the indices on the left-hand side of the equation.	$4^{3x} \times 4^{6-2x} = 4^4$ $4^{x+6} = 4^4$
3. As the bases are the same, equate the indices and solve the equation.	$x + 6 = 4$ $x = -2$
b. 1. Rearrange the equation.	b. $7^{x-3} - 3 = 0$ $7^{x-3} = 3$
2. Take the logarithm of both sides to base 7 and simplify.	$\log_7(7^{x-3}) = \log_7(3)$ $x - 3 = \log_7(3)$
3. Solve the equation.	$x = \log_7(3) + 3$
c. 1. Apply the Null Factor Law to solve each bracket.	c. $(5^x - 25)(5^x + 1) = 0$ $5^x - 25 = 0$ or $5^x + 1 = 0$ $5^x = 25$ $5^x = -1$

2. Convert 25 to base 5. $5^x > 0$, so there is no real solution for $5^x = -1$.

$$5^x = 5^2$$
$$x = 2$$

d. 1. Let $a = 3^x$ and substitute into the equation to create a quadratic to solve.

d. $3^{2x} - 9(3^x) + 14 = 0$

Let $a = 3^x$.

$$a^2 - 9a + 14 = 0$$

2. Factorise the left-hand side.

$$(a-7)(a-2) = 0$$

3. Apply the Null Factor Law to solve each bracket for a.

$$a - 7 = 0 \quad \text{or} \quad a - 2 = 0$$
$$a = 7 \qquad\qquad a = 2$$

4. Substitute back in for a.

$$3^x = 7 \qquad\qquad 3^x = 2$$

5. Take the logarithm of both sides to base 3 and simplify.

$$\log_3(3^x) = \log_3(7) \qquad \log_3(3^x) = \log_3(2)$$
$$x = \log_3(7) \qquad\qquad x = \log_3(2)$$

TI | THINK	DISPLAY/WRITE	CASIO | THINK	DISPLAY/WRITE
d. 1. On a Calculator page, press MENU, then select: 3: Algebra 1: Solve Complete the entry line as: solve $\left(3^{2x} - 9\left(3^x\right) + 14 = 0, x\right)$ then press ENTER.	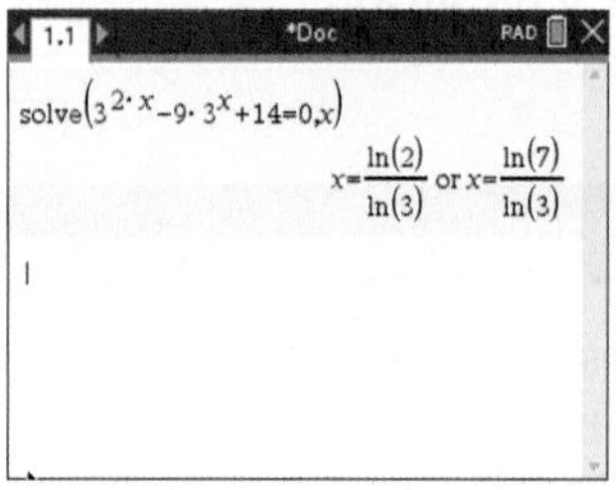	**d. 1.** On a Main screen, complete the entry line as: solve $\left(3^{2x} - 9\left(3^x\right) + 14 = 0, x\right)$ then press EXE.	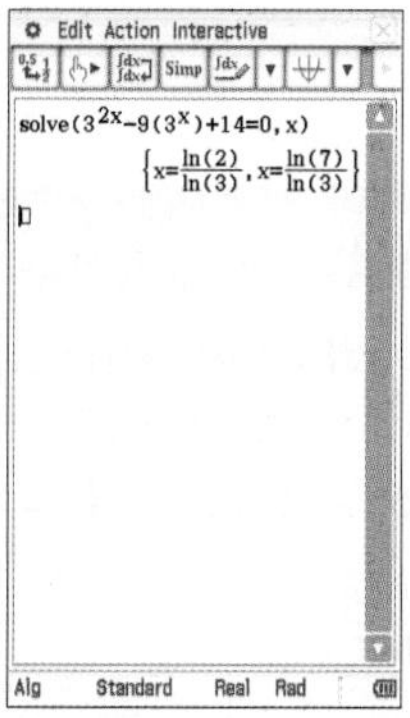
2. The answer appears on the screen.	$x = \dfrac{\ln(2)}{\ln(3)}$ or $x = \dfrac{\ln(7)}{\ln(3)}$	**2.** The answer appears on the screen.	$x = \dfrac{\ln(2)}{\ln(3)}$ or $x = \dfrac{\ln(7)}{\ln(3)}$

4.4 Exercise

Technology free

1. **WE6** Solve the following equations for x, giving your answers in exact form.

   a. $3^{2x+1} \times 27^{2-x} = 81$

   b. $10^{2x-1} - 5 = 0$

   c. $(4^x - 16)(4^x + 3) = 0$

   d. $2(10^{2x}) - 7(10^x) + 3 = 0$

2. Solve the following equations for x.

   a. $2^{x+3} - \dfrac{1}{64} = 0$

   b. $2^{2x} - 9 = 0$

   c. $3e^{2x} - 5e^x - 2 = 0$

   d. $e^{2x} - 5e^x = 0$

3. Solve the following equations for x.

a. $7^{2x-1}=5$
b. $(3^x-9)(3^x-1)=0$
c. $25^x-5^x-6=0$
d. $6(9^{2x})-19(9^x)+10=0$

4. Solve the following equations for x.

a. $16\times 2^{2x+3}=8^{-2x}$
b. $2\times 3^{x+1}=4$
c. $2(5^x)-12=-\frac{10}{5^x}$
d. $4^{x+1}=3^{1-x}$

5. Solve the following equations for x.

a. $2(2^{x-1}-3)+4=0$
b. $2(5^{1-2x})-3=7$

6. Solve the following equations for x.

a. $e^{x-2}-2=7$
b. $e^{\frac{x}{4}}+1=3$
c. $e^{2x}=3e^x$
d. $e^{x^2}+2=4$

7. Solve the following equations for x.

a. $e^{2x}=e^x+12$
b. $e^x=12-32e^{-x}$
c. $e^{2x}-4=2e^x$
d. $e^x-12=-\frac{5}{e^x}$

8. Solve $(\log_3(4m))^2=25n^2$ for m.

9. a. Simplify $x^{-1}-\dfrac{1}{1-\frac{1}{1+x^{-1}}}$

b. Solve $2^{3-4x}\times 3^{-4x+3}\times 6^{x^2}=1$ for x.

10. Solve the following for x.

a. $e^{m-kx}=2n$, where $k\in R\setminus\{0\}$ and $n\in R^+$
b. $8^{mx}\times 4^{2n}=16$, where $m\in R\setminus\{0\}$
c. $2e^{mx}=5+4e^{-mn}$, where $m\in R\setminus\{0\}$

Technology active

11. Solve the following for x, correct to 3 decimal places.

a. $2^x<0.3$
b. $(0.4)^x<2.$

12. If $y=m(10)^{nx}$, $y=20$ when $x=2$ and $y=200$ when $x=4$, determine the values of the constants m and n.

13. If $y=ae^{-kx}$, $y=3.033$ when $x=2$ and $y=1.1157$ when $x=6$, determine the values of the constants a and k. Give answers correct to 2 decimal places.

14. The compound interest formula $A=Pe^{rt}$ is an indicial equation, where A is the amount of interest, P is the principal, r is the annual interest rate and t is the number of years. If a principal amount of money, P, is invested for 5 years, the interest earned is \$12 840.25, but if this same amount is invested for 7 years, the interest earned is \$14 190.66. Determine the integer rate of interest and the principal amount of money invested, to the nearest dollar.

4.4 Exam questions

Question 1 (1 mark) TECH-ACTIVE

Source: *VCE 2017, Mathematical Methods Exam 2, Section A, Q8; © VCAA.*

MC If $y=a^{b-4x}+2$, where $a>0$, then x is equal to

A. $\frac{1}{4}\left(b-\log_a(y-2)\right)$
B. $\frac{1}{4}\left(b-\log_a(y+2)\right)$
C. $b-\log_a\left(\frac{1}{4}(y+2)\right)$
D. $\frac{b}{4}-\log_a(y-2)$
E. $\frac{1}{4}\left(b+2-\log_a(y)\right)$

Question 2 (3 marks) TECH-FREE

Source: *VCE 2015, Mathematical Methods (CAS) Exam 1, Q7b; © VCAA.*

Solve $3e^t = 5 + 8e^{-t}$ for t.

Question 3 (2 marks) TECH-FREE

Source: *VCE 2014, Mathematical Methods (CAS) Exam 1, Q4; © VCAA.*

Solve the equation $2^{3x-3} = 8^{2-x}$ for x.

More exam questions are available online.

4.5 Logarithmic graphs

LEARNING INTENTION

At the end of this subtopic you should be able to:

- identify key features and properties of the graph of a logarithm and draw the graphs of these, clearly identifying their key features and properties, including any vertical asymptotes
- sketch by hand graphs of $y = \log_e(x)$ and $y = \log_{10}(x)$ and simple transformations of these
- describe the effect of transformations on the graphs of a logarithm.

4.5.1 The graph of $y = \log_a(x)$

The graph of the logarithmic function $f: R^+ \to R, f(x) = \log_a(x), a > 1$ has the following characteristics.

Graphing $y = \log_a(x)$

For $f(x) = \log_a(x), a > 1$:

- **the domain is $(0, \infty)$**
- **the range is R**
- **the graph is an increasing function**
- **the graph cuts the x-axis at $(1, 0)$**
- **as $x \to 0, y \to -\infty$, so the line $x = 0$ is an asymptote**
- **as a increases, the graph rises more steeply for $x \in (0, 1)$ and is flatter for $x \in (1, \infty)$.**

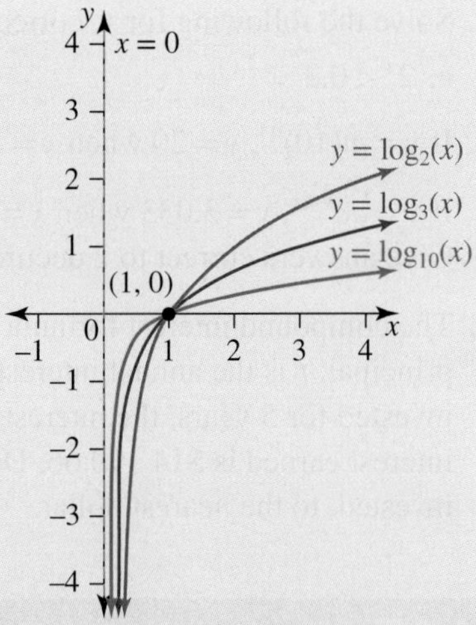

on Resources

Interactivity Logarithmic graphs (int-6418)

4.5.2 Transformations

Dilations: graphs of the form $y = n\log_a(x)$ and $y = \log_a(mx)$

The graph of $y = n\log_a(x)$ is the basic graph of $y = \log_a(x)$ dilated by factor n from the x axis or parallel to the y-axis. The graph of $y = \log_a(mx)$ is the basic graph of $y = \log_a(x)$ dilated by factor $\frac{1}{m}$ from the x axis or parallel to the y-axis. The line $x = 0$ or the y-axis remains the vertical asymptote, and the domain remains $(0, \infty)$.

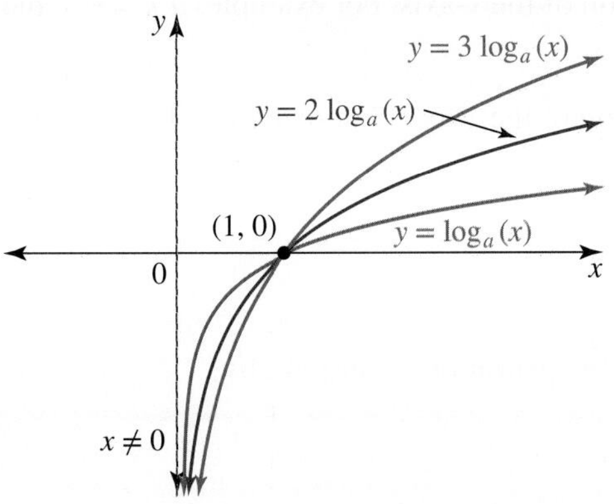

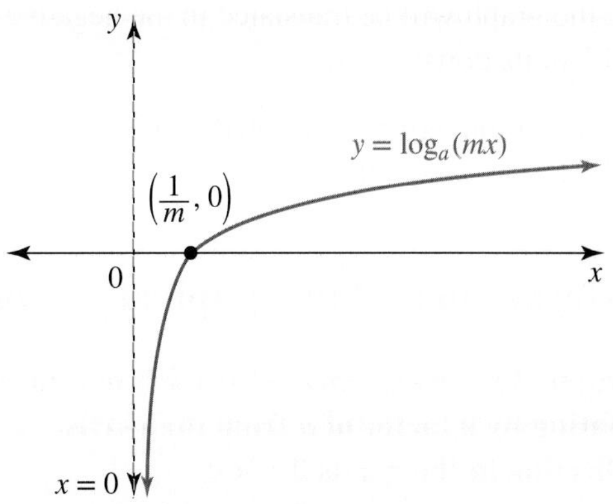

Reflections: graphs of the form $y = -\log_a(x)$ and $y = \log_a(-x)$

The graph of $y = -\log_a(x)$ is the basic graph of $y = \log_a(x)$ reflected in the x-axis. The line $x = 0$ or the y-axis remains the vertical asymptote, and the domain remains $(0, \infty)$.

It is worth noting that the graph of $\log_a\left(\frac{1}{x}\right) = \log_a(x^{-1}) = -\log_a(x)$.

The graph of $y = \log_a(-x)$ is the basic graph of $y = \log_a(x)$ reflected in the y-axis. The line $x = 0$ or the y-axis remains the vertical asymptote, but the domain changes to $(-\infty, 0)$.

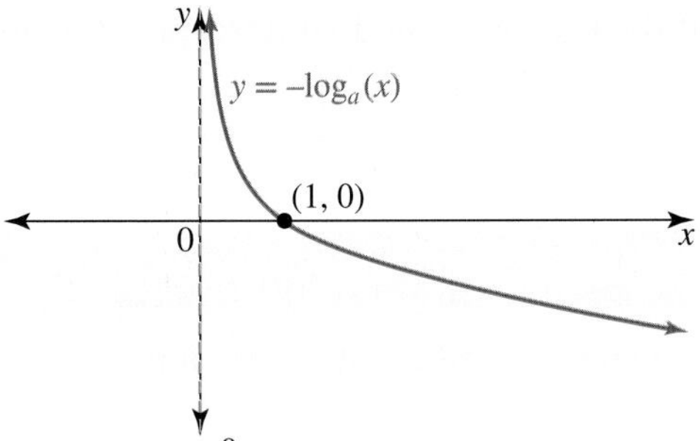

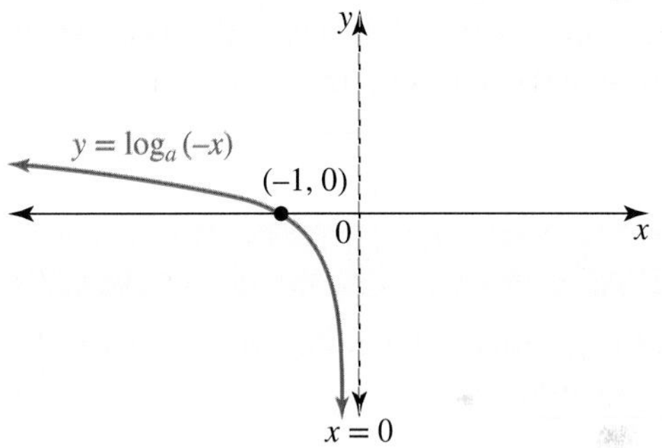

Translations: graphs of the form $y = \log_a(x) + k$ and $y = \log_a(x - h)$

For $k > 0$, the graph of $y = \log_a(x) + k$ is the basic graph of $y = \log_a(x)$ translated k units in the positive direction of the y-axis (i.e. vertically upwards). Thus, the line $x = 0$ or the y-axis remains the vertical asymptote, and the domain remains $(0, \infty)$.

For $h > 0$, the graph of $y = \log_a(x - h)$ is the basic graph of $y = \log_a(x)$ translated h units in the positive direction of the x-axis (i.e. horizontally right). Thus, the line $x = 0$ or the y-axis is no longer the vertical asymptote. The vertical asymptote is $x = h$ and the domain is (h, ∞).

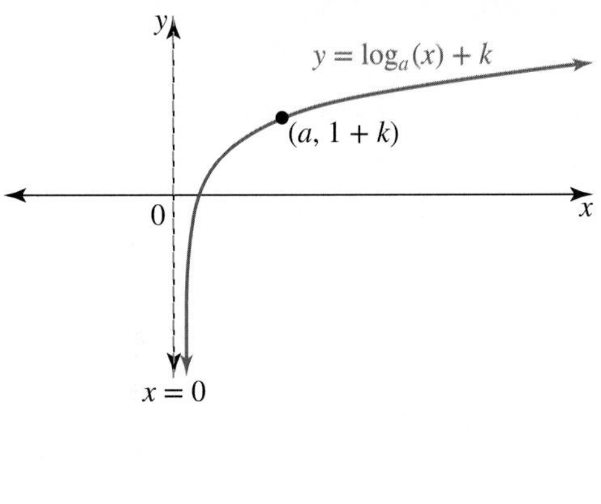

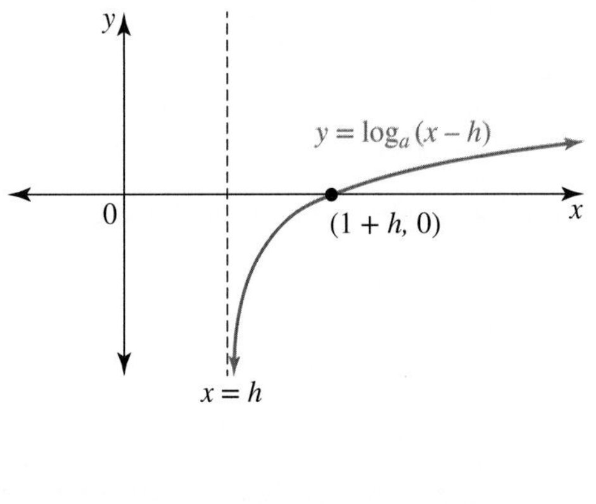

For $k < 0$, the graph will be translated in the negative direction of the y-axis. For example, if $k = -5$, the graph is translated 5 units down.

For $h < 0$, the graph will be translated in the negative direction of the x-axis. For example, if $h = -2$, the graph is translated 2 units to the left.

Transformations of the graph of $y = \log_a(x)$

The graph of $y = n \log_a(m(x - h)) + k$ is obtained from the graph of $y = \log_a(x)$ by:

- **dilating by a factor of n from the x-axis**
- **reflecting in the x-axis if $n < 0$**
- **dilating by a factor of $\frac{1}{m}$ from the y-axis**
- **reflecting in the y-axis if $m < 0$**
- **for $h > 0$, translating h units in the positive direction of the x-axis (h units right)**
- **for $k > 0$, translating k units in the positive direction of the y-axis (k units up).**

The graph will have:

- **an asymptote at $x = h$**
- **a domain of (h, ∞) for $m > 0$ or $(-\infty, h)$ for $m < 0$**
- **a range of R**
- **one x-intercept, obtained by letting $y = 0$**
- **either one or no y-intercept. The relative position of the asymptote and x-intercept will determine whether there is an y-intercept.**

WORKED EXAMPLE 7 Sketching transformed logarithmic functions

Sketch the graphs of the following, showing all important characteristics. State the domain and range for each graph.

a. $y = \log_e(x - 2)$

b. $y = \log_e(x + 1) + 2$

c. $y = \frac{1}{4}\log_e(2x)$

d. $y = -\log_e(-x)$

THINK	WRITE
a. 1. The basic graph of $y = \log_e(x)$ has been translated 2 units to the right, so $x = 2$ is the vertical asymptote.	a. $y = \log_e(x - 2)$ The domain is $(2, \infty)$. The range is R.
2. Find the x-intercept.	x-intercept, $y = 0$: $\log_e(x - 2) = 0$ $e^0 = x - 2$ $1 = x - 2$ $x = 3$
3. Determine another point through which the graph passes.	When $x = 4$, $y = \log_e(2)$. The point is $(4, \log_e(2))$.

4. Sketch the graph.
Note: When sketching the line, it must always appear to be moving closer to the asymptote but not touching it.

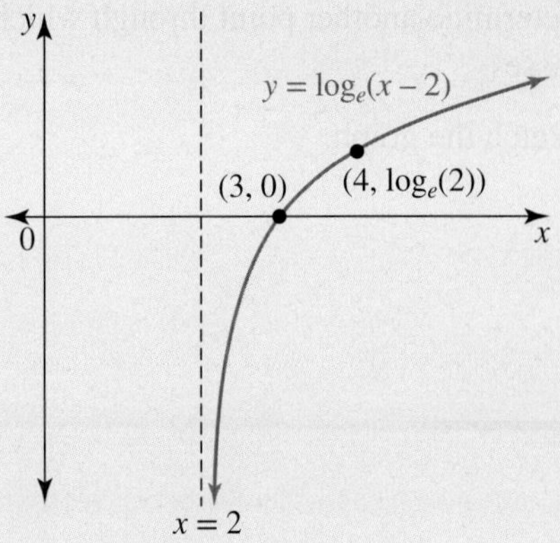

b. 1. The basic graph of $y = \log_e(x)$ has been translated up 2 units and 1 unit to the left, so $x = -1$ is the vertical asymptote.

b. $y = \log_e(x+1) + 2$
The domain is $(-1, \infty)$.
The range is R.

2. Find the x-intercept.

The graph cuts the x-axis where $y = 0$.

$$\log_e(x+1) + 2 = 0$$
$$\log_e(x+1) = -2$$
$$e^{-2} = x + 1$$
$$x = e^{-2} - 1$$

3. Find the y-intercept.

The graph cuts the y-axis where $x = 0$.

$$y = \log_e(1) + 2$$
$$= 2$$

4. Sketch the graph.

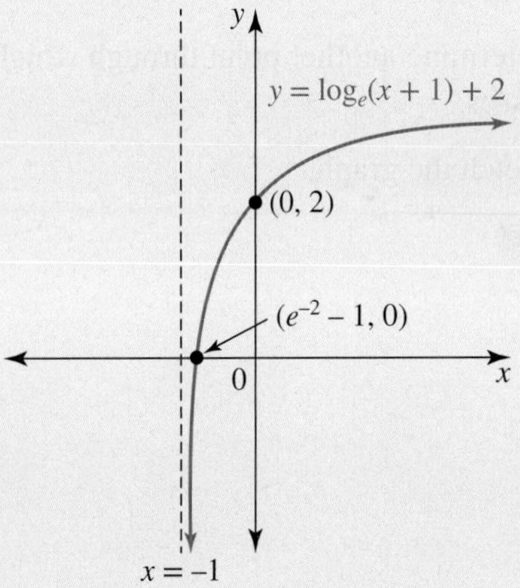

c. 1. The basic graph of $y = \log_e(x)$ has been dilated by factor $\frac{1}{4}$ from the x-axis and by factor $\frac{1}{2}$ from the y-axis. The vertical asymptote remains $x = 0$.

c. $y = \frac{1}{4}\log_e(2x)$
The domain is $(0, \infty)$.
The range is R.

2. Find the x-intercept.

x-intercept, $y = 0$:

$$\frac{1}{4}\log_e(2x) = 0$$
$$\log_e(2x) = 0$$
$$e^0 = 2x$$
$$1 = 2x$$
$$x = \frac{1}{2}$$

3. Determine another point through which the graph passes.

When $x = 1, y = \log_e(2)$.
The point is $(1, \log_e(2))$.

4. Sketch the graph.

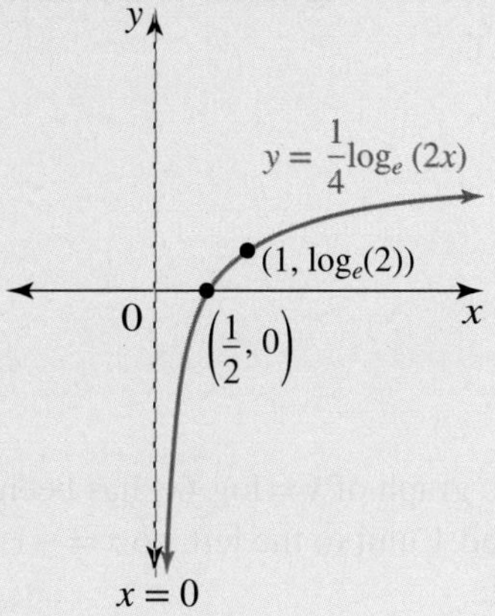

d. 1. The basic graph of $y = \log_e(x)$ has been reflected in both axes. The vertical asymptote remains $x = 0$.

d. $y = -\log_e(-x)$
The domain is $(-\infty, 0)$.
The range is R.

2. Find the x-intercept.

x-intercept, $y = 0$:

$$-\log_e(-x) = 0$$
$$\log_e(-x) = 0$$
$$e^0 = -x$$
$$x = -1$$

3. Determine another point through which the graph passes.

When $x = -2, \ y = -\log_e(2)$.
The point is $(-2, \ -\log_e(2))$.

4. Sketch the graph.

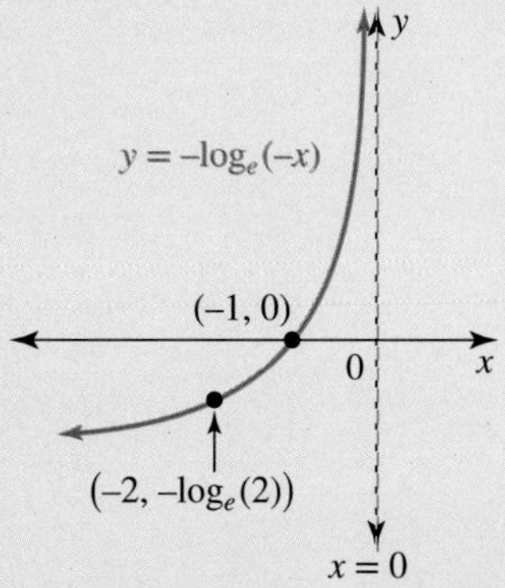

TI | THINK	**DISPLAY/WRITE**	**CASIO | THINK**	**DISPLAY/WRITE**
d. 1. On a Graphs page, complete the entry line for function 1 as: $f1(x) = -\ln(-x)$ then press ENTER.	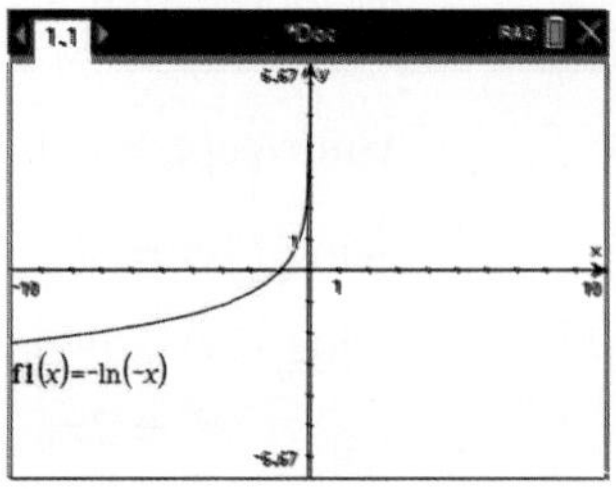	**d. 1.** On a Graph & Table screen, complete the entry line for $y1$ as: $y1 = -\ln(-x)$ then press EXE. Select the Graph icon.	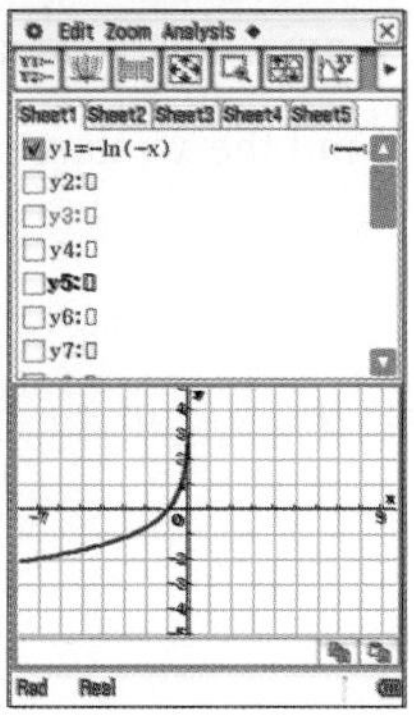

2. To find the x-intercept, press MENU, then select:
6: Analyze Graph
1: Zero
Move the cursor to the left of the x-intercept when prompted for the lower bound, then press ENTER. Move the cursor to the right of the x-intercept when prompted for the upper bound, then press ENTER.

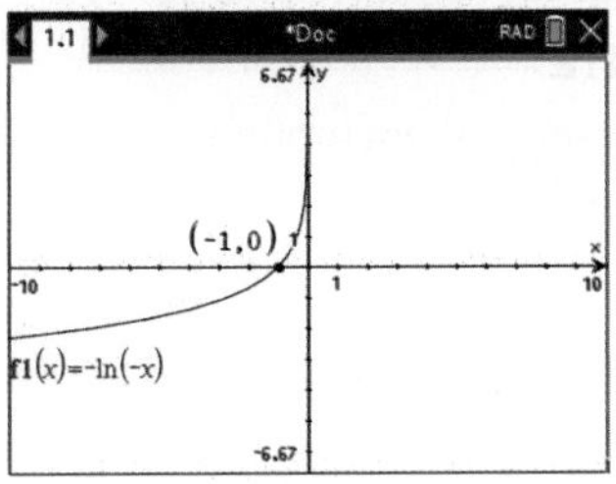

3. To find a second point, press MENU, then select:
5: Trace
1: Graph Trace
Type '−2', then press ENTER twice.
Note: The calculator will give decimal coordinates, not exact values.

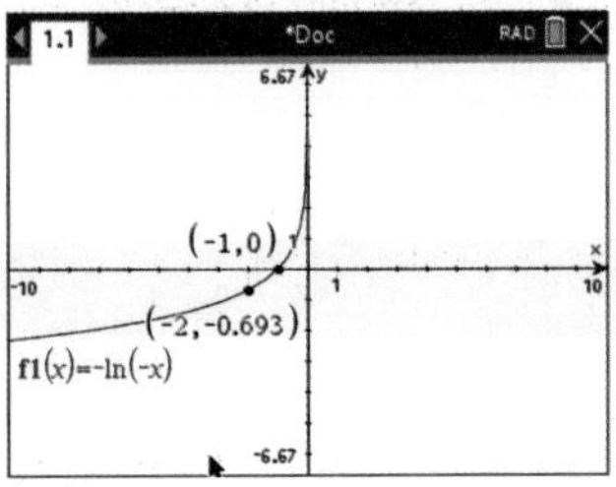

2. To find the x intercept, select:
- Analysis
- G-Solve
- Root

then press EXE.

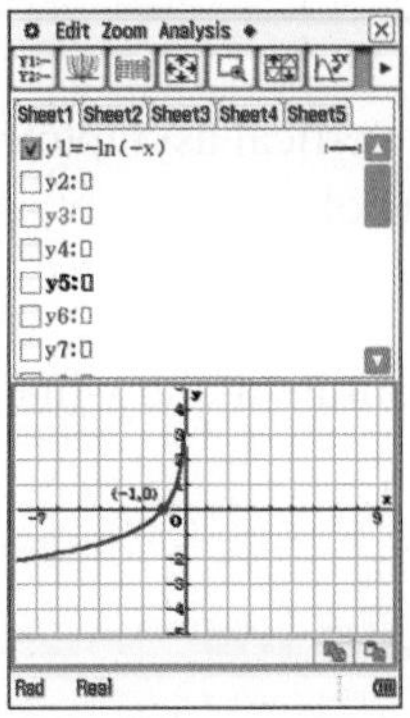

3. To find a second point, select:
- Analysis
- Trace

Type '−2', then select OK and press EXE.
Note: The calculator will give decimal coordinates, not exact values.

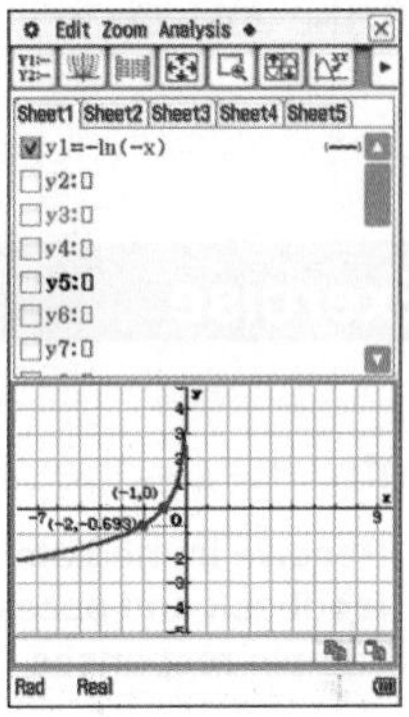

4.5.3 Determining the rule of a logarithmic function

The situation may arise where you are given the graph of a translated logarithmic function and you are required to find the rule. Information that could be provided to you is the equation of the asymptote, the intercepts and/or other points on the graph. As a rule, the number of pieces of information is equivalent to the number of unknowns in the equation.

WORKED EXAMPLE 8 Determining the rule of a logarithm

The rule for the function shown is of the form $y = \log_e(x - a) + b$. Determine the values of the constants a and b.

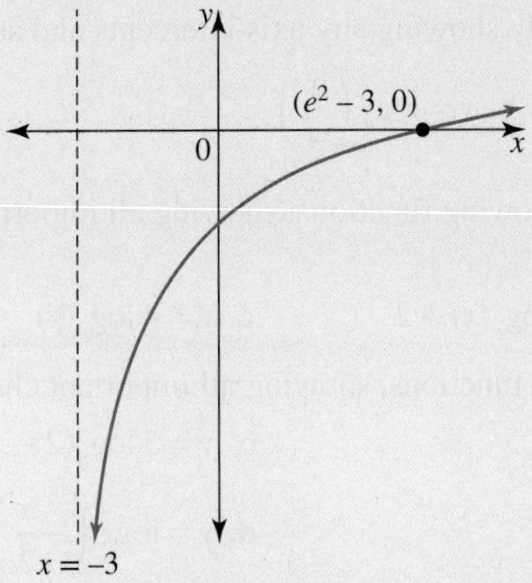

THINK	WRITE
1. The vertical asymptote corresponds to the value of a.	The vertical asymptote is $x = -3$; therefore, a must be -3. So $y = \log_e(x+3) + b$.
2. Substitute in the x-intercept to find b.	The graph cuts the x-axis at $(e^2 - 3,\ 0)$. $0 = \log_e(e^2 - 3 + 3) + b$ $-b = \log_e(e^2)$ $-b = 2$ $b = -2$ So $y = \log_e(x+3) - 2$.
3. Write the answer.	$a = -3, b = -2$.

4.5 Exercise

Technology free

1. Sketch the following graphs, clearly showing any axis intercepts and asymptotes.

   a. $y = \log_e(x) + 3$ b. $y = \log_e(x) - 5$ c. $y = \log_e(x) + 0.5$

2. Sketch the following graphs, clearly showing any axis intercepts and asymptotes.

   a. $y = \log_e(x - 4)$ b. $y = \log_e(x + 2)$ c. $y = \log_e(x + 0.5)$

3. Sketch the following graphs, clearly showing any axis intercepts and asymptotes.

   a. $y = \frac{1}{4}\log_e(x)$ b. $y = 3\log_e(x)$ c. $y = 6\log_e(x)$

4. Sketch the following graphs, clearly showing any axis intercepts and asymptotes.

   a. $y = \log_e(3x)$ b. $y = \log_e\left(\frac{x}{4}\right)$ c. $y = \log_e(4x)$

5. **WE7** Sketch the graphs of the following functions, showing all important characteristics. State the domain and range for each graph.

   a. $y = \log_e(x + 4)$ b. $y = \log_e(x) + 2$ c. $y = 4\log_e(x)$ d. $y = -\log_e(x - 4)$

6. Sketch the graphs of the following functions, showing all important characteristics.

   a. $y = \log_3(x + 2) - 3$ b. $y = 3\log_5(2 - x)$

   c. $y = 2\log_{10}(x + 1)$ d. $y = \log_2\left(-\frac{x}{2}\right)$

7. Sketch the following graphs, clearly showing any axis intercepts and asymptotes.

   a. $y = 1 - 2\log_e(x - 1)$ b. $y = \log_e(2x + 4)$ c. $y = \frac{1}{2}\log_e\left(\frac{x}{4}\right) + 1$

8. **WE8** The rule for the function shown is $y = \log_e(x - m) + n$. Determine the values of the constants m and n.

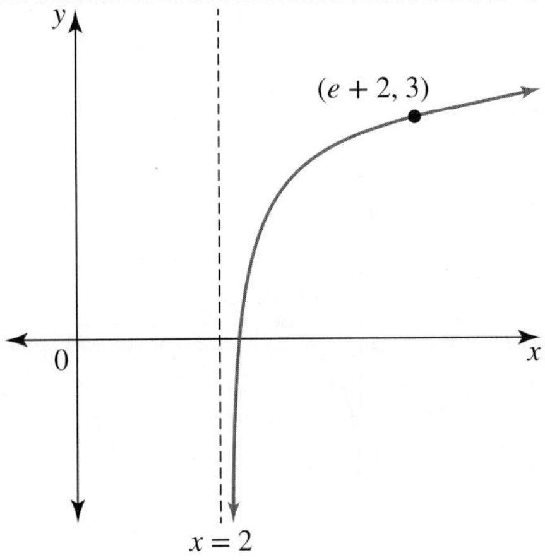

9. The graph of a logarithmic function of the form $y = a\log_e(x - h) + k$ is shown. Determine the values of a, h and k.

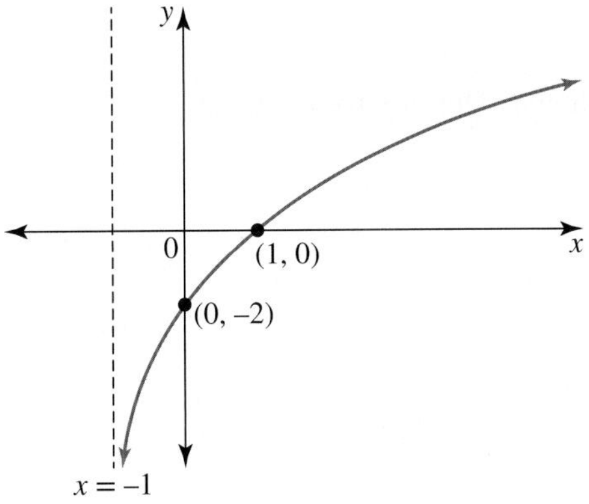

Technology active

10. **MC** The logarithmic function with the rule of the form $y = p\log_e(x - q)$ passes through the points $(0, 0)$ and $(1, -0.35)$. The values of the constants p and q are:

A. $p = -\dfrac{7}{20\log_e(2)},\ q = 1$

B. $p = 0,\ q = -1$

C. $p = -\dfrac{20}{7\log_e(2)},\ q = -1$

D. $p = -\dfrac{7}{20\log_e(2)},\ q = -1$

E. $p = \dfrac{7}{20\log_e(2)},\ q = 1$

11. The equation $y = a\log_e(bx)$ relates x to y. The table below shows values for x and y.

x	1	2	3
y	$\log_e(2)$	0	w

a. Determine the integer values of the constants a and b.
b. Determine the value of w correct to 4 decimal places.

12. The graph of $y = m\log_2(nx)$ passes through the points $(-2, 3)$ and $\left(-\frac{1}{2}, \frac{1}{2}\right)$. Show that the values of m and n are 1.25 and $-2\frac{7}{5}$ respectively.

4.5 Exam questions

Question 1 (3 marks) TECH-FREE

The graph of the equation $y = f(x)$ is shown.

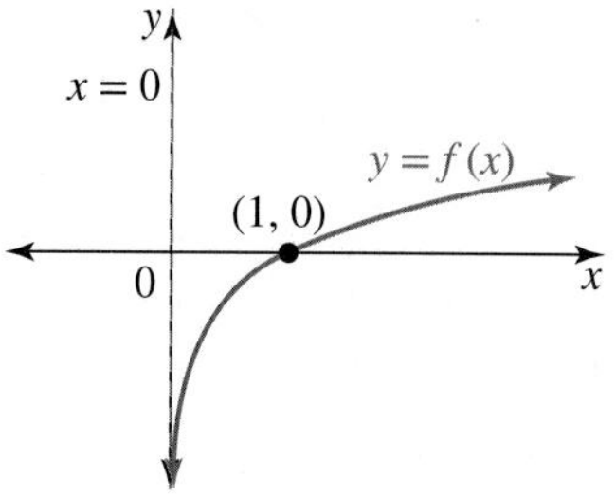

Sketch the graph of:

a. $y = f(-x)$ **(1 mark)**

b. $y = f(x - 1)$ **(1 mark)**

c. $y = f\left(\dfrac{x}{2}\right)$ **(1 mark)**

Question 2 (1 mark) TECH-ACTIVE

MC The graph of the function f is shown. The rule for f is most likely to be:

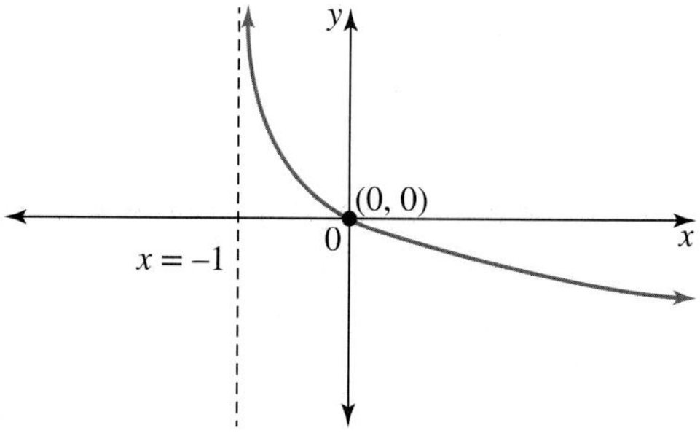

A. $f(x) = 1 + \log_e(x)$

B. $f(x) = \log_e(x) - 1$

C. $f(x) = 1 - \log_e(x)$

D. $f(x) = \log_e(x + 1)$

E. $f(x) = -\log_e(x + 1)$

Question 3 (1 mark) TECH-ACTIVE

MC If the equation of the graph shown is $y = \log_e(ax + b)$, then

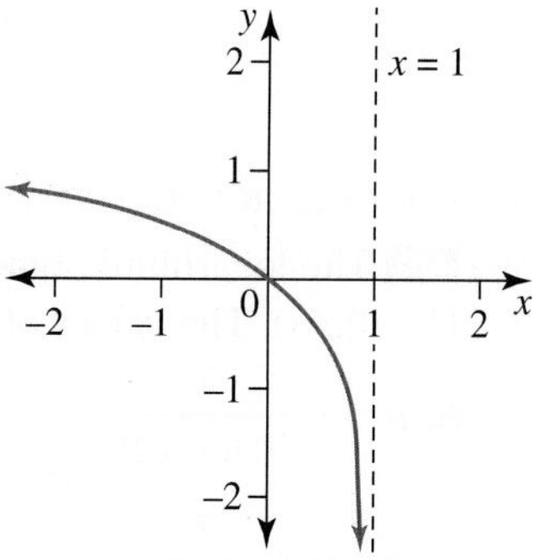

A. $a = 1$ and $b = -1$

B. $a = 1$ and $b = 1$

C. $a = -1$ and $b = -1$

D. $a = -1$ and $b = 1$

E. $a = -1$ and $b = 0$

More exam questions are available online.

4.6 Exponential graphs

LEARNING INTENTION

At the end of this subtopic you should be able to:

- identify key features and properties of the graph of an exponential and draw the graphs of these, clearly identifying their key features and properties, including any vertical asymptotes
- sketch by hand graphs of $y = e^x$ and $y = a^x$ using key points $\left(-1, \dfrac{1}{a}\right)$, $(0, 1)$ and $(1, a)$
- describe the effect of transformations on the graphs of an exponential
- determine the rule of the inverse of an exponential function and give its domain and range.

4.6.1 The graph of $y = a^x$

Exponential functions are those of the form $f: R \to R, f(x) = a^x$, where the base $a \in R^+ \setminus \{1\}$.

The index law $a^0 = 1$ explains why the graph of $y = a^x$ must contain the point (0, 1). The graph of $y = 2^x$ would also contain the point (1, 2), whereas the graph of $y = 3^x$ would contain the point (1, 3).

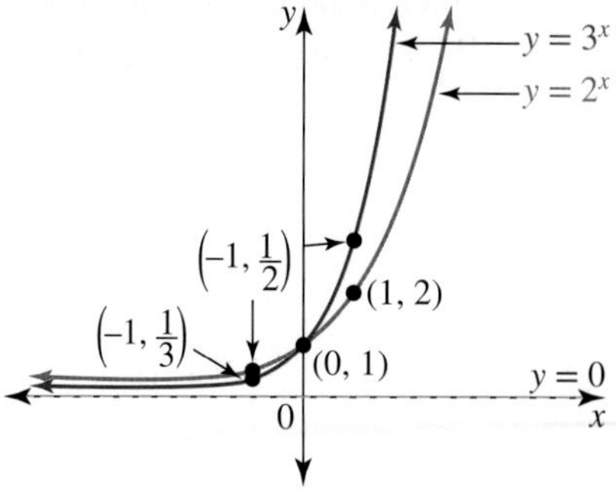

As the base becomes larger, exponential functions increase more quickly. This can be seen in the diagram comparing the graphs of $y = 2^x$ and $y = 3^x$.

The index law $a^{-x} = \frac{1}{a^x}$ explains why for negative values of x the graphs of $y = 2^x$ and $y = 3^x$ approach the x-axis but always lie above the x-axis. The x-axis is a horizontal asymptote for both of their graphs and for any graph of the form $y = a^x$.

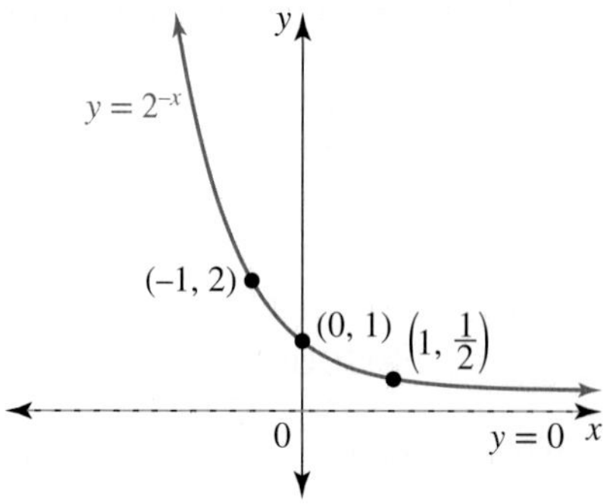

Exponential functions of the form $f: R \to R, f(x) = a^{-x}$, where $a \in R^+ \setminus \{1\}$, have base $\frac{1}{a}$. This is again explained by index laws, as $a^{-x} = \frac{1}{a^x} = \left(\frac{1}{a}\right)^x$. However, it is often preferable to write $y = 2^{-x}$ rather than $y = \left(\frac{1}{2}\right)^x$.

The graph of $y = 2^{-x}$ or $\left(\frac{1}{2}\right)^x$ must contain the point (0, 1), and other points on this graph include (−1, 2) and $\left(1, \frac{1}{2}\right)$.

The graph of $y = 2^x$ illustrates a 'growth' form, whereas the graph of $y = 2^{-x}$ takes a 'decay' form. The two graphs are reflections of each other in the y-axis.

WORKED EXAMPLE 9 Sketching reflections of an exponential function

Consider the function $f(x) = -5^x$.

a. Evaluate $f(2)$.

b. On the same set of axes, sketch the graphs of $y = 5^x$, $y = -5^x$ and $y = 5^{-x}$.

c. Express $y = 5^{-x}$ in an equivalent form.

THINK	WRITE
a. Calculate the required value. *Note:* $-5^2 \neq (-5)^2$	a. $f(x) = -5^x$ $f(2) = -5^2$ $= -25$
b. 1. Identify points on each curve.	b. $y = 5^x$ contains the points (0, 1) and (1, 5). $y = -5^x$ contains the points (0, −1) and (1, −5). $y = 5^{-x}$ contains the points (0, 1) and (−1, 5).

2. Sketch the graphs on the same axes.
Note: When sketching the line, it must always appear to be moving closer to the asymptote but not touching it.

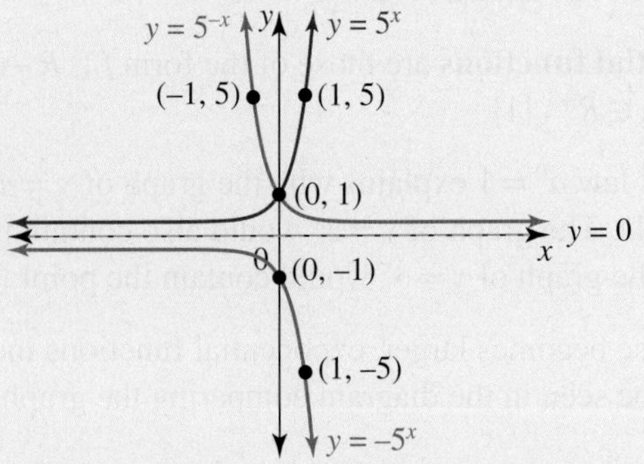

c. Write an equivalent form for the given rule.

c. Since $5^{-x} = \left(\frac{1}{5}\right)^x$, an alternative form for the rule is $y = \left(\frac{1}{5}\right)^x$ or $y = 0.2^x$.

on Resources

Interactivity Exponential functions (int-5959)

4.6.2 The exponential function $y = e^x$

Remember that $e = 2.718\,281...$

As $2 < e < 3$, the graph of $y = e^x$ lies between those of $y = 2^x$ and $y = 3^x$, and has much the same shape.

The graph of $y = e^x$

Key features:

- **The points $(0, 1)$ and $(1, e)$ lie on the graph.**
- **There is a horizontal asymptote at $y = 0$.**
- **The domain is R.**
- **The range is R^+.**
- **The function has one-to-one correspondence.**
- **As $x \to \infty, y \to \infty$, and as $x \to -\infty, y \to 0^+$.**

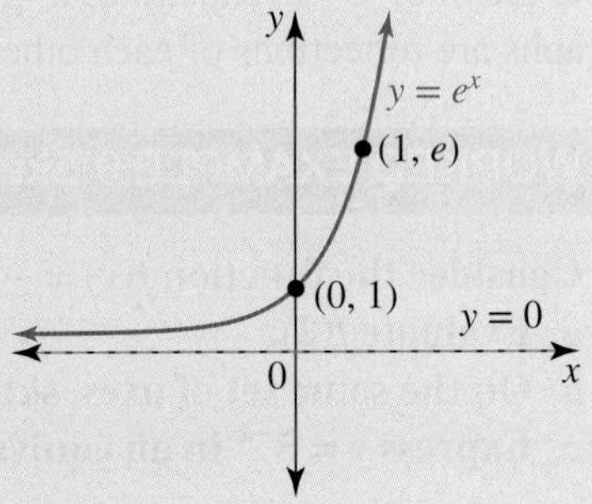

The graph shows an 'exponential growth' shape. Mathematical models of such phenomena, for example population **growth**, usually involve the exponential function $y = e^x$. Exponential **decay** models usually involve the function $y = e^{-x}$.

The graph of $y = e^{-x}$

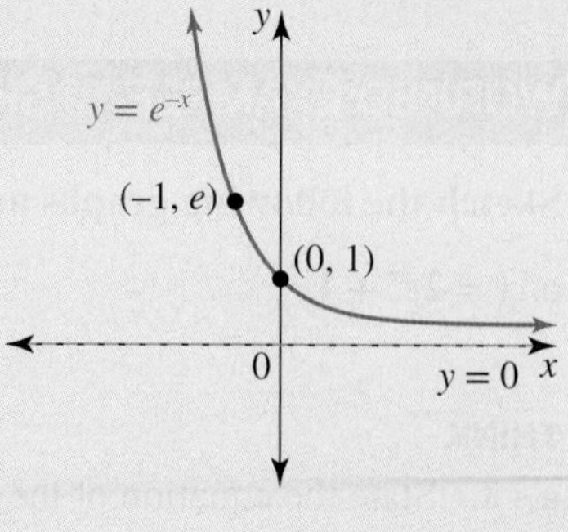

Key features:

- **The points $(0, 1)$ and $(-1, e)$ lie on the graph.**
- **There is a horizontal asymptote at $y = 0$.**
- **The domain is R.**
- **The range is R^+.**
- **The function has one-to-one correspondence.**
- **As $x \to -\infty, y \to \infty$, and as $x \to \infty, y \to 0^+$.**
- **The graph is a reflection of $y = e^x$ in the y-axis.**

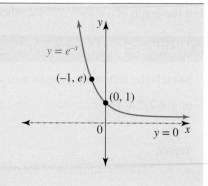

Sketching the graph of $y = ae^{nx} + k$

A vertical translation affects the position of the horizontal asymptote of an exponential graph in the same way it does for a hyperbola or truncus. The graph of $y = e^x + k$ has a horizontal asymptote with equation $y = k$. If $k < 0$, then $y = e^x + k$ will cut through the x-axis and its x-intercept will need to be calculated.

To sketch the graph of an exponential function:

- identify the equation of its asymptote
- calculate its y-intercept
- calculate its x-intercept if there is one.

If the function has no x-intercept, it may be necessary to obtain the coordinates of another point on its graph.

The graph of $y = ae^{nx} + k$

Key features:

- **There is a horizontal asymptote at $y = k$.**
- **There is one y-intercept, obtained by letting $x = 0$.**
- **There is either one or no x-intercept. The relative position of the asymptote and y-intercept will determine whether there is an x-intercept.**
- **If $a > 0$, the range is (k, ∞).**
- **If $a < 0$, the range is $(-\infty, k)$.**

The values of a and n in the equation $y = ae^{nx} + k$ are related to dilation factors, and their signs will affect the orientation of the graph. The possibilities are shown in the following table.

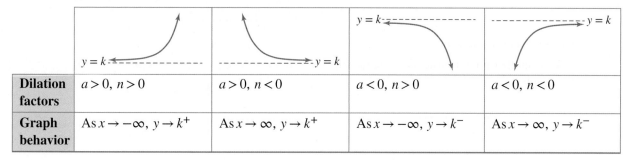

	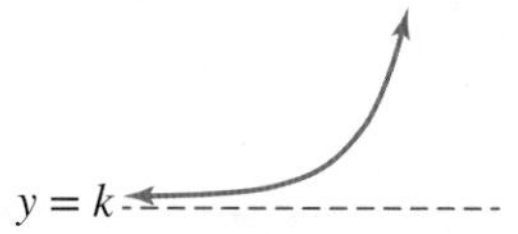	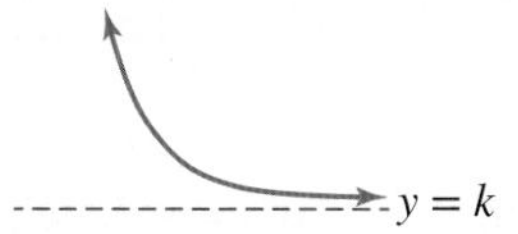	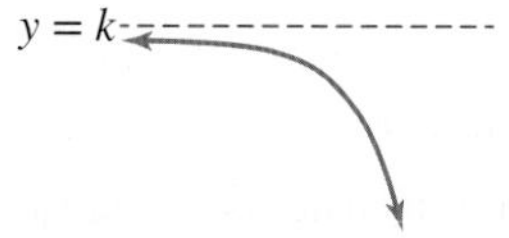	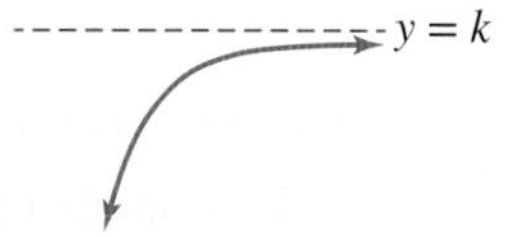
Dilation factors	$a > 0, n > 0$	$a > 0, n < 0$	$a < 0, n > 0$	$a < 0, n < 0$
Graph behavior	As $x \to -\infty$, $y \to k^+$	As $x \to \infty$, $y \to k^+$	As $x \to -\infty$, $y \to k^-$	As $x \to \infty$, $y \to k^-$

Sketching the graph of $y = ae^{n(x-h)} + k$

Under a horizontal translation of h units, the point $(0, 1)$ on $y = e^x$ is translated to the point $(h, 1)$ on the graph of $y = e^{x-h}$. The y-intercept is no longer $(0, 1)$, so it will need to be calculated.

By letting $x = h$ for the horizontal translation, the index for the exponential will be zero. This simplifies the calculation to obtain another point on the graph.

For the graph of $y = ae^{n(x-h)} + k$ when $x = h$, $y = ae^0 + k \Rightarrow y = a + k$.

WORKED EXAMPLE 10 Sketching transformed exponential functions

Sketch the following graphs and state the domain and range of each graph.

a. $y = 2e^x + 1$ **b.** $y = 3 - 3e^{-\frac{x}{2}}$ **c.** $y = -\frac{1}{4}e^{x+1}$

THINK	WRITE
a. 1. State the equation of the asymptote.	$y = 2e^x + 1$ The asymptote is $y = 1$.
2. Calculate the y-intercept.	y-intercept: let $x = 0$. $y = 2e^0 + 1$ $y = 2 + 1$ $y = 3$ The y-intercept is $(0, 3)$.
3. Calculate any x-intercepts.	As the y-intercept is above the positive asymptote, there is no x-intercept.
4. Locate another point if necessary and sketch the graph.	Growth shape 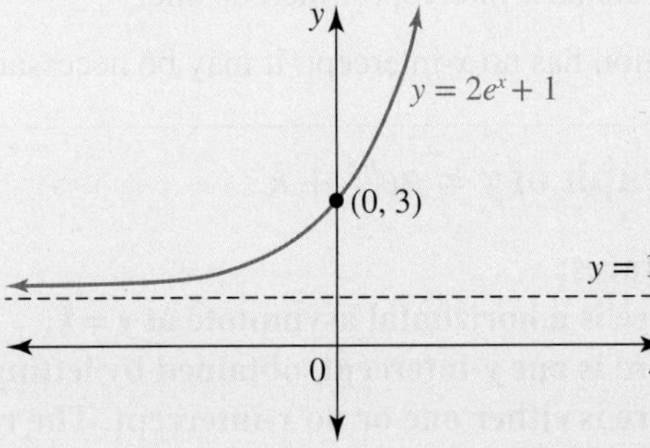
5. State the domain and range.	The domain is R and the range is $(1, \infty)$.
b. 1. State the equation of the asymptote.	$y = 3 - 3e^{-\frac{x}{2}}$ The asymptote is $y = 3$.
2. Calculate the y-intercept.	y-intercept: let $x = 0$. $y = 3 - 3e^0$ $y = 0$ The y-intercept is $(0, 0)$.
3. Calculate any x-intercepts.	$(0, 0)$ is also the x-intercept.
4. Locate another point if necessary and sketch the graph.	If $x = -2$, then $y = 3 - 3e < 0$. 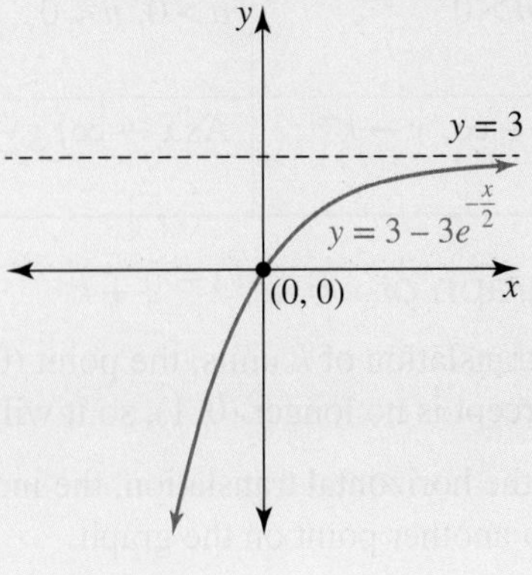

5. State the domain and range.	The domain is R and the range is $(-\infty, 3)$.
c. 1. State the equation of the asymptote.	$y = -\frac{1}{4}e^{x+1}$ The asymptote is $y = 0$.
2. Calculate the y-intercept.	y-intercept: let $x = 0$. $y = -\frac{1}{4}e^{0+1}$ $= -\frac{1}{4}e$ The y-intercept is $\left(0, -\frac{e}{4}\right)$.
3. Calculate any x-intercepts.	There are no x-intercepts as the x-axis is an asymptote.
4. Locate another point if necessary and sketch the graph.	Let $x = -1$. $y = -\frac{1}{4}e^{0}$ $= -\frac{1}{4}e$ Another point on the graph is $\left(-1, -\frac{1}{4}\right)$.

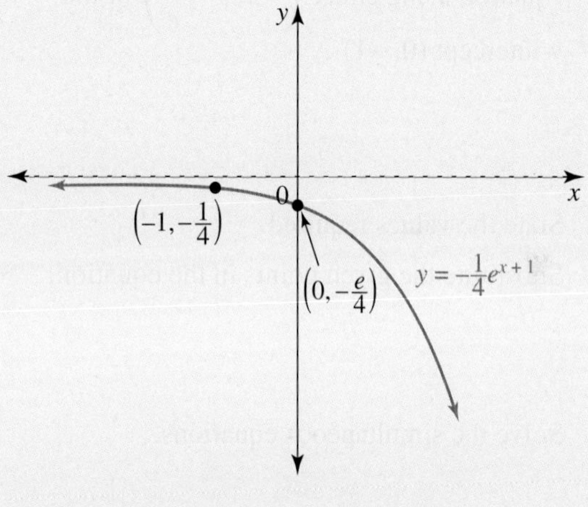

5. State the domain and range.	The domain is R and the range is R^{-}.

4.6.3 Determining the equation of an exponential function

The form of the equation is usually specified along with the given information needed to determine the equation. This is necessary because it could be difficult to decide whether the base is e or some other value. The number of pieces of information given will also need to match the number of parameters or unknown constants in the equation.

The asymptote is a key piece of information to obtain. If a graph is given, the equation of the asymptote will be apparent. Insert this value into the equation and then substitute coordinates of known points on the graph. Simultaneous equations may be required to calculate all the parameters in the equation.

WORKED EXAMPLE 11 Determining the rule of a exponential graph

a. The diagram shows the graph of $y = ae^x + b$. Determine the values of a and b.

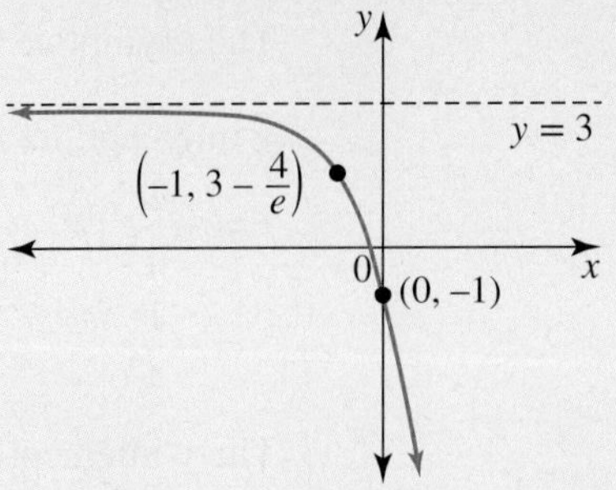

b. The graph of $y = a \times 10^{kx}$ contains the points (2, 30) and (4, 300). Form its equation.

THINK	WRITE
a. 1. Insert the equation of the asymptote into the equation of the graph.	**a.** $y = ae^x + b$ The asymptote is $y = 3$. This means $b = 3$. The equation becomes $y = ae^x + 3$.
2. Use a known point on the graph to fully determine the equation. *Note:* It is possible to get the correct equation using either $\left(-1, 3 - \frac{4}{e}\right)$ or the y-intercept $(0, -1)$	The graph passes through the point $\left(-1, 3 - \frac{4}{e}\right)$. Substitute into the equation. $3 - \frac{4}{e} = ae^{-1} + 3$ $-\frac{4}{e} = \frac{a}{e}$ $a = -4$ The equation is $y = -4e^x + 3$.
3. State the values required.	$a = -4, b = 3$
b. 1. Substitute the given points in the equation.	**b.** $y = a \times 10^{kx}$ $(2, 30) \Rightarrow 30 = a \times 10^{2k}$ $(4, 300) \Rightarrow 300 = a \times 10^{4k}$
2. Solve the simultaneous equations.	$a \times 10^{2k} = 30$ [1] $a \times 10^{4k} = 300$ [2] Divide equation [2] by equation [1]: $\frac{\not{a} \times 10^{4k}}{\not{a} \times 10^{2k}} = \frac{300}{30}$ $10^{4k-2k} = 10^1$ $10^{2k} = 10$ $2k = 1$ $k = \frac{1}{2}$ Substitute $k = \frac{1}{2}$ in equation [1]: $a \times 10^1 = 30$ $a = 3$
3. State the equation.	The equation is $y = 3 \times 10^{\frac{x}{2}}$.

TI | THINK | **DISPLAY/WRITE**

a. 1. On a Lists and Spreadsheet page, label the first column x and the third column y. Enter the x-coordinates of the two known points in the first column and the corresponding y-coordinates in the third column.

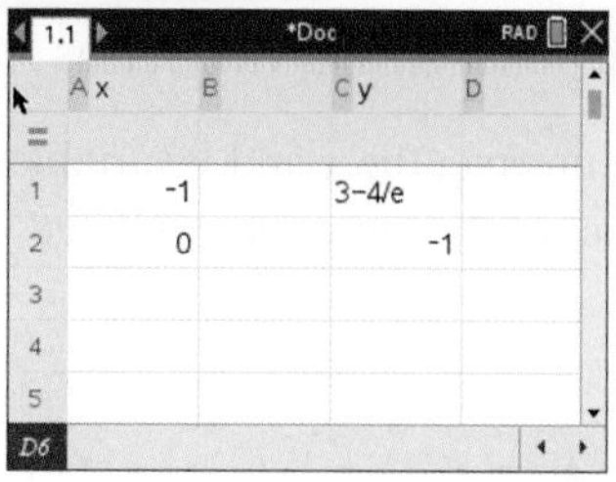

2. Label the second column *expx*. Select the function, entry line for *expx* and complete the entry line as e^x, then press ENTER. Select Variable Reference for x when prompted, then select OK.

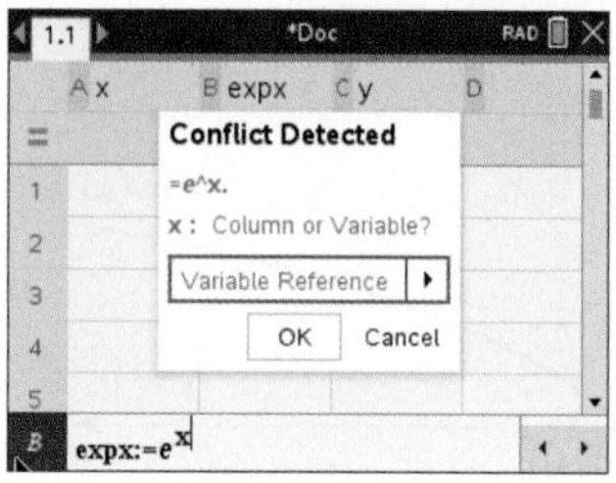

3. Press MENU then select:
6: Statistics
1: Stat Calculations
3: Linear Regression (mx + b)...
Complete the fields as:
X List: expx
Y List: y
1st Result Column: d[]
then select OK.

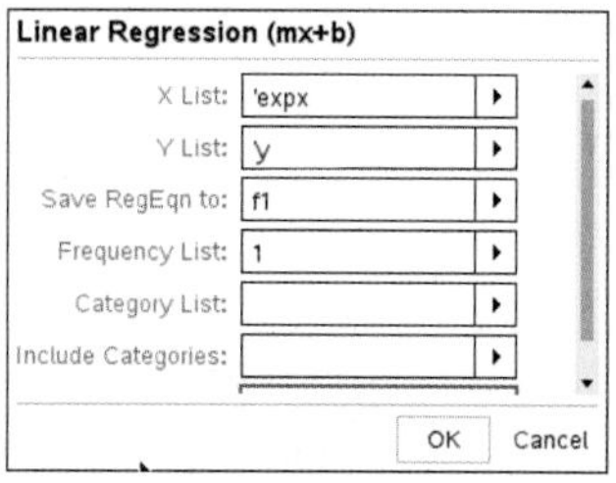

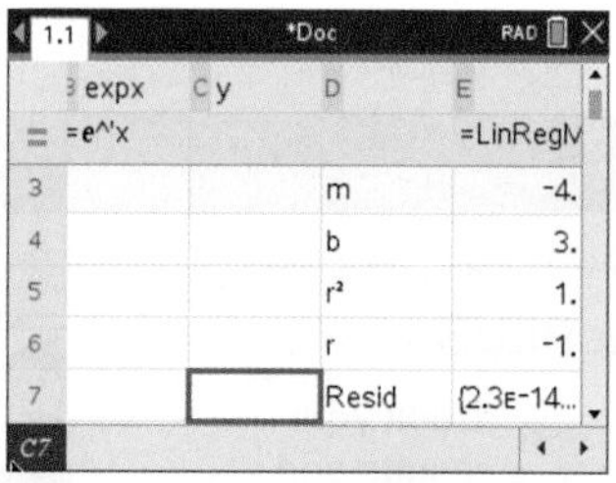

4. The answer appears on the screen.

The answer is of the form $y = mx + b$, where $m = -4$, $x = e^x$, and $b = 3$. The equation is $y = -4e^x + 3$.

CASIO | THINK | **DISPLAY/WRITE**

a. 1. On a Statistics screen, relabel list 1 as x and list 3 as y. Enter the x-coordinates of the two known points in the first column and the corresponding y-coordinates in the third column.

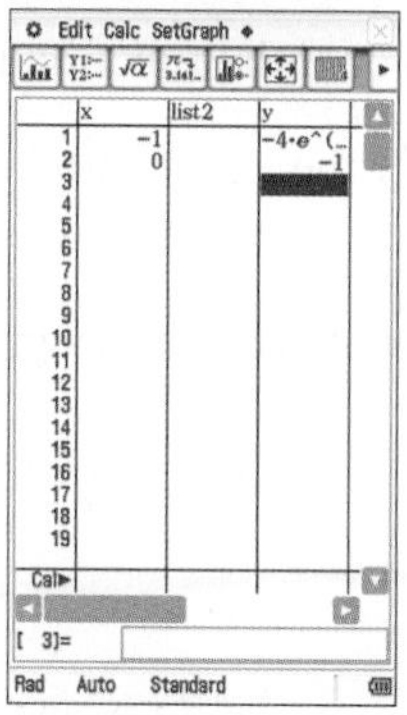

2. Relabel list 2 as *expx*. Select the function entry line for *expx* at the bottom of the list and complete the entry line as e^x, then press EXE.

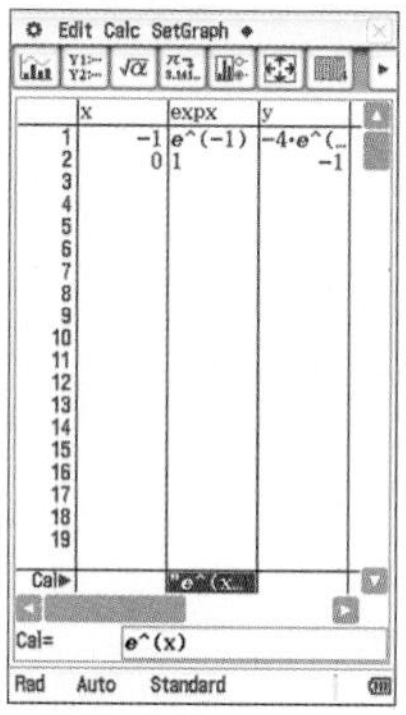

3. Select:

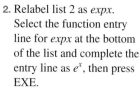

- Calc
- Regression
- Linear Reg

Complete the fields as:
XList: main\expx
YList: main\y
then select OK.

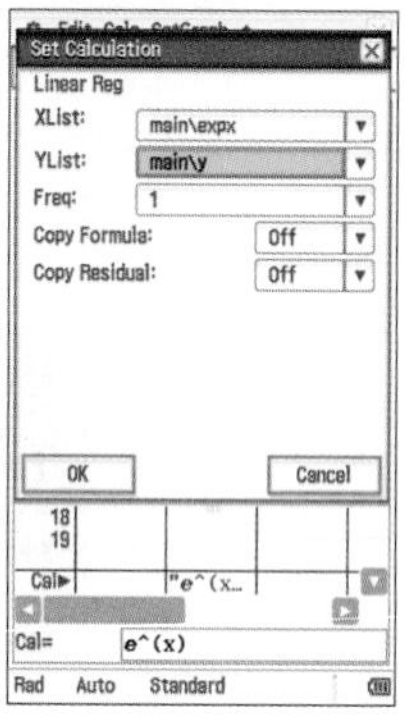

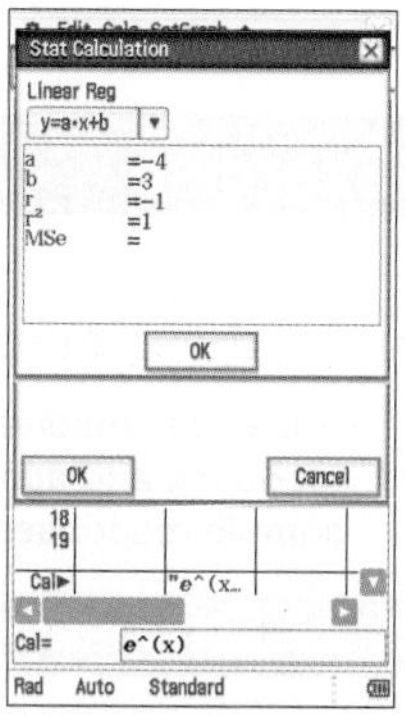

4. The answer appears on the screen.

The answer is of the form $y = ax + b$, where $a = -4$, $x = e^x$, and $b = 3$. The equation is $y = -4e^x + 3$.

4.6.4 Inverse functions

The exponential function is the inverse of the logarithmic function and vice versa.

Consider the exponential function $y = e^x$. To achieve the inverse, the x and y variables are interchanged. Therefore, $y = e^x$ becomes $x = e^y$. If we make y the subject of the equation, we have $y = \log_e(x)$.

This can also be shown graphically.

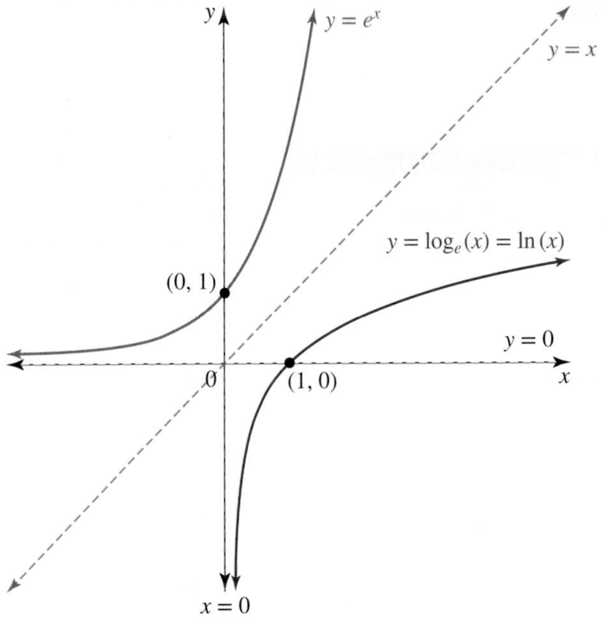

Rule	$y = e^x$	$y = \log_e(x) = \ln(x)$
Type of mapping	One-to-one	One-to-one
Domain	R	$(0, \infty)$
Range	$(0, \infty)$	R

4.6 Exercise

Technology free

1. WE9 Consider the function $f(x) = -10^x$.
   a. Evaluate $f(2)$.
   b. On the same set of axes, sketch the graphs of $y = 10^x$, $y = -10^x$ and $y = 10^{-x}$.
   c. Express $y = 10^{-x}$ in an equivalent form.
2. WE10 Sketch the following graphs and state the domain and range of each graph.
   a. $y = -2e^x - 3$
   b. $y = 4e^{-3x} - 4$
   c. $y = 5e^{x-2}$

3. Sketch the graph of each of the following exponential functions and state their long-term behaviour as $x \to \infty$.

a. $y = \frac{4}{5} \times 10^x$
b. $y = 3 \times 4^{-x}$
c. $y = -5 \times 3^{-\frac{x}{2}}$
d. $y = -\left(\frac{2}{3}\right)^{-x}$

4. For each of the following functions, sketch the graph, state the range and identify the exact position of any intercepts the graph makes with the coordinate axes.

a. $y = e^x - 3$
b. $y = -2e^{2x} - 1$
c. $y = \frac{1}{2}e^{-4x} + 3$
d. $y = 4 - e^{2x}$
e. $y = 4e^{2x-6} + 2$
f. $y = 1 - e^{-\frac{x+1}{2}}$

5. a. Sketch the graph of $y = 2e^{1-3x} - 4$, labelling any intercepts with the coordinate axes with their exact coordinates.
b. Sketch the graph of $y = 3 \times 2^x - 24$ and state its domain and range.

6. **WE11** a. The diagram shows the graph of $y = ae^x + b$. Determine the values of a and b.

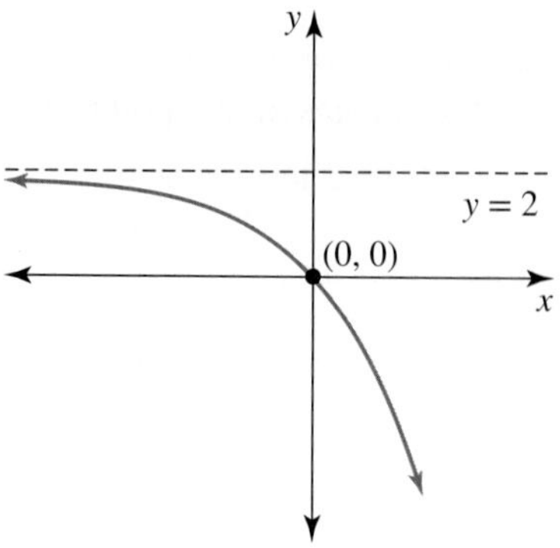

b. The graph of $y = a \times 10^{kx}$ contains the points $(4, -20)$ and $(8, -200)$. Form its equation.

Technology active

7. **MC** The graph of $y = a \times e^{kx}$ contains the points $(2, 36)$ and $(3, 108)$. The exact values of a and k are:

A. $a = 2, k = \log_e(3)$
B. $a = 4, k = 3$
C. $a = 4, k = \log_e(3)$
D. $a = 6, k = \log_e(3)$
E. $a = 6, k = 3$

8. a. The graph shown is of the function $f(x) = ae^x + b$. Determine the values of a and b, and write the function as a mapping.

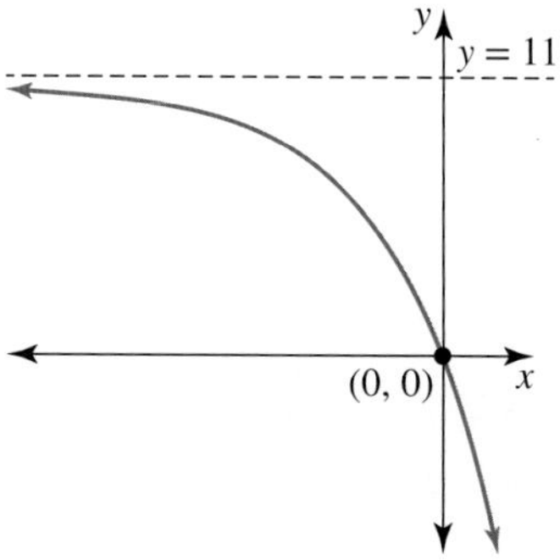

b. The graph shown has an equation of the form $y = Ae^{nx} + k$. Determine its equation.

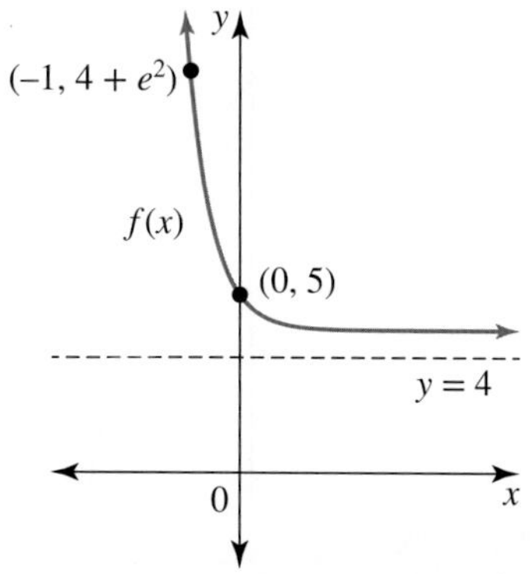

c. The graph of $y = 2^{x-b} + c$ contains the points $(0, -5)$ and $(3, 9)$.

i. Calculate the values of b and c.

ii. State the range of the graph.

d. The graph of $y = Ae^{x-2} + B$ contains the point $(2, 10)$. As $x \to -\infty$, $y \to -2$.

i. Calculate the values of A and B.

ii. The graph passes through the point $\left(a, 2\left(\frac{6}{e} - 1\right)\right)$. Determine the value of a.

9. For each of the following functions, state the domain and range. Define the inverse function, f^{-1}, and state the domain and range in each case.

a. $f(x) = 2\log_e(3x + 3)$

b. $f(x) = \log_e(2(x - 1)) + 2$

c. $f(x) = 2\log_e(1 - x) - 2$

10. For each of the following functions, sketch the graphs of f and f^{-1} on the same set of axes. Give the coordinates of any points of intersection, correct to 2 decimal places.

a. $f(x) = 2\log_e(3x + 3)$

b. $f(x) = \log_e(2(x - 1)) + 2$

c. $f(x) = 2\log_e(1 - x) - 2$

11. For the functions defined by $f(x) = 2^x$ and $g(x) = 2^{-x}$, sketch the graph of the difference function $y = (f - g)(x)$ and state its domain, range and rule.

12. Use addition of ordinates to sketch the graph of $y = e^{-x} + e^x$.

4.6 Exam questions

Question 1 (1 mark) TECH-ACTIVE

Source: *VCE 2018, Mathematical Methods Exam 2, Section A, Q7; © VCAA.*

MC Let $f: R^+ \to R, f(x) = k\log_2(x), k \in R$.

Given that $f^{-1}(1) = 8$, the value of k is

A. 0 **B.** $\frac{1}{3}$ **C.** 3 **D.** 8 **E.** 12

Question 2 (1 mark) TECH-ACTIVE

Source: *VCE 2013, Mathematical Methods (CAS) Exam 2, Section 1, Q4; © VCAA.*

MC Part of the graph of $y = f(x)$, where $f: R \to R, f(x) = 3 - e^x$, is shown below.

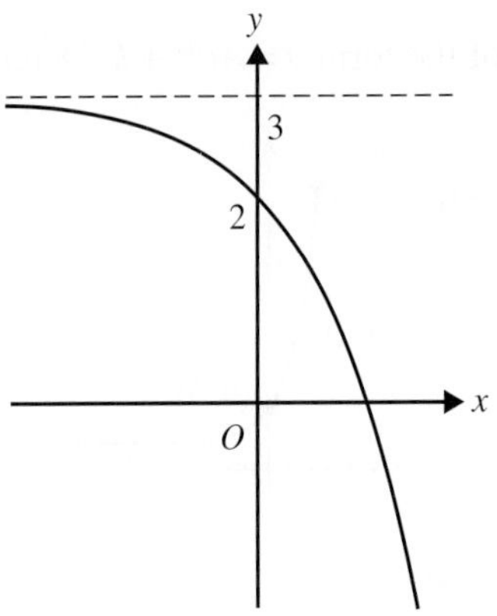

Which one of the following could be the graph of $y = f^{-1}(x)$, where f^{-1} is the inverse of f?

A.

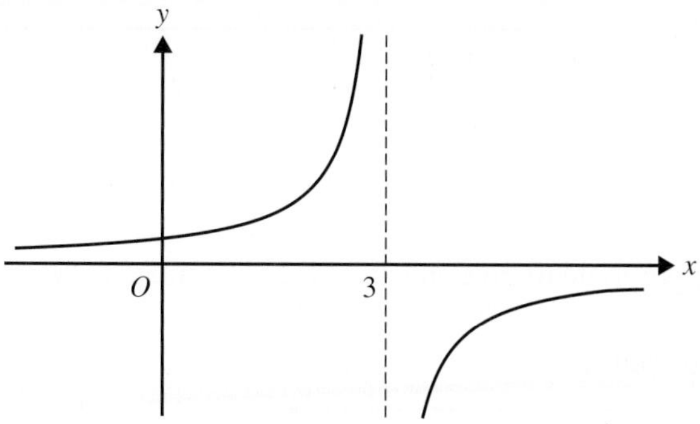

B.

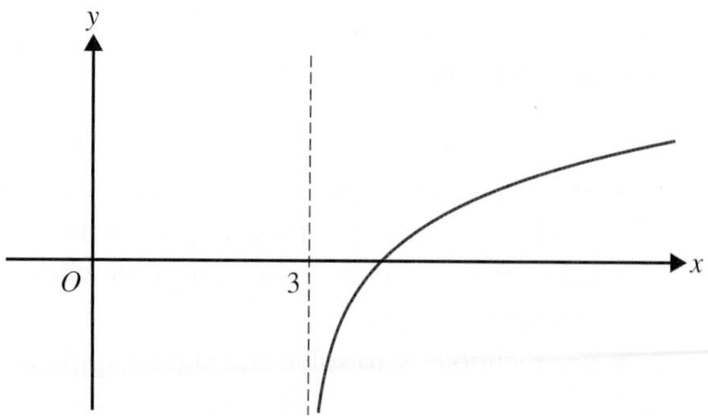

C.

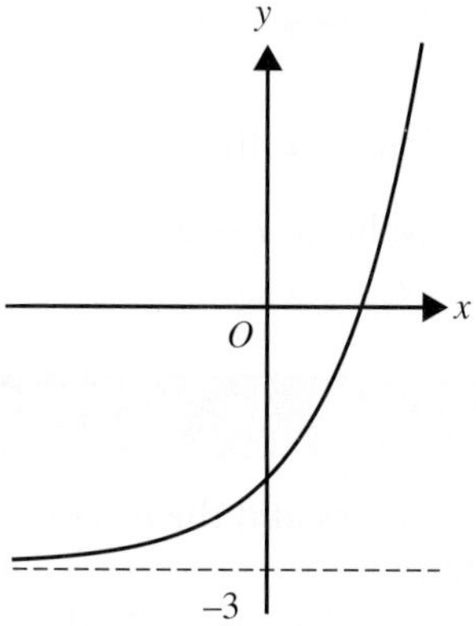

D.

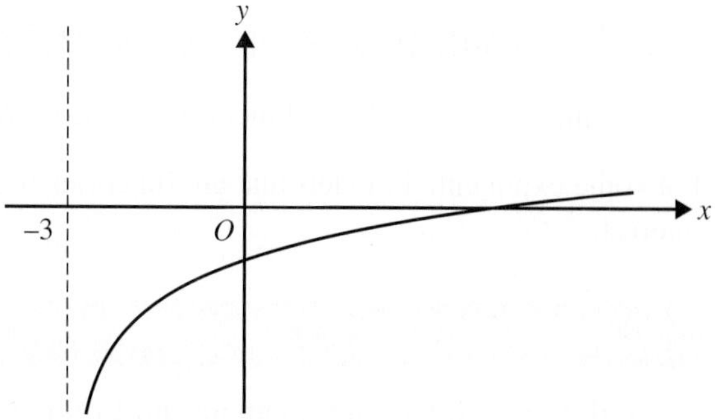

E.

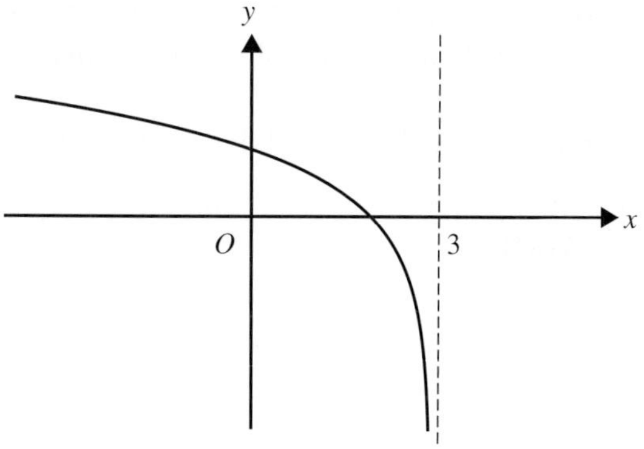

Question 3 (1 mark) TECH-ACTIVE

MC Consider the function $f: R \to R$ where $f(x) = 2 - e^{-x}$.

Select the false statement from the following.

A. The range is $(-\infty, 2)$.
B. The domain is R.
C. The line $y = 2$ is a horizontal asymptote.
D. The graph crosses the x-axis at $\left(\log_e(2), 0\right)$.
E. The graph crosses the y-axis at $(0, 1)$.

More exam questions are available online.

4.7 Applications

LEARNING INTENTION

At the end of this subtopic you should be able to:

- apply these concepts in non-routine contexts involving exponentials, requiring problem-solving and modelling to analyse these applications
- know from the context of a question when it is appropriate to give an exact or approximate answer using technology
- use technology to solve and sketch application problems.

4.7.1 Applications of exponential models

Logarithmic and exponential functions can be used to model many real-life situations directly.

For some exponential models that are functions of time, the behaviour or limiting value as $t \to \infty$ may be of interest.

WORKED EXAMPLE 12 Applications of exponential growth

If P dollars is invested into an account that earns interest at a rate of r for t years and the interest is compounded continuously, then $A = Pe^{rt}$, where A is the accumulated dollars.
A deposit of \$6000 is invested at the Western Bank, and \$9000 is invested at the Common Bank at the same time. Western offers compound interest continuously at a nominal rate of 6% per annum, whereas the Common Bank offers compound interest continuously at a nominal rate of 5% per annum.
Determine how many years it will take for the two investments to be the same. Give your answer to the nearest month.

THINK	WRITE
1. Write the compound interest equation for each of the two investments.	$A = Pe^{rt}$ Western Bank: $A = 6000e^{0.06t}$ Common Bank: $A = 9000e^{0.05t}$
2. Equate the two equations and solve for t. *Note:* If a question does not specify the number of decimal places to round to, an exact value must be given.	$6000e^{0.06t} = 9000e^{0.05t}$ $\frac{e^{0.06t}}{e^{0.05t}} = \frac{9000}{6000}$ $e^{0.01t} = \frac{3}{2}$ $0.01t = \log_e\left(\frac{3}{2}\right)$
3. Use CAS technology to find the decimal answer.	$t = 100\log_e\left(\frac{3}{2}\right)$ $t = 40.547$ years
4. Convert the decimal to a whole number of months. Do this by multiplying the decimal 0.547 by 12.	$0.547 \times 12 \sim 7$ $t = 40$ years and 7 months

WORKED EXAMPLE 13 Application of population growth

The population of foxes on the outskirts of a city is starting to increase. Data collected suggests that a model for the number of foxes is given by $N(t) = 480 - 320e^{-0.3t}$, $t \geq 0$, where N is the number of foxes t years after the observations began.

a. **State how many foxes were present at the start of the observations.**
b. **Calculate how many the population of foxes had grown by at the end of the first year of observations.**
c. **Determine when the model predicts the number of foxes would double its initial population.**
d. **Sketch the graph of N versus t.**
e. **Explain why this model does not predict the population of foxes will grow to 600.**

THINK	WRITE
a. Calculate the initial number.	a. $N(t) = 480 - 320e^{-0.3t}$ When $t = 0$, $N(0) = 480 - 320e^{0}$ $= 480 - 320$ $= 160$ There were 160 foxes present initially.
b. 1. Calculate the number after 1 year. *Note:* There cannot be a fraction of a fox, so give an integer answer.	b. When $t = 1$, $N(1) = 480 - 320e^{-0.3}$ ≈ 242.94 After the first year, approximately 243 foxes were present.
2. Express the change over the first year in context.	Over the first year the population grew from 160 to 243, an increase of 83 foxes.
c. 1. Calculate the required value of t. *Note:* An algebraic method requiring logarithms has been used here. CAS technology could also be used to solve the equation.	c. Let $N = 2 \times 160 = 320$. $320 = 480 - 320e^{-0.3t}$ $320e^{-0.3t} = 160$ $e^{-0.3t} = \frac{1}{2}$ $-0.3t = \log_e\left(\frac{1}{2}\right)$ $t = \frac{1}{0.3}\log_e\left(\frac{1}{2}\right)$ $t \approx 2.31$
2. Answer the question.	$0.31 \times 12 \approx 4$ The population doubles after 2 years and 4 months.

d. Sketch the graph.

d. $N(t) = 480 - 320e^{-0.3t}$

The horizontal asymptote is $N = 480$.

The y-intercept is (0, 160).

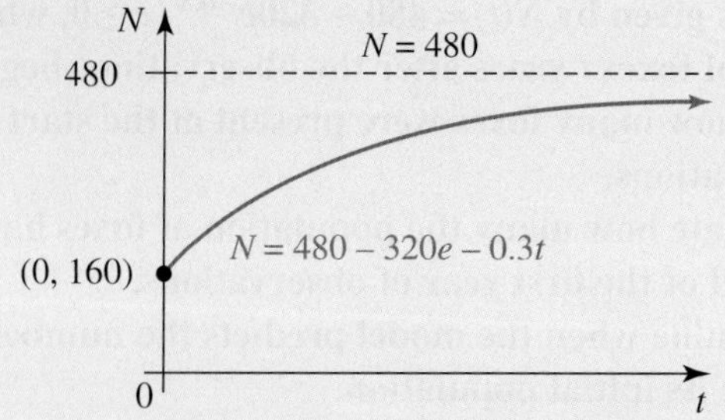

e. Give an explanation for the claim.

e. The presence of an asymptote on the graph shows that as $t \to \infty$, $N \to 480$. Hence, N can never reach 600. The population will never exceed 480 according to this model.

TI | THINK	WRITE/DISPALY
a. 1. If using a function more than once in an application, it is a good idea to first define the function. On a Calculator page, press MENU, then select: 1: Actions 1: Define	**a.** 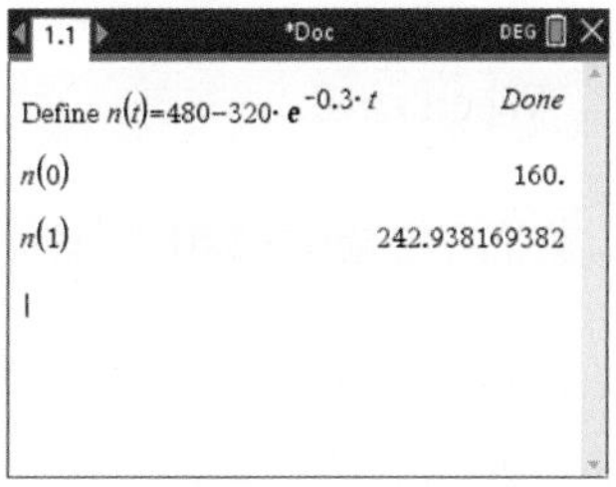 Initially there were 160 foxes.
2. The initial mean is $t = 0$. Substitute in 0 and round the answer to the nearest whole number.	
b. After one year the mean is $t = 1$. Substitute in 1 and round to the nearest whole number.	**b.** After one year there were 243 foxes.
c. 1. Double the number of foxes: $n(t) = 2 \times 160$. On a Calculator page, press MENU, then select: 3: Algebra 1: Solve	**c.** 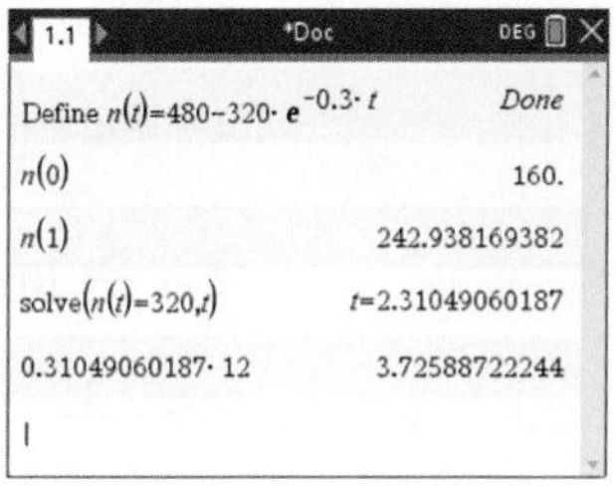 The population doubles in 2 years and 4 months.
2. Multiply the decimal to convert to months, rounding the answer up.	

CASIO | THINK	WRITE/DISPALY
a. 1. If using a function more than once in an application, it is a good idea to first define the function. In Main, select: • Interactive • Define	**a.** 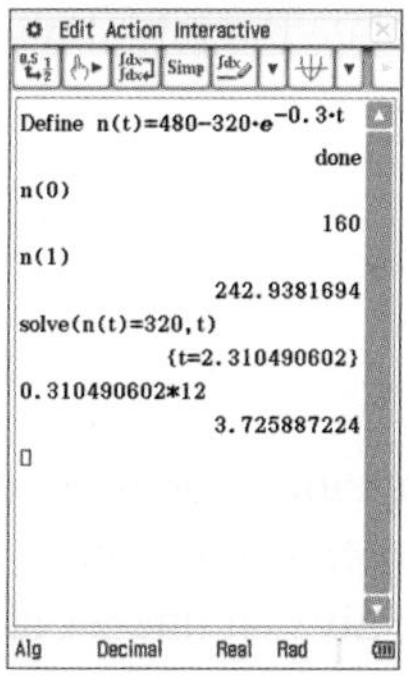 Initially there were 160 foxes.
2. The initial mean is $t = 0$. Substitute in 0 and round the answer to the nearest whole number.	
b. After one year the mean is $t = 1$. Substitute in 1 and round to the nearest whole number.	**b.** After one year there were 243 foxes.
c. 1. Double the number of foxes: $n(t) = 2 \times 160$. In Main, select: • Action • Advanced • Solve	**c.** The population doubles in 2 years and 4 months.
2. Multiply the decimal to convert to months, rounding the answer up.	

d. 1. Press ctrl, then doc to add a graphs page.

d. 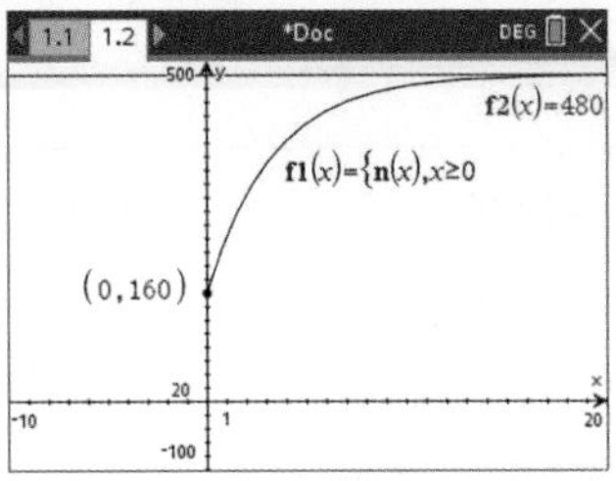

2. Enter the function as: $f1(x) = n(x) | x \geq 0$

3. Change the window to fit the graph. We already know that the graph passes through (0, 160) and (2.31, 320), which helps change the window settings. This can be done by double-clicking the values at the end of each axis.

e. The graph will not pass the asymptote of $n(t) = 480$.
Another way to see this is to use limits.
On a Calculator page, press MENU, then select:
4: Calculus
4: Limit
This can tell us what value the function approaches as $t \to \infty$.

e. 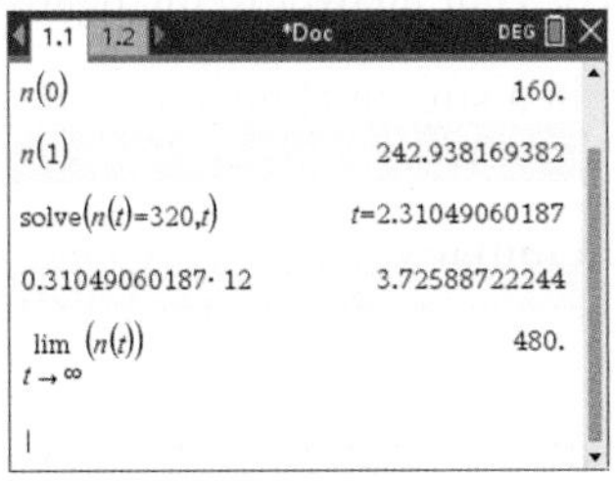

As $t \to \infty$, $n(t) \to 480$, so $n(t)$ will not grow to 600.

d. 1. On a Graph & Table screen, complete the entry line for y1 as:
$y1 = n(x) | x \geq 0$
then press EXE.
Select the Graph icon.

d. 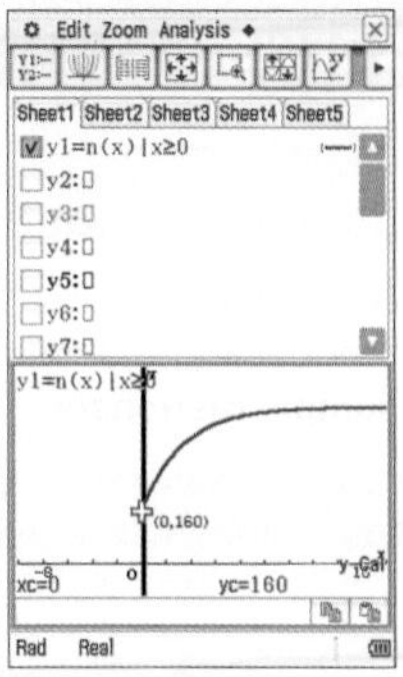

2. Change the window to fit the graph using the icon with four arrows. We already know that the graph passes through (0, 160) and (2.31, 320), which helps change the window settings.

e. The graph will not pass the asymptote of $n(t) = 480$.
Another way to see this is to use limits.
In Main, bring up the keyboard.
In Math2, select the limit icon.
The ∞ button can be found in the Trig menu on the keyboard.
This can tell us what value the function approaches as $t \to \infty$.

e. 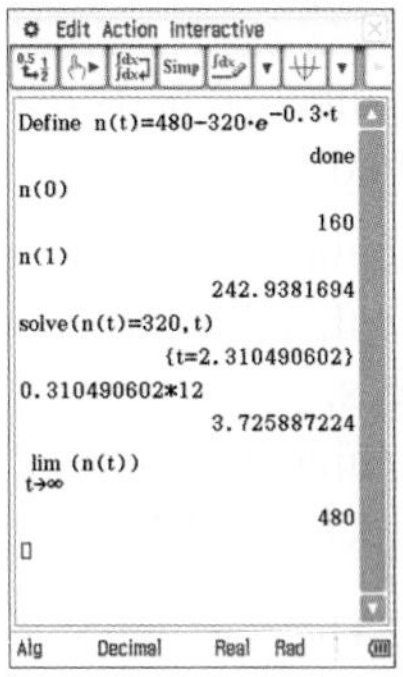

As $t \to \infty$, $n(t) \to 480$, so $n(t)$ will not grow to 600.

4.7 Exercise

Technology active

1. WE12 A deposit of \$4200 is invested at the Western Bank, and \$5500 is invested at the Common Bank at the same time. Western Bank offers compound interest continuously at a nominal rate of 5% per annum, whereas the Common bank offers compound interest continuously at a nominal rate of 4.5% per annum. Determine how many years it will take for the two investments to be the same. Give your answer to the nearest year.

2. If \$1000 is invested for 10 years at 5% per annum interest compounded continuously, find how much money will have accumulated after the 10 years.

3. a. An investment triples in 15 years. Determine the interest rate that this investment earns if it is compounded continuously. Give your answer correct to 2 decimal places.
   b. An investment of \$2000 earns 4.5% per annum interest compounded continuously. Find how long it will take for the investment to have grown to \$9000. Give your answer to the nearest month.

4. The number of parts per million, n, of a fungal bloom in a stream t hours after it was detected can be modelled by $n(t) = \log_e(t + e^2),\ t \geq 0$.
   a. Calculate how many parts per million were detected initially.
   b. Calculate how many parts of fungal bloom are in the stream after 12 hours. Give your answer to 2 decimal places.
   c. Calculate how long it will take before there are 4 parts per million of the fungal bloom. Give your answer correct to 1 decimal place.

5. WE13 The population of possums in an inner city suburb is starting to increase. Observations of the numbers present suggest a model for the number of possums in the suburb given by $P(t) = 83 - 65e^{-0.2t}, t \geq 0$, where P is the number of possums observed and t is the time in months since observations began.
   a. State how many possums were present at the start of the observations.
   b. Calculate how many possums the population had grown by at the end of the first month of observations.
   c. Determine when the model predicts the number of possums would double its initial population.
   d. Sketch the graph of P versus t.
   e. Explain why this model does not predict the population of possums will grow to 100.

6. Let $P(t) = 200^{kt} + 1000$ represent the number of bacteria present in a petri dish after t hours. Suppose the number of bacteria trebles every 8 hours. Determine the value of the constant k correct to 4 decimal places.

7. An epidemiologist studying the progression of a flu epidemic decides that the function

$$P(t) = \frac{3}{4}(1 - e^{-kt}), k > 0$$

will be a good model for the proportion of the earth's population that will contract the flu after t months. If after 3 months $\frac{1}{1500}$ of the earth's population has the flu, determine the value of the constant k, correct to 4 decimal places.

8. Carbon-14 dating works by measuring the amount of carbon-14, a radioactive element, that is present in a fossil. All living things have a constant level of carbon-14 in them. Once an organism dies, the carbon-14 in its body starts to decay according to the rule

$$Q = Q_0 e^{-0.000\,124t}$$

where t is the time in years since death, Q_0 is the amount of carbon-14 in milligrams present at death and Q is the quantity of carbon-14 in milligrams present after t years.

a. If it is known that a particular fossil initially had 100 milligrams of carbon-14, calculate how much carbon-14, in milligrams, will be present after 1000 years. Give your answer correct to 1 decimal place.
b. Calculate how long it will take before the amount of carbon-14 in the fossil is halved. Give your answer correct to the nearest year.

9. Glottochronology is a method of dating a language at a particular stage, based on the theory that over a long period of time linguistic changes take place at a fairly constant rate. Suppose a particular language originally has W_0 basic words and that at time t, measured in millennia, the number, $W(t)$, of basic words in use is given by $W(t) = W_0(0.805)^t$.

a. Calculate the percentage of basic words lost after ten millennia.
b. Calculate the length of time it would take for the number of basic words lost to be one-third of the original number of basic words. Give your answer correct to 2 decimal places.

10. Andrew believes that his fitness level can be modelled by the function

$$F(t) = 10 + 2\log_e(t + 2)$$

where $F(t)$ is his fitness level and t is the time in weeks since he started training.

a. State Andrew's level of fitness before he started training.
b. Calculate Andrew's level of fitness after 4 weeks of training.
c. Calculate how long it will take for Andrew's level of fitness to reach 15.

11. In 1947 a cave with beautiful prehistoric paintings was discovered in Lascaux, France. Some charcoal found in the cave contained 20% of the carbon-14 that would be expected in living trees. Determine the age of the paintings to the nearest whole number if

$$Q = Q_0 e^{-0.000\,124t}$$

where Q_0 is the amount of carbon-14 originally and t is the time in years since the death of the prehistoric material. Give your answer correct to the nearest year.

12. The sales revenue, R dollars, that a manufacturer receives for selling x units of a certain product can be modelled by the function

$$R(x) = 800 \log_e \left(2 + \frac{x}{250}\right).$$

Furthermore, each unit costs the manufacturer 2 dollars to produce, and the initial cost of adjusting the machinery for production is \$300, so the total cost in dollars, C, of production is

$$C(x) = 300 + 2x.$$

a. Write the profit, $P(x)$ dollars, obtained by the production and sale of x units.
b. Determine the number of units that need to be produced and sold to break even, that is, $P(x) = 0$. Give your answer correct to the nearest integer.

13. The value of a certain number of shares, $\$V$, can be modelled by the equation

$$V = ke^{mt}$$

where t is the time in months. The original value of the shares was \$10 000, and after one year the value of the shares was \$13 500.

a. Calculate the values of the constants k and m, giving answers correct to 3 decimal places where appropriate.
b. Determine the value of the shares to the nearest dollar after 18 months.
c. After t months, the shares are sold for 1.375 times their value at the time. Find an equation relating the profit made, P, over the time the shares were owned.
d. If the shares were kept for 2 years, calculate the profit made on selling the shares at that time.

14. The population, P, of trout at a trout farm is declining due to deaths of a large number of fish from fungal infections. The population is modelled by the function

$$P = a \log_e(t) + c$$

where t represents the time in weeks since the infection started. The population of trout was 10 000 after 1 week and 6000 after 4 weeks.

a. Calculate the values of the constants a and c. Give your answers correct to 1 decimal place where appropriate.
b. Determine the number of trout, correct to the nearest whole trout, after 8 weeks.
c. If the infection remains untreated, determine how long it will take for the population of trout to be fewer than 1000. Give your answer correct to 1 decimal place.

15. The mass, M grams, of a radioactive element, is modelled by the rule

$$M = a - \log_e(t + b)$$

where t is the time in years. The initial mass is 7.8948 grams, and after 80 years the mass is 7.3070 grams.

a. Determine the equation of the mass remaining after t years. Give a correct to 1 decimal place and b as an integer.
b. Calculate the mass remaining after 90 years.

4.7 Exam questions

Question 1 (4 marks) TECH-ACTIVE

Source: *VCE 2014, Mathematical Methods (CAS) Exam 2, Section 2, Q3; © VCAA.*

In a controlled experiment, Juan took some medicine at 8 pm. The concentration of medicine in his blood was then measured at regular intervals. The concentration of medicine in Juan's blood is modelled by the function $c(t) = \frac{5}{2}te^{-\frac{3t}{2}}, t \geq 0$, where c is the concentration of medicine in his blood, in milligrams per litre, t hours after 8 pm. Part of the graph of the function c is shown below.

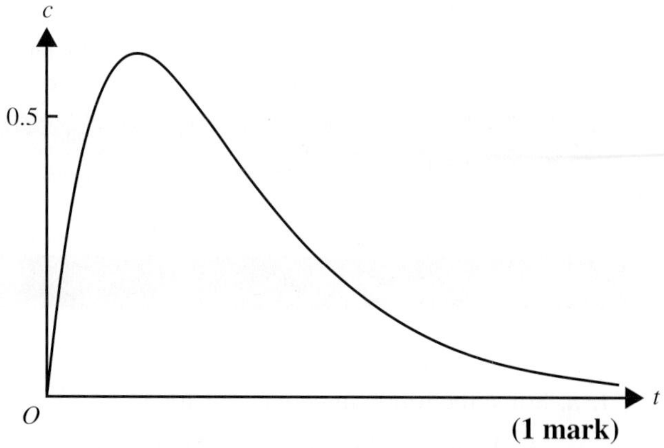

a. What was the maximum value of the concentration of medicine in Juan's blood, in milligrams per litre, correct to two decimal places? **(1 mark)**

b. Find the value of t, in hours, correct to two decimal places, when the concentration of medicine in Juan's blood first reached 0.5 milligrams per litre. **(1 mark)**

c. Find the length of time that the concentration of medicine in Juan's blood was above 0.5 milligrams per litre. Express the answer in hours, correct to two decimal places. **(2 marks)**

Question 2 (4 marks) TECH-ACTIVE

In her chemistry class, Hei is preparing a special solution for an experiment that she has to complete. The concentration of the solution can be modelled by the rule

$$C = A\ \log_e(kt)$$

where C is the concentration in moles per litre (M) and t represents the time of mixing in seconds. The concentration of the solution after 30 seconds of mixing is 4 M, and the concentration of the solution after 2 seconds of mixing was 0.1 M.

a. Calculate the values of the constants A and k, giving your answers correct to 3 decimal places. **(2 marks)**

b. Determine the concentration of the solution after 15 seconds of mixing. **(1 mark)**

c. Determine how long it will take, in minutes and seconds, for the concentration of the solution to reach 10 M. **(1 mark)**

Question 3 (7 marks) TECH-ACTIVE

Manoj pours himself a mug of coffee but gets distracted by a phone call before he can drink the coffee. The temperature of the cooling mug of coffee is given by $T = 20 + 75e^{-0.062t}$, where T is the temperature of the coffee t minutes after it was initially poured into the mug.

a. Calculate the initial temperature of the coffee when it was first poured. **(1 mark)**

b. Determine the temperature to which the coffee will cool if left unattended. **(1 mark)**

c. Determine how long it will take for the coffee to reach a temperature of 65 °C. Give your answer correct to 2 decimal places. **(1 mark)**

d. Manoj returns to the coffee when it has reached 65 °C and decides to reheat the coffee in a microwave. The temperature of the coffee in this warming stage is $T = A + Be^{-0.05t}$.
Given that the temperature of the reheated coffee cannot exceed 85 °C, calculate the values of A and B. **(2 marks)**

e. Sketch a graph showing the temperature of the coffee during its cooling and warming stages. **(2 marks)**

More exam questions are available online.

4.8 Review

4.8.1 Summary

doc-37033

Hey students! Now that it's time to revise this topic, go online to:

Access the topic summary | Review your results | Watch teacher-led videos | Practise VCAA exam questions

Find all this and MORE in jacPLUS

4.8 Exercise

Technology free: short answer

1. a. Solve the following equations for x.

   i. $2\log_e(x) - \log_e(x-1) = \log_e(x-4)$
   ii. $2\log_e(x+2) - \log_e(x) = \log_e 3(x-1)$
   iii. $2\left(\log_4(x)\right)^2 = 3 - \log_4(x^5)$

   b. Express y in terms of x for the following equations, giving any restrictions for x.

   i. $\log_2(y) = 2\log_2(x) - 3$
   ii. $\log_3(9x) - \log_3(x^4 y) = 2$

   c. If $\log_4(p) = x$ and $\log_4(q) = y$, show that $\log_4\left(\dfrac{64q^2}{p^3\sqrt{q}}\right) = 3 - 3x + \dfrac{3y}{2}$.

2. The pH of a substance is a value that defines the acidity or alkalinity of that substance. It depends on the concentration of the hydrogen ion, $\left[H^+\right]$ in moles/litre, and is calculated according to the formula

$$\text{pH} = -\log_{10}\left[H^+\right].$$

   Solutions with a pH less than 7 are acidic, solutions with a pH greater than 7 are basic, and solutions with a pH of 7 are neutral.

   a. For each of the following, find the pH and state whether the solution is acidic, basic or neutral.

   i. Vinegar has a hydrogen ion concentration of 0.01 moles/litre.
   ii. Ammonia has a hydrogen ion concentration of 10^{-11} moles/litre.

   b. Determine the hydrogen ion concentration for each of the following.

   i. Apples have a pH of 3.
   ii. Sodium hydroxide has a pH of 14.

3. Solve the following equations for x.

   a. $e^{2x} - 8e^x + 15 = 0$
   b. $2e^{2x} - 35 = 9e^x$
   c. $2^x + 18 \times 2^{-x} = 11$

4. Solve the following equations for x.

   a. $3^{kx} \times 9^{2m} = 27$, where $k \in R \setminus \{0\}$ and $m \in R$
   b. $3e^{kx} - 4 = 6e^{-kx}$, where $k \in R \setminus \{0\}$

5. Sketch the graphs of each of the following, showing any axis intercepts and the asymptote(s). State the domain and range in each case.

   a. $y = \log_e(x-1) + 3$
   b. $y = \log_e(x+3) - 1$
   c. $y = 2\log_e(-x)$
   d. $y = -\log_e(x-4)$

6. Sketch each of the following over the implied domain and state the range.

a. $y = 6 \times 2^{x-1} - 12$

b. $y = \frac{1}{2}\left(e^{3-x} + 5\right)$

7. The loudness of plant machinery at a manufacturing business is modelled by the equation $L = 10\log_{10}\left(\frac{I}{I_0}\right)$, where L is the loudness in decibels (dB), I is the intensity in watt/m^2 and $I_0 = 10^{-12}$ watt/m^2.

a. If the loudness of the plant machinery at this business is known to be 90 dB, calculate the intensity for this situation.

b. Calculate the loudness of the plant machinery if the intensity is 10^{-6} watt/m^2.

Technology active: multiple choice

8. **MC** Simplifying $3\log_e(5) + 2\log_e(2) - \log_e(20)$ gives:

A. $\log_e\left(\frac{19}{20}\right)$

B. $\log_e(109)$

C. $\log_e(480)$

D. $2\log_e(5)$

E. $6\log_e\left(\frac{1}{2}\right)$

9. **MC** The table below gives values for x and y that relate to the equation $y = a\log_e(bx)$. Calculate the exact values of a, b and m.

x	1	2	3
y	$-3\log_e(2)$	0	m

A. $a = -3,\ b = 2,\ m = \log_e\left(\frac{9}{4}\right)$

B. $a = -3,\ b = \frac{1}{2},\ m = \log_e\left(\frac{27}{4}\right)$

C. $a = -3,\ b = \frac{1}{2},\ m = 3\log_e\left(\frac{3}{2}\right)$

D. $a = 3,\ b = \frac{1}{2},\ m = \log_e\left(\frac{3}{2}\right)$

E. $a = 3,\ b = \frac{1}{2},\ m = 3\log_e\left(\frac{3}{2}\right)$

10. **MC** If $5\log_{10}(x) - \log_{10}(x^2) = 1 + \log_{10}(y)$, then x is equal to:

A. y

B. $10y$

C. $\sqrt[3]{10y}$

D. $\frac{10}{y}$

E. $\sqrt[3]{\frac{10}{y}}$

11. **MC** The exact solution of the equation $3^{2x+1} - 4 \times 3^x + 1 = 0$ is:

A. $x = 0,\ x = -1$

B. $x = 0,\ x = 1$

C. $x = -1,\ x = 1$

D. $x = \frac{1}{3},\ x = 1$

E. $x = \frac{1}{3},\ x = -1$

12. **MC** $e^{(3\log_e(x) - \log_e(3x))}$ is equal to:

A. $3\log_e\left(\frac{1}{3}\right)$

B. 0

C. 1

D. $x^3 - 3x$

E. $\frac{x^2}{3}$

13. **MC** The expression $\log_n(m) + \log_m(p) + \log_p(n)$ is equal to:

A. $\frac{1}{\log_n(m)} + \frac{1}{\log_m(p)} + \frac{1}{\log_p(n)}$

B. $\frac{1}{\log_m(n)} + \frac{1}{\log_p(m)} + \frac{1}{\log_n(p)}$

C. $-\frac{1}{\log_n(m)} - \frac{1}{\log_m(p)} - \frac{1}{\log_p(n)}$

D. $\frac{1}{\log_m(m)} + \frac{1}{\log_p(p)} + \frac{1}{\log_n(n)}$

E. $\frac{1}{\log_n(mp)} + \frac{1}{\log_p(mn)} + \frac{1}{\log_m(np)}$

14. **MC** The function h has the rule $h(x) = a\log_e(x - m) + k$, where m and k are positive constants and a is a negative constant. The maximal domain of h is:

A. R^+
B. $R\backslash\{m\}$
C. $R\backslash\{n\}$
D. (m, n)
E. (m, ∞)

15. **MC** If $7e^{ax} = 3$, then x equals:

A. $\frac{3}{7}\log_e a$
B. $a\log_e\left(\frac{3}{7}\right)$
C. $\frac{\log_e\left(\frac{3}{7}\right)}{a}$
D. $\frac{\log_e(3)}{a\log_e(7)}$
E. $\frac{a\log_e(3)}{\log_e(7)}$

16. **MC** $\log_e(4e^{3x})$ is equal to:

A. $12x$
B. $\log_e(12) + x$
C. $\log_e(e^{12x})$
D. $3x\log_e(4)$
E. $\log_e(4) + 3x$

17. **MC** A possible equation for the graph shown is:

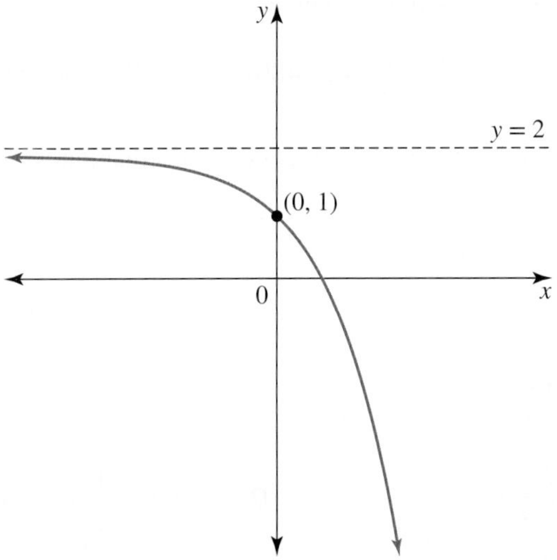

A. $y = 2 - e^x$
B. $y = 2 - e^{-x}$
C. $y = 2 + e^x$
D. $y = 2 + e^{-x}$
E. $y = e^{-x} - 2$

Technology active: extended response

18. a. The graph of the function $f: (-5, \infty) \to R, f(x) = \log_e(x + 5) + 1$ is shown.

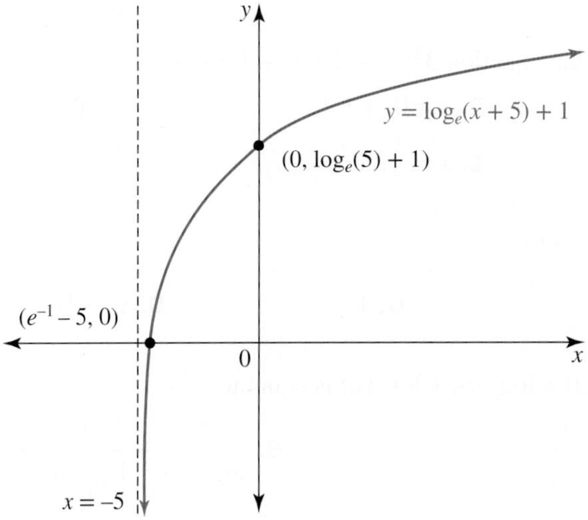

i. Determine the rule and domain of f^{-1}, the inverse function of f.
ii. On the same set of axes, sketch the graph of f^{-1}. Label the axis intercepts with their exact values.
iii. Determine the coordinates of the point(s) of intersection correct to 3 decimal places.

b. The graph below has the rule $g(x) = \log_e(x - h) + k$, where h and k are constants.

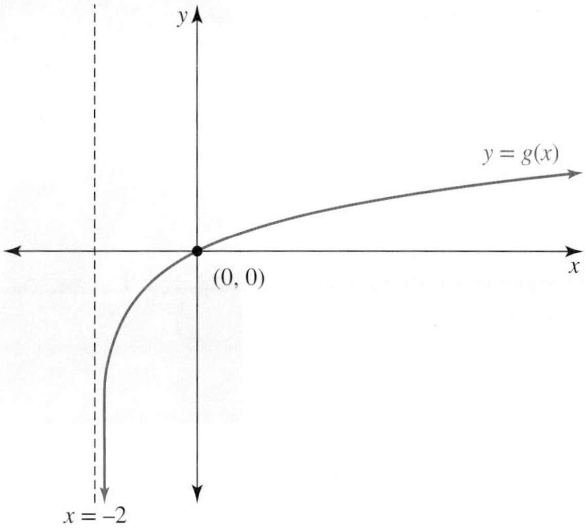

i. State the value of h.
ii. Show that $k = -\log_e(2)$.
iii. Hence, rewrite the rule in the form $g(x) = \log_e\left(\frac{x-h}{c}\right)$, where c is a constant.

19. Carbon-14 dating measures the amount of radioactive carbon-14 in fossils. This can be modelled by the relationship

$$Q = Q_0 e^{-0.000124t}$$

where Q is the amount, in milligrams, of carbon-14 currently present in the fossil of an organism, t is the time in years since the organism's death, and Q_0 is the initial amount, in milligrams, of carbon-14 present.

a. A fossil shell initially has 150 milligrams of carbon-14 present. Calculate how much carbon-14 will be present after 2000 years. Give your answer correct to 3 decimal places.
b. Determine the number of years it will take for the carbon-14 in the shell to be halved. Give your answer correct to the nearest year.
c. Suppose the amount of carbon-14 in the shell is $\frac{Q_0}{n}$. Determine an equation relating n to t. Hence, find how long it will be before the amount of carbon-14 in the fossil shell is $\frac{Q_0}{10}$. Give your answer to the nearest year.

20. The population of quokkas in a small corner of south-western Western Australia is currently described as vulnerable. The once-plentiful population of quokkas was drastically reduced after dingoes, foxes and wild pigs found their way to Australia.

Conservation efforts and dingo, fox and wild pig control programs have seen quokka populations recovering in some areas. In the Northern Jarrah forest, one of the areas where these conservation practices occur, there were known to be about 150 quokkas in 2008. Conservationists produced a model for the increase in population, which was given by

$$P = a\ \log_e(t) + b$$

where t is the time in years since 2007 and a and b are constants. There were estimated to be about 6000 quokkas present in the forest in 2013.

a. Determine the values of a and b. Give your answers correct to the nearest integer.
b. Calculate the number of quokkas that is expected to be present in 2020. Give your answer correct to the nearest integer.
c. Given that quokkas have a life expectancy of about 10 years, the model for the actual population is revised to

$$P_R = P - 0.25P = 0.75P$$

where P_R is the revised population.

i. Determine the equation relating P_R to t, the number of years since 2007.
ii. Calculate the revised population prediction for 2020. Give your answer correct to the nearest integer.

4.8 Exam questions

Question 1 (1 mark) TECH-ACTIVE

Source: *VCE 2021, Mathematical Methods Exam 2, Q2; © VCAA.*

MC The graph of $y = \log_e(x) + \log_e(2x)$, where $x > 0$, is identical, over the same domain, to the graph of

A. $y = 2\log_e\left(\frac{1}{2}x\right)$
B. $y = 2\log_e(2x)$
C. $y = \log_e(2x^2)$
D. $y = \log_e(3x)$
E. $y = \log_e(4x)$

Question 2 (7 marks) TECH-ACTIVE

A kettle was filled with water and the water heated to 98 °C. The kettle was then switched off at 1 pm and the water began to cool. By 3 pm, the temperature of the water was 58 °C. The temperature, T °C, of the water t hours after 1 pm is modelled by the rule $T = Ae^{-kt} + 18$.

a. Calculate the value of A and k. **(2 marks)**
b. Calculate the temperature of the water at 11 pm. Give your answer to 1 decimal place. **(1 mark)**
c. Sketch the graph of T versus t. **(2 marks)**
d. Find the time after which the temperature of the water will be less than 22 °C. **(1 mark)**
e. Explain what happens to the temperature in the long term. **(1 mark)**

Question 3 (5 marks) TECH-ACTIVE

A biologist conducts an experiment to determine conditions that affect the growth of bacteria. Her initial experiment finds the growth of the population of bacteria is modelled by the rule $N = 22 \times 2^t$, where N is the number of bacteria present after t days.

a. Calculate how long it will take for the number of bacteria to reach 2816. **(1 mark)**

b. Explain what will happen to the number of bacteria in the long term according to this model. **(1 mark)**

c. The biologist changes the conditions of her experiment and starts with a new batch of bacteria. She finds that under the changed conditions the growth of the population of bacteria is modelled by the rule $N = \dfrac{66}{1 + 2e^{-0.2t}}$.

i. Show that in both of her experiments the biologist used the same initial number of bacteria. **(2 marks)**

ii. Explain what will happen to the number of bacteria in the long term according to her second model. **(1 mark)**

Question 4 (5 marks) TECH-ACTIVE

In some parts of the world there have been measles (rubella) epidemics. For one such epidemic in Wales, the number of people in the population infected was modelled by the rule

$$P(t) = Ae^{kt}$$

where t was the number of days after the epidemic began. At the beginning of the epidemic, 200 cases were reported to authorities, but 30 days later there were 1000 cases.

a. Calculate the values of the constants A and k. Give k correct to 4 decimal places. **(2 marks)**

b. Calculate the expected number of cases after 60 days. Give your answer correct to the nearest integer. **(1 mark)**

c. Calculate how long it would have taken for the number of cases to reach 6000. Give your answer correct to 1 decimal place. **(1 mark)**

d. Thirty-eight thousand young people in Wales were at risk of contracting measles because they had not been immunised against rubella. If the epidemic went unchecked, find how long would it have taken for all these young people to be infected. Give your answer correct to 1 decimal place. **(1 mark)**

Question 5 (2 marks) TECH-FREE

Find all values of x for the equation $e^{2x} - 3e^x + 2 = 0$.

More exam questions are available online.

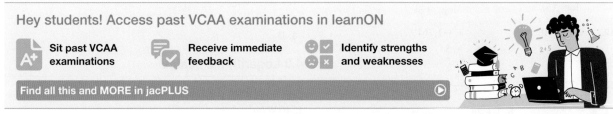

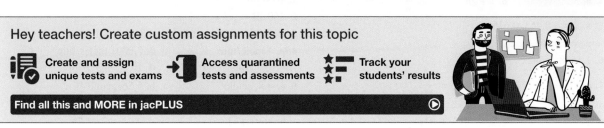

Answers

Topic 4 Exponential and logarithmic functions

4.2 Logarithm laws and equations

4.2 Exercise

1. a. $\log_6(216) = 3$ b. $\log_2(256) = 8$
   c. $\log_3(81) = 4$ d. $\log_{10}(0.0001) = -4$
   e. $\log_5(0.008) = -3$ f. $\log_7(7) = 1$

2. a. 4   b. −5   c. $\frac{1}{2}$ d. −1

3. a. 0
   b. $3\log_7(2x-1)$
   c. $\log_{10}(x-1)$

4. a. 7   b. −5   c. $-\frac{1}{2}$ d. −2   e. $\frac{5}{12}$ f. $\frac{2}{5}$

5. a. $3\log_3(x-4)$ b. $\log_e(2x+3)$
   c. 0   d. $2\log_4(5x+1)$

6. a. 3   b. $\frac{4}{3}$ c. $2^{-\frac{1}{3}}, 4$ d. $\frac{7}{2}$

7. a. 243   b. $\frac{47}{10}$

8. a. 4   b. −3   c. 11   d. −128

9. a. $\frac{1}{2}(e^{-3}+1)$ b. e^{-3} c. 7
   d. 15   e. 10   f. $-\frac{6}{5}$

10. a. 6, 1   b. $\frac{5}{4}$ c. $\frac{5}{4}$
    d. $x = 10, 10^{\frac{3}{2}}$ e. $x = e^2, \frac{1}{e}$ f. 4

11. a. $\frac{\log_{10}(9)}{\log_{10}(5)}$ b. $\frac{\log_{10}(12)}{\log_{10}\left(\frac{1}{2}\right)}$

12. a. i. 1.2770   ii. −1.2619
    b. i. 2×3^z ii. $\frac{3}{z}$

13. a. 1.7712   b. −6.9189

14. a. 5^{n+1} b. $1+2n$ c. $\frac{4}{n}$

15. a. $x = -0.463, 0.675$
    b. $x = 0.451, 1$

16. 1.5518, 1.4422

17. a. $y = \frac{4}{x^3}$ b. $y = \frac{x^2}{16}$

18. a. $y = \frac{9}{x}$ b. $y = 64x$

19. 16, $\frac{1}{16}$

20. a. $x = 3m$
    b. $\log_{10}(m) = x$, so $10^x = m$, and $\log_{10}(n) = y$, so $10^y = n$.

$$\begin{aligned}\log_{10}\left(\frac{100n^2}{m^5\sqrt{n}}\right) &= \log_{10}\left(\frac{100(10^y)^2}{(10^x)^5(10^y)^{\frac{1}{2}}}\right)\\ &= \log_{10}\left(\frac{10^2\times 10^{2y}}{10^{5x}\times 10^{\frac{y}{2}}}\right)\\ &= \log_{10}\left(\frac{10^2\times 10^{\frac{3y}{2}}}{10^{5x}}\right)\\ &= \log_{10}\left(10^{2+\frac{3y}{2}-5x}\right)\\ &= \left(2+\frac{3y}{2}-5x\right)\log_{10}(10)\\ &= 2+\frac{3y}{2}-5x\end{aligned}$$

21. a. $x < 4.0956$
    b. $x = 0.8653$

22. a. $f^{-1}(x) = \frac{1}{2}\log_e(x-2)$
    The domain of $f^{-1}(x)$ is the range of $f(x)$, which is $[2, \infty)$.
    b. 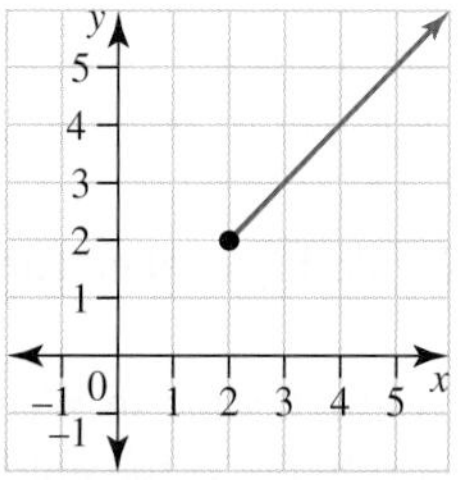

    c. $\frac{4x-3}{2x-2}$

4.2 Exam questions

Note: Mark allocations are available with the fully worked solutions online.

1. $x = -1$ only
2. B
3. D

4.3 Logarithmic scales

4.3 Exercise

1. 10 W/m^2
2. 160 dB
3. Lemon is acidic with a pH of 3.
4. a. 1 mole/litre   b. 0.0001 moles/litre
   c. 10^{-8} moles/litre   d. 10^{-12} moles/litre
5. a. 4.8, acidic   b. 5.56, acidic
6. a. 61 808 J
   b. 3691.17 J
   c. The magnitude 6.4 earthquake is 1.41 times stronger than the magnitude 6.3 earthquake.

7. The 500 W/m^2 amplifier is 13.98 dB louder.

8. a.
$$\begin{aligned} 0.5N_0 &= N_0 e^{-mt} \\ \frac{1}{2} &= e^{-mt} \\ \log_e\left(\frac{1}{2}\right) &= -mt \\ \log_e(2^{-1}) &= -mt \\ -\log_e(2) &= -mt \\ \log_e(2) &= mt \\ t &= \frac{\log_e(2)}{m} \end{aligned}$$

b. 9988 years

9. −437.97

4.3 Exam questions

Note: Mark allocations are available with the fully worked solutions online.

1. 361 cents
2. 133.98 dB, so protection should be worn.
3. The magnitude of the South American earthquake was 8.7.

4.4 Indicial equations

4.4 Exercise

1. a. 3 b. $\frac{1}{2}\log_{10}(5) + \frac{1}{2}$
   c. 2 d. $\log_{10}(3)$, $\log_{10}\left(\frac{1}{2}\right)$
2. a. −9 b. $\frac{1}{2}\log_2(9)$
   c. $\log_e(2)$ d. $\log_e(5)$
3. a. $\frac{1}{2}\log_7(5) + \frac{1}{2}$ b. 0, 2
   c. $\log_5(3)$ d. $\log_9\left(\frac{2}{3}\right)$ or $\log_9\left(\frac{5}{2}\right)$
4. a. $-\frac{7}{8}$ b. $\log_3(2) - 1$
   c. 0, 1 d. $\frac{\log_e\left(\frac{3}{4}\right)}{\log_e(12)}$
5. a. 1 b. 0
6. a. $2\log_e(3) + 2$ b. $4\log_e(2)$
   c. $\log_e(3)$ d. $-\sqrt{\log_e(2)}$, $\sqrt{\log_e(2)}$
7. a. $2\log_e(2)$
   b. $2\log_e(2)$, $3\log_e(2)$
   c. $\log_e\left(\sqrt{5}+1\right)$
   d. $\log_e\left(6-\sqrt{31}\right)$, $\log_e\left(\sqrt{31}+6\right)$
8. $\frac{3^{5n}}{4}$, $\frac{1}{4\times 3^{5n}}$
9. a. $\frac{1}{x} - x - 1$ b. 1 or 3
10. a. $\frac{m}{k} - \frac{1}{k}\log_e(2n)$, $k \in R\setminus\{0\}$ and $n \in R^+$
    b. $\frac{4-4n}{3m}$, $m \in R\setminus\{0\}$
    c. $\frac{1}{m}\log_e\left(\frac{\sqrt{57}+5}{4}\right)$, $m \in R\setminus\{0\}$
11. a. $x < -1.737$ b. $x > -0.756$
12. $m = 2$ and $n = \frac{1}{2}$
13. $a = 5$, $k = 0.25$
14. $P = \$10\,000$, $r = 5\%$

4.4 Exam questions

Note: Mark allocations are available with the fully worked solutions online.

1. A
2. $t = \log_e\left(\frac{8}{3}\right)$
3. $x = \frac{3}{2}$

4.5 Logarithmic graphs

4.5 Exercise

1. a.

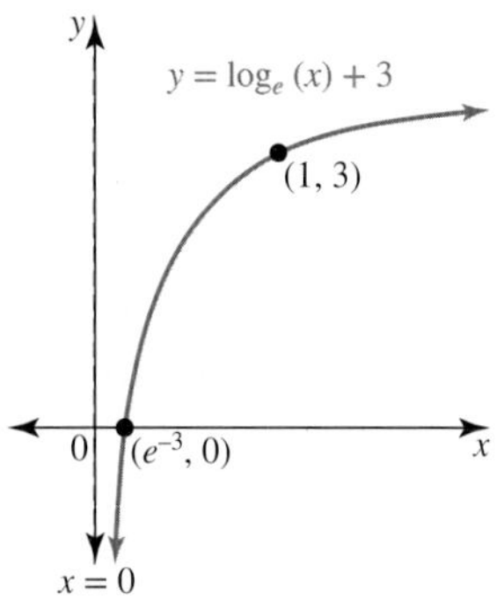

b.

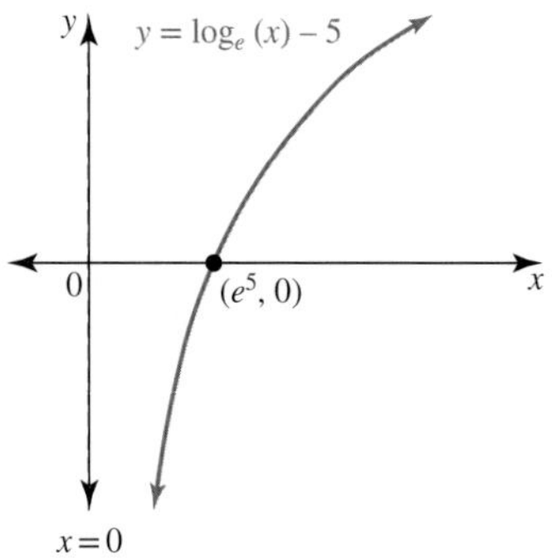

c.

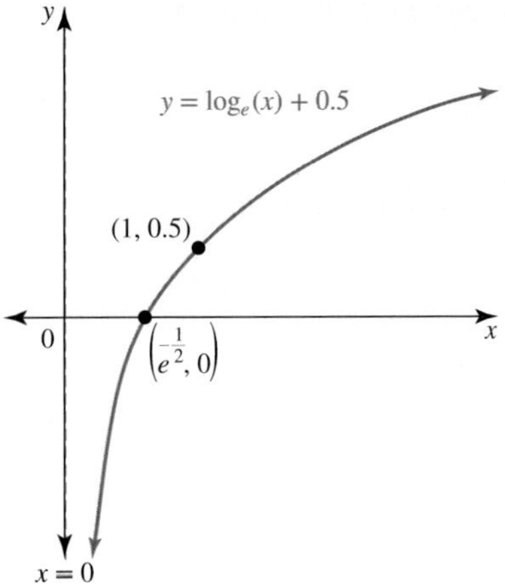

2. a.

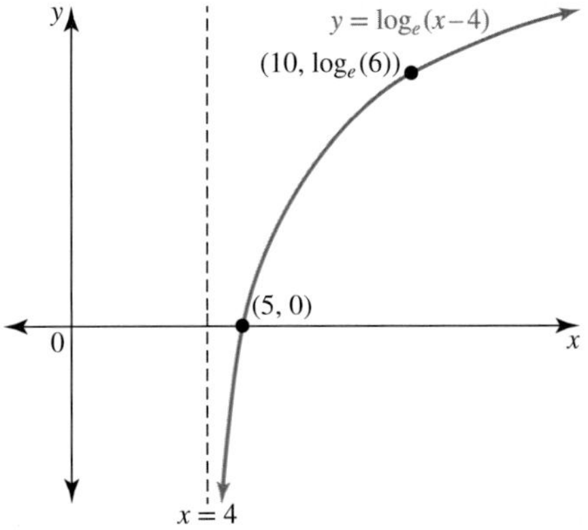

b.

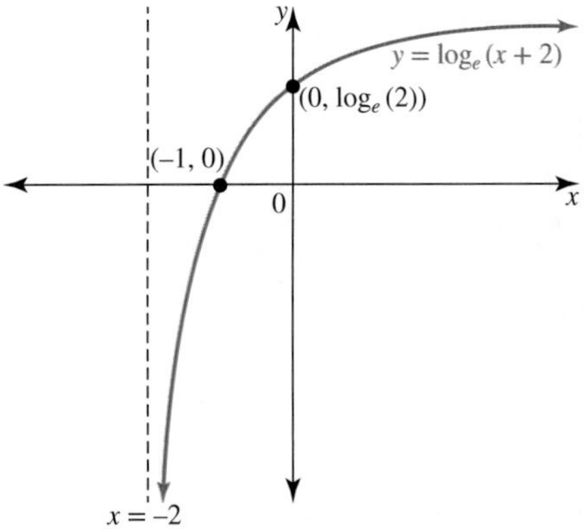

c.

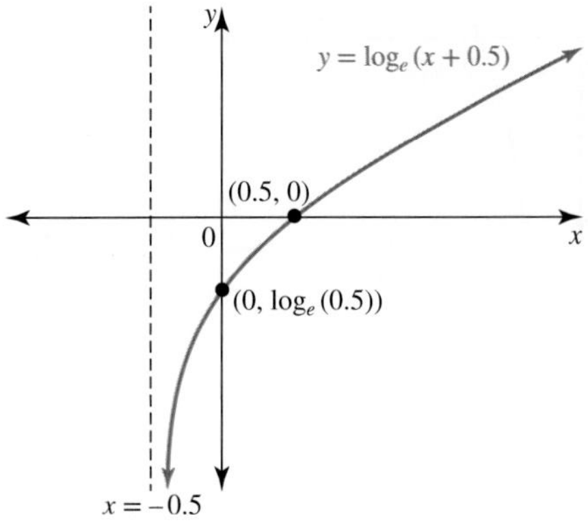

3. a. $y = \frac{1}{4}\log_e(x)$

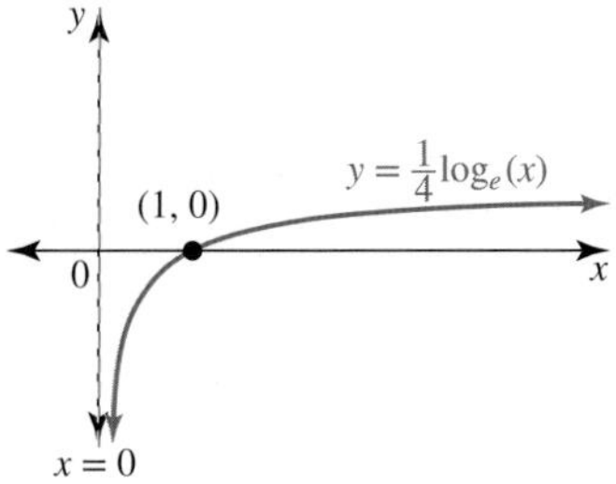

b. $y = 3\log_e(x)$

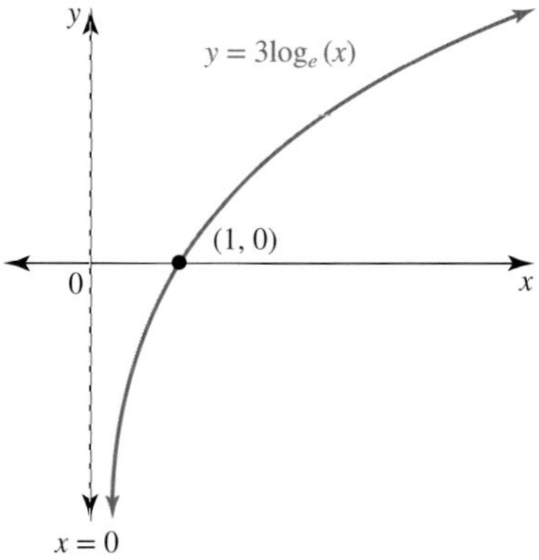

c. $y = 6\log_e(x)$

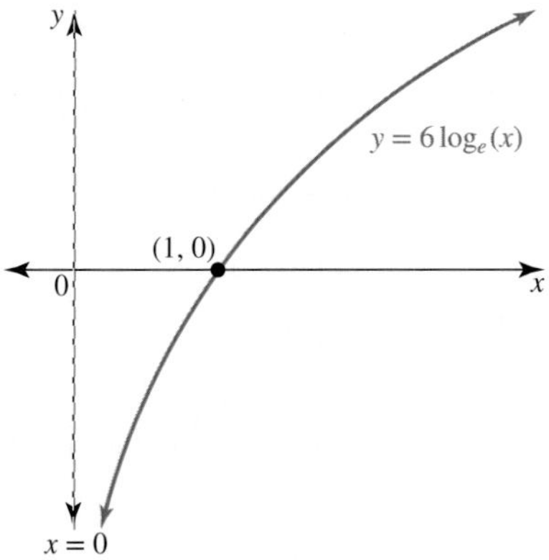

4. a. $y = \log_e(3x)$

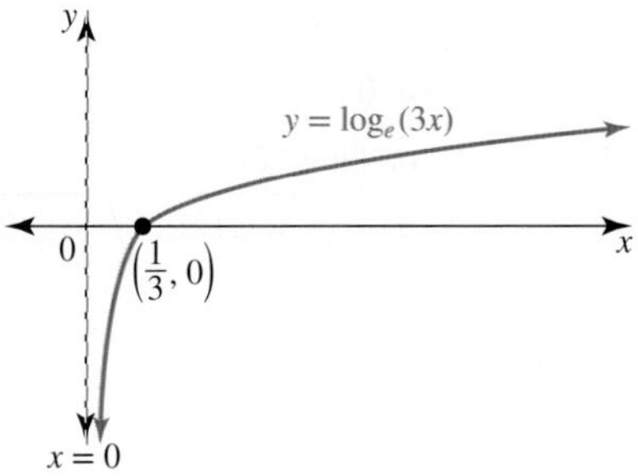

b. $y = \log_e\left(\frac{x}{4}\right)$

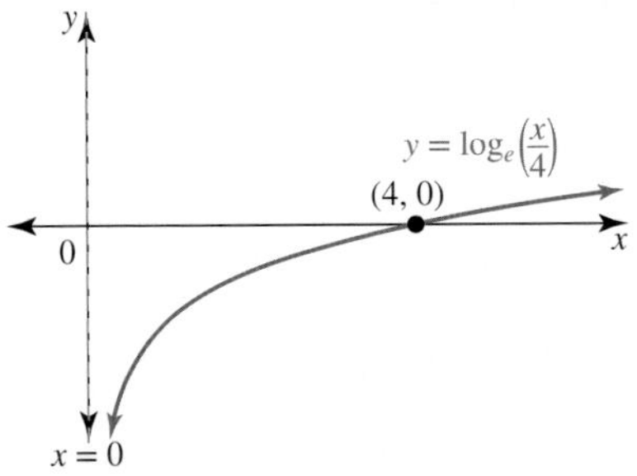

c. $y = \log_e(4x)$

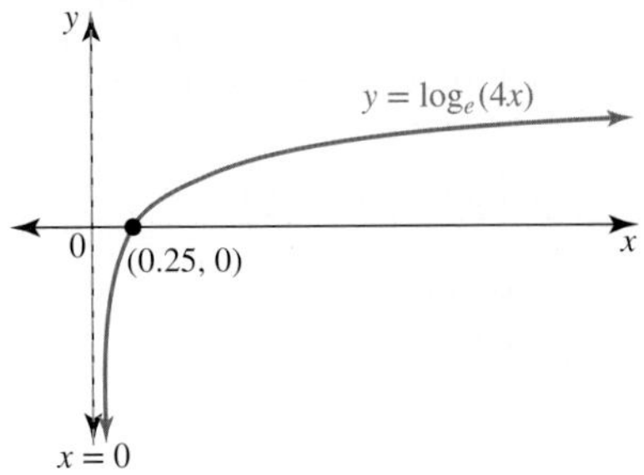

5. a.

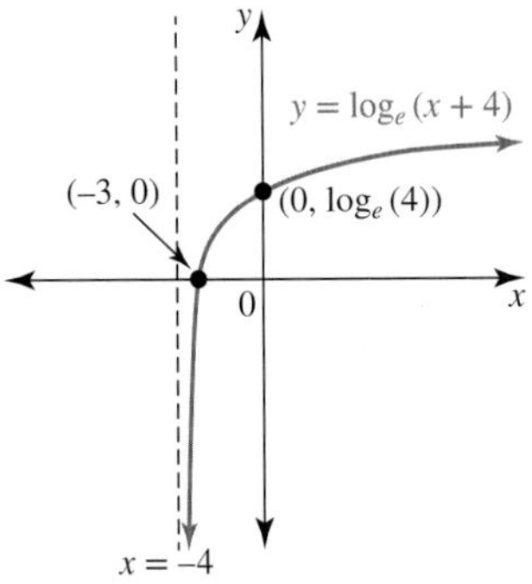

Domain $= (-4, \infty)$, range $= R$

b.

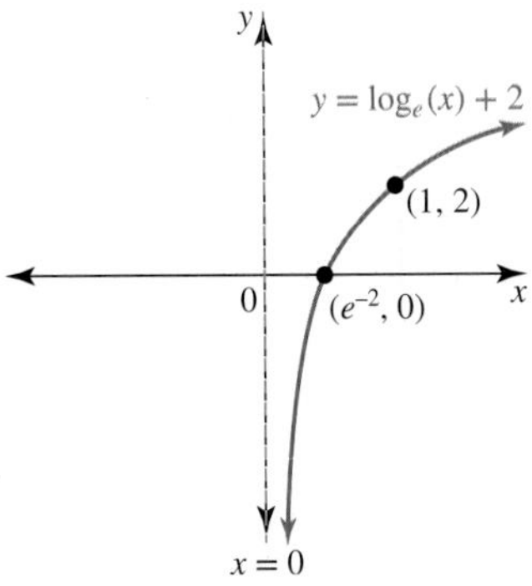

Domain $= (0, \infty)$, range $= R$

c.

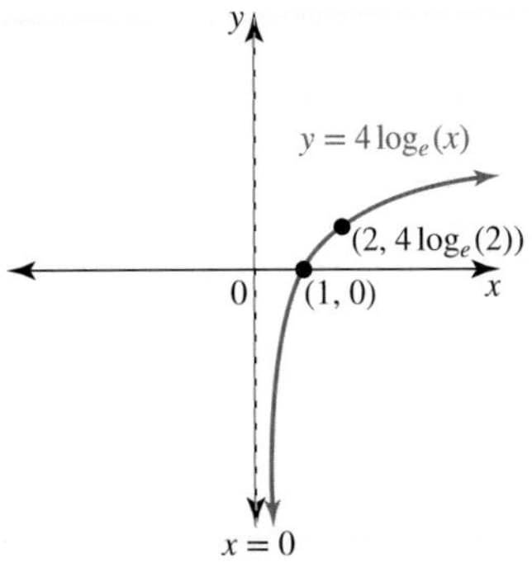

Domain $= (0, \infty)$, range $= R$

d.

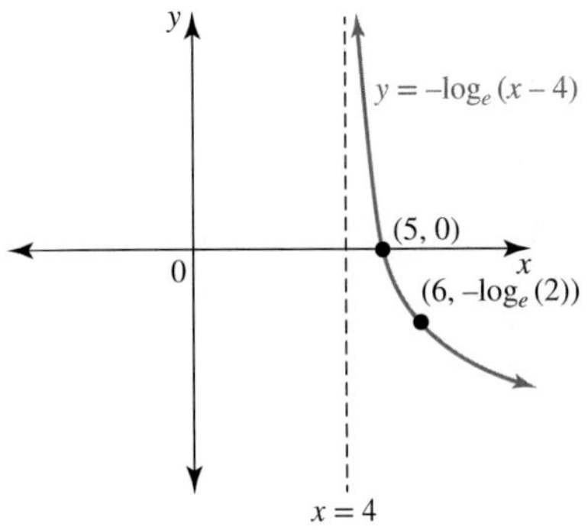

Domain $= (4, \infty)$, range $= R$

6. a.

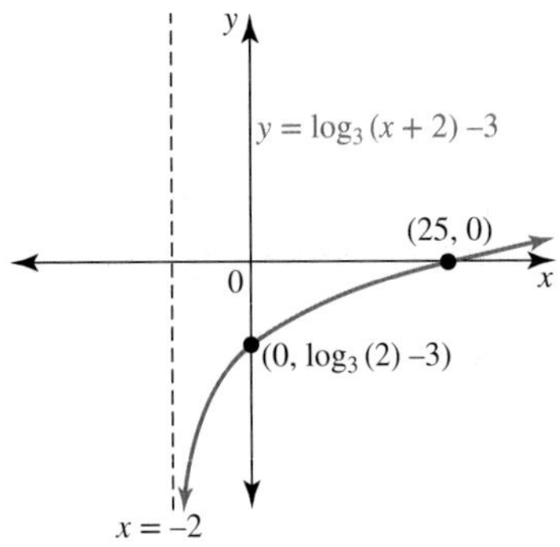

b.

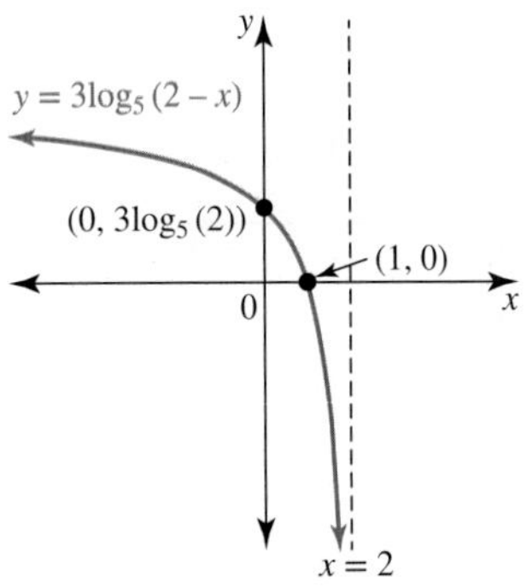

c.

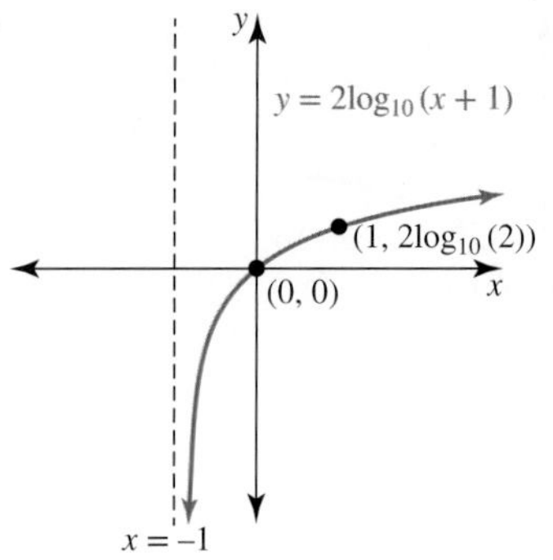

d.

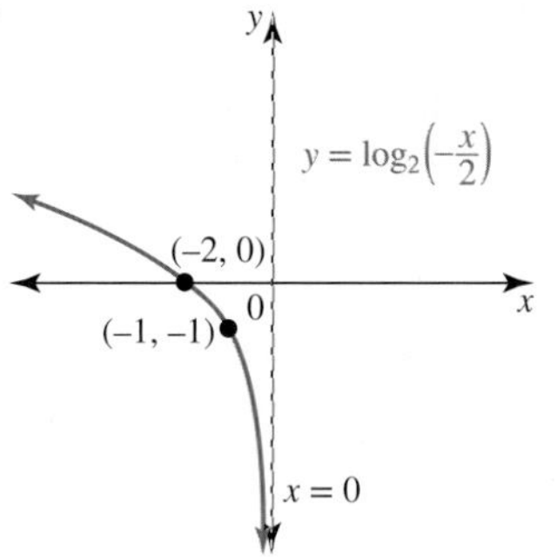

7. a. $y = 1 - 2\log_e(x - 1)$

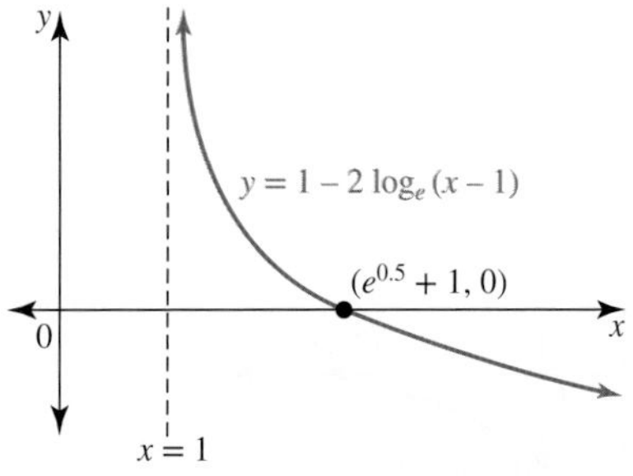

b. $y = \log_e(2x + 4)$

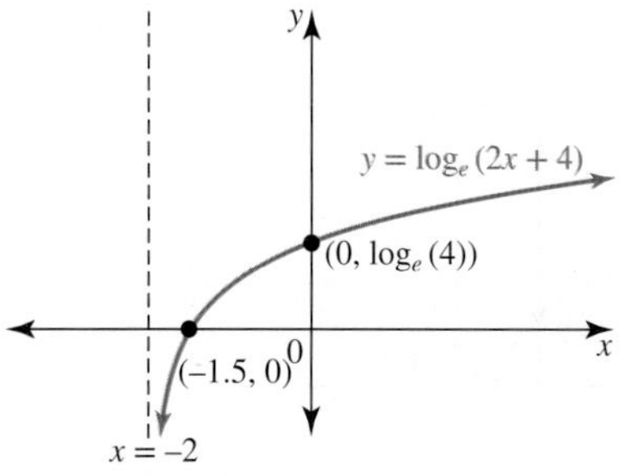

c. $y = \frac{1}{2}\log_e\left(\frac{x}{4}\right) + 1$

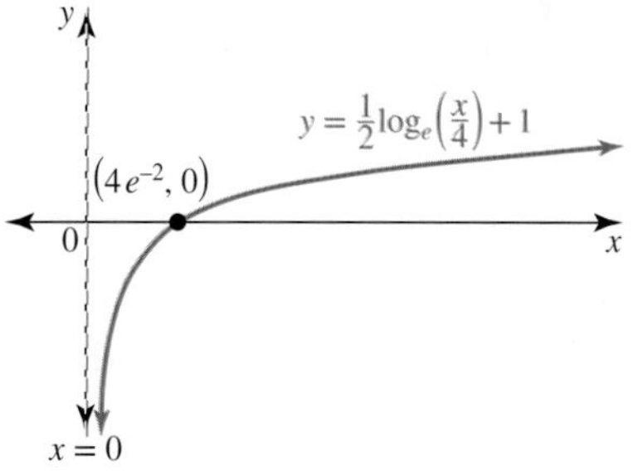

8. $m = 2, n = 2$

9. $a = \dfrac{2}{\log_e(2)},\quad h = -1,\quad k = -2$

10. D

11. a. $a = -1, b = \dfrac{1}{2}$ b. -0.4055

12. $(-2, 3) \Rightarrow 3 = m\log_2(-2n)$ [1]

$\left(-\frac{1}{2}, \frac{1}{2}\right) \Rightarrow \frac{1}{2} = m\log_2\left(-\frac{n}{2}\right)$ [2]

[1] – [2]:

$$\begin{aligned}
3 - \frac{1}{2} &= m\log_2(-2n) - m\log_2\left(-\frac{n}{2}\right) \\
\frac{5}{2} &= m\left(\log_2(-2n) - \log_2\left(-\frac{n}{2}\right)\right) \\
&= m\left(\log_2\left(\frac{-2n}{-\frac{n}{2}}\right)\right) \\
&= m\log_2(4) \\
&= m\log_2 2^2 \\
&= 2m \\
m &= \frac{5}{4}
\end{aligned}$$

Substitute $m = \frac{5}{4}$ into [1]:

$$\begin{aligned}
3 &= \frac{5}{4}\log_2(-2n) \\
\frac{12}{5} &= \log_2(-2n) \\
2^{\frac{12}{5}} &= -2n \\
n &= 2^{\frac{12}{5}} \div -2 \\
&= -2^{\frac{7}{5}}
\end{aligned}$$

4.5 Exam questions

Note: Mark allocations are available with the fully worked solutions online.

1. a.

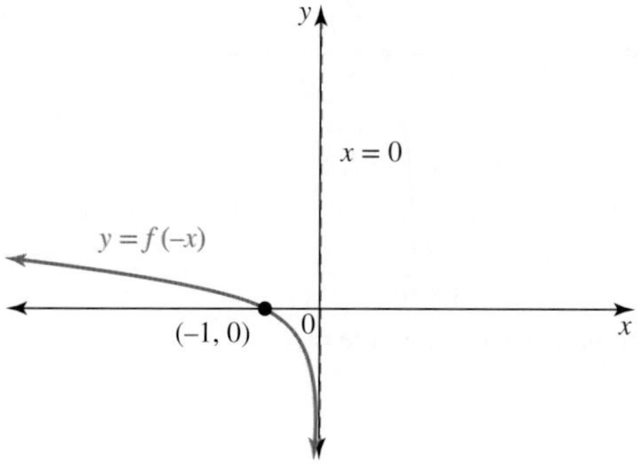

b.

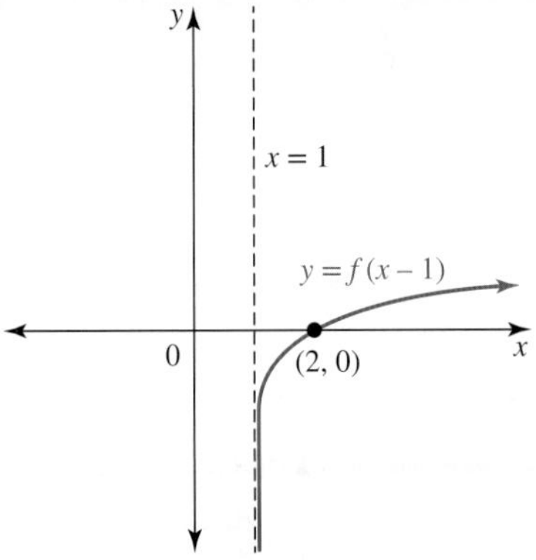

c.

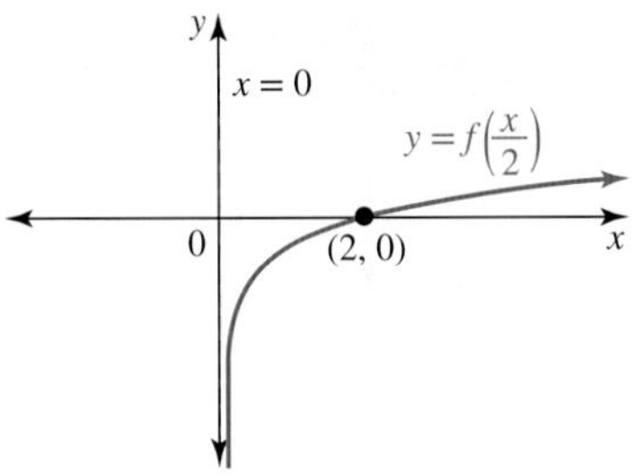

2. E
3. D

4.6 Exponential graphs

4.6 Exercise

1. a. -100

b.

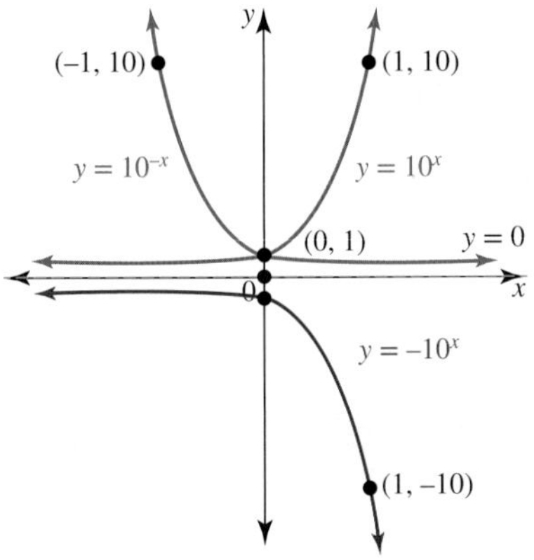

c. $y = \left(\dfrac{1}{10}\right)^x$ or $y = 0.1^x$.

2. a.

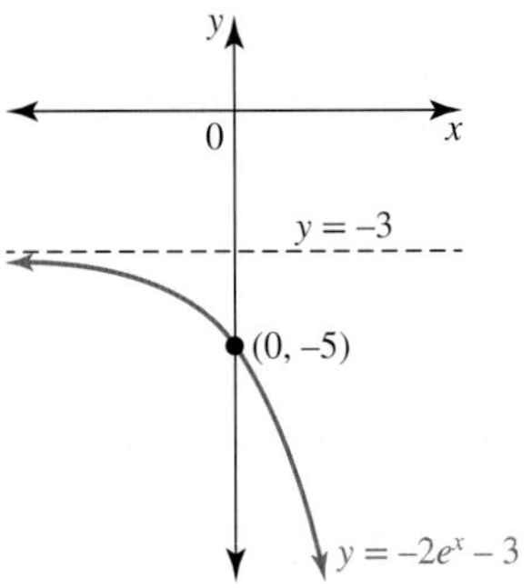

Domain R, range $(-\infty, -3)$

b.

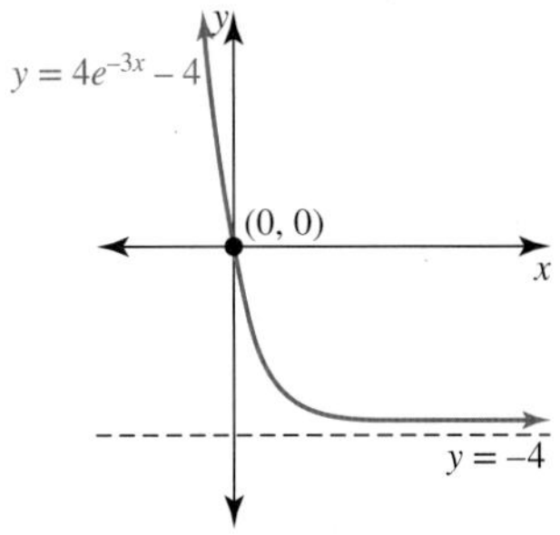

Domain R, range $(-4, \infty)$

c.

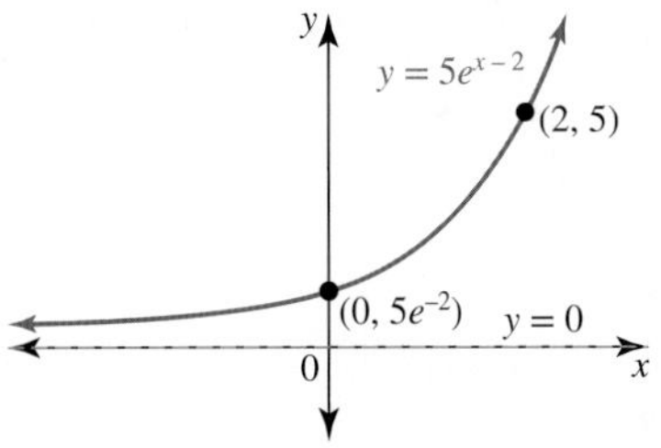

Domain R, range R^+

3. a.

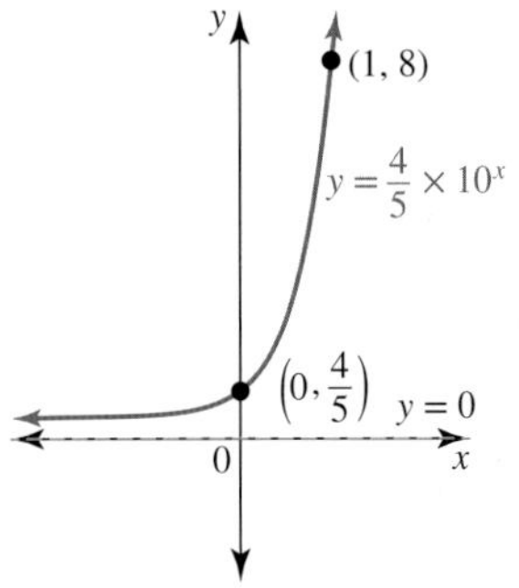

$x \to \infty, y \to \infty$

b.

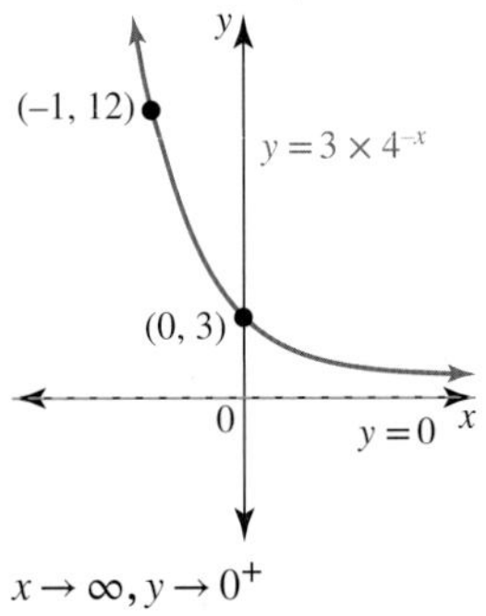

$x \to \infty, y \to 0^+$

c.

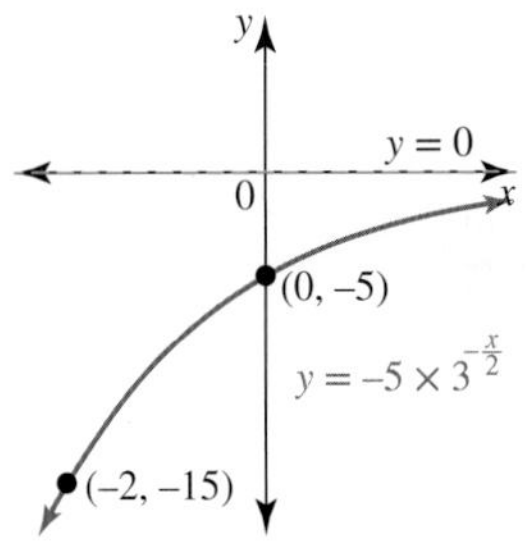

$x \to \infty, y \to 0^-$

d.

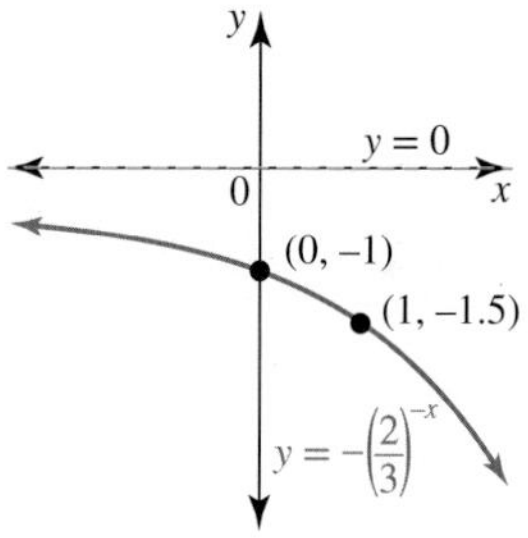

$x \to \infty, y \to -\infty$

4. a.

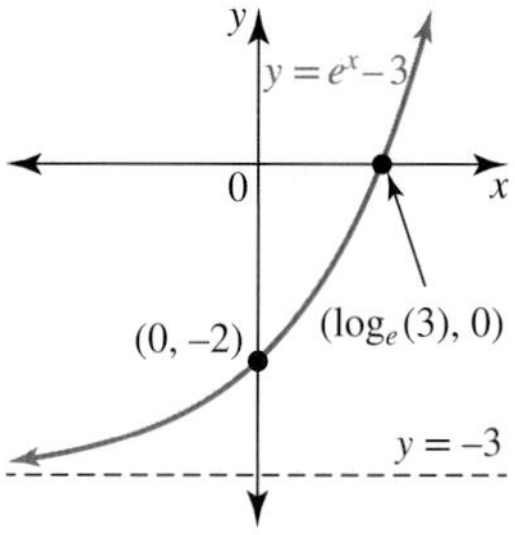

Range $(-3, \infty)$

b.

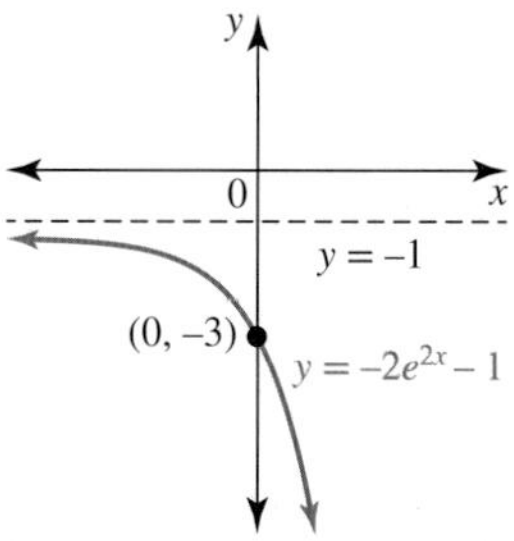

Range $(-\infty, -1)$

c.

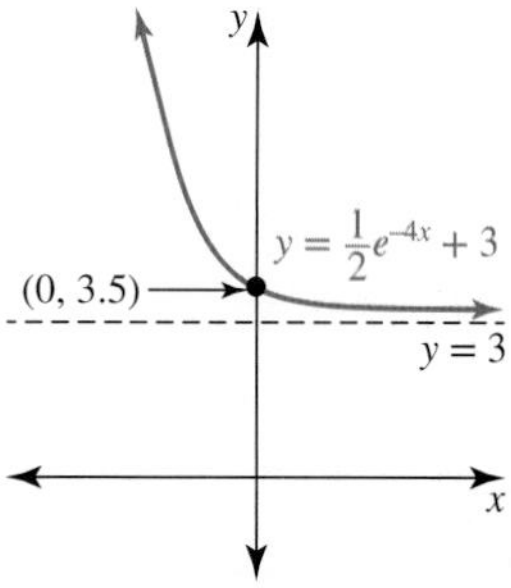

Range $(3, \infty)$

d.

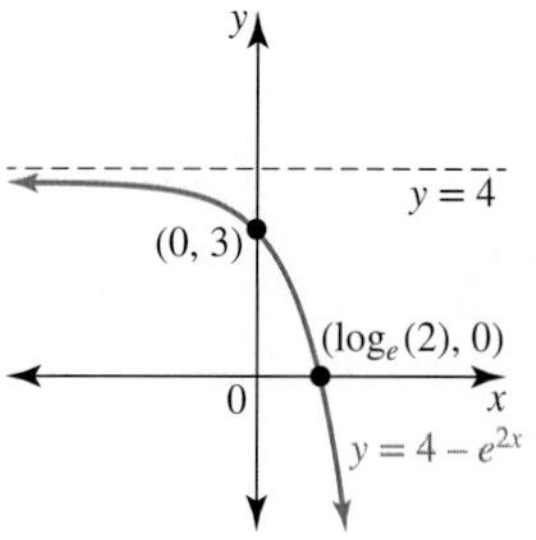

Range $(-\infty, 4)$

e.

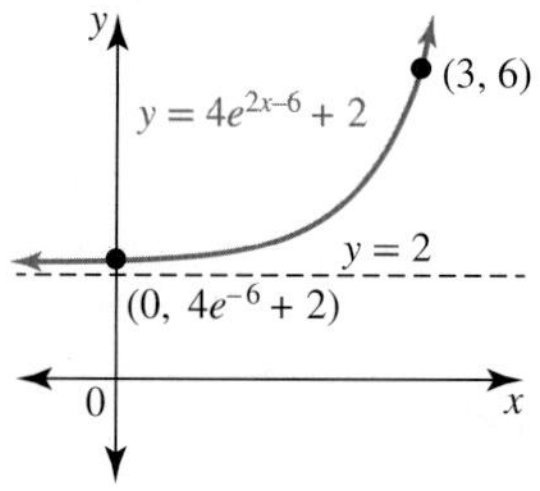

Range $(2, \infty)$

f.

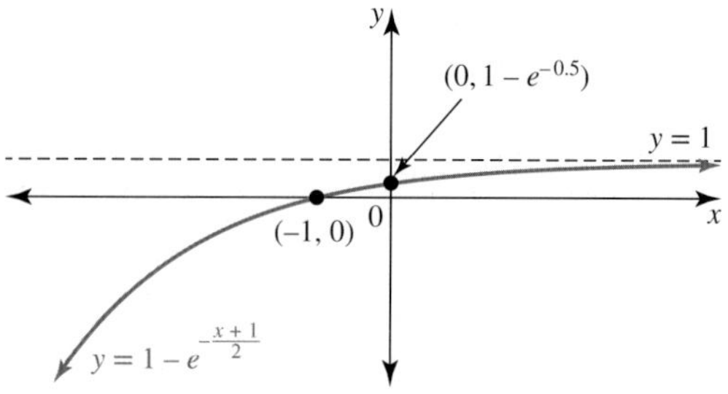

Range $(-\infty, 1)$

5. a.

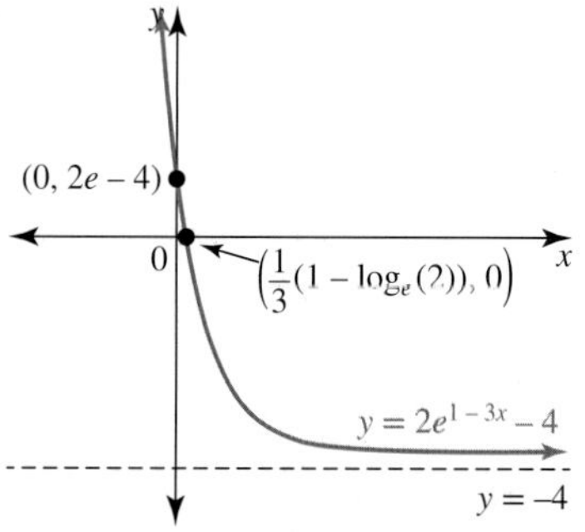

b.

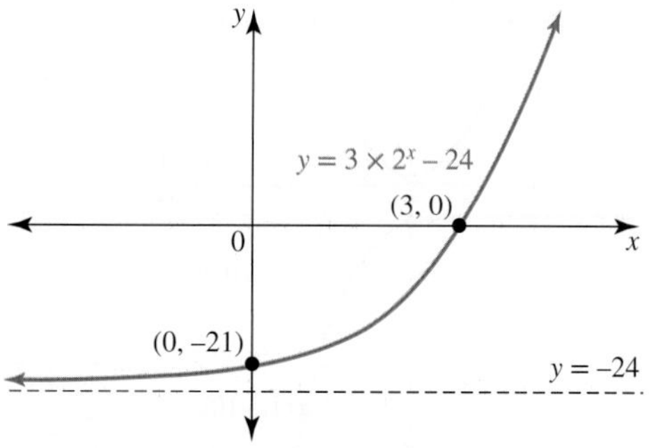

Domain R, range $(-24, \infty)$

6. a. $a = -2, b = 2$

b. $y = -2 \times 10^{\frac{x}{4}}$

7. C

8. a. $a = -11, b = 11, f: R \to R, f(x) = -11e^x + 11$

b. $a = 1,\ n = -2,\ y = e^{-2x} + 4$

c. i. $b = -1, c = -7$ **ii.** $(-7, \infty)$

d. i. $A = 12, B = -2$ **ii.** $a = 1$

9. a. $f(x) = 2\log_e(3(x+1))$, domain $= (-1, \infty)$ and range $= R$

$f^{-1}(x) = \frac{1}{3}e^{\frac{x}{2}} - 1$, domain $= R$ and range $= (-1, \infty)$

b. $f(x) = \log_e(2(x-1)) + 2$, domain $= (1, \infty)$ and range $= R$

$f^{-1}(x) = \frac{1}{2}e^{x-2} + 1$, domain $= R$ and range $= (1, \infty)$

c. $f(x) = 2\log_e(1 - x) - 2$, domain $= (-\infty, 1)$ and range $= R$

$f^{-1}(x) = 1 - e^{\frac{1}{2}(x+2)}$, domain $= R$ and range $= (-\infty, 1)$

10. a.

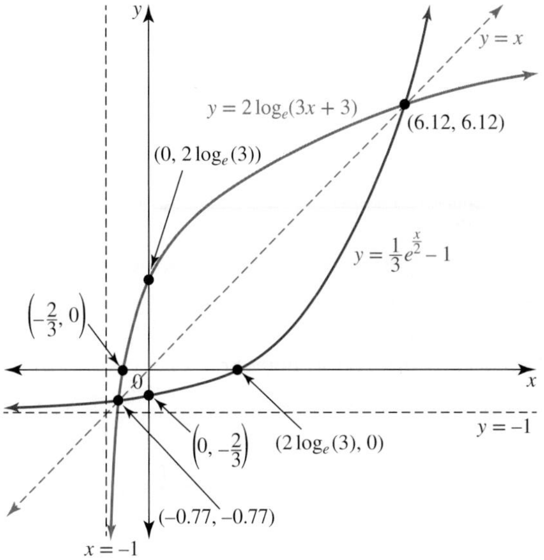

b.

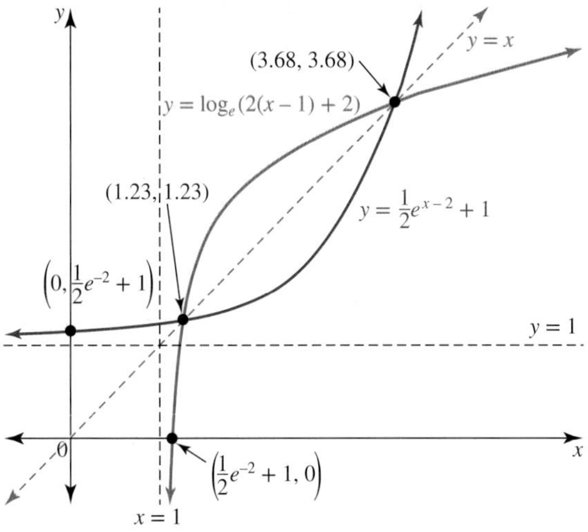

c.

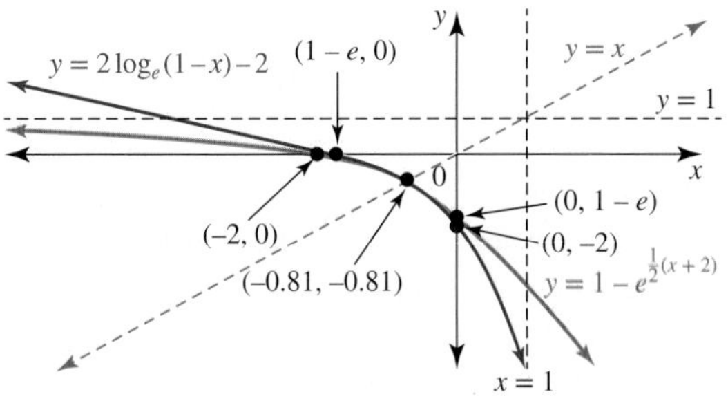

11. $y = 2^x - 2^{-x}$, domain R, range R

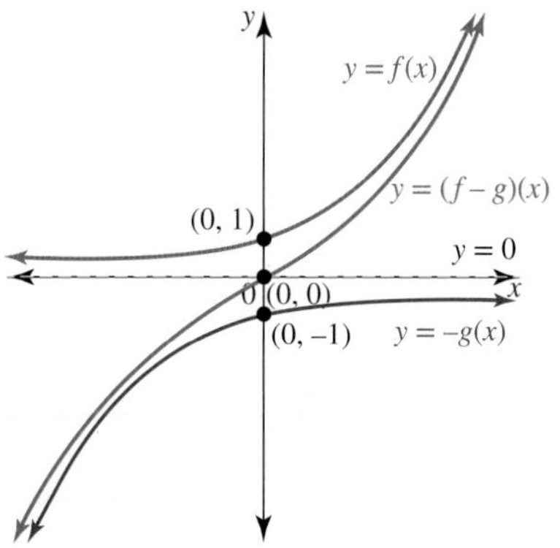

12.

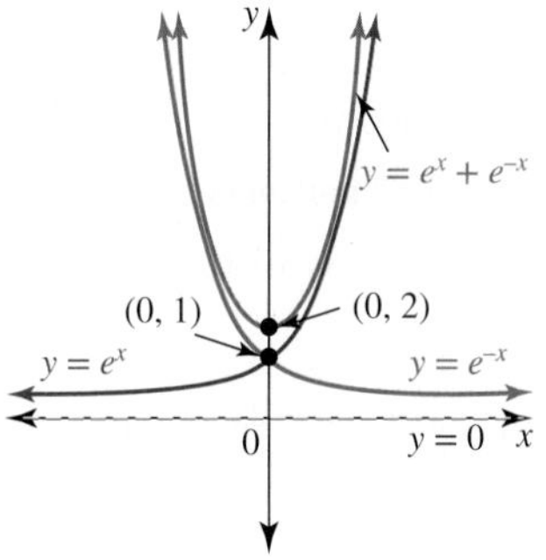

4.6 Exam questions

Note: Mark allocations are available with the fully worked solutions online.

1. B
2. E
3. D

4.7 Applications

4.7 Exercise

1. 54 years
2. \$1648.72
3. a. 7.32%
   b. 33 years 5 months
4. a. 2 parts per million
   b. 2.96 parts per million
   c. 47.2 hours
5. a. 18
   b. 12
   c. 1.62 months

d.

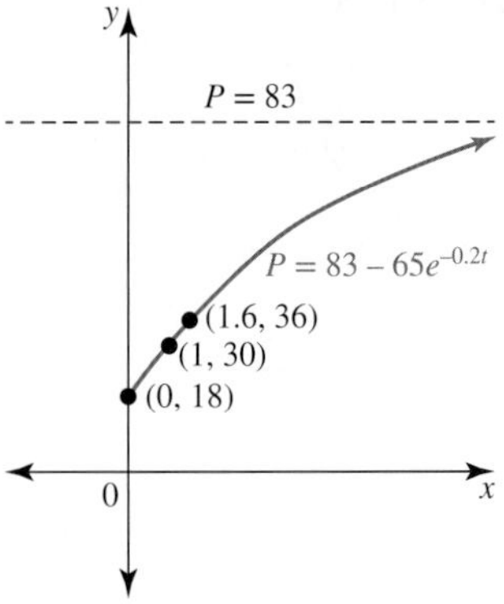

e. The population cannot exceed 83.

6. 0.1793

7. 0.0003

8. a. 88.3 milligrams b. 5590 years

9. a. 88.57% lost b. 1.87 millennia

10. a. 11.3863 b. 13.5835 c. 10.18 weeks

11. 12 979 years

12. a. $P(x) = 800\log_e\left(2 + \frac{x}{250}\right) - 300 - 2x$

b. 330

13. a. $k = 10\,000$, $m = 0.025$

b. \$15 685.58

c. $P = 13\,750e^{0.025t} - 10\,000$

d. \$15 059.38

14. a. $a = -2885.4$, $c = 10\,000$

b. 4000

c. 22.6 weeks

15. a. $a = 12.5$, $b = 100$

b. 7.253 g

4.7 Exam questions

Note: Mark allocations are available with the fully worked solutions online.

1. a. $c = \left(\frac{2}{3}\right) = \frac{5}{3}e^{-1} = 0.61$ mg/L

b. 0.33 hours

c. 0.86 hours

2. a. $A = 1.439$, $k = 0.536$

b. 2.999 M

c. 32 minutes 25 seconds

3. a. 95 °C

b. The temperature approaches 20 °C.

c. 8.24 minutes

d. $A = 85$, $B \approx -30$

e.

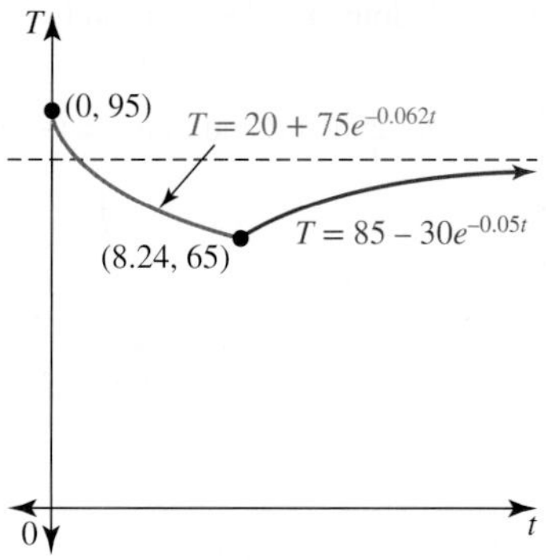

4.8 Review

4.8 Exercise

Technology free: short answer

1. a. i. $\frac{4}{5}$ ii. 4 iii. $2, \frac{1}{64}$

b. i. $y = \frac{x^2}{8}\ (x > 0)$ ii. $y = \frac{1}{x^3}\ (x > 0)$

c. $\log_4\left(\frac{64q^2}{p^3\sqrt{q}}\right) = \log_4\left(\frac{64q^{\frac{3}{2}}}{p^3}\right)$

$= \log_4(64) + \log_4\left(q^{\frac{3}{2}}\right) - \log_4\left(p^3\right)$

$= \log_4\left(4^3\right) + \frac{3}{2}\log_4(q) - 3\log_4(p)$

$= 3\log_4(4) + \frac{3}{2}\log_4(q) - 3\log_4(p)$

$= 3 + \frac{3}{2}\log_4(q) - 3\log_4(p)$

Substitute $x = \log_4(p)$ and $y = \log_4(q)$:

$$\begin{aligned}\log_4\left(\frac{64q^2}{p^3\sqrt{q}}\right) &= 3 + \frac{3}{2}\log_4(q) - 3\log_4(p)\\ &= 3 + \frac{3y}{2} - 3x\\ &= 3 - 3x + \frac{3y}{2}\end{aligned}$$

2. a. i. 2, acidic ii. 11, basic

b. i. 0.001 moles/litre ii. 10^{-14} moles/litre

3. a. $x = \log_e(3)$, $x = \log_e(5)$

b. $x = \log_e(7)$

c. $x = 1, \log_2(9)$

4. a. $x = \frac{3 - 4m}{k}$, where $k \in R \setminus \{0\}$ and $m \in R$

b. $x = \frac{1}{k}\log_e\left(\frac{2 + \sqrt{22}}{3}\right)$, where $k \in R \setminus \{0\}$

5. a. $y = \log_e(x - 1) + 3$

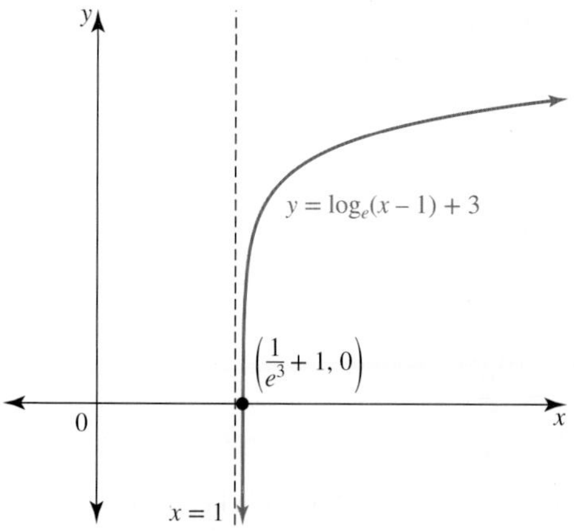

Domain $= (1, \infty)$, range $= R$

b. $y = \log_e(x + 3) - 1$

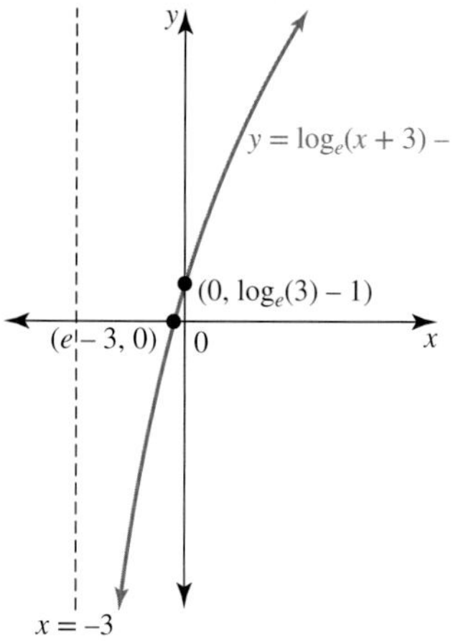

Domain $= (-3, \infty)$, range $= R$

c. $y = 2\log_e(-x)$

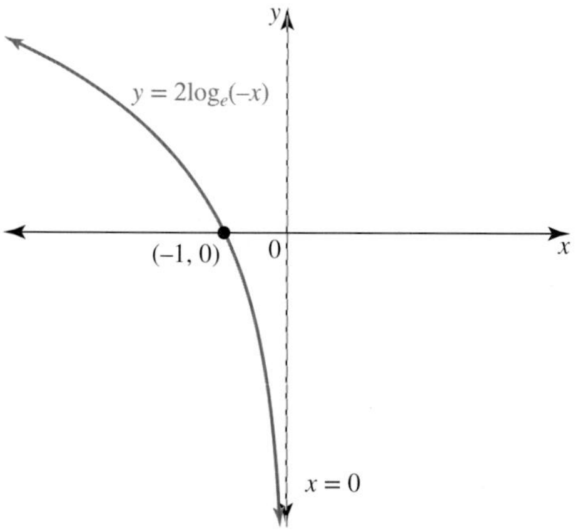

Domain $= (-\infty, 0)$, range $= R$

d. $y = -\log_e(x - 4)$

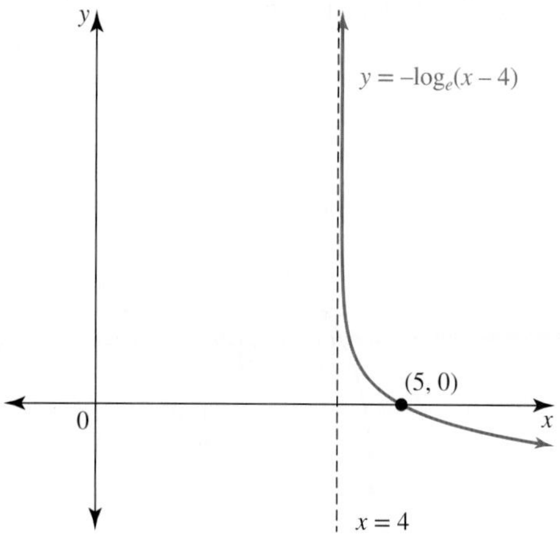

Domain $= (4, \infty)$, range $= R$

6. a. Range $(-12, \infty)$

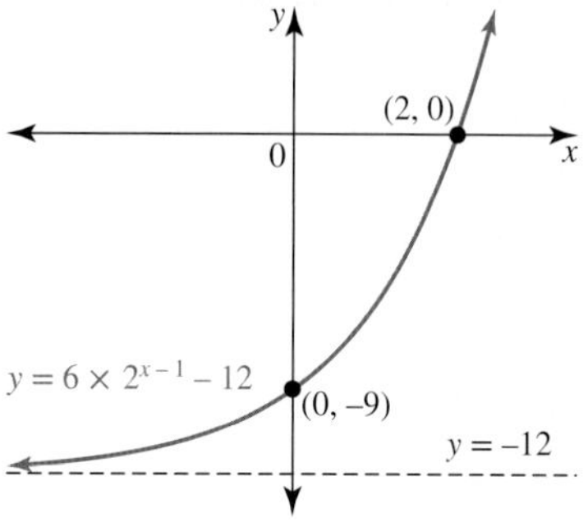

b. Range $(2.5, \infty)$

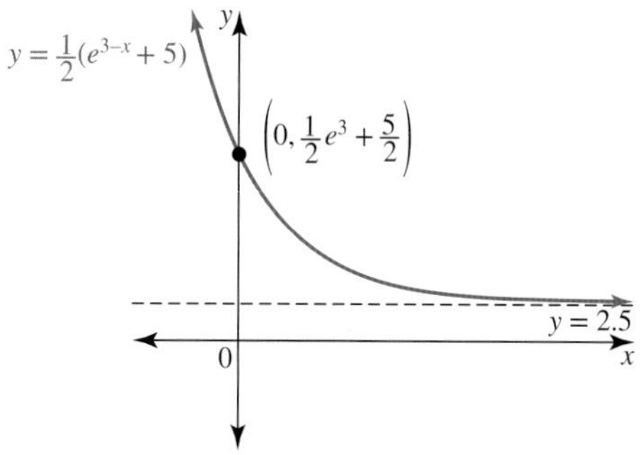

7. a. 0.001 watt/m^2 b. 60 dB

Technology active: multiple choice

8. D
9. E
10. C
11. A
12. E
13. B
14. E
15. C
16. E
17. A

Technology active: extended response

18. a. i. $y = e^{x-1} - 5$, domain $= R$

ii. f^{-1}: $\left(\log_e(5) + 1, 0\right)$, $\left(0, \frac{1}{e} - 5\right)$

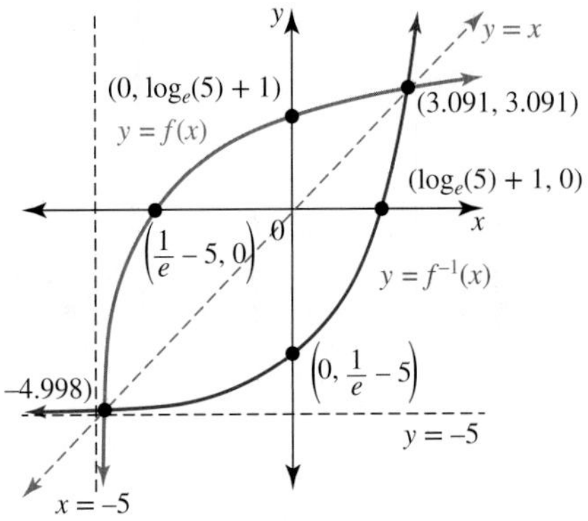

iii. $(-4.998, -4.998), (3.091, 3.091)$

b. i. $h = -2$

ii. $y = \log_e(x + 2) + k$

Substitute $(0, 0)$:

$0 = \log_e(2) + k$

$k = -\log_e(2)$

iii. $g(x) = \log_e\left(\frac{x + 2}{2}\right)$

19. a. 117.054 milligrams

b. 5590 years

c. $n = e^{0.000\,124t}$, 18 569 years

20. a. $a = 3265$, $b = 150$

b. 8525 quokkas

c. i. $P_R = 2448.75 \log_e(t) + 112.5$

ii. 6393 quokkas

4.8 Exam questions

Note: Mark allocations are available with the fully worked solutions online.

1. C

2. a. $A = 80$, $k = -\frac{1}{2}\log_e\left(\frac{1}{2}\right) \approx -0.3466$

b. 20.5 °C

c.

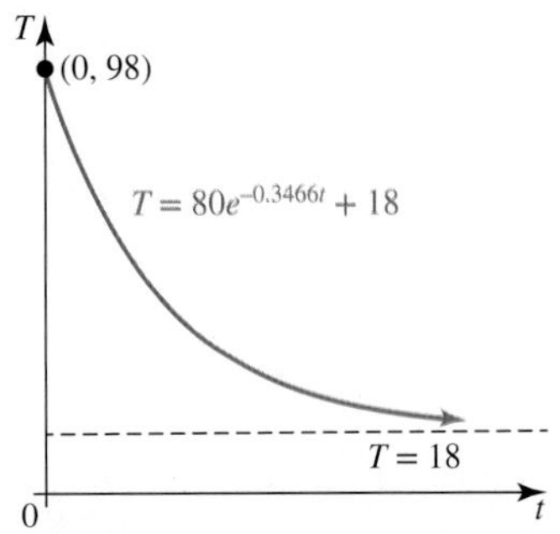

d. After approximately 9:38 pm

e. The temperature drops to no less than 18 °C.

3. a. 7 days

b. As $t \to \infty, N \to \infty$.

c. i. $N = 22 \times 2^t, t = 0$

$$\begin{aligned} N &= 22 \times 2^0 \\ &= 22 \times 1 \\ &= 22 \end{aligned}$$

$$\begin{aligned} N &= \frac{66}{1 + 2e^{-0.2t}}, t = 0 \\ &= \frac{66}{1 + 2e^{-0.2 \times 0}} \\ &= \frac{66}{3} \\ &= 22 \end{aligned}$$

Initially there are 22 bacteria in each model.

ii. The population will never exceed 66.

4. a. $P(t) = 200e^{0.0536t}$

b. 4986 cases

c. 63.5 days

d. 97.9 days

5. $x = \log_e(2)$ or $x = 0$

5 Differentiation

LEARNING SEQUENCE

Fully worked solutions for this topic are available online.

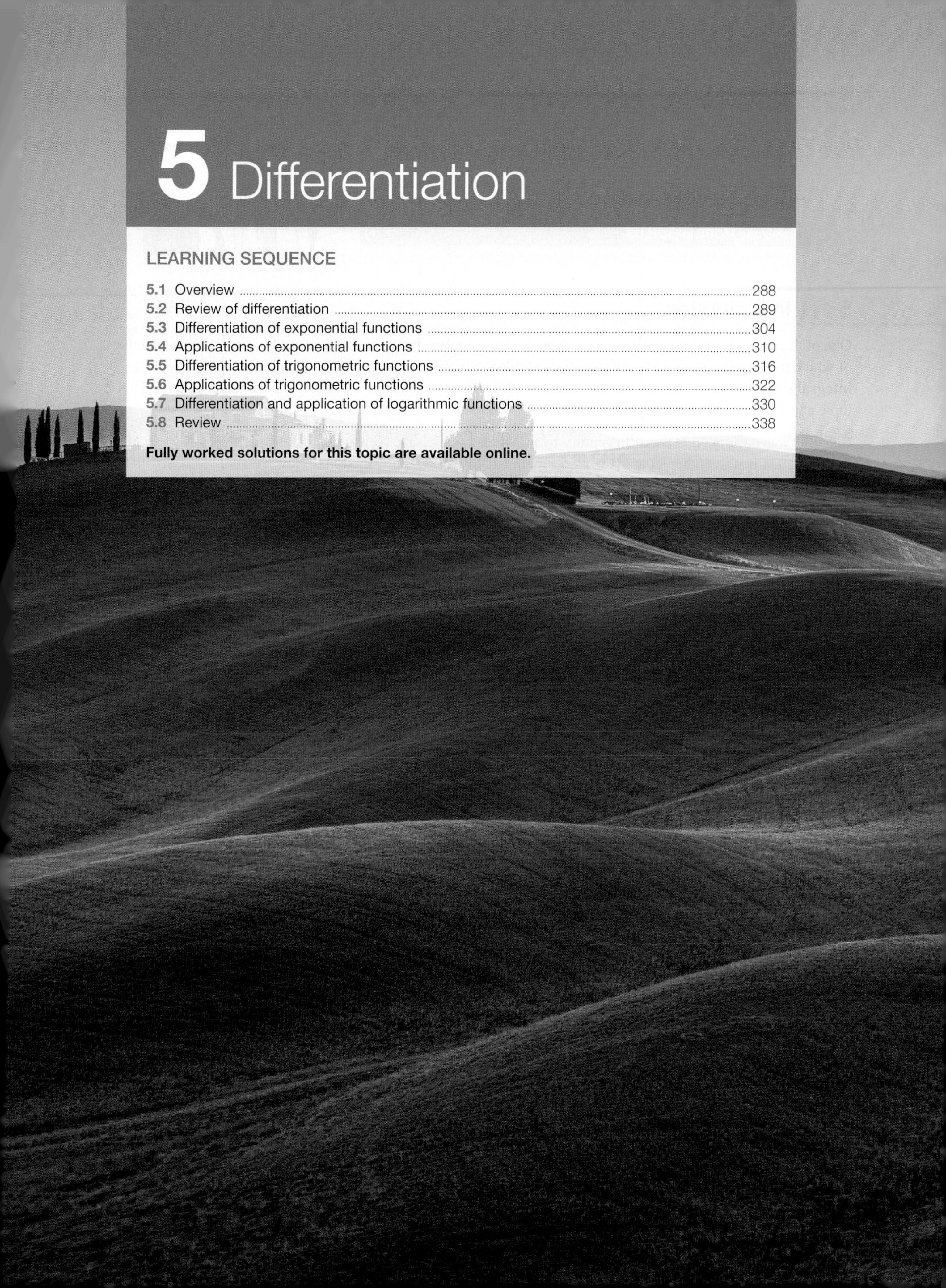

5.1 Overview

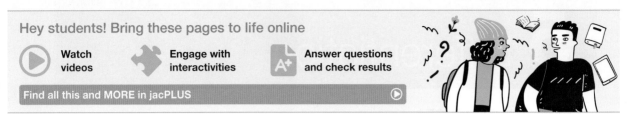

Hey students! Bring these pages to life online

Watch videos

Engage with interactivities

Answer questions and check results

Find all this and MORE in jacPLUS

5.1.1 Introduction

One of the most important branches of mathematics is **calculus**, the mathematical study of continuous change, of which there are two subfields: **differential calculus**, which deals with gradients and rates of change, and **integral calculus**, which deals with finding original functions from rates and finding areas under graphs. This topic and Topic 6 are concerned with differential calculus.

Calculus was developed in the 17th century by mathematicians Isaac Newton and Gottfried Leibniz. Each of them claimed that the other stole his work. Leibniz had his results published first; however, Newton derived his first. Today, the consensus is that they both independently contributed to the invention of calculus — Leibniz about 8 years after Newton. Leibniz is also given credit for introducing the calculus notations that are used today, including the $\frac{dy}{dx}$ notation and the integral symbol.

KEY CONCEPTS

This topic covers the following key concepts from the VCE Mathematics Study Design:

- deducing the graph of the derivative function from the graph of a given function and deducing the graph of an anti-derivative function from the graph of a given function
- derivatives of x^n for $n \in Q$, e^x, $\log_e(x)$, $\sin(x)$, $\cos(x)$ and $\tan(x)$.

Note: *Concepts shown in grey are covered in other topics.*

5.2 Review of differentiation

LEARNING INTENTION

At the end of this subtopic you should be able to:
- find derivatives of basic functions and apply differentiation to curve sketching
- determine the tangent to a curve at a given point.

5.2.1 The derivative of a function

The gradient of a curve is the instantaneous **rate of change** of the function at any point. The gradient at any point (x, y) can be found using the **central difference approximation**.

As shown in the diagram below left, the aim is to find the gradient of the tangent line at point P. The central difference approximation involves finding the coordinates of two points equidistant horizontally from P. These points are represented by R $(x-h,\ f(x-h))$ and Q $(x+h,\ f(x+h))$. This is shown on the diagram below right.

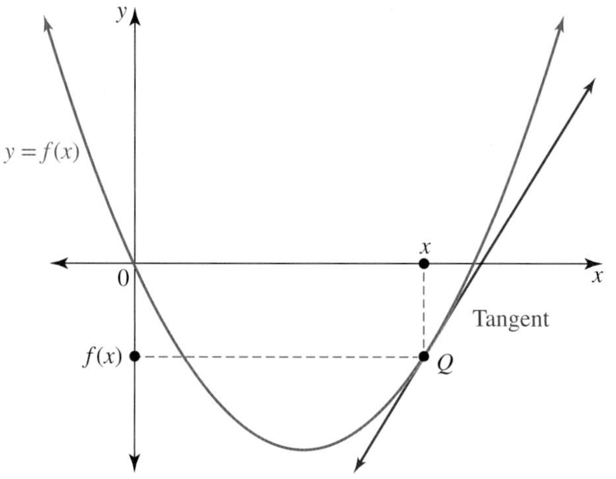

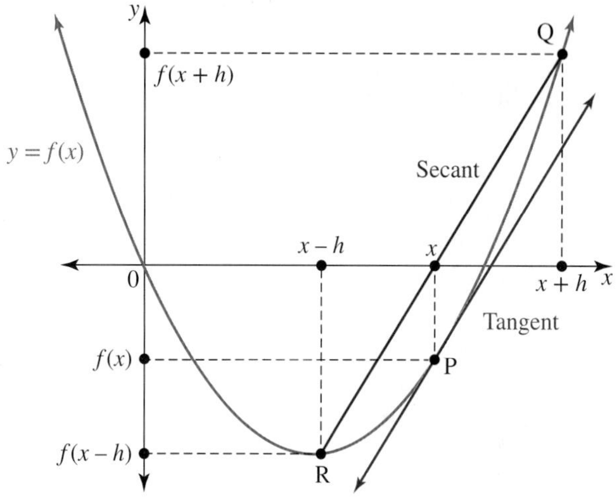

The gradient of the secant RQ can be calculated by:

$$\text{Gradient} = \frac{\text{rise}}{\text{run}} = \frac{f(x+h)-f(x-h)}{(x+h)-(x-h)}$$

$$= \frac{f(x+h)-f(x-h)}{2h}$$

This formula can be used to find the gradient function or the gradient at a particular point. As h gets smaller, the points R and Q get closer and closer to P, and the line RQ better approximates the tangent at P.

Central difference approximation

$$f'(x) \approx \frac{f(x+h)-f(x-h)}{2h}$$

where h is a small change in the x-coordinate.

In this notation, $f'(x)$ is the **derivative** of the function, or the gradient of the tangent to the curve at the point P $(x, f(x))$. $f'(x)$ is also the **gradient function** of $f(x)$, and $\frac{dy}{dx}$ is the gradient equation for y with respect to x.

WORKED EXAMPLE 1 Approximating $f'(x)$ by the central difference method

Consider the function $f(x) = -x^2 - x + 2$.

a. Determine the central difference approximation, $f'(x)$.

b. Hence, evaluate the gradient of the graph of $f(x)$ at the point (–2, 0).

THINK	WRITE
a. 1. Form an expression for $f(x+h)$.	a. $f(x+h) = -(x+h)^2 - (x+h) + 2$ $= -(x^2 + 2xh + h^2) - x - h + 2$ $= -x^2 - 2xh - h^2 - x - h + 2$
2. Form an expression for $f(x-h)$.	$f(x-h) = -(x-h)^2 - (x-h) + 2$ $= -(x^2 - 2xh + h^2) - x + h + 2$ $= -x^2 + 2xh - h^2 - x + h + 2$
3. Substitute the expressions into the central difference approximation formula, then expand and simplify.	$f'(x) \approx \frac{-x^2 - 2xh - h^2 - x - h + 2 - \left(-x^2 + 2xh - h^2 - x + h + 2\right)}{2h}$ $\approx \frac{-x^2 - 2xh - h^2 - x - h + 2 + x^2 - 2xh + h^2 + x - h - 2}{2h}$ $\approx \frac{-4xh - 2h}{2h}$ $\approx \frac{2h(-2x - 1)}{2h}$ $\approx -2x - 1$
b. Calculate the gradient at the given point.	b. For the point $(-2, 0)$, $x = -2$. The gradient at this point is $f'(-2)$. $f'(-2) = -(-2) - 1$ $= 3$ Therefore, the gradient at the point $(-2, 0)$ is 3.

5.2.2 The derivative of x^n

Differentiating by the central difference approximation is quite a tedious method, but there are rules to shortcut the process, depending on the function. Units 1 and 2 of Mathematical Methods covered differentiation of a polynomial function when $f(x) = x^n$. This general formula can also be used when $n \in R$.

Differentiation of $f(x) = x^n$

If $f(x) = ax^n$, $f'(x) = nax^{n-1}$, where $n \in R$ and $a \in R$

and

if $f(x) = g(x) \pm h(x)$, $f'(x) = g'(x) \pm h'(x)$.

WORKED EXAMPLE 2 Differentiation using $f(x) = x^n$

Differentiate the following.

a. $f(x) = x^3 - \dfrac{1}{2x} + 4$

b. $y = \dfrac{\sqrt{x} - 3x^3}{4x^2}$

THINK	WRITE
a. 1. Rewrite the equation with negative indices.	**a.** $f(x) = x^3 - \dfrac{1}{2x} + 4$ $= x^3 - \dfrac{1}{2}x^{-1} + 4$
2. Differentiate each term separately.	$f'(x) = 3x^2 + \dfrac{1}{2}x^{-2}$
3. Write the answer with positive indices.	$f'(x) = 3x^2 + \dfrac{1}{2x^2}$
b. 1. Split the fraction into two terms and rewrite $\sqrt{x}$ using a fractional index.	**b.** $y = \dfrac{\sqrt{x} - 3x^3}{4x^2}$ $= \dfrac{x^{\frac{1}{2}}}{4x^2} - \dfrac{3x^3}{4x^2}$
2. Simplify each term by applying the index laws.	$= \dfrac{x^{-\frac{3}{2}}}{4} - \dfrac{3}{4}x$
3. Differentiate each term separately.	$\dfrac{dy}{dx} = -\dfrac{3}{2} \times \dfrac{x^{-\frac{5}{2}}}{4} - \dfrac{3}{4}$
4. Simplify and write the answer with positive indices.	$= -\dfrac{3}{8x^{\frac{5}{2}}} - \dfrac{3}{4}$

TI | THINK	DISPLAY/WRITE
a. 1. On a Calculator page, press MENU, then select: 4: Calculus 1: Derivative Complete the entry line as: $\dfrac{dy}{dx}\left(x^3 - \dfrac{1}{2x} + 4\right)$ then press ENTER.	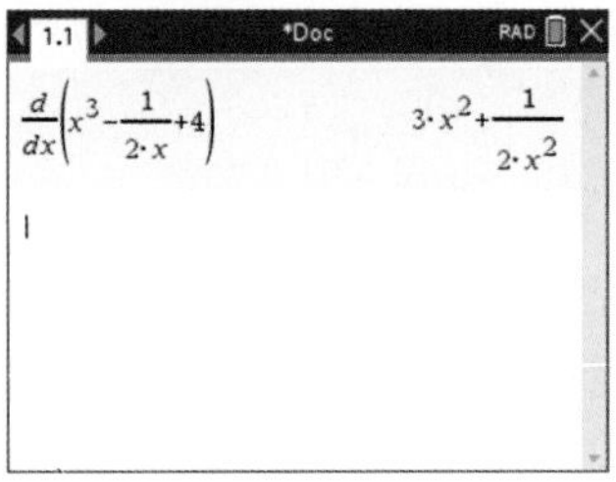
2. The answer appears on the screen.	$\dfrac{dy}{dx}\left(x^3 - \dfrac{1}{2x} + 4\right) = 3x^2 + \dfrac{1}{2x^2}$

CASIO | THINK	DISPLAY/WRITE
a. 1. On a Main screen, select: • Action • Calculation • diff Complete the entry line as: $\text{diff}\left(x^3 - \dfrac{1}{2x} + 4\right)$ then press EXE. Click the 'Simp' icon to simplify the answer.	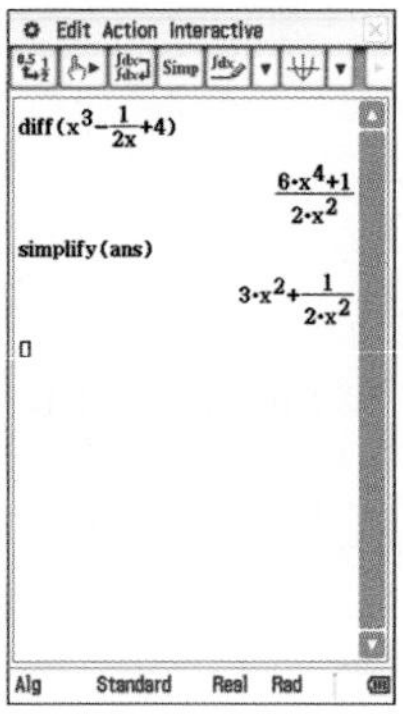
2. The answer appears on the screen.	$\dfrac{dy}{dx}\left(x^3 - \dfrac{1}{2x} + 4\right) =$ $3x^2 + \dfrac{1}{2x^2}$

WORKED EXAMPLE 3 Finding the gradient of a curve

a. If $f(x) = \frac{3}{x} - x^2$, calculate the gradient of the curve when $x = -2$.

b. If $f(x) = 2\sqrt{x} - 4$, determine the coordinates of the point where the gradient is 2.

THINK	WRITE
a. 1. Rewrite the equation with negative indices and differentiate each term.	a. $f(x) = \frac{3}{x} - x^2$ $= 3x^{-1} - x^2$ $f'(x) = -3x^{-2} - 2x$ $= -\frac{3}{x^2} - 2x$
2. The gradient of the curve when $x = -2$ is (2, 0).	$f'(-2) = -\frac{3}{(-2)^2} - 2 \times -2$ $= -\frac{3}{4} + 4$ $= \frac{13}{4}$
b. 1. Rewrite $\sqrt{x}$ with a fractional index and differentiate each term.	b. $f(x) = 2\sqrt{x} - 4$ $= 2x^{\frac{1}{2}} - 4$ $f'(x) = x^{-\frac{1}{2}}$ $= \frac{1}{\sqrt{x}}$
2. Finding where the gradient is 2 means solving $f'(x) = 2$.	$2 = \frac{1}{\sqrt{x}}$ $2\sqrt{x} = 1$ $\sqrt{x} = \frac{1}{2}$ $x = \frac{1}{4}$
3. Find $f\left(\frac{1}{4}\right)$ to determine the y-value where the gradient is 2.	$f\left(\frac{1}{4}\right) = 2\sqrt{\frac{1}{4}} - 4$ $= 2 \times \frac{1}{2} - 4$ $= -3$
4. Write the answer.	The gradient is 2 at the point $\left(\frac{1}{4}, -3\right)$.

TI | THINK

a. 1. On a Calculator page, press MENU, then select:
4: Calculus
2: Derivative at a Point…
Complete the fields as:
Variable: x
Value: -2
Derivative: 1st Derivative
then select OK.
Complete the entry line as:
$\dfrac{d}{dx}\left(\dfrac{3}{x}-x^2\right)|x=-2$
then press ENTER.

DISPLAY/WRITE

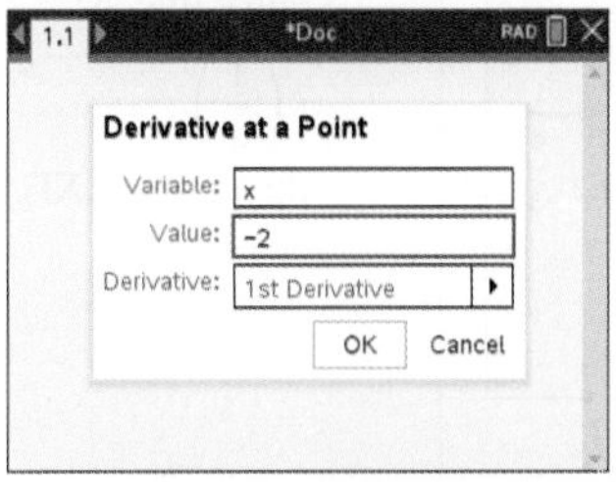

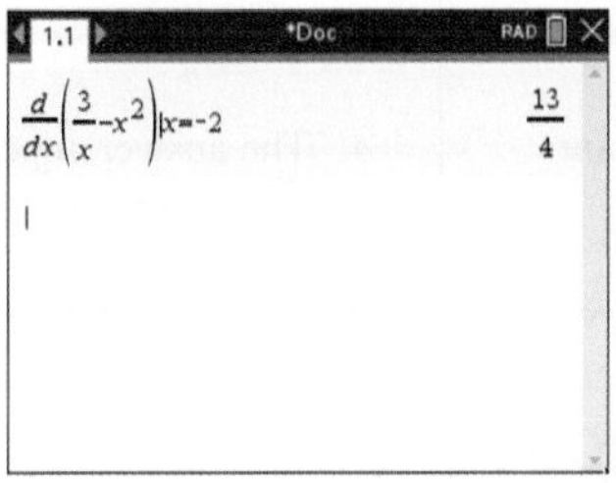

2. The answer appears on the screen.

When $x=-2$, $\dfrac{d}{dx}\left(\dfrac{3}{x}-x^2\right)=\dfrac{13}{4}$.

b. 1. On a Calculator page, press MENU, then select:
1: Actions
1: Define
Complete the entry line as:
Define $f1(x)=2\sqrt{x}-4$
then press ENTER.

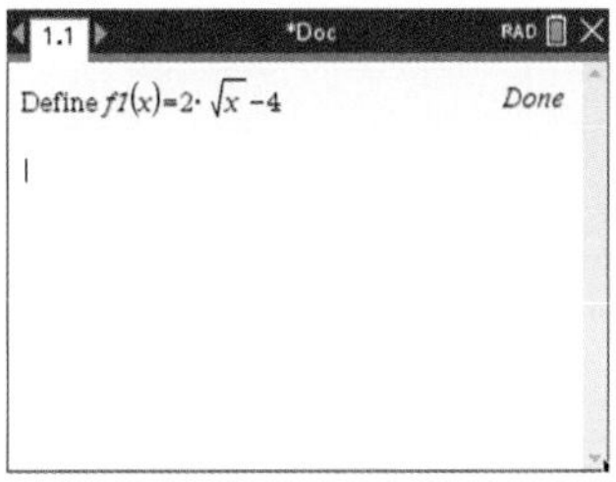

2. Press MENU, then select:
3: Algebra
1: Solve
Press MENU, then select:
4: Calculus
1: Derivative
Complete the entry line as:
$\text{solve}\left(\dfrac{d}{dx}(f1(x))=2,x\right)$
then press ENTER.

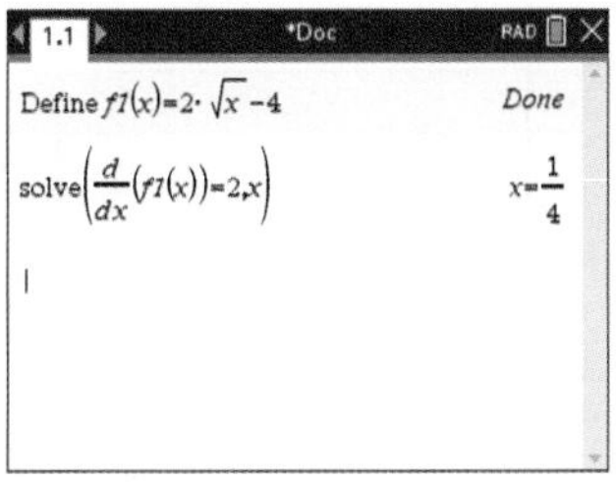

CASIO | THINK

a. 1. On a Main screen, select:
- Interactive
- Calculation
- diff

Select 'Derivative at value', then complete the fields as:
Expression: $\dfrac{3}{x}-x^2$
Variable: x
Order: 1
Value: -2
Select OK.

DISPLAY/WRITE

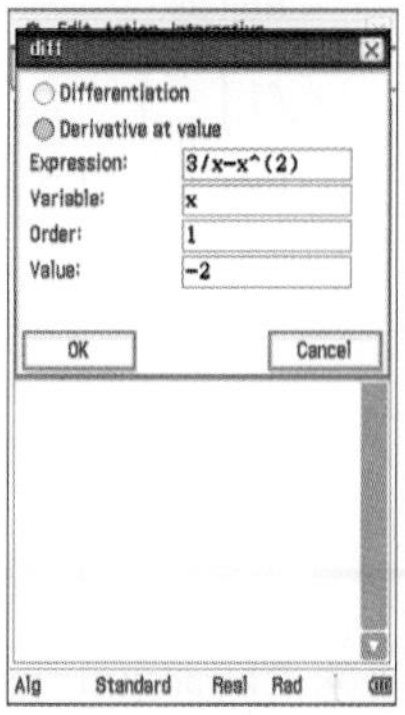

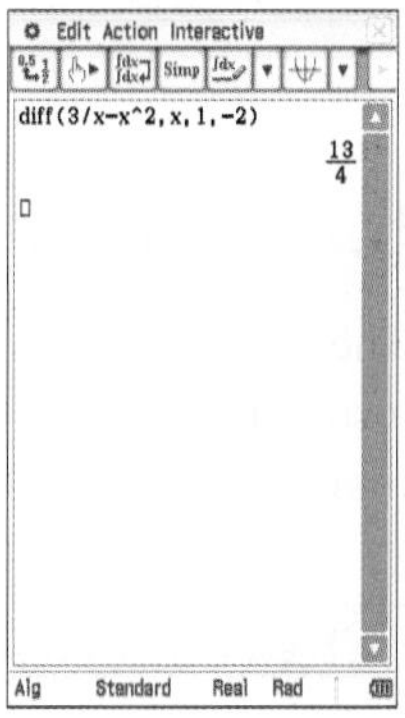

2. The answer appears on the screen.

When $x=-2$, $\dfrac{d}{dx}\left(\dfrac{3}{x}-x^2\right)=\dfrac{13}{4}$.

b. 1. On a Main screen, select:
- Interactive
- Define

Complete the fields as:
Func name: f
Variable/s: x
Expression: $2\sqrt{x}-4$
then select OK.

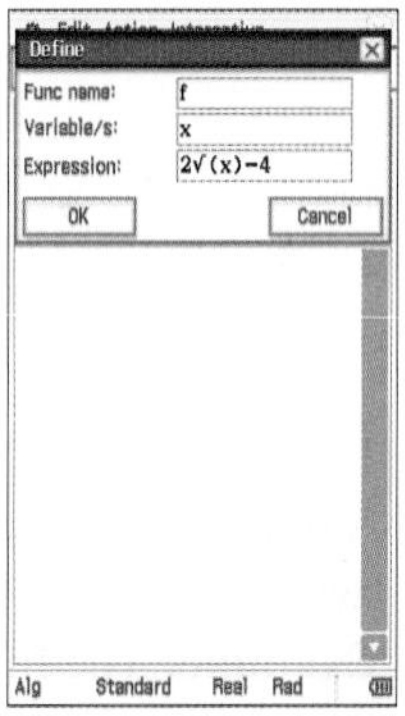

2. Complete the next entry line as:
solve
$(\text{diff}(f(x))=2,x)$
then press EXE.

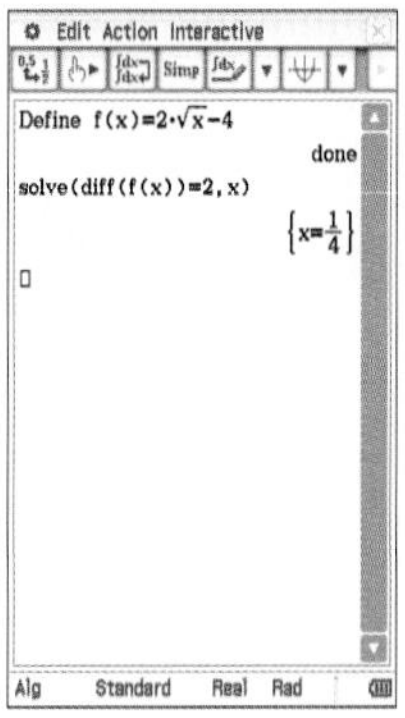

3. Complete the next entry line as:
$f1\left(\frac{1}{4}\right)$
then press ENTER.

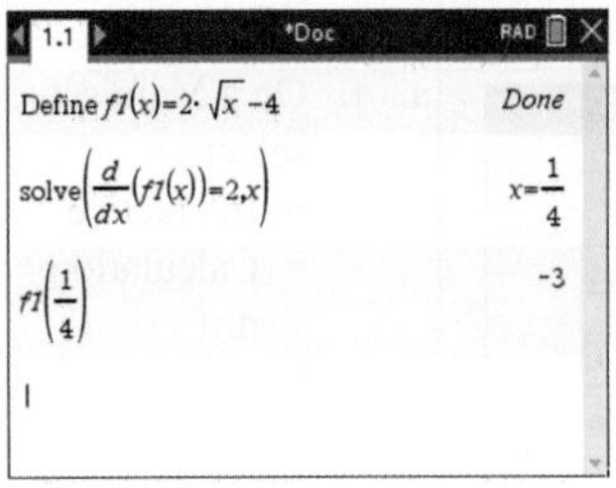

4. The answer appears on the screen.

The gradient is 2 at the point $\left(\frac{1}{4}, -3\right)$.

3. Complete the next entry line as:
$f\left(\frac{1}{4}\right)$
then press EXE.

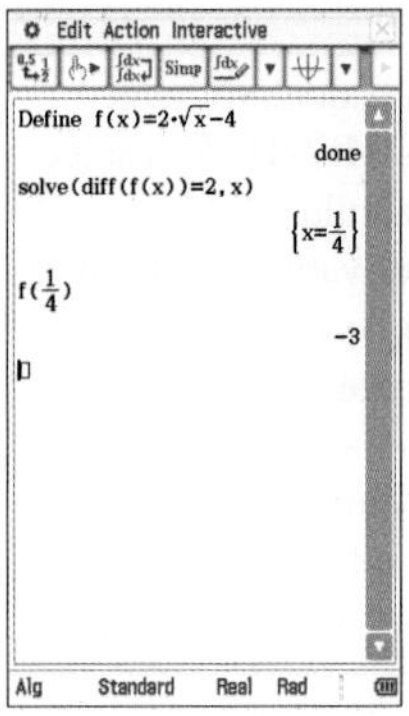

4. The answer appears on the screen.

The gradient is 2 at the point $\left(\frac{1}{4}, -3\right)$.

5.2.3 Graphs of the gradient function

The previous section shows that the derivative of the function $f(x) = x^n$ is one degree lower: $f'(x) = nx^{n-1}$. This also applies to the gradient graphs of these functions. For example, if $f(x)$ is a quadratic graph, $f'(x)$ will be a linear graph; if $f(x)$ is a cubic graph, $f'(x)$ will be a quadratic graph, and so on.

<table>
<tr><th>Given function $f(x)$</th><th>Gradient function $f'(x)$</th></tr>
<tr><td>A line of the form $y = mx + c$ is degree one, and the gradient is m.
Example:
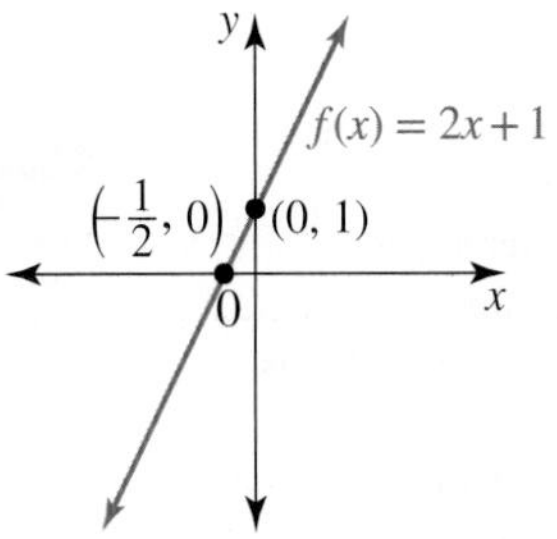
</td><td>The gradient is a constant value, so the gradient graph is a line parallel to the x-axis, $y = m$, degree zero.
Example:
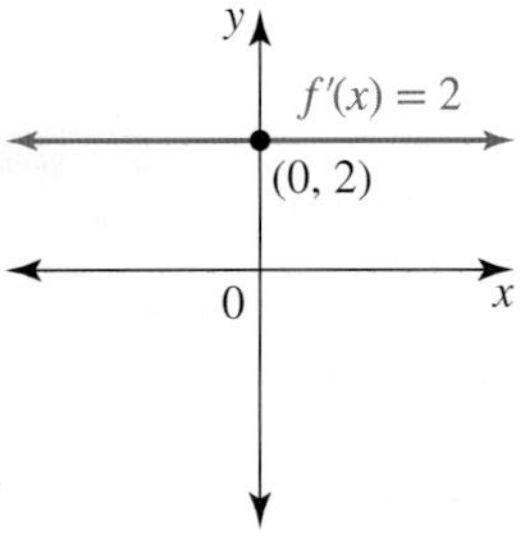
</td></tr>
<tr><td>A quadratic of the form $y = ax^2 + bx + c$ is degree two.
Example: The function shown has a local minimum at $x = -\frac{1}{2}$.
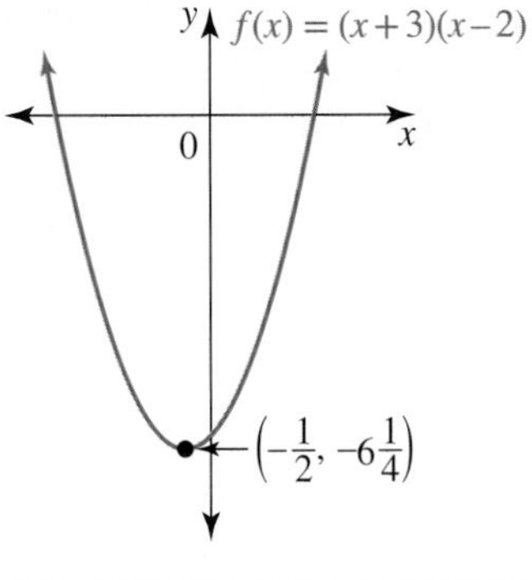
</td><td>A line of the form $y = mx + c$ is degree one.
Example: The line shown has an x-intercept at $x = -\frac{1}{2}$.
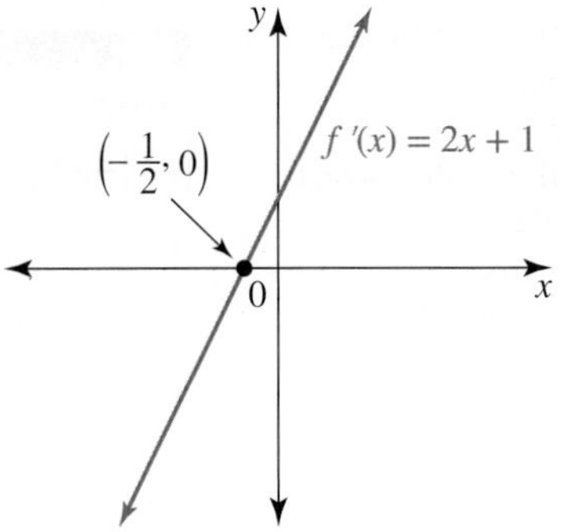
</td></tr>
</table>

A cubic of the form $y = ax^3 + bx^2 + cx + d$ is degree three. **Example:** The function shown has turning points at $x \simeq -1.8$ and $x \simeq 1.1$. 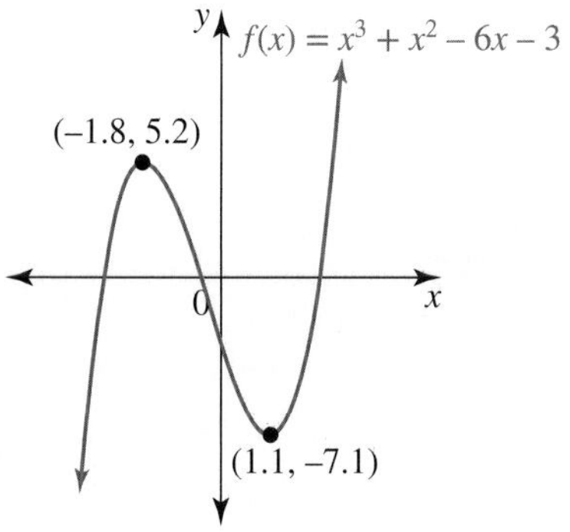 	A quadratic of the form $y = ax^2 + bx + c$ is degree two. **Example:** The curve shown has x-intercepts at $x \simeq -1.8$ and $x \simeq 1.1$. 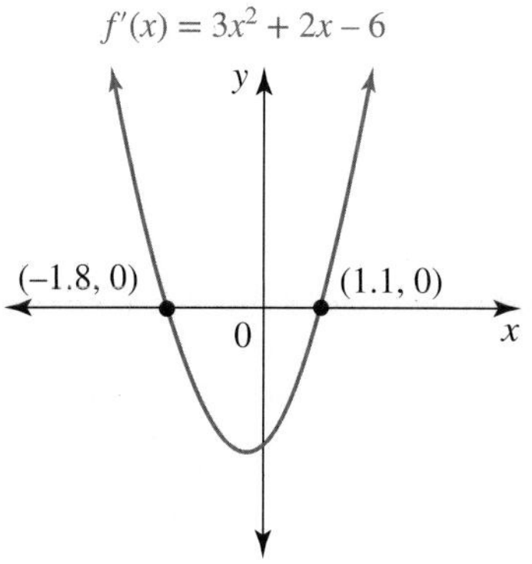

Sometimes $f(x)$ may not be a known function, so the features of the graph of $f(x)$ need to be studied carefully in order to sketch the gradient graph.

Summary of features for derivative graphs

- **Turning points on the graph of $f(x)$ represent x-intercepts on the graph of $f'(x)$.**
- **Where the graph of $f(x)$ has a positive gradient, the graph of $f'(x)$ is above the x-axis.**
- **Where the graph of $f(x)$ has a negative gradient, the graph of $f'(x)$ is below the x-axis.**

Existence of a gradient

The gradient of a function only exists where the graph is smooth and continuous. That is, a single tangent must be able to be drawn at $x = a$ for $f'(a)$ to exist.

WORKED EXAMPLE 4 Sketching a derivative graph and its domain

a. **The graph of the cubic function $f(x)$ is shown. Sketch the derivative function, $f'(x)$.**

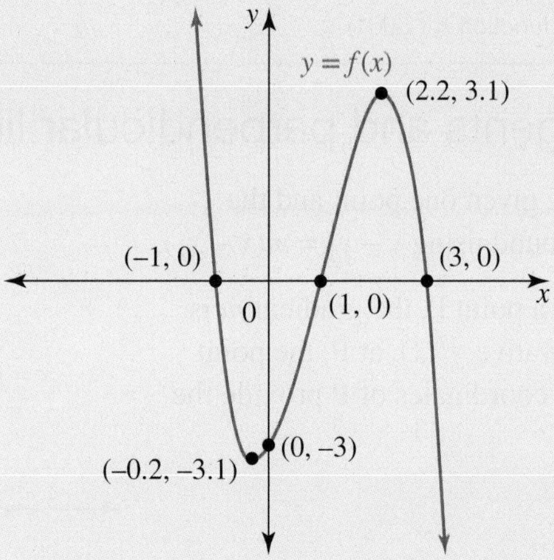

b. State the domain of the gradient function, $f'(x)$, for the function shown.

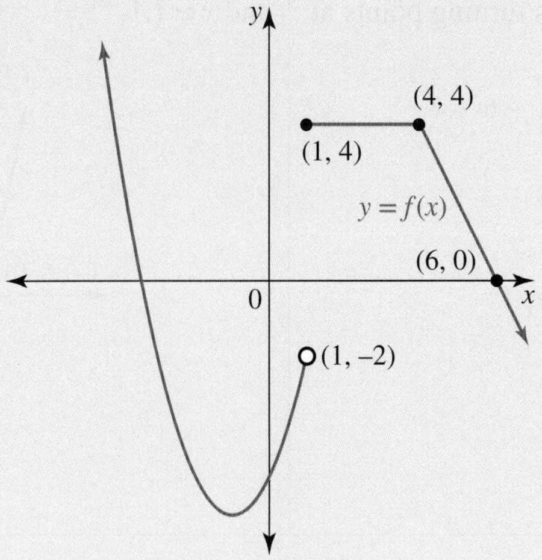

THINK

a. The turning points are points of zero gradient, so these points will correspond to x-intercepts on $f'(x)$.
$f'(x)$ lies above the x-axis where the gradient is positive.
Where the gradient is negative is where $f'(x)$ lies below the x-axis.
Use this information to sketch $f'(x)$.

WRITE

a.

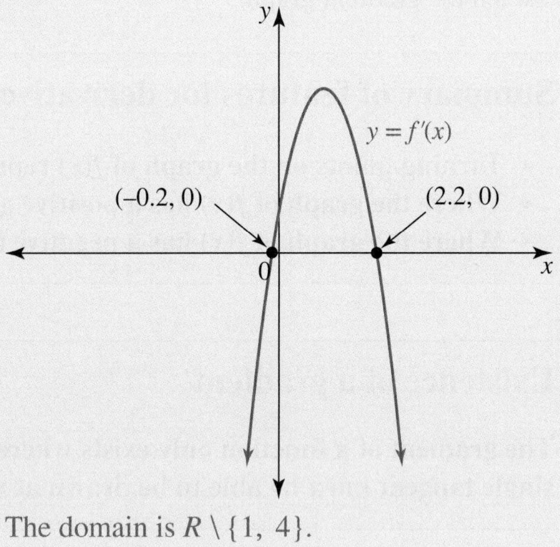

b. For the gradient to exist, the graph must be smooth and continuous. The gradient doesn't exist at $x = 1$ or at $x = 4$.

b. The domain is $R \setminus \{1, 4\}$.

on Resources

Interactivity Graph of a derivative function (int-5961)

5.2.4 Equations of tangents and perpendicular lines

A tangent is a straight line. If we are given one point and the gradient value, its equation can be found using $y - y_1 = m(x - x_1)$.

For the tangent to a curve $y = f(x)$ at a point P, the gradient m is found by evaluating the curve's derivative, $f'(x)$, at P, the point of contact or point of tangency. The coordinates of P provide the point (x_1, y_1) on the line.

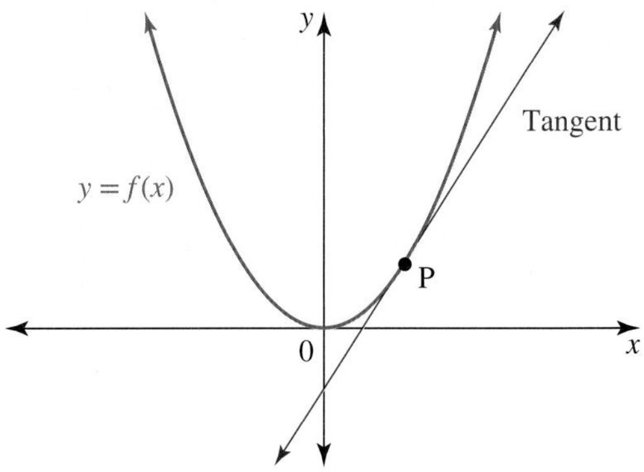

Important features of tangent lines

- **The angle of inclination of the tangent to the horizontal can be calculated using $m = \tan(\theta)$.**
- **Tangents that are parallel to each other have the same gradient.**
- **The gradient of a line perpendicular to the tangent is found using $m_T m_P = -1$. That is, if the gradient of a tangent is m_T, then the gradient of a perpendicular line, m_P, is $-\dfrac{1}{m_T}$.**
- **The gradient of a horizontal tangent is zero.**
- **The gradient of a vertical tangent is undefined.**

WORKED EXAMPLE 5 Finding equations of tangent and perpendicular lines

Consider the function $f(x) = (1 - x)(x - 3)(x - 6)$. The graph of this function is shown.

a. Determine the equation of the tangent to the curve at the point (4, 6).

b. Determine the equation of the line perpendicular to the tangent at the point (4, 6).

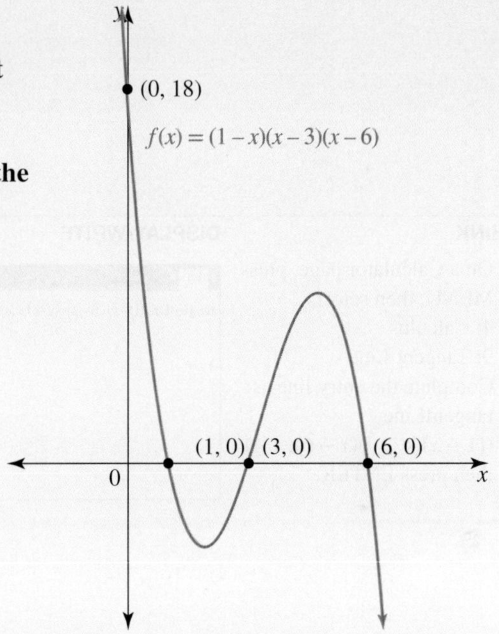

THINK	WRITE
a. 1. Expand $f(x)$.	a. $f(x) = (1 - x)(x - 3)(x - 6)$ $= (1 - x)(x^2 - 9x + 18)$ $= -x^3 + 10x^2 - 27x + 18$
2. Find the derivative of $f(x)$.	$f'(x) = -3x^2 + 20x - 27$
3. Find the gradient at $x = 4$.	$f'(4) = -3(4)^2 + 20(4) - 27$ $= -48 + 80 - 27$ $= 5$
4. Substitute the appropriate values into the formula $y - y_1 = m(x - x_1)$.	$m = 5$ and $(x_1, y_1) = (4, 6)$ $y - y_1 = m(x - x_1)$ $y - 6 = 5(x - 4)$ $y - 6 = 5x - 20$ $y = 5x - 14$

b. 1. Find the gradient of the line perpendicular to the tangent.

b. $m_P = -\dfrac{1}{m_T}$

$= -\dfrac{1}{5}$

2. Find the equation of the perpendicular line.

$m = -\dfrac{1}{5}$ and $(x_1, y_1) = (4, 6)$

$y - y_1 = m(x - x_1)$

$y - 6 = -\dfrac{1}{5}(x - 4)$

$y - 6 = -\dfrac{1}{5}x + \dfrac{4}{5}$

$y = -\dfrac{1}{5}x + \dfrac{4}{5} + \dfrac{30}{5}$

$y = -\dfrac{1}{5}x + \dfrac{34}{5}$

or

$5y = -x + 34$

$x + 5y = 34$

TI | THINK — **DISPLAY/WRITE**

a. 1. On a Calculator page, press MENU, then select:
4: Calculus
9: Tangent Line
Complete the entry line as:
tangentLine
$((1 - x)(x - 3)(x - 6), x, 4)$
then press ENTER.

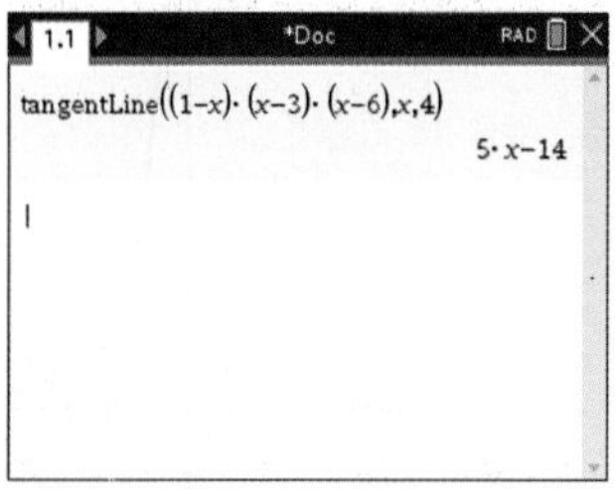

2. The answer appears on the screen.

The equation of the tangent to the curve at $x = 4$ is $y = 5x - 14$.

CASIO | THINK — **DISPLAY/WRITE**

a. 1. On a Main screen, select:
- Interactive
- Calculation
- line
- tanLine

Complete the fields as:
Expression:
$(1 - x)(x - 3)(x - 6)$
Variable: x
Point: 4
then select OK.

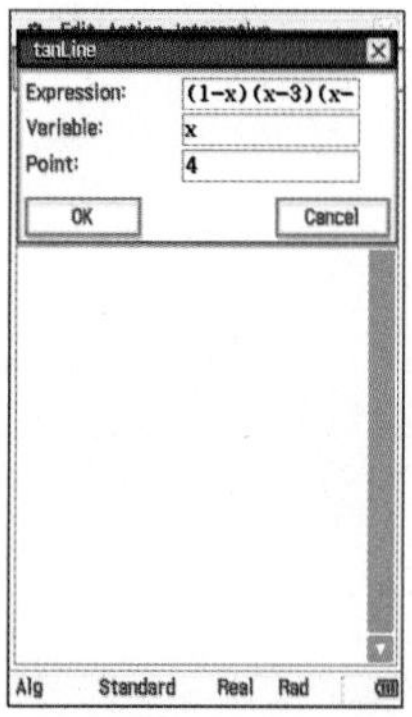

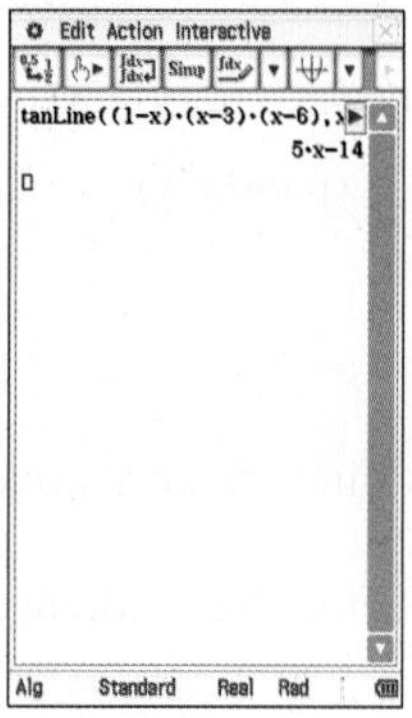

2. The answer appears on the screen.

The equation of the tangent to the curve at $x = 4$ is $y = 5x - 14$.

b. 1. On a Calculator page, press MENU, then select:
4: Calculus
A: Normal Line
Complete the entry line as:
normal Line
$((1-x)(x-3)(x-6), x, 4)$
then press ENTER.

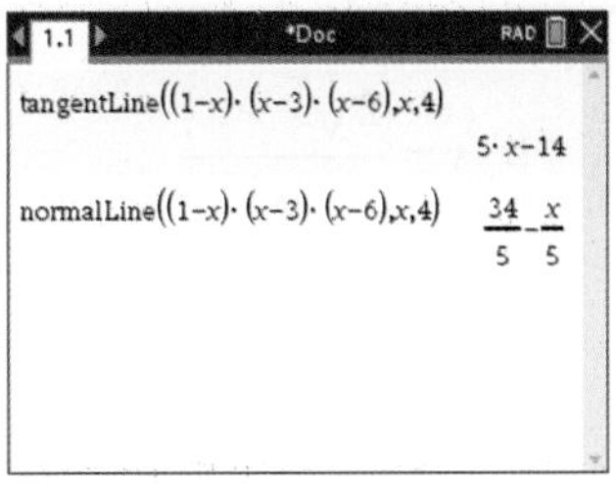

2. The answer appears on the screen.

The equation of the normal to the curve at $x = 4$ is $y = \frac{34}{5} - \frac{x}{5}$.

b. 1. On a Main screen, select:
- Interactive
- Calculation
- line
- normal

Complete the fields as:
Expression: $(1-x)(x-3)(x-6)$
Variable: x
Point: 4
then select OK.

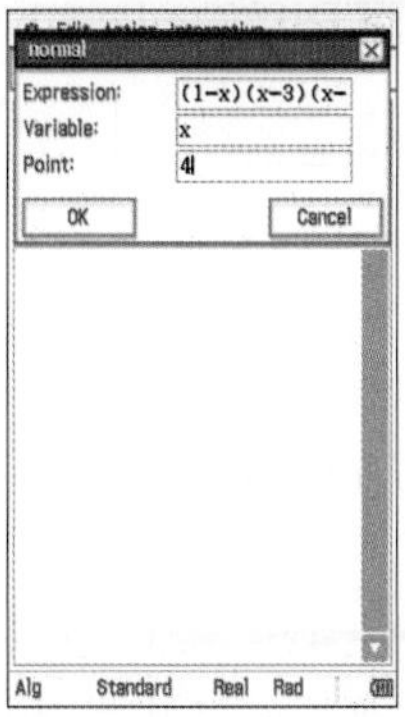

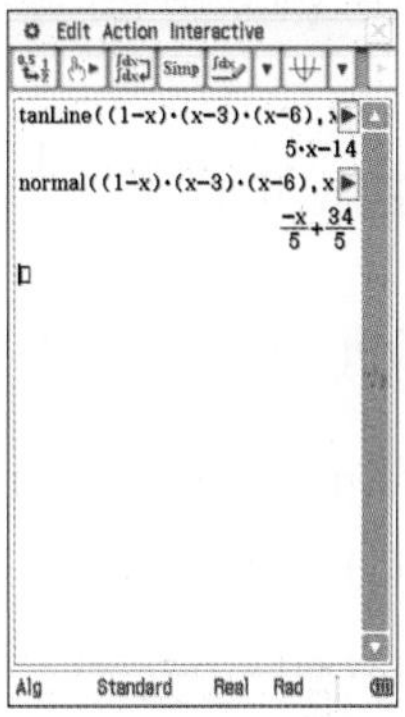

2. The answer appears on the screen.

The equation of the normal to the curve at $x = 4$ is $y = -\frac{x}{5} + \frac{34}{5}$.

5.2 Exercise

Technology free

1. **WE1** Consider the function $f(x) = (2-x)^2 + 1$.
   a. Determine the central difference approximation, $f'(x)$.
   b. Hence, evaluate the gradient of the curve at the point $(1, 2)$.

2. For the following functions, determine the central difference approximation, $f'(x)$.
   a. $f(x) = 12 - x$
   b. $f(x) = 3x^2 - 2x - 21$

3. WE2 Differentiate the following.

a. $f(x) = 4x^3 + \frac{1}{3x^2} + \frac{1}{2}$

b. $y = \frac{2\sqrt{x} - x^4}{5x^3}$

c. $f(x) = (x + 3)(x^2 + 1)$

d. $y = \frac{4 - \sqrt{x}}{\sqrt{x^3}}$

4. Differentiate the following.

a. $y = \frac{3}{4x^5} - \frac{1}{2x} + 4$

b. $f(x) = \frac{10x - 2x^3 + 1}{x^4}$

c. $y = \sqrt{x} - \frac{1}{2\sqrt{x}}$

d. $f(x) = \frac{(3 - x)^3}{2x}$

5. WE3 a. If $f(x) = -\frac{1}{x^2} + 2x$, calculate the gradient of the curve when $x = -\frac{1}{2}$.

b. If $f(x) = \frac{2x - 4}{x}$, determine the coordinates of the point where the gradient is 1.

6. Determine the gradient of the tangent to each of the following curves at the specified point.

a. $f(x) = x^2 - 3$ at $x = 2$

b. $f(x) = (3 - x)(x - 4)$ at $x = 1$

c. $f(x) = (x - 2)^3$ at $x = 4$

d. $f(x) = \sqrt{x} - \frac{3}{x} + 2x$ at $x = 4$

7. If $y = (x - a)(x^2 - 1)$, calculate the gradient of the curve when $x = -2$ in terms of a.

8. WE4 a. The graph of $f(x)$ is shown. Sketch the graph of the derivative function, $f'(x)$.

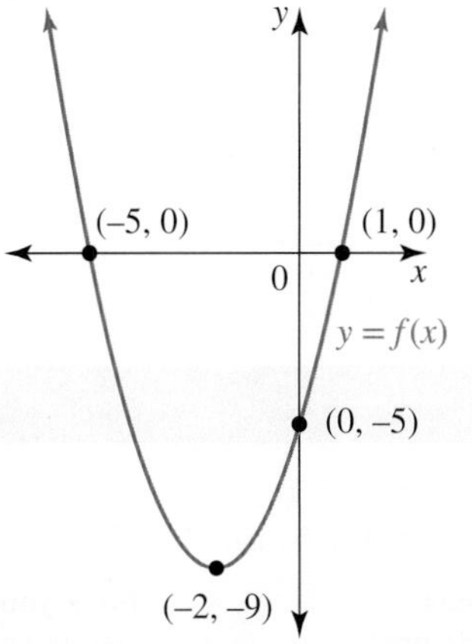

b. State the domain of the gradient function, $f'(x)$, for the function shown.

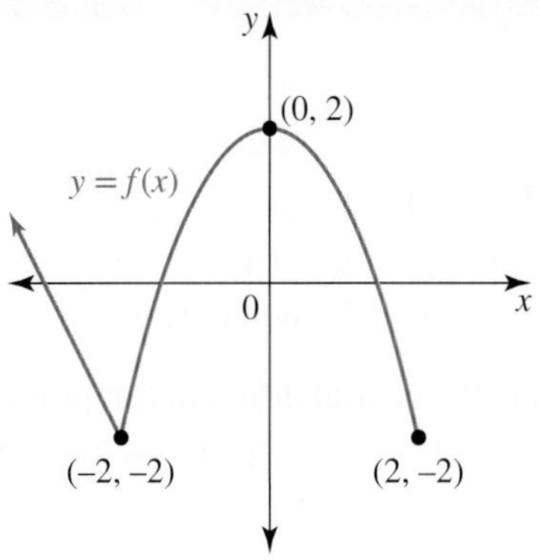

9. The graph of $f(x)$ is shown. Analyse this function and sketch the graph of $f'(x)$.

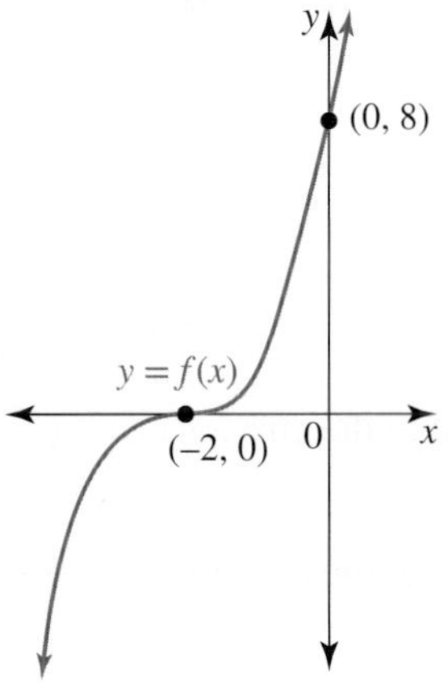

10. For the following graphs:
   i. state the domain of the gradient function, $f'(x)$
   ii. sketch the graph of $f'(x)$.

a.

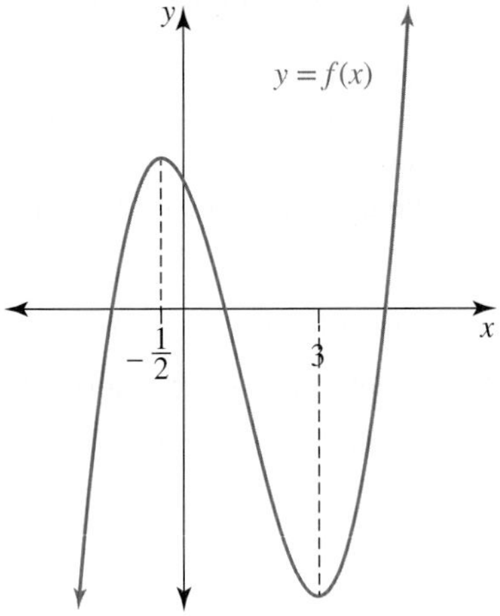

b.

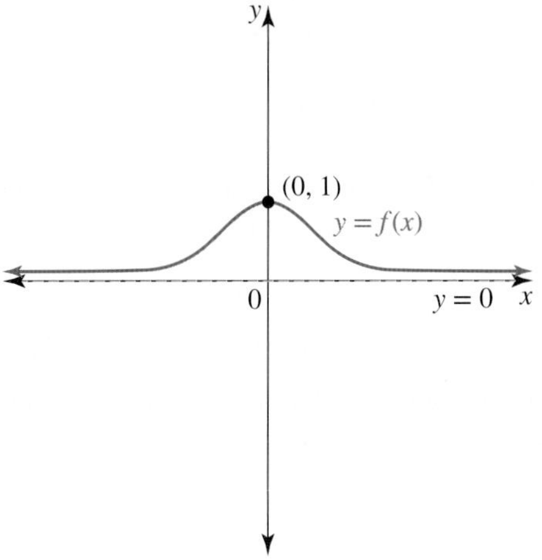

c.

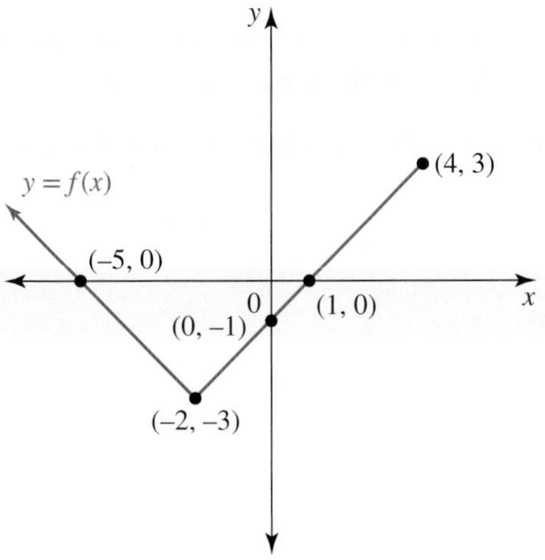

d.

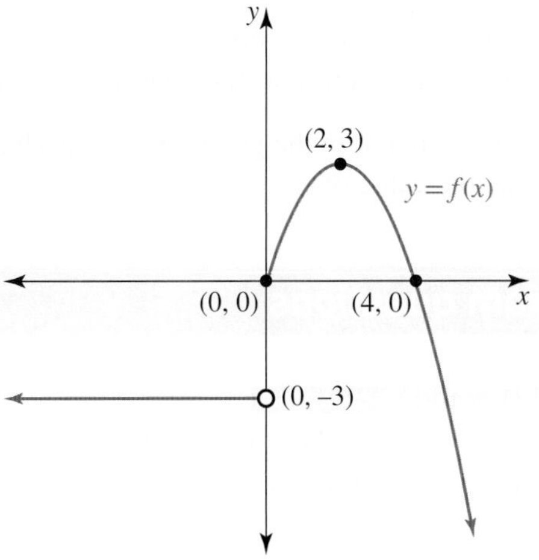

11. **WE5** a. Determine the equation of the tangent to the curve with equation $y = x(x-2)^2(x-4)$ at the point $(3, -3)$.
   b. Determine the equation of the line perpendicular to the tangent at the point (3, –3).

12. Determine the equations of the tangents to the following curves at the specified points.

a. $f(x) = (x+1)(x+3)$ at $x = -5$
b. $f(x) = 8 - x^3$ at $x = a$
c. $f(x) = 2\sqrt{x} - 5$ at $x = 3$
d. $f(x) = -\frac{2}{x} - 4x$ at $x = -2$

13. Determine the equation of the line perpendicular to the tangent for each of the following functions.

a. $f(x) = (x+1)(x+3)$ at $x = -5$
b. $f(x) = 8 - x^3$ at $x = a$
c. $f(x) = 2\sqrt{x} - 5$ at $x = 3$
d. $f(x) = -\frac{2}{x} - 4x$ at $x = -2$

14. a. Determine the equation of the tangent to the curve $f(x) = -(x-2)^2 + 3$ that is parallel to the line $y = 3x + 4$.

b. Determine the equation of the tangent to the curve $f(x) = -\frac{2}{x^2} + 1$ that is perpendicular to the line $2y - 2 = -4x$.

Technology active

15. a. Sketch the graphs of $y = (x+2)(2-x)$ and $y = x^2(4-x)$ on the one set of axes.
b. Determine the point(s) of intersection of the two curves, giving coordinates correct to 2 decimal places where appropriate.
c. If P is the point of intersection where $x \in Z$, use first principles to calculate the gradient of the tangents to each of the curves at this point.

16. a. Use CAS technology to sketch $y = x(x-2)(x+3)$ and $y = (2-x)(x+3)(x-3)$ on the same set of axes.
b. Determine the coordinates of the point of intersection between the graphs where $1 < x < 2$.
c. Determine the equation of the tangent and the line perpendicular to the tangent at the point defined in part **b** for the cubic function defined by $y = x(x-2)(x+3)$.

17. The equation of a tangent to a given parabola is $y = -2x + 5$. The equation of the line perpendicular to this tangent is $y = \frac{1}{2}x + \frac{5}{2}$. The parabola also has a stationary point at $(0, 4)$. Determine the equation of the parabola and hence sketch the parabola, the tangent and the line perpendicular to the tangent, on the one set of axes.

18. The tangent to a parabolic curve at $x = 4$ has the equation $y = -x + 6$. The curve also passes through the points $(0, -10)$ and $(2, 0)$. Determine the equation of the curve.

19. The tangent to a cubic function at the point $x = 2$ has a rule defined by $y = 11x - 16$. The cubic passes through the origin as well as the point $(-1, 0)$. Determine the equation of the cubic function.

20. A line perpendicular to the graph of $y = 2\sqrt{x}$ has the equation $y = -2x + m$, where m is a real constant. Determine the value of m.

5.2 Exam questions

Question 1 (1 mark) TECH-ACTIVE

Source: *VCE 2021, Mathematical Methods Exam 2, Q7; © VCAA.*

MC The tangent to the graph of $y = x^3 - ax^2 + 1$ at $x = 1$ passes through the origin.

The value of a is

A. $\frac{1}{2}$ **B.** 1 **C.** $\frac{3}{2}$ **D.** 2 **E.** $\frac{5}{2}$

Question 2 (1 mark) TECH-ACTIVE

Source: *VCE 2019, Mathematical Methods Exam 2, Section A, Q16; © VCAA.*

MC Part of the graph of $y = f(x)$ is shown below.

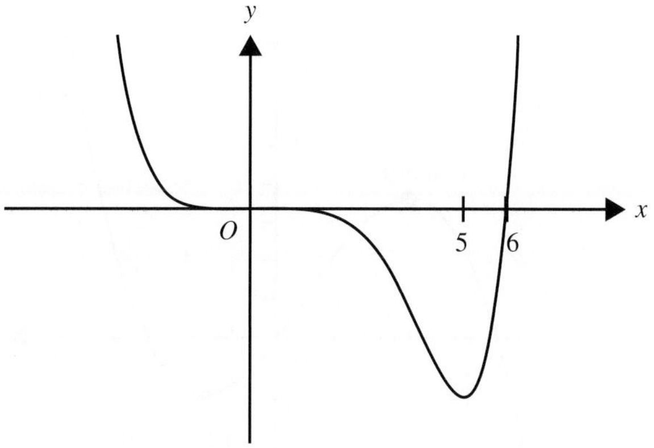

The corresponding part of the graph of $y = f'(x)$ is best represented by

A.

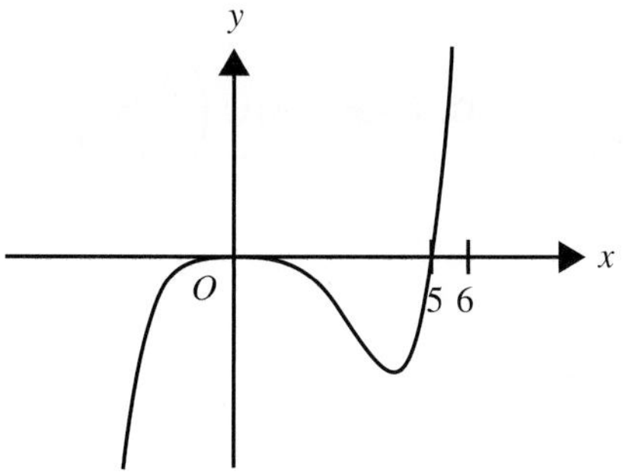

B.

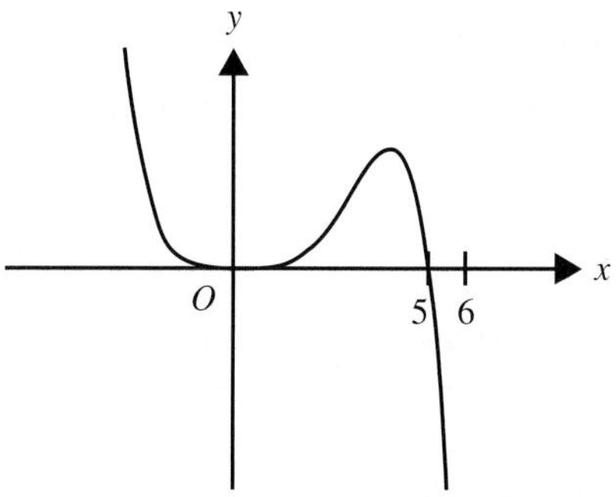

C.

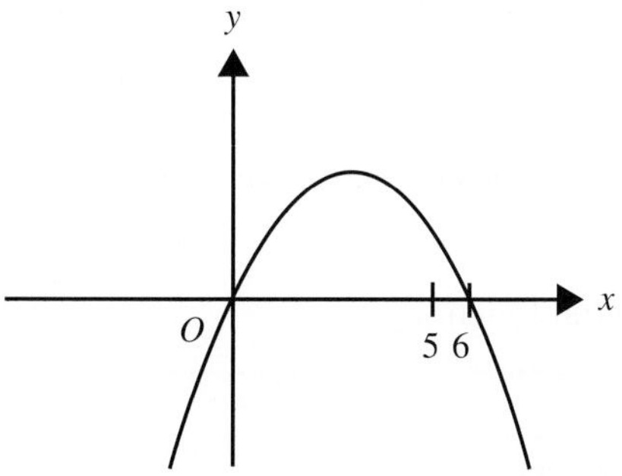

D.

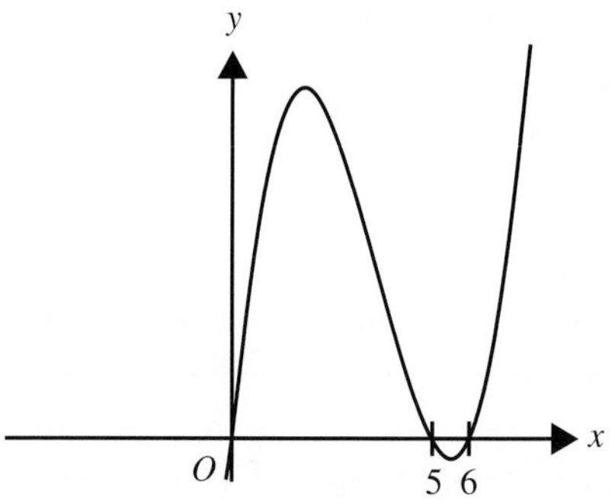

E.

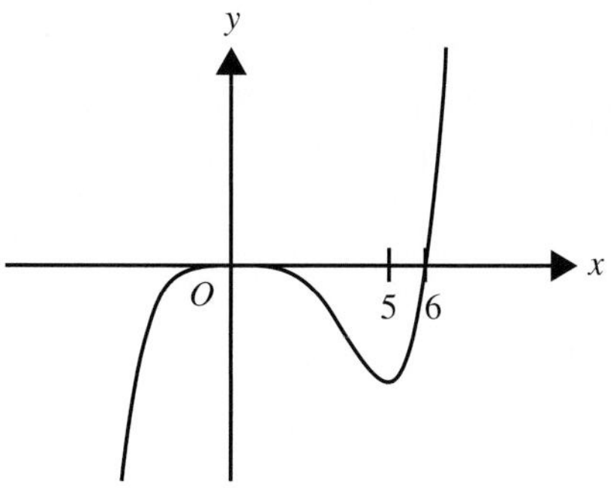

Question 3 (1 mark) TECH-ACTIVE

Source: *VCE 2017, Mathematical Methods Exam 2, Section A, Q2; © VCAA.*

MC Part of the graph of a cubic polynomial function f and the coordinates of its stationary points are shown below.

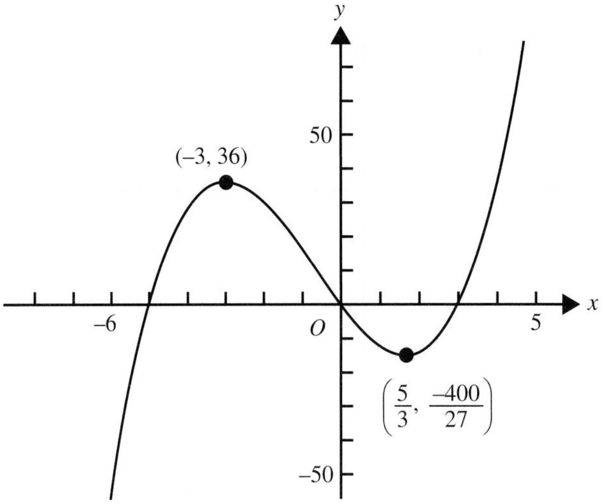

$f'(x) < 0$ for the interval

A. $(0, 3)$

B. $(-\infty, -5) \cup (0, 3)$

C. $(-\infty, -3) \cup \left(\frac{5}{3}, \infty\right)$

D. $\left(-3, \frac{5}{3}\right)$

E. $\left(\frac{-400}{27}, 36\right)$

More exam questions are available online.

5.3 Differentiation of exponential functions

LEARNING INTENTION

At the end of this subtopic you should be able to:
- calculate derivative of an exponential function
- determine the gradient and equation of a tangent to an exponential function.

5.3.1 The derivative of the exponential function

We can find the derivative of the exponential function as follows.

$$f'(x) = \lim_{h \to 0} \frac{f(x+h) - f(x)}{h}$$

$$f(x) = e^x$$

$$f(x+h) = e^{x+h}$$

$$\text{If } f'(x) = \lim_{h \to 0} \frac{f(x+h) - f(x)}{h}$$

$$\text{then } f'(x) = \lim_{h \to 0} \frac{e^{x+h} - e^x}{h}$$

$$f'(x) = \lim_{h \to 0} \frac{e^x(e^h - 1)}{h}$$

$$f'(x) = e^x \lim_{h \to 0} \frac{e^h - 1}{h}$$

We don't know the value of $\lim\limits_{h \to 0} \dfrac{e^h - 1}{h}$, but we can investigate by substituting different values for h and looking at what happens to the **limit** as the value of h approaches zero.

$$\begin{aligned}
\text{If } h = 1, &\quad \frac{e^h - 1}{h} = 1.7183 \\
h = 0.1, &\quad \frac{e^h - 1}{h} = 1.0517 \\
h = 0.01, &\quad \frac{e^h - 1}{h} = 1.0050 \\
h = 0.001, &\quad \frac{e^h - 1}{h} = 1.0005 \\
h = 0.001, &\quad \frac{e^h - 1}{h} = 1.000\,05
\end{aligned}$$

From these results, we can see that as the value of h gets smaller and approaches zero, the value of $\dfrac{e^h - 1}{h}$ approaches 1:

$$f'(x) = e^x \lim_{h \to 0} \frac{e^h - 1}{h}$$

Therefore, as $\lim\limits_{h \to 0} \dfrac{e^h - 1}{h} = 1$, $f'(x) = e^x$.

Derivative of e^x

If $f(x) = e^x$, then $f'(x) = e^x$.

The derivative of the exponential function is itself.

***Note*: This rule only applies to exponential functions of base *e*.**

The following can be shown using the **chain rule**, which will be introduced in the next topic.

Derivative of e^{kx} and $e^{g(x)}$

If $f(x) = e^{kx}$, then $f'(x) = ke^{kx}$

and

if $f(x) = e^{g(x)}$, then $f'(x) = g'(x)e^{g(x)}$.

WORKED EXAMPLE 6 Determining derivatives of exponential functions

Determine the derivative of each of the following functions with respect to x.

a. $y = e^{-\frac{1}{2}x}$ **b.** $y = \frac{1}{4}e^{2x} + e^{x^2}$ **c.** $y = \frac{e^{2x} + 3e^x - 1}{e^{2x}}$ **d.** $y = (e^x - 2)^2$

THINK	WRITE
a. 1. Write the equation to be differentiated.	**a.** $y = e^{-\frac{1}{2}x}$
2. Apply the rule for $\frac{d}{dx}(e^{kx})$ with $k = -\frac{1}{2}$.	$\frac{dy}{dx} = -\frac{1}{2}e^{-\frac{1}{2}x}$
b. 1. Write the equation to be differentiated.	**b.** $y = \frac{1}{4}e^{2x} + e^{x^2}$
2. Apply the rule $\frac{d}{dx}\left(e^{g(x)}\right) = g'(x)e^{g(x)}$ and differentiate each term separately.	$\frac{dy}{dx} = \frac{1}{4} \times 2e^{2x} + 2xe^{x^2}$ $= \frac{1}{2}e^{2x} + 2xe^{x^2}$
c. 1. Write the equation to be differentiated.	**c.** $y = \frac{e^{2x} + 3e^x - 1}{e^{2x}}$
2. Split the right-hand side into three separate terms and divide through by e^{2x}.	$= \frac{e^{2x}}{e^{2x}} + \frac{3e^x}{e^{2x}} - \frac{1}{e^{2x}}$ $= 1 + 3e^{-x} - e^{-2x}$
3. Apply the rule $\frac{d}{dx}\left(e^{kx}\right) = ke^{kx}$ and differentiate each term separately.	$\frac{dy}{dx} = -3e^{-x} + 2e^{-2x}$ $= -\frac{3}{e^x} + \frac{2}{e^{2x}}$
d. 1. Write the equation to be differentiated.	**d.** $y = (e^x - 2)^2$
2. Expand the right-hand side.	$= e^{2x} - 4e^x + 4$
3. Differentiate each term separately.	$\frac{dy}{dx} = 2e^{2x} - 4e^x$

WORKED EXAMPLE 7 Determining equations of tangents of exponential functions

a. Determine the gradient of the tangent to the curve with equation $y = e^{-x}$ at the point where $x = 1$.
b. i. Determine the equation of the tangent to the curve $y = e^{-x}$ at the point where $x = 1$.
ii. Hence, determine the equation of the line perpendicular to this tangent.

THINK	WRITE
a. 1. The gradient of the tangent is given by $\frac{dy}{dx}$.	**a.** $y = e^{-x}$ $\frac{dy}{dx} = -e^{-x}$

2. Substitute $x = 1$.	$\frac{dy}{dx} = -e^{-1}$ $= -\frac{1}{e}$ The gradient of the curve is $-\frac{1}{e}$.
b. i. 1. We have the gradient but we need a point. Determine the corresponding y-value when $x = 1$.	**b.** $x = 1$: $y = e^{-1}$ $= \frac{1}{e}$
2. Use $y - y_1 = m(x - x_1)$ to find the equation of the tangent.	If $(x_1, y_1) = \left(1, \frac{1}{e}\right)$ and $m = -\frac{1}{e}$. $y - \frac{1}{e} = -\frac{1}{e}(x - 1)$ $y - \frac{1}{e} = -\frac{1}{e}x + \frac{1}{e}$ $y = -\frac{1}{e}x + \frac{2}{e}$ $= -\frac{1}{e}(x - 2)$ The equation of the tangent is $y = -\frac{1}{e}(x - 2)$.
ii. 3. A line perpendicular to a tangent has a gradient that is the negative reciprocal of the gradient of the tangent: $m_P = -\frac{1}{m_T}$.	$m_P = -(-e)$ $= e$
4. The perpendicular line passes through the same point, $\left(1, \frac{1}{e}\right)$. Use $y - y_1 = m(x - x_1)$ to find the equation of the perpendicular line.	If $(x_1, y_1) = \left(1, \frac{1}{e}\right)$ and $m = e$: $y - \frac{1}{e} = e(x - 1)$ $y - \frac{1}{e} = ex - e$ $y = ex - e + \frac{1}{e}$ The equation of the perpendicular line is $y = ex - e + \frac{1}{e}$.

5.3 Exercise

Students, these questions are even better in jacPLUS

- **Receive immediate feedback and access sample responses**
- **Access additional questions**
- **Track your results and progress**

Find all this and MORE in jacPLUS

Technology free

1. **WE6** Determine the derivative of each of the following functions with respect to x.

   a. $e^{-\frac{1}{3}x}$
   b. $3x^4 - e^{-2x^2}$
   c. $y = \dfrac{4e^x - e^{-x} + 2}{3e^{3x}}$
   d. $y = \left(e^{2x} - 3\right)^2$

2. Differentiate the following with respect to x.

   a. $5e^{-4x} + 2e$
   b. $e^{-\frac{x}{2}} + \dfrac{1}{3}x^3$
   c. $4e^{3x} - \dfrac{1}{2}e^{6\sqrt{x}} - 3e^{-3x+2}$
   d. $\dfrac{e^{5x} - e^{-x} + 2}{e^{2x}}$
   e. $\dfrac{e^x(2 - e^{-3x})}{e^{-x}}$
   f. $(e^{2x} + 3)(e^{-x} - 1)$

3. Consider the function defined by the rule

$$f(x) = \frac{1}{2}e^{3x} + e^{-x}.$$

   Determine the gradient of the curve when $x = 0$.

4. Calculate the exact gradients of the tangents to the given functions at the specified points.

   a. $y = 2e^{-x}$ at $x = 0$
   b. $y = \dfrac{4}{e^{2x}}$ at $x = \dfrac{1}{2}$
   c. $y = \dfrac{1}{2}e^{3x}$ at $x = \dfrac{1}{3}$
   d. $y = 2x - e^{x}$ at $x = 0$

5. **WE7** a. Determine the gradient of the tangent to the curve with equation $y = e^{2x}$ at the point where $x = 0$.
   b. i. Determine the equation of the tangent to the curve $y = e^{2x}$ at the point where $x = 0$.
      ii. Hence, determine the equation of the line perpendicular to this tangent.

6. Determine the equations of the tangent and the line perpendicular to the curve with equation $y = e^{-3x} + 4$ at the point where $x = 0$.

7. Determine the equations of the tangent and the line perpendicular to the curve $y = e^{-3x} - 2$ at the point where $x = 0$.

8. Determine the equations of the tangent and the line perpendicular to the curve $y = e^{\sqrt{x}} + 1$ at the point where $x = 3$.

9. Determine the equation of the tangent to the curve $y = e^{x^2+3x-4}$ at the point where $x = 1$.

Technology active

10. **MC** The equation of the tangent to the curve with equation $y = e^{-2x}$ at the point where $x = -\dfrac{1}{2}$ is:

   A. $y = -2ex + e$
   B. $y = 2ex$
   C. $y = 2ex + 1$
   D. $y = -2ex - x$
   E. $y = -2ex$

11. a. The graphs of the equations $y = 2e^{-2x} + 1$ and $y = x^3 - 3x$ are shown. Determine the coordinates of the point of intersection, giving your answer correct to 2 decimal places.
b. Calculate the gradient of the tangent to the cubic at this point.

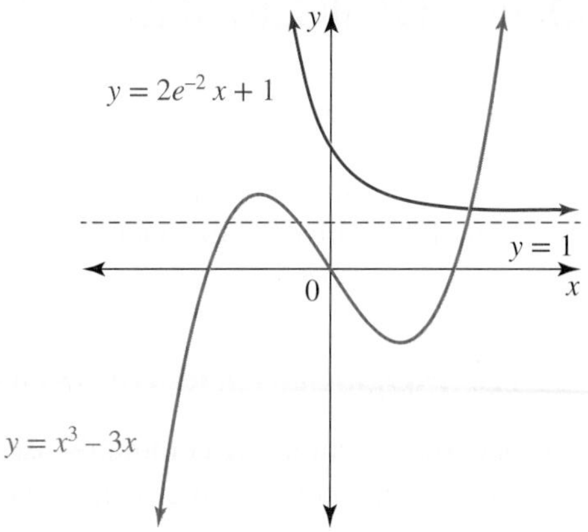

12. Determine the exact value for $f'(2)$ if $f(x) = 3^{2x-4}$.

13. Determine the derivative of the function $f(x) = e^{-2x+3} - 4$ and hence calculate:

a. $f'(-2)$ in exact form
b. $\{x : f'(x) = -2\}$.

14. Determine the derivative of the function $f(x) = \dfrac{e^{3x} + 2}{e^x}$ and hence calculate:

a. $f'(1)$ in exact form
b. $\{x : f'(x) = 0\}$.

15. The curve with the rule $A = A_0 e^{-0.69t}$ passes through the point (0, 2).

a. Determine the value of A_0.
b. Determine $\dfrac{dA}{dt}$ when $t = 0$.

16. For the function with the rule $f(x) = Ae^x + Be^{-3x}$, where A and B are non-zero real constants, determine $f'(x)$ and show that $f'(x) = 0$ when $e^{4x} = \dfrac{3B}{A}$.

5.3 Exam questions

Question 1 (1 mark) TECH-FREE

Source: *VCE 2021, Mathematical Methods Exam 1, Q1a; © VCAA.*

Differentiate $y = 2e^{-3x}$ with respect to x.

Question 2 (1 mark) TECH-ACTIVE

Source: *VCE 2013, Mathematical Methods (CAS) Exam 2, Section 1, Q11; © VCAA.*

MC If the tangent to the graph of $y = e^{ax}$, $a \neq 0$, at $x = c$ passes through the origin, then c is equal to

A. 0
B. $\dfrac{1}{a}$
C. 1
D. a
E. $-\dfrac{1}{a}$

Question 3 (2 marks) TECH-FREE

Determine the derivative of $y = \dfrac{e^{2x} + e^{-2x}}{e^x}$.

More exam questions are available online.

5.4 Applications of exponential functions

LEARNING INTENTION

At the end of this subtopic you should be able to:
- find derivatives of exponential functions and apply differentiation to real-world applications.

5.4.1 Applications of exponential functions

Exponential functions are commonly used to model a number of real-world applications, including Newton's Law of Cooling, population growth and decay, cell growth and decay, and radioactive decay.

A general equation to represent exponential growth and decay is given by

$$A = A_0 e^{kt}$$

where A_0 is the initial amount and k is a constant.

If the equation represents growth, then k is a positive value. If the equation represents decay, then k is a negative value.

WORKED EXAMPLE 8 Application of exponential functions

The number of bacterial cells on a culture plate, N, can be defined by the rule

$$N(t) = 2000e^{0.3t},\ t \geq 0$$

where t is the time in seconds the culture has been multiplying.

a. **Determine how many bacterial cells are initially present.**

b. **Determine how many bacterial cells, to the nearest whole number, are present after 10 seconds.**

c. **Determine the rate at which the population of bacteria is multiplying after 10 seconds. Give your answer correct to the nearest whole number.**

THINK	WRITE
a. 1. Initially $t = 0$, so substitute this value into the rule.	a. $N(0) = 2000e^{0.3(0)}$ $= 2000$
2. Write the answer.	Initially there are 2000 bacterial cells present.
b. 1. Substitute $t = 10$.	b. $N(10) = 2000e^{0.3(10)}$ $= 2000e^3$ $= 40\ 171$
2. Write the answer.	After 10 seconds there are 40 171 bacterial cells present.
c. 1. $\dfrac{dN}{dt}$ represents the required rate.	c. $\dfrac{dN}{dt} = 600e^{0.3t}$
2. Substitute $t = 10$.	$\dfrac{dN}{dt} = 600e^{0.3(10)}$ $= 600e^3$ $= 12\ 051$
3. Write the answer with the correct units.	After 10 seconds the bacterial cells are growing at a rate of 12 051 per second.

5.4 Exercise

Technology free

1. The graph shown is that of the function $f: R \to R, f(x) = e^{2x} + qe^{x} + 3$, where q is a constant.

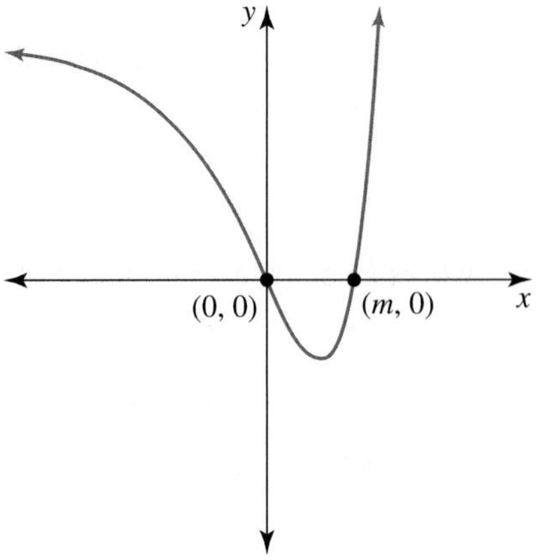

   a. Determine the value of q.
   b. Calculate the exact value of m, where m is a constant and $(m, 0)$ are the coordinates of the point where the function intersects the x-axis.
   c. Determine the derivative function, $f'(x)$.
   d. Calculate the gradient of the curve where it intersects the y-axis.

2. The graph shown is that of the function $f: R \to R, f(x) = e^{-2x} + ze^{-x} + 2$, where z is a constant.

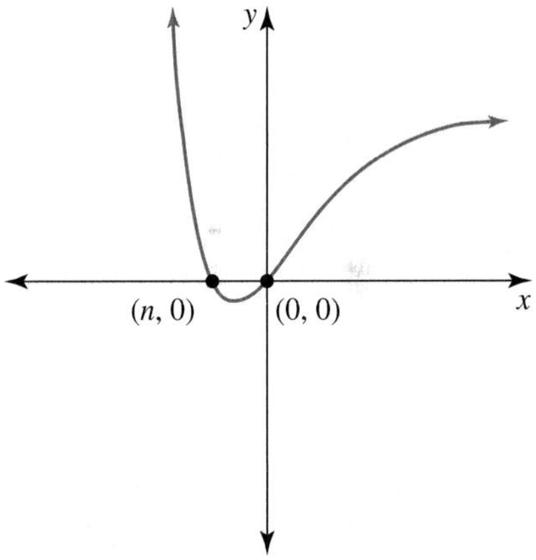

   a. Determine the value of z.
   b. Calculate the exact value of n, where n is a constant and $(n, 0)$ are the coordinates of the point where the graph intersects the x-axis.
   c. Determine the derivative function, $f'(x)$.
   d. Calculate the gradient of the curve where it passes through the origin.

Technology active

3. WE8 The mass, M grams, of a radioactive substance is initially 20 grams; 30 years later its mass is 19.4 grams. If the mass in any year is given by

$$M = M_0 e^{-0.00152t}$$

where t is the time in years and M_0 is a constant, determine:

   a. the value of M_0
   b. the annual rate of decay
   c. the rate of decay after 30 years.

4. The intensity of light decreases as it passes through water. The phenomenon can be modelled by the equation

$$I = I_0 e^{-0.0022d}$$

where I_0 is the intensity of light at the surface of the water and I is the intensity of light at a depth of d metres below the surface of the water.
   a. Calculate the intensity of light at a depth of 315 metres.
   b. Determine the rate at which the intensity of light is decreasing at 315 metres.

5. An unstable gas decomposes in such a way that the amount present, A units, at time t minutes is given by the equation

$$A = A_0 e^{-kt}$$

where k and A_0 are constants. It was known that initially there were 120 units of unstable gas.
   a. Determine the value of A_0.
   b. Show that $\frac{dA}{dt}$ is proportional to A.
   c. After 2 minutes there were 90 units of the gas left. Calculate the value of k.
   d. Determine the rate at which the gas is decomposing when $t = 5$. Give your answer correct to 3 decimal places.
   e. Explain whether there will ever be no gas left.

6. The bilby is an endangered species that can be found in the Kimberley in Western Australia as well as some parts of South Australia, the Northern Territory and Queensland. The gestation time for a bilby is 2 to 3 weeks and when they are born, they are only about 11 mm in length. The growth of a typical bilby can be modelled by the rule

$$L = L_0 e^{0.599t}$$

where L_0 is its length in millimetres at birth and L is the length of the bilby in millimetres t months after its birth.
   a. Determine the value of L_0.
   b. Determine the rate of change of length of the bilby at time t months.
   c. Determine the rate at which the bilby is growing when it is 3 months old. Give your answer correct to 3 decimal places.

7. The population of Australia since 1950 can be modelled by the rule

$$P = P_0 e^{0.016t}$$

where P_0 is the population in millions at the beginning of 1950 and P is the population in millions t years after 1950. It is known that there were 8.2 million people in Australia at the beginning of 1950.

a. Calculate the population in millions at the beginning of 2015, correct to 1 decimal place.
b. Determine the year and month in which the population reaches 20 million.
c. Determine the rate of change of population at the turn of the century, namely the year 2000, correct to 2 decimal places.
d. Determine the year in which the rate of increase of the population is predicted to exceed 400 000 people per year.

8. An entrance to a local suburban park has a series of posts connected with heavy chains as shown.

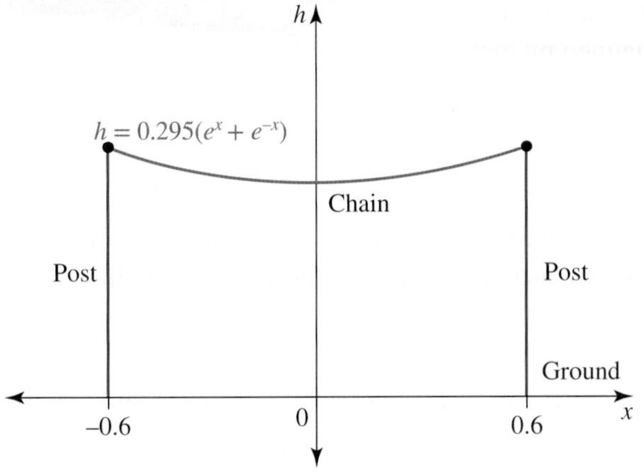

The chain between any two posts can be modelled by the curve defined by

$$h = 0.295(e^{x} + e^{-x}), -0.6 \leq x \leq 0.6$$

where h metres is the height of the chain above the ground and x metres is the horizontal distance from the midpoint of the posts. The x-axis represents the ground. The posts are positioned at $x = -0.6$ and $x = 0.6$.

a. Calculate the amount of sag in the chain (i.e. the difference in height between the highest points of the chain and the lowest point of the chain). Give your answer in centimetres.
b. Calculate the angle the chain makes with the post positioned on the right-hand side of the structure (i.e. at $x = 0.6$). Give your answer correct to 1 decimal place.

9. The graph of the function $f: R \to R$, $f(x) = e^{-x} - 0.5e^{-2x}$ is shown.

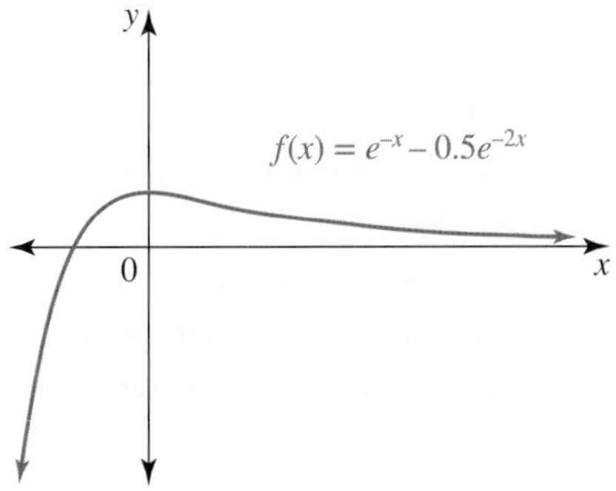

a. Calculate the coordinates of the point at which the graph crosses the y-axis.
b. Determine $f'(x)$.
c. Calculate the coordinates of the point at which the gradient is equal to zero.
d. Calculate the angle, correct to 1 decimal place, that the graph makes with the positive direction of the x-axis if it is known that the graph cuts the x-axis at $\left(-\log_e(2), 0\right)$.
e. Determine the equation of the tangent to the curve when $x = 1$.
f. Determine the equation of the line perpendicular to the curve when $x = 1$.

10. Consider the curve with equation $y = \dfrac{x^2 - 5}{2e^{x^2}}$.

Using CAS:

a. calculate $\dfrac{dy}{dx}$
b. determine the exact coordinates of the points on the curve where the gradient is equal to zero
c. calculate the gradient of the tangent to the curve at $x = \dfrac{3}{2}$, giving your answer correct to 3 decimal places.

11. The cane toad, originally from South America, is an invasive species in Australia. Cane toads were introduced to Australia from Hawaii in June 1935 in an attempt to control cane beetles, though this proved to be ineffective.

In a controlled experiment at a particular waterhole, it was observed that at the beginning of the experiment there were an estimated 30 000 tadpoles (future cane toads) in the water. The number of tadpoles increased by about 60 000 a day over the period of a week. This growth pattern can be defined by the equation

$$T = T_0 e^{kt}$$

where T_0 is the initial number of cane toad tadpoles (in thousands) at the waterhole during the time of the experiment, T is the number of cane toad tadpoles (in thousands) at the waterhole t days into the experiment, and k is a constant.

a. Calculate the value of T_0.
b. Calculate how many cane toad tadpoles are in the waterhole after a week if it is known that $k = 0.387$. Give your answer to the nearest thousand.
c. Determine the rate at which the cane toad tadpole numbers are increasing after 3 days.

After a week, no more tadpoles could be supported by the habitat. In favourable conditions, tadpoles take about 2 weeks to develop into small cane toads, at which point they leave the water. Once the small cane toads emerged, meat ants were introduced into their environment. This caused 90% of the cane toads to be killed off over a period of a week. The growth and decline of the tadpoles/cane toads is shown.
The decline in the number of young cane toads can be defined by the equation

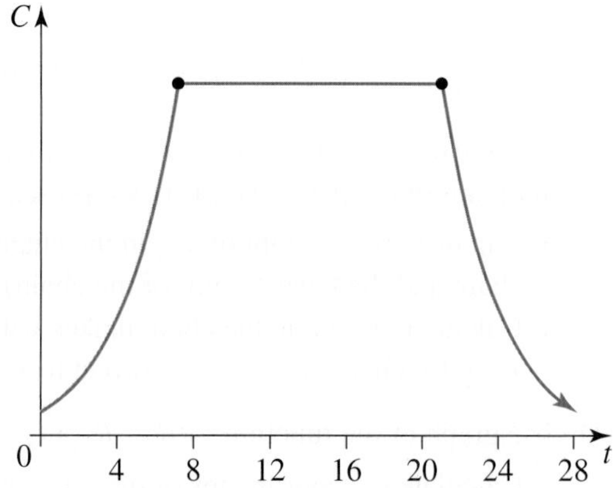

$$C = C_0 e^{mt}$$

where C_0 is the number of young cane toads (in thousands) just before the meat ants were introduced, C is the number of young cane toads (in thousands) t days after the meat ants were introduced, and m is a constant.

d. Determine the value of C_0.
e. Determine how many young cane toads still survived a week after the meat ants were introduced.
f. Determine m and the rate of decline in the number of cane toads after 4 days.

5.4 Exam questions

Question 1 (2 marks) TECH-ACTIVE

A body that is at a higher temperature than its surroundings cools according to Newton's Law of Cooling, which states that

$$T = T_0 e^{-zt}$$

where T_0 is the original *excess* of temperature, T is the excess of temperature in degrees centigrade after t minutes, and z is a constant.

a. The original temperature of the body was 95 °C and the temperature of the surroundings was 20 °C. Determine the value of T_0. **(1 mark)**

b. Determine the rate at which the temperature is decreasing after a quarter of an hour if it is known that $z = 0.034$. Give your answer correct to 3 decimal places. **(1 mark)**

Question 2 (4 marks) TECH-ACTIVE

The graph of $y = Ae^{-x^2}$, where A is a constant, is shown. Answer the following questions correct to 2 decimal places where appropriate.

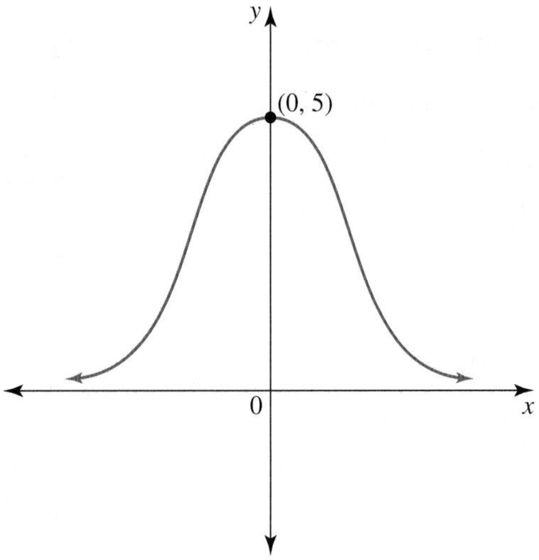

a. If the graph goes through (0, 5), determine the value of A. **(1 mark)**

b. Calculate $\frac{dy}{dx}$. **(1 mark)**

c. Determine the gradient of the tangent to the curve at the point where:

i. $x = -0.5$ **(1 mark)**

ii. $x = 1$ **(1 mark)**

Question 3 (3 marks) TECH-ACTIVE

The pressure of the atmosphere, P cm of mercury, decreases with the height, h km above sea level, according to the law

$$P = P_0 e^{-kh}$$

where P_0 is the pressure of the atmosphere at sea level and k is a constant. At 500 m above sea level, the pressure is 66.7 cm of mercury, and at 1500 m above sea level, the pressure is 52.3 cm of mercury.

a. Determine the values of P_0 and k, correct to 2 decimal places. **(2 marks)**

b. Determine the rate at which the pressure is falling when the height above sea level is 5 km. Give your answer correct to 2 decimal places. **(1 mark)**

More exam questions are available online.

5.5 Differentiation of trigonometric functions

LEARNING INTENTION

At the end of this subtopic you should be able to:

- find derivatives of trigonometric functions sine, cosine and tangent
- determine equations of the tangent to trigonometric functions.

5.5.1 The derivatives of trigonometric functions

The derivatives of sin(*x*) and cos(*x*)

The derivative of sin(*x*) can be determined as follows.

Consider $f: R \to R$, $f(x) = \sin(x)$, where *x* is an angle measurement in radians.

$$f'(x) = \lim_{h \to 0} \frac{f(x+h) - f(x)}{h}$$

$$f(x) = \sin(x)$$

$$f(x+h) = \sin(x+h)$$

$$\frac{f(x+h) - f(x)}{h} = \frac{\sin(x+h) - \sin(x)}{h}$$

$$f'(x) = \lim_{h \to 0} \frac{\sin(x+h) - \sin(x)}{h}$$

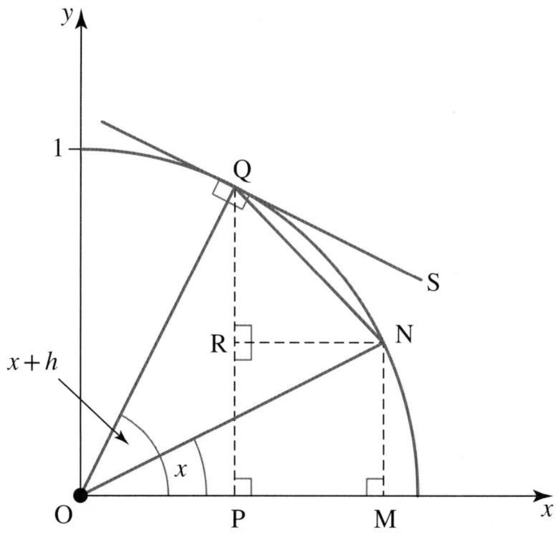

To evaluate this limit, we must look at the unit circle.

$$\angle \text{NOM} = x, \ \angle \text{QOM} = x + h$$

$$\angle \text{PQO} = \frac{\pi}{2} - (x + h)$$

$$\angle \text{RQS} = \frac{\pi}{2} - \left(\frac{\pi}{2} - (x + h)\right)$$

$$= x + h$$

By definition,

$$\sin(x) = MN$$

$$\sin(x+h) = PQ$$

$$\sin(x+h) - \sin(x) = PQ - MN = QR$$

$$\frac{\sin(x+h) - \sin(x)}{h} = \frac{QR}{h}$$

From the diagram, it can be seen that $\angle \text{RQS} = x + h$ and the arc QN has length *h*.
As $h \to 0$, $\angle$RQS approaches $\angle$RQN, which approaches *x*. Furthermore, the arc QN approaches the chord QN.

Consequently, $\frac{QR}{h} \to \frac{QR}{QN}$, but by definition, $\frac{QR}{QN} = \cos(x)$.

Hence,

$$\begin{aligned} f'(x) &= \lim_{h \to 0} \frac{\sin(x+h) - \sin(x)}{h} \\ &= \lim_{h \to 0} \frac{QR}{h} \\ &= \cos(x) \end{aligned}$$

Derivative of sin(x)

If $f(x) = \sin(x)$, then $f'(x) = \cos(x)$.

The following properties can be shown using the chain rule, which will be introduced in the next topic.

Derivatives of sin(kx) and sin (g(x))

If $f(x) = \sin(kx)$, then $f'(x) = k \cos(kx)$, where k is a constant

and

if $f(x) = \sin(g(x))$, then $f'(x) = g'(x) \cos(g(x))$.

The derivative of cos(x) can also be investigated geometrically, using the same method as shown for sin(x) and yielding the following result.

Derivative of cos(x)

If $f(x) = \cos(x)$, then $f'(x) = -\sin(x)$.

The following can also be shown using the chain rule.

Derivatives of cos(kx) and cos (g(x))

If $f(x) = \cos(kx)$, then $f'(x) = -k \sin(kx)$, where k is a constant

and

if $f(x) = \cos(g(x))$, then $f'(x) = -g'(x) \sin(g(x))$.

The derivative of tan(x)

Consider the function $f: R \to R,\ f(x) = \tan(x)$.

Derivative of tan(x)

If $f(x) = \tan(x)$, then $f'(x) = \dfrac{1}{\cos^2(x)} = \sec^2(x)$.

In order to prove this differentiation, we would use the trigonometric identity $\tan(x) = \dfrac{\sin(x)}{\cos(x)}$ in conjunction with the quotient rule, which will also be introduced in the next topic.

The following properties can be shown using the chain rule.

Derivatives of tan(*kx*) and tan(*g*(*x*))

If $f(x) = \tan(kx)$, then $f'(x) = \dfrac{k}{\cos^2(kx)} = k\sec^2(kx)$, where k is a constant

and

if $f(x) = \tan(g(x))$, then $f'(x) = \dfrac{g'(x)}{\cos^2(g(x))} = g'(x)\sec^2(g(x))$.

Remember that these rules only apply if the angle x is measured in radians.

WORKED EXAMPLE 9 Finding derivatives of trigonometrical functions and combinations

Differentiate each of the following functions with respect to x.

a. $\sin(8x) + x^4$

b. $\tan(5x) + 2\cos(x^2)$

c. $\dfrac{1 - \sin^2(x)}{\cos(x)}$

d. $\sin(6x^\circ)$

THINK	WRITE
a. Apply the rule $\dfrac{d}{dx}(\sin(kx)) = k\cos(kx)$ and differentiate each term separately.	a. $y = \sin(8x) + x^4$ $\dfrac{dy}{dx} = 8\cos(8x) + 4x^3$
b. Apply the rules $\dfrac{d}{dx}(\cos(g(x))) = -g'(x)\sin(g(x))$ and $\dfrac{d}{dx}(\tan(kx)) = \dfrac{k}{\cos^2(x)}$.	b. $y = \tan(5x) + 2\cos(x^2)$ $\dfrac{dy}{dx} = \dfrac{5}{\cos^2(5x)} - 2 \times 2x\sin(x^2)$ $= \dfrac{5}{\cos^2(5x)} - 4x\sin(x^2)$
c. 1. Remember the trigonometric identity $\sin^2(\theta) + \cos^2(\theta) = 1$. Use this to simplify the equation.	c. $y = \dfrac{1 - \sin^2(x)}{\cos(x)}$ $= \dfrac{\cos^2(x)}{\cos(x)}$ $= \cos(x), \cos(x) \neq 0$
2. Differentiate the simplified function.	$\dfrac{dy}{dx} = -\sin(x)$

d. 1. The function sin($6x°$) cannot be differentiated as the angle is not measured in radians. Convert the angle to radian measures by multiplying by $\frac{\pi}{180}$, as $1° = \frac{\pi}{180}^c$.

d. $\sin(6x°) = \sin\left(6 \times \frac{\pi}{180} x\right)$

$= \sin\left(\frac{\pi x}{30}\right)$

2. Differentiate the resultant function by applying the rule $\frac{d}{dx}(\sin(kx)) = k\cos(kx)$.

$y = \sin\left(\frac{\pi x}{30}\right)$

$\frac{dy}{dx} = \frac{\pi}{30}\cos\left(\frac{\pi x}{30}\right)$

WORKED EXAMPLE 10 Finding the equation of tangent of a trigonometrical function

Determine the equation of the tangent to the curve $y = \sin(3x) + 1$ at the point where $x = \frac{\pi}{3}$.

THINK

WRITE

1. First find the coordinates of the point; that is, determine the y-value when $x = \frac{\pi}{3}$.

When $x = \frac{\pi}{3}$,

$y = \sin\left(3 \times \frac{\pi}{3}\right) + 1$

$= \sin(\pi) + 1$

$= 0 + 1$

$= 1$

The point is $\left(\frac{\pi}{3}, 1\right)$.

2. Find the derivative of the function.

$\frac{dy}{dx} = 3\cos(3x)$

3. Determine the gradient at the point where $x = \frac{\pi}{3}$.

$x = \frac{\pi}{3}, \frac{dy}{dx} = 3\cos\left(3 \times \frac{\pi}{3}\right)$

$= 3\cos(\pi)$

$= 3(-1)$

$= -3$

4. Substitute the appropriate values into the rule $y - y_1 = m(x - x_1)$ to find the equation of the tangent.

$m = -3, (x_1, y_1) = \left(\frac{\pi}{3}, 1\right)$

$y - y_1 = m(x - x_1)$

$y - 1 = -3\left(x - \frac{\pi}{3}\right)$

$y - 1 = -3x + \pi$

$y = -3x + \pi + 1$

The equation of the tangent is $y = 1 + \pi - 3x$.

5.5 Exercise

Students, these questions are even better in jacPLUS

Receive immediate feedback and access sample responses

Access additional questions

Track your results and progress

Find all this and MORE in jacPLUS

Technology free

1. **WE9** Differentiate each of the following functions with respect to x.

a. $5x + 3\cos(x) + 5\sin(x)$

b. $\sin(3x + 2) - \cos(3x^2)$

c. $\frac{1}{3}\sin(9x)$

d. $5\tan(2x) - 2x^5$

e. $8\tan\left(\frac{x}{4}\right)$

f. $\tan(9x°)$

2. For each of the following functions, determine $\frac{dy}{dx}$.

a. $y = 2\cos(3x)$

b. $y = 3\cos\left(\frac{\pi}{2} - x\right)$

c. $y = -4\sin\left(\frac{x}{3}\right)$

d. $y = 2\sin\left(\frac{\pi}{2} + 3x\right)$

e. $y = -\frac{1}{2}\tan(5x^2)$

f. $y = \tan(20x)$

3. Simplify and then differentiate $\frac{\sin(x)\cos^2(2x) - \sin(x)}{\sin(x)\sin(2x)}$ with respect to x.

4. **WE10** Determine the equation of the tangent to the curve $y = -\cos(x)$ at the point where $x = \frac{\pi}{2}$.

5. Determine the equation to the tangent to the curve $y = \tan(2x)$ at the point where $x = -\frac{\pi}{8}$.

6. Determine the equation of the tangent to the curve with equation $y = 3\cos(x)$ at the point where $x = \frac{\pi}{6}$.

7. Determine the equation of the tangent to the curve with equation $y = 2\tan(x)$ at the point where $x = \frac{\pi}{4}$.

8. Determine the equations of the tangent and the line perpendicular to each of the following graphs at the points indicated.

a. $y = \sin(3x)$ at $\left(\frac{2\pi}{3}, 0\right)$

b. $y = \cos\left(\frac{x}{2}\right)$ at $(\pi, 0)$

9. a. Use both or either of the trigonometric identities $\sin^2(\theta) + \cos^2(\theta) = 1$ and $\tan(\theta) = \frac{\sin(\theta)}{\cos(\theta)}$ to simplify $\frac{\sin(x)\cos(x) + \sin^2(x)}{\sin(x)\cos(x) + \cos^2(x)}$.

b. Hence, calculate $\frac{d}{dx}\left(\frac{\sin(x)\cos(x) + \sin^2(x)}{\sin(x)\cos(x) + \cos^2(x)}\right)$.

Technology active

10. **MC** A point on the curve with equation $y = -2\sin\left(\frac{x}{2}\right)$, $x \in [0, 2\pi]$ where the gradient is equal to $\frac{1}{2}$ is:

A. $\left(\frac{4\pi}{3}, -\sqrt{3}\right)$ B. $\left(\frac{\pi}{3}, -\sqrt{3}\right)$ C. $\left(\frac{4\pi}{3}, \frac{1}{2}\right)$ D. $\left(\frac{4\pi}{3}, -\frac{1}{2}\right)$ E. $\left(\frac{\pi}{3}, \sqrt{3}\right)$

11. Consider the function $f: [0, 2\pi] \to R$, $f(x) = \sin(x) - \cos(x)$. Determine:

a. $f(0)$ b. $\{x : f(x) = 0\}$ c. $f'(x)$ d. $\{x : f'(x) = 0\}$.

12. Consider the function $f: [-\pi, \pi] \to R$, $f(x) = \sqrt{3}\cos(x) + \sin(x)$. Determine:

a. $f(0)$ b. $\{x : f(x) = 0\}$ c. $f'(x)$ d. $\{x : f'(x) = 0\}$.

13. For the function $f(x) = 2x + \cos(3x)$, $0 \le x \le \frac{\pi}{2}$, determine the point(s) where the gradient is zero. Give your answer correct to 3 decimal places.

14. For the function $f(x) = x - \sin(2x)$, $-\frac{\pi}{2} \le x \le \frac{\pi}{2}$, determine the point(s) where the gradient is zero. Give your answer correct to 3 decimal places.

15. Determine the angle that the curve with equation $y = \sin(2x)$ makes with the positive direction of the x-axis the first time it intersects the x-axis when $x > 0$. Give your answer correct to 1 decimal place.

16. Determine the x-values over the domain $x \in [-\pi, \pi]$ for which the gradients of the functions $f(x) = \sin(2x)$ and $f(x) = \cos(2x)$ are equal.

5.5 Exam questions

Question 1 (1 mark) TECH-ACTIVE

MC Consider the following function.

$$f(x) = \begin{cases} \sin(x) & x \ge 0 \\ 0 & x < 0 \end{cases}$$

Select the true statement from the following.

A. The function is continuous at $x = 0$ and differentiable at $x = 0$.
B. The function is continuous at $x = 0$ and not differentiable at $x = 0$.
C. The function is not continuous at $x = 0$ and differentiable at $x = 0$.
D. The function is not continuous at $x = 0$ and not differentiable at $x = 0$.
E. The $\lim_{x \to 0} f(x)$ exists and $f(0) = 0$. Therefore the function is differentiable at $x = 0$.

Question 2 (1 mark)

MC The derivative of $h(\cos(2x))$ is

A. $2h'(\cos(2x))\cos(2x)$
B. $h'(\cos(2x))\cos(2x)$
C. $-2h'(\cos(2x))(\cos(2x))$
D. $-2h'(\cos(2x))(\sin(2x))$
E. $2h'(\cos(2x))(\sin(2x))$

Question 3 (1 mark)

MC If $f(x) = e^{\sin(2x)}$, then $f'\left(\frac{\pi}{2}\right)$ is equal to

A. $-\frac{1}{e}$ B. 0 C. -1 D. -2 E. 2

More exam questions are available online.

5.6 Applications of trigonometric functions

LEARNING INTENTION

At the end of this subtopic you should be able to:
- apply differentiation of trigonometric functions to real-world problems.

5.6.1 Problem solving with trigonometric functions

Trigonometric functions can be used to model geometric scenarios in real-life situations where cyclic (or periodic) phenomena are being investigated.

WORKED EXAMPLE 11 Geometric application of trigonometric functions

The circle shown has a radius of 1 unit.

a. Show that the area of the triangle OQP is equal to $A = \frac{1}{2}\cos(\theta)$, where $\angle QOX = \theta$ and θ is in radian measure.

b. Evaluate $\frac{dA}{d\theta}$ when $\theta = \frac{\pi}{6}$.

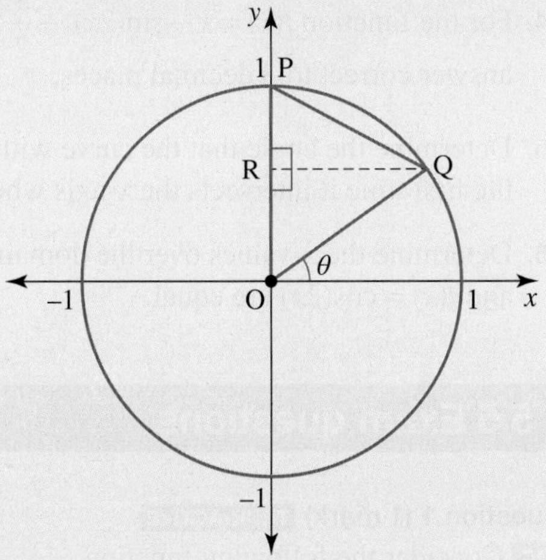

THINK	WRITE
a. 1. Determine the magnitude of $\angle RQO$.	a. As RQ is parallel to the x-axis, $\angle RQO = \theta$ because it is alternate to $\angle QOX$.
2. Determine the length of RQ and apply the formula for the area of a triangle, $A = \frac{1}{2} \times \text{base} \times \text{height}$.	$\cos(\theta) = \frac{RQ}{OQ} = \frac{RQ}{1}$ $\cos(\theta) = RQ$ and $OP = 1$ $\text{Area} = \frac{1}{2} \times OP \times RQ = \frac{1}{2} \times 1 \times \cos(\theta) = \frac{1}{2}\cos(\theta)$ (as required)
b. 1. Evaluate $\frac{dA}{d\theta}$.	b. $\frac{dA}{d\theta} = -\frac{1}{2}\sin(\theta)$

2. Substitute $\theta = \frac{\pi}{6}$.

$$\theta = \frac{\pi}{6}, \frac{dA}{d\theta} = -\frac{1}{2}\sin\left(\frac{\pi}{6}\right)$$
$$= -\frac{1}{2} \times \frac{1}{2}$$
$$= -\frac{1}{4}$$

The previous example involved a geometric application question, but everyday application questions can also be solved using trigonometric functions, as shown in the example below.

WORKED EXAMPLE 12 Problem solving using a trigonometric function

The temperature on a particular day can be modelled by the function

$$T(t) = -3\cos\left(\frac{\pi t}{9}\right) + 18,\ 0 \le t \le 18$$

where t is the time in hours after 5:00 am and T is the temperature in degrees Celsius.
For the remaining 6 hours of the 24-hour period, the temperature remains constant.

a. **Calculate the temperature at 8:00 am.**
b. **Determine the time(s) of the day at which the temperature is 20 °C. Give your answer correct to the nearest minute.**
c. **Determine $\frac{dT}{dt}$.**
d. **Determine the rate of change of temperature at the time(s) found in part b, correct to 2 decimal places.**

THINK	WRITE
a. At 8:00 am, $t = 3$. Substitute this value into the equation.	a. $T(3) = -3\cos\left(\frac{3\pi}{9}\right) + 18$ $= -3\cos\left(\frac{\pi}{3}\right) + 18$ $= -3 \times \frac{1}{2} + 18$ $= -1.5 + 18$ $= 16.5$ °C
b. 1. Substitute $T = 20$ into the equation.	b. $20 = -3\cos\left(\frac{\pi t}{9}\right) + 18$
2. Solve the equation for $0 \le t \le 18$ using CAS.	$20 = -3\cos\left(\frac{\pi t}{9}\right) + 18$ $t = 6.6,\ 11.4$
3. Interpret your answers and convert the t values to times of the day.	$t = 6.6 \Rightarrow$ 11:36 am $t = 11.4 \Rightarrow$ 4:24 pm

4. Write the answer.

The temperature is 20 °C at 11:36 am and 4:24 pm.

c. Determine $\frac{dT}{dt}$.

c. $\frac{dT}{dt} = 3 \times \frac{\pi}{9} \sin\left(\frac{\pi t}{9}\right)$

$= \frac{\pi}{3} \sin\left(\frac{\pi t}{9}\right)$

d. 1. Substitute $t = 6.6$ (11:36 am) and $t = 11.4$ (4:24 pm) into $\frac{dT}{dt}$.

d. When $t = 6.6$ (11:36 am),

$\frac{dT}{dt} = \frac{\pi}{3} \sin\left(\frac{6.6 \times \pi}{9}\right)$

$= 0.78$

When $t = 11.4$ (4:24 pm),

$\frac{dT}{dt} = \frac{\pi}{3} \sin\left(\frac{11.4 \times \pi}{9}\right)$

$= -0.78$

2. Write the answer.

At 11:36 am, the temperature is increasing at a rate of 0.78 °C per hour.
At 4:24 pm, the temperature is decreasing at a rate of 0.78 °C per hour.

5.6 Exercise

Technology free

1. WE11 Consider the triangle shown.

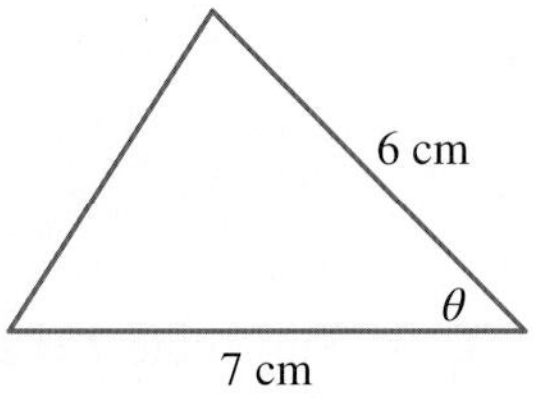

   a. Show that the area, A cm^2, is given by $A = 21 \sin(\theta)$.

   b. Determine $\frac{dA}{d\theta}$ when $\theta = \frac{\pi}{3}$.

2. A mass oscillates up and down at the end of a metal spring. The length of the spring, L cm after time t seconds, is modelled by the function $L(t) = 2\ \sin(\pi t) + 10$ for $t \geq 0$.

   a. Determine the length of the spring when the mass is not oscillating, that is, when it is at the mean position, P.

   b. Determine $\frac{dL}{dt}$.

   c. Evaluate the exact value of $\frac{dL}{dt}$ after 1 second.

3. **WE12** Between 6:00 am and 6:00 pm on a given day the height, H metres, of the tide in a harbour is given by

$$H(t) = 1.5 + 0.5 \sin\left(\frac{\pi t}{6}\right),\ 0 \le t \le 12.$$

a. State the period of the function.
b. Determine the value of H at low tide and the time when low tide occurs.
c. Determine $\frac{dH}{dt}$.
d. Determine the exact value of $\frac{dH}{dt}$ at 7:30 am.
e. Determine the second time during the given time interval that $\frac{dH}{dt}$ equals the value found in part **d**.

Technology active

4. Given that $f: \left[0, \frac{\pi}{2}\right] \to R,\ f(x) = 2\sin(4x) + 1$, determine:
   a. the values of x for which $f(x) = 0.5$, giving your answer correct to 3 decimal places
   b. the coordinates where the gradient of the function is zero
   c. the value of $f'(x)$ when $x = \frac{\pi}{4}$
   d. the interval over which the gradient is positive.

5. The diagram shows a garden bed bordered by wooden sleepers. BDC is a triangular herb garden and ABDE is a rectangular garden for vegetables.
   a. Determine BD and CD in terms of a and θ, where a is a constant, θ is $\angle BCD$ as shown and $0 < \theta < \frac{\pi}{2}$.
   b. Calculate the total length, L metres, of sleepers required to surround the garden bed. (This should include BD as well as the sleepers defining the perimeter.)
   c. Determine $\frac{dL}{d\theta}$ in terms of θ and a.
   d. Let $a = 2$ and use CAS to sketch $\frac{dL}{d\theta}$ for $0 < \theta < \frac{\pi}{2}$. Hence, determine when $\frac{dL}{d\theta} = 0$, correct to 1 decimal place.

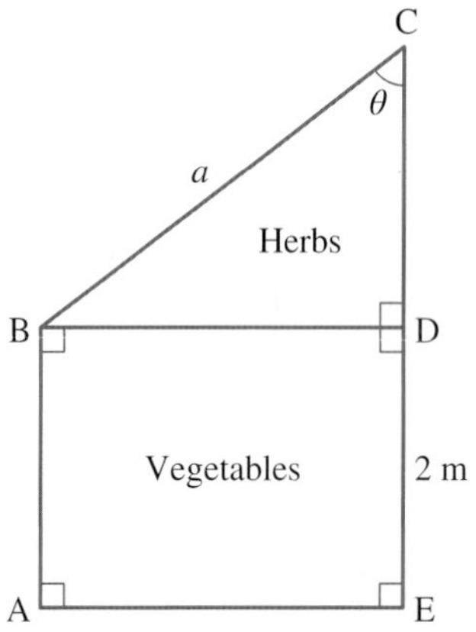

6. A wire frame is shaped as shown in the diagram. The diagonals shown are 100 cm long, and each diagonal makes an angle of θ with the horizontal.
   a. Show that the length of wire required to form the shape is given by

   $$L = 300\cos(\theta) + 400\sin(\theta) + 200,\ 0 \le \theta \le \frac{\pi}{2}$$

   where L is the total length of wire in centimetres and θ is the angle shown in radians.
   b. Determine $\frac{dL}{d\theta}$.
   c. Use CAS to sketch the graph of L. Determine the maximum length of the wire required and the value of θ, correct to 2 decimal places, for which this occurs.

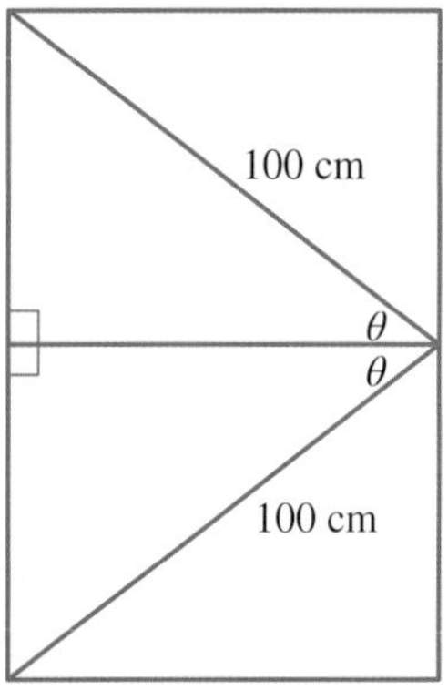

7. The figure shows a circular running track with centre O. The track has a radius of 200 metres. An athlete at a morning training session completes an obstacle course from N to P at a rate of 2 m/s and then a series of hurdles from P to M along the running track at a rate of 5 m/s.

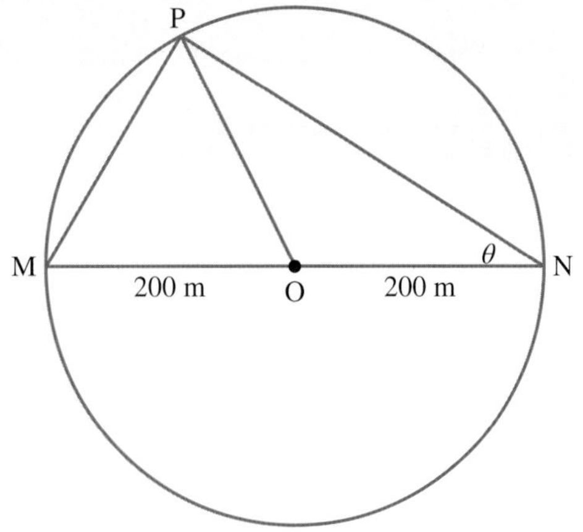

a. If $\angle MNP = \theta$ radians and the total time taken to complete the total course is T seconds, show that

$$T = 40\left(5\cos(\theta) + 2\theta\right),\ 0 < \theta \le \frac{\pi}{2}$$

b. Determine the value of θ when $\frac{dT}{d\theta} = 0$.

c. Determine the maximum time taken to complete the whole course. Give your answer in minutes and seconds.

8. A section of a water slide at a local aquatic complex is shown.

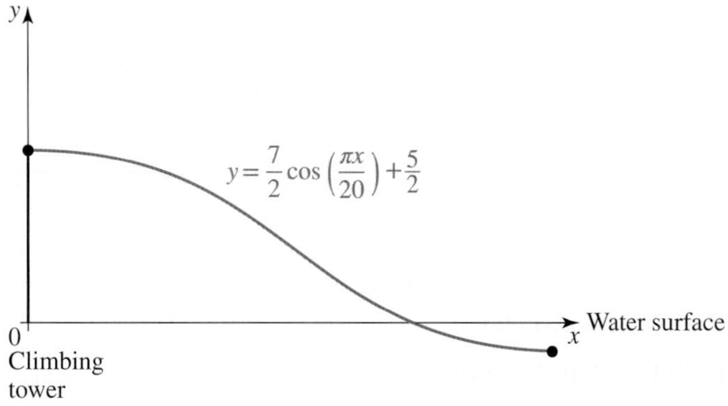

The water slide can be defined by the rule

$$y = \frac{7}{2}\cos\left(\frac{\pi x}{20}\right) + \frac{5}{2},\ 0 \le x \le 20$$

where y is the height in metres of the water slide above the water surface and x is the horizontal distance in metres between the start of the slide and the end of the slide. (*Note:* The x-axis represents the water surface.)

a. Determine how high a person must climb in order to reach the top of the water slide.

b. Determine $\frac{dy}{dx}$.

c. Calculate the exact gradient of the water slide:

i. when $x = 5$ ii. when $x = 10$.

d. i. Determine how far, to the nearest whole metre, from the climbing tower does the slide come into contact with the water surface.

ii. Determine the angle that the slide makes with the water surface at this point. Give your answer correct to 2 decimal places.

9. The diagram represents the cross-section of a waterfall feature in an Australian native garden. It consists of an undulating surface of corrugated plastic with vertical posts at each end. The relationship that defines this surface can be expressed by

$$h(x) = 10\cos\left(\frac{7x}{2}\right) - 5x + 90,\ 0 \leq x \leq 4.5$$

where h centimetres represents the vertical height of the water feature and x metres is the horizontal distance from the left-hand post. The posts supporting the undulating surface over which the water falls are situated at the points $x = 0$ and $x = 4.5$, as shown.

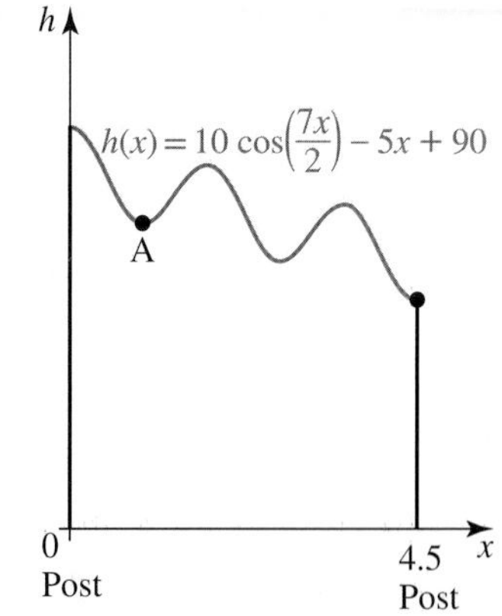

a. Determine the coordinates of the end points of the undulating surface. Give your answers correct to the nearest centimetre.

b. Determine the coordinates of point A, the first point in the interval [0, 4.5] where the gradient of the undulating surface is zero. Give your answer correct to 2 decimal places.

c. Determine the slope of the undulating surface at $x = 0.4$. Give your answer correct to 1 decimal place.

10. At a skateboard park, a new skateboard ramp has been constructed. A cross-section of the ramp is shown. The equation that approximately defines this curve is given by

$$h(x) = 2.5 - 2.5\cos\left(\frac{x}{4}\right),\ -5 \leq x \leq 5$$

where h is the height in metres above the ground level and x is the horizontal distance in metres from the lowest point of the ramp to each end of the ramp.

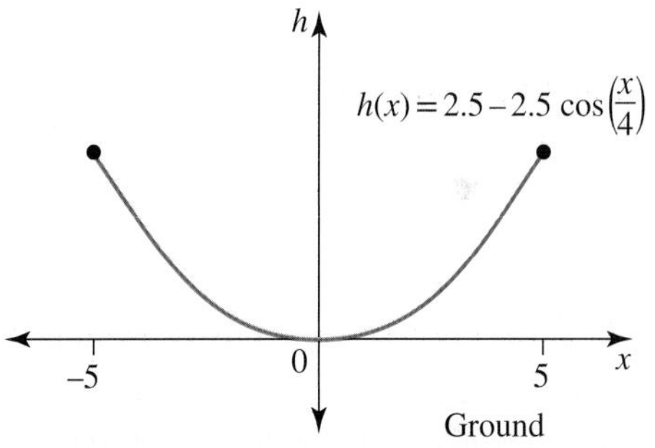

a. Determine the maximum depth of the skateboard ramp, giving your answer correct to 1 decimal place.

b. Calculate the gradient of the ramp, $\frac{dh}{dx}$.

c. Determine $\frac{dh}{dx}$ when $x = 3$, giving your answer correct to 3 decimal places.

d. Determine where $\frac{dh}{dx} = 0.58$, giving your answer correct to 3 decimal places.

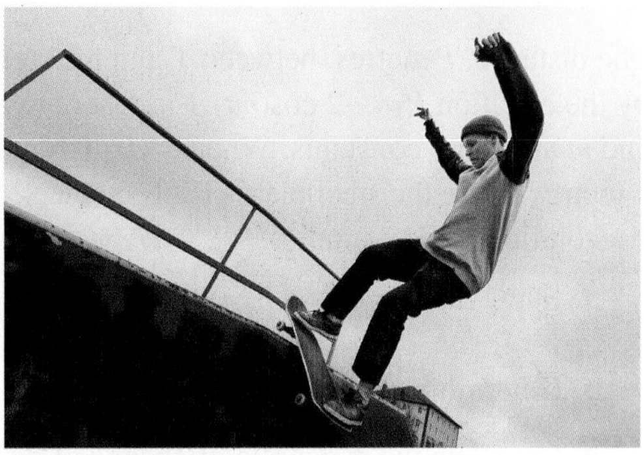

5.6 Exam questions

Question 1 (4 marks) TECH-ACTIVE

The triangle XYZ is inscribed by a circle with radius, r cm. The actual placement of the triangle is dependent on the size of the angle XZY, θ radians, and the length of ZM, where M is the midpoint of XY.

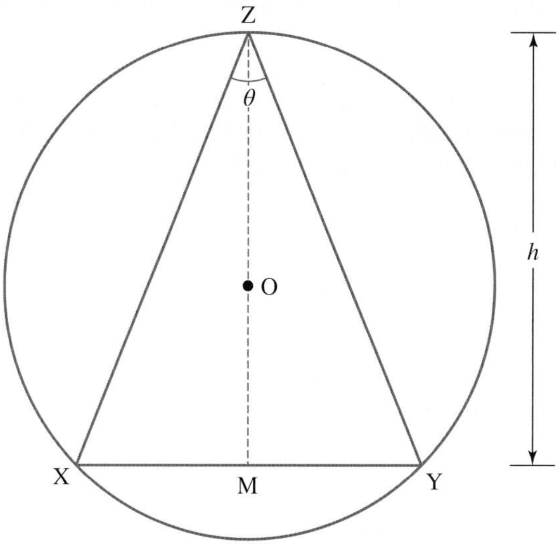

a. Show that $\angle XOM = \theta$. **(1 mark)**

b. Show that the relationship between θ, r and h, where $h = d\left(\overline{ZM}\right)$, is given by $\frac{h}{r} = \cos(\theta) + 1$. **(1 mark)**

c. If the radius of the circle is 3 cm, determine $\frac{dh}{d\theta}$. **(1 mark)**

d. Determine the exact value of $\frac{dh}{d\theta}$ when $\theta = \frac{\pi}{6}$. **(1 mark)**

Question 2 (4 marks) TECH-ACTIVE

A mechanism for crushing rock is shown. Rocks are placed on a steel platform, S, and a device raises and lowers a heavy mallet, H. The wheel, W, rotates, causing the upper block, B, to move up and down. The other wheel, V, is attached to the block, B, and rotates independently, causing the mallet to move up and down. T is the top of the block B.

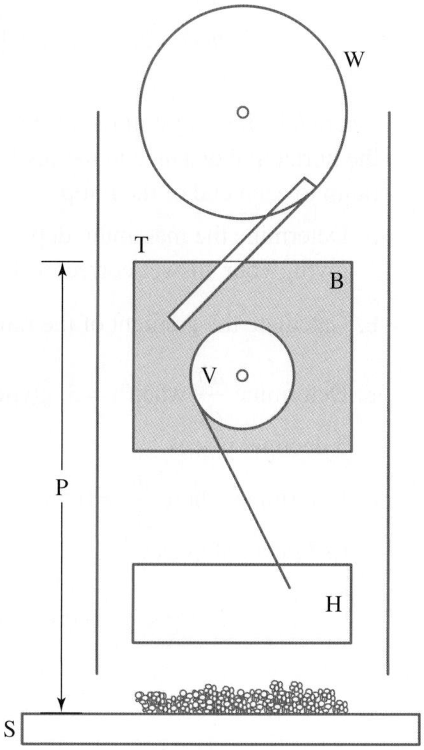

The distance, P metres, between T and the steel platform, is modelled by the equation $P = -2\cos(mt) + n$, where t is the time in minutes and m and n are constants. When $t = 0$, T is at its lowest point, 4 metres above the steel platform. The wheel, W, rotates at a rate of 1 revolution per 1.5 minutes.

a. Show that $n = 6$ and $m = \frac{4\pi}{3}$. **(2 marks)**

b. Determine $\frac{dP}{dt}$. **(1 mark)**

c. Calculate the exact rate of change of distance when $t = 0.375$ minutes. **(1 mark)**

Question 3 (7 marks) TECH-ACTIVE

An industrial process is known to cause the production of two separate gases that are released into the atmosphere. At a factory where this industrial process occurs, the technicians work a 12-hour day from 6:00 am until 6:00 pm.

The emission of the gas X can be modelled by the rule

$$x(t) = 1.5 \sin\left(\frac{\pi t}{3}\right) + 1.5,\ 0 \le t \le 12$$

and the emission of the gas Y can be modelled by the rule

$$y(t) = 2.0 - 2.0 \cos\left(\frac{\pi t}{3}\right),\ 0 \le t \le 12.$$

The graphs of these two functions are shown.

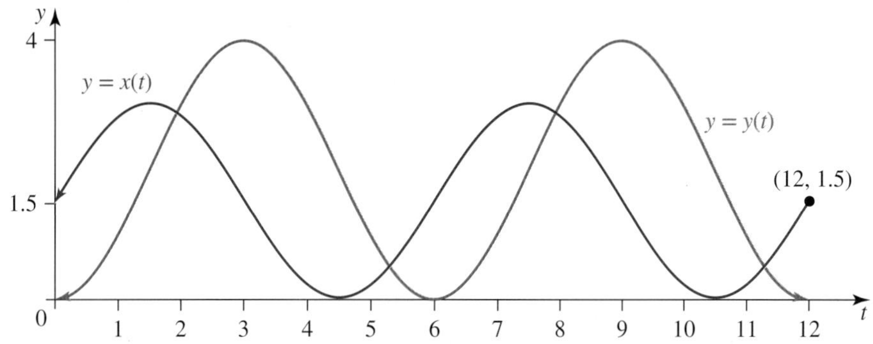

a. Determine the time of the day at which the emissions are the same for the first time and how many units of each gas are emitted at that time. Give your answer correct to 2 decimal places, and remember to note whether the time is am or pm. **(2 marks)**

b. The Environment Protection Authority (EPA) has strict rules about the emissions of gases. The total emission of gases for this particular industrial process is given by

$$T(t) = x(t) + y(t)$$

i. Sketch the graph of the function $T(t)$. **(2 marks)**

ii. Determine the maximum and minimum emissions in a 12-hour working day and the times at which these occur. **(2 marks)**

c. If the EPA rules state that all emissions from any one company must lie within the range 0 to 7 units at any one time, indicate whether this company works within the guidelines. **(1 mark)**

More exam questions are available online.

5.7 Differentiation and application of logarithmic functions

LEARNING INTENTION

At the end of this subtopic you should be able to:

- differentiate logarithmic functions
- determine the equation of the tangent to a logarithmic function.

5.7.1 The derivative of $f(x) = \log_e(x)$

The proof for the derivative of $y = \log_e(x)$ relies heavily on its link to its inverse function, $y = e^x$.

If $y = \log_e(x)$, then, by the definition of a logarithm, $e^y = x$.

If $x = e^y$, then applying the exponential derivative rule gives $\frac{dx}{dy} = e^y$.

However, it is known that

$$\frac{dy}{dx} = \frac{1}{\frac{dx}{dy}}$$

$$\text{so } \frac{dy}{dx} = \frac{1}{e^y}.$$

$$\text{But } e^y = x$$

$$\text{so therefore, } \frac{dy}{dx} = \frac{1}{x}.$$

In summary, $\frac{d}{dx}\left(\log_e(x)\right) = \frac{1}{x}$.

It is worth noting that $\frac{d}{dx}\left(\log_e(kx)\right) = \frac{1}{x}$ also.

This can be shown by applying the chain rule, which will be introduced in the next topic.

It can also be shown that

$$\frac{d}{dx}\left(\log_e(g(x))\right) = \frac{g'(x)}{g(x)}.$$

In summary, the following rules apply for derivatives of logarithmic functions.

Derivatives of logarithmic functions

$$\frac{d}{dx}(\log_e(x)) = \frac{1}{x}$$

$$\frac{d}{dx}(\log_e(kx)) = \frac{1}{x}$$

$$\frac{d}{dx}(\log_e(g(x))) = \frac{g'(x)}{g(x)}$$

Note: The above rules are only applicable for logarithmic functions of base *e*.

WORKED EXAMPLE 13 Derivatives of logarithmic functions

Determine the derivative of each of the following.

a. $3\log_e(2x)$

b. $3\log_e\left(\sqrt{x}\right)$

c. 8:

d. $\log_e(x^2 - 7x + 6)$

THINK	WRITE
a. 1. Use the rule $\frac{d}{dx}\left(\log_e(kx)\right) = \frac{1}{x}$ to differentiate the function.	a. $\frac{d}{dx}\left(3\log_e(2x)\right) = 3 \times \frac{1}{x}$
2. Simplify the answer.	$= \frac{3}{x}$
b. 1. Rewrite the function using $\sqrt{x} = x^{\frac{1}{2}}$.	b. $3\log_e\left(\sqrt{x}\right) = 3\log_e\left(x^{\frac{1}{2}}\right)$
2. Simplify the function by applying log laws.	$= 3 \times \frac{1}{2}\log_e(x)$ $= \frac{3}{2}\log_e(x)$
3. Differentiate the function and simplify.	$\frac{d}{dx}\left(\frac{3}{2}\log_e(x)\right) = \frac{3}{2} \times \frac{1}{x}$ $= \frac{3}{2x}$
c. 1. Use the rule $\frac{d}{dx}\left(\log_e(g(x))\right) = \frac{g'(x)}{g(x)}$ to differentiate the function. State $g(x)$ and $g'(x)$.	c. If $g(x) = \sin(x)$, $g'(x) = \cos(x)$
2. Substitute $g(x)$ and $g'(x)$ into the derivative rule.	$\frac{d}{dx}(\log_e(\sin(x))) = \frac{\cos(x)}{\sin(x)}$ or $\frac{1}{\tan(x)}$
d. 1. Use the rule $\frac{d}{dx}\left(\log_e(g(x))\right) = \frac{g'(x)}{g(x)}$ to differentiate the function. State $g(x)$ and $g'(x)$.	d. If $g(x) = x^2 - 7x + 6$, $g'(x) = 2x - 7$
2. Substitute $g(x)$ and $g'(x)$ into the derivative rule.	$\frac{d}{dx}(x^2 - 7x + 6) = \frac{2x - 7}{x^2 - 7x + 6}$

TI | THINK | **DISPLAY/WRITE**

d. 1. On a Calculator page, press MENU, then select:
4: Calculus
1: Derivative
Complete the entry line as:
$\frac{d}{dx}\left(\ln\left(x^2-7x+6\right)\right)$
then press ENTER.

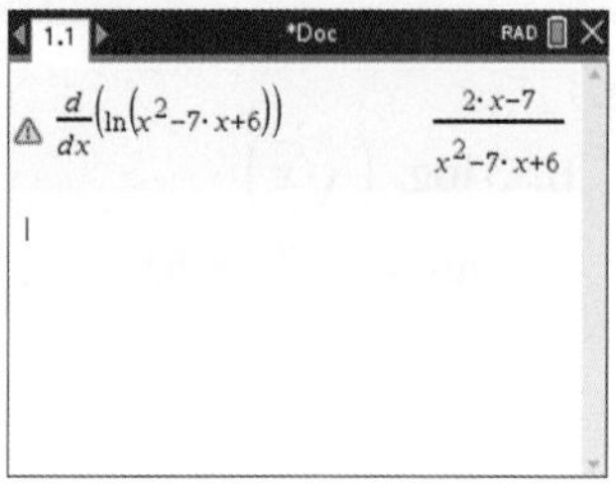

2. The answer appears on the screen.

$\frac{d}{dx}\left(\ln\left(x^2-7x+6\right)\right)$
$=\frac{2x-7}{x^2-7x+6}$

CASIO | THINK | **DISPLAY/WRITE**

d. 1. On a Main screen, select:
- Action
- Calculation
- diff

Complete the entry line as:
$\text{diff}\left(\ln\left(x^2-7x+6\right)\right)$
then press EXE.

2. The answer appears on the screen.

$\frac{d}{dx}\left(\ln\left(x^2-7x+6\right)\right)$
$=\frac{2x-7}{x^2-7x+6}$

5.7.2 Applications of the logarithmic derivative

Questions may also involve the differentiation of logarithmic functions to find the gradient of a curve at a given point or to find the equations of the tangent (and the perpendicular line) at a given point.

WORKED EXAMPLE 14 Determining equations of tangents and perpendicular lines

The graph of the function $f(x) = 0.5\log_e(x-1)$ is shown.

a. State the domain and range of f.

b. Calculate the value of the constant a given that $(a, 0)$ is the x-axis intercept.

c. Determine the equation of the tangent at $(a, 0)$.

d. Determine the equation of the line perpendicular to the curve at $(a, 0)$.

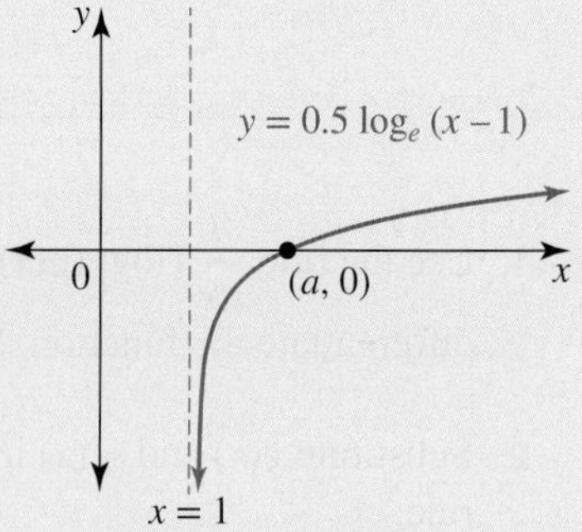

THINK	WRITE
a. State the domain and range of the function.	a. Domain $=(1,\infty)$ Range $=R$
b. 1. To find the x-intercept, let $f(x)=0$.	b. $0.5\log_e(x-1)=0$
2. Solve $0.5\log_e(x-1)=0$ for x.	$\log_e(x-1)=0$ $e^0=x-1$ $1=x-1$ $x=2$
3. Answer the question.	$(a,0)\equiv(2,0)$ $\therefore a=2$

c. 1. Determine the derivative of the function.

c. $f(x) = 0.5\log_e(x-1)$

$$f'(x) = \frac{1}{2} \times \frac{1}{x-1}$$
$$= \frac{1}{2(x-1)}$$

2. Substitute $x = 2$ into the derivative to find the gradient at this point.

$$f'(2) = \frac{1}{2(2-1)}$$
$$= \frac{1}{2}$$

The gradient at $x = 2$ is $\frac{1}{2}$.

3. State the general equation for a tangent.

The equation of the tangent is $y - y_1 = m_T(x - x_1)$.

4. State the known information.

The gradient of the tangent at $(x_1, y_1) = (2, 0)$ is $m_T = \frac{1}{2}$.

5. Substitute the values into the general equation.

$$y - 0 = \frac{1}{2}(x-2)$$

6. Simplify.

$$y = \frac{1}{2}x - 1$$

d. 1. A perpendicular line has a gradient of $m_p = -\frac{1}{m_T}$.

d. $m_P = -\frac{1}{\frac{1}{2}}$

$m_P = -2$

2. State the general equation for a line perpendicular to the tangent.

$y - y_1 = m_P(x - x_1)$

3. State the known information.

$m_P = -2,\ (x_1, y_1) = (2, 0)$

4. Substitute the values into the general equation.

$y - 0 = -2(x - 2)$

5. Simplify.

$$y = -2x + 4$$

TI | THINK — **DISPLAY/WRITE**

b. 1. On a Graphs page, complete the entry line for function 1 as: $f1(x) = 0.5\ln(x-1)$ then press ENTER.

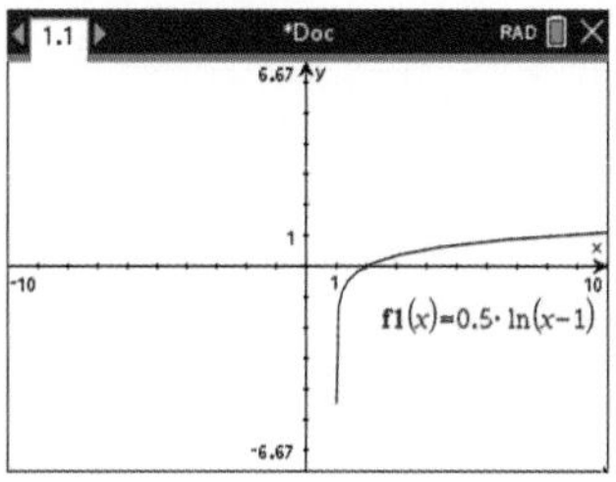

CASIO | THINK — **DISPLAY/WRITE**

b. 1. On a Graph & Table screen, complete the entry line for y1 as: $y1 = 0.5\ln(x-1)$ then press EXE. Select the 'Graph' icon to draw the graph.

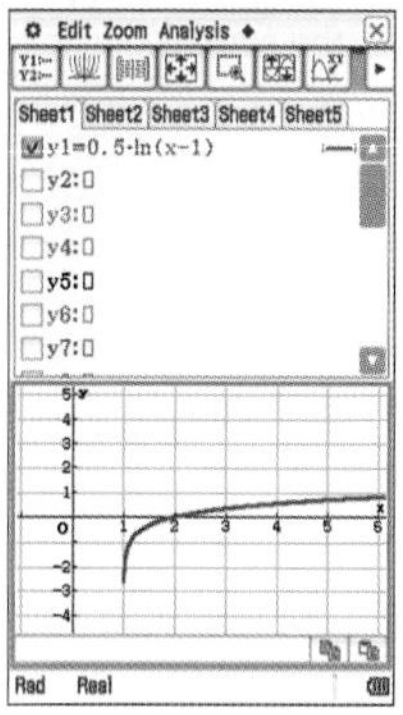

2. To find the x-intercept, press MENU, then select:
6: Analyze Graph
1: Zero
Move the cursor to the left of the x-intercept when prompted for the lower bound, then press ENTER. Move the cursor to the right of the x-intercept when prompted for the upper bound, then press ENTER.

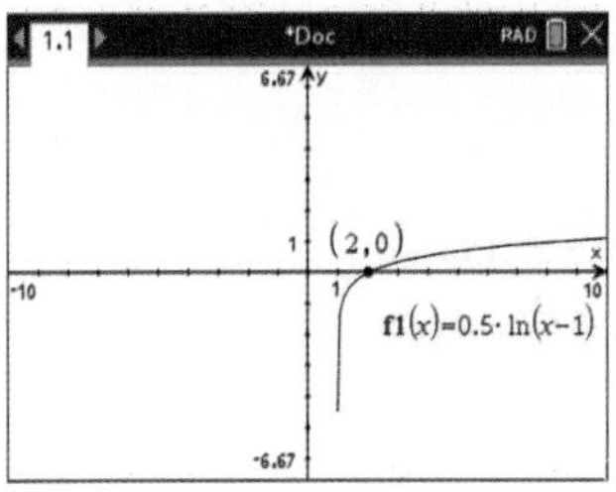

3. The answer appears on the screen. $(2, 0)$

c. 1. To find the gradient at $x = 2$, press MENU, then select:
6: Analyze
6: $\frac{dy}{dx}$
Click on the x-intercept then press ENTER.

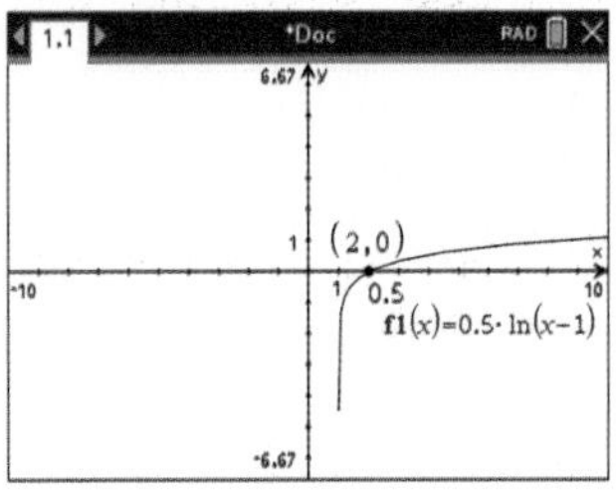

2. The answer appears on the screen. The gradient at $x = 2$ is 0.5.

3. To find the equation of the tangent at $x = 2$, press MENU, then select:
8: Geometry
1: Points & Lines
7: Tangent
Click on the x-intercept then press ENTER.

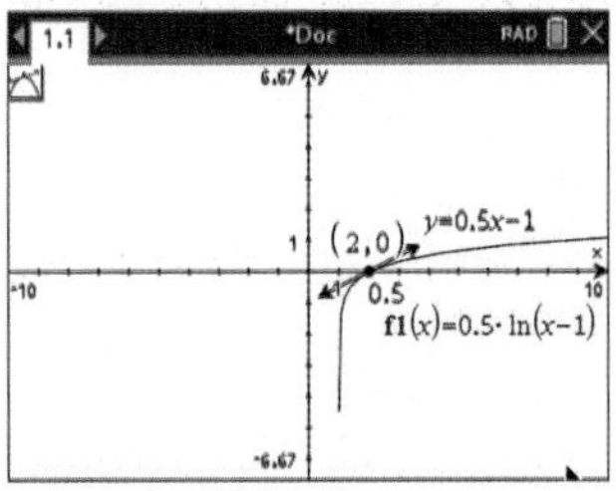

4. The answer appears on the screen. The equation of the tangent at $x = 2$ is $y = 0.5x - 1$.

2. To find the x-intercept, select:
- Analysis
- G-Solve
- Root

then press EXE.

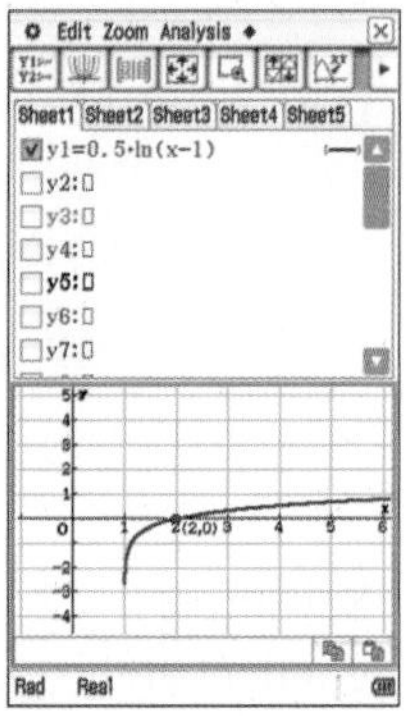

3. The answer appears on the screen. $(2, 0)$

c. 1. To find the gradient at $x = 2$, select:
- Analysis
- Sketch
- Tangent

Type ‘2’, select OK, then press EXE.

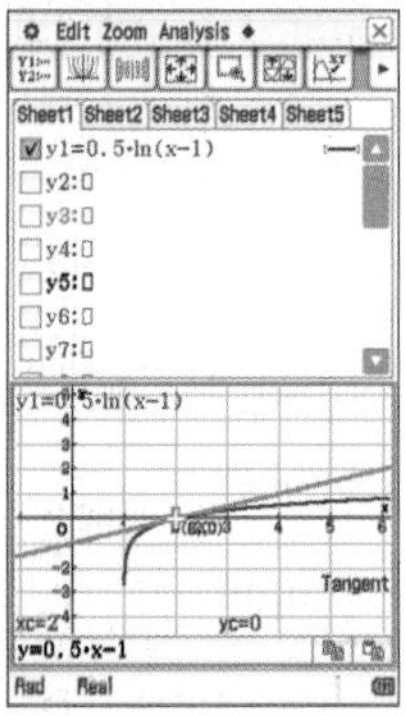

2. The equation of the tangent appears at the bottom of the screen. The gradient is the coefficient of the x term. The equation of the tangent at $x = 2$ is $y = 0.5x - 1$ and the gradient is 0.5.

d. 1. To find the equation of the perpendicular to the tangent at $x = 2$, press MENU, then select:
8: Geometry
4: Construction
1: Perpendicular
and choose the tangent line.
Note: To get the equation of the perpendicular line, press MENU, then select:
1: Actions
8: Coordinates and Equation

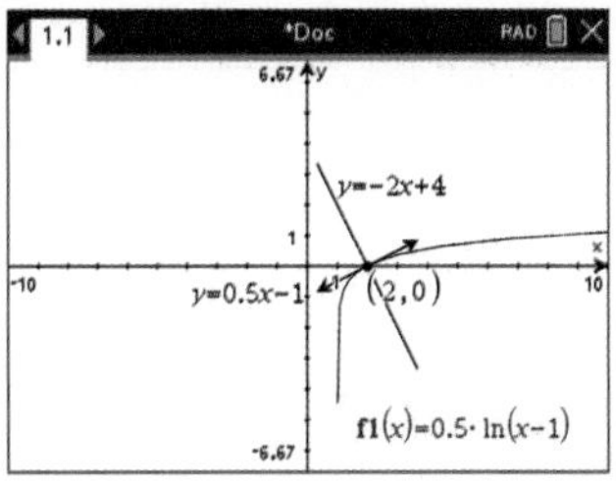

2. The answer appears on the screen.

The equation of the perpendicular line at $x = 2$ is $y = -2x + 4$.

d. 1. To find the equation of the perpendicular line at $x = 2$, select:
- Analysis
- Sketch
- Normal

Type ‘ 2’, select OK, then press EXE.

2. The answer appears on the screen.

The equation of the perpendicular line at $x = 2$ is $y = -2x + 4$.

5.7 Exercise

Technology free

1. WE13 Determine the derivative of each of the following.

a. $7\log_e\left(\dfrac{x}{3}\right)$

b. $2\log_e(x^3 + 2x^2 - 1)$

c. $3\log_e(e^{2x} - e^{-x})$

d. $\log_e(x^3 - 3x^2 + 7x - 1)$

2. Determine the derivative of each of the following.

a. $4\log_e\left(\dfrac{x}{2}\right)$

b. $-6\log_e(\cos(x))$

3. Differentiate the following functions with respect to x and state any restrictions on x.

a. $y = -5\log_e(2x)$

b. $y = \log_e\left(\dfrac{1}{x-2}\right)$

c. $y = \log_e\left(\dfrac{x+3}{x+1}\right)$

d. $y = \log_e(x^2 - x - 6)$

4. **WE14** The graph of the function $f: (2, \infty) \to R,\ f(x) = 2\log_e(x-2)$ is shown.

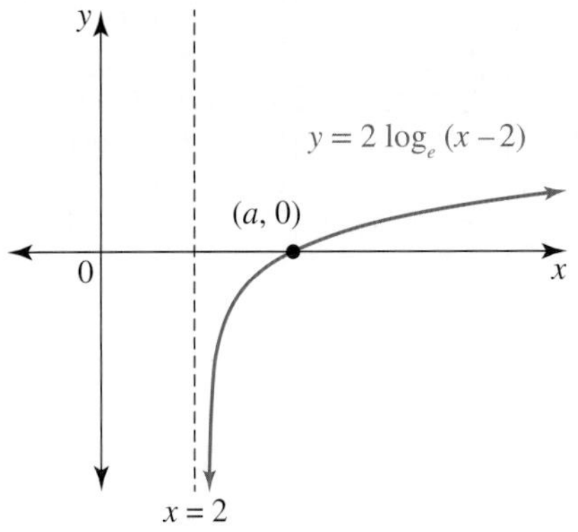

a. State the domain and range of f.
b. Calculate the value of the constant a given that $(a, 0)$ is the x-axis intercept.
c. Determine the equation of the tangent at $(a, 0)$.
d. Determine the equation of the line perpendicular to the curve at $(a, 0)$.

5. Determine the equation of the tangent to each of the given curves at the specified point.

a. $y = \log_e(2x-2)$ at $\left(\frac{3}{2}, 0\right)$
b. $y = 3\log_e(x)$ at $(e, 3)$

6. Obtain the equation of the tangent to the curve $y = 4\log_e(3x-1)$ at the point where the tangent is parallel to the line $6x - y + 2 = 0$.

7. The graph of the function defined by the rule $y = 2\log_e(2x)$ is shown.

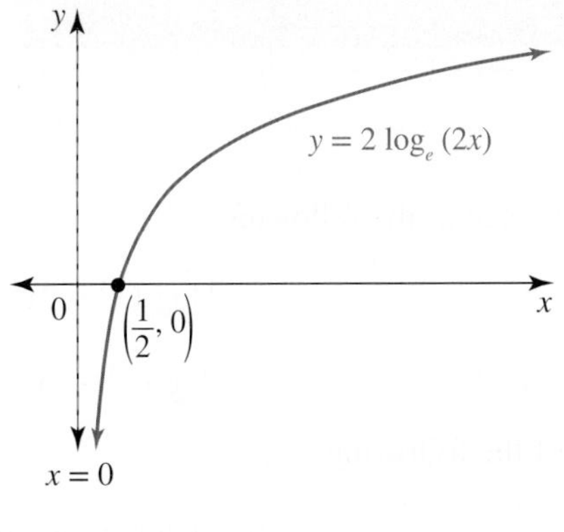

a. Determine the derivative of y with respect to x.
b. Determine the equation of the tangent at $\left(\frac{e}{2}, e\right)$.

Technology active

8. **MC** The line $y = x$ is a tangent to the curve $y = \log_e(x-1) + b$, where b is a constant. The value of b is:

A. 2
B. 1
C. $\frac{1}{2}$
D. -1
E. 0

9. The equation of a line perpendicular to the curve $y = \log_e(2(x-1))$ has the equation $y = -2x + k$, where k is a constant. Calculate the value of k, correct to 1 decimal place.

10. Calculate the gradient of each of the following functions at the specified point.

a. $y = 2\log_5(x);\ x = 5$

b. $y = \frac{1}{3}\log_3(x+1);\ x = 2$

c. $y = \log_6(x^2 - 3);\ x = 3$

d. $y = \frac{1}{2}\log_e\left(x^2\right);\ x = e$

11. The tangent to the curve $y = \log_e(2x-1)$ at $x = n$ intersects the x-axis at $x = 0.3521$. Calculate the value of the integer constant n.

5.7 Exam questions

Question 1 (1 mark) TECH-ACTIVE

Source: *VCE 2020, Mathematical Methods Exam 2, Section A, Q17; © VCAA.*

MC Let $f(x) = -\log_e(x+2)$.

A tangent to the graph of f has a vertical axis intercept at $(0, c)$.

The maximum value of c is

A. -1
B. $-1 + \log_e(2)$
C. $-\log_e(2)$
D. $-1 - \log_e(2)$
E. $\log_e(2)$

Question 2 (1 mark) TECH-ACTIVE

Source: *VCE 2018, Mathematical Methods Exam 2, Section A, Q9; © VCAA.*

MC A tangent to the graph of $y = \log_e(2x)$ has a gradient of 2.

A. 0
B. -0.5
C. -1
D. $-1 - \log_e(2)$
E. $-2\log_e(2)$

Question 3 (2 marks) TECH-FREE

If a function $f(x) = \log_e(2x)$, determine $f'(1)$.

More exam questions are available online.

5.8 Review

5.8.1 Summary

doc-37034

Hey students! Now that it's time to revise this topic, go online to:

Access the topic summary | Review your results | Watch teacher-led videos | Practise VCAA exam questions

Find all this and MORE in jacPLUS

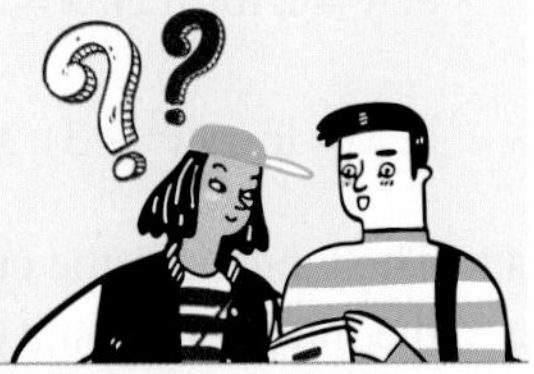

5.8 Exercise

Technology free: short answer

1. a. The graph of $y = f(x)$ is shown. Sketch the graph of its gradient function.

   b. Differentiate the following functions with respect to x and state any restrictions on x.

   i. $\log_e\left(\dfrac{x+2}{x-3}\right)$

   ii. $\log_e(x+2)^2$

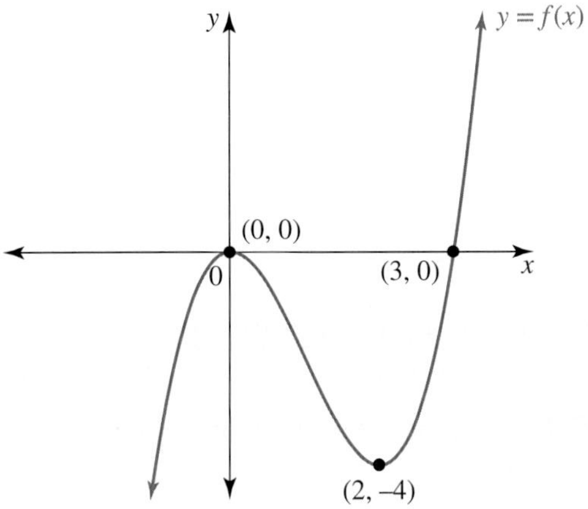

2. Given $f: \left[0, \dfrac{\pi}{2}\right] \rightarrow R,\ f(x) = 3\sin(2x) - 4$, determine:

   a. $f'(x)$

   b. the value of $f'(x)$ when $x = \dfrac{\pi}{3}$

   c. the domain over which the gradient is positive.

3. For the graph whose equation is given by $y = -3\cos\left(\dfrac{\pi x}{6}\right) + 7$, determine the equations of the tangent and the line perpendicular to the curve at the point where $x = 3$.

4. Determine the equation of the tangent to the curve $f(x) = (x-1)^2 + 2$ that is perpendicular to the line $x - 2y = 5$.

5. a. Sketch the graph of the function $f: (-\pi, \pi) \rightarrow R,\ f(x) = \tan\left(\dfrac{x}{2}\right) - 1$.

   b. Calculate $f'(x)$ and hence determine $f'\left(\dfrac{\pi}{3}\right)$.

6. A function has the equation $f(x) = ke^{x^2}$.

   a. Given $f'(4) = 4e^{16}$, show that $k = \dfrac{1}{2}$.

   b. Hence, determine $f'(2)$.

   c. Determine the equation of the line perpendicular to the curve at $x = 2$.

Technology active: multiple choice

7. **MC** If $f(x) = m\cos(3x)$ where m is a constant and $f'\left(\dfrac{\pi}{6}\right) = -3$, then m is equal to:

   A. 1 **B.** 3 **C.** 0 **D.** −3 **E.** −1

8. **MC** If $y = e^{-x} - 3$, then the gradient of the tangent to the curve when $x = 0$ is:

A. $-e$ B. -4 C. -1 D. 0 E. $e - 3$

9. **MC** If $y = \dfrac{\sqrt[3]{x} - 4x^2}{2x^3}$, then $\dfrac{dy}{dx}$ is:

A. $\dfrac{3x-8}{6x}$ B. $-\dfrac{4}{3x^{\frac{5}{3}}} + 2$ C. $-\dfrac{4}{3x^{\frac{11}{3}}} + \dfrac{2}{x^2}$

D. $-\dfrac{4}{3x^{\frac{13}{3}}} + \dfrac{2}{x^2}$ E. $\dfrac{3x-8}{2x^2}$

10. **MC** The derivative of $\dfrac{\sin^3(\theta) + \sin(\theta)\cos^2(\theta)}{\cos(\theta)}$ with respect to x is:

A. $\dfrac{3\sin^2(\theta) - 2\cos(\theta)\sin(\theta)}{-\sin(\theta)}$ B. $\cos^2(\theta)$ C. $-\dfrac{\cos(\theta)}{\sin(\theta)}$

D. $\dfrac{1}{\sin^2(\theta)}$ E. $\dfrac{1}{\cos^2(\theta)}$

11. **MC** If $y = \log_e\left(\dfrac{2}{x}\right)$, then $\dfrac{dy}{dx}$ is equal to:

A. x B. $\dfrac{1}{x}$ C. $\log_e\left(\dfrac{2}{x}\right)$ D. $\log_e\left(\dfrac{x}{2}\right)$ E. $-\dfrac{1}{x}$

12. **MC** If $y = 2\tan(3x)$, then $\dfrac{dy}{dx}$ equals:

A. $\dfrac{1}{\cos^2(3x)}$ B. $\dfrac{2}{\cos^2(3x)}$ C. $\dfrac{6}{\cos^2(3x)}$ D. $\dfrac{6}{\sin^2(3x)}$ E. $\dfrac{6}{\cos^2(x)}$

***The following information relates to Questions* 13 *and* 14.**

The depth of water, d metres, in a canal changes with the tides according to the rule

$$d = 6 + 2\cos\left(\frac{\pi t}{6}\right)$$

where t is the time in hours after high tide. On a particular day, high tide was at 10:00 am.

13. **MC** The water depths at high tide and low tide respectively were:

A. 6 m, 6 m B. 8 m, 6 m C. 8 m, 4 m D. 6 m, 4 m E. 4 m, 8 m

14. **MC** The rate of change of the depth of water with respect to time at 2:00 pm is equal to:

A. $-\dfrac{\sqrt{3}\pi}{6}$ m/h B. $-\dfrac{\pi}{6}$ m/h C. $-\dfrac{\pi}{3}$ m/h D. $\dfrac{\pi}{6}$ m/h E. $\dfrac{\sqrt{3}\pi}{6}$ m/h

15. **MC** If $y = 2x - e^{-2x}$, the equation of the tangent to the curve at $x = 0$ is given by:

A. $y = -4x + 1$ B. $y = 4x + 1$ C. $y = -4x - 1$

D. $y = 4x - 1$ E. $y = -\dfrac{1}{4}x - 1$

16. **MC** If $y = \log_e(x + 5)$, then the equation of the tangent to the curve at the point where $x = e - 5$ is given by:

A. $ey = x - 5$ B. $ey = x + 5$ C. $y = x + \dfrac{5}{e}$

D. $y = \dfrac{x}{e} + 5$ E. $y = x + 5$

Technology active: extended response

17. The profile of water waves produced by a wave machine in a scientific laboratory is modelled by the trigonometric function defined by
$f(x) = a\sin(x) + b\cos(x)$.

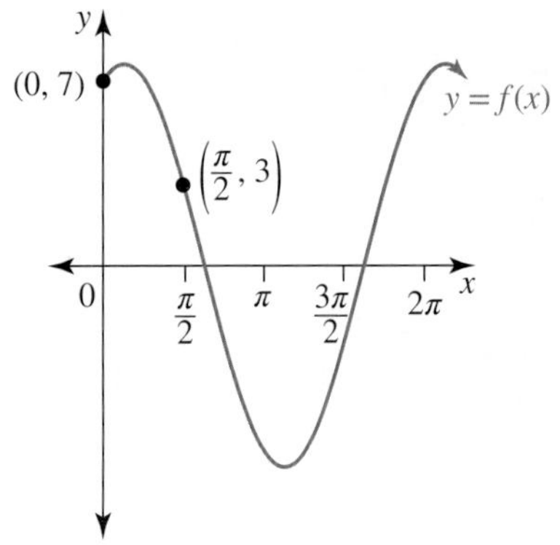

a. Given that the graph of the wave profile passes through the points (0, 7) and $\left(\frac{\pi}{2}, 3\right)$, determine the constants a and b.
b. Determine the maximum and minimum swells for the wave profile, correct to 1 decimal place. Hence, state the range of the function.
c. Determine $\{x : f(x) = 0,\ 0 \le x \le 2\pi\}$, giving your answers correct to 4 decimal places.
d. Determine $f'(x)$ and evaluate the gradient at the x-values found in part c. Give your answers correct to 3 decimal places.

18. The diagram shows a property in the country that includes a farmhouse that is rented out for holidays.

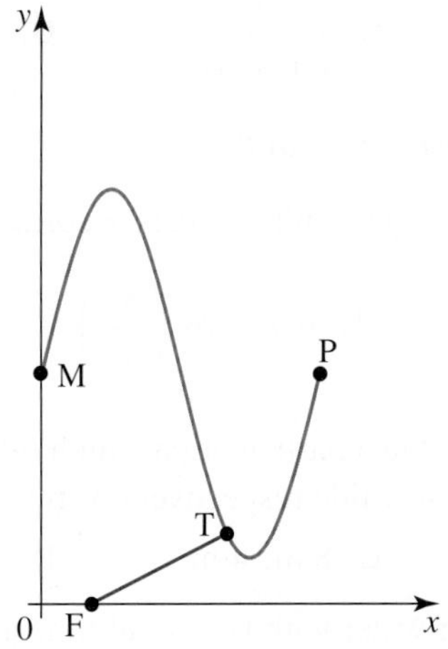

F represents the farmhouse, MP is a road, and TF is a track from the farmhouse to the road. MP is modelled by the equation

$$y = 2\sin(2x) + \frac{5}{2}$$

a. Write the coordinates of the points M and P.
b. Calculate $\frac{dy}{dx}$.
c. T is the point $\left(\frac{2\pi}{3}, t\right)$. Determine the value of t.
d. Given that the track, TF, lies perpendicular to the road at point T, determine the gradient of the track.
e. Determine the equation of the track, TF.

19. Scientists have been studying common microbes in soil. One such group is the *Streptomyces* genus of bacteria. These bacteria produce an organic compound called geosmin, which is one of the substances that causes the distinctive smell of damp topsoil. These microbes are dangerous to humans, but they are important because they can tolerate dry conditions that many others microbes cannot.

Scientists are interested in the microbial biomass of soil (the amount of microbes in the soil) because better management of the biomass leads to the improved use of the soil.

It is known that the relationship between the growth in the biomass and the breakdown of organic matter over time is given by

$$B(t) = B_0 e^{kt},\ 0 \le t \le 10$$

where B_0 is the initial biomass, B is the biomass t days after the start of the experiment, and k is a constant. A particular soil sample is being tested. The sample has an initial biomass of 1.5 units.

a. Determine the value of B_0, the initial biomass of the soil sample.

b. It is known that the rate of increase in the biomass of the soil after 4 days is 0.55 units/day. Calculate the value of k, correct to 4 decimal places.

The biomass reaches a maximum of 9 units after 10 days of incubation. After this, the biomass of the soil slowly reduces. This breakdown in the biomass of the soil can be described by

$$B(t) = A_0 e^{ht}$$

where B is the biomass of the soil t days after the beginning of the breakdown of the soil, and A_0 and h are constants. It was observed that 4 days after the biomass had started to break down, the biomass of the soil was 7 units. The graph depicting the increase and decrease in the biomass of the organic matter is shown.

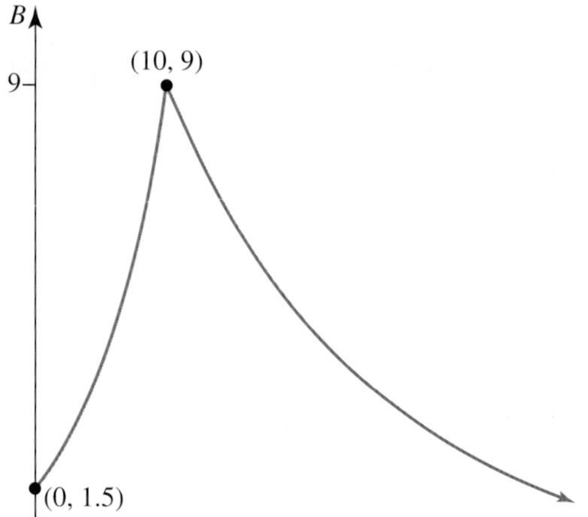

c. Determine the values of A_0 and h, correct to 4 decimal places.

d. Calculate the rate of decrease of the biomass after 4 days since the start of the breakdown, correct to 2 decimal places.

20. Microbiologists have been working with a certain type of bacteria that continues to thrive providing it has a favourable growth medium. However, it is known that once the medium is exhausted, the bacteria growth stagnates and then declines at a fast rate. The graph showing this growth, stagnation and decay is shown.

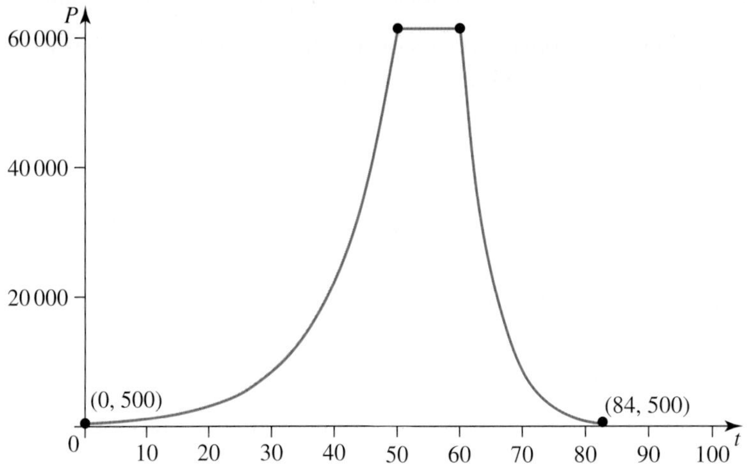

For a particular experiment, microbiologists started with 500 bacterial cells and observed that the population doubles every 7.2 hours. The relationship between the number of bacterial cells, P, and the time, t hours since the bacteria started multiplying, is given by

$$P = P_0 e^{kt}$$

where k is a constant.

a. State the value of P_0 and determine the value of k, correct to 4 decimal places.

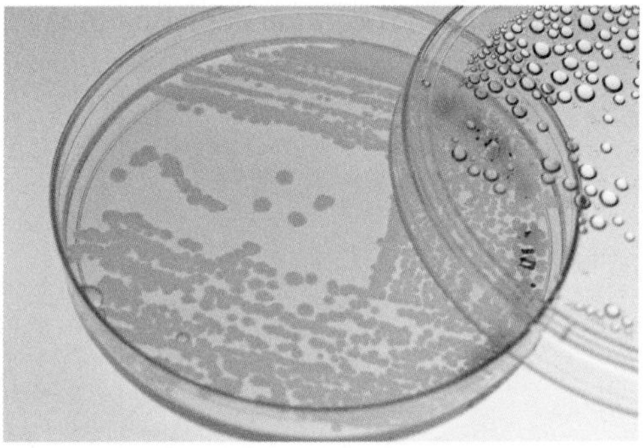

This growth phase lasts for 50 hours, after which the size of the colony stagnates (stays the same) for 10 hours.

b. How many bacterial cells are present in the colony after 50 hours? Give your answer correct to the nearest thousand.

c. Determine the rate of increase in the colony size after 30 hours. Give your answer correct to 1 decimal place.

The colony population declines over the next 24 hours. It is known that over this time the number of bacterial cellsdrops to just 500. It is also known that the number ofbacterial cells present in the colony can be defined by

$$P = Q_0 e^{mt},\ 0 \le t \le 24$$

where t is the number of hours since the population started to decline, and m is a constant.

d. Determine the value of Q_0 to the nearest thousand.

e. Determine the rate of decrease in the size of the colony after 10 hours if it is known that $m = -0.2008$. Give your answer correct to the nearest bacterial cells.

5.8 Exam questions

Question 1 (8 marks) TECH-FREE

Source: *VCE 2020, Mathematical Methods Exam 1, Q7; © VCAA.*

Consider the function $f(x) = x^2 + 3x + 5$ and the point $P(1, 0)$. Part of the graph of $y = f(x)$ is shown below.

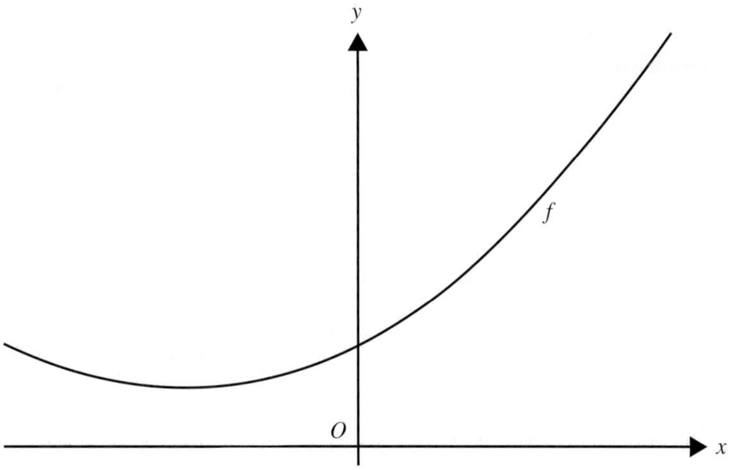

a. Show that point P is not on the graph of $y = f(x)$. **(1 mark)**

b. Consider a point $Q(a, f(a))$ to be a point on the graph of f.

i. Find the slope of the line connecting points P and Q in terms of a. **(1 mark)**

ii. Find the slope of the tangent to the graph of f at point Q in terms of a. **(1 mark)**

iii. Let the tangent to the graph of f at $x = a$ pass through point P. Find the values of a. **(2 marks)**

iv. Give the equation of one of the lines passing through point P that is tangent to the graph of f. **(1 mark)**

c. Find the value, k, that gives the shortest possible distance between the graph of the function of $y = f(x - k)$ and point P. **(2 marks)**

Question 2 (1 mark) TECH-ACTIVE

Source: *VCE 2016, Mathematical Methods Exam 2, Section A, Q3; © VCAA.*

MC Part of the graph $y = f(x)$ of the polynomial function f is shown below.

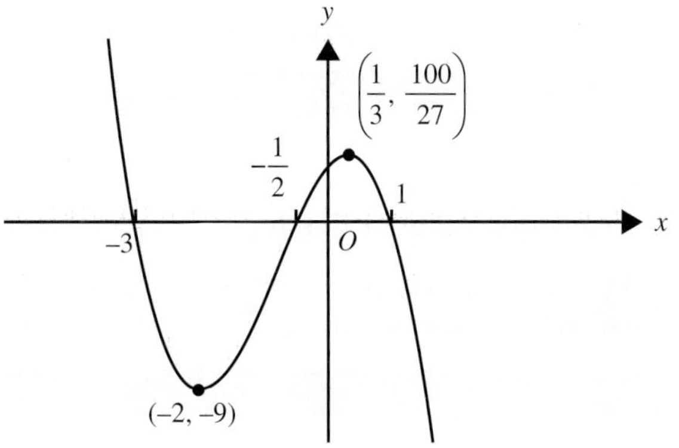

$f'(x) < 0$ for

A. $x \in (-2, 0) \cup \left(\frac{1}{3}, \infty\right)$

B. $x \in \left(-9, \frac{100}{27}\right)$

C. $x \in (-\infty, -2) \cup \left(\frac{1}{3}, \infty\right)$

D. $x \in \left(-2, \frac{1}{3}\right)$

E. $x \in (-\infty, -2] \cup (1, \infty)$

Question 3 (1 mark) TECH-ACTIVE

Source: *VCE 2016, Mathematical Methods Exam 2, Section A, Q10; © VCAA.*

MC For the curve $y = x^2 - 5$, the tangent to the curve will be parallel to the line connecting the positive x-intercept and the y-intercept when x is equal to

A. $\sqrt{5}$
B. 5
C. -5
D. $\dfrac{\sqrt{5}}{2}$
E. $\dfrac{1}{\sqrt{5}}$

Question 4 (8 marks) TECH-ACTIVE

Source: *Adapted from VCE 2016, Mathematical Methods Exam 2, Section B, Q1; © VCAA.*

Let $f: [0, 8\pi] \to R, f(x) = 2\cos\left(\dfrac{x}{2}\right) + \pi$.

a. Find the period and range of f. **(2 marks)**
b. State the rule for the derivative function f'. **(1 mark)**
c. Find the equation of the tangent to the graph of f at $x = \pi$. **(1 mark)**
d. Find the equations of the tangents to the graph of $f: [0, 8\pi] \to R, f(x) = 2\cos\left(\dfrac{x}{2}\right) + \pi$ that have a gradient of 1. **(2 marks)**
e. Find the values of x, $0 \le x \le 8\pi$, such that $f(x) = 2f'(x) + \pi$. **(2 marks)**

Question 5 (1 mark) TECH-ACTIVE

Source: *VCE 2015, Mathematical Methods (CAS) Exam 2, Section 1, Q4; © VCAA.*

MC Consider the tangent to the graph of $y = x^2$ at the point $(2, 4)$.

Determine which of the following points lies on this tangent.

A. $(1, -4)$
B. $(3, 8)$
C. $(-2, 6)$
D. $(1, 8)$
E. $(4, -4)$

More exam questions are available online.

Answers

Topic 5 Differentiation

5.2 Review of differentiation

5.2 Exercise

1. a. $f'(x) = 2x - 4$
   b. -2
2. a. $f'(x) = -1$ b. $f'(x) = 6x - 2$
3. a. $12x^2 - \dfrac{2}{3x^3}$ b. $-\dfrac{1}{x^{\frac{7}{2}}} - \dfrac{1}{5}$
   c. $3x^2 + 6x + 1$ d. $-\dfrac{6}{x^{\frac{5}{2}}} + \dfrac{1}{x^2}$
4. a. $\dfrac{dy}{dx} = -\dfrac{15}{4x^6} + \dfrac{1}{2x^2}$ b. $f'(x) = -\dfrac{30}{x^4} + \dfrac{2}{x^2} - \dfrac{4}{x^5}$
   c. $\dfrac{dy}{dx} = \dfrac{1}{2\sqrt{x}} + \dfrac{1}{4x^{\frac{3}{2}}}$ d. $f'(x) = -\dfrac{27}{2x^2} - x + \dfrac{9}{2}$
5. a. -14 b. $(2,\ 0)$ and $(-2,\ 4)$
6. a. 4 b. 5 c. 12 d. $\dfrac{39}{16}$
7. $11 + 4a$
8. a.

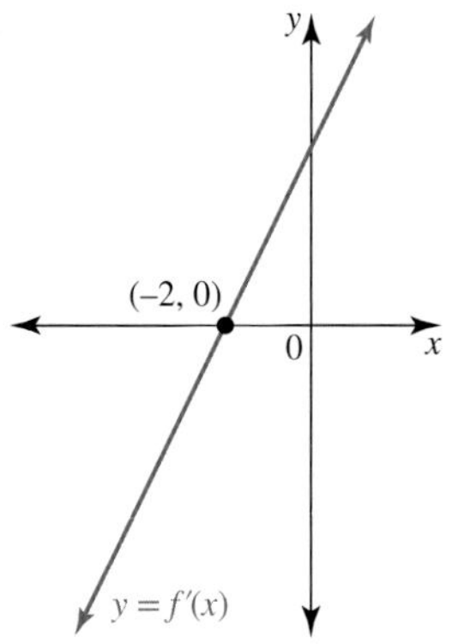

   b. $x \in (-\infty,\ 2) \backslash \{-2\}$
9.

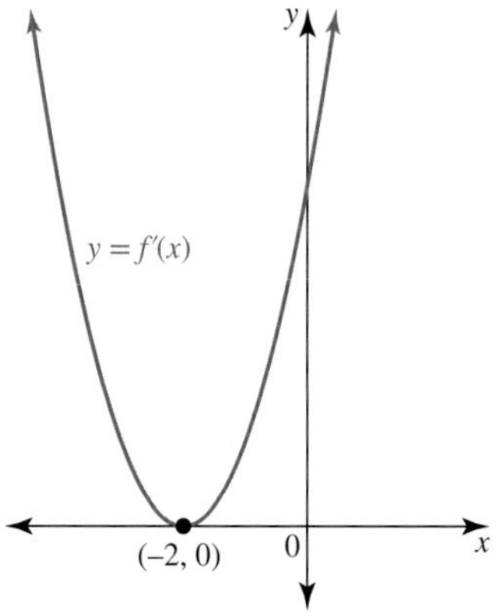

10. a. i. Domain $= R$
    ii.

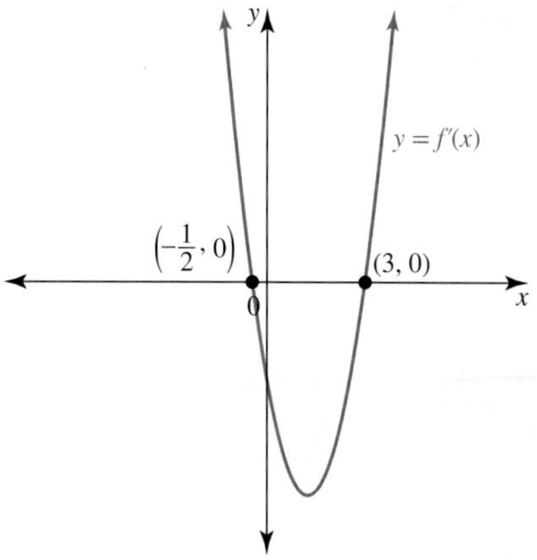

   b. i. Domain $= R$
    ii.

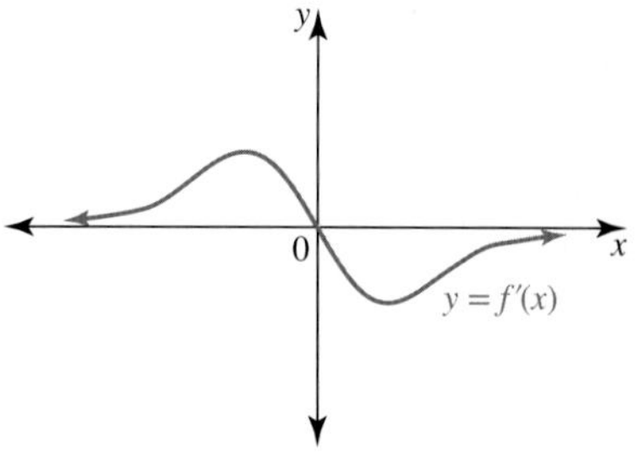

   c. i. Domain $= (-\infty, 4) \backslash \{-2\}$
    ii.

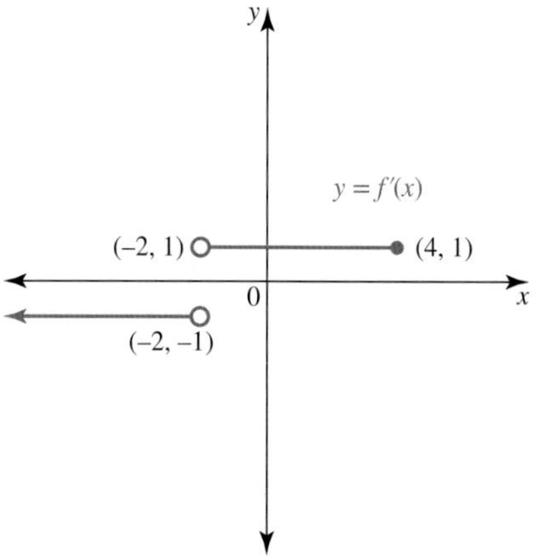

   d. i. Domain $= R \backslash \{0\}$
    ii.

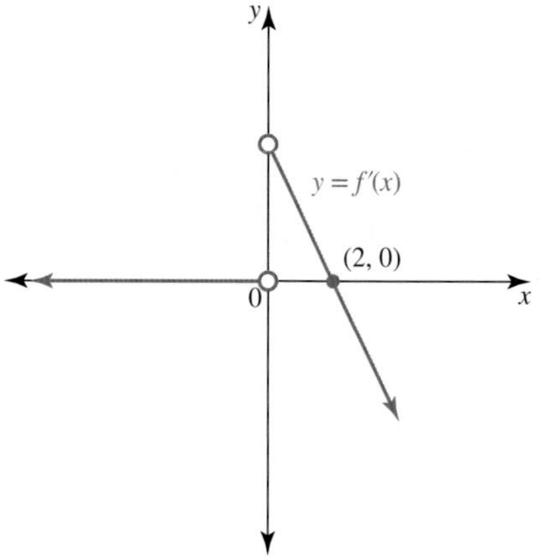

11. a. $y = -4x + 9$ b. $y = \dfrac{1}{4}x - \dfrac{15}{4}$

12. a. $y = -6x - 22$

b. $y = -3a^2x + 2a^3 + 8$

c. $y = \dfrac{\sqrt{3}}{3}x + \sqrt{3} - 5$

d. $y = -\dfrac{7}{2}x + 2$

13. a. $y = \dfrac{1}{6}x + \dfrac{53}{6}$

b. $y = \dfrac{1}{3a^2}x + 8 - a^3 - \dfrac{1}{3a}$

c. $y = -\sqrt{3}x + 5\sqrt{3} - 5$

d. $y = \dfrac{2}{7}x + \dfrac{67}{7}$

14. a. $y = 3x - \dfrac{3}{4}$ b. $y = \dfrac{1}{2}x - \dfrac{1}{2}$

15. a and b

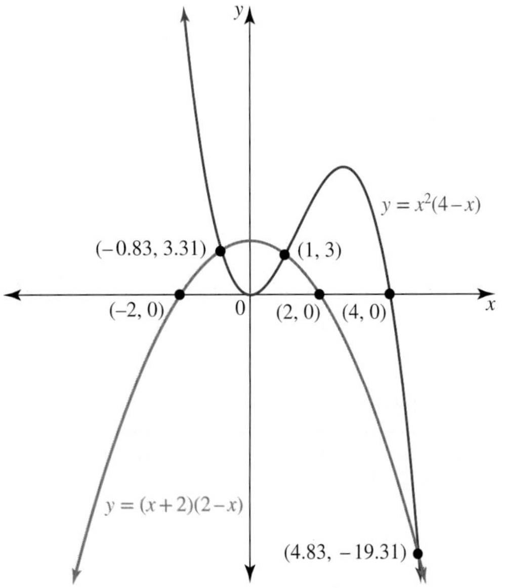

c. The gradient of $y = (x + 2)(2 - x)$ when $x = 1$ is -2.
The gradient of $y = x^2(4 - x)$ when $x = 1$ is 5.

16. a.

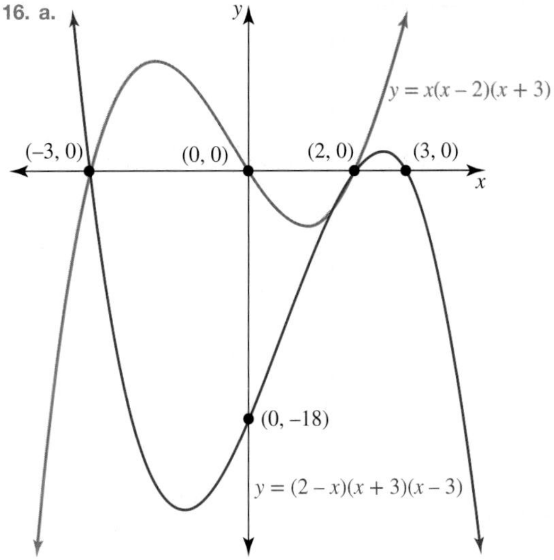

b. $\left(1\dfrac{1}{2}, -3\dfrac{3}{8}\right)$

c. Tangent: $y = \dfrac{15}{4}x - 9$

Line perpendicular to tangent: $y = -\dfrac{4}{15}x - \dfrac{119}{40}$

17. $y = 4 - x^2$

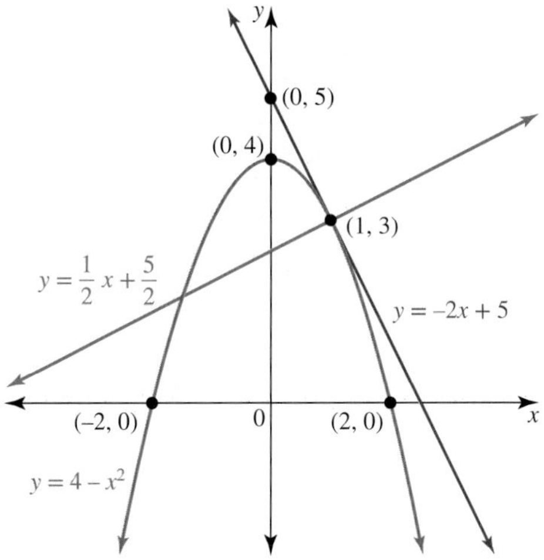

18. $y = -x^2 + 7x - 10$

19. $y = x^3 - x$

20. $m = 12$

5.2 Exam questions

Note: Mark allocations are available with the fully worked solutions online.

1. B
2. A
3. D

5.3 Differentiation of exponential functions

5.3 Exercise

1. a. $-\dfrac{1}{3}e^{-\frac{1}{3}x}$

b. $12x^3 + 4xe^{-2x^2}$

c. $-\dfrac{8}{3}e^{-2x} + \dfrac{4}{3}e^{-4x} - 2e^{-3x}$

d. $4e^{4x} - 12e^{2x}$

2. a. $-20e^{-4x}$

b. $-\dfrac{1}{2}e^{-\frac{1}{2}x} + x^2$

c. $12e^{3x} - \dfrac{3e^{6\sqrt{x}}}{2\sqrt{x}} + 9e^{-3x+2}$

d. $3e^{3x} + 3e^{-3x} - 4e^{-2x}$

e. $4e^{2x} + e^{-x}$

f. $e^x - 2e^{2x} - 3e^{-x}$

3. $\dfrac{1}{2}$

4. a. -2 b. $-\dfrac{8}{e}$ c. $\dfrac{3e}{2}$ d. 1

5. a. 2

b. i. $y = 2x + 1$ ii. $y = -\frac{1}{2}x + 1$

6. $y_T = -3x + 5,\ y_P = \frac{1}{3}x + 5$

7. $y_T = -3x - 1,\ y_P = \frac{1}{3}x - 1$

8. $y_T = \frac{e^{\sqrt{3}}}{2\sqrt{3}}x + e^{\sqrt{3}} + 1 - \frac{3e^{\sqrt{3}}}{2\sqrt{3}}$

$y_P = -\frac{2\sqrt{3}}{e^{\sqrt{3}}}x + e^{\sqrt{3}} + 1 + \frac{6\sqrt{3}}{e^{\sqrt{3}}}$

9. $y = 5x - 4$

10. E

11. a. $(1.89,\ 1.05)$ b. 7.66

12. $2\log_e(3)$

13. a. $-2e^7$ b. $\frac{3}{2}$

14. a. $2e^2 - \frac{2}{e}$ b. 0

15. a. 2 b. -1.38

16. $f'(x) = Ae^x - 3Be^{-3x}$

$0 = Ae^x - 3Be^{-3x}$

$= e^{-3x}\left(Ae^{4x} - 3B\right)$

$e^{-3x} \neq 0$, no real solution

So, $0 = Ae^{4x} - 3B$

$3B = Ae^{4x}$

$e^{4x} = \frac{3B}{A}$

5.3 Exam questions

Note: Mark allocations are available with the fully worked solutions online.

1. $-6e^{-3x}$
2. B
3. $\frac{dy}{dx} = e^x - 3e^{-3x}$

5.4 Applications of exponential functions

5.4 Exercise

1. a. $q = -4$ b. $(m,\ 0) = (\log_e 3,\ 0)$
c. $f'(x) = 2e^{2x} - 4e^x$ d. $f'(0) = -2$

2. a. $z = -3$ b. $(n,\ 0) = (-\log_e(2),\ 0)$
c. $f'(x) = -2e^{-2x} + 3e^{-x}$ d. 1

3. a. 20
b. $-0.0304e^{-0.00152t}$
c. 0.0291 g/year

4. a. $0.5I_0$ b. $0.0011I_0$

5. a. 120
b. $\frac{dA}{dt} = -120ke^{-kt}$
$= -k \times 120e^{-kt}$
$= -k \times A$
$\propto A$
c. $k = \frac{1}{2}\log_e\left(\frac{4}{3}\right)$
d. 8.408 units/min
e. As $t \to \infty, A \to 0$. Technically, the graph approaches the line $A = 0$ (with asymptotic behaviour, so it never reaches $A = 0$ exactly). However, the value of A would be so small that in effect, after a long period of time, there will be no gas left.

6. a. 11
b. $\frac{dL}{dt} = 6.589e^{0.599t}$
c. 39.742 mm/month

7. a. 23.2 million b. 2005, September
c. 0.29 million/year d. 2019

8. a. 10.94 cm b. 69.4°

9. a. $(0,\ 0.5)$ b. $f'(x) = -e^{-x} + e^{-2x}$
c. $(0,\ 0.5)$ d. 63.4°
e. $y = -0.2325x + 0.5327$ f. $y = 4.3011x - 4.0009$

10. a. $\frac{x^3 - 6x}{e^{x^2}}$
b. $\left(\pm\sqrt{6},\ \frac{1}{2e^6}\right)$ and $\left(0,\ -\frac{5}{2}\right)$
c. 0.593

11. a. 30
b. 450 000
c. 37 072.2/day
d. 450
e. 45 000
f. $m = -\frac{1}{7}\log_e(10)$, 39 711/day

5.4 Exam questions

Note: Mark allocations are available with the fully worked solutions online.

1. a. $T_0 = 75$ b. 1.531 °C/min

2. a. 5
b. $-10xe^{-x^2}$
c. i. 3.89
ii. -3.68

3. a. $P_0 = 75.32$ cm of mercury, $k = 0.24$
b. 5.45 cm of mercury/km

5.5 Differentiation of trigonometric functions

5.5 Exercise

1. a. $5 - 3\sin(x) + 5\cos(x)$
b. $3\cos(3x + 2) + 6x\sin(3x^2)$
c. $3\cos(9x)$
d. $10\sec^2(2x) - 10x^4$

e. $2\sec^2\left(\frac{x}{4}\right)$

f. $\frac{\pi}{20}\sec^2\left(\frac{\pi x}{20}\right)$

2. a. $-6\sin(3x)$ b. $3\sin\left(\frac{\pi}{2}-x\right)$

c. $-\frac{4}{3}\cos\left(\frac{x}{3}\right)$ d. $6\cos\left(\frac{\pi}{2}+3x\right)$

e. $-5x\sec^2(5x^2)$ f. $20\sec^2(20x)$

3. $-2\cos(2x)$

4. $y = x - \frac{\pi}{2}$

5. $y = 4x + \frac{\pi}{2} - 1$

6. $y = -\frac{3}{2}x + \frac{\pi}{4} + \frac{3\sqrt{3}}{2}$

7. $y = 4x + 2 - \pi$

8. a. $y_T = 3x - 2\pi,\ y_P = -\frac{1}{3}x + \frac{2\pi}{9}$

b. $y_T = -\frac{1}{2}x + \frac{\pi}{2},\ y_P = 2x - 2\pi$

9. a. $\tan(x)$ b. $\sec^2(x)$

10. A

11. a. -1 b. $\frac{\pi}{4}, \frac{5\pi}{4}$

c. $\cos(x) + \sin(x)$ d. $\frac{3\pi}{4}, \frac{7\pi}{4}$

12. a. $\sqrt{3}$ b. $-\frac{\pi}{3}, \frac{2\pi}{3}$

c. $-\sqrt{3}\sin(x) + \cos(x)$ d. $-\frac{5\pi}{6}, \frac{\pi}{6}$

13. (0.243, 1.232), (0.804, 0.863)

14. (−0.524, 0.342), (0.524, −0.342)

15. 116.6°

16. $x = -\frac{5\pi}{8}, -\frac{\pi}{8}, \frac{3\pi}{8}, \frac{7\pi}{8}$

5.5 Exam questions

Note: Mark allocations are available with the fully worked solutions online.

1. B
2. D
3. D

5.6 Applications of trigonometric functions

5.6 Exercise

1. a. $A = \frac{1}{2}ab\sin(c)$

$= \frac{1}{2} \times 6 \times 7\cos(\theta)$

$= 21\sin(\theta)$

b. 10.5 cm^2/radian

2. a. 10 b. $2\pi\cos(\pi t)$ c. -2π cm/s

3. a. 12 hours

b. Low tide = 1 metre at 3:00 pm

c. $\frac{\pi}{12}\cos\left(\frac{\pi t}{6}\right)$

d. $\frac{\sqrt{2}\pi}{24}$

e. 4:30 pm

4. a. 0.849, 1.508

b. $\left(\frac{\pi}{8}, 3\right), \left(\frac{3\pi}{8}, -1\right)$

c. -8

d. $\left[0, \frac{\pi}{8}\right) \cup \left(\frac{3\pi}{8}, \frac{\pi}{2}\right]$

5. a. $BD = a\sin(\theta)$, $CD = a\cos(\theta)$

b. $L = a + 2a\sin(\theta) + a\cos(\theta) + 4$

c. $2a\cos(\theta) - a\sin(\theta)$

d. $\theta = 1.1^c$

6. a. $L = 3 \times 100\cos(\theta) + 4 \times 100\sin(\theta) + 2 \times 100$ $L = 300\cos(\theta) + 400\sin(\theta) + 200$ as required

b. $-300\sin(\theta) + 400\cos(\theta)$

c. 700 cm, $\theta = 0.93^c$

7. a.

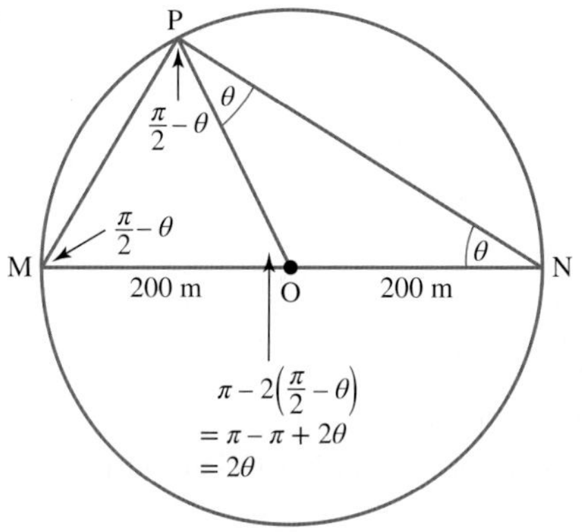

Distance ÷ time = velocity
Distance = velocity × time
Distance ÷ velocity = time
So $d(PM) = 400\cos(\theta)$.

$T_{obstacles} = \frac{400\cos(\theta)}{2}$

$T_{obstacles} = 200\cos(\theta)$

$d_{PM} = 200 \times 2\theta$

$d_{PM} = 400\theta$

$T_{hurdles} = \frac{400\theta}{5}$

$= 80\theta$

$T_{total} = T_{obstacles} + T_{hurdles}$

$T_{total} = 200\cos(\theta) + 80\theta$

$T_{total} = 40(5\cos(\theta) + 2\theta)$

b. 0.4115

c. T_{max} = 3 min 36 s

8. a. 6 m

b. $-\frac{7\pi}{40}\sin\left(\frac{\pi x}{20}\right)$

c. i. $-\frac{7\pi}{40}\sin\left(\frac{\pi}{4}\right)=-0.3888$

ii. $-\frac{7\pi}{40}\sin\left(\frac{\pi}{2}\right)=-0.5498$

d. i. 15 m ii. 158.95°

9. a. (0, 100), (4.5, 57.5) b. (0.94, 75.41)

c. −39.5

10. a. 1.7 m b. $0.625\sin\left(\frac{x}{4}\right)$

c. 0.426 d. 4.756 m

5.6 Exam questions

Note: Mark allocations are available with the fully worked solutions online.

1. a.

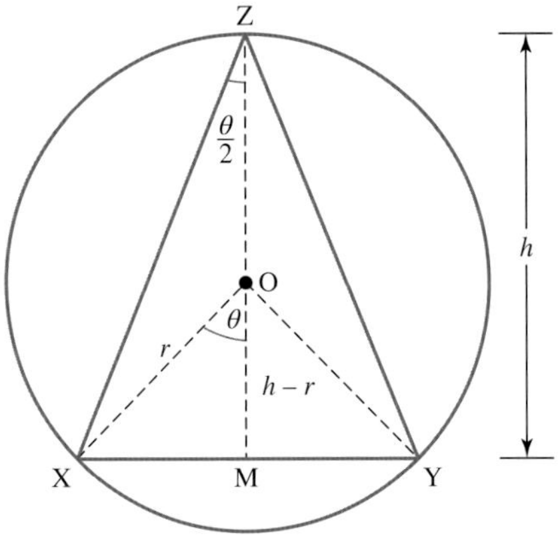

∠XOY = 2θ because the angle at the centre of the circle is twice the angle at the circumference.

$\angle XOM = \angle YOM = \frac{1}{2}\times 2\theta$

∠XOM = θ as required

b.
$$XM = r\sin(\theta)$$
$$\frac{XM}{h-r}=\tan(\theta)$$
$$\frac{r\sin(\theta)}{h-r}=\frac{\sin(\theta)}{\cos(\theta)}$$
$$\frac{r}{h-r}=\frac{1}{\cos(\theta)}$$
$$\frac{h-r}{r}=\cos(\theta)$$
$$\frac{h}{r}-1=\cos(\theta)$$
$$\frac{h}{r}=\cos(\theta)+1$$

c. $-3\sin(\theta)$

d. $-\frac{3}{2}$

2. a. $P=-2\cos(mt)+n$

When $t=0$, $P=4$:
$$4=-2\cos(0)+n$$
$$4+2=n$$
$$6=n$$
Period: $\frac{3}{2}=\frac{2\pi}{m}$
$$3m=4\pi$$
$$m=\frac{4\pi}{3}$$

b. $\frac{8\pi}{3}\sin\left(\frac{4\pi t}{3}\right)$

c. $\frac{8\pi}{3}$ m/min

3. a. 2.86 units at 7:55 am ($t=1.92$)

b. i.

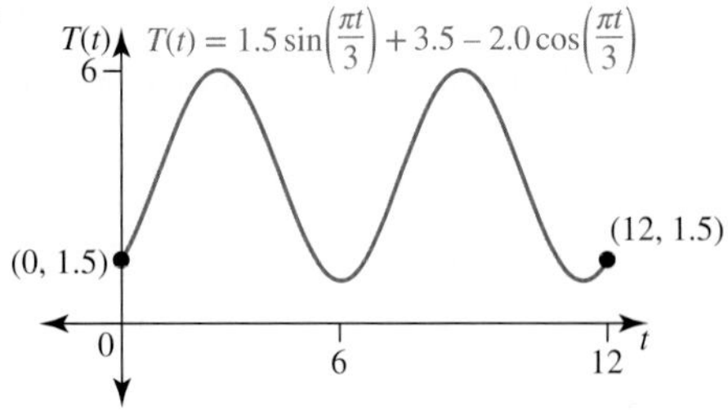

ii. Minimum 1 unit at 11:23 am ($t=5.39$) and 5:23 pm ($t=11.39$)
Maximum 6 units at 8:23 am ($t=2.39$) and 2:23 pm ($t=8.39$)

c. Emissions of 1 unit and 6 units lie within the guidelines.

5.7 Differentiation and application of logarithmic functions

5.7 Exercise

1. a. $\frac{7}{x}$

b. $\frac{2(3x^2+4x)}{x^3+2x^2-1}$

c. $\frac{3(2e^{3x}+1)}{e^{3x}-1}$

d. $\frac{3x^2-6x+7}{x^3-3x^2+7x-1}$

2. a. $\frac{4}{x}$

d. $6\tan(x)$

3. a. $\frac{dy}{dx}=-\frac{5}{x}, x\in(0, \infty)$

b. $\frac{dy}{dx}=-\frac{1}{x-2}, \; x\in(2, \infty)$

c. $\frac{dy}{dx}=-\frac{2}{(x+3)(x+1)}, x\in(-\infty, -3)\cup(-1, \infty)$

d. $\frac{dy}{dx}=\frac{2x-1}{x^2-x-6}, x\in(-\infty, -2)\cup(3, \infty)$

4. a. Domain = $(2, \infty)$, range = R

b. $a=3$

c. $y=2x-6$

d. $y=-\frac{1}{2}x+\frac{3}{2}$

5. a. $y = 2x - 3$ b. $y = \frac{3}{e}x$

6. Tangent: $y = 6x + 4\log_e(2) - 6$

7. a. $\frac{dy}{dx} = \frac{2}{x}$ b. $y = \frac{4}{e}x + e - 2$

8. A

9. $k = 7.4$

10. a. $\frac{2}{5}\log_5(e)$ b. $\frac{1}{9}\log_3(e)$

c. $\log_6(e)$ d. $\frac{1}{e}$

11. $n = 2$

5.7 Exam questions

Note: Mark allocations are available with the fully worked solutions online.

1. C
2. C
3. 1

5.8 Review

5.8 Exercise

Technology free: short answer

1. a.

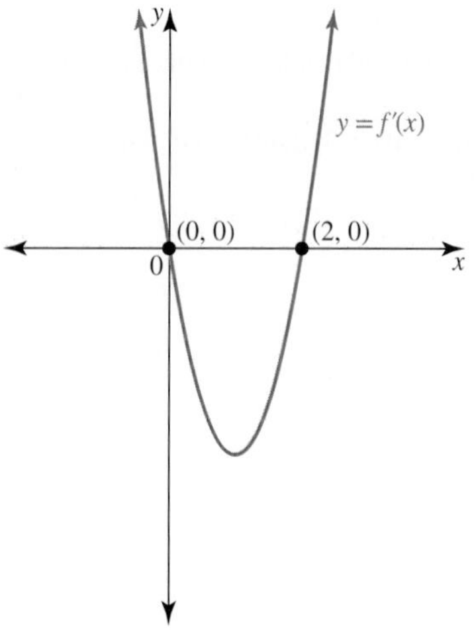

b. i. $\frac{-5}{(x-3)(x+2)}, x \in (-\infty, -2) \cup (3, \infty)$

ii. $\frac{2}{x+2}, x \in R \setminus \{-2\}$

2. a. $6\cos(2x)$

b. -3

c. $x \in \left[0, \frac{\pi}{4}\right)$

3. $y_T = \frac{\pi}{2}x - \frac{3\pi}{2} + 7, \ y_P = -\frac{2}{\pi}x + \frac{6}{\pi} + 7$

4. $y = -2x + 3$

5. a.

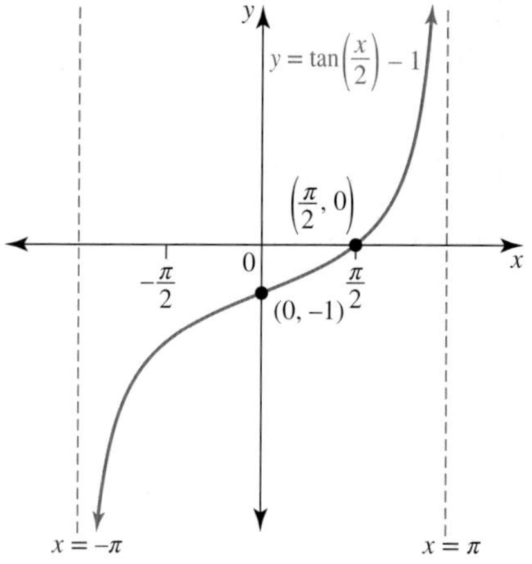

b. $f'(x) = \frac{1}{2}\sec^2\left(\frac{x}{2}\right) = \frac{1}{2\cos^2\left(\frac{x}{2}\right)} \ f'\left(\frac{\pi}{3}\right) = \frac{2}{3}$

6. a.
$$f(x) = ke^{x^2}$$
$$f'(x) = 2xke^{x^2}$$
$$f'(4) = 8ke^{16}$$
$$\therefore 8ke^{16} = 4e^{16}$$
$$\therefore k = \frac{1}{2}$$

b. $f'(2) = 2e^4$

c. $y = -\frac{x}{2e^4} + \frac{1}{e^4} + \frac{e^4}{2}$

Technology active: multiple choice

7. A
8. C
9. C
10. E
11. E
12. C
13. C
14. A
15. D
16. B

Technology active: extended response

17. a. $a = 3, b = 7$

b. Max = 7.6, min = −7.6; range = $[-7.6, 7.6]$

c. 1.9757, 5.1173

d. −7.616, 7.616

18. a. $M(0, 2.5)$, $P(\pi, 2.5)$

b. $\frac{dy}{dx} = 4\cos(2x)$

c. $t = \frac{5}{2} - \sqrt{3}$

d. $\frac{1}{2}$

e. $y = \frac{1}{2}x + \frac{5}{2} - \sqrt{3} - \frac{\pi}{3}$

19. a. $B_0 = 1.5$

b. $k = 0.1791$

c. $A_0 = 16.8696,\ h = -0.0628$

d. 0.44 units/day

20. a. $P_0 = 500,\ k = 0.0963$

b. 62 000 bacterial cells

c. 865.5 bacterial cells/hour

d. 62 000 bacterial cells

e. 1671 bacterial cells/hour

5.8 Exam questions

Note: Mark allocations are available with the fully worked solutions online.

1. a. $f(x) = x^2 + 3x + 5$
   $f(1) = 1 + 3 + 5 = 9 \neq 0$, so the point $P(1, 0)$ is not on the graph of $y = f(x)$

   b. i. $\dfrac{a^2 + 3a + 5}{a - 1}$

   ii. $f'(a) = 2a + 3$

   iii. $a = -2, 4$

   iv. $y = 11x - 11$ or $y = 1 - x$

2. C

3. D

4. a. Period $= 4\pi$, range $[-2 + \pi, 2 + \pi]$

   b. $f'(x) = -\sin\left(\dfrac{x}{2}\right)$

   c. $y = -x + 2\pi$

   d. $y = x - 2\pi,\ y = x - 6\pi$

   e. $x = \dfrac{3\pi}{2}, \dfrac{7\pi}{2}, \dfrac{11\pi}{2}, \dfrac{15\pi}{2}$

5. B

6 Further differentiation and applications

LEARNING SEQUENCE

Fully worked solutions for this topic are available online.

6.1 Overview

Hey students! Bring these pages to life online

Watch videos

Engage with interactivities

Answer questions and check results

Find all this and MORE in jacPLUS

6.1.1 Introduction

Differential calculus is one of the two subfields of calculus, the other being integral calculus. There are many and varied applications of differential calculus, some of which will be investigated in this topic.

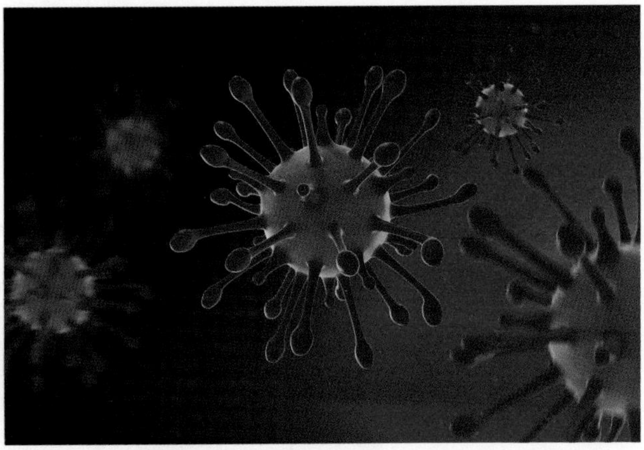

In physics, the derivative of the position of a moving object with respect to time is equal to the velocity of that object, and the derivative of the velocity of a moving object with respect to time is equal to the acceleration. The equation associated with Newton's second law of motion, $F = ma$, comes from a derivative statement as well.

The study of infectious diseases such as COVID-19 utilises calculus to determine how far and how fast it might be spreading, as well as where it might have come from. Calculations take into account factors such as people who are vulnerable to the disease, those who are already infected, and those that have had the disease and recovered. These types of situations change rapidly; therefore, the calculus equations must be dynamic so as to form new models with every new piece of information.

Weather prediction involves the use of calculus modelling through algorithms that take into account factors such as wind speed, moisture level and temperature.

KEY CONCEPTS

This topic covers the following key concepts from the VCE Mathematics Study Design:

- derivatives of $f(x) \pm g(x)$, $f(x) \times g(x)$, $\dfrac{f(x)}{g(x)}$ and $(f \circ g)(x)$ where f and g are polynomial functions, exponential, circular, or power functions and transformations or simple combinations of these functions
- application of differentiation to graph sketching and identification of key features of graphs, including stationary points and points of inflection, and intervals over which a function is strictly increasing or strictly decreasing
- identification of local maximum/minimum values over an interval and application to solving optimisation problems in context, including identification of interval endpoint maximum and minimum values.

6.2 The chain rule

LEARNING INTENTION

At the end of this subtopic you should be able to:
- use the chain rule to find the derivative of functions and equations of their tangents.

6.2.1 The proof of the chain rule

A composite function, also known as a function of a function, consists of two or more functions nested within each other. Consider the functions $g(x) = x^4$ and $h(x) = 2x + 1$.

If $f(x) = g(h(x))$, we are actually determining the rule for $g(2x + 1)$, so

$$f(x) = g(h(x)) = g(2x + 1) = (2x + 1)^4.$$

It is worth noting, however, that $g(h(x))$ is not necessarily equal to $h(g(x))$. In this instance, $h(g(x)) = h\left(x^4\right) = 2x^4 + 1$.

The **chain rule** for differentiation is another name for the derivative of a composite function.

Consider again $f(x) = (2x + 1)^4$. If this is expanded, it is possible to find the derivative.

$$\begin{aligned} f(x) &= (2x+1)^4 \\ &= (2x)^4 + 4(2x)^3(1) + 6(2x)^2(1)^2 + 4(2x)(1)^3 + 1^4 \\ &= 16x^4 + 32x^3 + 24x^2 + 8x + 1 \end{aligned}$$

Therefore, the derivative can be given by

$$\begin{aligned} f'(x) &= 64x^3 + 96x^2 + 48x + 8 \\ &= 8(8x^3 + 12x^2 + 6x + 1) \\ &= 8(2x+1)^3 \end{aligned}$$

The chain rule allows us to reach this same outcome without having to expand the function first.

The proof of the chain rule is as follows.

If $f(x) = m(n(x))$,

then $f(x + h) = m(n(x + h))$.

Therefore, $\dfrac{f(x+h) - f(x)}{h} = \dfrac{m(n(x+h)) - m(n(x))}{h}$.

Multiply the numerator and the denominator by $n(x + h) - n(x)$, as it is expected that at some stage $n'(x)$ will appear somewhere in the rule.

$$\frac{f(x+h) - f(x)}{h} = \frac{n(x+h) - n(x)}{h} \times \frac{m(n(x+h)) - m(n(x))}{n(x+h) - n(x)}$$

$$\begin{aligned} f'(x) &= \lim_{h \to 0} \left[\frac{m(n(x+h)) - m(n(x))}{n(x+h) - n(x)} \times \frac{n(x+h) - n(x)}{h} \right] \\ &= \lim_{h \to 0} \left[\frac{m(n(x+h)) - m(n(x))}{n(x+h) - n(x)} \right] \times \lim_{h \to 0} \left[\frac{n(x+h) - n(x)}{h} \right] \end{aligned}$$

By definition, $n'(x) = \lim_{h \to 0} \dfrac{n(x+h) - n(x)}{h}$. Also, if we let $n(x) = A$ and $n(x+h) = A + B$, then $n(x+h) - n(x) = A + B - A$, so that

$$\frac{m(n(x+h)) - m(n(x))}{n(x+h) - n(x)} = \frac{m(A+B) - n(A)}{B}.$$

Also, as $h \to 0,\ B \to 0$.

Consequently, $\lim_{B \to 0} \dfrac{m(A+B) - m(A)}{B} = m'(A)$.

Therefore, $\lim_{h \to 0} \left[\dfrac{m(n(x+h)) - m(n(x))}{n(x+h) - n(x)}\right] = m'(n(x))$.

Bringing this all together, we can state the rule as follows.

The chain rule

If $f(x) = m(n(x))$,

$f'(x) = m'(n(x)) \times n'(x)$.

Using Leibniz notation, this becomes

$\dfrac{dy}{dx} = \dfrac{dy}{du} \times \dfrac{du}{dx}$, where $y = f(u)$ and u is a function of x.

Consider again $y = f(x) = (2x+1)^4$. The chain rule can be used to find the derivative of this function.

Let $u = 2x + 1$; therefore, $\dfrac{du}{dx} = 2$.

Let $y = u^4$; therefore, $\dfrac{dy}{du} = 4u^3$.

By the chain rule, $\dfrac{dy}{dx} = \dfrac{dy}{du} \times \dfrac{du}{dx}$

$= 4u^3 \times 2$

$= 8u^3$

Since $u = 2x + 1$, $\dfrac{dy}{dx} = 8(2x+1)^3$.

In Topic 5, the following derivatives were given. We are now in a position to derive these using the chain rule.

$$\frac{d}{dx}(\sin(kx)) = k\cos(kx)$$
$$\frac{d}{dx}(\cos(kx)) = -k\sin(kx)$$
$$\frac{d}{dx}(\tan(kx)) = k\sec^2(kx) = \frac{k}{\cos^2(kx)}$$

We will prove the last of these facts.

If $y = \tan(kx)$,

let $u = kx$; therefore, $\dfrac{du}{dx} = k$.

Let $y = \tan(u)$; therefore, $\dfrac{dy}{du} = \sec^2(u)$

$$\frac{dy}{dx} = \frac{dy}{du} \times \frac{du}{dx}$$
$$\frac{dy}{dx} = \sec^2(u) \times k$$
$$\frac{dy}{dx} = k\sec^2(u)$$
$$\text{But } u = kx;\ \therefore\ \frac{dy}{dx} = k\sec^2(kx) = \frac{k}{\cos^2(kx)}$$

The other derivatives can be shown in a similar way.

The chain rule can also be applied without showing so much detail. Suppose you are asked to differentiate $y = \sin(e^{2x})$. First, differentiate the inner function.

$$\frac{d}{dx}\left(e^{2x}\right) = 2e^{2x}$$

Then differentiate the outer function.

$$\frac{d}{dx}(\sin(u)) = \cos(u), \text{ where } u = e^{2x}$$

The required derivative is the product of the two derivatives.

$$\frac{dy}{dx} = 2e^{2x} \times \cos\left(e^{2x}\right) = 2e^{2x}\cos\left(e^{2x}\right)$$

WORKED EXAMPLE 1 The chain rule

Use the chain rule to determine the derivative of $y = \left(x^3 + 2x^2 - x^{-2}\right)^{-7}$.

THINK	WRITE
1. Write the function to be derived.	$y = \left(x^3 + 2x^2 - x^{-2}\right)^{-7}$
2. Let u equal the inner function.	Let $u = x^3 + 2x^2 - x^{-2}$.
3. Differentiate to find $\dfrac{du}{dx}$.	$\dfrac{du}{dx} = 3x^2 + 4x + 2x^{-3}$
4. State the equation relating y and u.	$y = u^{-7}$
5. Differentiate to find $\dfrac{dy}{du}$.	$\dfrac{dy}{du} = -7u^{-8}$
6. Apply the chain rule.	$\dfrac{dy}{dx} = \dfrac{dy}{du} \times \dfrac{du}{dx}$ $= -7u^{-8} \times \left(3x^2 + 4x + 2x^{-3}\right)$
7. Substitute back in for u.	$\dfrac{dy}{dx} = -7\left(3x^2 + 4x + 2x^{-3}\right)\left(x^3 + 2x^2 - x^{-2}\right)^{-8}$

The chain rule must often be applied first before application problems involving the derivative can be solved.

WORKED EXAMPLE 2 Further use of the chain rule

If $y = \cos^2(e^x)$, evaluate the derivative when $x = 0$, giving your answer correct to 4 decimal places.

THINK	WRITE
1. Write the function to be derived.	$y = \cos^2(e^x) = [\cos(e^x)]^2$
2. Consider first the inner function.	$\frac{d}{dx}(\cos(e^x))$
3. Use the chain rule to differentiate this inner function.	$\frac{d}{dx}(\cos(e^x)) = -e^x \sin(e^x)$
4. Consider the outer function.	$\frac{d}{dx}(u^2)$, where $u = \cos(e^x)$
5. Apply the chain rule to differentiate this function.	$\frac{d}{dx}(u^2) = 2u$
6. The required derivative is the product of the two previously found derivatives.	$\frac{dy}{dx} = 2\cos(e^x) \times -e^x \sin(e^x)$ $= -2e^x \cos(e^x)\sin(e^x)$
7. Evaluate $\frac{dy}{dx}$ when $x = 0$.	Let $x = 0$. $\frac{dy}{dx} = -2e^0 \cos(e^0)\sin(e^0)$ $= -2(1)\cos(1)\sin(1)$ $= -0.9093$

WORKED EXAMPLE 3 Equations of the tangents using the chain rule

For the function with the rule $y = (x-1)^{\frac{2}{3}}$, determine:

a. $\frac{dy}{dx}$

b. **the equations of the tangents at (2, 1) and (0, 1).**

THINK	WRITE
a. 1. Write the function to be derived.	a. $y = (x-1)^{\frac{2}{3}}$
2. Apply the chain rule to find the derivative. Multiply the derivative of the outer function with the derivative of the inner function. Write the answer in surd form.	$\frac{dy}{dx} = \frac{2}{3}(x-1)^{-\frac{1}{3}} \times (1)$ $= \frac{2}{3\sqrt[3]{(x-1)}}$
b. 1. Find the gradient at $x = 2$.	b. When $x = 2$: $\frac{dy}{dx} = \frac{2}{3\sqrt[3]{2-1}}$ $= \frac{2}{3}$

2. Find the equation of the tangent at $x = 2, y = 1$.	If $m_T = \frac{2}{3}$, $(x_1, y_1) = (2, 1)$. $y - y_1 = m(x - x_1)$ $y - 1 = \frac{2}{3}(x - 2)$ $y - 1 = \frac{2}{3}x - \frac{4}{3}$ $y = \frac{2}{3}x - \frac{1}{3}$ or $2x - 3y = 1$
3. Find the gradient at $x = 0$.	When $x = 0$: $\frac{dy}{dx} = \frac{2}{3\sqrt[3]{0 - 1}}$ $= -\frac{2}{3}$
4. Find the equation of the tangent at $x = 0$, $y = 1$.	If $m_T = -\frac{2}{3}$, $(x_1, y_1) = (0, 1)$. $y - y_1 = m(x - x_1)$ $y - 1 = -\frac{2}{3}(x - 0)$ $y - 1 = -\frac{2}{3}x$ $y = -\frac{2}{3}x + 1$ or $2x + 3y = 3$

TI | THINK

a. 1. On a Calculator page, press MENU, then select:
1: Actions
1: Define
Complete the entry line as:
Define $f1(x) = (x - 1)^{\frac{2}{3}}$
then press ENTER.

DISPLAY/WRITE

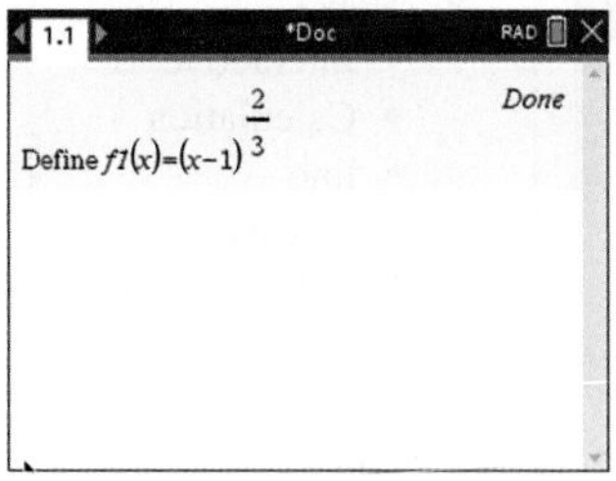

CASIO | THINK

a. 1. On a Main screen, select:
- Interactive
- Define

Complete the fields as
Func name: f
Variable/s: x
Expression: $(x - 1)^{\frac{2}{3}}$
then select OK.

DISPLAY/WRITE

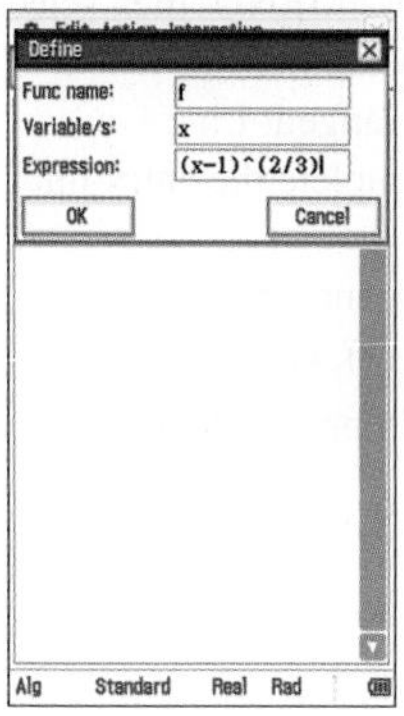

2. Press MENU, then select:
4: Calculus
1: Derivative
Complete the entry line as:
$\frac{d}{dx}(f1(x))$
then press ENTER.

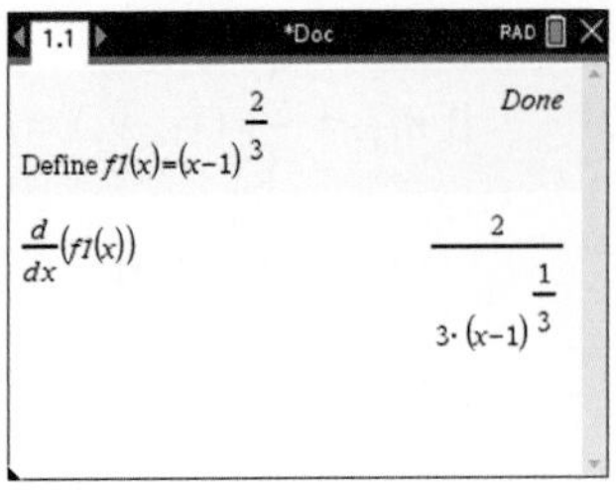

3. The answer appears on the screen.

$$\frac{d}{dx}\left((x-1)^{\frac{2}{3}}\right)=\frac{2}{3(x-1)^{\frac{1}{3}}}$$

b. 1. Press MENU, then select:
4: Calculus
9: Tangent Line
Complete the entry line as:
tangentLine
$(f1(x), x, 2)$
then press ENTER.

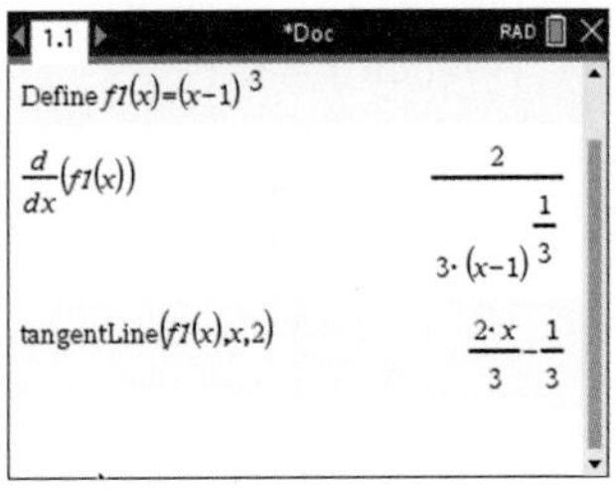

2. Press MENU, then select:
4: Calculus
9: Tangent Line
Complete the entry line as:
tangentLine
$(f1(x), x, 0)$
then press ENTER.

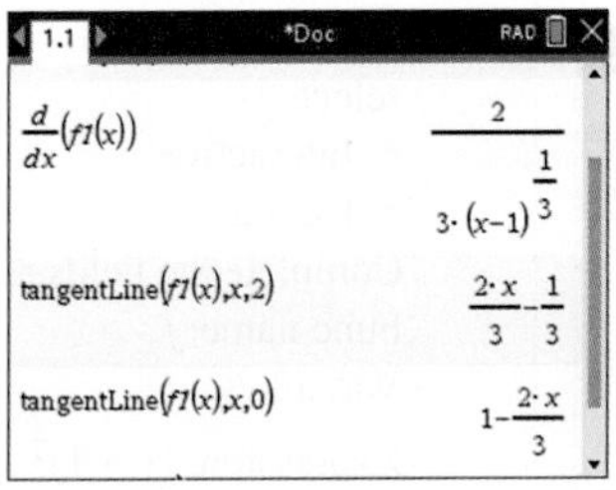

2. Select:
- Action
- Calculation
- diff

Complete the entry line as:
diff $(f(x))$
then press EXE.

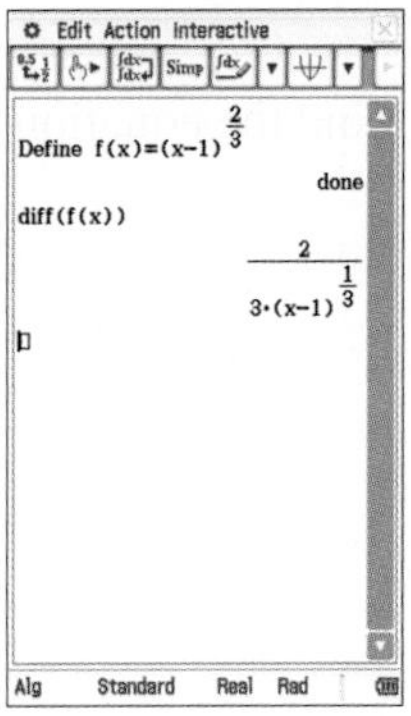

3. The answer appears on the screen.

$$\frac{d}{dx}\left((x-1)^{\frac{2}{3}}\right)=\frac{2}{3(x-1)^{\frac{1}{3}}}$$

b. 1. Select:
- Interactive
- Calculation
- line
- tanLine

Complete the fields as:
Expression: $f(x)$
Variable: x
Point: 2
then select OK.

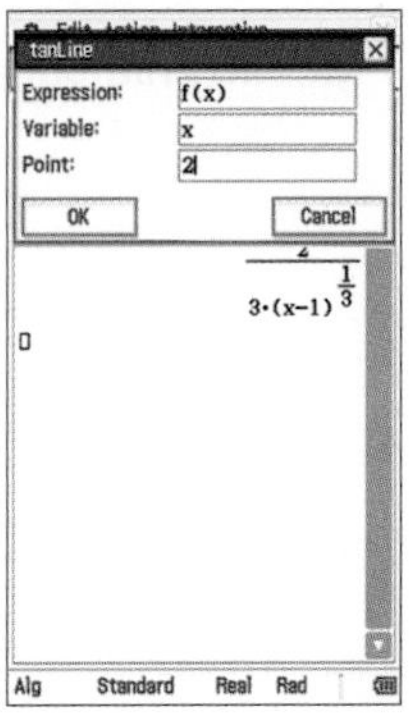

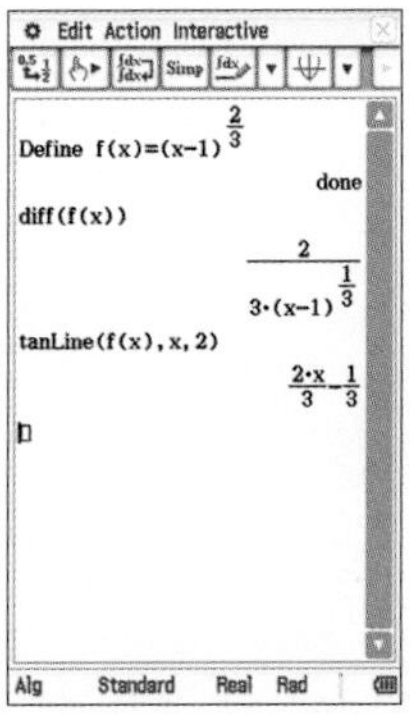

2. Select:
- Interactive
- Calculation
- line
- tanLine

Complete the fields as:
Expression: $f(x)$
Variable: x
Point: 0
then select OK.

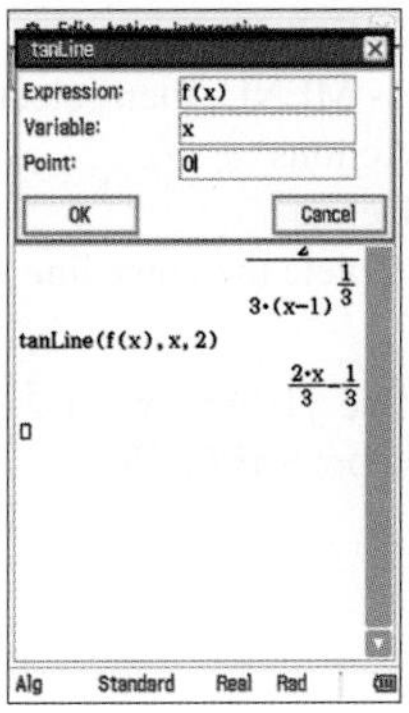

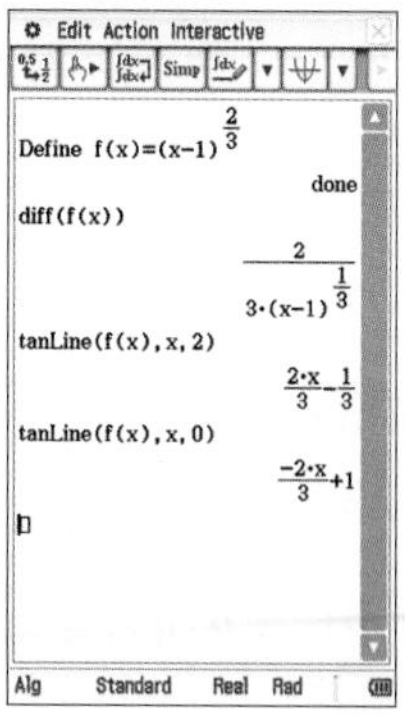

3. The answers appear on the screen.	The tangent to the curve at (2, 1) is $y=\frac{2x}{3}-\frac{1}{3}$ and the tangent to the curve at (0, 1) is $y=1-\frac{2x}{3}$.	**3.** The answers appear on the screen.	The tangent to the curve at (2, 1) is $y=\frac{2x}{3}-\frac{1}{3}$ and the tangent to the curve at (0, 1) is $y=-\frac{2x}{3}+1$.

6.2 Exercise

Technology free

1. **WE1** Use the chain rule to determine the derivatives of the following.

a. $y=\sqrt{x^2-7x+1}$ **b.** $y=\left(3x^2+2x-1\right)^3$ **c.** $y=\sin^2(x)$ **d.** $y=e^{\cos(3x)}$

2. State the derivatives of the following functions.

a. $g(x)=3\left(x^2+1\right)^{-1}$ **b.** $g(x)=e^{\cos(x)}$ **c.** $g(x)=\sqrt{(x+1)^2+2}$

d. $g(x)=\dfrac{1}{\sin^2(x)}$ **e.** $f(x)=\sqrt{x^2-4x+5}$ **f.** $f(x)=3\cos\left(x^2-1\right)$

3. State the derivatives of the following functions.

a. $f(x)=5e^{3x^2-1}$ **b.** $f(x)=\left(x^3-\dfrac{2}{x^2}\right)^{-2}$ **c.** $f(x)=\dfrac{\sqrt{2-x}}{2-x}$ **d.** $f(x)=\cos^3(2x+1)$

4. Determine the derivative of each of the following.

a. $\sqrt{\log_e(3-2x)}$ **b.** $\dfrac{1}{2}\log_e\left(\sqrt{x-2}\right)$

5. **WE2** If $y=\sin^3(x)$, evaluate the derivative when $x=\dfrac{\pi}{3}$.

6. Determine the derivatives of the following functions, and hence calculate the gradients at the given x-values.

a. $f(x)=\tan(4x+\pi)$; evaluate $f'\left(\dfrac{\pi}{4}\right)$. **b.** $f(x)=(2-x)^{-2}$; evaluate $f'\left(\dfrac{1}{2}\right)$.

c. $f(x)=e^{2x^2}$; evaluate $f'(-1)$. **d.** $f(x)=\sqrt[3]{\left(3x^2-2\right)^4}$; evaluate $f'(1)$.

7. a. If $f(x) = e^{\sin^2(x)}$, calculate $f'\left(\frac{\pi}{4}\right)$.

b. If $f(x) = (\cos(3x) - 1)^5$, calculate $f'\left(\frac{\pi}{2}\right)$.

8. Determine $f'(x)$ for each of the following.

a. $f(x) = g[\cos(x)]$

b. $f(x) = g\left(2x^3\right)$

c. $f(x) = g\left(3e^{2x+1}\right)$

d. $f(x) = g\left(\sqrt{2x^2 - x}\right)$

9. Determine $f'(x)$ for each of the following.

a. $f(x) = [h(x)]^{-2}$

b. $f(x) = \sin^2[h(x)]$

c. $f(x) = \sqrt[3]{2h(x) + 3}$

d. $f(x) = -2e^{h(x)+4}$

10. Let $f: R^+ \to R, f(x) = \frac{1}{x^2}$. Determine $g(x) = f(f(x))$ and hence state $g'(x)$.

11. **WE3** For the function $y = \frac{1}{(2x-1)^2}$, determine:

a. $\frac{dy}{dx}$

b. the equation of the tangent to the curve at the point where $x = 1$.

12. Let $f(x) = (x-1)^3$ and $g(x) = e^x$.

a. Write the rule for $f(g(x))$.

b. Determine $h'(x)$ where $h(x) = f(g(x))$.

c. Determine the equation of the tangent of $y = h(x)$ at the origin.

13. If $f(x) = \sin^2(2x)$, calculate the points where $f'(x) = 0$ for $x \in [0, \pi]$.

Technology active

14. If $z = 4y^2 - 5$ and $y = \sin(3x)$, find $\frac{dz}{dx}$.

15. For the functions $f(x) = 2\sin(x)$ and $h(x) = e^x$:

a. state the rule for:

i. $m(x) = f(h(x))$

ii. $n(x) = h(f(x))$

b. determine when $m'(x) = n'(x)$ over the interval $x \in [0, 3]$, correct to 3 decimal places.

16. For the functions $m(x) = 3^x$ and $n(x) = x^2 + 4x - 5$:

a. state the rule of $m(n(x))$

b. calculate the gradient of the function at the point where $x = 1$.

17. For the functions with the rules $f(x) = \sqrt[3]{x^2}$ and $h(x) = 2x - 1$:

a. define the rule for $g(x) = f(h(x))$

b. determine $g'(x)$

c. determine the equations of the tangents at the points $(1, 1)$ and $(0, 1)$

d. determine the coordinates of the point of intersection of these two tangents.

18. The function f is defined by $f: [0, \infty) \to R, f(x) = \left(2\log_e(x)\right)^2$, and the function g is defined by $g: [0, \infty) \to R, f(x) = 2\log_e(x)$.

a. Calculate the coordinates of the points of intersection between f and g.

b. Calculate the gradient of each graph at the point $(1, 0)$.

c. Sketch both graphs on the same set of axes.

d. Determine the x-values for which $2\log_e x > \left(2\log_e(x)\right)^2$.

19. The function h has a rule $h(x) = \sqrt{x^2 - 16}$ and the function g has the rule $g(x) = x - 3$.

a. Determine the integers m and n such that $h(g(x)) = \sqrt{(x+m)(x+n)}$.

b. State the maximal domain of $h(g(x))$.

c. Determine the derivative of $h(g(x))$.
d. Calculate the gradient of the function $h(g(x))$ at the point when $x=-2$.

20. The line perpendicular to the graph $y=g(f(x))$ where $f(x)=\frac{1}{x}$ and $g(x)=x-\frac{1}{x^2}$ is given by $y=-x+a$, where a is a real constant. Calculate the possible value(s) of a.

6.2 Exam questions

Question 1 (1 mark) TECH-ACTIVE

Source: *VCE 2020, Mathematical Methods Exam 2, Section A, Q7; © VCAA.*

MC If $f(x)=e^{g(x^2)}$, where g is a differentiable function, then $f'(x)$ is equal to

A. $2xe^{g(x^2)}$
B. $2xg(x^2)\,e^{g(x^2)}$
C. $2xg'(x^2)\,e^{g(x^2)}$
D. $2xg'(2x)\,e^{g(x^2)}$
E. $2xg'(x^2)\,e^{g(2x)}$

Question 2 (1 mark) TECH-FREE

Source: *VCE 2015, Mathematical Methods (CAS) Exam 1, Q1a; © VCAA.*

Let $y=(5x+1)^7$.

Find $\frac{dy}{dx}$.

Question 3 (3 marks) TECH-FREE

Source: *VCE 2014, Mathematical Methods (CAS) Exam 1, Q1b; © VCAA.*

If $f(x)=\sqrt{x^2+3}$, find $f'(1)$.

More exam questions are available online.

6.3 The product rule

LEARNING INTENTION

At the end of this subtopic you should be able to:

- use the product rule to find the derivative of functions, equations of their tangents and determine stationary points.

6.3.1 The proof of the product rule

There are many functions that have rules which are the product of two simpler functions, such as $x\sin(x)$ or $e^x(2x+1)$. In order to differentiate such functions, we need to apply the **product rule**.

The product rule

If $f(x)=g(x)\times h(x)$, then

$$f'(x)=g(x)\times h'(x)+h(x)\times g'(x)$$

Alternatively, if $y=uv$, then

$$\frac{dy}{dx}=u\frac{dv}{dx}+v\frac{du}{dx}$$

This rule can be proven as follows.

Let $f(x) = u(x)v(x)$

so $f(x+h) = u(x+h)v(x+h)$.

$$\frac{f(x+h) - f(x)}{h} = \frac{u(x+h)v(x+h) - u(x)v(x)}{h}$$

Add and subtract $u(x)v(x+h)$, as it is expected that at some stage $v'(x)$ will appear somewhere in the rule.

$$\begin{aligned}\frac{f(x+h) - f(x)}{h} &= \frac{u(x+h)v(x+h) - u(x)v(x+h) + u(x)v(x+h) - u(x)v(x)}{h} \\ &= \frac{[u(x+h) - u(x)]v(x+h) + u(x)[v(x+h) - v(x)]}{h}\end{aligned}$$

$$\begin{aligned}f'(x) &= \lim_{h\to 0} \frac{f(x+h) - f(x)}{h} \\ &= \lim_{h\to 0} \frac{[u(x+h) - u(x)]v(x+h) + u(x)[v(x+h) - v(x)]}{h} \\ &= \lim_{h\to 0} \left[\frac{u(x+h) - u(x)}{h} \times v(x+h)\right] + \lim_{h\to 0} \left[\frac{v(x+h) - v(x)}{h} \times u(x)\right] \\ &= \lim_{h\to 0} \frac{u(x+h) - u(x)}{h} \times \lim_{h\to 0} v(x+h) + \lim_{h\to 0} \frac{v(x+h) - v(x)}{h} \times \lim_{h\to 0} u(x) \\ &= u'(x)v(x) + v'(x)u(x) \\ &= u(x)v'(x) + v(x)u'(x)\end{aligned}$$

The Leibniz notation states that if $y = uv$,

$$\frac{dy}{dx} = u\frac{dv}{dx} + v\frac{du}{dx}$$

Apply this to differentiate $y = x\sin(x)$.

Let $u = x$; therefore, $\frac{du}{dx} = 1$.

Let $v = \sin(x)$, so $\frac{dv}{dx} = \cos(x)$.

$$\begin{aligned}\frac{dy}{dx} &= u\frac{dv}{dx} + v\frac{du}{dx} \\ \frac{dy}{dx} &= x \times \cos(x) + \sin(x) \times 1 \\ &= x\cos(x) + \sin(x)\end{aligned}$$

For another example, differentiate $y = e^x(2x+1)$.

Let $u = e^x$; therefore, $\frac{du}{dx} = e^x$

Let $v = 2x + 1$, so $\frac{dv}{dx} = 2$.

$$\begin{aligned}\frac{dy}{dx} &= u\frac{dv}{dx} + v\frac{du}{dx} \\ &= e^x \times 2 + (2x+1) \times e^x \\ &= 2e^x + 2xe^x + e^x \\ &= 3e^x + 2xe^x\end{aligned}$$

WORKED EXAMPLE 4 The product rule

If $g(x) = x^3 \sin(3x)$, determine $g'(x)$.

THINK	WRITE
1. Define u and v as functions of x.	$g(x) = x^3 \sin(3x)$ Let $u(x) = x^3$ and $v(x) = \sin(3x)$.
2. Differentiate u and v with respect to x.	$u'(x) = 3x^2 \quad v'(x) = 3\cos(3x)$
3. Apply the product rule to determine $g'(x)$.	$g'(x) = u(x)v'(x) + v(x)u'(x)$ $= x^3 \times 3\cos(3x) + \sin(3x) \times 3x^2$ $= 3x^3 \cos(3x) + 3x^2 \sin(3x)$

The product rule may have to be used first before an application problem can be solved.

WORKED EXAMPLE 5 Equation of tangents using the product rule

Given that $y = e^{2x}(x+1)^2$, determine $\frac{dy}{dx}$ and hence calculate the equation of the tangent to the curve at the point (0, 1).

THINK	WRITE
1. Define u and v as functions of x.	$y = e^{2x}(x+1)^2$ Let $u = e^{2x}$ and $v = (x+1)^2$.
2. Differentiate u and v with respect to x.	$\frac{du}{dx} = 2e^{2x} \quad \frac{dv}{dx} = 2(x+1)$
3. Apply the product rule to determine $\frac{dy}{dx}$ and simplify.	$\frac{dy}{dx} = u\frac{dv}{dx} + v\frac{du}{dx}$ $\frac{dy}{dx} = e^{2x} \times 2(x+1) + (x+1)^2 \times 2e^{2x}$ $= 2e^{2x}(x+1) + 2e^{2x}(x+1)^2$ $= 2e^{2x}(x+1)(1+x+1)$ $= 2e^{2x}(x+1)(x+2)$
4. Evaluate $\frac{dy}{dx}$ when $x = 0$.	When $x = 0$, then $\frac{dy}{dx} = 2e^0(0+1)(0+2)$ $= 4$
5. Find the equation of the tangent.	If $m = 4$ and $(x_1, y_1) = (0, 1)$, $y - y_1 = m(x - x_1)$ $y - 1 = 4(x - 0)$ $y - 1 = 4x$ $y = 4x + 1$

Frequently, problems may involve graphs of a function being given so that aspects of the function can be investigated.

WORKED EXAMPLE 6 Using the product rule to determine stationary points

The graph of $f: R \to R, f(x) = x^2e^x$ is shown. Using calculus, calculate the coordinates where $f'(x) = 0$.

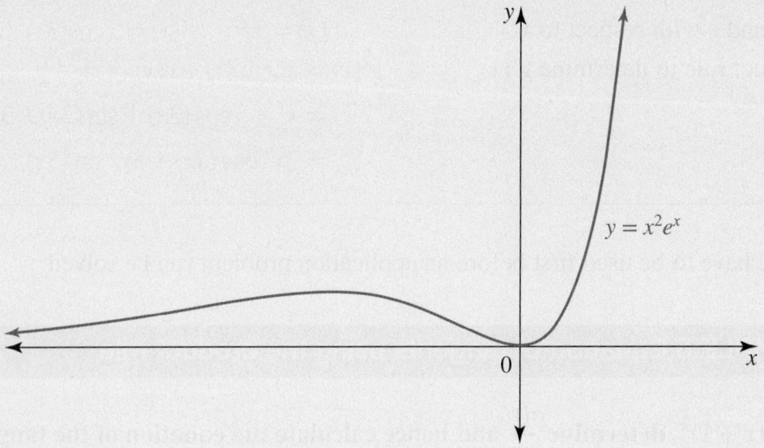

THINK	WRITE
1. Define u and v as functions of x.	$f(x) = x^2e^x$ Let $u(x) = x^2$ and $v(x) = e^x$.
2. Differentiate u and v with respect to x.	$u'(x) = 2x \quad v'(x) = e^x$
3. Apply the product rule to determine $f'(x)$.	$f'(x) = u(x)v'(x) + v(x)u'(x)$ $= x^2 \times e^x + e^x \times 2x$ $= x^2e^x + 2xe^x$
4. Solve $f'(x) = 0$.	$x^2e^x + 2xe^x = 0$ $e^x x(x+2) = 0$ $e^x > 0$ for all values of x. Either $x = 0$ or $x + 2 = 0$. $\therefore x = 0, -2$
5. Substitute the x-values to find the corresponding y-values.	When $x = -2$, $y = (-2)^2e^{-2}$ $= 4e^{-2}$ When $x = 0$, $y = (0)^2e^0$ $= 0$
6. Write the answer.	The coordinates where the gradient is zero are $(0, 0)$ and $(-2, 4e^{-2})$.

6.3 Exercise

Students, these questions are even better in jacPLUS

Receive immediate feedback and access sample responses

Access additional questions

Track your results and progress

Find all this and MORE in jacPLUS

Technology free

1. **WE4** For each of the following functions, determine $f'(x)$.

   a. $f(x) = \sin(3x)\cos(3x)$
   b. $f(x) = x^2 e^{3x}$
   c. $f(x) = \left(x^2 + 3x - 5\right)e^{5x}$
   d. $f(x) = \sqrt{x^2 - 1}\,\tan(3x)$

2. Differentiate the following.

   a. $x^2 e^{5x}$
   b. $e^{2x+1}\tan(2x)$
   c. $x^{-2}(2x+1)^3$
   d. $x\cos(x)$
   e. $2\sqrt{x}\,(4-x)$
   f. $\sin(2x-\pi)e^{-3x}$

3. Differentiate the following.

   a. $3x^{-2}e^{x^2}$
   b. $e^{2x}\sqrt{4x^2-1}$
   c. $x^2\sin^3(2x)$
   d. $(x-1)^4(3-x)^{-2}$
   e. $(3x-2)^2 g(x)$
   f. $-e^{5x}g\left(\sqrt{x}\right)$

4. Determine the derivative of each of the following. State any restrictions on x.

   a. $(x^2 - 3x + 7)\log_e(2x-1)$
   b. $\sin(x)\log_e(x^2)$

5. If $f(x) = 2x^4\cos(2x)$, calculate $f'\left(\frac{\pi}{2}\right)$.

6. Determine the derivatives of the following functions, and hence calculate the gradient at the given points.

   a. $f(x) = xe^x$; find $f'(-1)$.
   b. $f(x) = x\left(x^2 + x\right)^4$; find $f'(1)$.
   c. $f(x) = (1-x)\tan^2(x)$; find $f'\left(\frac{\pi}{3}\right)$.
   d. $f(x) = \sqrt{x}\sin\left(2x^2\right)$; find $f'\left(\sqrt{\pi}\right)$.

7. a. Given the function $f(x) = (x+1)\sin(x)$, determine $f'(x)$ and hence evaluate the gradient of the function when $x = 0$.

   b. Let $y = 2x\tan(2x)$. Evaluate $\frac{dy}{dx}$ when $x = \frac{\pi}{12}$, giving your answer in exact form.

Technology active

8. Differentiate $y = 2^x\sin(x)$ and evaluate the value of $\frac{dy}{dx}$ when $x = \frac{\pi}{2}$. Give your answer correct to 2 decimal places.

9. **MC** Given that $f: R \to R$, $f(x) = x^4 e^{-3x}$ and that $f'(x)$ may be written in the form $f'(x) = e^{-3x}\left(ax^3 + bx^4\right)$, the constants a and b are respectively:

   A. -4 and 3
   B. 4 and 3
   C. 4 and -3
   D. 12 and 3
   E. 12 and -3

10. **WE6** Given $f(x) = 2x^2(1-x)^3$, use calculus to determine the coordinates where $f'(x) = 0$.

11. The graph of $f: R^+ \to R, f(x) = e^{-\frac{x}{2}} \sin(x)$ is shown.

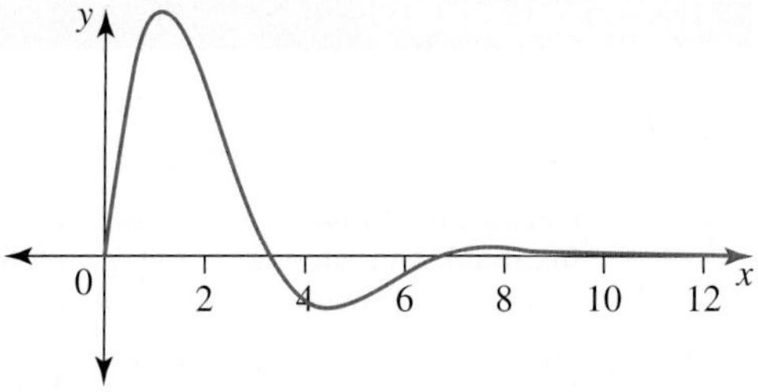

a. Calculate the values of x when $f(x) = 0$ for $x \in [0, 3\pi]$.
b. Use calculus to determine the values of x when $f'(x) = 0$ for $x \in [0, 3\pi]$. Give your answers correct to 2 decimal places.

12. Determine the derivative of $y = -\cos(x)\tan(x)$ by:
a. simplifying the expression first
b. applying the product rule and then simplifying.

13. a. If $f(x) = (x - a)^2 g(x)$, determine the derivative of f.
b. If $f(x) = g(x)\sin(2x)$ and $f'\left(\frac{\pi}{2}\right) = -3\pi$, calculate the constant a if $g(x) = ax^2$.

14. **WE5** Given that $y = \left(x^2 + 1\right) e^{3x}$, determine $\frac{dy}{dx}$ and hence calculate the equation of the tangent to the curve at $x = 0$.

15. For the function with the rule $y = xe^x$, determine the equations of the tangent and the line perpendicular to the curve at the point where $x = 1$.

16. The graph of $y = e^{-x^2}(1 - x)$ is shown.

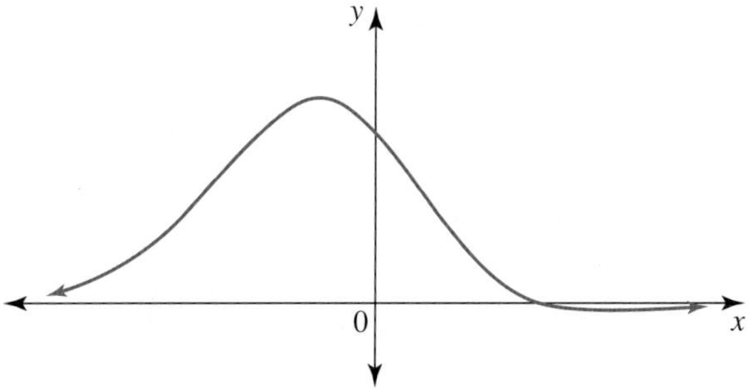

a. Evaluate the coordinates of the points where the graph cuts the x- and y-axes.
b. Calculate the coordinates of the points where the gradient is zero, giving your answers correct to 3 decimal places.
c. Determine the equation of the tangent to the curve at the point where the curve intersects the x-axis.
d. Determine the equation of the line perpendicular to the curve where the curve crosses the y-axis.
e. Determine where the tangent and the perpendicular line from parts **c** and **d** intersect. Give your answer correct to 2 decimal places.

17. A country town has decided to construct a new road. The x-axis is also the position of the railway line that connects Sydney with Brisbane. The road can be approximated by the equation$y = \left(4x^2 - 5x\right) e^x$.

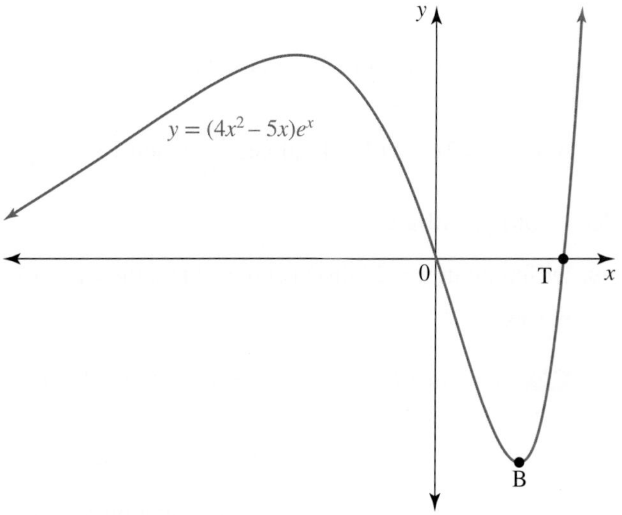

a. The post office for the town is positioned at $(-2, 3.5)$. They want the new road to be adjacent to the post office. Explain whether they have made a sensible decision regarding the placement of the road.
b. Calculate the coordinates of the point T where the road crosses the railway line.
c. Use calculus to find the coordinates of the point B. Give your answer correct to 3 decimal places.

18. The graph of the function $f: R \to R,\ f(x) = 3x^3 e^{-2x}$ is shown. The derivative may be written as $f'(x) = ae^{-2x}\left(bx^2 + cx^3\right)$, where a, b and c are constants.

   a. Calculate the exact values of a, b and c.

   b. Calculate the exact coordinates where $f'(x) = 0$.

   c. Determine the equation of the tangent to the curve at $x = 1$.

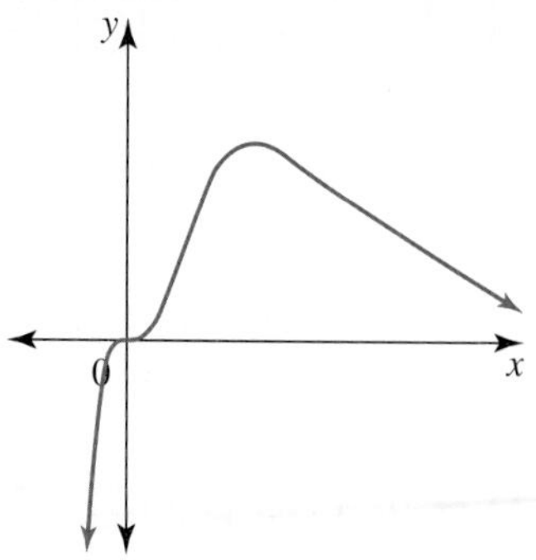

19. An artist has been commissioned to produce a sculpture for an art gallery. The artist intends to construct a Perspex square-based pyramid as shown.
She also plans to have an animal-themed sculpture inside the pyramid. Each face of the pyramid is an isosceles triangle as shown.
D is the midpoint of AB. Angles CAB and CBA are each θ radians.

   a. i. Determine CD in terms of θ.
   ii. Determine BD in terms of θ.

   b. Show that the total surface area, $S\,\text{m}^2$, of the pyramid, including the base, is given by $S = 36\left(\cos^2(\theta) + \cos(\theta)\sin(\theta)\right)$.

   c. Determine $\dfrac{dS}{d\theta}$.

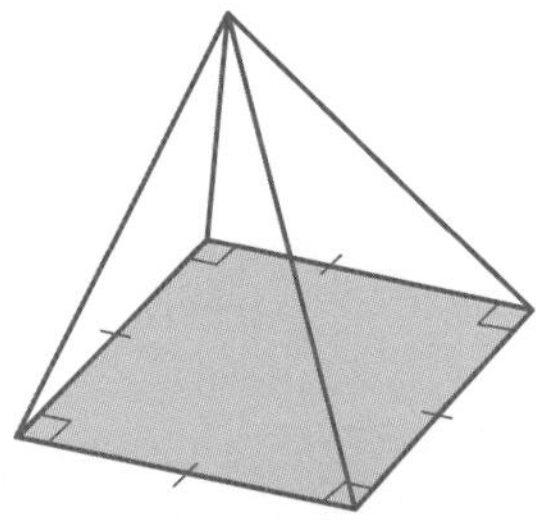

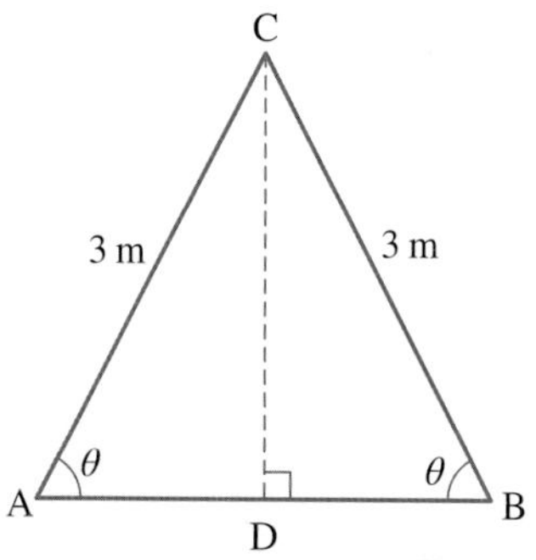

6.3 Exam questions

Question 1 (1 mark) TECH-FREE

Source: *VCE 2020, Mathematical Methods Exam 1, Q1a; © VCAA.*

Let $y = x^2 \sin(x)$.

Find $\dfrac{dy}{dx}$.

Question 2 (2 marks) TECH-FREE

Source: *VCE 2016, Mathematical Methods Exam 1, Q1b; © VCAA.*

Let $f(x) = x^2 e^{5x}$.

Evaluate $f'(1)$.

Question 3 (2 marks) TECH-FREE

Source: *VCE 2013, Mathematical Methods (CAS) Exam 1, Q1a; © VCAA.*

If $y = x^2 \log_e(x)$, find $\dfrac{dy}{dx}$

More exam questions are available online.

6.4 The quotient rule

LEARNING INTENTION

At the end of this subtopic you should be able to:
- use the quotient rule to find the derivative of functions and the gradient at a given point.

6.4.1 The proof of the quotient rule

When one function is divided by a second function, for example $f(x) = \dfrac{x}{x^2 - 1}$ or $f(x) = \dfrac{e^x}{\cos(x)}$, we have the quotient of the two functions. For such functions, there is a rule for finding the derivative. It is called the **quotient rule**.

The quotient rule

If $f(x) = \dfrac{u(x)}{v(x)}$, where $v(x) \neq 0$, then

$$f'(x) = \frac{v(x)u'(x) - u(x)v'(x)}{[v(x)]^2}$$

Alternatively, if $y = \dfrac{u}{v}$, then

$$\frac{dy}{dx} = \frac{v\frac{du}{dx} - u\frac{dv}{dx}}{v^2}$$

This rule can be proven as follows by using the product rule.

$$\text{If } f(x) = \frac{u(x)}{v(x)}, \text{ then } f(x) = u(x) \times [v(x)]^{-1}$$

$$\begin{aligned} f'(x) &= u(x) \times -1 \times [v(x)]^{-2} \times v'(x) + [v(x)]^{-1} \times u'(x) \\ &= -\frac{u(x)\,v'(x)}{[v(x)]^2} + \frac{u'(x)}{[v(x)]} \\ &= \frac{u'(x)\,v(x)}{[v(x)]^2} - \frac{u(x)\,v'(x)}{[v(x)]^2} \\ &= \frac{v(x)\,u'(x) - u(x)\,v'(x)}{[v(x)]^2} \end{aligned}$$

The Leibnitz notation for the quotient rule states that if $y = \dfrac{u}{v}$, then $\dfrac{dy}{dx} = \dfrac{v\frac{du}{dx} - u\frac{dv}{dx}}{v^2}$.

For example, we can differentiate $y = \dfrac{x}{x^2 - 1}$ as follows.

Let $u = x$ so that $\frac{du}{dx} = 1$ and let $v = x^2 - 1$ so that $\frac{dv}{dx} = 2x$.

$$\begin{aligned}\frac{dy}{dx} &= \frac{v\frac{du}{dx} - u\frac{dv}{dx}}{v^2} \\ &= \frac{\left(x^2-1\right)(1) - x(2x)}{\left(x^2-1\right)^2} \\ &= \frac{x^2 - 1 - 2x^2}{\left(x^2-1\right)^2} \\ &= \frac{-x^2-1}{\left(x^2-1\right)^2} \\ &= -\frac{x^2+1}{\left(x^2-1\right)^2}\end{aligned}$$

Note that although the numerator has been factorised and simplified, it is more common not to expand the denominator.

Always check that the quotient rule is the best method to use to differentiate the function. For example, $y = \frac{x-2}{\sqrt{x-2}}$ can be broken down to $y = \sqrt{x-2}$; therefore, the chain rule should be used. Also, $y = \frac{5x^2 - 2x}{\sqrt{x}}$ can be split into separate fractions, and each term can be differentiated using the basic differentiation rule. Before applying the quotient rule, always check if the function can be simplified first.

WORKED EXAMPLE 7 The quotient rule

Determine the derivative of $y = \frac{\sin(2t)}{t^2}$ with respect to t.

THINK	WRITE
1. Define u and v as functions of t.	$y = \frac{\sin(2t)}{t^2}$ Let $u = \sin(2t)$ and $v = t^2$.
2. Differentiate u and v with respect to t.	$\frac{du}{dt} = 2\cos(2t)$ $\frac{dv}{dt} = 2t$
3. Apply the product rule to determine $\frac{dy}{dt}$ and simplify.	$\frac{dy}{dt} = \frac{v\frac{du}{dt} - u\frac{dv}{dt}}{v^2}$ $= \frac{t^2 \times 2\cos(2t) - \sin(2t) \times 2t}{\left(t^2\right)^2}$ $= \frac{2t(t\cos(2t) - \sin(2t))}{t^4}$ $= \frac{2(t\cos(2t) - \sin(2t))}{t^3}$

WORKED EXAMPLE 8 Further use of the quotient rule

Determine the derivative of $f(x) = \dfrac{\cos(3x)}{2e^x - x}$ and hence calculate the gradient at the point where $x = 0$.

THINK	WRITE
1. Define u and v as functions of x.	$f(x) = \dfrac{\cos(3x)}{2e^x - x}$ Let $u(x) = \cos(3x)$ and $v(x) = 2e^x - x$.
2. Differentiate u and v with respect to x.	$u'(x) = -3\sin(3x)$ $v'(x) = 2e^x - 1$
3. Apply the product rule to determine $\dfrac{dy}{dx}$ and simplify.	$f'(x) = \dfrac{v(x)u'(x) - u(x)v'(x)}{[v(x)]^2}$ $= \dfrac{(2e^x - x) \times -3\sin(3x) - \cos(3x) \times (2e^x - 1)}{(2e^x - x)^2}$ $= \dfrac{-3(2e^x - x)\sin(3x) - (2e^x - 1)\cos(3x)}{(2e^x - x)^2}$
4. Evaluate $f'(0)$.	$f'(0) = \dfrac{-3\left(2e^0 - 0\right)\sin(0) - \left(2e^0 - 1\right)\cos(0)}{\left(2e^0 - 0\right)^2}$ $= \dfrac{0 - 1}{4}$ $= -\dfrac{1}{4}$

6.4 Exercise

Technology free

1. WE7 Use the quotient rule to determine the derivatives of:

   a. $\dfrac{e^{2x}}{e^x + 1}$

   b. $\dfrac{\cos(3t)}{t^3}$.

2. Determine the derivative of $\dfrac{x + 1}{x^2 - 1}$.

3. Differentiate the following.

   a. $\dfrac{\sin(x)}{\sqrt{x}}$

   b. $\dfrac{\tan(2x)}{e^x}$

   c. $f(x) = \dfrac{(5 - x)^2}{\sqrt{5 - x}}$

   d. $y = \dfrac{\sin^2\left(x^2\right)}{x}$

   e. $y = \dfrac{3x - 1}{2x^2 - 3}$

   f. $f(x) = \dfrac{x - 4x^2}{2\sqrt{x}}$

4. Differentiate the following.

a. $\dfrac{e^x}{\cos(2x+1)}$
b. $\dfrac{e^{-x}}{x-1}$
c. $y=\dfrac{3\sqrt{x}}{x+2}$
d. $y=\dfrac{\cos(3x)}{\sin(3x)}$
e. $\dfrac{x-2}{2x^2-x-6}$
f. $\dfrac{1-e^{2x}}{1+e^{2x}}$

5. Determine the derivative of each of the following.

a. $\dfrac{\log_e(x^2)}{2x-1}$
b. $\dfrac{2\log_e(2x)}{e^{2x}+1}$

6. Differentiate the following functions.

a. $f(x)=\dfrac{x+2}{\sin(g(x))}$
b. $f(x)=\dfrac{g\left(e^{-2x}\right)}{e^x}$

7. **WE8** If $y=\dfrac{\sin(x)}{e^{2x}}$, determine the gradient and hence calculate the gradient of the function at the point where $x=0$.

8. Evaluate the gradient at the stated point for each of the following functions.

a. $y=\dfrac{2x}{x^2+1}$, $x=1$
b. $y=\dfrac{\sin(2x+\pi)}{\cos(2x+\pi)}$, $x=\dfrac{\pi}{2}$
c. $y=\dfrac{x+1}{\sqrt{3x+1}}$, $x=5$
d. $y=\dfrac{5-x^2}{e^x}$, $x=0$

9. a. Calculate the gradient of the tangent to the curve with equation $y=\dfrac{2x}{(3x+1)^{\frac{3}{2}}}$ at the point where $x=1$.

b. Determine the equation of the tangent to the curve $y=\dfrac{e^x}{x^2+2}$ when $x=0$.

Technology active

10. Consider the curve defined by the rule $y=\dfrac{2x-1}{3x^2+1}$.

a. Determine the rule for the gradient.
b. Determine the value(s) of x for which the gradient is equal to 0.875. Give your answers correct to 4 decimal places.

11. Let $f:[-2,2]\to R, f(x)=\dfrac{\sin(2x-3)}{e^x}$. The graph of this function is shown.

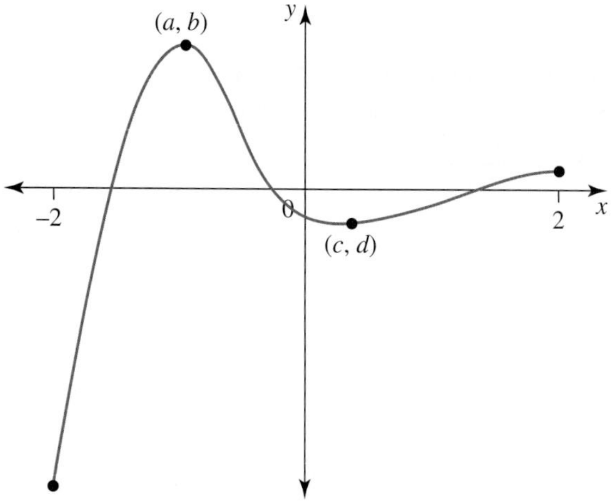

a. The stationary points occur at (a, b) and (c, d). Calculate the values of a, b, c and d, giving your answers correct to 3 decimal places.
b. Calculate the gradient of the tangent to the curve at the point where $x=1$, correct to 3 decimal places.

12. Show that $\frac{d}{dx}\left(\frac{1+\cos(x)}{1-\cos(x)}\right) = -\frac{-2\sin(x)}{(\cos(x)-1)^2}$.

13. **MC** Given that $f(x) = \frac{\sqrt{2x-1}}{\sqrt{2x+1}}$, the value of m such that $f'(m) = \frac{2}{5\sqrt{15}}$ is:

A. $m = \pm 2$
B. $m = 2$
C. $m = -1$
D. $m = \sqrt{2}$
E. $m = \pm\sqrt{2}$

14. If $\frac{d}{dx}\left(\frac{e^{-3x}}{e^{2x}+1}\right) = \frac{e^{-x}\left(a+be^{-2x}\right)}{\left(e^{2x}+1\right)^2}$, calculate the exact values of a and b.

15. Determine the derivative of the function $f : R \to R,\ f(x) = \frac{10x}{x^2+1}$ and calculate when the gradient is negative.

16. For the curve with the rule $y = \frac{x-5}{x^2+5x-14}$:

a. state when the function is undefined
b. determine the coordinates when the gradient is zero
c. determine the equation of the tangent to the curve at the point where $x = 1$.

6.4 Exam questions

Question 1 (2 marks) TECH-FREE

Source: *VCE 2018, Mathematical Methods Exam 1, Q1b; © VCAA.*

Let $f(x) = \frac{e^x}{\cos(x)}$.

Evaluate $f'(\pi)$.

Question 2 (2 marks) TECH-FREE

Source: *VCE 2017, Mathematical Methods Exam 1, Q1a; © VCAA.*

Let $f : (-2, \infty) \to R, f(x) = \frac{x}{x+2}$.

Differentiate f with respect to x.

Question 3 (2 marks) TECH-FREE

Source: *VCE 2016, Mathematical Methods Exam 1, Q1a; © VCAA.*

Let $y = \frac{\cos(x)}{x^2+2}$.

Find $\frac{dy}{dx}$.

More exam questions are available online.

6.5 Curve sketching

LEARNING INTENTION

At the end of this subtopic you should be able to:

- graph functions and use differentiation to identify their stationary points and their nature
- identify local maximum/minimum values over an interval
- identify interval endpoint maximum and minimum values.

At the points where a differentiable function is neither increasing nor decreasing, the function is stationary and its gradient is zero. Identifying such **stationary points** provides information that assists curve sketching.

6.5.1 Stationary points

Stationary points

There are three types of stationary points:

- **(local) minimum turning points**
- **(local) maximum turning points**
- **stationary points of inflection.**

For a curve $y = f(x)$ at a stationary point, $f'(x) = 0$.

The word 'local' means that the point is a minimum or a maximum in a particular locality or neighbourhood. Beyond this section of the graph, there could be other points on the graph that are lower than the local minimum or higher than the local maximum. Our purpose for the time being is simply to identify the turning points and their nature, so we shall continue to refer to them just as minimum or maximum turning points.

Nature of a stationary point

At each of the three types of stationary points, $f'(x) = 0$. This means that the tangents to the curve at these points are horizontal. By examining the slope of the tangent to the curve immediately before and immediately after the stationary point, the nature or type of stationary point can be determined.

	Minimum turning point	**Maximum turning point**	**Stationary point of inflection**
Stationary point	 For a minimum turning point, the behaviour of the function changes from decreasing just before the point, to stationary at the point, to increasing just after the point.	 For a maximum turning point, the behaviour of the function changes from increasing just before the point, to stationary at the point, to decreasing just after the point.	 or  For a stationary point of inflection, the behaviour of the function remains either increasing or decreasing before and after the point and stationary at the point.

Slope of tangent	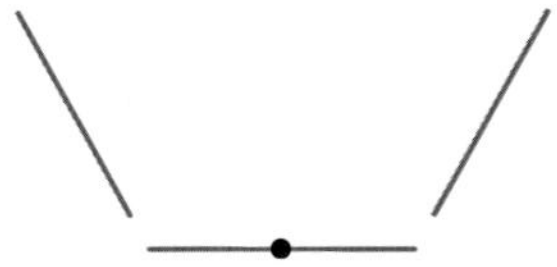 The slope of the tangent changes from negative to zero to positive.	 The slope of the tangent changes from positive to zero to negative.	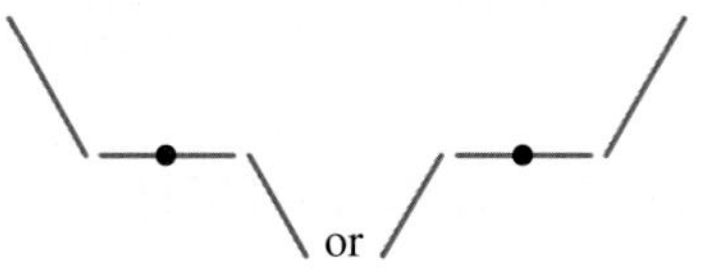 The slope of the tangent is zero at the point but does not change sign either side of the point.

Identifying stationary points

To identify stationary points and their nature:

- **establish where $f'(x) = 0$**
- **determine the nature by testing the slope of the tangent at selected points either side of, and in the neighbourhood of, the stationary point.**

WORKED EXAMPLE 9 Stationary points and their nature

a. **Determine the stationary points of $f(x) = 2 + 4x - 2x^2 - x^3$ and justify their nature.**
b. **The curve $y = ax^2 + bx - 24$ has a stationary point at $(-1, -25)$. Calculate the values of a and b.**

THINK	WRITE
a. 1. Derive the function.	a. $f(x) = 2 + 4x - 2x^2 - x^3$ $f'(x) = 4 - 4x - 3x^2$
2. Calculate the x-coordinates of the stationary points by solving $f'(x) = 0$. *Note:* Always include the reason why $f'(x) = 0$.	At stationary points, $f'(x) = 0$. $4 - 4x - 3x^2 = 0$ $(2 - 3x)(2 + x) = 0$ $x = \frac{2}{3}$ or $x = -2$
3. Calculate the corresponding y-coordinates.	When $x = \frac{2}{3}$, $f\left(\frac{2}{3}\right) = 2 + 4\left(\frac{2}{3}\right) - 2\left(\frac{2}{3}\right)^2 - \left(\frac{2}{3}\right)^3$ $= \frac{94}{27}$ When $x = -2$, $f(-2) = 2 + 4(-2) - 2(-2)^2 - (-2)^3$ $= -6$
4. Write the answer.	The stationary points are $\left(\frac{2}{3}, \frac{94}{27}\right)$, $(-2, -6)$.
5. To justify the nature of the stationary points, draw a table to show the gradient of the curve either side of the stationary points. *Note:* The shape of the cubic graph would suggest the nature of the stationary points.	(see table below)

x	-3	-2	0	$\frac{2}{3}$	1
$f'(x)$	-11	0	4	0	-3
Slope	\	—	/	—	\

6. Identify the nature of each stationary point by examining the sign of the gradient before and after each point.

At $x = -2$, the gradient changes from negative to positive, so $(-2, -6)$ is a minimum turning point.

At $x = \frac{2}{3}$, the gradient changes from positive to negative, so $\left(\frac{2}{3}, \frac{94}{27}\right)$ is a maximum turning point.

b. 1. Use the coordinates of the given point to form an equation.

Note: As there are two unknowns to determine, two pieces of information are needed to form two equations in the two unknowns.

b. $y = ax^2 + bx - 24$

The point $(-1, -25)$ lies on the curve.

$-25 = a(-1)^2 + b(-1) - 24$

$a - b = -1 \quad [1]$

2. Use the other information given about the point to form a second equation.

The point $(-1, -25)$ is a stationary point, so $\frac{dy}{dx} = 0$ at this point.

$y = ax^2 + bx - 24$

$\frac{dy}{dx} = 2ax + b$

At$(-1, -25)$,

$\frac{dy}{dx} = 2a(-1) + b$

$= -2a + b$

3. Solve the simultaneous equations and state the answer.

$-2a + b = 0 \quad [2]$

$a - b = -1 \quad [1]$

Add the equations:

$-a = -1$

$\therefore a = 1$

Substitute $a = 1$ in equation [2]:

$-2 + b = 0$

$\therefore b = 2$

The values are $a = 1$ and $b = 2$.

TI | THINK

b. 1. On a Calculator page, press MENU, then select:
1: Actions
1: Define
Complete the entry line as:
Define $f1(x) = a \times x^2 + b \times x - 24$
then press ENTER.
Note: Remember to include the multiplication operator between a and x, and between b and x.

DISPLAY/WRITE

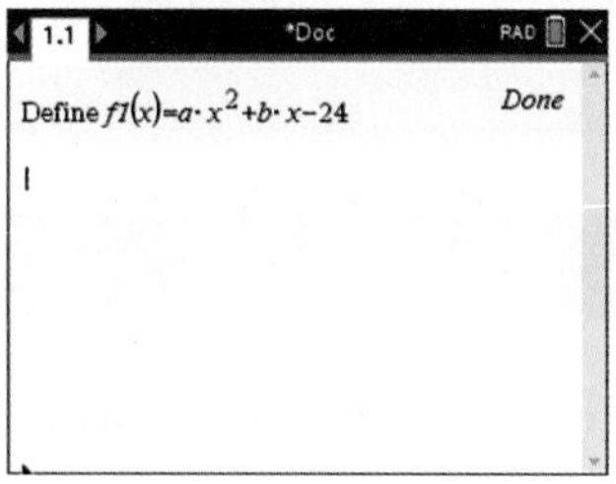

CASIO | THINK

b. 1. On a Main screen, select:
- Interactive
- Define

Complete the fields as:
Func name: f
Variable/s: x
Expression: $a \times x^2 + b \times x - 24$
then select OK.
Note: Remember to include the multiplication operator between a and x, and between b and x.

DISPLAY/WRITE

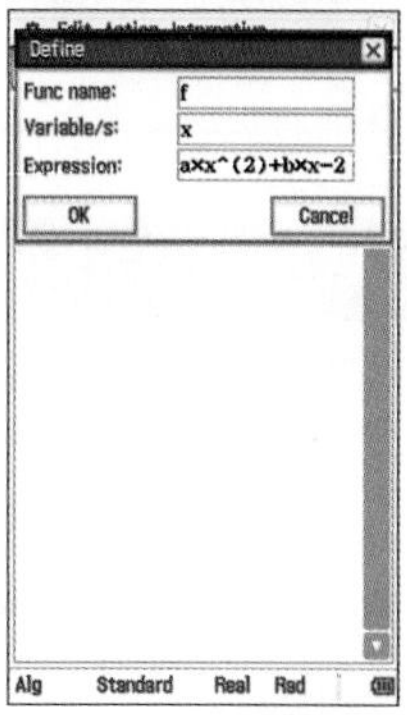

2. Press MENU, then select:
3: Algebra
1: Solve
Complete the entry line as:
solve $\left(f1(-1)=-25 \text{ and } \left(\frac{d}{dx}(f1(x))|x=-1\right)=0, a, b\right)$
Press ENTER.
Note: Press MENU, then select:
4: Calculus
2: Derivative at a point
to find the derivative template.

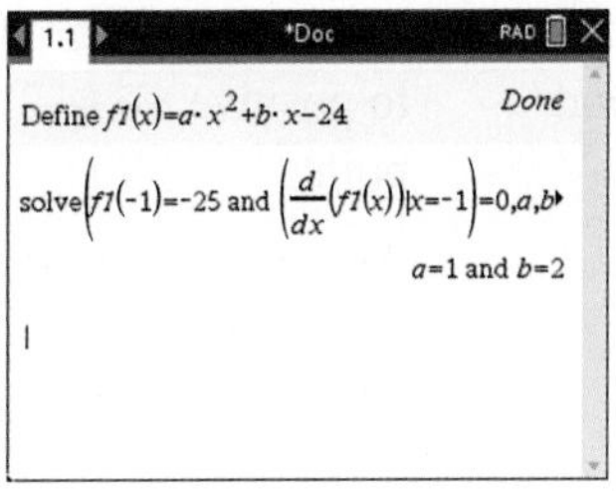

3. The answer appears on the screen.

$a=1$ and $b=2$.

2. Complete the next entry line as:
solve($\{f(-1)=-25,$
$\text{diff}(f(x), x, 1, -1)=0\},$
$\{a, b\}$)
then press EXE.

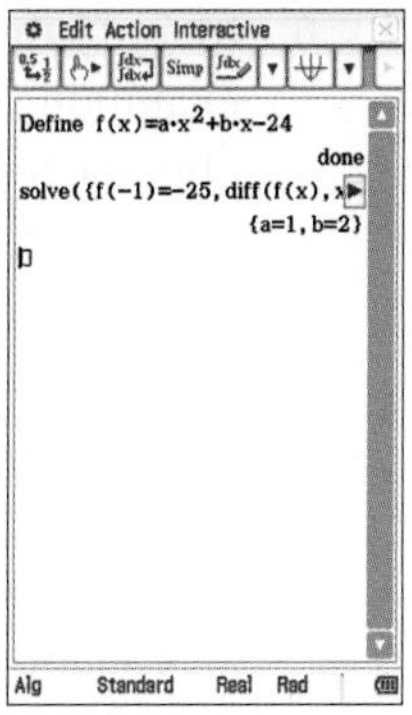

3. The answer appears on the screen.

$a=1$ and $b=2$.

on Resources

Interactivity Stationary points (int-5963)

6.5.2 Curve sketching

To sketch the graph of any function $y=f(x)$, perform the following steps:

- Obtain the y-intercept by evaluating $f(0)$.
- Obtain any x-intercepts by solving, if possible, $f(x)=0$. This may require the use of factorisation techniques including the factor theorem.
- Calculate the x-coordinates of the stationary points by solving $f'(x)=0$. Use the equation of the curve to obtain the corresponding y-coordinates.
- Identify the nature of the stationary points.
- Calculate the coordinates of the end points of the domain where appropriate.
- Identify any other key features of the graph where appropriate.

Conditions for strictly increasing and decreasing

A function is strictly increasing on an interval $x \in [a, b]$ if, for every value in that interval, $f(b) > f(a)$.

Similarly, a function is strictly decreasing on an interval $x \in [a, b]$ if, for every value in that interval, $f(b) < f(a)$.

WORKED EXAMPLE 10 Curve sketching

a. Sketch the function $y=\frac{1}{2}x^3-3x^2+6x-8$. Locate any intercepts with the coordinate axes and any stationary points, and justify their nature.

b. State the domain over which the function is strictly increasing.

THINK	WRITE
a. 1. State the y-intercept.	a. $y=\frac{1}{2}x^3-3x^2+6x-8$ y-intercept: $(0,-8)$

2. Calculate any x-intercepts. For a polynomial of degree 3, try to identify a factor by looking at the equation.	x-intercepts: When $y = 0$, $\frac{1}{2}x^3 - 3x^2 + 6x - 8 = 0$ $x^3 - 6x^2 + 12x - 16 = 0$ Let $P(x) = x^3 - 6x^2 + 12x - 16$. $P(4) = 64 - 96 + 48 - 16$ $= 0$ $\therefore (x - 4)$ is a factor. $0 = x^3 - 6x^2 + 12x - 16$ $0 = (x - 4)\left(x^2 - 2x + 4\right)$ $\therefore x = 4$ or $x^2 - 2x + 4 = 0$ The discriminant of $x^2 - 2x + 4$ is $\Delta = 4 - 16 < 0$. Therefore, there is only one x-intercept, $(4, 0)$.
3. Obtain the derivative in order to locate any stationary points.	Stationary points: $y = \frac{1}{2}x^3 - 3x^2 + 6x - 8$ $\frac{dy}{dx} = \frac{3}{2}x^2 - 6x + 6$ At stationary points, $\frac{dy}{dx} = 0$. $\frac{3}{2}x^2 - 6x + 6 = 0$ $\frac{3}{2}(x^2 - 4x + 4) = 0$ $\frac{3}{2}(x - 2)^2 = 0$ $x = 2$ Substitute $x = 2$ into the function's equation: $y = \frac{1}{2}(2)^3 - 3(2)^2 + 6(2) - 8$ $= -4$ The stationary point is $(2, -4)$.
4. Identify the type of stationary point by evaluating the slope either side of the stationary point.	(see table below) The point $(2, -4)$ is a stationary point of inflection.

x	0	2	4
$\frac{dy}{dx}$	6	0	6
Slope	/	—	/

5. Sketch the curve, showing the intercepts with the axes and the stationary point.

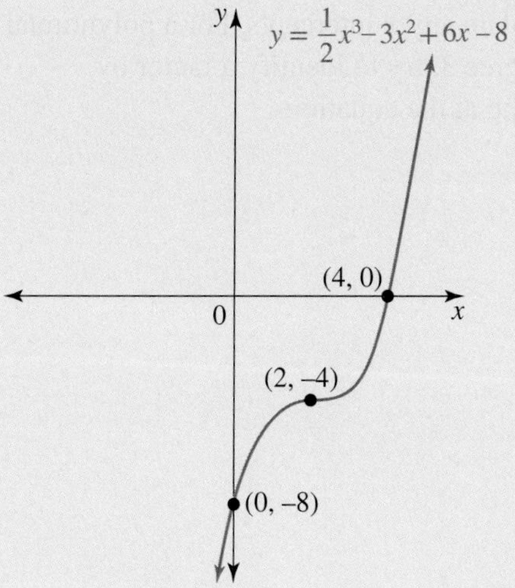

b. Identify the domain over which $f(b) > f(a)$, where $b > a$.

b. $x \in R$

TI | THINK — **DISPLAY/WRITE**

a. 1. On a Graphs page, complete the entry line for function 1 as:
$f1(x) = \frac{1}{2}x^3 - 3x^2 + 6x - 8$
then press ENTER.

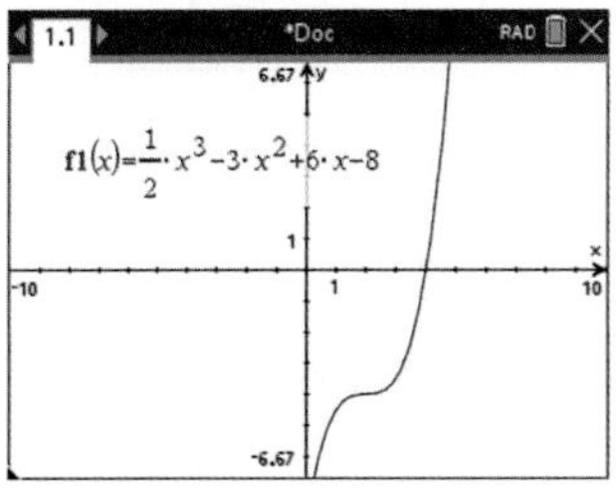

2. To find the x-intercept, press MENU, then select:
6: Analyze Graph
1: Zero
Move the cursor to the left of the x-intercept when prompted for the lower bound, then press ENTER. Move the cursor to the right of the x-intercept when prompted for the upper bound, then press ENTER.

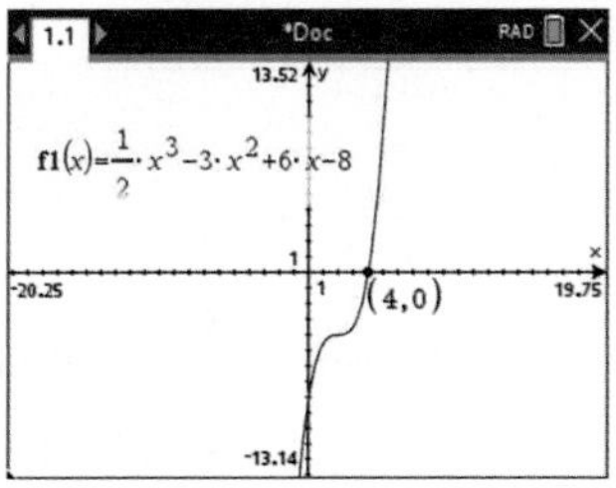

3. To find the y-intercept, press MENU, then select:
5: Trace
1: Graph Trace
Type '0', then press ENTER twice.

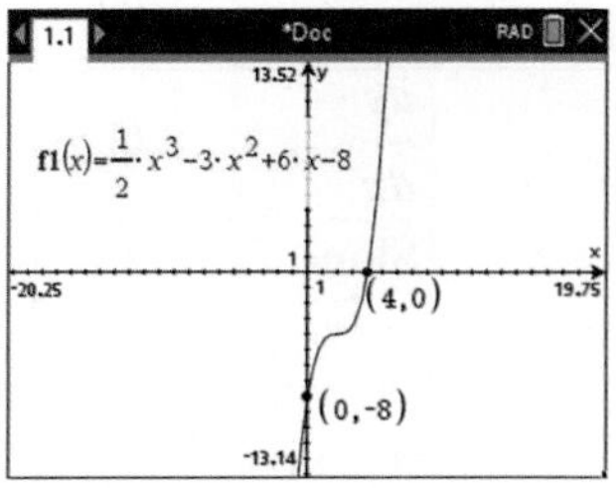

CASIO | THINK — **DISPLAY/WRITE**

a. 1. On a Graph & Table screen, complete the entry line for $y1$ as:
$y1 = \frac{1}{2}x^3 - 3x^2 + 6x - 8$
then press EXE.
Select the 'Graph' icon to draw the graph.

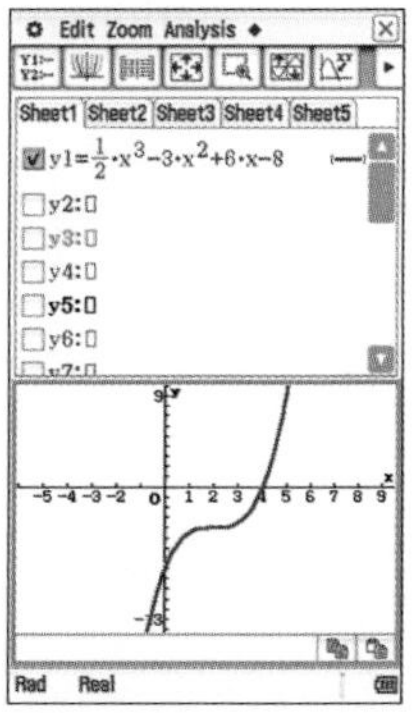

2. To find the x-intercept, select:
- Analysis
- G-solve
- Root

then press EXE.

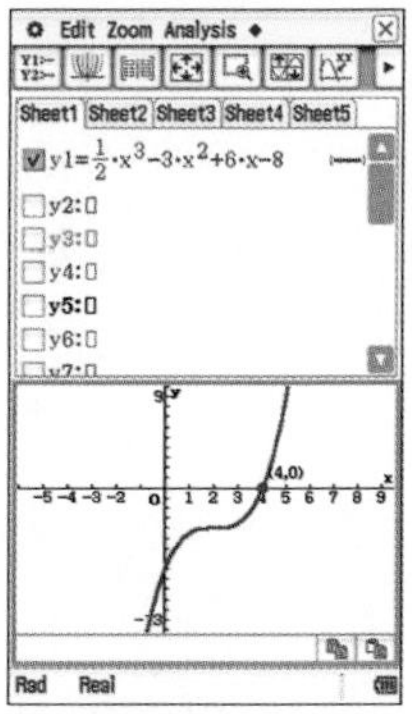

3. To find the y-intercept, select:
- Analysis
- G-Solve
- Y Intercept

then press EXE.

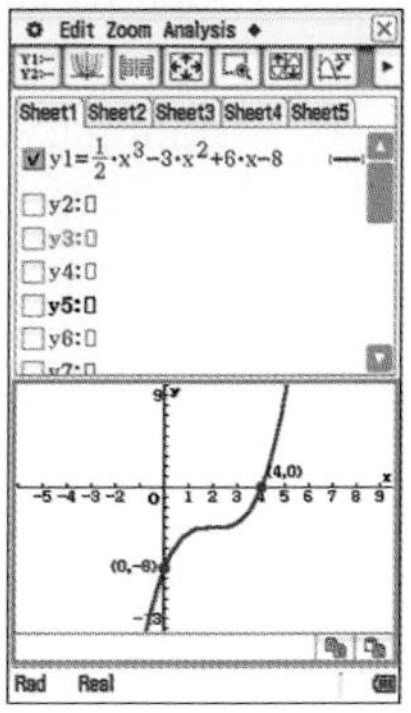

4. To find the point of inflection, press MENU, then select:
6: Analyze Graph
5: Inflection
Move the cursor to the left of the point of inflection when prompted for the lower bound, then press ENTER. Move the cursor to the right of the point of inflection when prompted for the upper bound, then press ENTER.

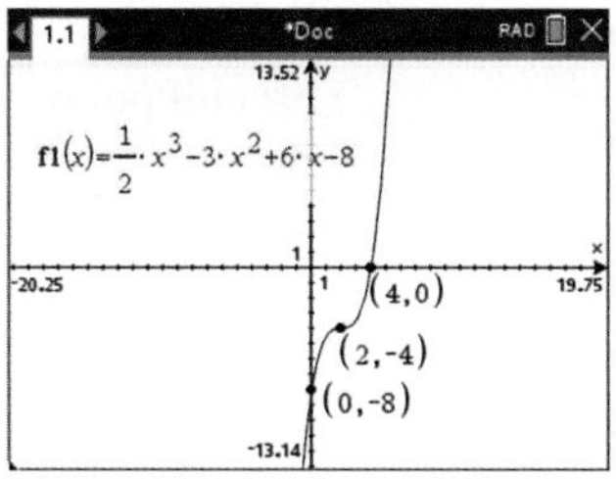

4. To find the point of inflection, select:
- Analysis
- G-Solve
- inflection

then press EXE.

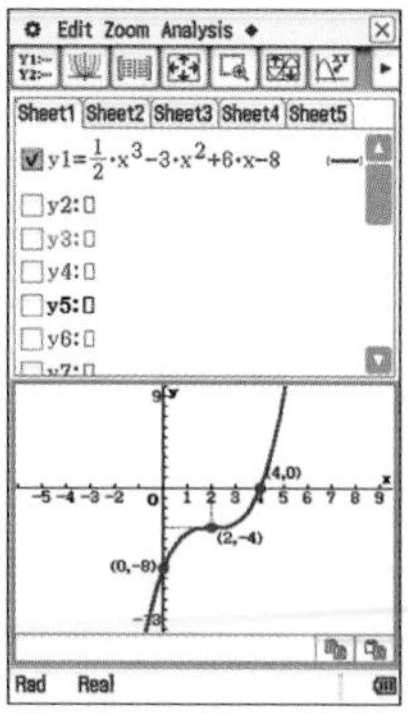

6.5.3 Local and absolute maxima and minima

The diagram shows the graph of a function sketched over a domain with end points D and E.

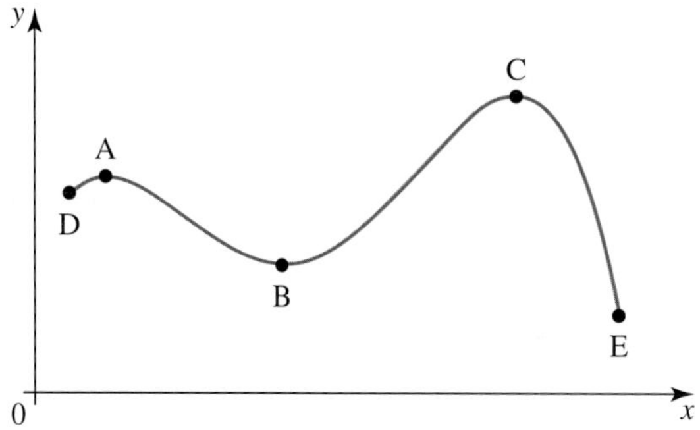

There are three turning points: A and C are maximum turning points, and B is a minimum turning point.

The y-coordinate of point A is greater than those of its neighbours, so A is a local maximum point. However, the y-coordinate of point C is not only greater than those of its neighbours; it is greater than that of any other point on the graph. For this reason, C is called the **absolute maximum** point.

The **absolute minimum** point is the point whose y-coordinate is less than any others on the graph. For this function, point E, an end point of the domain, is the absolute minimum point. Point B is a local minimum point; it is not the absolute minimum point.

Absolute maximums and minimums may not exist for all functions. For example, a cubic function on its maximal domain may have one local maximum turning point and one local minimum turning point, but there is neither an absolute maximum nor an absolute minimum point, because as $x \to \pm\infty$, $y \to \pm\infty$ (assuming a positive coefficient of x^3).

If a differentiable function has an absolute maximum or an absolute minimum value, then this will occur at either a turning point or an end point of the domain. The y-coordinate of such a point gives the value of the absolute maximum or the absolute minimum.

WORKED EXAMPLE 11 Absolute maximum and minimum values

A function defined on a restricted domain has the rule $y = \frac{x}{2} + \frac{2}{x}, x \in \left[\frac{1}{4}, 4\right]$.

a. **Specify the coordinates of the end points of the domain.**
b. **Obtain the coordinates of any stationary point and determine its nature.**
c. **Sketch the graph of the function.**
d. **State the absolute maximum and minimum values of the function, if they exist.**

THINK	WRITE
a. Use the given domain to calculate the coordinates of the end points.	a. $y = \frac{x}{2} + \frac{2}{x}$ for the domain $\frac{1}{4} \le x \le 4$. Substitute each of the end values of the domain in the function's rule.

Left end point: when $x = \frac{1}{4}$,

$$y = \frac{x}{2} + \frac{2}{x}$$

$$= \frac{1}{8} + 8$$

$$= 8\frac{1}{8}$$

Right end point: when $x = 4$,

$$y = 2 + \frac{1}{2}$$

$$= 2\frac{1}{2}$$

The end points are $\left(\frac{1}{4}, \frac{65}{8}\right)$ and $\left(4, \frac{5}{2}\right)$.

b. 1. Calculate the derivative of the function.

b. $y = \frac{x}{2} + \frac{2}{x}$

$$y = \frac{x}{2} + 2x^{-1}$$

$$\frac{dy}{dx} = \frac{1}{2} - 2x^{-2}$$

$$\frac{dy}{dx} = \frac{1}{2} - \frac{2}{x^2}$$

2. Calculate the coordinates of any stationary point.

At a stationary point, $\frac{dy}{dx} = 0$.

$$\frac{1}{2} - \frac{2}{x^2} = 0$$

$$\frac{1}{2} = \frac{2}{x^2}$$

$$x^2 = 4$$

$$x = \pm 2$$

$$x = 2,\ x \in \left[\frac{1}{4}, 4\right]$$

When $x = 2, y = \frac{2}{2} + \frac{2}{2}$

$$= 2$$

(2, 2) is a stationary point.

3. Test the gradient at two selected points either side of the stationary point.

$\frac{dy}{dx}$	$\frac{1}{2} - \frac{2}{1} = -\frac{3}{2}$	0	$\frac{1}{2} - \frac{2}{9} = \frac{5}{18}$
Slope	\	—	/

4. State the nature of the stationary point. The gradient changes from negative to zero to positive about the stationary point.

The point (2, 2) is a minimum turning point.

c. 1. Calculate any intercepts with the coordinate axes.

c. There is no y-intercept, since $x = 0$ is not in the given domain, nor is $y = \frac{x}{2} + \frac{2}{x}$ defined at $x = 0$.
There is no x-intercept, since the end points and the minimum turning point all have positive y-coordinates, and there are no other turning points.

2. Sketch the graph using the three known points.

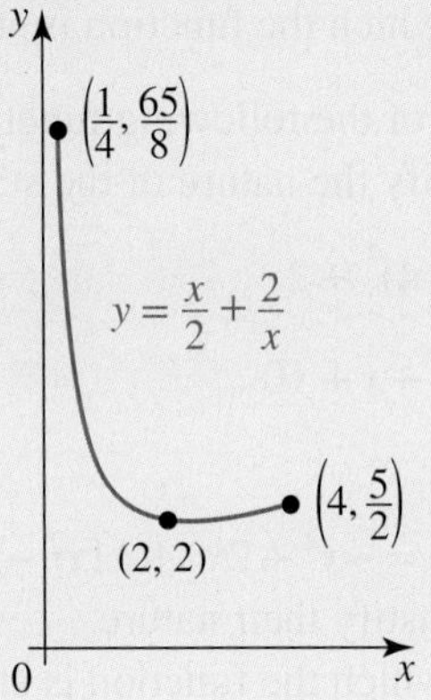

d. Examine the graph and the y-coordinates to identify the absolute extremes.

d. The function has an absolute maximum of $\frac{65}{8}$ at the left end point and an absolute minimum, and local minimum, of 2 at its turning point.

6.5 Exercise

Technology free

1. **WE9** a. Determine the stationary points of $f(x) = \frac{2x^3}{3} + \frac{3x^2}{2} - 2x + 4$ and justify their nature.

   b. The curve $y = ax^2 + bx + c$ passes through the point $(0, -8)$ and has a stationary point at $(-1, -5)$. Calculate the values of a, b and c.

2. Obtain any stationary points of the following curves and justify their nature.

   a. $y = x(x + 2)^2$

   b. $y = x^3 + 3x^2 - 24x + 5$

   c. $y = \frac{x^2}{x + 1}$

   d. $y = (x - 1)e^{-x}$

3. Consider the function defined by $f(x) = 16x^2 - x^4$.

a. Show that $(2\sqrt{2}, 64)$ is a stationary point of the function.
b. Determine the nature of this stationary point.
c. State the coordinates of any other stationary points and state their nature.

4. The curve $y = x^3 + ax^2 + bx - 11$ has stationary points when $x = 1$ and $x = \frac{5}{3}$.

a. Calculate a and b.
b. Determine the coordinates of the stationary points and their nature.

5. **WE10** a. Sketch the function $f(x) = 2x^3 - x^2$. Locate any intercepts with the coordinate axes and any stationary points, and justify their nature.

b. State the domain over which the function is strictly decreasing.

6. Sketch the graphs of each of the following functions. Label any intercepts and any stationary points with their coordinates, and justify the nature of the stationary points.

a. $f: R \to R, f(x) = -\frac{1}{4}(x-4)^3 + 2$
b. $g(x) = 2x^3 - x^2, x \in [-1, 1]$
c. $h: R \to R, h(x) = x^3 - x^2 - x + 10$
d. $f(x) = x^4 - 6x^2 + 8$
e. $f(x) = (x+3)^3(4-x)$
f. $f(x) = x^3 - 4x^2 - 3x + 12$

7. a. Sketch the function $f(x) = -x^4 + 2x^3 + 11x^2 - 12x$. Locate any intercepts with the coordinate axes and any stationary points, and justify their nature.
b. State the domain over which the function is strictly increasing.

Technology active

8. a. Sketch the graph of $f(x) = \frac{1}{2}(2x-3)^4(x+1)^5$, showing all intercepts and stationary points.
b. State the domain over which the function is strictly decreasing.

9. The graph of the function $f: R^+ \to R, f(x) = \frac{1}{10x} + \log_e(x)$ is shown.
Use calculus to determine the coordinates of the minimum turning point.

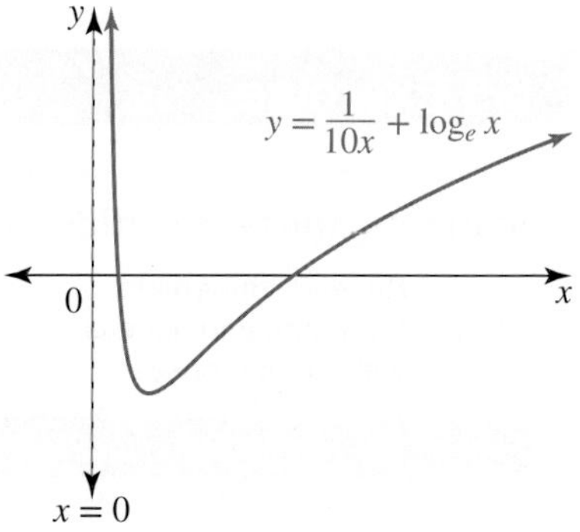

10. Use calculus to determine the exact local maximum or minimum value of the function defined by:

a. $f(x) = 2x\log_e(x), x > 0$
b. $f(x) = \frac{\log_e(2x)}{x}, x > 0$
c. $f(x) = x\log_e\left(\frac{3}{x}\right), x > 0$
d. $f(x) = x - \log_e\left(\sqrt{x-2}\right)$

In each case, investigate the nature of the turning point to determine whether it is a maximum or a minimum.

11. Sketch a possible graph of the function $y = f(x)$ for which:
a. $f'(-1) = 0, f'(3) = 0, f(3) = 0, f'(x) > 0$ for $x < -1, f'(x) < 0$ for $-1 < x < 3$ and $f'(x) > 0$ for $x > 3$
b. $f'(-1) = 0, f'(x) < 0$ for $x \in R \setminus \{-1\}$ and $f(-1) = 2$.

12. a. The point $(2, -8)$ is a stationary point of the curve $y = x^3 + bx + c$. Calculate the values of b and c.
b. The point $(1.5, 6)$ is a stationary point of the curve $y = ax^2 + bx + 15$. Calculate the values of a and b.
c. A curve has equation $y = x^3 + bx^2 + cx + d$. The curve has a stationary point at $(-3, -10)$ and passes through the point $(1, 6)$. Determine the values of b, c and d.

13. WE11 A function defined on a restricted domain has the rule $f(x) = \frac{1}{4x} + x, x \in \left[-2, -\frac{1}{4}\right]$.

a. Specify the coordinates of the end points of the domain.
b. Obtain the coordinates of any stationary point and determine its nature.
c. Sketch the graph of the function.
d. State the absolute maximum and minimum values of the function, if they exist.

14. Calculate, if possible, the absolute maximum and minimum values of the function $f(x) = 2x^3 - 8x$ over the domain $\{x : x \le 2\}$.

15. Sketch the graphs and state the absolute maximum and minimum values over the given domain for each of the following functions.

a. $y = \frac{1}{2}(x-1)^2 - 2, 0 \le x \le 5$ b. $y = x^3 - 2x^2, -2 \le x \le 3$ c. $y = 4 - x^3, x \le 2$

16. The graph of $f(x) = 2\sqrt{x} + \frac{1}{x}$, $0.25 \le x \le 5$ is shown.

a. Determine the coordinates of the end points A and C and the stationary point B.
b. Determine the point at which the absolute maximum occurs.
c. State the absolute maximum and minimum values.

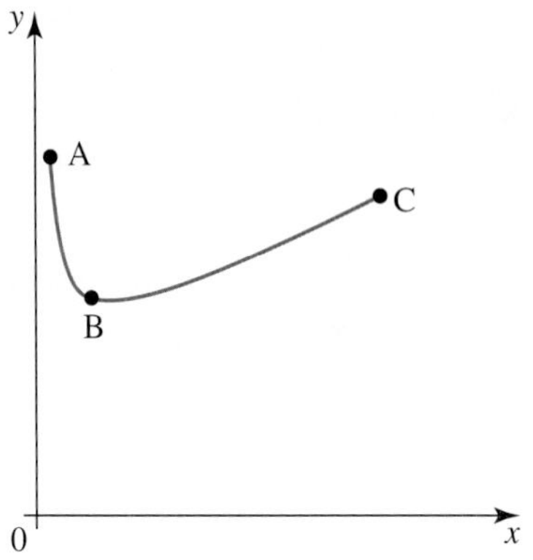

17. Let $f(x) = xe^x$.

a. Determine any stationary points and state their nature.
b. Determine $\{x : f'(x) > 0\}$.
c. State, if possible, the absolute maximum and minimum values.

18. a. Sketch the graph of $y = e^{-x^2}$.
b. If $y = 20e^{-2x^2-4x+1}$, use calculus to calculate the values of x for which the function is strictly increasing.

19. The graph of the function $f: R^- \to R, f(x) = \frac{1}{x^2} - 2\log_e(x+3)$ is shown.

a. Calculate the coordinates of the x-intercepts, correct to 3 decimal places.
b. Determine the equations of the tangents at the x-axis intercepts.
c. Determine the coordinates of the minimum turning point. Give your answer correct to 4 decimal places.

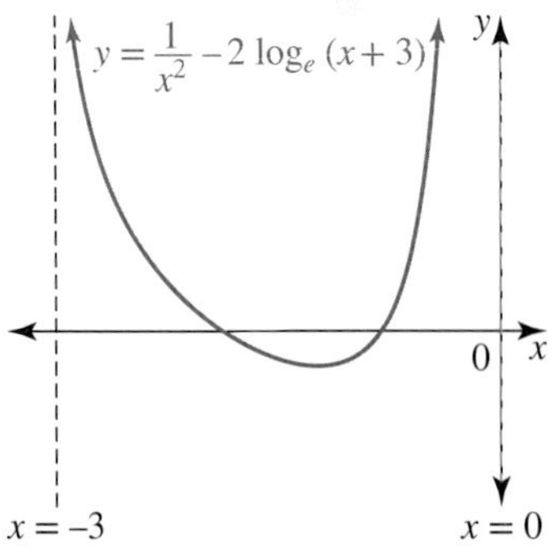

20. The graph of the function $f: R^+ \to R, f(x) = -\frac{1}{x^2} - 8\log_e(x)$ is shown.

a. Calculate the coordinates of the x-intercepts, giving your answers correct to 4 decimal places.
b. Calculate the gradient of the curve at the points found in part a, giving your answers correct to 2 decimal places.
c. Determine the equation of the tangent at $(1, -1)$ and the equation of the line perpendicular to the curve at $(1, -1)$.
d. Show that the coordinates of the maximum turning point are $\left(\frac{1}{2}, -4 + 8\log_e(2)\right)$.

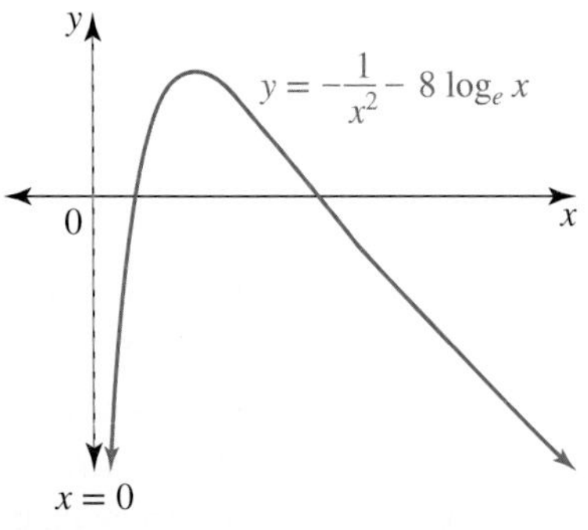

21. Consider the function $f(x) = (a - x)^2(x - 2)$ where $a > 2$.

a. Locate or calculate the coordinates of the stationary points.
b. State the nature of the stationary points.
c. Calculate the value of a if the graph of $y = f(x)$ has a turning point at $(3, 4)$.

22. Consider the function $f(x) = (x - a)(x - b)^3$, where $a > 0$, $b > 0$ and $a < b$.

a. Determine the x-intercepts
b. Locate or calculate the coordinates of the stationary points.
c. State the nature of the stationary points.
d. If one of the stationary points has coordinates $(3, -27)$, calculate a and b.

6.5 Exam questions

Question 1 (1 mark) TECH-ACTIVE

Source: *VCE 2017, Mathematical Methods Exam 2, Section A, Q11; © VCAA.*

MC The function $f: R \to R$, $f(x) = x^3 + ax^2 + bx$ has a local maximum at $x = -1$ and a local minimum at $x = 3$. The values of a and b are respectively

A. -2 and -3
B. 2 and 1
C. 3 and -9
D. -3 and -9
E. -6 and -15

Question 2 (1 mark) TECH-ACTIVE

Source: *VCE 2013, Mathematical Methods (CAS) Exam 2, Section 1, Q21; © VCAA.*

MC The cubic function $R \to R$, $f(x) = ax^3 - bx^2 + cx$, where a, b and c are positive constants, has no stationary points when

A. $c > \dfrac{b^2}{4a}$
B. $c < \dfrac{b^2}{4a}$
C. $c < 4b^2a$
D. $c > \dfrac{b^2}{3a}$
E. $c < \dfrac{b^2}{3a}$

Question 3 (5 marks) TECH-ACTIVE

Source: *VCE 2017, Mathematical Methods Exam 2, Section B, Q1a, b; © VCAA.*

Let $f: R \to R$, $f(x) = x^3 - 5x$. Part of the graph of f is shown below.

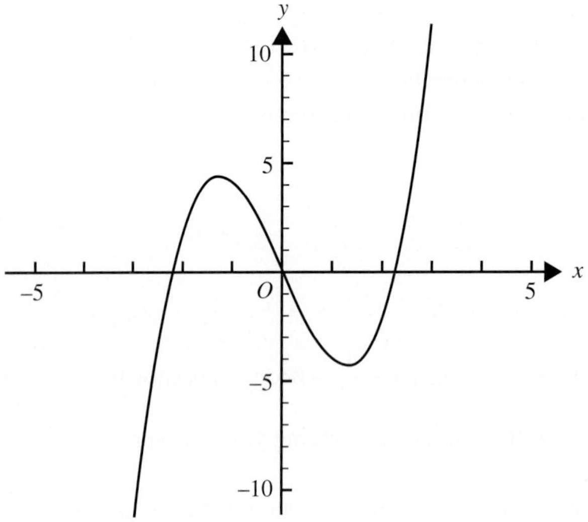

a. Find the coordinates of the turning points. **(2 marks)**
b. $A(-1, f(-1))$ and $B(1, f(1))$ are two points on the graph of f.
 i. Find the equation of the straight line through A and B. **(2 marks)**
 ii. Find the distance AB. **(1 mark)**

More exam questions are available online.

6.6 Maximum and minimum problems

LEARNING INTENTION

At the end of this subtopic you should be able to:
- solving problems involving maximum and minimum values.

6.6.1 How to solve maximum and minimum problems

In many practical situations, it is necessary to find the maximum or minimum value of the function that describes it. For example, if you were running your own business, you would always want to minimise the production costs while maximising the profits.

To solve maximum or minimum problems, apply the following steps.
- Draw a diagram if possible and label it with as few variables as possible.
- If there is more than one variable, find a connection between the variables from the information given. For instance, if you are finding the area of a rectangle, where the dimensions are length, l, and width, w, the width needs to be expressed in terms of the length or vice versa, so that the area can be expressed in terms of one variable only. Pythagoras' theorem, trigonometry, similar triangles, standard formulas, or given information could be required to express one variable in terms of another.
- Find an expression for the quantity to be maximised or minimised in terms of the one nominated variable.
- Differentiate the expression, equate to zero and solve for the unknown variable.
- Reject any unrealistic solutions.
- Check the nature of the stationary point.
- Check whether the answer is the absolute maximum or minimum by evaluating the end points of the domain.
- Always sketch the shape of the graph.
- Answer the actual question.

WORKED EXAMPLE 12 Maximum and minimum problems

The new owner of an apartment wants to install a window in the shape of a rectangle surmounted by a semicircle in order to allow more light into the apartment.
The owner has 336 cm of wood for a frame to surround the window. They want to determine the dimensions of the window that will allow as much light into the apartment as possible.

a. **Show that the area, A in cm^2, of the window is $A = 336x - \frac{1}{2}(4+\pi)x^2$.**
b. **Hence, determine, to the nearest cm, the width and the height of the window for which the area is greatest.**
c. **Structural limitations mean that the width of the window should not exceed 84 cm. Determine the required dimensions of the window of maximum area.**

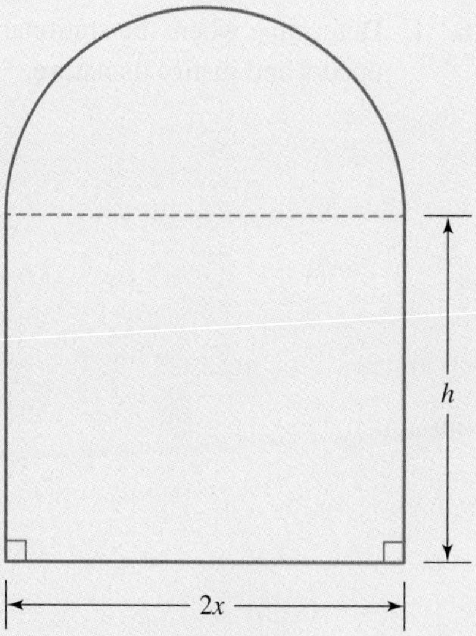

THINK | **WRITE**

a. 1. Form an expression for the total area. The total area is the sum of the areas of the rectangle and semicircle.
Note: This expression involves more than one variable.

a. Rectangle: length $2x$ cm, width h cm

$\therefore A_{\text{rectangle}} = 2xh$

Semicircle: diameter $2x$ cm, radius x cm

$\therefore A_{\text{semicircle}} = \frac{1}{2}\pi x^2$

The total area of the window is $A = 2xh + \frac{1}{2}\pi x^2$.

2. Use the perimeter information to form an expression connecting the two variables.

$P_{\text{window}} = 336$ cm

$C_{\text{semicircle}} = \frac{1}{2}(2\pi x)$

$\therefore P_{\text{shape}} = h + 2x + h + \frac{1}{2}(2\pi x)$

Hence, $2h + 2x + \pi x = 336$.

3. Express one appropriately chosen variable in terms of the other. The required expression for the area is in terms of x, so express h in terms of x.

$$2h = 336 - 2x - \pi x$$
$$h = \frac{1}{2}(336 - 2x - \pi x)$$

4. Write the area as a function of x by substituting for h.

$$\begin{aligned} A &= x(2h) + \frac{1}{2}\pi x^2 \\ &= x(336 - 2x - \pi x) + \frac{1}{2}\pi x^2 \\ &= 336x - 2x^2 - \pi x^2 + \frac{1}{2}\pi x^2 \\ &= 336x - \left(2 + \frac{1}{2}\pi\right)x^2 \end{aligned}$$

$\therefore A = 336x - \frac{1}{2}(4 + \pi)x^2$ as required.

b. 1. Determine where the stationary point occurs and justify its nature.

b. At the stationary point, $\frac{dA}{dx} = 0$.

$$\frac{dA}{dx} = 336 - (4 + \pi)x$$
$$0 = 336 - (4 + \pi)x$$
$$x = \frac{336}{4 + \pi}$$
$$x = 47.05$$

x	40	$\frac{336}{4+\pi}$	50
$\frac{dA}{dx}$	$176 - 40\pi$	0	$136 - 50\pi$
Slope	/	—	\

The maximum area is obtained when $x = \frac{336}{4 + \pi}$ cm.

2. State the values of both variables.

When $x = \dfrac{336}{4+\pi}$,

$$2h = 336 - 2 \times \left(\frac{336}{4+\pi}\right) - \pi \times \left(\frac{336}{4+\pi}\right)$$

$\therefore h = 47.05$ cm

3. Calculate the required dimensions and state the answer.

The width of the window is $2x \simeq 94$ cm.
The total height of the window is $h + x \simeq 94$ cm.
Therefore, the area of the window will be greatest if its width is 94 cm and its height is 94 cm.

c. 1. Give the restricted domain of the area function.

c. If the width is not to exceed 84 cm, then
$2x \leq 84$.
$\therefore x \leq 42$
With the restriction, the domain of the area function is $[0, 42]$.

2. Determine where the function is greatest. As the stationary point occurs when $x \approx 47$, for the domain $[0, 42]$ there is no stationary point, so the greatest area must occur at an end point of the domain.

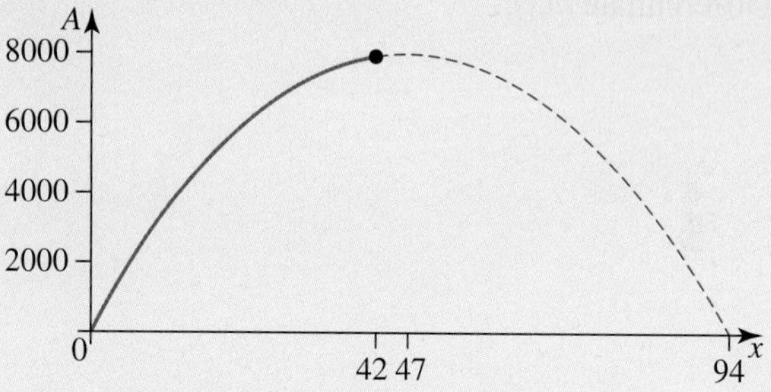

The maximum occurs when $x = 42$.

3. Calculate the required dimensions and state the answer.

When $x = 42$,

$$h = \frac{1}{2}(336 - 84 - 42\pi)$$

$\therefore h = 60.03$ cm
≈ 60 cm

The width of the window is $2x = 84$ cm.
The height of the window is $h + x = 102$ cm.
With the restriction, the area of the window will be greatest if its width is 84 cm and its height is 102 cm.

WORKED EXAMPLE 13 Determining minimum distance

Calculate the minimum distance from the curve $y = 2x^2$ to the point (4, 0), correct to 2 decimal places. You do not need to justify your answer.

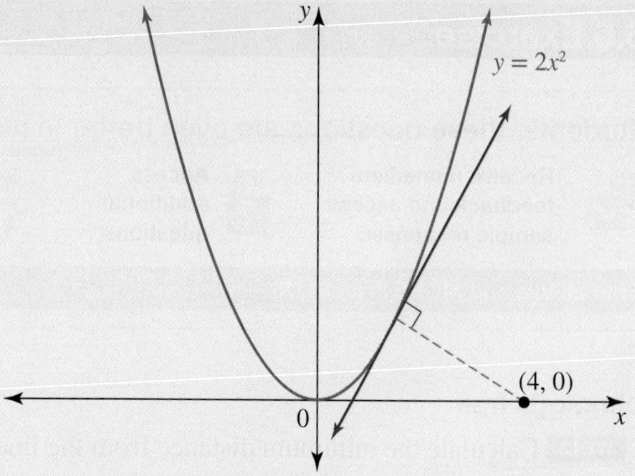

THINK	WRITE
1. Let P be the point on the curve such that the distance from P to the point $(4, 0)$ is a minimum.	$P = (x, y)$
2. Write the formula for the distance between two points.	$d(x) = \sqrt{(x_2 - x_1)^2 + (y_2 - y_1)^2}$ $= \sqrt{(x-4)^2 + (y-0)^2}$ $= \sqrt{(x-4)^2 + y^2}$
3. Express the distance between the two points as a function of x only.	$y = 2x^2$ $\therefore d(x) = \sqrt{(x-4)^2 + (2x^2)^2}$ $= \left(x^2 - 8x + 16 + 4x^4\right)^{\frac{1}{2}}$
4. Differentiate $d(x)$.	$d'(x) = \frac{1}{2} \times \left(4x^4 + x^2 - 8x + 16\right)^{-\frac{1}{2}} \times \left(16x^3 + 2x - 8\right)$ $= \frac{16x^3 + 2x - 8}{2\sqrt{4x^4 + x^2 - 8x + 16}}$ $= \frac{8x^3 + x - 4}{\sqrt{4x^4 + x^2 - 8x + 16}}$
5. Solve $d'(x) = 0$ using CAS.	$0 = \frac{8x^3 + x - 4}{\sqrt{4x^4 + x^2 - 8x + 16}}$ $0 = 8x^3 + x - 4$ $x = 0.741$
6. Evaluate $d(0.741)$.	$d(0.741) = \sqrt{(0.741)^2 - 8(0.741) + 16 + 4(0.741)^4}$ $= 3.439$
7. Write the answer.	The minimum distance is 3.44 units.

Note: When finding the minimum distance between two points, one of which is on a curve, the line joining the points is always perpendicular to the curve. This fact can also be used to determine the minimum distance between two points.

6.6 Exercise

Technology free

1. **WE13** Calculate the minimum distance from the line $y = 2x - 5$ to the origin. You do not need to justify your answer.

2. Calculate the minimum distance from the line $y = 2\sqrt{x}$ to the point $(5, 0)$.

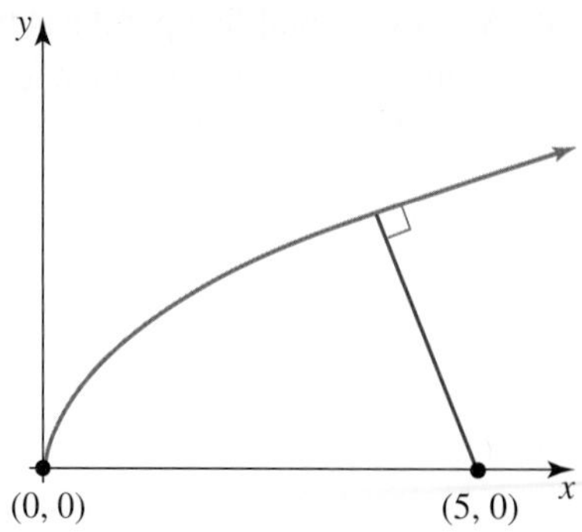

3. The sum of two positive numbers is 32. Determine the numbers if their product is a maximum.

4. A rectangular box with an open top is to be constructed from a rectangular sheet of cardboard measuring 16 cm by 10 cm. The box will be made by cutting equal squares of side length x cm out of the four corners and folding the flaps up.
   a. Express the volume as a function of x.
   b. Determine the dimensions of the box with greatest volume and give this maximum volume.

Technology active

5. **WE12** The owner of an apartment wants to create a stained glass feature in the shape of a rectangle surmounted by an isosceles triangle of height equal to half its base. This will be adjacent to a door opening on to a balcony.
   The owner has 150 cm of plastic edging to place around the perimeter of the figure and wants to determine the dimensions of the figure with the greatest area.
   a. Show that the area, A in cm^2, of the stained glass figure is $A = 150x - \left(2\sqrt{2} + 1\right)x^2$.
   b. Determine, correct to 1 decimal place, the width and the height of the figure for which the area is greatest.
   c. Due to structural limitations, the width of the figure should not exceed 30 cm. Determine the required dimensions of the stained glass figure of maximum area.

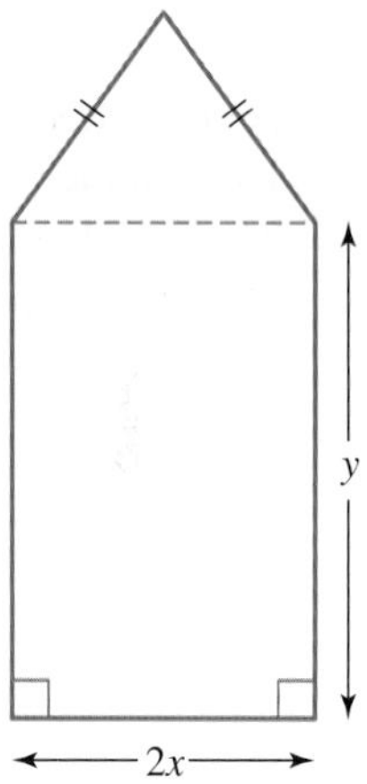

6. A pen for holding farm animals has dimensions $l \times w$ metres. This pen is to be partitioned so that there are four spaces of equal area as shown.

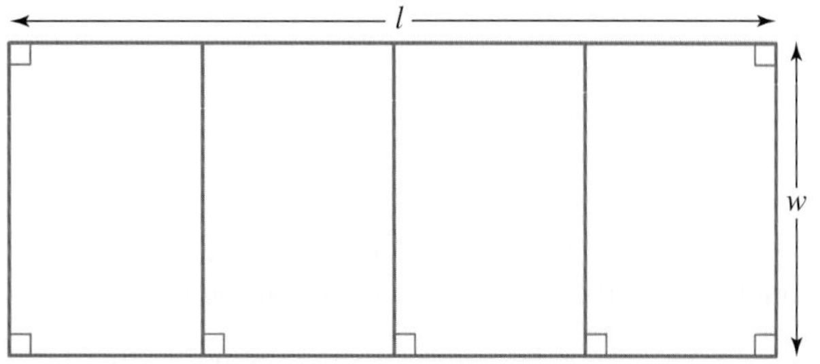

   The farmer has 550 metres of fencing material to construct this pen.
   a. Calculate the required length and width in order to maximise the area of the pen.
   b. Calculate the maximum area.

7. A playground is being constructed by the local council. The shape of the playground is shown. All measurements are in metres.
   The perimeter of the playground is known to be 96 metres.
   a. Determine the values of a and b that give a maximum area for the playground.
   b. Calculate the maximum area.

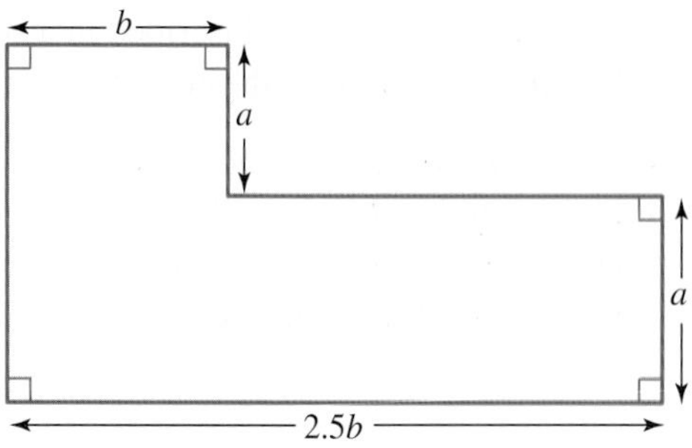

8. An ornamental fish pond has the shape shown.
A plastic cover is being made for the pond for the winter months. If the surface area of the pond, A, is a constant, show that the perimeter of the pond is a minimum when both R and l are equal to $\sqrt{\dfrac{2A}{\pi+4}}$.

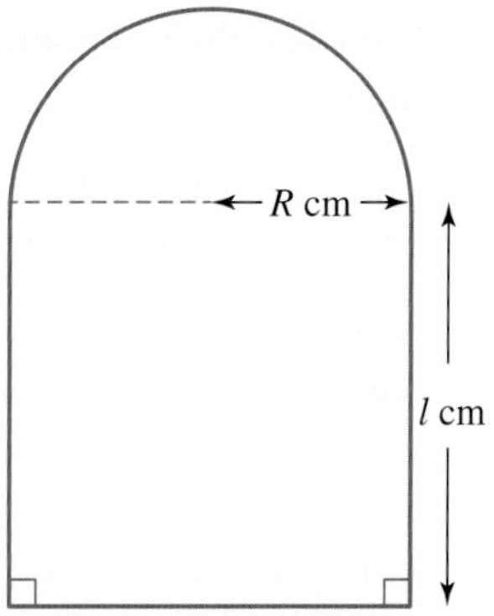

9. A colony of blue wrens, also known as superb fairy wrens, survives in a national park in Sassafras Victoria because the wooded areas have rich undergrowth and a plentiful supply of insects, the wrens' main food source. Breeding begins in spring and continues until late summer.
The population of the colony at any time t months after 1 September can be modelled by the function

$$P(t) = 200te^{\frac{-t}{4}} + 400, 0 \leq t \leq 12$$

where P is the number of birds in the colony.
Determine:
a. the initial population of the birds
b. when the largest number of birds is reached
c. the maximum number of birds, to the nearest bird.

10. The amount of money in a savings account t years after the account was opened on 1 January 2016 is given by the equation

$$A(t) = 1000 - 12te^{\frac{4-t^3}{8}} \text{ for } t \in [0, 6].$$

a. Calculate how much money was in the account when the account was first opened.
b. Determine the least amount of money in the account.
c. Determine when the account contained its lowest amount. Give the year and month.
d. Calculate how much money was in the account at the end of the 6 years.

11. A cylinder has a surface area of 220π cm^2. Calculate the height and radius of each end of the cylinder so that the volume of the cylinder is maximised, and determine the maximum volume for the cylinder. Give answers correct to 2 decimal places.

12. Calculate the volume of the largest cone, correct to the nearest cubic centimetre, that can be inscribed in a sphere of radius 12 centimetres. Let the base radius of the cone be r cm and the vertical height h cm.

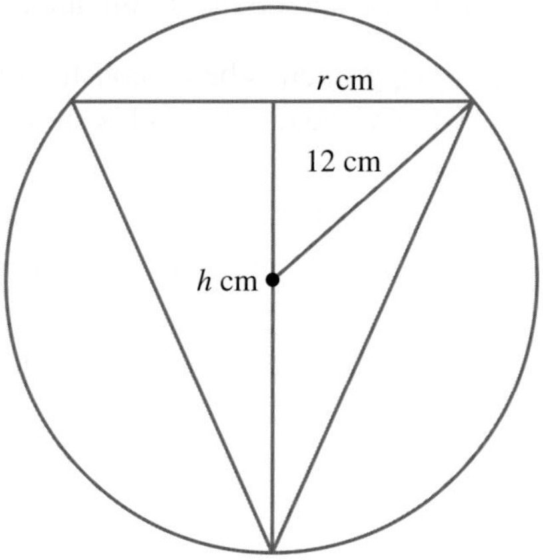

13. A cone is 10 cm high and has a base radius of 8 cm. Calculate the radius and height of a cylinder that is inscribed in the cone such that the volume of the cylinder is a maximum. Determine the maximum volume of the cylinder, correct to the nearest cubic centimetre.

14. A rower is in a boat 4 km from the nearest point, O, on a straight beach. His destination is 8 km along the beach from O. If he is able to row at 5 km/h and walk at 8 km/h, determine the point on the beach he should row to in order to reach his destination in the least possible time. Give your answer correct to 1 decimal place.

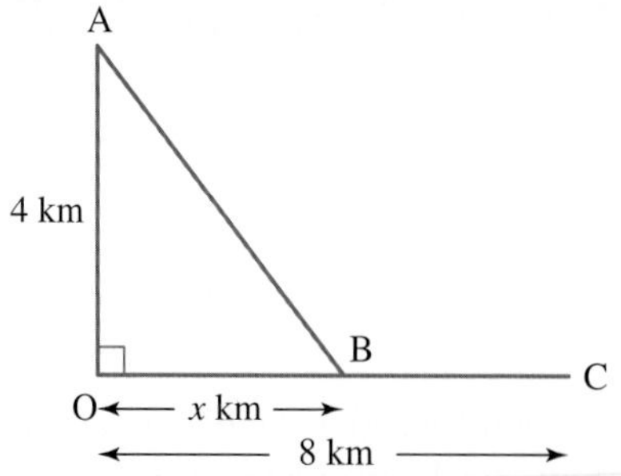

6.6 Exam questions

Question 1 (1 mark) TECH-ACTIVE

Source: *VCE 2016, Mathematical Methods Exam 2, Section A, Q14; © VCAA.*

MC A rectangle is formed by using part of the coordinate axes and a point (u, v), where $u > 0$ on the parabola $y = 4 - x^2$.

Which one of the following is the maximum area of the rectangle?

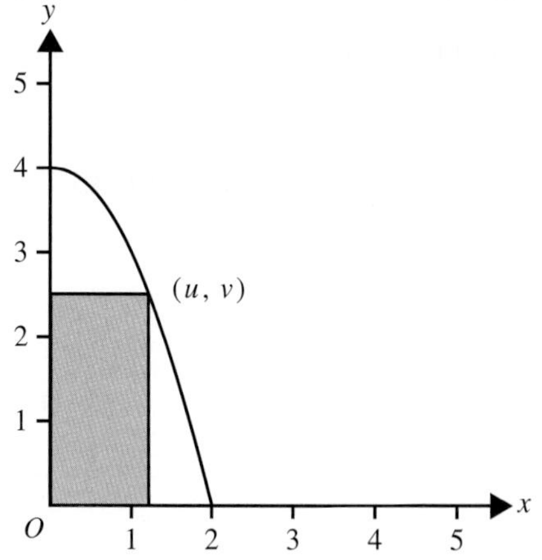

A. 4

B. $\dfrac{2\sqrt{3}}{3}$

C. $\dfrac{8\sqrt{3}-4}{3}$

D. $\dfrac{8}{3}$

E. $\dfrac{16\sqrt{3}}{9}$

Question 2 (5 marks) TECH-FREE

Source: *VCE 2018, Mathematical Methods Exam 1, Q7; © VCAA.*

Let P be a point on the straight line $y = 2x - 4$ such that the length of OP, the line segment from the origin O to P, is a minimum.

a. Find the coordinates of P. **(3 marks)**

b. Find the distance OP. Express your answer in the form $\dfrac{a\sqrt{b}}{b}$, where a and b are positive integers. **(2 marks)**

Question 3 (1 mark) TECH-ACTIVE

Source: *VCE 2014, Mathematical Methods (CAS) Exam 2, Section 1, Q21; © VCAA.*

MC The trapezium $ABCD$ is shown below. The sides AB, BC and DA are of equal length, p. The size of the acute angle BCD is x radians.

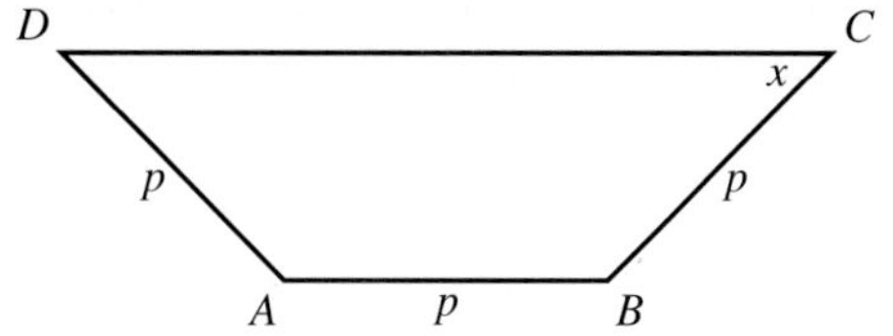

The area of the trapezium is a maximum when the value of x is

A. $\dfrac{\pi}{12}$ **B.** $\dfrac{\pi}{6}$ **C.** $\dfrac{\pi}{4}$ **D.** $\dfrac{\pi}{3}$ **E.** $\dfrac{5\pi}{12}$

More exam questions are available online.

6.7 Rates of change

LEARNING INTENTION

At the end of this subtopic you should be able to:
- solve problems involving instantaneous and average rates of change.

Calculus enables the behaviour of a quantity that changes to be analysed. Many topics of interest in the biological, physical and social sciences involve the study of rates of change. In this section we consider the application of calculus to rates of change in general contexts and then as applied to the motion of a moving object.

6.7.1 Rates of change

The **average rate of change** of a function, f, over the interval x_1 to x_2 is calculated by finding the gradient of the line connecting the two points. The **instantaneous rate of change**, or rate of change, of the function $y = f(x)$ is given by the derivative, $\frac{dy}{dx}$ or $f'(x)$.

Rate of change

$$\textbf{Average rate of change} = \frac{f(x_2) - f(x_1)}{x_2 - x_1}$$

$$\textbf{Instantaneous rate of change} = \frac{dy}{dx} \text{ or } f'(x)$$

For example, the derivative $\frac{dV}{dt}$ could be the rate of change of volume with respect to time, with possible units being litres per minute; the rate of change of volume with respect to height would be $\frac{dV}{dh}$, with possible units being litres per cm. To calculate these rates, V would need to be expressed as a function of one independent variable, either time or height. Similar methods to those encountered in optimisation problems are often required to connect variables together in order to obtain this function of one variable.

To solve rates of change problems, apply the following steps.
- Draw a diagram of the situation where appropriate.
- Identify the rate of change required and define the variables involved.
- Express the quantity that is changing as a function of one independent variable, the variable the rate is measured with respect to.
- Calculate the derivative that measures the rate of change.
- To obtain the rate at a given value or instant, substitute the given value into the derivative expression.
- Remember that a negative value for the rate of change means the quantity is decreasing (negative gradient), whereas a positive value for the rate of change means the quantity is increasing (positive gradient).

WORKED EXAMPLE 14 Rates of change applications

A container in the shape of an inverted right cone of radius 2 cm and depth 5 cm is being filled with water. When the depth of water is *h* cm, the radius of the water level is *r* cm.

a. Use similar triangles to express *r* in terms of *h*.

b. Express the volume of the water as a function of *h*.

c. Determine the rate with respect to the depth of water at which the volume of water is changing when its depth is 1 cm.

THINK	WRITE
a. 1. Draw a diagram of the situation.	a. 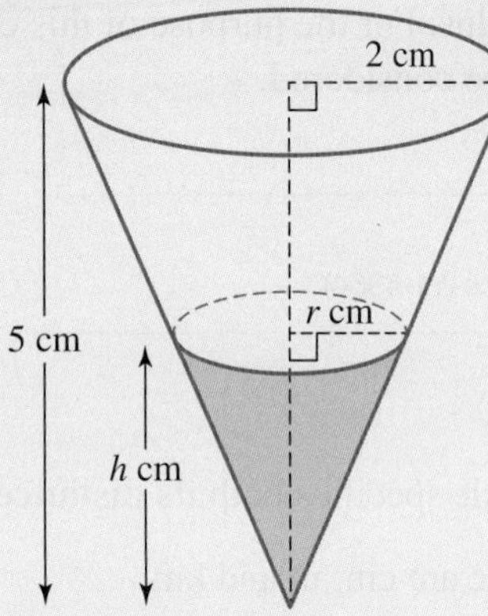
2. Obtain the required relationship between the variables using similar triangles.	$\frac{2}{r} = \frac{5}{h}$ $\therefore r = \frac{2h}{5}$
b. Express the function in the required form.	b. $V_{\text{cone}} = \frac{1}{3}\pi r^2 h$ Therefore, the volume of water is $V = \frac{1}{3}\pi r^2 h$. Substitute $r = \frac{2h}{5}$: $V = \frac{1}{3}\pi\left(\frac{2h}{5}\right)^2 h$ $= \frac{4\pi h^3}{75}$ $\therefore V = \frac{4\pi}{75}h^3$
c. 1. Calculate the derivative of the function. The derivative gives the rate of change at any depth.	c. $\frac{dV}{dh} = \frac{4\pi}{75} \times 3h^2$ $= \frac{4\pi}{25}h^2$
2. Evaluate the derivative at the given value.	When $h = 1$, $\frac{dV}{dh} = \frac{4\pi}{25}$ $= 0.16\pi$
3. Write the answer in context, with the appropriate units.	At the instant the depth is 1 cm, the volume of water is increasing at the rate of 0.16π cm^3/cm.

Resources

Interactivity Rates of change (int-5960)

6.7.2 Kinematics

Many quantities change over time, so many rates measure that change with respect to time. Motion is one such quantity. The study of the motion of a particle without considering the causes of the motion is called **kinematics**. Analysing motion requires interpretation of the **position, velocity** and **acceleration**, and this analysis depends on calculus. For the purpose of this course, only motion in a straight line, also called **rectilinear motion**, will be considered.

Resources

Interactivity Kinematics (int-5964)

Position

The position, x, of a particle specifies both its **distance** and its direction from a fixed origin.

Common units for distance are cm, m and km.

The commonly used conventions for motion along a horizontal straight line are:

- if $x > 0$, the particle is to the right of the origin
- if $x < 0$, the particle is to the left of the origin
- if $x = 0$, the particle is at the origin.

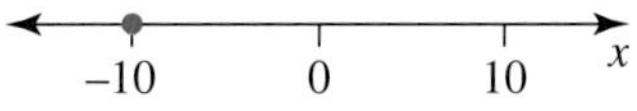

For example, if $x = -10$, this means the particle is 10 units to the left of origin O.

Distance is not concerned with the direction of motion. This can have implications if there is a change of direction in a particle's motion. For example, suppose a particle that is initially 3 cm to the right of the origin travels 2 cm further to the right and then 2 cm to the left, thus returning to where it started. Its change in position (displacement) is zero, but the distance it has travelled is 4 cm.

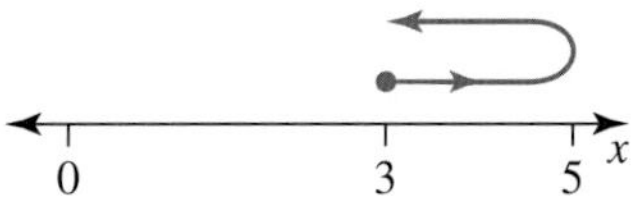

Velocity

Velocity, v, measures the rate of change of position which means that $v = \frac{dx}{dt}$.

For a particle moving in a horizontal straight line, the sign of the velocity indicates that:

- if $v > 0$, the particle is moving to the right
- if $v < 0$, the particle is moving to the left
- if $v = 0$, the particle is stationary (instantaneously at rest).

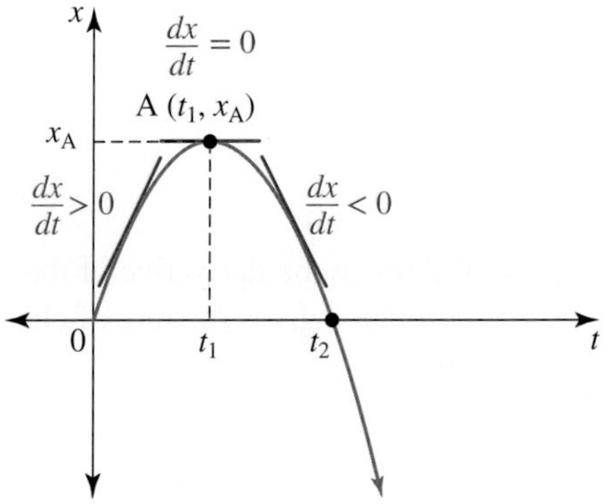

Common units for velocity and **speed** include m/s and km/h.

Just as for distance, speed is not concerned with the direction the particle travels and is never negative. A velocity of -10 m/s means the particle is travelling at 10 m/s to the left. Its speed, however, is 10 m/s, regardless of whether the particle is moving to the left or to the right; that is, the speed is 10 m/s for $v = \pm 10$ m/s.

The position, x, of a particle can be plotted against time t to create a position–time graph, $x = f(t)$. Because $v = \frac{dx}{dt}$, the gradient of the tangent to the curve $f(t)$ at any point represents the velocity of the particle at that point: $v = \frac{dx}{dt} = f'(t)$.

This position–time graph shows the position of a particle that starts at the origin and initially moves to the right, as the gradient of the graph, that is the velocity, is positive.

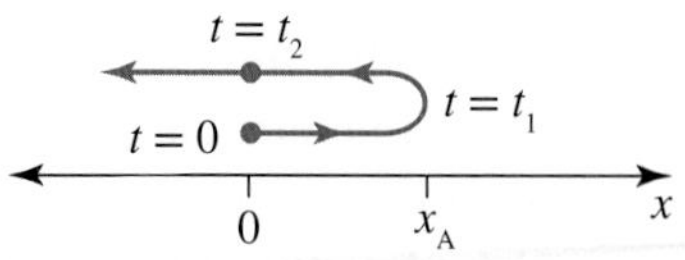

At the point A the tangent is horizontal and the velocity is zero, indicating the particle changes its direction of motion at that point.

The particle then starts to move to the left as indicated by the gradient of the graph, that is the velocity, having a negative sign. The particle returns to the origin and continues to move to the left, so its position becomes negative.

Average velocity is the average rate of change of the position over time. It is measured by the gradient of the chord joining two points on the position–time graph. It must be evaluated using coordinate geometry, not calculus.

Average velocity and speed

- **Average velocity** $= \dfrac{\textbf{change in position}}{\textbf{change in time}} = \dfrac{x_2 - x_1}{t_2 - t_1}$
- **Average speed** $= \dfrac{\textbf{distance travelled}}{\textbf{time taken}}$

Acceleration

Acceleration, a, measures the rate of change of velocity; thus, $a = \frac{dv}{dt}$.

Common units for acceleration include m/s^2.

Position, velocity and acceleration are linked by calculus. Differentiation enables us to obtain the velocity function from the position function, and to obtain the acceleration function from the velocity function.

Relationship between x, v, and a

$$x \xrightarrow{\text{derive}} v \xrightarrow{\text{derive}} a$$

$$x \longrightarrow \frac{dx}{dt} \longrightarrow \frac{dv}{dt}$$

Acceleration acts tangentially to the velocity–time graph, whereas average acceleration measures the gradient of the chord joining two points on the velocity–time graph.

Average acceleration

$$\textbf{Average acceleration} = \frac{\textbf{change in velocity}}{\textbf{change in time}}$$

WORKED EXAMPLE 15 Kinematics applications

A particle moves in a straight line such that its position, x metres, from a fixed origin at time t seconds is modelled by $x = t^2 - 4t - 12$, $t \geq 0$.

a. **Identify its initial position.**

b. **Obtain its velocity and hence state its initial velocity and describe its initial motion.**

c. **Determine the time and position at which the particle is momentarily at rest.**

d. **Show the particle is at the origin when $t = 6$ and calculate the distance it has travelled to reach the origin.**

e. **Calculate the average speed over the first 6 seconds.**

f. **Calculate the average velocity over the first 6 seconds.**

THINK	WRITE
a. Calculate the value of x when $t = 0$.	a. $x = t^2 - 4t - 12,\ t \geq 0$ When $t = 0$, $x = -12$. Initially the particle is 12 metres to the left of the origin.
b. 1. Calculate the rate of change required.	b. $v = \frac{dx}{dt}$ $v = 2t - 4$
2. Calculate the value of v at the given instant.	When $t = 0$, $v = -4$. The initial velocity is −4 m/s.
3. Describe the initial motion.	Since the initial velocity is negative, the particle starts to move to the left with an initial speed of 4 m/s.
c. 1. Calculate when the particle is momentarily at rest. *Note*: This usually represents a change of direction of motion.	c. The particle is momentarily at rest when its velocity is zero. When $v = 0$, $2t - 4 = 0$ $t = 2$ The particle is at rest after 2 seconds.
2. Calculate where the particle is momentarily at rest.	The position of the particle when $t = 2$ is $x = (2)^2 - 4(2) - 12$ $= -16$ Therefore, the particle is momentarily at rest after 2 seconds at the position 16 metres to the left of the origin.
d. 1. Calculate the position to show the particle is at the origin at the given time.	d. When $t = 6$, $x = 36 - 24 - 12$ $= 0$ The particle is at the origin when $t = 6$.
2. Track the position over the first 6 seconds and calculate the required distance.	The motion of the particle for the first 6 seconds is shown. 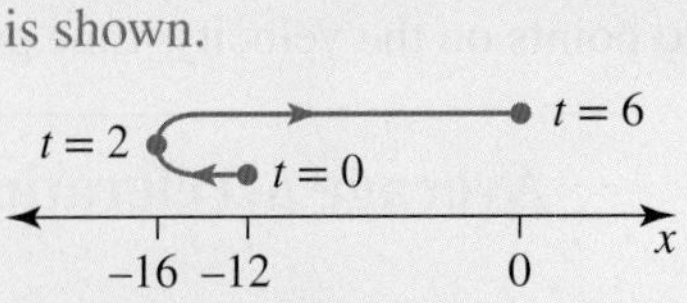 The distances travelled are 4 metres to the left, then 16 metres to the right.

The total distance travelled is the sum of the distances in each direction.
The particle has travelled a total distance of 20 metres.

e. Calculate the value required.

e. Average speed $= \dfrac{\text{distance travelled}}{\text{time taken}}$

$$= \frac{20}{6}$$

$$= 3\frac{1}{3}$$

The average speed is $3\frac{1}{3}$ m/s.

f. Calculate the average rate of change required.
Note: As there is a change of direction, the average velocity will not be the same as the average speed.

f. Average velocity is the average rate of change of position.
For the first 6 seconds,
$(t_1, x_1) = (0, -12), \ (t_2, x_2) = (6, 0)$.

Average velocity $= \dfrac{x_2 - x_1}{t_2 - t_1}$

$$= \frac{0 - (-12)}{6 - 0}$$

$$= 2$$

The average velocity is 2 m/s.

6.7 Exercise

Technology free

1. **WE14** A container in the shape of an inverted right cone of radius 4 cm and depth 12 cm is being filled with water. When the depth of water is h cm, the radius of the water level is r cm.
   a. Use similar triangles to express r in terms of h.
   b. Express the volume of the water as a function of h.
   c. Determine the rate with respect to the depth of water at which the volume of water is changing when its depth is 5 cm.

2. a. A spherical balloon of radius r is expanding. Calculate the rate of change of the volume with respect to the radius when the radius is 10 cm.
   b. An ice cube melts in such a way as to maintain its shape as a cube. Calculate the rate at which its surface area is changing with respect to its side length at the instant the side length is 6 mm.

3. Water is being poured into a vase. The volume, V mL, of water in the vase after t seconds is given by

$$V = \frac{2}{3}t^2(15 - t),\ 0 \le t \le 10.$$

a. Calculate the volume after 10 seconds.
b. Determine the rate at which the water is flowing into the vase at t seconds.
c. Determine the rate of flow after 3 seconds.
d. Determine the time at which the rate of flow is the greatest, and the rate of flow at this time.

4. **WE15** A particle moves in a straight line such that its position, x metres, from a fixed origin at time t seconds is given by $x = 2t^2 - 8t,\ t \ge 0$.

a. Identify its initial position.
b. Obtain its velocity and hence state its initial velocity and describe its initial motion.
c. Determine the time and position at which the particle is momentarily at rest.
d. Show that the particle is at the origin when $t = 4$ and calculate the distance it has travelled to reach the origin.
e. Calculate the average speed over the first 4 seconds.
f. Calculate the average velocity over the first 4 seconds.

5. The position, in metres, of a particle after t seconds is given by $x(t) = -\frac{1}{3}t^3 + t^2 + 8t + 1,\ t \ge 0$.

a. Find its initial position and initial velocity.
b. Calculate the distance travelled before it changes its direction of motion.
c. Determine its acceleration at the instant it changes direction.

Technology active

6. The number of rabbits on a farm is modelled by $N = \frac{110}{t},\ t > 0$, where N is the number of rabbits present after t months.

a. Determine the rate at which the population of rabbits is changing after 5 months.
b. Calculate the average rate of change of the population over the interval $t \in [1, 5]$.
c. Describe what will happen to the population of rabbits in the long term.

7. The volume of water, V litres, in a bath t minutes after the plug is removed is given by $V = 0.4(8 - t)^3,\ 0 \le t \le 8$.

a. Determine the rate at which the water is leaving the bath after 3 minutes.
b. Determine the average rate of change of the volume for the first 3 minutes.
c. Determine when the rate of water leaving the bath is the greatest.

8. A ball is thrown vertically upwards into the air so that after t seconds its height h metres above the ground is $h = 50t - 4t^2$.

a. Calculate the rate at which its height is changing after 3 seconds.
b. Calculate its velocity when $t = 5$.
c. Determine the time at which its velocity is -12 m/s and the direction the ball is then travelling.
d. Determine when its velocity is zero.
e. Calculate the greatest height the ball reaches.
f. Determine the time and speed at which the ball strikes the ground.

9. A colony of viruses can be modelled by the rule

$$N(t) = \frac{2t}{(t+0.5)^2} + 0.5$$

where N hundred thousand is the number of viruses on a nutrient plate t hours after they started multiplying.

a. Calculate how many viruses were present initially.
b. Find $N'(t)$.
c. Determine the maximum number of viruses and the time at which this maximum will occur.
d. Determine the rate at which the virus numbers were changing after 10 hours.

10. A population of butterflies in an enclosure at a zoo is modelled by

$$N = 220 - \frac{150}{t+1}, \ t \geq 0$$

where N is the number of butterflies t years after observations of the butterflies commenced.

a. Determine how long it took for the butterfly population to reach 190 butterflies and the rate at which the population was growing at that time.
b. Determine the time at which the growth rate was 12 butterflies per year. Give your answer correct to 2 decimal places.
c. Sketch the graphs of population versus time and rate of growth versus time, and explain what happens to each as $t \to \infty$.

11. A particle moves in a straight line so that at time t seconds its position, x metres, from a fixed origin O is given by $x(t) = 2t^2 - 16t - 18, \ t \geq 0$.

a. Calculate the distance the particle is from O after 2 seconds.
b. Calculate the speed at which it is travelling after 2 seconds.
c. Calculate the average velocity of the particle over the first 2 seconds of motion.
d. Determine the time and velocity when it reaches O.

12. The position, x m, relative to a fixed origin of a particle moving in a straight line at time t seconds is $x = \frac{2}{3}t^3 - 4t^2, \ t \geq 0$.

a. Show the particle starts at the origin from rest.
b. Determine the time and position at which the particle is next at rest.
c. Determine when the particle returns to the origin.
d. Determine the particle's speed and acceleration when it returns to the origin.

13. A cone has a slant height of 20 cm. The diameter of its circular base is increased in such a way that the cone maintains its slant height of 20 cm while its perpendicular height decreases. When the base radius is r cm, the perpendicular height of the cone is h cm.

a. Use Pythagoras's theorem to express r in terms of h.
b. Express the volume of the cone as a function of h.
c. Calculate the rate of change of the volume with respect to the perpendicular height when the height is 8 cm.

14. A tent in the shape of a square-based right pyramid has perpendicular height h metres, base side length x metres and volume $\frac{1}{3}Ah$, where A is the area of its base.

a. Express the length of the diagonal of the square base in terms of x.
b. If the slant height of the pyramid is 12 metres, show that $x^2 = 288 - 2h^2$ and hence express the volume of air inside the tent in terms of h.
c. Calculate the rate of change of the volume with respect to height when the height is $3\sqrt{3}$ metres.

15. A container in the shape of an inverted right circular cone is being filled with water. The cone has a height of 15 cm and a radius of 6 cm. Calculate the rate at whcih the volume of water is changing with respect to the depth of water when:

a. the depth of water reaches half the height of the cone
b. the container is one-third full.

16. A veterinarian has administered a painkiller by injection to a sick horse.
The concentration of painkiller in the blood, y mg/L, can be defined by the rule

$$y = \frac{3t}{\left(4 + t^2\right)}$$

where t is the number of hours since the medication was administered.

a. Determine $\frac{dy}{dt}$.
b. Determine the maximum concentration of painkiller in the blood and the time at which this is achieved.
c. The effect of the painkiller is considerably reduced once the concentration falls below 0.5 mg/L, when a second dose needs to be given to the horse. Determine when this occurs.
d. Calculate the rate of change of concentration of painkiller in the blood after one hour. Give your answer correct to 2 decimal places.
e. Determine when the rate of change of painkiller in the blood equal to -0.06 mg/L/h. Give your answer correct to 2 decimal places.

6.7 Exam questions

Question 1 (1 mark) TECH-ACTIVE

Source: *VCE 2019, Mathematical Methods Exam 2, Section A, Q3; © VCAA.*

MC Let $f: R\backslash\{4\} \to R,\ f(x) = \frac{a}{x-4}$, where $a > 0$.

The average rate of change of f from $x = 6$ to $x = 8$ is

A. $a \log_e(2)$ **B.** $\frac{a}{2}\log_e(2)$ **C.** $2a$ **D.** $-\frac{a}{4}$ **E.** $-\frac{a}{8}$

Question 2 (1 mark) TECH-ACTIVE

Source: *VCE 2017, Mathematical Methods Exam 2, Section A, Q9; © VCAA.*

MC The average rate of change of the function with the rule $f(x) = x^2 - 2x$ over the interval $[1, a]$, where $a > 1$, is 8.

The value of a is

A. 9 **B.** 8 **C.** 7 **D.** 4 **E.** $1 + \sqrt{2}$

Question 3 (1 mark) TECH-ACTIVE

MC For $y = 2e^{-x+1}\sin(x-1)$, the rate of change of y with respect to x when $x = 1$ is

A. -1 **B.** 1 **C.** 4 **D.** 2 **E.** -2

More exam questions are available online.

6.8 Newton's method

LEARNING INTENTION

At the end of this subtopic you should be able to:
- use Newton's method to calculate roots of an equation.

6.8.1 Approximate solutions to equations using tangents

Newton's method, also known as the Newton–Raphson method, is a numerical method for solving equations iteratively. It involves using the x-intercepts of tangent lines to best approximate the roots or solutions of a curve.

Consider the curve $y = f(x)$ as shown. We want to find the value of p, the solution to the equation $f(x) = 0$.

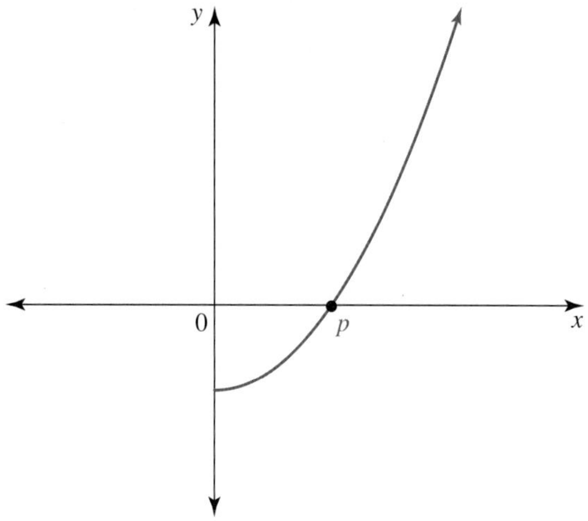

Step 1: Choose a starting value, x_0, that is an estimate of the root of the equation. In this example, there is only one solution visible, but $f(x)$ may have more than one solution, so it is important to choose the starting value carefully.

Step 2: At the point $(x_0, f(x_0))$ the tangent is drawn, and the x-intercept of this tangent is found. This point is called x_1 and is a better estimate for the root of the equation than x_0.

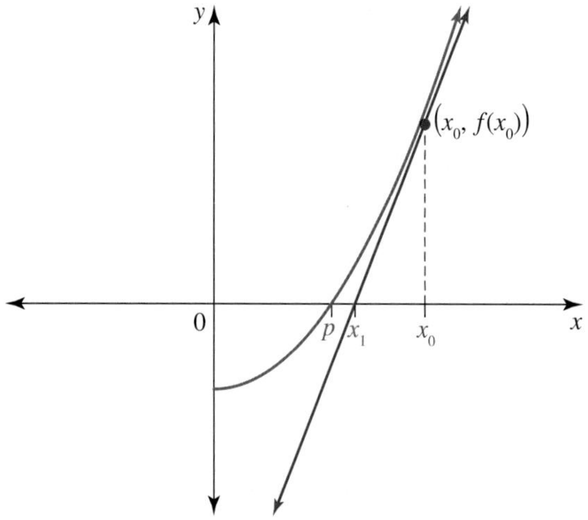

Step 3: At the point $(x_1, f(x_1))$ the tangent is drawn, and the x-intercept of this tangent is found. This point is called x_2 and is a better estimate for the root of the equation than x_1.

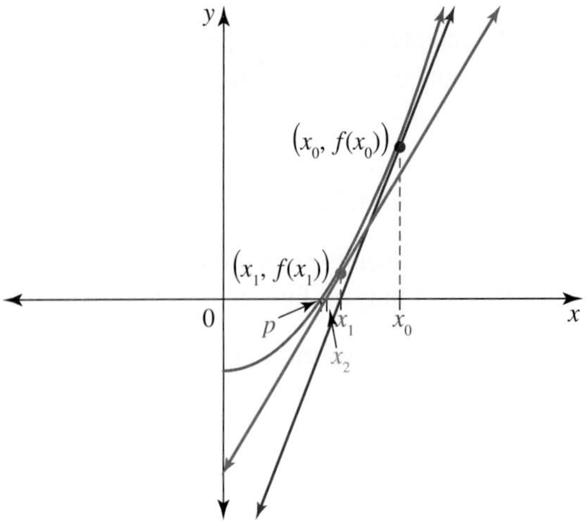

This process is repeated until the specified degree of accuracy is obtained.

The iterative formula

Going back to step 2, the equation of the tangent needs to be determined.

The point is $(x_0, f(x_0))$, and the gradient at this point is $f'(x_0)$.

The equation of the tangent is:

$$y - f(x_0) = f'(x_0)(x - x_0)$$
$$y = f(x_0) + f'(x_0)(x - x_0)$$

The value of x_1 is the x-intercept of this tangent. Let $y = 0$.

$$0 = f(x_0) + f'(x_0)(x_1 - x_0)$$
$$-f(x_0) = f'(x_0)(x_1 - x_0)$$
$$x_1 - x_0 = -\frac{f(x_0)}{f'(x_0)}$$
$$x_1 = x_0 - \frac{f(x_0)}{f'(x_0)}$$

In step 3 the process is repeated and the value of x_2 is found by

$$x_2 = x_1 - \frac{f(x_1)}{f'(x_1)}$$

We can generalise the rule as follows:

Iterative formula for Newton's method

$$x_{n+1} = x_n - \frac{f(x_n)}{f'(x_n)}, \text{ where } n = 0, 1, 2...$$

WORKED EXAMPLE 16 Using Newton's method

Use Newton's method to calculate the root of the equation $x^3 - x^2 - 2 = 0$ with starting value $x_0 = 2$. Express your answer to an accuracy of 4 decimal places.

THINK	WRITE
1. Define $f(x)$ and state $f'(x)$.	$f(x) = x^3 - x^2 - 2$ $f'(x) = 3x^2 - 2x$
2. Using the formula for Newton's method, calculate x_1 from x_0.	$x_1 = 2 - \frac{f(2)}{f'(2)}$ $f(2) = 2^3 - 2^2 - 2$ $= 2$ $f'(2) = 3(2)^2 - 2(2)$ $= 8$ $\therefore x_1 = 2 - \frac{2}{8}$ $= 1.75$
3. Now calculate x_2 from x_1.	$f(1.75) = (1.75)^3 - (1.75)^2 - 2$ $= 0.296\,875$ $f'(1.75) = 3(1.75)^2 - 2(1.75)$ $= 5.6875$ $\therefore x_2 = 1.75 - \frac{0.296\,875}{5.6875}$ $= 1.697\,802\,2$
4. It may be more efficient to define the function and derivative in your calculator, and use the Ans function to carry the previous answer into the next iteration.	$\therefore x_3 = \text{Ans} - \frac{f(\text{Ans})}{f'(\text{Ans})}$ $= 1.695\,624\,5$
5. Continue the process until the 4th decimal place is constant.	$\therefore x_4 = \text{Ans} - \frac{f(\text{Ans})}{f'(\text{Ans})}$ $= 1.695\,620\,8$ Both x_3 and x_4 are the same value for the required accuracy of 4 decimal places.
6. State the solution.	To 4 decimal places, the solution to $x^3 - x^2 - 2 = 0$ is $x = 1.6956$.

Newton's method does not always work

1. If the initial estimate chosen, x_0, is a stationary point, then the tangent at this point will be horizontal; therefore, it will not produce an x-intercept or x_1 value.
2. If the initial estimate chosen is on the other side of a stationary point to where the root is located, Newton's Method may (a) find a different solution or (b) move further away and not locate the root.

a.

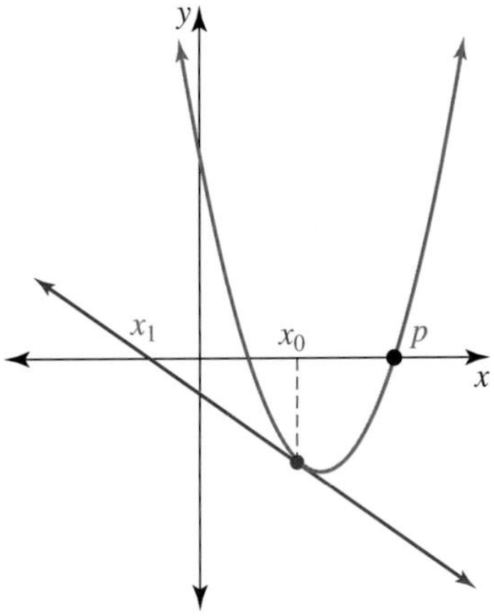

b.

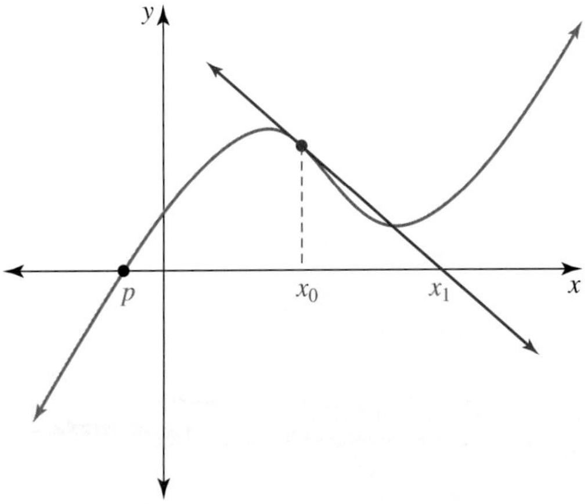

3. There is an oscillating sequence where the x-intercepts of the tangents oscillate between x_1, x_0, x_1, x_0

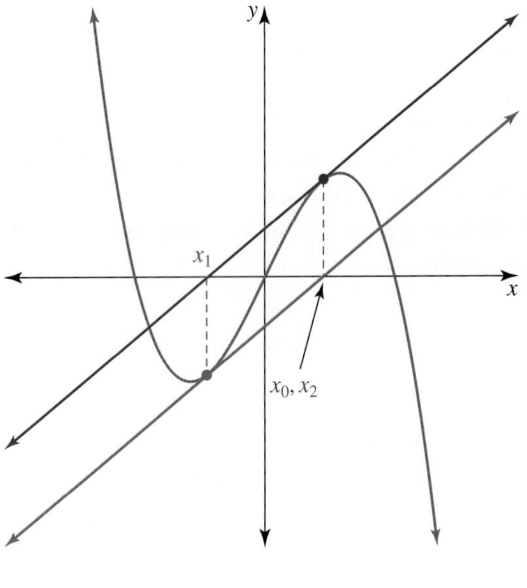

6.8 Exercise

Technology active

1. WE16 Use Newton's method to calculate the root of the equation $x^3 + x - 5 = 0$ with a starting value, $x_0 = 1$. Express your answer to an accuracy of 4 decimal places.
2. The equation $0 = x^3 - 6x - 12$ has only one solution. Determine between which two integer values the solution lies, then calculate the solution to an accuracy of 3 decimal places.
3. Determine the root of $0 = 2\log_e(x-2) + 1$, accurate to 3 decimal places.
4. Use a suitable equation to calculate $\sqrt[3]{10}$, accurate to 2 decimal places.
5. MC If Newton's method is used to find the solutions to $\cos(3x) = \sin(x)$, and $x_0 = 0$, then x_3 will be:
   - A. $-0.900\,524$
   - B. $-0.716\,125$
   - C. $0.392\,699$
   - D. $1.184\,399$
   - E. $-0.785\,398$
6. Determine the root of $2x^2 + 5 = e^x$ in the interval $[3, 4]$, to an accuracy of 3 decimal places.

6.8 Exam questions

Question 1 (2 marks) TECH-ACTIVE

Determine the root of $2 - x^2 = \sin(x)$ in the interval $[1, 2]$ to an accuracy of 3 decimal places.

Question 2 (3 marks) TECH-ACTIVE

Use a suitable equation to calculate $\sqrt[4]{12}$, accurate to 2 decimal places.

Question 3 (3 marks) TECH-ACTIVE

The equation $0 = x^4 - 5x - 8$ has only one negative solution. Determine between which two integer values the solution lies, then calculate the solution to an accuracy of 3 decimal places.

More exam questions are available online.

6.9 Review

6.9.1 Summary

doc-37035

Hey students! Now that it's time to revise this topic, go online to:

Access the topic summary | Review your results | Watch teacher-led videos | Practise VCAA exam questions

Find all this and MORE in jacPLUS

6.9 Exercise

Technology free: short answer

1. A curve is represented by the equation $y = ax\cos(3x)$, where a is a constant.
   a. If $\frac{dy}{dx} = -5$ when $x = \pi$, calculate the value of a.
   b. Determine the equation of the line perpendicular to the curve at $x = \frac{\pi}{3}$.
2. Let $f: [0, 2] \to R$, $f(x) = 4x^3 - 6x^2 + 3$.
   a. Calculate the coordinates of the end points of this function.
   b. Determine the coordinates of all stationary points and specify their nature.
   c. Sketch a graph of this function over the specified domain.
   d. State the absolute maximum and minimum of this function (if they exist).
3. The graph shown is of $y = x^4 - 2x^2 + 5,\ x \in [-1, 2]$.

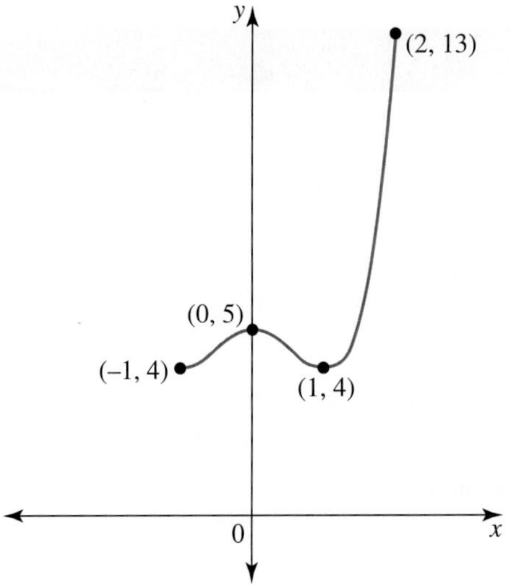

   a. Calculate the average rate of change over the interval $x \in [0, 2]$.
   b. Determine the domain over which the function is strictly increasing.
4. The function $f: R \to R, f(x) = 6\log_e(x^2 - 4x + 8)$ has one stationary point.
   a. Use calculus to determine the coordinates of this stationary point.
   b. Determine the nature of this stationary point.

5. a. If the function f has a rule $f(x) = \sqrt{x^2 - 1}$ and the function g has the rule $g(x) = x + 3$, calculate the integers m and n such that $f(g(x)) = \sqrt{(x+m)(x+n)}$, where $m > n$.

b. If $h(x) = f(g(x))$, determine $h'(x)$.

6. a. Consider the function $f: R \rightarrow R, f(x) = x^4 e^{-3x}$. The derivative $f'(x)$ may be written in the form $f'(x) = e^{-3x}\left(mx^4 + nx^3\right)$, where m and n are real constants.
Calculate the exact values of m and n.

b. The graph of f is shown. Locate the coordinates of the stationary points.

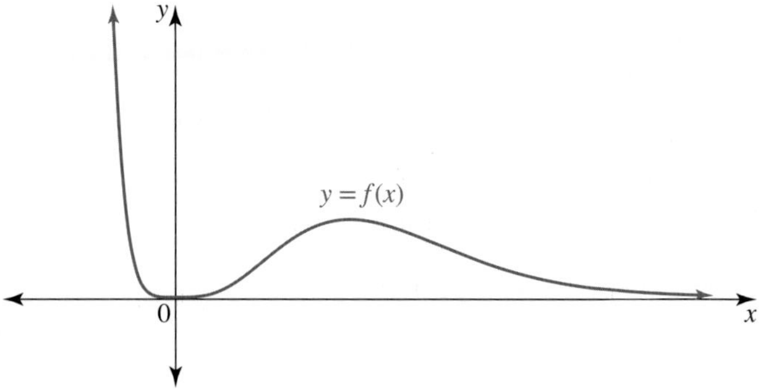

Technology active: multiple choice

7. **MC** If $y = e^{3\cos(5x)}$, then $\dfrac{dy}{dx}$ is:

A. $15\sin(5x)e^{3\cos(5x)}$
B. $e^{3\cos(5x)}$
C. $-15\sin(5x)e^{3\cos(5x)}$
D. $e^{-15\sin(5x)}$
E. $-15\cos(5x)e^{-3\sin(5x)}$

8. **MC** If $y = e^{ax}\sin(bx)$, where a and b are constants, then $\dfrac{dy}{dx}$ is:

A. $ae^{ax}\sin(bx) + be^{ax}\cos(bx)$
B. $ae^{ax}\sin(bx) - be^{ax}\cos(bx)$
C. $abe^{ax}\cos(bx)$
D. $e^{ax}\sin(bx) + be^{ax}\cos(bx)$
E. $\dfrac{1}{ab}e^{ax}\cos(bx)$

9. **MC** The derivative of $\dfrac{\cos(7t)}{t^2}$ with respect to t is:

A. $\dfrac{-7t^2\sin(7t) - 2t\cos(7t)}{t^4}$
B. $\dfrac{-7\sin(7t) - 2\cos(7t)}{t^2}$
C. $\dfrac{7t^2\sin(7t) - 2t\cos(7t)}{t^4}$
D. $\dfrac{2t\cos(7t) + 2t^2\sin(7t)}{t^4}$
E. $\dfrac{-t^2\sin(7t) - 2t\cos(7t)}{t^4}$

10. **MC** If $y = x\log_e(5x)$, then $\dfrac{dy}{dx}$ is equal to:

A. $\dfrac{1}{5}$
B. $\dfrac{1}{5x}$
C. $1 + \log_e(5x)$
D. $5 + \log_e(5x)$
E. $\dfrac{1}{5} + \log_e(5x)$

11. **MC** Let $f: R \rightarrow R$ be a differentiable function. For all real values of x, the derivative of $f(e^{4x})$ with respect to x will be:

A. $4e^{4x}f'(x)$
B. $e^{4x}f'(x)$
C. $4e^{4x}f'\left(e^{4x}\right)$
D. $4f'\left(e^{4x}\right)$
E. $f'\left(e^{4x}\right)$

12. **MC** For $y = \sqrt{7 - 2f(x)}$, $\dfrac{dy}{dx}$ is equal to:

A. $\dfrac{2f'(x)}{\sqrt{7 - 2f(x)}}$
B. $\dfrac{-1}{2\sqrt{7 - 2f(x)}}$
C. $\dfrac{1}{2}\sqrt{7 - 2f'(x)}$
D. $\dfrac{7}{2(7 - 2f'(x))}$
E. $\dfrac{-f'(x)}{\sqrt{7 - 2f'(x)}}$

13. **MC** Consider the function $f(x) = x^3 + 2x^2 - 15x + 7$. It has stationary points with coordinates:

A. $(-3, 43)$ and $\left(\frac{5}{3}, -\frac{211}{27}\right)$
B. $(0, 7)$
C. $(-3, 43)$ and $(5, 107)$
D. $(-3, 0)$ and $\left(\frac{5}{3}, 0\right)$
E. $(3, 43)$ and $\left(\frac{5}{3}, \frac{211}{27}\right)$

14. **MC** The graph of $f(x) = \frac{1}{5}(1 - x)(x - 3)(x - 5)$ is shown.

The absolute maximum is:

A. -3
B. -0.616
C. 3
D. 0.616
E. There is no absolute maximum.

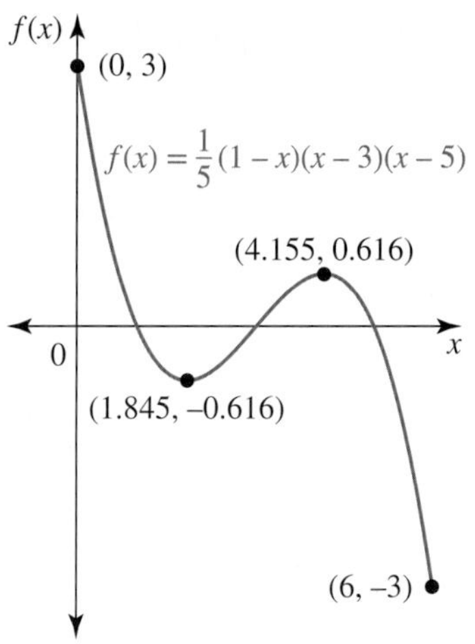

15. **MC** If $f(x) = (x - a)^3 g(x)$, then the derivative of $f(x)$ is equal to:

A. $3(x - a)g(x)$
B. $3(x - a)g'(x)$
C. $3(x - a)^2 g'(x)$
D. $3g'(x)$
E. $3(x - a)^2 g(x) + (x - a)^3 g'(x)$

16. **MC** If Newton's method is used to find the solutions to $f(x) = x\cos(x) - x^2$, and $x_0 = 1$, then x_2 will be:

A. 0.943 861
B. 0.800 232
C. 0.739 085
D. 0.744 094
E. 0.762 334

Technology active: extended response

17. An amusement park has constructed a new super roller-coaster ride. The profile of the ride is shown.

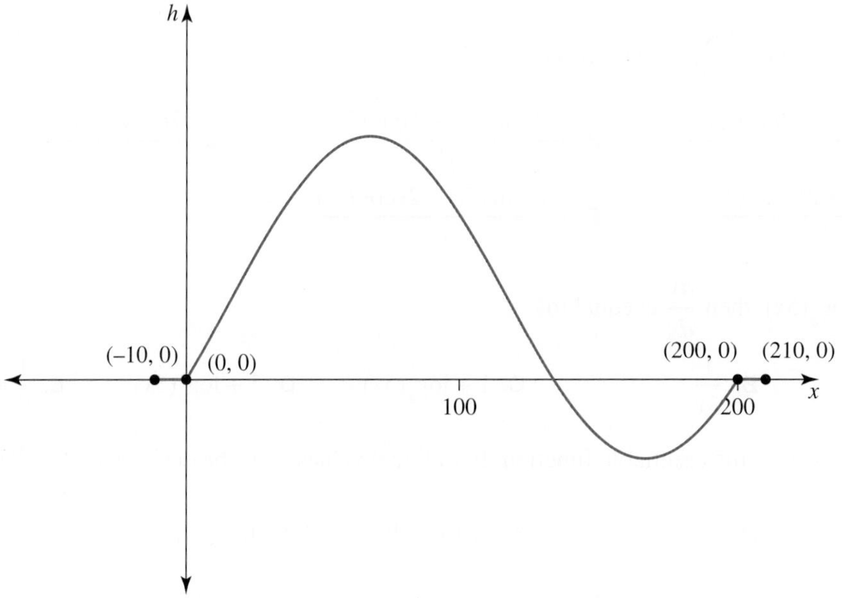

The thrill ride can be modelled by the equation

$$h = 10\cos\left(\frac{\pi(x-67)}{100}\right) + 5,\ 0 \le x \le 200$$

where h metres is the height of the track above or below the platform where riders climb on or off the ride, and x metres is the horizontal distance from the start of the ride. There are flat sections at the beginning and end of the ride so that riders can climb on or off.

a. Determine an expression for the gradient of the track at any point during the ride.
b. Determine the greatest and least heights above the platform that the car reaches during the duration of the ride.
c. Calculate the gradient of the track when x is equal to:
   i. 50 metres
   ii. 100 metres.

   Give your answers correct to 2 decimal places.

18. Metal box guttering has to be formed on a common wall between two adjacent town houses. The cross section of the box guttering is shown.
For the most efficient elimination of rain water, this box guttering needs to have a maximum cross-sectional area within the given dimensions.

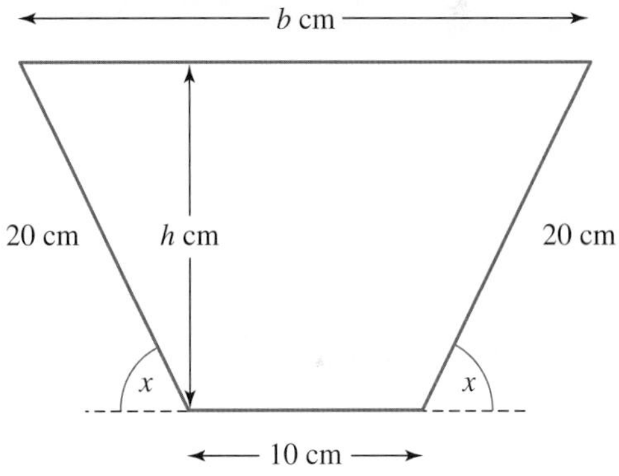

a. Determine an expression for h, the height of the trapezium, in terms of the angle x in radians, as shown.
b. Determine an expression for b, the base length of the trapezium, in terms of x.
c. Show that the cross-sectional area of the box guttering, A cm^2, is given by $A = 200\sin(x)(2\cos(x) + 1)$.
d. Calculate, correct to 3 decimal places, the value of x that gives maximum cross-sectional area, and find this maximum area correct to the nearest cm^2.

19. A vessel is leaking its liquid contents out very slowly through a small hole. The volume of liquid, V mL, remaining in the vessel after t hours is modelled by the equation

$$V = -0.5t^2 - t + 1.5,\ t \in [0, 1].$$

a. Determine the rate at which liquid is leaking from the vessel after 0.2 hours.
b. Determine the time at which the vessel has lost half its initial volume. Give your answer correct to 2 decimal places.
c. Calculate the average rate of change in the volume over this time. Give your answer correct to 2 decimal places.
d. Determine when the rate of flow is greatest.

20. A swimming pool consists of a rectangle and a semicircle as shown. The radius of the semi-circular section is r metres, and the length of the rectangular section is L metres. The total surface area of the pool is 145 m^2.

a. Show that the equation representing the perimeter of the pool, P metres, is $P = \dfrac{290 - \pi r^2}{2r} + (2 + \pi)r$.

b. Determine an expression for r that will cause this perimeter to be a minimum.

c. Calculate the minimum perimeter and the values of r and L that produce this minimum. Give all answers correct to 1 decimal place.

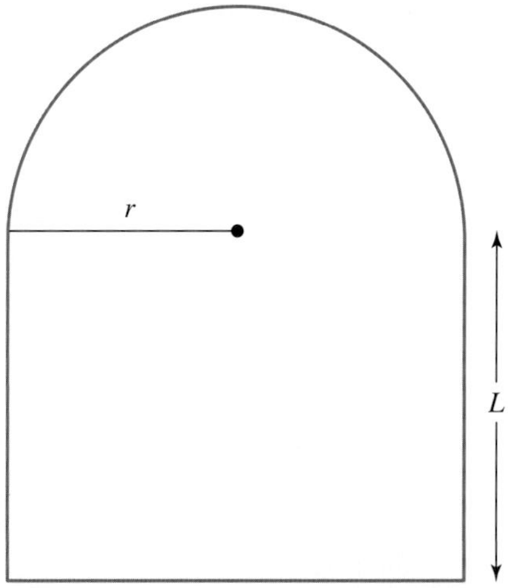

6.9 Exam questions

Question 1 (1 mark) TECH-FREE

Source: *VCE 2018, Mathematical Methods Exam 1, Q1a; © VCAA.*

If $y = \left(-3x^3 + x^2 - 64\right)^3$, find $\dfrac{dy}{dx}$.

Question 2 (2 marks) TECH-FREE

Source: *VCE 2014, Mathematical Methods (CAS) Exam 1, Q1a; © VCAA.*

If $y = x^2 \sin(x)$, find $\dfrac{dy}{dx}$.

Question 3 (1 mark) TECH-ACTIVE

Source: *VCE 2020, Mathematical Methods Exam 2, Section A, Q16; © VCAA.*

MC A right-angled triangle, OBC, is formed using the horizontal axis and the point $C(m, 9 - m^2)$ $m \in (0, 3)$, where, on the parabola $y = 9 - x^2$, as shown below.

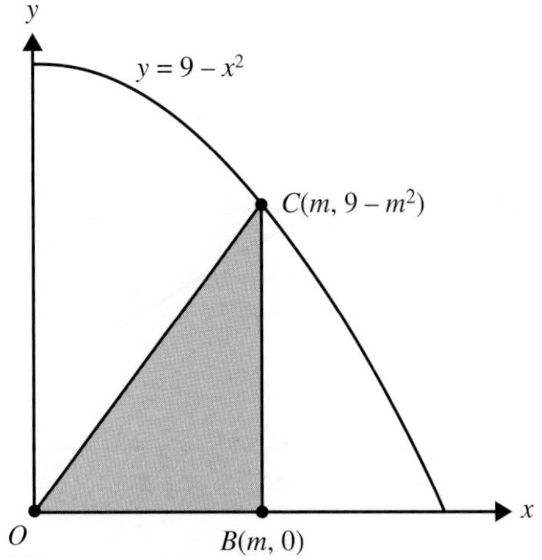

The maximum area of the triangle OBC is

A. $\dfrac{\sqrt{3}}{3}$ **B.** $\dfrac{2\sqrt{3}}{3}$ **C.** $\sqrt{3}$ **D.** $3\sqrt{3}$ **E.** $9\sqrt{3}$

Question 4 (1 mark) TECH-ACTIVE

Source: *VCE 2016, Mathematical Methods Exam 2, Section A, Q4; © VCAA.*

MC The average rate of change of the function f with rule $f(x) = 3x^2 - 2\sqrt{x+1}$, between $x = 0$ and $x = 3$, is

A. 8 **B.** 25 **C.** $\dfrac{53}{9}$ **D.** $\dfrac{25}{3}$ **E.** $\dfrac{13}{9}$

Question 5 (11 marks) TECH-ACTIVE

Source: *VCE 2019, Mathematical Methods Exam 2, Section B, Q2; © VCAA.*

An amusement park is planning to build a zip-line above a hill on its property.

The hill is modelled by $y = \dfrac{3x(x-30)^2}{2000}$, $x \in [0, 30]$, where x is the horizontal distance, in metres, from an origin and y is the height, in metres, above this origin, as shown in the graph below.

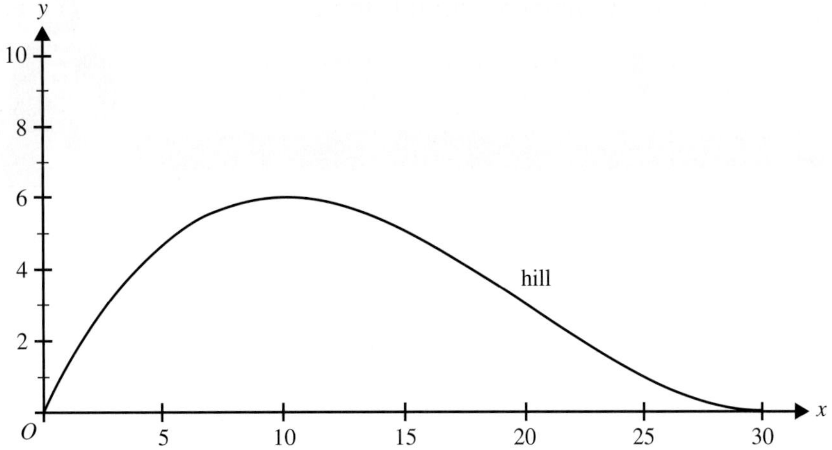

a. Find $\dfrac{dy}{dx}$. **(1 mark)**

b. State the set of values for which the gradient of the hill is strictly decreasing. **(1 mark)**

The cable for the zip-line is connected to a pole at the origin at a height of 10 m and is straight for $0 \le x \le a$, where $10 \le a \le 20$. The straight section joins the curved section at $A(a, b)$. The cable is then exactly 3 m vertically above the hill from $a \le x \le 30$, as shown in the graph below.

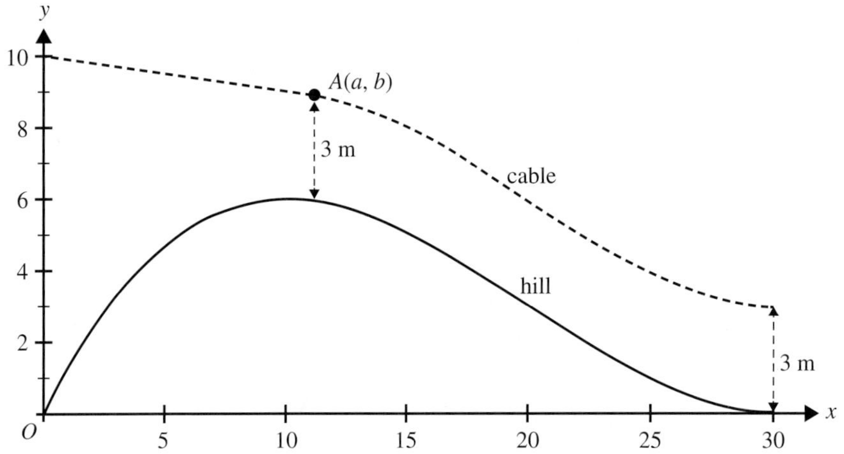

c. State the rule, in terms of x, for the height of the cable above the horizontal axis for x $[a, 30]$. **(1 mark)**

d. Find the values of x for which the gradient of the cable is equal to the average gradient of the hill for $x \in [10, 30]$ **(3 marks)**

The gradients of the straight and curved sections of the cable approach the same value at $x = a$, so there is a continuous and smooth join at A.

e. **i.** State the gradient of the cable at A, in terms of a. **(1 mark)**

ii. Find the coordinates of A, with each value correct to two decimal places. **(3 marks)**

iii. Find the value of the gradient at A, correct to one decimal place. **(1 mark)**

More exam questions are available online.

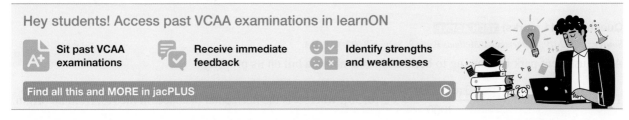

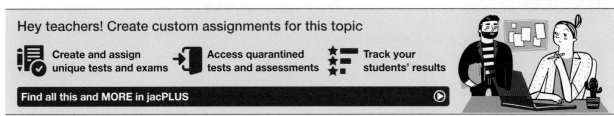

Answers

Topic 6 Further differentiation and applications

6.2 The chain rule

6.2 Exercise

1. a. $\dfrac{2x-7}{2\sqrt{x^2-7x+1}}$ b. $6(3x+1)\left(3x^2+2x-1\right)^2$
   c. $2\cos(x)\sin(x)$ d. $-3\sin(3x)e^{\cos(3x)}$
2. a. $-\dfrac{6x}{(x^2+1)^2}$ b. $-\sin(x)e^{\cos(x)}$
   c. $\dfrac{x+1}{\sqrt{x^2+2x+3}}$ d. $\dfrac{-2\cos(x)}{\sin^3(x)}$
   e. $\dfrac{x-2}{\sqrt{x^2-4x+5}}$ f. $-6x\sin\left(x^2-1\right)$
3. a. $30xe^{3x^2-1}$
   b. $-\dfrac{6x^5+8}{x^3\left(x^3-\frac{2}{x^2}\right)^3}$
   c. $\dfrac{1}{2(2-x)^{\frac{3}{2}}}$
   d. $-6\sin(2x+1)\cos^2(2x+1)$
4. a. $\dfrac{1}{(2x-3)\sqrt{\log_e(3-2x)}}$
   b. $\dfrac{1}{4(x-2)}$
5. $\dfrac{dy}{dx}=3\cos(x)\sin^2(x)$; $\dfrac{dy}{dx}=\dfrac{9}{8}$
6. a. $f'(x)=\dfrac{4}{\cos^2(4x+\pi)}$, $f'\left(\dfrac{\pi}{4}\right)=4$
   b. $f'(x)=\dfrac{2}{(2-x)^3}$, $f'\left(\dfrac{1}{2}\right)=\dfrac{16}{27}$
   c. $f'(x)=4xe^{2x^2}$, $f'(-1)=-4e^2$
   d. $f'(x)=8x\sqrt[3]{3x^2-2}$, $f'(1)=8$
7. a. $\dfrac{dy}{dx}=2\cos(x)\sin(x)e^{\sin^2(x)}$; $\dfrac{dy}{dx}=e^{\frac{1}{2}}=\sqrt{e}$
   b. $f'(x)=-15\sin(3x)(\cos(3x)-1))^4$; $f'\left(\dfrac{\pi}{2}\right)=15$
8. a. $f'(x)=-\sin(x)g'[\cos(x)]$
   b. $f'(x)=6x^2g'\left(2x^3\right)$
   c. $f'(x)=6e^{2x+1}g'\left(3e^{2x+1}\right)$
   d. $f'(x)=\dfrac{(4x-1)g'\left(\sqrt{2x^2-x}\right)}{2\sqrt{2x^2-x}}$
9. a. $f'(x)=-2h'(x)[h(x)]^{-3}$
   b. $f'(x)=2h'(x)\sin[h(x)]$
   c. $f'(x)=\dfrac{2h'(x)}{3(2h(x)+3)^{\frac{2}{3}}}$
   d. $f'(x)=-2h'(x)e^{h(x)+4}$
10. $f(f(x))=x^4$, $g'(x)=4x^3$
11. a. $\dfrac{dy}{dx}=-\dfrac{4}{(2x-1)^3}$ b. $y=-4x+5$
12. a. $f(g(x))=(e^x-1)^3$ b. $h'(x)=3e^x(e^x-1)^2$
    c. $y=0$
13. $(0,0)$, $\left(\dfrac{\pi}{4},1\right)$, $\left(\dfrac{\pi}{2},0\right)$, $\left(\dfrac{3\pi}{4},1\right)$, $(\pi,0)$
14. $24\cos(3x)\sin(3x)$
15. a. i. $f(h(x))=2\sin(e^x)$
    ii. $h(f(x))=e^{2\sin(x)}$
    b. 1.555, 2.105, 2.372
16. a. 3^{x^2+4x-5} b. 6.5916
17. a. $g(x)=\sqrt[3]{(2x-1)^2}$
    b. $g'(x)=\dfrac{4}{3\sqrt[3]{2x-1}}$
    c. At $(1,1)$: $y=\dfrac{4}{3}x-\dfrac{1}{3}$
    At $(0,1)$: $y=-\dfrac{4}{3}x+1$
    d. $\left(\dfrac{1}{2},\dfrac{1}{3}\right)$
18. a. $(1,0)$ and $(e^{0.5},1)$
    b. For f: at $(1,0)$, $\dfrac{dy}{dx}=0$.
    For g: at $(1,0)$, $\dfrac{dy}{dx}=2$.
    c.

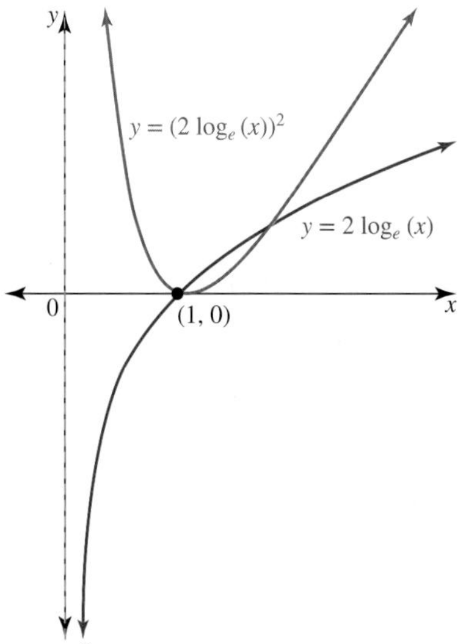

    d. $\{x : 1<x<e^{0.5}\}$

19. a. $m = -7, \ n = 1$

b. $\{x : x \leq -1\} \cup \{x : x \geq 7\}$

c. $\dfrac{x-3}{\sqrt{x^2-6x-7}}$

d. $-\dfrac{5}{3}$

20. $a = -3$

6.2 Exam questions

Note: Mark allocations are available with the fully worked solutions online.

1. C
2. $35(5x+1)^6$
3. $\dfrac{1}{2}$

6.3 The product rule

6.3 Exercise

1. a. $f'(x) = 3\cos^2(3x) - 3\sin^2(3x)$

b. $f'(x) = 3x^2e^{3x} + 2xe^{3x}$

c. $f'(x) = \left(5x^2 + 17x - 22\right)e^{5x}$

d. $f'(x) = 3\sqrt{x^2-1}\sec^2(3x) + \dfrac{x\tan(3x)}{\sqrt{x^2-1}}$

2. a. $5x^2e^{5x} + 2xe^{5x}$

b. $2e^{2x+1}\sec^2(2x) + 2e^{2x+1}\tan(2x)$

c. $\dfrac{2(x-1)(2x+1)^2}{x^3}$

d. $-x\sin(x) + \cos(x)$

e. $\dfrac{4-3x}{\sqrt{x}}$

f. $-3e^{-3x}\sin(2x-\pi) + 2e^{-3x}\cos(2x-\pi)$

3. a. $\dfrac{6e^{x^2}\left(x^2-1\right)}{x^3}$

b. $\dfrac{2e^{2x}\left(4x^2+2x-1\right)}{\sqrt{4x^2-1}}$

c. $2x\sin^2(2x)[3x\cos(2x) + \sin(2x)]$

d. $\dfrac{2(x-5)(x-1)^3}{(x-3)^3}$

e. $(3x-2)((3x-2)g'(x) + 6g(x))$

f. $\dfrac{-e^{5x}\left(g'\left(\sqrt{x}\right) + 10\sqrt{x}g\left(\sqrt{x}\right)\right)}{2\sqrt{x}}$

4. a. $(2x-3)\log_e(2x-1) + \dfrac{2(x^2-3x+7)}{2x-1}, x \in \left(\dfrac{1}{2}, \infty\right)$

b. $\dfrac{x\cos(x)\log_e(x^2) + 2\sin(x)}{x}, \ x \in (0, \infty)$

5. $f'(x) = 8x^3\cos(2x) - 4x^4\sin(2x); f'\left(\dfrac{\pi}{2}\right) = -\pi^3$

6. a. $f'(x) = e^x(x+1), \ f'(-1) = 0$

b. $f'(x) = \left(x^2+x\right)^3\left(9x^2+5x\right), \ f'(1) = 112$

c. $f'(x) = \dfrac{2(1-x)\tan(x)}{\cos^2(x)} - \tan^2(x),$

$f'\left(\dfrac{\pi}{3}\right) = 8\sqrt{3}\left(1-\dfrac{\pi}{3}\right) - 3$

d. $f'(x) = \dfrac{8x^2\cos\left(2x^2\right)\sin\left(2x^2\right) + \sin\left(2x^2\right)}{2\sqrt{x}},$

$f'\left(\sqrt{\pi}\right) = 0$

7. a. $f'(x) = \sin(x) + (x+1)\cos(x); f'(0) = 1$

b. $\dfrac{4\pi + 6\sqrt{3}}{9}$

8. 2.06
9. C
10. $(0, 0), (1, 0), \left(\dfrac{2}{5}, \dfrac{216}{3125}\right)$

11. a. $x = 0, \pi, 2\pi, 3\pi$

b. $x = 1.11, \ 4.25, \ 7.39$

12. a. Simplified $y = -\sin(x)$; $\dfrac{dy}{dx} = -\cos(x)$

b. $-\cos(x)$

13. a. $f'(x) = (x-a)^2g'(x) + 2(x-a)g(x)$

b. $\dfrac{6}{\pi}$

14. $\dfrac{dy}{dx} = 2xe^{3x} + 3\left(x^2+1\right)e^{3x}$; $\dfrac{dy}{dx} = 3$. The equation of the tangent is $y = 3x + 1$.

15. Tangent $y = 2ex - e$; perpendicular line $y = -\dfrac{x}{2e} + \left(\dfrac{1}{2e} + e\right)$

16. a. $(0, 1)$ and $(1, 0)$

b. $(1.366, -0.057)$ and $(-0.366, 1.195)$

c. $y = -\dfrac{1}{e}x + \dfrac{1}{e}$

d. $y = x + 1$

e. $(-0.46, 0.54)$

17. a. The decision is appropriate because the road goes through the point.

b. $\left(\dfrac{5}{4}, 0\right)$

c. $(0.804, \ -3.205)$

18. a. $a = 3, \ b = 3, \ c = -2$

b. $(0, 0)$ and $\left(\dfrac{3}{2}, \dfrac{81}{8e^3}\right)$

c. $y = \dfrac{3}{e^2}x$

19. a. i. $CD = 3\sin(\theta)$

ii. $BD = 3\cos(\theta)$

b. S = area of 4 triangles + area of square base

$S = 4 \times \dfrac{1}{2} \times 6\cos(\theta) \times 3\sin(\theta) + (6\cos(\theta))^2$

$= 36\cos(\theta)\sin(\theta) + 36\cos^2(\theta)$

$= 36\left(\cos^2(\theta) + \cos(\theta)\sin(\theta)\right)$ as required

c. $\dfrac{dS}{d\theta} = 72\cos^2(\theta) - 72\sin(\theta)\cos(\theta) - 36$

6.3 Exam questions

Note: Mark allocations are available with the fully worked solutions online.

1. $x(x\cos(x) + 2\sin(x))$
2. $7e^5$
3. $x\left(1 + 2\log_e(x)\right)$

6.4 The quotient rule

6.4 Exercise

1. a. $\dfrac{e^{3x} + 2e^{2x}}{(e^x+1)^2}$ b. $\dfrac{-3(t\sin(3t) + \cos(3t))}{t^4}$
2. $-\dfrac{1}{(x-1)^2}$
3. a. $\dfrac{2x\cos(x) - \sin(x)}{2x\sqrt{x}}$
   b. $\dfrac{2 - \sin(2x)\cos(2x)}{e^x\cos^2(2x)}$
   c. $-\dfrac{3\sqrt{5-x}}{2}$
   d. $\dfrac{4x^2\sin\left(x^2\right)\cos\left(x^2\right) - \sin^2\left(x^2\right)}{x^2}$
   e. $\dfrac{-6x^2 + 4x - 9}{(2x^2-3)^2}$
   f. $\dfrac{1}{4\sqrt{x}} - 3\sqrt{x}$
4. a. $\dfrac{e^x\cos(2x+1) + 2e^x\sin(2x+1)}{\cos^2(2x+1)}$
   b. $-\dfrac{xe^{-x}}{(x-1)^2}$
   c. $\dfrac{6-3x}{2\sqrt{x}(x+2)^2}$
   d. $-\dfrac{3}{\sin^2(3x)}$
   e. $-\dfrac{2}{(2x+3)^2}$
   f. $\dfrac{-4e^{2x}}{(1+e^{2x})^2}$
5. a. $\dfrac{2(2x-1) - 2x\log_e\left(x^2\right)}{x(2x-1)^2}$
   b. $\dfrac{2e^{2x} + 2 - 4e^{2x}x\log_e(2x)}{x(e^{2x}+1)^2}$
6. a. $f'(x) = \dfrac{\sin[g(x)] - (x+2)g'(x)\cos[g(x)]}{\sin^2[g(x)]}$
   b. $f'(x) = \dfrac{-2e^{-2x}g'\left(e^{-2x}\right) - g\left(e^{-2x}\right)}{e^x}$
7. $\dfrac{dy}{dx} = \dfrac{\cos(x) - 2\sin(x)}{e^{2x}}$; $x = 0$, $\dfrac{dy}{dx} = 1$
8. a. 0 b. 2 c. $\dfrac{7}{64}$ d. -5
9. a. $-\dfrac{1}{32}$ b. $y = \dfrac{1}{2}x + \dfrac{1}{2}$
10. a. $\dfrac{-6x^2 + 6x + 2}{(3x^2+1)^2}$
    b. $x = -0.1466, 0.5746$
11. a. $a = -1.088, b = 2.655, c = 0.483, d = -0.552$
    b. 0.707
12. $y = \dfrac{1+\cos(x)}{1-\cos(x)}$, $u = 1 + \cos(x)$, $v = 1 - \cos(x)$,

$$\frac{du}{dx} = -\sin(x),$$

$$\frac{dv}{dx} = \sin(x)$$

$$\frac{dy}{dx} = \frac{v\frac{du}{dx} - u\frac{dv}{dx}}{v^2}$$

$$= \frac{(1-\cos(x)) \times -\sin(x) - (1+\cos(x)) \times \sin(x)}{(1-\cos(x))^2}$$

$$= \frac{-\sin(x)(1-\cos(x)+1+\cos(x))}{(1-\cos(x))^2}$$

$$= \frac{-2\sin(x)}{(1-\cos(x))^2}$$

$$= \frac{-2\sin(x)}{(-(\cos(x)-1))^2}$$

$$= \frac{-2\sin(x)}{(\cos(x)-1)^2}$$

13. B
14. $a = -5, b = -3$
15. $\dfrac{dy}{dx} = \dfrac{10\left(1-x^2\right)}{(x^2+1)^2}$; negative gradient when $x \in (-\infty, -1) \cup (1, \infty)$
16. a. Undefined function when $x = 2, -7$
    b. $\left(-1, \dfrac{1}{3}\right)$ and $\left(11, \dfrac{1}{27}\right)$
    c. $y = \dfrac{5}{16}x + \dfrac{3}{16}$

6.4 Exam questions

Note: Mark allocations are available with the fully worked solutions online.

1. $f'(x) = -e^{\pi}$
2. $f'(x) = \dfrac{2}{(x+2)^2}$, for $x > -2$
3. $\dfrac{dy}{dx} = \dfrac{-\left(x^2+2\right)\sin(x) - 2x\cos(x)}{(x^2+2)^2}$

6.5 Curve sketching

6.5 Exercise

1. a. $\left(\frac{1}{2}, \frac{83}{24}\right)$, minimum turning point; $\left(-2, \frac{26}{3}\right)$, maximum turning point

   b. $a = -3, b = -6, c = -8$

2. a. $(-2, 0)$ maximum turning point, $\left(-\frac{2}{3}, -\frac{32}{27}\right)$ minimum turning point

   b. $(-4, 85)$ maximum turning point, $(2, -23)$ minimum turning point

   c. $(-2, -4)$ maximum turning point, $(0, 0)$ minimum turning point

   d. $\left(2, e^{-2}\right)$ maximum turning point

3. a. $f'(x) = 32x - 4x^3$

   $$\begin{aligned} 0 &= 32 - 4x^3 \\ &= 4x\left(8 - x^2\right) \\ &= 4x(2\sqrt{2} - x)(2\sqrt{2} + x) \\ x &= 0, \pm 2\sqrt{2} \end{aligned}$$

   $$\begin{aligned} f(2\sqrt{2}) &= 16(2\sqrt{2})^2 - (2\sqrt{2})^4 \\ &= 16 \times 8 - 16 \times 4 \\ &= 64 \end{aligned}$$

   $\therefore \left(2\sqrt{2}, 64\right)$ is a stationary point.

   b. Maximum

   c. $\left(-2\sqrt{2}, 64\right)$ maximum turning point, $(0, 0)$ minimum turning point

4. a. $a = -4, b = 5$

   b. $(1, -9)$, maximum turning point

   $\left(\frac{5}{3}, \frac{-247}{27}\right)$, minimum turning point

5. a.

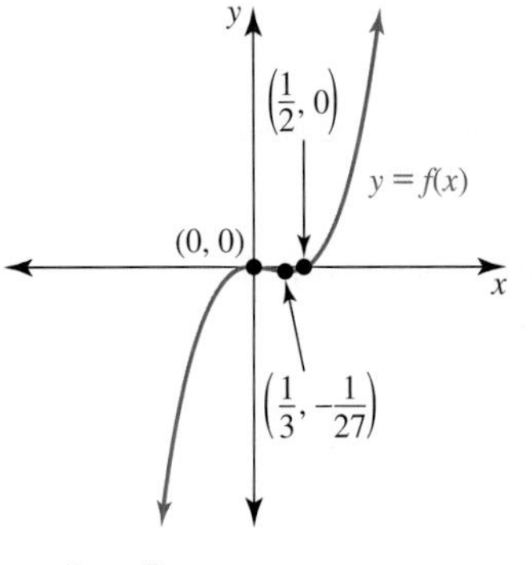

   b. $x \in \left[0, \frac{1}{3}\right]$

6. a.

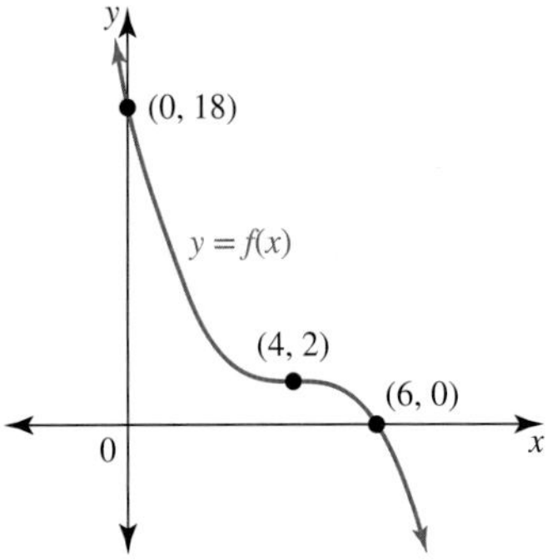

   b.

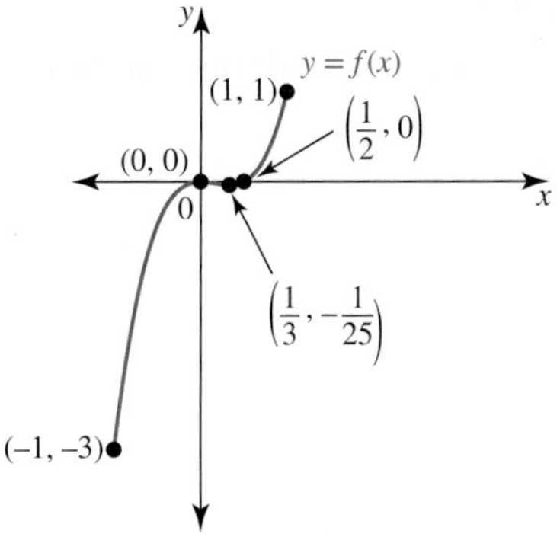

   c.

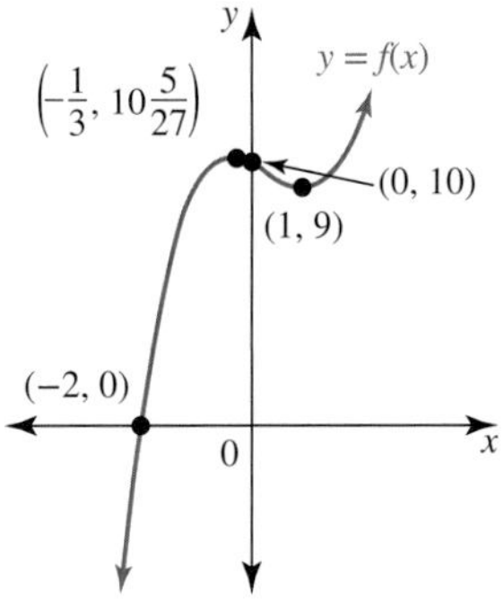

   d.

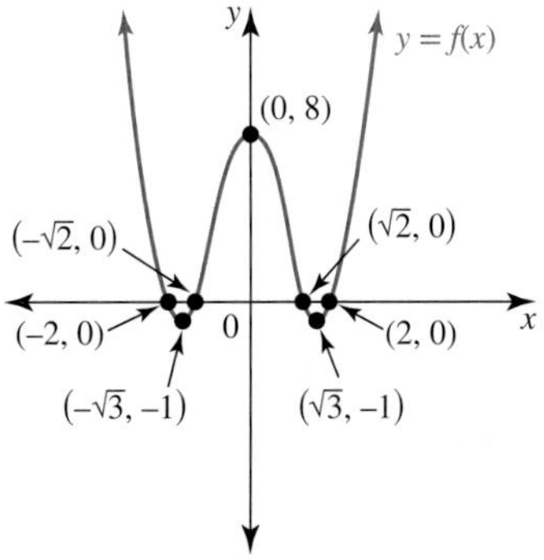

   e.

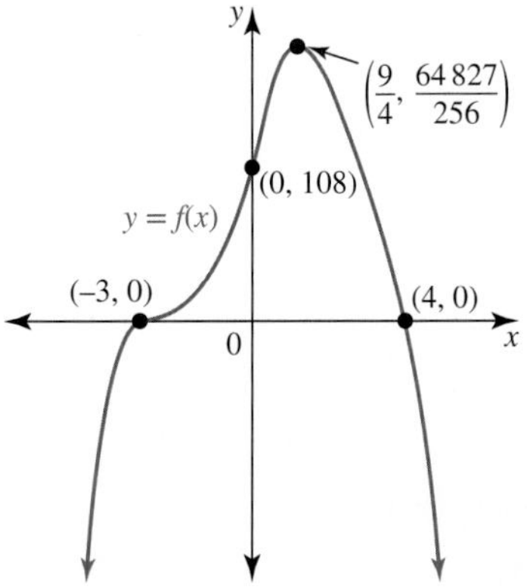

f.

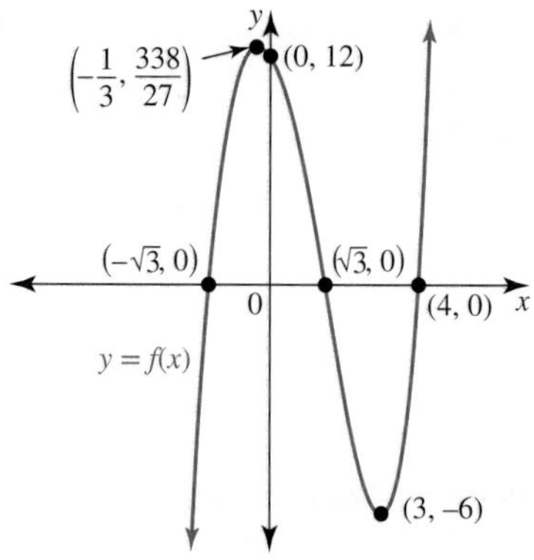

7. a.

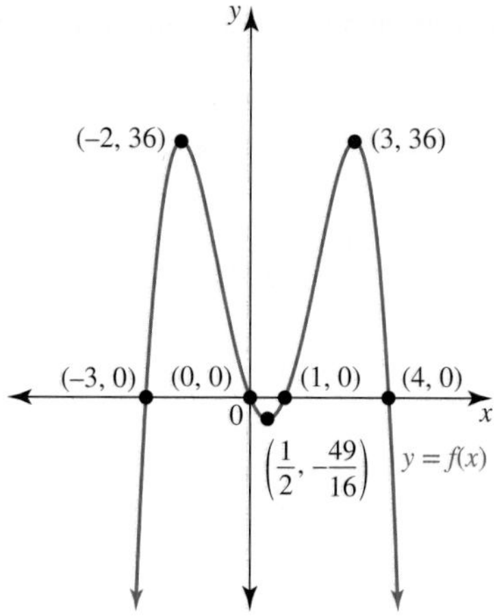

b. $x \in (-\infty, -2] \cup \left[\frac{1}{2}, 3\right]$

8. a.

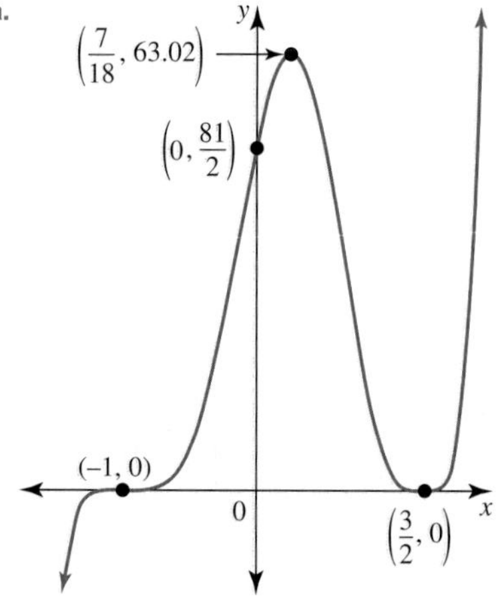

b. $x \in \left[\frac{7}{18}, \frac{3}{2}\right]$

9. Minimum turning point at $(0.1, 1 - \log_e(10))$

10. a. Local minimum at $\left(\frac{1}{e}, -\frac{2}{e}\right)$

b. Local maximum at $\left(\frac{e}{2}, \frac{2}{e}\right)$

c. Local maximum at $\left(\frac{3}{e}, \frac{3}{e}\right)$

d. Local minimum at $\left(\frac{5}{2}, \frac{5}{2} + \frac{1}{2}\log_e(2)\right)$

11. a.

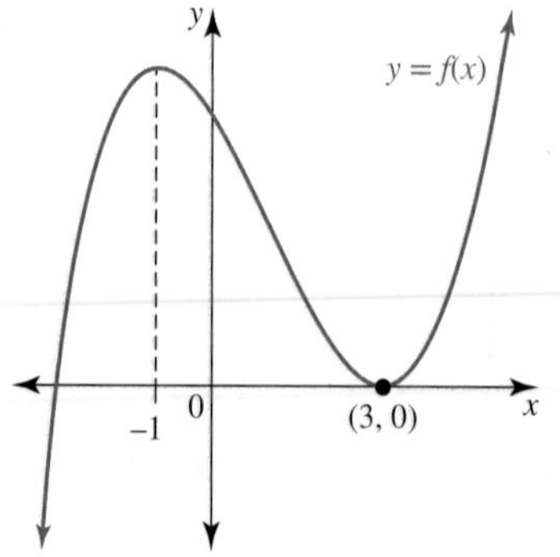

b.

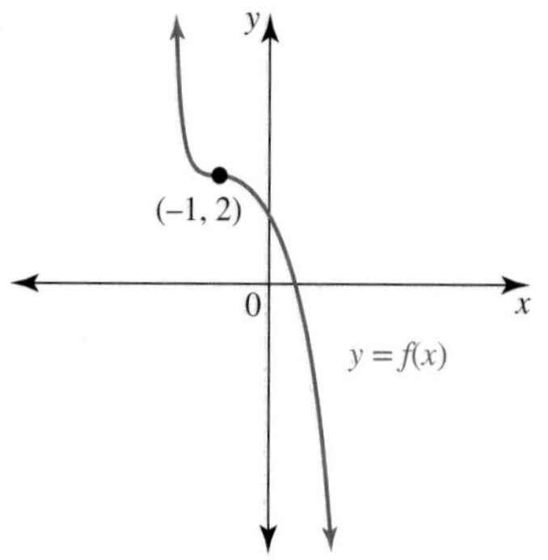

12. a. $b = -12,\ c = 8$ **b.** $a = 4,\ b = -12$

c. $b = 6,\ c = 9,\ d = -10$

13. a. $\left(-2, -\frac{17}{8}\right), \left(-\frac{1}{4}, -\frac{5}{4}\right)$

b. $\left(-\frac{1}{2}, -1\right)$, maximum turning point

c.

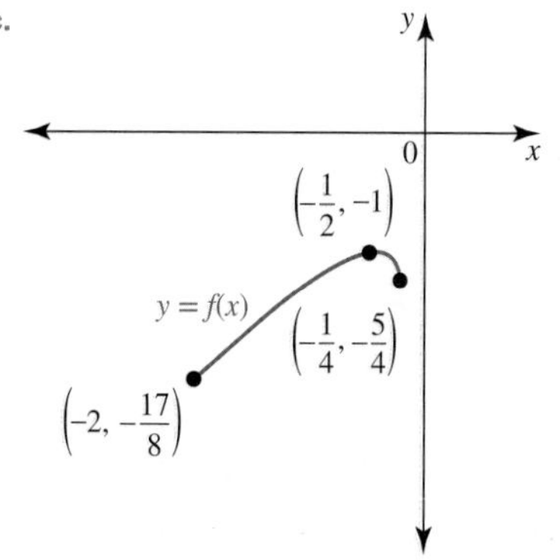

d. Absolute minimum $-\frac{17}{8}$; absolute maximum is -1.

14. No absolute minimum; absolute maximum is $\frac{32}{3\sqrt{3}}$.

15. a.

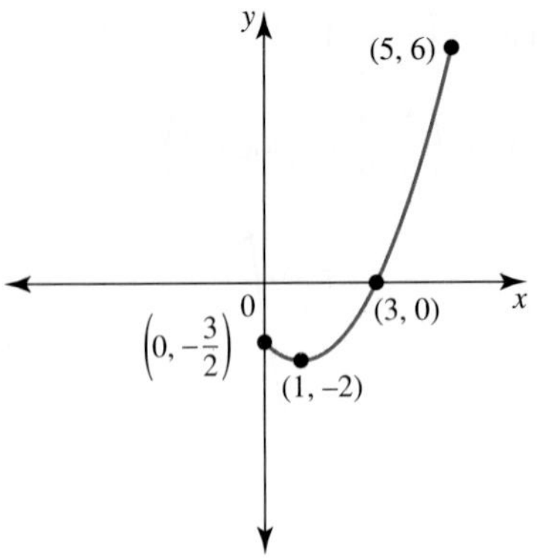

Absolute maximum = 6, absolute minimum = −2

b.

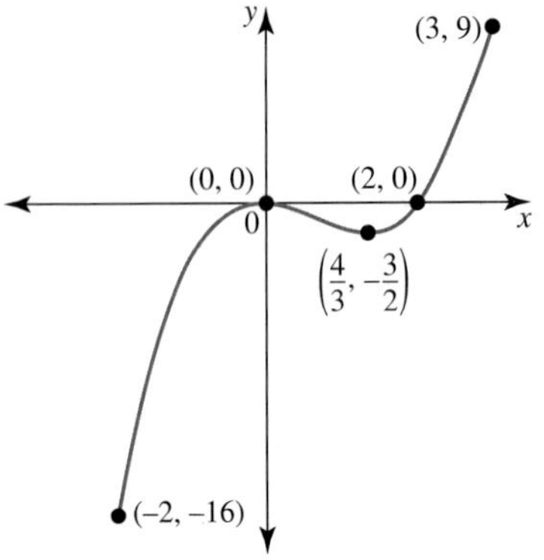

Absolute maximum = 9, absolute minimum = −16

c.

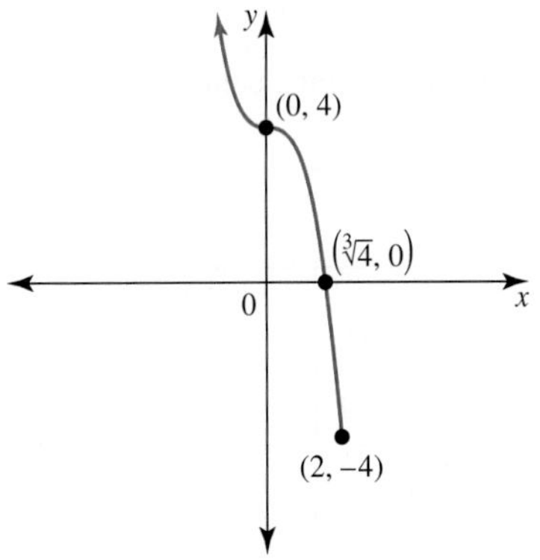

No absolute maximum; absolute minimum = −4

16. a. A (0, 25, 5), B (1, 3), C $\left(5,\ 2\sqrt{5}+0.2\right)$

b. A

c. Absolute minimum = 3, absolute maximum = 5

17. a. $\left(-1, -\frac{1}{e}\right)$, minimum turning point

b. $x \in (-1, \infty)$

c. Absolute minimum $= -\frac{1}{e}$, no absolute maximum

18. a.

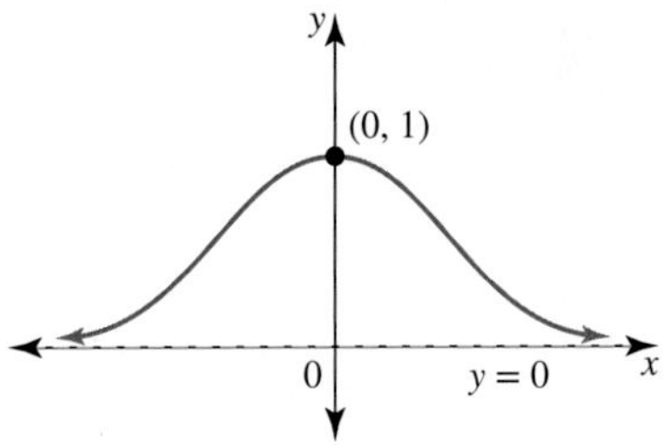

b. $x \in (-\infty, -1]$

19. a. $x = -1.841, -0.795$

b. At $(-1.841, 0)$, $y = -1.1989x - 2.2072$.
At $(-0.795, 0)$, $y = 3.0735x + 2.4434$.

c. Minimum turning point $= (-1.2134, -0.4814)$.

20. a. $x = 0.3407, 0.8364$

b. At $(0.3407, 0)$, $\frac{dy}{dx} = 27.09$; at $(0.8364, 0)$, $\frac{dy}{dx} = -6.15$.

c. Tangent: $y = -6x + 5$
Perpendicular line: $y = \frac{1}{6}x - \frac{7}{6}$ or $x - 6y = 7$

d. The turning point occurs where $\frac{dy}{dx} = 0$.

$$\begin{aligned}\frac{2-8x^2}{x^3} &= 0\\ 2-8x^2 &= 0\\ 1-4x^2 &= 0\\ (1-2x)(1+2x) &= 0\\ x = \frac{1}{2}, -\frac{1}{2}&\text{ but } x > 0\\ x = \frac{1}{2}, y = -\frac{1}{\left(\frac{1}{2}\right)^2} &- 8\log_e\left(\frac{1}{2}\right)\\ &= -\frac{1}{\frac{1}{2}} - 8\log_e\left(2^{-1}\right)\\ &= -4 + 8\log_e(2)\end{aligned}$$

The maximum turning point is at $\left(\frac{1}{2}, -4 + 8\log_e(2)\right)$.

21. a. Stationary points $(a, 0)$, $\left(\frac{a+4}{3}, \frac{4(a-2)^3}{27}\right)$

b. Minimum turning point $(a, 0)$, maximum turning point $\left(\frac{a+4}{3}, \frac{4(a-2)^3}{27}\right)$

c. $a = 5$

22. a. $(a, 0)$, $(b, 0)$

b. $(b, 0)$, $\left(\frac{3a+b}{4}, \frac{-27(a-b)^4}{256}\right)$

c. $(b, 0)$ is a stationary point of inflection; $\left(\frac{3a+b}{4}, \frac{-27(a-b)^4}{256}\right)$ is a minimum turning point.

d. $a = 2$, $b = 6$

6.5 Exam questions

Note: Mark allocations are available with the fully worked solutions online.

1. D
2. D
3. a. $\left(\frac{\sqrt{15}}{3}, -\frac{10\sqrt{15}}{9}\right), \left(-\frac{\sqrt{15}}{3}, \frac{10\sqrt{15}}{9}\right)$
   b. i. $y = -4x$
      ii. $2\sqrt{17}$

6.6 Maximum and minimum problems

6.6 Exercise

1. The point on the curve is $(2, -1)$. The minimum distance is $\sqrt{5}$ units.
2. The point on the line is $\left(3, 2\sqrt{3}\right)$. Minimum distance is 4 units.
3. Both numbers are 16.
4. a. $V = x(16 - 2x)(10 - 2x)$
   b. $x = 2$; therefore, height is 2 cm, length is 12 cm and width is 6 cm. Volume is 144 cm^3.
5. a.

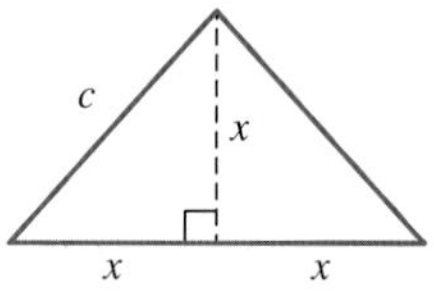

Using Pythagoras' theorem, $c^2 = x^2 + x^2$, so the sloping sides have lengths $\sqrt{2}x$ cm.
Since the perimeter of the figure is 150 cm,

$$2x + 2y + 2\sqrt{2}x = 150$$
$$\therefore y = 75 - x - \sqrt{2}x$$

The area of the figure is the sum of the areas of the rectangle and the triangle with base $2x$, height x.

$$A = 2xy + \frac{1}{2}(2x)x$$
$$\therefore A = 2x\left(75 - x - \sqrt{2}x\right) + x^2$$
$$= 150 - 2x^2 - 2\sqrt{2}x^2 + x^2$$
$$\therefore A = 150x - \left(1 + 2\sqrt{2}\right)x^2$$

   b. Width = 39.2 cm and height = 47.3 cm
   c. Width = 30 cm and height = 53.8 cm
6. a. $l = 137.5$ m, $w = 55$ m
   b. $A_{max} = 7562.5\text{ m}^2$
7. a. $a = 12, \quad b = 9.6$ b. $A_{max} = 403.2\text{ m}^2$
8.
$$P = 2l + 2R + \pi R$$
$$A = 2Rl + \frac{\pi R^2}{2}$$
$$A - \frac{\pi R^2}{2} = 2Rl$$
$$l = \frac{A}{2R} - \frac{\pi R}{4}$$

Substitute for l into the perimeter formula:

$$P = 2\left(\frac{A}{2R} - \frac{\pi R}{4}\right) + 2R + \pi R$$
$$= \frac{A}{R} - \frac{\pi R}{2} + 2R + \pi R$$
$$= \frac{A}{R} + \frac{\pi R}{2} + 2R$$

For minimum perimeter, solve $\frac{dP}{dR} = 0$.

$$\frac{dP}{dR} = -\frac{A}{R^2} + \frac{\pi}{2} + 2$$
$$0 = -\frac{A}{R^2} + \frac{\pi + 4}{2}$$
$$\frac{A}{R^2} = \frac{\pi + 4}{2}$$
$$R^2 = \frac{2A}{\pi + 4}$$
$$R = \pm\sqrt{\frac{2A}{\pi + 4}}$$
$$= \sqrt{\frac{2A}{\pi + 4}}, R > 0$$

Substitute R into the length equation to determine l:

$$l = \frac{A}{2R} - \frac{\pi R}{4}$$
$$= \frac{2A - \pi R^2}{4R}$$
$$= \frac{2A - \frac{2A\pi}{\pi+4}}{4\sqrt{\frac{2A}{\pi+4}}}$$
$$= \frac{\frac{2A(\pi+4) - 2A\pi}{\pi+4}}{4\sqrt{\frac{2A}{\pi+4}}}$$
$$= \frac{\frac{8A\pi}{\pi+4}}{4\sqrt{\frac{2A}{\pi+4}}}$$
$$= \frac{4 \times \frac{2A\pi}{\pi+4}}{4\sqrt{\frac{2A}{\pi+4}}}$$
$$= \sqrt{\frac{2A}{\pi + 4}}$$

If I and R are both $\sqrt{\frac{2A}{\pi + 4}}$ m, the perimeter is a minimum (as required).

9. a. 400
   b. $t = 4$, so after 4 months (i.e. end of December)
   c. 694 birds
10. a. \$1000 b. $A_{least} = \$980.34$
    c. May 2017 d. \$1000

11. $r = 6.06$ cm, $h = 12.11$ cm and $V_{max} = 1395.04\text{ cm}^3$

12. $h = 16$ cm, $V_{max} = 2145\text{ cm}^3$

13. $h = \frac{10}{3}$ cm, $r = \frac{16}{3}$ cm and $V_{max} = 298\text{ cm}^3$

14. Row to a point that is 3.2 km to the right of O.

6.6 Exam questions

Note: Mark allocations are available with the fully worked solutions online.

1. E

2. a. $\left(\frac{8}{5}, -\frac{4}{5}\right)$

   b. $\frac{4\sqrt{5}}{5}$

3. D

6.7 Rates of change

6.7 Exercise

1. a. $r = \frac{h}{3}$ b. $V = \frac{\pi h^3}{27}$

   c. $\frac{dV}{dh} = \frac{25\pi}{9}\text{ cm}^3/\text{cm}$

2. a. $400\pi\text{ cm}^3/\text{cm}$ b. $-72\text{ mm}^2/\text{mm}$

3. a. $V = 333\frac{1}{3}$ mL   b. $\frac{dV}{dt} = 20t - 2t^2$

   c. $\frac{dV}{dt} = 42$ mL/s   d. $t = 5$ s, $\frac{dV}{dt} = 50$ mL/s

4. a. 0 m

   b. $v = 4t - 8$; the initial velocity is -8 m/s, so the object is moving left at a speed of 8 m/s.

   c. After 2 seconds, 8 m to the left of the origin

   d. $2t^2 - 8t = 0$
   $2t(t-4) = 0$
   $t = 0, 4$
   The distance travelled is 16 m.

   e. 4 m/s

   f. 0 m/s

5. a. 1 m right, 8 m/s

   b. $27\frac{2}{3}$ m

   c. -6 m/s^2

6. a. -4.4 rabbits/month

   b. -22 rabbits/month

   c. $t \to \infty$, $N \to 0$. Effectively, the population of rabbits will be zero in the long term.

7. a. $\frac{dV}{dt} = 30$ L/min   b. -51.6 L/min

   c. $t = 0$

8. a. 26 m/s

   b. 10 m/s

   c. 7.75 seconds, travelling down towards the ground

   d. 6.25 seconds

   e. 156.25 metres

   f. 12.5 seconds, 50 m/s

9. a. 0.5 hundred thousand or 50 000

   b. $N'(t) = \frac{-2t^2 + 0.5}{(t + 0.5)^4}$

   c. $N_{max} = 1.5$ hundred thousand or 150 000 after half an hour.

   d. -1641 viruses/hour

10. a. 4 years, 6 butterflies per year

    b. 2.54 years

    c. As $t \to \infty$, $N \to 220$ and $\frac{dN}{dt} \to 0$.

11. a. 42 metres to the left   b. 8 m/s

    c. -12 m/s   d. 9 seconds, 20 m/s

12. a. $t = 0$, $x = \frac{2}{3}(0)^3 - 4(0)^2$
    $= 0$ m
    $v = \frac{dx}{dt} = 2t^2 - 8t$
    $t = 0, v = 2(0)^2 - 8(0)$
    $= 0$ m/s

    b. 4 seconds, $21\frac{1}{3}$ metres left of origin

    c. 6 seconds

    d. 24 m/s, 16 m/s^2

13. a. $r = \sqrt{400 - h^2}$ b. $V = \frac{400\pi h}{3} - \frac{\pi h^3}{3}$

    c. $\frac{dV}{dh} = \frac{208\pi}{3}\text{ cm}^3/\text{cm}$

14. a. $\sqrt{2}x$ metres

    b. Right-angled triangle with lengths h, $\frac{\sqrt{2}x}{2}$ and hypotenuse 12

    $$h^2 + \left(\frac{\sqrt{2}x}{2}\right)^2 = 144$$
    $$\frac{2x^2}{4} = 144 - h^2$$
    $$x^2 = 288 - 2h^2$$
    (as required)
    $$V = \frac{1}{3}\left(288h - 2h^3\right)$$

    c. $42\text{ m}^3/\text{m}$

15. a. $9\pi\text{ cm}^3/\text{cm}$ b. $12 \times 3^{\frac{1}{3}}\pi\text{ cm}^3/\text{cm}$

16. a. $\frac{3\left(4 - t^2\right)}{(4 + t^2)^2}$

    b. $y_{max} = 0.75$ mg/L after 2 hours

    c. Next dose after 5.24 hours

    d. 0.36 mg/L/h

    e. $t = 2.45$ h and $t = 6$ h

6.7 Exam questions

Note: Mark allocations are available with the fully worked solutions online.

1. E   2. D   3. A

6.8 Newton's method

6.8 Exercise

1. $x = 1.5160$
2. $x = 3.135$
3. $x = 2.607$
4. $\sqrt[3]{10} = 2.15$
5. B
6. $x = 3.276$

6.8 Exam questions

Note: Mark allocations are available with the fully worked solutions online.

1. $x = 1.062$
2. $\sqrt[4]{12} = 1.86$
3. $x = -1.194$

6.9 Review

6.9 Exercise

Technology free: short answer

1. a. 5   b. $y = \frac{1}{5}x - \frac{26\pi}{15}$
2. a. $(0, 3)$ and $(2, 11)$
   b. There is a minimum turning point at $(1, 1)$ and a maximum turning point at $(0, 3)$.
   c. 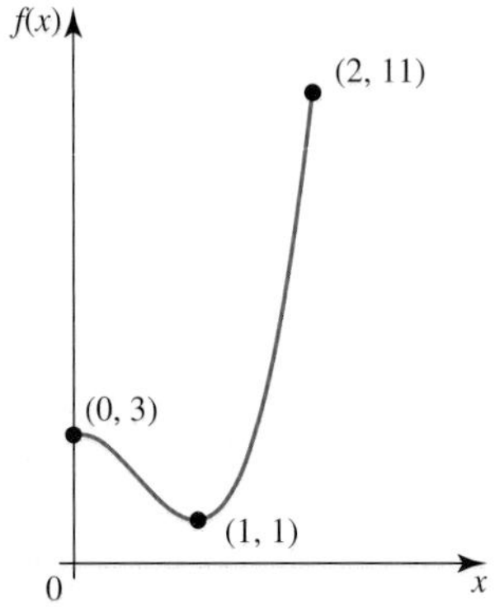

   d. Absolute maximum = 11, absolute minimum = 1.
3. a. 4   b. $x \in [-1, 0] \cup [1, 2]$
4. a. $\left(2, 12\log_e(2)\right)$ b. Local minimum
5. a. $m = 2, n = 4$ b. $h'(x) = \frac{x+3}{\sqrt{(x+2)(x+4)}}$
6. a. $m = -3,\ n = 4$ b. $(0,\ 0)$ and $\left(\frac{4}{3}, \frac{256}{81e^4}\right)$

Technology active: multiple choice

7. C   8. A   9. A   10. C   11. C
12. E   13. A   14. C   15. E   16. D

Technology active: extended response

17. a. $\frac{dh}{dx} = -\frac{\pi}{10}\sin\left(\frac{\pi(x-67)}{100}\right)$
    b. 5 m below the platform, 15 m above the platform
    c. i. 0.16   ii. -0.27
18. a. $h = 20\sin(x)$
    b. $b = 40\cos(x) + 10$
    c. $A = \frac{1}{2}(40\cos(x) + 10 + 10) \times 20\sin(x)$
       $A = 10\sin(x)(40\cos(x) + 20)$
       $A = 200\sin(x)(2\cos(x) + 1)$ as required
    d. 352 cm^2 when $x = 0.936$
19. a. 1.2 mL/h   b. 0.58 h
    c. -1.29 mL/h   d. After 1 hour
20. a. $145 = 2Lr + \frac{1}{2}\pi r^2$
       $145 - \frac{1}{2}\pi r^2 = 2Lr$
       $290 - \pi r^2 = 4Lr$
       $\frac{290 - \pi r^2}{4r} = L$
       $P = 2L + 2r + \frac{1}{2} \times 2\pi r$
       $P = 2L + 2r + \pi r$
       $P = 2\left(\frac{290 - \pi r^2}{4r}\right) + (2 + \pi)r$
       $P = \frac{290 - \pi r^2}{2r} + (2 + \pi)r$ as required
    b. $0 = 2 + \frac{\pi}{2} - \frac{145}{r^2}$
    c. $P_{min} = 45.5$ m when $r = 6.4$ m and $L = 6.4$ m

6.9 Exam questions

Note: Mark allocations are available with the fully worked solutions online.

1. $-3\left(9x^2 - 2x\right)\left(3x^3 - x^2 + 64\right)^2$
2. $\frac{dy}{dx} = x(x\cos(x) + 2\sin(x))$
3. D
4. D
5. a. $\frac{dy}{dx} = \frac{9(x-10)(x-30)}{2000}$
   b. $x \in (0, 20]$
   c. $\frac{3x(x-30)^2}{2000} + 3$
   d. $x = \frac{10}{3}\left(6 \pm \sqrt{3}\right)$
   e. i. $\frac{9(a-10)(a-30)}{2000}$
      ii. $A(11.12, 8.95)$
      iii. -0.1

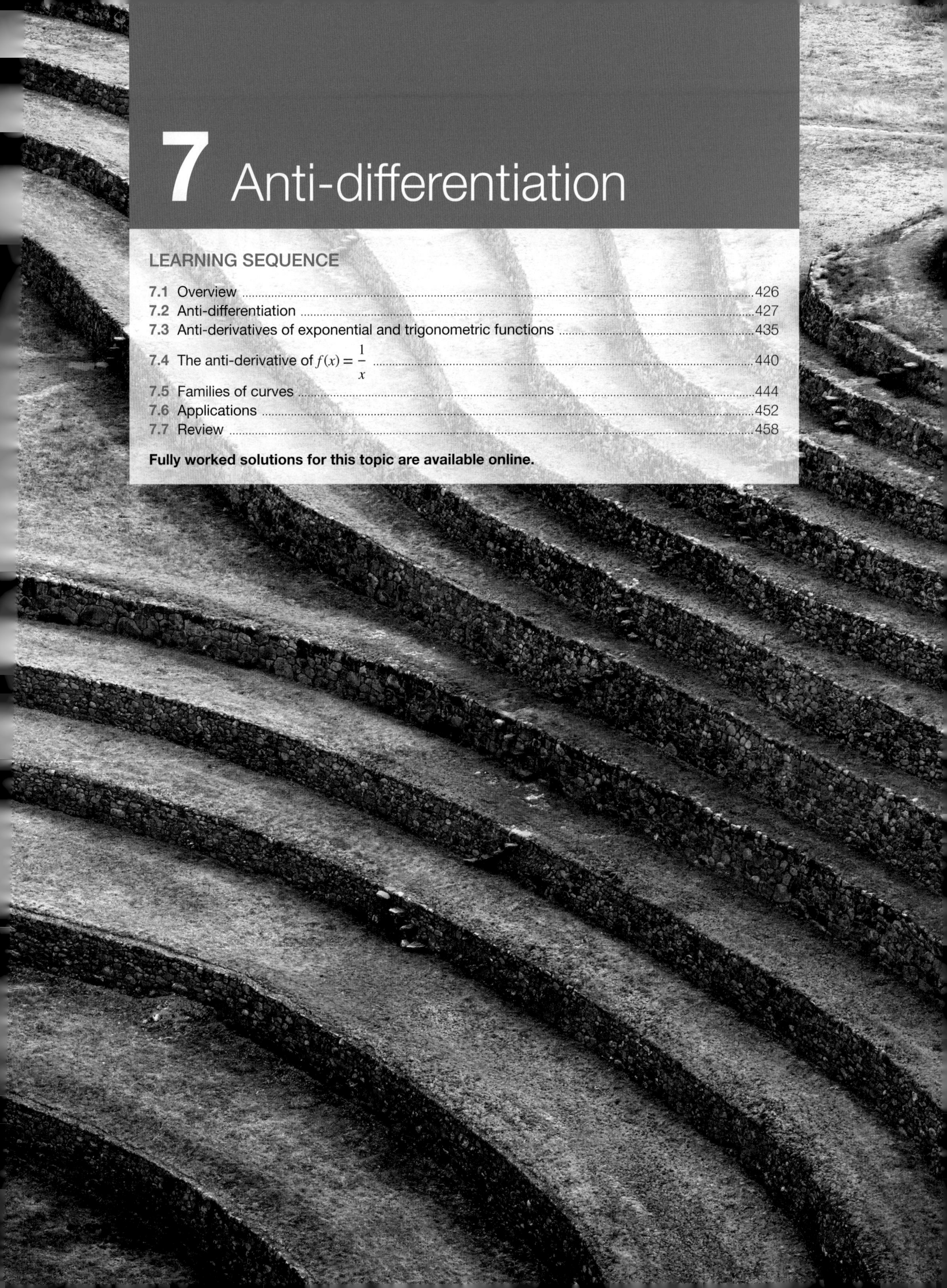

7 Anti-differentiation

LEARNING SEQUENCE

Fully worked solutions for this topic are available online.

7.1 Overview

Hey students! Bring these pages to life online

Watch videos

Engage with interactivities

Answer questions and check results

Find all this and MORE in jacPLUS

7.1.1 Introduction

Anti-differentiation or integration is the opposite of differentiation. These are the two main operations of calculus. This topic introduces the concept and the basic rules of anti-differentiating particular functions. Topic 8 extends this further to cover applications.

We have to go right back to around 350 BCE for the first documented system of determining integrals. This was developed by Eudoxus, an ancient Greek astronomer. A similar method was developed independently in China around 300 CE. But further significant advances would not occur after this until the 17th century via the individual work of Cavalieri and Fermat.

Newton was the one who first created symbols for integration — a small vertical bar above the variable or a box around the variable. However, these symbols were not widely used as they were too easily confused with other mathematical symbols already in use, such as the dot above the variable, $\dot{x}$.

Leibniz, in 1675, created the symbol for integration that we know today, $\int$.
He based it on the letter *ſ* (long s) as a representation of standing the Latin word *summa*, meaning 'sum' or 'total'.

KEY CONCEPTS

This topic covers the following key concepts from the VCE Mathematics Study Design:

- deducing the graph of the derivative function from the graph of a given function and deducing the graph of an anti-derivative function from the graph of a given function
- anti-derivatives of polynomial functions and functions of the form $f(ax+b)$ where f is x^n, for $n \in Q, e^x, \sin(x), \cos(x)$, and linear combinations of these
- anti-differentiation by recognition that $F'(x) = f(x)$ implies $\int f(x)dx = F(x) + c$ and the informal treatment of the fundamental theorem of calculus, $\int_a^b f(x)dx = F(b) - F(a)$
- properties of anti-derivatives and definite integrals
- application of integration to problems involving finding a function from a known rate of change given a boundary condition, calculation of the area of a region under a curve and simple cases of areas between curves, average value of a function and other situations.

Note: *Concepts shown in grey are covered in other topics.*

7.2 Anti-differentiation

7.2.1 Introduction

LEARNING INTENTION

At the end of this subtopic you should be able to:
- anti-differentiate functions
- recognise patterns between differentiation and anti-differentiation.

Anti-differentiation, also known as **integration**, is the reverse process of differentiation. Anti-differentiation allows us to find $f(x)$ when we are given $f'(x)$.

In Topic 5, you learned that $\frac{d}{dx}\left(x^2\right) = 2x$. Alternatively, this can be expressed in function notation: if $f(x) = x^2$, then $f'(x) = 2x$. So if you were given $f'(x) = 2x$ and asked to find $f(x)$, you might expect that $f(x) = x^2$.

However, this is not quite as simple as it first appears. Consider each of the following derivatives.

$$\frac{d}{dx}\left(x^2 + 7\right) = 2x$$
$$\frac{d}{dx}\left(x^2 + 2\right) = 2x$$
$$\frac{d}{dx}\left(x^2 - 1\right) = 2x$$
$$\frac{d}{dx}\left(x^2 - 5\right) = 2x$$

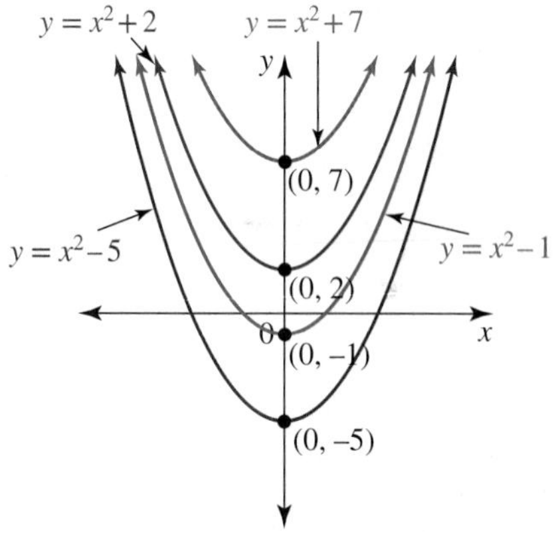

If we are asked to find $f(x)$ given that $f'(x) = 2x$, how do we know which of the equations above is the correct answer? To give a totally correct answer, additional information about the function must be given.

If $f'(x) = 2x$, then $f(x) = x^2 + c$, where c is an arbitrary constant. This means we have a family of curves that fit the criteria for the function f.

To know which specific curve matches f, we must know additional information such as a point through which the curve passes.

Notation

The notation that is commonly used for anti-differentiation was introduced by the German mathematician Gottfried Leibniz (1646–1716). An example of this notation is:

$$\int 2x\,dx = x^2 + c$$

This equation indicates that the **anti-derivative** of $2x$ with respect to x is equal to x^2 plus an unknown constant, c. $\int f(x)dx$ is known as the indefinite integral. It is read as 'the integral of $f(x)$ with respect to x'.

The $\int$ tells us to anti-differentiate and the dx tells us that x is the variable.

Remembering that the reverse process of differentiation is integration, consider the following:

$$\frac{d}{dx}\left(x^3\right) = 3x^2 \quad \therefore \int 3x^2 dx = x^3 + c$$
$$\frac{d}{dx}\left(x^4\right) = 4x^3 \quad \therefore \int 4x^3 dx = x^4 + c$$
$$\frac{d}{dx}\left(2x^5\right) = 10x^4 \quad \therefore \int 10x^4 dx = 2x^5 + c$$
$$\frac{d}{dx}\left(3x^6\right) = 18x^5 \quad \therefore \int 18x^5 dx = 3x^6 + c$$

From this series of derivatives and integrals, two important observations can be made.

Anti-derivatives of x^n and ax^n

$$\int x^n dx = \frac{x^{n+1}}{n+1} + c,\ n \neq -1$$

$$\int ax^n dx = a\int x^n dx = \frac{ax^{n+1}}{n+1} + c,\ n \neq -1$$

WORKED EXAMPLE 1 Anti-derivative of ax^n

Determine:

a. $\int 3x^4\, dx$ b. $\int -4\sqrt[3]{x}\, dx$ c. $\int \frac{5}{x^3} dx$

THINK	WRITE
a. 1. Apply the rule $\int ax^n dx = \frac{ax^{n+1}}{n+1} + c$	$\int 3x^4 dx = \frac{3x^{4+1}}{5} + c$
2. Simplify the answer.	$= \frac{3x^5}{5} + c$
b. 1. Write the integral in the form $\int ax^n dx$.	$\int -4\sqrt[3]{x} dx = \int -4x^{\frac{1}{3}} dx$
2. Apply the rule $\int ax^n dx = \frac{ax^{n+1}}{n+1} + c$.	$= -\frac{4x^{\frac{1}{3}+1}}{\frac{4}{3}} + c$
3. Simplify the answer.	$= -3x^{\frac{4}{3}} + c$
c. 1. Write the integral in the form $\int ax^n dx$.	$\int \frac{5}{x^3} dx = \int 5x^{-3} dx$
2. Apply the rule $\int ax^n dx = \frac{ax^{n+1}}{n+1} + c$.	$= \frac{5x^{-3+1}}{-2} + c$ $= -\frac{5x^{-2}}{2} + c$
3. Simplify the answer.	$= -\frac{5}{2x^2} + c$

on Resources

Interactivity Integration of ax^n (int-6419)

7.2.2 Properties of integrals

As differentiation is a linear operation, so too is anti-differentiation.

Properties of anti-differentiation

$$\int (f(x) \pm g(x))\,dx = \int f(x)\,dx \pm \int g(x)\,dx$$

$$\int af(x)\,dx = a\int f(x)\,dx, \text{ where } a \text{ is a constant.}$$

That is, we can anti-differentiate the separate components of an expression. For example:

$$\begin{aligned}\int \left(4x^3 + 6x^2 - 9x + 7\right) dx &= \frac{4x^{3+1}}{4} + \frac{6x^{2+1}}{3} - \frac{9x^{1+1}}{2} + 7x + c \\ &= x^4 + 2x^3 - \frac{9}{2}x^2 + 7x + c\end{aligned}$$

To check your anti-derivative is correct, it is always good to differentiate your answer to see if the derivative matches the original expression.

Note: If you are asked to find 'the' anti-derivative of an expression, then the '$+c$' component must be part of the answer. However, if you are asked to find 'an' anti-derivative, then you can choose what the value of c is. The convention when finding 'an' anti-derivative is to let c equal 0. So the example above would have an anti-derivative of $x^4 + 2x^3 - \frac{9}{2}x^2 + 7x$.

WORKED EXAMPLE 2 Applying anti-derivative properties

a. Evaluate $\int (3x^4 - x^3 + 2)\,dx$.

b. If $f'(x) = 2\sqrt{x} + \frac{1}{x^2} - 7$, determine the rule for f.

c. Determine an anti-derivative of $(2x - 3)(4 - x)$.

d. Determine $\int \frac{x^4 - 2x^3 + 5}{x^3}\,dx$.

THINK	WRITE
a. Anti-differentiate each term separately by applying the rule.	**a.** $\int (3x^4 - x^3 + 2)\,dx = \frac{3x^5}{5} - \frac{x^4}{4} + 2x + c$
b. 1. First rewrite any surds as fractional powers, and rewrite any powers in the denominator as negative powers. That is, write each term in the form ax^n.	**b.** $f(x) = \int \left(2\sqrt{x} + \frac{1}{x^2} - 7\right) dx$ $= \int \left(2x^{\frac{1}{2}} + x^{-2} - 7\right) dx$

2. Anti-differentiate each term separately by applying the rule.

$$= \frac{2x^{\frac{3}{2}}}{\frac{3}{2}} + \frac{x^{-1}}{-1} - 7x + c$$

3. Simplify.

$$= \frac{4x^{\frac{3}{2}}}{3} - \frac{1}{x} - 7x + c$$

c. 1. First expand the expression.

c. $(2x-3)(4-x) = 8x - 2x^2 - 12 + 3x$

$$= -2x^2 + 11x - 12$$

2. Anti-differentiate each term separately by applying the rule.
Note: The '+ *c*' is not needed as the question asked for 'an' anti-derivative.

$$\int(-2x^2 + 11x - 12)\,dx = -\frac{2x^3}{3} + \frac{11x^2}{2} - 12x$$

d. 1. Rewrite the expression as separate fractions.

d. $\int \frac{x^4 - 2x^3 + 5}{x^3}\,dx = \int\left(\frac{x^4}{x^3} - \frac{2x^3}{x^3} + \frac{5}{x^3}\right)dx$

2. Anti-differentiate each term separately by applying the rule.

$$= \int\left(x - 2 + 5x^{-3}\right)dx$$

3. Simplify.
Note: Simplest form usually assumes positive indices only.

$$= \frac{x^2}{2} - 2x + \frac{5x^{-2}}{-2} + c$$

$$= \frac{x^2}{2} - 2x - \frac{5}{2x^2} + c$$

TI | THINK

d. 1. On a Calculator page, press MENU, then select:
4: Calculus
3: Integral
Complete the entry line as:
$\int \frac{x^4 - 2x^3 + 5}{x^3}\,dx$
then press ENTER.

DISPLAY/WRITE

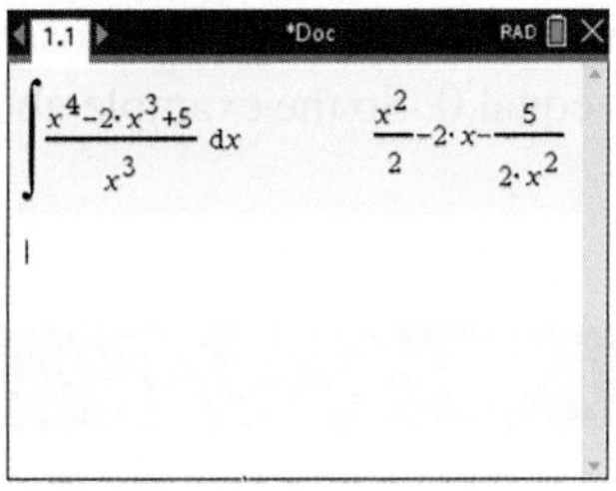

2. The answer appears on the screen.

$\int \frac{x^4 - 2x^3 + 5}{x^3}\,dx = \frac{x^2}{2} - 2x - \frac{5}{2x^2} + c$

Note: Remember to include the constant, *c*, as the calculator will not include it.

CASIO | THINK

d. 1. On a Main screen, select:
- Action
- Calculation
- $\int$

Complete the entry line as:
$\int \frac{x^4 - 2x^3 + 5}{x^3}\,dx$
then press EXE.

DISPLAY/WRITE

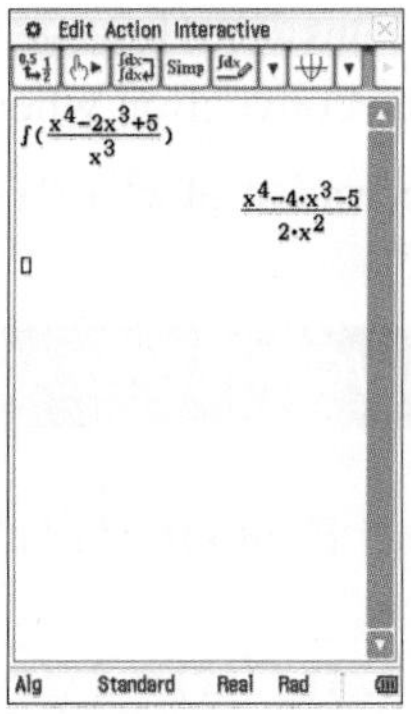

2. The answer appears on the screen.

$\int \frac{x^4 - 2x^3 + 5}{x^3}\,dx$
$= \frac{x^4 - 4x^3 - 5}{2x^2} + c$

Note: Remember to include the constant, *c*, as the calculator will not include it.

Note: It is extremely useful to differentiate the answer of an anti-derivative in order to check its validity.

on Resources

Interactivity Properties of integrals (int-6420)

7.2.3 Integrals of the form $\int f(ax+b)\,dx, n \neq 1$

Consider the function $f: R \to R,\ f(x) = (ax+b)^n$.

Using the chain rule, $f'(x) = an(ax+b)^{n-1}$.

Hence,

$$\int an(ax+b)^{n-1}dx = (ax+b)^n$$

$$an\int (ax+b)^{n-1}dx = (ax+b)^n$$

$$\int (ax+b)^{n-1}dx = \frac{1}{an}(ax+b)^n + c$$

Thus we obtain the general rule that follows.

Anti-derivative of $(ax+b)^n$

$$\int (ax+b)^n dx = \frac{(ax+b)^{n+1}}{a(n+1)} + c,\ n \neq -1$$

WORKED EXAMPLE 3 Anti-derivative of $(ax+b)^n$

Anti-differentiate:

a. $(2x+3)^5$

b. $2(3x-1)^{-2}$

THINK	WRITE
a. Apply the rule $\int (ax+b)^n dx = \frac{(ax+b)^{n+1}}{a(n+1)} + c$.	**a.** $\int (2x+3)^5 dx$ $= \frac{(2x+3)^6}{2(6)} + c$ $= \frac{(2x+3)^6}{12} + c$
b. 1. Take 2 out as a factor.	**b.** $\int 2(3x-1)^{-2}dx$ $= 2\int (3x-1)^{-2}dx$
2. Apply the rule $\int (ax+b)^n dx = \frac{(ax+b)^{n+1}}{a(n+1)} + c$.	$= \frac{2(3x-1)^{-1}}{3(-1)} + c$ $= \frac{2(3x-1)^{-1}}{-3} + c$ $= -\frac{2}{3(3x-1)} + c$

Note: The rules described above only apply if the expression inside the brackets is linear. If the expression is of any other kind, it must be expanded, if possible, before integrating, or you must use CAS to integrate the expression.

WORKED EXAMPLE 4 Expansion and anti-differentiation

Determine $\int \left(x+\frac{1}{x}\right)^2 dx$.

THINK	WRITE
1. As the inner function is not linear, there is no anti-differentiation rule we can apply, so the expression must first be expanded.	$\int \left(x+\frac{1}{x}\right)^2 dx$ $=\int \left(x^2+2x\left(\frac{1}{x}\right)+\left(\frac{1}{x}\right)^2\right) dx$ $=\int \left(x^2+2+\frac{1}{x^2}\right) dx$
2. Write all terms in the form ax^n.	$=\int (x^2+2+x^{-2})\, dx$
3. Apply the rules for anti-differentiation.	$=\frac{x^3}{3}+2x-\frac{x^{-1}}{1}+c$ $=\frac{1}{3}x^3+2x-\frac{1}{x}+c$

7.2.4 Integration by recognition

Sometimes you may be required to find an anti-derivative of a very complex function. In order to complete this task, you will first be given a function to differentiate. The technique is then to recognise the patterns between the derivative you have found and the function you have been given to anti-differentiate.

In general, if $f(x) = g'(x)$, then $\int g'(x)\, dx = f(x) + c$.

WORKED EXAMPLE 5 Integration by recognition

If $y = (3x^2 + 4x - 7)^5$**, find** $\frac{dy}{dx}$**. Hence, determine an anti-derivative of** $20(3x + 2)(3x^2 + 4x - 7)^4$**.**

THINK	WRITE
1. Use the chain rule to differentiate the given function.	$y = (3x^2+4x-7)^5$ $\frac{dy}{dx} = 5(6x+4)(3x^2+4x-7)^4$
2. Remove 2 as a factor from the linear bracket.	$= 10(3x+2)(3x^2+4x-7)^4$
3. Rewrite the result as an integral.	$\int 10(3x+2)(3x^2+4x-7)^4 dx = (3x^2+4x-7)^5$

4. Adjust the left-hand side so that it matches the expression to be integrated. In this case, multiply both sides of the equation by 2.

$$2\int 10(3x+2)(3x^2+4x-7)^4\,dx = 2(3x^2+4x-7)^5$$

5. Write the answer.
Note: c is not required because 'an' anti-derivative was required.

$$\int 20(3x+2)(3x^2+4x-7)^4\,dx = 2(3x^2+4x-7)^5$$

7.2 Exercise

Students, these questions are even better in jacPLUS

Receive immediate feedback and access sample responses

Access additional questions

Track your results and progress

Find all this and MORE in jacPLUS

Technology free

1. **WE1** Determine:

a. $\int -2x^3\,dx$

b. $\int \frac{1}{2}\sqrt[4]{x}\,dx$

c. $\int -\frac{3}{2x^2}\,dx$

d. $\int \frac{4}{\sqrt[3]{x^2}}\,dx$

2. **WE2** Determine:

a. $f(x)$ if $f'(x) = \frac{3}{2}x - 4x^2 + 2x^3$

b. an anti-derivative of $\frac{3}{\sqrt{x}} - 4x^3 + \frac{2}{5x^3}$

c. $\int x(x-3)(2x+5)\,dx$

d. $\int \frac{3x^3 - x}{2\sqrt{x}}\,dx$

3. Determine:

a. $\int \left(\frac{2}{\sqrt{x}} + \frac{3}{x^2} - \frac{1}{2x^3}\right) dx$

b. $\int (x+1)(2x^2-3x+4)\,dx$

4. Given that $f'(x) = x^2 - \frac{1}{x^2}$, express the rule for f.

5. Determine:

a. $\int x^3\,dx$

b. $\int 7x^2 - \frac{2}{5x^3}\,dx$

c. $\int (4x^3 - 7x^2 + 2x - 1)\,dx$

d. $\int \left(2\sqrt{x}\right)^3 dx$

6. **WE3** Anti-differentiate:

a. $(3x-5)^5$

b. $\frac{1}{(2x-3)^{\frac{5}{2}}}$

c. $(2x+3)^4$

d. $(1-2x)^{-5}$

Technology active

7. MC The gradient function for a particular curve is given by $\frac{dy}{dx}=x^3-3\sqrt{x}$. The general rule for the function, y, is:

A. $y=3x^2-\frac{3}{2\sqrt{x}}+c$
B. $y=\frac{1}{4}x^4-2x\sqrt{x}+c$
C. $y=\frac{1}{3}x^4-3x+c$
D. $y=\frac{1}{4}x^4-\frac{9}{2}x\sqrt{x}+c$
E. $y=\frac{1}{4}x^4-\frac{2}{9}x\sqrt{x}+c$

8. MC The general equation of the curve, whose gradient at any point is given by $\frac{x^3+3x^2-3}{x^2}$, is:

A. $y=x+3-3x^{-2}+c$
B. $y=1+\frac{6}{x^3}+c$
C. $y=2x^2+3x-\frac{3}{x}+c$
D. $y=\frac{1}{2}x^2+3x+\frac{3}{x}+c$
E. $y=x^2+3x+\frac{6}{x}+c$

9. Determine the indefinite integral of each of the following functions.

a. $(3x-1)^3$
b. $\frac{1}{4x^3}$
c. $x^{\frac{5}{2}}-3x^{\frac{2}{5}}$
d. $\frac{x^4-2x}{x^3}$
e. $\sqrt{x}(2x-\sqrt{x})$
f. $\sqrt{4-x}$

10. MC The general equation of the curve, whose gradient at any point on the curve is given by $\sqrt{x}+\frac{1}{\sqrt{x}}$, is:

A. $y=\frac{2}{3}x\sqrt{x}+\frac{1}{2}\sqrt{x}+c$
B. $y=\frac{1}{2\sqrt{x}}-\frac{1}{2x\sqrt{x}}+c$
C. $y=\frac{3}{2}x\sqrt{x}+2\sqrt{x}+c$
D. $y=\frac{2}{3}x\sqrt{x}+\sqrt{x}+c$
E. $y=\frac{2}{3}x\sqrt{x}+2\sqrt{x}+c$

11. WE4 Determine $\int\left(2x^2+\frac{1}{x}\right)^3 dx$.

12. Determine:

a. $\int\left(\sqrt{x}-x\right)^2 dx$
b. $\int\left(\sqrt{x}+\frac{1}{\sqrt{x}}\right)^3 dx$

13. Determine an anti-derivative for each of the following functions.

a. $(2x+3)(3x-2)$
b. $\frac{x^3+x^2+1}{x^2}$
c. $2\sqrt{x}-\frac{4}{\sqrt{x}}$
d. $\left(x^3-\frac{2}{x^3}\right)^2$
e. $2(1-4x)^{-3}$
f. $\frac{2}{(2x-3)^{\frac{5}{2}}}$

14. Determine $\int\frac{x^2}{\sqrt{x^3+1}}dx$.

15. Determine $\int 2(3x+5)^{\frac{1}{2}}(7x^2+4x-1)\,dx$.

16. WE5 If $y=(3x^2+2x-4)^3$, find $\frac{dy}{dx}$. Hence, find an anti-derivative of $(3x+1)(3x^2+2x-4)^2$.

17. If $y=\left(7x+\sqrt{x}-\frac{1}{\sqrt{x}}\right)^4$, find $\frac{dy}{dx}$. Hence, find an anti-derivative of $y=\left(7+\frac{1}{2\sqrt{x}}+\frac{1}{2\sqrt{x^3}}\right)\left(7x+\sqrt{x}-\frac{1}{\sqrt{x}}\right)^3$.

18. If $y=\sqrt{x^2+1}$, find $\frac{dy}{dx}$ and hence find the anti-derivative of $\frac{5x}{\sqrt{x^2+1}}$.

19. If $y=(5x^2+2x-1)^4$, find $\frac{dy}{dx}$ and hence find an anti-derivative of $16(5x+1)(5x^2+2x-1)^3$.

20. If $y=\sqrt{5x^3+4x^2}$, find $\frac{dy}{dx}$ and hence find an anti-derivative of $\frac{15x^2+8x}{\sqrt{5x^3+4x^2}}$.

7.2 Exam questions

Question 1 (2 marks) TECH-FREE

Source: *VCE 2013, Mathematical Methods (CAS) Exam 1, Q2; © VCAA.*

Find an anti-derivative of $(4-2x)^{-5}$ with respect to x.

Question 2 (1 mark) TECH-ACTIVE

Source: *VCE 2015, Mathematical Methods (CAS) Exam 2, Section 1, Q16; © VCAA.*

MC Let $f(x)=ax^m$ and $g(x)=bx^n$, where a, b, m and n are positive integers. The domain of f = domain of $g=R$. If $f'(x)$ is an anti-derivative of $g(x)$, then which one of the following must be true?

A. $\frac{m}{n}$ is an integer

B. $\frac{n}{m}$ is an integer

C. $\frac{a}{b}$ is an integer

D. $\frac{b}{a}$ is an integer

E. $n-m=2$

Question 3 (1 mark) TECH-FREE

Find $\int\left(3x^4-\frac{2}{x^2}\right)dx$.

More exam questions are available online.

7.3 Anti-derivatives of exponential and trigonometric functions

LEARNING INTENTION

At the end of this subtopic you should be able to:

- anti-differentiate exponential functions
- anti-differentiate trigonometric functions.

7.3.1 The anti-derivative of e^x

As we have seen in Topic 5, $\frac{d}{dx}(e^x)=e^x$ and $\frac{d}{dx}(e^{kx})=ke^{kx}$.

Therefore, the following properties hold.

Anti-derivative of e^x

$$\int e^x dx = e^x + c$$

and

$$\int e^{kx} dx = \frac{1}{k}e^{kx} + c$$

WORKED EXAMPLE 6 Anti-derivative of e^x

Determine:

a. $\int (x^7 - e^{3x})\, dx$

b. $\int 8e^{2x} dx$

THINK	WRITE
a. Integrate each term separately.	a. $\int (x^7 - e^{3x})\, dx$ $= \frac{1}{8}x^8 - \frac{1}{3}e^{3x} + c$
b. 1. Apply the rule $\int e^{kx} dx = \frac{1}{k}e^{kx} + c$.	b. $\int 8e^{2x} dx$ $= 8\int e^{2x} dx$ $= 8 \times \frac{1}{2}e^{2x} + c$
2. Simplify.	$= 4e^{2x} + c$

Algebraic expansion may also be necessary for questions involving e^{kx}.

WORKED EXAMPLE 7 Further anti-differentiation

Determine y if it is known that $\frac{dy}{dx} = (e^x + e^{-x})^3$.

THINK	WRITE
1. Expand the brackets.	$\frac{dy}{dx} = (e^x + e^{-x})^3$ $= (e^x)^3 + 3(e^x)^2(e^{-x}) + 3(e^x)(e^{-x})^2 + (e^{-x})^3$ $= e^{3x} + 3e^{2x}e^{-x} + 3e^x e^{-2x} + e^{-3x}$ $= e^{3x} + 3e^x + 3e^{-x} + e^{-3x}$
2. Anti-differentiate each term separately.	$y = \int (e^{3x} + 3e^x + 3e^{-x} + e^{-3x})dx$ $= \frac{1}{3}e^{3x} + 3e^x - 3e^{-x} - \frac{1}{3}e^{-3x} + c$

7.3.2 Anti-derivatives of $\sin(x)$ and $\cos(x)$

We know that:

- if $y = \sin(x)$, then $\dfrac{dy}{dx} = \cos(x)$
- if $y = \sin(ax + b)$, then $\dfrac{dy}{dx} = a\cos(ax + b)$
- if $y = \cos(x)$, then $\dfrac{dy}{dx} = -\sin(x)$
- if $y = \cos(ax + b)$, then $\dfrac{dy}{dx} = -a\sin(ax + b)$.

Hence, we can state the following properties.

Anti-derivative of $\sin(x)$

$$\int \sin(x)\,dx = -\cos(x) + c$$

and

$$\int \sin(ax + b)\,dx = -\frac{1}{a}\cos(ax + b) + c$$

Anti-derivative of $\cos(x)$

$$\int \cos(x)\,dx = \sin(x) + c$$

and

$$\int \cos(ax + b)\,dx = \frac{1}{a}\sin(ax + b) + c$$

WORKED EXAMPLE 8 Anti-derivatives of $\sin(x)$ and $\cos(x)$

Determine an anti-derivative of $f(x) = 2\sin(5x) + 3\cos\left(\dfrac{x}{3}\right)$.

THINK	WRITE
1. Separate the two terms.	$\displaystyle\int \left(2\sin(5x) + 3\cos\left(\frac{x}{3}\right)\right) dx$ $\displaystyle= \int 2\sin(5x)\,dx + \int 3\cos\left(\frac{x}{3}\right) dx$
2. Take out 2 and 3 as factors.	$\displaystyle= 2\int \sin(5x)\,dx + 3\int \cos\left(\frac{x}{3}\right) dx$
3. Apply the anti-differentiation rules for sin and cos.	$\displaystyle= -\frac{2}{5}\cos(5x) + 9\sin\left(\frac{x}{3}\right)$

For particularly difficult anti-differentiation problems, you may first be asked to differentiate a function so that you can use this result to carry out the anti-differentiation. This technique is called integration by recognition.

Recall that if $f(x) = g'(x)$, then $\displaystyle\int g'(x)\,dx = f(x) + c$.

WORKED EXAMPLE 9 Integration by recognition

Given that $y = e^{x^2}$, find $\frac{dy}{dx}$ and hence find an anti-derivative of xe^{x^2}.

THINK	WRITE
1. Use the chain rule to differentiate the given function.	$y = e^{x^2}$ $\frac{dy}{dx} = 2xe^{x^2}$
2. Rewrite the result as an integral.	$\int 2xe^{x^2}\,dx = e^{x^2}$
3. Adjust the left-hand side so that it matches the expression to be integrated. In this case, both sides need to multiplied by $\frac{1}{2}$.	$2\int xe^{x^2}\,dx = e^{x^2}$ $\frac{1}{2} \times 2\int xe^{x^2}\,dx = \frac{1}{2} \times e^{x^2}$
4. Write the answer.	$\int xe^{x^2}\,dx = \frac{1}{2}e^{x^2}$

7.3 Exercise

Technology free

1. WE6 Determine:

a. $\int (x^4 - e^{-4x})dx$

b. $\int \left(\frac{1}{2}e^{2x} - \frac{2}{3}e^{-\frac{1}{2}x}\right) dx$

2. WE7 Determine $\int (e^{2x} - e^{-3x})^3\,dx$.

3. Determine the indefinite integral of $\left(e^{\frac{x}{2}} - \frac{1}{e^x}\right)^2$.

4. WE8 a. Determine the indefinite integral of $\frac{1}{2}\cos(3x+4) - 4\sin\left(\frac{x}{2}\right)$.

b. Determine an anti-derivative of $\cos\left(\frac{2x}{3}\right) - \frac{1}{4}\sin(5-2x)$.

5. a. Determine $\int \left(\sin\left(\frac{x}{2}\right) - 3\cos\left(\frac{x}{2}\right)\right) dx$.

b. If $f'(x) = 7\cos(2x) - \sin(3x)$, determine a general rule for f.

6. Express the indefinite integral of:

a. $e^{\frac{x}{3}} + \sin\left(\frac{x}{3}\right) + \frac{x}{3}$

b. $\cos(4x) + 3e^{-3x}$

7. Determine an anti-derivative of $\frac{1}{4x^2} + \sin\left(\frac{3\pi x}{2}\right)$.

8. The gradient of a tangent to a curve is given by $\frac{dy}{dx} = \cos(2x) - e^{-3x}$. Determine a possible general rule for the curve y.

9. Determine the following.

a. $\int (2e^{3x} - \sin(2x))\, dx$

b. $\int \frac{e^{2x} + 3e^{-5x}}{2e^x}\, dx$

c. $\int (0.5\cos(2x+5) - e^{-x})\, dx$

d. $\int (e^x - e^{2x})^2 dx$

Technology active

10. Determine $\int \frac{e^{2x} + e^x - 1}{e^x + 1}\, dx$.

11. If it is known that $\int ae^{bx} dx = -3e^{3x} + c$, calculate the values of the constants a and b.

12. If $f'(x) = a\sin(mx) - be^{nx}$ and $f(x) = \cos(2x) - 2e^{-2x} + 3$, calculate the value of the constants a, b, m and n.

13. Heat escapes from a storage tank at a rate of kilojoules per day. This rate can be modelled by

$$\frac{dH}{dt} = 1 + \frac{\pi^2}{9}\sin\left(\frac{\pi t}{45}\right),\ 0 \le t \le 100$$

where $H(t)$ is the total accumulated heat loss in kilojoules, t days after June 1.

a. Determine $H(t)$.

b. Calculate the total accumulated heat loss after 15 days. Give your answer correct to 3 decimal places.

14. Calculate $\{x : \int e^x \sin(x)\, dx = \int e^x \cos(x)\, dx,\ 0 \le x \le 2\pi\}$.

15. If $x(t) = 20 + \cos\left(\frac{\pi t}{4}\right)$ and $\frac{dy}{dt} = \frac{\pi}{20}x(t) - \pi$, determine a possible rule for y in terms of t.

16. **WE9** Find the derivative of $e^{\cos^2(x)}$ and hence find an anti-derivative of $\sin(x)\cos(x)\, e^{\cos^2(x)}$.

17. Find the derivative of $e^{(x+1)^3}$ and hence find $\int 9(x+1)^2 e^{(x+1)^3} dx$.

18. **MC** Differentiate $y = 2xe^{3x}$. Hence, an anti-derivative of xe^{3x} is:

A. $\frac{1}{3}xe^{3x} - \frac{1}{9}e^{3x}$

B. $\frac{1}{9}xe^{3x} - \frac{1}{3}e^{3x}$

C. $\frac{1}{3}e^{3x} - \frac{1}{9}e^{3x}$

D. $-\frac{1}{3}xe^{3x} - \frac{1}{9}e^{3x}$

E. $\frac{1}{3}xe^{3x} + \frac{1}{9}e^{3x}$

19. Given that $y = e^{2x^2+3x-1}$, find $\frac{dy}{dx}$ and hence find an anti-derivative of $2(4x+3)e^{2x^2+3x-1}$.

20. Find $\frac{d}{dx}(x\cos(x))$ and hence find an anti-derivative of $x\sin(x)$.

7.3 Exam questions

Question 1 (1 mark) TECH-ACTIVE

Source: *VCE 2016, Mathematical Methods Exam 2, Section A, Q9; © VCAA.*

MC Given that $\frac{d\left(xe^{kx}\right)}{dx} = (kx+1)e^{kx}$, then $\int xe^{kx}dx$ is equal to

A. $\frac{xe^{kx}}{kx+1}+c$

B. $\left(\frac{kx+1}{k}\right)e^{kx}+c$

C. $\frac{1}{k}\int e^{kx}dx$

D. $\frac{1}{k}\left(xe^{kx}-\int e^{kx}dx\right)+c$

E. $\frac{1}{k^2}\left(xe^{kx}-e^{kx}\right)+c$

Question 2 (1 mark) TECH-ACTIVE

MC The gradient of a curve is given by $2\sin(2x)-4e^{-2x}$. The curve passes through the origin. The equation of the curve is given by

A. $2e^{-3x}-1-\cos(2x)$

B. $1+\cos(2x)-2e^{-2x}$

C. $\cos(2x)+2e^{-2x}-3$

D. $4\cos(2x)-8e^{-2x}+4$

E. $-4\cos(2x)+8e^{-2x}-4$

Question 3 (1 mark) TECH-ACTIVE

MC If $\int ae^{bx}\,dx = -2e^{2x}+c$, then

A. $a=4$ and $b=-2$

B. $a=-2$ and $b=2$

C. $a=-1$ and $b=2$

D. $a=-4$ and $b=2$

E. $a=-4$ and $b=-2$

More exam questions are available online.

7.4 The anti-derivative of $f(x)=\frac{1}{x}$

LEARNING INTENTION

At the end of this subtopic you should be able to:

- anti-differentiate logarithmic functions.

7.4.1 The anti-derivative of $\frac{1}{x}$

In Topic 5, we found that if $y=\log_e(x)$, then $\frac{dy}{dx}=\frac{1}{x}$.

Therefore, we can find the anti-derivative of $f(x)=\frac{1}{x}$ as follows.

Anti-derivative of $f(x)=\frac{1}{x}$

$$\int \frac{1}{x}\,dx = \log_e(x)+c,\ x>0$$

and

$$\int \frac{1}{ax+b}\,dx = \frac{1}{a}\log_e(ax+b)+c,\ x>-\frac{b}{a}$$

WORKED EXAMPLE 10 The anti-derivative of $f(x) = \frac{1}{x}$

Determine:

a. $\int \frac{3}{2x}\, dx$

b. $\int \frac{4}{2x+1}\, dx$

THINK	WRITE
a. 1. Remove $\frac{3}{2}$ as a factor.	**a.** $\int \frac{3}{2x}\, dx = \frac{3}{2}\int \frac{1}{x}\, dx$
2. Apply the integration rule.	$= \frac{3}{2}\log_e(x) + c, x > 0$
b. 1. Remove 4 as a factor.	**b.** $\int \frac{4}{2x+1}\, dx = 4\int \frac{1}{2x+1}\, dx$
2. Apply the integration rule.	$= 4 \times \frac{1}{2}\log_e(2x+1),\ x > -\frac{1}{2}$ $= 2\log_e(2x+1),\ x > -\frac{1}{2}$

TI | THINK

a. 1. On a Calculator page, press MENU, then select:
4: Calculus
3: Integral
Complete the entry line as:
$\int \frac{3}{2x} dx$
then press ENTER.

DISPLAY/WRITE

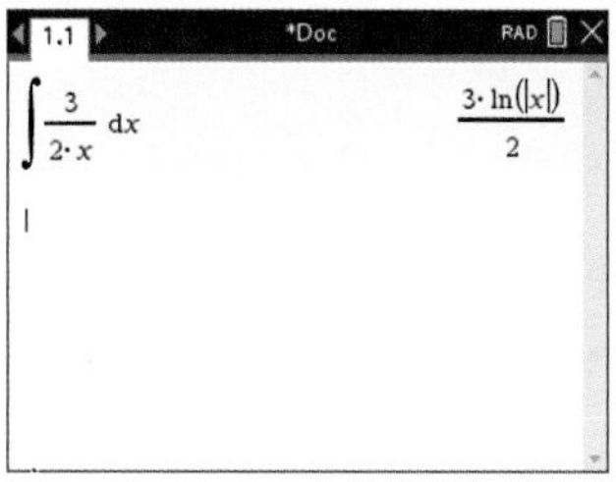

2. The answer appears on the screen.
Note: Remember to include the constant, c, as the calculator will not include it.

$\int \frac{3}{2x}\, dx = \frac{3\log_e(x)}{2} + c, x > 0$

CASIO | THINK

a. 1. On a Main screen, select:
- Interactive
- Calculation
- $\int$

Select 'Indefinite integral', then complete the fields as:
Expression: $\frac{3}{2x}$
Variable: x
then select OK.

DISPLAY/WRITE

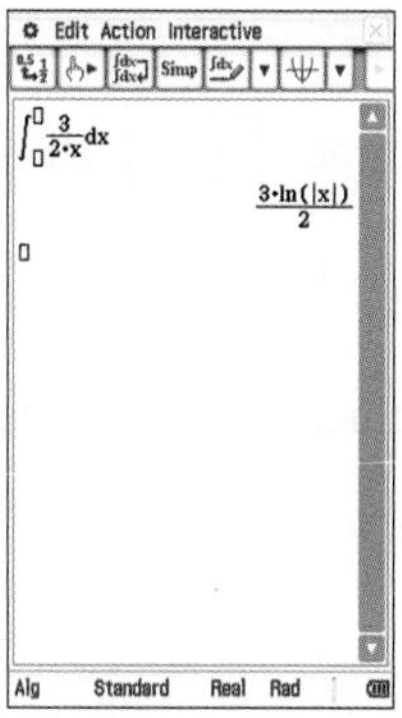

2. The answer appears on the screen.
Note: Remember to include the constant, c, as the calculator will not include it.

$\int \frac{3}{2x} dx = \frac{3\log_e(x)}{2} + c, x > 0$

7.4.2 Integration by recognition

Integration by recognition is used when we want to anti-differentiate more complex functions that we don't have an anti-derivative rule for. This method involves finding the derivative of a related function and using this derivative to find the anti-derivative.

WORKED EXAMPLE 11 Integration by recognition

Differentiate $y = \frac{x^2}{4}\log_e(x)$ and hence determine $\int x\log_e(x)\,dx$.

THINK	WRITE
1. Use the product rule to differentiate the given function.	$y = \frac{x^2}{4}\log_e(x)$ Let $u = \frac{x^2}{4}$, so $\frac{du}{dx} = \frac{x}{2}$. Let $v = \log_e(x)$, so $\frac{dv}{dx} = \frac{1}{x}$. $\frac{dy}{dx} = u\frac{dv}{dx} + v\frac{du}{dx}$ $= \frac{x^2}{4} \times \frac{1}{x} + \log_e(x) \times \frac{x}{2}$ $= \frac{x}{4} + \frac{x}{2}\log_e(x)$
2. Express the answer in integral form. *Note:* There is no need to include $+c$ as the question asked for 'an' anti-derivative.	$\int\left(\frac{x}{4} + \frac{x}{2}\log_e(x)\right)dx = \frac{x^2}{4}\log_e(x),\ x > 0$
3. Separate the two parts of the integral.	$\int\left(\frac{x}{4}\right)dx + \int\left(\frac{x}{2}\log_e(x)\right)dx = \frac{x^2}{4}\log_e(x)$
4. Subtract $\int\left(\frac{x}{4}\right)dx$ from both sides to make $\int\frac{x}{2}\log_e(x)\,dx$ the subject. (Remember we are determining $\int x\log_e(x)\,dx$.)	$\int\frac{x}{2}\log_e(x)dx = \frac{x^2}{4}\log_e(x) - \int\left(\frac{x}{4}\right)dx$
5. Anti-differentiate the function on the right-hand side of the equation, $\frac{x}{4}$.	$\int\frac{x}{2}\log_e(x)dx = \frac{x^2}{4}\log_e(x) - \frac{x^2}{8}$
6. Remove $\frac{1}{2}$ as a factor so that the function to be integrated matches the one in the question.	$\frac{1}{2}\int x\log_e(x)dx = \frac{x^2}{4}\log_e(x) - \frac{x^2}{8}$

7. Multiply the equation through by 2. $2 \times \frac{1}{2}\int x\log_e(x)dx = 2\left(\frac{x^2}{4}\log_e(x) - \frac{x^2}{8}\right)$

8. State the answer. $\int x\log_e(x)dx = \frac{x^2}{2}\log_e(x) - \frac{x^2}{4},\ x > 0$

7.4 Exercise

Technology free

1. **WE10** Determine:

a. $\int \frac{2}{5x}\,dx$

b. $\int \frac{3}{4x-1}\,dx$

2. Anti-differentiate the following.

a. $-\frac{4}{x}$

b. $\frac{3}{4x+7}$

c. $\frac{x^3+2x^2+3x-1}{x^2}$

d. $\frac{3}{2-x}+\cos(4x)$

3. Determine:

a. $\int \frac{3}{1-2x}dx$

b. $\int \frac{2}{x+4}dx$

4. **WE11** Differentiate $y = 2\log_e(\cos(2x))$ and hence determine an anti-derivative of $\tan(2x)$ with respect to x.

5. If $f(x) = 2x\log_e(mx)$, state $f'(x)$ and hence determine $\int \log_e(mx)dx$, where m is a constant.

6. Differentiate $3x\log_e(x)$ and hence find an anti-derivative for $2\log_e(x)$.

Technology active

7. If $\int \frac{6}{4-2x}dx = a\log_e(b-x)+c$, determine the values of the constants a and b.

8. **MC** $\int \frac{2}{2x+3}+3\sin(4x+1)\,dx$ is:

A. $2\log_e(2x+3) - \frac{3}{4}\cos(4x+1)+c$

B. $\log_e(2x+3) - \frac{3}{4}\cos(4x+1)+c$

C. $\frac{1}{2}\log_e(2x+3) - \frac{3}{4}\cos(4x+1)+c$

D. $\log_e(2x+3) + \frac{3}{4}\cos(4x+1)+c$

E. $2\log_e(2x+3) + \frac{3}{4}\cos(4x+1)+c$

9. **MC** Differentiate $y = (\log_e(x))^2$. Hence, an anti-derivative for $\int \frac{4\log_e(x)}{x}dx$ is:

A. $2\log_e(x)$

B. $\frac{1}{2}(\log_e(x))^2$

C. $\frac{2}{x}$

D. $2(\log_e(x))^2$

E. $4\log_e(x)$

10. If $\int \frac{d - 2x^b + x}{ax} dx = \frac{5}{4}\log_e(x) - \frac{x^2}{4} + \frac{x}{4} + c$, determine the values of the constants a, b and d.

7.4 Exam questions

Question 1 (1 mark) TECH-FREE

Determine $\int \frac{x^2 + 2x - 3}{x^2} dx$.

Question 2 (1 mark) TECH-ACTIVE

MC $\int \frac{6}{x+5} dx$ is equal to

A. $6\int \frac{1}{x+5} dx$

B. $\int 6dx \int \frac{1}{x+5} dx$

C. $\int 6dx + \int \frac{1}{x+5} dx$

D. $\frac{\int 6dx}{\int (x+5)dx}$

E. $\frac{6}{\int (x+5)dx}$

Question 3 (3 marks) TECH-FREE

If $f(x) = x\log_e(x)$, find $f'(x)$ and hence $\int \log_e(x)dx$.

More exam questions are available online.

7.5 Families of curves

LEARNING INTENTION

At the end of this subtopic you should be able to:
- use anti-differentiation to graph $f(x)$ from $f'(x)$
- recognise a family of curves
- calculate the value of c when given extra information.

7.5.1 Initial conditions

Suppose we are asked to investigate $\frac{dy}{dx} = 2e^{2x}$. Because $y = e^{2x} + c$, this is a series of an infinite number of exponential functions. We call this a family of curves. The functions with c values of 1, 0, −2 and −4, as shown in the diagram, are four of the possible functions for $y = e^{2x} + c$.

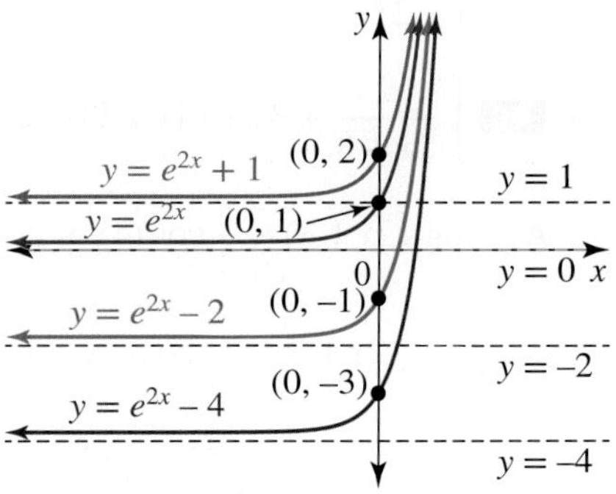

A specific function can only be found if we are given some additional information to allow us to evaluate the constant, c. For example, we might be told that the curve passes through the origin. This lets us know that when $x = 0$, $y = 0$.

Hence,

$$\begin{aligned} y &= e^{2x} + c \\ (0,0) \Rightarrow 0 &= e^{2(0)} + c \\ 0 &= 1 + c \\ c &= -1 \end{aligned}$$

Therefore, $y = e^{2x} - 1$.

This additional information is referred to as an initial condition. The question could have been given as follows: 'If $f'(x) = 2e^{2x}$, find f given that $f(0) = 0$.'

WORKED EXAMPLE 12 Sketching a family of curves

a. Sketch a family of curves that have the derivative function $f'(x) = 2\cos(2x)$ for $0 \le x \le 2\pi$.

b. Determine the specific rule for this function if $f(\pi) = 2$.

THINK

a. 1. Apply the rule $\int \cos(ax)\,dx = \frac{1}{a}\sin(ax) + c$ to anti-differentiate the function.

2. Graph the function, first with $c = 0$. That is, sketch $f(x) = \sin(2x)$. Then translate this graph up or down to sketch the graphs with different c values. Any c values may be used.

b. 1. Substitute the known point into the equation.

2. Simplify and determine the value for c.

3. State the rule for $f(x)$.

WRITE

a. $f(x) = \int 2\cos(2x)dx$
$= \sin(2x) + c$

$f(x) = \sin(2x) + c$

Amplitude = 1 and period $= \frac{2\pi}{2} = \pi$

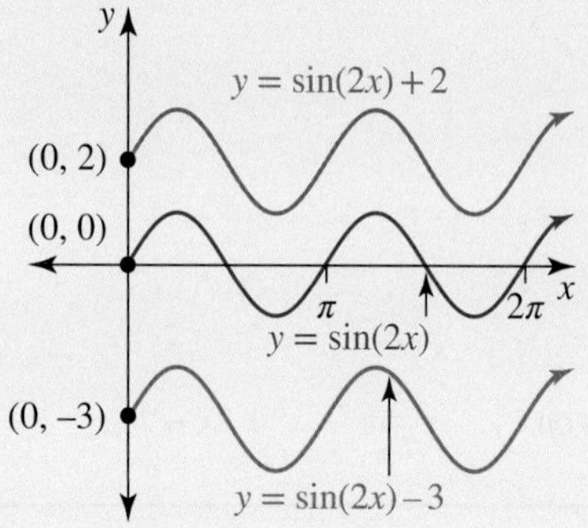

b. $f(x) = \sin(2x) + c$
$f(\pi) = 2 \Rightarrow 2 = \sin(2\pi) + c$

$2 = 0 + c$
$c = 2$

$f(x) = \sin(2x) + 2$

WORKED EXAMPLE 13 Using initial conditions to find c

Determine the equation of the curve that passes through the point (1, 0) if the gradient is given by $\frac{dy}{dx} = 3x^2 - 2x + 2$.

THINK

1. Write the gradient rule and anti-differentiate to find y.

2. Substitute the known point into the equation.

3. State the rule for y.

WRITE

$\frac{dy}{dx} = 3x^2 - 2x + 2$

$y = \int (3x^2 - 2x + 2)\,dx$
$= x^3 - x^2 + 2x + c$

When $x = 1$, $y = 0$:
$0 = 1 - 1 + 2 + c$
$c = -2$

$y = x^3 - x^2 + 2x - 2$

TI | THINK	DISPLAY/WRITE
1. On a Calculator page, press MENU, then select: 4: Calculus D: Differential Equation Solve Complete the fields as: Equation: $y\prime = 3x^2 - 2x + 2$ Independent Var: x Dependent Var: y Condition: $y(1) = 0$ then press OK and then ENTER.	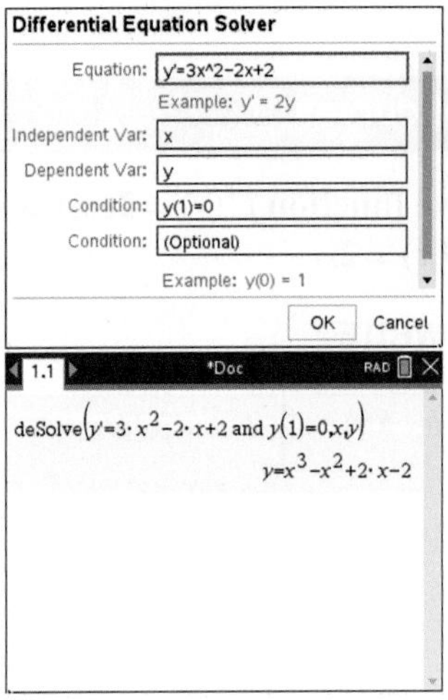
2. The answer appears on the screen.	$y = x^3 - x^2 + 2x - 2$

CASIO | THINK	DISPLAY/WRITE
1. On a Main screen, select: • Interactive • Advanced • dSolve Select 'Include condition', then complete the fields as: Equation: $y\prime = 3x^2 - 2x + 2$ Inde var: x Depe var: y Condition: $y(1) = 0$ Select OK.	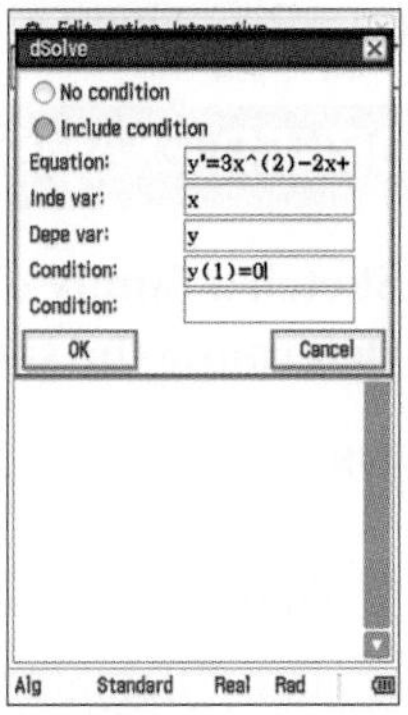 
2. The answer appears on the screen.	$y = x^3 - x^2 + 2x - 2$

on Resources

Interactivity Families of curves (int-6421)

7.5.2 Sketching the graph of a function given the graph of $f'(x)$

When $f'(x)$ represents the equation of a polynomial function, the graph of $f(x)$ can be drawn by raising the degree of $f'(x)$ by one.

Note: When we sketch $f(x)$ from the gradient function, the '$+ c$' component is unknown, so there is not just one single answer. Often we will choose c to be zero, but any vertical translation of this general graph is correct.

Gradient function, $f'(x)$	Original function, $f(x)$
A line parallel to the x-axis ($y = m$) is degree 0. **Example:** 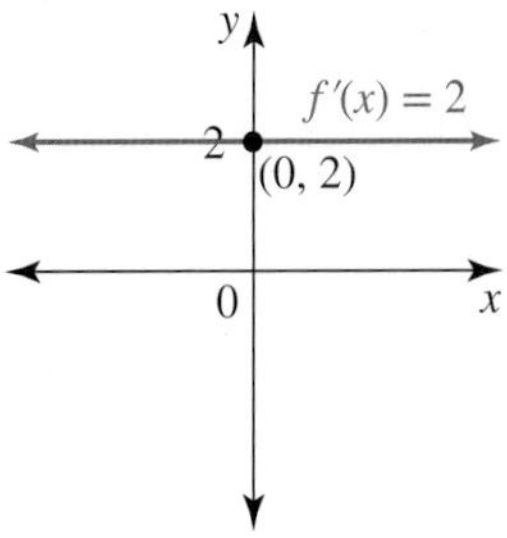	A line of the form $y = mx + c$ is degree 1 and its gradient is m. **Example:** 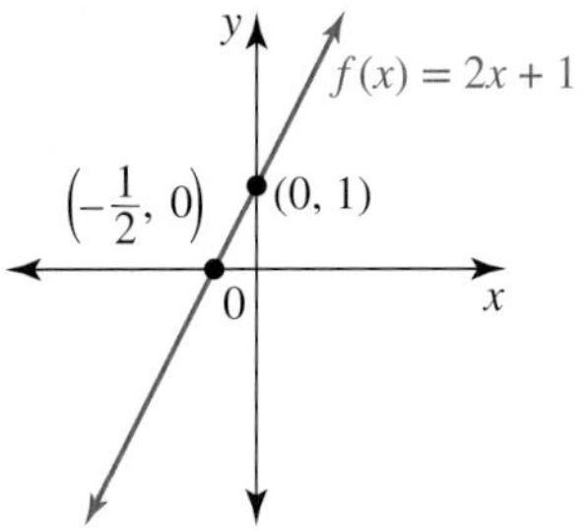

A line of the form $y = mx + c$ is degree 1. **Example:** The line shown has an x-intercept at $x = 3$. 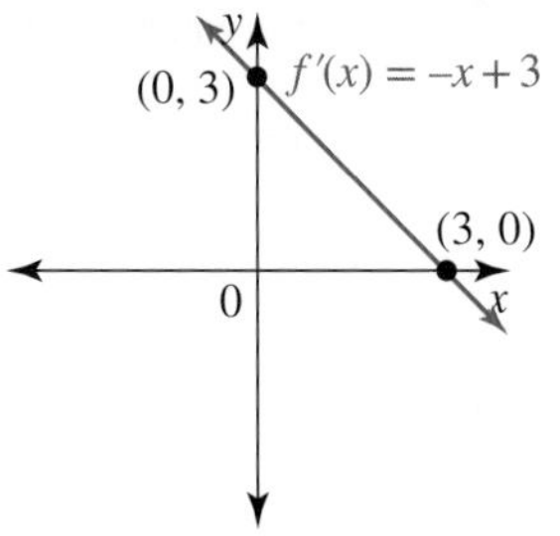	A quadratic of the form $y = ax^2 + bx + c$ is degree 2. **Example:** The function shown has a turning point at $x = 3$. 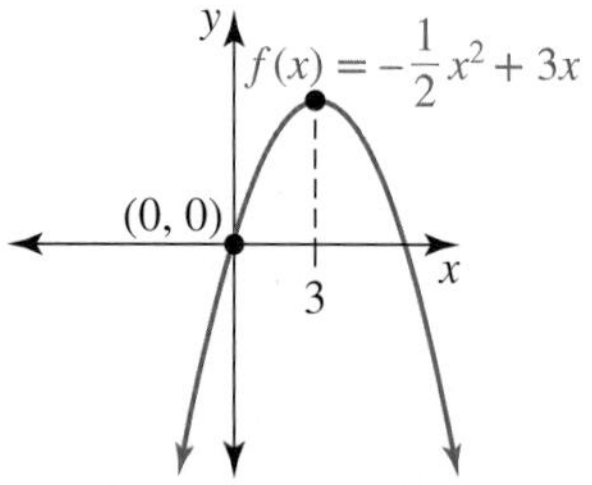
A quadratic of the form $y = ax^2 + bx + c$ is degree 2. **Example:** The graph shown has x-intercepts at $x = -1$ and $x = 5$. 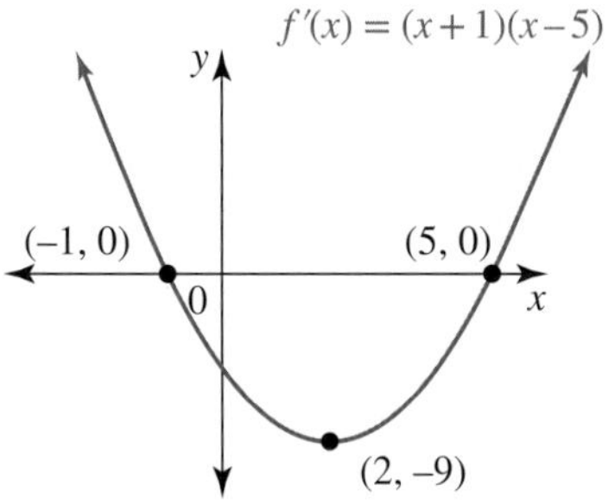	A cubic of the form $y = ax^3 + bx^2 + cx + d$ is degree 3. **Example:** The graph has turning points at $x = -1$ and $x = 5$. 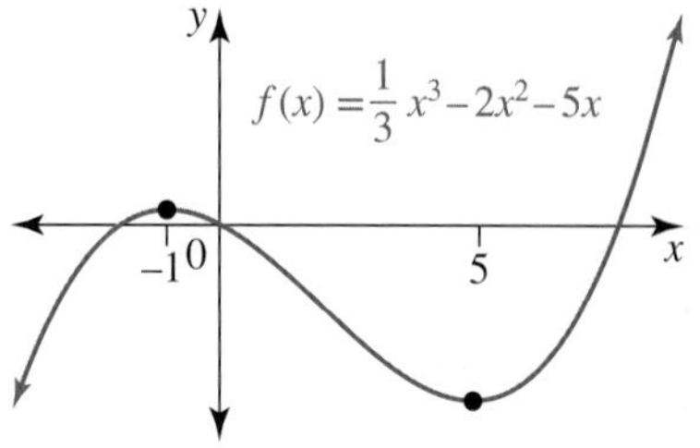

The derivative functions will not always be polynomial functions, so it is important to analyse the graph of the derivative carefully, as it will give you key information about the anti-derivative graph.

Features of derivative graphs

- **x-intercepts on the graph of $y = f'(x)$ give the x-coordinates of stationary points on the graph of $y = f(x)$, that is, points where the derivative is zero.**
- **When the graph of $y = f'(x)$ is above the x-axis, this indicates the graph of $y = f(x)$ has a positive gradient for these x-values.**
- **When the graph of $y = f'(x)$ is below the x-axis, this indicates the graph of $y = f(x)$ has a negative gradient for these x-values.**

WORKED EXAMPLE 14 Sketching $f(x)$ from $f'(x)$

The graph of the gradient function $y = f'(x)$ is shown. Analyse this derivative function and sketch a possible graph for $y = f(x)$.

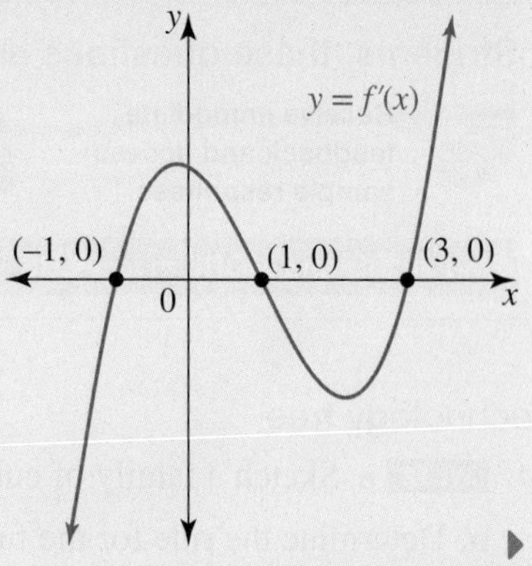

THINK | **WRITE**

1. Use the derivative graph to determine the key features of f — where f' lies above or below the x axis (this indicates where f has positive or negative gradient) and where the x-intercepts are located (this indicates the position of stationary points on f). As $y = f'(x)$ is a positive cubic, $y = f(x)$ will be a positive quartic.

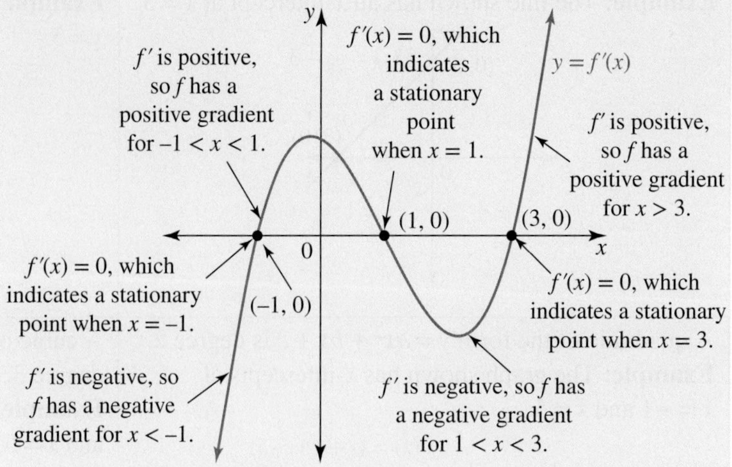

$y = f(x)$ has stationary points when $x = -1, 1, 3$.
The gradient of $y = f(x)$ is positive when $x \in (-1, 1) \cup (3, \infty)$.
The gradient of $y = f(x)$ is negative when $x \in (-\infty, -1) \cup (1, 3)$.

2. Use this knowledge to sketch the function $y = f(x)$.
Note: In this graph of $y = f(x)$, we have chosen $c = 0$. Your graph may have a different c value. However, it should be of the same basic shape, just translated vertically.

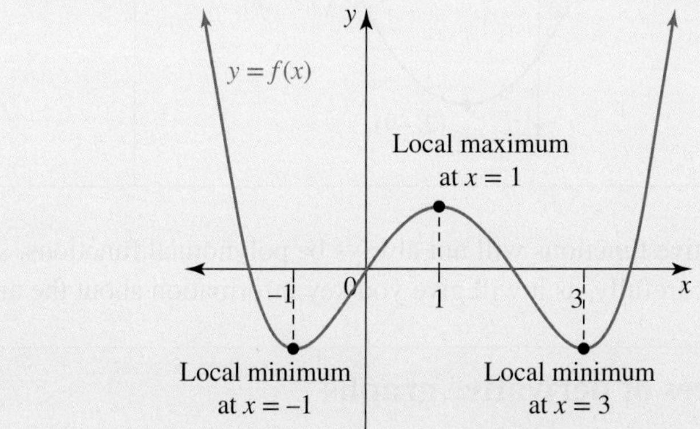

Resources

Interactivity Sketching the anti-derivative graph (int-5965)

7.5 Exercise

Technology free

1. **WE12** a. Sketch a family of curves that have the derivative function $f'(x) = 3x^2$.
   b. Determine the rule for the function that belongs to this family of curves and passes through the point $(2, 16)$.

2. a. Sketch a family of curves related to the derivative function $f'(x) = -2\cos(2x)$.

b. Determine the rule for the function that belongs to this family of curves and passes through the point $\left(\frac{\pi}{2}, 4\right)$.

3. a. Sketch a family of curves with the derivative function $f'(x) = 3e^{-3x}$.

b. Determine the rule for the function that belongs to this family of curves and passes through the point $(0, 1)$.

4. **WE13** Determine the equation of the curve that passes through the point $(0, 3)$ if the gradient is given by $\frac{dy}{dx} = 2e^{2x} + e^{-x}$.

5. The gradient function of a particular curve is given by $f'(x) = \cos(2x) - \sin(2x)$. Express the rule for this function if it is known that the curve passes through the points $(\pi, 2)$.

6. Determine the anti-derivative of $\cos(2x) + 3e^{-3x}$ if $y = 4$ when $x = 0$.

7. Express the equation of the curve defined by $\frac{dy}{dx} = e^{\frac{1}{2}x}$, given that it passes through the point $(0, 5)$.

8. Express the equation of the curve defined by $f'(x) = \frac{1}{(1-x)^2}$, given that it passes through the point $(0, 4)$.

9. Determine $f(x)$ for each of the following.

a. $f'(x) = 5 - 2x$ and $f(1) = 4$

b. $f'(x) = \sin\left(\frac{x}{2}\right)$ and $f(\pi) = 3$

10. Determine $f(x)$ if $f'(x) = x^3 - \frac{1}{x}$ and $f(1) = \frac{1}{4}$.

11. a. Given that $\frac{dy}{dx} = \frac{5}{2x+4}$ and $y = 3$ when $x = -\frac{3}{2}$, obtain an expression for y in terms of x.

b. Given that $\frac{dy}{dx} = \frac{3}{2-5x}$ and $y = 1$ when $x = \frac{1}{5}$, obtain an expression for y in terms of x.

12. **WE14** The graph of the gradient function $y = f'(x)$ is shown. Analyse this derivative function and sketch the given function $y = f(x)$.

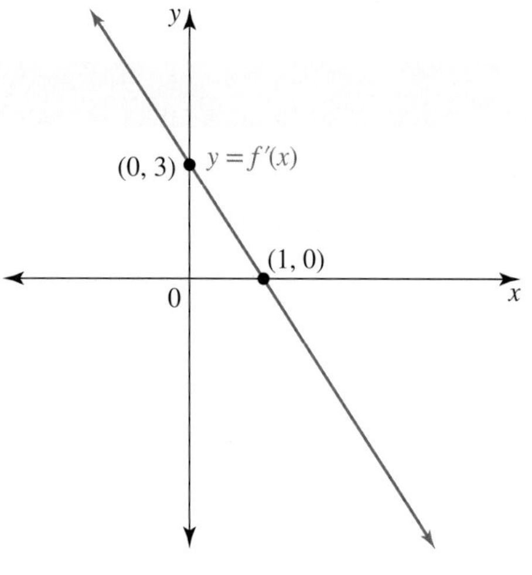

13. The graph of the gradient function $y=f'(x)$ is shown. Analyse this derivative function and sketch the given function $y=f(x)$.

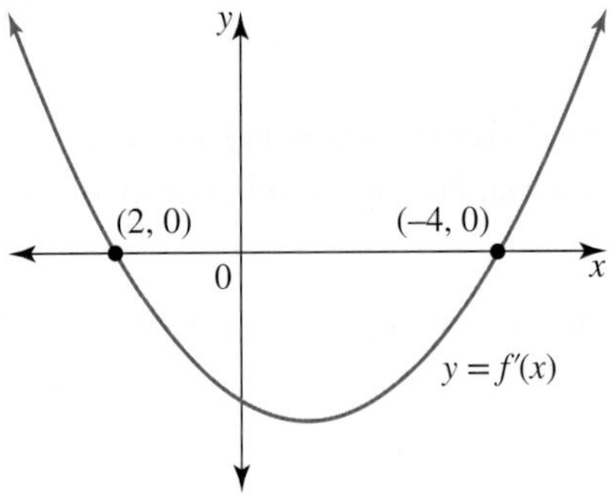

14. The graphs of some gradient functions are shown. Sketch the corresponding anti-derivative graphs.

a.

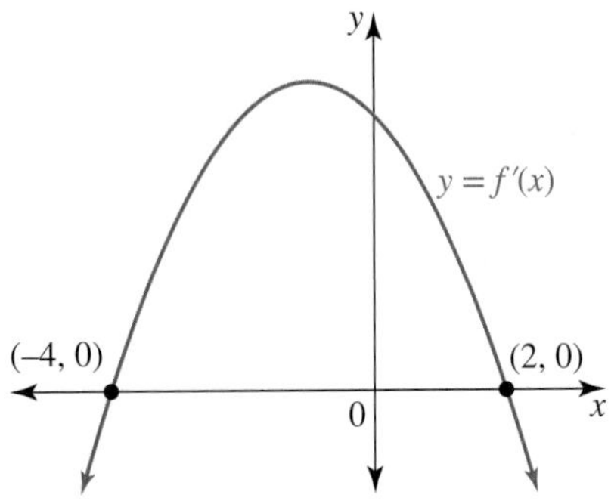

b.

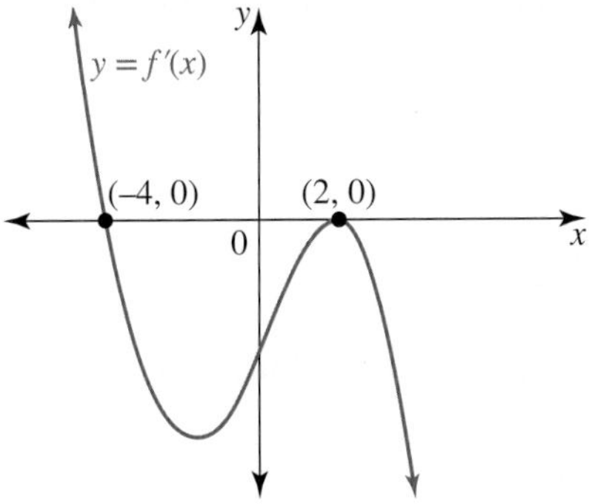

c.

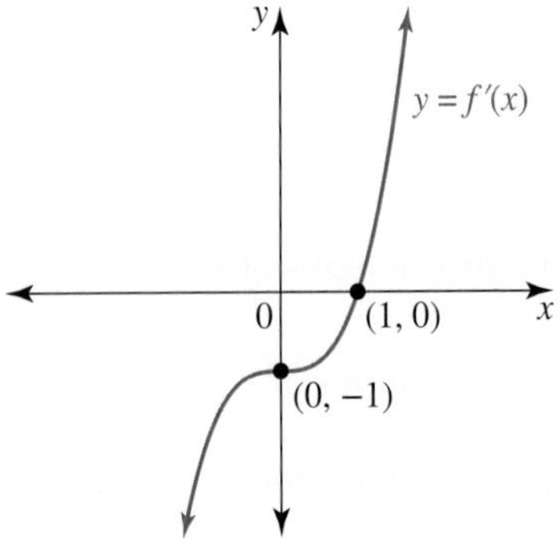

d.

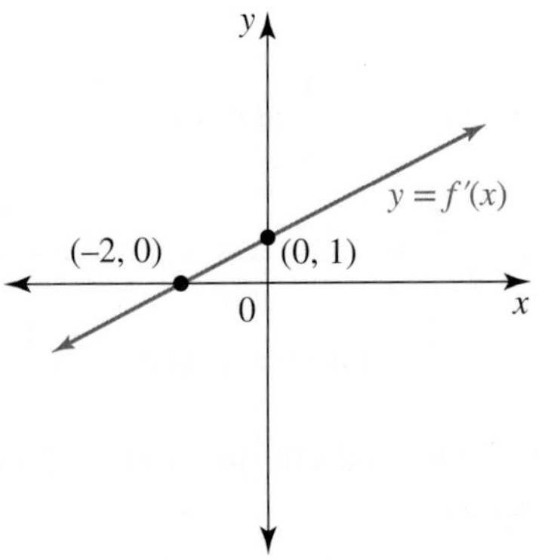

7.5 Exam questions

Question 1 (1 mark) TECH-ACTIVE

Source: *VCE 2020, Mathematical Methods Exam 2, Section A, Q3; © VCAA.*

MC Let $f'(x)=\dfrac{2}{\sqrt{2x-3}}$.

If $f(6)=4$, then

A. $f(x)=2\sqrt{2x-3}$

B. $f(x)=\sqrt{2x-3}-2$

C. $f(x)=2\sqrt{2x-3}-2$

D. $f(x)=\sqrt{2x-3}+2$

E. $f(x)=\sqrt{2x-3}$

Question 2 (1 mark) TECH-ACTIVE

Source: *VCE 2020, Mathematical Methods Exam 2, Section A, Q6; © VCAA.*

MC Part of the graph of $y = f'(x)$ is shown below.

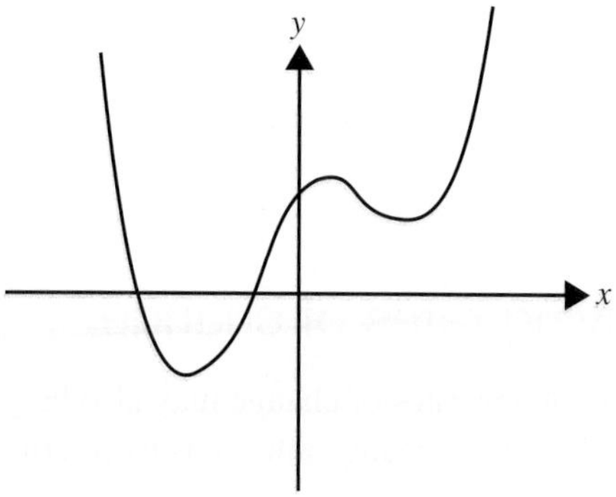

The corresponding part of the graph of $y = f(x)$ is best represented by

A.

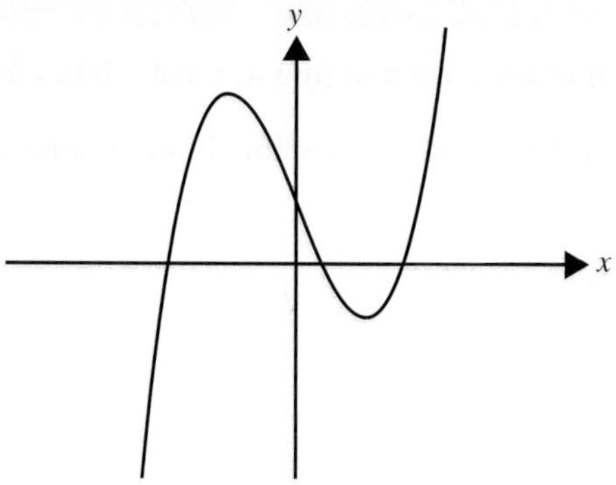

B.

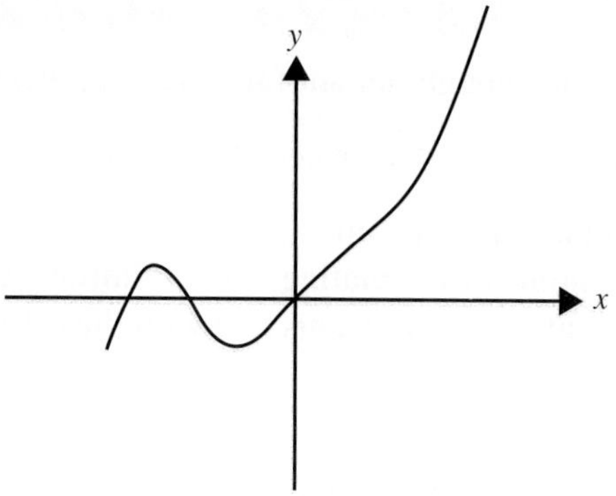

C.

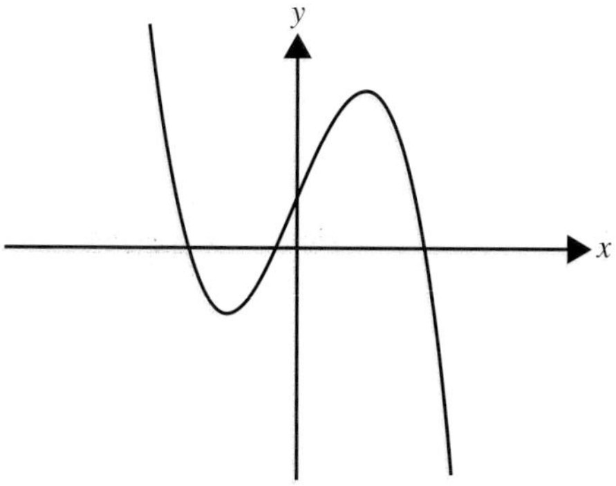

D.

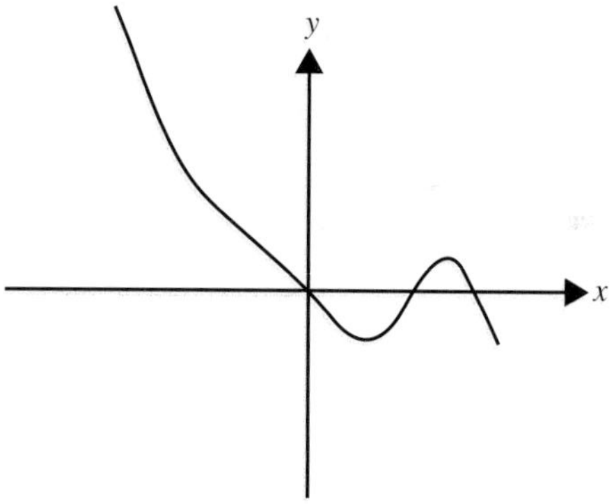

E.

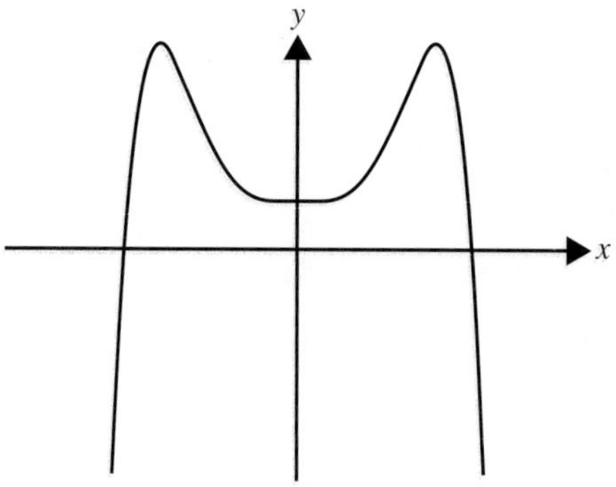

Question 3 (3 marks) TECH-FREE

Source: *VCE 2014, Mathematical Methods (CAS) Exam 1, Q7; © VCAA.*

If $f'(x) = 2\cos(x) - \sin(2x)$ and $f\left(\frac{\pi}{2}\right) = \frac{1}{2}$, find $f(x)$.

More exam questions are available online.

7.6 Applications

LEARNING INTENTION

At the end of this subtopic you should be able to:
- use anti-differentiation to solve problems involving finding a function from a known rate of change given a boundary condition.

7.6.1 Applications involving rates of change

Application questions such as those involving rates of change may also be given in terms of the derivative function. Integrating the equation for the rate of change allows us to determine the original function.

WORKED EXAMPLE 15 Rates of change application

A young boy bought an ant farm. It is known that the ant population is changing at a rate defined by $\frac{dN}{dt} = 20e^{0.2t}$, $0 \le t \le 20$, where N is the number of ants in the colony and t is the time in days since the ant farm has been set up.

a. Determine a rule relating N to t if initially there were 50 ants.

b. Calculate how many ants make up the colony after 8 days.

THINK	WRITE
a. 1. Write the rate rule and anti-differentiate to find the function for N.	a. $\frac{dN}{dt} = 20e^{0.2t}$ $N = \int (20e^{0.2t})\, dt$ $= \frac{20}{0.2}e^{0.2t} + c$ $= 100e^{0.2t} + c$
2. Use the initial condition to determine the value of c.	When $t = 0$, $N = 50$: $50 = 100e^{0.2\times 0} + c$ $50 = 100 + c$ $c = -50$
3. State the equation for N.	$N = 100e^{0.2t} - 50$
b. 1. Substitute $t = 8$ into the population equation.	b. When $t = 8$: $N = 100e^{0.2\times 8} - 50$ $= 100e^{1.6} - 50$ $= 445.3$
2. Answer the question. Round to the nearest whole number.	There are 445 ants after 8 days.

7.6.2 Linear motion

From Topic 6 we know that the study of the motion of a particle in a straight line is called kinematics. When this motion is only in a straight line, it is referred to as **rectilinear motion**.

Position–velocity–acceleration relationship

Because velocity is the derivative of position with respect to time, it follows that position is the anti-derivative of velocity.

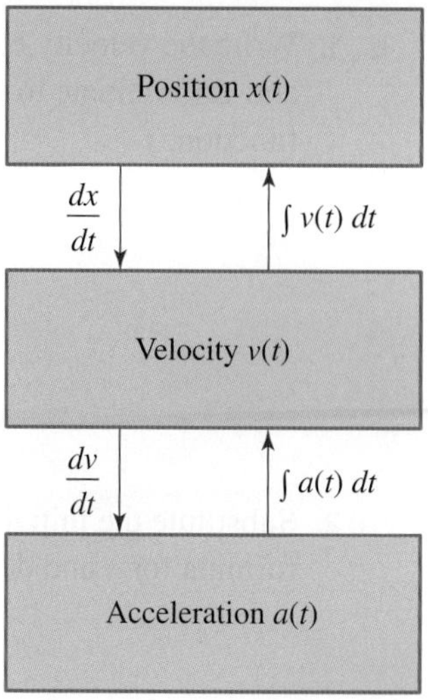

Consider a particle whose position, x metres, from the origin at time t seconds is defined by

$$x(t) = t^2 - 5t - 6,\ t \geq 0$$

Initially, at $t = 0$, the particle is 6 metres to the left of the origin. The velocity of the particle can be defined as $v = x'(t) = 2t - 5$ metres/second.

The initial velocity of the particle is -5 metres/second.

This same situation could have been approached in the following way. A particle has an instantaneous velocity defined by $v = x'(t) = 2t - 5$ metres/second.

If it is known that the particle is initially 6 metres to the left of the origin, then the position can be given by:

$$x = \int (2t - 5)dt$$

$$= t^2 - 5t + c$$

When $t = 0$, $x = -6$: $\quad -6 = 0 + c$

$$c = -6$$

$$\therefore x = t^2 - 5t - 6$$

WORKED EXAMPLE 16 Determining position from velocity

In each of the following cases, express the position as a function of t if initially the particle is at the origin.

a. $v = t^3 - t$ **b.** $v = (2t - 3)^3$ **c.** $v = \dfrac{1}{(t-1)^2}$

THINK	WRITE
a. 1. Write the velocity equation and anti-differentiate to find the position function, x.	**a.** $v = \dfrac{dx}{dt} = t^3 - t$ $x = \int (t^3 - t)\,dt$ $x = \frac{1}{4}t^4 - \frac{1}{2}t^2 + c$
2. Substitute the initial condition into the formula for x and determine c.	When $t = 0$, $x = 0$: $0 = 0 + c$ $c = 0$
3. State the rule.	$x = \frac{1}{4}t^4 - \frac{1}{2}t^2$

b. 1. Write the velocity equation and anti-differentiate to find the position function, x.

b. $v = \dfrac{dx}{dt}$

$= (2t-3)^3$

$x = \int (2t-3)^3\,dt$

$= \dfrac{(2t-3)^4}{2(4)} + c$

$= \dfrac{1}{8}(2t-3)^4 + c$

2. Substitute the initial condition into the formula for x and determine c.

When $t = 0, x = 0$:

$0 = \dfrac{1}{8}(-3)^4 + c$

$= \dfrac{81}{8} + c$

$c = -\dfrac{81}{8}$

3. State the rule.

$x = \dfrac{1}{8}(2t-3)^4 - \dfrac{81}{8}$

c. 1. Write the velocity equation and anti-differentiate to find the position function, x.

c. $v = \dfrac{dx}{dt}$

$= \dfrac{1}{(t-1)^2}$

$= (t-1)^{-2}$

$x = \int (t-1)^{-2}\,dt$

$= \dfrac{(t-1)^{-1}}{-1} + c$

$= -(t-1)^{-1} + c$

$= -\dfrac{1}{(t-1)} + c$

2. Substitute the initial condition into the formula for x and determine c.

When $t = 0, x = 0$:

$0 = -\dfrac{1}{(-1)} + c$

$0 = 1 + c$

$c = -1$

3. State the rule.

$x = -\dfrac{1}{(t-1)} - 1$

WORKED EXAMPLE 17 Linear motion

The velocity of a particle moving in a straight line along the x-axis is given by $v = \dfrac{dx}{dt} = 9 - 9e^{-3t}$ where t is the time in seconds and x is the position in metres.

a. Show that the particle is initially at rest.

b. Determine the equation relating x to t if it is known that initially the particle was 3 metres to the left of the origin.

THINK	WRITE
a. 1. Substitute $t=0$ and evaluate.	a. $v=9-9e^{-3t}$ $t=0 \Rightarrow v=9-9e^{0}$ $=9-9\times 1$ $=0\text{ m/s}$
2. Answer the question.	Initially the particle is at rest as its velocity is 0 m/s.
b. 1. Write the velocity equation and anti-differentiate to find the position equation, x.	b. $v=\dfrac{dx}{dt}$ $=9-9e^{-3t}$ $x=\int(9-9e^{-3t})\,dt$ $=9t+3e^{-3t}+c$
2. Substitute the initial condition to determine c. Remember, left of the origin means the position is negative.	When $t=0, x=-3$: $-3=9\times 0+3e^{0}+c$ $-3=3+c$ $c=-6$
3. State the equation.	$x=9t+3e^{-3t}-6$

7.6 Exercise

Technology free

1. WE16 In each of the following cases, express the position as a function of t if initially the particle is at the origin.

   a. $v=(3t+1)^{\frac{1}{2}}$
   b. $v=\dfrac{1}{(t+2)^2}$
   c. $v=(2t+1)^3$
   d. $v=e^{(3t-1)}$
   e. $v=-\sin(2t+3)$.
   f. $v=2\cos(3t)$

2. A particle moves in a straight line so that its velocity, in metres per second, can be defined by the rule $v=3t^2+7t,\ t\geq 0$. Determine the rule relating the position of the particle, x metres, to t, if it is known that the particle started from the origin.

3. A particle starting at the origin moves in a straight line with a velocity of $\dfrac{12}{(t-1)^2}+6$ metres per second after t seconds.

   a. Determine the rule relating the position of the particle, x metres, to t.
   b. Calculate the position of the particle after 3 seconds.

4. WE17 A particle is oscillating so that its velocity, v cm/s, can be defined by

$$v = \frac{dx}{dt} = \sin(2t) + \cos(2t)$$

where t is the time in seconds and x centimetres is its position.

a. Show that initially the particle is moving at 1 cm/s.
b. Determine the equation relating x to t if it is known that initially the particle was at the origin.

Technology active

5. A particle attached to a spring moves up and down in a straight line so that at time t seconds its velocity, v metres per second, is given by

$$v = 3\pi \sin\left(\frac{\pi t}{8}\right), t \geq 0$$

Initially the particle is stationary. Determine:

a. the rule relating the position of the particle, x centimetres, to t
b. the maximum position of the particle
c. where the particle is, relative to the stationary position, after 4 seconds.

6. When a bus travels along a straight road in heavy traffic from one stop to another stop, the velocity at time t seconds is given by $v = 0.25t(50 - t)$, where v is the velocity in m/s.

a. Calculate the greatest velocity reached by the bus.
b. Determine the rule for the position of the bus, x metres, in terms of t.

7. If $v = 2t\cos(t)$ metres per second, determine a rule relating the position x metres to t if it is known the particle starts from rest at the origin.

8. WE15 The rate of change of volume of a balloon as it is being blown up can be modelled by

$$\frac{dV}{dr} = \pi r^2$$

where V cm^3 is the volume of the balloon and r cm is the radius of the balloon.

a. Determine the rule for the volume of the balloon.
b. Calculate the volume of the balloon when its radius is 4 cm.

9. A chemical factory has permission from the Environment Protection Authority to release particular toxic gases into the atmosphere for a period of 20 seconds no more than once every 3 hours. This maintains safe levels of the gases in the atmosphere. This rate of emission is given by

$$\frac{dV}{dt} = 20t^2 - t^3 \text{ cm}^3/\text{s}$$

where $0 \leq t \leq 20$ and V cm^3 is the total volume of toxic gases released over t seconds. Calculate the total volume of toxic gases released during a 20-second release period.

10. A population of sea lions on a distant island is growing according to the model

$$\frac{dP}{dt} = 30e^{0.3t},\ 0 \le t \le 10$$

where P is the number of sea lions present after t years.

a. If initially there were 50 sea lions on the island, determine the rule for the number of sea lions present, P, after t years.

b. Calculate the number of sea lions on the island after 10 years. Give your answer correct to the nearest whole sea lion.

11. A newly established suburban area of Perth is growing at a rate modelled by the rule

$$\frac{dN}{dt} = 400 + 1000\sqrt{t},\ 0 \le t \le 10$$

where N is the number of families living in the suburb t years after the suburb was established in 2015.

a. Determine a rule relating N and t if initially there were 40 families living in this suburb.

b. Calculate how many families will be living in the suburb 5 years after its establishment. Give your answer correct to the nearest number of families.

12. The rate of change of the depth of water in a canal is modelled by the rule

$$\frac{dh}{dt} = \frac{\pi}{2}\cos\left(\frac{\pi t}{4}\right)$$

where h is the depth of the water in metres and t is the number of hours since 6 am.

a. Determine an equation for h in terms of t if the water is 3 metres deep at 6 am.

b. Determine the maximum and minimum depths of the water.

c. Determine how many hours a day the water level is 4 metres or more.

7.6 Exam questions

Question 1 (3 marks) TECH-ACTIVE

A particle moves in a straight line so that its velocity, in metres per second, can be defined by the rule $v = 2t^3 - t + 1,\ t \ge 0$

a. Determine the rule for the position of the particle, if it is known the particle started 1 m to the left of the origin. **(1 mark)**

b. Determine the velocity and position of the particle after 3 seconds. **(2 marks)**

Question 2 (2 marks) TECH-ACTIVE

A particle moves in a straight line so that its acceleration, after t seconds, is given by $a = 4 - 2t, t \ge 0$. If $v = 0$ and $x = 3$ when $t = 0$, calculate the position and velocity of the particle when $t = 2$.

Question 3 (2 marks) TECH-ACTIVE

The rate of growth of bacteria in a petri dish in a laboratory can be modelled by $\frac{db}{dt} = 100t^{\frac{3}{2}}$, where b is the number of bacteria after t hours. Initially there were 80 bacteria cells.

a. Determine an equation for b in terms of t. **(1 mark)**

b. Determine how long, to the nearest minute, it would take the number of bacteria to reach 500. **(1 mark)**

More exam questions are available online.

7.7 Review

7.7.1 Summary

doc-37036

Hey students! Now that it's time to revise this topic, go online to:

Access the topic summary | Review your results | Watch teacher-led videos | Practise VCAA exam questions

Find all this and MORE in jacPLUS

7.7 Exercise

Technology free: short answer

1. Determine:

a. $\int \frac{4x^4 - 4}{x^2}\, dx$

b. $\int (e^x - e^{-x})^2\, dx$

c. $\int \frac{e^{2x} + 3e^x + 2}{e^x + 1}\, dx$

d. $\int \left(-\frac{1}{3}\cos(4x - 3) - \sin\left(\frac{x}{2}\right)\right) dx$

e. $\int \left(\frac{4}{3 - x} + \sin(x)\right) dx$

f. $\int \frac{x^2 + 4x - 1}{x^2}\, dx$

2. The rate of change of temperature of a kettle can be modelled by

$$\frac{dT}{dt} = abe^{-bt}$$

where T is the temperature in degrees Celsius and t is the time in minutes since 12 noon.

a. Determine the equation for T in terms of t, where c is an arbitrary constant.

b. If $a = 100$, $b = 0.08$ and $c = 195$, calculate the temperature of the kettle at 12 noon.

3. a. Given that $f'(x) = (e^{2x} + 1)^2$ and $f(0) = -\frac{5}{4}$, determine a rule for f.

b. If $\frac{dy}{dx} = \sqrt{3x - 5}$ and $y = 0$ when $x = 2$, determine the rule for y.

c. Determine $f(x)$ if $f'(x) = \frac{x^3 + 4x}{x^2}$ and $f(1) = \log_e(2) + \frac{1}{2}$.

4. a. Sketch the family of curves related to each of the following derivative functions.

i. $f'(x) = x^2 - 2x$

ii. $f'(x) = -6e^{-3x}$

b. Each of the following graphs shows a gradient function, $y = f'(x)$. Sketch the function $y = f(x)$ that corresponds to each given gradient function.

i.

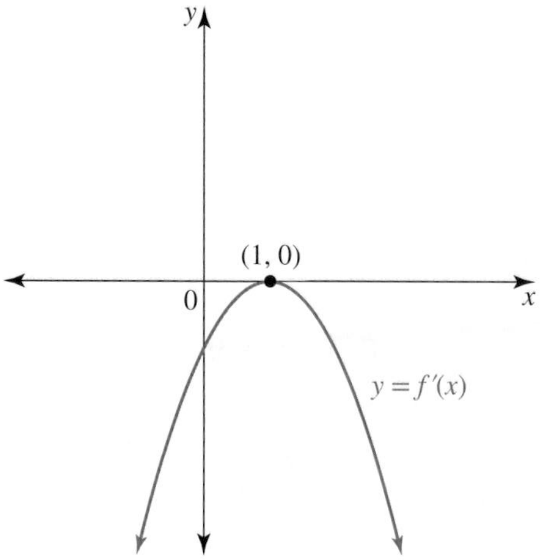

ii.

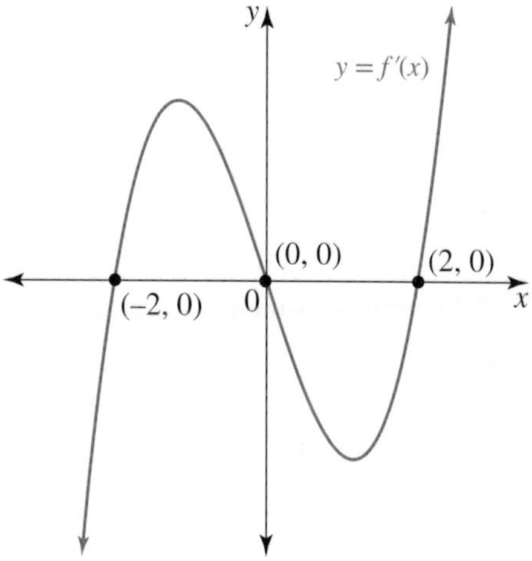

5. An athlete inhales oxygen at a rate of $3.5 - (t - 0.5)^{-2}$ litres per minute. Determine how much oxygen the athlete inhales after t minutes.

6. a. Given that $y = e^{3x^2 - 2x + 1}$, find $\frac{dy}{dx}$ and hence find an anti-derivative of $(3x - 1)e^{3x^2 - 2x + 1}$.

b. Given that $y = x\sin(2x)$, find $\frac{dy}{dx}$ and hence find an anti-derivative of $x\cos(2x)$.

Technology active: multiple choice

7. **MC** $\int \frac{5x^2 - x + 2}{4\sqrt{x}}\, dx$ is equal to:

A. $\frac{25x^{\frac{5}{2}}}{8} - \frac{3x^{\frac{3}{2}}}{8} + \frac{\sqrt{x}}{4} + c$

B. $\frac{15\sqrt{x}}{8} - \frac{1}{8\sqrt{x}} - \frac{1}{4\sqrt[3]{x}} + c$

C. $\frac{x^{\frac{5}{2}}}{2} + \frac{x^{\frac{3}{2}}}{6} - \sqrt{x} + c$

D. $\frac{x^{\frac{5}{2}}}{2} - \frac{x^{\frac{3}{2}}}{6} + \sqrt{x} + c$

E. $\frac{5x^{\frac{5}{2}}}{6} - \frac{x^{\frac{3}{2}}}{2} - \sqrt{x} + c$

8. **MC** An anti-derivative of $\frac{1}{(4x-1)^{\frac{3}{2}}}$ is equal to:

A. $-\frac{1}{2(4x-1)^{\frac{5}{2}}}$

B. $-\frac{1}{10(4x-1)^{\frac{5}{2}}}$

C. $-\frac{2}{\sqrt{4x-1}}$

D. $-\frac{1}{8\sqrt{4x-1}}$

E. $-\frac{1}{2\sqrt{4x-1}}$

9. **MC** An anti-derivative of $\cos\left(\frac{x}{5}\right) + 3e^{-3x}$ is:

A. $5\sin\left(\frac{x}{5}\right) - e^{-3x}$

B. $-5\sin\left(\frac{x}{5}\right) - 9e^{-3x}$

C. $-\frac{1}{5}\sin\left(\frac{x}{5}\right) - e^{-3x}$

D. $5\sin\left(\frac{x}{5}\right) + e^{-3x}$

E. $5\sin\left(\frac{x}{5}\right) + 9e^{-3x}$

10. **MC** If $f'(x) = g'(x) + 2$, $f(0) = 3$ and $g(0) = 1$, then $f(x)$ is given by:

A. $f(x) = g(x) + 2x + 2$
B. $f(x) = g'(x) + 2$
C. $f(x) = g(x) + 2x$
D. $f(x) = 2$
E. $f(x) = g(x) + 2$

11. **MC** If $\int ae^{bx}dx = -3e^{4x} + c$, then:

A. $a = 12, b = -4$
B. $a = -3, b = 4$
C. $a = -4, b = 12$
D. $a = -12, b = 4$
E. $a = 3, b = -4$

12. **MC** A particle starts from the origin. If the velocity of the particle, v m/s, is defined by $v = 2t + 3$, then the position of the particle, x metres from the origin, is given by:

A. $x = 2t + 3$
B. $x = t^2 + 3$
C. $x = t^2 + 3t$
D. $x = 2$
E. $x = t^2 + 3t + 2$

13. **MC** The velocity, v cm/s, of a particle moving up and down can be defined by $v = -\sin(2t)$, where t is measured in seconds. Initially the particle is 2 cm above its resting position. The rule relating its position, x cm, to t could be given by:

A. $x = \frac{1}{2}\cos(2t) + \frac{3}{2}$
B. $x = \frac{1}{2}\cos(2t)$
C. $x = \frac{1}{2}\cos(2t) + \frac{1}{2}$
D. $x = 2\cos(2t) + \frac{3}{2}$
E. $x = -2\cos(2t) + \frac{1}{2}$

14. **MC** If $f'(x) = \frac{4x}{\pi^2} + \sin(2x)$ and $f\left(\frac{\pi}{4}\right) = \frac{1}{4}$, then $f(x)$ equals:

A. $\frac{2x^2}{\pi^2} - \frac{1}{2}\cos(2x)$
B. $\frac{2x^2}{\pi^2} - \frac{1}{2}\cos(2x) + \frac{1}{8}$
C. $\frac{8x^2}{\pi^2} - \frac{1}{2}\cos(2x)$
D. $\frac{8x^2}{\pi^2} - \frac{1}{2}\cos(2x) + \frac{1}{8}$
E. $\frac{2x^2}{\pi^2} + \frac{1}{2}\cos(2x) + \frac{1}{8}$

15. **MC** The equation of the curve that passes through the point $(0, 0)$ and has $\frac{dy}{dx} = 4\cos(2x) + 2\sin(2x)$ is given by:

A. $y = 2\sin(2x) - \cos(2x) + 1$
B. $y = 2\sin(2x) + \cos(2x) - 1$
C. $y = 8\sin(2x) + 4\cos(2x) - 4$
D. $y = 2\sin(2x) - \cos(2x) - 1$
E. $y = 2\sin(2x) - \cos(2x) - 2$

16. **MC** If $f'(x) = \cos(3x) - e^{-3x}$ and $f(0) = \frac{2}{3}$, then $f(x)$ is equal to:

A. $\sin(3x) + e^{-3x} + 1$
B. $\sin(3x) - e^{-3x} - 1$
C. $\frac{1}{3}\left(\sin(3x) - e^{-3x} - 1\right)$
D. $\frac{1}{3}\left(\sin(3x) + e^{-3x} + 1\right)$
E. $\frac{1}{3}\left(\sin(3x) + e^{-3x}\right)$

Technology active: extended response

17. The velocity of a particle is given by $v = 3\sin(3t) + 1$, where x is the position in metres at time t seconds. Initially the particle is at the origin.
    a. Show that the position of the particle is given by $x = -\cos(3t) + t + 1$.
    b. Determine how long it takes for the particle to first come to rest. Give your answer correct to 3 decimal places.
    c. Determine the maximum velocity of the particle.
    d. Calculate where the particle is relative to the origin after 3 seconds. Give your answer correct to 3 decimal places.

18. George went to his favourite coffee shop and ordered a large plunger of his favourite South American black coffee.

The initial temperature of the coffee was 98 °C. George estimated the coffee was cooling at a rate of 8 °C per minute, but after 10 minutes its rate of cooling was 4 °C per minute. If T °C is the temperature of the coffee t minutes after the coffee was made, then a model for the rate of change of temperature is

$$\frac{dT}{dt} = me^{nt}$$

a. Show that $m = -8$.
b. Show that $n = \frac{-\log_e(2)}{10}$.
c. Determine a relationship between T and t.
d. Calculate the temperature of the coffee 5 minutes after it was made. Give your answer correct to the nearest whole degree.

19. An investor has just bought six townhouses that they intend to rent out. Part of the sales pitch from their real estate agent was that the experience of the previous owner suggested the rate of increase of the maintenance cost for each townhouse could be approximated by

$$\frac{dM}{da} = 48a^2 + 250$$

where a is the age of the townhouse in years and M is the total accumulated cost of the maintenance over the a years.

a. The townhouses are currently 5 years old. State what the investor should expect as an increase in the cost of maintenance for each townhouse for the next year.
b. Determine a relationship between M and a if it was known that the total maintenance cost for each townhouse at the end of the first year was \$350.
c. Determine the total accumulated maintenance cost for one townhouse during the next 5 years.

20. An oil tanker off the coast of Queensland has sprung a leak, and the oil is threatening the marine life of the Great Barrier Reef. The oil leak has produced a circular oil slick that is spreading at a rate approximated by

$$\frac{dr}{dt} = \frac{30}{\sqrt{t+1}}, t \geq 0$$

where r metres is the radius of the oil slick and t is the time in hours since the spill started.

a. Determine a relationship between r and t if the radius of the circular oil slick was 45 metres after 3 hours.
b. Calculate the radius of the oil slick to the nearest metre after 5 hours.
c. The captain and crew of the tanker are working furiously to plug the hole to stop the oil leaking. They aim to have the hole plugged so that the radius of the circular slick does not exceed 75 metres. Determine how long they expect it will take them to plug the hole.

7.7 Exam questions

Question 1 (1 mark) TECH-ACTIVE

Source: *VCE 2019, Mathematical Methods Exam 2, Section A, Q5; © VCAA.*

MC Let $f'(x) = 3x^2 - 2x$ such that $f(4) = 0$.

The rule of f is

A. $f(x) = x^3 - x^2$

B. $f(x) = x^3 - x^2 + 48$

C. $f(x) = x^3 - x^2 - 48$

D. $f(x) = 6x - 2$

E. $f(x) = 6x - 24$

Question 2 (1 mark) TECH-ACTIVE

MC The graph of the function $y = f(x)$ is shown.

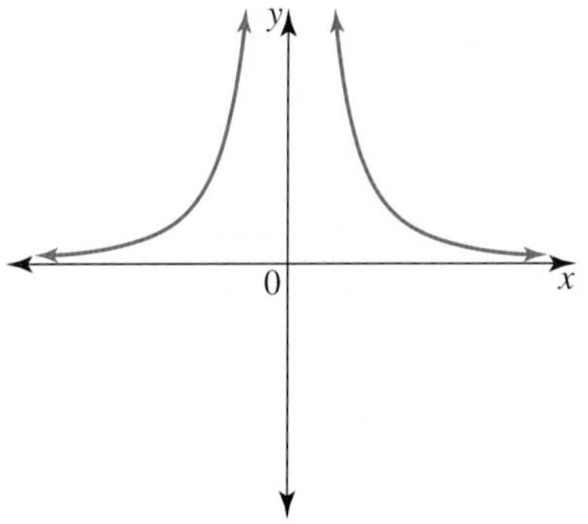

Select the graph from the following that could be the graph of the anti-derivative.

A.

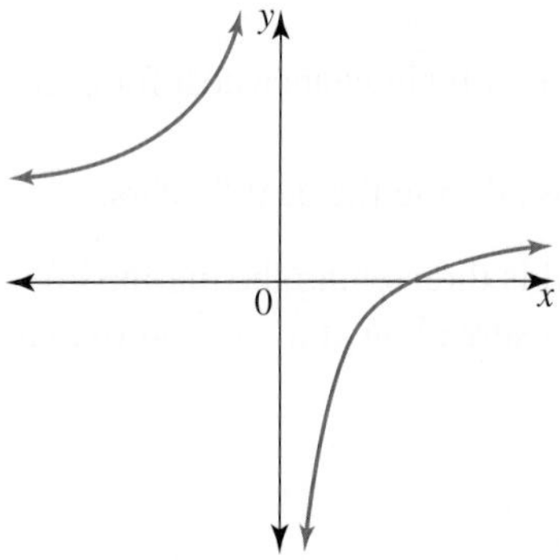

B.

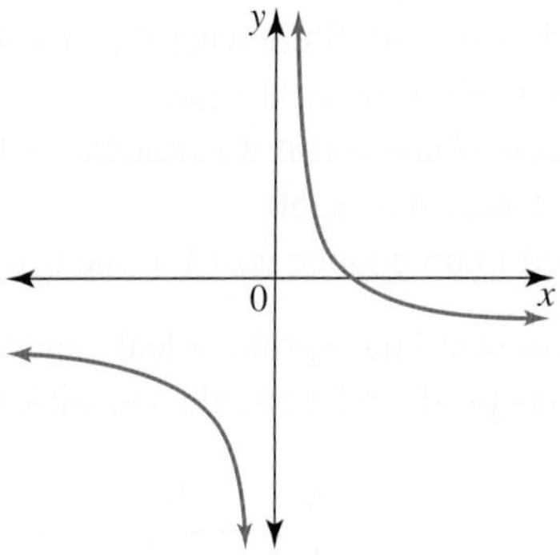

C.

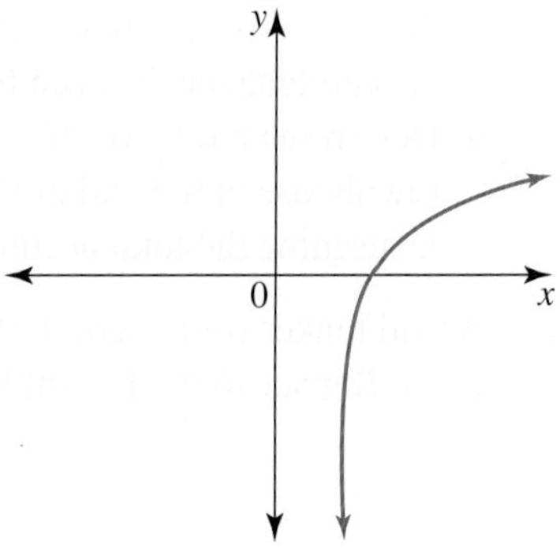

D.

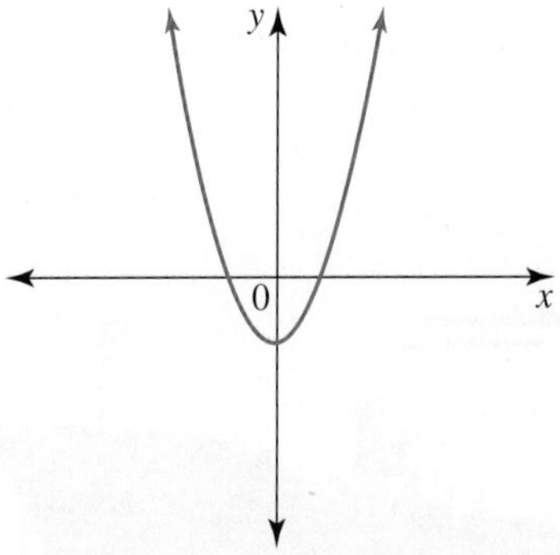

E.

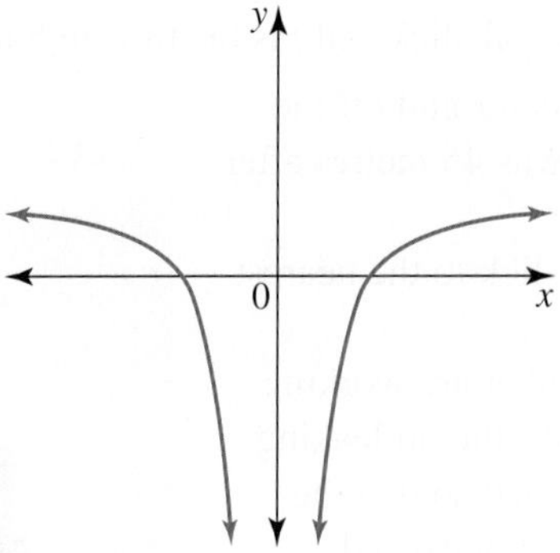

Question 3 (1 mark) TECH-ACTIVE

MC An anti-derivative of $\int \frac{1}{(3x-4)^{\frac{5}{2}}} dx$ is

A. $\frac{1}{3(x-4)^{\frac{3}{2}}}$

B. $\frac{-3}{2(3x-4)^{\frac{3}{2}}}$

C. $\frac{-9}{2(3x-4)^{\frac{3}{2}}}$

D. $\frac{3}{(3x-4)^{\frac{3}{2}}}$

E. $\frac{-2}{9(3x-4)^{\frac{3}{2}}}$

Question 4 (1 mark) TECH-ACTIVE

MC The gradient of a curve is given by $6x - 1$. The curve passes through the point $(2, 1)$. The equation of the curve is given by

A. $y = 3x^2 - x$

B. $y = 3x^2 - x - 9$

C. $y = 3x^2 - 11$

D. $y = 6$

E. $y = 2x^2 - x - 3$

Question 5 (3 marks) TECH-FREE

Find the derivative of $x \sin(x)$ and hence find an anti-derivative of $x \cos(x)$.

More exam questions are available online.

Answers

Topic 7 Anti-differentiation

7.2 Anti-differentiation

7.2 Exercise

1. a. $-\frac{x^4}{2}+c$ b. $\frac{2}{5}x^{\frac{5}{4}}+c$

c. $\frac{3}{2x}+c$ d. $12x^{\frac{1}{3}}+c$

2. a. $\frac{3x^2}{4}-\frac{4x^3}{3}+\frac{x^4}{2}+c$ b. $6\sqrt{x}-x^4-\frac{1}{5x^2}$

c. $\frac{x^4}{2}-\frac{x^3}{3}-\frac{15x^2}{2}+c$ d. $\frac{3x^{\frac{7}{2}}}{7}-\frac{x^{\frac{3}{2}}}{3}+c$

3. a. $4\sqrt{x}-\frac{3}{x}+\frac{1}{4x^2}+c$

b. $\frac{1}{2}x^4-\frac{1}{3}x^3+\frac{1}{2}x^2+4x+c$

4. $f(x)=\frac{1}{3}x^3+\frac{1}{x}+c$

5. a. $\frac{1}{4}x^4+c$ b. $\frac{7}{3}x^3+\frac{1}{5x^2}+c$

c. $x^4-\frac{7}{3}x^3+x^2-x+c$ d. $\frac{16}{5}x^2\sqrt{x}+c$

6. a. $\frac{(3x-5)^6}{18}+c$ b. $-\frac{1}{3(2x-3)^{\frac{3}{2}}}+c$

c. $\frac{(2x+3)^5}{10}+c$ d. $\frac{1}{8(1-2x)^4}+c$

7. B

8. D

9. a. $\frac{27}{4}x^4-9x^3+\frac{9}{2}x^2-x+c$

b. $-\frac{1}{8x^2}+c$

c. $\frac{2}{7}x^{\frac{7}{2}}-\frac{15}{7}x^{\frac{7}{5}}+c$

d. $\frac{1}{2}x^2+\frac{2}{x}+c$

e. $\frac{4x^{\frac{5}{2}}}{5}-\frac{1}{2}x^2+c$

f. $-\frac{2(4-x)^{\frac{3}{2}}}{3}+c$

10. E

11. $\frac{8x^7}{7}+3x^4+6x-\frac{1}{2x^2}+c$

12. a. $\frac{x^3}{3}-\frac{4x^{\frac{5}{2}}}{5}+\frac{x^2}{2}+c$

b. $\frac{2}{5}x^2\sqrt{x}+2x\sqrt{x}+6\sqrt{x}-\frac{2}{\sqrt{x}}+c$

13. a. $2x^3+\frac{5}{2}x^2-6x$ b. $\frac{1}{2}x^2+x-\frac{1}{x}$

c. $\frac{4}{3}x^{\frac{3}{2}}-8x^{\frac{1}{2}}$ d. $\frac{1}{7}x^7-4x-\frac{4}{5x^5}$

e. $\frac{1}{4(1-4x)^2}$ f. $-\frac{2}{3(2x-3)^{\frac{3}{2}}}$

14. $\frac{2}{3}\sqrt{x^3+1}+c$

15. $\frac{4}{405}(3x+5)^{\frac{3}{2}}(135x^2-72x+35)+c$

16. $\frac{dy}{dx}=6(3x+1)(3x^2+2x-4)^2$

$\int(3x+1)(3x^2+2x-4)^2=\frac{1}{6}(3x^2+2x-4)^3$

17. $\frac{dy}{dx}=4\left(7+\frac{1}{2\sqrt{x}}+\frac{1}{2\sqrt{x^3}}\right)\left(7x+\sqrt{x}-\frac{1}{\sqrt{x}}\right)^3$

$\int\left(7+\frac{1}{2\sqrt{x}}+\frac{1}{2\sqrt{x^3}}\right)\left(7x+\sqrt{x}-\frac{1}{\sqrt{x}}\right)^3dx$

$=\frac{1}{4}\left(7x+\sqrt{x}-\frac{1}{\sqrt{x}}\right)^4$

18. $\frac{dy}{dx}=\frac{x}{\sqrt{x^2+1}}$

$\int\frac{5x}{\sqrt{x^2+1}}=5\sqrt{x^2+1}+c$

19. $\frac{dy}{dx}=8(5x+1)(5x^2+2x-1)^3$

$\int 16(5x+1)(5x^2+2x-1)^3dx=2(5x^2+2x-1)^4$

20. $\frac{dy}{dx}=\frac{15x^2+8x}{2\sqrt{5x^3+4x^2}}$

$\int\frac{15x^2+8x}{\sqrt{5x^3+4x^2}}dx=2\sqrt{5x^3+4x^2}$

7.2 Exam questions

Note: Mark allocations are available with the fully worked solutions online.

1. $\frac{1}{8(4-2x)^4}+c$

2. $\frac{3x^5}{5}+\frac{2}{x}+c$

3. D

7.3 Anti-derivatives of exponential and trigonometric functions

7.3 Exercise

1. a. $\frac{x^5}{5}+\frac{1}{4}e^{-4x}+c$ b. $\frac{1}{4}e^{2x}+\frac{4}{3}e^{-\frac{1}{2}x}+c$

2. $\frac{1}{6}e^{6x}-3e^x-\frac{3}{4}e^{-4x}+\frac{1}{9}e^{-9x}+c$

3. $e^x + 4e^{-\frac{x}{2}} - \frac{1}{2}e^{-2x} + c$

4. a. $\frac{1}{6}\sin(3x+4) + 8\cos\left(\frac{x}{2}\right) + c$

b. $\frac{3}{2}\sin\left(\frac{2x}{3}\right) - \frac{1}{8}\cos(5-2x)$

5. a. $-2\cos\left(\frac{x}{2}\right) - 6\sin\left(\frac{x}{2}\right) + c$

b. $f(x) = \frac{7}{2}\sin(2x) + \frac{1}{3}\cos(3x) + c$

6. a. $3e^{\frac{x}{2}} - 3\cos\left(\frac{x}{3}\right) + \frac{x^2}{6} + c$

b. $\frac{1}{4}\sin(4x) - e^{-3x} + c$

7. $-\frac{1}{4x} - \frac{2}{3\pi}\cos\left(\frac{3\pi x}{2}\right)$

8. $y = \frac{1}{2}\sin(2x) + \frac{e^{-3x}}{3} + c$

9. a. $\frac{2}{3}e^{3x} + \frac{1}{2}\cos(2x) + c$

b. $\frac{1}{2}e^x - \frac{1}{4}e^{-6x} + c$

c. $\frac{1}{4}\sin(2x+5) + e^{-x} + c$

d. $\frac{1}{2}e^{2x} - \frac{2}{3}e^{3x} + \frac{1}{4}e^{4x} + c$

10. $e^x - x + \log_e(e^x + 1) + c$

11. $a = -9, b = 3$

12. $a = -2, b = -4, m = 2, n = -2$

13. a. $H(t) = t - 5\pi\cos\left(\frac{\pi t}{45}\right)$

b. $H = 7.146$ kilojoules

14. $x = \frac{\pi}{2}, \frac{3\pi}{2}$

15. $\frac{dy}{dt} = \frac{\pi}{20}\cos\left(\frac{\pi t}{4}\right)$ $\quad y = \frac{1}{5}\sin\left(\frac{\pi t}{4}\right) + c$

16. $\frac{dy}{dx} = -2\sin(x)\cos(x)e^{\cos^2(x)}$

$\int \sin(x)\cos(x)e^{\cos^2(x)} = -\frac{1}{2}e^{\cos^2(x)}$

17. $\frac{dy}{dx} = 3(x+1)^2e^{(x+1)^3}$

$\int 9(x+1)^2e^{(x+1)^3}\,dx = 3e^{(x+1)^3} + c$

18. A

19. $\frac{dy}{dx} = (4x+3)e^{2x^2+3x-1}$

$\int 2(4x+3)e^{2x^2+3x-1}dx = 2e^{2x^2+3x-1}$

20. $\frac{dy}{dx} = -x\sin(x) + \cos(x)$

$\int x\sin(x)\,dx = \sin(x) - x\cos(x)$

7.3 Exam questions

Note: Mark allocations are available with the fully worked solutions online.

1. D
2. A
3. D

7.4 The anti-derivative of $f(x) = \frac{1}{x}$

7.4 Exercise

1. a. $\frac{2}{5}\log_e(x) + c, x > 0$ $\quad$ b. $\frac{3}{4}\log_e(4x-1) + c$

2. a. $-4\log_e(x) + c, x > 0$

b. $\frac{3}{4}\log_e(4x+7) + c, x > -\frac{7}{4}$

c. $\frac{1}{2}x^2 + 2x + 3\log_e(x) + \frac{1}{x} + c, x > 0$

d. $-3\log_e(2-x) + \frac{1}{4}\sin(4x) + c, x < 2$

3. a. $-\frac{3}{2}\log_e(1-2x) + c$ $\quad$ b. $2\log_e(x+4) + c$

4. $\int \tan(2x)\,dx = -\frac{1}{2}\log_e(\cos(2x))$

5. $f'(x) = 2\log_e(mx) + 2$ and

$\int \log_e(mx)\,dx = x\log_e(mx) - x + c$

6. $f'(x) = 3 + 3\log_3(x)$ and $\int 2\log_e(x)\,dx = 2x\log_e(x) - 2x$

7. $a = -3, b = 2$

8. B

9. D

10. $a = 4, b = 2, d = 5$

7.4 Exam questions

Note: Mark allocations are available with the fully worked solutions online.

1. $x + 2\log_e(x) + \frac{3}{x} + c, x > 0$
2. A
3. $x\log_e(x) - x$

7.5 Families of curves

7.5 Exercise

1. a. $f(x) = x^3 + c$

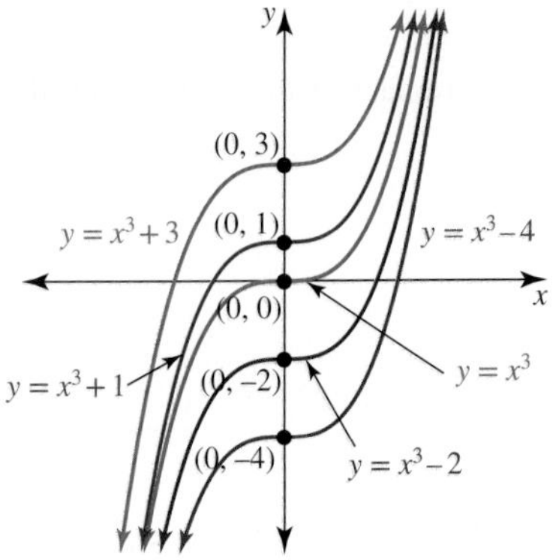

b. $f(x) = x^3 + 8$

2. a. $f(x) = -\sin(2x) + c$

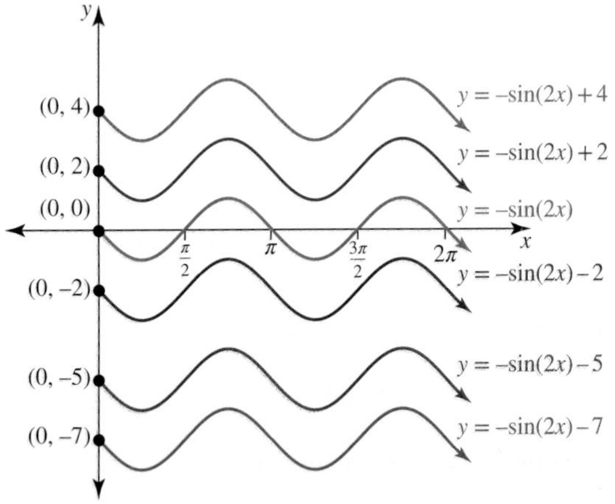

b. $f(x) = 4 - \sin(2x)$

3. a. $f(x) = -e^{-3x} + c$

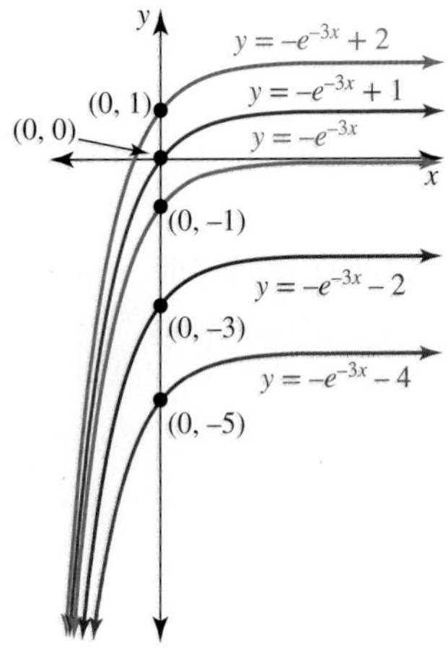

b. $f(x) = 2 - e^{-3x}$

4. $e^{2x} - e^{-x} + 3$

5. $\frac{1}{2}\sin(2x) + \frac{1}{2}\cos(2x) + \frac{3}{2}$

6. $y = \frac{1}{2}\sin(2x) - e^{-3x} + 5$

7. $y = 2e^{\frac{x}{2}} + 3$

8. $f(x) = \frac{1}{1-x} + 3$

9. a. $f(x) = 5x - x^2$

b. $f(x) = 3 - 2\cos\left(\frac{x}{2}\right)$

10. $f(x) = \frac{1}{4}x^4 - \log_e(x), x > 0$

11. a. $y = \frac{5}{2}\log_e(2(x+2)) + 3$

b. $y = -\frac{3}{5}\log_e(2 - 5x) + 1$

12.

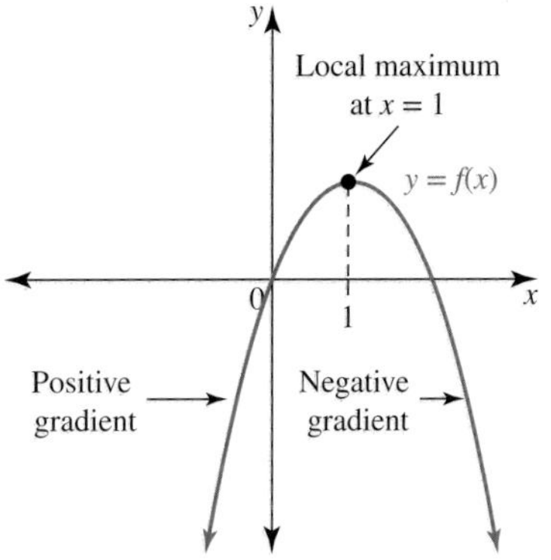

13.

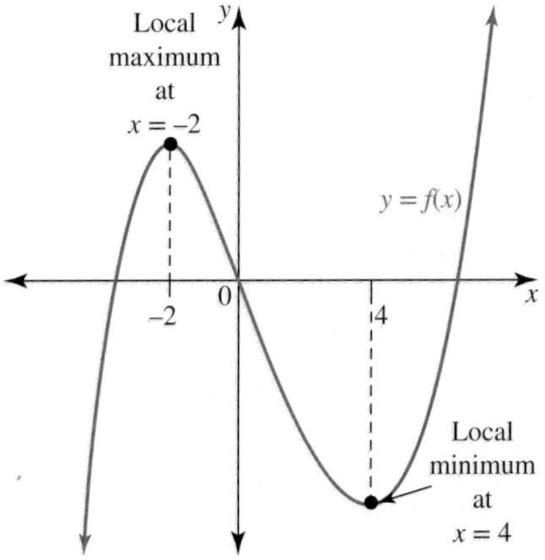

14. a.

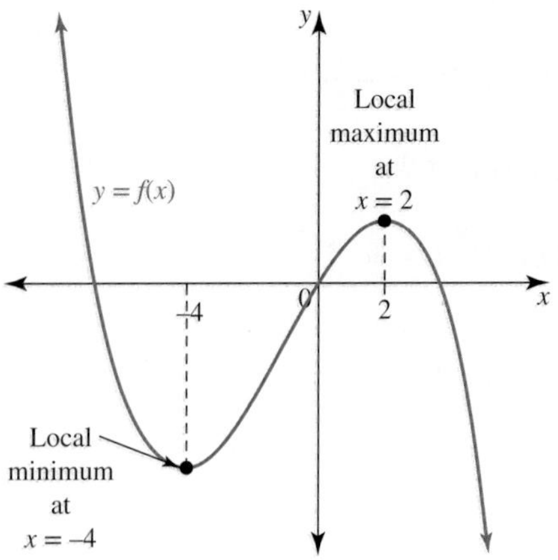

b.

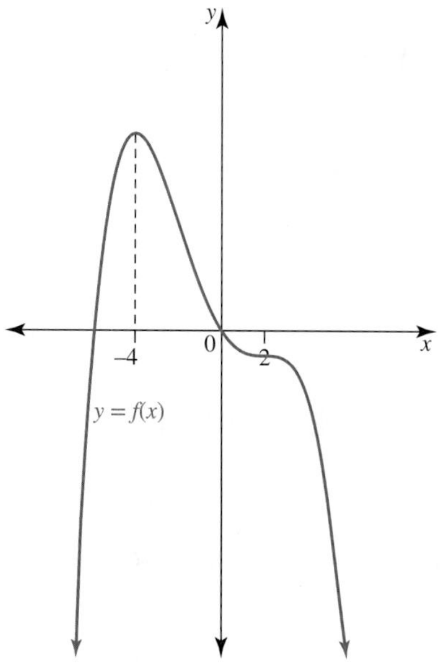

c.

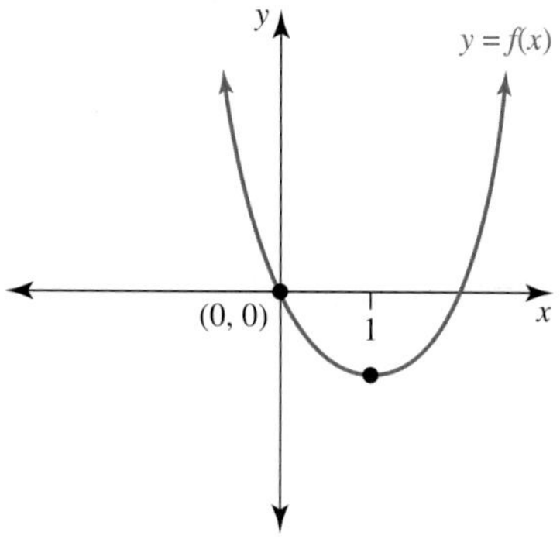

d.

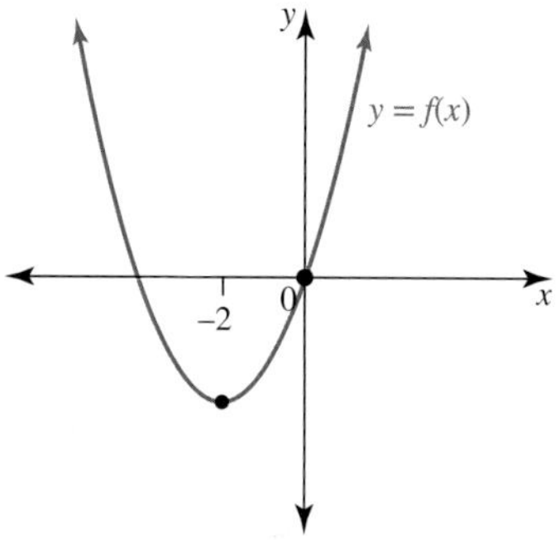

7.5 Exam questions

Note: Mark allocations are available with the fully worked solutions online.

1. C
2. B
3. $f(x) = 2\sin(x) + \frac{1}{2}\cos(2x) - 1$

7.6 Applications

7.6 Exercise

1. **a.** $x = \frac{2}{9}\sqrt{(3t+1)^3} - \frac{2}{9}$

   b. $x = \frac{1}{2} - \frac{1}{t+2}$

   c. $x = \frac{1}{8}(2t+1)^4 - \frac{1}{8}$

   d. $x = \frac{1}{3}e^{3t-1} - \frac{1}{3e}$

   e. $x = \frac{1}{2}\cos(2t+3) - \frac{1}{2}\cos(3)$

   f. $x = \frac{2}{3}\sin(3t)$

2. $x = t^3 + \frac{7}{2}t^2$

3. **a.** $x = 6t - \frac{12}{(t-1)} - 12$

   b. At the origin

4. **a.** $v = \sin(2t) + \cos(2t)$

   $t = 0 \Rightarrow v = \sin(0) + \cos(0)$
   $= 0 + 1$
   $= 1$ cm/s

   b. $x = -\frac{1}{2}\cos(2t) + \frac{1}{2}\sin(2t) + \frac{1}{2}$

5. **a.** $x = 24 - 24\cos\left(\frac{\pi t}{8}\right)$

   b. Maximum position = 48 metres

   c. After 4 seconds the particle is 24 metres above the stationary position.

6. **a.** 156.25 m/s **b.** $x = 6.25t^2 - \frac{1}{12}t^3$

7. $x = 2t\sin(t) + 2\cos(t) - 2$

8. **a.** $V = \frac{\pi}{3}r^3$ **b.** $\frac{64\pi}{3}\text{cm}^3$

9. $V = 13\,333\frac{1}{3}\text{ cm}^3$

10. **a.** $P = 100e^{0.3t} - 50$

    b. 1959 seals

11. **a.** $N = 400t + \frac{2000}{3}\sqrt{t^3} + 40$

    b. 9494 families

12. **a.** $h = 2\sin\left(\frac{\pi t}{4}\right) + 3$

    b. The minimum depth is 1 metre and the maximum depth is 5 metres.

    c. 8 hours a day

7.6 Exam questions

Note: Mark allocations are available with the fully worked solutions online.

1. **a.** $x = \frac{1}{2}t^4 - \frac{1}{2}t^2 + t - 1$

   b. 34 m to the right of the origin, velocity 52 m/s

2. $\frac{25}{3}$ m to the right of the origin, velocity 4 m/s

3. **a.** $b = 40t^{\frac{5}{2}} + 80$

   b. 2 hours, 34 minutes

7.7 Review

7.7 Exercise

Technology free: short answer

1. a. $\dfrac{4x^3}{3} + \dfrac{4}{x} + c$

   b. $\dfrac{e^{2x}}{2} - \dfrac{1}{2e^{2x}} - 2x + c$

   c. $e^x + 2x + c$

   d. $2\cos\left(\dfrac{x}{2}\right) - \dfrac{1}{12}\sin(4x - 3) + c$

   e. $-4\log_e(x - 3) - \cos(x) + c$

   f. $4\log_e(x) + x + \dfrac{1}{x} + c$

2. a. $T = -ae^{-bt} + c$ b. 95 °C

3. a. $f(x) = \dfrac{1}{4}e^{4x} + e^{2x} + x - \dfrac{5}{2}$

   b. $y = \dfrac{2}{9}\sqrt{(3x-5)^3} - \dfrac{2}{9}$

   c. $f(x) = 4\log_e(x) + \dfrac{x^2}{2} + \log_e(2)$

4. a. i.

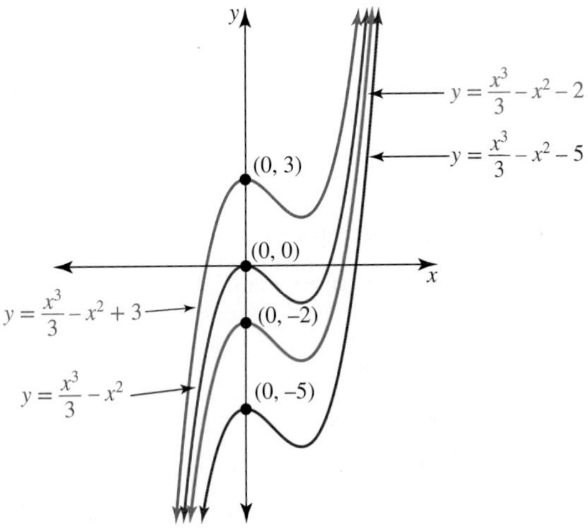

   ii.

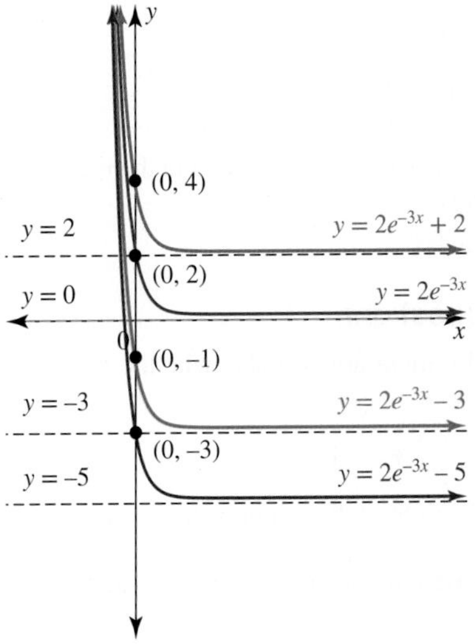

   b. i.

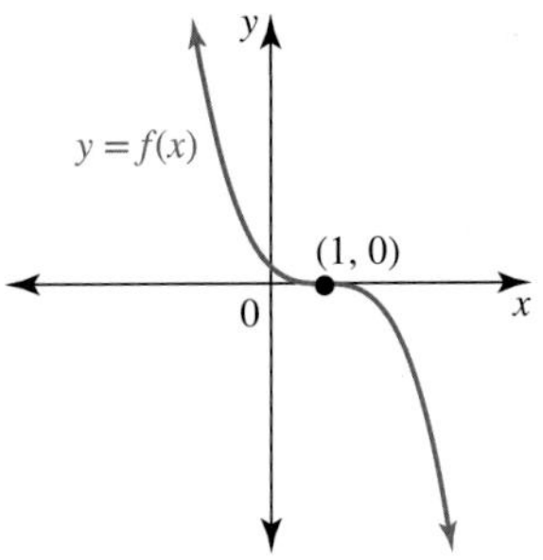

   ii.

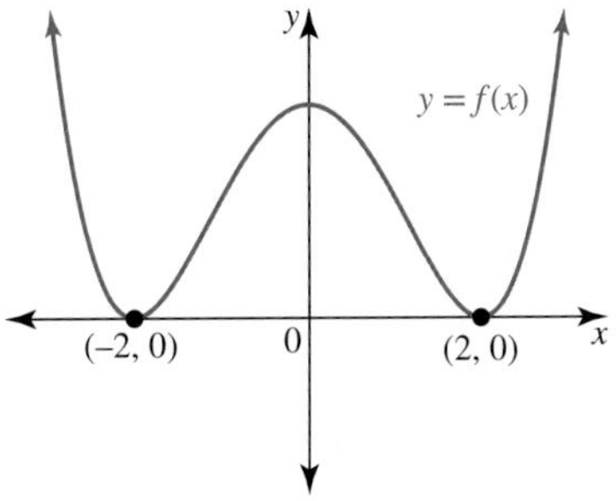

5. $V = 3.5t + \dfrac{1}{(t - 0.5)} + 2$

6. a. $\dfrac{dy}{dx} = 2(3x - 1)e^{3x^2 - 2x + 1}$,

   $\displaystyle\int (3x - 1)e^{3x^2 - 2x + 1}\,dx = \frac{1}{2}e^{3x^2 - 2x + 1}$

   b. $\dfrac{dy}{dx} = 2x\cos(2x) + \sin(2x)$,

   $\displaystyle\int x\cos(2x)\,dx = \frac{x\sin(2x)}{2} + \frac{\cos(2x)}{4}$

Technology active: multiple choice

7. D
8. E
9. A
10. A
11. D
12. C
13. A
14. B
15. A
16. D

Technology active: extended response

17. a. $v = \dfrac{dx}{dt} = 3\sin(3t) + 1$

   $x = \displaystyle\int (3\sin(3t) + 1)\,dt$

   $= -\cos(3t) + t + c$

   When $t = 0, x = 0$:

   $0 = -\cos(0) + 0 + c$

   $0 = -1 + c$

   $c = 1$

   $x = -\cos(3t) + t + 1$ as required.

   b. 1.160 s

c. 4 m/s

d. 4.911 m from the origin

18. a. When $t = 0$, $\frac{dT}{dt} = -8$.

$$\begin{aligned} -8 &= me^0 \\ m \times 1 &= -8 \\ m &= -8 \end{aligned}$$

b. $\frac{dT}{dt} = -8e^{nt}$

When $t = 10$, $\frac{dT}{dt} = -4$:

$$\begin{aligned} -4 &= -8e^{10n} \\ e^{10n} &= \frac{-4}{-8} \\ e^{10n} &= \frac{1}{2} \\ 10n &= \log_e\left(\frac{1}{2}\right) \\ 10n &= \log_e(2^{-1}) \\ n &= \frac{-\log_e(2)}{10} \end{aligned}$$

c. $T = \frac{80e^{\frac{-t\log_e(2)}{10}}}{\log_e(2)} + 98 - \frac{80}{\log_e(2)}$

d. 64 °C

19. a. \$1978

b. $M = 16a^3 + 250a + 84$

c. \$15 250

20. a. $r = 60\sqrt{t+1} - 75$

b. 72 metres

c. 5 hours and 15 minutes

7.7 Exam questions

Note: Mark allocations are available with the fully worked solutions online.

1. C
2. A
3. E
4. B
5. $\frac{d}{dx}(x\sin(x)) = \sin(x) + x\cos(x)$

$$\begin{aligned} \int (x\cos(x))\,dx &= x\sin(x) - \int \sin(x)\,dx \\ &= x\sin(x) + \cos(x) \end{aligned}$$

8 Integral calculus

LEARNING SEQUENCE

Fully worked solutions for this topic are available online.

8.1 Overview

Hey students! Bring these pages to life online

Watch videos | Engage with interactivities | Answer questions and check results

Find all this and MORE in jacPLUS

8.1.1 Introduction

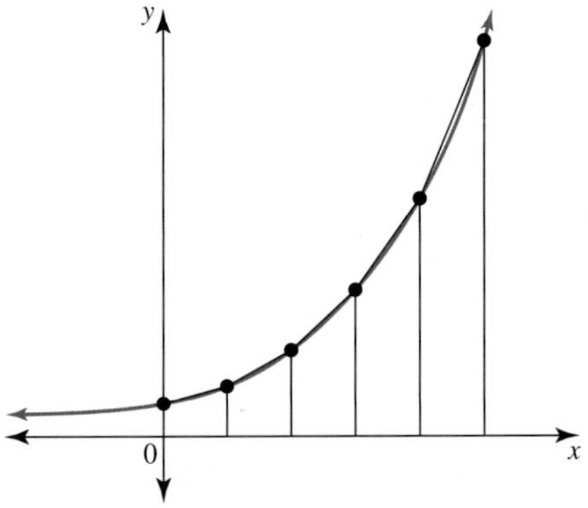

Integrals, in particular definite integrals, can be used to describe concepts such as area and volume.

Isaac Newton and Gottfried Leibniz both independently discovered the fundamental theorem of calculus in the late seventeenth century. They thought of the integral as an infinite sum of rectangles of infinitesimal width. The theorem also demonstrates the connection between integration and differentiation.

Following on from this, Riemann formalised integration using limits in 1854. Since then, it has been shown that his definition does not apply to all functions; however, he was the first to introduce rigour to the approach of integration.

Integrals are used extensively in mathematics, notably in the field of probability, where integrals are needed to calculate the probability of a value falling within a certain range. Volumes of three-dimensional objects with a curved boundary can be calculated, as well as the centre of mass of an object. Integration is also important in physics, as it is used to find position, velocity, time and work.

KEY CONCEPTS

This topic covers the following key concepts from the VCE Mathematics Study Design:

- informal consideration of the definite integral as a limiting value of a sum involving quantities such as area under a curve and approximation of definite integrals using the trapezium rule
- anti-differentiation by recognition that $F'(x) = f(x)$ implies $\int f(x)dx = F(x) + c$ and informal treatment of the fundamental theorem of calculus, $\int_a^b f(x)dx = F(b) - F(a)$
- properties of anti-derivatives and definite integrals
- application of integration to problems involving finding a function from a known rate of change given a boundary condition, calculation of the area of a region under a curve and simple cases of areas between curves, average value of a function and other situations.

8.2 The fundamental theorem of integral calculus

LEARNING INTENTION

At the end of this subtopic you should be able to:

- use the trapezium rule to approximate definite integrals
- calculate definite integrals using the fundamental theorem of calculus $\int_a^b f(x)\,dx = F(b) - F(a)$.

8.2.1 Estimation of the area under a curve

There are several different ways to approximate or estimate the area between a curve and the x-axis. This section will cover the **trapezium rule**.

Consider the curve defined by the rule $f: R \to R$, $f(x) = x^2 + 2$. Suppose we wish to know the area between this curve and the x-axis from $x = 0$ to $x = 3$. This can be achieved by constructing trapeziums of width 1 unit.

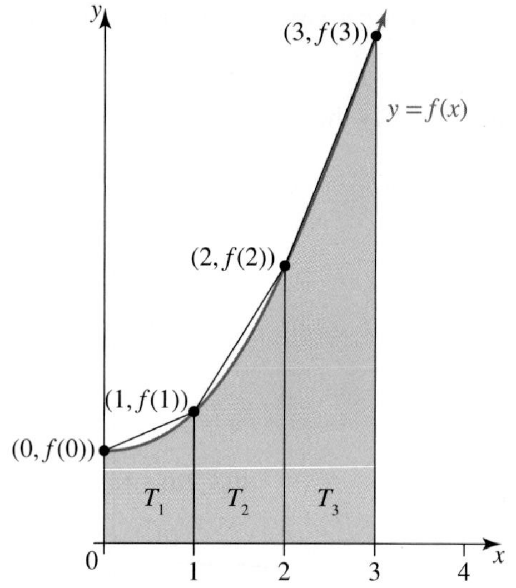

Remember that the formula for the area of a trapezium is $A = \frac{1}{2}(a + b)h$, where a and b are the lengths of the two parallel sides. The value of h is constant.

In the example, $h = 1$. The values of a and b are calculated by evaluating $f(0)$ and $f(1)$ for the first trapezium, $f(1)$ and $f(2)$ for the second trapezium, and $f(2)$ and $f(3)$ for the third trapezium.

$$
\begin{aligned}
A &= T_1 + T_2 + T_3 \\
&= \frac{1}{2}(f(0) + f(1)) \times 1 + \frac{1}{2}(f(1) + f(2)) \times 1 + \frac{1}{2}(f(2) + f(3)) \times 1 \\
&= \frac{1}{2}(2 + 3) \times 1 + \frac{1}{2}(3 + 6) \times 1 + \frac{1}{2}(6 + 11) \times 1 \\
&= \frac{5}{2} + \frac{9}{2} + \frac{17}{2} \\
&= \frac{31}{2} \\
&= 15.5 \text{ units}^2
\end{aligned}
$$

WORKED EXAMPLE 1 Approximating areas under curves

The graph of the function defined by the rule $f(x) = e^x$ is shown.

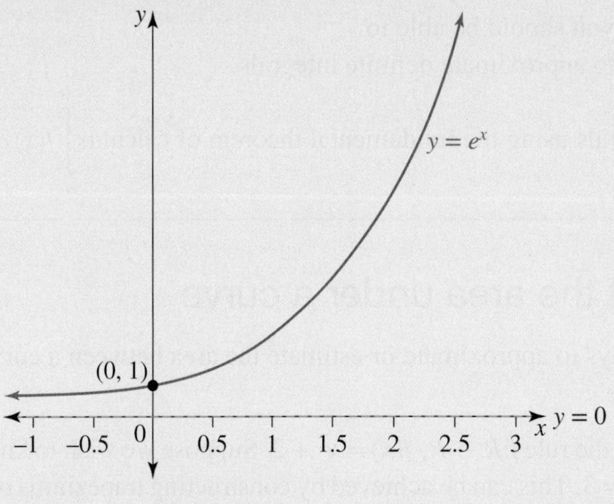

Use the trapezium rule with interval widths of 0.5 to approximate the area bound by the curve and the x-axis from $x = 0.5$ to $x = 2.5$. Give your answer correct to 2 decimal places.

THINK	WRITE
1. Draw the trapeziums on the graph. Identify the key points needed for the trapeziums.	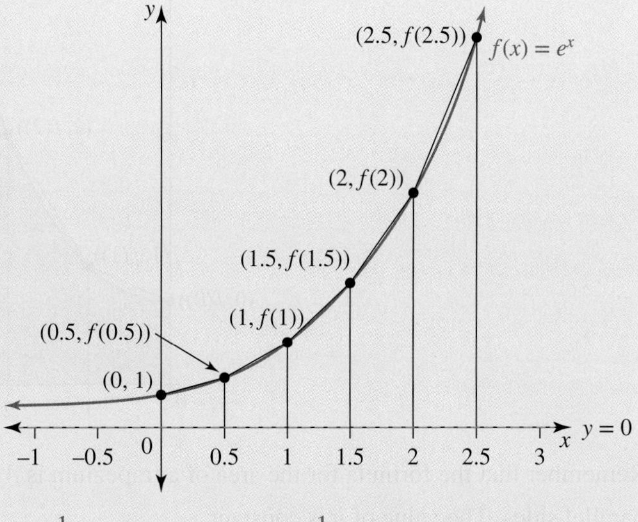
2. Determine the total approximate area by adding the areas of all the trapeziums.	$A = \frac{1}{2}(f(0.5) + f(1)) \times 0.5 + \frac{1}{2}(f(1) + f(1.5)) \times 0.5$ $+ \frac{1}{2}(f(1.5) + f(2)) \times 0.5 + \frac{1}{2}(f(2) + f(2.5)) \times 0.5$ $= \frac{1}{2}\left(e^{0.5} + e^{1}\right) \times 0.5 + \frac{1}{2}\left(e^{1} + e^{1.5}\right) \times 0.5$ $+ \frac{1}{2}\left(e^{1.5} + e^{2}\right) \times 0.5 + \frac{1}{2}\left(e^{2} + e^{2.5}\right) \times 0.5$ $= 10.75 \text{ units}^2$

8.2.2 The definite integral

The **definite integral**, $\int_a^b f(x)\,dx$, is similar to the indefinite integral, $\int f(x)\,dx$, except that it has end points, or terminals, a and b. The **indefinite integral** involves finding only an anti-derivative of f, but the presence of the end points means that the definite integral requires further calculation involving these values. In fact, the end points a and b indicate the range of the values of x over which the integral is taken.

The definite integral

$$\begin{aligned}\int_a^b f(x)\,dx &= [F(x)]_a^b \\ &= F(b) - F(a)\end{aligned}$$

Consider

$$\begin{aligned}\int_{-1}^{1} \left(1 - x^2\right) dx &= \left[x - \frac{1}{3}x^3\right]_{-1}^{1} \\ &= \left(1 - \frac{1}{3}(1)^3\right) - \left(-1 - \frac{1}{3}(-1)^3\right) \\ &= 1 - \frac{1}{3} + 1 - \frac{1}{3} \\ &= \frac{4}{3}\end{aligned}$$

Note: For the definite integral, no arbitrary constant is required for the anti-differentiation, as this would only be eliminated once the end points were used in the calculation.

$$\begin{aligned}\int_{-1}^{1} \left(1 - x^2\right) dx &= \left[x - \frac{1}{3}x^3 + c\right]_{-1}^{1} \\ &= \left(1 - \frac{1}{3}(1)^3 + c\right) - \left(-1 - \frac{1}{3}(-1)^3 + c\right) \\ &= 1 - \frac{1}{3} + c + 1 - \frac{1}{3} - c \\ &= 2 - \frac{2}{3} \\ &= \frac{4}{3}\end{aligned}$$

WORKED EXAMPLE 2 Calculating definite integrals

Evaluate:

a. $\displaystyle\int_0^{\frac{\pi}{2}} \cos(x)\, dx$ **b.** $\displaystyle\int_0^{2} \left(e^{-x}+2\right) dx$

THINK	WRITE
a. 1. Anti-differentiate the given function and specify the end points for the calculation using square brackets.	**a.** $\displaystyle\int_0^{\frac{\pi}{2}} \cos(x)\, dx = [\sin(x)]_0^{\frac{\pi}{2}}$
2. Substitute the upper and lower end points into the anti-derivative and calculate the difference between the two values.	$= \sin\left(\frac{\pi}{2}\right) - \sin(0)$ $= 1$
b. 1. Anti-differentiate the given function and specify the end points for the calculation using square brackets.	**b.** $\displaystyle\int_0^{2} (e^{-x}+2)\, dx = [-e^{-x}+2x]_0^2$
2. Substitute the upper and lower end points into the anti-derivative and calculate the difference between the two values.	$= \left(-e^{-2}+2(2)\right) - \left(-e^{0}+2(0)\right)$ $= -\frac{1}{e^2}+5$

TI | THINK

a. 1. On a Calculator page, press MENU, then select:
4: Calculus
3: Integral
Complete the entry line as:
$\displaystyle\int_0^{\frac{\pi}{2}} \cos(x)dx$
then press ENTER.
Note: Make sure your calculator is in Radians (RAD) mode.

DISPLAY/WRITE

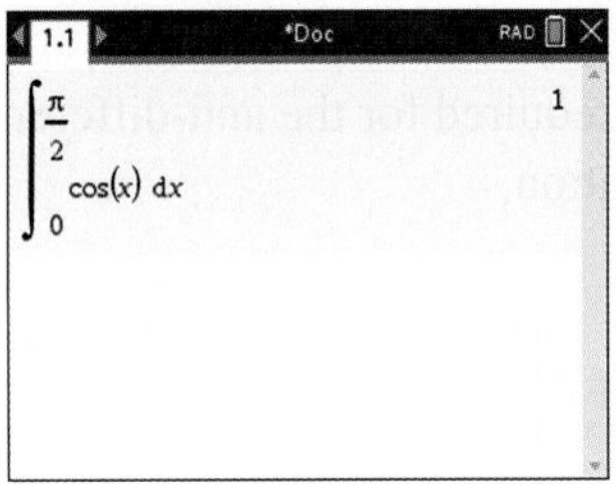

2. The answer appears on the screen.

$\displaystyle\int_0^{\frac{\pi}{2}} \cos(x)\, dx = 1$

CASIO | THINK

a. 1. On a Main screen, select:
- Interactive
- Calculation
- $\int$

Select 'Definite', then complete the fields as:
Expression: cos(x)
Variable: x
Lower: 0
Upper: $\pi/2$
and select OK.

DISPLAY/WRITE

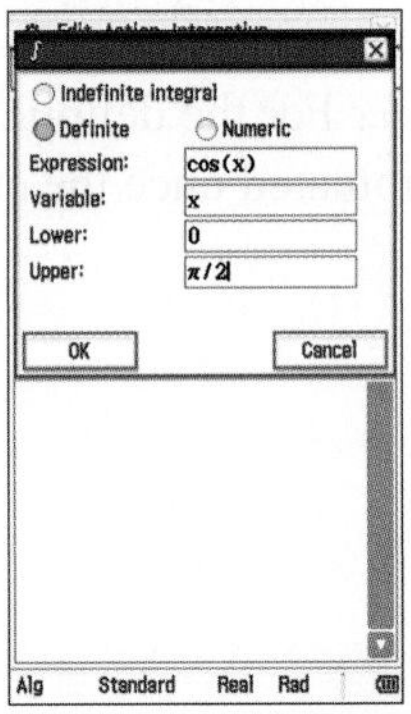

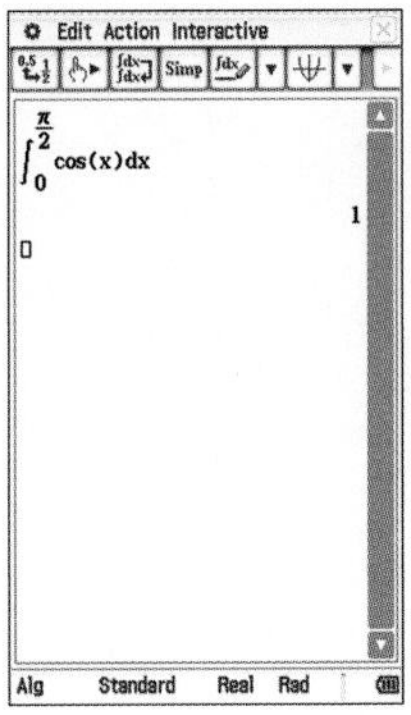

2. The answer appears on the screen.

$\displaystyle\int_0^{\frac{\pi}{2}} \cos(x)\, dx = 1$

b. 1. On a Calculator page, press MENU, then select:
4: Calculus
3: Integral
Complete the entry line as:

$$\int_0^2 (e^{-x} + 2)\,dx$$

then press ENTER.

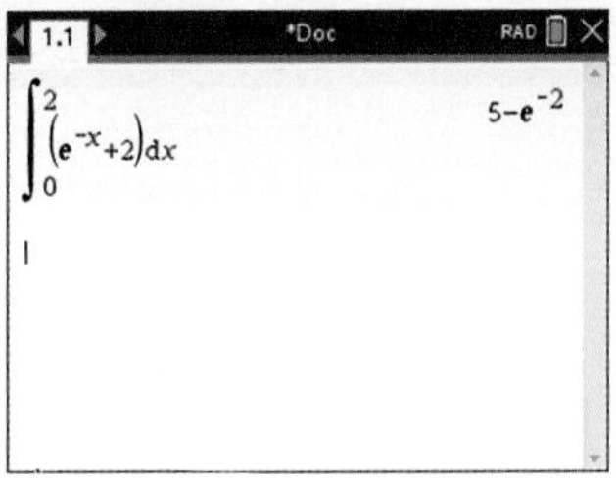

2. The answer appears on the screen.

$$\int_0^2 (e^{-x} + 2)\,dx = 5 - e^{-2}$$

b. 1. On a Main screen, select:

- Interactive
- Calculation
- $\int$

Select 'Definite', then complete the fields as:
Expression: $e^{-x} + 2$
Variable: x
Lower: 0
Upper: 2
and select OK.

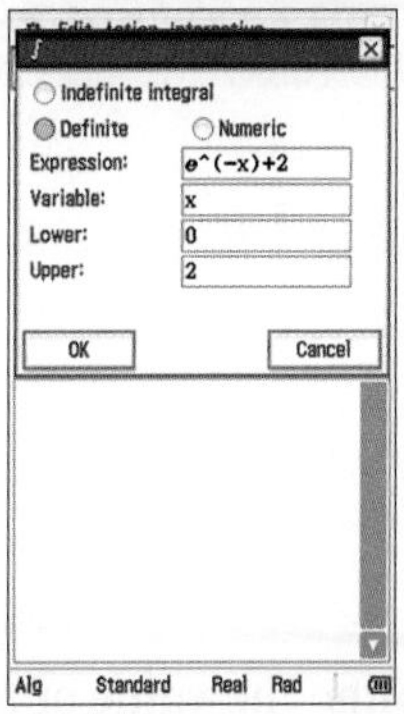

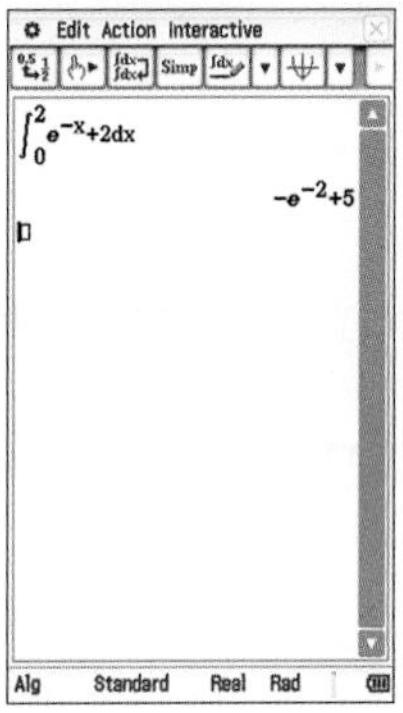

2. The answer appears on the screen.

$$\int_0^2 (e^{-x} + 2)\,dx = -e^{-2} + 5$$

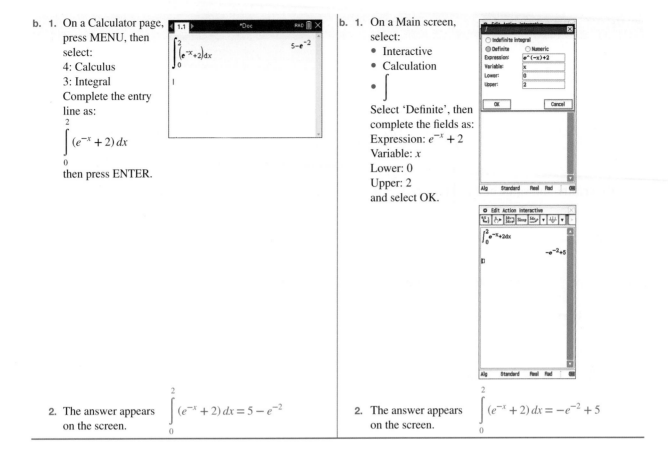

8.2.3 Properties of the definite integral

If f and g are continuous functions on an interval where $a < x < b$ and k is a constant, then the following rules apply.

Properties of the definite integral

$$\int_a^a f(x)\,dx = 0$$

$$\int_a^b f(x)\,dx = -\int_b^a f(x)\,dx$$

$$\int_a^b kf(x)\,dx = k\int_a^b f(x)\,dx$$

$$\int_a^b (f(x) \pm g(x))\,dx = \int_a^b f(x)\,dx \pm \int_a^b g(x)\,dx$$

$$\int_a^b f(x)\,dx = \int_a^c f(x)\,dx + \int_c^b f(x)\,dx, \text{ providing } a < c < b$$

Sometimes, definite integral questions take more of a theoretical approach to problem solving. Even if the function is unknown, we can use the properties of definite integrals to find the values of related integrals.

WORKED EXAMPLE 3 Definite integral properties

a. Given that $\int_1^3 f(x)\,dx = 8$, evaluate:

i. $\int_1^3 2f(x)\,dx$ ii. $\int_1^3 (f(x)+1)\,dx$ iii. $\int_3^1 f(x)\,dx$ iv. $\int_1^3 (f(x)-x)\,dx$.

b. Determine the value of k if $\int_1^k (x+2)\,dx = 0$.

THINK	WRITE
a. i. Apply the definite integral property $\int_a^b kf(x)\,dx = k\int_a^b f(x)\,dx$.	a. i. $\int_1^3 2f(x)\,dx = 2\int_1^3 f(x)\,dx$ $= 2\times 8$ $= 16$
ii. 1. Apply the definite integral property $\int_a^b (f(x) \pm g(x))\,dx = \int_a^b f(x)\,dx \pm \int_a^b g(x)\,dx$.	ii. $\int_1^3 (f(x)+1)\,dx = \int_1^3 f(x)\,dx + \int_1^3 1\,dx$
2. Integrate the second function and evaluate.	$= 8 + [x]_1^3$ $= 8 + (3-1)$ $= 10$
iii. Apply the definite integral property $\int_a^b f(x)\,dx = -\int_b^a f(x)\,dx$	iii. $\int_3^1 f(x)\,dx = -\int_1^3 f(x)\,dx$ $= -8$
iv. 1. Apply the definite integral property $\int_a^b (f(x) \pm g(x))\,dx = \int_a^b f(x)\,dx \pm \int_a^b g(x)\,dx$	iv. $\int_1^3 (f(x)-x)\,dx = \int_1^3 f(x)\,dx - \int_1^3 x\,dx$
2. Integrate the second function and evaluate.	$= 8 - \left[\frac{1}{2}x^2\right]_1^3$ $= 8 - \left(\frac{1}{2}(3)^2 - \frac{1}{2}(1)^2\right)$ $= 8 - \left(\frac{9}{2} - \frac{1}{2}\right)$ $= 8 - 4$ $= 4$

b. 1. Anti-differentiate and substitute the values of 1 and k.

b. $0 = \int_1^k (x+2)\,dx$

$0 = \left[\frac{1}{2}x^2 + 2x\right]_1^k$

$0 = \left(\frac{1}{2}k^2 + 2k\right) - \left(\frac{1}{2}(1)^2 + 2(1)\right)$

2. Simplify and solve for k.

$0 = \frac{1}{2}k^2 + 2k - \frac{5}{2}$

$0 = k^2 + 4k - 5$

$0 = (k+5)(k-1)$

$k = -5 \text{ or } k = 1$

3. Write the answer.

$k = 1, -5$

The fundamental theorem of integral calculus

In this section, the variable t is used and the function f is defined as a continuous function on the interval $[a, b]$ where $x \in [a, b]$. $A(x)$ is defined as

$$A\,(x) = \int_a^x f(t)\,dt$$

where $A(x)$ is the area between the curve $y = f(x)$ and the t-axis from $t = a$ to $t = x$. $A(x + \delta x)$ represents the area between the curve $y = f(t)$ and the t-axis from $t = a$ to $t = x + \delta x$.

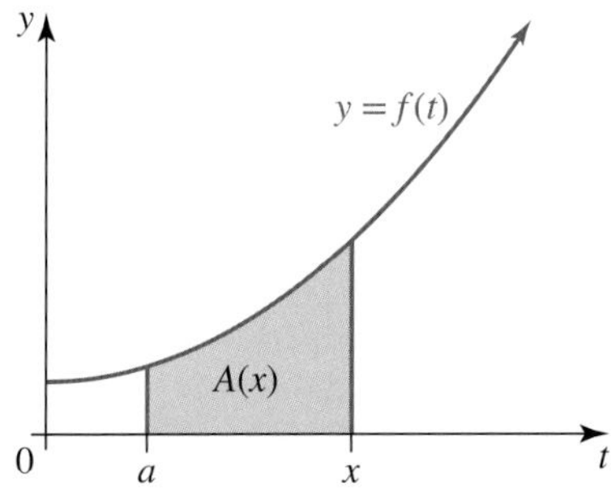

Area between the curve, the t-axis and the lines $t = a$ and $t = x$

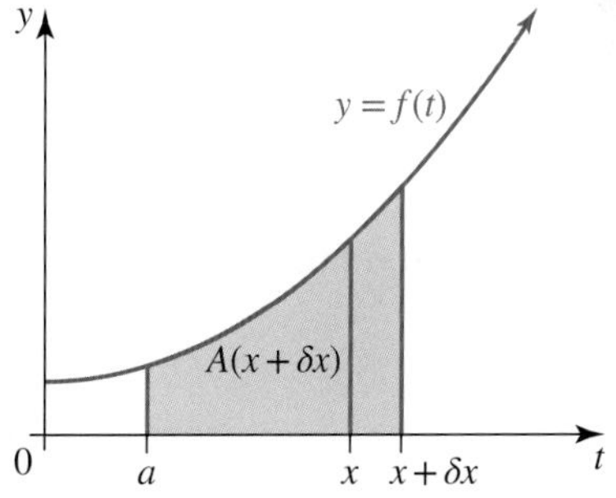

Area between the curve, the t-axis and the lines $t = a$ and $t = x + \delta x$

The difference between the areas is therefore $A(x + \delta x) - A(x)$. As $\delta x \to 0$, this results in an increasingly accurate area calculation between the curve and the t-axis. This concept is a limiting situation.

By definition,

$$\lim_{\delta x \to 0} \frac{A\,(x + \delta x) - A\,(x)}{\delta x} = \frac{d}{dx}(A\,(x))$$

and as $\delta x \to 0$, then $f(x + \delta x) \to f(x)$.

Consequently, we can say that

$$\frac{d}{dx}(A\ (x)) = f(x)$$

If we then integrate both sides with respect to x, we have

$$\int \frac{d}{dx}(A(x))\ dx = \int f(x)\ dx \qquad \text{or} \qquad A(x) = \int f(x)\ dx$$

To further investigate this theorem, we will let F be any anti-derivative of f, and A be the special anti-derivative defined as $\int_a^x f(t)\ dt$.

From our knowledge of anti-differentiation,

$$A(x) - F(x) = c \text{ where } c \text{ is a number.}$$

Therefore, $\int_a^x f(t)\ dt - F(x) = c$.

If we let $x = a$, then $\int_a^a f(t)\ dt = 0$, so

$$0 - F(a) = c \quad \text{or} \quad -F(a) = c.$$

Therefore, $\int_a^x f(t)\ dt - F(x) = -F(a)$.

If we now let $x = b$, then

$$\int_a^b f(t)\ dt - F(b) = -F(a) \quad \text{or} \quad \int_a^b f(t)\ dt = F(b) - F(a).$$

It is customary that $F(b) - F(a)$ is represented by $[F(x)]_a^b$.

Therefore,

$$\begin{aligned}\int_a^b f(t)dt &= [F(x)]_a^b \\ &= F(b) - F(a)\end{aligned}$$

as stated previously in section 8.2.2.

on Resources

Interactivity The fundamental theorem of integral calculus (int-6423)

8.2 Exercise

Technology free

1. **WE1** The graph of $y = \frac{1}{x}, x > 0$ is shown.

   Use the trapezium rule with interval widths of 0.5 to approximate the area bound by the curve and the x-axis from $x = 0.5$ to $x = 2.5$.

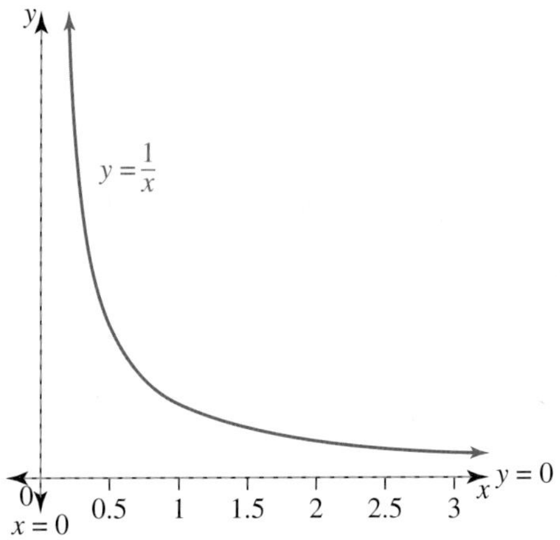

2. Given the function $f: [0, 4] \to R, f(x) = -(x - 1)^2 + 9$, use the trapezium rule to approximate the area between the curve and the x-axis from $x = 0$ to $x = 4$. Use interval widths of 1 unit.

3. The graph of the function $f(x) = 4 - \frac{1}{4}x^2$ is shown.

   Approximate the area bound by the curve and the x-axis using the trapezium rule with interval widths of 1 unit.

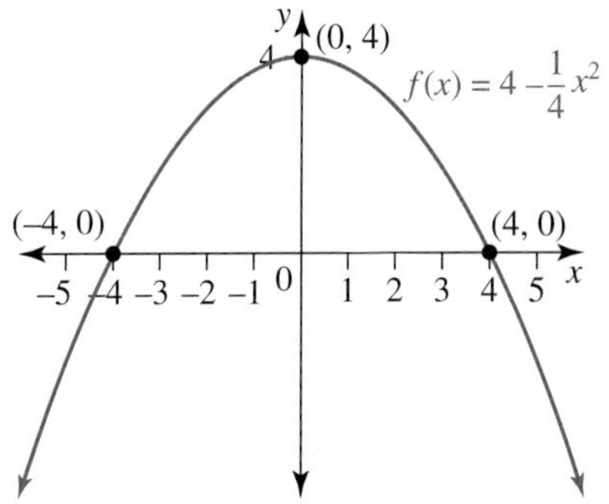

4. **WE2** Evaluate:

   a. $\int_0^1 (4x^3 + 3x^2 + 2x + 1)\, dx$
   b. $\int_{-\pi}^{\pi} (\cos(x) + \sin(x))\, dx$.

5. Evaluate:

   a. $\int_{-3}^{2} (x + 1)^3 dx$
   b. $\int_0^1 (e^x + e^{-x})^2\, dx$.

6. Evaluate the following.

   a. $\int_0^3 (3x^2 - 2x + 3)\, dx$
   b. $\int_1^2 \frac{2x^3 + 3x^2}{x}\, dx$
   c. $\int_{-1}^{1} \left(e^{2x} - e^{-2x}\right) dx$
   d. $\int_{2\pi}^{4\pi} \sin\left(\frac{x}{3}\right) dx$
   e. $\int_{-3}^{-1} \frac{2}{\sqrt{1 - 3x}}\, dx$
   f. $\int_{-\frac{\pi}{3}}^{\frac{\pi}{2}} \left[\cos(2x) - \sin\left(\frac{x}{2}\right)\right] dx$.

7. Differentiate $\log_e(3x^3 - 4)$ and hence evaluate $\int_2^3 \frac{x^2}{3x^3 - 4}\, dx$.

Technology active

8. **MC** Consider the function defined by the rule
$f: R \to R,\ f(x) = -0.01x^3(x-5)(x+5),\ x \geq 0.$
The graph of the function is shown.
Use the trapezium rule with interval widths of 1 unit to approximate the area bound by the curve and the x-axis.
The area is:
A. 9.12 units^2
B. 12 units^2
C. 11.52 units^2
D. 12.48 units^2
E. 13 units^2

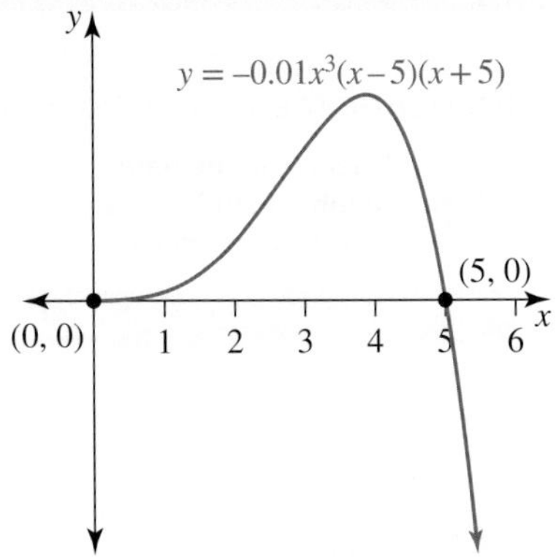

9. The graph of $f(x) = \sqrt{x}(4-x)$ for $x \in [0, a]$ is shown.
   a. The graph intersects the x-axis at the point $(a, 0)$ as shown. Find the value of the constant a.
   b. Use the trapezium rule to determine the approximate area between the curve and the x-axis from $x = 0$ to $x = a$. Use width intervals of 1 unit and give your answer correct to 2 decimal places.

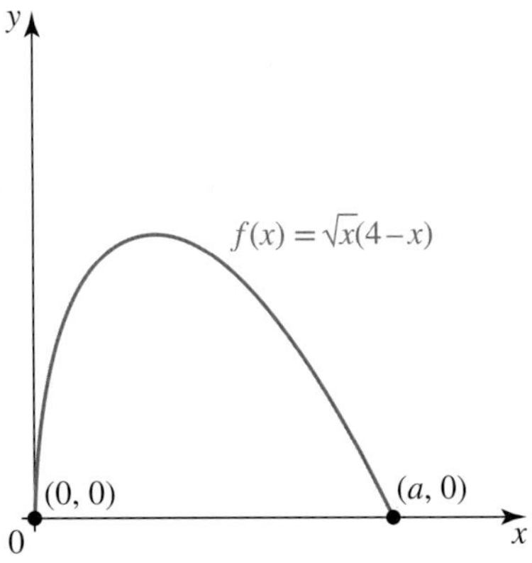

10. **WE3** a. Given that $\int_2^5 m(x)\,dx = 7$ and $\int_2^5 n(x)\,dx = 3$, evaluate:

   i. $\int_2^5 3m(x)\,dx$
   ii. $\int_2^5 (2m(x) - 1)\,dx$
   iii. $\int_5^2 (m(x) + 3)\,dx$
   iv. $\int_2^5 (2m(x) + n(x) - 3)\,dx.$

   b. Determine the value of k if $\int_k^1 \left(4x^3 - 3x^2 + 1\right) dx = 0.$

11. Given that $\int_0^5 f(x)\,dx = 7.5$ and $\int_0^5 g(x)\,dx = 12.5$, find:

   a. $\int_0^5 -2f(x)\,dx$
   b. $\int_5^0 g(x)\,dx$
   c. $\int_0^5 (3f(x) + 2)\,dx$
   d. $\int_0^5 (g(x) + f(x))\,dx$
   e. $\int_0^5 (8g(x) - 10f(x))\,dx$
   f. $\int_0^3 g(x)\,dx + \int_3^5 g(x)\,dx.$

12. **MC** If $\int_1^h \frac{3}{x^2}\,dx = -\frac{12}{5}$, the value of h is:

   A. $\frac{9}{5}$
   B. $\frac{5}{81}$
   C. 3
   D. $\frac{5}{9}$
   E. $\frac{81}{5}$

13. a. Determine a if $\int_0^a e^{-2x}dx = \frac{1}{2}\left(1 - \frac{1}{e^8}\right).$
   b. If $\int_1^k (2x - 3)\,dx = 7 - 3\sqrt{5}$, find k, given $k > 1$.

14. The graph of the function $f: R \to R, f(x) = x^3 - 8x^2 + 21x - 14$ is shown.

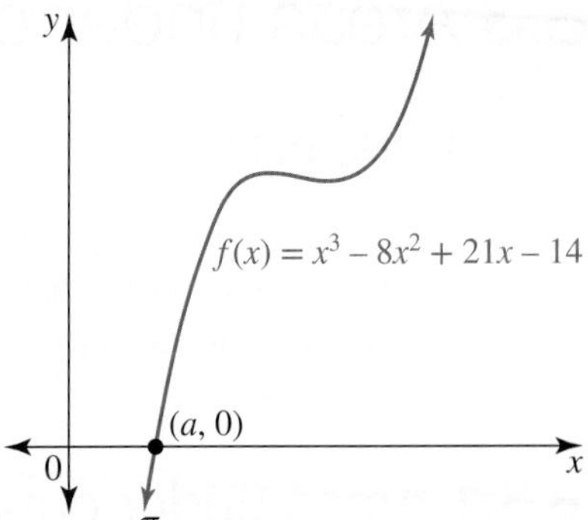

a. The graph cuts the x-axis at the point $(a, 0)$. Determine the value of the constant a.

b. Evaluate $\int_a^5 \left(x^3 - 8x^2 + 21x - 14\right) dx$.

15. Determine $\int_{-2}^{0} \frac{1 + e^{2x} - 2xe^{2x}}{\left(e^{2x}+1\right)^2} dx$, correct to 3 decimal places.

16. a. If $y = x\sin(x)$, find $\frac{dy}{dx}$.

b. Hence, find the value of $\int_{-\pi}^{\frac{\pi}{2}} 2x\cos(x)\, dx$.

17. a. If $y = e^{x^3 - 3x^2} + 2$, find $\frac{dy}{dx}$.

b. Hence, find the value of $\int_0^1 \left(x^2 - 2x\right) e^{x^3 - 3x^2} dx$.

18. **MC** Differentiate $y = \left(\log_e(x)\right)^2$. Hence, the value of $\int_1^e \frac{4\log_e(x)}{x} dx$ is:

A. 4 B. 1 C. −1 D. 2 E. 0

19. Differentiate $\log_e(e^x + 1)^2$ and hence find $\int_1^5 \frac{e^x}{e^x + 1} dx$, correct to 4 decimal places.

8.2 Exam questions

Question 1 (1 mark) TECH-ACTIVE

Source: VCE 2020, Mathematical Methods Exam 2, Section A, Q9; © VCAA.

MC If $\int_4^8 f(x)\,dx = 5$, then $\int_0^2 f(2(x+2))\,dx$ is equal to

A. 12 B. 10 C. 8 D. $\frac{1}{2}$ E. $\frac{5}{2}$

Question 2 (1 mark) TECH-ACTIVE

Source: VCE 2018, Mathematical Methods Exam 2, Section A, Q8; © VCAA.

MC If $\int_1^{12} g(x)\,dx = 5$ and $\int_{12}^{5} g(x)\,dx = -6$, then $\int_1^5 g(x)\,dx$ is equal to

A. −11 B. −1 C. 1 D. 3 E. 11

Question 3 (2 marks) TECH-FREE

Source: VCE 2015, Mathematical Methods (CAS) Exam 1, Q3; © VCAA.

Evaluate $\int_1^4 \left(\frac{1}{\sqrt{x}}\right) dx$.

More exam questions are available online.

8.3 Areas under curves

LEARNING INTENTION

At the end of this subtopic you should be able to:
- identify the relationship between a definite integral and area under a curve
- apply definite integrals to calculate the area under a curve over a specified interval.

8.3.1 Areas under curves

If we are interested in the area between a curve that is a continuous function, $y = f(x)$, and the x-axis between $x = a$ and $x = b$, then the following graph shows us exactly what we require.

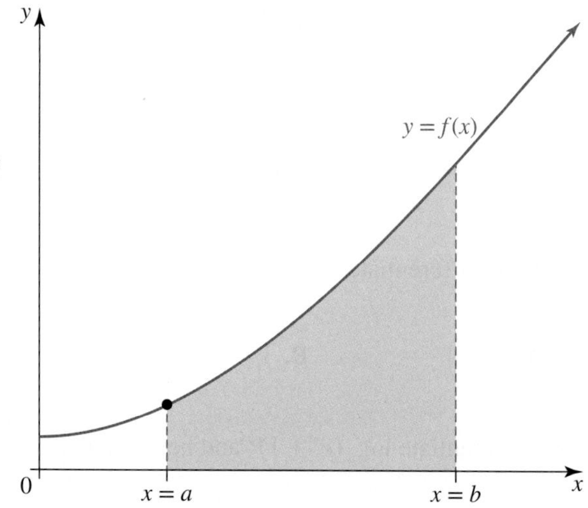

This area can be approximated by dividing it into a series of thin vertical strips or rectangles. The approximate value of the area is the sum of the areas of all the rectangles.

Suppose A represents the sum of the areas of all the rectangular strips between $x = a$ and $x = b$, where each strip has a width of δx.

Providing there is a very large number of rectangular strips so that δx is extremely small,

$$A = \lim_{\delta x \to 0} \sum_{x=a}^{x=b} y\,\delta x \qquad [1]$$

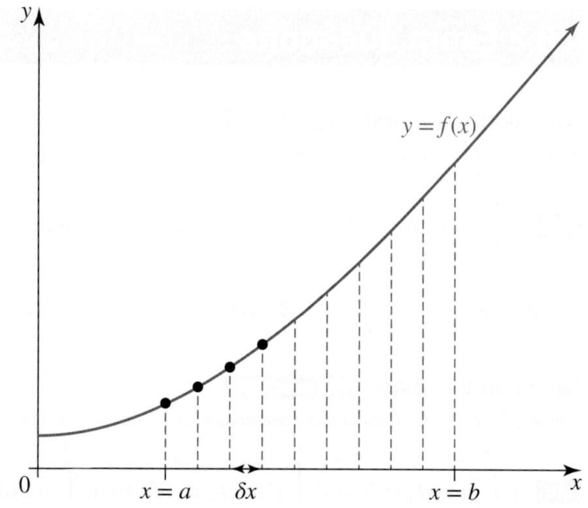

where $\sum_{x=a}^{x=b}$ means 'the sum from $x = a$ to $x = b$'.

Also, since each strip can have its area defined as $\delta A \simeq \delta x \times y$,

$$\frac{\delta A}{\delta x} \simeq y$$

Therefore, if the area under the curve is divided into a very large number of strips, then

$$\lim_{\delta x \to 0} \frac{\delta A}{\delta x} = y$$

$$\frac{dA}{dx} = y$$

This leads to the statement that

$$A = \int y\,dx$$

But since $x = a$ and $x = b$ are the boundary points or end points, then

$$A = \int_a^b y\,dx \qquad [2]$$

Equating [1] and [2], we have the following rule for the area under a curve.

Area under the curve

$$A = \lim_{\delta x \to 0} \sum_{x=a}^{x=b} y\,\delta x = \int_a^b y\,dx$$

This statement allows us to calculate the area between a curve and the x-axis from $x = a$ to $x = b$.

Consider the function defined by the rule $f: [0, 4] \to R, f(x) = -(x-1)^2 + 9$.

$$\begin{aligned} A &= \int_0^4 -(x-1)^2 + 9\,dx \\ &= \left[-\frac{(x-1)^3}{3} + 9x\right]_0^4 \\ &= \left(-\frac{(3)^3}{3} + 9(4)\right) - \left(-\frac{(-1)^3}{3} + 9(0)\right) \\ &= -9 + 36 - \frac{1}{3} + 0 \\ &= 27 - \frac{1}{3} \\ &= 26\frac{2}{3} \text{ units}^2 \end{aligned}$$

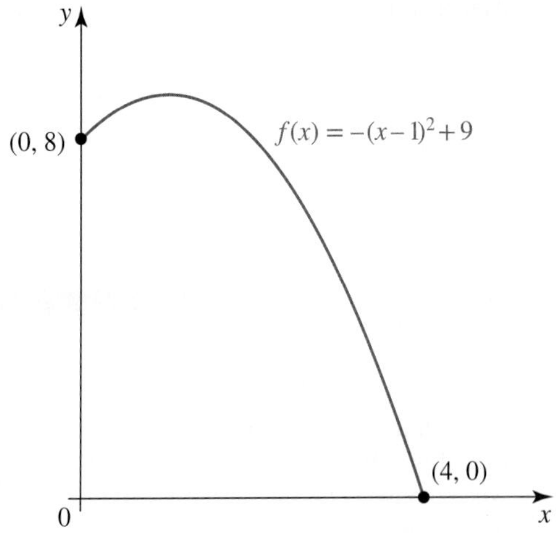

WORKED EXAMPLE 4 Determining areas under curves

Determine the area bound by the curve defined by the rule $y = e^{-x} + 3$ and the x-axis from $x = 0$ to $x = 3$.

THINK

1. Sketch the graph of the given function and shade the required area.

WRITE

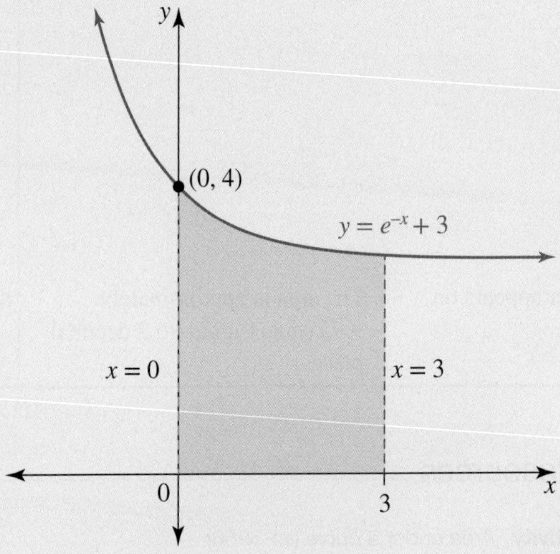

2. Write the integral needed to find the area. $A = \int_0^3 (e^{-x} + 3)\,dx$

3. Anti-differentiate the function and evaluate.
$$A = [-e^{-x} + 3x]_0^3$$
$$= \left(-e^{-3} + 3(3)\right) - \left(-e^{0} + 3(0)\right)$$
$$= -e^{-3} + 9 + 1$$
$$= -e^{-3} + 10$$

4. Write the answer. The area is $-e^{-3} + 10$ square units.

TI | THINK

1. On a Graphs page, complete the entry line for function 1 as:
$f1(x) = e^{-x} + 3$
then press ENTER.

DISPLAY/WRITE

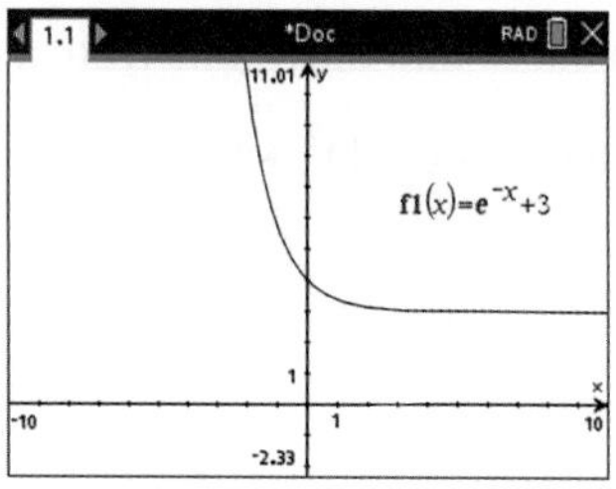

2. Press MENU, then select:
6: Analyze Graph
7: Integral
Type '0' when prompted for the lower bound, then press ENTER. Type '3' when prompted for the upper bound, then press ENTER.

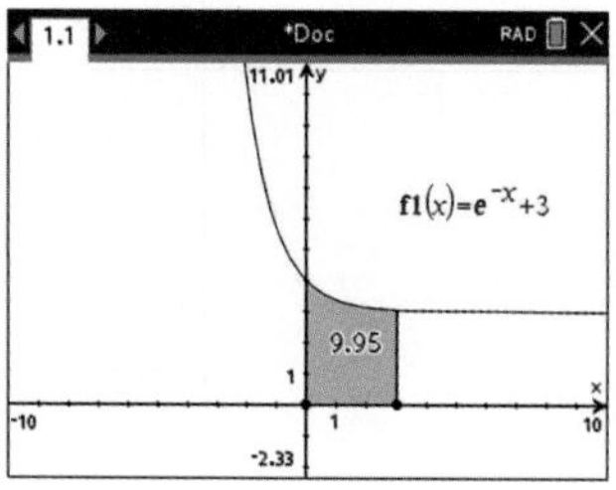

3. The answer appears on the screen.

The area is approximately 9.95 square units (to 2 decimal places).

CASIO | THINK

1. On a Graph & Table screen, complete the entry line for y1 as:
$y1 = e^{-x} + 3$
then press EXE.
Select the 'Graph' icon to draw the graph.

DISPLAY/WRITE

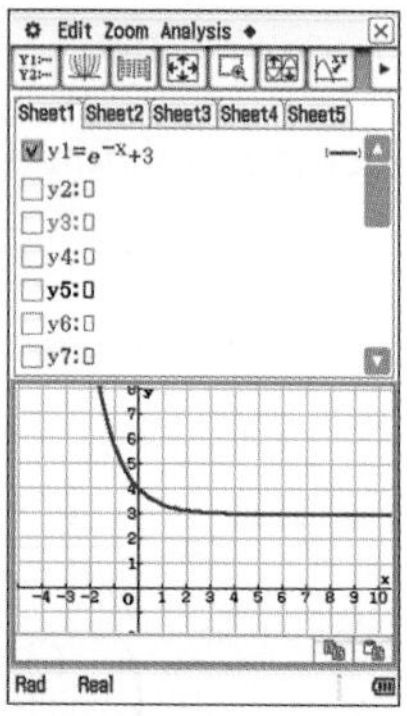

2. Select:
- Analysis
- G-Solve
- Integral
- $\int dx$

Press '0' to bring up the dialogue box, then complete the fields as:
Lower: 0
Upper: 3
and select OK.

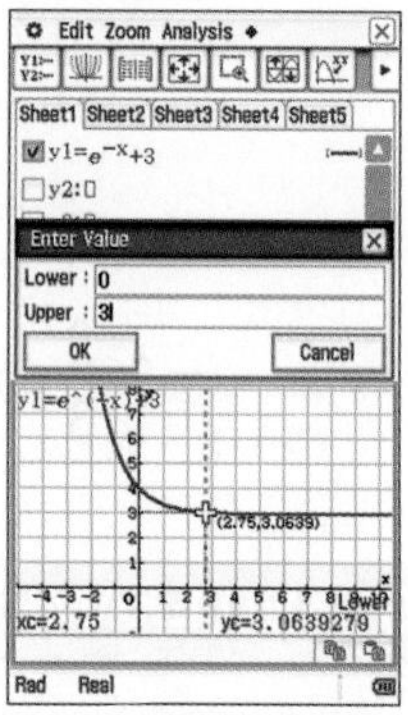

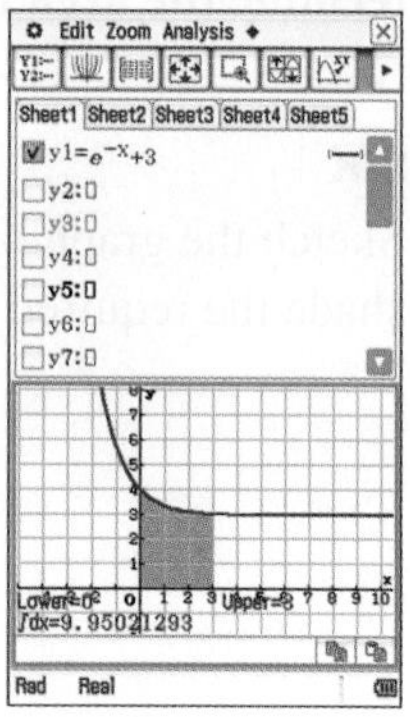

3. The answer appears on the screen.

The area is approximately 9.95 square units (to 2 decimal places).

on Resources

Interactivity Area under a curve (int-5966)

8.3.2 Signed areas

When we calculate the area between a graph $y = f(x)$ and the x-axis from $x = a$ to $x = b$ using the definite integral $\int_a^b f(x)\, dx$, the result can either be positive or negative.

Consider the function defined by the rule $f: R \rightarrow R$, $f(x) = x(3 - x)(x + 2)$, which is shown.

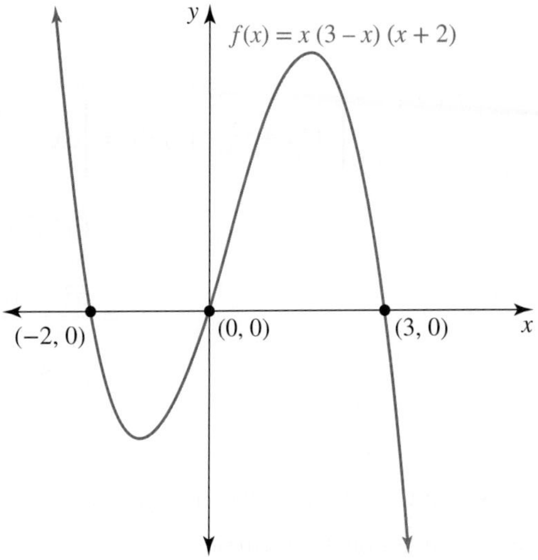

We can evaluate the area between $x = -2$ and $x = 0$ as follows.

$$
\begin{aligned}
\int_{-2}^{0} x(3-x)(x+2)\sim dx &= \int_{-2}^{0} \left(6x + x^2 - x^3\right) dx \\
&= \left[3x^2 + \frac{1}{3}x^3 - \frac{1}{4}x^4\right]_{-2}^{0} \\
&= 0 - \left(3(-2)^2 + \frac{1}{3}(-2)^3 - \frac{1}{4}(-2)^4\right) \\
&= 0 - \left(12 - \frac{8}{3} - 4\right) \\
&= -\left(8 - \frac{8}{3}\right) \\
&= -\frac{16}{3}
\end{aligned}
$$

This area is negative because the region lies below the x-axis.

$$
\begin{aligned}
\text{However, } \int_{0}^{3} x(3-x)(x+2)\, dx &= \int_{0}^{3} \left(6x + x^2 - x^3\right) dx \\
&= \left[3x^2 + \frac{1}{3}x^3 - \frac{1}{4}x^4\right]_{0}^{3} \\
&= \left(3(3)^2 + \frac{1}{3}(3)^3 + \frac{1}{4}(3)^4\right) - 0 \\
&= \left(27 + 9 - \frac{81}{4}\right) - 0 \\
&= \frac{63}{4}
\end{aligned}
$$

This area is positive as the region lies above the x-axis.

If we want an accurate answer for the area bound by the curve from $x = -2$ to $x = 3$, we counteract the negative region by subtracting it from the positive region. By subtracting the negative area, we are actually adding the area.

$$
\begin{aligned}
A &= \int_{0}^{3} x(3-x)(x+2)\, dx - \int_{-2}^{0} x(3-x)(x+2)\, dx \\
&= \frac{63}{4} - \left(-\frac{16}{3}\right) \\
&= \frac{253}{12} \\
&= 21\frac{1}{12} \text{ units}^2
\end{aligned}
$$

The total area bound by the curve, the x-axis and the lines $x=-2$ and $x=3$ is $21\frac{1}{12}$ square units.

This confirms the theory that if $f(x)>0$, then the region above the x-axis has a positive area, but if $f(x)<0$, then the region below the x-axis has a negative area.

Had we not broken up the interval and calculated $\int_{-2}^{3}\left(6x+x^2-x^3\right)dx$, the result would have been

$$\begin{aligned}\int_{-2}^{3}\left(6x+x^2-x^3\right)dx &= \left[3x^2+\frac{1}{3}x^3-\frac{1}{4}x^4\right]_{-2}^{3}\\ &= \left((3)^2+\frac{1}{3}(3)^3+\frac{1}{4}(3)^4\right)-\left(3(-2)^2+\frac{1}{3}(-2)^3-\frac{1}{4}(-2)^4\right)\\ &= \frac{63}{4}-\frac{16}{3}\\ &= 10\frac{5}{12}\end{aligned}$$

This result would not have given us the required area. The value of $10\frac{5}{12}$ is the value of the definite integral but not the area under the curve.

This shows that it is imperative to have a 'picture' of the function to determine when $f(x)>0$ and when $f(x)<0$; otherwise, we are just evaluating the definite integral rather than finding the necessary area.

The other method to account for the negative area is to switch the terminals within the integral for the negative region.

Calculation of area under a curve

The total area between the function $y=f(x)$ and the x-axis from $x=a$ to $x=b$ is given by:

$$\begin{aligned}A_{\text{total}} &= \int_{c}^{b} f(x)dx-\int_{a}^{c} f(x)dx\\ &= A_2-A_1\end{aligned}$$

or

$$\begin{aligned}A_{\text{total}} &= \int_{c}^{b} f(x)dx+\int_{c}^{a} f(x)dx\\ &= A_2+A_1\end{aligned}$$

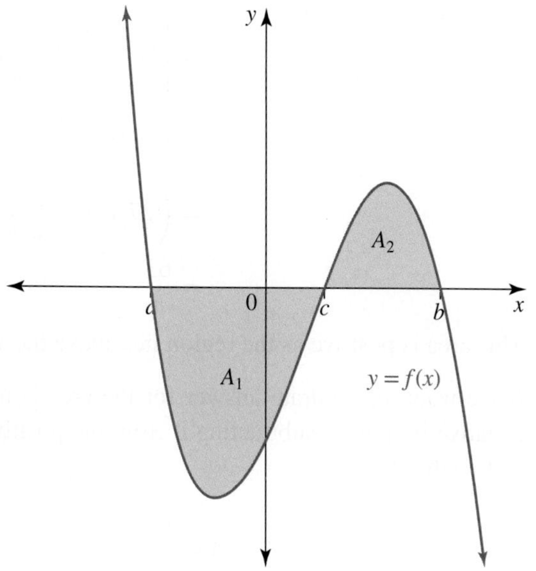

WORKED EXAMPLE 5 Calculating areas under curves

Determine the area bound by the curve $y = \left(x^2 - 1\right)\left(x^2 - 4\right)$ and the x-axis from $x = -2$ to $x = 2$.

THINK

WRITE

1. Make a careful sketch of the given function. Shade the required region.

The graph cuts the y-axis where $x = 0$.
$\therefore$ the y-intercept is (0, 4).
The graph cuts the x-axis where $y = 0$:

$$\left(x^2 - 1\right)(x-4)^2 = 0$$
$$(x-1)(x+1)(x-2)(x+2) = 0$$
$$x = \pm 1, x = \pm 2$$

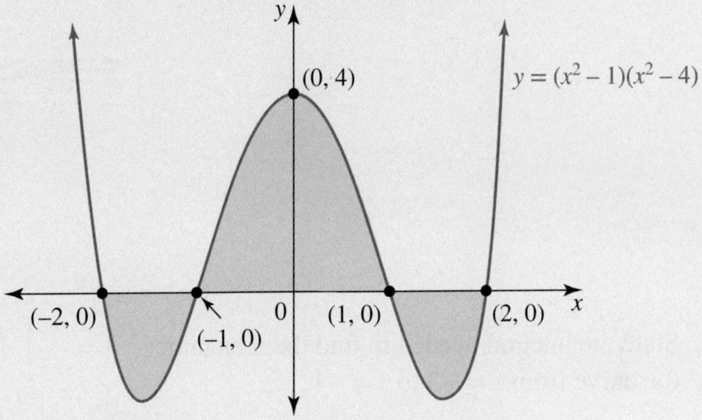

2. Express the area using definite integrals. Account for the negative regions by subtracting these from the positive areas. Note that the region from $x = -2$ to $x = -1$ is the same as the region from $x = 1$ to $x = 2$ due to the symmetry of the graph.

$$A = \int_{-1}^{1} \left(x^2 - 1\right)\left(x^2 - 4\right) dx - 2\int_{1}^{2} \left(x^2 - 1\right)\left(x^2 - 4\right) dx$$
$$= \int_{-1}^{1} \left(x^4 - 5x^2 + 4\right) dx - 2\int_{1}^{2} \left(x^4 - 5x^2 + 4\right) dx$$

3. Anti-differentiate and evaluate.

$$= \left[\frac{1}{5}x^5 - \frac{5}{3}x^3 + 4x\right]_{-1}^{1} - 2\left[\frac{1}{5}x^5 - \frac{5}{3}x^3 + 4x\right]_{1}^{2}$$
$$= \left(\frac{1}{5}(1)^5 - \frac{5}{3}(1)^3 + 4(1)\right) - \left(\frac{1}{5}(-1)^5 - \frac{5}{3}(-1) + 4(-1)\right)$$
$$- 2\left[\left(\frac{1}{5}(2)^5 - \frac{5}{3}(2)^3 + 4(2)\right) - \left(\frac{1}{5}(1)^5 - \frac{5}{3}(1)^3 + 4(1)\right)\right]$$
$$= \left(\frac{1}{5} - \frac{5}{3} + 4\right) - \left(-\frac{1}{5} + \frac{5}{3} - 4\right)$$
$$- 2\left[\left(\frac{32}{5} - \frac{40}{3} + 8\right) - \left(\frac{1}{5} - \frac{5}{3} + 4\right)\right]$$
$$= \frac{1}{5} - \frac{5}{3} + 4 + \frac{1}{5} - \frac{5}{3} + 4 - 2\left(\frac{32}{5} - \frac{40}{3} + 8 - \frac{1}{5} + \frac{5}{3} - 4\right)$$
$$= -\frac{60}{5} + \frac{60}{3}$$
$$= 8$$

4. Write the answer.

The area is 8 units^2.

WORKED EXAMPLE 6 Determining the area under a curve

Using calculus, calculate the area enclosed between the curve $y = \frac{2}{x}$, the x-axis and the lines $x = -3$ and $x = -1$.

THINK	WRITE
1. Sketch a graph of the required area.	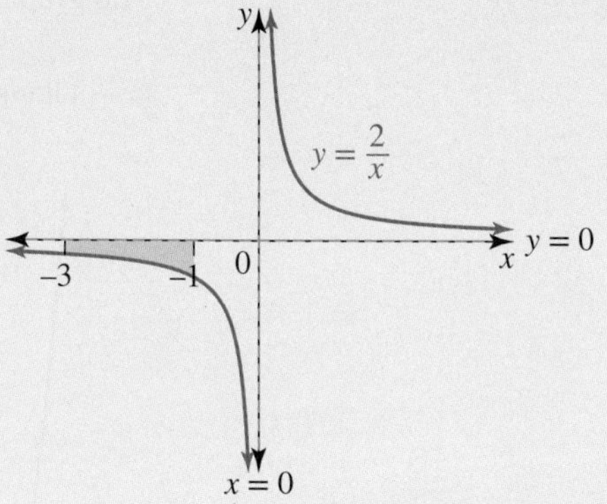
2. State the integral needed to find the area under the curve from $x = -3$ to $x = -1$.	$A = -\int_{-3}^{-1} \left(\frac{2}{x}\right) dx$
3. The integral of $y = \frac{1}{x}$ is $y = \log_e(x)$. Negative values cannot be substituted, so symmetry must be used to find the area.	$A = \int_{1}^{3} \left(\frac{2}{x}\right) dx$
4. Anti-differentiate and evaluate.	$= \left[2\log_e(x)\right]_1^3$ $= 2\log_e(3) - 2\log_e(1)$ $= 2\log_e(3)$ units^2

8.3 Exercise

Students, these questions are even better in jacPLUS

Receive immediate feedback and access sample responses

Access additional questions

Track your results and progress

Find all this and MORE in jacPLUS

Technology free

1. WE4 Determine the area bound by the curve defined by the rule $y = 2\sqrt{x}$, $x \geq 0$ and the x-axis from $x = 0$ to $x = 25$.

2. Determine the area bounded by the curve $y = 2\sin(2x) + 3$, the x-axis and the lines $x = 0$ and $x = \pi$.

3. The graph of $y = \frac{1}{x^2}$, $x < 0$ is shown.

Calculate the area of the shaded region (i.e. for $-2.5 \le x \le -0.5$).

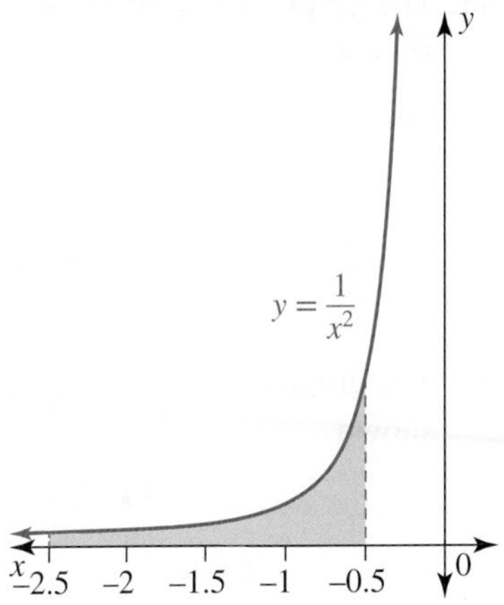

4. **WE5** Determine the area bound by the curve $y = 1 - e^{-x}$ and the x-axis from $x = -1$ to $x = 1$.

5. Sketch the graph of $y = \sqrt[3]{x}$ and hence calculate the area between the curve and the x-axis from $x = -8$ to $x = 8$.

6. **WE6** Using calculus, calculate the area enclosed between the curve $y = \frac{1}{x-1} + 2$, the x-axis and the lines $x = -2$ and $x = -1$.

Technology active

7. Consider the function defined by the rule $f: R \setminus \{0\} \to R, f(x) = \frac{1}{\sqrt{x}}$.
   a. Sketch the graph of f for $x > 0$.
   b. Using calculus, determine the area enclosed by the function, the lines $x = 1$ and $x = 3$, and the x-axis.

8. Consider the function $f: R \to R, f(x) = -\left(x^2 - 1\right)\left(x^2 - 9\right)$.
   a. Sketch the graph of f, showing the axis intercepts and turning points.
   b. Using calculus, calculate the area enclosed by the function, the lines $x = -3$ and $x = 3$, and the x-axis, correct to 2 decimal places.

9. **MC** Using calculus, the area, in units^2, enclosed between the curve $y = \frac{1}{x}$, the x-axis and the lines $x = -4$ and $x = -2$ is:

   A. $-\log_e(2)$
   B. 1
   C. $\log_e(2)$
   D. $\log_e(8)$
   E. undefined

10. The graph of the function $y = 2\sin(x) + 3\cos(x)$ is shown. Using calculus, determine the area between the curve and the x-axis from $x = 0$ to $x = 4\pi$.

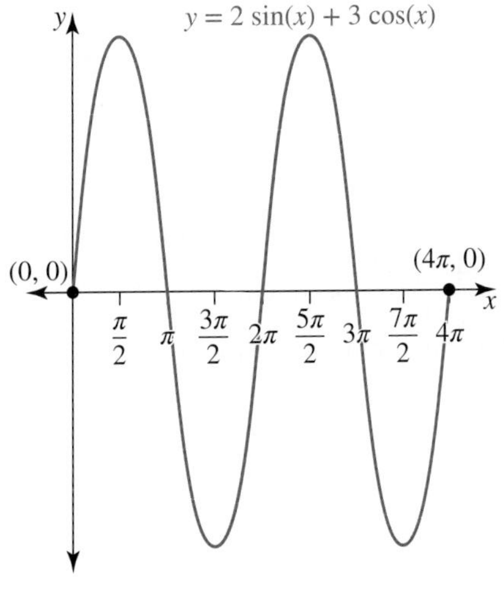

11. The graph of $y=-0.5(x+2)(x+1)(x-2)(x-3)$ is shown.

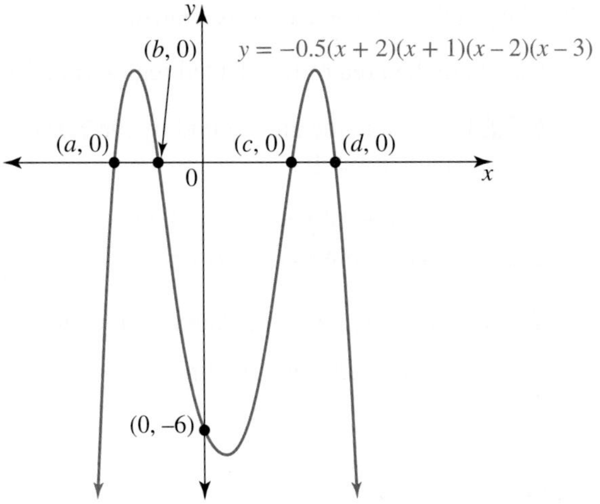

a. The graph intersects the x-axis at $(a,0)$, $(b,0)$, $(c,0)$ and $(d,0)$. Determine the values of the constants a, b, c and d.

b. Calculate the area between the curve and the x-axis from $x=a$ to $x=d$, correct to 2 decimal places.

12. The graph of $y=2\sin(x)+\cos(x)$ for $0\leq x\leq \pi$ is shown.

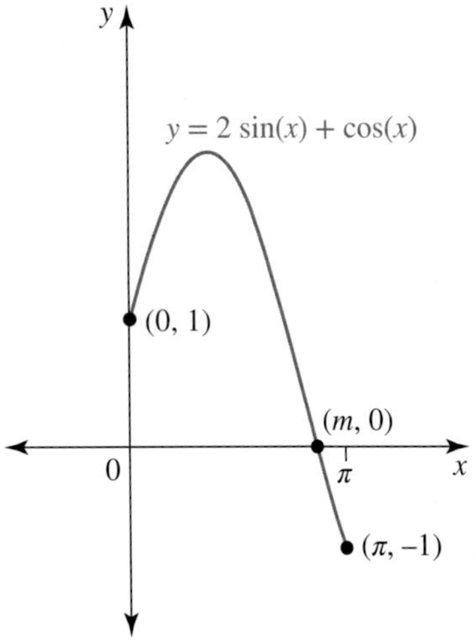

a. The graph intersects the x-axis at $(m,0)$. Determine the value of the constant m, correct to 4 decimal places.

b. Calculate $\int_0^m (2\sin(x)+\cos(x))\,dx$, correct to 4 decimal places.

13. a. The graph of $y=e^{-x^2}$ is shown. Determine the area between the curve and the x-axis from $x=-2$ to $x=2$, giving your answer correct to 4 decimal places.

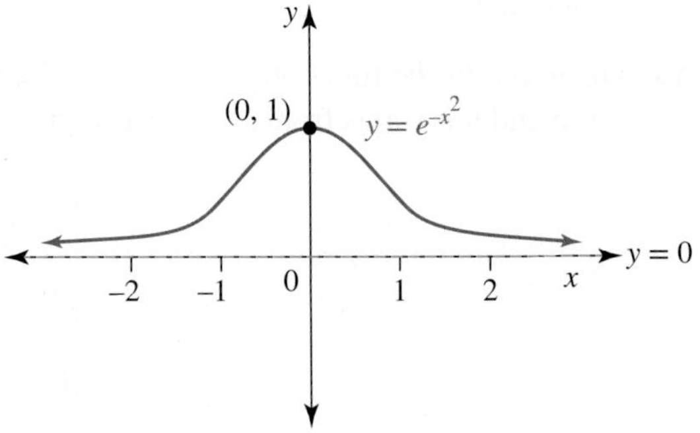

b. The graph of the function $y=\dfrac{x^2+3x-4}{x^2+1}$ is shown. Calculate the area, correct to 3 decimal places, between the curve and the x-axis from $x=-2$ to $x=3$.

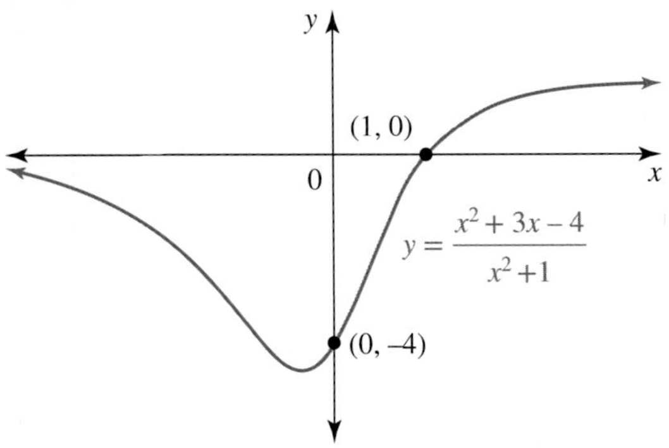

14. The graph of $y = 2xe^{x^2}$ is shown.

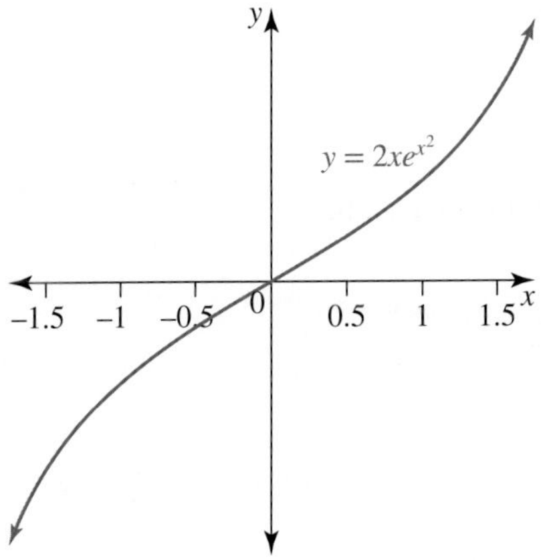

a. Calculate $\frac{d}{dx}\left(e^{x^2}\right)$.

b. Hence, find the exact area between the curve $y = 2xe^{x^2}$ and the x-axis from $x = -1$ to $x = 1$.

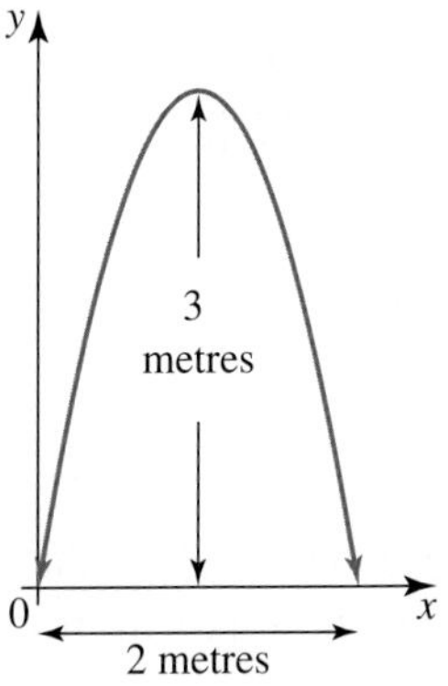

15. The 'Octagon Digital' store on the corner of two main roads in the north-eastern suburbs of a large Australian city has two very distinctive parabolic windows, each one facing one of the main roads. In the early hours of a Sunday morning, a motorist smashed through one of the windows. The owner decided it would be beneficial to replace both windows with strongly reinforced and quite heavily tinted glass. Each window has the dimensions shown in the diagram.

a. Determine the equation of the parabola that defines the shape of each window.

b. Calculate the area of glass required to replace each window.

c. If the cost per square metre of the replacement reinforced and tinted glass is \$55, evaluate the cost of replacing the two windows.

16. The graph of $f: R \to R, f(x) = 3x^3$ is shown.

a. Determine the area bounded by the curve and the x-axis from $x = 0$ to $x = 1$.

b. Hence, or otherwise, calculate the area of the shaded region.

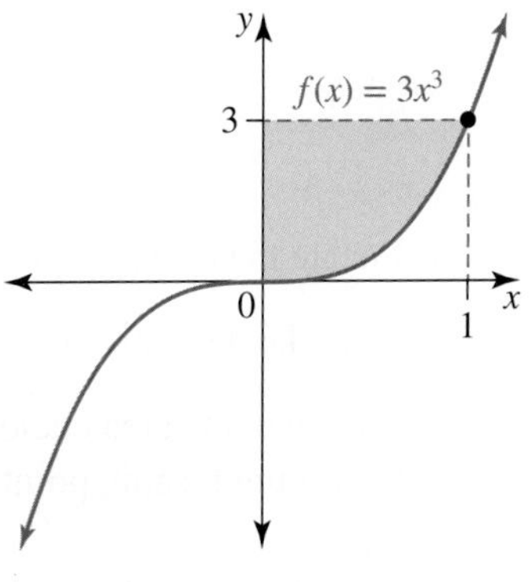

17. The graph of $y = 2\sin(x), -\frac{3\pi}{2} \le x \le \frac{3\pi}{2}$ is shown.

a. Calculate $\int_0^{\frac{\pi}{2}} 2\sin(x)\,dx$.

b. Hence, or otherwise, determine the area of the shaded region.

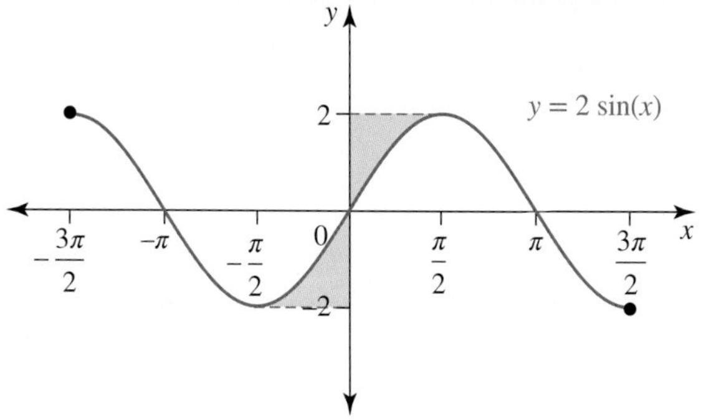

18. The graph of the function $f: (1, \infty) \to R, f(x) = 2\log_e(x-1)$ is shown.

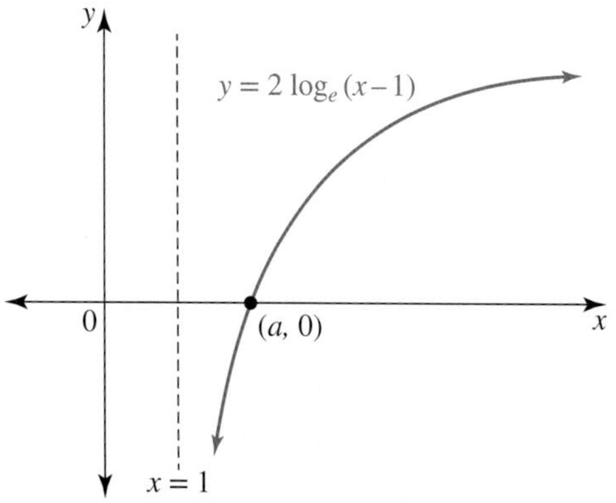

a. State the domain and range of f.

b. Calculate the value of the constant a, given that $(a, 0)$ is the x-axis intercept.

c. Determine the area between the curve and the x-axis from $x = a$ to $x = 5$, correct to 4 decimal places.

19. The graph of $y = \frac{10x}{5 + x^2}$ is shown.

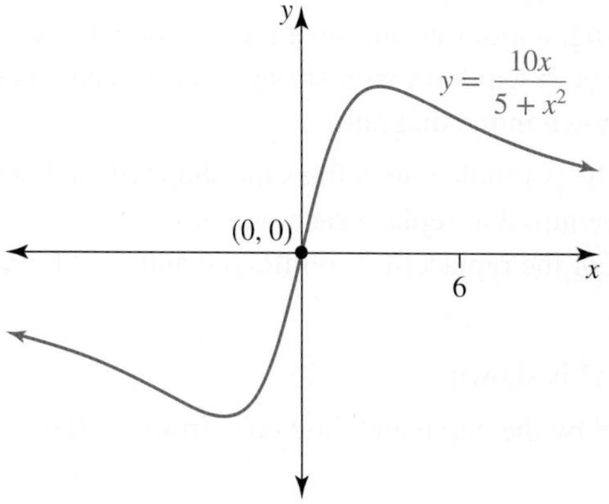

a. Locate the exact coordinates of the minimum and maximum turning points.

b. State the derivative of $\log_e(5 + x^2)$ and hence determine an anti-derivative for $\frac{10x}{5 + x^2}$.

c. Calculate the area enclosed between the curve, the x-axis, the line where x equals the x-coordinate of the maximum turning point, and the line $x = 6$.

20. The graph of the function $f: (3, \infty) \to R, f(x) = 5\log_e(x - 3)$ is shown.

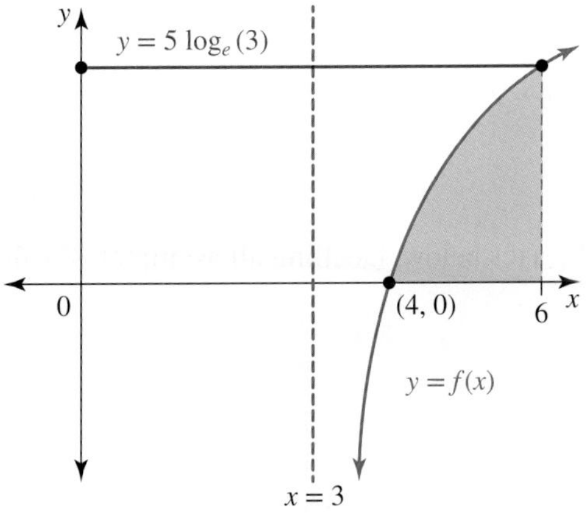

a. Calculate the area of the shaded region, correct to 3 decimal places.
b. Determine the rule for the inverse function, $y = f^{-1}(x)$.
c. Verify your answer to part a by calculating the area enclosed between the curve $y = f^{-1}(x)$, the y-axis and the line $y = 6$.

21. The Red Fish Restaurant is a new restaurant about to open. The owners commissioned a graphic artist to design a logo that will be seen on the menus and on advertisements for the restaurant, and will also be etched into the front window of the restaurant. The logo is shown in Figure 1.

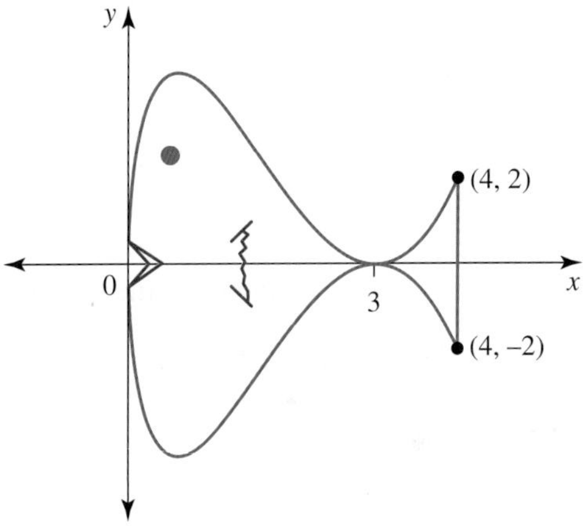

As the logo is to appear in a number of different scenarios, the owners need to know the area of the original to allow for enlargement or diminishing processes. The graphic artist formed the shape by using the rule

$$y = \sqrt{x}(x - 3)^2, 0 \le x \le 4$$

for the upper part of the fish and

$$y = -\sqrt{x}(x - 3)^2, 0 \le x \le 4$$

for the lower part of the fish.

The original outline is shown in Figure 2.

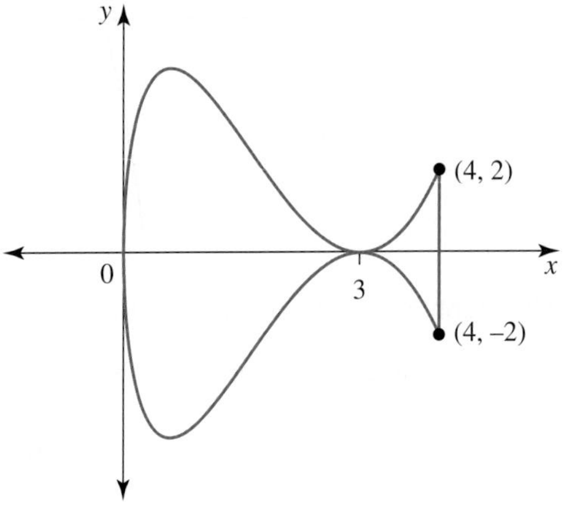

a. Calculate the area between the upper curve and the x-axis from $x = 0$ to $x = 4$, correct to 4 decimal places.
b. Calculate the area of the entire fish logo. (All measurements are in centimetres.) Give your answer correct to 1 decimal place.
c. The etched fish on the front window of the restaurant has an area of 0.348 75 m^2. Determine the scale factor, to the nearest integer, that was used to enlarge the fish motif.

8.3 Exam questions

Question 1 (5 marks) TECH-FREE

Source: *VCE 2019, Mathematical Methods Exam 1, Q5; © VCAA.*

Let $f: R\backslash\{1\} \to R, f(x) = \dfrac{2}{(x-1)^2} + 1$.

a. i. Evaluate $f(-1)$. **(1 mark)**

ii. Sketch the graph of f on the axes below, labelling all asymptotes with their equations. **(2 marks)**

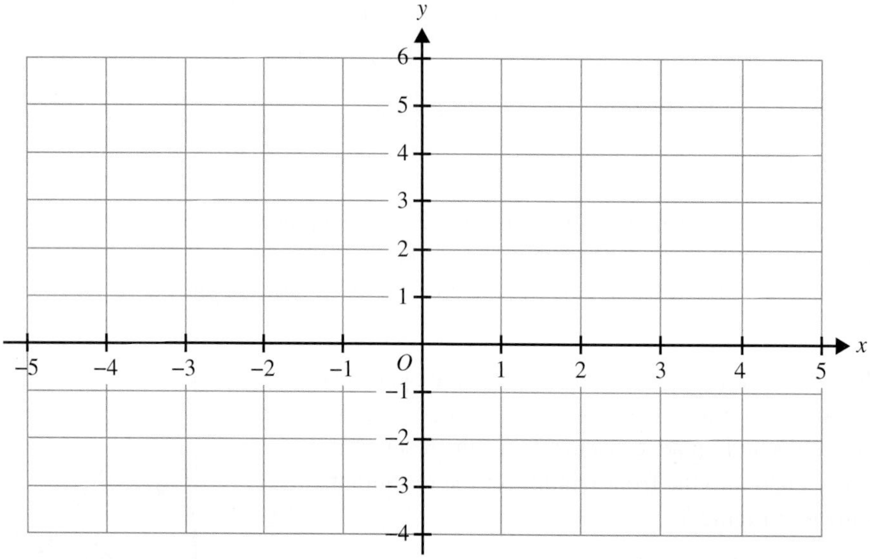

b. Find the area bounded by the graph of f, the x-axis, and the line $x = -1$ and the line $x = 0$. **(2 marks)**

Question 2 (1 mark) TECH-ACTIVE

Source: *VCE 2017, Mathematical Methods Exam 2, Section A, Q17; © VCAA.*

MC The graph of a function f, where $f(-x) = f(x)$, is shown below.

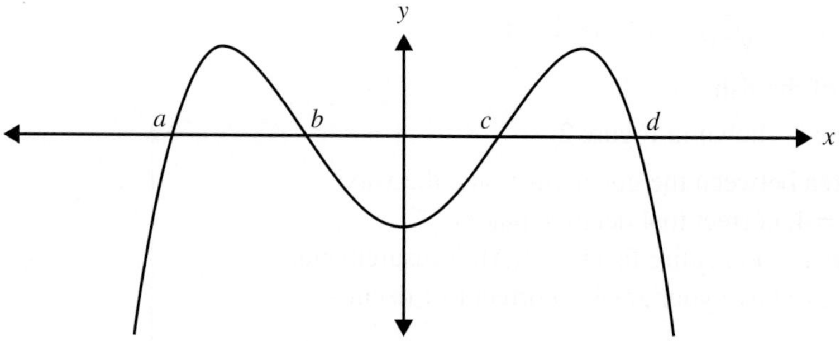

The graph has x-intercepts at $(a, 0)$, $(b, 0)$, $(c, 0)$ and $(d, 0)$ only.

The area bound by the curve and the x-axis on the interval $[a, d]$ is

A. $\displaystyle\int_a^d f(x)dx$

B. $\displaystyle\int_a^b f(x)dx - \int_c^b f(x)dx + \int_c^d f(x)dx$

C. $\displaystyle 2\int_a^b f(x)dx + \int_b^c f(x)dx$

D. $\displaystyle 2\int_a^b f(x)dx - 2\int_b^{b+c} f(x)dx$

E. $\displaystyle\int_a^b f(x)dx + \int_c^b f(x)dx + \int_d^c f(x)dx$

Question 3 (1 mark) TECH-ACTIVE

Source: *VCE 2017, Mathematical Methods Exam 2, Section A, Q20; © VCAA.*

MC The graphs of $f: \left[0, \frac{\pi}{2}\right] \to R, f(x) = \cos(x)$ and $g: \left[0, \frac{\pi}{2}\right] \to R, g(x) = \sqrt{3}\sin(x)$ are shown below.

The graphs intersect at B.

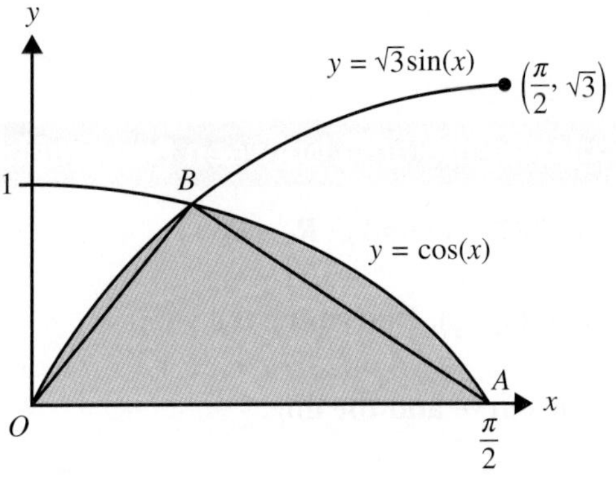

The ratio of the area of the shaded region to the area of triangle OAB is

A. $9 : 8$

B. $\sqrt{3} - 1 : \frac{\sqrt{3}\pi}{8}$

C. $8\sqrt{3} - 3 : 3\pi$

D. $\sqrt{3} - 1 : \frac{\sqrt{3}\pi}{4}$

E. $1 : \frac{\sqrt{3}\pi}{8}$

More exam questions are available online.

8.4 Areas between curves and average values

LEARNING INTENTION

At the end of this subtopic you should be able to:
- apply definite integrals to calculate the area between curves over a specified interval
- calculate the average or mean value of a function.

8.4.1 Areas between curves

Consider the functions f and g, which are both continuous on the interval $[a, e]$. Sometimes $f > g$ and on other occasions $f < g$. It is absolutely critical to know when $f > g$ or $f < g$, so a graphic representation of the situation is essential, particularly to show the points of intersection of the graphs.

We can find the area between the curves, providing we take each section one at a time. Within each section, the area is found by subtracting the lower function from the higher function. As we are finding the area between two curves, we don't need to worry about whether the region is above or below the x-axis.

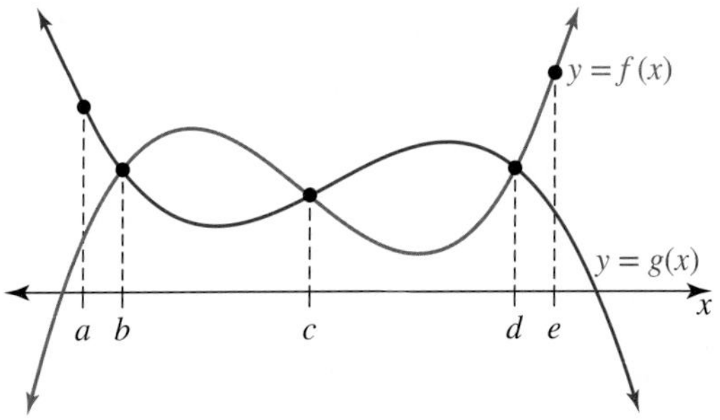

Calculation of areas between curves

$$A = \int_a^b (g(x) - f(x))\,dx + \int_b^c (f(x) - g(x))\,dx + \int_c^d (g(x) - f(x))\,dx + \int_d^e (f(x) - g(x))\,dx$$

$f < g$ $\quad f > g \quad$ $f < g$ $\quad f > g$

WORKED EXAMPLE 7 Calculating the area between two graphs

The functions $f: R \to R, f(x) = (x+2)(2-x)$ and $g: R \to R$, $g(x) = x + 2$ are shown.

a. The graphs intersect at $(m, 0)$ and (p, q). Determine the values of the constants m, p and q.

b. Calculate the area bound by the curve and the line.

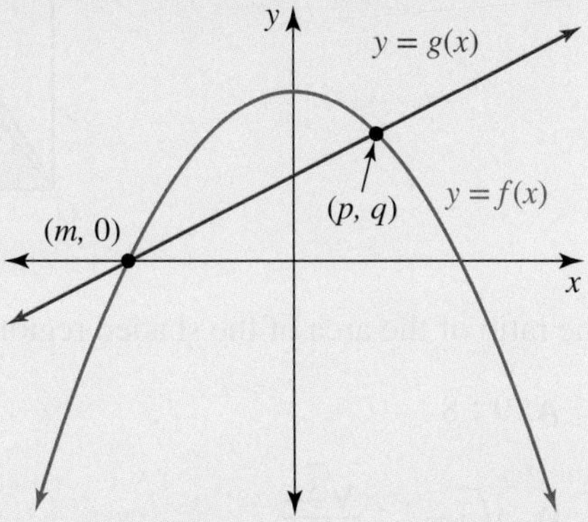

THINK	WRITE
a. 1. Points of intersection are found by solving the equations simultaneously, so equate the equations and solve for x.	**a.** $x+2 = (x+2)(2-x)$ $x+2 = 4-x^2$ $x^2+x-2 = 0$ $(x+2)(x-1) = 0$ $x = -2, x = 1$
2. Find the corresponding y-values.	When $x = -2, y = -2+2 = 0$. When $x = 1, y = 1+2 = 3$.
3. State the solution.	$m = -2, p = 1, q = 3$
b. 1. Determine whether $f > g$ or $f < g$.	**b.** As $f(x) = 4 - x^2$ lies above $g(x) = x+2, f > g$.
2. Express the area in definite integral notation and simplify the expression within the integral.	$A = \int_{-2}^{1} (f(x) - g(x))\,dx$ $= \int_{-2}^{1} \left(4 - x^2 - (x+2)\right)dx$ $= \int_{-2}^{1} \left(-x^2 - x + 2\right)dx$

3. Anti-differentiate and evaluate.

$$= \left[-\frac{1}{3}x^3 - \frac{1}{2}x^2 + 2x\right]_{-2}^{1}$$

$$= \left(-\frac{1}{3}(1)^3 - \frac{1}{2}(1)^2 + 2(1)\right) - \left(-\frac{1}{3}(-2)^3 - \frac{1}{2}(-2)^2 + 2(-2)\right)$$

$$= -\frac{1}{3} - \frac{1}{2} + 2 - \frac{8}{3} + 2 + 4$$

$$= -3 - \frac{1}{2} + 8$$

$$= 4\frac{1}{2}$$

4. Write the answer. The area is 4.5 units^2.

TI | THINK

DISPLAY/WRITE

a. 1. On a Graphs page, complete the entry line for function 1 as:
$f1(x) = (x + 2)(2 - x)$
then press ENTER.

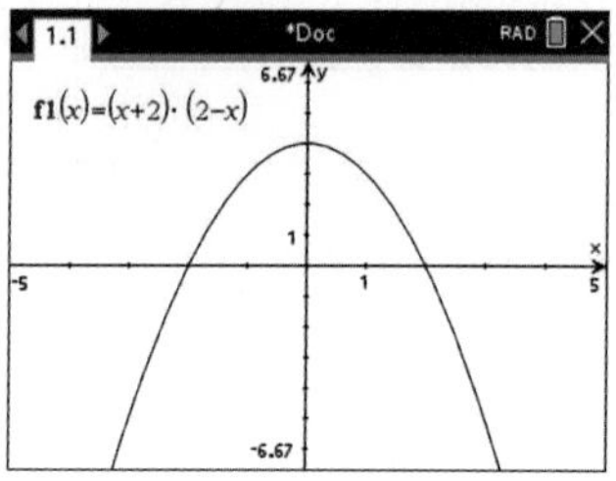

2. Complete the entry line for function 2 as:
$f2(x) = x + 2$
then press ENTER.

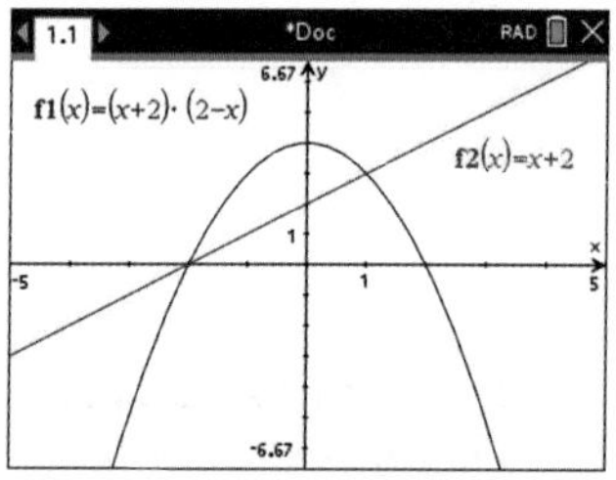

3. To find the points of intersection, press MENU, then select:
6: Analyze Graph
4: Intersection
When prompted, select a point before and after an intersection point to identify coordinates. Repeat this process to find the other point of intersection.

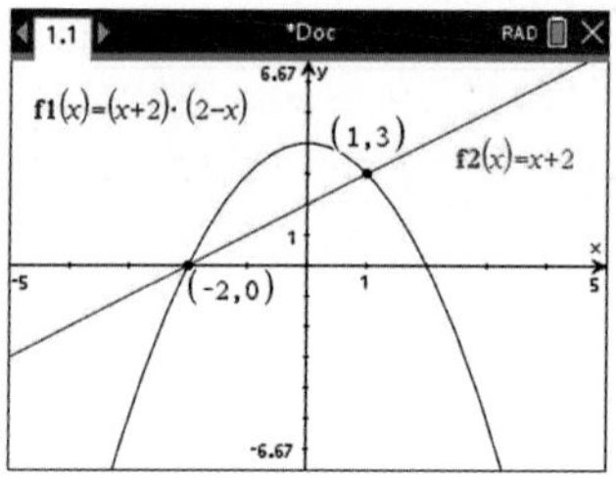

CASIO | THINK

DISPLAY/WRITE

a. 1. On a Graph & Table screen, complete the entry line for $y1$ as:
$y1 = (x + 2)(2 - x)$
then press EXE.
Complete the entry line for $y2$ as:
$y2 = x + 2$
then press EXE. Select the 'Graph' icon to draw the graphs.

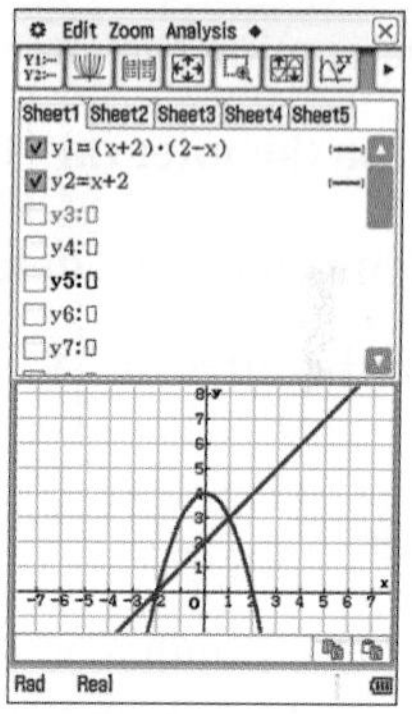

2. To find the points of intersection, select:
- Analysis
- G-Solve
- Intersection

With the cursor on the first point of intersection, press EXE. Use the left/right arrows to move to the other point of intersection, then press EXE.

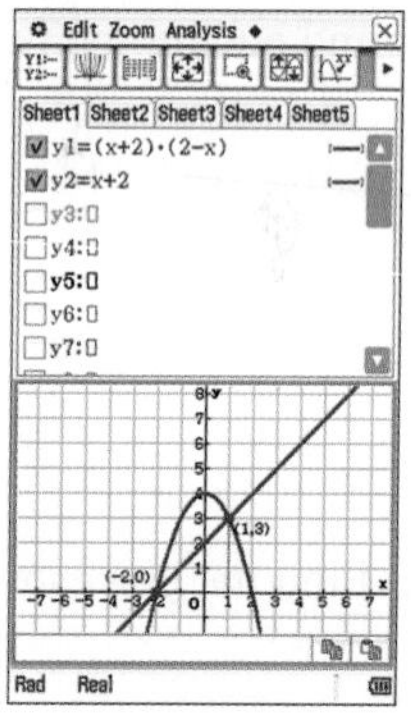

b. 1. To find the area bounded by the two curves, select:
- Analysis
- G-Solve
- Integral
- $\int$ dx Intersection

With the cursor on the first point of intersection, press EXE. Use the left/right arrows to move to the other point of intersection, then press EXE.

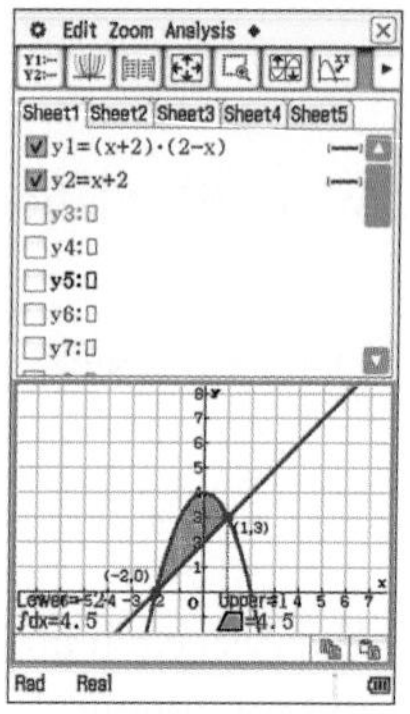

b. 1. To find the area bounded by the two curves, press MENU, then select:
6: Analyze Graph
8: Bounded Area
Click on the left-most point of intersection, then click on the right-most point of intersection.

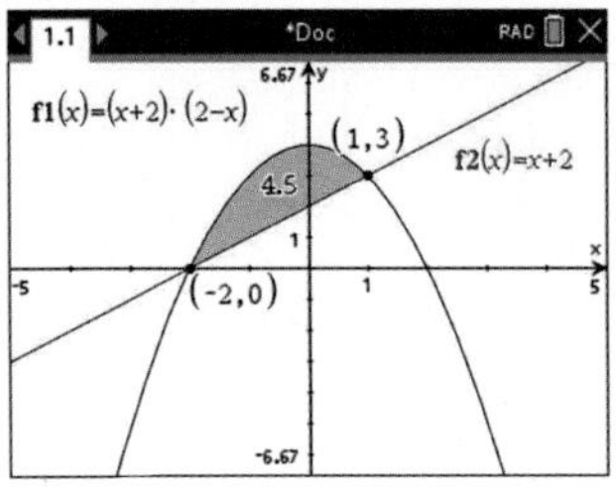

2. The answer appears on the screen. — The area is 4.5 square units.

2. The answer appears on the screen. — The area is 4.5 square units.

WORKED EXAMPLE 8 Calculating the area between two curves

The graphs of $f(x) = 3\sin(2x)$ and $g(x) = 3\cos(2x)$ are shown for $x \in [0, \pi]$.

a. Determine the coordinates of the point(s) of intersection of f and g for the interval $\left[0, \frac{\pi}{2}\right]$.

b. Using calculus, determine the area enclosed between the curves on the interval $\left[0, \frac{\pi}{2}\right]$.

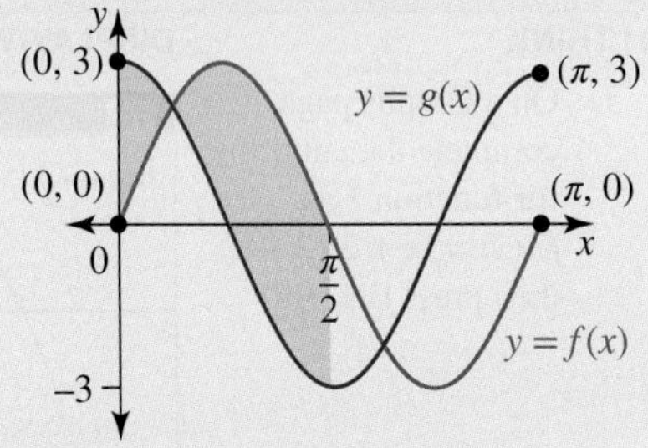

THINK / **WRITE**

a. 1. Use simultaneous equations to find where the graphs intersect, and equate the two equations.

a. $3\sin(2x) = 3\cos(2x)$

$$\frac{3\sin(2x)}{3\cos(2x)} = 1,\ 0 \le x \le \frac{\pi}{2}$$

2. Solve for $2x$ by changing the domain.

$$\tan(2x) = 1,\ 0 \le 2x \le \pi$$
$$2x = \frac{\pi}{4}$$
$$\therefore x = \frac{\pi}{8}$$

3. Find the corresponding y-value.

$$f\left(\frac{\pi}{8}\right) = 3\sin\left(\frac{\pi}{4}\right) = \frac{3\sqrt{2}}{2}$$

4. Write the solution.

The coordinates are $\left(\frac{\pi}{8}, \frac{3\sqrt{2}}{2}\right)$.

b. 1. Determine when $f > g$ and $f < g$.

b. When $0 < x < \frac{\pi}{8}$, $g > f$.

When $\frac{\pi}{8} < x < \frac{\pi}{2}$, $f > g$.

2. Express each area individually in definite integral notation.

The area is equal to:

$$A = \int_0^{\frac{\pi}{8}} (3\cos(2x) - 3\sin(2x))\,dx + \int_{\frac{\pi}{8}}^{\frac{\pi}{2}} (3\sin(2x) - 3\cos(2x))\,dx$$

3. Use calculus to anti-differentiate and evaluate.

$$= \left[\frac{3}{2}\sin(2x) + \frac{3}{2}\cos(2x)\right]_0^{\frac{\pi}{8}} + \left[-\frac{3}{2}\cos(2x) - \frac{3}{2}\sin(2x)\right]_{\frac{\pi}{8}}^{\frac{\pi}{2}}$$

$$= \frac{3}{2}\sin\left(\frac{\pi}{4}\right) + \frac{3}{2}\cos\left(\frac{\pi}{4}\right) - \left(\frac{3}{2}\sin(0) + \frac{3}{2}\cos(0)\right)$$

$$+ -\frac{3}{2}\cos(\pi) - \frac{3}{2}\sin(\pi) - \left(-\frac{3}{2}\cos\left(\frac{\pi}{4}\right) - \frac{3}{2}\sin\left(\frac{\pi}{4}\right)\right)$$

$$= \frac{3}{2} \times \frac{\sqrt{2}}{2} + \frac{3}{2} \times \frac{\sqrt{2}}{2} - 0 - \frac{3}{2} + \frac{3}{2} - 0 + \frac{3}{2} \times \frac{\sqrt{2}}{2} + \frac{3}{2} \times \frac{\sqrt{2}}{2}$$

$$= \frac{3\sqrt{2}}{4} + \frac{3\sqrt{2}}{4} + \frac{3\sqrt{2}}{4} + \frac{3\sqrt{2}}{4}$$

$$= 3\sqrt{2}$$

4. Write the answer.

The area is $3\sqrt{2}$ square units.

on Resources

Interactivity Areas between curves (int-6425)

8.4.2 The average or mean value of a function

Geometrically, the **average** or **mean value** of a function is the height of a rectangle, y_{avg}, with a width of $(b - a)$, that has the same area as the area under the curve $y = f(x)$ for the interval $[a, b]$.

The average or mean value

$$\text{Average} = \frac{1}{b-a}\int_a^b f(x)dx$$

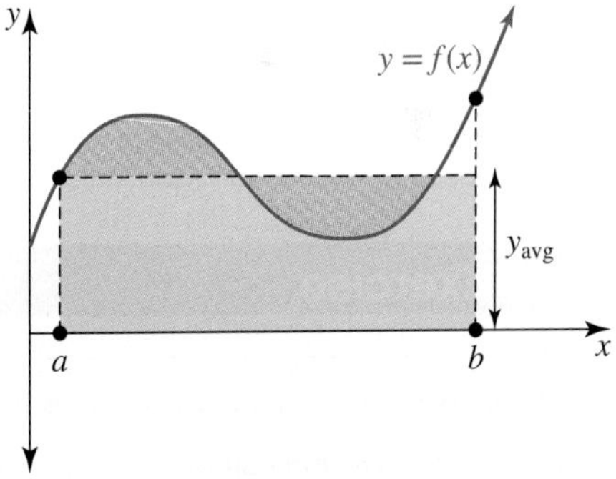

WORKED EXAMPLE 9 Calculating the average value

Calculate the average value for the function defined by $f(x) = \sin(2x)$ for the interval $x \in \left[\frac{\pi}{8}, \frac{3\pi}{8}\right]$.

THINK

WRITE

1. Write the rule for the average or mean value of a function.

$$\text{Average} = \frac{1}{b-a}\int_a^b f(x)\,dx$$

2. Substitute the appropriate values into the rule.

$$= \frac{1}{\left(\frac{3\pi}{8} - \frac{\pi}{8}\right)} \int_{\frac{\pi}{8}}^{\frac{3\pi}{8}} \sin(2x)\, dx$$

$$= \frac{4}{\pi} \int_{\frac{\pi}{8}}^{\frac{3\pi}{8}} \sin(2x)\, dx$$

3. Anti-differentiate and evaluate.

$$= \frac{4}{\pi}\left[-\frac{1}{2}\cos(2x)\right]_{\frac{\pi}{8}}^{\frac{3\pi}{8}}$$

$$= \frac{4}{\pi}\left(-\frac{1}{2}\cos\left(2\times\frac{3\pi}{8}\right) + \frac{1}{2}\cos\left(2\times\frac{\pi}{8}\right)\right)$$

$$= \frac{4}{\pi}\left(-\frac{1}{2}\times -\frac{\sqrt{2}}{2} + \frac{1}{2}\times\frac{\sqrt{2}}{2}\right)$$

$$= \frac{4}{\pi}\left(\frac{\sqrt{2}}{4} + \frac{\sqrt{2}}{4}\right)$$

$$= \frac{4}{\pi}\times\frac{\sqrt{2}}{2}$$

$$= \frac{2\sqrt{2}}{\pi}$$

on Resources

Interactivity Average value of a function (int-6424)

8.4 Exercise

Students, these questions are even better in jacPLUS

Receive immediate feedback and access sample responses

Access additional questions

Track your results and progress

Find all this and MORE in jacPLUS

Technology free

1. **WE7** The graphs of $g(x) = \sqrt{x}$ and the line $f(x) = 4$ are shown.

   a. Determine the coordinates of the point(s) of intersection between f and g.

   b. Calculate the area bound by the two graphs.

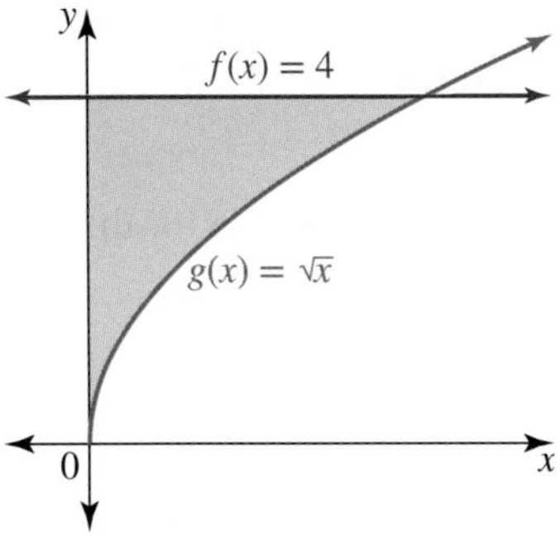

2. Calculate the area enclosed between the curve $f(x) = (x - 3)^2$ and the line $g(x) = 9 - x$.

3. **WE8** a. Determine the coordinates of the point(s) of intersection of $f(x) = \sin(x)$ and $g(x) = -\cos(x)$ for the interval $[0, \pi]$.

   b. Using calculus, determine the area enclosed between the curves on the interval $[0, \pi]$.

4. **WE9** Calculate the average value of the function defined by the rule $f(x) = e^{3x}$ for $x \in \left[0, \frac{1}{3}\right]$.

5. Calculate the average value of the function defined by the rule $f(x) = x^2 - 2x$ for $x \in [0.5, 1]$.

Technology active

6. The graphs of $f(x) = 4e^{-x}$ and $g(x) = 4 - x$ are shown.
   Using calculus, determine the area enclosed between $y = f(x)$, $y = g(x)$ and the lines $x = 0$ and $x = 5$. Give your answer correct to 4 decimal places.

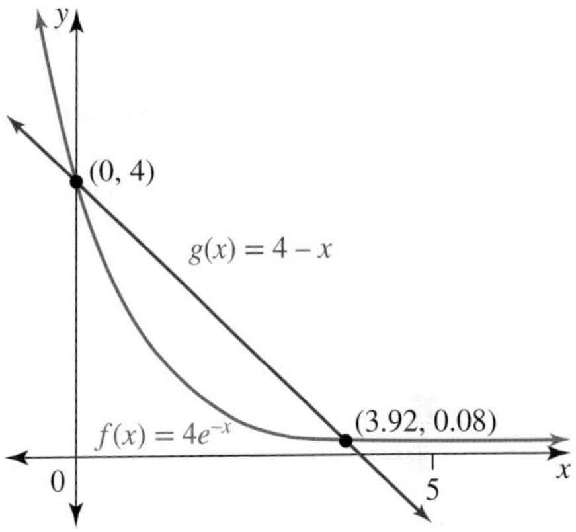

7. The graphs of $f(x) = 0.5e^x$ and $g(x) = \cos(x)$ are shown. The graphs intersect at $(-1.5, 0.12)$ and $(0.5, 0.86)$.
   Using calculus, calculate the area enclosed between the curves from $x = -1.5$ to $x = 0.5$. Give your answer correct to 4 decimal places.

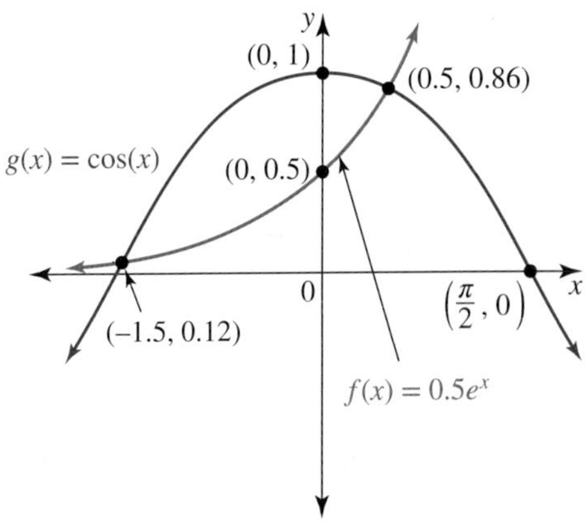

8. **MC** The graphs of $y = 0.25x^2(x - 3)(x + 3)$ and $y = -0.5x^2(x - 3)(x + 3)$ are shown.
   The area of the region enclosed between the curves and the lines $x = -3$ and $x = 3$ is:
   A. 60.75 units^2
   B. 48.6 units^2
   C. 121.5 units^2
   D. 24.3 units^2
   E. 97.2 units^2

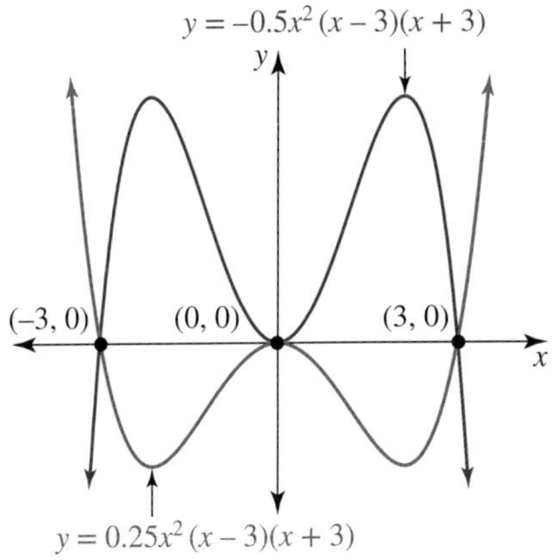

9. a. Sketch the graphs of $y=0.5(x+4)(x-1)(x-3)$ and $y=(3-x)(x+4)$ on the one set of axes.
b. Show that the three coordinate pairs of the points of intersection of the two graphs are $(-4,0)$, $(-1,12)$ and $(3,0)$.
c. Calculate the area, correct to 2 decimal places, enclosed between the curves from $x=-4$ to $x=3$.

10. Consider $f(x)=\dfrac{1}{x+2}-1, x>-2$.
a. Calculate the value of the constant a, where $(a,0)$ is the x-axis intercept.
b. Calculate the area between the curve and the x-axis from $x=a$ to $x=2$.
c. A straight line given by $y=-\dfrac{1}{2}x+\dfrac{1}{4}$ intersects $y=f(x)$ in two places. Determine the coordinates of the points of intersection.
d. Use calculus to evaluate the area between the curve and the line.

11. The graph of $y=\dfrac{5x}{x^2+1}$ is shown. The tangent to the curve at $x=-0.5$ is also shown.

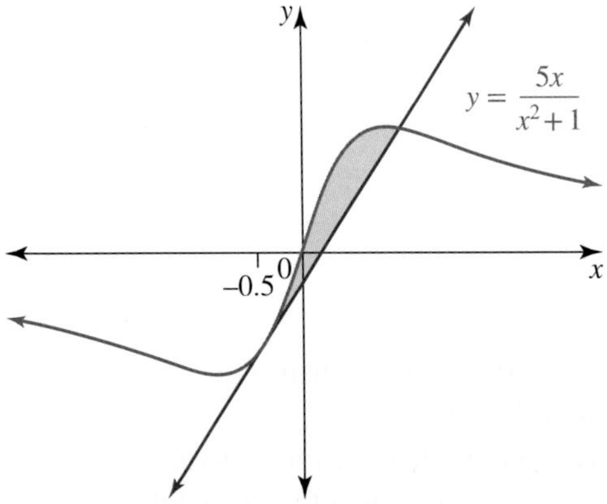

a. Determine the equation of the tangent to the curve at $x=-0.5$.
b. State the derivative of $\log_e(x^2+1)$ and hence determine an anti-derivative for $\dfrac{5x}{x^2+1}$.
c. Using calculus, calculate the area of the shaded region. Give your answer correct to 4 decimal places.

12. The graph of the function $y=\dfrac{1}{x}+x^3-4$ is shown. The tangent to the curve at $x=1$ is also shown.

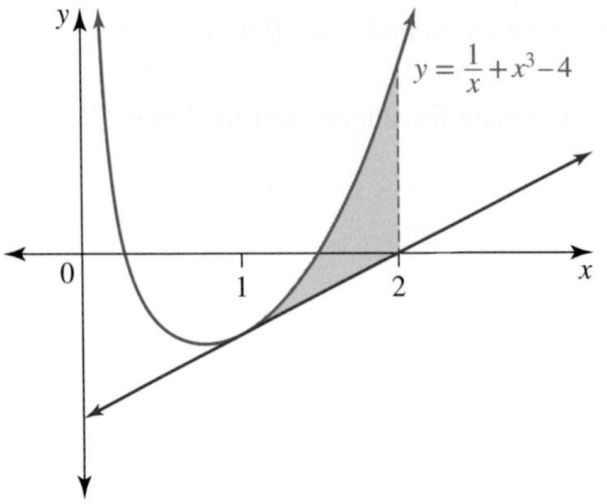

a. Determine the equation of the tangent to the curve at $x=1$.
b. Calculate the area of the shaded region.

13. The graphs of $f(x)=3x^3-x^4$ and $g(x)=-x+3$ are shown. The graphs intersect at the points (a, b) and $(c, 0)$.

a. Find the constants a, b and c.

b. Determine the area enclosed between the curves from $x=a$ to $x=c$.

c. Calculate the average value or mean value of the function $f(x)=3x^3-x^4$ for $x \in [1, 2.5]$. Give your answer correct to 3 decimal places.

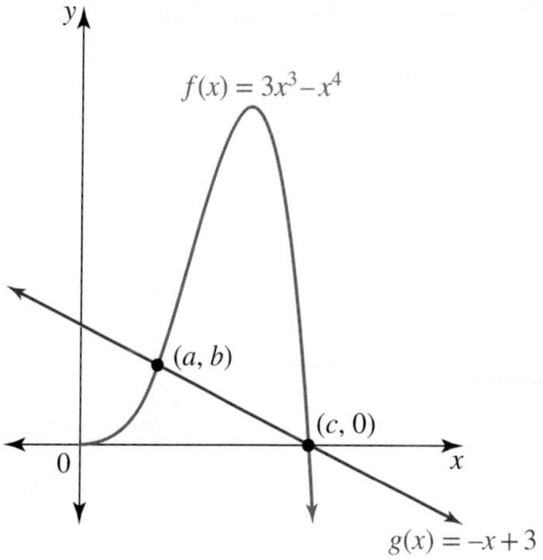

14. Consider the functions $f(x)=\sin^2(x)$ and $g(x)=\cos^2(x)$.

a. Sketch the graphs on the same set of axes for $0 \le x \le \pi$.

b. Determine the area between the curves for $0 \le x \le \pi$.

15. The graphs of $y^2=4-x$ and $y=x-2$ are shown.

a. Determine the points of intersection of the two graphs.

b. Calculate the blue shaded area.

c. Calculate the pink shaded area.

d. Hence, determine the area enclosed between the two graphs that is represented by the blue and pink shaded regions.

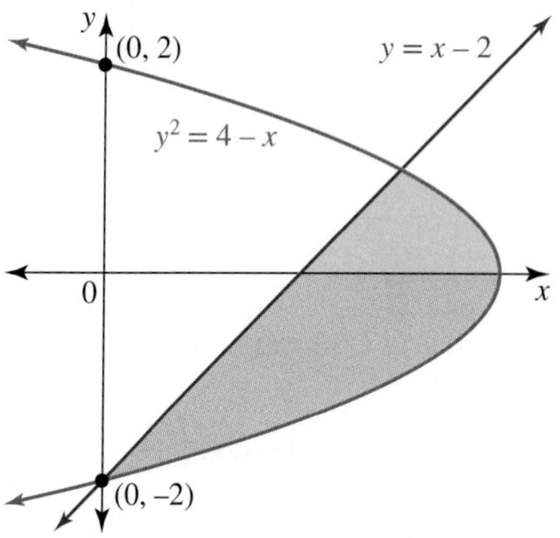

8.4 Exam questions

Question 1 (1 mark) TECH-ACTIVE

Source: *VCE 2020, Mathematical Methods Exam 2, Section A, Q15; © VCAA.*

MC Part of the graph of a function f, where $a > 0$, is shown below.

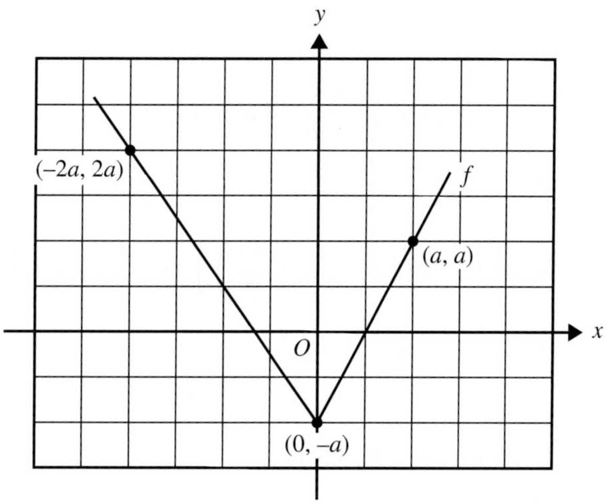

The average value of the function f over the interval $[-2a, a]$ is

A. 0 **B.** $\frac{a}{3}$ **C.** $\frac{a}{2}$ **D.** $\frac{3a}{4}$ **E.** a

Question 2 (1 mark) TECH-ACTIVE

Source: *VCE 2018, Mathematical Methods Exam 2, Section A, Q19; © VCAA.*

MC The graphs $f: R \to R, f(x) = \cos\left(\frac{\pi x}{2}\right)$ and $g: R \to R, g(x) = \sin(\pi x)$ are shown in the diagram below.

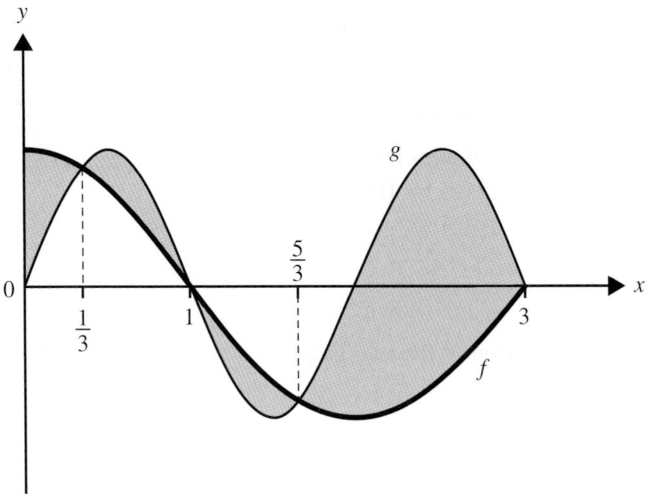

An integral expression that gives the total area of the shaded regions is

A. $\int_0^3 \left(\sin(\pi x) - \cos\left(\frac{\pi x}{2}\right)\right) dx$

B. $2\int_{\frac{5}{3}}^3 \left(\sin(\pi x) - \cos\left(\frac{\pi x}{2}\right)\right) dx$

C. $\int_0^{\frac{1}{3}} \left(\cos\left(\frac{\pi x}{2}\right) - \sin(\pi x)\right) dx - 2\int_{\frac{1}{3}}^1 \left(\cos\left(\frac{\pi x}{2}\right) - \sin(\pi x)\right) dx - \int_{\frac{5}{3}}^3 \left(\cos\left(\frac{\pi x}{2}\right) - \sin(\pi x)\right) dx$

D. $2\int_1^{\frac{5}{3}} \left(\cos\left(\frac{\pi x}{2}\right) - \sin(\pi x)\right) dx - 2\int_{\frac{5}{3}}^3 \left(\cos\left(\frac{\pi x}{2}\right) - \sin(\pi x)\right) dx$

E. $\int_0^{\frac{1}{3}} \left(\cos\left(\frac{\pi x}{2}\right) - \sin(\pi x)\right) dx + 2\int_{\frac{1}{3}}^1 \left(\sin(\pi x) - \cos\left(\frac{\pi x}{2}\right)\right) dx + \int_{\frac{5}{3}}^3 \left(\cos\left(\frac{\pi x}{2}\right) - \sin(\pi x)\right) dx$

Question 3 (7 marks) TECH-FREE

Source: *VCE 2018, Mathematical Methods Exam 1, Q8; © VCAA.*

Let $f: R \to R, f(x) = x^2e^{kx}$, where k is a positive real constant.

a. Show that $f'(x) = xe^{kx}(kx + 2)$. **(1 mark)**

b. Find the value of k for which the graphs of $y = f(x)$ and $y = f'(x)$ have exactly one point of intersection. **(2 marks)**

Let $g(x) = -\frac{2xe^{kx}}{k}$. The diagram below shows sections of the graphs of f and g for $x \geq 0$.

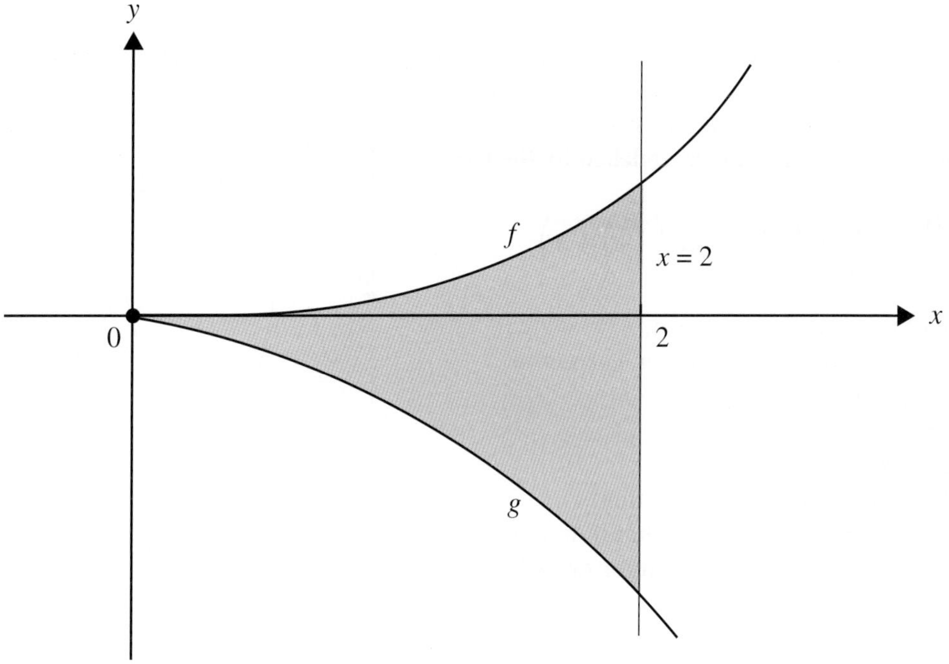

Let A be the area of the region bounded by the curves $y = f(x)$, $y = g(x)$ and the line $x = 2$.

c. Write down a definite integral that gives the value of A. **(1 mark)**

d. Using your result from part **a**, or otherwise, find the value of k such that $A = \frac{16}{k}$. **(3 marks)**

More exam questions are available online.

8.5 Applications

LEARNING INTENTION

At the end of this subtopic you should be able to:

- apply integration to real world problems.

8.5.1 Total change as the integral of instantaneous change

If we are given the equation for the rate of change and we want to find the amount that has changed over a particular time period, we would integrate the rate of change equation using the starting and finishing times as the terminals.

For example, if we know the rate of water flowing, $\frac{dV}{dt}$ in L/min, and we want to find the amount of liquid that has flowed in the first 30 minutes, we would evaluate $\int_0^{30} \frac{dV}{dt}\, dt$.

WORKED EXAMPLE 10 Applications of areas under curves

It is common practice to include heating in concrete slabs when new residential homes or units are being constructed, because it is more economical than installing heating later. A typical reinforced concrete slab, 10–15 centimetres thick, has tubing installed on top of the reinforcement, then concrete is poured on top. When the system is complete, hot water runs through the tubing. The concrete slab absorbs the heat from the water and releases it into the area above.

The number of litres/minute of water flowing through the tubing over t minutes can be modelled by the rule.

$$\frac{dV}{dt} = 2\left(\cos\left(\frac{\pi t}{3}\right) + \sin\left(\frac{\pi t}{9}\right) + 3\right)$$

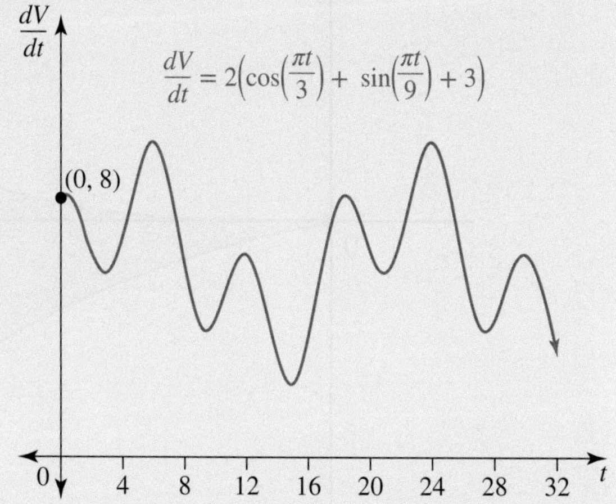

The graph of this function is shown.

a. Calculate the rate of flow of water, correct to 2 decimal places, at:
  i. 4 minutes
  ii. 8 minutes.

b. State the period of the given function.

c. Determine the volume of water that flows through the tubing during the time period for one whole cycle.

THINK	WRITE
a. i. Substitute $t = 4$ into the given equation and evaluate.	**a. i.** $\frac{dV}{dt} = 2\left(\cos\left(\frac{\pi t}{3}\right) + \sin\left(\frac{\pi t}{9}\right) + 3\right)$ When $t = 4$, $\frac{dV}{dt} = 2\left(\cos\left(\frac{4\pi}{3}\right) + \sin\left(\frac{4\pi}{9}\right) + 3\right)$ $= 6.97$ The rate at 4 minutes is 6.97 litres/minute.
ii. Substitute $t = 8$ into the given equation and evaluate.	**ii.** When $t = 8$, $\frac{dV}{dt} = 2\left(\cos\left(\frac{8\pi}{3}\right) + \sin\left(\frac{8\pi}{9}\right) + 3\right)$ $= 5.68$ The rate at 8 minutes is 5.68 litres/minute.
b. Determine the cycle for the function by analysing the shape of the graph.	**b.** A complete cycle for the function occurs between $t = 6$ and $t = 24$, so the period is $24 - 6 = 18$ minutes.
c. 1. The area under the curve of the equation of the rate of flow gives the total volume that has flowed through the tubing.	**c.** $A = \int_6^{24} 2\left(\cos\left(\frac{\pi t}{3}\right) + \sin\left(\frac{\pi t}{9}\right) + 3\right) dt$ $= 2\int_6^{24} \left(\cos\left(\frac{\pi t}{3}\right) + \sin\left(\frac{\pi t}{9}\right) + 3\right) dt$ $= 2\left[\frac{3}{\pi}\sin\left(\frac{\pi t}{3}\right) - \frac{9}{\pi}\cos\left(\frac{\pi t}{9}\right) + 3t\right]_6^{24}$

2. Anti-differentiate and evaluate.

$$= 2\left(\left(\frac{3}{\pi}\sin(8\pi) - \frac{9}{\pi}\cos\left(\frac{8\pi}{3}\right) + 72\right) - \left(\frac{3}{\pi}\sin(2\pi) - \frac{9}{\pi}\cos\left(\frac{2\pi}{3}\right) + 18\right)\right)$$

$$= 2\left(-\frac{9}{\pi}\cos\left(\frac{2\pi}{3}\right) + 72 + \frac{9}{\pi}\cos\left(\frac{2\pi}{3}\right) - 18\right)$$

$$= 2 \times 54$$

$$= 108$$

3. Write the answer.

The volume of water that passes through the tubing during one cycle is 108 litres.

8.5.2 Kinematics

You are already aware of the relationships between position, velocity and acceleration.

However, our knowledge about the definite integral and the area under curves now gives us additional skills for the calculation of facts related to kinematics.

Applications using kinematics

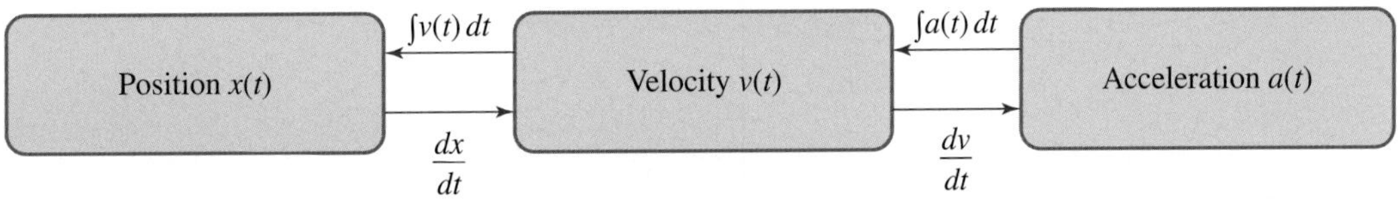

WORKED EXAMPLE 11 Applications using kinematics

A particle starting from rest accelerates according to the rule $a = 3t(2 - t)$.

a. Determine a relationship between the velocity of the particle, v metres/second, and the time, t seconds.

b. Calculate the displacement of the particle over the first 4 seconds.

c. Sketch the graph of velocity versus time for the first 4 seconds of the motion.

d. Calculate the distance travelled by the particle in the first 4 seconds.

THINK

a. 1. Anti-differentiate the acceleration equation to find the velocity equation.

WRITE

a. $v = \int a(t)\, dt$

$$= \int (3t(2 - t))\, dt$$

$$= \int \left(6t - 3t^2\right) dt$$

$$= 3t^2 - t^3 + c$$

2. Apply the initial conditions to find v in terms of t.

When $t = 0$, $v = 0$, so $c = 0$.

$\therefore v = 3t^2 - t^3$

b. 1. Integrate v between $t = 0$ and $t = 4$. As we are finding displacement, there is no need to sketch the graph.	**b.** $x = \int_0^4 \left(3t^2 - t^3\right) dt$ $= \left[t^3 - \frac{1}{4}t^4\right]_0^4$ $= \left(4^3 - \frac{1}{4}(4)^4\right) - \left(0^3 - \frac{1}{4}(0)^4\right)$ $= 0$
2. Write the answer.	The displacement over the first 4 seconds is zero.
c. Sketch a graph of v versus t.	**c.** y-intercept: (0, 0) t-intercepts: $0 = 3t^2 - t^3$ $= t^2(3 - t)$ $t = 0, 3$ When $t = 4$, $v = 3 \times 4^2 - 4^3$ $= -16.$ 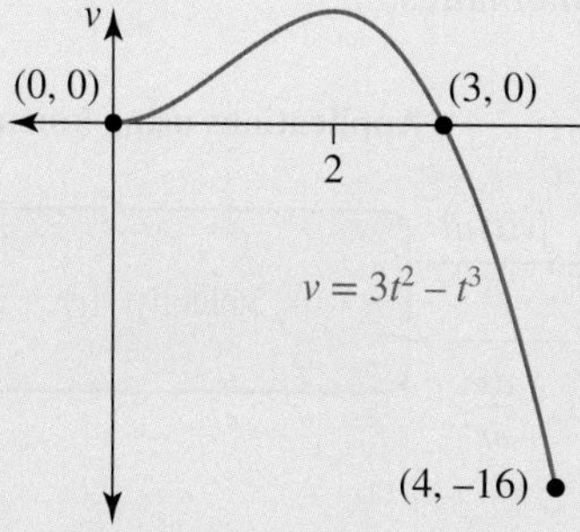
d. 1. The area under the curve of a velocity-time graph gives the distance covered. Set up the integrals and subtract the negative region.	**d.** $D = \int_0^3 \left(3t^2 - t^3\right) dt - \int_3^4 \left(3t^2 - t^3\right) dt$
2. Anti-differentiate and evaluate.	$= \left[t^3 - \frac{1}{4}t^4\right]_0^3 - \left[t^3 - \frac{1}{4}t^4\right]_3^4$ $= \left(3^3 - \frac{3^4}{4}\right) - \left(0^3 - \frac{0^4}{4}\right) - \left[\left(\left(4^3 - \frac{4^4}{4}\right) - \left(3^3 - \frac{3^4}{4}\right)\right)\right.$ $= 27 - \frac{81}{4} - 0 - 64 + 64 + 27 - \frac{81}{4}$ $= 54 - \frac{162}{4}$ $= 13.5$
3. Write the answer.	The distance travelled by the particle in 4 seconds is 13.5 metres.

8.5.3 Using the inverse function

Application problems can involve real-life applications of logarithms. We cannot anti-differentiate a logarithmic function without technology, so if we want to find the area under a logarithmic curve, we require another method.

One option is to use integration by recognition. Another is to link the areas bound by the curve of the inverse of the required function and the axes.

To find the inverse of a function, all components relating to x of the original function will relate to y of the inverse. Similarly, all components relating to y of the original function will relate to x of the inverse. This is also true for areas bound by the curve and the axes.

If $f(x) = \log_e(x)$, $f^{-1}(x) = e^x$.

The area bound by the curve of $f(x)$ and the x-axis from $x = 1$ to $x = 3$ is shown. This area is equivalent to the area bound by the curve of $f^{-1}(x)$ and the y-axis from $y = 1$ to $y = 3$.

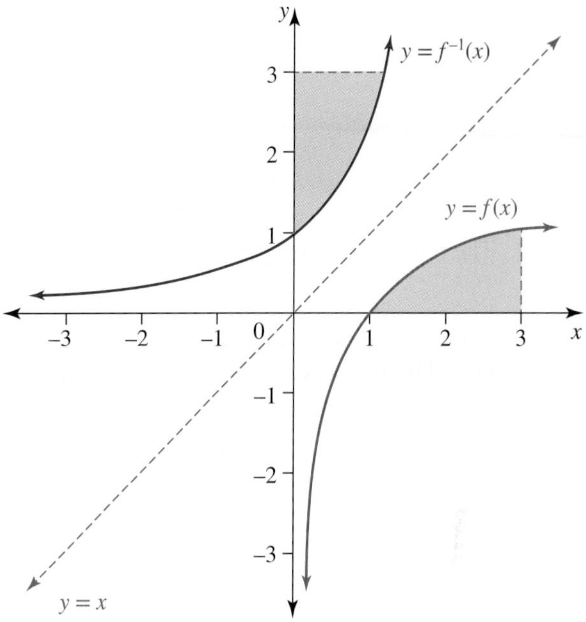

WORKED EXAMPLE 12 Using the inverse function to calculate areas

The graph of the function $f: \left[\frac{1}{3}, \infty\right) \to R, f(x) = \log_e(3x)$ is shown.

a. **Determine $f^{-1}(x)$.**

b. **Calculate $\int_0^1 f^{-1}(x)\,dx$.**

c. **Hence, calculate the exact area of the shaded region.**

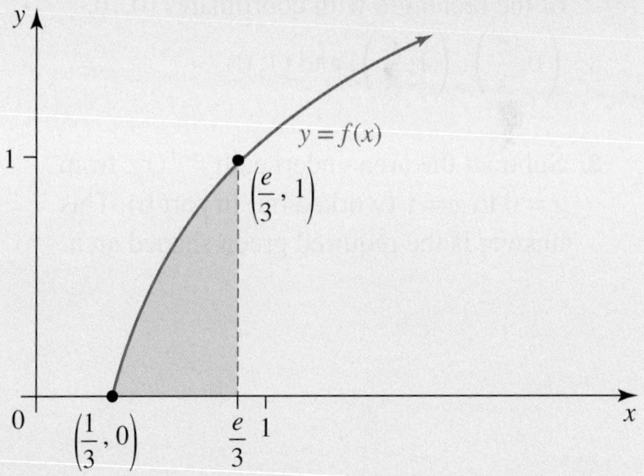

THINK

a. To find the inverse, swap x and y, and solve for y.

WRITE

a. Let $y = f(x)$.

Swap x and y:

$$\Rightarrow x = \log_e(3y)$$
$$e^x = 3y$$
$$y = \frac{1}{3}e^x$$
$$\therefore f^{-1}(x) = \frac{1}{3}e^x$$

b. 1. Set up the appropriate integral and anti-differentiate.

b. $\int_0^1 f^{-1}(x)\,dx = \int_0^1 \frac{1}{3}e^x dx$

$= \left[\frac{1}{3}e^x\right]_0^1$

2. Evaluate.

$= \frac{1}{3}e^1 - \frac{1}{3}e^0$

$= \frac{e}{3} - \frac{1}{3}$

c. 1. The required shaded area is $\int_{\frac{1}{3}}^{\frac{e}{3}} f(x)\,dx$, the blue area. This is equivalent to the area bound by the curve of $f^{-1}(x)$ and the y-axis from $y = \frac{1}{3}$ to $y = \frac{e}{3}$.

c.

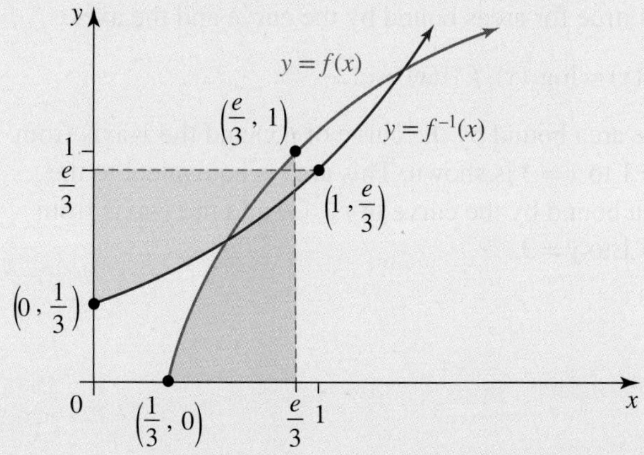

2. To find the area bound by the y-axis, the green shaded area, we need to find the area of the rectangle with coordinates $(0, 0)$, $\left(0, \frac{e}{3}\right)$, $\left(1, \frac{e}{3}\right)$, and $(1, 0)$.

$A_{\text{rectangle}} = 1 \times \frac{e}{3}$

$= \frac{e}{3}$

3. Subtract the area underneath $f^{-1}(x)$, from $x = 0$ to $x = 1$ (worked out in part **b**). This answer is the required green shaded area.

$A = A_{\text{rectangle}} - \int_0^1 f^{-1}(x)\,dx$

$= \frac{e}{3} - \left(\frac{e}{3} - \frac{1}{3}\right)$

$= \frac{1}{3}$

4. State the answer.

$\int_{\frac{1}{3}}^{\frac{e}{3}} f(x)\,dx = \frac{1}{3}$ units^2

on Resources

Interactivity Area under a curve (int-6426)

8.5 Exercise

Students, these questions are even better in jacPLUS

Receive immediate feedback and access sample responses

Access additional questions

Track your results and progress

Find all this and MORE in jacPLUS

Technology free

1. WE10 A number of apprentice bricklayers are competing in a competition in which they are required to build a fence. The competitors must produce a fence that is straight, neatly constructed and level. The winner will also be judged on how many bricks they have laid during a 30-minute period. The winner laid bricks at a rate defined by the rule

$$\frac{dN}{dt} = 0.8t + 2$$

where N is the number of bricks laid after t minutes.

a. Sketch the graph of the given function for $0 \le t \le 30$.
b. Shade the region defined by $10 \le t \le 20$.
c. Determine how many bricks in total the winner laid in the 10-minute period defined by $10 \le t \le 20$.

2. WE11 A particle moves in a line so that its velocity v metres/second from a fixed point, O, is defined by $v = 1 + 3\sqrt{t+1}$, where t is the time in seconds.

a. Determine the initial velocity of the particle.
b. Calculate the acceleration of the particle when:
   i. $t = 0$
   ii. $t = 8$
c. Sketch the graph of v versus t for the first 10 seconds.
d. Calculate the distance covered by the particle in the first 8 seconds.

3. A stone footbridge over a creek is shown along with the mathematical profile of the bridge.

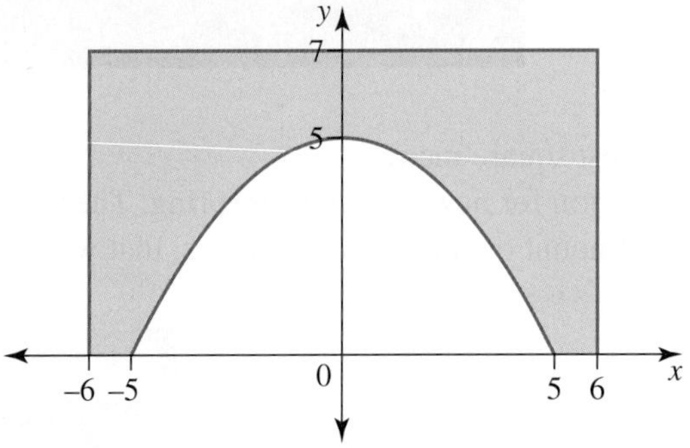

The arch of the footbridge can be modelled by a quadratic function for $x \in [-5, 5]$, with all measurements in metres.

a. Determine the equation for the arch of the bridge
b. Calculate the area between the curve and the x-axis from $x = -5$ to $x = 5$.
c. Calculate the area of the side of the bridge represented by the shaded area.
d. The width of the footbridge is 3 metres. Determine the volume of stones used in the construction of the footbridge.

4. **WE12** The graph of the function $f: \left[\frac{1}{4}, \infty\right] \to R, f(x) = 2\log_e(4x)$ is shown.

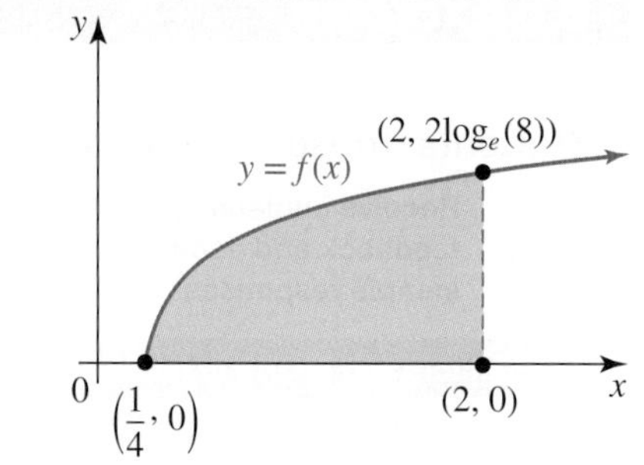

a. Determine $f^{-1}(x)$.

b. Calculate $\displaystyle\int_0^{2\log_e(8)} f^{-1}(x)\,dx$.

c. Hence, calculate the exact area of the shaded region.

5. Part of the graph of the function $g: \left[\frac{1}{5}, \infty\right) \to R, g(x) = -\log_e(5x)$. is shown.

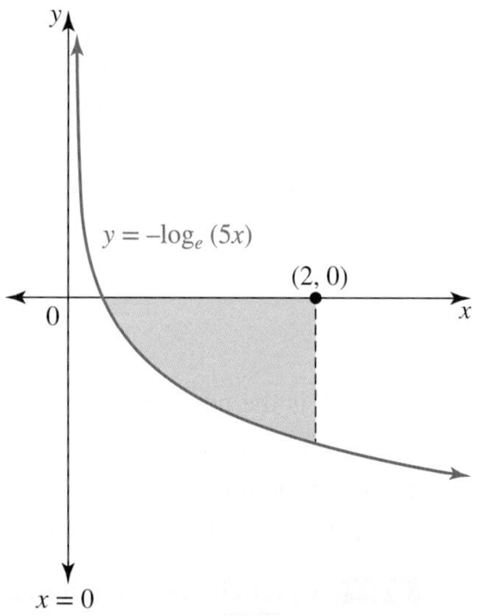

a. Calculate the coordinates of the point where the graph intersects the x-axis.

b. If $y = -x\log_e(5x) + x$, determine $\frac{dy}{dx}$.

c. Use your result from part **b** to calculate the area of the shaded region.

Technology active

6. The edge of a garden bed can be modelled by the rule

$$y = 0.5\sin\left(\frac{x}{2}\right) + 2$$

The bed has edges defined by $y = 0$, $x = 0$ and $x = 4\pi$. All measurements are in metres.

a. Sketch the graph of $y = 0.5\sin\left(\frac{x}{2}\right) + 2$ along with $y = 0$, $x = 0$ and $x = 4\pi$ as edges to show the shape of the garden bed.

b. Calculate the area of the garden bed, correct to the nearest square metre.

c. Topsoil is going to be used on the garden bed in preparation for new planting for spring. The topsoil is to be spread so that it is uniformly 50 cm thick. Find the amount of soil, in cubic metres, that will be needed for the garden bed.

7. The average rate of increase, in cm/month, in the length of a baby boy from birth until age 36 months is given by the rule

$$\frac{dL}{dt} = \frac{4}{\sqrt{t}}$$

where t is the time in months since birth and L is the length in centimetres. Calculate the average total increase in length of a baby boy from 6 months of age until 36 months of age. Give your answer correct to 1 decimal place.

8. The rate of growth of mobile phone subscribers with a particular company in the Australia can be modelled by the rule

$$\frac{dN}{dt} = 0.853e^{0.1333t}$$

where N million is the number of subscribers with the company since 2004 and t is the number of years since 2004, the year the company was established. Determine how many millions of mobile phone subscribers have joined the company between 2004 and 2021, correct to 1 decimal place.

9. The maintenance costs for a car increase as the car gets older. It has been suggested that the increase in maintenance costs of dollars per year could be modelled by

$$\frac{dC}{dt} = 15t^2 + 250$$

where t is the age of the car in years and C is the total accumulated cost of maintenance for t years.

a. Sketch the graph of the given function for $0 \le t \le 10$.
b. Determine the total accumulated cost of maintenance for $t = 5$ to $t = 10$ years.

10. Part of the graph of the function $h: (-5, \infty) \to R$, $h(x) = 2\log_e(x + 5) + 1$ is shown.

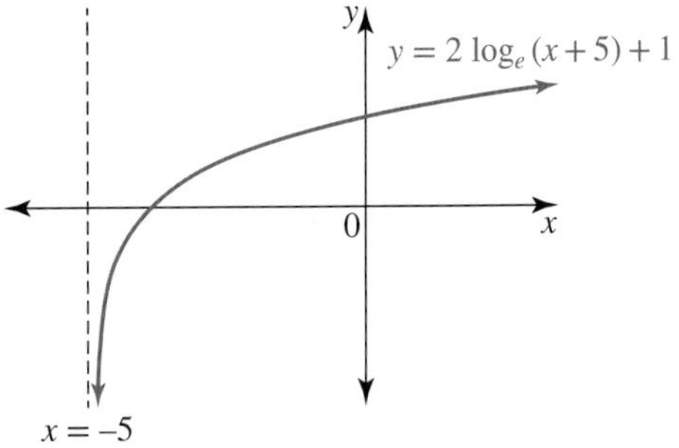

a. Calculate the coordinates of the axial intercepts.
b. State the rule and domain for h^{-1}, the inverse of h.
c. On the one set of axes, sketch the graphs of $y = h(x)$ and $y = h^{-1}(x)$. Clearly label the axial intercepts with exact values and any asymptotes.
d. Calculate the values of x, correct to 4 decimal places, for which $h(x) = h^{-1}(x)$.
e. Calculate the area of the region enclosed by the graphs of h and h^{-1}. Give your answer correct to 4 decimal places.

11. a. If $y = x\log_e(x)$, find $\frac{dy}{dx}$. Hence, find the exact value of $\int_1^{e^2} \log_e(x)\, dx$.

b. If $y = x\left(\log_e(x)\right)^m$ where m is a positive integer, determine $\frac{dy}{dx}$.

c. Let $I_m = \int_1^{e^2} \left(\log_e(x)\right)^m dx$ for $m > 1$. Show that $I_m + mI_{m-1} = 2^m e^2$.

d. Hence, calculate the value of $\int_1^{e^2} \left(\log_e(x)\right)^3 dx$.

12. The shaded area in the diagram is the plan of a mine site. All distances are in kilometres.

Two of the boundaries of the mine site are in the shape of graphs defined by the functions with equations $f: R \to R, f(x) = 2e^x$ and $g: R^+ \to R, g(x) = \log_e\left(\frac{x}{2}\right)$, where $g(x)$ is the inverse function of $f(x)$.

Calculate the area of the region bounded by the graphs of f and g, the y-axis and the lines $x = 1$ and $y = -4$. Give your answer correct to 1 decimal place.

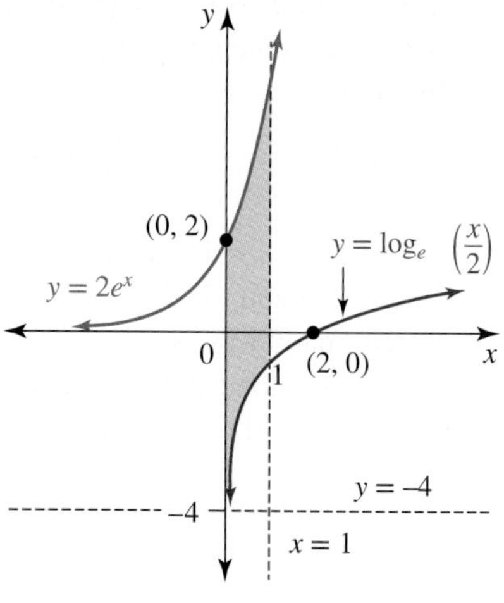

13. A patient has just had a medical procedure that required a general anaesthetic. Five minutes after the end of the procedure was completed, the patient starts to show signs of awakening. The alertness, A, of the patient t minutes after the completion of the procedure can be modelled by the rule $A = 4.6 \log_e(t - 4)$.

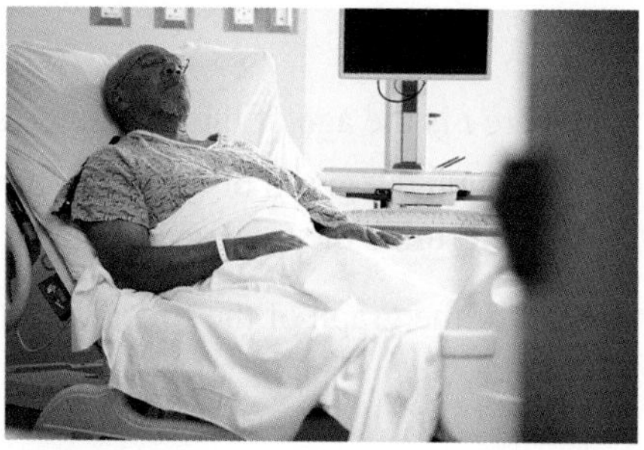

The graph of the function is shown.

a. Calculate the value of the constant a, given that $(a, 0)$ is the x-axis intercept.
b. When the patient has an alertness of 15, they are allowed to have water to sip, and 15 minutes later they can be given a warm drink and something to eat. Calculate how long it takes for the patient to reach an alertness of 15. Give your answer correct to the nearest minute.
c. Calculate the rate at which the alertness of the patient is changing 10 minutes after the completion of the medical procedure.
d. Use the inverse function of A to determine the total change of alertness for 30 minutes after the completion of the medical procedure. That is, calculate the area between the curve and the t axis from $t = 5$ to $t = 30$.

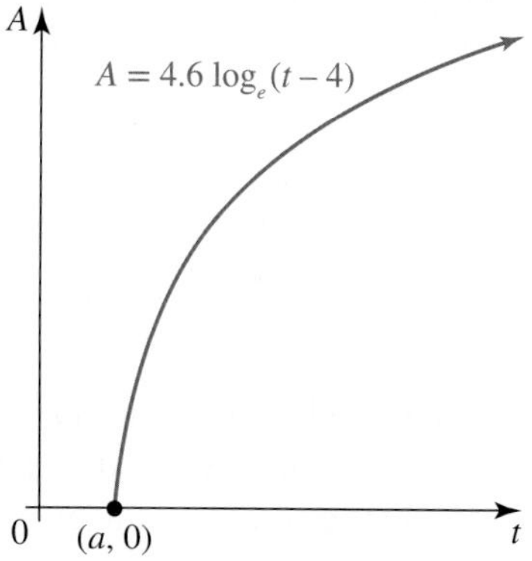

14. An object travels in a line so that its velocity, v metres/second, at time t seconds is given by

$$v = 3\cos\left(\frac{t}{2} - \frac{\pi}{4}\right), t \geq 0.$$

Initially the object is $-3\sqrt{2}$ metres from the origin.

a. Determine the relationship between the position of the object, x metres, and time, t seconds.
b. Calculate the position of the object when time is equal to 3π seconds.
c. Sketch the graph of v versus t for $0 \leq t \leq 4\pi$.
d. Calculate the distance travelled by the object after 3π seconds. Give your answer in metres, correct to 2 decimal places.
e. Determine a relationship between the acceleration of the object, a metres/second^2, and time, t seconds.
f. Calculate acceleration of the object when $t = 3\pi$ seconds.

15. A particle moves in a straight line. At time t seconds its velocity, v metres per second, is defined by the rule $v = e^{-0.5t} - 0.5$, $t \geq 0$. The graph of the motion is shown.

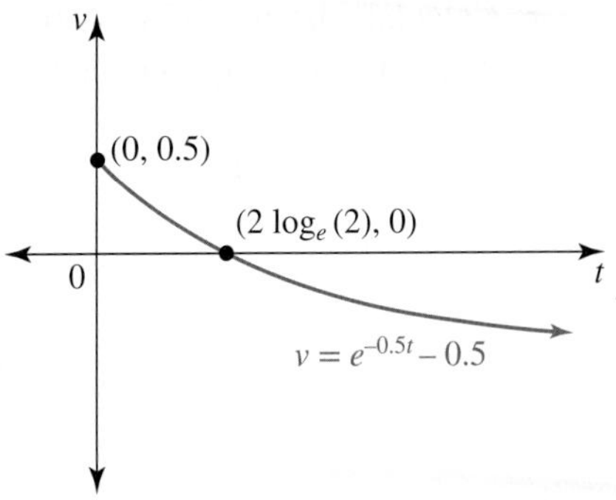

a. Determine the acceleration of the particle, a m/s^2, in terms of t.
b. Determine the position of the particle, x m, in terms of t, if $x = 0$ when $t = 0$.
c. Calculate the position of the particle after 4 seconds, correct to 4 decimal places.
d. Determine the distance covered by the particle in the fourth second. Give your answer correct to 4 decimal places.

16. The cross-section of a waterway is parabolic. Its depth is 3 metres, and the width across the top of the waterway is 4 metres. Determine the depth of the water in metres, to 2 decimal places, when the waterway is one-third full.

17. At the Royal Botanical Gardens, a new area of garden is being prepared for native Australian plants.

The area of garden has two curved walking paths as borders.

One of the paths can be modelled by the rule $f(x) = e^{\frac{1}{2}(x-1)} + 3$.

a. The other curved walking path is defined by the rule for the inverse of f, f^{-1}. State the rule for f^{-1}.

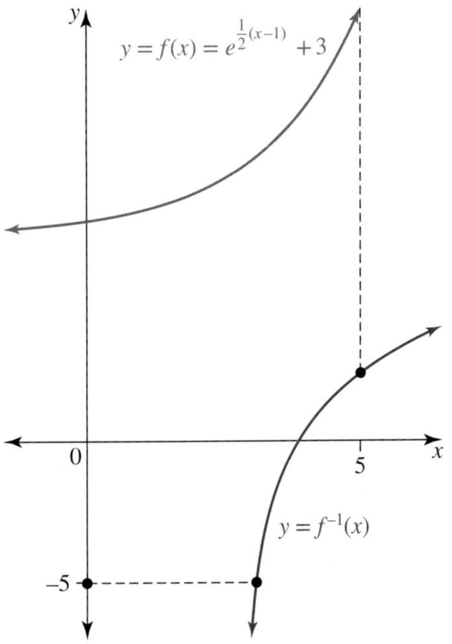

The other borders are given by $x = 5$ and $y = -5$ as shown. The remaining border is formed by the y-axis, as shown. All measurements are in metres.

b. Determine the respective axis intercepts of the graphs of f and f^{-1}.

c. Calculate the area of the garden above the x-axis, as shown in the diagram below, by calculating

$$\int_0^5 \left(e^{\frac{1}{2}(x-1)}+3\right)\,dx - \int_{e^{-0.5}+3}^{5} (2\log_e(x-3)+1)\,dx.$$

Give your answer correct to 2 decimal places.

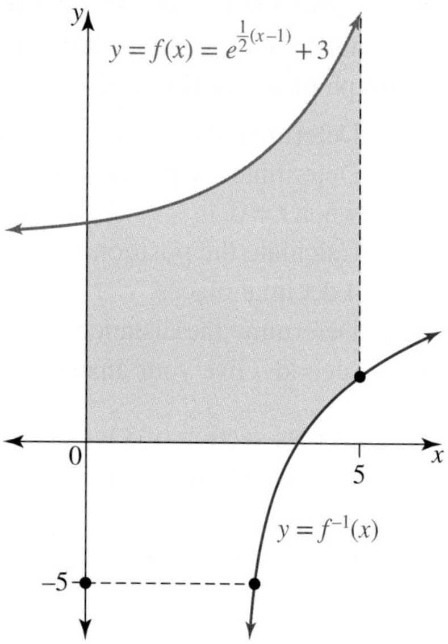

d. Calculate the area of the garden below the x-axis, as shown in the diagram below, correct to 2 decimal places.

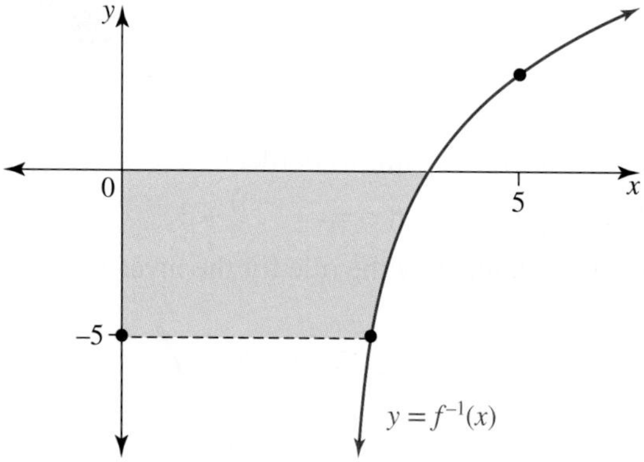

e. Hence, determine the total area of the garden correct to 1 decimal place.

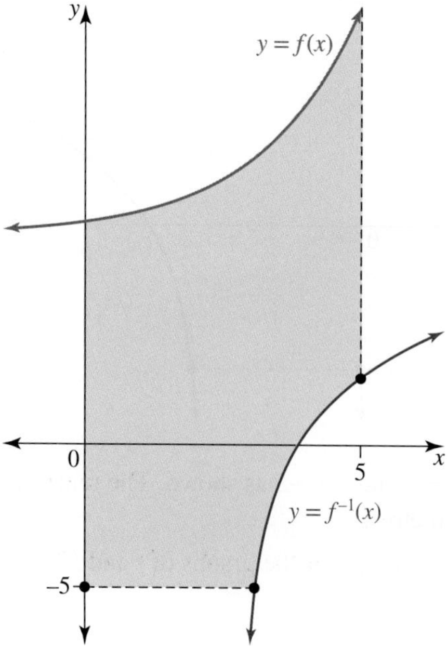

8.5 Exam questions

Question 1 (11 marks) TECH-ACTIVE

Source: *Adapted from VCE 2020, Mathematical Methods Exam 2, Section B, Q2; © VCAA.*

An area of parkland has a river running through it, as shown below. The river is shown shaded.

The north bank of the river is modelled by the function $f_1 : [0, 200] \to R, f_1(x) = 20\cos\left(\frac{\pi x}{100}\right) + 40$

The south bank of the river is modelled by the function $f_2 : [0, 200] \to R, f_2(x) = 20\cos\left(\frac{\pi x}{100}\right) + 30$

The horizontal axis points east and the vertical axis points north.

All distances are measured in metres.

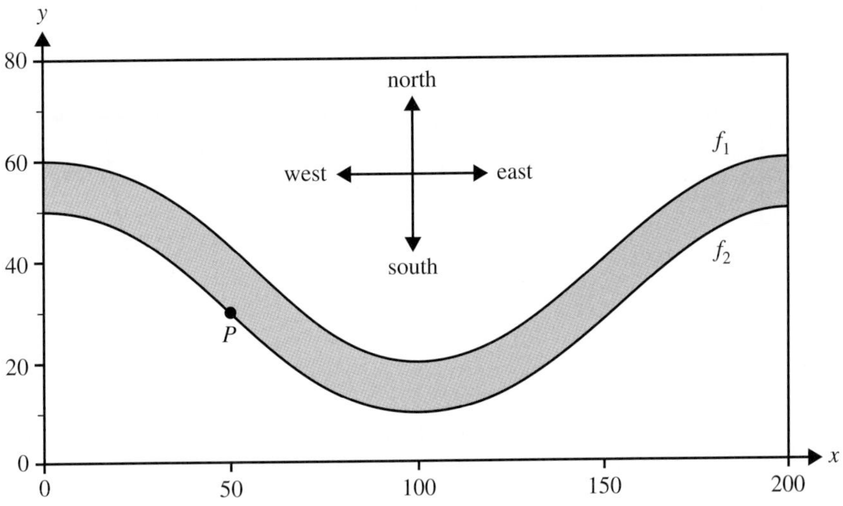

A swimmer always starts at point ***P***, which has coordinates (50, 30).

Assume that no movement of water in the river affects the motion or path of the swimmer, which is always a straight line.

a. The swimmer swims north from point P.

Find the distance, in metres, that the swimmer needs to swim to get to the north bank of the river. **(1 mark)**

b. The swimmer swims east from point P.

Find the distance, in metres, that the swimmer needs to swim to get to the north bank of the river. **(2 marks)**

c. On another occasion, the swimmer swims the minimum distance from point P to the north bank of the river.

Find this minimum distance. Give your answer in metres, correct to one decimal place. **(2 marks)**

d. Calculate the surface area of the section of the river shown on the graph above in square metres. **(1 mark)**

e. A horizontal line is drawn through point P. The section of the river that is south of the line is declared a 'no swimming' zone.

Find the area of the 'no swimming' zone, correct to the nearest square metre. **(3 marks)**

f. Scientists observe that the north bank of the river is changing over time. It is moving further north from its current position. They model its predicted new location using the function with rule $y = kf_1(x)$, where $k \geq 1$.

Find the values of k for which the distance **north** across the river, for all parts of the river, is strictly less than 20 m. **(2 marks)**

Question 2 (7 marks) TECH-ACTIVE

Source: *Adapted from VCE 2019, Mathematical Methods Exam 2, Section B, Q3; © VCAA.*

During a telephone call, a phone uses a dual-tone frequency electrical signal to communicate with the telephone exchange.

The strength, f, of a simple dual-tone frequency signal is given by the function $f(t) = \sin\left(\frac{\pi t}{3}\right) + \sin\left(\frac{\pi t}{6}\right)$, where t is a measure of time and $t \geq 0$.

Part of the graph of $y = f(t)$ is shown below.

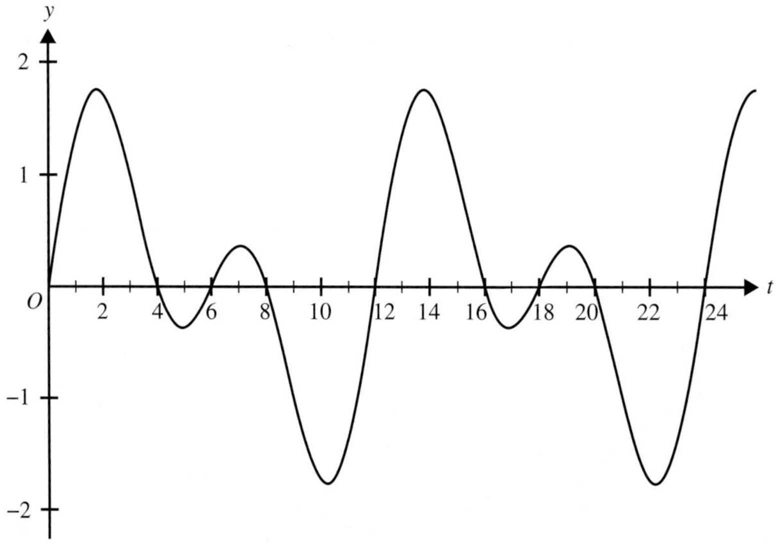

a. State the period of the function. **(1 mark)**

b. Find the values of t where $f(t) = 0$ for the interval $t \in [0, 6]$. **(1 mark)**

c. Find the maximum strength of the dual-tone frequency signal, correct to two decimal places. **(1 mark)**

d. Find the area between the graph of f and the horizontal axis for $t \in [0, 6]$. **(2 marks)**

e. The rectangle bounded by the line $y = k, k \in R^+$, the horizontal axis, and the lines $x = 0$ and $x = 12$ has the same area as the area between the graph of f and the horizontal axis for one period of the dual-tone frequency signal.

Find the value of k. **(2 marks)**

Question 3 (14 marks) TECH-ACTIVE

Source: *VCE 2015, Mathematical Methods (CAS) Exam 2, Section 2, Q2; © VCAA.*

A city is located on a river that runs through a gorge.

The gorge is 80 m across, 40 m high on one side and 30 m high on the other side.

A bridge is to be built that crosses the river and the gorge.

A diagram for the design of the bridge is shown below.

The main frame of the bridge has the shape of a parabola. The parabolic frame is modelled by $y = 60 - \frac{3}{80}x^2$ and is connected to concrete pads at $B(40, 0)$ and $A(-40, 0)$.

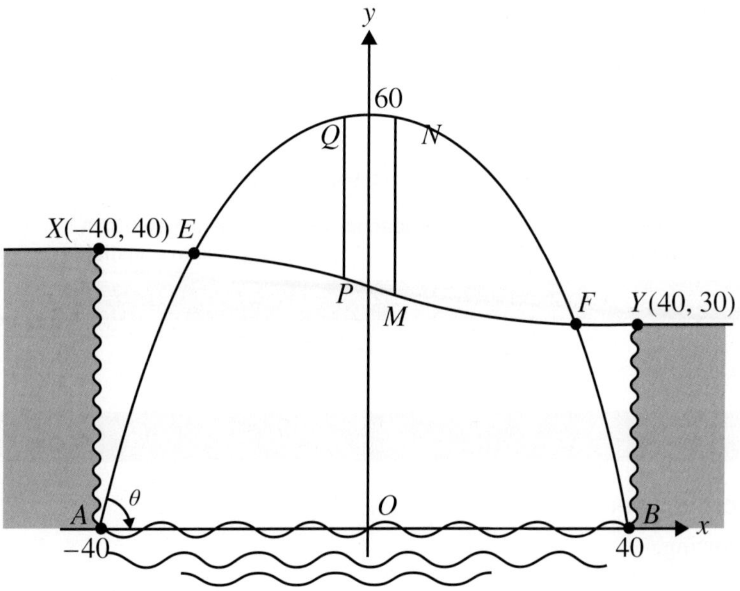

The road across the gorge is modelled by a cubic polynomial function.

a. Find the angle, θ, between the tangent to the parabolic frame and the horizontal at the point $A(-40, 0)$ to the nearest degree. **(2 marks)**

The road from X to Y across the gorge has gradient zero at $X(-40, 40)$ and at $Y(40, 30)$, and has equation

$$y = \frac{x^3}{25\,600} - \frac{3x}{16} + 35.$$

b. Find the maximum downwards slope of the road. Give your answer in the form $-\frac{m}{n}$ where m and n are positive integers. **(2 marks)**

Two vertical supporting columns, MN and PQ, connect the road with the parabolic frame.

The supporting column, MN, is at the point where the vertical distance between the road and the parabolic frame is a maximum.

c. Find the coordinates (u, v) of the point M, stating your answers correct to two decimal places. **(3 marks)**

The second supporting column, PQ, has its lowest point at $P(-u, w)$.

d. Find, correct to two decimal places, the value of w and the lengths of the supporting columns MN and PQ. **(3 marks)**

For the opening of the bridge, a banner is erected on the bridge, as shown by the shaded region in the diagram at right.

e. Find the x-coordinates, correct to two decimal places, of E and F, the points at which the road meets the parabolic frame of the bridge. **(3 marks)**

f. Find the area of the banner (shaded region), giving your answer to the nearest square metre. **(1 mark)**

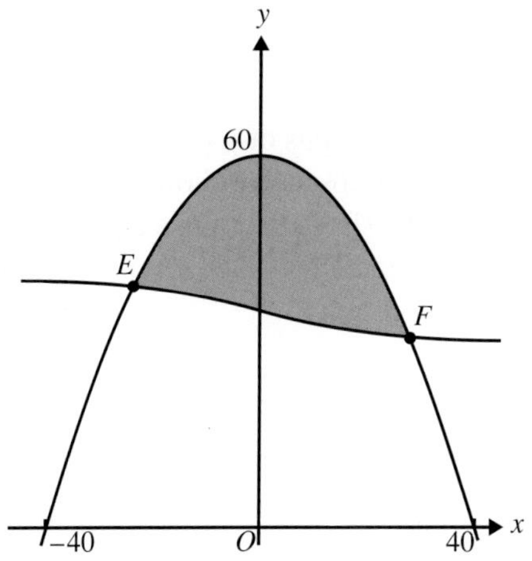

More exam questions are available online.

8.6 Review

8.6.1 Summary

doc-37037

Hey students! Now that it's time to revise this topic, go online to:

Access the topic summary | Review your results | Watch teacher-led videos | Practise VCAA exam questions

Find all this and MORE in jacPLUS

8.6 Exercise

Technology free: short answer

1. a. Evaluate the following.

   i. $\displaystyle\int_{-\frac{\pi}{6}}^{\frac{\pi}{2}} (4\sin(2x) + \cos(3x))\,dx$ ii. $\displaystyle\int_{0}^{2} \left(3x + 6\sqrt{x} + 1\right) dx$ iii. $\displaystyle\int_{0}^{\frac{1}{2}} (e^x + 1)(e^x - 1)\,dx$

   b. Given that $\displaystyle\int_{1}^{5} f(x)dx = 4$ and $\displaystyle\int_{1}^{5} g(x)dx = 3$, determine:

   i. $\displaystyle\int_{1}^{5} (4f(x) + 1)\,dx$ ii. $\displaystyle\int_{1}^{5} (2f(x) - g(x))\,dx$ iii. $\displaystyle\int_{1}^{5} (3f(x) + 2g(x) - 5)\,dx$

2. a. Evaluate the following.

   i. $\displaystyle\int_{0}^{2} \frac{x+2}{x+1}\,dx$ ii. $\displaystyle\int_{-2}^{-1} \frac{3}{2x+1}\,dx$

   b. If $\displaystyle\int_{3}^{a} \frac{1}{3(x-1)}dx = 1$ and $a > 3$, calculate the exact value of a.

3. a. Determine m if $\displaystyle\int_{\frac{1}{2}}^{m} 6(2x-1)^2\ dx = 1$.

   b. The graphs of $y = 2\sin(x) + k$ and $y = k\cos(x)$ are shown.
   The shaded region is equal to $(3\pi + 4)$ square units. Determine the value of the constant k.

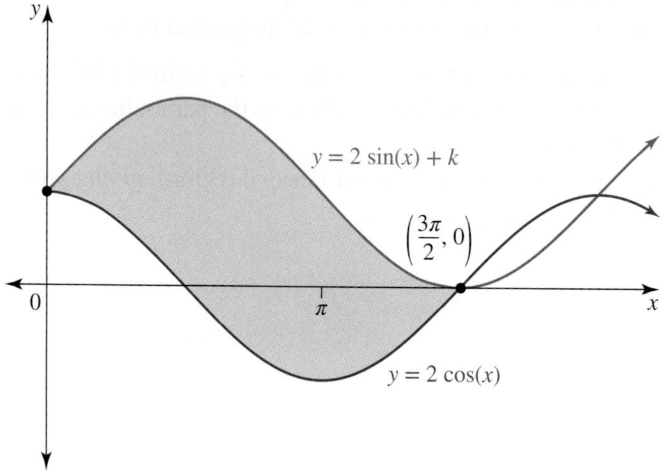

4. The graphs with equations $y = \sqrt{x}$ and $y = 2 - x$ are shown.

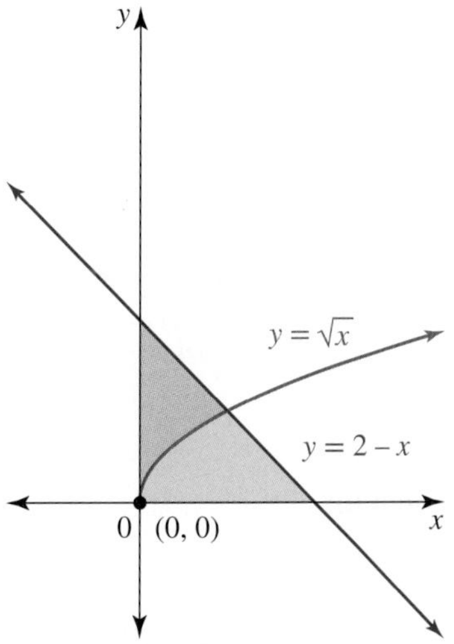

a. Calculate the average value of the function $y = \sqrt{x}$ for $x \in [0, 2]$.
b. Determine the solution to the equation $\sqrt{x} = 2 - x$.
c. Determine the area of the blue shaded region (the lower shaded region).
d. Hence or otherwise, calculate the area of the pink shaded region (the upper shaded region).

5. The graphs of $y = 4x^2$ and $y = \dfrac{1}{4x^2}$ are shown.

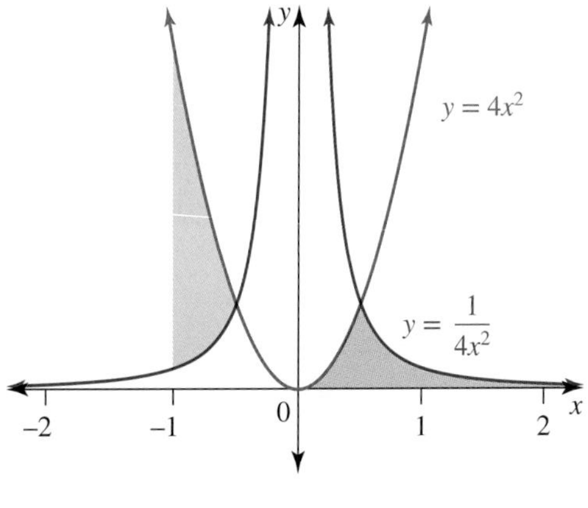

a. Determine the coordinates of the points of intersection of the two curves.
b. Calculate the area of the pink shaded region.
c. Calculate the area of the blue shaded region.

6. a. Determine any point(s) of intersection between the two curves $f(x) = x^3 - 3x + 2$ and $g(x) = x + 2$.
b. Sketch $f(x)$ and $g(x)$ on the same set of axes. Label the point(s) of intersection and any x- and y-intercepts.
c. Evaluate the area between the two curves.

Technology active: multiple choice

7. **MC** Using the trapezium rule with interval widths of 1 unit, the area of the region bounded by the x-axis, the y-axis, the line $x = -3$ and the curve whose equation is $y = e^{-x}$ is approximated by:

A. $e^3 + 2e^2 + 2e^1 + 1$
B. $\frac{1}{2}e^3 + e^2 + e^1 + \frac{1}{2}$
C. $\frac{1}{2}e^3 + e^2 + e^1$
D. $e^3 + 2e^2 + 2e^1$
E. $\frac{1}{2}e^3 + e^2 + e^1$

8. **MC** The graph of $y = (4 - x)\left(x^2 - x + 1\right)$ is shown. The approximate area between the curve and the x- and y-axes using four trapeziums of equal width is:

A. 36 units^2
B. 14.5 units^2
C. 29 units^2
D. 18 units^2
E. 20.5 units^2

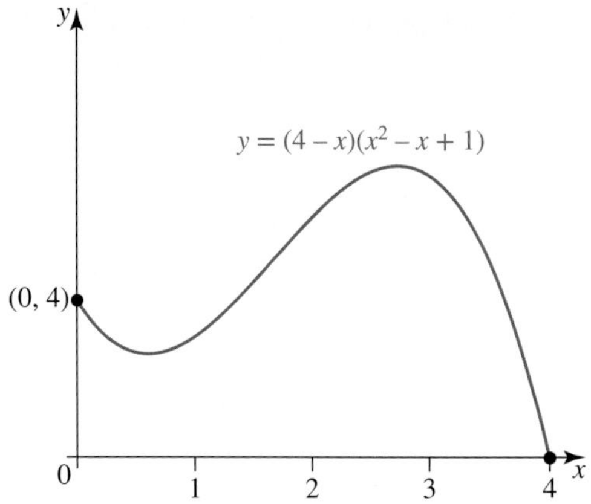

9. **MC** $\displaystyle\int_{-\frac{\pi}{2}}^{\frac{\pi}{3}} \left(\cos\left(\frac{x}{2}\right) - \sin(2x)\right) dx$ is equal to:

A. $\sqrt{2} + \dfrac{5}{4}$
B. $\sqrt{2} + \dfrac{3}{4}$
C. $\dfrac{3\sqrt{3}}{4} + \sqrt{2} + \dfrac{1}{2}$
D. $\sqrt{2} - \dfrac{3}{4}$
E. $\dfrac{1}{4} - \sqrt{2}$

10. **MC** The graph of $y = \sqrt{x - 0.5}$ is shown.
The approximate area under the curve from $x = 0.5$ to $x = 3.5$, using the trapezium method with interval widths of 1 unit, is:

A. 2.8 units^2
B. 6.6 units^2
C. 3.3 units^2
D. 4.1 units^2
E. 5.6 units^2

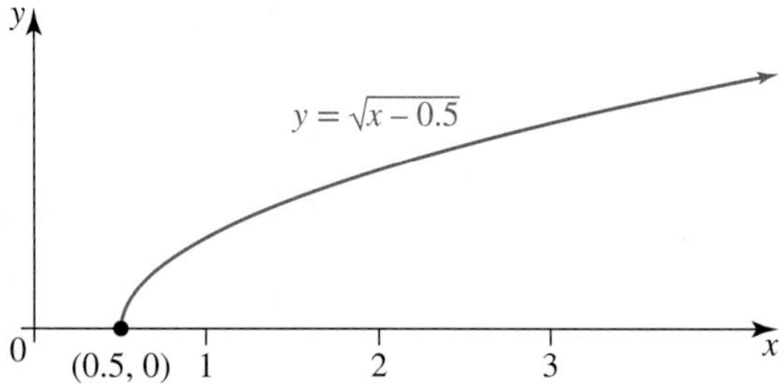

11. **MC** The graphs of $y = f(x)$ and $y = g(x)$ are shown.
The area of the shaded region is given by:

A. $\displaystyle\int_0^n (g(x) - f(x))\, dx$

B. $\displaystyle\int_0^n (f(x) - g(x))\, dx$

C. $\displaystyle\int_0^m (f(x) - g(x))\, dx + \int_m^n (g(x) - f(x))\, dx$

D. $\displaystyle\int_0^m (f(x) - g(x))\, dx - \int_m^n (g(x) - f(x))\, dx$

E. $\displaystyle\int_0^n (f(x) + g(x))\, dx$

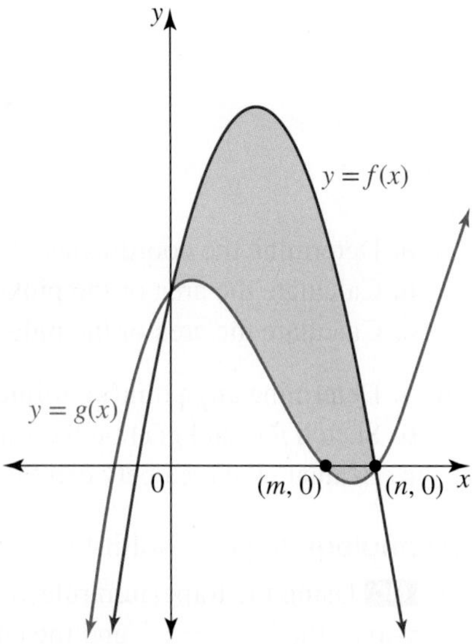

12. **MC** The graph with equation $y = k\left(x^2 + 2\right)$ is shown.
The area of the shaded region is equal to 5 square units. The value of k is:

A. $\frac{1}{5}$ B. $\frac{1}{2}$ C. 1 D. $\frac{1}{3}$ E. $\frac{2}{3}$

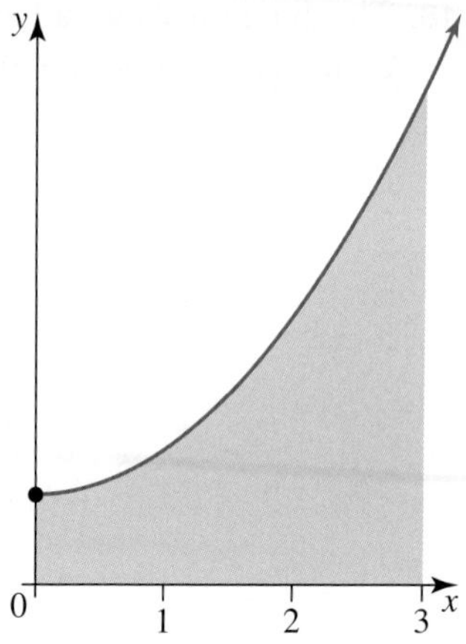

13. **MC** The rate of flow of water from a hose nozzle can be defined by $\frac{dV}{dt} = 3e^{-0.2t}$, where V is the volume in litres and t is the time in minutes that the hose has been turned on. The number of litres that has flowed out in 2 minutes is:

A. 10.05 B. 0.2 C. 2.01 D. 5 E. 4.95

***The following information refers to Questions* 14 *and* 15.**

A particle starts at the origin and travels in a straight line with a velocity, v m/s, modelled by $v = t^2 - t - 2$.

14. **MC** The position of the particle after 3 seconds is:

A. 1.5 m B. 4.5 m C. −4.5 m D. 12 m E. −1.5 m

15. **MC** The distance covered by the particle in the first 2 seconds is:

A. $1\frac{5}{6}$ m B. 3 m C. $8\frac{2}{3}$ m D. $3\frac{1}{3}$ m E. 1 m

16. **MC** If $k = \int_1^3 \frac{1}{x}\, dx$, then e^k is equal to:

A. $\log_e(3)$ B. 1 C. 3 D. e E. e^3

Technology active: extended response

17. The graph of the curve $f(x) = x^2 + 3$ is shown. Let A represent the area enclosed by the curve, the x- and y-axes, and the line $x = 2$.

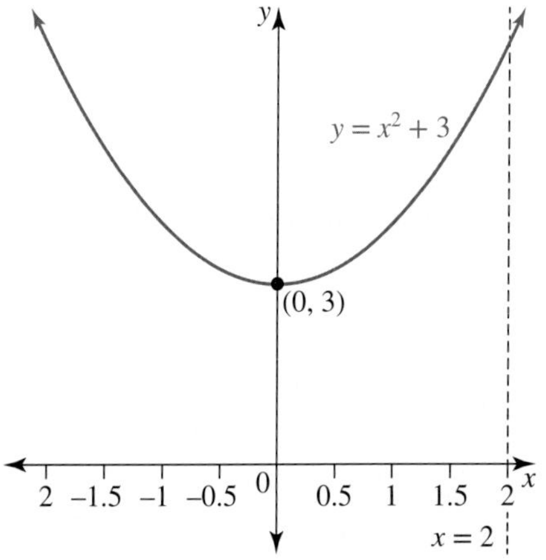

a. Determine an approximation for the area, A, using four trapeziums of width 0.5.

b. Find the exact value of A, that is $\int_0^2 \left(x^2 + 3\right) dx$.

18. The diagram shows a feeding trough for horses, with dimensions given in centimetres. The parabola is a cross-sectional profile of each end of the trough.

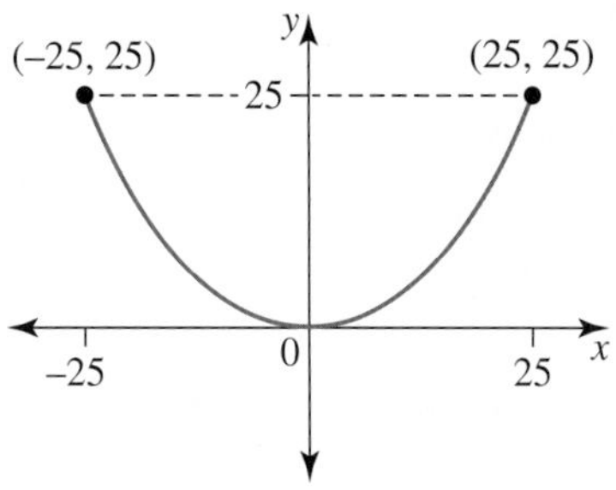

a. Determine the equation of the parabola.
b. Calculate the area of the parabolic cross-section.
c. Hence, calculate the volume of the trough if it is known to be 120 cm long.

19. The graph of $y = m\log_e(n(x+p))$ is shown.

a. Determine the values of the non-zero constants m, n and p.
b. Describe the transformations that have been applied to $y = \log_e(x)$ to achieve this function.
c. On one set of axes, sketch the graph of $y = f(x)$ and its inverse function, $y = f^{-1}(x)$. State the domain and range of each function, and give the equations of any asymptotes.
d. Calculate the point where $f(x) = f^{-1}(x)$.
e. Calculate the area between the two functions from $x = -1.5$ to $x = 0$. Give your answer correct to 4 decimal places.

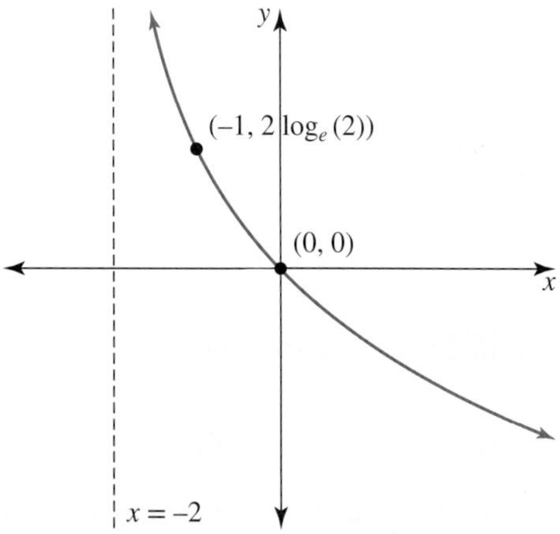

20. a. Let $f: R^+ \to R, f(x) = e^{-\frac{x}{3}}(m\cos(x) + n\sin(x))$. Given that $f'(x) = e^{-\frac{x}{3}}\sin(x)$, determine the constants m and n.

b. Calculate $\int_0^{2\pi} e^{-\frac{x}{3}}\sin(x)\,dx$.

c. The graph of $y = e^{-\frac{x}{3}}\sin(x)$ is shown. Calculate the area of the shaded region, correct to 4 decimal places.

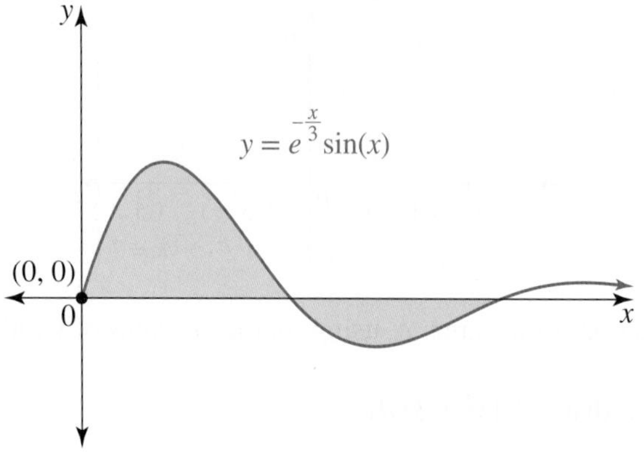

8.6 Exam questions

Question 1 (1 mark) TECH-ACTIVE

Source: *VCE 2021, Mathematical Methods Exam 2, Section A, Q11; © VCAA.*

MC If $\int_0^a f(x)\,dx = k$, then $\int_0^a (3f(x)+2)\,dx$ is

A. $3k + 2a$
B. $3k$
C. $k + 2a$
D. $k + 2$
E. $3k + 2$

Question 2 (4 marks) TECH-FREE

Source: *VCE 2017, Mathematical Methods Exam 1, Q2; © VCAA.*

Let $y = x\log_e(3x)$.

a. Find $\frac{dy}{dx}$. **(2 marks)**

b. Hence, calculate $\int_1^2 (\log_e(3x)+1)dx$. Express your answer in the form $\log_e(a)$, where a is a positive integer. **(2 marks)**

Question 3 (5 marks) TECH-FREE

Source: *VCE 2016, Mathematical Methods Exam 1, Q3; © VCAA.*

Let $f: R\backslash\{1\} \to R$, where $f(x) = 2 + \frac{3}{x-1}$.

a. Sketch the graph of f. Label the axis intercepts with their coordinates and label any asymptotes with the appropriate equation. **(3 marks)**

b. Find the area enclosed by the graph of f, the lines $x = 2$ and $x = 4$, and the x-axis. **(2 marks)**

Question 4 (1 mark) TECH-ACTIVE

Source: *VCE 2019, Mathematical Methods Exam 2, Section A, Q12; © VCAA.*

MC If $\int_1^4 f(x)\,dx = 4$ and $\int_2^4 f(x)\,dx = -2$, then $\int_1^2 (f(x)+x)$ is equal to

A. 2
B. 6
C. 8
D. $\frac{7}{2}$
E. $\frac{15}{2}$

Question 5 (1 mark) TECH-ACTIVE

Source: VCE 2013, Mathematical Methods (CAS) Exam 2, Section 1, Q15; © VCAA.

MC Let h be a function with an average value of 2 over the interval [0, 6].

The graph of h over this interval could be

A.

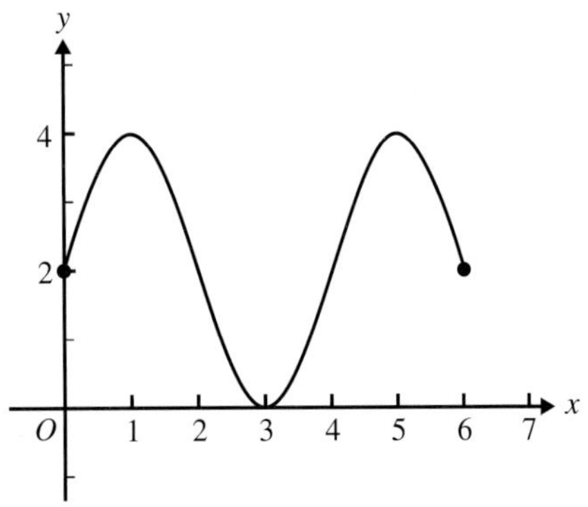

B.

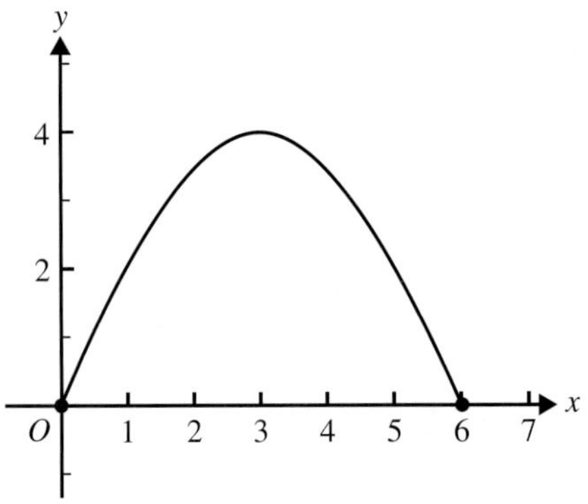

C.

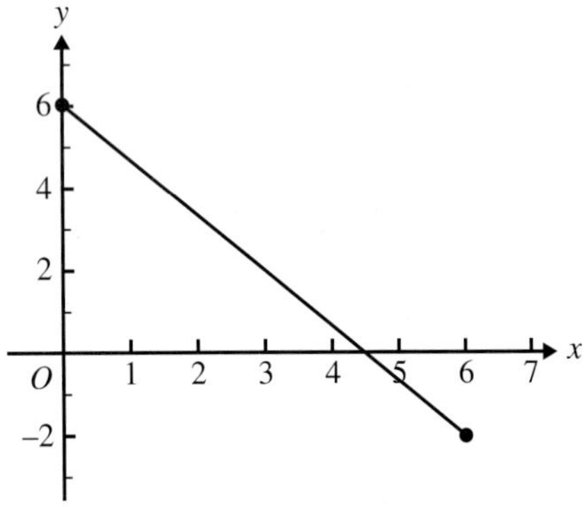

D.

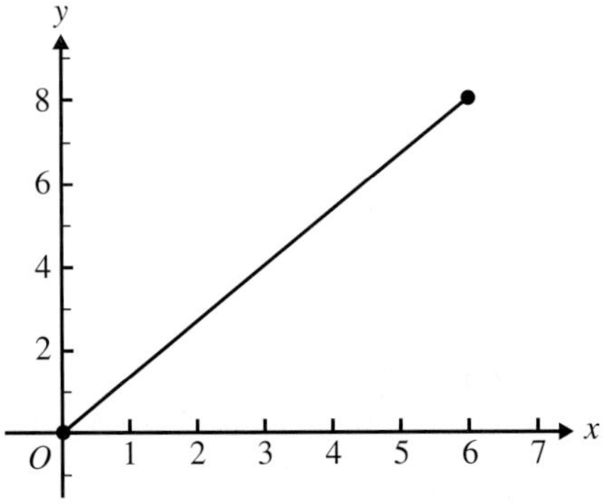

E.

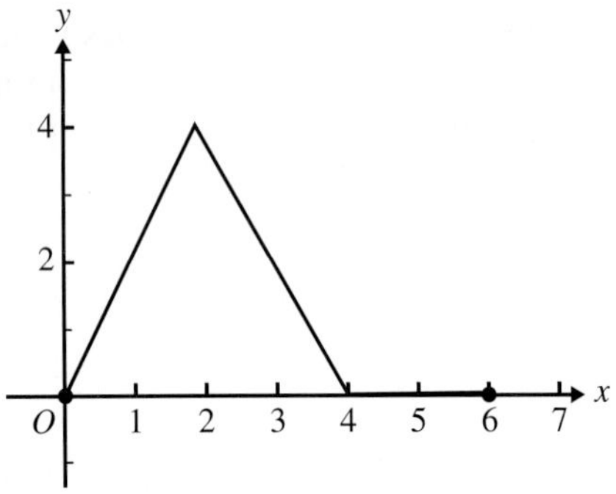

More exam questions are available online.

Answers

Topic 8 Integral calculus

8.2 The fundamental theorem of integral calculus

8.2 Exercise

1. $\frac{51}{30}$ units^2
2. 26 units^2
3. 21 units^2
4. a. 4 b. 0
5. a. $\frac{65}{4}$ b. $2 + 0.5e^2 - 0.5e^{-2}$
6. a. 27 b. $\frac{55}{6}$
   c. 0 d. 0
   e. $\frac{4}{3}\left(\sqrt{10} - 2\right)$ f. $\sqrt{2} - \frac{3\sqrt{3}}{4}$
7. $\frac{dy}{dx} = \frac{9x}{3x^3 - 4}$ and $\int_2^3 \frac{x^2}{3x^3 - 4}\,dx = \frac{1}{9}\log_3\left(\frac{77}{20}\right)$
8. B
9. a. $a = 4$ b. 7.56 units^2
10. a. i. 21 ii. 11
    iii. -16 iv. 8
    b. $k = \pm 1$
11. a. -15 b. -12.5
    c. 32.5 d. 20
    e. 25 f. 12.5
12. D
13. a. $a = 4$ b. $k = \sqrt{5}$
14. a. $a = 1$ b. $21\frac{1}{3}$ units^2
15. 1.964
16. a. $\frac{dy}{dx} = x\cos(x) + \sin(x)$ b. $\pi + 2$
17. a. $\frac{dy}{dx} = 3\left(x^2 - 2x\right)e^{x^3 - 3x^2}$
    b. $\frac{1}{3}\left(e^{-2} - 1\right)$
18. D
19. $\frac{dy}{dx} = \frac{2e^x}{(e^x + 1)}$ and $\int_1^5 \frac{e^x}{e^x + 1}\,dx = 3.6935$

8.2 Exam questions

Note: Mark allocations are available with the fully worked solutions online.

1. E
2. B
3. 2

8.3 Areas under curves

8.3 Exercise

1. $166\frac{2}{3}$ units^2
2. 3π units^2
3. 1.6 units^2
4. $e + e^{-1} - 2$ units^2
5. 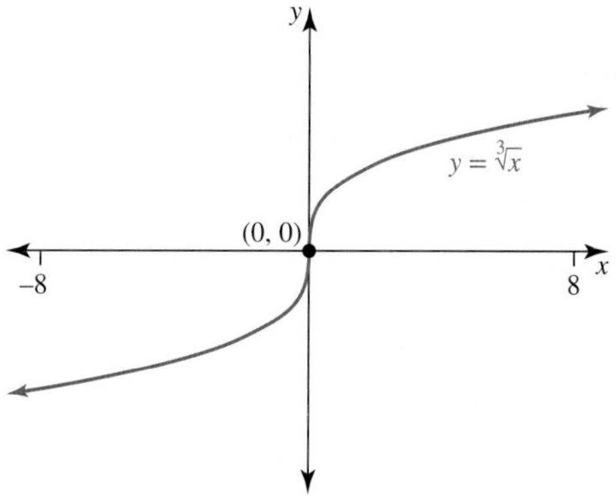

   24 units^2
6. $\log_e\left(\frac{2}{3}\right) + 2$
7. a. 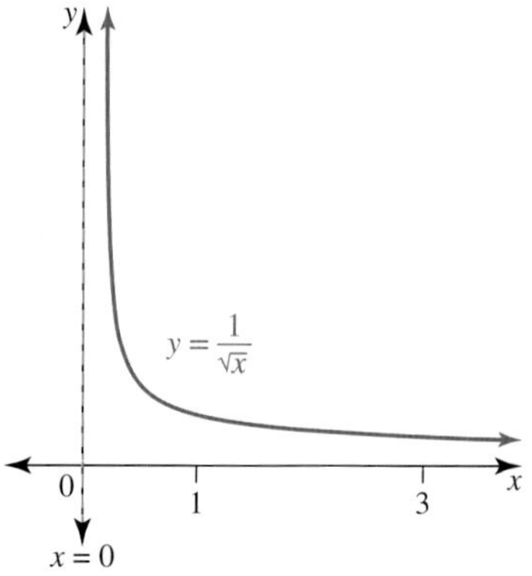

   b. $2\sqrt{3} - 2$ units^2
8. a. 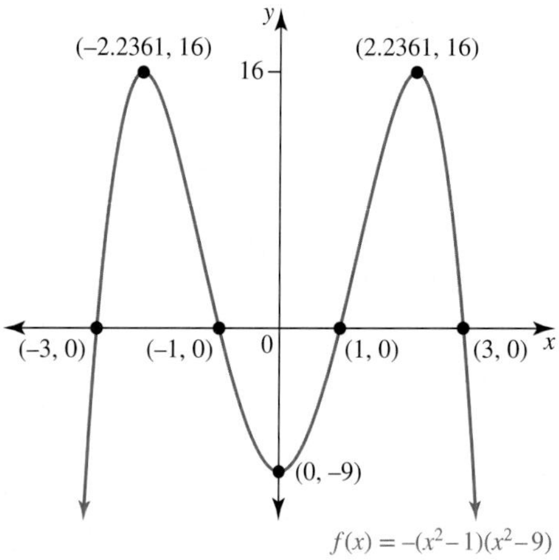

   b. 52.27 units^2
9. C
10. 16 units^2

11. a. $a = -2,\ b = -1,\ c = 2,\ d = 3$
   b. 15.68 units^2
12. a. $m = 2.6779$ b. 4.2361 units^2
13. a. 1.7642 units^2 b. 9.933 units^2
14. a. $\dfrac{dy}{dx} = 2xe^{x^2}$ b. $2(e-1)$ units^2
15. a. $y = -3x^2 + 6x$
   b. 4 m^2
   c. \$440
16. a. $\dfrac{3}{4}$ units^2 b. $2\dfrac{1}{4}$ units^2
17. a. 2 b. $2(\pi - 2)$ units^2
18. a. Domain $= (1, \infty)$, range $= R$
   b. $a = 2$
   c. 5.0904 units^2
19. a. Maximum turning point $= \left(\sqrt{5}, \sqrt{5}\right)$,
   minimum turning point $= \left(-\sqrt{5}, -\sqrt{5}\right)$
   b. $5\log_e(5 + x^2)$
   c. $5\log_e\left(\dfrac{41}{10}\right)$ units^2
20. a. 6.479 units^2
   b. $f^{-1}(x) = e^{\frac{x}{5}} + 3, x \in R$
   c. 6.479 units^2
21. a. 7.7714 cm^2 b. 15.5 cm^2 c. Scale factor = 224

8.3 Exam questions

Note: Mark allocations are available with the fully worked solutions online.

1. a. i. $\dfrac{3}{2}$
   ii.

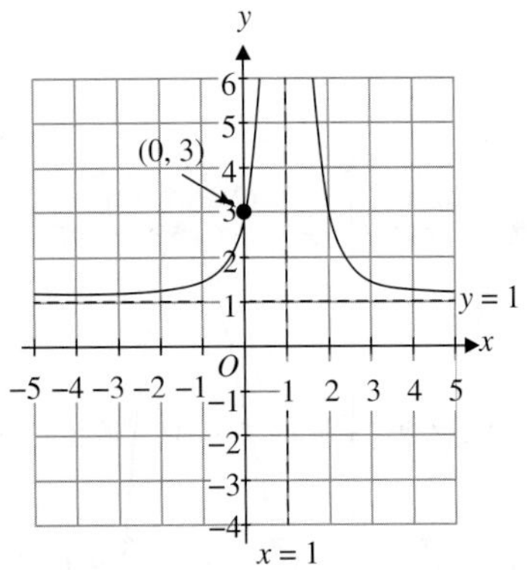

   b. $A = 2$ units^2
2. D
3. B

8.4 Areas between curves and average values

8.4 Exercise

1. a. Point of intersection = (16, 4)
   b. Area $= 21\dfrac{1}{3}$ units^2
2. Area $= 20\dfrac{5}{6}$ units^2
3. a. $\left(\dfrac{3\pi}{4}, \dfrac{\sqrt{2}}{2}\right)$ b. $2\sqrt{2}$ units^2
4. $e - 1$
5. $-\dfrac{11}{12}$
6. 4.6254 units^2
7. 0.7641 units^2
8. B
9. a.

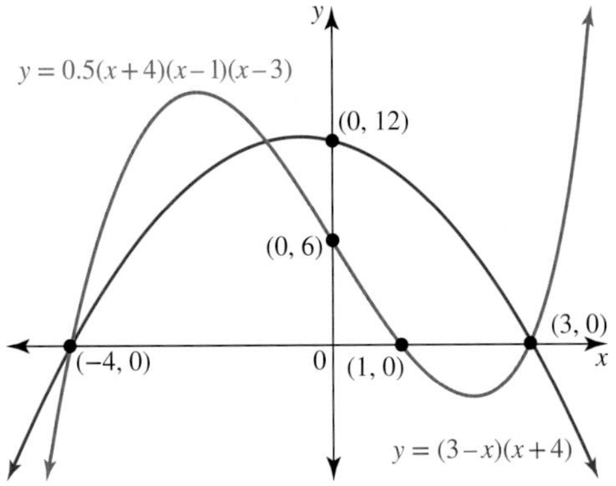

   b. $y = 0.5(x+4)(x-1)(x-3)$ [1]
   $y = (3-x)(x+4)$ [2]
   [1] = [2]:
   $0.5(x+4)(x-1)(x-3) = (3-x)(x+4)$
   $0.5(x+4)(x-1)(x-3) - (3-x)(x+4) = 0$
   $0.5(x+4)(x-1)(x-3) + (x-3)(x+4) = 0$
   $(x-3)(x+4)(0.5(x-1)+1) = 0$
   $(x-3)(x+4)(0.5x+0.5) = 0$
   $x-3 = 0,\ x+4 = 0$ or $0.5x + 0.5 = 0$
   $x = 3\ x = -4\ x = -1$
   When $x = -4$, $y = (3+4)(-4+4) = 0$.
   When $x = -1$, $y = (3+1)(-1+4) = 12$.
   When $x = 3$, $y = (3-1)(3+4) = 0$.
   Therefore, the coordinates are $(-4, 0)$, $(-1, 12)$ and $(3, 0)$.
   c. 39.04 units^2
10. a. $a = -1$ b. $\log_e(4) - 3$
   c. $\left(-\dfrac{3}{2}, 1\right), \left(2, -\dfrac{3}{4}\right)$ d. $\dfrac{63}{16} - 3\log_e(2)$
11. a. $y = \dfrac{12}{5}x - \dfrac{4}{5}$ or $12x - 5y = 4$
   b. $\dfrac{5}{2}\log_e(1 + x^2)$
   c. 1.629 units^2
12. a. $y = 2x - 4$
   b. $\dfrac{3}{4} + \log_e(2)$ units^2
13. a. $a = 1,\ b = 2,\ c = 3$
   b. 9.6 units^2
   c. 6.144

14. a. 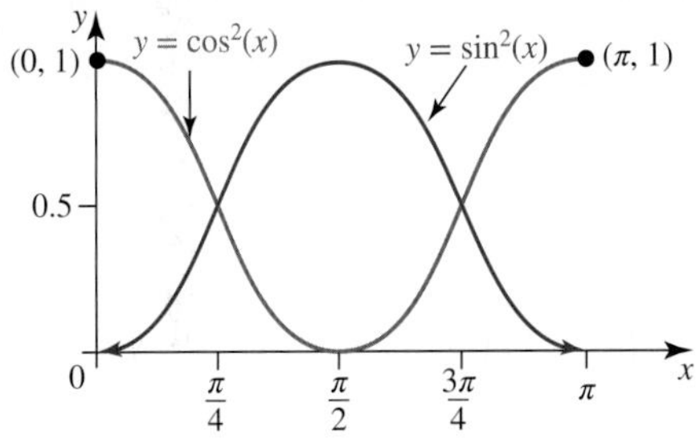

b. 2 units^2

15. a. $(0, -2), (3, 1)$

b. $\frac{7}{6}$ units^2

c. $3\frac{1}{3}$ units^2

d. 4.5 units^2

8.4 Exam questions

Note: Mark allocations are available with the fully worked solutions online.

1. B
2. C
3. a. $f'(x) = xe^{kx}(kx + 2)$

b. $k = 1$

c. $A = \int_0^2 \left(x^2 e^{kx} + \frac{2xe^{kx}}{k} \right) dx$

d. $k = \log_e(2)$

8.5 Applications

8.5 Exercise

1. a and b.

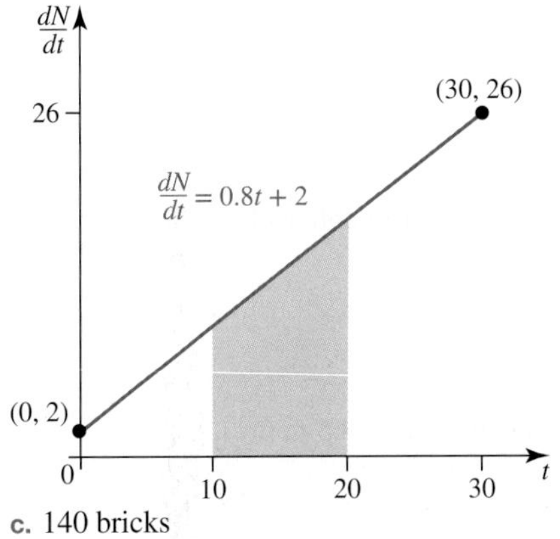

c. 140 bricks

2. a. 4 m/s

b. $a = \frac{3}{2\sqrt{t+1}}$

i. 1.5 m/s^2

ii. 0.5 m/s^2

c. 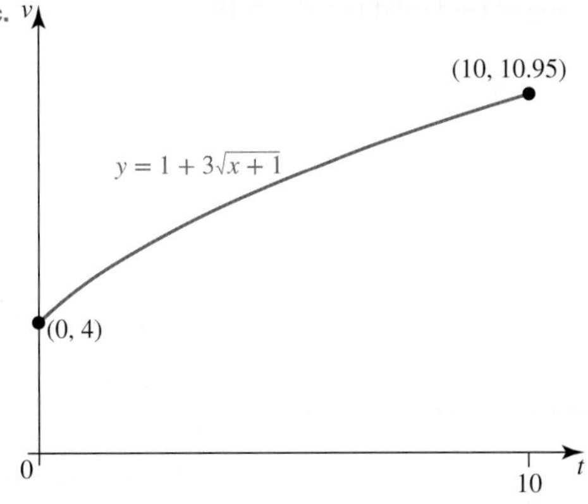

d. 60 m

3. a. $y = 5 - 0.2x^2$

b. $33\frac{1}{3}$ m^2

c. $50\frac{2}{3}$ m^2

d. 152 m^3

4. a. $f^{-1}(x) = \frac{1}{4}e^{\frac{x}{2}}$

b. $\frac{7}{2}$ units^2

c. $4\log_e(8) - \frac{7}{2}$ units^2

5. a. $\left(\frac{1}{5}, 0\right)$

b. $\frac{dy}{dx} = -\log_e(5x)$

c. $2\log_e(10) - \frac{9}{5}$ units^2

6. a. 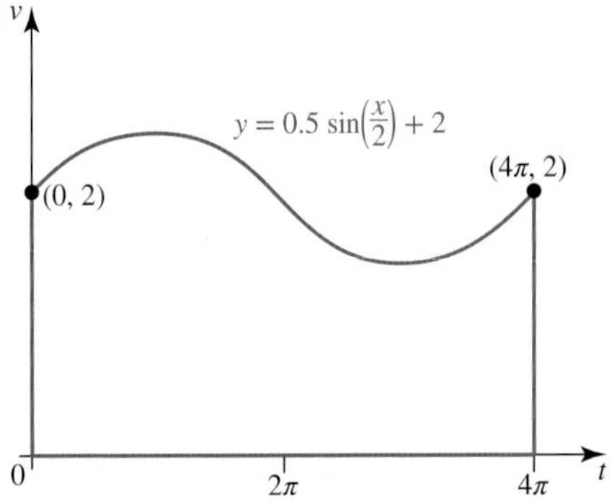

b. 25 m^2

c. 12.5 m^3

7. 28.4 cm
8. 55.3 million
9. a. 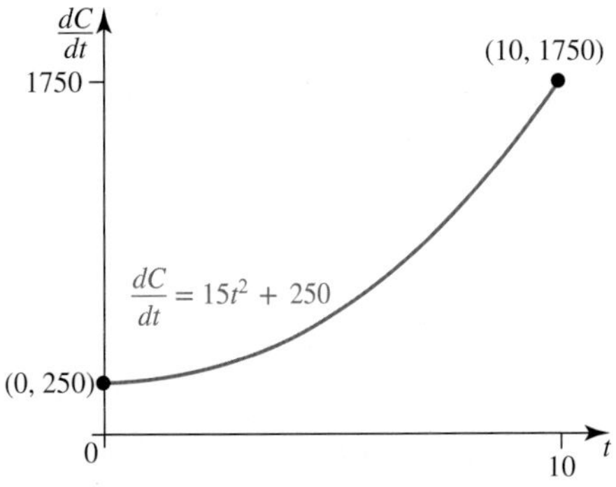

b. \$5625

10. a. $(0, 2\log_e(5)+1)$ and $\left(e^{-0.5}-5, 0\right)$

b. $h^{-1}: R \to R, h^{-1}(x) = e^{\frac{1}{2}(x-1)} - 5.$

c.

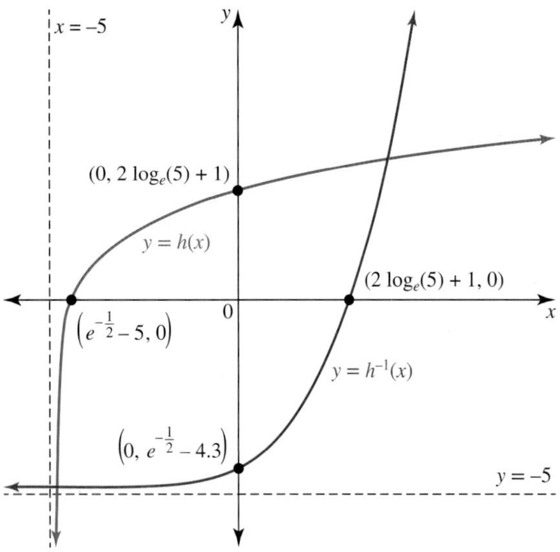

d. $x = -4.9489, 5.7498$

e. 72.7601 units^2

11. a. $\dfrac{dy}{dx} = \log_e(x) + 1$ and $\displaystyle\int_1^{e^2} \log_e(x)\,dx = e^2 + 1$

b. $\dfrac{dy}{dx} = \left(\log_e(x)\right)^m + m\left(\log_e(x)\right)^{m-1}$

c. Consequentially

$$\int_1^{e^2} \left((\log_e(x))^m + m(\log_e(x))^{m-1}\right) dx = \left[x(\log_e(x))^m\right]_1^{e^2}$$

$$\int_1^{e^2} (\log_e(x))^m dx + \int_1^{e^2} m(\log_e(x))^{m-1}\,dx = \left[x(\log_e(x))^m\right]_1^{e^2}$$

$$\int_1^{e^2} (\log_e(x))^m dx + \int_1^{e^2} m(\log_e(x))^{m-1}\,dx = e^2(\log_e(e^2))^m - (1)(\log_e(1))^m$$

$$\int_1^{e^2} (\log_e(x))^m dx + \int_1^{e^2} m(\log_e(x))^{m-1}\,dx = e^2(2\log_e(e))^m - 0$$

$$\int_1^{e^2} (\log_e(x))^m dx + \int_1^{e^2} m(\log_e(x))^{m-1}\,dx = 2^m e^2$$

If $I_m = \displaystyle\int_1^{e^2} (\log_e(x))^m dx$, then

$I_m + mI_{m-1} = 2^m\, e^2$ as required

d. $\displaystyle\int_1^{e^2} \left(\log e(x)\right)^3 dx = 2e^2 + 6$

12. 5.1 km^2

13. a. $a = 5$

b. 30 minutes

c. $\dfrac{23}{30}$ units/minute

d. 274.6683 units

14. a. $x = 6\sin\left(\dfrac{t}{2} - \dfrac{\pi}{4}\right)$

b. $-3\sqrt{2}$ m

c.

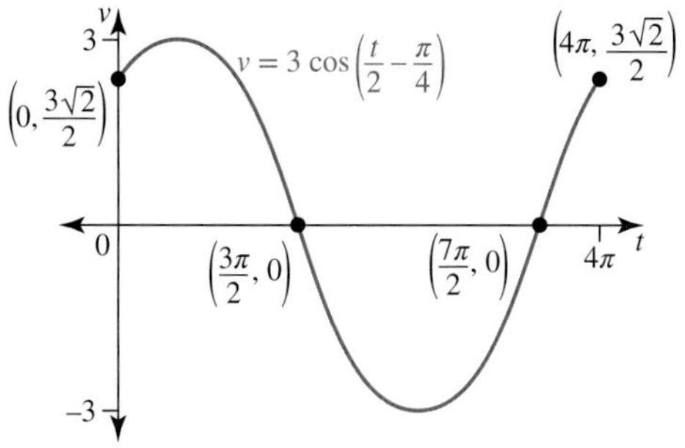

d. 20.49 m

e. $a = -\dfrac{3}{2}\sin\left(\dfrac{t}{2} - \dfrac{\pi}{4}\right)$

f. $\dfrac{3\sqrt{2}}{4}$ m/s^2

15. a. $a = -0.5e^{-0.5t}$

b. $x = -2e^{-0.5t} - 0.5t + 2$

c. -0.2707 m

d. 0.3244 m

16. 1.44 m

17. a. $f^{-1}(x) = 2\log_e(x-3) + 1$

b. $(0, e^{-\frac{1}{2}} + 3)$ and $(e^{-\frac{1}{2}} + 3, 0)$

c. 26.58 m^2

d. 16.11 m^2

e. 42.69 m^2

8.5 Exam questions

Note: Mark allocations are available with the fully worked solutions online.

1. a. 10 m

b. $\dfrac{50}{3}$

c. The minimum distance is 8.5 m (to 1 d.p.).

d. 2000 m^2

e. 837 m^2

f. Given that $k \geq 1$, $k \in \left[1, \dfrac{7}{6}\right)$.

2. a. 12

b. $t = 0, 4, 6$

c. 1.76

d. $\dfrac{15}{\pi}$

e. $k = \dfrac{5}{2\pi}$

3. a. 72°

b. The maximum slope is $-\frac{3}{16}$, that is $\frac{3}{16}$ downwards.

c. $M(2.49, 34.53)$

d. $w = 35.47$ m, $PQ = 24.30$ m, $MN = 25.23$ m

e. E: $x = -23.71$
F: $x = 28.00$

f. 870 m^2

8.6 Review

8.6 Exercise

Technology free: short answer

1. a. i. 3   ii. $8\left(\sqrt{2}+1\right)$ iii. $\frac{e}{2}-1$

b. i. 20   ii. 5   iii. −2

2. a. i. $\log_e(3)+2$

ii. $-\frac{3}{2}\log_e(3)$

b. $a = 2e^3 + 1$

3. a. $m = 1$ b. $k = 2$

4. a. $f_{avg} = \frac{2\sqrt{2}}{3}$ b. $x = 1$

c. $1\frac{1}{6}$ units^2 d. $\frac{5}{6}$ units^2

5. a. $\left(-\frac{1}{2}, 1\right), \left(\frac{1}{2}, 1\right)$ b. $\frac{13}{24}$ units^2

c. $\frac{11}{12}$ units^2

6. a. (−2, 0), (0, 2), (2, 4)

b.

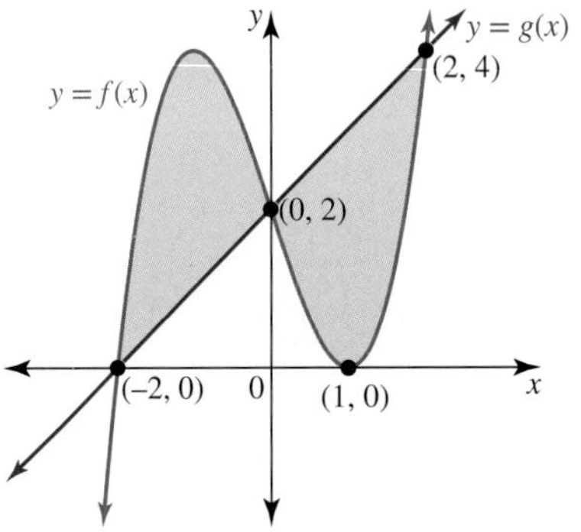

c. 8 units^2

Technology active: multiple choice

7. B   8. D   9. A   10. C
11. B   12. D   13. E   14. E
15. D   16. C

Technology active: extended response

17. a. 8.75 units^2 b. $\frac{26}{3}$ units^2

18. a. $y = \frac{1}{25}x^2$ b. $\frac{1250}{3}$ cm^2

c. 50 000 cm^3

19. a. $m = -2$, $n = \frac{1}{2}$ and $p = 2$

b. $y = \log_e x$ is reflected in the x-axis, dilated by a factor of 2 parallel to the y-axis or from the x-axis and dilated by a factor of 2 parallel to the x-axis or from the y-axis, and then translated 2 units to the left.

c. $f^{-1}: y = 2e^{-\frac{x}{2}} - 2$

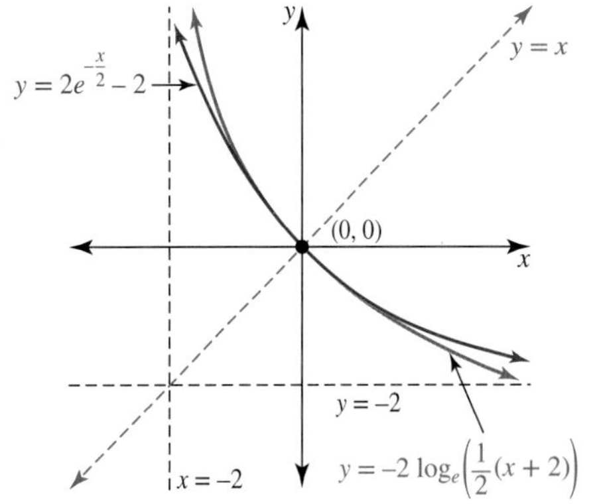

For $f: y = -2\log_e\left(\frac{1}{2}(x+2)\right)$, domain = $(-2, \infty)$ and range = R.

For $f^{-1}: y = 2e^{-\frac{x}{2}} - 2$, domain = R and range = $(-2, \infty)$.

d. (0, 0)

e. 0.1457 units^2

20. a. $m = -\frac{9}{10}$, $n = -\frac{3}{10}$ b. $\frac{9}{10}\left(1 - e^{\frac{-2\pi}{3}}\right)$

c. 1.6425 units^2

8.6 Exam questions

Note: Mark allocations are available with the fully worked solutions online.

1. A

2. a. $\log_e(3x) + 1$ b. $\log_e(12)$

3. a.

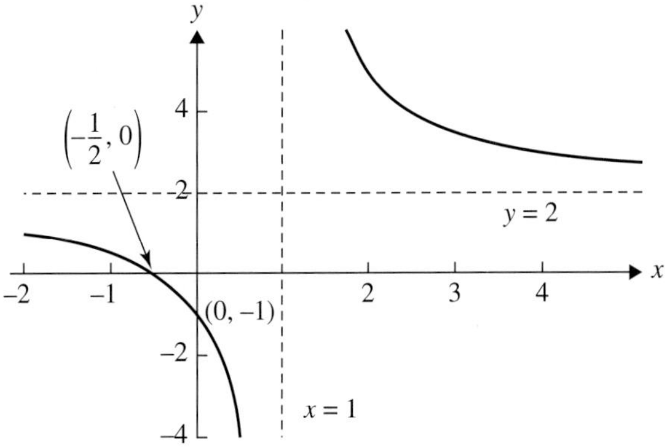

b. $4 + 3\log_e(3)$

4. E

5. C

9 Discrete random variables

LEARNING SEQUENCE

Fully worked solutions for this topic are available online.

9.1 Overview

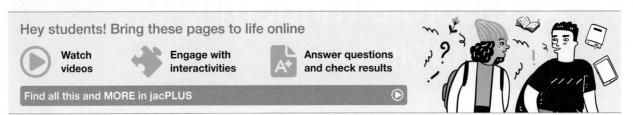

Hey students! Bring these pages to life online

Watch videos

Engage with interactivities

Answer questions and check results

Find all this and MORE in jacPLUS

9.1.1 Introduction

A random variable is a variable whose value is determined by the outcome of a random procedure.

Continuous random variables deal with values that are obtained by measuring, whereas discrete random variables are values that are obtained by counting. For example, the number of people attending a series of concerts is a discrete random variable. A discrete random variable may assume only certain values within a given range.

There are many different types of discrete distributions. Topic 10 covers one particular type, the binomial distribution. Others are the Poisson distribution, the geometric distribution, the hypergeometric distribution and the multinomial distribution. The Poisson distribution was named after French mathematician Simeon Denis Poisson. It gives the probability of a given number of events occurring over a fixed time period. For example, if a restaurant monitors its customers every day and for a particular week the average number each day is 200, this can be used to predict the probability of a particular day having more diners. Businesses use the Poisson distribution to make forecasts about the number of customers or sales so that they can accurately stock their shelves without too much overstock or understock, or hotels can prepare for a predicted influx of visitors by employing more staff.

KEY CONCEPTS

This topic covers the following key concepts from the VCE Mathematics Study Design:

- random variables, including the concept of a random variable as a real function defined on a sample space and examples of discrete random variables
- discrete random variables:
  - specification of probability distributions for discrete random variables using graphs, tables and probability mass functions
  - calculation and interpretation of mean, μ, variance, σ^2 and standard deviation of a discrete random variable and their use
  - calculation of probabilities for specific values of a random variable and intervals defined in terms of a random variable, including conditional probability

9.2 Probability review

LEARNING INTENTION

At the end of this subtopic you should be able to:
- use probability language and determine probabilities of events
- use Venn diagrams to represent two or more events
- use the additional rule of probability to solve problems
- determine mutually exclusive events.

9.2.1 Probability events

This section will review the key concepts and skills studied in Units 1 and 2 of Mathematical Methods. This is assumed knowledge for Units 3 and 4.

The **sample space** is the set of all possible outcomes for an experiment, and is denoted by the symbol ξ. For example, if rolling a standard die, $\xi = \{1, 2, 3, 4, 5, 6\}$. The sample space can also be written as a tree diagram or table.

The **probability** of an **event** is the likelihood of that event occurring. For example, the probability of rolling a 6 for a standard die is $\frac{1}{6}$: one desired outcome from 6 possible outcomes. Probability values range between 0 (impossible) to 1 (certain). The higher the probability value of an event, the more likely it is to occur.

The **complement** of an event, A, is everything other than the event occurring; that is, 'not' A. So, for the example of event A above, rolling a 6, $A' = \{1, 2, 3, 4, 5\}$.

$$\Pr(A') = 1 - \Pr(A)$$

Probability rules

- $\Pr(A) = \dfrac{\text{number of favourable outcomes corresponding to event } A}{\text{total number of possible outcomes}}$
- $0 \leq \Pr(A) \leq 1$
- **The sum of all probabilities of the outcomes for an experiment is 1.**
- $\Pr(\varnothing) = 0$. **$\varnothing$ is the null set, which is a set with no elements in it.**
- $\Pr(A') = 1 - \Pr(A)$

WORKED EXAMPLE 1 Determining probabilities

A bag contains 6 blue discs, 2 red discs and 7 green discs. A disc is drawn randomly from the bag. Determine the probability that the disc drawn is:

a. blue **b. not red** **c. either green or red.**

THINK	WRITE
a. 1. Define the event.	Let B = a blue disc.
2. Substitute the values into the probability rule: $\Pr(B) = \dfrac{\text{number of blue discs}}{\text{total number of discs}}$.	$\Pr(B) = \dfrac{6}{15}$

3. Simplify. $= \frac{2}{5}$

b. 1. Define the event. Let R' = not a red disc.

2. State the complementary rule. $\Pr(R') = 1 - \Pr(R)$

3. Substitute values into the probability rule. $= 1 - \frac{2}{15}$

4. Evaluate. $= \frac{13}{15}$

c. 1. Define the events. Let R = a red disc and let G = a green disc.

2. Substitute values into the probability rule. Add the total number of red and green discs. $\Pr(R \text{ or } G) = \frac{2+7}{15}$

3. Evaluate. $= \frac{9}{15}$

4. Simplify. $= \frac{3}{5}$

9.2.2 Venn diagrams

A **Venn diagram** is a visual way of representing the sample space, generally involving two events. It can also be used to represent the probabilities for each event.

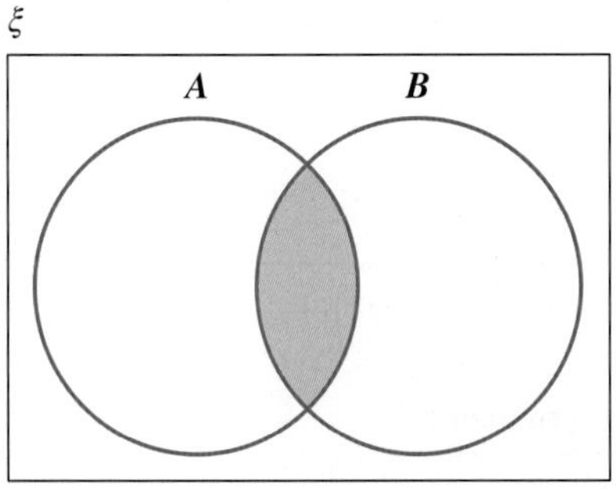

$A \cap B$

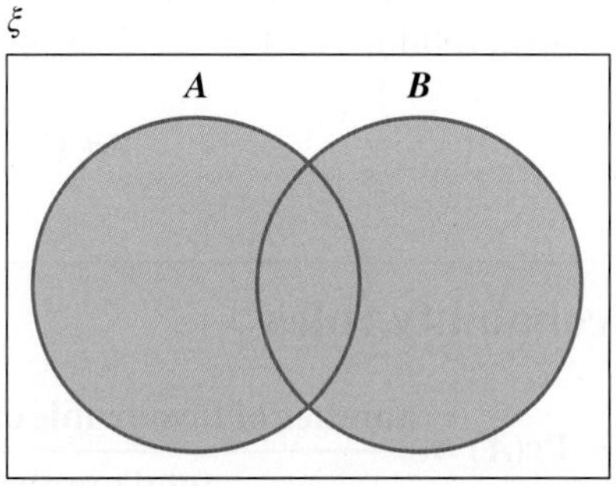

$A \cup B$

The intersection of events A and B is denoted by $A \cap B$. This represents the elements that are common to both; that is, the elements that are in both A **and** B.

The union of events A and B is denoted by $A \cup B$. This represents the total elements in either A or B; that is, the elements that are in A **or** B **or** both.

The addition rule of probability

$$\Pr(A \cup B) = \Pr(A) + \Pr(B) - \Pr(A \cap B)$$

Conditional probability occurs when additional information is known, such as the probability of event A occurring when it is known that another event, B, has already occurred. The notation for this is $\Pr(A \mid B)$.

The conditional probability rule

$$\Pr(A \mid B) = \frac{\Pr(A \cap B)}{\Pr(B)} \quad \text{if } \Pr(B) \neq 0$$

or

$$\Pr(A \mid B) = \frac{n(A \cap B)}{n(B)} \quad \text{if } n(B) \neq 0.$$

This can be rearranged to the multiplication rule of probability:

$$\Pr(A \cap B) = \Pr(A \mid B) \times \Pr(B)$$

WORKED EXAMPLE 2 Venn diagrams

A group of 40 students were surveyed as to whether they studied Maths and/or History in Year 12. Five people said they studied both subjects, 18 said they studied Maths and 8 studied neither.

a. **Represent the information in a Venn diagram.**

b. **Determine the probability of selecting a student who:**
   - i. **studies History**
   - ii. **studies Maths and History**
   - iii. **studies Maths given they study History.**

THINK	WRITE
a. First place 5 in the intersection region. Maths only section = 18–5 = 13. Place 8 in the region outside the circles. Subtract the values from 40. Place the remaining 14 in the History only area.	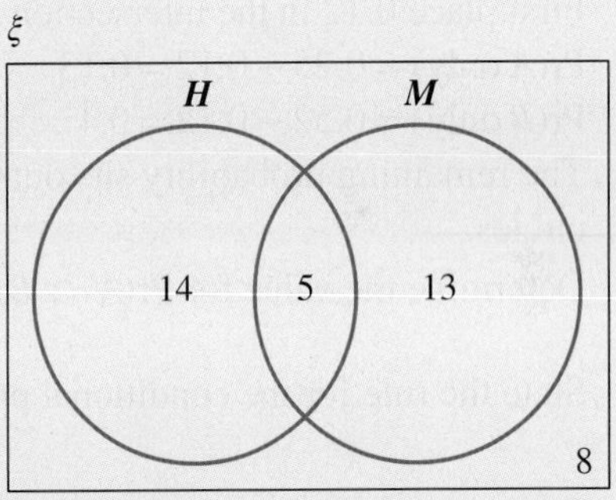
b. i. 1. Substitute the values into the probability rule: $\Pr(H) = \frac{\text{number of people who study History}}{\text{total number of students}}$.	$\Pr(H) = \frac{19}{40}$
ii. 1. Study History **and** Maths is denoted by the intersection, $H \cap M$. Substitute the values into the probability rule.	$\Pr(H \cap M) = \frac{5}{40}$
2. Simplify.	$= \frac{1}{8}$
iii. 1. This is a conditional probability. Given that the Venn diagram is written in terms of numbers of people, not probabilities, choose the appropriate rule.	$\Pr(M \mid H) = \frac{n(M \cap H)}{n(H)}$
2. Substitute the appropriate values and evaluate.	$\Pr(M \mid H) = \frac{5}{19}$

WORKED EXAMPLE 3 The addition rule

If $\Pr(A) = 0.25$, $\Pr(B) = 0.52$ and $\Pr(A \cap B) = 0.12$, find:

a. **$\Pr(A \cup B)$**

b. **$\Pr(A \mid B)$**

c. **$\Pr(A' \mid B)$**

THINK	WRITE
a. 1. State the addition rule.	$\Pr(A \cup B) = \Pr(A) + \Pr(B) - \Pr(A \cap B)$
2. Substitute the known probabilities and evaluate $\Pr(A \cup B)$.	$= 0.25 + 0.52 - 0.12$ $= 0.65$
b. 1. State the conditional probability rule.	$\Pr(A \mid B) = \dfrac{\Pr(A \cap B)}{\Pr(B)}$
2. Substitute the known probabilities and evaluate $\Pr(A \mid B)$.	$= \dfrac{0.12}{0.52}$ $= \dfrac{12}{52}$ $= \dfrac{3}{13}$
c. 1. $\Pr(A' \cap B)$ needs to be determined for the conditional probability rule. A Venn diagram may be helpful to determine $\Pr(A' \cap B)$. First place 0.12 in the intersection region. $\Pr(A \text{ only}) = 0.25 - 0.12 = 0.13$ $\Pr(B \text{ only}) = 0.52 - 0.12 = 0.4$ The remaining probability sits outside the circles.	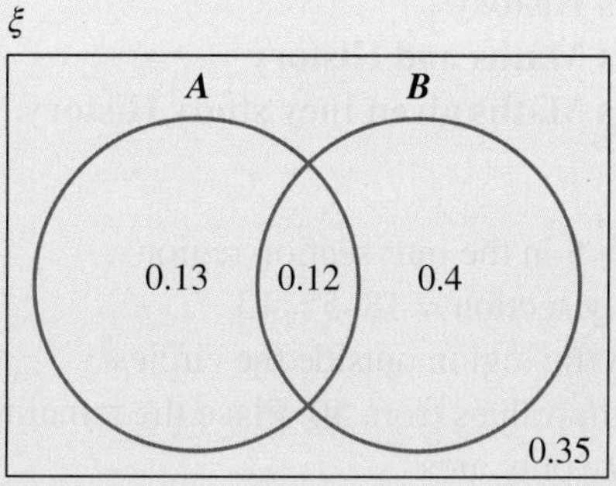
2. Determine the value for $\Pr(A' \cap B)$.	$\Pr(A' \cap B) = 0.4$
3. State the rule for the conditional probability.	$\Pr(A' \mid B) = \dfrac{\Pr(A' \cap B)}{\Pr(B)}$
4. Substitute the known probabilities and evaluate $\Pr(A' \mid B)$.	$= \dfrac{0.4}{0.52}$ $= \dfrac{40}{52}$ $= \dfrac{10}{13}$

9.2.3 Independent events

Two events are independent if the occurrence of one of the events has no impact on the chance of the other event occurring.

Independence rule

If two events A and B are independent, then $\Pr(A \cap B) = \Pr(A) \times \Pr(B)$

and

$$\Pr(A \mid B) = \Pr(A)$$

Mutually exclusive events

If events A and B have no elements in common, for example if event A is rolling an even number and event B is rolling an odd number, then they are said to be mutually exclusive.

$$A \cap B = \varnothing \quad \text{or} \quad \Pr(A \cap B) = 0$$

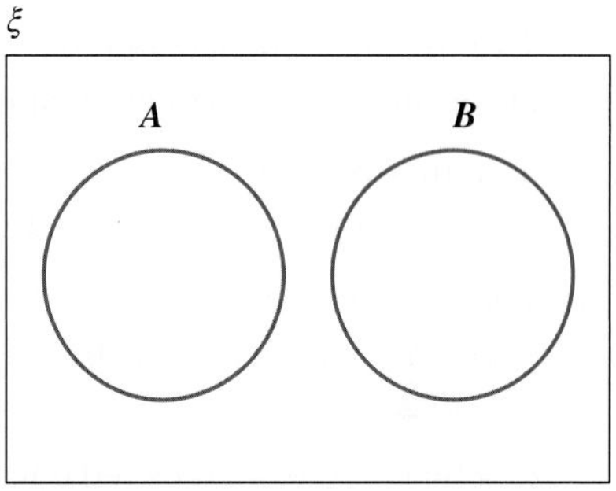

WORKED EXAMPLE 4 Independent events

For two events A and B, $\Pr(A) = 0.3$, $\Pr(B) = 0.4$ and $\Pr(A \cup B) = 0.58$.

a. Determine whether A and B are mutually exclusive events.

b. Determine whether A and B are independent events.

THINK	WRITE
a. 1. Determine $\Pr(A \cap B)$ via the addition rule.	$\Pr(A \cup B) = \Pr(A) + \Pr(B) - \Pr(A \cap B)$ $0.58 = 0.3 + 0.4 - \Pr(A \cap B)$ $\Pr(A \cap B) = 0.7 - 0.58$ $= 0.12$
2. Use the definition for mutually exclusive events that $\Pr(A \cap B) = 0$ to answer the question.	$\Pr(A \cap B) \neq 0$ Therefore, events A and B are not mutually exclusive.
b. 1. State the rule for independence.	If independent, $\Pr(A \cap B) = \Pr(A) \times \Pr(B)$.
2. Determine the right-hand side separately.	$\Pr(A) \times \Pr(B) = 0.3 \times 0.4$ $= 0.12$
3. Check if the independence rule holds.	$= \Pr(A \cap B)$
4. Write the final statement.	Therefore, events A and B are independent.

9.2 Exercise

Technology free

1. **WE1** A bag contains 4 blue discs, 8 red discs and 10 green discs. A disc is drawn randomly from the bag. Determine the probability that the disc drawn is:
   a. blue
   b. not red
   c. either green or red.
2. Twenty balls labelled with the numbers 1 to 20 are placed into a box and a ball is selected at random. Determine the probability that the number of the ball selected is:
   a. a multiple of 4
   b. less than 16
   c. greater than 5 but no more than 12.
3. There are 4 blue, 10 red and 6 green counters in a bag. A counter is selected, the colour is noted, and the counter is returned to the bag. Three counters are selected in total. Determine the probability of selecting:
   a. 3 blue counters
   b. a blue, then a green, then a red counter
   c. a blue, a green and a red counter.
4. **WE2** A group of 60 students were surveyed as to whether they studied Maths and/or History in Year 12. Eighteen people said they studied both subjects, 32 said they studied Maths and 10 studied neither.
   a. Represent the information in a Venn diagram
   b. Determine the probability of selecting a student who:
      i. studies History
      ii. studies Maths and History
      iii. studies Maths given they study History.
5. **WE3** If $\Pr(A) = 0.35$, $\Pr(B) = 0.48$ and $\Pr(A \cap B) = 0.15$, calculate:
   a. $\Pr(A \cup B)$
   b. $\Pr(A \mid B)$
   c. $\Pr(A' \mid B)$
6. For two events A and B, $\Pr(A) = 0.7$, $\Pr(B) = 0.5$ and $\Pr(A \mid B) = 0.64$. Calculate:
   a. $\Pr(A \cap B)$
   b. $\Pr(A \cup B)$
   c. $\Pr(B \mid A)$
7. In a certain town, the probability of it raining the following day if it has been raining today is 0.7. If it has been sunny today, the chance of it being sunny the next day is 0.5. If it is raining today, determine the probability that:
   a. it rains the next two days
   b. it rains tomorrow, given that it will be sunny in two days' time.
8. There are 6 white plates and 4 blue plates in a cupboard. Two are removed at random. Determine the probability of obtaining:
   a. two blue plates
   b. one plate of each colour.
9. **WE4** For two events A and B, $\Pr(A) = 0.5$, $\Pr(B) = 0.36$ and $\Pr(A \cup B) = 0.75$.
   a. Determine whether A and B are mutually exclusive events.
   b. Determine whether A and B are independent events.
10. Two fair dice are rolled. M represents the event of rolling a number less than 3 on the die and N represents the event of rolling a number greater than 5 on the second die. Determine if:
    a. M and N are mutually exclusive events
    b. M and N are independent events.

11. If $\Pr(A) = 0.44$ and $\Pr(B) = 0.6$, determine $\Pr(A \mid B)$ if:

a. A and B are mutually exclusive

b. A and B are independent.

12. For two events, M and N, $\Pr(M \cap N) = 0.18$, $\Pr(N) = k$, and $\Pr(M') = 0.7 - k$. Determine the value of k if M and N are independent.

9.2 Exam questions

Question 1 (3 marks) TECH-FREE

Source: *VCAA 2020, Mathematical Methods Exam 1, Q2*

A car manufacturer is reviewing the performance of its car model X. It is known that at any given six-month service, the probability of model X requiring an oil change is $\frac{17}{20}$, the probability of model X requiring an air filter change is $\frac{3}{20}$ and the probability of model X requiring both is $\frac{1}{20}$.

a. State the probability that at any given six-month service model X will require an air filter change without an oil change. **(1 mark)**

b. The car manufacturer is developing a new model, Y. The production goals are that the probability of model Y requiring an oil change at any given six-month service will be $\frac{m}{m+n}$, the probability of model Y requiring an air filter change will be $\frac{n}{m+n}$ and the probability of model Y requiring both will be $\frac{1}{m+n}$, where $m, n \in Z^+$.

Determine m in terms of n if the probability of model Y requiring an air filter change without an oil change at any given six-month service is 0.05. **(2 marks)**

Question 2 (1 mark) TECH-ACTIVE

Source: *VCAA 2019, Mathematical Methods Exam 2, Section A, Q17*

MC A box contains n marbles that are identical in every way except colour, of which k marbles are coloured red and the remainder of the marbles are coloured green. Two marbles are drawn randomly from the box.

If the first marble is **not** replaced into the box before the second marble is drawn, then the probability that the two marbles drawn are the same colour is

A. $\dfrac{k^2 + (n-k)^2}{n^2}$

B. $\dfrac{k^2 + (n-k-1)^2}{n^2}$

C. $\dfrac{2k(n-k-1)}{n(n-1)}$

D. $\dfrac{k(n-1) + (n-k)(n-k-1)}{n(n-1)}$

E. ${}^nC_2 \left(\frac{k}{n}\right)^2 \left(1 - \frac{k}{n}\right)^{n-2}$

Question 3 (1 mark) TECH-ACTIVE

Source: *VCAA 2018, Mathematical Methods Exam 2, Section A, Q14*

MC Two events, A and B, are independent, where $\Pr(B) = 2\Pr(A)$ and $\Pr(A \cup B) = 0.52$. $\Pr(A)$ is equal to

A. 0.1 **B.** 0.2 **C.** 0.3 **D.** 0.4 **E.** 0.5

More exam questions are available online.

9.3 Discrete random variables

LEARNING INTENTION

At the end of this subtopic you should be able to:
- define a discrete random variable
- construct a discrete probability distribution table
- calculate probabilities over specific intervals, including conditional probability restrictions.

9.3.1 Random variables

The numerical value of a **random variable** is determined by conducting a random experiment. Random variables are represented by uppercase letters of the alphabet. Lowercase letters of the alphabet are used for the associated values of the random variable. For example, $\Pr(X = x)$ is interpreted as the probability that the random variable X will equal x.

Consider tossing three unbiased coins, where the number of Tails obtained is recorded. X is defined as the number of Tails obtained; therefore, x can be 0, 1, 2 or 3 (the different number of Tails that can be obtained from three tosses). In order to determine the associated probabilities for this random experiment, where each of the outcomes is equally likely, we need to list the sample space.

$$\xi = \{\text{TTT, TTH, THT, HTT, THH, HTH, HHT, HHH}\}$$

- $\Pr(X = 0) = \Pr(\text{HHH})$
$$= \frac{1}{2} \times \frac{1}{2} \times \frac{1}{2}$$
$$= \frac{1}{8}$$
- $\Pr(X = 1) = \Pr(\text{HHT}) + \Pr(\text{HTH}) + \Pr(\text{THH})$
$$= \left(\frac{1}{2} \times \frac{1}{2} \times \frac{1}{2}\right) + \left(\frac{1}{2} \times \frac{1}{2} \times \frac{1}{2}\right) + \left(\frac{1}{2} \times \frac{1}{2} \times \frac{1}{2}\right)$$
$$= \frac{3}{8}$$
- $\Pr(X = 2) = \Pr(\text{TTH}) + \Pr(\text{THT}) + \Pr(\text{HTT})$
$$= \left(\frac{1}{2} \times \frac{1}{2} \times \frac{1}{2}\right) + \left(\frac{1}{2} \times \frac{1}{2} \times \frac{1}{2}\right) + \left(\frac{1}{2} \times \frac{1}{2} \times \frac{1}{2}\right)$$
$$= \frac{3}{8}$$
- $\Pr(X = 3) = \Pr(\text{TTT})$
$$= \frac{1}{2} \times \frac{1}{2} \times \frac{1}{2}$$
$$= \frac{1}{8}$$

This is an example of a **discrete random variable**. A discrete random variable can have only countable numbers or integer values. For the tossing of the three coins, $X = \{0, 1, 2, 3\}$.

Other examples of discrete random variables include the number of pups in a litter, the number of soft-centred chocolates in a box of mixed chocolates, the number of rainy days in the month of March, the number of blue smarties in a standard 15-gram packet and the number of traffic accidents at a main intersection over the period of three months.

The **probability distribution** of a discrete random variable defines the probabilities associated with each value the random variable can assume. For the experiment of the tossing of three coins, the distribution can be displayed in a table.

x	0	1	2	3
$\Pr(X=x)$	$\frac{1}{8}$	$\frac{3}{8}$	$\frac{3}{8}$	$\frac{1}{8}$

The probability distribution can also be represented graphically, with probability on the vertical axis and the possible x-values on the horizontal axis. For the probability distribution shown in the table below, the graph would appear as shown.

x	0	1	2	3
$\Pr(X=x)$	$\frac{1}{10}$	$\frac{4}{10}$	$\frac{4}{10}$	$\frac{1}{10}$

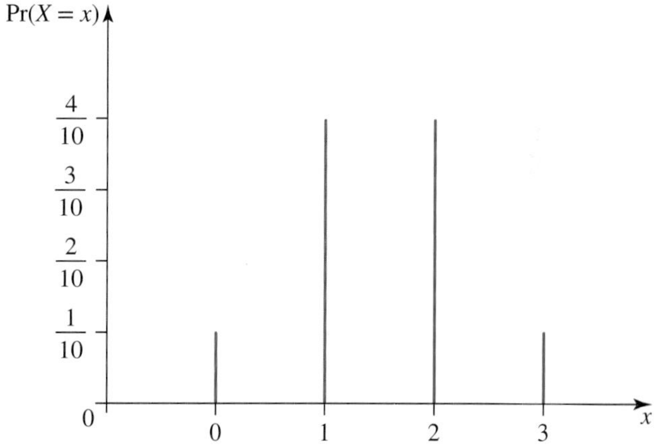

WORKED EXAMPLE 5 Creating a discrete probability distribution

A motorist travels along a main road in Melbourne. In doing so they must travel through three intersections with traffic lights over a stretch of two kilometres. The probability that the motorist will have to stop because of a red light at any of the intersections is $\frac{2}{5}$.

Let X be the number of red lights encountered by the motorist.

a. **Use a tree diagram to produce a sample space for this situation.**

b. **Determine the probability of each outcome.**

c. **State the probability distribution for this random variable.**

THINK	WRITE
a. 1. Set up a tree diagram to show the sample space. *Note:* $\Pr(R) = \frac{2}{5}$, $\Pr(G) = \frac{3}{5}$.	**a.** Let R = a red light and G = a green light.

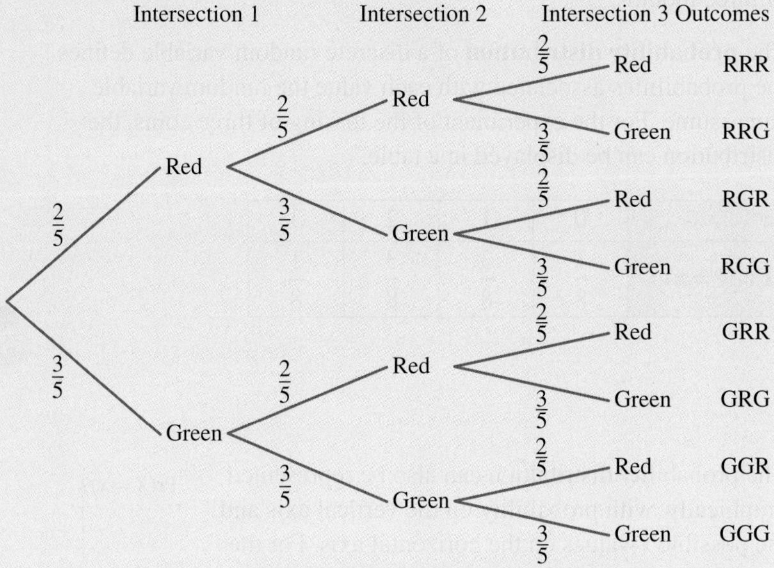

THINK	WRITE
2. List the event or sample space.	$\xi = \{RRR, RRG, RGR, RGG, GRR, GRG, GGR, GGG\}$
b. Calculate the probability of each outcome.	**b.** $\Pr(RRR) = \frac{2}{5} \times \frac{2}{5} \times \frac{2}{5} = \frac{8}{125}$ $\Pr(RRG) = \frac{2}{5} \times \frac{2}{5} \times \frac{3}{5} = \frac{12}{125}$ $\Pr(RGR) = \frac{2}{5} \times \frac{3}{5} \times \frac{2}{5} = \frac{12}{125}$ $\Pr(RGG) = \frac{2}{5} \times \frac{3}{5} \times \frac{3}{5} = \frac{18}{125}$ $\Pr(GRR) = \frac{3}{5} \times \frac{2}{5} \times \frac{2}{5} = \frac{12}{125}$ $\Pr(GRG) = \frac{3}{5} \times \frac{2}{5} \times \frac{3}{5} = \frac{18}{125}$ $\Pr(GGR) = \frac{3}{5} \times \frac{3}{5} \times \frac{2}{5} = \frac{18}{125}$ $\Pr(GGG) = \frac{3}{5} \times \frac{3}{5} \times \frac{3}{5} = \frac{27}{125}$
c. 1. Set up the probability distribution by combining the outcomes related to each possible value of *x*.	**c.** $\Pr(X = 0) = \Pr(GGG) = \frac{27}{125}$ $\Pr(X = 1) = \Pr(RGG) + \Pr(GRG) + \Pr(GGR)$ $= \frac{18}{125} + \frac{18}{125} + \frac{18}{125}$ $= \frac{54}{125}$

$$\Pr(X=2) = \Pr(RRG) + \Pr(RGR) + \Pr(GRR)$$
$$= \frac{12}{125} + \frac{12}{125} + \frac{12}{125}$$
$$= \frac{36}{125}$$
$$\Pr(X=3) = \Pr(RRR) = \frac{8}{125}$$

2. Enter the combined results into a table.

X = number of red lights

x	0	1	2	3
$\Pr(X=x)$	$\frac{27}{125}$	$\frac{54}{125}$	$\frac{36}{125}$	$\frac{8}{125}$

9.3.2 Conditions for a discrete probability distribute

Properties of a discrete probability distribution

1. **The probability of each outcome is restricted to a value from 0 to 1; that is, $0 \le \Pr(X=x) \le 1$.**
2. **The sum of the probabilities of each outcome add up to 1; that is, $\sum_{\text{all } x} \Pr(X=x) = 1$.**

WORKED EXAMPLE 6 Conditions for a discrete probability distribution

a. State, giving reasons, whether each of the following represents a discrete probability distribution.

i.

x	0	2	4	6
$\Pr(X=x)$	−0.1	0.3	0.4	0.2

ii.

x	−3	−1	4	6
$\Pr(X=x)$	0.01	0.32	0.52	0.15

iii.

x	−1	0	1	2
$\Pr(X=x)$	0.2	0.1	0.2	0.3

b. A random variable, X, has the following probability distribution.

x	1	2	3	4	5
$\Pr(X=x)$	b	$2b$	$0.5b$	$0.5b$	b

Find the value of the constant b.

THINK

a. i. 1. Check that each probability is a value from 0 to 1.

WRITE

a. i. Each probability does not meet the requirement $0 \le \Pr(X=x) \le 1$, as $\Pr(X=0) = -0.1$.

2. If this condition is satisfied, add the probabilities together to see if they add to 1.

As one of the probabilities is a negative value, there is no point checking the sum of the probabilities.

3. Answer the question.

This is not a discrete probability distribution.

ii. 1. Check that each probability is a value from 0 to 1.

ii. Each probability does meet the requirement $0 \le \Pr(X=x) \le 1$.

2. If this condition is satisfied, add the probabilities together to see if they add to 1.

$$\begin{aligned}\sum \Pr(X=x) &= 0.01+0.32+0.52+0.15\\ &= 1\end{aligned}$$

3. Answer the question.

Yes, this is a discrete probability function, as both of the conditions have been satisfied.

iii. 1. Check that each probability is a value from 0 to 1.

iii. Each probability does meet the requirement $0 \le \Pr(X=x) \le 1$.

2. If this condition is satisfied, add the probabilities together to see if they add to 1.

$$\begin{aligned}\sum \Pr(X=x) &= 0.2+0.1+0.2+0.3\\ &= 0.8\end{aligned}$$

3. Answer the question.

As the sum of the probabilities is not equal to 1, this is not a discrete probability distribution.

b. 1. As we know this is a probability distribution, we can equate the probabilities to 1.

b. $\sum \Pr(X=x) = 1$

$b + 2b + 0.5b + 0.5b + b = 1$

2. Simplify.

$5b = 1$

3. Solve for b.

$b = \dfrac{1}{5}$

The tossing of an unbiased die 3 times to see how many sixes are obtained is an example of a **uniform distribution**, because all of the outcomes are equally likely. Another example is seeing how many Heads are obtained when a single coin is tossed n times. However, a non-uniform distribution exists when a biased coin is used, because all of the outcomes are not equally likely.

WORKED EXAMPLE 7 Non-uniform distribution

A coin is biased so that there are twice as many chances of it landing with Heads up. The coin is tossed 3 times.

a. List the sample space and calculate the associated probabilities for each of the possible outcomes.

b. Construct a probability distribution table.

c. Calculate $\Pr(X > 1)$.

THINK | **WRITE**

a. 1. Determine the probability for each event.

a. If a Head is twice as likely to happen, then $\Pr(\text{H}) = \dfrac{2}{3}$ and $\Pr(\text{T}) = \dfrac{1}{3}$.

2. List the sample space.

$\xi = \{\text{HHH, HHT, HTH, THH, HTT, THT, TTH, TTT}\}$

3. Calculate the individual probabilities.

$$\Pr(\text{HHH}) = \frac{2}{3} \times \frac{2}{3} \times \frac{2}{3} = \frac{8}{27}$$
$$\Pr(\text{HHT}) = \frac{2}{3} \times \frac{2}{3} \times \frac{1}{3} = \frac{4}{27}$$
$$\Pr(\text{HTH}) = \frac{2}{3} \times \frac{1}{3} \times \frac{2}{3} = \frac{4}{27}$$
$$\Pr(\text{HTT}) = \frac{2}{3} \times \frac{1}{3} \times \frac{1}{3} = \frac{2}{27}$$
$$\Pr(\text{THH}) = \frac{1}{3} \times \frac{2}{3} \times \frac{2}{3} = \frac{4}{27}$$
$$\Pr(\text{THT}) = \frac{1}{3} \times \frac{2}{3} \times \frac{1}{3} = \frac{2}{27}$$
$$\Pr(\text{TTH}) = \frac{1}{3} \times \frac{1}{3} \times \frac{2}{3} = \frac{2}{27}$$
$$\Pr(\text{TTT}) = \frac{1}{3} \times \frac{1}{3} \times \frac{1}{3} = \frac{1}{27}$$

b. 1. Group the outcomes that contain the same number of Heads.

b. Let X be the number of Heads.

$$\Pr(X = 0) = \Pr(\text{TTT}) = \frac{8}{27}$$
$$\Pr(X = 1) = \Pr(\text{HHT}) + \Pr(\text{HTH}) + \Pr(\text{THH})$$
$$= 3 \times \frac{4}{27}$$
$$= \frac{4}{9}$$
$$\Pr(X = 2) = \Pr(\text{HTT}) + \Pr(\text{TTH}) + \Pr(\text{THT})$$
$$= 3 \times \frac{2}{27}$$
$$= \frac{2}{9}$$
$$\Pr(X = 3) = \Pr(\text{HHH}) = \frac{1}{27}$$

2. Check that the probabilities add to 1.

Check: $\frac{1}{27} + \frac{6}{27} + \frac{12}{27} + \frac{8}{27} = \frac{27}{27}$

$= 1$

3. Write the answer.

x	0	1	2	3
$\Pr(X = x)$	$\frac{1}{27}$	$\frac{2}{9}$	$\frac{4}{9}$	$\frac{8}{27}$

c. 1. Define what probabilities are included in this inequality.

c $\Pr(X > 1) = \Pr(X = 2) + \Pr(X = 3)$

2. Add the probabilities.

$$\Pr(X > 1) = \frac{4}{9} + \frac{8}{27}$$
$$= \frac{12}{27} + \frac{8}{27}$$

3. Write the answer.

$$\Pr(X > 1) = \frac{20}{27}$$

9.3 Exercise

Technology free

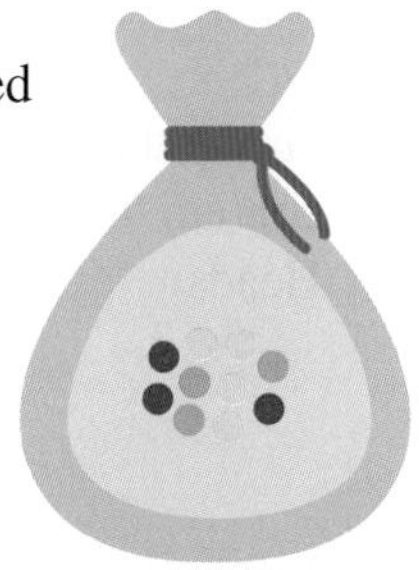

1. WE5 A bag contains 3 red, 3 green and 4 yellow balls. A ball is withdrawn from the bag, its colour is noted, and then the ball is returned to the bag. This process is repeated on two more occasions. Let Y be the number of green balls obtained.
   a. Use a tree diagram to produce the sample space for the experiment.
   b. Determine the probability of each outcome.
   c. State the probability distribution for this random variable.

2. An unbiased die is tossed twice. Let the random variable X be the number of sixes obtained. Determine the probability distribution for this discrete random variable.

3. WE6 State, giving reasons, whether each of the following represents a discrete probability distribution.

a.

y	3	6	9	12
$\Pr(Y=y)$	0.2	0.3	0.3	0.2

b.

y	−2	−1	0	1	2
$\Pr(Y=y)$	0.15	0.2	0.3	0.2	0.15

4. State, giving reasons, whether each of the following represents a discrete probability distribution.

a.

y	5	10	15	20
$\Pr(Y=y)$	0.15	0.35	0.35	0.05

b.

y	0	1	2	3	4
$\Pr(Y=y)$	−0.2	−0.1	0.1	0.2	0.3

5. State, with reasons, which of the following could be a probability distribution.

a.

x	−3	−1	1	3
$\Pr(X=x)$	0.1	0.5	0.5	0.1

b.

y	1	2	3	4	5
$\Pr(Y=y)$	−0.2	−0.1	0.4	0.1	0.2

c.

z	−2	0	2	4
$\Pr(Z=z)$	0.25	0.15	0.45	0.25

d.

x	1	2	3	4	5
$\Pr(X=x)$	0.1	0.25	0.3	0.25	0.1

6. State, with reasons, whether the following are discrete probability distributions.
   a. $p(x)=\dfrac{1}{7}(5-x),\ x\in\{1,3,4\}$
   b. $p(x)=\dfrac{x^2-x}{40},\ x\in\{-1,1,2,3,4,5\}$
   c. $p(x)=\dfrac{1}{15}\sqrt{x},\ x\in\{1,4,9,16,25\}$

7. Calculate the value(s) of k if the tables represents discrete probability distributions.

a.

x	2	3	4	5	6
$\Pr(X=x)$	$5k$	$3k-0.1$	$2k$	k	$0.6-3k$

b.

x	1	2	3	4	5
$\Pr(X=x)$	$0.5k^2$	$0.5k$	$0.25(k+1)$	0.5	$0.5k^2$

8. Each of the following tables shows a discrete probability distribution. Determine the unknown value in each case. (Assume the unknown value is not zero.)

a.

x	2	4	6	8	10
$\Pr(X=x)$	$3d$	$0.5-3d$	$2d$	$0.4-2d$	$d-0.05$

b.

y	-6	-3	0	3	6
$\Pr(Y=y)$	$0.5k$	$1.5k$	$2k$	$1.5k$	$0.5k$

c.

z	1	3	5	7
$\Pr(Z=z)$	$\frac{1}{3}-a^2$	$\frac{1}{3}-a^2$	$\frac{1}{3}-a^2$	a

Technology active

9. **MC** The following is a discrete probability function.

$$p(x) = \frac{1}{a}(15-3x), x \in \{1, 2, 3, 4, 5\}$$

The value of a is:

A. 20 B. 60 C. 15 D. 10 E. 30

10. **WE7** Two dice are weighted so that $\Pr(2) = 0.2$, $\Pr(1) = \Pr(3) = \Pr(5) = 0.1$ and $\Pr(4) = \Pr(6) = 0.25$. They are both rolled at the same time. Let Z be the number of even numbers obtained.

a. List the sample space.
b. Construct a probability distribution table.
c. Calculate $\Pr(Z = 1)$.

11. Samara and Simon are going to play tennis together. Samara has been playing tennis for longer than Simon, and the probability that Samara wins a set is 0.6. They intend to play 3 sets of tennis. Let X be the number of sets that Simon wins.

a. List the sample space.
b. List the possible values of X and construct a probability distribution table.
c. Calculate $\Pr(X \le 2)$.

12. A mature British Blue female cat has just given birth to 4 kittens. Assume that there is an equally likely chance of a kitten being of either sex.

a. Use a tree diagram to list the sample space for the possible number of males and females in the litter.
b. Let X be the number of females in the litter. Construct a probability distribution table for the gender of the kittens.
c. Calculate the probability that 4 females will be born.
d. Calculate the probability that at least 1 female will be born.
e. Calculate the probability that at most 2 females will be born.

13. Matthew likes to collect differently shaped dice. Currently he has two tetrahedrons (4 sides), an icosahedron (20 sides), two dodecahedrons (12 sides) and an octahedron (8 sides) as well as two standard 6-sided cubes. Matthew has decided to play a game of chance using the octahedral die (with sides numbered 1 to 8) and one dodecahedral die (with sides numbered 1 to 12). The dice are tossed simultaneously and Matthew notes the number showing uppermost on both dice.

a. List the sample space for the simultaneous tossing of the two dice.

b. Let X be the number of primes obtained as a result of a toss. Determine $\Pr(X = 0)$, $\Pr(x = 1)$ and $\Pr(X = 2)$.

c. This particular game of chance involves tossing the two dice simultaneously on three occasions. The winner of the game must obtain two primes with each of the three tosses. Calculate the probability of being a winner. Give your answer correct to 3 decimal places.

14. A card game has the following rules. A card is chosen at random from a standard deck of 52 cards. Each card is awarded a numerical score. The premium cards — aces, kings and queens (P) — are each awarded 10 points. Each jack (J) is awarded 5 points, and each standard card (S) is awarded 2 points. The game is played twice.

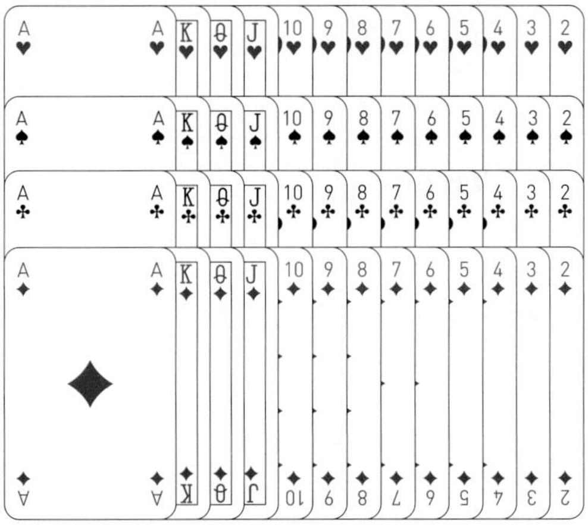

a. List the possible total points scored when two games are played.

b. If X is the total points scored when two games are played, construct the probability distribution.

c. Calculate:

i. $\Pr(X = 15)$

ii. $\Pr(X \geq 12)$

iii. $\Pr(X = 15 \mid X \geq 12)$.

15. Diabetes is the name of a group of diseases that affect how the body uses blood glucose. If you have diabetes, it means that you have too much glucose in your blood. This can lead to serious health problems. Treatment for type 2 diabetes primarily involves monitoring your blood sugar level along with medications, insulin or both.

A new diabetes medication is to be trialled by 5 patients. From experiments that have been performed with mice, the success rate of the new medication is about 60%.

a. Let X denote the number of patients who improve their health with the new medication. Construct the probability distribution.

b. The new medication will be considered a success if 68% or more of the patients improve their health. Determine $\Pr(X = 3) + \Pr(X = 4) + \Pr(X = 5)$ and comment on the success of the new medication.

16. A biased coin is tossed 6 times.

a. If $\Pr(\text{H}) = \frac{2}{3}$ and X defines the number of Tails obtained, construct a probability distribution for this discrete probability experiment. Give your answers correct to 4 decimal places.

b. Calculate, correct to 4 decimal places:

i. the probability of more than 2 Heads

ii. the probability of more than 2 Heads, given that fewer than 5 Heads come up.

17. A discrete random variable, X, can take the values $-5, -1, 0, 1$ and 5. The probability distribution is defined in the following manner.

$\Pr(X = -1) = \Pr(X = 0) = \Pr(X = 1) = m$

$\Pr(X = -5) = \Pr(X = 5) = n$

$3\Pr(X = 0) = \Pr(X = 5)$

a. Determine the values of m and n.
b. Calculate:

i. $\Pr(X \geq 0)$ ii. $\Pr(X = 1 \mid X \geq 0)$.

18. A discrete random variable has the following probability distribution.

y	1	2	3	4	5
$\Pr(Y=y)$	$0.5k^2$	$0.3-0.2k$	0.1	$0.5k^2$	0.3

Determine the value(s) of k, correct to 4 decimal places, that meet the criteria for this to be a valid probability distribution function.

9.3 Exam questions

Question 1 (1 mark) TECH-ACTIVE

Source: *VCE 2016 Mathematical Methods Exam 2, Section A, Q7; © VCAA.*

MC The number of pets, X, owned by each student in a large school is a random variable with the following discrete probability distribution.

x	0	1	2	3
$\Pr(X=x)$	0.5	0.25	0.2	0.05

If two students are selected at random, the probability that they own the same number of pets is

A. 0.3 **B.** 0.305 **C.** 0.355 **D.** 0.405 **E.** 0.8

Question 2 (1 mark) TECH-ACTIVE

MC For the table below to represent a probability function, the value(s) of p must be

x	0	1	2	3	4
$\Pr(X=x)$	0.2	$0.6p^2$	0.1	$1-p$	0.1

A. $p=0$ **B.** $p=\frac{4}{5}$ **C.** $p=\frac{2}{3}$ **D.** $p=\frac{2}{3}$ or $p=1$ **E.** $p=1$

Question 3 (5 marks) TECH-ACTIVE

A game is played using a spinner that has been loaded so that it is more likely to land on the red side. In fact, $\Pr(\text{red})=\frac{2}{5}$, and $\Pr(\text{blue})=\Pr(\text{green})=\Pr(\text{yellow})=\frac{1}{5}$.

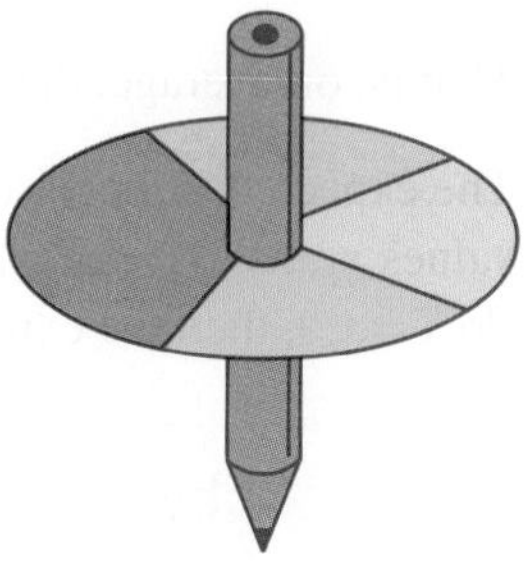

Each player pays \$2 to play. The player spins the spinner a total of 3 times; however, once the spinner lands on the red side the game is over. If a player has a combination of any 3 colours, they win \$1, but if the player has a combination of 3 colours that are all the same, they win \$10. There are a total of 40 different outcomes for the game.

a. List the possible ways in which the game could end. **(1 mark)**
b. List the possible ways in which the player could win \$10. **(1 mark)**
c. Suppose X equals the amount of money won by playing the game, excluding the amount the person pays to play, so $X=\{0, 1, 10\}$. Construct the probability distribution. Give your answers correct to 4 decimal places. **(3 marks)**

More exam questions are available online.

9.4 Measures of centre and spread

LEARNING INTENTION

At the end of this subtopic you should be able to:
- calculate the mean (expected value) of a discrete probability distribution
- calculate the variance and standard deviation of a discrete probability distribution.

The mean, variance and standard deviation are common statistical measurements that give us insight about sets of data, including discrete random variable distributions.

9.4.1 The expected value or mean

The **expected value** or **mean** of a discrete probability function represents the 'average' outcome for the random experiment. When we find the mean, we are not stating the actual outcome; we are stating the outcome that we expect to happen.

Consider again a weighted spinner where $\Pr(\text{red}) = \frac{2}{5}$ and $\Pr(\text{blue}) = \Pr(\text{green}) = \Pr(\text{yellow}) = \frac{1}{5}$.

The player spins the spinner 3 times, although the game is over if the spinner lands on its red side. If a player obtains a combination of 3 colours, they win \$1, and if a player obtains a combination of 3 colours that are the same, they win \$10. The game costs \$2 to play. The calculated probabilities are:
$\Pr(\text{Win}) = \frac{3}{125}$, $\Pr(\text{Win } \$1) = \frac{24}{125}$ and $\Pr(\text{Game over}) = \frac{98}{125}$.

If we consider the profit made by the person conducting the game, then they can lose \$8 (the player pays \$2 to play but wins \$10), gain \$1 (the player pays \$2 but wins \$1) or gain \$2 (the player pays \$2 and wins nothing). If Y represents the profit made by the person conducting the game, the following table would represent this situation.

y	$-\$8$	$\$1$	$\$2$
$\Pr(Y=y)$	$\frac{3}{125}$	$\frac{24}{125}$	$\frac{98}{125}$

$$\begin{aligned}\text{The expected profit} &= -8\times\frac{3}{125} + 1\times\frac{24}{125} + 2\times\frac{98}{125}\\ &= -\frac{24}{125} + \frac{24}{125} + \frac{196}{125}\\ &= \$1.57\end{aligned}$$

That is, on average, the person conducting the game makes a profit of \$1.57 per game.

The expected value of a random variable, X, is denoted by $\mathrm{E}(X)$ or μ (mu). If a random variable assumes the values $x_1, x_2, x_3 \ldots x_{n-1}, x_n$ with associated probabilities $\Pr(X=x_1), \Pr(X=x_2), \Pr(X=x_3) \ldots \Pr(X=x_{n-1})$, $\Pr(X=x_n)$, then the expected value is as follows.

The expected value

$$\begin{aligned}\mathrm{E}(X) &= x_1\Pr(X=x_1) + x_2\Pr(X=x_2) + \ldots + x_{n-1}\Pr(X=x_{n-1}) + x_n\Pr(X=x_n)\\ &= \sum_{x=1}^{x=n} x_n \Pr(X=x_n)\end{aligned}$$

WORKED EXAMPLE 8 Calculating the expected value

Calculate the expected value of the random variable with the following probability distribution.

x	10	20	30	40	50
$\Pr(X=x)$	0.42	0.34	0.16	0.07	0.01

THINK	WRITE
1. Write the rule to find the expected value.	$E(X) = \sum_{x=1}^{x=n} x_n \Pr(X = x_n)$
2. Substitute the appropriate values into the rule.	$E(X) = 10\,(0.42) + 20\,(0.34) + 30\,(0.16) + 40\,(0.07) + 50\,(0.01)$
3. Simplify.	$= 4.2 + 6.8 + 4.8 + 2.8 + 0.5$ $= 19.1$

on Resources

Interactivity Expected value or mean (int-6428)

9.4.2 Linear properties of the expected value

Sometimes we may be required to find the expected value of a linear function $aX + b$.

Linear properties of E(X)

$$E(aX + b) = E(aX) + E(b)$$
$$= aE(X) + b$$
$$\text{Also, } E(X + Y) = E(X) + E(Y)$$

where X and Y are discrete random variables and a and b are constants.

Note that the above properties are linear in nature, so

$$E(X^2) \neq [E(X)]^2$$

WORKED EXAMPLE 9 Applying the linear properties of E(X)

A discrete random variable, X, has the following probability distribution.

x	$5-d$	$3-d$	$-d$	$3+d$
$\Pr(X=x)$	$\frac{7}{20}$	$\frac{9}{20}$	$\frac{1}{10}$	$\frac{1}{10}$

If $E(X) = 1$, determine:

a. **the value of the constant d**

b. i. **$E(7X)$** ii. **$E(5X + 3)$** iii. **$E(3X - 2)$.**

THINK	WRITE
a. 1. Write the rule to find the expected value.	**a.** $\text{E}(X) = \sum_{x=1}^{x=n} x_n \Pr(X = x_n)$
2. Substitute the appropriate values into the rule.	$\text{E}(X) = \frac{7}{20}(5-d) + \frac{9}{20}(3-d) - \frac{2d}{20} + \frac{2}{20}(3+d)$
3. Simplify.	$\text{E}(X) = \frac{35}{20} - \frac{7d}{20} + \frac{27}{20} - \frac{9d}{20} - \frac{2d}{20} + \frac{6}{20} + \frac{2d}{20}$ $\text{E}(X) = \frac{68-16d}{20}$
4. Substitute $\text{E}(X) = 1$ and solve for d.	$1 = \frac{68-16d}{20}$ $20 = 68 - 16d$ $16d = 48$ $d = 3$
b. i. 1. Apply the linear property of $\text{E}(X)$: $\text{E}(aX+b) = a\text{E}(X) + b$.	**b. i.** $\text{E}(7X) = 7\text{E}(X)$
2. Substitute in the value of $\text{E}(X)$ and evaluate.	$\text{E}(7X) = 7 \times 1$ $= 7$
ii. 1. Apply the linear property of $\text{E}(X)$: $\text{E}(aX+b) = a\text{E}(X) + b$.	**ii.** $\text{E}(5X+3) = 5\text{E}(X) + 3$
2. Substitute in the value of $\text{E}(X)$ and evaluate.	$\text{E}(5X+3) = 5 \times 1 + 3$ $= 8$
iii. 1. Apply the linear property of $\text{E}(X)$: $\text{E}(aX+b) = a\text{E}(X) + b$.	**iii.** $\text{E}(3X-2) = 3\text{E}(X) - 2$
2. Substitute in the value of $\text{E}(X)$ and evaluate.	$\text{E}(3X-2) = 3 \times 1 - 2$ $= 1$

9.4.3 The variance and standard deviation

The measure of spread of a random variable distribution tells us how the data is dispersed. The measure of spread is called the **variance**, and the square root of the variance gives the **standard deviation**. The variance is denoted by $\text{Var}(X)$ or σ^2 (sigma squared) and is defined as follows.

Variance rule

$$\textbf{Var}(X) = \sigma^2 = \textbf{E}\left(X^2\right) - [\textbf{E}(X)]^2$$

This may also be written as

$$\text{Var}(X) = \sigma^2 = \text{E}\left(X^2\right) - \mu^2, \text{ where } \mu = \text{E}(X).$$

The derivation of this rule is follows:

$$\begin{aligned}\text{Var}(X) &= \text{E}(X-\mu)^2 \\ &= \text{E}\left(X^2 - 2X\mu + \mu^2\right) \\ &= \text{E}\left(X^2\right) - \text{E}(2X\mu) + \text{E}\left(\mu^2\right) \\ &= \text{E}\left(X^2\right) - 2\mu\text{E}(X) + \mu^2 \\ &= \text{E}\left(X^2\right) - 2\mu^2 + \mu^2\end{aligned}$$

Since $\text{E}(X) = \mu$,

$$\begin{aligned}\text{Var}(X) &= \text{E}\left(X^2\right) - \mu^2 \\ \text{Var}(X) &= \text{E}\left(X^2\right) - [\text{E}(X)]^2\end{aligned}$$

The standard deviation of X is the square root of the variance of X and is denoted by $\text{SD}(X)$ or σ.

Standard deviation rule

$$\textbf{SD}(X) = \sigma = \sqrt{\textbf{Var}(X)}$$

If the standard deviation is large, the spread of the data is large. If the standard deviation is small, the data is clumped together, close to the mean.

WORKED EXAMPLE 10 Variance and standard deviation

A discrete random variable, X, has the following probability distribution.

x	1	2	3	4	5
$\text{Pr}(X=x)$	0.15	0.25	0.3	0.2	0.1

Calculate:

a. **$\text{E}(X)$**
b. **$\text{Var}(X)$**
c. **$\text{SD}(X)$, correct to 4 decimal places.**

THINK	WRITE
a. 1. Write the rule to find the expected value.	a. $\text{E}(X) = \sum_{\text{all } x} x\,\text{Pr}(X=x)$
2. Substitute the appropriate values into the rule.	$\text{E}(X) = 1(0.15) + 2(0.25) + 3(0.3) + 4(0.2) + 5(0.1)$
3. Simplify.	$\text{E}(X) = 0.15 + 0.5 + 0.9 + 0.8 + 0.5$ $= 2.85$
b. 1. Evaluate $\text{E}(X^2)$.	b. $\text{E}\left(X^2\right) = \sum_{\text{all } x} x^2\,\text{Pr}(X=x)$ $\text{E}\left(X^2\right) = 1^2(0.15) + 2^2(0.25) + 3^2(0.3) + 4^2(0.2) + 5^2(0.1)$ $= 0.15 + 1 + 2.7 + 3.2 + 2.5$ $= 9.55$

2. Write the rule for the variance. $\text{Var}(X) = \text{E}(X^2) - [\text{E}(X)]^2$

3. Substitute in the appropriate values and evaluate.

$$\begin{aligned}\text{Var}(X) &= 9.55 - (2.85)^2 \\ &= 9.55 - 8.1225 \\ &= 1.4275\end{aligned}$$

c. 1. Write the rule for the standard deviation. **c.** $\text{SD}(X) = \sqrt{\text{Var}(X)}$

2. Substitute in the variance and evaluate.

$$\begin{aligned}\text{SD}(X) &= \sqrt{1.4275} \\ &= 1.1948\end{aligned}$$

TI | THINK — **DISPLAY/WRITE**

a. 1. On a Lists & Spreadsheet page, label the first column as x and the second column as p.
Enter the given x-values in the first column and their probabilities in the second column.

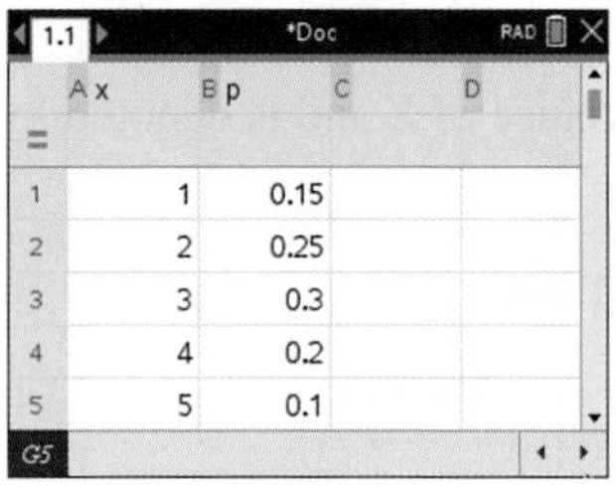

2. On a Calculator page, complete the entry line as:
sum($x \times p$)
then press ENTER.

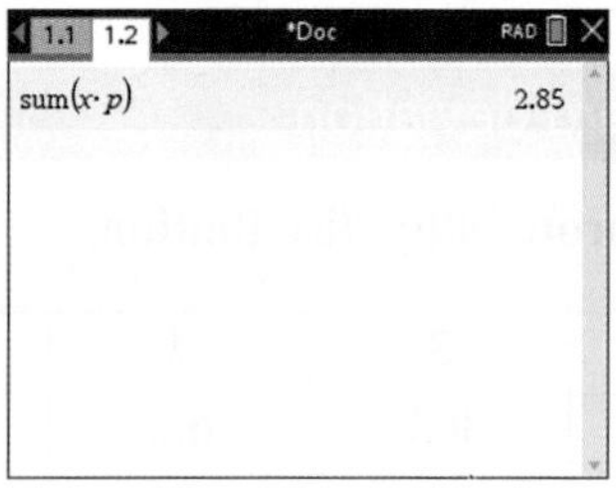

3. The answer appears on the screen. $E(X) = 2.85$

b. 1. Complete the next entry line as:
sum $(x^2 \times p)\left(\text{sum}(x \times p)\right)^2$
then press ENTER.

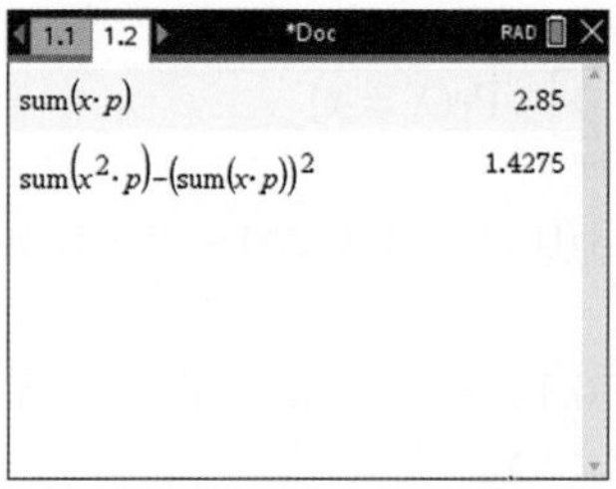

2. The answer appears on the screen. $\text{Var}(X) = 1.4275$

CASIO | THINK — **DISPLAY/WRITE**

a. 1. On a Statistics screen, relabel list 1 as x and list 2 as p.
Enter the given x-values in the first column and their probabilities in the second column.

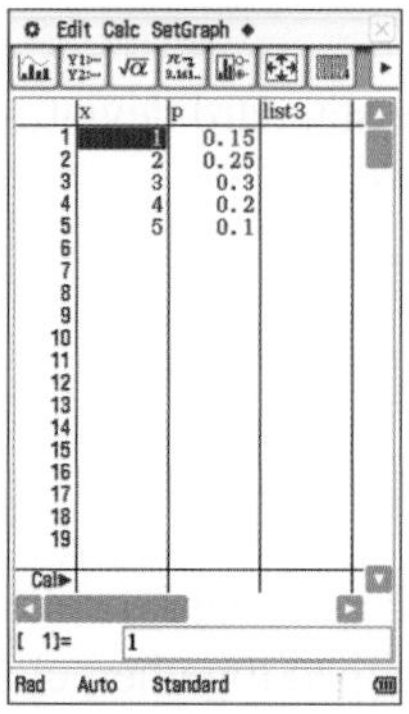

2. On a Main screen, complete the entry line as:
sum($x \times p$)
then press EXE.

3. The answer appears on the screen. $E(X) = 2.85$

b. 1. Complete the next entry line as:
sum $(x^2 \times p) - \left(\text{sum}(x \times p)\right)^2$
then press EXE.

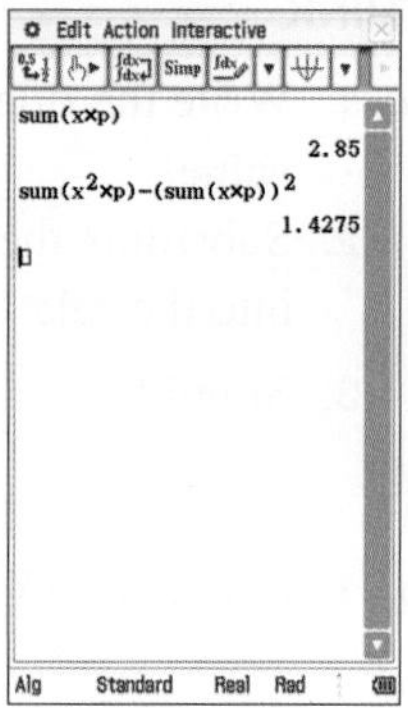

2. The answer appears on the screen. $\text{Var}(X) = 1.4275$

c. 1. Complete the next entry line as:
$\sqrt{ans}$
then press ENTER.

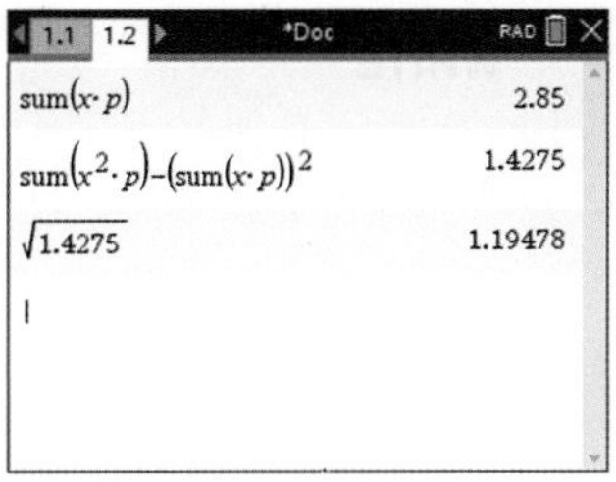

2. The answer appears on the screen.

SD(X) = 1.1948

c. 1. Complete the next entry line as:
$\sqrt{ans}$
then press EXE.

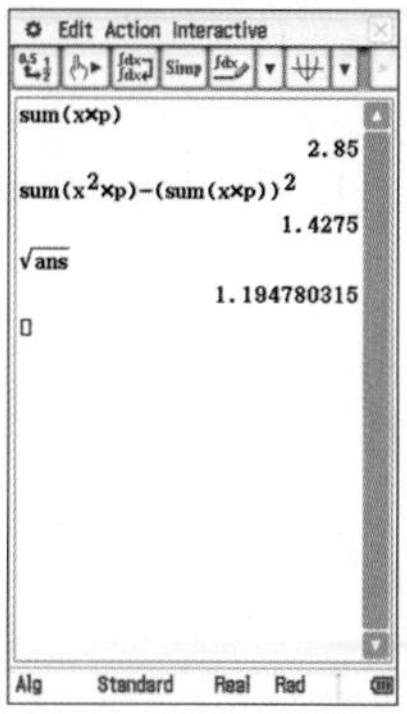

2. The answer appears on the screen.

SD(X) = 1.1948

on Resources

Interactivity Variance and standard deviation (int-6429)

9.4.4 Properties of the variance

The variance of a linear function has rules similar to those for the expectation of a linear function.

$$\text{Var}(aX+b)=a^2\text{Var}(X)$$

This can be proved in the following manner.

$$\begin{aligned}\text{Var}(aX+b)&=\text{E}(aX+b)^2-[\text{E}(aX+b)]^2\\&=\text{E}\left(a^2X^2+2abX+b^2\right)-[a\text{E}(X)+b]^2\\&=\text{E}\left(a^2X^2\right)+\text{E}(2abX)+\text{E}\left(b^2\right)-\left(a^2\,[\text{E}(X)]^2-2ab\text{E}(X)+b^2\right)\\&=a^2\text{E}(X^2)+2ab\text{E}(X)+b^2-a^2\,[\text{E}(X)]^2-2ab\text{E}(X)-b^2\\&=a^2\left(\text{E}\left(X^2\right)-[\text{E}(X)]^2\right)\end{aligned}$$

But $\text{Var}(X)=\text{E}\left(X^2\right)-[\text{E}(X)]^2$, so

$$\text{Var}(aX+b)=a^2\text{Var}(X).$$

Properties of the variance

$$\mathbf{Var}(aX+b)=a^2\mathbf{Var}(X)$$

WORKED EXAMPLE 11 Properties of the variance

A discrete probability function is defined by the rule $p(y)=\dfrac{1}{12}(10-3y), y\in\{1, 2, 3\}$.

a. Show that the sum of the probabilities is equal to one.

b. Calculate:

i. E(Y) **ii. Var(Y).**

c. Calculate:

i. Var($3Y-1$) **ii. Var($4-5Y$).**

THINK	WRITE
a. 1. Evaluate the probabilities for the given values of y.	**a.** $p(y) = \frac{1}{12}(10 - 3y), y \in \{1, 2, 3\}$ $p(1) = \frac{1}{12}(10 - 3(1)) = \frac{7}{12}$ $p(2) = \frac{1}{12}(10 - 3(2)) = \frac{4}{12} = \frac{1}{3}$ $p(3) = \frac{1}{12}(10 - 3(3)) = \frac{1}{12}$
2. Add the probabilities.	$\Pr(Y=1) + \Pr(Y=2) + \Pr(Y=3)$ $= \frac{7}{12} + \frac{4}{12} + \frac{1}{12}$ $= \frac{12}{12}$ $= 1$
b. i. 1. Write the rule to find the expected value.	**b. i.** $E(Y) = \sum_{\text{all } y} y \Pr(Y = y)$
2. Substitute the appropriate values into the rule.	$E(Y) = 1\left(\frac{7}{12}\right) + 2\left(\frac{4}{12}\right) + 3\left(\frac{1}{12}\right)$
3. Simplify.	$= \frac{7}{12} + \frac{8}{12} + \frac{3}{12}$ $= \frac{18}{12}$ $= \frac{3}{2}$
ii. 1. Evaluate $E(Y^2)$.	**ii.** $E(Y^2) = 1^2\left(\frac{7}{12}\right) + 2^2\left(\frac{4}{12}\right) + 3^2\left(\frac{1}{12}\right)$ $= \frac{7}{12} + \frac{16}{12} + \frac{9}{12}$ $= \frac{32}{12}$ $= \frac{8}{3}$
2. Write the rule for the variance.	$\text{Var}(Y) = E(Y^2) - [E(Y)]^2$
3. Substitute in the appropriate values and evaluate.	$\text{Var}(Y) = \frac{32}{12} - \left(\frac{3}{2}\right)^2$ $= \frac{32}{12} - \frac{9}{4}$ $= \frac{32 - 27}{12}$ $= \frac{5}{12}$

c. i. 1. Apply the property of the variance: $\text{Var}(aY+b)=a^2\text{Var}(Y)$.	**c. i.** $\text{Var}(3Y-1)=3^2\text{Var}(Y)$
2. Substitute in the value of Var(Y) and evaluate.	$\text{Var}(3Y-1)=9\times\frac{5}{12}$ $=\frac{15}{4}$
ii. 1. Apply the property of the variance: $\text{Var}(aY+b)=a^2\text{Var}(Y)$.	**ii.** $\text{Var}(4-5Y)=(-5)^2\,\text{Var}(Y)$
2. Substitute in the value of Var(Y) and evaluate.	$\text{Var}(4-5Y)=25\times\frac{5}{12}$ $=\frac{125}{12}$

9.4 Exercise

Students, these questions are even better in jacPLUS

Receive immediate feedback and access sample responses

Access additional questions

Track your results and progress

Find all this and MORE in jacPLUS

Technology free

1. **WE8** Calculate the expected value of the random variable for each of the following probability distributions.

a.

x	−3	−2	−1	0	1	2	3
$\Pr(X=x)$	$\frac{1}{9}$	$\frac{1}{9}$	$\frac{1}{9}$	$\frac{2}{9}$	$\frac{2}{9}$	$\frac{1}{9}$	$\frac{1}{9}$

b.

z	1	2	3	4	5	6
$\Pr(Z=z)$	$\frac{1}{12}$	$\frac{1}{4}$	$\frac{1}{3}$	$\frac{1}{6}$	$\frac{1}{12}$	$\frac{1}{12}$

2. A discrete random variable is defined by the function $p(x)=\frac{1}{16}(2x-1), x\in\{1, 2, 3, 4\}$.
   a. Construct a probability distribution table for this function.
   b. Find the expected value of the function.

3. Recently the large supermarket chains have been waging a price war on bread.
   On a particular Tuesday, a standard loaf of bread was purchased from a number of outlets of different chains. The following table shows the probability distribution for the price of the bread, X.

x	\$1	\$2	\$3	\$4	\$5
$\Pr(X=x)$	0.3	0.15	0.4	0.1	0.05

   Calculate the expected cost of a loaf of bread on that given Tuesday.

4. A game of chance is played with a spinner.

Each sector represents $\frac{1}{8}$ of the circular spinner. If the pointer lands on yellow, the player receives nothing. If the pointer lands on green, the player receives \$5. If the pointer lands on red, the player receives \$8, and if the pointer lands on blue, the player receives \$10. The game costs \$2 to play. Let X represent the net profit made by the player.

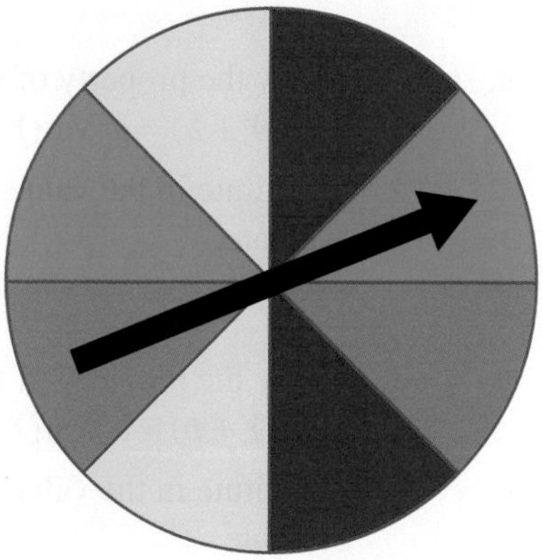

a. Construct a probability distribution table for the net profit.
b. Calculate the expected net profit in dollars for any player.

5. A discrete random variable, X, has the following probability distribution.

x	-2	0	2	4	6
$\Pr(X=x)$	k	k	$2k$	$3k$	$3k$

a. Determine the value of the constant k.
b. Determine the expected value of X.

6. **WE9** The discrete random variable, Y, has the following probability distribution.

y	-5	0	5	d	25
$\Pr(Y=y)$	$\frac{1}{10}$	$\frac{3}{10}$	$\frac{1}{5}$	$\frac{3}{10}$	$\frac{1}{10}$

If $E(Y)=7.5$, determine:

a. the value of the constant d

b. i. $E(2Y+3)$ ii. $E(5-Y)$ iii. $E(-2Y)$.

7. Given that $E(X)=4.5$, determine:

a. $E(2X-1)$ b. $E(5-X)$ c. $E(3X+1)$.

Technology active

8. **MC** A discrete random variable, Z, is defined by $p(z)=\frac{1}{38}\left(z^2-4\right), 2\le z\le 5$.

The mean, correct to 2 decimal places, is:

A. 4.02 B. 4.42 C. 4.47 D. 3.16 E. 1.66

9. **WE10** A discrete random variable, y, has the following probability distribution.

y	1	4	7	10	13
$\Pr(Y=y)$	0.15	0.2	0.3	0.2	0.15

Calculate:

a. $E(X)$
b. $Var(X)$
c. $SD(X)$, correct to 4 decimal places.

10. **WE11** A discrete probability function is defined by $p(x)=\frac{x^2}{30}, x=1, 2, 3, 4$.

Where appropriate, give your answers to the following to 2 decimal places.

a. Construct a probability distribution table and show that $\sum_{\text{all } x}\Pr(X=x)=1$.

b. Calculate:

i. $E(X)$ ii. $Var(X)$.

c. Calculate:

i. $Var(4X+3)$ ii. $Var(2-3X)$.

11. a. Determine the value of the constant m if the discrete random variable Z has the probability distribution shown and $E(Z) = 14.94$.

z	-7	m	23	31
$\Pr(Z = z)$	0.21	0.34	0.33	0.12

b. Determine $\text{Var}(Z)$ and hence calculate $\text{Var}(2(Z-1))$ and $\text{Var}(3-Z)$, correct to 2 decimal places.

12. **MC** Given that $\text{SD}(X) = \sigma = 2.5$, $\text{Var}(2X + 3)$ is:

A. 28 B. 6.32 C. 25 D. 15.5 E. 12.5

13. A discrete probability function is defined by the rule $p(x) = h(3-x)(x+1)$, $x = 0, 1, 2$.

a. Show that the value of h is $\frac{1}{10}$.

b. Hence, calculate the mean, variance and standard deviation of X. Where appropriate, give your answers to 4 decimal places.

14. Two octahedral dice (with faces numbered 1 to 8) are rolled simultaneously and the two numbers are recorded.

a. List the probability or event space and find $n(\xi)$.

Let Z be the larger of the two numbers on the two dice.

b. State the probability distribution for Z.

c. Calculate the expected value and standard deviation of Z, correct to 4 decimal places.

15. At a beginner's archery competition, each archer has two arrows to shoot at the target. A target is marked with ten evenly spaced concentric rings. The following is a summary of the scoring for the beginner's competition.

Yellow: 10 points
Red: 7 points
Blue: 5 points
Black: 3 points
White: 1 point

Let X be the total score after a beginner shoots two arrows.

a. List the possible score totals.

The probability of a beginner hitting each of the rings has been calculated as follows:
$\Pr(\text{yellow}) = 0.1$, $\Pr(\text{red}) = 0.2$, $\Pr(\text{blue}) = 0.3$, $\Pr(\text{black}) = 0.2$ and $\Pr(\text{white}) = 0.2$.

b. Construct a probability distribution table for the total score achieved by a beginner archer.

c. Calculate the expected score and the standard deviation for a beginner. Where appropriate, give your answers correct to 4 decimal places.

16. A random variable, Y, has the following probability distribution.

y	-1	1	3	5	7
$\Pr(Y = y)$	$1-2c$	c^2	c^2	c^2	$1-2c$

a. Determine the value of the constant c.

b. Calculate $E(Y)$, the mean of Y.

c. Calculate $\text{Var}(Y)$ and hence determine the standard deviation of Y, correct to 2 decimal places.

17. A discrete random variable, X, has the following probability distribution.

x	-2	-1	0	1	2	3	4
$\Pr(X = x)$	$0.5k^2$	$0.5k^2$	$k + k^2$	$4k$	$2k$	$2k + k^2$	$7k^2$

a. Determine the value of the constant k.

b. Calculate the expected value of X.

c. Calculate the standard deviation of X, correct to 4 decimal places.

18. A discrete probability function has the following distribution.

x	1	2	3	4	5
$\Pr(X = x)$	a	0.2	0.3	b	0.1

The expected value of the function is 2.5.

a. Calculate the values of the constants a and b.

b. Hence, evaluate the variance and standard deviation of X. Where appropriate, give your answers to 4 decimal places.

19. For a given discrete random variable, X, it is known that $E(X) = a$ and $Var(X) = 2a - 2$, where a is a constant that is greater than zero.

a. Calculate $E(X^2)$ in terms of a.

b. If $E(X^2)$ is known to be 6, determine $E(X)$ and $Var(X)$.

20. For a discrete random variable, Y, the probability function is defined by

$$p(y) = \begin{cases} ny, & y \in \{1, 2, 3, 4\} \\ n(7 - y), & y \in \{5, 6\} \end{cases}$$

a. Calculate the value of the constant n.

b. Determine the expected value, the variance and the standard deviation of Y, correct to 4 decimal places.

9.4 Exam questions

Question 1 (1 mark) TECH-ACTIVE

Source: VCE 2018, Mathematical Methods Exam 2, Section A, Q12; © VCAA.

MC The discrete random variable X has the following probability distribution.

x	0	1	2	3	6
$\Pr(X = x)$	$\frac{1}{4}$	$\frac{9}{20}$	$\frac{1}{10}$	$\frac{1}{20}$	$\frac{3}{20}$

Let μ be the mean of X.

$\Pr(X < \mu)$ is

A. $\frac{1}{2}$ B. $\frac{1}{4}$ C. $\frac{17}{20}$ D. $\frac{4}{5}$ E. $\frac{7}{10}$

Question 2 (1 mark) TECH-ACTIVE

Source: VCE 2017, Mathematical Methods Exam 2, Section A, Q14; © VCAA.

MC The random variable X has the following probability distribution, where $0 < p < \frac{1}{3}$.

x	-1	0	1
$\Pr(X = x)$	p	$2p$	$1 - 3p$

The variance of X is

A. $2p(1 - 3p)$ B. $1 - 4p$ C. $(1 - 3p)^2$ D. $6p - 16p^2$ E. $p(5 - 9p)$

Question 3 (10 marks) TECH-FREE

A dart competition at a local sports centre allows each player to throw one dart at the board, which has a radius of 20 centimetres. The board consists of five concentric circles, each with the same width.

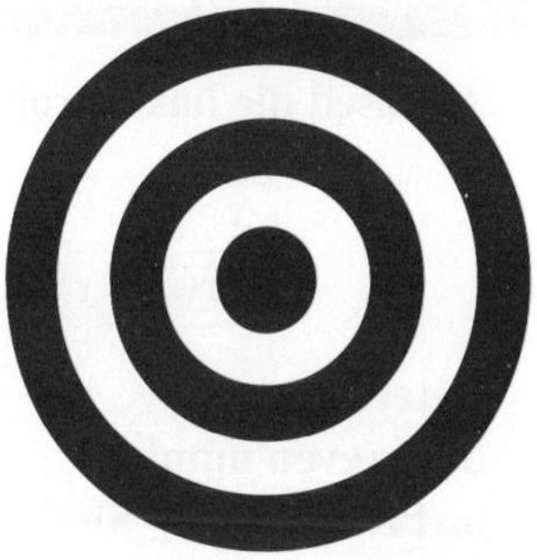

The inner circle has a radius of 4 cm. The probability of landing on each band is determined by the area of that band available on the board.

The outer red band is called band E, the next white band is called band D and so on until you get to the inner red circle, which is band A.

a. Calculate the probability of landing on each of the bands. **(5 marks)**

The competition costs \$1 to enter and the prizes are as follows:

If a dart hits band E, the player receives nothing.
If a dart hits band D, the player receives \$1.
If a dart hits band C, the player receives \$2.
If a dart hits band B, the player receives \$5.
If a dart hits band A, the player receives \$10.

b. If X is a discrete random variable that represents the profit in dollars for the player, construct a probability distribution table for this game. **(1 mark)**

c. Calculate:

i. the expected profit a player could make in dollars **(2 marks)**

ii. the standard deviation. **(2 marks)**

More exam questions are available online.

9.5 Applications

LEARNING INTENTION

At the end of this subtopic you should be able to:

- determine probabilities within 2 standard deviations of the mean
- determine probabilities for different applications.

9.5.1 Applications of discrete random variables

One important application of the expected value and standard deviation of a random variable is that approximately 95% of the distribution lies within two standard deviations of the mean.

Probability within 2 standard deviations of the mean

$$\Pr(\mu - 2\sigma \le X \le \mu + 2\sigma) \approx 0.95$$

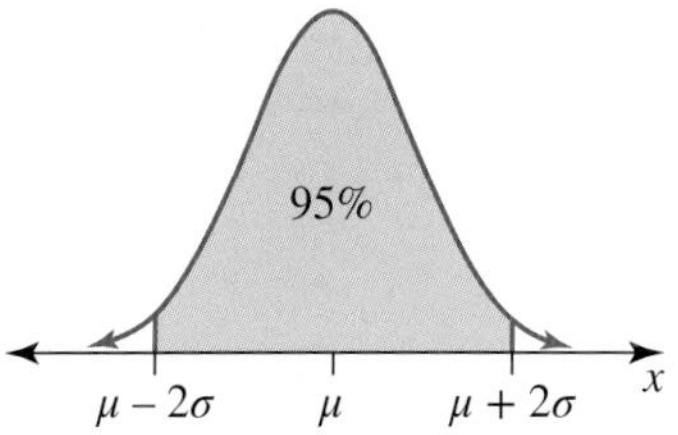

This can be illustrated by the normal distribution curve. This type of distribution is covered in Topic 12.

WORKED EXAMPLE 12 Applications of discrete random variables

A biased die has a probability distribution for the outcome of the die being rolled as follows.

x	1	2	3	4	5	6
$\Pr(X = x)$	0.1	0.1	0.2	0.25	0.25	0.1

Determine:

a. **Pr(even number)**
b. **$\Pr(X \geq 3 \mid X \leq 5)$**
c. **the expected value.**

THINK

a. 1. State the probabilities to be added.

2. Substitute the values and simplify.

b. 1. Define the rule.

2. Find $\Pr(X \geq 3 \cap X \leq 5)$.

3. Calculate $\Pr(X \leq 5)$.

4. Substitute the appropriate values into the formula.

5. Evaluate and simplify.

c. Calculate the expected value.

WRITE

a. $\Pr(\text{even number}) = \Pr(X=2) + \Pr(X=4) + \Pr(X=6)$
$= 0.1 + 0.25 + 0.1$
$= 0.45$

b. $\Pr(X \geq 3 \mid X \leq 5) = \dfrac{\Pr(X \geq 3 \cap X \leq 5)}{\Pr(X \leq 5)}$

$\Pr(X \geq 3 \cap X \leq 5) = \Pr(3 \leq X \leq 5)$
$= \Pr(X=3) + \Pr(X=4) + \Pr(X=5)$
$= 0.2 + 0.25 + 0.25$
$= 0.7$

$\Pr(X \leq 5) = 1 - \Pr(X=6)$
$= 1 - 0.1$
$= 0.9$

$\Pr(X \geq 3 \mid X \leq 5) = \dfrac{\Pr(X \geq 3 \cap X \leq 5)}{\Pr(X \leq 5)}$
$= \dfrac{\Pr(3 \leq X \leq 5)}{\Pr(X \leq 5)}$
$= \dfrac{0.7}{0.9}$

$= \dfrac{7}{9}$

c. $E(X) = 1(0.1) + 2(0.1) + 3(0.2) + 4(0.25) + 5(0.25) + 6(0.1)$
$= 0.1 + 0.2 + 0.6 + 1 + 1.25 + 0.6$
$= 3.75$

WORKED EXAMPLE 13 Further applications

Let Y be a discrete random variable with the following probability distribution.

y	0	1	2	3	4
$\Pr(Y = y)$	0.08	0.34	0.38	0.17	0.03

a. Determine the expected value of Y.
b. Calculate the standard deviation of Y, correct to 4 decimal places.
c. Calculate $\Pr(\mu - 2\sigma \le Y \le \mu + 2\sigma)$.

THINK	WRITE
a. 1. Write the rule to find the expected value.	**a.** $E(Y) = \sum_{\text{all } y} y \Pr(Y = y)$
2. Substitute the appropriate values into the rule.	$E(Y) = 0(0.08) + 1(0.34) + 2(0.38) + 3(0.17) + 4(0.03)$
3. Simplify.	$E(Y) = 0 + 0.34 + 0.76 + 0.51 + 0.12$ $= 1.73$
b. 1. Find $E(Y^2)$.	**b.** $E(Y^2) = 0^2(0.08) + 1^2(0.34) + 2^2(0.38) + 3^2(0.17) + 4^2(0.03)$ $= 0 + 0.34 + 1.52 + 1.53 + 0.48$ $= 3.87$
2. Write the rule for the variance.	$\text{Var}(Y) = E(Y^2) - [E(Y)]^2$
3. Substitute in the appropriate values and evaluate.	$\text{Var}(Y) = 3.87 - 1.73^2$ $= 0.8771$
4. Write the rule for the standard deviation.	$SD(Y) = \sqrt{\text{Var}(Y)}$
5. Substitute in the variance and evaluate.	$SD(Y) = \sqrt{0.8771}$ $= 0.9365$
c. 1. Find $\mu - 2\sigma$.	**c.** $\mu - 2\sigma = 1.73 - 2(0.9365)$ $= -0.143$
2. Find $\mu + 2\sigma$.	$\mu + 2\sigma = 1.73 + 2(0.9365)$ $= 3.603$
3. Substitute the values into $\Pr(\mu - 2\sigma \le Y \le \mu + 2\sigma)$.	$\Pr(\mu - 2\sigma \le Y \le \mu + 2\sigma)$ $= \Pr(-0.143 \le Y \le 3.603)$
4. Interpret this interval in the context of a discrete distribution. The smallest y-value in the distribution table is 0, so -0.173 is rounded up to 0. The largest y-value in the distribution table that is smaller than 3.573 is 3.	$\Pr(\mu - 2\sigma \le Y \le \mu + 2\sigma)$ $= \Pr(0 \le Y \le 3)$ $= 0.08 + 0.34 + 0.38 + 0.17$ $= 0.97$ *Note:* This is very close to the estimated value of 0.95.

9.5 Exercise

Technology free

1. Let X be a discrete random variable with the following probability distribution.

x	0	1	2	3	4
$\Pr(X = x)$	0.2	0.1	0.3	0.3	0.1

a. Calculate the expected value of X

b. If the standard deviation of X is 1.3, evaluate $\Pr(\mu - 2\sigma \le X \le \mu + 2\sigma)$.

2. A discrete random variable, Y, has the following probability distribution.

y	1	2	d	8
$\Pr(Y = y)$	0.3	0.2	0.4	0.1

a. Calculate the value of the constant d if it is known that $E(Y) = 3.5$.

b. Determine $\Pr(Y \ge 2 \mid Y \le d)$.

3. **WE12** A financial adviser for a large company has put forward a number of options to improve the company's profitability, X (measured in hundreds of thousands of dollars). The decision to implement the options will be based on the cost of the options as well as their profitability. The company stands to make an extra profit of 1 million dollars with a probability of 0.1, an extra profit of \$750 000 with a probability of 0.3, an extra profit of \$500 000 with a probability of 0.3, an extra profit of \$250 000 with a probability of 0.2 and an extra profit of \$100 000 with a probability of 0.1.
Determine:

a. $\Pr(X \le \$500\,000)$

b. $\Pr(X \ge \$250\,000 \mid X \le \$750\,000)$

c. the expected profit.

4. A discrete random variable, Z, has the following probability distribution.

z	1	3	5	7	9
$\Pr(Z = z)$	0.2	0.15	a	b	0.05

The expected value of Z is known to be equal to 4.6.
Calculate the values of the constants a and b.

Technology active

5. **WE13** A discrete random variable, X, has the following probability distribution.

x	5	10	15	20	25
$\Pr(X = x)$	0.05	0.25	0.4	0.25	0.05

a. Determine the expected value of X.

b. Calculate the standard deviation of X, correct to 4 decimal places.

c. Calculate $\Pr(\mu - 2\sigma \le X \le \mu + 2\sigma)$.

6. The number of Tails, X, when a fair coin is tossed six times has the following probability distribution.

x	0	1	2	3	4	5	6
$\Pr(X=x)$	0.012	0.093	0.243	0.315	0.214	0.1	0.023

Evaluate $\Pr(\mu-2\sigma \le X \le \mu+2\sigma)$.

7. A probability function is defined as $p(x)=\frac{1}{9}(4-x), x \in \{0, 1, 2\}$.

a. Construct a probability distribution table.
b. Calculate, correct to 4 decimal places where appropriate:

i. $E(X)$ ii. $\text{Var}(X)$ iii. $\text{SD}(X)$.

c. Calculate $\Pr(\mu-2\sigma \le X \le \mu+2\sigma)$.

8. The probability distribution table for the discrete random variable, Z, is as follows.

z	1	3	5
$\Pr(Z=z)$	$\frac{k^2}{7}$	$\frac{5-2k}{7}$	$\frac{8-3k}{7}$

a. Determine the value(s) of the constant k.
b. Determine, correct to 4 decimal places:

i. $E(Z)$ ii. $\text{Var}(Z)$ iii. $\text{SD}(Z)$.

c. Evaluate $\Pr(\mu-2\sigma \le Z \le \mu+2\sigma)$.

9. A discrete random variable, Z, can take the values 0, 1, 2, 3, 4 and 5. The probability distribution of Z is as follows:
$\Pr(Z=0)=\Pr(Z=1)=\Pr(Z=2)=m$,
$\Pr(Z=3)=\Pr(Z=4)=\Pr(Z=5)=n$,
and $\Pr(Z<2)=3\Pr(Z>4)$ where m and n are constants.

a. Determine the values of m and n.
b. Show that the expected value of Z is $\frac{11}{5}$, and determine the variance and standard deviation for Z, correct to 4 decimal places.
c. Calculate $\Pr(\mu-2\sigma \le Z \le \mu+2\sigma)$.

10. A probability distribution is such that
$\Pr(Z=0)=\Pr(Z=1)=\Pr(Z=2)=\Pr(Z=3)=m$,
$\Pr(Z=3)=\Pr(Z=5)=n$,
and $\Pr(Z \le 3)=\Pr(Z \ge 4)$.

a. Calculate the values of the constants m and n.
b. Determine:

i. $E(Z)$ ii. $\text{Var}(Z)$.

c. Calculate $\Pr(\mu-2\sigma \le Z \le \mu+2\sigma)$.

11. The number of customers, X, waiting in line at a bank just before closing time has a probability distribution as follows.

x	0	1	2	3
$\Pr(X=x)$	$\frac{k^2}{4}$	$\frac{5k-1}{12}$	$\frac{3k-1}{12}$	$\frac{4k-1}{12}$

a. Determine the value of the constant k.
b. Determine the expected number of customers waiting in line just before closing time.
c. Calculate the probability that the number of customers waiting in line just before closing time is no greater than $E(X)$.

12. A television game show features 26 cases with various amounts of money ranging from 50 cents to \$200 000. The contestant chooses one case and then proceeds to open the other cases. At the end of each round, the banker makes an offer to end the game. The game ends when the contestant accepts the offer or when all the other 25 cases have been opened; in the latter event, the contestant receives the amount of money in the case they first chose.
Suppose a contestant has five cases left and the amounts of \$200 000, \$100 000, \$50 000, \$15 000 and \$1000 are still to be found.
   a. Determine the expected amount that the banker should offer the contestant to end the game.
   b. The contestant turned down the offer and opened a case containing \$100 000. State what you would expect the banker to offer the contestant at this stage.
13. A bookstore sells both new and secondhand books. A particular new autobiography costs \$65, a good-quality used autobiography costs \$30 and a worn autobiography costs \$12. A new cookbook costs \$54, a good-quality used cookbook costs \$25 and a worn cookbook costs \$15. Let X denote the total cost of buying two books (an autobiography and a cookbook). Assume that the purchases are independent of one another.
   a. Construct a probability distribution table for the cost of the two textbooks if the following probabilities apply.
      - The probability of buying a new autobiography is 0.4.
      - The probability of buying a good-quality used autobiography is 0.3.
      - The probability of buying a worn used autobiography is 0.3.
      - The probability of buying a new cookbook is 0.4.
      - The probability of buying a good-quality used cookbook is 0.25.
      - The probability of buying a worn used cookbook is 0.35.
   b. Calculate the expected cost of the two books.
14. Let X be the number of dining suites sold by the dining suite department of a large furniture outlet on any given day. The probability function for this discrete random variable is as follows.

x	0	1	2	3
Pr($X = x$)	0.3	0.4	0.2	0.1

The dining suite department receives a profit of \$350 for every dining setting sold. The daily running costs for the sales operation of the department are \$120. The net profit per day is a function of the random variable such that $y(x) = 350x - 120$ dollars.
   a. Set up a probability distribution table for the net profit, \$$Y$, per day.
   b. Determine the expected daily profit for the dining suite department.
   c. Determine $\Pr(\mu - 2\sigma \leq Y \leq \mu + 2\sigma)$.
15. A loaded 6-sided die and a biased coin are tossed simultaneously. The coin is biased such that the probability of obtaining a Head is 3 times the probability of obtaining a Tail. The loaded die has the following probabilities for each of the numbers 1 to 6.

$$\Pr(1) = \Pr(2) = \Pr(5) = \frac{1}{12}$$

$$\Pr(3) = \Pr(4) = \Pr(6) = \frac{1}{4}$$

When a player tosses the coin and die simultaneously, they receive the following points as a result of the outcome.

$$\overbrace{\text{1T 2T 5T}}^{\text{10 points}} \qquad \overbrace{\text{1H 2H 5H}}^{\text{5 points}} \qquad \overbrace{\text{All other outcomes}}^{\text{1 point}}$$

Let X be the number of points scored from a simultaneous toss.
   a. Construct a probability distribution table for the number of points scored.
   b. Determine the expected points received from a single toss, correct to 1 decimal place.

c. If 25 simultaneous tosses occurred, determine the expected score, correct to 1 decimal place.
d. Determine the minimum number of simultaneous tosses that would have to occur for the expected total to be a score of 100.

16. In a random experiment the events M and N are independent events where $\Pr(M) = 0.45$ and $\Pr(N) = 0.48$.
  a. Determine the probability that both M and N occur.
  b. Calculate the probability that neither M nor N occur.

  Let Y be the discrete random variable that defines the number of times M and N occur.
  $Y = 0$ if neither M nor M occurs.
  $Y = 1$ if only one of M and N occurs.
  $Y = 2$ if both M and N occur.

  c. Specify the probability distribution for Y.
  d. Determine, correct to 4 decimal places where appropriate:
    i. $E(Y)$ ii. $\text{Var}(Y)$ iii. $SD(Y)$.

9.5 Exam questions

Question 1 (2 marks) TECH-FREE

X is a discrete random variable representing the number of touchscreen tablets sold on a particular day. Given $E(X) = 15$ and $E(X^2) = 250$, determine the values of x_1 and x_2 in the interval $[x_1, x_2]$. The interval represents a probability of 95% that the number of touchscreen tablets sold on a particular day falls within this range.

Question 2 (1 mark) TECH-ACTIVE

MC Over a twenty-day period, Pete recorded the number of minutes that he spent waiting for the lift to go from the ground floor to a classroom on the sixth floor.

Number of minutes spent waiting x	0	1	2	3	4
Proportion of days on which x minutes were spent waiting	$\frac{5}{20}$	$\frac{3}{20}$	$\frac{4}{20}$	$\frac{6}{20}$	$\frac{2}{20}$

During this twenty-day period, the mean number of minutes that Pete spent waiting for the lift was

A. 1.85 **B.** 0.75 **C.** 18.75 **D.** 37 **E.** 2.1

Question 3 (8 marks) TECH-FREE

In a certain random experiment the events V and W are independent events.

a. If $\Pr(V \cup W) = 0.7725$ and $\Pr(V \cap W) = 0.7725$, calculate $\Pr(V)$ and $\Pr(W)$, given $\Pr(V) < \Pr(W)$. **(3 marks)**
b. Determine the probability that neither V nor W occur. **(1 mark)**

Let X be the discrete random variable that defines the number of times events V and W occur.

$X = 0$ if neither V nor W occurs.
$X = 1$ if only one of V and W occurs.
$X = 2$ if both V and W occur.

c. Specify the probability distribution for X. **(1 mark)**
d. Determine, correct to 4 decimal places where appropriate

i. $E(X)$ **(1 mark)**
ii. $\text{Var}(X)$ **(1 mark)**
iii. $SD(X)$. **(1 mark)**

More exam questions are available online.

9.6 Review

9.6.1 Summary

doc-37039

Hey students! Now that it's time to revise this topic, go online to:

Access the topic summary | Review your results | Watch teacher-led videos | Practise exam questions

Find all this and MORE in jacPLUS

9.6 Exercise

Technology free: short answer

1. The random variable X has a probability distribution as follows.

x	0	1	2	3	4
$\Pr(X = x)$	0.1	0.3	0.4	0.1	0.1

   a. Calculate $\Pr(X \geq 2 \mid X \leq 3)$.
   b. Determine the expected value of X.

2. On any given day, the number of text messages, Y, received by Garish is a discrete random variable with a distribution as follows.

y	0	2	4	6	8	10
$\Pr(Y = y)$	0.05	0.4	0.2	0.15	0.15	0.05

   a. Calculate the expected value of Y.
   b. Determine the probability that Garish receives no texts on four consecutive days.
   c. Garish received text messages on Thursday and Friday. Determine the probability that he received 10 text messages over these two days.

3. A biased coin is tossed four times. The probability of a Head from a toss is a where $0 < a < 1$.
   a. Calculate, in terms of a, the probability of obtaining:
      i. four Tails from four tosses
      ii. one Head and three Tails from four tosses.
   b. If the probability of obtaining four Heads is the same as the probability of obtaining one Head and three Tails, evaluate a.

4. X has a probability function given by $p(x) = \frac{1}{5}$, $x = 1, 3, 5, 7, 9$.

   Determine:

   a. $E(X)$
   b. $\text{Var}(X)$
   c. $E(3X + 1)$
   d. $\text{Var}(5X + 2)$

5. Alicia and Harry have devised a game where a biased spinner is spun. There are 5 colours on the wheel and the sectors are of varying sizes.

   $\Pr(\text{red}) = \frac{1}{20}$, $\Pr(\text{blue}) = \Pr(\text{green}) = 2 \times \Pr(\text{red})$, and the other sector colour is yellow.

   Players have to pay \$2.00 to play. If yellow is spun up, players receive nothing. If green is spun up or blue is spun up, players get their money back. If the spinner lands on red, players win \$5.

   a. Determine the probability distribution for the amount of money a person can win.
   b. Determine the expected amount of money that a player will win each game.

6. A discrete random variable, X, has a standard deviation of 1.1. If $E(X^2)$ is equal to 2.02, determine:

a. the variance of X

b. the expected value of X

c. $E(2X-4)$

d. $Var(2X-4)$.

Technology active: multiple choice

7. **MC** Consider the discrete probability function with the following distribution.

x	2	4	6	8	10
$Pr(X=x)$	$2a$	$3a$	$4a$	$5a$	$6a$

The value of the constant a is:

A. 20 B. $\frac{1}{20}$ C. $\frac{1}{2}$ D. $\frac{1}{14}$ E. $\frac{1}{19}$

8. **MC** State which of the following random variables is **not** discrete.

A. The number of goals scored at a football match
B. The number of T-shirts owned by a student
C. The number of kittens in a litter
D. The number of customers at a department store sale
E. The volume of soft drink consumed by a family over the period of a week

***The following information refers to Questions* 9, 10 *and* 11.**

The discrete random variable Z has a probability distribution as follows.

z	1	2	3	4	5
$Pr(Z=z)$	0.1	0.25	0.35	0.25	0.05

9. **MC** The expected value of Z is equal to:

A. 2.9 B. 9.5 C. 1.09 D. 1.044 E. 3.082

10. **MC** The variance of Z is equal to:

A. 2.9 B. 9.5 C. 1.09 D. 1.044 E. 3.082

11. **MC** The standard deviation of Z is equal to:

A. 2.9 B. 9.5 C. 1.09 D. 1.044 E. 3.082

12. **MC** The random variable X has the following probability distribution.

x	0	2	4
$Pr(X=x)$	m	n	0.7

If the mean of X is 3.2, then m is equal to:

A. 0.3 B. 0.2 C. 0.1 D. 0 E. 0.25

13. **MC** Maya constructed a spinner that will fall onto one of the numbers 1 to 5 with the following probability.

Number	1	2	3	4	5
Probability	0.1	0.3	0.3	0.2	0.1

The mean and standard deviation, correct to 2 decimal places, respectively are:

A. 3 and 1 B. 3 and 1.29 C. 3 and 1.13
D. 2.9 and 1.29 E. 2.9 and 1.14

14. **MC** For a discrete random variable X with a mean of 2.1 and a variance of 1.3, the values of $E(2X+1)$ and $Var(2X+1)$ are respectively:

A. 4.2 and 5.2 B. 5.2 and 6.2 C. 5.2 and 5.2 D. 4.2 and 6.2 E. 5.2 and 4.2

15. **MC** The random variable Y has the following probability distribution.

y	-2	0	2
$\Pr(Y = y)$	$2p$	$3p$	$1 - 5p$

The mean of Y is:

A. $2 - 11p$
B. $2 - 14p$
C. $1 - 3p$
D. $14p$
E. $2 + 14p$

16. **MC** The probability distribution for the random variable X is as follows.

x	-1	0	1	2
$\Pr(X = x)$	m	$m + n$	$3m$	$m - n$

If $E(X) = 0.4$, then m and n are equal to:

A. $m = \frac{1}{6}, n = \frac{1}{5}$
B. $m = \frac{1}{5}, n = \frac{1}{6}$
C. $m = \frac{1}{6}, n = \frac{2}{15}$
D. $m = \frac{2}{15}, n = \frac{1}{6}$
E. $m = \frac{1}{6}, n = \frac{1}{15}$

Technology active: extended response

17. A discrete random variable, Z, has a probability distribution as shown.

z	1	2	3	4	5	6
$\Pr(Z = z)$	$\frac{m}{5}$	$\frac{1}{6}$	$\frac{1}{6}$	$\frac{1}{6}$	$\frac{2m}{5}$	$\frac{1}{10}(5 - 6m)$

This random variable describes the outcome of tossing a loaded die. The die is thrown twice.

a. Prove that the chance of throwing a total of 11 is $\frac{10m - 12m^2}{25}$.

b. Determine the value of m that makes this chance a maximum, and find the maximum probability.

c. Using the value of m from part **b**, calculate:

i. the expected value of Z and the standard deviation of Z

ii. $\Pr(\mu - 2\sigma \leq Z \leq \mu + 2\sigma)$.

18. A random variable, X, represents the number of televisions serviced per week by a television serviceman. The probability distribution is as follows.

x	10	11	12	13	14	15	16	17	18	19	20
$\Pr(X = x)$	0.07	0.12	0.12	0.1	0.1	0.1	0.1	0.08	0.08	0.08	0.05

a. Determine the mean number of televisions serviced per week.

b. Determine the variance and standard deviation of X, correct to 4 decimal places.

The serviceman is paid a bonus depending on how many televisions he services a week. The bonuses are as follows:

- If less than 13 televisions are serviced, there is no bonus.
- If 13–16 televisions are serviced, he receives a bonus of \$120.
- If more than 16 televisions are serviced, he receives a bonus of \$250.

Let B be the bonus paid to the serviceman.

c. Construct a probability distribution for B.

d. Calculate the expected bonus to be paid.

19. The number of passengers per car, X, entering Melbourne on a freeway on a workday morning is as follows.

x	0	1	2	3	4	5
$\Pr(X = x)$	0.37	0.22	0.21	0.1	0.05	0.05

a. Determine the mean number of passengers per car.
b. Determine the variance and standard deviation for X.
c. Determine $\Pr(\mu - 2\sigma \leq X \leq \mu + 2\sigma)$.
The fees for cars at a toll booth on the freeway are as follows.
- Cars carrying no passengers: \$2.50
- Cars carrying 1 or 2 passengers: \$1.00
- Cars carrying more than 2 passengers: no fee

Let T be the toll paid by a randomly selected car on the freeway.
d. Construct a probability distribution for T.
e. Calculate the expected value of the toll per car.
f. Calculate $\Pr(\mu - 2\sigma \leq T \leq \mu + 2\sigma)$.

20. The discrete random variable Z can only take the values 0, 1, 2, 3, 4, 5 and 6. The probability distribution for Z is given by the following:
$\Pr(Z = 0) = \Pr(Z = 2) = \Pr(Z = 4) = \Pr(Z = 6) = m$,
$\Pr(Z = 1) = \Pr(Z = 3) = \Pr(Z = 5) = n$,
and $2\Pr(0 < Z < 2) = \Pr(3 < Z \leq 6)$, where m and n are constants.
a. Determine the values of m and n.
b. Calculate the expected value of Z.
c. Calculate the standard deviation of Z.
d. Evaluate:
i. $E(2 - 3Z)$ ii. $\text{Var}(2Z - 3)$.
e. Evaluate $\Pr(\mu - 2\sigma \leq Z \leq \mu + 2\sigma)$.

9.6 Exam questions

Question 1 (1 mark) TECH-ACTIVE
Source: *VCE 2021 Mathematical Methods Exam 2, Q20; © VCAA.*

MC Let A and B be two independent events from a sample space.

If $\Pr(A) = p$, $\Pr(B) = p^2$ and $\Pr(A) + \Pr(B) = 1$, then $\Pr(A' \cup B)$ is equal to
A. $1 - p - p^2$
B. $p^2 - p^3$
C. $p - p^3$
D. $1 - p + p^3$
E. $1 - p - p^2 + p^3$

Question 2 (3 marks) TECH-FREE
Source: *VCE 2019 Mathematical Methods Exam 1, Q3; © VCAA.*

The only possible outcomes when a coin is tossed are a head or a tail. When an unbiased coin is tossed, the probability of tossing a head is the same as the probability of tossing a tail. Jo has three coins in her pocket; two are unbiased and one is biased. When the biased coin is tossed, the probability of tossing a head is $\frac{1}{3}$.

Jo randomly selects a coin from her pocket and tosses it.
a. Find the probability that she tosses a head. **(2 marks)**
b. Find the probability the she selected an unbiased coin, given that she tossed a head. **(1 mark)**

Question 3 (1 mark) TECH-ACTIVE

Source: *VCE 2016 Mathematical Methods Exam 2, Section A, Q19; © VCAA.*

MC Consider the discrete probability distribution with random variable X shown in the table below.

x	-1	0	b	$2b$	4
$\Pr(X=x)$	a	b	b	$2b$	0.2

The smallest and largest possible values of $E(X)$ are respectively

A. -0.8 and 1
B. -0.8 and 1.6
C. 0 and 2.4
D. 0.2125 and 1
E. 0 and 1

Question 4 (1 mark) TECH-ACTIVE

Source: *VCE 2015 Mathematical Methods (CAS) Exam 2, Section 1, Q14; © VCAA.*

MC Consider the following discrete probability distribution for the random variable X.

x	1	2	3	4	5
$\Pr(X=x)$	p	$2p$	$3p$	$4p$	$5p$

The mean of this distribution is

A. 2 **B.** 3 **C.** $\frac{7}{2}$ **D.** $\frac{11}{3}$ **E.** 4

Question 5 (1 mark) TECH-ACTIVE

Source: *VCE 2014 Mathematical Methods (CAS) Exam 2, Section 1, Q14; © VCAA.*

MC If X is a random variable such that $\Pr(X>5)=a$ and $\Pr(X>8)=b$, then $\Pr(X<5 \mid X<8)$ is

A. $\frac{a}{b}$ **B.** $\frac{a-b}{1-b}$ **C.** $\frac{1-b}{1-a}$ **D.** $\frac{ab}{1-b}$ **E.** $\frac{a-1}{b-1}$

More exam questions are available online.

Answers

Topic 9 Discrete random variables

9.2 Probability review

9.2 Exercise

1. a. $\frac{2}{11}$ b. $\frac{7}{11}$ c. $\frac{9}{11}$

2. a. $\frac{1}{4}$ b. $\frac{3}{4}$ c. $\frac{7}{20}$

3. a. $\frac{1}{125}$ b. $\frac{3}{100}$ c. $\frac{9}{50}$

4. a. ξ

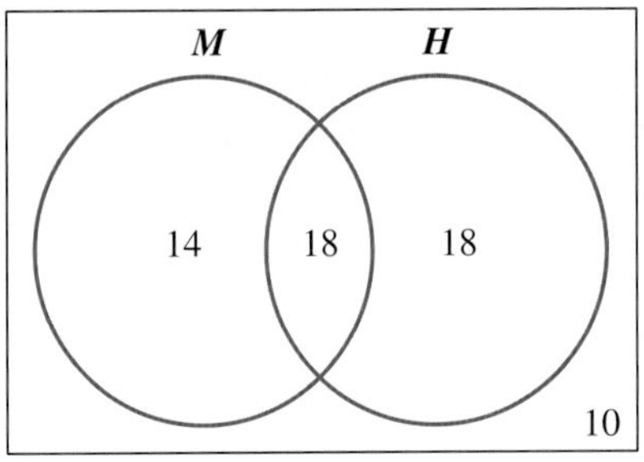

b. i. $\frac{3}{5}$

ii. $\frac{3}{10}$

iii. $\frac{1}{2}$

5. a. 0.68 b. $\frac{5}{16}$ c. $\frac{11}{16}$

6. a. 0.32 b. 0.88 c. $\frac{16}{35}$

7. a. 0.49 b. $\frac{7}{12}$

8. a. $\frac{2}{15}$ b. $\frac{8}{15}$

9. a. Not mutually exclusive events
   b. Not independent events

10. a. Not mutually exclusive events
    b. Yes, independent events

11. a. 0 b. 0.44

12. 0.3

9.2 Exam questions

Note: Mark allocations are available with the fully worked solutions online.

1. a. 0.1 b. $19n - 20$

2. D

3. B

9.3 Discrete random variables

9.3 Exercise

1. a.

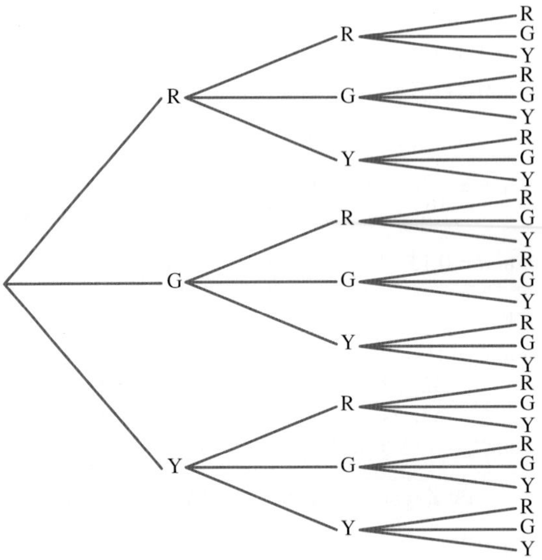

ξ = {RRR, RRG, RRY, RGR, RGG, RGY, RYR, RYG, RYY, GRR, GRG, GRY, GGR, GGG GGY, GYR, GYG, GYY, YRR, YRG, YRY, YGR, YGG, YGY, YYR, YYG, YYY,}

b. $\Pr(Y=3) = \frac{27}{1000}$, $\Pr(Y=2) = \frac{189}{1000}$, $\Pr(Y=1) = \frac{441}{1000}$, $\Pr(Y=0) = \frac{343}{1000}$

c.

y	0	1	2	3
$\Pr(Y=y)$	$\frac{343}{1000}$	$\frac{441}{1000}$	$\frac{189}{1000}$	$\frac{27}{1000}$

2.

x	0	1	2
$\Pr(X=x)$	$\frac{25}{36}$	$\frac{10}{36} = \frac{5}{18}$	$\frac{1}{36}$

3. a. $0 \le \Pr(Y=y) \le 1$ for all y and $\sum_{\text{all } y} \Pr(Y-y) = 1$.
   Yes, this is a discrete probability function.
   b. $0 \le \Pr(Y=y) \le 1$ for all y and $\sum_{\text{all } y} \Pr(Y-y) = 1$.
   Yes, this is a discrete probability function.

4. a. $0 \le \Pr(Y=y) \le 1$ for all y and $\sum_{\text{all } y} \Pr(Y-y) = 0.9$.
   No, this is not a discrete probability function.
   b. Probabilities cannot have negative values. No, this is not a discrete probability function.

5. a. $0 \le \Pr(X=x) \le 1$ for all x but $\sum_{\text{all } x} \Pr(X=x) \neq 1$.
   This is not a probability distribution.
   b. Probabilities cannot have negative values. This is not a probability distribution.
   c. $0 \le \Pr(Z=z) \le 1$ for all z but $\sum_{\text{all } x} \Pr(Z=z) = 1.1$.
   This is not a probability distribution.
   d. $0 \le \Pr(Z=z) \le 1$ for all x and $\sum_{\text{all } x} \Pr(X=x) = 1$.
   This is a probability distribution.

6. a. $0 \leq \Pr(X = x) \leq 1$ for all x and $\sum_{\text{all } x} \Pr(X = x) = 1$.
This is a discrete probability distribution.
b. $0 \leq \Pr(Z = z) \leq 1$ for all x but $\sum_{\text{all } x} \Pr(X = x) \neq 1$.
This is not a discrete probability distribution.
c. $0 \leq \Pr(Z = z) \leq 1$ for all x and $\sum_{\text{all } x} \Pr(X = x) = 1$.
This is a discrete probability distribution.

7. a. $k = \frac{1}{16}$ b. $k = \frac{1}{4}$

8. a. $d = 0.15$ b. $k = \frac{1}{6}$ c. $a = \frac{1}{3}$

9. E

10. a. $\xi = \{11, 12, 13, 14, 15, 16, 21, 22, 23, 24, 25, 26, 31, 32, 33, 34, 35, 36, 41, 42, 43, 44, 45, 46, 51, 52, 53, 54, 55, 56, 61, 62, 63, 64, 65, 66\}$

b.

z	0	1	2
Pr($Z = z$)	0.09	0.42	0.49

c. 0.42

11. a. $\xi = \{\text{SSS, SSA, SAS, SAA, ASS, ASA, AAS, AAA}\}$

b.

x	0	1	2	3
Pr($X = x$)	0.216	0.432	0.288	0.064

c. 0.936

12. a. $\xi = \{\text{FFFF, FFFM, FFMF, FFMM, FMFF, FMFM, FMMF, FMMM, MFFF, MFFM, MFMF, MFMM, MMFF, MMFM, MMMF, MMMM}\}$

b.

x	0	1	2	3	4
Pr($X = x$)	$\frac{1}{16}$	$\frac{4}{16} = \frac{1}{4}$	$\frac{6}{16} = \frac{3}{8}$	$\frac{4}{16} = \frac{1}{4}$	$\frac{1}{16}$

c. $\frac{1}{16}$

d. $\frac{15}{16}$

e. $\frac{11}{16}$

13. a. $\xi = \{11, 12, 13, 14, 15, 16, 17, 18, 19, 110, 111, 112, 21, 22, 23, 24, 25, 26, 27, 28, 29, 210, 211, 212, 31, 32, 33, 34, 35, 36, 37, 38, 39, 310, 311, 312, 41, 42, 43, 44, 45, 46, 47, 48, 49, 410, 411, 412, 51, 52, 53, 54, 55, 56, 57, 58, 59, 510, 511, 512, 61, 62, 63, 64, 65, 66, 67, 68, 69, 610, 611, 612, 71, 72, 73, 74, 75, 76, 77, 78, 79, 710, 711, 712, 81, 82, 83, 84, 85, 86, 87, 88, 89, 810, 811, 812\}$

b. $\Pr(X = 0) = \frac{28}{96}, \Pr(X = 1) = \frac{48}{96}, \Pr(X = 2) = \frac{20}{96}$

c. $\Pr(\text{win}) = 0.009$

14. a. Possible scores = 4, 7, 10, 12, 15 and 20 points
b. See the table at the bottom of the page.*
c. i. $\frac{6}{169}$ ii. $\frac{69}{169}$ iii. $\frac{2}{23}$

15. a. See the table at the bottom of the page.*
b. 0.6826. It is a success, helping 3 or more patients.

16. a. See the table at the bottom of the page.*
i. 0.8999 ii. 0.8457

17. a. $m = \frac{1}{9}, \ n = \frac{1}{3}$
b. i. $\frac{5}{9}$ ii. $\frac{1}{5}$

18. i. $k = 0.6568$ or $k = -0.4568$

9.3 Exam questions

Note: Mark allocations are available with the fully worked solutions online.

1. C
2. D
3. a. The game can end as follows:
Throw 1 red — game over.
Throw 1 blue, green or yellow, throw 2 red — game over.
Throws 1 and 2 combinations of blue, green and yellow, throw 3 red — game over.
b. Wins \$10 with BBB, GGG or YYY
c.

x	\$0	\$1	\$10
Pr($X = x$)	$\frac{98}{125}$	$\frac{24}{125}$	$\frac{3}{125}$

*14. b.

x	4	7	10	12	15	20
Pr($X = x$)	$\frac{81}{169}$	$\frac{18}{169}$	$\frac{1}{169}$	$\frac{54}{169}$	$\frac{6}{169}$	$\frac{9}{169}$

*15. a.

x	0	1	2	3	4	5
Pr($X = x$)	0.0102	0.0768	0.2304	0.3456	0.2592	0.0778

*16. a.

x	0	1	2	3	4	5	6
Pr($X = x$)	0.0014	0.0165	0.0823	0.2195	0.3292	0.2634	0.0878

9.4 Measures of centre and spread

9.4 Exercise

1. a. $E(X) = \frac{1}{9}$ b. $E(Z) = \frac{19}{6}$

2. a.

y	1	2	3	4
$\Pr(X = x)$	$\frac{1}{16}$	$\frac{3}{16}$	$\frac{5}{16}$	$\frac{7}{16}$

b. 3.125

3. \$2.45

4. a.

x	−\$2	\$3	\$6	\$8
$\Pr(X = x)$	$\frac{1}{4}$	$\frac{1}{4}$	$\frac{1}{4}$	$\frac{1}{4}$

b. \$3.75

5. a. $k = \frac{1}{10}$ b. 3.2

6. a. $d = 15$
b. i. 18 ii. −2.5 iii. −15

7. a. 8 b. 0.5 c. 14.5

8. B

9. a. 7 b. 14.4 c. 3.797

10. a. See the table at the bottom of the page*
b. i. $\frac{10}{3}$ ii. $\frac{31}{45} = 0.69$
c. i. 11.02 ii. 6.2

11. a. $m = 15$
b. $\text{Var}(Z) = 153.48$, $\text{Var}(2(Z - 1)) = 613.91$, $\text{Var}(3 - Z) = 153.48$

12. C

13. a. $p(x) = h(3 - x)(x + 1)$
$p(0) = h(3)(1) = 3h$
$p(1) = h(3 - 1)(1 + 1) = 4h$
$p(2) = h(3 - 2)(2 + 1) = 3h$
$3h + 4h + 3h = 1$
$10h = 1$
$h = \frac{1}{10}$
b. $E(X) = 1$, $\text{Var}(X) = 0.6$, $\text{SD}(X) = 0.7746$

14. a. $\xi = \{11, 12, 13, 14, 15, 16, 17, 18, 21, 22, 24, 25, 26, 27, 28, 31, 32, 33, 34, 35, 36, 37, 38, 41, 42, 43, 44, 45, 46, 47, 48, 51, 52, 53, 54, 55, 56, 57, 58, 61, 62, 63, 64, 65, 66, 67, 68, 71, 72, 73, 74, 75, 76, 77, 78, 81, 82, 83, 84, 85, 86, 87, 88\}$
b. See the table at the bottom of the page.*
c. $E(Z) = 5.8125$, $\text{SD}(Z) = 1.8781$

15. a. Score totals = $\{1 + 1 = 2,\ 1 + 3 = 4,\ 1 + 5 = 6,\ 1 + 7 = 8,\ 1 + 10 = 11,\ 3 + 1 = 4,\ 3 + 3 = 6,\ 3 + 5 = 8,\ 3 + 7 = 10,\ 3 + 10 = 13, 5 + 1 = 6,\ 5 + 3 = 8,\ 5 + 5 = 10,\ 5 + 7 = 12,\ 5 + 10 = 15,\ 7 + 1 = 8,\ 7 + 3 = 10,\ 7 + 5 = 12,\ 7 + 7 = 14,\ 7 + 10 = 17,\ 10 + 1 = 11,\ 10 + 3 = 13,\ 10 + 5 = 15,\ 10 + 7 = 17,\ 10 + 10 = 20\}$
b. See the table at the bottom of the page.*
c. $E(X) = 9.4$, $\text{SD}(X) = 3.7974$

16. a. $c = \frac{1}{3}$
b. $E(Y) = 3$
c. $\text{Var}(Y) = 11.56$, $\text{SD}(Y) = 3.40$

17. a. $k = 0.1$
b. $E(X) = 1.695$
c. $\text{SD}(X) = 1.1670$

18. a. $a = 0.3$, $b = 0.1$
b. $\text{Var}(X) = 1.65$, $\text{SD}(X) = 1.2845$

19. a. $E(X^2) = a^2 + 2a - 2$
b. $E(X) = a = 2$, $\text{Var}(X) = 2a - 2 = 2$

20. a. $n = \frac{1}{13}$
b. $E(Y) = \frac{46}{13} = 3.5385$, $\text{Var}(Y) = 1.7870$, $\text{SD}(Y) = 1.3368$

9.4 Exam questions

Note: Mark allocations are available with the fully worked solutions online.

1. E
2. D

*10. a.

x	1	2	3	4
$\Pr(X = x)$	$\frac{1}{30}$	$\frac{4}{30} = \frac{2}{15}$	$\frac{9}{30} = \frac{3}{10}$	$\frac{16}{30} = \frac{8}{15}$

*14. b.

z	1	2	3	4	5	6	7	8
$\Pr(Z = z)$	$\frac{1}{64}$	$\frac{3}{64}$	$\frac{5}{64}$	$\frac{7}{64}$	$\frac{9}{64}$	$\frac{11}{64}$	$\frac{13}{64}$	$\frac{15}{64}$

*15. b.

x	2	4	6	8	10	11	12	13	14	15	17	20
$\Pr(X = x)$	0.04	0.08	0.16	0.2	0.17	0.04	0.12	0.04	0.04	0.06	0.04	0.01

3. a. $\Pr(B \text{ and } A) = \frac{1}{25}$, $\Pr(B \text{ and } B) = \frac{3}{25}$,

$\Pr(B \text{ and } C) = \frac{5}{25}$, $\Pr(B \text{ and } D) = \frac{7}{25}$,

$\Pr(B \text{ and } E) = \frac{9}{25}$

b.

x	−\$1	\$0	\$1	\$4	\$9
$\Pr(X = x)$	$\frac{9}{25}$	$\frac{7}{25}$	$\frac{5}{25} = \frac{1}{5}$	$\frac{3}{25}$	$\frac{1}{25}$

c. i. $E(X) = 0.68$ cents ii. $SD(X) = \$2.29$

9.5 Applications

9.5 Exercise

1. a. 2.0 b. 1
2. a. $d = 5$ b. $\frac{2}{3}$
3. a. 0.6 b. $\frac{8}{9}$ c. \$535 000
4. $a = 0.35, b = 0.25$
5. a. 15 b. 4.7434 c. 0.9
6. 0.965
7. a.

x	0	1	2
$\Pr(X = x)$	$\frac{4}{9}$	$\frac{3}{9} = \frac{1}{3}$	$\frac{2}{9}$

b. i. $\frac{7}{9}$ ii. $\frac{50}{81} = 0.6173$ iii. 0.7857

c. 1

8. a. $k = 2$

b. i. $E(Z) = 2.4286$

ii. $Var(Z) = 3.1019$

iii. $SD(Z) = 1.7613$

c. $\Pr(\mu - 2\sigma \le Z \le \mu + 2\sigma) = 1$

9. a. $m = \frac{1}{5}$, $n = \frac{2}{15}$

b. $Var(Z) = 2.8267$, $SD(Z) = 1.6813$

c. 1

10. a. $m = \frac{1}{8}, n = \frac{1}{4}$

b. i. $E(Z) = 3$ ii. $Var(Z) = 3$

c. $\Pr(\mu - 2\sigma \le Z \le \mu + 2\sigma) = 1$

11. a. $k = 1$ b. 1.4 c. $\frac{7}{12}$
12. a. \$73 200 b. \$66 500
13. a. See the table at the bottom of the page.*

b. \$71.70

14. a.

y	−\$120	\$230	\$580	\$930
$\Pr(Y = y)$	0.3	0.4	0.2	0.1

b. $E(Y) = +\$265$

c. $\Pr(\mu - 2\sigma \le Y \le \mu + 2\sigma) = 0.9$

15. a.

x	1	5	10
$\Pr(X = x)$	$\frac{12}{16} = \frac{3}{4}$	$\frac{3}{16}$	$\frac{1}{16}$

b. 2.3

c. 57.8

d. 44 tosses

16. a. 0.216

b. 0.286

c.

y	0	1	2
$\Pr(Y = y)$	0.289	0.498	0.216

d. i. $E(Y) = 0.93$

ii. $Var(Y) = 0.4971$

iii. $SD(Y) = 0.7050$

9.5 Exam questions

Note: Mark allocations are available with the fully worked solutions online.

1. $[5, 25]$
2. A
3. a. $\Pr(V) = 0.35$, $\Pr(W) = 0.65$

b. 0.2275

c.

x	0	1	2
$\Pr(X = x)$	0.2275	0.5450	0.2275

d. i. $E(X) = 1$

ii. $Var(X) = 0.455$

iii. $SD(X) = 0.6745$

9.6 Review

9.6 Exercise

Technology free: short answer

1. a. $\frac{5}{9}$

b. $E(X) = 1.8$, $Var(X) = 1.16$

2. a. $E(Y) = 4.2$ b. 0.000 006 25 c. $\Pr(10) = 0.185$
3. a. i. $\Pr(TTTT) = (1 - a)^4$

ii. $\Pr(3 \text{ Tails and a Head}) = 4a(1 - a)^3$

b. $a = \frac{1}{5}$

4. a. $E(X) = 5$ c. 16

b. $Var(X) = 8$ d. 200

*13. a.

x	119	90	84	80	66	55	45	37	27
$\Pr(X = x)$	0.16	0.1	0.12	0.14	0.12	0.075	0.105	0.075	0.105

5. a. X = amount of money a player can win

x	0	2	5
$\Pr(X = x)$	$\frac{3}{4}$	$\frac{1}{5}$	$\frac{1}{20}$

b. $E(X) = \$0.65$

6. a. $\text{Var}(X) = 1.21$
b. $E(X) = 0.9$
c. -2.2
d. 4.84

Technology active: multiple choice

7. B
8. E
9. A
10. C
11. D
12. C
13. E
14. C
15. B
16. C

Technology active: extended response

17. a. $\Pr(\text{total of } 11)$

$$= \Pr(5, 6) + \Pr(6, 5)$$
$$= \frac{2m}{5} \times \frac{1}{10}(5 - 6m) + \frac{1}{10}(5 - 6m) \times \frac{2m}{5}$$
$$= \frac{2m}{25}(5 - 6m)$$
$$= \frac{10m - 12m^2}{25} \text{ as required}$$

b. $m = \frac{5}{12}$, $\Pr(\text{total of } 11) = \frac{1}{12}$

c. i. $E(Z) = \frac{47}{12}$, $SD(Z) = 1.6562$

ii. $\Pr(\mu - 2\sigma \le Z \le \mu + 2\sigma) = 1$

18. a. 14.58

b. $\text{Var}(X) = 8.7436$, $SD(X) = 2.9570$

c.

b	\$0	\$120	\$250
$\Pr(B = b)$	0.31	0.4	0.29

d. $E(B) = \$120.50$

19. a. 1.39

b. $\text{Var}(X) = 2.0779$, $SD(X) = 1.4415$

c. $\Pr(\mu - 2\sigma \le X \le \mu + 2\sigma) = 0.95$

d.

t	\$0	\$1	\$2.50
$\Pr(T = t)$	0.2	0.43	0.37

e. $E(T) = \$1.36$

f. $\Pr(\mu - 2\sigma \le T \le \mu + 2\sigma) = 1$

20. a. $m = \frac{1}{10}$, $n = \frac{1}{5}$

b. $E(Z) = 3$

c. $SD(Z) = 1.8974$

d. i. -7
ii. 14.4

e. $\Pr(\mu - 2\sigma \le Z \le \mu + 2\sigma) = 1$

9.6 Exam questions

Note: Mark allocations are available with the fully worked solutions online.

1. D
2. a. $\frac{4}{9}$
b. $\frac{3}{4}$
3. E
4. D
5. E

10 The binomial distribution

LEARNING SEQUENCE

Fully worked solutions for this topic are available online.

10.1 Overview

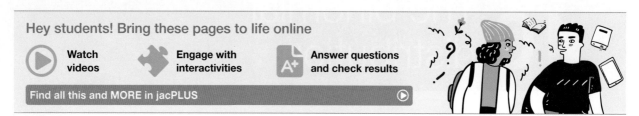

Hey students! Bring these pages to life online

Watch videos

Engage with interactivities

Answer questions and check results

Find all this and MORE in jacPLUS

10.1.1 Introduction

The Bernoulli or binomial distribution is named after the Swiss mathematician Jakob Bernoulli (1654–1705). This distribution is a type of discrete probability distribution and involves Bernoulli trials where there are only two outcomes, success or failure. Each trial is independent of the previous one, and the probability of success remains fixed for each trial.

Researchers may use the binomial distribution when interested in the occurrence of an event, rather than the effect of an event. For example, in the public health sector, medical researchers might focus on whether a drug cures a particular illness or not (success or failure), rather than whether the illness reduces in severity. The binomial distribution can also be used in the manufacturing industry, for example in analysing the probability of a batch containing a defective item. Other areas that utilise the binomial distribution are computing (the distribution of IP addresses), economics (prediction of the way the economy will perform over the next couple of years) and engineering (calculation of time and costs of big projects).

KEY CONCEPTS

This topic covers the following key concepts from the VCE Mathematics Study Design:

- discrete random variables:
  - calculation and interpretation of mean, μ, variance, σ^2, and standard deviation of a discrete random variable and their use
  - Bernoulli trials and the binomial distribution, $\text{Bi}(n, p)$, as an example of a probability distribution for a discrete random variable
  - effect of variation in the value/s of defining parameters on the graph of a given probability mass function for a discrete random variable
  - calculation of probabilities for specific values of a random variable and intervals defined in terms of a random variable, including conditional probability.

Source: VCE Mathematics Study Design (2023–2027) extracts © VCAA; reproduced by permission.

10.2 Bernoulli trials

LEARNING INTENTION

At the end of this subtopic you should be able to:

- define a Bernoulli trial
- calculate the mean, variance and standard deviation of a Bernoulli distribution.

10.2.1 The Bernoulli distribution

In probability theory, the **Bernoulli distribution** is a discrete probability distribution of the simplest kind. The term '**Bernoulli trial**' refers to a single event that has only 2 possible outcomes, a success or a failure, with each outcome having a fixed probability. The following are examples of Bernoulli trials.

- Will a coin land Heads up?
- Are a random person's eyes blue or not?
- Will a person vote for a particular candidate at the next local council elections or not?
- Will you pass or fail an examination?

The Bernoulli distribution has only one controlling parameter: the probability of success, p. The alternative to success is failure, which is denoted by $1-p$ (and can also be denoted by q).

The Bernoulli distribution

For a discrete probability distribution that has a Bernoulli random variable, X:

$$\Pr(X=x)=\begin{cases} p & \text{if } x=1 \\ 1-p & \text{if } x=0 \end{cases}$$

The mean and variance for a Bernoulli distribution

If X is a Bernoulli random variable with the following distribution,

x	0	1
$\Pr(X=x)$	$1-p$	p

then

$$\begin{aligned} \mathrm{E}(X) &= \sum_{\text{all } x} x\Pr(X=x) \\ &= 0(1-p)+1\times p \\ &= p \\ \mathrm{E}\left(X^2\right) &= 0^2(1-p)+1^2\times p \\ &= p \\ \mathrm{Var}(X) &= \mathrm{E}\left(X^2\right)-[\mathrm{E}(X)]^2 \\ &= p-p^2 \\ &= p(1-p) \end{aligned}$$

Mean and variance formulas

For a Bernoulli distribution:

$$E(X) = \mu = p$$
$$Var(X) = \sigma^2 = p(1-p)$$

WORKED EXAMPLE 1 Identifying Bernoulli trials

Determine which of the following can be defined as Bernoulli trials.

a. Interviewing a random person to see if they have had a flu injection this year
b. Rolling a die in an attempt to obtain an even number
c. Choosing a ball from a bag that contains 3 red balls, 5 blue balls and 4 yellow balls

THINK	WRITE
a. Check for the characteristics of a Bernoulli trial.	a. Yes, this is a Bernoulli trial, as there are 2 possible outcomes. A person either has or has not had a flu injection this year.
b. Check for the characteristics of a Bernoulli trial.	b. Yes, this is a Bernoulli trial, as there are 2 possible outcomes. The die will show either an odd number or an even number.
c. Check for the characteristics of a Bernoulli trial.	c. No, this is not a Bernoulli trial, as success has not been defined.

WORKED EXAMPLE 2 Calculations of a Bernoulli distribution

A new cream has been developed for the treatment of dermatitis. In laboratory trials the cream was found to be effective in 72% of the cases. Hang's doctor has prescribed the cream for her. Let X be the effectiveness of the cream.

a. Construct a probability distribution table for X.
b. Determine $E(X)$.
c. Calculate the variance and the standard deviation of X, correct to 4 decimal places.

THINK

a. Construct a probability distribution table and clearly state the value of p.

WRITE

a. p = success with cream = 0.72

x	0	1
$Pr(X = x)$	0.28	0.72

b. 1. State the rule for the expected value.

b. $E(X) = \sum_{\text{all } x} x\,Pr(X = x)$

2. Substitute the appropriate values and evaluate.

$$E(X) = 0 \times 0.28 + 1 \times 0.72 = 0.72$$

c. 1. Find $E(X^2)$.

c. $E(X^2) = 0^2 \times 0.28 + 1^2 \times 0.72$
$= 0.72$

2. Calculate the variance.

$\text{Var}(X) = E(X^2) - [E(X)]^2$
$\text{Var}(X) = 0.72 - (0.72)^2$
$= 0.2016$

3. Calculate the standard deviation.

$SD(X) = \sqrt{0.2016}$
$= 0.4490$

on Resources

Interactivity The Bernoulli distribution (int-6430)

10.2 Exercise

Students, these questions are even better in jacPLUS

Receive immediate feedback and access sample responses

Access additional questions

Track your results and progress

Find all this and MORE in jacPLUS

Technology free

1. WE1 Determine which of the following can be defined as Bernoulli trials.

a. Spinning a spinner with 3 coloured sections
b. A golfer is at the tee of the first hole of a golf course. As she is an experienced golfer, the chance of her getting a hole in one is 0.15. We want to know if she will get a hole in one at this first hole.
c. A card is drawn from a standard pack of 52 cards. We are interested in the chance of drawing an ace.
d. A new drug for arthritis is said to have a success rate of 63%. Jingjing has just been prescribed the drug to treat her arthritis, and her doctor is interested in whether her symptoms improve or not.
e. Juanita has just given birth to a baby and we are interested in the gender of the baby, in particular whether the baby is a girl.
f. You are asked what your favourite colour is.

2. State clearly why the following are not Bernoulli trials.

a. A bag contains 12 balls, 5 of which are black, 3 of which are white and 4 of which are red. Paul has just drawn a ball from the bag without returning it. Now it is Alice's turn to draw a ball from the bag. We want to know if she gets a red one.
b. A die is tossed and the outcome is recorded.
c. A little penguin colony at Phillip Island in Victoria is being studied by an ecologist. We want to know if the habitat will be able to sustain the colony in the future.

3. A discrete random variable, Z, has a Bernoulli distribution as follows.

z	0	1
$\Pr(Z = z)$	0.4	0.6

a. Determine $E(Z)$.
b. Calculate $\text{Var}(Z)$.

4. Prisha is sitting for a Mathematics examination. She has just started question 1, which is a multiple choice question with 5 possible answer choices. Prisha plans to randomly guess the answer to the question. Let X be the discrete random variable that Prisha answers the question correctly.

a. Construct a probability distribution table for X.
b. Determine $E(X)$.
c. In total the test has 5 multiple choice questions to be answered. Determine the probability that Prisha answers all five questions correctly.

Technology active

5. **WE2** Azra is playing basketball. The chance that Azra scores a goal is 0.42. The ball has just been passed to her and she shoots for a goal. Let X be the random variable that defines Azra getting a goal. (Assume X obeys the Bernoulli distribution.)

a. Construct a probability distribution for X.
b. Determine $E(X)$.
c. Calculate the variance and standard deviation of X, correct to 4 decimal places.

6. **MC** Eli and Siti are about to play a game of chess. As Eli is a much more experienced chess player, the chance that they win is 0.68. Let Y be the discrete random variable that defines the fact that Eli wins.
$\text{Var}(Y)$ is:

A. 1
B. 0.2176
C. 0.32
D. 0.68
E. 0.4665

7. During the wet season, the probability that it rains on any given day in Cairns in northern Queensland is 0.89. You are going to Cairns tomorrow and it is the wet season. Let X be the chance that it rains on any given day during the wet season.

a. Construct a probability distribution table for X.
b. Evaluate:

i. $E(X)$
ii. $\text{Var}(X)$, correct to 4 decimal places
iii. $\text{SD}(X)$, correct to 4 decimal places.

c. Determine $\Pr(\mu - 2\sigma \leq X \leq \mu + 2\sigma)$.

8. A manufacturer of sweets reassures their customers that when they buy a box of their 'All Sorts' chocolates there is a 33% chance that the box will contain one or more toffees. Kasper bought a box of 'All Sorts' and selected one. Let Y be the discrete random variable that Kasper chose a toffee.

   a. Construct a probability distribution table for Y.
   b. Determine $\mathrm{E}(Y)$.
   c. Calculate $\Pr(\mu - 2\sigma \leq Y \leq \mu + 2\sigma)$.

9. X is a discrete random variable that has a Bernoulli distribution. It is known that the variance for this distribution is 0.21.
   a. Calculate the probability of success, p, where $p > 1 - p$.
   b. Determine $\mathrm{E}(X)$.

10. Z is a discrete random variable that has a Bernoulli distribution. It is known that the variance of Z is 0.1075.
   a. Calculate the probability of success, correct to 4 decimal places, if Pr(success) > Pr(failure).
   b. Construct a probability distribution table for Z.
   c. Evaluate the expected value of Z.

10.2 Exam questions

Question 1 (4 marks) TECH-ACTIVE

It has been found that when a breast ultrasound is combined with a common mammogram, the rate in which breast cancer is detected in a group of women is 7.2 per 1000. Noa is due for her two-yearly mammography testing, which will involve an ultrasound combined with a mammogram. Let Z be the discrete random variable that breast cancer is detected.

   a. Calculate the probability that Noa has breast cancer detected at this next test. **(1 mark)**
   b. Construct a probability distribution table for Z. **(1 mark)**
   c. Calculate $\Pr(\mu - 2\sigma \leq Z \leq \mu + 2\sigma)$. **(2 marks)**

Question 2 (3 marks) TECH-ACTIVE

Y is a discrete random variable that has a Bernoulli distribution. It is known that the standard deviation for this distribution is 0.4936.

   a. Calculate the variance of Y correct to 4 decimal places. **(1 mark)**
   b. Calculate the probability of success, p, if $p > 1 - p$. **(1 mark)**
   c. Determine $\mathrm{E}(Y)$. **(1 mark)**

Question 3 (3 marks) TECH-ACTIVE

Y is a discrete random variable that has a Bernoulli distribution. It is known that the standard deviation of Y is 0.3316.

   a. Calculate the variance correct to 2 decimal places. **(1 mark)**
   b. Calculate the probability of success correct to 4 decimal places if Pr(success) > Pr(failure). **(2 marks)**

More exam questions are available online.

10.3 The binomial distribution

LEARNING INTENTION

At the end of this subtopic you should be able to:

- identify and define a binomial probability distribution
- calculate the mean or expected value of a binomial probability distribution
- calculate the variance and standard deviation of a binomial probability distribution.

10.3.1 Introduction to the binomial distribution

When a Bernoulli trial is repeated a number of times, we have a **binomial distribution**. A binomial distribution is characterised by the following rules:

- It is made up of n Bernoulli trials or n identical trials.
- Each trial is an independent trial.
- There are two possible outcomes for each trial, a success, p, and a failure, $1-p$.

Consider again Question **4** from Exercise 10.2. Prisha has 5 multiple choice questions to answer on her mathematics examination. Each question has 5 different choices for the correct answer, and she plans to randomly guess every question. Prisha can get all 5 questions incorrect; 1 correct and 4 incorrect; 2 correct and 3 incorrect; 3 correct and 2 incorrect; 4 correct and 1 incorrect; or all 5 correct. This situation represents a binomial distribution and can be analysed as follows.

If X represents the number of questions answered correctly, then $p=\frac{1}{5}$ and $1-p=\frac{4}{5}$.

Let I = an incorrect answer and C = a correct answer.

0 correct answers: outcome = IIIII

$$\begin{aligned}\Pr(X=0) &= \left(\frac{4}{5}\right)^5\\ &= \frac{1024}{3125}\\ &= 0.3277\end{aligned}$$

1 correct answer: outcomes = IIIIC, IIICI, IICII, ICIII, CIIII

$$\begin{aligned}\Pr(X=1) &= 5\times\left(\frac{4}{5}\right)^4\left(\frac{1}{5}\right)\\ &= \frac{1280}{3125}\\ &= 0.4096\end{aligned}$$

2 correct answers: outcomes = IIICC, IICIC, ICIIC, CIIIC, IICCI, ICICI, CIICI, ICCII, CICII, CCIII

$$\begin{aligned}\Pr(X=2) &= 10\times\left(\frac{4}{5}\right)^3\left(\frac{1}{5}\right)^2\\ &= \frac{640}{3125}\\ &= 0.2048\end{aligned}$$

3 correct answers: outcomes = IICCC, ICICC, CIICC, ICCIC, CICIC, CCIIC, ICCCI, CICCI, CCICI, CCCII

$$\begin{aligned}\Pr(X=3) &= 10 \times \left(\frac{4}{5}\right)^2 \left(\frac{1}{5}\right)^3 \\ &= \frac{160}{3125} \\ &= 0.0512\end{aligned}$$

4 correct answers: outcomes = ICCCC, CICCC, CCICC, CCCIC, CCCCI

$$\begin{aligned}\Pr(X=4) &= 5 \times \left(\frac{4}{5}\right) \left(\frac{1}{5}\right)^4 \\ &= \frac{20}{3125} \\ &= 0.0064\end{aligned}$$

5 correct answers: outcome = CCCCC

$$\begin{aligned}\Pr(X=5) &= \left(\frac{1}{5}\right)^5 \\ &= \frac{1}{3125} \\ &= 0.0003\end{aligned}$$

This can then be represented in a table:

x	0	1	2	3	4	5
$\Pr(X=x)$	0.3277	0.4096	0.2048	0.0512	0.0064	0.0003

where $\sum_{\text{all } x} \Pr(X=x) = 1$.

It can be quite tedious to work out all the possible outcomes, especially when the number of trials is large. However, we usually just want to know how many different ways there are of obtaining each number of correct answers, not the actual specific order of the incorrect and correct answers.

From your study of probability in Units 1 and 2, you will recall that the number of ways of obtaining x successes from n independent trials is given by ${}^nC_x$, which can also be written as $\binom{n}{x}$.

$${}^nC_x = \frac{n!}{(n-x)!\ x!}$$

Binomial distribution definition

If a discrete random variable, X, has a binomial distribution, we say that

$$X \sim \text{Bi}(n, p)$$

where n is the number of independent trials and p is the probability of success.

If $X \sim \text{Bi}(n, p)$, then $\Pr(X=x) = {}^nC_x\,(1-p)^{n-x}\,p^x$, where $x = 0, 1, 2, 3 \ldots n$.

Note: If the order is specified for a particular scenario, then the binomial probability distribution rule cannot be used. The probabilities need to be multiplied in the given order.

Graphing the binomial distribution

The probability distribution for the previous example, where Prisha answered 5 questions on a mathematics exam, can be graphed as follows.

x	0	1	2	3	4	5
$\Pr(X = x)$	0.3277	0.4096	0.2048	0.0512	0.0064	0.0003

The shape of this graph (Figure 1) is positively skewed. It indicates that the probability of success is low (in this case $p = 0.2$), as the larger x-values (number of successful outcomes) have corresponding low probabilities.

FIGURE 1

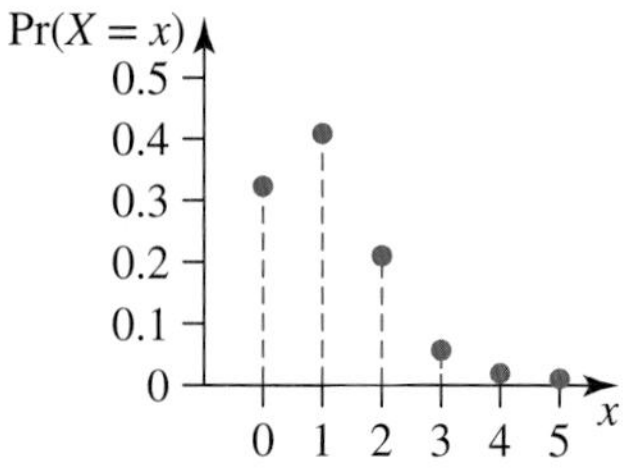

If the value of p was higher, for example if Prisha was 80% sure of getting a question right ($p = 0.8$), then the graph would look like Figure 2. The shape of this graph is negatively skewed. It indicates that the probability of success is high (in this case $p = 0.8$), as the larger x-values (number of successful outcomes) have corresponding high probabilities.

FIGURE 2

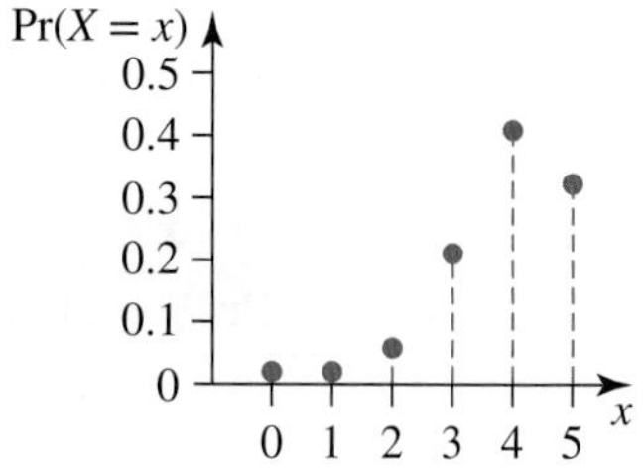

If the value of p was 0.5, for example if Prisha was 50% sure of getting a question right, then the graph would look like Figure 3. The shape of this graph is symmetrical. It indicates that the probability of success is equal to the probability of failure. If the number of trials increased, the graph would approach the shape of a bell-shaped curve.

FIGURE 3

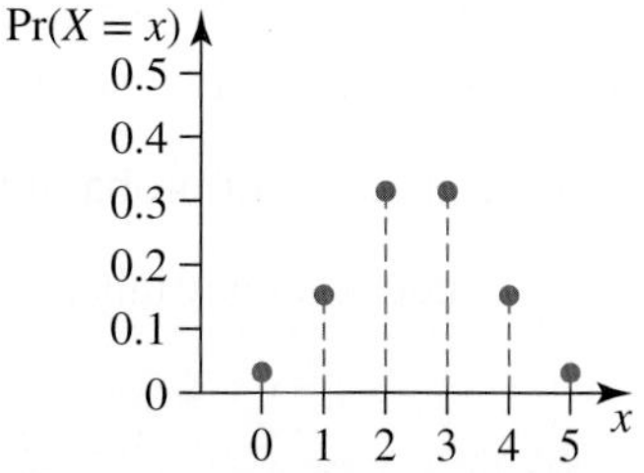

WORKED EXAMPLE 3 Determining the probabilities of a binomial distribution

It is known that 52% of the population participates in sport on a regular basis. Five random individuals are interviewed and asked whether they participate in sport on a regular basis. Let X be the number of people who regularly participate in sport. Give answers correct to 4 decimal places for the following.

a. **Construct a probability distribution table for X.**
b. **Calculate the probability that 3 people or less play sport.**
c. **Determine the probability that at least one person plays sport, given that no more than 3 people play sport.**
d. **Calculate the probability that the first person interviewed plays sport but the next 2 do not.**

THINK	WRITE
a. 1. Write the rule for the probabilities of the binomial distribution.	**a.** $X \sim \text{Bi}(5, 0.52)$ $\Pr(X=x) = {}^nC_x\,(1-p)^{n-x}\,p^x$
2. Substitute $x=0$ into the rule and simplify.	$\Pr(X=0) = {}^5C_0\,(0.48)^5$ $= 0.025\,5$
3. Substitute $x=1$ into the rule and simplify.	$\Pr(X=1) = {}^5C_1\,(0.48)^4\,(0.52)$ $= 0.138\,0$
4. Substitute $x=2$ into the rule and simplify.	$\Pr(X=2) = {}^5C_2\,(0.48)^3\,(0.52)^2$ $= 0.299\,0$
5. Substitute $x=3$ into the rule and simplify.	$\Pr(X=3) = {}^5C_3\,(0.48)^2\,(0.52)^3$ $= 0.324\,0$
6. Substitute $x=4$ into the rule and simplify.	$\Pr(X=4) = {}^5C_4(0.48)\,(0.52)^4$ $= 0.175\,5$
7. Substitute $x=5$ into the rule and simplify.	$\Pr(X=5) = {}^5C_5\,(0.52)^5$ $= 0.038\,0$
8. Construct a probability distribution table and check that $\sum_{\text{all } x} \Pr(X=x) = 1$.	See table below.

x	$\Pr(X=x)$
0	0.025 5
1	0.138 0
2	0.299 0
3	0.324 0
4	0.175 5
5	0.038 0

$$\sum_{\text{all } x} \Pr(X=x) = 1$$

THINK	WRITE
b. 1. Interpret the question and write the probability to be found.	**b.** $\Pr(X \le 3)$
2. State the probabilities included in $\Pr(X \le 3)$.	$\Pr(X \le 3) = \Pr(X=0) + \Pr(X=1) + \Pr(X=2) + \Pr(X=3)$ $\Pr(X \le 3) = 1 - (\Pr(X=4) + \Pr(X=5))$
3. Substitute the appropriate probabilities and evaluate.	$\Pr(X \le 3) = 1 - (0.175\,5 + 0.038\,0)$ $\Pr(X \le 3) = 0.7865$ *Note*: CAS technology can be used to add up multiple probabilities.
c. 1. State the rule for conditional probability.	**c.** $\Pr(X \ge 1 \mid X \le 3) = \dfrac{\Pr(X \ge 1 \cap X \le 3)}{\Pr(X \le 3)}$ $= \dfrac{\Pr(1 \le X \le 3)}{\Pr(X \le 3)}$
2. Evaluate $\Pr(X \ge 1 \cap X \le 3)$.	$\Pr(X \ge 1 \cap X \le 3) = \Pr(X=1) + \Pr(X=2) + \Pr(X=3)$ $= 0.138\,0 + 0.299\,0 + 0.324\,0$ $= 0.761\,0$

3. Substitute the appropriate values into the rule.

$$\Pr(X \geq 1 \mid X \leq 3) = \frac{\Pr(X \geq 1 \cap X \leq 3)}{\Pr(X \leq 3)}$$
$$= \frac{0.761\,0}{0.7865}$$

4. Simplify.

$$\Pr(X \geq 1 \mid X \leq 3) = 0.9676$$

d. 1. Order has been specified for this question. Therefore, the binomial probability distribution rule cannot be used. The probabilities must be multiplied together in order.

d. S = plays sport, N = doesn't play sport
$\Pr(SNN) = \Pr(S) \times \Pr(N) \times \Pr(N)$

2. Substitute the appropriate values and evaluate.

$$\Pr(SNN) = 0.52 \times 0.48 \times 0.48$$
$$= 0.1198$$

TI | THINK / **DISPLAY/WRITE**

a. 1. On a Calculator page, press MENU, then select:
5: Probability
5: Distributions
A: Binomial Pdf …
Complete the fields as:
Num Trials, n: 5
Prob Success, p: 0.52
then select OK.

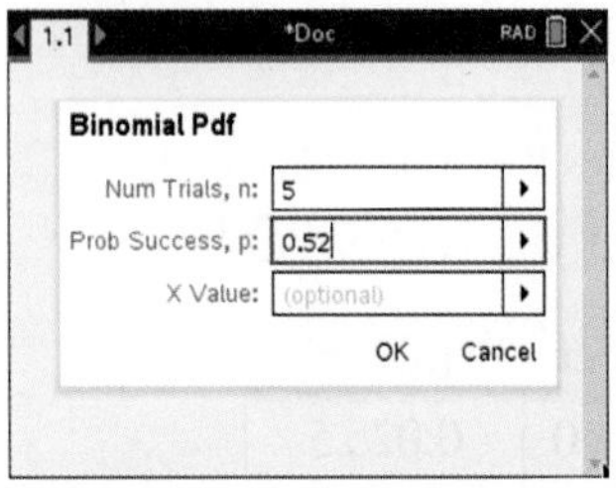

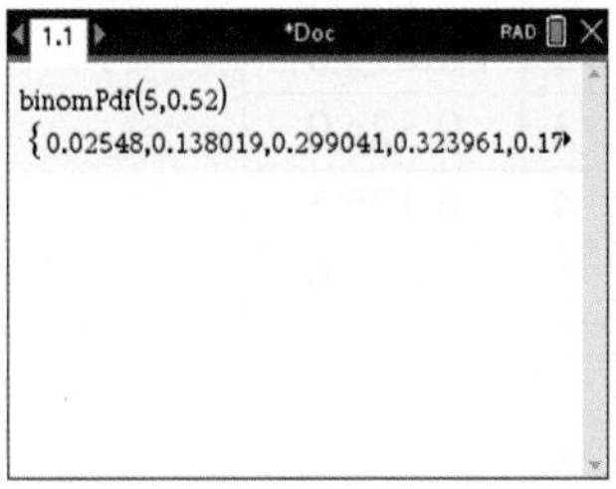

2. The probability distribution appears on the screen.

A probability distribution table can be drawn from the probabilities listed on the screen.

x	$\Pr(X = x)$
0	0.025 5
1	0.138 0
2	0.299 0
3	0.324 0
4	0.175 5
5	0.038 0

CASIO | THINK / **DISPLAY/WRITE**

a. 1. On a Statistics screen, select:
- Calc
- Distribution

Select 'Distribution' and 'Binomial PD' from the drop-down menus, then select NEXT.
Complete the fields as:
x: 0
Numtrial: 5
pos: 0.52
then select NEXT.

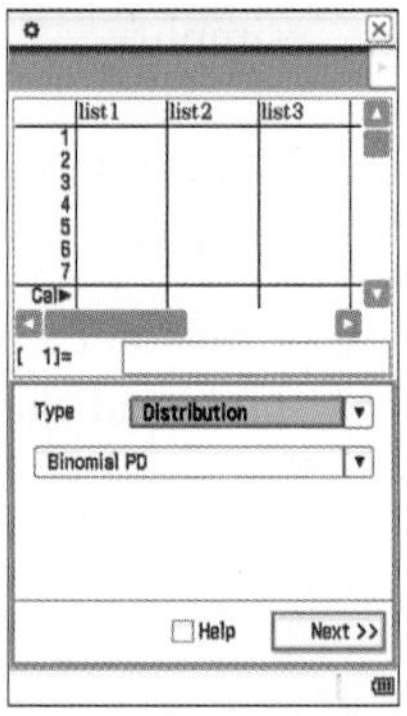

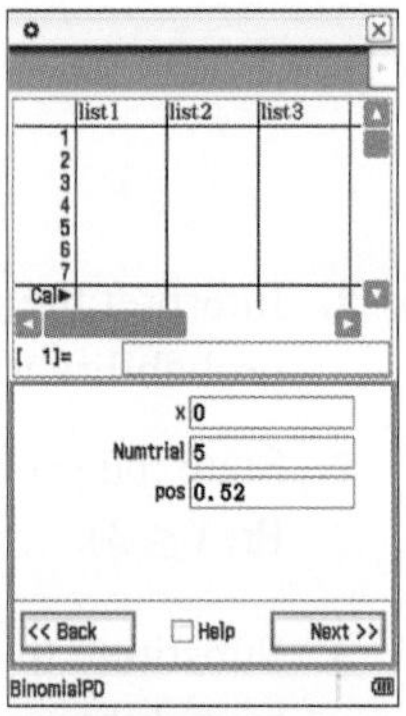

2. Select the 'Graph' icon to view the probability distribution. Use the left/right arrows to view the probability for each X.

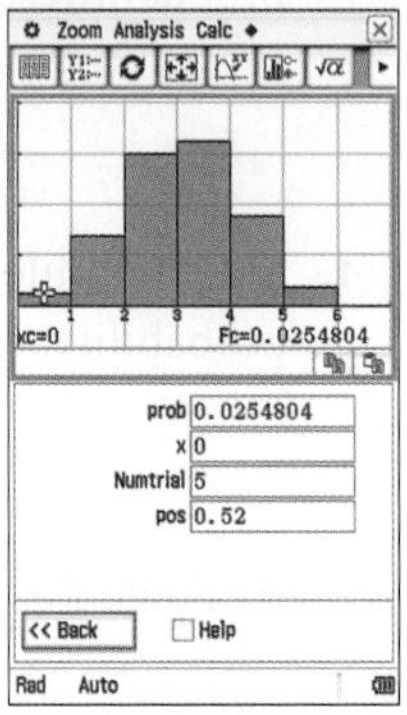

b. 1. On a Calculator page, press MENU, then select
5: Probability
5: Distributions
B: Binomial Cdf …
Complete the fields as:
Num Trials, n: 5
Prob Success, p: 0.52
Lower Bound: 0
Upper Bound: 3
then select OK.

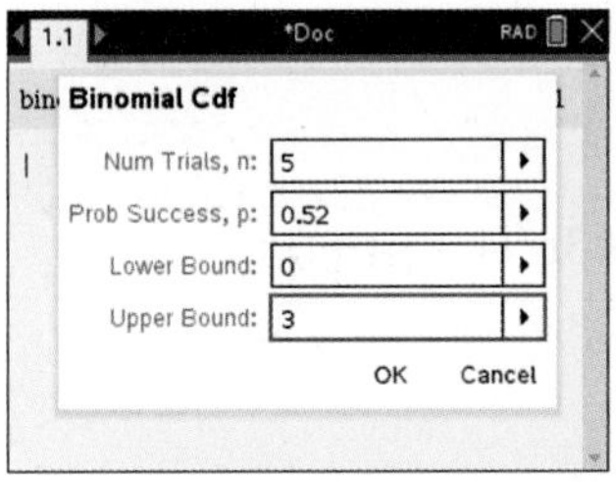

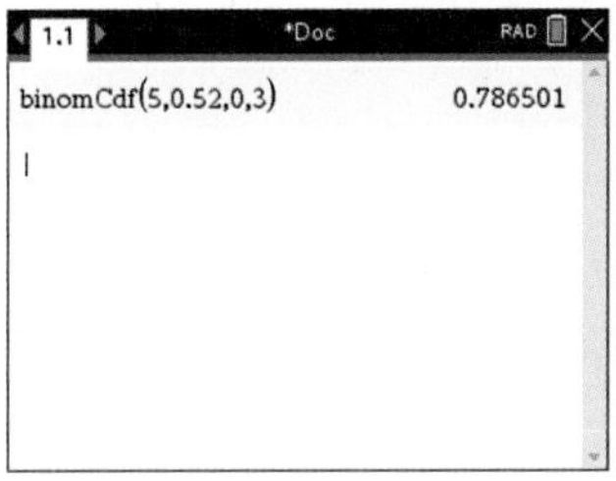

2. The answer appears on the screen.

$\Pr(X \le 3) = 0.7865$

3. The probability distribution appears on the screen.

A probability distribution table can be drawn from the probabilities in the graph.

x	$\Pr(X = x)$
0	0.025 5
1	0.138 0
2	0.299 0
3	0.324 0
4	0.175 5
5	0.038 0

b. 1. On a Main screen, select:
- Interactive
- Distribution/Inv. Dist
- Discrete
- binomial CDf

Complete the fields as:
Lower: 0
Upper: 3
NumTrial: 5
pos: 0.52
then select OK.

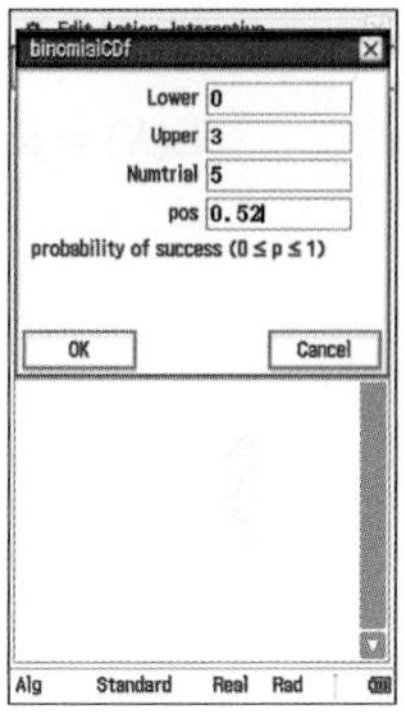

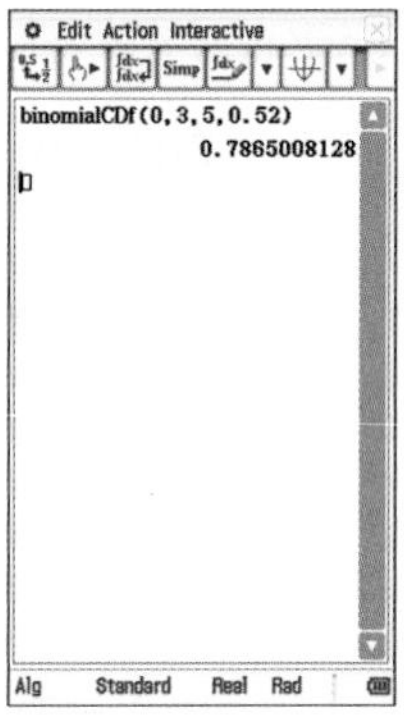

2. The answer appears on the screen.

$\Pr(X \le 3) = 0.7865$

on Resources

Interactivity Graphing the binomial distribution (int-6431)

10.3.2 The mean and variance of the binomial distribution

If $X \sim \text{Bi}(n,\ p)$, then $\Pr(X=x) = {}^n\text{C}_x(1-p)^{n-x}p^x$, where $x = 0, 1, 2, 3 \ldots n$.

Suppose $n = 3$:

$$
\begin{aligned}
\Pr(X=0) &= {}^3\text{C}_0\,(1-p)^3\,p^0 \\
&= (1-p)^3 \\
\Pr(X=1) &= {}^3\text{C}_1\,(1-p)^2\,p \\
&= 3p\,(1-p)^2 \\
\Pr(X=2) &= {}^3\text{C}_2(1-p)p^2 \\
&= 3p^2(1-p) \\
\Pr(X=3) &= {}^3\text{C}_3\,(1-p)^0\,p^3 \\
&= p^3
\end{aligned}
$$

$$
\begin{aligned}
\text{E}(X) = \mu &= \sum_{\text{all } x} x\Pr(X=x) \\
&= 0(1-p)^3 + 1\times 3p(1-p)^2 + 2\times 3p^2(1-p) + 3p^3 \\
&= 0 + 3p(1-p)^2 + 6p^2(1-p) + 3p^3 \\
&= 3p\left[(1-p)^2 + 2p(1-p) + p^2\right] \\
&= 3p\left[(1-p)+p\right]^2 \\
&= 3p(1-p+p)^2 \\
&= 3p
\end{aligned}
$$

Suppose now $n = 4$:

$$
\begin{aligned}
\Pr(X=0) &= {}^4\text{C}_0(1-p)^4p^0 \\
&= (1-p)^4 \\
\Pr(X=1) &= {}^4\text{C}_1(1-p)^3p \\
&= 4p(1-p)^3 \\
\Pr(X=2) &= {}^4\text{C}_2(1-p)^2p^2 \\
&= 6p^2(1-p)^2 \\
\Pr(X=3) &= {}^4\text{C}_3(1-p)p^3 \\
&= 4p^3(1-p) \\
\Pr(X=4) &= {}^4\text{C}_4(1-p)^0p^4 \\
&= p^4
\end{aligned}
$$

$$
\begin{aligned}
\text{E}(X) = \mu &= \sum_{\text{all} x} x\Pr(X=x) \\
&= 0(1-p)^4 + 1\times 4p(1-p)^3 + 2\times 6p^2(1-p)^2 + 3\times 4p^3(1-p) + 4p^4 \\
&= 0 + 4p(1-p)^3 + 12p^2(1-p)^2 + 12p^3(1-p) + 4p^4 \\
&= 4p\,\left[(1-p)^3 + 3p(1-p)^2 + 3p^2(1-p) + p^3\right] \\
&= 4p\left[(1-p)+p\right]^3 \\
&= 4p(1-p+p)^3 \\
&= 4p
\end{aligned}
$$

In both cases the expected value of X is the number chosen for n multiplied by p. The same result can be achieved for any value of n.

The expected value

If $X \sim \text{Bi}(n,p)$, then $\text{E}(X) = \mu = np$.

Consider again $n = 3$:

$$\begin{aligned}\text{E}\left(X^2\right) &= 0^2(1-p)^3 + 1^2 \times 3p(1-p)^2 + 2^2 \times 3p^2(1-p) + 3^2p^3 \\ &= 0 + 3p(1-p)^2 + 12p^2(1-p) + 9p^3 \\ &= 3p(1-p)^2 + 12p^2(1-p) + 9p^3\end{aligned}$$

$$\begin{aligned}\text{Var}(X) &= \text{E}\left(X^2\right) - [\text{E}(X)]^2 \\ &= 3p(1-p)^2 + 12p^2(1-p) + 9p^3 - (3p)^2 \\ &= 3p(1-p)^2 + 12p^2(1-p) + 9p^3 - 9p^2 \\ &= 3p(1-p)^2 + 12p^2(1-p) - 9p^2(1-p) \\ &= 3p(1-p)^2 + 3p^2(1-p) \\ &= 3p(1-p)(1-p+p) \\ &= 3p(1-p)\end{aligned}$$

Consider again $n = 4$:

$$\begin{aligned}\text{E}\left(X^2\right) &= 0^2(1-p)^4 + 1^2 \times 4p(1-p)^3 + 2^2 \times 6p^2(1-p)^2 + 3^2 \times 4p^3(1-p) + 4^2p^4 \\ &= 0 + 4p(1-p)^3 + 24p^2(1-p)^2 + 36p^3(1-p) + 16p^4 \\ &= 4p(1-p)^3 + 24p^2(1-p)^2 + 36p^3(1-p) + 16p^4\end{aligned}$$

$$\begin{aligned}\text{Var}(X) &= \text{E}\left(X^2\right) - [\text{E}(X)]^2 \\ &= 4p(1-p)^3 + 24p^2(1-p)^2 + 36p^3(1-p) + 16p^4 - (4p)^2 \\ &= 4p(1-p)^3 + 24p^2(1-p)^2 + 36p^3(1-p) + 16p^4 - 16p^2 \\ &= 4p(1-p)^3 + 24p^2(1-p)^2 + 36p^3(1-p) + 16p^2\left(p^2-1\right) \\ &= 4p(1-p)^3 + 24p^2(1-p)^2 + 36p^3(1-p) - 16p^2\left(1-p^2\right) \\ &= 4p(1-p)^3 + 24p^2(1-p)^2 + 36p^3(1-p) - 16p^2(1-p)(1+p) \\ &= 4p(1-p)\left[(1-p)^2 + 6p(1-p) + 9p^2 - 4p(1+p)\right] \\ &= 4p(1-p)\left(1 - 2p + p^2 + 6p - 6p^2 + 9p^2 - 4p - 4p^2\right) \\ &= 4p(1-p)\end{aligned}$$

Again, in both cases, the variance of X is the number chosen for n times $p(1-p)$. This same result can be shown for any value of n.

The variance and standard deviation

If $X \sim \text{Bi}(n,p)$, then $\text{Var}(X) = \sigma^2 = np(1-p)$

and

$\text{SD}(X) = \sigma = \sqrt{np(1-p)}$.

WORKED EXAMPLE 4 Expected value and variance

A test consists of 20 multiple choice questions, each with 5 alternatives for the answer. A student has not studied for the test so she chooses the answers at random. Let X be the discrete random variable that describes the number of correct answers. Calculate:

a. the expected number of correct questions answered

b. the variance of the correct number of questions answered.

THINK	WRITE
a. 1. Write the rule for the expected value.	a. $\mu = np$
2 Substitute the appropriate values and simplify.	$n = 20, p = \frac{1}{5}$ $\mu = np$ $= 20 \times \frac{1}{5}$ $= 4$
3. Write the answer.	The expected number of questions correct is 4.
b. 1. Write the rule for the variance.	b. $\sigma^2 = np(1-p)$
2. Substitute the appropriate values and evaluate.	$\text{Var}(X) = 20 \times \frac{1}{5} \times \frac{4}{5}$ $= \frac{16}{5}$ $= 3.2$

WORKED EXAMPLE 5 Determining n and p

A binomial random variable, Z, has a mean of 8.4 and a variance of 3.696. Calculate:

a. the probability of success, p

b. the number of trials, n.

THINK	WRITE
a. 1. Write the rules for the variance and expected value.	a. $\mu = np$ $\text{Var}(Z) = np(1-p)$
2. Substitute the known information and label the two equations.	$8.4 = np$ [1] $\text{Var}(Z) = np(1-p)$ $3.696 = np(1-p)$ [2]
3. To cancel out the n, divide equation [2] by equation [1].	$[2] \div [1]: \frac{np(1-p)}{np} = \frac{3.696}{8.4}$
4. Simplify.	$1 - p = 0.44$ $p = 0.56$
5. Write the answer.	The probability of success is 0.56.
b. 1. Substitute $p = 0.56$ into $\text{E}(Z) = np$ and solve for n.	b. $\mu = np$ $8.4 = n \times 0.56$ $n = 15$
2. Write the answer.	There are 15 trials.

WORKED EXAMPLE 6 Determining n

The probability of an Olympic archer hitting the centre of the target is 0.7. Determine the smallest number of arrows he must shoot to ensure that the probability he hits the centre at least once is more than 0.9.

THINK	WRITE
1. Write the rule for the probabilities of the binomial distribution.	$X \sim \text{Bi}(n, 0.7)$ $\Pr(X \geq 1) > 0.9$
2. The upper limit of successes is unknown, because n is unknown. Therefore, $\Pr(X \geq 1)$ cannot be found by adding up the probabilities. However, the required probability can be found by subtracting from 1 the only probability not included in $\Pr(X \geq 1)$.	$\Pr(X \geq 1) = 1 - \Pr(X = 0)$
3. Substitute in the appropriate values and simplify.	$\Pr(X \geq 1) = 1 - \Pr(X = 0)$ $1 - \Pr(X = 0) > 0.9$ $1 - {}^nC_x (1-p)^{n-x} p^x > 0.9$ $1 - {}^nC_0 (0.3)^n (0.7)^0 > 0.9$ $1 - 1 \times (0.3)^n \times 1 > 0.9$ $1 - (0.3)^n > 0.9$
4. Solve for n using CAS.	$n > 1.91249$
5. Interpret the result and answer the question.	$n = 2$ (as n must be an integer) The smallest number of arrows the archer needs to shoot in order to guarantee a probability of 0.9 of hitting the centre is 2.

TI | THINK	DISPLAY/WRITE
1. Rewrite the probability statement.	$\Pr(X \geq 1) > 0.9$ $1 - \Pr(X = 0) > 0.9$ $\Pr(X = 0) < 0.1$
2. On a Calculator page, press MENU, then select: 5: Probability 5: Distributions D: Inverse Binomial N … Complete the fields as: Cumulative Prob: 0.1 Prob Success, p: 0.7 Successes, x: 0 then select OK.	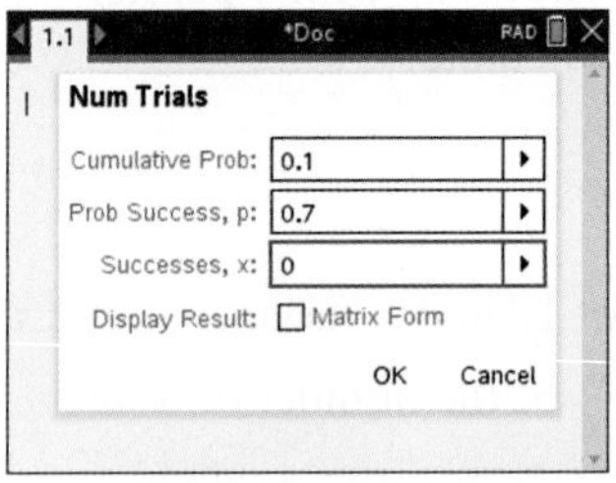 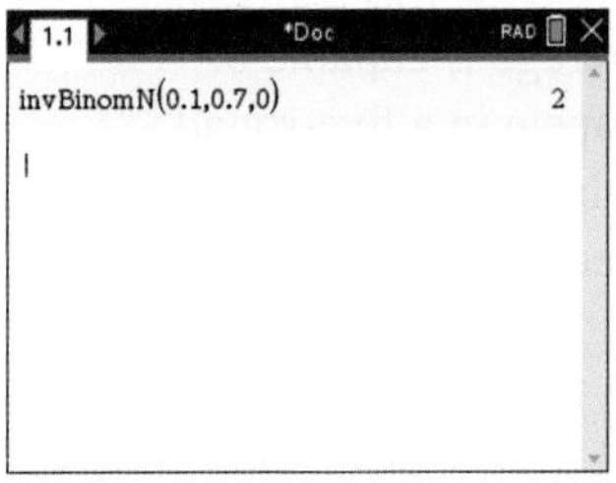
3. The answer appears on the screen.	$n = 2$.

CASIO | THINK	DISPLAY/WRITE
1. Rewrite the probability statement.	$\Pr(X \geq 1) > 0.9$ $1 - \Pr(X = 0) > 0.9$ $\Pr(X = 0) < 0.1$ ${}^nC_0 \times (0.3)^n (0.7)^0 < 0.1$
2. On a Main screen, complete the entry line as: solve(nCr(n, 0) × $(0.3)^n (0.7)^0 < 0.1, n$) then press EXE. *Note:* The ${}^nC_r$ template can be found in the Advance tab of the Keyboard menu.	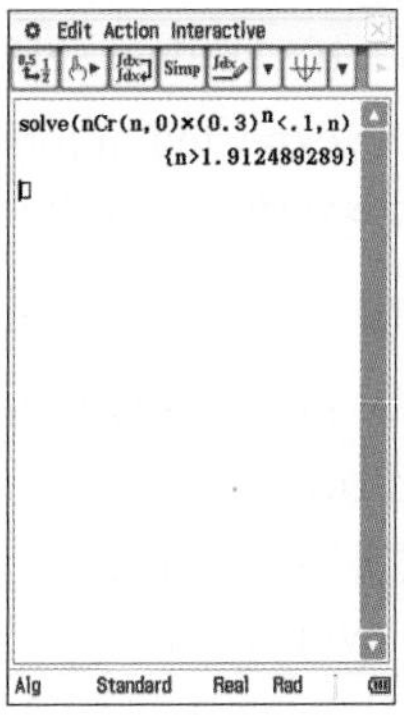
3. The answer appears on the screen.	$n = 2$.

on Resources

Interactivity Effects of n and p on the binomial distribution (int-6432)

10.3 Exercise

Students, these questions are even better in jacPLUS

Receive immediate feedback and access sample responses

Access additional questions

Track your results and progress

Find all this and MORE in jacPLUS

Technology free

1. Chiara has a 0.5 chance of being late to school each day. For the next week, calculate the probability that Chiara is:
   a. late 3 out of 5 days
   b. late less than 2 days
   c. late the first 2 days and on time the next 3 days.

2. Surveys show that 25% of adults read a daily newspaper. If 4 people are selected at random, calculate the probability that:
   a. all 4 of them read the daily newspaper
   b. at least one person reads a newspaper every day
   c. two people read the newspaper daily, given that at least one person does.

3. Consider a woman with the genotype XX and a man with the genotype XY. Their offspring have an equal chance of inheriting one of these genotypes. Calculate the probability that 6 of their 7 offspring have the genotype XY.

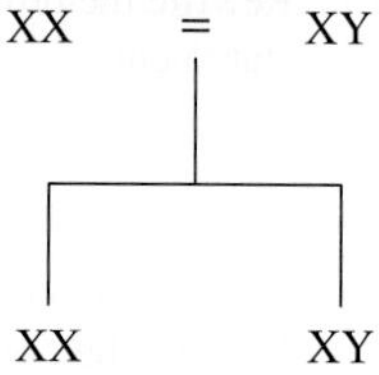

4. **WE4** A science test consists of 60 multiple choice questions with 4 possible options, A–D. If a student selected each answer at random, calculate:
   a. the expected number of correct answers
   b. the variance of the number of correct answers.

5. **WE5** A binomial random variable, X, has a mean of 12 and a variance of 3. Calculate:
   a. the probability of success, p
   b. the number of trials, n.

Technology active

6. **WE3** Jack is an enthusiastic darts player and on average is capable of achieving a bullseye 3 out of 7 times. Jack will compete in a five-round tournament. Let Y be the discrete random variable that defines the number of bullseyes Jack achieves. Give your answers correct to 4 decimal places for the following.
   a. Construct a probability distribution table for Y.
   b. Calculate the probability that Jack will score at most 3 bullseyes.
   c. Determine the probability that Jack will score at least 1 bullseye, given that he scored at most 3 bullseyes.
   d. Calculate the probability that his first shot missed, his second shot was a bullseye and then his next 2 shots missed.

7. At a poultry farm, eggs are collected daily and classified as large or medium. Then they are packed into cartons containing 12 eggs of the same classification. Experience has enabled the director of the poultry farm to know that 42% of all eggs produced at the farm are considered to be large. Ten eggs are randomly chosen from a conveyor belt on which the eggs are to be classified. Let Z be the discrete random variable that gives the number of large eggs. Give your answers correct to 4 decimal places for the following.

   a. Calculate $\Pr(Z=0)$, $\Pr(Z=1) \ldots \Pr(Z=9)$, $\Pr(Z=10)$ for this binomial distribution.
   b. Construct a probability distribution table for Z.
   c. Calculate $\Pr(Z \geq 5 \mid Z \leq 8)$.

8. A particular medication used by asthma sufferers has been found to be beneficial if used 3 times a day. In a trial of the medication it was found to be successful in 63% of the cases. Eight random asthma sufferers have had the medication prescribed for them. Give your answers correct to places for the following.

   a. Construct a probability distribution table for the number of sufferers who have benefits from the medication, X.
   b. Determine the probability that no more than 7 people will benefit from the medication.
   c. Calculate the probability that at least 3 people will benefit from the medication, given that no more than 7 will.
   d. Calculate the probability that the first person won't benefit from the medication, but the next 5 will.

9. If X has a binomial distribution so that $n=15$ and $p=0.62$, calculate, correct to 4 decimal places:
   a. $\Pr(X=10)$
   b. $\Pr(X \geq 10)$
   c. $\Pr(X<4 \mid X \leq 8)$

10. Wan is selling raffle tickets for cancer research outside her local supermarket. As people pass her table, there is a probability of 0.45 that they will stop and buy a ticket. During the course of 15 minutes, 15 people walked past her table. Let X be the binomial random variable for the number of people who stopped and bought a ticket. Determine:
    a. the expected value for the number of people who will stop and buy a ticket
    b. the probability that 4 people will stop and buy a ticket, correct to 4 decimal places
    c. the probability that no more than 8 people will buy a ticket, correct to 4 decimal places
    d. the probability that the first 2 people will buy a ticket but the next 2 won't, correct to 4 decimal places

11. The executive committee for an independent school consists of 12 members. Calculate, correct to 4 decimal places, the probability that there are 8 or more women on the executive committee if:
    a. it is equally likely that a man or a woman is chosen for the executive position
    b. women have a 58% chance of being chosen for an executive position.

12. For each of the following binomial random variables, calculate:
    i. the expected value
    ii. the variance.

    a. $X \sim \text{Bi}(45,\ 0.72)$
    b. $Y \sim \text{Bi}\left(100,\ \frac{1}{5}\right)$
    c. $Z \sim \text{Bi}\left(72,\ \frac{2}{9}\right)$

13. A fair die is tossed 25 times. Let X be the discrete random variable that represents the number of ones achieved. Calculate, correct to 4 decimal places:

a. the expected number of ones achieved
b. the standard deviation of the number of ones achieved.

14. At midday at the local supermarket, three checkouts are in operation. The probability that a customer can walk up to a register without queuing is 0.32. Milo visits the supermarket at noon on 7 different occasions. Give your answers correct to 4 decimal places for the following.

a. Construct a probability distribution table for the number of times Milo doesn't have to queue to pay for his purchases, Z.
b. Determine $E(Z)$ and $Var(Z)$.
c. Calculate $Pr(\mu - 2\sigma \le Z \le \mu + 2\sigma)$.

15. A large distributor of white goods has found that 1 in 7 people who buy goods from them do so by using their lay-by purchasing system. On one busy Saturday morning, 10 customers bought white goods. Let X be the number of people who use the lay-by purchasing system to buy their goods. Calculate, correct to 4 decimal places:

a. $E(X)$ and $Var(X)$
b. $Pr(\mu - 2\sigma \le X \le \mu + 2\sigma)$.

16. A binomial random variable, Z, has a mean of 32.535 and a variance of 9.021 95.

a. Calculate the probability of success, p, correct to 3 decimal places.
b. Calculate the number of trials, n.

17. A binomial random variable has an expected value of 9.12 and a variance of 5.6544.

a. Calculate the probability of success, p.
b. Calculate the number of trials, n.

18. **MC** A binomial random variable has an expected value of 3.8325 and a variance of 3.412 842 15. The probability of success, p, is:

A. 1.1230
B. 0.8905
C. 0.0286
D. 0.7644
E. 0.1095

19. **MC** A binomial experiment is completed 16 times and has an expected value of 10.16. The standard deviation is closest to:

A. 0.635
B. 2.4886
C. 1.9257
D. 3.7084
E. 0.365

20. The tram that stops outside Maia's house is late 20% of the time. If there are 12 times during the day that the tram stops outside Maia's house, determine, correct to 4 decimal places:

a. the probability that the tram is late 3 times
b. the probability that the tram is late 3 times for at least 6 out of the next 14 days.

21. WE6 The probability of winning a prize in a particular competition is 0.2. Calculate how many tickets someone would need to buy in order to guarantee them a probability of at least 0.85 of winning a prize.

22. Lizzie and Matt enjoy playing card games. The probability that Lizzie will beat Matt is 0.67. Calculate how many games they need to play so that the probability of Matt winning at least one game is more than 0.9.

10.3 Exam questions

Question 1 (1 mark) TECH-ACTIVE

Source: *VCE 2021, Mathematical Methods Exam 2, Section A, Q17; © VCAA.*

MC A discrete random variable X has a binomial distribution with a probability of success of $p = 0.1$ for n trials, where $n > 2$.

If the probability of obtaining at least two successes after n trials is at least 0.5, then the smallest possible value of n is

A. 15 **B.** 16 **C.** 17 **D.** 18 **E.** 19

Question 2 (1 mark) TECH-ACTIVE

Source: *VCE 2017, Mathematical Methods Exam 2, Section A, Q18; © VCAA.*

MC Let X be a discrete random variable with binomial distribution $X \sim \text{Bi}(n, p)$. The mean and the standard deviation of this distribution are equal.

Given that $0 < p < 1$, the smallest number of trials, n, such that $p \leq 0.01$ is

A. 37 **B.** 49 **C.** 98 **D.** 99 **E.** 101

Question 3 (1 mark) TECH-ACTIVE

Source: *VCE 2015, Mathematical Methods (CAS) Exam 2, Section 1, Q10; © VCAA.*

MC The binomial random variable, X, has $\text{E}(X) = 2$ and $\text{Var}(X) = \frac{4}{3}$.

$\Pr(X = 1)$ is equal to

A. $\left(\frac{1}{3}\right)^6$

B. $\left(\frac{2}{3}\right)^6$

C. $\frac{1}{3} \times \left(\frac{2}{3}\right)^2$

D. $6 \times \frac{1}{3} \times \left(\frac{2}{3}\right)^5$

E. $6 \times \frac{2}{3} \times \left(\frac{1}{3}\right)^5$

More exam questions are available online.

10.4 Applications

LEARNING INTENTION

At the end of this subtopic you should be able to:
- apply the binomial distribution properties to real-world problems and calculate specific probabilities and values.

10.4.1 Applications of the binomial distribution

The binomial distribution has important applications in medical research, quality control, simulation and genetics. In this section we will explore some of these areas.

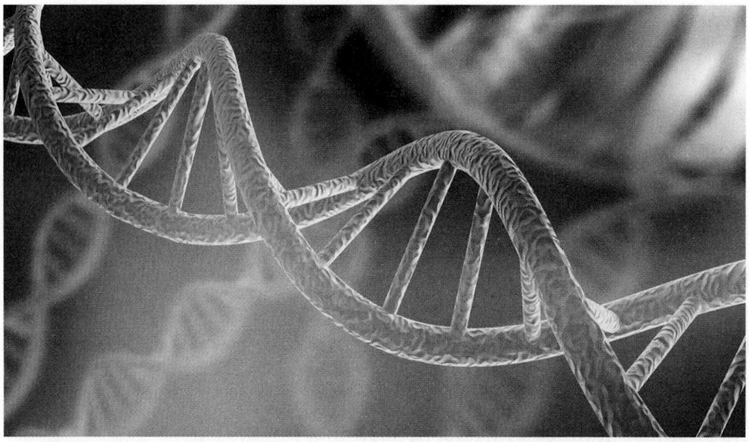

WORKED EXAMPLE 7 Applying the binomial distribution

It has been found that 9% of the population have diabetes. A sample of 15 people were tested for diabetes. Let X be the random variable that gives the number of people who have diabetes. Calculate, correct to 4 decimal places:

a. **$\Pr(X \leq 5)$**

b. **$E(X)$ and $SD(X)$.**

THINK	WRITE
a. 1. Summarise the information using binomial notation.	a. $X \sim \text{Bi}(15, 0.09)$
2. Use CAS technology to add up the required probabilities.	$\Pr(X \leq 5) = 0.9987$
b. 1. State the rule for the expected value.	b. $E(X) = np$
2. Substitute the appropriate values and simplify.	$E(X) = 15 \times 0.09$ $= 1.35$
3. Determine the variance.	$\text{Var}(X) = np(1-p)$ $= 15 \times 0.09 \times 0.91$ $= 1.2285$
4. Determine the standard deviation.	$SD(X) = \sqrt{\text{Var}(X)}$ $= \sqrt{1.2285}$ $= 1.1084$

10.4 Exercise

Students, these questions are even better in jacPLUS

Receive immediate feedback and access sample responses

Access additional questions

Track your results and progress

Find all this and MORE in jacPLUS

Technology active

1. WE7 It is thought that about 30% of teenagers receive their spending money from part-time jobs. Ten random teenagers were interviewed about their spending money and how they obtained it. Let Y be the random variable that defines the number of teenagers who obtain their spending money by having a part-time job. Calculate, correct to 4 decimal places:
   a. $\Pr(Y \geq 7)$
   b. $E(Y)$ and $SD(Y)$.

2. A mobile phone manufacturer has a relatively simple but important quality test at the end of the manufacturing, which is that the phone should be dropped onto a hard surface. If the phone cracks or breaks in any way it is rejected and destroyed. The probability that a phone is rejected and destroyed is 0.01. Let Z be the number of mobile phones that are dropped and broken when 5 mobile phones are tested.
   a. Calculate $\Pr(Z \leq 3)$.
   b. Determine:
      i. $E(Z)$
      ii. $SD(Z)$, correct to 4 decimal places.
   c. Determine $\Pr(\mu - 2\sigma \leq Z \leq \mu + 2\sigma)$.
3. In Australia, it is estimated that 30% of the population over the age of 25 have hypertension. A statistician wishes to investigate this, so he arranges for 15 random adults over the age of 25 to be tested to see if they have high blood pressure. Let X be the random variable that defines the number of adults over the age of 25 with hypertension. Calculate, correct to 4 decimal places where appropriate:
   a. $\Pr(X \leq 5)$.
   b. i. $E(X)$
      ii. $SD(X)$.
4. It is estimated that about 8% of men and 1% of women have colour blindness. Six men and six women are checked for any signs of colour blindness. Let Y be the discrete random variable that defines the number of men who have colour blindness, and let Z be the discrete random variable that defines the number of women who have colour blindness.
   a. Calculate, correct to 4 decimal places where appropriate:
      i. $E(Y)$
      ii. $SD(Y)$
      iii. $\Pr(\mu - 2\sigma \leq Y \leq \mu + 2\sigma)$
      iv. $E(Z)$
      v. $SD(Z)$
      vi. $\Pr(\mu - 2\sigma \leq Z \leq \mu + 2\sigma)$.
   b. Compare the two distributions.

5. Suppose that 85% of adults with allergies report systematic relief with a new medication that has just been released. The medication has just been given to 12 patients who suffer from allergies. Let Z be the discrete random variable that defines the number of patients who get systematic relief from allergies with the new medication.

a. Evaluate the probability that no more than 8 people get relief from allergies.
b. Given that no more than 8 people get relief from allergies after taking the medication, evaluate the probability that at least 5 people do.
c. Calculate:
   i. $\mathrm{E}(Z)$
   ii. $\mathrm{SD}(Z)$, correct to 4 decimal places.

6. Silicon chips are tested at the completion of the fabrication process. Chips either pass or fail the inspection, and if they fail they are destroyed. The probability, correct to 4 decimal places, that a chip fails an inspection is 0.02. Calculate the probability, correct to 4 decimal places, that in a manufacturing run of 250 chips only 7 will fail the inspection.

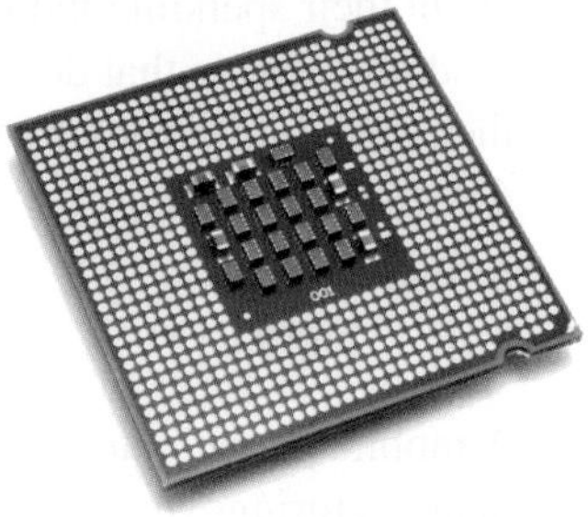

7. The probability of a person in Australia suffering anaemia is 1.3%. A group of 100 different Australians of differing ages were tested for anaemia. Calculate, correct to 4 decimal places:
a. the probability that at least 5 of the 100 Australians suffer from anaemia.
b. the probability that 4 of the 100 Australians suffer from anaemia, given that less than 10 do.
c. the value of $\Pr(\mu - 2\sigma \le X \le \mu + 2\sigma)$ and interpret this value.

8. Haruto is completing a multiple choice test of 20 questions. Each question has 5 possible answers.

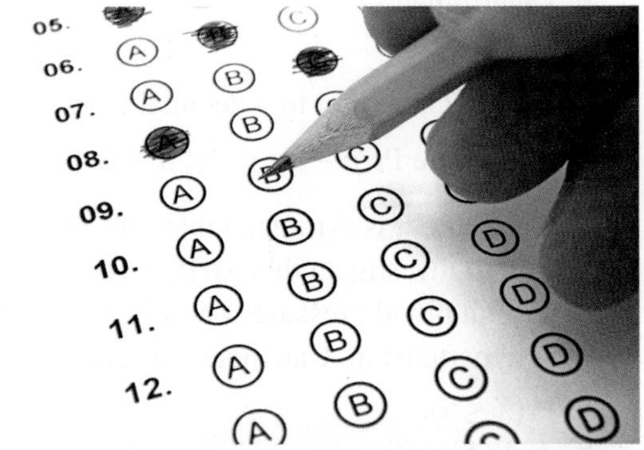

a. If Haruto randomly guesses every question, calculate the probability, correct to 4 decimal places, that he correctly answers 10 or more questions.
b. If Haruto knows the answers to the first 4 questions but must randomly guess the answers to the other questions, find the probability that he correctly answers 10 or more questions. Give your answer correct to 4 decimal places.

9. Six footballers are chosen at random and asked to kick a football. The probability of a footballer being able to kick at least 50 m is 0.7.

a. Determine the probability, correct to 4 decimal places, that:
   i. only the first three footballers chosen kick the ball at least 50 m
   ii. exactly three of the footballers chosen kick the ball at least 50 m
   iii. at least three of the footballers chosen kick the ball at least 50 m, given that the first footballer chosen kicks it at least 50 m.
b. Determine the minimum number of footballers required to ensure that the probability that at least one of them can kick the ball 50 m is at least 0.95.

10. A manufacturer of electric kettles has a process of randomly testing the kettles as they leave the assembly line to see if they are defective. For every 50 kettles produced, 3 are selected and tested for any defects. Let X be the binomial random variable that is the number of kettles that are defective so that $X \sim \text{Bi}(3, p)$.

a. Construct a probability distribution table for X, giving your probabilities in terms of p.
b. Assuming $\Pr(X = 0) = \Pr(X = 1)$, determine the value of p where $0 < p < 1$.
c. Determine:

i. μ

ii. σ.

11. Lee Yin is a goal shooter for her netball team. The probability of her scoring a goal is 0.85. In one particular game, Lee Yin had 12 shots at goal. Determine the probability, correct to 4 decimal places, that:

a. she scored more than 9 goals
b. only her last 9 shots were goals
c. she scored 10 goals, given that her last 9 shots were goals.

12. The chance of winning a prize in the local raffle is 0.08. Determine the least number of tickets Siena needs to purchase so that the chance of both her and her sister each winning at least one prize is more than 0.8.

10.4 Exam questions

Question 1 (1 mark) TECH-ACTIVE

Source: VCE 2021, Mathematical Methods Exam 2, Section A, Q6; © VCAA.

MC The probability of winning a game is 0.25.

The probability of winning a game is independent of winning any other game.

If Ben plays 10 games, the probability that he will win exactly four times is closest to

A. 0.1460 **B.** 0.2241 **C.** 0.9219 **D.** 0.0781 **E.** 0.7759

Question 2 (1 mark) TECH-ACTIVE

Source: VCE 2019, Mathematical Methods Exam 2, Section A, Q8; © VCAA.

MC An archer can successfully hit a target with a probability of 0.9. The archer attempts to hit the target 80 times. The outcome of each attempt is independent of any other attempt.

Given that the archer successfully hits the target at least 70 times, the probability that the archer successfully hits the target exactly 74 times, correct to four decimal places, is

A. 0.3635 **B.** 0.8266 **C.** 0.1494 **D.** 0.3005 **E.** 0.1701

Question 3 (1 mark) TECH-ACTIVE

Source: VCE 2013, Mathematical Methods (CAS) Exam 2, Section 1, Q9; © VCAA.

MC Harry is a soccer player who practises penalty kicks many times each day.

Each time Harry takes a penalty kick, the probability that he scores a goal is 0.7, independent of any other penalty kick. One day Harry took 20 penalty kicks.

Given that he scored at least 12 goals, the probability that Harry scored exactly 15 goals is closest to

A. 0.1789 **B.** 0.8867 **C.** 0.8 **D.** 0.6396 **E.** 0.2017

More exam questions are available online.

10.5 Review

10.5.1 Summary

doc-37040

Hey students! Now that it's time to revise this topic, go online to:

Access the topic summary | Review your results | Watch teacher-led videos | Practise VCAA exam questions

Find all this and MORE in jacPLUS

10.5 Exercise

Technology free: short answer

1. One-third of all customers at a particular bookstore buy non-fiction books. If 5 customers purchase a book on a particular day, calculate the probability that 3 of them purchased a non-fiction book.
2. Aki is a goal shooter for his local netball team. During his matches, he has many attempts at scoring a goal. Assume that each attempt at scoring a goal is independent of any other attempt. In the long run his scoring rate has been shown to be 80%. Determine:
   a. the probability that his first 4 attempts at scoring a goal in a match are successful
   b. the probability that exactly 3 out of his first 4 attempts at scoring a goal in a match are successful
   c. the probability that his first shot is a goal, his next 2 shots miss, and his fourth is a goal.

3. It is known that 50% of customers who enter a restaurant order a cup of coffee. Assume that what any customer orders is independent of what any other customer orders.
   a. If 4 customers enter the restaurant, calculate the probability that more than 1 of these customers will order coffee.
   b. If 12 customers enter the restaurant, calculate the expected number of customers who will order a cup of coffee.
4. A particular binomial random variable has a mean of 24 and a variance of 6. Calculate the probability of success, p, and the number of independent trials, n.
5. The probability that Dineesha will answer any question independently of her answer to any other question is p where $p > 0$. Let the random variable X be the number of questions that Dineesha answers correctly in any set of 20 questions.
   a. If $\Pr(X > 18) = 6\Pr(X = 20)$, show that the value of p is $\frac{4}{5}$.
   b. Evaluate $\mathrm{E}(X)$ and $\mathrm{Var}(X)$.
6. A barrel contains 100 balls, some of which have a zebra pattern painted on them. Five balls are randomly selected from the barrel with replacement after each ball has been withdrawn. Let p be the proportion of zebra-patterned balls in the barrel such that $0 < p < 1$.
   a. Write an expression for the probability that exactly 1 of the 5 balls selected has a zebra pattern.
   b. Use calculus to find the exact value of p for which this probability will be a maximum.

Technology active: multiple choice

7. **MC** A Bernoulli random variable, X, has a probability of failure of 0.35. The expected value and variance of X are respectively:

   A. 0.35 and 0.2275
   B. 0.35 and 0.65
   C. 0.65 and 0.2275
   D. 0.65 and 0.35
   E. 0.65 and 0.65

8. **MC** On Hugo's drive towards the city, he initially has to go through a set of traffic lights at his local shopping centre. As the lights are for people to cross the main road from one side of shops to the other, they are frequently red. Hugo estimates that they are red 80% of the time. Over a period of a week, he drives through these lights twice a day. The probability that these lights are red on exactly 13 occasions is:

   A. ${}^{14}C_{13}(0.2)(0.8)^{13}$
   B. ${}^{14}C_{13}(0.8)(0.2)^{13}$
   C. ${}^{14}C_{13}(0.2)(0.8)^{13} + (0.8)^{14}$
   D. ${}^{14}C_{13}(0.8)(0.2)^{13} + (0.2)^{14}$
   E. $1 - (0.8)^{14}$

9. **MC** Joop takes the bus to and from school each school day over a period of a week. The probability that the bus is late on 4 occasions is given by ${}^{10}C_4(0.7)^6(0.3)^4$.

   The mean and variance of the number of times that Joop finds the bus is late respectively are:

   A. 3 and 0.21 B. 7 and 0.21 C. 0.3 and 2.1
   D. 7 and 2.1 E. 3 and 2.1

10. **MC** A ticket inspector at the Australian Tennis Open has observed that in the long run 65% of all tickets viewed are full-price tickets and the remaining 35% are concession tickets. The ticket inspector takes note of the next 25 tickets he views. The probability that the sample contains exactly 14 full-price tickets is equal to:

    A. 1 B. ${}^{25}C_{14}(0.35)^{11}(0.65)^{14}$ C. ${}^{25}C_{14}(0.65)^{11}(0.35)^{14}$
    D. $(0.35)^{14}(0.65)^{11}$ E. $(0.65)^{14}(0.35)^{11}$

11. **MC** A soccer player is practising his goal kicking. He has a probability of $\frac{4}{7}$ of scoring a goal with each independent attempt. He has 8 attempts at kicking a goal. The probability that the number of goals he scores is less than 4 is:

    A. 0.4717
    B. 0.2518
    C. 0.2199
    D. 0.4728
    E. 0.2210

12. **MC** A fair coin is tossed 7 times. The probability, correct to 4 decimal places, that at most 3 Heads are obtained is:

A. 0.2734
B. 0.2266
C. 0.2188
D. 0.5000
E. 0.0625

13. **MC** Let X be a discrete random variable with a binomial distribution. The mean is 3.5 and the variance is 1.05. The values of n (the number of independent trials) and p (the probability of success) are:

A. $n = 4,\ p = 0.3$
B. $n = 4,\ p = 0.7$
C. $n = 3,\ p = 0.5$
D. $n = 5,\ p = 0.3$
E. $n = 5,\ p = 0.7$

14. **MC** During a holiday Mateo and Airi play a total of n games of golf. The probability that Airi wins any independent game is 0.45. No games are drawn. If the probability that Airi wins no games is 0.0152, then the total number of games played is:

A. 1
B. 2
C. 5
D. 7
E. 10

15. **MC** A random variable, Z, has the following probability distribution, where $0 < p < 1$.

z	0	1
Pr$(Z = z)$	$1 - p$	p

The standard deviation of Z is:

A. $p - 1$
B. $\sqrt{p(1 - p)}$
C. $1 - p$
D. $\sqrt{p(p - 1)}$
E. $p(1 - p)$

16. **MC** The number of defective computer parts in a box of computer parts ready for sale is a random variable with a binomial distribution that has a mean of 12 and a standard deviation of 2. The probability of success, p, is equal to

A. $\frac{1}{3}$
B. $\frac{1}{6}$
C. $\frac{1}{2}$
D. $\frac{5}{6}$
E. $\frac{2}{3}$

Technology active: extended response

17. It has been found in the past that 8% of the computers produced in a certain factory are defective. A sample of 12 is drawn randomly from each hour's production and the number of defectives in the sample is noted. When a computer is chosen, it is tested for defectiveness, and it is returned to the production line if it is not defective.
    a. Calculate, correct to 2 decimal places, the percentage of these hourly samples in which there would be at least 2 defective computers.
    b. Calculate the mean and standard deviation of the number of defective computers noted and calculate $\mu \pm 2\sigma$. Give your answers correct to 4 decimal places.
    c. A particular sample is found to contain 3 defectives. Explain whether this would cause doubts about the production process.

18. For a large population of people, it is fair to assume that the dates of birth are evenly distributed throughout the year, so that the probability of having your birthday in a particular month is $\frac{1}{12}$.

    a. Determine the probability that of 8 people randomly chosen, exactly 3 will have their birthday in August. Give your answer correct to 4 decimal places.
    b. Determine the probability that of the 8 randomly chosen people, at least 1 will have their birthday in November. Give your answer correct to 4 decimal places.
    c. If n people are chosen at random, find the smallest value of n so that the probability that at least 1 will have their birthday in March is greater than 0.9.

19. Australia has the highest incidence of melanoma in the world. One in 17 Australians will be diagnosed with melanoma before the age of 85. In fact, 1 in 14 males are at risk of being diagnosed with a melanoma before the age of 85, because men are 2.5 times more likely to die from melanoma compared to women. Fifty randomly chosen men were asked if they had ever been diagnosed with a melanoma.
    a. State the expected number of melanoma cases for this sample if X is the random variable that defines the number of melanoma cases.
    b. Determine, correct to 4 decimal places:
        i. the probability that fewer than 6 men are diagnosed with melanoma
        ii. the probability that at least 3 men are diagnosed with melanoma, given that fewer than 6 are
        iii. the variance and standard deviation for X.

20. The game of two-up is thought to have originated in the United Kingdom in the 18th century, but it became historically well-known because it was played by Australian soldiers during World War I. The game is now played at casinos because it is illegal to play it anywhere else in Australia, except on Anzac Day, when it can be played in all states except Queensland.

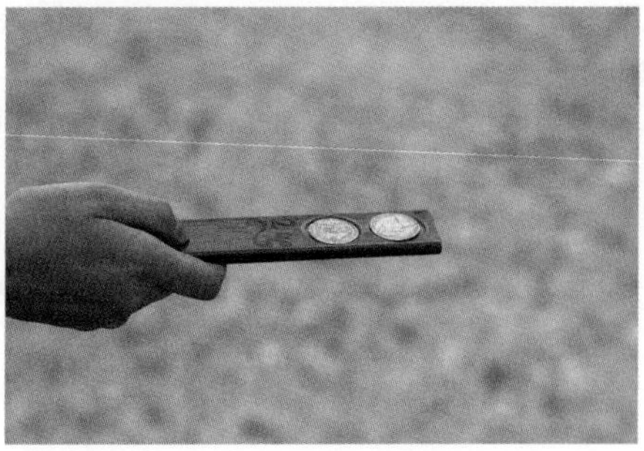

    The usual rules for two-up are quite straightforward. The ringkeeper controls the spinner and the conduct of the game. The spinner places two pennies on a wooden plank called a kip and tosses them.

The pennies must have spun at least 2 metres over the head of the spinner. Players bet on either two Heads or two Tails coming uppermost when the pennies land. To make the outcomes perfectly clear, the Tails side of the pennies is marked with a white or yellow cross.

If the pennies show two Heads, the result is Heads. If the pennies show two Tails, the result is Tails. If the pennies show one Tail and one Head, the result is 'odds'.

A spinner bet means a wager by the spinner. If the spinner nominates to spin for Heads, they win if they spin Heads 3 times without spinning any Tails and without spinning 5 consecutive odds.

Gus has just accepted the roll of the spinner and has nominated Heads to win. He now has 5 spins to try to achieve Heads 3 times.

a. Draw a tree diagram of the possible outcomes of the two coins from one spin. State the accompanying probabilities.
b. Let X be the random variable of both coins showing Heads when a spin occurs. State the values of p and n.
c. Gus successfully completes his 5 spins and did not achieve Heads in 3 consecutive spins. Construct a probability distribution table for X. Give your answers correct to 4 decimal places.
d. Determine μ and σ, and hence calculate $\Pr(\mu - 2\sigma \leq X \leq \mu + 2\sigma)$. Give your answers correct to 4 decimal places where appropriate.

10.5 Exam questions

Question 1 (3 marks) TECH-FREE

Source: *VCE 2016, Mathematical Methods Exam 1, Q4; © VCAA.*

A paddock contains 10 tagged sheep and 20 untagged sheep. Four times each day, one sheep is selected at random from the paddock, placed in an observation area and studied, and then returned to the paddock.

a. What is the probability that the number of tagged sheep selected on a given day is zero? **(1 mark)**
b. What is the probability that at least one tagged sheep is selected on a given day? **(1 mark)**
c. What is the probability that no tagged sheep are selected on each of six consecutive days? Express your answer in the form $\left(\frac{a}{b}\right)^c$ where a, b and c are positive integers. **(1 mark)**

Question 2 (4 marks) TECH-FREE

Source: *VCE 2020, Mathematical Methods Exam 1, Q5; © VCAA.*

For a certain population the probability of a person being born with the specific gene SPGE1 is $\frac{3}{5}$.

The probability of a person having this gene is independent of any other person in the population having this gene.

a. In a randomly selected group of four people, what is the probability that three or more people have the SPGE1 gene? **(2 marks)**
b. In a randomly selected group of four people, what is the probability that exactly two people have the SPGE1 gene, given that at least one of those people has the SPGE1 gene? Express your answer in the form $\frac{a^3}{b^4 - c^4}$, where $a, b, c \in Z^+$. **(2 marks)**

Question 3 (1 mark) TECH-ACTIVE

Source: *VCE 2020, Mathematical Methods Exam 2, Section A, Q19; © VCAA.*

MC Shown below is the graph of p, which is the probability function for the number of times, x, that a '6' is rolled on a fair six-sided die in 20 trials.

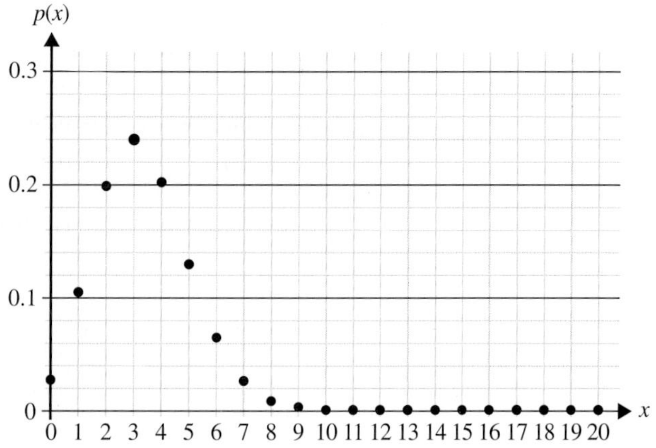

Let q be the probability function for the number of times, w, that a '6' is **not** rolled on a fair six-sided die in 20 trials. $q(w)$ is given by

A. $p(20-w)$ **B.** $p\left(1-\frac{w}{20}\right)$ **C.** $p\left(\frac{w}{20}\right)$ **D.** $p(w-20)$ **E.** $1-p(w)$

Question 4 (1 mark) TECH-ACTIVE

Source: *VCE 2020, Mathematical Methods Exam 2, Section A, Q8; © VCAA.*

MC Items are packed in boxes of 25 and the mean number of defective items per box is 1.4.

Assuming that the probability of an item being defective is binomially distributed, the probability that a box contains more than three defective items, correct to three decimal places, is

A. 0.037 **B.** 0.048 **C.** 0.056 **D.** 0.114 **E.** 0.162

Question 5 (1 mark) TECH-ACTIVE

MC In a shipment of electrical conductors, it is found that 8% are defective.

A sample of 30 conductors is selected at random.

The standard deviation of the number of defective conductors selected is closest to

A. 4.80 **B.** 2.4 **C.** 2.19 **D.** 2.21 **E.** 1.49

More exam questions are available online.

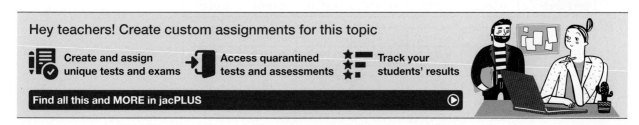

Answers

Topic 10 The binomial distribution

10.2 Bernoulli trials

10.2 Exercise

1. a. No b. Yes c. Yes
   d. Yes e. Yes f. No
2. a. No replacement of ball
   b. There are 6 outcomes, not 2.
   c. Success unknown
3. a. 0.6 b. 0.24
4. a.

x	0	1
$\Pr(X = x)$	$\frac{4}{5}$	$\frac{1}{5}$

   b. $\frac{1}{5}$
   c. 0.000 32
5. a.

x	0	1
$\Pr(X = x)$	0.58	0.42

   b. 0.42
   c. $\text{Var}(X) = 0.2436$, $\text{SD}(X) = 0.4936$
6. B
7. a.

x	0	1
$\Pr(X = x)$	0.11	0.89

   b. i. 0.89 ii. 0.0979 iii. 0.3129
   c. 0.89
8. a.

y	0	1
$\Pr(Y = y)$	0.67	0.33

   b. 0.33
   c. 1
9. a. 0.7 b. 0.7
10. a. 0.8775
    b.

z	0	1
$\Pr(Z = z)$	0.1225	0.8775

    c. 0.8775

10.2 Exam questions

Note: Mark allocations are available with the fully worked solutions online.

1. a. 0.0072
   b.

z	0	1
$\Pr(Z = z)$	0.9928	0.0072

   c. 0.9928
2. a. 0.2436 b. 0.58 c. 0.58
3. a. 0.11 b. 0.8742

10.3 The binomial distribution

10.3 Exercise

1. a. $\frac{5}{16}$ b. $\frac{3}{16}$ c. $\frac{1}{32}$
2. a. $\frac{1}{256}$ b. $\frac{175}{256}$ c. $\frac{54}{175}$
3. $\frac{7}{128}$
4. a. 15 b. 11.25
5. a. $\frac{3}{4}$ b. 16
6. a. See the table at the bottom of the page.*
   b. 0.8891
   c. 0.6745
   d. 0.0800
7. a. $\Pr(Z = 0) = 0.0043$,
   $\Pr(Z = 1) = 0.0312$,
   $\Pr(Z = 2) = 0.1017$,
   $\Pr(Z = 3) = 0.1963$,
   $\Pr(Z = 4) = 0.2488$,
   $\Pr(Z = 5) = 0.2162$,
   $\Pr(Z = 6) = 0.1304$,
   $\Pr(Z = 7) = 0.0540$,
   $\Pr(Z = 8) = 0.0147$,
   $\Pr(Z = 9) = 0.0024$,
   $\Pr(Z = 10) = 0.0002$
   b. See the table at the bottom of the page.*
   c. 0.4164

---

*6. a.

y	0	1	2	3	4	5
$\Pr(Y = y)$	0.0609	0.2285	0.3427	0.2570	0.0964	0.0145

*7. b.

z	0	1	2	3	4	5
$\Pr(Z = z)$	0.0043	0.0312	0.1017	0.1963	0.2488	0.2162

z	6	7	8	9	10
$\Pr(Z = z)$	0.1304	0.0540	0.0147	0.0024	0.0002

8. a. See the table at the bottom of the page.*

b. 0.9752

c. 0.9655

d. 0.0367

9. a. 0.1997 b. 0.4665 c. 0.0034

10. a. $E(X) = 6.75$ b. 0.0780

c. 0.8182 d. 0.0613

11. a. 0.1938 b. 0.3825

12. a. i. 32.4 ii. 9.072

b. i. 20 ii. 16

c. i. 16 ii. $12.\dot{4}$

13. a. 4.1667 b. 1.8634

14. a. See the table at the bottom of the page.*

b. $E(Z) = 2.24$, $Var(Z) = 1.5232$

c. 0.9620

15. a. $E(X) = 1.4286$, $Var(X) = 1.2245$

b. 0.9574

16. a. 0.723 b. 45

17. a. 0.38 b. 24

18. E

19. C

20. a. 0.2362 b. 0.0890

21. 9 tickets

22. 6 games

10.3 Exam questions

Note: Mark allocations are available with the fully worked solutions online.

1. C 2. D 3. D

10.4 Applications

10.4 Exercise

1. a. 0.0106

b. $E(Y) = 3$, $SD(Y) = 1.4491$

2. a. 1

b. i. 0.05

ii. $Var(Z) = 0.0495$, $SD(Z) = 0.2225$

c. 0.9510

3. a. 0.7216

b. i. 4.5 ii. 1.7748

4. a. i. 0.48 ii. 0.6645

iii. 0.9227 iv. 0.06

v. 0.2437 vi. 0.9415

b. There is a probability of 0.9228 that a maximum of 1 male will have colour blindness, whereas there is a probability of 0.9415 that no females will have colour blindness.

5. a. 0.0922

b. 0.9992

c. i. 10.2 ii. 1.2369

6. 0.1051

7. a. 0.0101

b. 0.0319

c. 0.9580. This means there is a probability of 0.9580 that a maximum of 3 people will suffer from anaemia per 100.

8. a. 0.0026 b. 0.0817

9. a. i. 0.0093 ii. 0.1852 iii. 0.9692

b. 3

10. a.

x	0	1	2	3
Pr($X = x$)	$(1-p)^3$	$3(1-p)^2p$	$3(1-p)p^2$	p^3

b. $p = \frac{1}{4}$

c. i. $\frac{3}{4}$ ii. $\frac{3}{4}$

11. a. 0.7358 b. 0.0008 c. 0.0574

12. 37

10.4 Exam questions

Note: Mark allocations are available with the fully worked solutions online.

1. A

2. C

3. E

10.5 Review

10.5 Exercise

Technology free: short answer

1. $\frac{40}{243}$

2. a. $\frac{256}{625}$ b. $\frac{256}{625}$ c. $\frac{16}{625}$

3. a. $\frac{11}{16}$ b. 6

4. $p = \frac{3}{4}$, $n = 32$

*8. a.

x	0	1	2	3	4	5	6	7	8
Pr($X = x$)	0.0004	0.0048	0.0285	0.0971	0.2067	0.2815	0.2397	0.1166	0.0248

*14. a.

z	0	1	2	3	4	5	6	7
Pr($Z = z$)	0.0672	0.2215	0.3127	0.2452	0.1154	0.0326	0.0051	0.0003

5. a. $X \sim \text{Bi}(20, p)$

$\Pr(X > 18) = \Pr(X = 19) + \Pr(X = 20)$

$\Pr(X > 18) = {}^{20}C_{19}(1-p)p^{19} + p^{20}$

$\Pr(X > 18) = 20(1-p)p^{19} + p^{20}$

$\Pr(X = 20) = p^{20}$

$\Pr(X = 20) = 6\Pr(X > 18)$

$$20(1-p)p^{19} + p^{20} = 6p^{20}$$
$$0 = 6p^{20} - p^{20} - 20(1-p)p^{19}$$
$$0 = 5p^{20} - 20p^{19} + 20p^{20}$$
$$0 = 25p^{20} - 20p^{19}$$
$$0 = 5p^{19}(5p - 4)$$
$$p = 0 \text{ or } 5p - 4 = 0$$
$$5p = 4$$
$$p = \frac{4}{5}$$

As $p > 0$, $p = \frac{4}{5}$.

b. $E(X) = 16$, $\text{Var}(X) = 3.2$

6. a. $5p(1-p)^4$ b. $p = \frac{1}{5}$

Technology active: multiple choice

7. C
8. A
9. E
10. B
11. E
12. D
13. E
14. D
15. B
16. E

Technology active: extended response

17. a. 24.87%

b. $\mu = 0.96$, $\sigma = 0.9398$, $\mu - 2\sigma = -0.9196$, $\mu + 2\sigma = 2.8396$

c. The result lies outside the 95% limit, so there would be concerns about the production process.

18. a. 0.0210 b. 0.5015 c. 27

19. a. $\frac{25}{7}$

b. 0.8554

c. 0.6523

d. $\text{Var}(X) = 3.3163$, $\text{SD}(X) = 1.8211$

20. a. $\Pr(HH) = \frac{1}{4}$, $\Pr(HT) = \frac{1}{4}$, $\Pr(TH) = \frac{1}{4}$, $\Pr(TT) = \frac{1}{4}$

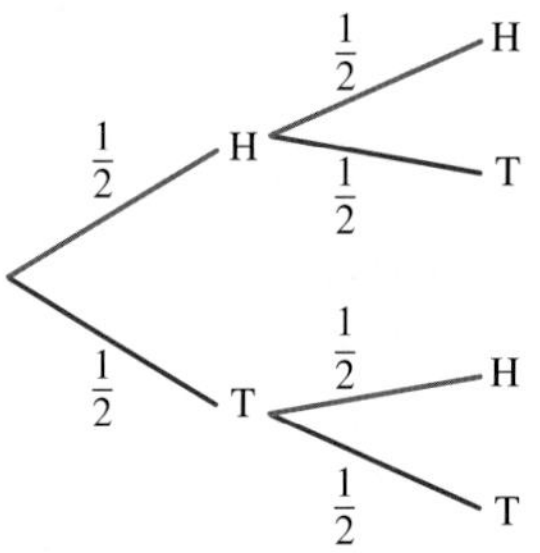

b. $X \sim \text{Bi}\left(5, \frac{1}{4}\right)$

$\therefore n = 5$, $p = \frac{1}{4}$

c. See the table at the bottom of the page.*

d. $\mu = 1.25$, $\sigma = 0.9682$,

$\Pr(\mu - 2\sigma \le X \le \mu + 2\sigma) = 0.9844$

10.5 Exam questions

Note: Mark allocations are available with the fully worked solutions online.

1. a. $\frac{16}{81}$ b. $\frac{65}{81}$ c. $\left(\frac{16}{81}\right)^6$

2. a. $\frac{297}{625}$

b. $\frac{6^3}{5^4 - 2^4}$, $a = 6$, $b = 5$, $c = 2$

3. A
4. B
5. E

*20. c.

x	0	1	2	3	4	5
$\Pr(X = x)$	0.2373	0.3955	0.2637	0.0879	0.0146	0.0010

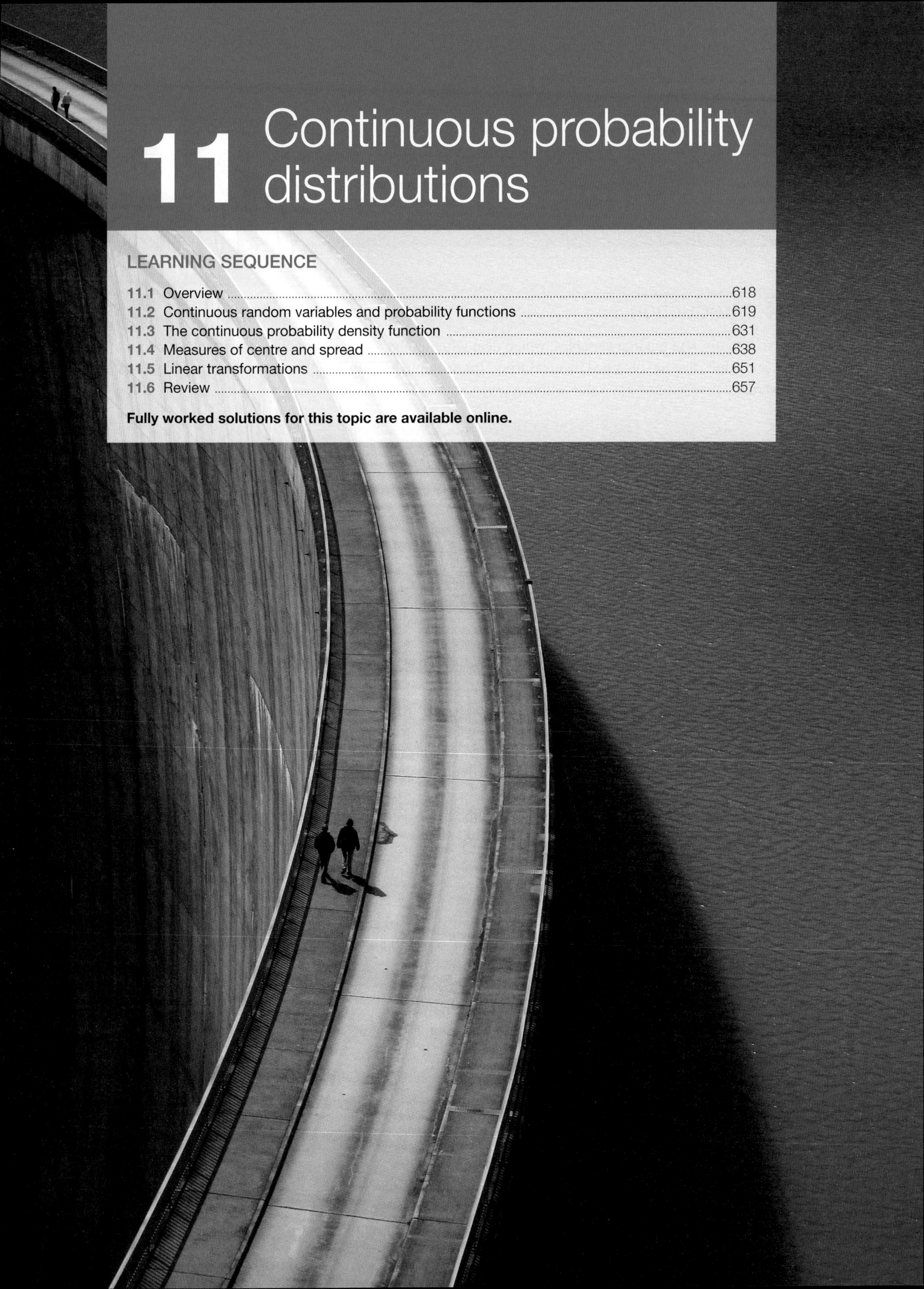

11 Continuous probability distributions

LEARNING SEQUENCE

Fully worked solutions for this topic are available online.

11.1 Overview

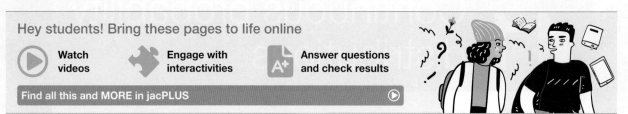

Hey students! Bring these pages to life online

Watch videos

Engage with interactivities

Answer questions and check results

Find all this and MORE in jacPLUS

11.1.1 Introduction

In contrast with discrete random variables, which deal with numbers that are counted, continuous random variables are values that are obtained by measuring, for example heights, weights or time taken. A continuous random variable may assume any value within a given range.

A probability density function is such that the area under the curve is 1 and $f(x) \geq 0$ for all x-values. Some functions are quite basic, but others are more elaborate and have particular applications, such as the Kumaraswamy distribution, which is used in relation to the storage volume of a reservoir. The Birnbaum–Saunders distribution, also called the fatigue life distribution, is used in reliability applications to model failure times. The Erlang distribution was developed with the purpose of examining the number of telephone calls that could be made simultaneously to switchboard operators. Erlang's work has now been expanded to encompass queue waiting times.

KEY CONCEPTS

This topic covers the following key concepts from the VCE Mathematics Study Design:

- random variables, including the concept of a random variable as a real function defined on a sample space and examples of continuous random variables
- continuous random variables:
  - construction of probability density functions from non-negative functions of a real variable
  - specification of probability distributions for continuous random variables using probability density functions
  - calculation and interpretation of mean, μ, variance, σ^2, and standard deviation of a continuous random variable and their use
  - calculation of probabilities for intervals defined in terms of a random variable, including conditional probability.

11.2 Continuous random variables and probability functions

LEARNING INTENTION

At the end of this subtopic you should be able to:
- define and identify probability density functions
- determine unknown values of given probability density functions.

11.2.1 Probability density functions

Discrete data is data that is finite or countable, such as the number of soft-centred chocolates in a box of soft- and hard-centred chocolates.

A **continuous random variable** assumes an uncountable or infinite number of possible outcomes between two values. That is, the variable can assume any value within a given range. For example, the birth weights of babies and the number of millimetres of rain that falls in a night are continuous random variables. In these examples, the measurements come from an interval of possible outcomes. If a newborn boy is weighed at 4.46 kilograms, that is just what the weight scale's output said. In reality, he may have weighed 4.463 279 ... kilograms. Therefore, a possible range of outcomes is valid within an interval that depends on the precision of the scale.

Consider an Australian health study that was conducted. The study targeted young people aged 5 to 17 years old. They were asked to estimate the average number of hours of physical activity they participated in each week. The results of this study are shown in the following histogram.

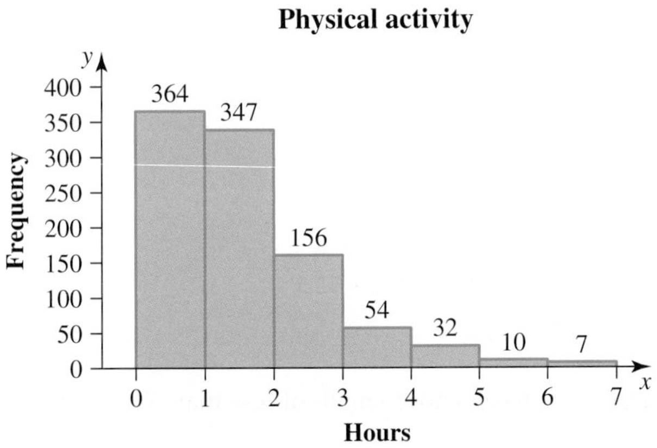

Remember, continuous data has no limit to the accuracy with which it is measured. In this case, for example, $0 \leq x < 1$ means from 0 seconds to 59 minutes and 59 seconds, and so on, because x is not restricted to integer values. In the physical activity study, x taking on a particular value is equivalent to x taking on a value in an appropriate interval. For instance,

$$\Pr(X = 0.5) = \Pr(0 \leq X < 1)$$
$$\Pr(X = 1.5) = \Pr(1 \leq X < 2)$$

and so on.

From the histogram,

$$\begin{aligned}\Pr(X=2.5) &= \Pr(2 \le X < 3)\\ &= \frac{156}{(364+347+156+54+32+10+7)}\\ &= \frac{156}{970}\end{aligned}$$

In another study, the nose lengths, X millimetres, of 75 adults were measured. This data is continuous because the results are measurements. The result of the study is shown in the table and accompanying histogram.

Nose length	Frequency
$27.5 < X \le 32.5$	2
$32.5 < X \le 37.5$	5
$37.5 < X \le 42.5$	17
$42.5 < X \le 47.5$	21
$47.5 < X \le 52.5$	11
$52.5 < X \le 57.51$	7
$57.5 < X \le 62.5$	6
$62.5 < X \le 67.5$	5
$67.5 < X \le 72.5$	1

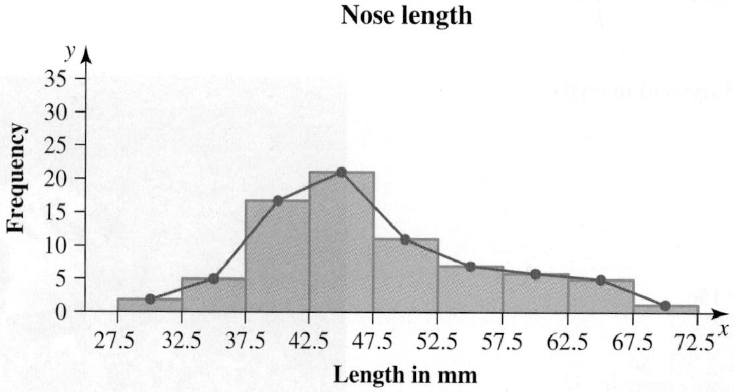

It is possible to use the histogram to find the number of people who have a nose length of less than 47.5 mm.

$$\begin{aligned}\Pr(\text{nose length is} < 47.5) &= \frac{2+5+17+21}{75}\\ &= \frac{45}{75}\\ &= \frac{3}{5}\end{aligned}$$

It is worth noting that we cannot find the probability that a person has a nose length that is less than 45 mm, as this is not the end point of any interval. However, if we had a mathematical formula to approximate the shape of the graph, then the formula could give us the answer to this important question.

In the histogram, the midpoints at the top of each bar have been connected by line segments. If the class intervals were much smaller, say 1 mm or even less, these line segments would take on the appearance of a smooth curve.

This smooth curve is of considerable importance for continuous random variables, because it represents the **probability density function** for the continuous data.

This problem for a continuous random variable can be addressed by using calculus.

Probability density function conditions

For any continuous random variable, X, the probability density function is such that

$$\Pr(a < X < b) = \int_a^b f(x)dx$$

which is the area under the curve from $x = a$ to $x = b$.

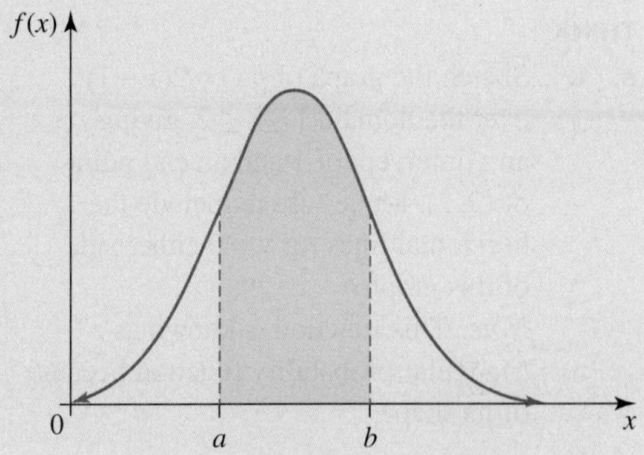

A probability density function must satisfy the following conditions:

- $f(x) \geq 0$ **for all** $x \in [a, b]$
- $\int_a^b f(x)\,dx = 1$**; this is absolutely critical.**

Other properties are:

- $\Pr(X = x) = 0$, **where** $x \in [a, b]$
- $\Pr(a < X < b) = \Pr(a \leq X < b) = \Pr(a < X \leq b) = \Pr(a \leq X \leq b) = \int_a^b f(x)\,dx$
- $\Pr(X < c) = \Pr(X \leq c) = \int_a^c f(x)\,dx$ **when** $x \in [a, b]$ **and** $a < c < b$.

In theory, the domain of a continuous probability density function is R, so that

$$\int_{-\infty}^{\infty} f(x)dx = 1.$$

However, if we must address the condition that

$$\int_a^b f(x)dx = 1,$$

then the function must be zero everywhere else.

on Resources

Interactivity Continuous random variables (int-6433)

WORKED EXAMPLE 1 Probability density functions

Sketch the graph of each of the following functions and state whether each function is a probability density function.

a. $f(x) = \begin{cases} 2(x-1), & 1 \le x \le 2 \\ 0, & \text{elsewhere} \end{cases}$ b. $f(x) = \begin{cases} 0.5, & 2 \le x \le 4 \\ 0, & \text{elsewhere} \end{cases}$ c. $f(x) = \begin{cases} 2e^{-x} & 0 \le x \le 2 \\ 0, & \text{elsewhere} \end{cases}$

THINK

a. 1. Sketch the graph of $f(x) = 2(x-1)$ over the domain $1 \le x \le 2$, giving an x-intercept of 1 and an end point of $(2, 2)$. Make sure to include the horizontal lines for $y = 0$ either side of this graph.
Note: This function is known as a triangular probability function because of its shape.

WRITE

a.

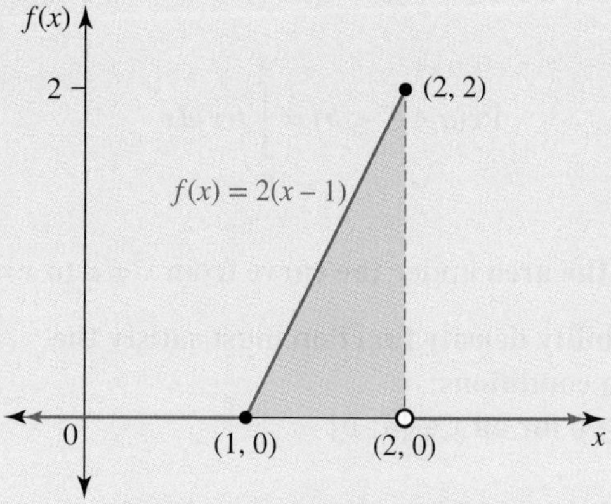

2. Inspect the graph to determine if the function is always positive or zero, that is, $f(x) \ge 0$ for all $x \in [a, b]$.

Yes, $f(x) \ge 0$ for all x-values.

3. Calculate the area of the shaded region to determine if $\int_1^2 2(x-1)\,dx - 1$.

Method 1: Using the area of triangles

$$\begin{aligned}\text{Area of shaded region} &= \frac{1}{2} \times \text{base} \times \text{height} \\ &= \frac{1}{2} \times 1 \times 2 \\ &= 1\end{aligned}$$

Method 2: Using calculus

$$\begin{aligned}\text{Area of shaded region} &= \int_1^2 2(x-1)\,dx \\ &= \int_1^2 (2x-2)\,dx \\ &= [x^2 - 2x]_1^2 \\ &= (2^2 - 2(2)) - (1^2 - 2(1)) \\ &= 0 - 1 + 2 \\ &= 1\end{aligned}$$

4. Interpret the results.

$f(x) \ge 0$ for all values, and the area under the curve $= 1$. Therefore, this is a probability density function.

b. 1. Sketch the graph of $f(x) = 0.5$ for $2 \leq x \leq 4$. This gives a horizontal line, with end points of $(2, 0.5)$ and $(4, 0.5)$. Make sure to include the horizontal lines for $y = 0$ on either side of this graph.
Note: This function is known as a uniform or rectangular probability density function because of its rectangular shape.

b.

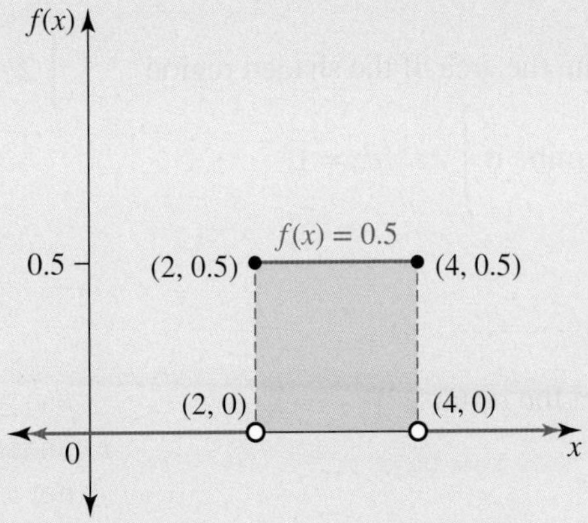

2. Inspect the graph to determine if the function is always positive or zero, that is, $f(x) \geq 0$ for all $x \in [a, b]$.

Yes, $f(x) \geq 0$ for all x-values.

3. Calculate the area of the shaded region to determine if $\int_2^4 0.5\, dx = 1$.

Again, it is not necessary to use calculus to find the area.

Method 1: Using the area of a rectangle

$$\begin{aligned}\text{Area of shaded region} &= \text{length} \times \text{width}\\ &= 2 \times 0.5\\ &= 1\end{aligned}$$

Method 2: Using calculus

$$\begin{aligned}\text{Area of shaded region} &= \int_2^4 0.5\, dx\\ &= [0.5x]_2^4\\ &= 0.5(4) - 0.5(2)\\ &= 2 - 1\\ &= 1\end{aligned}$$

4. Interpret the results.

$f(x) \geq 0$ for all values, and the area under the curve $= 1$. Therefore, this is a probability density function.

c. 1. Sketch the graph of $f(x) = 2e^{-x}$ for $0 \leq x \leq 2$. End points will be $(0, 2)$ and $(2,\ 2e^{-2})$. Make sure to include the horizontal lines for $y = 0$ on either side of this graph.

c.

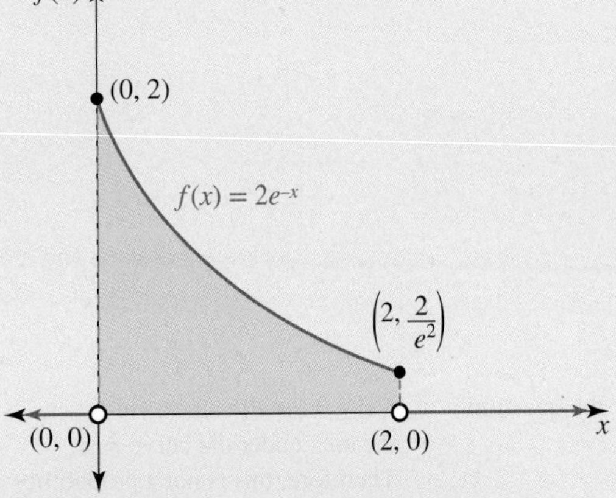

2. Inspect the graph to determine if the function is always positive or zero, that is, $f(x) \geq 0$ for all $x \in [a, b]$.

Yes, $f(x) \geq 0$ for all x-values.

3. Calculate the area of the shaded region to determine if $\int_0^2 2e^{-x}dx = 1$.

$$\int_0^2 2e^{-x}dx = 2\int_0^2 e^{-x}dx$$
$$= 2[-e^{-x}]_0^2$$
$$= 2(-e^{-2} + e^0)$$
$$= 2(-e^{-2} + 1)$$
$$= 1.7293$$

4. Interpret the results.

$f(x) \geq 0$ for all values. However, the area under the curve $\neq 1$. Therefore, this is not a probability density function.

TI | THINK — **DISPLAY/WRITE**

c. 1. On a Graphs page, complete the entry line for function 1 as: $f1(x) = 2e^{-x}\ |0 \leq x \leq 2$ then press ENTER.

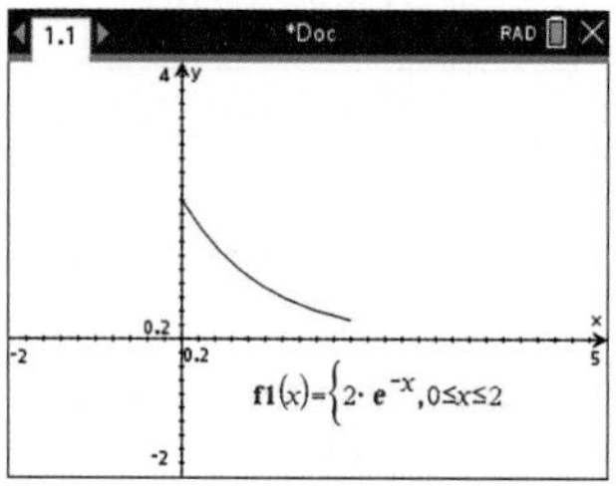

2. To find the area under the curve, press MENU, then select:
6: Analyze Graph
7: Integral
Type '0' when prompted for the lower bound, then press ENTER. Type '2' when prompted for the upper bound, then press ENTER.
The area appears on the screen.

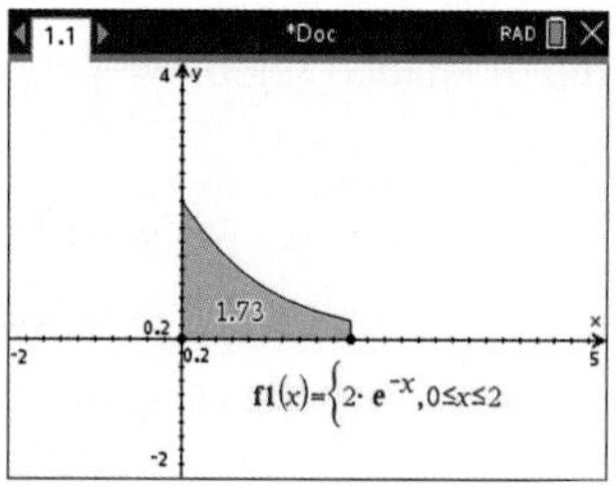

3. Answer the question.

$f(x) \geq 0$ for all values. However, the area under the curve $\neq 1$. Therefore, this is not a probability density function.

CASIO | THINK — **DISPLAY/WRITE**

c. 1. On a Graph & Table screen, complete the entry line for $y1$ as: $y1 = 2e^{-x}\ |0 \leq x \leq 2$ then press ENTER. Select the 'Graph' icon to draw the graph.

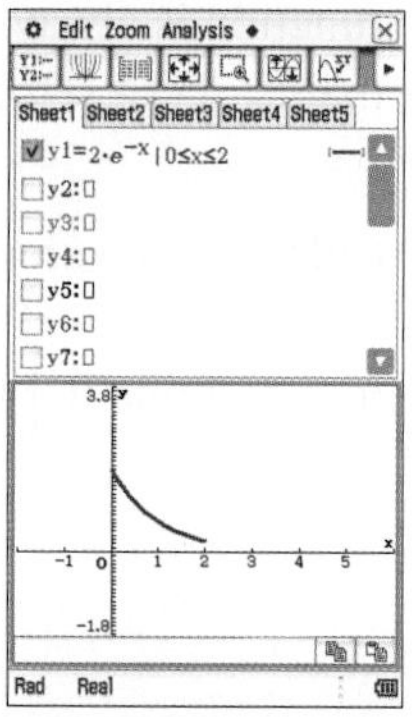

2. To find the area under the curve, select:
- Analysis
- G-Solve
- Integral
- $\int dx$

Type '0' to bring up the dialogue box, then complete the fields as:
Lower: 0
Upper: 2
Select OK.

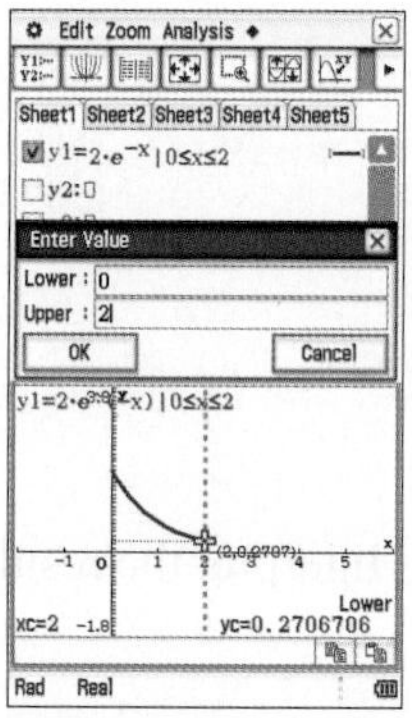

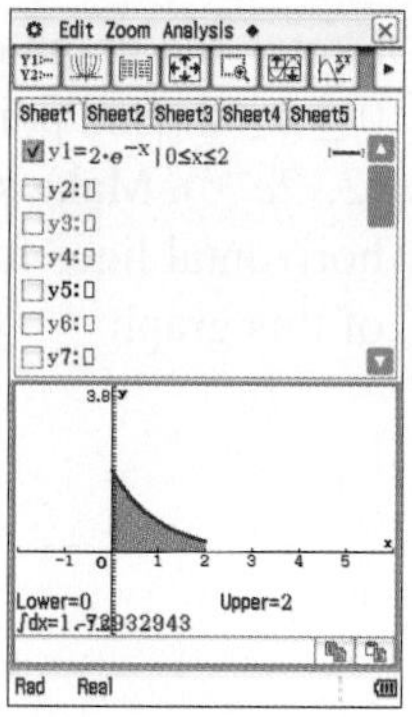

3. Answer the question.

$f(x) \geq 0$ for all values. However, the area under the curve $\neq 1$. Therefore, this is not a probability density function.

WORKED EXAMPLE 2 Determining unknown values in probability density functions

Given that the functions below are probability density functions, calculate the value of a in each function.

a. $f(x) = \begin{cases} a(x-1)^2, & 0 \le x \le 4 \\ 0, & \text{elsewhere} \end{cases}$

b. $f(x) = \begin{cases} ae^{-4x}, & x > 0 \\ 0, & \text{elsewhere} \end{cases}$

THINK	WRITE
a. 1. As the function has already been defined as a probability density function, this means that the area under the graph is definitely 1.	a. $\int_0^4 f(x)\,dx = 1$ $\int_0^4 a(x-1)^2\,dx = 1$
2. Remove a from the integral, as it is a constant.	$a\int_0^4 (x-1)^2\,dx = 1$
3. Anti-differentiate and substitute in the terminals.	$a\left[\frac{(x-1)^3}{3}\right]_0^4 = 1$ $a\left[\frac{3^3}{3} - \frac{(-1)^3}{3}\right] = 1$
4. Solve for a.	$a\left(9 + \frac{1}{3}\right) = 1$ $a \times \frac{28}{3} = 1$ $a = \frac{3}{28}$
b. 1. As the function has already been defined as a probability density function, this means that the area under the graph is definitely 1.	b. $\int_0^\infty f(x)\,dx = 1$ $\int_0^\infty ae^{-4x}\,dx = 1$
2. Remove a from the integral, as it is a constant.	$a\int_0^\infty e^{-4x}\,dx = 1$
3. To evaluate an integral containing infinity as one of the terminals, we find the appropriate limit.	$a \times \lim_{k\to\infty} \int_0^k e^{-4x}\,dx = 1$
4. Anti-differentiate and substitute in the terminals.	$a \times \lim_{k\to\infty}\left[-\frac{1}{4}e^{-4x}\right]_0^k = 1$ $a \times \lim_{k\to\infty}\left(-\frac{e^{-4k}}{4} + \frac{1}{4}\right) = 1$ $a \times \lim_{k\to\infty}\left(-\frac{1}{4e^{4k}} + \frac{1}{4}\right) = 1$

5. Solve for a. Remember that a number divided by an extremely large number is effectively zero, so $\lim_{k \to \infty} \left(\frac{1}{e^{4k}} \right) = 0$.

$$a\left(0 + \frac{1}{4}\right) = 1$$
$$\frac{a}{4} = 1$$
$$a = 4$$

Resources

Interactivity Probability density functions (int-6434)

11.2 Exercise

Technology free

1. A small car-hire firm keeps note of the age and kilometres covered by each of the cars in their fleet. Generally, cars are no longer used once they have either covered 350 000 kilometres or are more than five years old. The following information describes the ages of the cars in their current fleet.

Age	Frequency
$0 < x \leq 1$	10
$1 < x \leq 2$	26
$2 < x \leq 3$	28
$3 < x \leq 4$	20
$4 < x \leq 5$	11
$5 < x \leq 6$	4
$6 < x \leq 7$	1

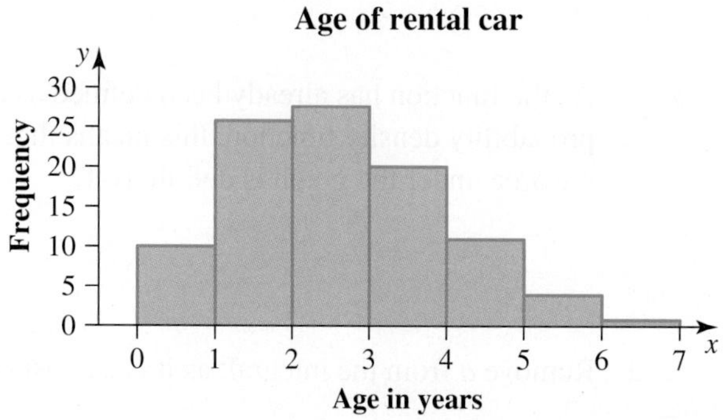

a. Determine:

i. $\Pr(X \leq 2)$

ii. $\Pr(X > 4)$.

b. Determine:

i. $\Pr(1 < X \leq 4)$

ii. $\Pr(X > 1 \mid X \leq 4)$.

2. The battery life for batteries in television remote controls was investigated in a study.

Hours of life	Frequency
$0 < x \le 15$	15
$15 < x \le 30$	33
$30 < x \le 45$	23
$45 < x \le 60$	26
$60 < x \le 75$	3

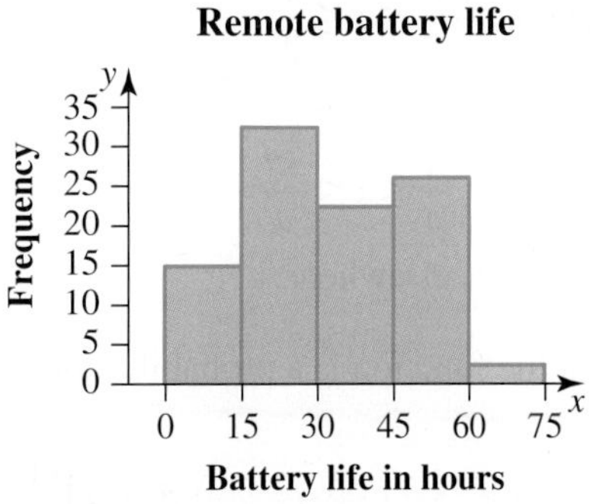

a. State how many remote control batteries were included in the study.
b. Calculate the probability that a battery will last more than 45 hours.
c. Calculate the probability that a battery will last between 15 and 60 hours.
d. A new battery producer is advocating that their batteries have a long life of 60 + hours. If it is known that this is just advertising hype because these batteries are no different from the batteries in the study, calculate the probability that these new batteries will have a life of 60 + hours.

3. A number of experienced shot-putters were asked to aim for a line 10 metres away.
After each of them put their shot, its distance from the 10-metre line was measured. All of the shots were on or between the 8- and 10-metre lines. The results of the measurements are shown, where X is the distance in metres from the 10-metre line.

Metres	Frequency
$0 < x \le 0.5$	75
$0.5 < x \le 1$	63
$1 < x \le 1.5$	45
$1.5 < x \le 2$	17

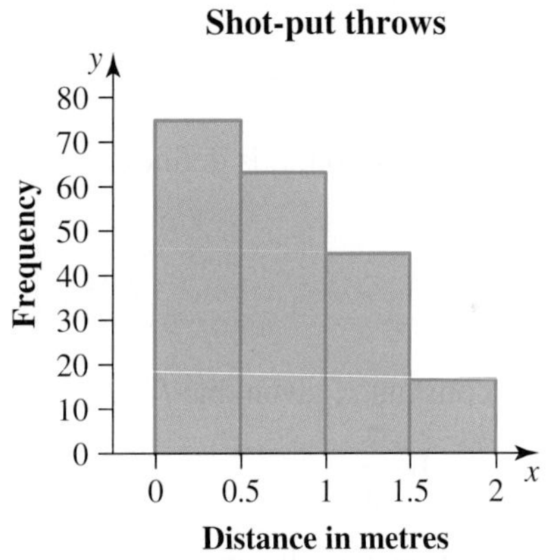

a. State how many shot-put throws were measured.
b. Calculate:

i. $\Pr(X > 0.5)$
ii. $\Pr(1 < X \le 2)$

c. A guest shot-putter is visiting the athletics club where the measurements are being conducted. His shot-putting ability is equivalent to the abilities of the club members. Determine the probability that he puts the shot within 50 cm of the 10-metre line if it is known that he put the shot within 1 metre of the 10-metre line.

4. **WE1** Sketch the graph of each of the following functions and state whether each function is a probability density function.

a. $f(x) = \begin{cases} \frac{1}{4}e^{2x}, & 0 \le x \le \log_e(3) \\ 0, & \text{elsewhere} \end{cases}$

b. $f(x) = \begin{cases} 0.25, & -2 \le x \le 2 \\ 0, & \text{elsewhere} \end{cases}$

5. Sketch each of the following functions and determine whether each one is a probability density function.

a. $f(x)=\begin{cases}\frac{1}{2}\cos(x), & -\frac{\pi}{2}\le x\le\frac{\pi}{2}\\ 0, & \text{elsewhere}\end{cases}$

b. $f(x)=\begin{cases}-\frac{1}{x}, & -e\le x\le -1\\ 0, & \text{elsewhere}\end{cases}$

c. $f(x)=\begin{cases}\cos(x)+1, & \frac{\pi}{4}\le x\le\frac{3\pi}{4}\\ 0, & \text{elsewhere}\end{cases}$

d. $f(x)=\begin{cases}\frac{1}{2}\sin(x), & 0\le x\le\pi\\ 0, & \text{elsewhere}\end{cases}$

6. **WE2** Given that the function is a probability density function, calculate the value of n.

$$f(x)=\begin{cases}n(x^3-1), & 1\le x\le 3\\ 0, & \text{elsewhere}\end{cases}$$

7. Given that the function is a probability density function, calculate the value of a.

$$f(x)=\begin{cases}-ax, & -2\le x<0\\ 2ax, & 0\le x\le 3\\ 0, & \text{elsewhere}\end{cases}$$

Technology active

8. Sketch each of the following functions and determine whether each function is a probability density function.

a. $f(x)=\begin{cases}\frac{1}{2\sqrt{x}}, & \frac{1}{2}\le x\le 4\\ 0, & \text{elsewhere}\end{cases}$

b. $f(x)=\begin{cases}\frac{1}{2\sqrt{x-1}}, & 1<x\le 2\\ 0, & \text{elsewhere}\end{cases}$

9. **MC** The rectangular function, f, is defined by the rule

$$f(x)=\begin{cases}c, & 0.25<x<1.65\\ 0, & \text{elsewhere}\end{cases}$$

The value of the constant c, given that f is a probability density function, is:

A. 1
B. $\frac{7}{5}$
C. $\frac{2}{3}$
D. $\frac{3}{4}$
E. $\frac{5}{7}$

10. The graph of a function, f, is shown.
If f is known to be a probability density function, show that the value of z is $\frac{1}{3}$.

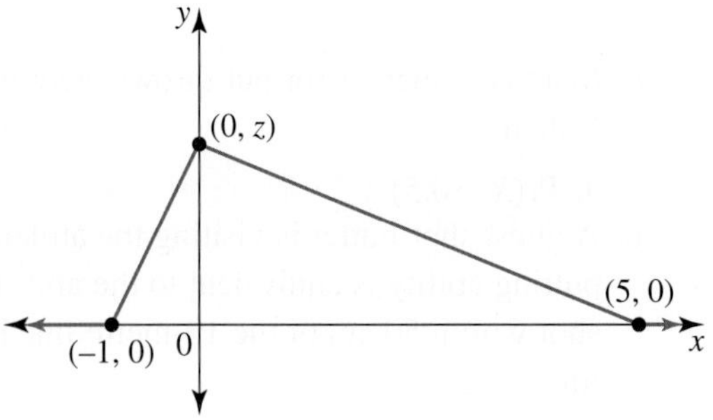

11. Calculate the value of the constant m in each of the following if each function is a probability density function.

a. $f(x)=\begin{cases}m(6-2x), & 0\le x\le 2\\ 0, & \text{elsewhere}\end{cases}$

b. $f(x)=\begin{cases}me^{-2x}, & x\ge 0\\ 0, & \text{elsewhere}\end{cases}$

c. $f(x)=\begin{cases}me^{2x}, & 0\le x\le \log_e(3)\\ 0, & \text{elsewhere}\end{cases}$

12. Let X be a continuous random variable with the probability density function

$$f(x) = \begin{cases} x^2 + 2kx + 1, & 0 \le x \le 3 \\ 0, & \text{elsewhere} \end{cases}$$

Show that the value of k is $-\frac{11}{9}$.

13. **MC** X is a continuous random variable such that

$$f(x) = \begin{cases} n\sin(3x)\cos(3x), & 0 < x < \frac{\pi}{12} \\ 0, & \text{elsewhere} \end{cases}$$

If f is known to be a probability density function, the value of the constant, n, is:

A. 12 **B.** $\frac{1}{6}$ **C.** 6 **D.** $\frac{1}{8}$ **E.** $\frac{1}{12}$

14. A function, f, is defined by the rule

$$f(x) = \begin{cases} \log_e(x), & x > 0 \\ 0, & \text{elsewhere} \end{cases}$$

a. If $\int_1^a f(x)\,dx = 1$, determine the value of the real constant a.

b. State whether this function defines a probability density function.

15. X is a continuous random variable such that

$$f(x) = \begin{cases} \frac{1}{2}\log_e\left(\frac{x}{2}\right), & 2 \le x \le a \\ 0, & \text{elsewhere} \end{cases}$$

and $\int_2^a f(x)\,dx = 1$. The graph of this function is shown.

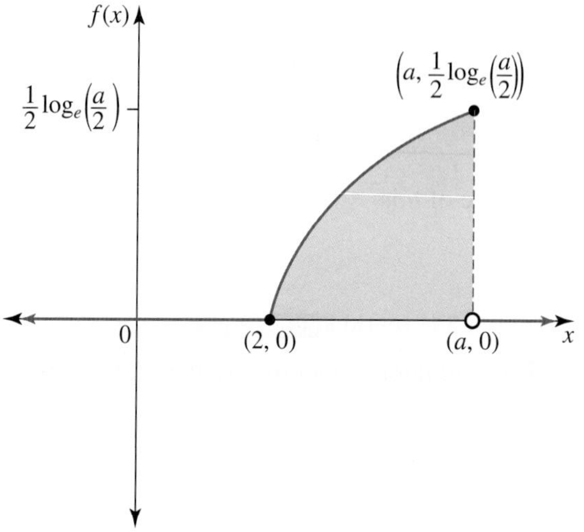

Determine the value of the constant a.

16. X is a continuous random variable such that

$$f(x) = \begin{cases} -x, & -1 \le x < 0 \\ x, & 0 \le x \le a \\ 0, & \text{elsewhere} \end{cases}$$

where a is a constant.

Y is another continuous random variable such that

$$f(y) = \begin{cases} \frac{1}{y}, & 1 \le y \le e \\ 0, & \text{elsewhere} \end{cases}$$

a. Sketch the graph of the function for X and determine $\int_{-1}^a f(x)\,dx$.

b. Sketch the graph of the function for Y and determine $\int_1^e f(y)\,dy$.

c. Calculate the value of the constant a if $\int_{-1}^{a} f(x)dx = \int_1^e f(y)dy$.

11.2 Exam questions

Question 1 (1 mark) TECH-ACTIVE

Source: *VCE 2017 Mathematical Methods Exam 2, Section A, Q19; © VCAA.*

MC A probability density function f is given by

$$f(x) = \begin{cases} \cos(x) + 1 & k < x < (k+1) \\ 0 & \text{elsewhere} \end{cases}$$

where $0 < k < 2$.

The value of k is

A. 1 **B.** $\frac{3\pi - 1}{2}$ **C.** $\pi - 1$ **D.** $\frac{\pi - 1}{2}$ **E.** $\frac{\pi}{2}$

Question 2 (1 mark) TECH-ACTIVE

Source: *VCE 2015 Mathematical Methods (CAS) Exam 2, Section 1, Q13; © VCAA.*

MC The function f is a probability density function with rule

$$f(x) = \begin{cases} ae^x & 0 \le x \le 1 \\ ae & 1 < x \le 2 \\ 0 & \text{otherwise} \end{cases}$$

The value of a is

A. 1 **B.** e **C.** $\frac{1}{e}$ **D.** $\frac{1}{2e}$ **E.** $\frac{1}{2e - 1}$

Question 3 (1 mark) TECH-ACTIVE

MC A continuous random variable, X, has a probability density function defined by

$$f(x) = \begin{cases} cx^2 & \text{for } 0 \le x \le 3 \\ 0 & \text{elsewhere} \end{cases}$$

Determine which of the following options is **correct**.

A. $c = \frac{3}{26}$ **B.** $c = \frac{26}{3}$ **C.** $c = \frac{1}{14}$

D. $c = \frac{1}{9}$ **E.** $c = 9$

More exam questions are available online.

11.3 The continuous probability density function

LEARNING INTENTION

At the end of this subtopic you should be able to:

- calculate various probabilities for a given probability distribution function, including conditional probability.

11.3.1 Continuous probability density functions

As stated in subtopic 11.2, if X is a continuous random variable, then

$$\Pr(a \le X \le b) = \int_a^b f(x)\,dx.$$

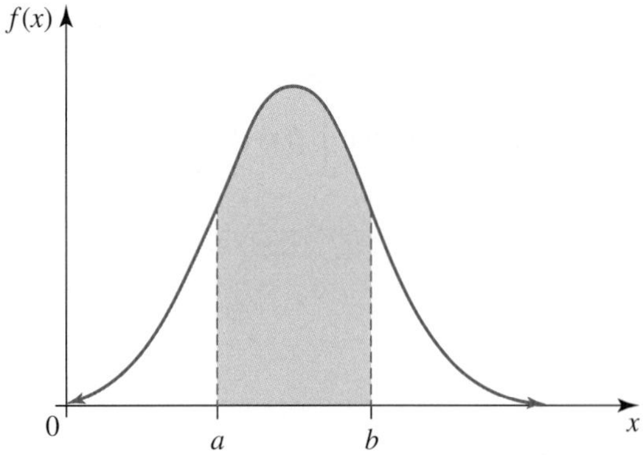

In other words, by finding the area between the curve of the continuous probability function, the x-axis, the line $x = a$ and the line $x = b$, providing $f(x) \ge 0$, then we are finding $\Pr(a \le X \le b)$. It is worth noting that because we are dealing with a continuous random variable, $\Pr(X = a) = 0$. Consequently, we can calculate probabilities as follows.

Calculating probabilities within a probability density function

$$\Pr(a \le X \le b) = \Pr(a < X \le b) = \Pr(a \le X < b) = \Pr(a < X < b)$$

$$\Pr(a \le X \le b) = \Pr(a \le X \le c) + \Pr(c < X \le b), \text{ where } a < c < b.$$

This property is particularly helpful when the probability density function is a hybrid function and the required probability encompasses two functions.

WORKED EXAMPLE 3 Calculating probabilities

A continuous random variable, X, has a probability density function, f, defined by

$$f(x) = \begin{cases} -ax, & -3 \le x \le 0 \\ ax, & 0 < x \le 3 \\ 0, & \text{elsewhere} \end{cases}$$

where a is a constant.

a. **Sketch the graph of f.**
b. **Calculate the value of the constant, a.**
c. **Determine $\Pr(1 \le X \le 3)$.**
d. **Determine $\Pr(X < 2 \mid X > -1)$.**

THINK | **WRITE**

a. The hybrid function contains three sections. The first graph, $f(x) = -ax$, is a straight line with end points of (0, 0) and (−3, 3a). The second graph is also a straight line and has end points of (0, 0) and (3, 3a). Don't forget to include the $f(x) = 0$ lines for $x > 3$ and $x < -3$.

a. $f(-3) = 3a$ and $f(3) = 3a$

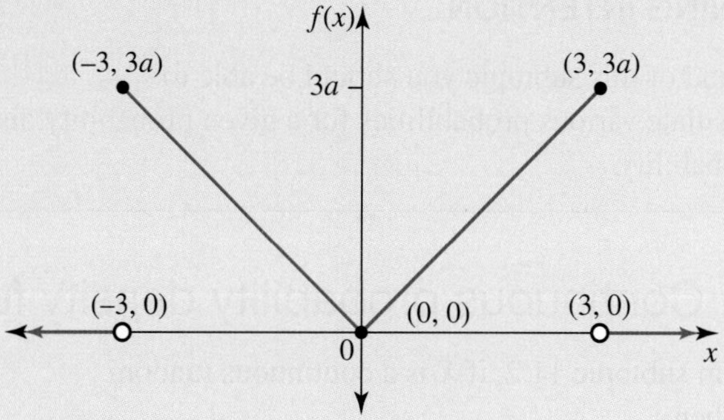

b. Use the fact that $\int_{-3}^{3} f(x)\,dx = 1$ to solve for a.

b. $\int_{-3}^{3} f(x)\,dx = 1$

Using the area of a triangle, we find:

$$\frac{1}{2} \times 3 \times 3a + \frac{1}{2} \times 3 \times 3a = 1$$

$$\frac{9a}{2} + \frac{9a}{2} = 1$$

$$9a = 1$$

$$a = \frac{1}{9}$$

c. Identify the part of the function that the required x-values sit within: the values $1 \le X \le 3$ are within the region where $f(x) = \frac{1}{9}x$.

c. $\Pr(1 \le X \le 3) = \int_{1}^{3} f(x)\,dx$

$$= \int_{1}^{3} \left(\frac{1}{9}x\right) dx$$

$$= \left[\frac{1}{18}x^2\right]_1^3$$

$$= \frac{1}{18}(3)^2 - \frac{1}{18}(1)^2$$

$$= \frac{8}{18}$$

$$= \frac{4}{9}$$

Note: The method of finding the area of a trapezium could also be used.

d. 1. State the rule for the conditional probability.

d. $\Pr(X < 2 \mid X > -1) = \dfrac{\Pr(X < 2 \cap X > -1)}{\Pr(X > -1)}$

$$= \frac{\Pr(-1 < X < 2)}{\Pr(X > -1)}$$

2. Find $\Pr(-1 < X < 2)$. As the interval is across two functions, the interval needs to be split.

$\Pr(-1 < X < 2) = \Pr(-1 < X < 0) + \Pr(0 \le X < 2)$

3. To find the probabilities we need to find the areas under the curve.

$$= \int_{-1}^{0} -\frac{1}{9}x\,dx + \int_{0}^{2} \frac{1}{9}x\,dx$$

$$= -\int_{-1}^{0} \frac{1}{9}x\,dx + \int_{0}^{2} \frac{1}{9}x\,dx$$

4. Anti-differentiate and evaluate after substituting the terminals.

$$= -\left[\frac{1}{18}x^2\right]_{-1}^{0} + \left[\frac{1}{18}x^2\right]_{0}^{2}$$

$$= -\left(\frac{1}{18}(0)^2 - \frac{1}{18}(-1)^2\right) + \frac{1}{18}(2)^2 - \frac{1}{18}(0)^2$$

$$= \frac{1}{18} + \frac{4}{18}$$

$$= \frac{5}{18}$$

5. Find $\Pr(X > -1)$. As the interval is across two functions, the interval needs to be split.

$$\Pr(X > -1) = \Pr(-1 < X < -1) + \Pr(0 \le X \le 3)$$

6. To find the probabilities we need to find the areas under the curve. As $\Pr(0 \le X \le 3)$ covers exactly half the area under the curve, $\Pr(0 \le X \le 3) = \frac{1}{2}$. (The entire area under the curve is always 1 for a probability density function.)

$$= \int_{-1}^{0} -\frac{1}{9}x\,dx + \int_{0}^{3} \frac{1}{9}x\,dx$$

$$= -\int_{-1}^{0} \frac{1}{9}x\,dx + \frac{1}{2}$$

7. Anti-differentiate and evaluate after substituting the terminals.

$$= -\left[\frac{1}{18}x^2\right]_{-1}^{0} + \frac{1}{2}$$

$$= -\left(\frac{1}{18}(0)^2 - \frac{1}{18}(-1)^2\right) + \frac{1}{2}$$

$$= \frac{1}{18} + \frac{9}{18}$$

$$= \frac{10}{18}$$

$$= \frac{5}{9}$$

8. Now substitute into the formula to find $\Pr(X < 2 \mid X > -1)$.

$$\Pr(X < 2 \mid X > -1) = \frac{\Pr(-1 < X < 2)}{\Pr(X > -1)}$$

$$= \frac{5}{18} \div \frac{5}{9}$$

$$= \frac{5}{18} \times \frac{9}{5}$$

$$= \frac{1}{2}$$

11.3 Exercise

Technology free

1. The continuous random variable X has a uniform rectangular probability density function defined by

$$f(x) = \begin{cases} \dfrac{1}{5}, & 1 \le x \le 6 \\ 0, & \text{elsewhere} \end{cases}$$

   a. Sketch the graph of f.
   b. Determine $\Pr(2 \le X \le 5)$.

2. The continuous random variable X has a probability density function given by

$$f(x) = \begin{cases} 4x^3, & 0 \le x \le a \\ 0, & \text{elsewhere} \end{cases}$$

   where a is a constant.

   a. Calculate the value of the constant a.
   b. Sketch the graph of f.
   c. Determine $\Pr(0.5 \le X \le 1)$.

3. **WE3** A continuous random variable, X, has a probability density function, f, defined by

$$f(x) = \begin{cases} k(2+x), & -2 \le x < 0 \\ k(2-x), & 0 \le x \le 2 \\ 0, & \text{elsewhere} \end{cases}$$

   where k is a constant.

   a. Sketch the graph of f.
   b. Show that the value of k is $\dfrac{1}{4}$.
   c. Determine $\Pr(-1 \le X \le 1)$.
   d. Determine $\Pr(X \ge -1 \mid X \le 1)$.

4. The continuous random variable Z has a probability density function given by

$$f(z) = \begin{cases} -z+1, & 0 \le z < 1 \\ z-1, & 1 \le z \le 2 \\ 0, & \text{elsewhere} \end{cases}$$

   a. Sketch the graph of f.
   b. Calculate $\Pr(Z < 0.75)$.
   c. Calculate $\Pr(Z > 0.5)$.

5. Let X be a continuous random variable with a probability density function defined by

$$f(x) = \begin{cases} \dfrac{1}{2}\sin(x), & 0 \le x \le \pi \\ 0, & \text{elsewhere} \end{cases}$$

a. Sketch the graph of f.

b. Calculate $\Pr\left(\frac{\pi}{4} < X < \frac{3\pi}{4}\right)$.

c. Calculate $\Pr\left(X > \frac{\pi}{4} \,\middle|\, X < \frac{3\pi}{4}\right)$.

6. The continuous random variable Z has a probability density function defined by

$$f(z) = \begin{cases} \frac{1}{2}\cos(z), & -\frac{\pi}{2} \le z \le \frac{\pi}{2} \\ 0, & \text{elsewhere} \end{cases}$$

a. Sketch the graph of f and verify that $y = f(z)$ is a probability density function.

b. Calculate $\Pr\left(-\frac{\pi}{6} \le Z \le \frac{\pi}{4}\right)$.

Technology active

7. The amount of petrol sold daily by a busy service station is a uniformly distributed probability density function. A minimum of 18 000 litres and a maximum of 30 000 litres are sold on any given day. The graph of the function is shown. Determine:

a. the value of the constant k

b. the probability that between 20 000 and 25 000 litres of petrol are sold on a given day

c. the probability that as much as 26 000 litres of petrol were sold on a particular day, given that it was known that at least 22 000 litres were sold.

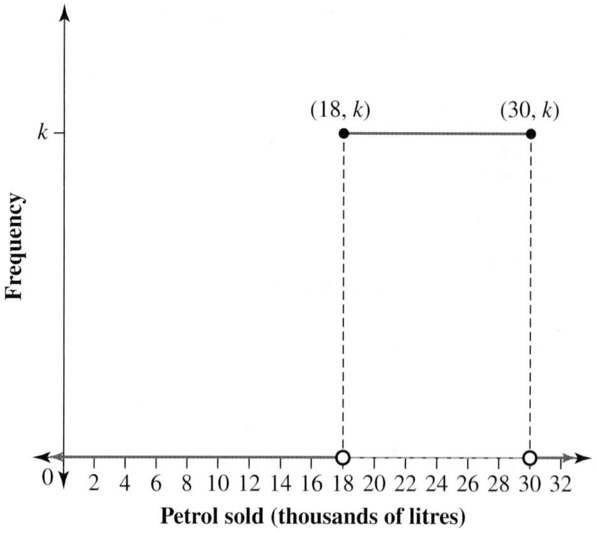

8. The continuous random variable U has a probability density function defined by

$$f(u) = \begin{cases} 1 - \frac{1}{4}(2u - 3u^2), & 0 \le u \le a \\ 0, & \text{elsewhere} \end{cases}$$

where a is a constant. Calculate:

a. the value of the constant a

b. $\Pr(U < 0.75)$

c. $\Pr(0.1 < U < 0.5)$

d. $\Pr(U = 0.8)$.

9. **MC** The continuous random variable X has a probability density function given as

$$f(x) = \begin{cases} 3e^{-3x}, & x \ge 0 \\ 0, & \text{elsewhere} \end{cases}$$

$\Pr(0 \le X \le 1)$, correct to 4 decimal places, is:

A. 0.3167 B. 0.9502 C. 1.8964 D. 0.0498 E. 0.6833

10. **MC** The continuous random variable X has a probability density function given as

$$f(x) = \begin{cases} 3e^{-3x}, & x \geq 0 \\ 0, & \text{elsewhere} \end{cases}$$

$\Pr(X > 2)$, correct to 4 decimal places, is:

A. 1 **B.** 0.0498 **C.** 0.0025 **D.** 0.0008 **E.** 0.4060

11. The continuous random variable X has a probability density function defined by

$$f(x) = \begin{cases} \log_e(x^2), & 1 \leq x \leq a \\ 0, & \text{elsewhere} \end{cases}$$

Calculate, correct to 4 decimal places:

a. the value of the constant a

b. $\Pr(1.25 \leq X \leq 2)$.

12. The continuous random variable X has a probability density function defined by

$$f(x) = \begin{cases} \frac{3}{8}x^2, & 0 \leq x \leq 2 \\ 0, & \text{elsewhere} \end{cases}$$

Calculate:

a. $\Pr(X > 1.2)$

b. $\Pr(X > 1 \mid X > 0.5)$

c. the value of n such that $\Pr(X \leq n) = 0.75$.

13. The graph of the probability function

$$f(z) = \frac{1}{\pi(z^2 + 1)}$$

is shown.

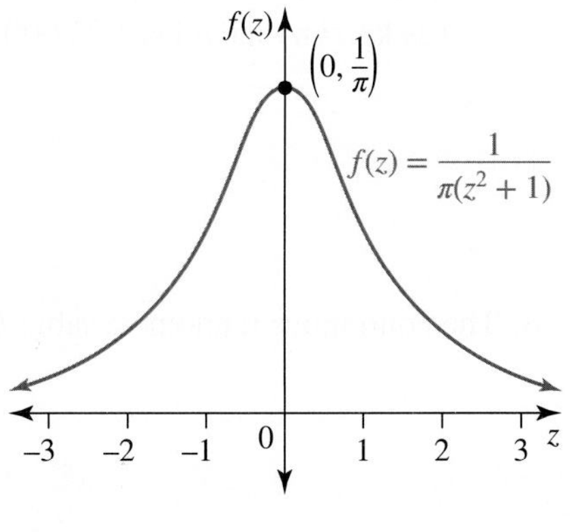

a. Calculate, correct to 4 decimal places, $\Pr(-0.25 < Z < 0.25)$.

b. Suppose another probability density function is defined as

$$f(x) = \begin{cases} \frac{1}{x^2 + 1}, & -a \leq x \leq a \\ 0, & \text{elsewhere} \end{cases}$$

Determine the value of the constant a. Give your answer correct to 2 decimal places.

14. The continuous random variable Z has a probability density function defined by

$$f(z) = \begin{cases} \frac{1}{2z}, & 1 \leq z \leq e^2 \\ 0, & \text{elsewhere} \end{cases}$$

a. Sketch the graph of f and shade the area that represents $\int_1^{e^2} f(z)\, dz$.

b. Determine $\int_1^{e^2} f(z)\, dz$. Explain your result.

The continuous random variable U has a probability density function defined by

$$f(u) = \begin{cases} e^{4u}, & 0 \le u \le a \\ 0, & \text{elsewhere} \end{cases}$$

c. Sketch the graph of f and shade the area that represents $\int_0^a f(u)\,du$, where a is a constant.

d. Determine the exact value of a.

11.3 Exam questions

Question 1 (1 mark) TECH-ACTIVE

Source: *VCE 2016 Mathematical Methods Exam 2, Section A, Q18; © VCAA.*

MC The continuous random variable, X, has a probability density function given by

$$f(x) = \begin{cases} \frac{1}{4}\cos\left(\frac{x}{2}\right) & 3\pi \le x \le 5\pi \\ 0 & \text{elsewhere} \end{cases}$$

The value of a such that $\Pr(X < a) = \frac{\sqrt{3}+2}{4}$ is

A. $\frac{19\pi}{6}$ **B.** $\frac{14\pi}{3}$ **C.** $\frac{10\pi}{3}$ **D.** $\frac{29\pi}{6}$ **E.** $\frac{17\pi}{3}$

Question 2 (4 marks) TECH-FREE

Source: *Adapted from VCE 2016 Mathematical Methods Exam 1, Q8; © VCAA.*

Let X be a continuous random variable with probability density function

$$f(x) = \begin{cases} -4x\log_e(x) & 0 < x \le 1 \\ 0 & \text{elsewhere} \end{cases}$$

Part of the graph of f is shown below. The graph has a turning point at $x = \frac{1}{e}$.

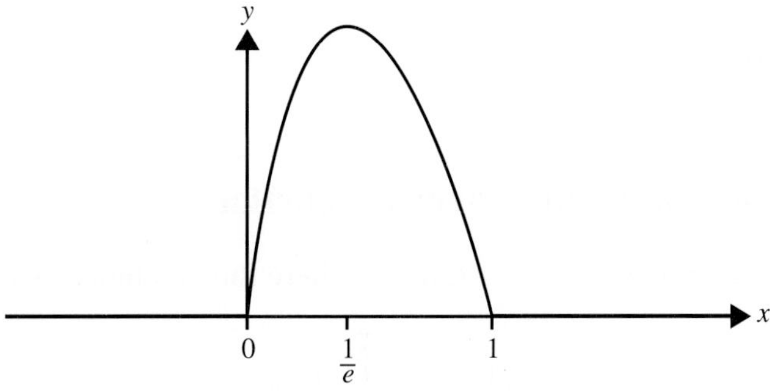

a. Show by differentiation that

$$\frac{x^k}{k^2}\left(k\log_e(x) - 1\right)$$

is an antiderivative of $x^{k-1}\log_e(x)$, where k is a positive real number. **(2 marks)**

b. Calculate $\Pr\left(X > \frac{1}{e}\right)$. **(2 marks)**

Question 3 (7 marks) TECH-ACTIVE

The continuous random variable Z has a probability density function defined by

$$f(z)=\begin{cases} e^{-\frac{z}{3}}, & 0 \le z \le a \\ 0, & \text{elsewhere} \end{cases}$$

where a is a constant. Determine:

a. the value of the constant a such that $\int_0^a f(z)\,dz = 1$ **(2 marks)**

b. $\Pr(0 < Z < 0.7)$, correct to 4 decimal places **(1 mark)**

c. $\Pr(Z < 0.7 \mid Z > 0.2)$, correct to 4 decimal places **(2 marks)**

d. the value of b, correct to 2 decimal places, such that $\Pr(Z \le b) = 0.54$. **(2 marks)**

More exam questions are available online.

11.4 Measures of centre and spread

LEARNING INTENTION

At the end of this subtopic you should be able to:

- calculate the mean of a probability density function
- calculate percentiles of a probability density function
- determine the variance and standard deviation of a probability density function.

11.4.1 Measures of central tendency

The mean

Remember that for a discrete random variable,

$$E(X) = \mu = \sum_{x=1}^{x=n} x_n \Pr(X = x_n)$$

This definition can also be applied to a continuous random variable.

We define $E(X) = \mu = \int_{-\infty}^{\infty} x f(x)\,dx$.

The mean of a probability density function

If $f(x) = 0$ everywhere except for $x \in [a, b]$, where the function is defined, then

$$E(X) = \mu = \int_a^b x\,f(x)dx$$

and the function of X, $g(x)$, has a mean defined by

$$E(g(x)) = \mu = \int_{-\infty}^{\infty} g(x)f(x)dx.$$

Consider the continuous random variable, X, which has a probability density function defined by

$$f(x) = \begin{cases} x^2, & 0 \le x \le 1 \\ 0, & \text{elsewhere} \end{cases}$$

For this function,

$$\begin{aligned} \mathrm{E}(X) = \mu &= \int_0^1 x f(x)\,dx \\ &= \int_0^1 x(x^2)\,dx \\ &= \int_0^1 x^3\,dx \\ &= \left[\frac{x^4}{4}\right]_0^1 \\ &= \frac{1^4}{4} - 0 \\ &= \frac{1}{4} \end{aligned}$$

Similarly, if the continuous random variable X has a probability density function of

$$f(x) = \begin{cases} 7e^{-7x}, & x \ge 0 \\ 0, & \text{elsewhere} \end{cases}$$

then

$$\begin{aligned} \mathrm{E}(X) = \mu &= \int_0^\infty x f(x)\,dx \\ &= \lim_{k \to \infty} \int_0^k 7xe^{-7x}\,dx \\ &= 0.1429 \end{aligned}$$

where CAS technology is required to determine the integral.

If we again consider

$$f(x) = \begin{cases} x^2, & 0 \le x \le 1 \\ 0, & \text{elsewhere} \end{cases}$$

then

$$\begin{aligned}
E(X)^2 &= \int_0^1 x^2 f(x)dx \\
&= \int_0^1 x^4 dx \\
&= \left[\frac{x^5}{5}\right]_0^1 \\
&= \frac{1^5}{5} - 0 \\
&= \frac{1}{5}
\end{aligned}$$

This definition is important when we investigate the variance of a continuous random variable.

Percentiles

If a person wishes to know the value below which a certain percentage of the values lie, this is equivalent to working out a percentile. For example, the 60th **percentile** is the value below which 60% of the data lie.

Percentiles

The percentile value, p, is given by

$$\Pr(X \le p) = q$$

or

$$\int_{-\infty}^{p} f(x)dx = q$$

where q is the percentile amount, such as 0.7 for the 70th percentile.

Other percentiles that are frequently calculated are the 25th percentile or lower **quartile**, Q_1, and the 75th percentile or upper quartile, Q_3.

The interquartile range

The **interquartile range** is calculated as:

$$\text{IQR} = Q_3 - Q_1$$

Consider the continuous random variable X, which has a probability density function of

$$f(x) = \begin{cases} \dfrac{x^3}{4}, & 0 \le x \le 2 \\ 0, & \text{elsewhere} \end{cases}$$

The 50th percentile is given by $\Pr(0 \leq x \leq p) = 0.5$:

$$\int_0^p \frac{x^3}{4} dx = 0.5$$

$$\left[\frac{x^4}{16}\right]_0^p = \frac{1}{2}$$

$$\frac{p^4}{16} - 0 = \frac{1}{2}$$

$$p^4 = 8$$

$$p = \pm\sqrt[4]{8}$$

$$p = 1.6818 \ (0 \leq m \leq 2)$$

To find the lower quartile, we make the area under the curve equal to 0.25. Thus, the lower quartile is given by $\Pr(0 \leq x \leq a) = 0.25$:

$$\int_0^a \frac{x^3}{4} dx = 0.25$$

$$\left[\frac{x^4}{16}\right]_0^a = \frac{1}{4}$$

$$\frac{a^4}{16} - 0 = \frac{1}{4}$$

$$a^4 = 4$$

$$a = \pm\sqrt[4]{4}$$

$$a = Q_1 = 1.4142 \ (0 \leq a \leq m)$$

Similarly, to find the upper quartile, we make the area under the curve equal to 0.75. Thus, the upper quartile is given by $\Pr(0 \leq x \leq n) = 0.75$:

$$\int_0^n \frac{x^3}{4} dx = 0.75$$

$$\left[\frac{x^4}{16}\right]_0^n = \frac{3}{4}$$

$$\frac{n^4}{16} - 0 = \frac{3}{4}$$

$$n^4 = 12$$

$$n = \pm\sqrt[4]{12}$$

$$n = Q_3 = 1.8612 \ (m \leq x \leq 2)$$

So the interquartile range is given by

$$\begin{aligned} Q_3 - Q_1 &= 1.8612 - 1.4142 \\ &= 0.4470 \end{aligned}$$

WORKED EXAMPLE 4 The mean of a probability density function

A continuous random variable, Y, has a probability density function, f, defined by

$$f(y) = \begin{cases} ky, & 0 \le y \le 1 \\ 0, & \text{elsewhere} \end{cases}$$

where k is a constant.

a. **Sketch the graph of f.**

b. **Calculate the value of the constant k.**

c. **Calculate the mean of y.**

THINK

a. The graph $f(y) = ky$ is a straight line with end points at $(0, 0)$ and $(1, k)$. Remember to include the lines $f(y) = 0$ for $y > 1$ and $y < 0$.

WRITE

a.

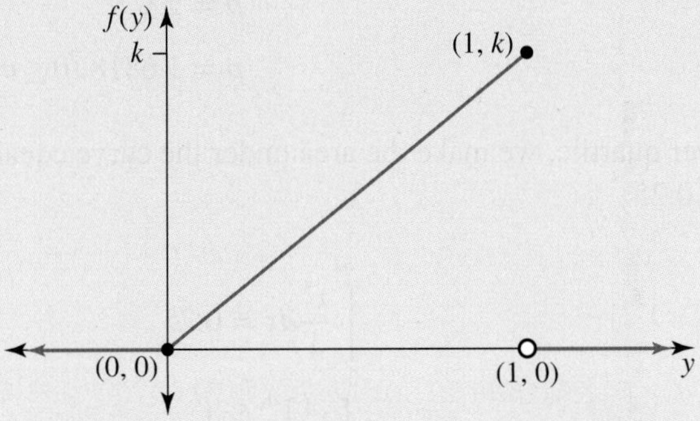

THINK

b. Solve $\int_0^1 ky\, dy = 1$ to find the value of k.

WRITE

b.

$$\int_0^1 ky\, dy = 1$$

$$k\int_0^1 y\, dy = 1$$

$$k\left[\frac{y^2}{2}\right]_0^1 = 1$$

$$\frac{k(1)^2}{2} - 0 = 1$$

$$\frac{k}{2} = 1$$

$$k = 2$$

Using the area of a triangle also enables you to find the value of k.

$$\frac{1}{2} \times 1 \times k = 1$$

$$\frac{k}{2} = 1$$

$$k = 2$$

THINK

c. 1. State the rule for the mean.

WRITE

c. i. $\mu = \int_0^1 y\,(2y)dy$

$$= \int_0^1 2y^2 dy$$

2. Anti-differentiate and simplify.

$$= \left[\frac{2y^3}{3}\right]_0^1$$

$$= \frac{2(1)^3}{3} - 0$$

$$= \frac{2}{3}$$

TI | THINK	DISPLAY/WRITE	CASIO | THINK	DISPLAY/WRITE
c. 1. On a Calculator page, complete the entry line as: $\int_0^1 (y \times 2y)\, dy$ then press ENTER. *Note:* The integral template can be found by pressing the 'Templates' button.	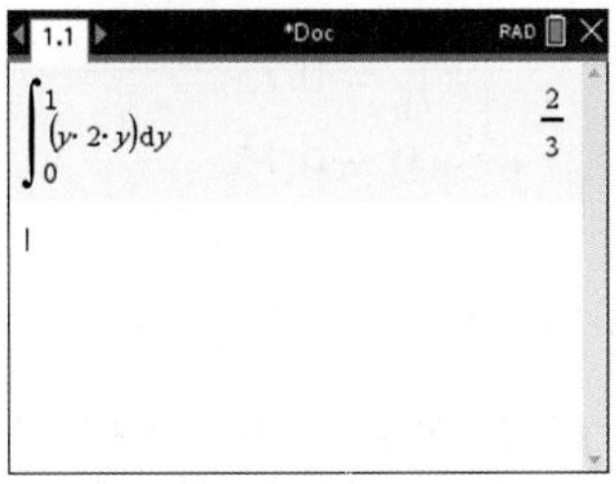	c. 1. On a Main screen, select: • Interactive • Calculation • $\int$ Select 'Definite', then complete the fields as: Expression: $y \times 2y$ Variable: y Lower: 0 Upper: 1 Select OK.	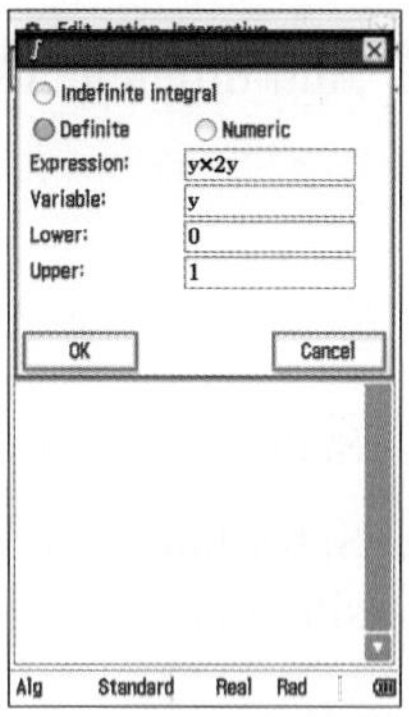 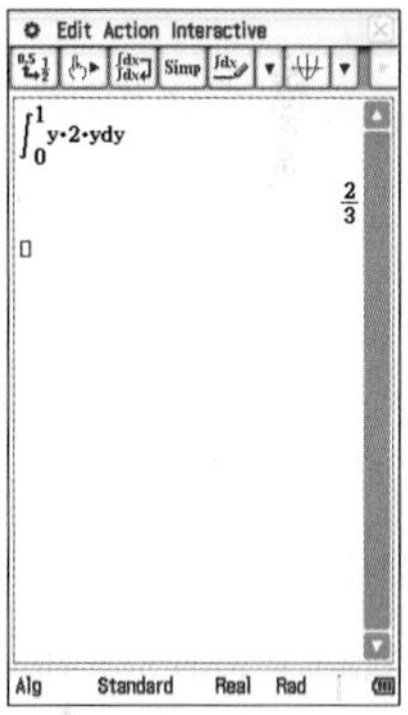
2. The answer appears on the screen.	$\int_0^1 (2y^2)\, dy = \frac{2}{3}$	2. The answer appears on the screen.	$\int_0^1 (2y^2)\, dy = \frac{2}{3}$

WORKED EXAMPLE 5 The interquartile range

For the function in Worked example 4, $f(y) = \begin{cases} 2y, & 0 \le y \le 1 \\ 0, & \text{elsewhere} \end{cases}$, calculate, correct to 4 decimal places:

a. the lower and upper quartiles of Y **b. the inter-quartile range of Y.**

THINK	WRITE
a. 1. State the rule for the lower quartile, Q_1.	a. $\int_0^a f(y)\, dy = 0.25$ $\int_0^a 2y\, dy = 0.25$
2. Anti-differentiate and solve for Q_1.	$[y^2]_0^a = 0.25$ $a^2 - 0 = 0.25$

$a = \pm\sqrt{0.25}$

$a = Q_1 = 0.5 \ (0 \le y \le 1)$

3. State the rule for the upper quartile, Q_3.

$$\int_0^n f(y)\,dy = 0.75$$

$$\int_0^n 2y\,dy = 0.75$$

4. Anti-differentiate and solve for Q_3.

$$\left[y^2\right]_0^n = 0.75$$

$$n^2 - 0 = 0.75$$

$$n = \pm\sqrt{0.75}$$

$$n = Q_3 = 0.8660 \ (0 \le y \le 1)$$

b. 1. State the rule for the interquartile range.

2. Substitute the appropriate values and simplify.

b. $\text{IQR} = Q_3 - Q_1$

$= 0.8660 - 0.5$

$= 0.3660$

on Resources

Interactivities Mean (int-6435)

Median and percentiles (int-6436)

11.4.2 Measures of spread

Variance, standard deviation and range

The variance and standard deviation are important measures of spread. From previous topics we know that

$$\text{Var}(X) = \text{E}(X^2) - [\text{E}(X)]^2$$

For continuous probability functions,

$$\begin{aligned}
\text{Var}(X) &= \int_{-\infty}^{\infty} (x-\mu)^2 f(x)dx \\
&= \int_{-\infty}^{\infty} \left(x^2 - 2x\mu + \mu^2\right) f(x)dx \\
&= \int_{-\infty}^{\infty} x^2 f(x)dx - \int_{-\infty}^{\infty} 2xf(x)\mu\ dx + \int_{-\infty}^{\infty} \mu^2 f(x)dx \\
&= \text{E}\left(X^2\right) - 2\mu \int_{-\infty}^{\infty} xf(x)dx + \mu^2 \int_{-\infty}^{\infty} 1f(x)dx \\
&= \text{E}\left(X^2\right) - 2\mu \times \text{E}(X) + \mu^2 \\
&= \text{E}\left(X^2\right) - 2\mu^2 + \mu^2 \\
&= \text{E}\left(X^2\right) - \mu^2 \\
&= \text{E}\left(X^2\right) - [\text{E}(X)]^2
\end{aligned}$$

Two important facts were used in this proof: $\int_{-\infty}^{\infty} f(x)\,dx = 1$ and $\int_{-\infty}^{\infty} xf(x)dx = \mu = \text{E}(X)$.

The variance and standard deviation of a probability distribution function

If $f(x) = 0$ everywhere except for $x \in [a, b]$, where the function is defined, then

$$\begin{aligned}\text{Var}(X) &= \text{E}(X^2) - [\text{E}(X)]^2 \\ &= \int_a^b x^2 f(x)dx - \left[\int_a^b x\,f(x)dx\right]^2\end{aligned}$$

and

$$\text{SD}(X) = \sqrt{\text{Var}(X)}.$$

The range is calculated as the highest value minus the lowest value, so for the probability density function given by $f(x) = \begin{cases} \frac{1}{5}, & 1 \le x \le 6 \\ 0, & \text{elsewhere} \end{cases}$, the highest possible x-value is 6 and the lowest is 1.

Therefore, the range for this function $= 6 - 1$
$= 5$

WORKED EXAMPLE 6 Variance and standard deviation

For a continuous random variable, X, with a probability density function, f, defined by

$$f(x) = \begin{cases} \frac{1}{2}x + 2, & -4 \le x \le -2 \\ 0, & \text{elsewhere} \end{cases}$$

find:

a. **the mean**

b. **the variance**

c. **the standard deviation, correct to 4 decimal places.**

THINK	WRITE
a. 1. State the rule for the mean and simplify.	a. $\mu = \int_{-4}^{-2} xf(x)\,dx$ $= \int_{-4}^{-2} x\left(\frac{1}{2}x + 2\right) dx$ $= \int_{-4}^{-2} \left(\frac{1}{2}x^2 + 2x\right) dx$

2. Anti-differentiate and evaluate.

$$= \left[\frac{1}{6}x^3 + x^2\right]_{-4}^{-2}$$

$$= \left(\frac{1}{6}(-2)^3 + (-2)^2\right) - \left(\frac{1}{6}(-4)^3 + (-4)^2\right)$$

$$= \frac{-4}{3} + 4 + \frac{32}{3} - 16$$

$$= -2\frac{2}{3}$$

b. 1. Write the rule for variance.

2. Find $E(X^2)$ first.

3. Substitute $E(X)$ and $E(X^2)$ into the rule for variance.

b. $\text{Var}(X)^2 = E(X^2) - [E(X)]^2$

$$E(X^2) = \int_a^b x^2 f(x)dx$$

$$= \int_{-4}^{-2} x^2\left(\frac{1}{2}x + 2\right)dx$$

$$= \int_{-4}^{-2}\left(\frac{1}{2}x^3 + 2x^2\right)dx$$

$$= \left[\frac{1}{8}x^4 + \frac{2}{3}x^3\right]_{-4}^{-2}$$

$$= \left(\frac{1}{8}(-2)^4 + \frac{2}{3}(-2)^3\right) - \left(\frac{1}{8}(-4)^4 + \frac{2}{3}(-4)^3\right)$$

$$= 2 - \frac{16}{3} - 32 + \frac{128}{3}$$

$$= -30 + \frac{112}{3}$$

$$= \frac{22}{3}$$

$$\text{Var}(X) = E(X^2) - [E(X)]^2$$

$$= \frac{22}{3} - \left(-\frac{8}{3}\right)^2$$

$$= \frac{22}{3} - \frac{64}{9}$$

$$= \frac{66}{9} - \frac{64}{9}$$

$$= \frac{2}{9}$$

c. 1. Write the rule for standard deviation.

2. Substitute the variance into the rule and evaluate.

c. $\text{SD}(X) = \sqrt{\text{Var}(X)}$

$$= \sqrt{\frac{2}{9}}$$

$$= 0.4714$$

on Resources

Interactivity Variance, standard deviation and range (int-6437)

11.4 Exercise

Students, these questions are even better in jacPLUS

Receive immediate feedback and access sample responses

Access additional questions

Track your results and progress

Find all this and MORE in jacPLUS

Technology free

1. Calculate the mean for each of the following probability density functions.

   a. $f(x) = \begin{cases} -\frac{1}{2}x + 1, & 0 \leq x \leq 2 \\ 0, & \text{elsewhere} \end{cases}$

   b. $f(x) = \begin{cases} \frac{1}{x^2}, & x \geq 1 \\ 0, & \text{elsewhere} \end{cases}$

2. Calculate the mean and variance for each of the following probability density functions.

   a. $f(x) = \begin{cases} 4x, & 0 \leq x \leq 0.5 \\ 0, & \text{elsewhere} \end{cases}$

   b. $f(x) = \begin{cases} 0.5, & 1 \leq x \leq 3 \\ 0, & \text{elsewhere} \end{cases}$

3. **WE4** The continuous random variable Z has a probability density function of

$$f(z) = \begin{cases} \frac{1}{\sqrt{z}}, & 1 \leq z \leq a \\ 0, & \text{elsewhere} \end{cases}$$

   where a is a constant.

   a. Calculate the value of the constant a.

   b. Calculate the mean of Z

4. The continuous random variable, Z, has a probability density function of

$$f(z) = \begin{cases} \frac{3}{z^2}, & 1 \leq z \leq a \\ 0, & \text{elsewhere} \end{cases}$$

   where a is a constant.

   a. Show that the value of a is $\frac{3}{2}$.

   b. Calculate the mean value of f.

   c. Calculate the value of p such that $\Pr(1 \leq Z \leq p) = 0.5$

5. Consider the continuous random variable X with a probability density function of

$$f(x) = \begin{cases} h(2-x), & 0 \le x \le 2 \\ h(x-2), & 2 < x \le 4 \\ 0, & \text{elsewhere} \end{cases}$$

where h is a constant. Calculate:

a. the value of the constant h
b. $\text{E}(X)$
c. $\text{Var}(X)$.

Technology active

6. Let X be a continuous random variable with a probability density function of

$$f(x) = \begin{cases} \dfrac{1}{2\sqrt{x}}, & 0 \le x \le 1 \\ 0, & \text{elsewhere} \end{cases}$$

a. Prove that f is a probability density function.
b. Calculate $\text{E}(X)$.

7. The continuous random variable, Y, has a probability density function of

$$f(y) = \begin{cases} \sqrt{y}, & 0 \le y \le a \\ 0, & \text{elsewhere} \end{cases}$$

where a is a constant.
Calculate, correct to 4 decimal places:

a. the value of the constant a
b. $\text{E}(Y)$
c. the value of k such that $\Pr(0 \le y \le k) = 0.6$

8. WE6 For the continuous random variable Z, the probability density function is

$$f(z) = \begin{cases} 2\log_e(2z), & \dfrac{1}{2} \le z \le \dfrac{e}{2} \\ 0, & \text{elsewhere} \end{cases}$$

Determine the mean, variance and standard deviation, correct to 4 decimal places.

9. The function

$$f(x) = \begin{cases} 3e^{-3x}, & x \ge 0 \\ 0, & \text{elsewhere} \end{cases}$$

defines the probability density function for the continuous random variable, X. Determine the mean, variance and standard deviation of X.

10. The time in minutes that an individual must wait in line to be served at the local bank branch is defined by

$$f(t) = 2e^{-2t}, \ t \ge 0$$

where T is a continuous random variable.

a. Calculate the mean waiting time for a customer in the queue, correct to 1 decimal place.
b. Calculate the standard deviation for the waiting time in the queue, correct to 1 decimal place.

11. WE5 The continuous random variable Y has a probability density function defined by

$$f(y) = \begin{cases} \dfrac{y^2}{3}, & 0 \le y \le \sqrt[3]{9} \\ 0, & \text{elsewhere} \end{cases}$$

Calculate, correct to 4 decimal places:

a. the lower and upper quartiles of Y

b. the interquartile range of Y.

12. The continuous random variable Z has a probability density function defined by

$$f(y) = \begin{cases} \dfrac{a}{z}, & 1 \le z \le 8 \\ 0, & \text{elsewhere} \end{cases}$$

where a is a constant.

a. Calculate the value, correct to 4 decimal places, of the constant a.
b. Calculate E(Z) correct to 4 decimal places.
c. Calculate Var(Z) and SD(Z), correct to 4 decimal places.
d. Determine the interquartile range for Z, correct to 4 decimal places.
e. Determine the range for Z.

13. X is a continuous random variable. The graph of the probability density function

$$f(x) = \frac{1}{\pi}(\sin(2x) + 1) \text{ for } 0 \le x \le \pi$$

is shown.

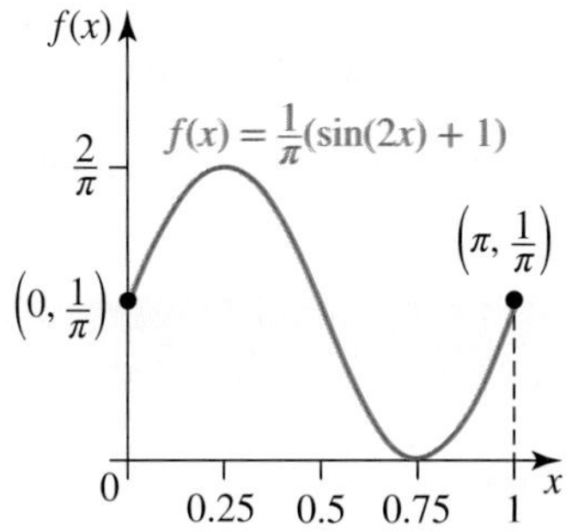

a. Show that $f(x)$ is a probability density function.
b. Calculate E(X) correct to 4 decimal places.
c. Calculate, correct to 4 decimal places:

i. Var(X)

ii. SD(X).

14. The continuous random variable Y has a probability density function

$$f(y) = \begin{cases} 0.2 \log_e \left(\dfrac{y}{2}\right), & 2 \le y \le 7.9344 \\ 0, & \text{elsewhere} \end{cases}$$

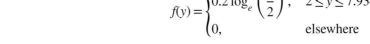

a. Verify that f is a probability density function.
b. Calculate E(Y) correct to 4 decimal places.
c. Calculate Var(Y) and SD(Y) correct to 4 decimal places.
d. State the range.

15. The continuous random variable Z has a probability density function

$$f(z) = \begin{cases} \sqrt{z-1}, & 1 \le z \le a \\ 0, & \text{elsewhere} \end{cases}$$

where a is a constant.

a. Calculate the value of the constant a correct to 4 decimal places.
b. Determine, correct to 4 decimal places:

i. E(Z)

ii. E(Z^2)

iii. Var(Z)

iv. SD(Z).

16. The continuous random variable X has a probability density function defined by

$$f(x) = \begin{cases} ax - bx^2, & 0 \le x \le 2 \\ 0, & \text{elsewhere} \end{cases}$$

Determine the values of the constants a and b if $E(X) = 1$.

17. a. State the derivative of $\sqrt{4 - x^2}$.

b. Hence, determine the mean value of the probability density function defined by

$$f(x) = \begin{cases} \dfrac{3}{\pi\sqrt{4 - x^2}}, & 0 \le x \le \sqrt{3} \\ 0, & \text{elsewhere} \end{cases}$$

18. Consider the continuous random variable X with a probability density function of

$$f(x) = \begin{cases} k, & a \le x \le b \\ 0, & \text{elsewhere} \end{cases}$$

where a, b and k are positive constants.

a. Sketch the graph of the function f.

b. Show that $k = \dfrac{1}{b - a}$.

c. Determine $E(X)$ in terms of a and b.

d. Determine $\text{Var}(X)$ in terms of a and b.

11.4 Exam questions

Question 1 (3 marks) TECH-FREE

Source: *VCE 2021 Mathematical Methods Exam 1, Q7; (C) VCAA.*

A random variable X had the probability density function f given by

$$f(x) = \begin{cases} \dfrac{k}{x^2}, & 1 \le x \le 2 \\ 0, & \text{elsewhere} \end{cases}$$

where k is a positive real number.

a. Show that $k = 2$. **(1 mark)**

b. Find $E(X)$. **(2 marks)**

Question 2 (1 mark) TECH-ACTIVE

Source: *VCE 2015 Mathematical Methods (CAS) Exam 2, Section 1, Q9; © VCAA.*

MC The graph of the probability density function of a continuous random variable, X, is shown below.

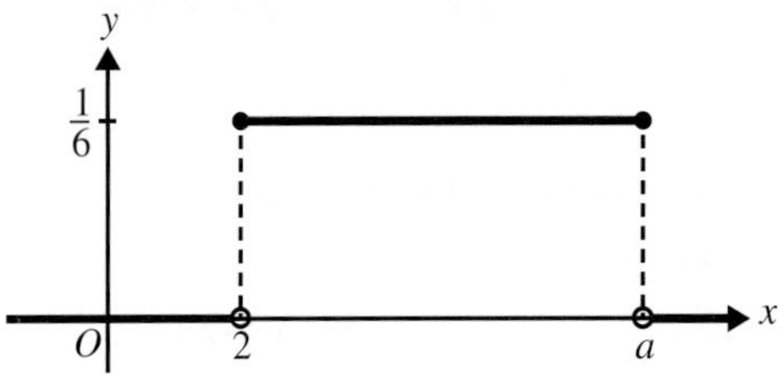

If $a > 2$, then $E(X)$ is equal to

A. 8 B. 5 C. 4 D. 3 E. 2

Question 3 (2 marks) TECH-FREE

Source: *VCE 2013 Mathematical Methods (CAS) Exam 1, Q8; © VCAA.*

A continuous random variable, X, has a probability density function

$$f(x) = \begin{cases} \frac{\pi}{4}\cos\left(\frac{\pi x}{4}\right) & \text{if } x \in [0,\ 2] \\ 0 & \text{otherwise} \end{cases}$$

Given that $\frac{d}{dx}\left(x\sin\left(\frac{\pi x}{4}\right)\right) = \frac{\pi x}{4}\cos\left(\frac{\pi x}{4}\right) + \sin\left(\frac{\pi x}{4}\right)$, find $\text{E}(X)$.

More exam questions are available online.

11.5 Linear transformations

LEARNING INTENTION

At the end of this subtopic you should be able to:

- calculate the mean and variance when linear transformations are applied to a probability density function.

11.5.1 Transformations

Sometimes it is necessary to apply transformations to a continuous random variable. A transformation is a change that is applied to the random variable. The change may consist of one or more operations that may involve adding or subtracting a constant or multiplying or dividing the variable by a constant.

Suppose a linear transformation is applied to the continuous random variable X to create a new continuous random variable, Y. For instance

$$Y = aX + b$$

It can be shown that $\text{E}(Y) = \text{E}(aX + b) = a\text{E}(X) + b$

and $\text{Var}(Y) = \text{Var}(aX + b) = a^2\text{Var}(X)$.

First let us show that $\text{E}(Y) = \text{E}(aX + b) = a\text{E}(X) + b$.

Since $\text{E}(X) = \int_{-\infty}^{\infty} xf(x)dx$,

then $\text{E}(aX + b) = \int_{-\infty}^{\infty} (ax + b)f(x)dx$.

Using the distributive law, it can be shown that this is equal to

$$\begin{aligned}\text{E}(aX + b) &= \int_{-\infty}^{\infty} axf(x)\,dx + \int_{-\infty}^{\infty} bf(x)\,dx \\ &= a\int_{-\infty}^{\infty} xf(x)\,dx + b\int_{-\infty}^{\infty} f(x)\,dx\end{aligned}$$

But $E(X)=\int_{-\infty}^{\infty} xf(x)dx$, so

$$E(aX+b)=aE(X)+b\int_{-\infty}^{\infty} f(x)\,dx$$

Also, $\int_{-\infty}^{\infty} f(x)dx=1$, so $E(aX+b)=aE(X)+b$.

Also note that $E(aX)=aE(X)$ and $E(b)=b$.

Now let us show that $\text{Var}(Y)=\text{Var}(aX+b)=a^2\text{Var}(X)$.

Since $\text{Var}(X)=E(X^2)-[E(X)]^2$,

$$\begin{aligned}\text{Var}(aX+b) &= E((aX+b)^2)-[E(aX+b)]^2\\ &=\int_{-\infty}^{\infty}(ax+b)^2f(x)dx-(aE(X)+b)^2\\ &=\int_{-\infty}^{\infty}(a^2x^2+2abx+b^2)f(x)dx-[a^2[E(X)]^2+2abE(X)+b^2]\end{aligned}$$

Using the distributive law to separate the first integral, we have

$$\begin{aligned}\text{Var}(aX+b) &= \int_{-\infty}^{\infty}a^2x^2f(x)dx+\int_{-\infty}^{\infty}2abxf(x)dx+\int_{-\infty}^{\infty}b^2f(x)dx-a^2[E(X)]^2-2abE(X)-b^2\\ &=a^2\int_{-\infty}^{\infty}x^2f(x)dx+2ab\int_{-\infty}^{\infty}xf(x)dx+b^2\int_{-\infty}^{\infty}f(x)dx-a^2[E(X)]^2-2abE(X)-b^2\end{aligned}$$

But $E(X)=\int_{-\infty}^{\infty}xf(x)\,dx$, $E(X^2)=\int_{-\infty}^{\infty}x^2f(x)\,dx$ and $\int_{-\infty}^{\infty}f(x)\,dx=1$ for a probability density function. Thus,

$$\begin{aligned}\text{Var}(aX+b) &= a^2E(X^2)+2abE(X)+b^2-a^2[E(X)]^2-2abE(X)-b^2\\ &=a^2E(X^2)-a^2[E(X)]^2\\ &=a^2(E(X)^2-[E(X)]^2)\\ &=a^2\text{Var}(X)\end{aligned}$$

Thus, we can calculate the mean and variance of linear transformations as follows.

Mean and variance of linear transformations

$$\mathbf{E(aX+b)=aE(X)+b}$$

and

$$\mathbf{Var(aX+b)=a^2Var(X)}$$

WORKED EXAMPLE 7 Mean and variance of linear transformations

A continuous random variable, X, has a mean of 3 and a variance of 2. Determine:

a. $\mathrm{E}(2X+1)$ **b.** $\mathrm{Var}(2X+1)$ **c.** $\mathrm{E}(X^2)$
d. $\mathrm{E}(3X^2)$ **e.** $\mathrm{E}(X^2-5)$.

THINK	WRITE
a. Use $\mathrm{E}(aX+b)=a\mathrm{E}(X)+b$ to find $\mathrm{E}(2X+1)$.	a. $\mathrm{E}(2X+1)=2\mathrm{E}(X)+1$ $=2(3)+1$ $=7$
b. Use $\mathrm{Var}(aX+b)=a^2\mathrm{Var}(X)$ to find $\mathrm{Var}(2X+1)$.	b. $\mathrm{Var}(2X+1)=2^2\mathrm{Var}(X)$ $=4\times 2$ $=8$
c. Use $\mathrm{Var}(X)=\mathrm{E}(X^2)-[\mathrm{E}(X)]^2$ to find $\mathrm{E}(X^2)$.	c. $\mathrm{Var}(X)=\mathrm{E}(X^2)-[\mathrm{E}(X)]^2$ $2=\mathrm{E}(X^2)-3^2$ $2=\mathrm{E}(X^2)-9$ $\mathrm{E}(X^2)=11$
d. Use $\mathrm{E}(aX^2)=a\mathrm{E}(X^2)$ to find $\mathrm{E}(3X^2)$.	d. $\mathrm{E}(3X^2)=3\mathrm{E}(X^2)$ $=3\times 11$ $=33$
e. Use $\mathrm{E}(aX^2+b)=a\mathrm{E}(X^2)+b$ to find $\mathrm{E}(X^2-5)$.	e. $\mathrm{E}(X^2-5)=\mathrm{E}(X)^2-5$ $=11-5$ $=6$

11.5 Exercise

Technology free

1. WE7 A continuous random variable, Y, has a mean of 4 and a variance of 3. Determine:
   a. $\mathrm{E}(2Y-3)$
   b. $\mathrm{Var}(2Y-3)$
   c. $\mathrm{E}(Y^2)$
   d. $\mathrm{E}(Y(Y-1))$.
2. For a continuous random variable Z, where $\mathrm{E}(Z)=5$ and $\mathrm{Var}(Z)=2$, determine:
   a. $\mathrm{E}(3Z-2)$
   b. $\mathrm{Var}(3Z-2)$
   c. $\mathrm{E}(Z^2)$
   d. $\mathrm{E}\left(\frac{1}{3}Z^2-1\right)$.

3. The continuous random variable X has a probability density function defined by

$$f(x) = \begin{cases} -kx, & -2 \le x \le 0 \\ kx, & 0 < x \le 2 \\ 0, & \text{elsewhere} \end{cases}$$

where k is a constant. The graph of the function is shown.

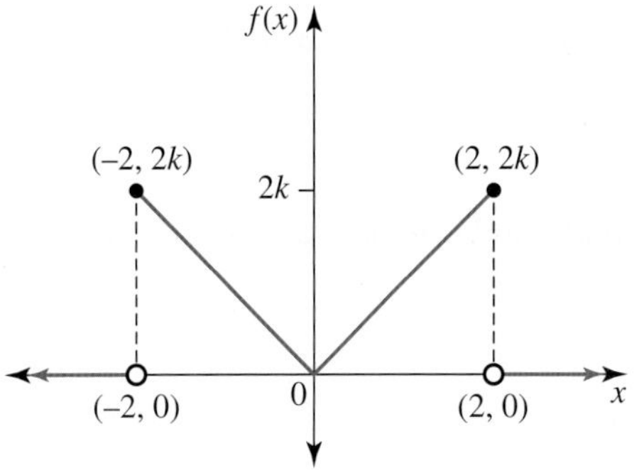

a. Calculate the value of the constant k.
b. Determine $E(X)$ and $\text{Var}(X)$.
c. Evaluate $E(5X + 3)$ and $\text{Var}(5X + 3)$.
d. Evaluate $E((3X - 2)^2)$.

4. The probability density function for the continuous random variable X is

$$f(x) = \begin{cases} mx(2 - x), & 0 \le x \le 2 \\ 0, & \text{elsewhere} \end{cases}$$

where m is a constant. Determine:

a. the value of the constant m
b. $E(X)$ and $\text{Var}(X)$
c. $E(5 - 2X)$ and $\text{Var}(5 - 2X)$.

Technology active

5. The continuous random variable Z has a probability density function given by

$$f(z) = \begin{cases} \dfrac{2}{z + 1}, & 0 \le z \le a \\ 0, & \text{elsewhere} \end{cases}$$

where a is a constant. Calculate, correct to 4 decimal places:

a. the value of the constant a
b. the mean and variance of Z
c. i. $E(3Z + 1)$ ii. $\text{Var}(3Z + 1)$ iii. $E(Z^2 + 2)$.

6. The continuous random variable Z has a probability density function defined by

$$f(z) = \begin{cases} \dfrac{5 \log_e(z)}{\sqrt{z}}, & 1 \le z \le a \\ 0, & \text{elsewhere} \end{cases}$$

where a is a constant. Given $a = 1.7755$, determine, correct to 4 decimal places:

a. $E(Z)$ and $\text{Var}(Z)$ b. $E(3 - 2Z)$ and $\text{Var}(3 - 2Z)$.

7. The continuous random variable Z has a probability density function given by

$$f(z)=\begin{cases}\dfrac{3}{\sqrt{z}}, & 1\le z\le a\\ 0, & \text{elsewhere}\end{cases}$$

where a is a constant.

a. Calculate the value of the constant a.
b. Calculate the mean and variance of Z correct to 4 decimal places.
c. Determine, correct to 4 decimal places:
i. $E(4-3Z)$
ii. $Var(4-3Z)$.

8. MC The length of time it takes for an electric kettle to come to the boil is a continuous random variable with a mean of 1.5 minutes and a standard deviation of 1.1 minutes.
If each time the kettle is brought to the boil is an independent event and the kettle is boiled five times a day, calculate the standard deviation of the total time taken for the kettle to boil during a day.

A. 30.25 minutes
B. 1.21 minutes
C. 7.5 minutes
D. 5.5 minutes
E. 10.1 minutes

9. Two continuous random variables, X and Y, are related such that $Y=aX+5$ where a is a positive integer and $E(aX+5)=Var(aX+5)$. The mean of X is 9 and the variance of X is 2. Determine:

a. the value of the constant a
b. $E(Y)$ and $Var(Y)$.

10. The continuous random variable X is transformed so that $Y=aX+3$ where a is a positive integer. If $E(X)=5$ and $Var(X)=2$, calculate the value of the constant a, given that $E(Y)=Var(Y)$. Then calculate both $E(Y)$ and $Var(Y)$ to verify this statement.

11. A continuous random variable, X, is transformed so that $Y=aX+1$, where a is a positive constant. If $E(X)=2$ and $Var(X)=7$, determine the value of the constant a, given $E(Y)=Var(Y)$. Then calculate both $E(Y)$ and $Var(Y)$ to verify this statement. Give your answers correct to 4 decimal places.

12. The mass, Y kilograms, of flour sold in bags labelled as 1 kilogram is known to have a probability density function given by

$$f(y)=\begin{cases}k(2y+1), & 0.9\le y\le 1.25\\ 0, & \text{elsewhere}\end{cases}$$

where k is a constant.

a. Determine the value of the constant k.
b. Calculate the expected mass of a bag of flour, correct to 3 decimal places.
c. On a particular day, the machinery packaging the bags of flour needed to be recalibrated and produced a batch which had a mass of Z kilograms, where the probability density function for Z was given by $Z=0.75Y+0.45$. Calculate the expected mass of a bag of flour for this particular batch, correct to 3 decimal places.

11.5 Exam questions

Question 1 (5 marks) TECH-FREE

The mean of the continuous random variable Y is known to be 3.5, and its standard deviation is 1.2. Determine:

a. $\text{E}(2-Y)$ **(1 mark)**

b. $\text{E}\left(\frac{Y}{2}\right)$ **(1 mark)**

c. $\text{Var}(Y)$ **(1 mark)**

d. $\text{Var}(2-Y)$ **(1 mark)**

e. $\text{Var}\left(\frac{Y}{2}\right)$. **(1 mark)**

Question 2 (7 marks) TECH-ACTIVE

The continuous random variable X has a probability density function defined by

$$f(x)=\begin{cases}-\cos(x), & \frac{\pi}{2}\le x\le\pi\\ 0, & \text{elsewhere}\end{cases}$$

a. Sketch the graph of f and verify that it is a probability density function. **(2 marks)**

b. Calculate $\text{E}(X)$ and $\text{Var}(X)$, correct to 4 decimal places. **(2 marks)**

c. Calculate $\text{E}(3X+1)$ and $\text{Var}(3X+1)$, correct to 4 decimal places. **(2 marks)**

d. Calculate $\text{E}((2X-1)(3X-2))$, correct to 4 decimal places. **(1 mark)**

Question 3 (5 marks) TECH-ACTIVE

The daily rainfall, X mm, in a particular Australian town has a probability density function defined by

$$f(x)=\begin{cases}\frac{x}{k\pi}\sin\left(\frac{x}{3}\right), & 0\le x\le 3\pi\\ 0, & \text{elsewhere}\end{cases}$$

where k is a constant.

a. Determine the value of the constant k. **(2 marks)**

b. Calculate the expected daily rainfall, correct to 2 decimal places. **(2 marks)**

c. During the winter the daily rainfall is better approximated by $W=2X-1$. Calculate the expected daily rainfall during winter, correct to 2 decimal places. **(1 mark)**

More exam questions are available online.

11.6 Review

11.6.1 Summary

doc-37041

Hey students! Now that it's time to revise this topic, go online to:

Access the topic summary | Review your results | Watch teacher-led videos | Practise exam questions

Find all this and MORE in jacPLUS

11.6 Exercise

Technology free: short answer

1. a. Sketch the graph of $f(x) = \begin{cases} \frac{1}{4}(1-x), & -1 \le x < 1 \\ \frac{1}{4}(x-1), & 1 \le x \le 3 \\ 0, & \text{elsewhere} \end{cases}$

   b. Show that $f(x)$ is a probability density function.

   c. Determine $E(X)$.

2. For a continuous random variable, Z, with $E(Z) = 3$ and $SD(Z) = 1.2$, determine:

   a. $E(Z^2)$ b. $E(3Z - 2)$ c. $Var(3Z - 2)$ d. $E(Z(Z + 2))$ e. $E((Z - 2)(Z + 1))$.

3. X is a random variable with a probability density function given by

$$f(x) = \begin{cases} 2\sin(4x), & 0 \le x \le \frac{\pi}{4} \\ 0, & \text{elsewhere} \end{cases}$$

   Calculate:

   a. $\Pr\left(X < \frac{\pi}{6}\right)$

   b. the mean

   c. $\Pr\left(X < \frac{\pi}{8} \,\middle|\, X < \frac{\pi}{6}\right)$.

4. The continuous random variable Y has a probability density function defined by

$$f(y) = \begin{cases} gy^2, & 0 \le y \le h \\ 0, & \text{elsewhere} \end{cases}$$

   where g and h are constants and $E(Y) = \frac{3}{2}$.

   a. Evaluate the values of the constants g and h.

   b. Determine $Var(Y)$.

   c. Determine:

   i. $E(2Y + 1)$ ii. $Var(2Y + 1)$ iii. $E(Y^2 - 1)$ iv. $E(Y(Y + 4))$.

5. The continuous random variable X has a probability density function of

$$f(x) = \begin{cases} mx - nx^2, & -3 \le x \le 0 \\ 0, & \text{elsewhere} \end{cases}$$

where m and n are constants. Determine the values of m and n if $E(X) = -\frac{3}{2}$.

6. a. Determine the derivative of $3x\cos(\pi x)$.

b. Hence, calculate the mean of the continuous random variable X with a probability density function of $f(x) = \pi\sin(\pi x)$ for $0 \le x \le \frac{1}{2}$ and zero elsewhere.

Technology active: multiple choice

The following information relates to Questions 7 and 8.

A survey was taken to determine the amount of time, X hours, that teenagers spend interacting with digital devices during a 24-hour period. The table of findings and histogram are shown.

Time in hours	Frequency
$0 \le x \le 1$	20
$1 < x \le 2$	50
$2 < x \le 3$	60
$3 < x \le 4$	20

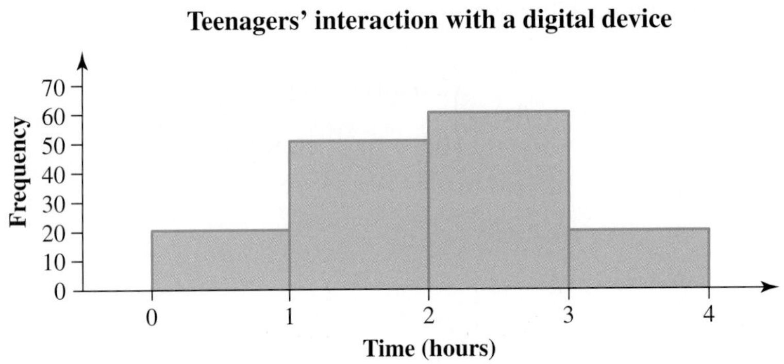

7. **MC** The number of teenagers in the survey was:

A. 100 B. 150 C. 280 D. 40 E. 80

8. **MC** $\Pr(X \le 3)$ is equal to:

A. $\frac{13}{15}$ B. $\frac{2}{5}$ C. $\frac{1}{4}$ D. $\frac{7}{15}$ E. $\frac{2}{15}$

9. **MC** A continuous random variable, X, has a probability density function defined by

$$f(x) = \begin{cases} \frac{1}{2}e^x, & 0 \le x \le m \\ 0, & \text{elsewhere} \end{cases}$$

The exact value of the constant m is:

A. $\frac{1}{3}\log_e(9)$ B. $\log_e(7)$ C. $\log_e(9)$ D. $\log_e(3)$ E. $\frac{1}{3}\log_e(7)$

10. **MC** The graph of a rectangular or uniform probability density function, $f(x)$, is shown.

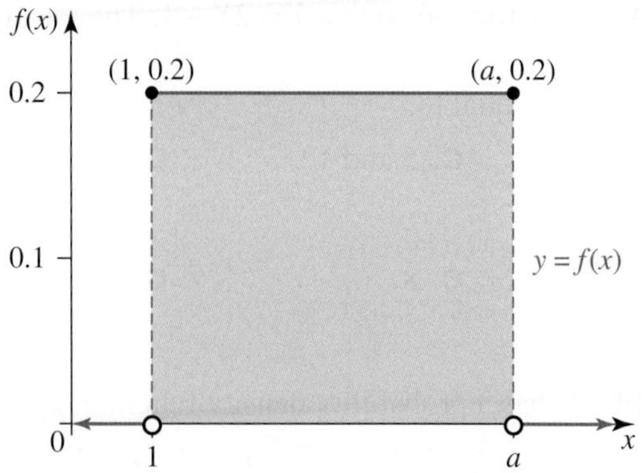

The value of the constant a is:

A. 5 **B.** 4 **C.** 6 **D.** 2 **E.** 3

11. **MC** A continuous probability density function is defined by

$$f(x) = \begin{cases} 2x, & 0 \le x \le 1 \\ 0, & \text{elsewhere} \end{cases}$$

The mean and variance of X are respectively:

A. $\frac{2}{3}$ and $\frac{1}{2}$ **B.** $\frac{1}{18}$ and $\frac{2}{3}$ **C.** $\frac{1}{2}$ and $\frac{1}{18}$

D. $\frac{2}{3}$ and $\frac{1}{18}$ **E.** $\frac{1}{2}$ and $\frac{2}{3}$

12. **MC** Y is a continuous random variable with a probability density function of

$$f(y) = \begin{cases} 3y^2, & 0 \le y \le 1 \\ 0, & \text{elsewhere} \end{cases}$$

$\Pr(0.2 < Y < 0.7)$ is equal to:

A. 0.5 **B.** 0.008 **C.** 0.343 **D.** 0.335 **E.** 0.3

***The following information relates to Questions* 13 *and* 14.**

The continuous random variable Z has a probability density function defined by

$$f(z) = \begin{cases} \frac{1}{\sqrt{3}}\cos(z), & -\frac{\pi}{3} < z < \frac{\pi}{3} \\ 0, & \text{elsewhere} \end{cases}$$

13. **MC** The mean of z is:

A. 0 **B.** 1.209 **C.** 1 **D.** $\frac{\pi}{6}$ **E.** −0.324

14. **MC** $\Pr\left(Z < \frac{\pi}{6}\right)$ is equal to:

A. $\frac{1}{2\sqrt{3}}$ **B.** $\frac{2}{\sqrt{3}}$ **C.** $\frac{1+\sqrt{3}}{2\sqrt{3}}$ **D.** $\frac{\sqrt{3}}{2}$ **E.** $\frac{2\sqrt{3}}{1+\sqrt{3}}$

***The following information relates to Questions* 15 *and* 16.**

The continuous random variable X is transformed so that $Y = 2X - 1$. The mean of X is 3 and the variance of X is 1.5.

15. **MC** E(Y) and Var(Y) are respectively equal to:

    A. 5 and 6 B. 2 and 12 C. 5 and 12 D. 2 and 6 E. 7 and 6

16. **MC** $E(X^2 - 1)$ is equal to:

    A. 3.75 B. 10.5 C. 8 D. 9 E. 9.5

Technology active: extended response

17. The continuous random variable, Y, has a probability density function defined by

$$f(y) = \begin{cases} 4 - \dfrac{1}{y^2}, & 0.5 \le y \le a \\ 0, & \text{elsewhere} \end{cases}$$

where a is a constant. Calculate:

a. the value of the constant a

b. i. E(Y) ii. Var(Y) iii. SD(Y)

c. the value of p such that $\Pr(0.5 \le y \le p) = 0.5$.

18. The continuous random variable, Z, has a probability density function defined by

$$f(z) = \begin{cases} \dfrac{1}{2}\sin\left(\dfrac{z}{4}\right), & 0 \le z \le k \\ 0, & \text{elsewhere} \end{cases}$$

where k is a constant. Determine:

a. the value of the constant k

b. E(Z) and Var(Z)

c. the interquartile range of Z.

19. The graph of the probability density function $f(x) = \dfrac{4x}{1 + x^2}$, $x \ge 0$ for the continuous random variable X is shown.

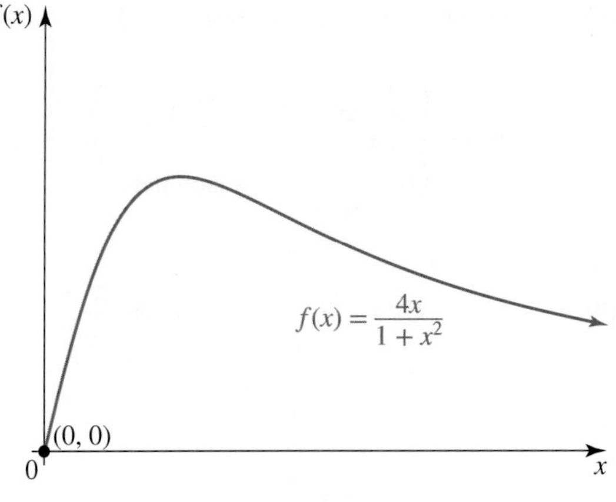

a. If X is a continuous random variable with a probability density function of $f(x) = \dfrac{4x}{1 + x^2}$ for $0 \le x \le a$ and zero elsewhere, calculate the value of the constant a.

b. Calculate the 30th percentile.

20. Beran has just spread lawn seed on his nature strip. With constant watering and plenty of sunshine, the time it takes for the lawn seed to germinate, T days after the seeding, can be determined by the probability density function

$$f(t) = \begin{cases} ke^{-0.15t}, & t \geq 0 \\ 0, & \text{elsewhere} \end{cases}$$

where k is a constant.

a. Determine the value of the constant k.
b. Calculate the expected period of time for the germination of the lawn seed. Give your answer correct to the nearest day.
c. Calculate the standard deviation for the germination time, correct to 2 decimal places.
d. Calculate the probability that the time it takes for the lawn seed to germinate is more than 7 days, given that the time taken is less than 10 days.

11.6 Exam questions

Question 1 (19 marks) TECH-ACTIVE

Source: *VCE 2017 Mathematical Methods Exam 2, Section B, Q3; © VCAA.*

The time Jennifer spends on her homework each day varies, but she does some homework every day.

The continuous random variable T, which models the time, t, in minutes, that Jennifer spends each day on her homework, has a probability density function f, where

$$f(t) = \begin{cases} \dfrac{1}{625}(t-20) & 20 \leq t < 45 \\ \dfrac{1}{625}(70-t) & 45 \leq t \leq 70 \\ 0 & \text{elsewhere} \end{cases}$$

a. Sketch the graph of f on the axes provided below. **(3 marks)**

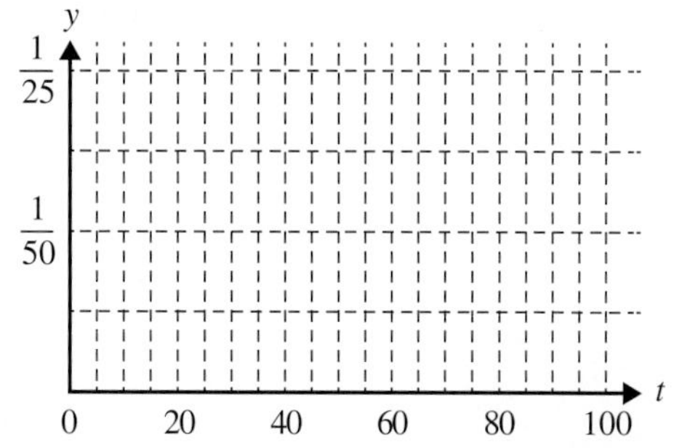

b. Find $\Pr(25 \le T \le 55)$. **(2 marks)**

c. Find $\Pr(T \le 25 \mid T \le 55)$. **(2 marks)**

d. Find a such that $\Pr(T \ge a) = 0.7$, correct to four decimal places. **(2 marks)**

e. The probability that Jennifer spends more than 50 minutes on her homework on any given day is $\frac{8}{25}$. Assume that the amount of time spent on her homework on any day is independent of the time spent on her homework on any other day.

i. Find the probability that Jennifer spends more than 50 minutes on her homework on more than three of seven randomly chosen days, correct to four decimal places. **(2 marks)**

ii. Find the probability that Jennifer spends more than 50 minutes on her homework on at least two of seven randomly chosen days, given that she spends more than 50 minutes on her homework on at least one of those days, correct to four decimal places. **(2 marks)**

f. Let p be the probability that on any given day Jennifer spends more than d minutes on her homework. Let q be the probability that on two or three days out of seven randomly chosen days she spends more than d minutes on her homework.
Express q as a polynomial in terms of p. **(2 marks)**

g. i. Find the maximum value of q, correct to four decimal places, and the value of p for which this maximum occurs, correct to four decimal places. **(2 marks)**

ii. Find the value of d for which the maximum found in **part g.i.** occurs, correct to the nearest minute. **(2 marks)**

Question 2 (1 mark) TECH-ACTIVE

MC A continuous random variable X, has a probability density function defined by

$$f(x) = \begin{cases} k\sin\left(\dfrac{\pi x}{8}\right) & \text{for } 0 \le x \le 8 \\ 0 & \text{elsewhere} \end{cases}$$

Determine which of the following options is **correct**.

A. $k = \dfrac{16}{\pi}$

B. $k = \dfrac{\pi}{16}$

C. $k = \dfrac{8}{\pi}$

D. $k = \dfrac{\pi}{8}$

E. $k = 8$

Question 3 (1 mark) TECH-ACTIVE

MC A continuous random variable X, has a probability density function defined by

$$f(x) = \begin{cases} 12(x-1)(x-2)^2 & \text{for } 1 \le x \le 2 \\ 0 & \text{elsewhere} \end{cases}$$

The mean is

A. $x = 1$ **B.** $x = \dfrac{7}{5}$ **C.** $x = \dfrac{4}{3}$ **D.** $x = \dfrac{8}{5}$ **E.** $x = 2$

Question 4 (3 marks) TECH-FREE

A probability density function is defined by

$$f(x)=\begin{cases}k\left(a^2-x^2\right) & -a<x<a\\ 0 & \text{otherwise}\end{cases}$$

Determine the exact value of a that gives a standard deviation of 2.

Question 5 (1 mark) TECH-ACTIVE

MC The time T spent waiting for a tram to arrive is a continuous random variable, and has a probability density function defined by

$$f(x)=\begin{cases}2e^{-2t} & \text{for } t\geq 0\\ 0 & \text{elsewhere}\end{cases}$$

The variance is equal to

A. $\frac{1}{2}$

B. $\frac{1}{\sqrt{2}}$

C. $\frac{1}{4}$

D. 0.26

E. 2

More exam questions are available online.

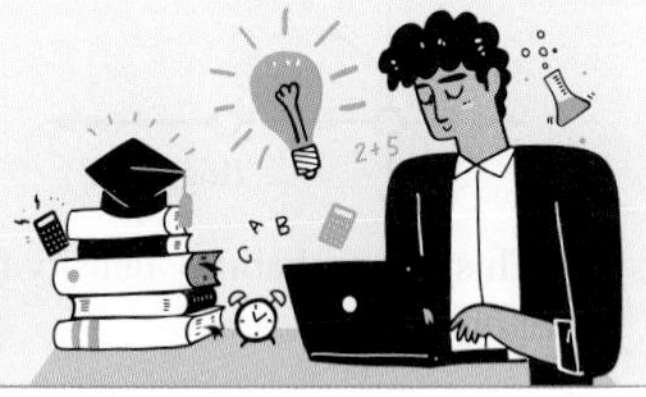

Answers

Topic 11 Continuous probability distributions

11.2 Continuous random variables and probability functions

11.2 Exercise

1. a. i. $\frac{9}{25}$ ii. $\frac{4}{25}$

   b. i. $\frac{37}{50}$ ii. $\frac{37}{42}$

2. a. 100 b. $\frac{29}{100}$

   c. $\frac{41}{50}$ d. $\frac{3}{100}$

3. a. 200

   b. i. $\frac{5}{8}$ ii. $\frac{31}{100}$

   c. $\frac{21}{46}$

4. a.

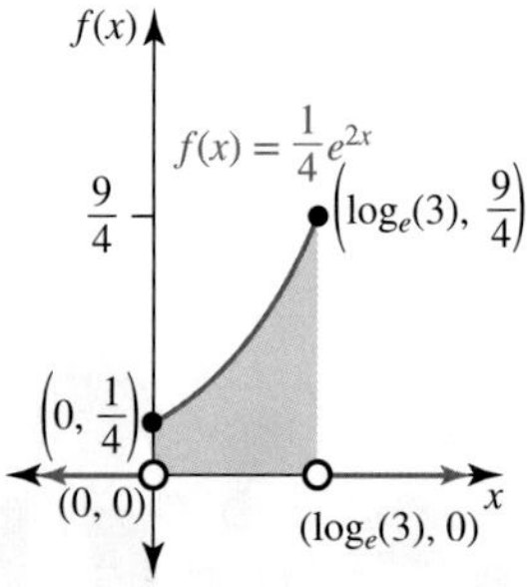

This is a probability density function as the area is 1 unit^2.

   b.

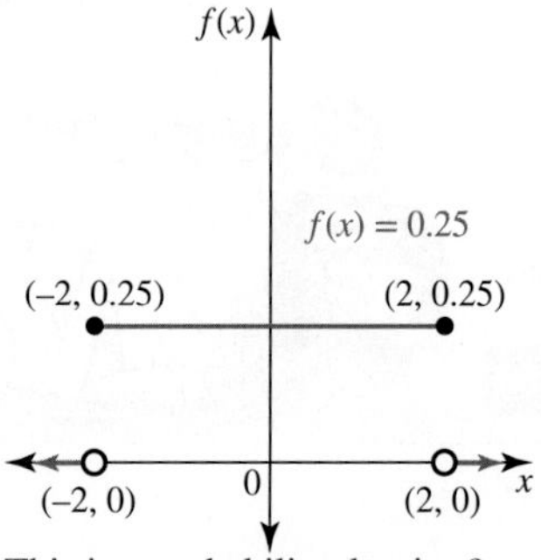

This is a probability density function as the area is 1 unit^2.

5. a.

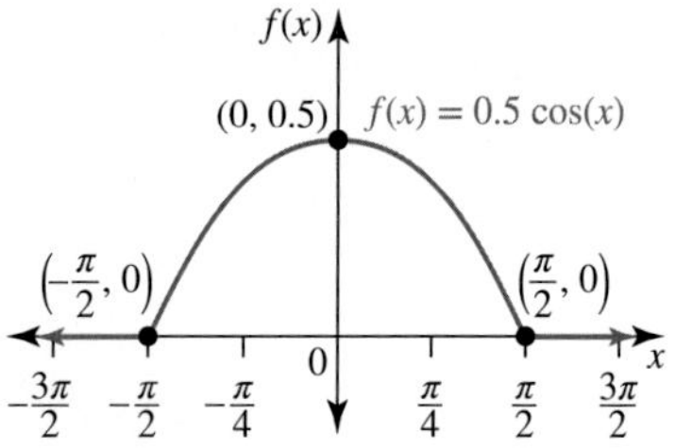

This is a probability density function as the area is 1 unit^2.

   b.

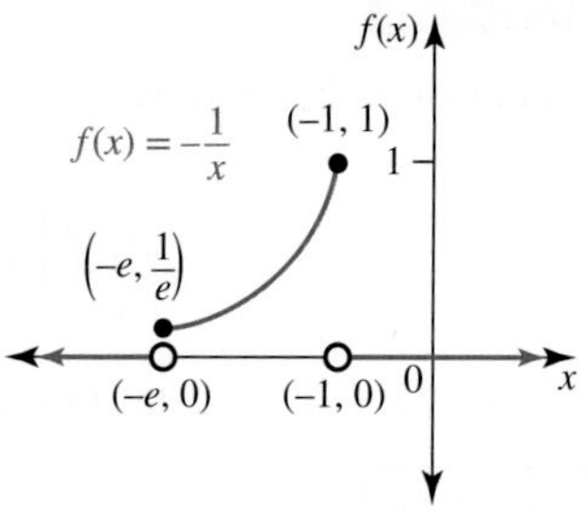

This is a probability density function as the area is 1 unit^2.

   c.

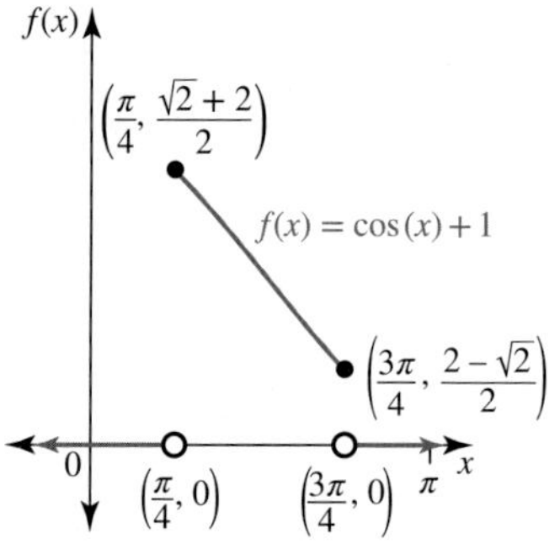

This is not a probability density function as the area is $\frac{\pi}{2}$ unit^2.

   d.

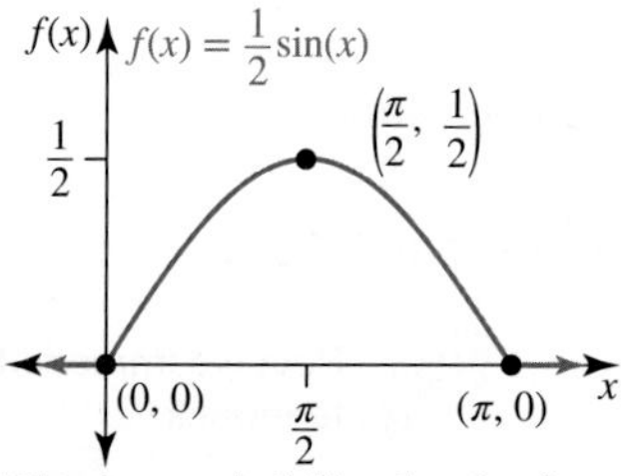

This is a probability density function as the area is 1 unit^2.

6. $n = \frac{1}{18}$

7. $a = \frac{1}{11}$

8. a.

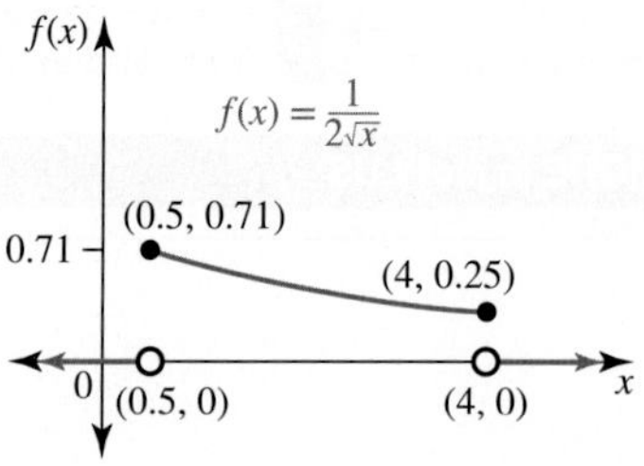

This is not a probability density function as the area is 1.2929 unit^2.

b.

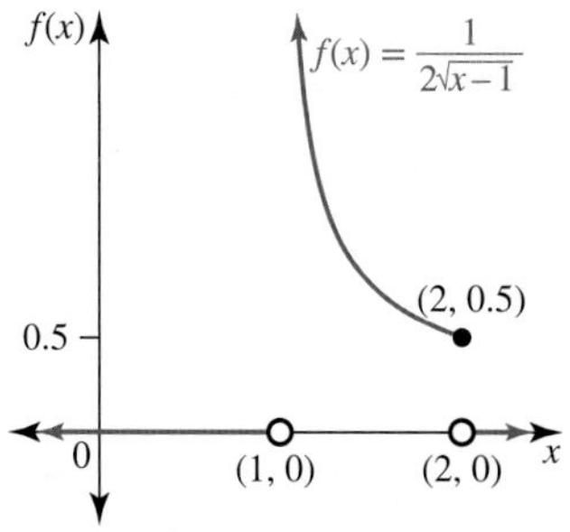

This is a probability density function as the area is 1 unit^2.

9. E

10. $\int_{-1}^{5} f(z)dz = 1$

$$A_{\text{triangle}} = 1$$

$$\frac{1}{2}bh = 1$$

$$\frac{1}{2} \times 6 \times z = 1$$

$$3z = 1$$

$$z = \frac{1}{3}$$

11. a. $m = \frac{1}{8}$ **b.** $m = 2$ **c.** $m = \frac{1}{4}$

12.

$$\int_{0}^{3} (x^2 + 2kx + 1)dx = 1$$

$$\left[\frac{1}{3}x^3 + kx^2 + x\right]_0^3 = 1$$

$$\left(\frac{1}{3}(3)^3 + k(3)^2 + 3\right) - 0 = 1$$

$$9 + 9k + 3 = 1$$

$$9k + 12 = 1$$

$$9k = -11$$

$$k = -\frac{11}{9}$$

13. $n = 12$

14. a. $a = e$

b. As $f(x) \geq 0$ and $\int_{1}^{e} f(x)\, dx = 1$, this is a probability density function.

15. $a = 2e$

16. a.

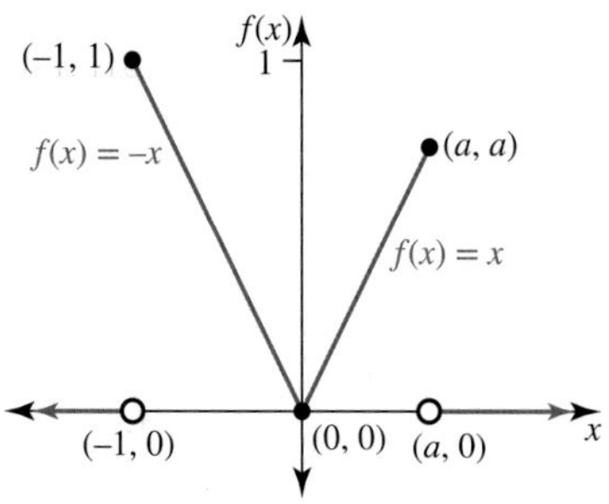

$$\int_{-1}^{0} -x\, dx + \int_{0}^{a} x\, dx = \frac{a^2 + 1}{2}$$

b.

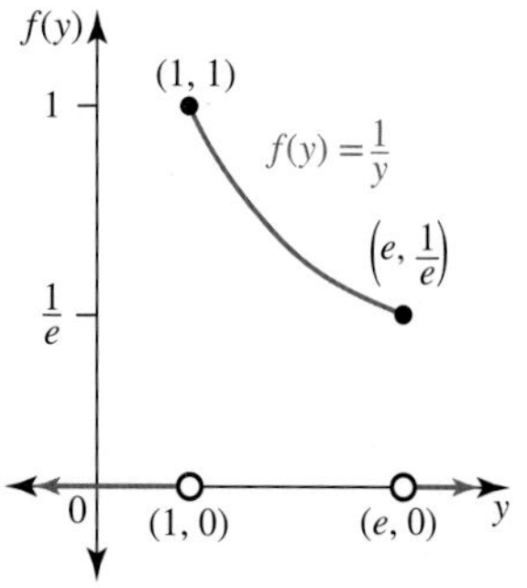

$$\int_{1}^{e} \frac{1}{y}\, dy = 1$$

c. $a = 1$

11.2 Exam questions

Note: Mark allocations are available with the fully worked solutions online.

1. D

2. E

3. D

11.3 The continuous probability density function

11.3 Exercise

1. a.

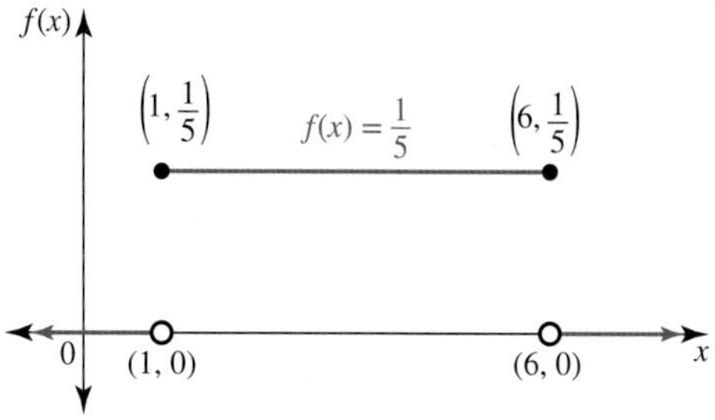

b. $\frac{3}{5}$

2. a. $a = 1$

b.

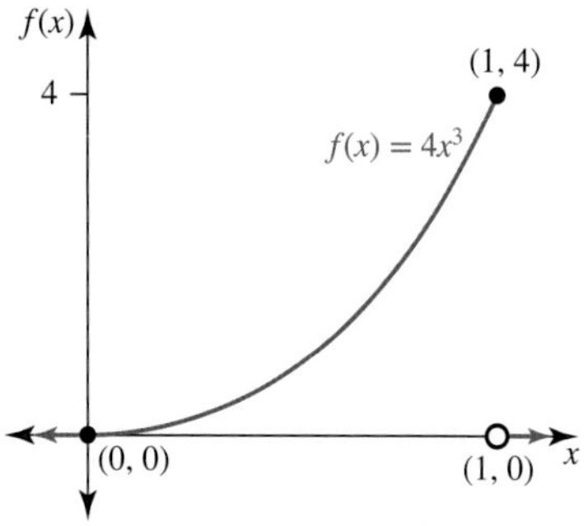

c. $\frac{15}{16}$

3. a.

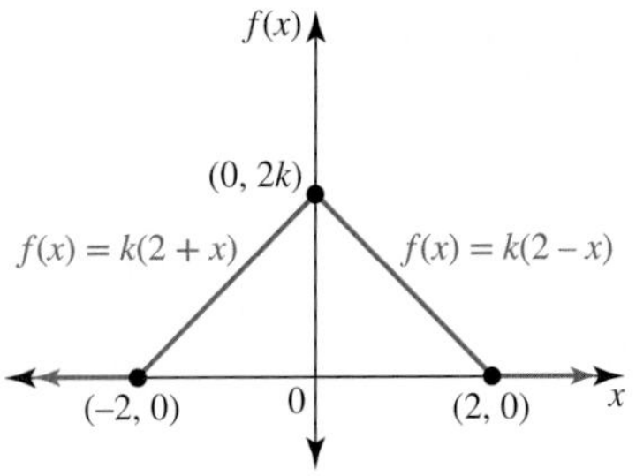

b. $A = \frac{1}{2}bh$

$1 = \frac{1}{2} \times 4 \times 2k$

$1 = 4k$

$k = \frac{1}{4}$

c. $\frac{3}{4}$

d. $\frac{6}{7}$

4. a.

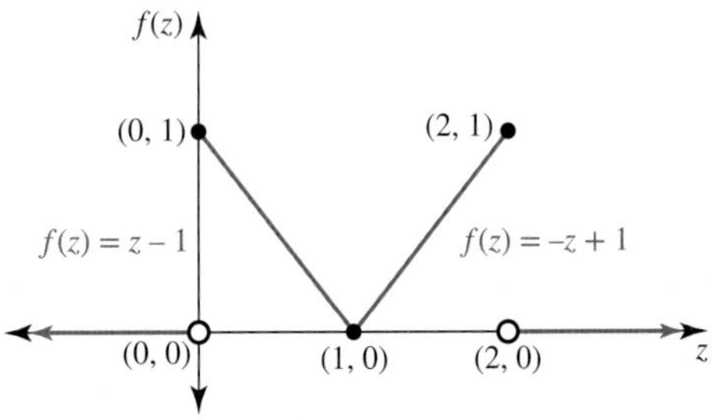

b. $\frac{15}{32}$

c. $\frac{5}{8}$

5. a.

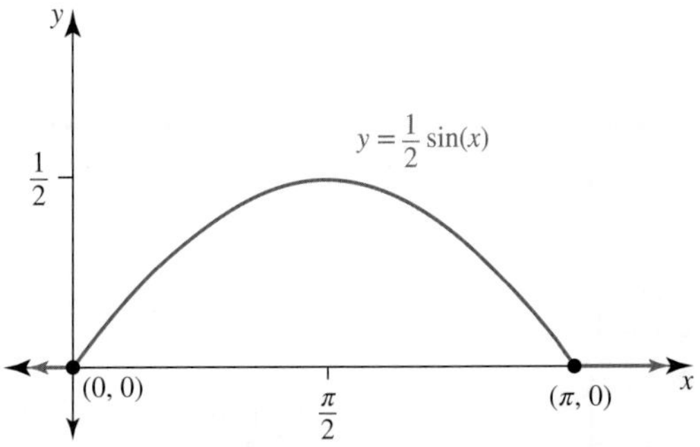

b. $\frac{\sqrt{2}}{2}$

c. $2\sqrt{2} - 2$

6. a.

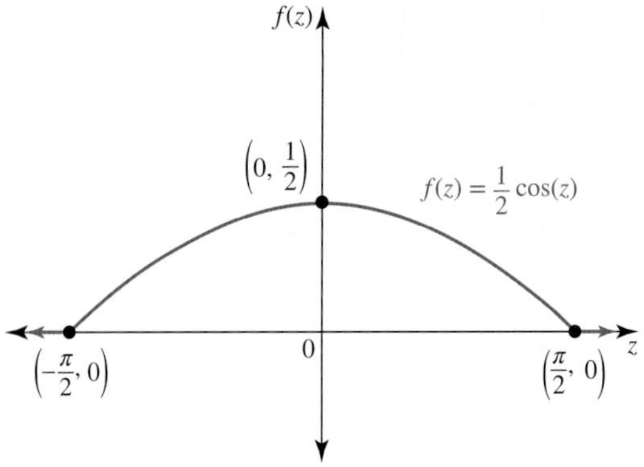

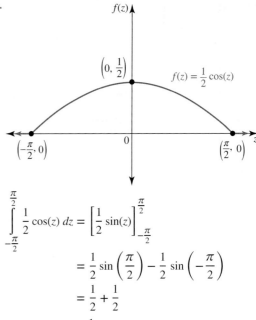

$$\int_{-\frac{\pi}{2}}^{\frac{\pi}{2}} \frac{1}{2}\cos(z)\,dz = \left[\frac{1}{2}\sin(z)\right]_{-\frac{\pi}{2}}^{\frac{\pi}{2}}$$

$$= \frac{1}{2}\sin\left(\frac{\pi}{2}\right) - \frac{1}{2}\sin\left(-\frac{\pi}{2}\right)$$

$$= \frac{1}{2} + \frac{1}{2}$$

$$= 1$$

This is a probability density function as the area under the curve is 1 and $f(z) \geq 0$ for all values of z.

b. $\frac{\sqrt{2}+1}{4}$

7. a. $\frac{1}{12}$ **b.** $\frac{5}{12}$ **c.** $\frac{1}{2}$

8. a. $a = 1$ **b.** $\frac{183}{256}$ **c.** 0.371 **d.** 0

9. B

10. C

11. a. $a = 2.1555$ **b.** 0.7147

12. a. $\frac{98}{125}$ **b.** $\frac{8}{9}$ **c.** $6^{\frac{1}{3}}$

13. a. 0.1560 **b.** $a = 0.55$

14. a.

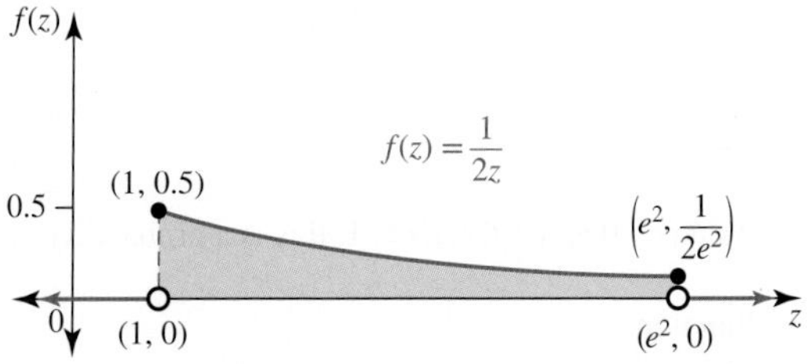

b. $\int_1^{e^2} \frac{1}{2z}\,dz = 1$. As $f(z) \geq 0$ and $\int_1^{e^2} f(z)\,dz = 1$, this is a probability density function.

c.

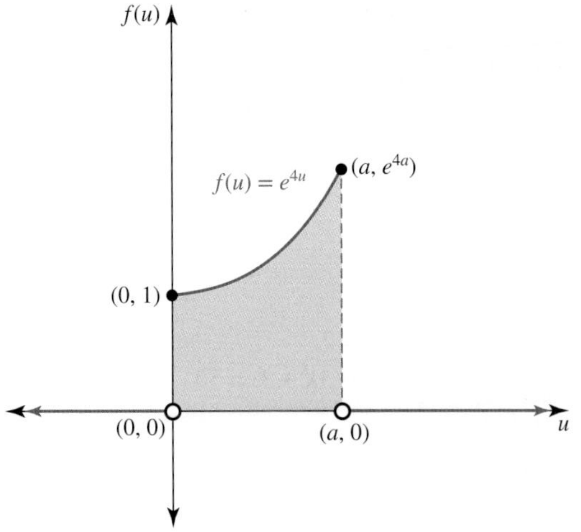

d. $a = \frac{1}{4}\log_e(5)$

11.3 Exam questions

Note: Mark allocations are available with the fully worked solutions online.

1. B
2. a. See the worked solution in the online resources.

 b. $1 - \frac{3}{e^2}$
3. a. $a = 3\log_e\left(\frac{3}{2}\right)$ b. 0.6243

 c. 0.5342 d. 0.60

11.4 Measures of centre and spread

11.4 Exercise

1. a. Mean $= \frac{2}{3}$ b. Mean $= \frac{1}{4}$
2. a. Mean $= \frac{1}{6}$ b. Mean = 2

 Variance $= \frac{5}{144}$ Variance $= \frac{1}{3}$
3. a. $a = \frac{9}{4}$ b. $\frac{19}{12}$
4. a.

$$\int_1^a \frac{3}{z^2}dz = 1$$
$$\int_1^a 3z^{-2}dz = 1$$
$$[-3z^{-1}]_1^a = 1$$
$$[-\frac{3}{z}]_1^a = 1$$
$$-\frac{3}{a} + \frac{3}{1} = 1$$
$$-\frac{3}{a} + 3 = 1$$
$$-\frac{3}{a} = -2$$
$$3 = 2a$$
$$a = \frac{3}{2}$$

 b. $E(Z) = 3\log_e\left(\frac{3}{2}\right)$

 c. $p = \frac{6}{5}$
5. a. $h = \frac{1}{4}$

 b. 2

 c. 2
6. a.

$$\int_0^1 \frac{1}{2\sqrt{x}}\,dx = \int_0^1 \frac{1}{2}x^{-\frac{1}{2}}dx$$
$$= \frac{1}{2}\int_0^1 x^{-\frac{1}{2}}dx$$
$$= \frac{1}{2}\left[2x^{\frac{1}{2}}\right]_0^1$$
$$= \frac{1}{2}(2\sqrt{1} - 2\sqrt{0})$$
$$= \frac{1}{2} \times 2$$
$$= 1$$

 As $f(x) \geq 0$ for all x-values and the area under the curve is 1, $f(x)$ is a probability density function.

 b. $\frac{1}{3}$
7. a. 1.3104

 b. 0.7863

 c. $k = 0.9322$
8. $E(Z) = 1.0486$, $Var(z) = 0.0440$, $SD(Z) = 0.2099$
9. $E(X) = \frac{1}{3}$, $Var(X) = \frac{1}{9}$, $SD(X) = \frac{1}{3}$
10. a. 0.5 min

 b. 0.5 min
11. a. $Q_1 = 1.3104,\ Q_3 = 1.8899$

 b. 0.5795
12. a. $a = 0.4809$

 b. 3.3663

 c. $Var(Z) = 3.8195,\ SD(Z) = 1.9571$

 d. 3.0751

 e. 7
13. a.

$$\int_0^\pi \frac{1}{\pi}(\sin(2x) + 1)\,dx = \frac{1}{\pi}\int_0^\pi (\sin(2x) + 1)\,dx$$
$$= \frac{1}{\pi}\left[-\frac{1}{2}\cos(2x) + x\right]_0^x$$
$$= \frac{1}{\pi}\left(\left(-\frac{1}{2}\cos(2\pi) + \pi\right) - \left(-\frac{1}{2}\cos(0) + 0\right)\right)$$
$$= \frac{1}{\pi}\left(-\frac{1}{2} + \pi + \frac{1}{2}\right)$$
$$= 1$$

 As $f(x) \geq 0$ for all x-values and the area under the curve is 1, $f(x)$ is a probability density function.

 b. 1.0708

c. i. 0.5725

ii. 0.7566

14. a. $\int_{2}^{7.9344} f(y)\,dy = \int_{2}^{7.9344} 0.2 \log_e\left(\frac{y}{2}\right) dy = 1$

b. 5.7278

c. $\text{Var}(Y) = 2.1600$, $\text{SD}(Y) = 1.4697$

d. 5.9344

15. a. 2.3104

b. i. 1.7863

ii. 3.3085

iii. 0.1176

iv. 0.3430

16. $a = \frac{3}{2}$, $b = \frac{3}{4}$

17. a. $-\frac{x}{\sqrt{4 - x^2}}$

b. $\frac{3}{\pi}$

18. a.

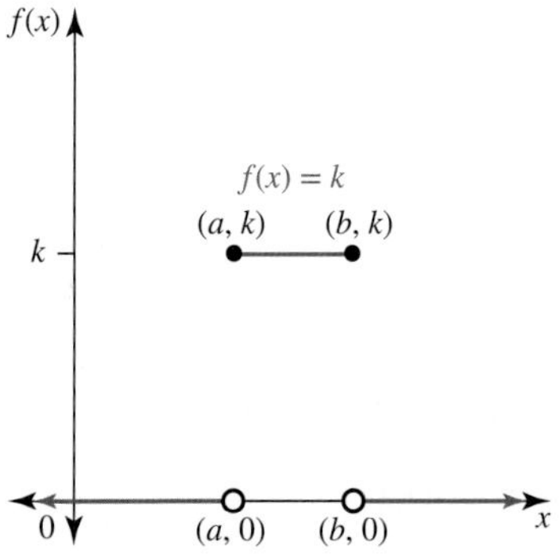

b. $\int_{a}^{b} k\,dx = 1$

$[kx]_a^b = 1$

$kb - ka = 1$

$k(b - a) = 1$

$k = \frac{1}{b - a}$

c. $\frac{b + a}{2}$

d. $\frac{(a - b)^2}{12}$

11.4 Exam questions

Note: Mark allocations are available with the fully worked solutions online.

1. a. See the worked solution in the online resources.

b. $\log_e(4)$

2. B

3. $2 - \frac{4}{\pi}$

11.5 Linear transformations

11.5 Exercise

1. a. 5 b. 12
c. 19 d. 15

2. a. 13 b. 18
c. 27 d. 8

3. a. $k = \frac{1}{4}$

b. $E(X) = 0$, $\text{Var}(X) = 2$

c. $E(5X + 3) = 3$, $\text{Var}(5X + 3) = 50$

d. $E((3X - 2)^2) = 22$

4. a. $m = \frac{3}{4}$

b. $E(X) = 1$, $\text{Var}(X) = 0.2$

c. $E(5 - 2X) = 3$, $\text{Var}(5 - 2X) = 0.8$

5. a. $a = 0.6487$

b. $E(Z) = 0.2974$, $\text{Var}(Z) = 0.0349$

c. i. 1.8922

ii. 0.3141

iii. 2.1234

6. a. $E(Z) = 1.4921$, $\text{Var}(Z) = 0.0361$

b. $E(3 - 2Z) = 0.0158$, $\text{Var}(3 - 2Z) = 0.1444$

7. a. $a = \frac{49}{36}$

b. $E(Z) = 1.1759$, $\text{Var}(Z) = 0.0109$

c. i. 0.4722 ii. 0.0978

8. D

9. a. $a = 5$

b. $E(Y) = 50$, $\text{Var}(Y) = 50$

10. $a = 3$, $E(Y) = 18$, $\text{Var}(Y) = 18$

11. $a = 0.5469$, $E(Y) = 2.0938$, $\text{Var}(Y) = 2.0938$

12. a. $k = \frac{400}{441}$ b. 1.081 kg c. 1.261 kg

11.5 Exam questions

Note: Mark allocations are available with the fully worked solutions online.

1. a. −1.5 b. 1.75
c. 1.44 d. 1.44
e. 0.36

2. a.

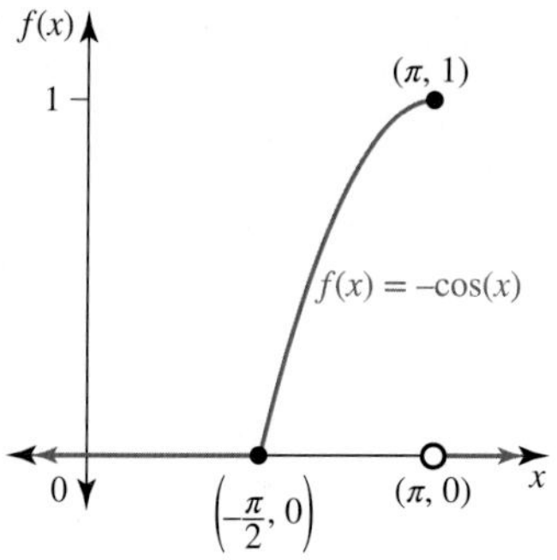

$$\int_{\frac{\pi}{2}}^{\pi}(-\cos(x))dx = \left[-\sin(x)\right]_{\frac{\pi}{2}}^{\pi}$$

$$= -\sin(x) + \sin\left(\frac{\pi}{2}\right)$$

$$= 0 + 1$$

$$= 1$$

As $f(x) \geq 0$ for all x-values and the area under the curve is 1, $f(x)$ is a probability density function.

b. $E(X) = 2.5708$, $Var(X) = 0.1416$

c. $E(3X + 1) = 8.7124$, $Var(3X + 1) = 1.2743$

d. $E((2X - 1)(3X - 2)) = 24.5079$

3. a. $k = 9$

b. 5.61 mm

c. 10.21 mm

11.6 Review

11.6 Exercise

Technology free: short answer

1. a.

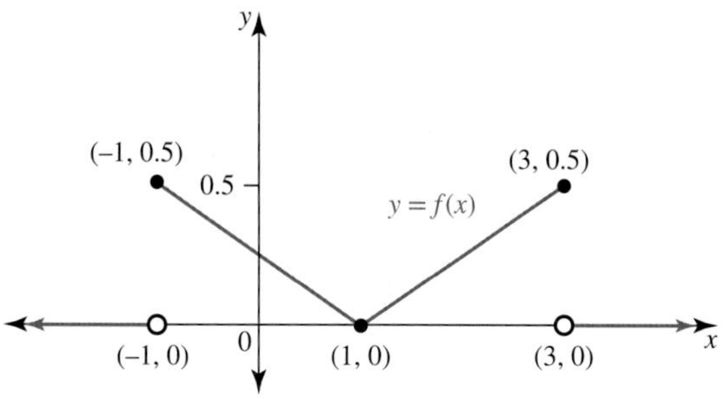

b. $A = \int_{-1}^{3} f(x)dx$

$$= \frac{1}{2} \times 2 \times \frac{1}{2} + \frac{1}{2} \times 2 \times \frac{1}{2}$$

$$= \frac{1}{2} + \frac{1}{2}$$

$$= 1$$

As $f(x) \geq 0$ for all x-values and the area under the curve is 1, $f(x)$ is a probability density function.

c. 1

2. a. 10.44 b. 7
c. 12.96 d. 16.44
e. 5.44

3. a. $\frac{3}{4}$ b. $\frac{\pi}{8}$
c. $\frac{2}{3}$

4. a. $g = \frac{3}{8}$, $h = 2$

b. $\frac{3}{20}$

c. i. 4 ii. $\frac{3}{5}$ iii. $\frac{7}{5}$ iv. $8\frac{2}{5}$

5. $m = -\frac{2}{3}$, $n = \frac{2}{9}$

6. a. $3\cos(\pi x) - 3\pi x \sin(\pi x)$

b. $\frac{1}{\pi}$

Technology active: multiple choice

7. B 8. A 9. D 10. C 11. D
12. D 13. A 14. C 15. A 16. E

Technology active: extended response

17. a. 1

b. i. 0.8069

ii. 0.0156

iii. 0.1248

c. $p = 0.8202$

18. a. $k = \frac{4\pi}{3}$

b. $E(Z) = 2.7394$,
$Var(Z) = 0.9912$

c. 1.56

19. a. 0.8054 b. 0.4023

20. a. 0.15 b. 7 days
c. 6.67 d. 0.1632

11.6 Exam questions

Note: Mark allocations are available with the fully worked solutions online.

1. a.

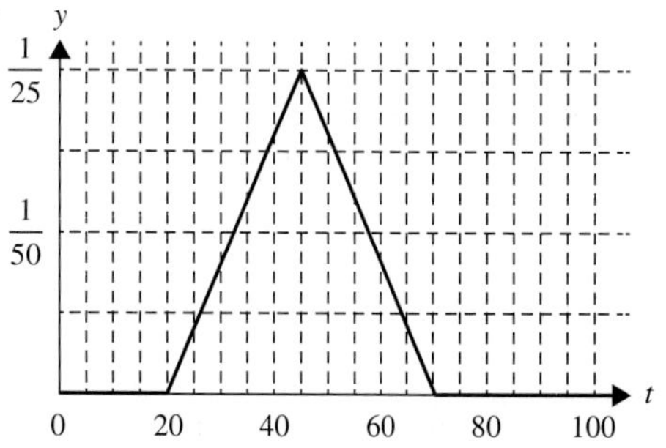

b. $\frac{4}{5}$

c. $\frac{1}{41}$

d. $a = 39.3649$

e. i. 0.1534

ii. 0.7626

f. $q(p) = 7p^2(p - 1)^4(2p + 3)$

g. i. 0.5665

ii. $d = 49$ minutes

2. B

3. B

4. $a = 2\sqrt{5}$

5. C

12 The normal distribution

LEARNING SEQUENCE

Fully worked solutions for this topic are available online.

12.1 Overview

12.1.1 Introduction

The normal distribution is sometimes referred to as the bell curve. It is also called the Gaussian curve, after mathematician Carl Friedrich Gauss, although he was not actually the first discover the normal distribution. That honour went to Abraham de Moivre, who was a statistician and gambling consultant in the eighteenth century. He noticed that when the number of events, such as flipping a coin, increased, the shape of the binomial distribution approached a smooth curve. He was able to find a mathematical expression for this curve, which we now know as the normal distribution. Gauss and Pierre-Simon Laplace were also significant contributors in the development of the normal distribution.

The importance of the normal curve stems from the fact that many phenomena have distributions that are at least approximately normally distributed, such as measurement errors, IQ scores, salaries, and the heights, weights and strength of people. The study scores for a VCE subject are also normally distributed with a mean of 30 and a standard deviation of approximately 7. This is why such a large percentage of students score around 30 and only a very small percentage score about 45.

KEY CONCEPTS

This topic covers the following key concepts from the VCE Mathematics Study Design:

- random variables, including the concept of a random variable as a real function defined on a sample space and examples of continuous random variables
- continuous random variables:
  - standard normal distribution, N(0, 1), and transformed normal distributions, $N\left(\mu, \sigma^2\right)$, as examples of a probability distribution for a continuous random variable
  - effect of variation in the value(s) of defining parameters on the graph of a given probability density function for a continuous random variable
  - calculation of probabilities for intervals defined in terms of a random variable, including conditional probability.

Source: VCE Mathematics Study Design (2023–2027) extracts © VCAA; reproduced by permission.

12.2 The normal distribution

LEARNING INTENTION

At the end of this subtopic you should be able to:

- define a normal distribution
- describe how the mean and standard deviation affect the curve of a normal distribution
- state the three key probabilities associated with the normal distribution.

12.2.1 Introduction

The **normal distribution** is arguably the most important distribution in statistics. It is characterised by the well-known bell-shaped curve, which is symmetrical about the mean (as well as the median and mode). Continuous random variables such as height, weight, time and other naturally occurring phenomena are frequently analysed with normal distribution calculations.

Normal distributions may vary depending on their means and standard deviations. The diagram shows three different normal distributions.

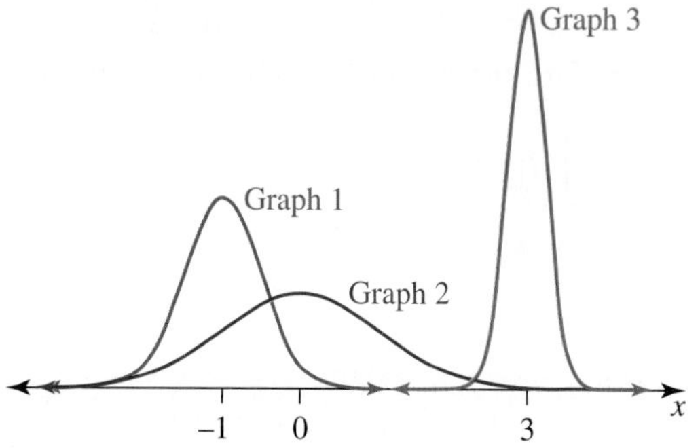

Graph 1 has mean of −1 and a standard deviation of 0.5.

Graph 2 has a mean of 0 and a standard deviation of 1.

Graph 3 has a mean of 3 and a standard deviation of 0.25.

The probability density function for the normal distribution is given by

$$f(x) = \frac{1}{\sigma\sqrt{2\pi}}\, e^{-\frac{1}{2}\left(\frac{x-\mu}{\sigma}\right)^2}$$

where the parameters μ and σ are the mean and standard deviation of the distribution, respectively.

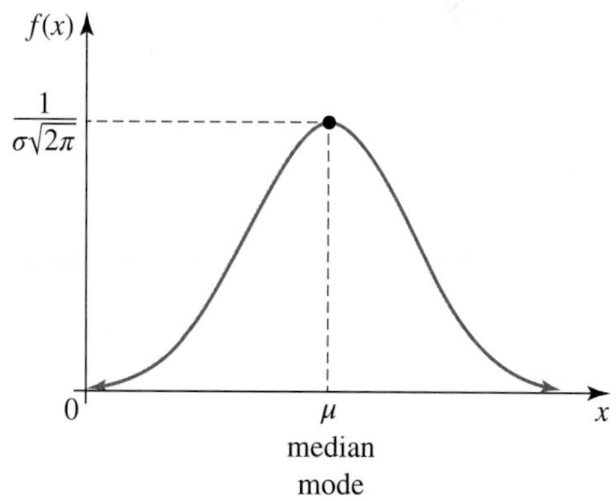

We say that

$$X \sim \mathrm{N}\left(\mu, \sigma^2\right)$$

meaning X is distributed normally with the mean and variance specified.

As the mean and standard deviation can vary, and the area under the graph must be constant and equal to 1, in effect, changing the mean and the standard deviation transforms the normal curve.

The normal distribution curve

Changing the standard deviation affects the normal curve twofold. The transformed curve will display:

- **dilation by a factor $\frac{1}{\sigma}$ from the x-axis**
- **dilation by a factor σ from the y-axis.**

Changing the mean has the effect of a translation parallel to the x-axis.

The normal probability density function has the following characteristics.

- $f(x) = \dfrac{1}{\sigma\sqrt{2\pi}} e^{-\frac{1}{2}\left(\frac{x-\mu}{\sigma}\right)^2}, x \in R$
- **The distribution is symmetrical about the mean.**
- **μ = median = mode**
- **The maximum value is $\dfrac{1}{\sigma\sqrt{2\pi}}$ when $x = \mu$.**
- **The curve continues infinitely in both directions.**
- $\displaystyle\int_{-\infty}^{\infty} f(x)dx = 1$

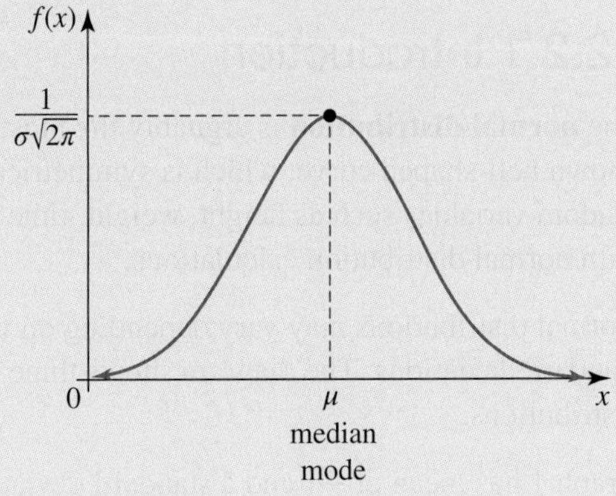

WORKED EXAMPLE 1 The normal distribution curve

A normal probability density function is defined by

$$f(x) = \frac{2}{\sqrt{2\pi}} e^{-\frac{1}{2}(2(x-1))^2}, \; x \in R.$$

a. State the mean and standard deviation of the distribution.
b. State what effect the mean and standard deviation have on the graph of the normal distribution.

THINK

a. Use $f(x) = \dfrac{1}{\sigma\sqrt{2\pi}} e^{-\frac{1}{2}\left(\frac{x-\mu}{\sigma}\right)^2}$ to determine μ and σ.

WRITE

a. $f(x) = \dfrac{1}{\sigma\sqrt{2\pi}} e^{-\frac{1}{2}\left(\frac{x-\mu}{\sigma}\right)^2}$

$= \dfrac{2}{\sqrt{2\pi}} e^{-\frac{1}{2}(2(x-1))^2}$

$\dfrac{1}{\sigma} = 2$, so $\sigma = \dfrac{1}{2}$ and $\mu = 1$.

THINK

b. The mean has the effect of a translation parallel to the x-axis. The standard deviation has the effect of dilations from both the x- and y-axes.

WRITE

b. $\sigma = \dfrac{1}{2}$ is a dilation of factor 2 from the x-axis and a dilation of factor $\dfrac{1}{2}$ from the y-axis.

$\mu = 1$ means a translation of 1 unit in the positive x-direction.

on Resources

Interactivity The normal distribution (int-6438)

12.2.2 Important intervals and their probabilities

Often we are required to find the proportion of a population for a given interval. Using the property that the symmetry of the normal distribution is about the mean, we are able to predict with certainty the following facts.

Probabilities within 1, 2, 3 standard deviations of the mean

- **Approximately 68% of the population will fall within 1 standard deviation of the mean:**

$$\Pr(\mu - \sigma \le X \le \mu + \sigma) \approx 0.68$$

- **Approximately 95% of the population will fall within 2 standard deviations of the mean:**

$$\Pr(\mu - 2\sigma \le X \le \mu + 2\sigma) \approx 0.95$$

We say that a randomly chosen member of the population will most probably be or is highly likely to be within 2 standard deviations of the mean.

- **Approximately 99.7% of the population will fall within 3 standard deviations of the mean:**

$$\Pr(\mu - 3\sigma \le X \le \mu + 3\sigma) \approx 0.997$$

We say that a randomly chosen member of the population will almost certainly be within 3 standard deviations of the mean.

This is shown in the following graphs.

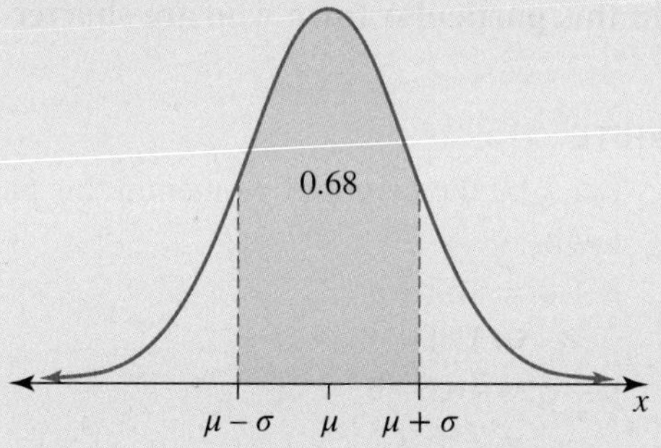

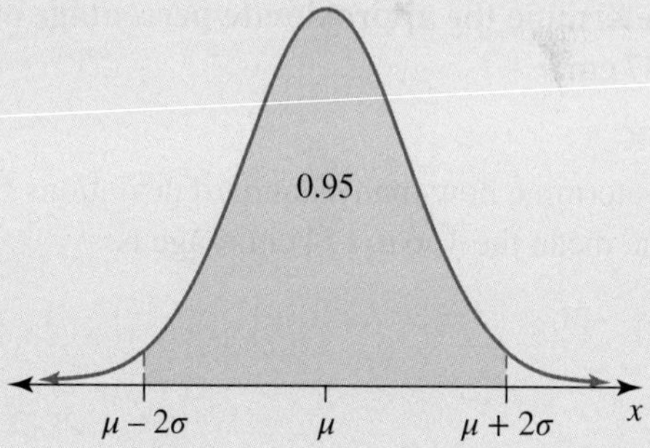

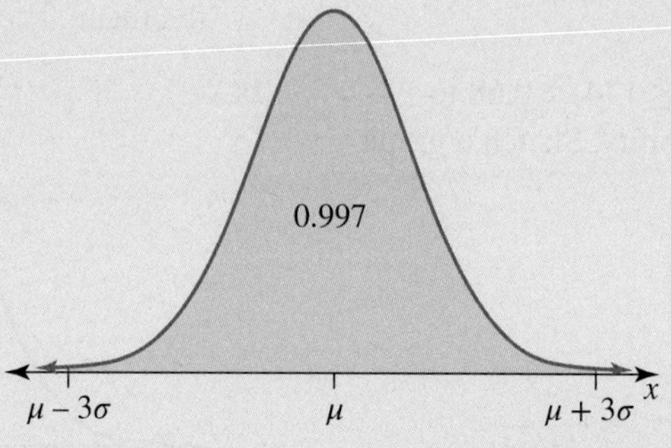

A more comprehensive breakdown of the proportion of the population for each standard deviation is shown on the following graph.

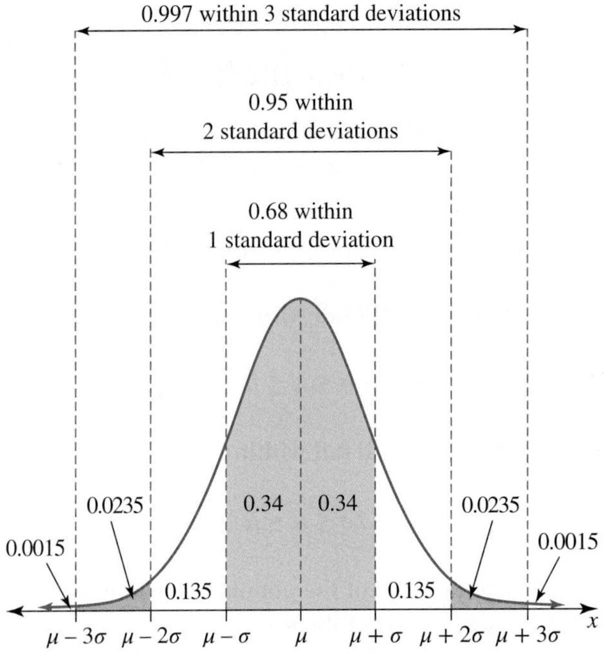

WORKED EXAMPLE 2 Important probabilities of the normal curve

The heights of the women in a particular town are normally distributed with a mean of 165 centimetres and a standard deviation of 9 centimetres.

a. **Determine the approximate probability that a woman chosen at random has a height that is between 156 and 174 cm.**

b. **Determine the approximate probability that a woman chosen at random is taller than 174 cm.**

c. **Determine the approximate percentage of the women in this particular town who are shorter than 147 cm.**

THINK	WRITE
a. Determine how many standard deviations from the mean the 156 to 174 cm range is.	a. Let X be the height of women in this particular town. $\mu+\sigma = 165+9$ $= 174$ $\mu-\sigma = 165-9$ $= 156$ Since the range is 1 standard deviation from the mean, $\Pr(156 \le X \le 174) \approx 0.68$.
b. Use the fact that $\Pr(156 \le X \le 174) \approx 0.68$ to calculate the required probability. Sketch a graph to help.	b. 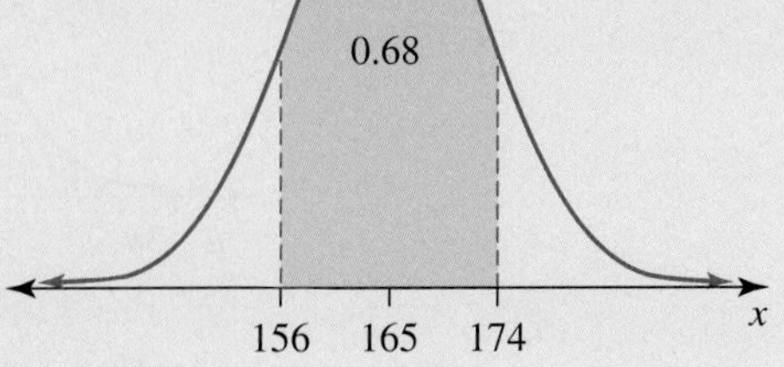

Since $\Pr(156 \le X \le 174) \approx 0.68$,

$\Pr(X < 156) \cup \Pr(X > 174) \approx 1 - 0.68$
$= 0.32$

Because of symmetry,

$$\Pr(X < 156) = \Pr(X > 174) = \frac{0.32}{2} = 0.16$$

Thus, $\Pr(X > 174) \approx 0.16$.

c. 1. Determine how many standard deviations 147 cm is from the mean.

c. $\mu - \sigma = 165 - 9 = 156$

$\mu - 2\sigma = 165 - 2 \times 9 = 147$

147 cm is 2 standard deviations from the mean.

The corresponding upper value is

$165 + 2 \times 9 = 183$

$\Pr(147 \le X \le 183) \approx 0.95$

2. Using symmetry, calculate $\Pr(X < 147)$.

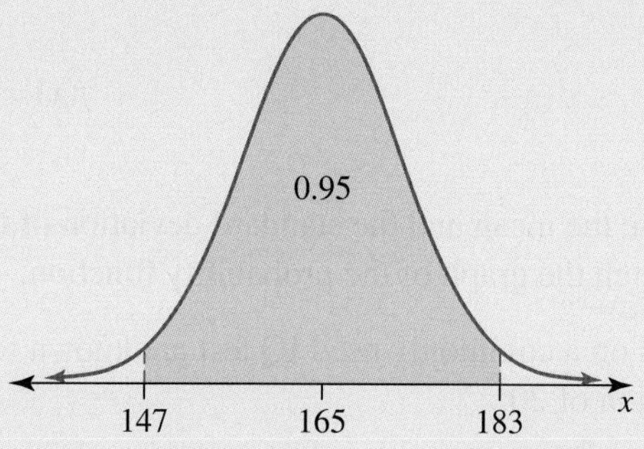

Thus,

$\Pr(X < 147) \cup \Pr(X > 183) \approx 1 - 0.95$
$= 0.05$

and by symmetry,

$$\Pr(X < 147) = \Pr(X > 183) \approx \frac{0.05}{2} = 0.025$$

Thus, approximately 2.5% of the population of women in this particular town are shorter than 147 cm.

on Resources

Interactivity The 68 – 95 – 99.7% rule (int-6439)

12.2 Exercise

Technology free

1. **WE1** A normal probability density function is defined by

$$f(x) = \frac{10}{3\sqrt{2\pi}} e^{-\frac{1}{2}\left(\frac{10(x-1)}{3}\right)^2}, x \in R.$$

   a. State the mean and standard deviation of the distribution.
   b. State what effect the mean and standard deviation have on the graph of the normal distribution.

2. The probability density function of a normal distribution is given by

$$f(x) = \frac{1}{3\sqrt{2\pi}} e^{-\frac{1}{2}\left(\frac{x-2}{3}\right)^2}$$

   a. State the mean and the standard deviation of the distribution.
   b. Sketch the graph of the probability function.

3. Scores on a commonly used IQ test are known to be normally distributed with a mean of 120 and a standard deviation of 20.
   a. Determine:
      i. $\mu \pm \sigma$
      ii. $\mu \pm 2\sigma$
      iii. $\mu \pm 3\sigma$.
   b. Determine:
      i. $\Pr(X < 80)$
      ii. $\Pr(X > 180)$.

4. **WE2** The results of a Mathematical Methods test are normally distributed with a mean of 72 and a standard deviation of 8.
   a. Determine the approximate probability that a student who sat the test has a score that is greater than 88.
   b. Determine the approximate proportion of the students who sat the test and had a score that was less than 48.
   c. Determine the approximate percentage of the students who sat the test and scored less than 80.

5. A continuous random variable, X, is known to be normally distributed with a mean of 15 and a standard deviation of 5. State the range between which approximately:
   a. 68% of the values lie
   b. 95% of the values lie
   c. 99.7% of the values lie.

6. A normal probability density function, X, has a mean of 24 and a standard deviation of 7. Determine the approximate values for:
   a. $\Pr(X < 31)$
   b. $\Pr(10 < X < 31)$
   c. $\Pr(X > 10 \mid X < 31)$.

7. The number of pears harvested from each tree in a large orchard is normally distributed with a mean of 230 and a standard deviation of 25. Determine the approximate probability that the number of pears harvested from a randomly selected tree is:

a. less than 280
b. between 180 and 280
c. is greater than 180, given that less than 280 pears were harvested.

Technology active

8. **MC** The length of pregnancy for a human is normally distributed with a mean of 275 days and a standard deviation of 14 days. A mother gave birth after less than 233 days. Select the approximate probability of this happening for the general population.

A. 0.0015
B. 0.05
C. 0.003
D. 0.0235
E. 0.025

9. **MC** The results of a Year 12 Biology examination are known to be normally distributed with a mean of 70 and a standard deviation of 6. Select the approximate percentage of students sitting for this examination who can be expected to achieve a score that is greater than 88.

A. 2.35%
B. 0.3%
C. 99.85%
D. 0.15%
E. 99.7%

10. In a particular area of Australia, the annual rainfall, X mm, is known to be normally distributed with a mean of 305 mm and a standard deviation of 50 mm.

a. Calculate the approximate value of $\Pr(205 < X < 355)$.
b. Evaluate k such that $\Pr(X < k) \approx 0.025$.
c. Evaluate h such that $\Pr(X < h) \approx 0.0015$.

11. A continuous random variable, X, is normally distributed with a mean of 72.5 and a standard deviation of 8.4. Determine the approximate values for:

a. $\Pr(64.1 < X < 89.3)$
b. $\Pr(X < 55.7)$
c. $\Pr(X > 47.3 \mid X < 55.7)$
d. m such that $\Pr(X > m) \approx 0.16$.

12. A normal distribution has a probability density function of

$$f(x) = \frac{1}{\sqrt{2\pi}} e^{-\frac{1}{2}(x-3)^2}.$$

a. Using CAS technology, verify that $\int_{-\infty}^{\infty} f(x)\,dx = 1$.
b. State μ and σ.
c. Sketch the graph of the probability function.

13. Consider the normal probability density function

$$f(x) = \frac{1}{4\sqrt{2\pi}} e^{-\frac{1}{2}\left(\frac{x+2}{4}\right)^2}, x \in R.$$

a. Using CAS technology, verify that $\int_{-\infty}^{\infty} f(x)\,dx = 1$

b. State μ.

14. A normal probability density function is given by

$$f(x) = \frac{1}{10\sqrt{2\pi}} e^{-\frac{1}{2}\left(\frac{x+4}{10}\right)^2}, x \in R.$$

a. State the values of μ and σ.

b. State what effect the mean and standard deviation have on the graph of the normal distribution.

c. Determine:

i. Var(X) ii. E$\left(X^2\right)$.

d. Verify that this is a probability density function.

15. $f(x) = \frac{5}{2\sqrt{2\pi}} e^{-\frac{1}{2}\left(\frac{5(x-2)}{2}\right)^2}$, $x \in R$ defines a normal probability density function.

a. State the values of μ and σ.

b. Calculate E$\left(X^2\right)$.

12.2 Exam questions

Question 1 (2 marks) TECH-FREE

Sketch the probability density curve for the random variable X, which is normally distributed with mean 12 and standard deviation of 1.5.

Question 2 (1 mark) TECH-ACTIVE

MC The diagram shows the graphs of two normal distributions curves, with means μ_1 and μ_2 and standard deviations σ_1 and σ_2 respectively.

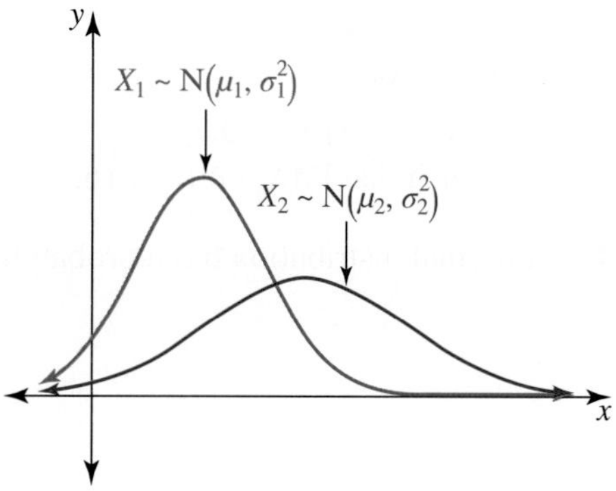

Select the **true** statement from the following.

A. $\mu_1 = \mu_2$ and $\sigma_1 = \sigma_2$

B. $\mu_1 < \mu_2$ and $\sigma_1 = \sigma_2$

C. $\mu_1 = \mu_2$ and $\sigma_1 < \sigma_2$

D. $\mu_1 < \mu_2$ and $\sigma_1 > \sigma_2$

E. $\mu_1 < \mu_2$ and $\sigma_1 < \sigma_2$

Question 3 (2 marks) TECH-FREE

The study score of a subject is normally distributed with a mean $\mu = 30$ and a standard deviation $\sigma = 7$.

a. Determine the approximate percentage of student scores that are between 23 and 37. **(1 mark)**

b. Determine the approximate percentage of student scores that are above 44. **(1 mark)**

More exam questions are available online.

12.3 Calculating probabilities and the standard normal distribution

LEARNING INTENTION

At the end of this subtopic you should be able to:
- calculate probabilities associated with normal distributions
- define the standard normal distribution
- standardise data values.

12.3.1 The standard normal distribution

Suppose we are comparing the results of two students on two similar IQ tests. Theo obtained 92 on one IQ test, for which the results were known to be normally distributed with a mean of 80 and a standard deviation of 6. Samara obtained 88 on a similar IQ test, for which the results were known to be normally distributed with a mean of 78 and a standard deviation of 10. Which student was the most successful?

This question is very difficult to answer unless we have some common ground for a comparison. This can be achieved by using a transformed or standardised form of the normal distribution called the **standard normal distribution**. The variable in a standard normal distribution is always denoted by Z, so that it is immediately understood that we are dealing with the standard normal distribution. The standard normal distribution always has a mean of 0 and a standard deviation of 1, so that Z indicates how many standard deviations the corresponding X-value is from the mean. To find the value of Z, we find the difference between the x-value and the mean, $x - \mu$. To find how many standard deviations this equals, we divide by the standard deviation, σ.

The standard normal distribution

- **The variable is denoted by Z.**
- **The mean is 0 and the standard deviation is 1.**
- **To standardise a data value, x, use the rule $z = \dfrac{x-\mu}{\sigma}$.**

Therefore, if $z = \dfrac{x-\mu}{\sigma}$, $\mu = 0$ and $\sigma = 1$, the probability density function is given by

$$f(z) = \frac{1}{\sqrt{2\pi}} e^{-\frac{1}{2}z^2},\ z \in R.$$

Remember that $\mu \pm 3\sigma$ encompasses approximately 99.7% of the data, so for the standard normal curve, these figures are $0 \pm 3 \times 1 = 0 \pm 3$. Therefore, approximately 99.7% of the data lies between −3 and 3.

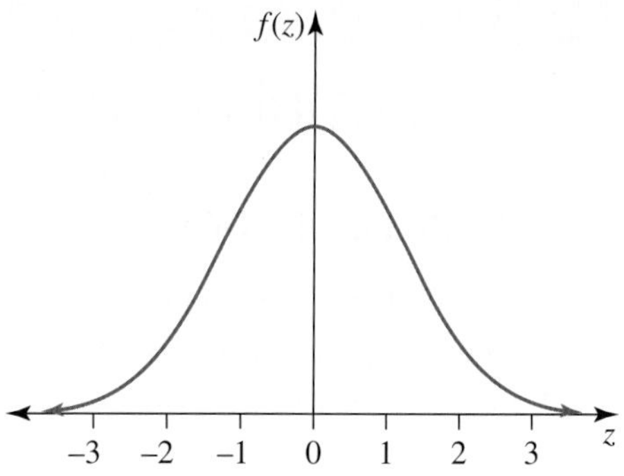

For the standard normal distribution, we say $Z \sim N(0, 1)$.

Let us return to the comparison between Theo and Samara.

For Theo: $X \sim N(80, 6^2)$, $z = \dfrac{x-\mu}{\sigma}$

$$= \frac{92-80}{6}$$
$$= \frac{12}{6}$$
$$= 2$$

For Samara: $X \sim N(78, 10^2)$, $z = \dfrac{x-\mu}{\sigma}$

$$= \frac{88-78}{10}$$
$$= \frac{10}{10}$$
$$= 1$$

Theo's mark lies within 2 standard deviations of the mean, so it lies in the top 2.5%, whereas Samara's mark is 1 standard deviation from the mean, so it is in the top 16%. Hence, Theo performed better than Samara.

Obviously, not all data values will lie exactly 1, 2 or 3 standard deviations from the mean. In these cases technology such as a CAS calculator is needed to calculate the required probability. CAS can be used to calculate probabilities associated with the normal distribution for any value of μ and σ.

WORKED EXAMPLE 3 Probabilities of the normal distribution

a. Calculate the values of the following probabilities correct to 4 decimal places.
   i. $Pr(Z < 2.5)$
   ii. $Pr(-1.25 \le Z \le 1.25)$
b. X is a normally distributed random variable such that $X \sim N(25, 3^2)$.
   i. Calculate $Pr(X > 27)$ correct to 4 decimal places.
   ii. Convert X to a standard normal variable, Z.

THINK	WRITE
a. i. 1. Sketch a graph to help understand the problem.	a. i. 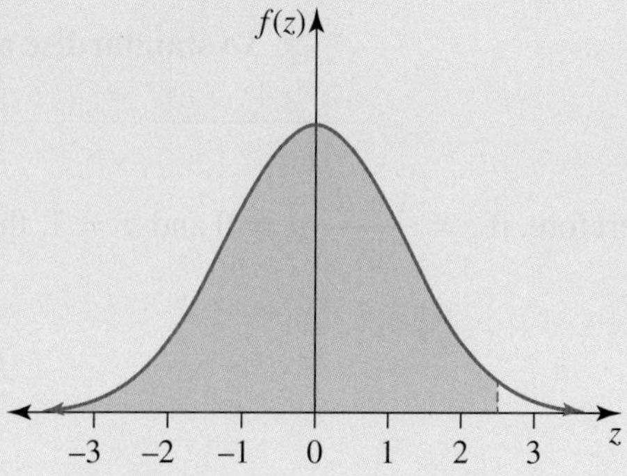

2. Use CAS to find the probability. The upper limit is 2.5 and the lower limit is $-\infty$. The mean is 0 and the standard deviation is 1.	$\Pr(Z < 2.5) = 0.9938$
ii. 1. Sketch a graph to help understand the problem.	**ii.** 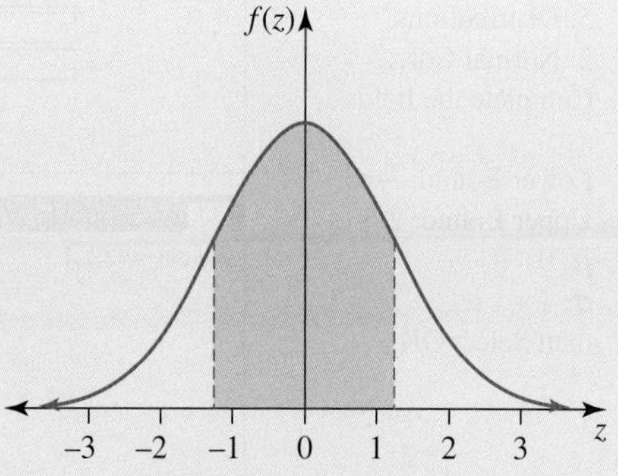
2. Use CAS to find the probability. The upper limit is 1.25 and the lower limit is -1.25.	$\Pr(-1.25 \le Z \le 1.25) = 0.7887$
b. i. 1. Sketch a graph to help understand the problem.	**b. i.** 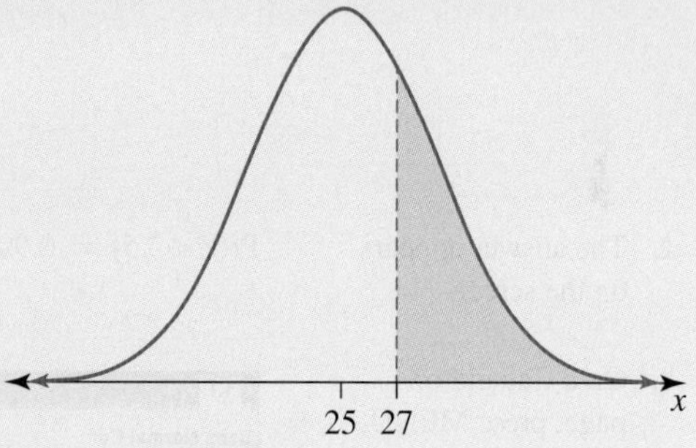
2. Use CAS to find the probability. The upper limit is ∞ and the lower limit is 27. The mean is 25 and the standard deviation is 3.	$\Pr(X > 27) = 0.2525$
ii. 1. Write the rule to standardise X.	**ii.** $z = \dfrac{x - \mu}{\sigma}$
2. Substitute the mean and standard deviation.	$z = \dfrac{27 - 25}{3}$ $= \dfrac{2}{3}$

TI | THINK — **DISPLAY/WRITE**

a. i. 1. On a Calculator page, press MENU, then select:
5: Probability
5: Distributions
2: Normal Cdf…
Complete the fields as:
Lower Bound: $-\infty$
Upper Bound: 2.5
μ: 0
σ: 1
then Select OK.

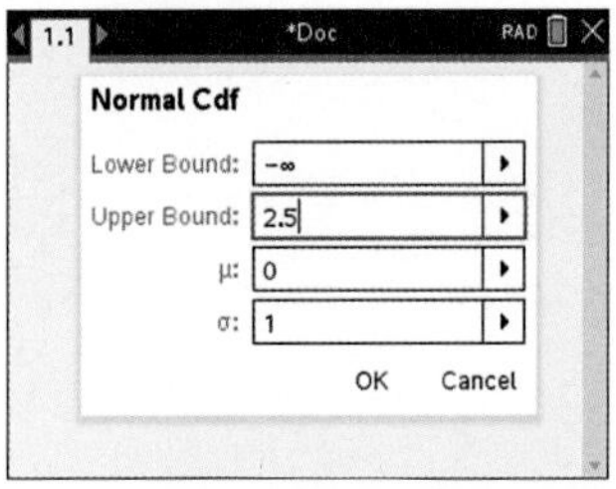

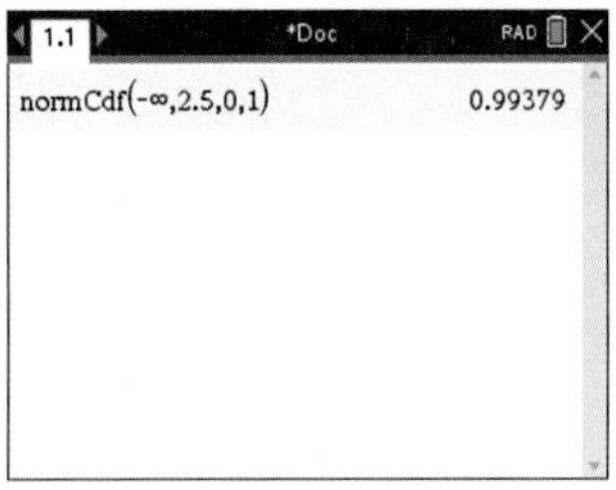

2. The answer appears on the screen.

$\Pr(Z < 2.5) = 0.9938$

a. ii. 1. On a Calculator page, press MENU, then select:
5: Probability
5: Distributions
2: Normal Cdf…
Complete the fields as:
Lower Bound: -1.25
Upper Bound: 1.25
μ: 0
σ: 1
then Select OK.

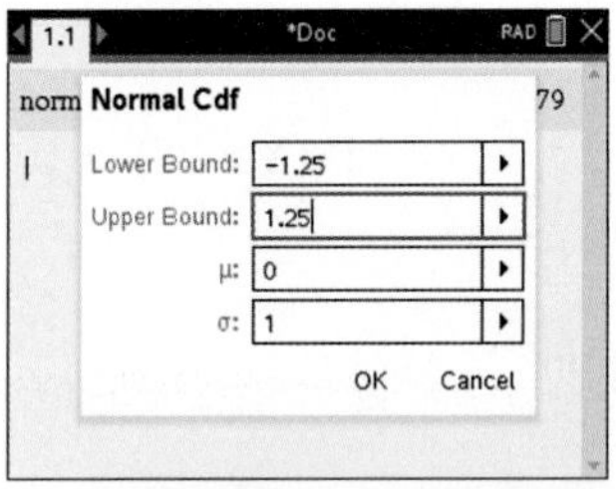

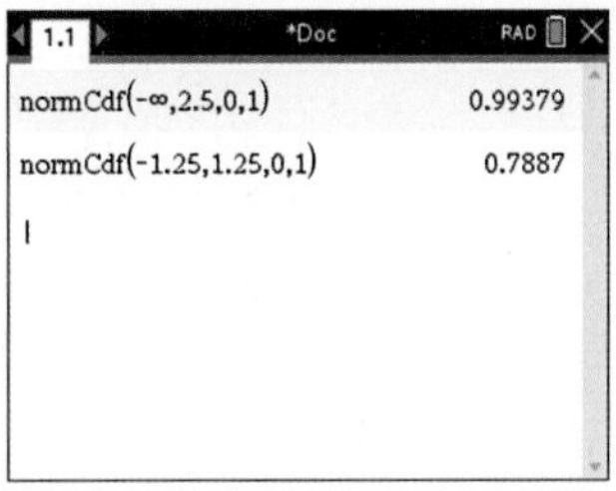

2. The answer appears on the screen.

$\Pr(-1.25 \leq Z \leq 1.25) = 0.7887$

CASIO | THINK — **DISPLAY/WRITE**

a. i. 1. On a Main screen, select:
- Interactive
- Distribution/ Inv. Dist
- Continuous
- normCDf

Complete the fields as:
Lower: $-\infty$
Upper: 2.5
σ: 1
μ: 0
then select OK.

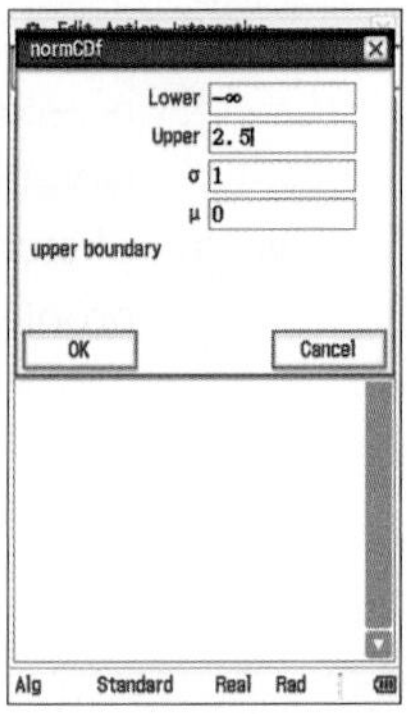

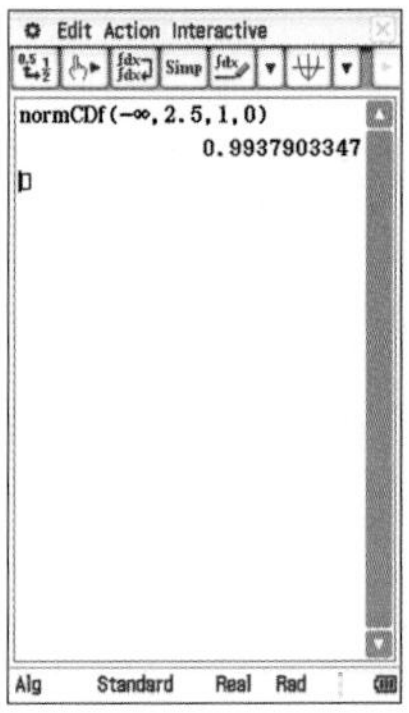

2. The answer appears on the screen.

$\Pr(Z < 2.5) = 0.9938$

a. ii. 1. On a Main screen, select:
- Interactive
- Distribution/ Inv.Dist
- Continuous
- normCDf

Complete the fields as:
Lower: -1.25
Upper: 2.5
σ: 1
μ: 0
then select OK.

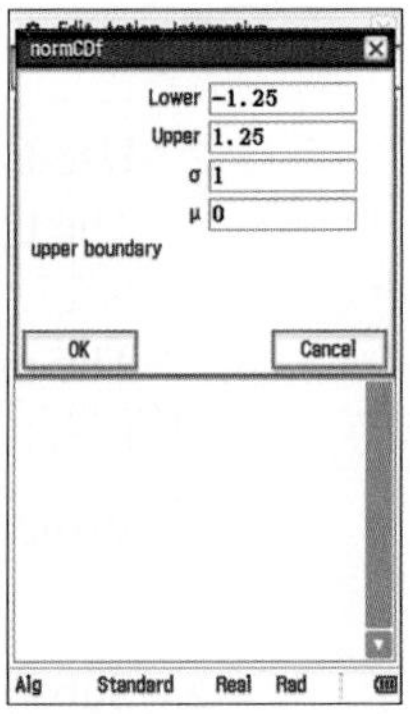

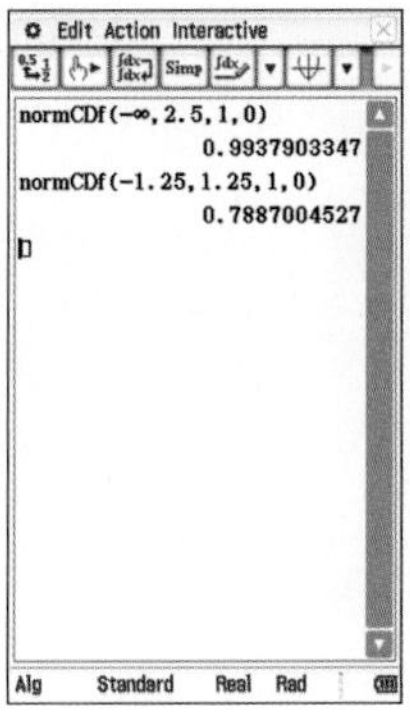

2. The answer appears on the screen.

$\Pr(-1.25 \leq Z \leq 1.25)$
$= 0.7887$

b. i. 1. On a Calculator page, press MENU, then select:
5: Probability
5: Distributions
2: Normal Cdf…
Complete the fields as:
Lower Bound: 27
Upper Bound: ∞
μ: 25
σ: 3
then Select OK.

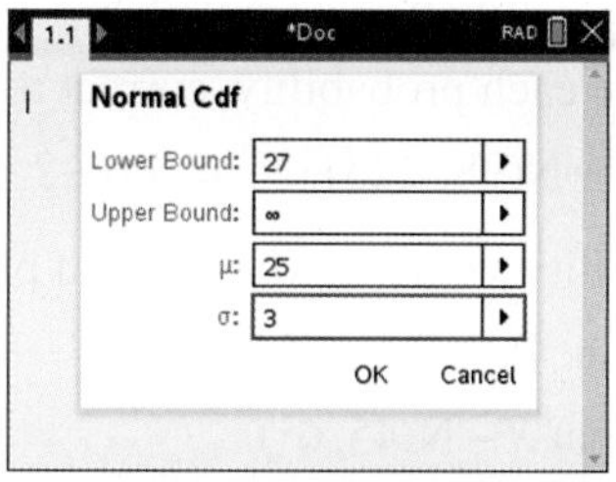

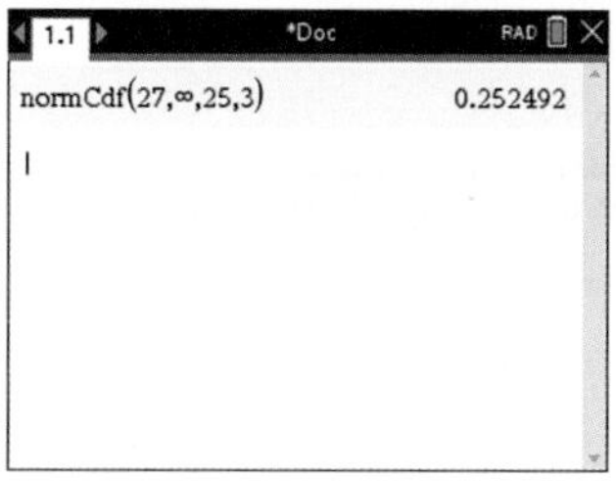

2. The answer appears on the screen.

$\Pr(X > 27) = 0.2525$

b. i. 1. On a Main screen, select:
- Interactive
- Distribution/Inv. Dist
- Continuous
- normCDf

Complete the fields as:
Lower: 27
Upper: ∞
σ: 3
μ: 25
then select OK.

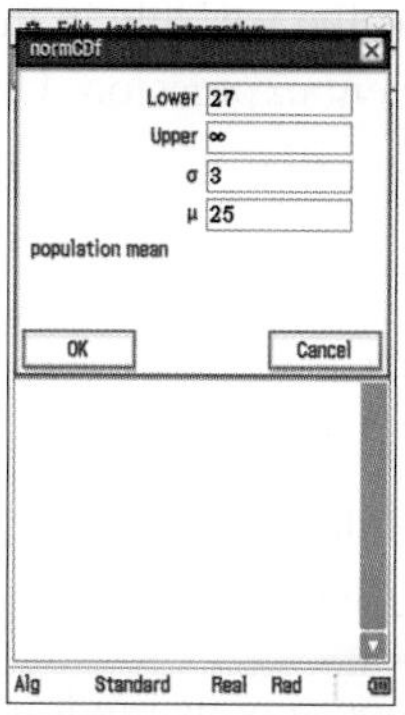

2. The answer appears on the screen.

$\Pr(X > 27) = 0.2525$

on Resources

Interactivities Calculation of probabilities (int-6440)
The standard normal distribution (int-6441)

12.3 Exercise

Technology active

1. For a particular type of laptop computer, the length of time, X hours, between charges of the battery is normally distributed such that

$$X \sim N(50, 15^2)$$

Calculate $\Pr(50 < X < 70)$, correct to 4 decimal places.

2. If $Z \sim N(0, 1)$, calculate, correct to 4 decimal places:

a. $\Pr(Z \le 2)$
b. $\Pr(Z \le -2)$
c. $\Pr(-2 < Z \le 2)$
d. $\Pr(Z > 1.95) \cup \Pr(Z < -1.95)$.

3. Convert the variable in the following expressions to a standard normal variable, Z, and use it to write an equivalent expression. Use your calculator to evaluate each probability, correct to 4 decimal places.

a. $\Pr(X < 61), X \sim N(65, 9)$ b. $\Pr(X \geq 110), X \sim N(98, 225)$ c. $\Pr(-2 < X \leq 5), X \sim N(2, 9)$.

4. WE3 a. Calculate the values of the following probabilities correct to 4 decimal places.

i. $\Pr(Z < 1.2)$ ii. $\Pr(-2.1 < Z < 0.8)$

b. X is a normally distributed random variable such that $X \sim N(45, 6^2)$.

i. Calculate $\Pr(X > 37)$ correct to 4 decimal places.

ii. Convert X to a standard normal variable, Z.

5. X is a continuous random variable and is known to be normally distributed.

a. If $\Pr(X < a) = 0.35$ and $\Pr(X < b) = 0.62$, evaluate:

i. $\Pr(X > a)$ ii. $\Pr(a < X < b)$.

b. If $\Pr(X < c) = 0.27$ and $\Pr(X < d) = 0.56$, evaluate:

i. $\Pr(c < X < d)$ ii. $\Pr(X > c \mid X < d)$.

6. A random variable, X, is normally distributed with a mean of 20 and a standard deviation of 5.

a. Determine k if $\Pr(X > 32) = \Pr(Z > k)$.

b. Determine k if $\Pr(X < 12) = \Pr(Z > k)$.

7. A speed camera is used to measure the speeds of cars on a freeway.

The speeds are normally distributed with a mean of 98 km/h and a standard deviation of 6 km/h. Calculate the probability, correct to 4 decimal places, that a car picked at random is travelling at:

a. more than 110 km/h

b. less than 90 km/h

c. a speed between 90 km/h and 110 km/h.

8. MC A large number of students took a test in Physics. Their final grades have a mean of 72 and a standard deviation of 12. If the distribution of these grades can be approximated by a normal distribution, calculate the percentage of students, correct to 2 decimal places, who:

a. gained a score of more than 95

A. 5.18% B. 97.24% C. 94.82% D. 4.13% E. 2.76%

b. should pass the test if grades greater than or equal to 55 are considered passes.

A. 89.13% B. 92.17% C. 93.25% D. 7.83% E. 10.87%

9. A salmon farm in Tasmania has a very large number of salmon in its ponds. It is known that the lengths of the salmon from this farm are normally distributed with a mean of 38 cm and a standard deviation of 2.4 cm. A randomly chosen fish from this farm was measured as 39.5 cm. If salmon with lengths in the top 15% are considered to be gourmet salmon, determine whether this particular fish can be classified as gourmet.

10. The labels on packets of sugar say the bags have a weight of 1 kg. The actual mean weight of the bags is 1.025 kg in order to minimise the number of bags which may be underweight. If the weight of the bags is normally distributed with a standard deviation of 10 g, calculate the percentage of bags, correct to 2 decimal places, that would be expected to weigh:

a. more than 1.04 kg
b. less than 996 g, the legal meaning of underweight.

11. Teresa has taken her pulse each day for a month after going for a brisk walk. Her pulse rate in beats per minute is known to be normally distributed with a mean of 80 beats per minute and a standard deviation of 5 beats per minute. After her most recent walk she took her pulse rate. Calculate the probability, correct to 4 decimal places, that her pulse rate was:

a. in excess of 85 beats per minute
b. equal to or less than 75 beats per minute
c. between 78 and 82 beats per minute, given that it was higher than 75 beats per minute.

12. Jingjing scored 85 on the mathematics section of a scholarship examination, the results of which were known to be normally distributed with a mean of 72 and a standard deviation of 9. Rani scored 18 on the mathematics section of a similar examination, the results of which were normally distributed with a mean of 15 and a standard deviation of 4. Assuming that both tests measure the same kind of ability, determine which student has the higher score.

13. Juan's results in Chemistry, Mathematical Methods and Physics are shown in the table below. The marks, X, the mean, μ, and standard deviation, σ, for each examination are given.

Subject	**Mark, X**	**Mean, μ**	**Standard deviation, σ**	**Standardised mark, Z**
Chemistry	72	68	5	
Maths Methods	75	69	7	
Physics	68	61	8	

Complete the table by finding Juan's standardised mark for each subject and use this to determine in which subject he did best when compared to his peers.

14. A continuous random variable, Z, has a probability density function defined by $f(z) = 0.025e^{-0.025z}$, $z \geq 0$. A second continuous random variable, Y, is distributed normally with a mean of 25 and a standard deviation of 3. In order to find k such that $\Pr(Z > k) = \Pr(Y < k)$,

$$\int_k^{\infty} f(z)\, dz = \int_{-\infty}^{k} g(y)dy$$

must be solved. Calculate the value of k, correct to 2 decimal places.

12.3 Exam questions

Question 1 (2 marks) TECH-FREE

Source: *VCE 2018 Mathematical Methods Exam 1, Q4; © VCAA.*

Let X be a normally distributed random variable with a mean of 6 and a variance of 4. Let Z be a random variable with the standard normal distribution

a. Find $\Pr(X > 6)$. **(1 mark)**

b. Find b such that $\Pr(X > 7) = \Pr(Z < b)$. **(1 mark)**

Question 2 (1 mark) TECH-ACTIVE

Source: *VCE 2016 Mathematical Methods Exam 2, Section A, Q16; © VCAA.*

MC The random variable, X, has a normal distribution with mean 12 and standard deviation 0.25.

If the random variable, Z, has the standard normal distribution, then the probability that X is greater than 12.5 is equal to

A. $\Pr(Z < -4)$
B. $\Pr(Z < -1.5)$
C. $\Pr(Z < 1)$
D. $\Pr(Z \geq 1.5)$
E. $\Pr(Z > 2)$

Question 3 (1 mark) TECH-ACTIVE

MC If Z has the standard normal distribution, and a and b are positive real numbers, state which of the following is **false**.

A. $\Pr(Z \geq -a) = 1 - \Pr(Z < -a)$
B. $\Pr(Z \leq a) = 0.5 + \Pr(0 < Z < a)$
C. $\Pr(-a \leq Z \leq b) = \Pr(-a < Z < 0) + \Pr(0 < Z < b)$
D. $\Pr(a \leq Z \leq b) = \Pr(Z < b) + \Pr(Z > a) - 0.5$
E. $\Pr(Z = a) = 0$

More exam questions are available online.

12.4 The inverse normal distribution

LEARNING INTENTION

At the end of this subtopic you should be able to:

- calculate values that correspond to particular probabilities
- calculate unknown means and standard deviations.

12.4.1 The inverse normal distribution

CAS technology provides an easy way to find a Z or X value, given a probability for a normal distribution. Suppose X is normally distributed with a mean of 32 and a standard deviation of 5. We wish to find $\Pr(X \leq a) = 0.72$.

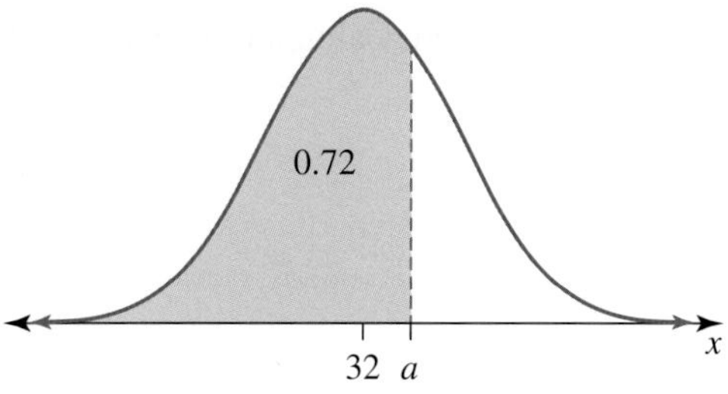

The key information to enter into your calculator is the known probability, that is, the area under the curve. It is essential to input the correct area so that your calculator knows if you are inputting the 'less than' area or the 'greater than' area.

WORKED EXAMPLE 4 Determining values that correspond to given probabilities

If X is a normally distributed random variable, determine, correct to 2 decimal places:

a. m **given that** $\Pr(X \le m) = 0.85$, $X \sim \mathrm{N}\left(15.2,\ 1.5^2\right)$

b. n **given that** $\Pr(X > n) = 0.37$, $X \sim \mathrm{N}\left(22, 2.75^2\right)$

c. p **given that** $\Pr(37.6 - p \le X \le 37.6 + p) = 0.65$, $X \sim \mathrm{N}\left(37.6,\ 12^2\right)$.

THINK	WRITE
a. Use the probability menus on the CAS calculator to find the required X value.	a. $\Pr(X \le m) = 0.85, \mu = 15.2, \sigma = 1.5$ $m = 16.75$
b. Use the probability menus on the CAS calculator to find the required X value. *Note:* It may be a requirement to input the 'less than' area, so $\Pr(X < n) = 1 - 0.37$ $= 0.63$	b. $\Pr(X > n) = 0.37, \mu = 22, \sigma = 2.75$ $n = 22.91$
c. 1. Sketch a graph to visualise the problem. Due to symmetry, the probabilities either side of the upper and lower limits can be calculated.	c. 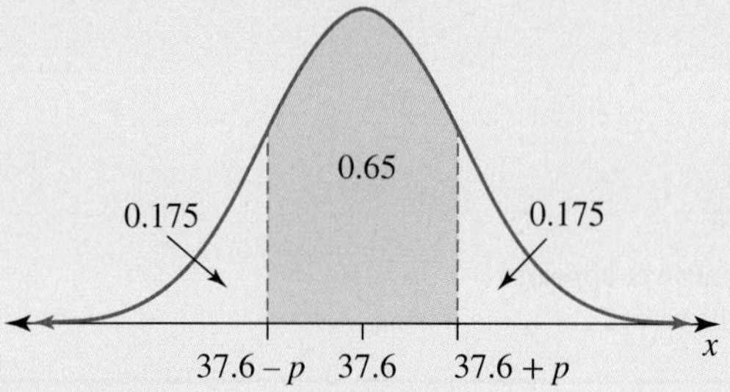 $1 - 0.65 = 0.35$ $\Pr(X < 37.6 - p) = \Pr(X > 37.6 + p)$ $= \frac{0.35}{2}$ $= 0.175$
2. Determine p by finding X given that $\Pr(X < 37.6 - p) = 0.175$. *Note:* p could also be found by using the upper limit.	$\Pr(X < 37.6 - p) = 0.175$ $37.6 - p = 26.38$ $p = 37.6 - 26.38$ $= 11.22$

TI | THINK

DISPLAY/WRITE

a. 1. On a Calculator page, press MENU, then select:
5: Probability
5: Distributions
3: Inverse Normal…
Complete the fields as:
Area: 0.85
μ: 15.2
σ: 1.5
then Select OK.

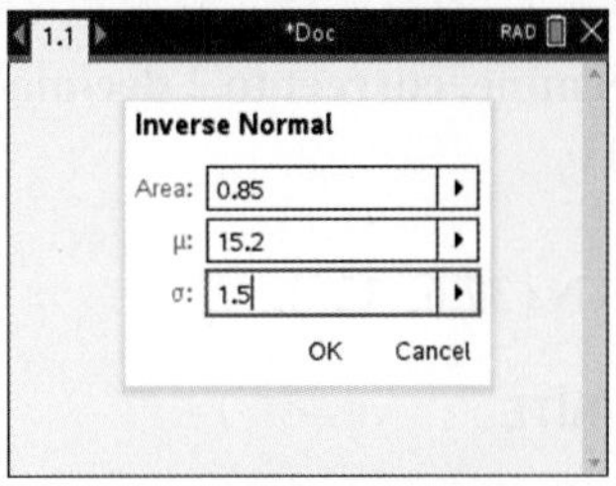

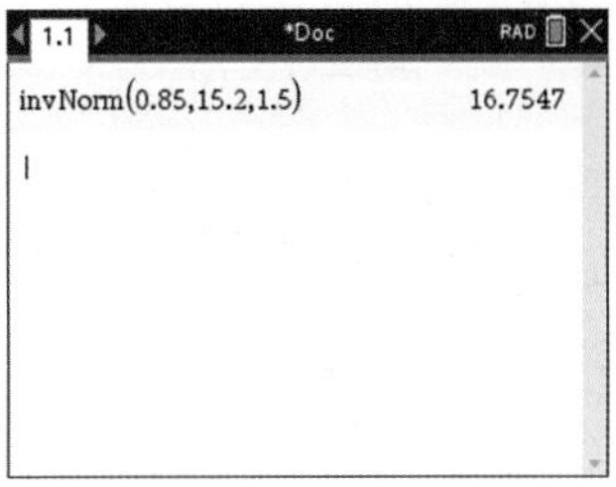

2. The answer appears on the screen.

$m = 16.75$

CASIO | THINK

DISPLAY/WRITE

a. 1. On a Main screen, select:
- Interactive
- Distribution/ Inv. Dist
- Inverse
- invNormCDf

Complete the fields as:
Tail setting: Left
prob: 0.85
σ : 1.5
μ : 15.2
then select OK.

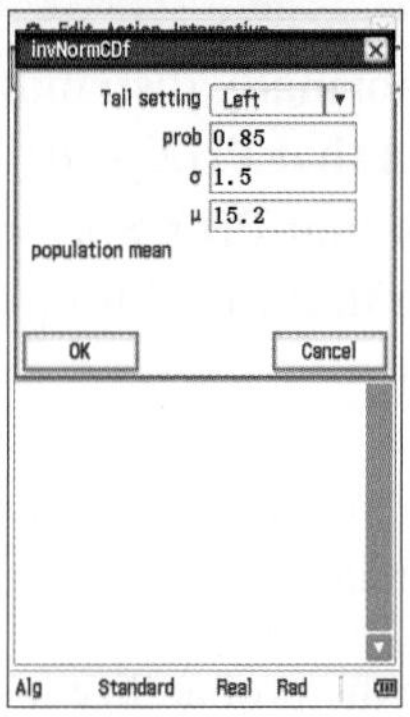

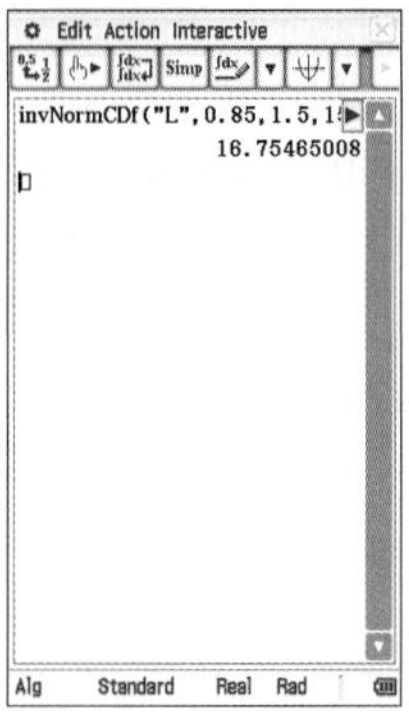

2. The answer appears on the screen.

$m = 16.75$

Quantiles and percentiles

Quantiles and percentiles are terms that enable us to convey information about a distribution. Quantiles refer to the value below which there is a specified probability that a randomly selected value will fall. For example, to find the 0.7 quantile of a standard normal distribution, we find a such that $\Pr(Z < a) = 0.7$.

Percentiles are very similar to quantiles. For the example of $\Pr(Z < a) = 0.7$, we could also be asked to find the 70th percentile for the standard normal distribution.

12.4.2 Calculating the mean or standard deviation

Steps for determining the mean or standard deviation

If the mean or standard deviation is unknown:

- **the known probability needs to be linked to the standard normal distribution and the corresponding z-value calculated via CAS**
- **once the z-value has been found, the missing mean or standard deviation can be calculated via the rule $z = \dfrac{x - \mu}{\sigma}$.**

WORKED EXAMPLE 5 Determining the mean and standard deviation

a. For the normally distributed variable X, the 0.15 quantile is 1.9227 and the mean is 2.7. Calculate the standard deviation of the distribution, correct to 2 decimal places.

b. X is normally distributed so that the 63rd percentile is 15.896 and the standard deviation is 2.7. Calculate the mean of X, correct to 2 decimal places.

THINK	WRITE
a. 1. Write the probability statement.	**a.** The 0.15 quantile is 1.9227. $\Pr(X < 1.9227) = 0.15$
2. Find the corresponding standardised value, Z, by using CAS.	$\Pr(Z < z) = 0.15$ $z = -1.0364$
3. Write the standardised formula connecting z and x.	$z = \dfrac{x - \mu}{\sigma}$
4. Substitute the appropriate values and solve for σ.	$-1.0364 = \dfrac{1.9227 - 2.7}{\sigma}$ $-1.0364\,\sigma = -0.7773$ $\sigma = 0.75$
b. 1. Write the probability statement.	**b.** The 63rd percentile is 15.896. $\Pr(X < 15.896) = 0.63$
2. Find the corresponding standardised value, Z, by using CAS.	$\Pr(Z < z) = 0.63$ $z = 0.3319$
3. Write the standardised formula connecting z and x.	$z = \dfrac{x - \mu}{\sigma}$
4. Substitute in the appropriate values and solve for μ.	$0.3319 = \dfrac{15.896 - \mu}{2.7}$ $0.8960 = 15.896 - \mu$ $\mu = 15.00$

12.4 Exercise

Technology active

1. **WE4** If X is a normally distributed random variable, determine the value of a, correct to 2 decimal places, given that:
   a. $\Pr(X \le a) = 0.16$, $X \sim \text{N}(41, 6.7^2)$
   b. $\Pr(X > a) = 0.21$, $X \sim \text{N}(12.5, 2.7^2)$
   c. $\Pr(15 - a \le X \le 15 + a) = 0.32$, $X \sim \text{N}(15, 4^2)$.
2. Calculate the values of m and n if X is normally distributed and $\Pr(m \le X \le n) = 0.92$ when $\mu = 27.3$ and $\sigma = 8.2$. The specified interval is symmetrical about the mean. Give your answers correct to 2 decimal places.
3. If $Z \sim \text{N}(0, 1)$, calculate the value of z correct to 4 decimal places given that:
   a. $\Pr(Z < z) = 0.39$
   b. $\Pr(Z \ge z) = 0.15$
   c. $\Pr(-z < Z < z) = 0.28$.

4. If $X \sim N(37.5,\ 8.62^2)$, determine a correct to 2 decimal places such that:

a. $\Pr(X < a) = 0.72$

b. $\Pr(X > a) = 0.32$

c. $\Pr(37.5 - a < X < 37.5 + a) = 0.88$.

5. For a standard normal distribution, determine, correct to 4 decimal places:

a. the 0.57 quantile

b. the 63rd percentile.

6. If X is distributed normally with $\mu = 43.5$ and $\sigma = 9.7$, determine, correct to 4 decimal places:

a. the 0.73 quantile

b. the 24th percentile.

7. WE5a X is distributed normally with a mean of 112, and the 42nd percentile is 108.87. Calculate the standard deviation of the distribution, correct to 1 decimal place.

8. X is a normally distributed random variable such that $X \sim N(\mu, 4.45^2)$. If the 0.11 quantile is 32.142, determine the value of μ, correct to 1 decimal place.

9. WE5b X is distributed normally with a standard deviation of 5.67, and $\Pr(X > 20.952) = 0.09$. Calculate the mean of X, giving your answer correct to 2 decimal places.

10. MC X is distributed normally with a standard deviation of 3.5, and $\Pr(X < 23.96) = 0.28$. The mean for X is closest to:

A. 26

B. 25

C. 22

D. −22

E. 21

11. $X \sim N(115,\ \sigma^2)$ and the 76th percentile is 122.42. Determine the value of σ, giving your answer correct to 1 decimal place.

12. MC X is distributed normally with $\mu = 41$ and $\Pr(X > 55.9636) = 0.11$. The standard deviation is:

A. 10.5

B. −12.2

C. −1.2

D. 12.2

E. 1.5

13. X is distributed normally and $\Pr(X < 33.711) = 0.36$, while $\Pr(X < 34.10) = 0.42$. Determine the mean and the standard deviation of X, giving your answer correct to 1 decimal place.

14. X is distributed normally such that $\Pr(X > 18.35) = 0.31$ and the 45th percentile is 15.09. Determine μ and σ for X, giving your answers correct to 1 decimal place.

15. X is distributed normally such that $\Pr(X < 39.9161) = 0.5789$ and $\Pr(X > 38.2491) = 0.4799$. Determine the mean and the standard deviation of X, giving your answers correct to 2 decimal places.

16. $\Pr(a < X < b) = 0.52$ and the specified interval is symmetrical about the mean. If X is normally distributed with a mean of 42.5 and a standard deviation of 10.3, determine $\Pr(X > a \mid X < b)$.

12.4 Exam questions

Question 1 (1 mark) TECH-ACTIVE

Source: *VCE 2020 Mathematical Methods Exam 2, Section A, Q11; © VCAA.*

MC The lengths of plastic pipes that are cut by a particular machine are a normally distributed random variable, X, with a mean of 250 mm.

Z is the standard normal random variable.

If $\Pr(X < 259) = 1 - \Pr(Z > 1.5)$, then the standard deviation of the lengths of plastic pipes, in millimetres, is

A. 1.5

B. 3

C. 6

D. 9

E. 12

Question 2 (1 mark) TECH-ACTIVE

Source: *VCE 2020 Mathematical Methods Exam 2, Section A, Q14; © VCAA.*

MC The random variable X is normally distributed.

The mean of X is twice the standard deviation of X.

If $\Pr(X > 5.2) = 0.9$, then the standard deviation of X is closest to

A. 7.238 **B.** 14.476 **C.** 3.327 **D.** 1.585 **E.** 3.169

Question 3 (1 mark) TECH-ACTIVE

Source: *VCE 2019 Mathematical Methods Exam 2, Section A, Q14; © VCAA.*

MC The weights of packets of lollies are normally distributed with a mean of 200 g.

If 97% of these packets of lollies have a weight of more than 190 g, then the standard deviation of the distribution, correct to one decimal place, is

A. 3.3 g **B.** 5.3 g **C.** 6.1 g **D.** 9.4 g **E.** 12.1 g

More exam questions are available online.

12.5 Mixed probability applications

LEARNING INTENTION

At the end of this subtopic you should be able to:

- apply your knowledge of normal distributions to a range of application problems
- answer mixed probability questions.

12.5.1 Applications

Application problems involving the normal distribution cover a wide range of topics. Such questions will not only incorporate theory associated with the normal distribution but may also include other areas of probability you have previously studied.

WORKED EXAMPLE 6 Mixed probability

The amount of instant porridge oats in packets packed by a particular machine is normally distributed with a mean of μ grams and a standard deviation of 6 grams. The advertised weight of a packet is 500 grams.

a. Calculate the proportion of packets that will be underweight (less than 500 grams) when $\mu = 505$ grams. Give your answer to 4 decimal places.

b. Calculate the value of μ required to ensure that only 1% of packets are underweight. Give your answer correct to 2 decimal places.

c. As a check on the setting of the machine, a random sample of 5 boxes is chosen and the setting is changed if more than one of them is underweight. Determine the probability, correct to 4 decimal places, that the setting on the machine is changed when $\mu = 505$ grams.

THINK	WRITE
a. 1. Rewrite the information in the question using appropriate notation.	**a.** X is the amount of instant porridge oats in a packet and $X \sim N(505, 6^2)$.
2. Use CAS to find $\Pr(X < 500)$.	$\Pr(X < 500) = 0.2023$
b. 1. State the known probability.	**b.** $\Pr(X < 500) = 0.01$
2. Find the corresponding standardised value, Z, by using CAS.	$\Pr(Z < z) = 0.01$ $z = -2.3263$

3. Write the standardised formula connecting z and x.

$$z = \frac{x - \mu}{\sigma}$$

4. Substitute the appropriate values and solve for μ.

$$-2.3263 = \frac{500 - \mu}{6}$$
$$-13.9581 = 500 - \mu$$
$$\mu = 513.96\text{ g}$$

c. 1. The wording of the question (sample of 5 boxes) indicates that this is now a binomial distribution. Rewrite the information in the question using appropriate notation.

c. Let $Y =$ the number of underweight packets.
$Y \sim \text{Bi}(5, 0.2023)$

2. Using CAS, calculate the probability.

$$\Pr(Y > 1) = 1 - \Pr(Y \leq 1)$$
$$= 1 - 0.7325$$
$$= 0.2675$$

12.5 Exercise

Technology active

1. **WE6** Packages of butter with a stated weight of 500 grams have an actual weight of W grams, which is normally distributed with a mean of 508 grams.

   a. If the standard deviation of W is 3.0 grams, calculate, correct to 4 decimal places:
      i. the proportion of packages that weigh less than 500 grams
      ii. the weight that is exceeded by 99% of the packages.
   b. If the probability that a package weighs less than 500 grams is not to exceed 0.01, determine the maximum allowable standard deviation of W. Give your answer correct to 4 decimal places.

2. Chocolate Surprise is a toy that is packed inside an egg-shaped chocolate. A certain manufacturer provides four different types of Chocolate Surprise toy — a car, an aeroplane, a ring and a doll — in the proportions given in the table.

Toy	Proportion
Car	$3k^2 + 2k$
Aeroplane	$6k^2 + 2k$
Ring	$k^2 + 2k$
Doll	$3k$

a. Show that k must be a solution to the equation $10k^2 + 9k - 1 = 0$.
b. Evaluate the value of k.

In response to customer demand, the settings on the machine that produce Chocolate Surprise have been changed so that 25% of all Chocolate Surprises produced contain rings. A sample of 8 Chocolate Surprises is randomly selected from a very large number produced by the machine.

c. Calculate the expected number of Chocolate Surprises that contain rings.
d. Calculate the probability, correct to 4 decimal places, that this sample has exactly 2 Chocolate Surprises that contain rings.
e. Determine the smallest sample size that should be taken so that the probability of selecting no Chocolate Surprise that contain a ring is at most 0.09.

A Chocolate Surprise is considered defective if it weighs less than 100 grams. The weight of a Chocolate Surprise is known to be normally distributed with a mean of 125 grams.

f. If 8.2% of the Chocolate Surprises produced are defective, determine, to the nearest gram, the standard deviation for the weight of the Chocolate Surprises.

3. A particular brand of car speedometer was tested for accuracy. The error measured is 0 km/h and a standard deviation of 0.76 km/h. Speedometers are considered unacceptable if the error is more than 1.5 km/h. Calculate the proportion of speedometers that are unacceptable. Give your answer correct to 4 decimal places.

4. The heights of adult males in Perth can be taken as normally distributed with a mean of 174 cm and a standard deviation of 8 cm. Suppose the Western Australian Police Force accepts recruits only if they are at least 180 cm tall.
   a. Calculate the percentage of Perth adult males who satisfy the height requirement for the Western Australian Police Force. Give your answer correct to 2 decimal places.
   b. Determine the minimum height, to the nearest centimetre, that the Western Australian Police Force would have to accept if it wanted a quarter of the Perth adult male population to satisfy the height requirement.

5. a. Farmer David grows avocados on a farm on Mount Tamborine, Queensland.
   The average weight of his avocados is known to be normally distributed with a mean weight of 410 grams and a standard deviation of 20 grams.
   Calculate, correct to 4 decimal places:
      i. the probability that an avocado chosen at random weighs less than 360 grams.
      ii. the probability that an avocado that weighs less than 360 grams weighs more than 340 grams.

   b. Farmer Jane grows avocados on a farm next to farmer David's. If Y represents the average weight of Jane's avocados, the weights of which are also normally distributed where $\Pr(Y < 400) = 0.4207$ and $\Pr(Y > 415) = 0.3446$, calculate the mean and standard deviation of the weights of Jane's avocados. Give answers correct to the nearest integer.

6. A manufacturer produces metal rods whose lengths are normally distributed with a mean of 145.0 cm and a standard deviation 1.4 cm.

a. Determine the probability, correct to 4 decimal places, that a randomly selected metal rod is longer than 146.5 cm.

b. A metal rod has a size fault if its length is not within d cm either side of the mean. The probability of a metal rod having a size fault is 0.15. Calculate the value of d, giving your answer correct to 1 decimal place.

c. A random sample of 12 metal rods is taken from a crate containing a very large number of metal rods. Determine the probability that there are exactly 2 metal rods with a size fault, giving your answer correct to 4 decimal places.

d. The sales manager is considering what price, x dollars, to sell each of the metal rods for, whether they are good or have some kind of fault. The materials cost is \$5 per rod. The metal rods are sorted into three bins. The staff know that 15% of the manufactured rods have a size fault and another 17% have some other fault. The profit, Y dollars, is a random variable whose probability distribution is shown in the following table.

Bin	Description	Profit (\$$y$)	$\Pr(Y = y)$
A	Good metal rods that are sold for x dollars each.	$x - 5$	a
B	Metal rods with a size fault — these are not sold but recycled.	0	0.15
C	Metal rods with other faults — these are sold at a discount of \$3 each.	$x - 8$	0.17

i. Determine the value of a.
ii. Determine the mean of Y in terms of x.
iii. Hence or otherwise, evaluate, the selling price of good rods so that the mean profit is zero.
iv. The metal rods are stored in the bins until a large number is ready to be sold. Determine what proportion of the rods ready to be sold are good rods.

7. A company sells two different products, X and Y, for \$5.00 and \$6.50 respectively. Regular markets exist for both products, with sales being normally distributed and averaging 2500 units (standard deviation 700) and 3000 units (standard deviation 550) respectively each week. It is company policy that if in any one week the market for a particular product falls below half the average, that product is advertised as a 'special' the following week.

a. Calculate the probability, correct to 4 decimal places, that product X will be advertised as a 'special' next week.

b. Calculate the probability, correct to 4 decimal places, that product Y will be advertised as a 'special' next week.

c. Calculate the probability, correct to 4 decimal places, that both products will be advertised as a 'special' next week.

d. If 40% of the company's product is product X and 60% is product Y, calculate the probability correct to 4 decimal places that:

i. one product is a 'special'
ii. if one product is advertised as 'special', then it is product X.

8. The height of plants sold at a garden nursery supplier are normally distributed with a mean of 18 cm and a standard deviation of 5 cm.

a. Complete the following table by determining the proportions for each of the three plant sizes, correct to 4 decimal places.

Description of plant	Plant size (cm)	Cost in $	Proportion of plants
Small	Less than 10 cm	2.00	
Medium	10–25 cm	3.50	
Large	Greater than 25 cm	5.00	

b. Evaluate the expected cost, to the nearest dollar, for 150 plants chosen at random from the garden nursery.

9. A fruit grower produces peaches whose weights are normally distributed with a mean of 185 grams and a standard deviation of 20 grams.

Peaches whose weights exceed 205 grams are sold to the cannery, yielding a profit of 60 cents per peach. Peaches whose weights are between 165 and 205 grams are sold to wholesale markets at a profit of 45 cents per peach. Peaches whose weights are less than 165 grams are sold for jam at a profit of 30 cents per peach.

a. Determine the percentage of peaches sold to the canneries correct to 2 decimal places.

b. Determine the percentage of peaches sold to the wholesale markets correct to 2 decimal places.

c. Calculate the mean profit per peach correct to the nearest cent.

10. The Lewin Tennis Ball Company makes tennis balls whose diameters are distributed normally with a mean of 70 mm and a standard deviation of 1.5 mm. The tennis balls are packed and sold in cylindrical tins that each hold five tennis balls. A tennis ball fits in the tin if the diameter is less than 71.5 mm.

a. Calculate the probability, correct to 4 decimal places, that a randomly chosen tennis ball produced by the Lewin company fits into the tin.

The Lewin management would like each ball produced to have a diameter between 68.6 mm and 71.4 mm.

b. Calculate the probability, correct to 4 decimal places, that a randomly chosen tennis ball produced by the Lewin company is in this range.

c. A tin of five balls is selected at random. Calculate the probability, correct to 4 decimal places, that at least one ball has a diameter outside the range of 68.6 mm to 71.4 mm.

d. Lewin management wants engineers to change the manufacturing process so that 99.5% of all balls produced have a diameter between 68.6 mm and 71.4 mm. The mean is to stay at 70 mm but the standard deviation is to be changed. Determine what should the new standard deviation be, correct to 4 decimal places.

11. The Apache Orchard grows a very juicy apple called the Fuji apple. Fuji apples are picked and then sorted by diameter in three categories:
- small — diameter less than 60 mm
- jumbo — the largest 15% of the apples
- standard — all other apples.

Diameters of Fuji apples are found to be normally distributed with a mean of 71 mm and a standard deviation of 12 mm.

a. A particular apple is the largest possible whose diameter lies within 2 standard deviations of the mean. Calculate the diameter. Give your answer correct to the nearest millimetre.
b. Calculate, correct to 4 decimal places, the probability that a Fuji apple, selected at random, has a diameter less than 85 mm.
c. Calculate the percentage of apples (to the nearest 1%) that is sorted into the small category.
d. Determine, correct to the nearest millimetre, the minimum diameter of a jumbo Fuji.
e. An apple is selected at random from a bin of jumbo apples. Calculate the probability, correct to 4 decimal places, that it has a diameter greater than 100 mm.
f. The Apache Orchard receives the following prices for Fuji apples:
   - small — 12 cents each
   - standard — 15 cents each
   - jumbo — 25 cents each.

   Calculate the orchard's expected income, correct to the nearest dollar, for a container of 2500 unsorted apples.
g. Some apples are selected before sorting and are packed into bags of six to be sold at the front gate of the orchard. Calculate the probability, correct to 4 decimal places, that one of these bags contains at least two jumbo apples.

12. A brand of disinfectant is sold in two sizes: standard and large. For each size, the contents, in litres, of a randomly chosen bottle is normally distributed with a mean and standard deviation as shown in the following table.

Bottle size	Mean	Standard deviation
Standard	0.765 L	0.007 L
Large	1.015 L	0.009 L

a. Calculate the probability, correct to 4 decimal places, that a randomly chosen standard bottle contains less than 0.75 litres.
b. Calculate the probability, correct to 4 decimal places, that a box of 12 randomly chosen large bottles contains at least 4 bottles whose contents are each less than 1 litre.

13. Amalie is gathering data on two particular species of yellow butterflies: the lemon emigrant and the yellow emigrant, which can be very difficult to tell apart as the intensity of the yellow can be confusing. Both species are equally likely to be caught in a particular area of Australia. One technique for telling them apart is by measuring the lengths of their antennae. For the lemon emigrant, the antennae are distributed normally with a mean of 22 mm and a standard deviation of 1.5 mm.

In the region where Amalie is hunting for yellow butterflies, 45% of the yellow butterflies are lemon emigrants and 55% are lemon emigrants.

a. Calculate the probability, correct to 4 decimal places, that a randomly chosen lemon emigrant butterfly will have antennae which are shorter than 18 mm.
b. Amalie knows that 8% of the yellow emigrants have antennae that are shorter than 15.5 mm, and 8% of yellow emigrant butterflies have antennae that are longer than 22.5 mm. Assuming that the antenna lengths are normally distributed, determine the mean and standard deviation of the antenna length of yellow emigrant butterflies, giving your answers correct to the nearest 0.1 mm.
c. Calculate the probability, correct to 4 decimal places, that a random sample of 12 butterflies from the region will contain 5 yellow emigrant butterflies.

14. The daily error (in seconds) of a particular brand of clock is known to be normally distributed. Only those clocks with an error of less than 3 seconds are acceptable. Determine, correct to 4 decimal places:

a. the mean and standard deviation of the distribution of error if 2.5% of the clocks are rejected for losing time and 2.5% of the clocks are rejected for gaining time
b. the probability that fewer than 2 clocks are rejected in a batch of 12 such clocks.

12.5 Exam questions

Question 1 (1 mark) TECH-ACTIVE

Source: *VCE 2013 Mathematical Methods (CAS) Exam 2, Section 1, Q22; © VCAA.*

MC Butterflies of a particular species die T days after hatching, where T is a normally distributed random variable with a mean of 120 days and a standard deviation of σ days.

If, from a population of 2000 newly hatched butterflies, 150 are expected to die in the first 90 days, then the value of σ is closest to

A. 7 days
B. 13 days
C. 17 days
D. 21 days
E. 37 days

Question 2 (8 marks) TECH-ACTIVE

Source: *VCE 2014 Mathematical Methods (CAS) Exam 2, Section 2, Q4a–e; © VCAA.*

Patricia is a gardener and she owns a garden nursery. She grows and sells basil plants and coriander plants.

The heights, in centimetres, of the basil plants that Patricia is selling are distributed normally with a mean of 14 cm and a standard deviation of 4 cm. There are 2000 basil plants in the nursery.

a. Patricia classifies the tallest 10 per cent of her basil plants as **super**.
What is the minimum height of a super basil plant, correct to the nearest millimetre? **(1 mark)**

b. Patricia decides that some of her basil plants are not growing quickly enough, so she plans to move them to a special greenhouse. She will move the basil plants that are less than 9 cm in height.

How many basil plants will Patricia move to the greenhouse, correct to the nearest whole number? **(2 marks)**

c. The heights of the coriander plants, x centimetres, follow the probability density function $h(x)$, where

$$h(x) = \begin{cases} \dfrac{\pi}{100}\sin\left(\dfrac{\pi x}{50}\right) & 0 < x < 50 \\ 0 & \text{otherwise} \end{cases}$$

State the mean height of the coriander plants. **(1 mark)**

d. Patricia thinks that the smallest 15 per cent of her coriander plants should be given a new type of plant food.
Find the maximum height, correct to the nearest millimetre, of a coriander plant if it is to be given the new type of plant food. **(2 marks)**

e. Patricia also grows and sells tomato plants that she classifies as either **tall** or **regular**. She finds that 20 per cent of her tomato plants are tall.
A customer, Jack, selects n tomato plants at random.
Let q be the probability that at least one of Jack's n tomato plants is tall.
Find the minimum value of n so that q is greater than 0.95. **(2 marks)**

Question 3 (11 marks) TECH-ACTIVE

Source: *VCE 2015 Mathematical Methods (CAS) Exam 2, Section 2, Q3; © VCAA.*

Mani is a fruit grower. After his oranges have been picked, they are sorted by a machine, according to size. Oranges classified as **medium** are sold to fruit shops and the remainder are made into orange juice.

The distribution of the diameter, in centimetres, of medium oranges is modelled by a continuous random variable, X, with probability density function

$$f(x) = \begin{cases} \frac{3}{4}(x-6)^2(8-x) & 6 \le x \le 8 \\ 0 & \text{otherwise} \end{cases}$$

a. i. Find the probability that a randomly selected medium orange has a diameter greater than 7 cm. **(2 marks)**

ii. Mani randomly selects three medium oranges.
Find the probability that exactly one of the oranges has a diameter greater than 7 cm.
Express the answer in the form $\frac{a}{b}$, where a and b are positive integers. **(2 marks)**

b. Find the mean diameter of medium oranges, in centimetres. **(1 mark)**

c. For oranges classified as large, the quantity of juice obtained from each orange is a normally distributed random variable with a mean of 74 mL and a standard deviation of 9 mL.

What is the probability, correct to three decimal places, that a randomly selected large orange produces less than 85 mL of juice, given that it produces more than 74 mL of juice? **(2 marks)**

d. Mani also grows lemons, which are sold to a food factory. When a truckload of lemons arrives at the food factory, the manager randomly selects and weighs four lemons from the load. If one or more of these lemons is underweight, the load is rejected. Otherwise it is accepted.

It is known that 3% of Mani's lemons are underweight.

i. Find the probability that a particular load of lemons will be rejected. Express the answer correct to four decimal places. **(2 marks)**

ii. Suppose that instead of selecting only four lemons, n lemons are selected at random from a particular load.

Find the smallest integer value of n such that the probability of at least one lemon being underweight exceeds 0.5. **(2 marks)**

More exam questions are available online.

12.6 Review

12.6.1 Summary

doc-37042

Hey students! Now that it's time to revise this topic, go online to:

Access the topic summary | Review your results | Watch teacher-led videos | Practise VCAA exam questions

Find all this and MORE in jacPLUS

12.6 Exercise

Technology free: short answer

1. Let X be normally distributed with a mean of 76 and a standard deviation of 6. Let Z be a standard normal distribution. Use the result that $\Pr(Z < 1) = 0.84$, correct to 2 decimal places, to determine:
   a. the probability that X is greater than 82
   b. the probability that $70 < X < 76$
   c. the probability that $X > 70$ given that $X < 76$.
2. A random variable, X, is normally distributed with a mean of 31 and a standard deviation of 6. State how the mean and standard deviation affect the shape of the normal distribution graph (in terms of transformations).
3. The time Lee Yin spends swimming each day is normally distributed with a mean of 35 minutes and a standard deviation of 4 minutes.

   Determine the approximate percentage of swims that are:
   a. between 31 and 39 minutes
   b. less than 43 minutes
   c. between 27 and 39 minutes.
4. If $\Pr(X > m) = 0.65$, evaluate:
   a. $\Pr(X < m)$
   b. $\Pr(X > -m)$.
5. By standardising, convert the following X-values to Z-values.
   a. $X \sim N(15, 2^2)$, $X = 22$
   b. $X \sim N(180, 25)$, $X = 192$

6. a. If $\Pr(m < X < n) = 0.54$ and $\Pr(X < n) = 0.72$, determine $\Pr(X \leq m)$.
   b. If $\Pr(m < X < n) = 0.75$ and $\Pr(X > n) = 0.18$, determine $\Pr(X \leq m)$.

Technology active: multiple choice

7. **MC** The diagram below shows two normal distributions with means μ_1 and μ_2 and variances σ_1^2 and σ_2^2.
   State which one of the following sets of statements is true.
   A. $\mu_1 < \mu_2$ and $\sigma_1 < \sigma_2$
   B. $\mu_1 > \mu_2$ and $\sigma_1 < \sigma_2$
   C. $\mu_1 < \mu_2$ and $\sigma_1 > \sigma_2$
   D. $\mu_1 > \mu_2$ and $\sigma_1 > \sigma_2$
   E. $\mu_1 > \mu_2$ and $\sigma_1 = \sigma_2$

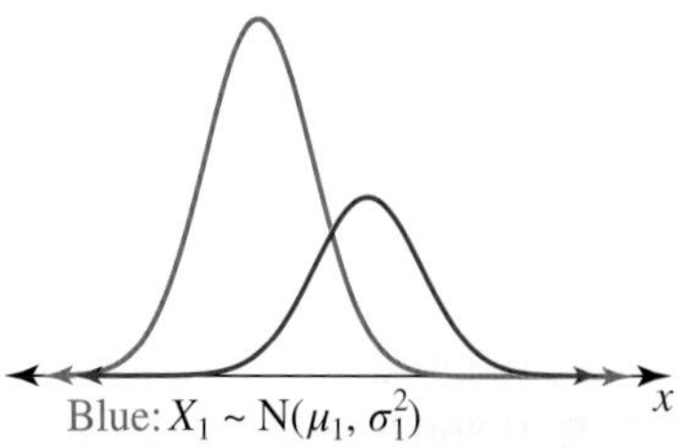

8. **MC** Chris carefully measures the quantity of orange juice contained in 1-litre cartons of a particular brand. He finds the actual quantity of orange juice is approximately normally distributed with a mean of 1.05 litres and a standard deviation of 0.05 litres.

   The proportion of cartons that contain more than 1 litre of orange juice is closest to:
   A. 0.16
   B. 0.84
   C. 0.90
   D. 0.10
   E. 0.05

9. **MC** If Z is a standard normal distribution, state which one of the following is **not** true.
   A. The mean, median and mode of Z are all the same.
   B. The mean is zero and the standard deviation is 1.
   C. $\Pr(Z < -1) = 1 - \Pr(Z > 1)$
   D. $\Pr(Z > 0) = 0.5$
   E. $\Pr(-3 < Z < 3) = 0.997$

10. **MC** Let X be normally distributed with a mean of μ and a standard deviation of σ. State which one of the following is not always true.
    A. $\Pr(X < \mu) = 0.5$
    B. $\Pr(\mu - \sigma < X < \mu + \sigma) \approx 0.68$
    C. $\Pr(X < a) = 1 - \Pr(X > a)$
    D. $\Pr(\mu - 2\sigma < X < \mu + 2\sigma) \approx 0.95$
    E. $\Pr(a < X < b) = \Pr(X < b) - \Pr(X > a)$

11. **MC** The random variable X has a normal distribution with a mean of 4.9 and a standard deviation of 1.4. If Z is a standard normal distribution, then the probability that X is less than 3.5 is equal to:
    A. $\Pr(Z < 1)$
    B. $\Pr(Z > 1)$
    C. $1 - \Pr(Z < -1)$
    D. $\Pr(Z > -1)$
    E. $\Pr(-1 < Z < 1)$

12. **MC** The eggs laid by a particular breed of chicken have a mass that is normally distributed with a mean of 63 grams and a standard deviation of 2.7 grams.

The probability, correct to 4 decimal places, that a single egg chosen at random has a mass between 60 grams and 65 grams is:

A. 0.1332
B. 0.7705
C. 0.6373
D. 0.2294
E. 0.8667

13. **MC** The continuous random variable X is normally distributed with a mean of 32 and a standard deviation of 7. For a given number a, $\Pr(X > a) = 0.20$. Correct to 2 decimal places, a is equal to:

A. 26.11 B. 36.72 C. 27.28 D. 37.89 E. 35.89

14. **MC** In an orchard of 3000 apple trees, it is found that 2570 trees have a height greater than 2.9 metres. The heights of the trees are normally distributed with a mean of μ and a standard deviation of 0.3 metres. The value of μ is closest to:

A. 3.220
B. 2.580
C. 3.315
D. 2.485
E. 2.850

15. **MC** The weights of bags of flour are normally distributed with a mean of 255 grams and a standard deviation of 14 grams. If 45% of the bags weigh more than x grams, then the maximum possible value of x is closest to:

A. 253.2
B. 260.4
C. 258.6
D. 249.6
E. 256.8

16. **MC** The 0.85 quantile for the standard normal distribution is approximately equal to:

A. −1.0364 B. 0.8416 C. −0.8416 D. 1.0364 E. 0

Technology active: extended response

17. X is a normally distributed variable for which $\Pr(X < 47) = 0.3694$ and $\Pr(X > 56) = 0.3385$.
    a. Calculate the mean and standard deviation of X. Give your answers to the nearest integer.
    b. Determine the integer value of a given that $\Pr(51 - a < X < 51 + a) = 0.7566$. Hence, calculate, correct to 4 decimal places, $\Pr(X > 51 - a \mid X < 51 + a)$.
    c. Calculate:
        i. $\text{Var}(X)$ ii. $E(X^2)$.

18. In a chocolate factory, the material for the making of each chocolate is sent to one of two machines, machine A and machine B.

The time, X seconds, taken to produce a chocolate by machine A is distributed normally with a mean of 2.5 seconds and a standard deviation of 0.7 seconds. The time, Y seconds, taken to produce a chocolate by machine B has the probability density function

$$f(y) = \begin{cases} 0, & y < 0 \\ \dfrac{y}{12}, & 0 \le y \le 3 \\ \dfrac{1}{3}e^{-0.5(y-3)}, & y > 3 \end{cases}.$$

a. Calculate, correct to 4 decimal places:
   i. $\Pr(2 \le X \le 5)$
   ii. $\Pr(2 \le Y \le 5)$.

b. Determine $E(Y)$.

c. It can be shown that $\Pr(Y \le 2) = \frac{1}{6}$. A random sample of 10 chocolates produced by machine B is chosen. Calculate the probability, correct to 4 decimal places, that exactly 4 of the 10 chocolates took 2 seconds or less to produce.

d. i. Determine $\Pr(X \le 2)$, correct to 4 decimal places
   ii. A sample of 10 chocolates produced by machine A is chosen at random. Determine the probability, correct to 4 decimal places, that exactly 4 of the 10 chocolates took 2 seconds or less to produce.

19. Pieter has a strawberry farm in the Dandenong Ranges in the outer suburbs of Melbourne. The average length of a strawberry is normally distributed with a mean of 3.5 cm and a standard deviation of 0.8 cm.

Strawberries that are longer than 4.5 cm are sold to a restaurant supplier for \$6.50 per kilogram. Strawberries that are between 2.5 cm and 4.5 cm long are sold to a supermarket supplier for \$4.50 per kilogram, and strawberries that are less than 2.5 cm long are sold to a manufacturer of jam for \$1.75 per kilogram.

a. Determine, correct to 2 decimal places:
   i. the percentage of strawberries that are sold to the restaurant supplier
   ii. the percentage of strawberries that are sold to the jam manufacturer
   iii. the mean profit for a kilogram of strawberries.

b. Pieter's neighbour, Marta, also has a strawberry farm. Y is the average length of a strawberry from Marta's farm and is distributed normally.
   Calculate the mean and standard deviation, correct to 1 decimal place, of the average length of Marta's strawberries if $\Pr(Y > 4.5) = 0.0316$ and $\Pr(Y < 2.5) = 0.1587$.

20. Merilyn is training to throw a javelin at the next Australian Athletics Championship. The 'A standard' throwing distance in an authorised competition is 65 metres. The current Olympic record for the women's javelin is 72.28 metres. If Merilyn is to be selected in an Olympic training squad, she must throw to the A standard. Merilyn feels confident that the distance in metres that she can throw the javelin from the marked throwing line is normally distributed with a mean of 64.5 metres and a standard deviation of 3.5 metres.

a. Complete the following table. Give probabilities correct to 4 decimal places.

Distance thrown	Probability
Less than the A standard	
Greater than the A standard but less than the Olympic record	
Greater than the Olympic record	

b. 90% of Merilyn's throws travel at least m metres. Calculate the value of m correct to 2 decimal places.
c. Merilyn throws a javelin that does not reach the Olympic record. Calculate the probability, correct to 4 decimal places, that it reaches the A standard.

Merilyn's sponsor offers her an incentive to perform her best in competitions. The cash rewards for each throw are shown in the following table.

Length of throw	Amount paid ($)
Less than the A standard	$500
Between the A standard and the Olympic record	$1000
Greater than the Olympic record	$5000

d. Calculate Merilyn's expected reward, correct to the nearest dollar, for each throw she completes in a competition.
e. In a particular competition Merilyn completes 5 throws. Calculate:
   i. the total reward she would expect to receive, correct to the nearest 10 dollars
   ii. the probability, correct to 4 decimal places, that at least 3 of her throws will be over the A standard
   iii. the expected number of times she will be over the A standard, correct to 2 decimal places.

12.6 Exam questions

Question 1 (3 marks) TECH-FREE

Source: *VCE 2015 Mathematical Methods (CAS) Exam 1, Q6; © VCAA.*

Let the random variable X be normally distributed with mean 2.5 and standard deviation 0.3.

Let Z be the standard normal random variable, such that $Z \sim N(0, 1)$.

a. Find b such that $\Pr(X > 3.1) = \Pr(Z < b)$. **(1 mark)**
b. Using the fact that, correct to two decimal places, $\Pr(Z < -1) = 0.16$, find $\Pr(X < 2.8 \mid X > 2.5$.
   Write the answer correct to two decimal places. **(2 marks)**

Question 2 (1 mark) TECH-ACTIVE

Source: *VCE 2014 Mathematical Methods (CAS) Exam 2, Section 1, Q5; © VCAA.*

MC The random variable X has a normal distribution with mean 12 and standard deviation 0.5. If Z has the standard normal distribution, then the probability that X is less than 11.5 is equal to

A. $\Pr(Z > -1)$ **B.** $\Pr(Z < -0.5)$ **C.** $\Pr(Z > 1)$ **D.** $\Pr(Z \geq 0.5)$ **E.** $\Pr(Z < 1)$

Question 3 (12 marks) TECH-ACTIVE

Source: *Adapted from VCE 2020 Mathematical Methods Exam 2, Section B, Q3; © VCAA.*

A transport company has detailed records of all its deliveries. The number of minutes a delivery is made before or after its scheduled delivery time can be modelled as a normally distributed random variable, T, with a mean of zero and a standard deviation of four minutes. A graph of the probability distribution of T is shown below.

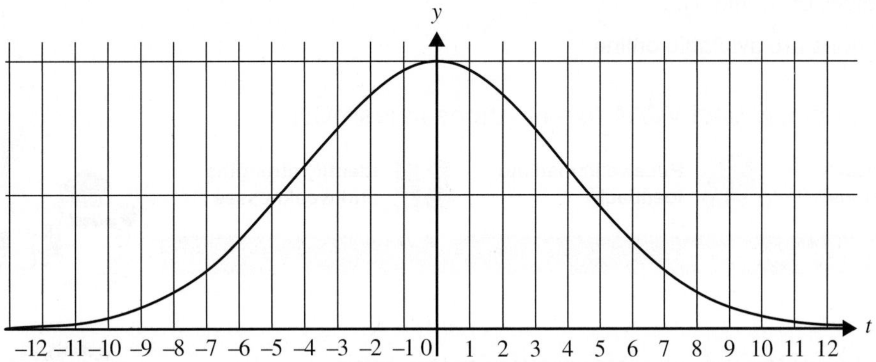

a. If $\Pr(T \leq a) = 0.6$, find a to the nearest minute. **(1 mark)**

b. Find the probability, correct to three decimal places, of a delivery being no later than three minutes after its scheduled delivery time, given that it arrives after its scheduled delivery time. **(2 marks)**

c. Using the model described above, the transport company can make 46.48% of its deliveries over the interval $-3 \leq t \leq 2$.

It has an improved delivery model with a mean of k and a standard deviation of four minutes.

Find the values of k, correct to one decimal place, so that 46.48% of the transport company's deliveries can be made over the interval $-4.5 \leq t \leq 0.5$. **(3 marks)**

d. A rival transport company claims that there is a 0.85 probability that each delivery it makes will arrive on time or earlier.

Assume that whether each delivery is on time or earlier is independent of other deliveries.

Assuming that the rival company's claim is true, find the probability that on a day in which the rival company makes eight deliveries, fewer than half of them arrive on time or earlier. Give your answer correct to three decimal places. **(2 marks)**

e. Assuming that the rival company's claim is true, consider a day in which it makes n deliveries.

i. Express, in terms of n, the probability that one or more deliveries will **not** arrive on time or earlier. **(1 mark)**

ii. Hence, or otherwise, find the minimum value of n such that there is at least a 0.95 probability that one or more deliveries will **not** arrive on time or earlier. **(1 mark)**

f. An analyst from a government department believes the rival transport company's claim is only true for deliveries made before 4 pm. For deliveries made after 4 pm, the analyst believes the probability of a delivery arriving on time or earlier is x, where $0.3 \leq x \leq 0.7$.

After observing a large number of the rival transport company's deliveries, the analyst believes that the overall probability that a delivery arrives on time or earlier is actually 0.75.

Let the probability that a delivery is made after 4 pm be y.

Assuming that the analyst's beliefs are true, find the minimum and maximum values of y. **(2 marks)**

Question 4 (1 mark) TECH-ACTIVE

MC For the standard normal distribution, state which one of the following is true.

A. $\Pr(Z > a) = \Pr(Z \le a)$
B. $\Pr(Z < -a) = \Pr(Z > -a)$
C. $\Pr(Z > a) = 1 - \Pr(Z \ge a)$
D. $\Pr(-a < Z < a) = 1 - 2\Pr(Z \le -a)$
E. $\Pr(-a < Z < a) = 1 - 2\Pr(Z \le a)$

Question 5 (2 marks) TECH-ACTIVE

X is a continuous random variable that is normally distributed, with a mean of 14.2 and a standard deviation of 2.1. The values a_1 and a_2 are evenly distributed either side of the mean such that $\Pr(a_1 < X < a_2) = 0.8$. Determine the values of a_1 and a_2.

More exam questions are available online.

Answers

Topic 12 The normal distribution

12.2 The normal distribution

12.2 Exercise

1. a. $\mu = 1, \sigma = 0.3$ or $\frac{3}{10}$

   b. Dilation factor $\frac{10}{3}$ parallel to the y-axis, dilation factor $\frac{3}{10}$ parallel to the x-axis, translation 1 unit in the positive x-direction

2. a. $\mu = 2, \sigma = 3$

   b.

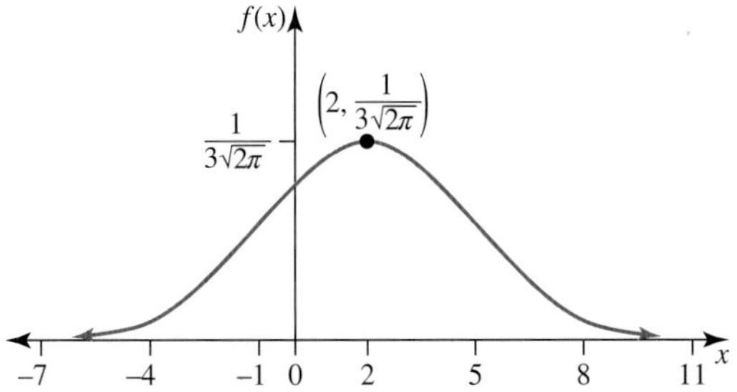

3. a. i. 100 and 140 ii. 80 and 160 iii. 60 and 180

   b. i. 0.025 ii. 0.0015

4. a. 0.025 b. 0.0015 c. 0.84
5. a. 10 and 20 b. 5 and 25 c. 0 and 30
6. a. 0.84 b. 0.815 c. 0.9702
7. a. 0.975 b. 0.95 c. 0.9744
8. A
9. D
10. a. 0.815 b. $k = 205$ c. $h = 155$
11. a. 0.815 b. 0.025 c. 0.94

    d. $m = 80.9$

12. a. Use CAS technology to verify the integral.

    b. $\mu = 3, \sigma = 1$

    c.

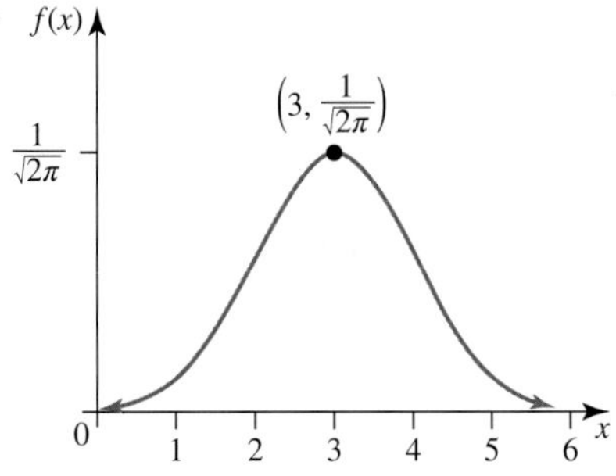

13. b. $\mu = -2$
14. a. $\mu = -4, \sigma = 10$

    b. Dilation factor $\frac{1}{10}$ from the x-axis, dilation factor 10 from the y-axis, translation 4 units in the negative x-direction

    c. i. 100 ii. 116

    d. $\displaystyle\int_{-\infty}^{\infty} \frac{1}{10\sqrt{2\pi}} e^{-\frac{1}{2}\left(\frac{x+4}{10}\right)^2} dx = 0.9999 \simeq 1$

    $f(x) \geq 0$ for all values of x, and the area under the curve is 1. Therefore, this function is a probability density function.

15. a. $\mu = 2, \sigma = \frac{2}{5}$ b. $\frac{104}{25} = 4.16$

12.2 Exam questions

Note: Mark allocations are available with the fully worked solutions online.

1.

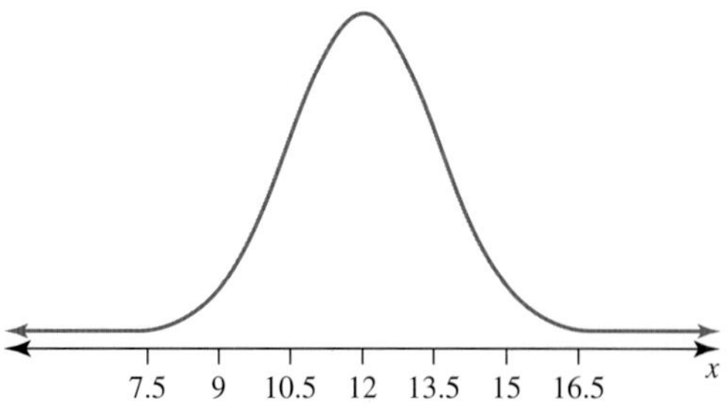

2. E
3. a. 68% b. 2.5%

12.3 Calculating probabilities and the standard normal distribution

12.3 Exercise

1. 0.4088
2. a. 0.9772 b. 0.0228 c. 0.9545 d. 0.0512
3. a. 0.0912 b. 0.2119 c. 0.7501
4. a. i. 0.8849 ii. 0.7703

   b. i. 0.9088 ii. $-\frac{4}{3}$

5. a. i. 0.65 ii. 0.27

   b. i. 0.29 ii. $\frac{29}{56}$

6. a. $k = 2.4$ b. $k = 1.6$
7. a. 0.0228 b. 0.0912 c. 0.8860
8. a. E b. B
9. The salmon is in the top 26.6%, so it is not gourmet.
10. a. 6.68% b. 0.19%
11. a. 0.1587 b. 0.1587 c. 0.3695
12. Jingjing
13. Chemistry 0.8, Maths Methods 0.86, Physics 0.875; Juan did best in Physics when compared to his peers.
14. $k = 25.24$

12.3 Exam questions

Note: Mark allocations are available with the fully worked solutions online.

1. a. 0.5 b. $-\frac{1}{2}$
2. E
3. D

12.4 The inverse normal distribution

12.4 Exercise

1. a. $a = 34.34$ b. $a = 14.68$ c. $a = 1.65$
2. $m = 12.94$, $n = 41.66$
3. a. -0.2793 b. 1.0364 c. 0.3585
4. a. $a = 42.52$ b. $a = 41.53$ c. $a = 13.40$
5. a. 0.1764 b. 0.3319
6. a. 49.4443 b. 36.6489
7. $\sigma = 15.5$
8. $\mu = 37.6$
9. $\mu = 13.35$
10. A
11. $\sigma = 10.5$
12. D
13. $\mu = 34.6$, $\sigma = 2.5$
14. $\mu = 15.8$, $\sigma = 5.2$
15. $\mu = 37.68$, $\sigma = 11.21$
16. 0.6842

12.4 Exam questions

Note: Mark allocations are available with the fully worked solutions online.

1. C
2. A
3. B

12.5 Mixed probability applications

12.5 Exercise

1. a. i. 0.0038 ii. 501.0210
   b. 3.4389 grams
2. a. $3k^2 + 2k + 6k^2 + 2k + k^2 + 2k + 3k = 1$
   $$10k^2 + 9k - 1 = 0$$
   b. $k = \frac{1}{10}$
   c. 2
   d. 0.3115
   e. 9
   f. $\sigma = 18$
3. 0.0484
4. a. 22.66% b. 179 cm
5. a. i. 0.0062 ii. 0.9625
   b. $\mu = 405$, $\sigma = 25$
6. a. 0.1420
   b. $d = 2.0$
   c. 0.2924
   d. i. $a = 0.68$ ii. $0.85x - 4.76$
   iii. \$5.60 iv. 80%
7. a. 0.0371
   b. 0.0032
   c. 0.0001
   d. i. 0.0167 ii. 0.8856
8. a. Small: 0.0548, medium: 0.8644, large: 0.0808
   b. \$531
9. a. 15.87% b. 68.27% c. 45 cents
10. a. 0.8413 b. 0.6494 c. 0.8845
    d. $\sigma = 0.4987$
11. a. 95 mm b. 0.8783 c. 18%
    d. 83 mm e. 0.0078 f. \$399
    g. 0.2236
12. a. 0.0161 b. 0.0019
13. a. 0.0038 b. $\mu = 19.0$ mm, $\sigma = 2.5$ mm
    c. 0.2225
14. a. $\mu = 0$, $\sigma = 1.5306$ b. 0.8816

12.5 Exam questions

Note: Mark allocations are available with the fully worked solutions online.

1. D
2. a. 19.1 cm b. 211 c. 25 cm
   d. 12.7 cm e. 14
3. a. i. $\frac{11}{16}$ or 0.6875 ii. $\frac{825}{4096}$
   b. 7.2
   c. 0.778
   d. i. 0.1147 ii. 23

12.6 Review

12.6 Exercise

Technology free: short answer

1. a. 0.16
   b. 0.34
   c. $\frac{17}{25}$
2. Dilation factor 6 from the x-axis, dilation factor $\frac{1}{6}$ from the y-axis, translation 31 units in the positive y-direction
3. a. 68% b. 97.5% c. 81.5%
4. a. 0.35 b. 0.35
5. a. 3.5 b. 2.4
6. a. 0.18 b. 0.07

Technology active: multiple choice

7. A
8. B
9. C
10. E
11. B
12. C
13. D
14. A
15. E
16. D

Technology active: extended response

17. a. $\mu = 51, \sigma = 12$
    b. $a = 14$, $\Pr(X > 51 - a \mid X < 51 + a) = 0.8615$
    c. i. 144 ii. 2745
18. a. i. 0.7623 ii. 0.6297
    b. 4.083
    c. 0.0543
    d. i. 0.2375 ii. 0.1313
19. a. i. 10.56% ii. 10.56% iii. $4.42/kg
    b. $\mu = 3.2, \sigma = 0.7$
20. a. $\Pr(X < 65) = 0.5568$, $\Pr(65 < X < 72.28) = 0.4301$, $\Pr(X > 72.28) = 0.0131$
    b. 60.01 m
    c. 0.4359
    d. $774
    e. i. $3870 ii. 0.3944 iii. 1.97

12.6 Exam questions

Note: Mark allocations are available with the fully worked solutions online.

1. a. −2 b. 0.68
2. C
3. a. 1 minute
   b. 0.547
   c. −2.5
   d. 0.003
   e. i. $1 - 0.85^n$ ii. 19
   f. Minimum y-value: $y = \frac{2}{11}$
      Maximum y-value: $y = \frac{2}{3}$
4. D
5. $a_1 = 11.5087$
   $a_2 = 16.8913$

13 Statistical inference

LEARNING SEQUENCE

Fully worked solutions for this topic are available online.

13.1 Overview

Hey students! Bring these pages to life online

Watch videos | Engage with interactivities | Answer questions and check results

Find all this and MORE in jacPLUS

13.1.1 Introduction

Statistical inference describes the process where conclusions are drawn about the characteristics of a population, based on the analysis of the sample drawn from it.

It is quite impractical, and in some cases impossible, to analyse the entire population for a particular situation. Therefore, investigating a sample is the most common practice. The sample needs to be representative of the entire population, and where possible, multiple samples should be analysed, especially if the sample size is small. This then leads to a sampling distribution.

Statistical inference, of which only the basic foundations are covered in this course, is used in many important industries. In the Future Cities area, mathematical models relating to crime, security and resilience are analysed for urban environments. This knowledge can then be applied to areas such as infrastructure and transport. Statistical inference can also be used in predictions of weather, or of voting preferences on election day. In medical research, it can be used to help predict what parts of the brain are active during an MRI scan while the patient is doing a basic task such as tapping their finger, or to analyse whether a new treatment program is reducing the progression of cancer.

KEY CONCEPTS

This topic covers the following key concepts from the VCE Mathematics Study Design:

- statistical inference, including definition and distribution of sample proportions, simulations and confidence intervals:
  - distinction between a population parameter and a sample statistic and the use of the sample statistic to estimate the population parameter
  - simulation of random sampling, for a variety of values of p and a range of sample sizes, to illustrate the distribution of $\hat{P}$ and variations in confidence intervals between samples
  - concept of the sample proportion $\hat{P} = \frac{X}{n}$ as a random variable whose value varies between samples, where X is a binomial random variable which is associated with the number of items that have a particular characteristic and n is the sample size
  - approximate normality of the distribution of $\hat{P}$ for large samples and, for such a situation, the mean p (the population proportion) and standard deviation $\sqrt{\frac{p(1-p)}{n}}$
  - determination and interpretation of, from a large sample, an approximate confidence interval $\left(\hat{p} - z\sqrt{\frac{\hat{p}(1-\hat{p})}{n}}, \hat{p} + z\sqrt{\frac{\hat{p}(1-\hat{p})}{n}}\right)$, for a population proportion where z is the appropriate quantile for the standard normal distribution, in particular the 95% confidence interval as an example of such an interval where $z \approx 1.96$ (the term standard error may be used but is not required).

13.2 Population parameters and sample statistics

LEARNING INTENTION

At the end of this subtopic you should be able to:
- identify populations and samples
- identify sample statistics and population parameters
- understand different types of sampling.

13.2.1 Samples and populations

Suppose you were interested in the percentage of Year 12 graduates who plan to study Mathematics once they complete school. It is probably not practical to question every student. There must be a way that we can ask a smaller group and then use this information to make generalisations about the whole group.

A **population** is a group that you want to know something about, and a **sample** is the group within the population that you collect the information from. Normally, a sample is smaller than the population; the exception is a census, where the whole population is the sample.

The number of members in a sample is called the **sample size** (symbol n), and the number of members of a population is called the **population size** (symbol N). Sometimes the population size is unknown.

WORKED EXAMPLE 1 Samples and populations

A total of 137 people volunteer to take part in a medical trial. Of these, 57 are identified as suitable candidates and are given the medication. Identify the population and sample size.

THINK	WRITE
1. 57 people are given the medication. This is the sample size.	$n = 57$
2. We are interested in the group of people who might receive the drug in the future. This is the population.	The population is unknown, as we don't know how many people may be given this drug in the future.

Using technology to select a sample

If you know the population size, it should also be possible to produce a list of population members. Assign each population member a number (from 1 to $n = N$). Use the random number generator on your calculator to generate a random number between 1 and N. The population member who was allocated that number becomes the first member of the sample. Continue generating random numbers until the required number of members has been picked for the sample. If the same random number is generated more than once, ignore it and continue selecting members until the required number has been chosen.

WORKED EXAMPLE 2 Samples and populations

Cameron has uploaded a popular YouTube video. He thinks that the 133 people in his year group at school have seen it, and he wants to know what they think. He decides to question 10 people. Identify the population and sample size.

THINK	WRITE
1. Cameron wants to know what the people in his year at school think. This is the population.	$N = 133$
2. He asks 10 people. This is the sample.	$n = 10$

TI | THINK — **DISPLAY/WRITE**

1. A random sample of people can be generated using technology.

If each of the 133 students in Cameron's year group is assigned a number from 1 to 133, the calculator can randomly select 10 integers between 1 and 133 to represent a random sample of 10 students.

2. On a Calculator page, press MENU, then select:
5: Probability
4: Random
2: Integer
Complete the entry line as:
randInt (1, 133, 10)
then press ENTER.

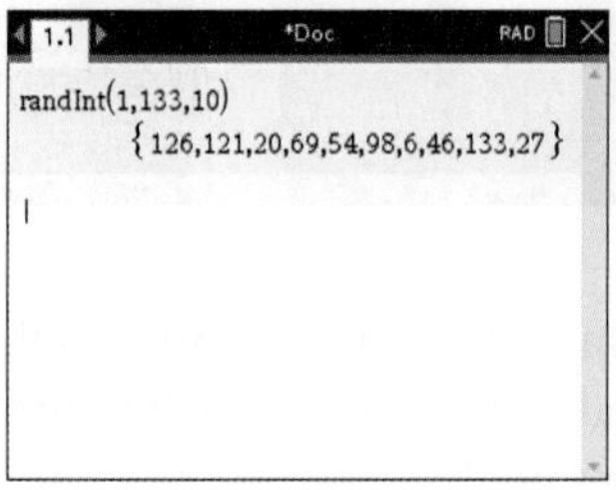

3. The numbers on the screen represent the students chosen to be included in the sample.

Students assigned numbers 126, 121, 20, 69, 54, 98, 6, 46, 133 and 27 are to be included in the sample.

CASIO | THINK — **DISPLAY/WRITE**

1. A random sample of people can be generated using technology.

If each of the 133 students in Cameron's year group is assigned a number from 1 to 133, the calculator can randomly select 10 integers between 1 and 133 to represent a random sample of 10 students.

2. On a Main screen, complete the entry line as:
randList (10, 1, 133)
then press EXE.
Note: randList can be found by using the down arrow and then selecting it from the catalog.

3. The numbers on the screen represent the students chosen to be included in the sample.

Students assigned numbers 92, 37, 122, 85, 78, 127, 33, 66, 14 and 77 are to be included in the sample.

13.2.2 Statistics and parameters

A **parameter** is a characteristic of a population, whereas a **statistic** is a characteristic of a sample. This means that a statistic is always known exactly (because it is measured from the sample that has been selected). A parameter is usually estimated from a sample statistic. (The exception is if the sample is a census, in which case the parameter is known exactly.)

In this unit, we will study binomial data (that means that each data point is either yes/no or success/failure) with special regard to the proportion of successes.

The relationship between populations and samples

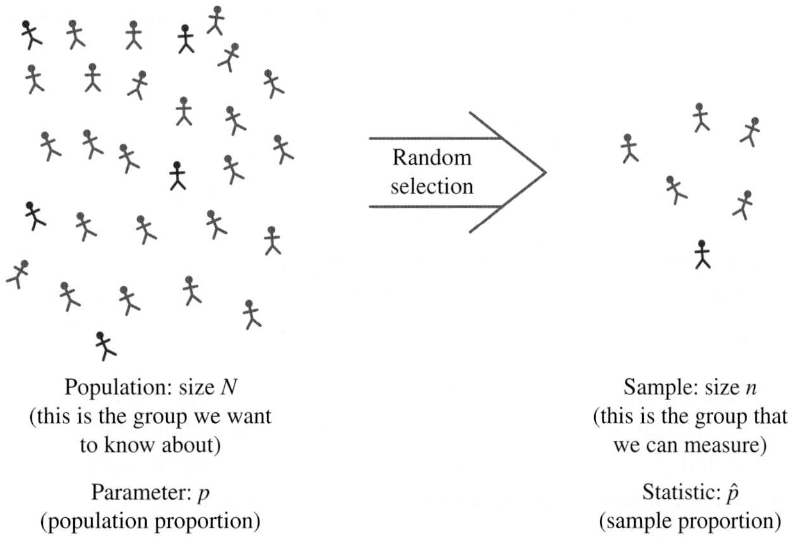

WORKED EXAMPLE 3 Statistics and parameters

Identify the following as either sample statistics or population parameters.

a. 43% of voters polled say that they are in favour of banning fast food.
b. According to Australian Bureau of Statistics census data, the average family has 1.7 children.
c. Between 18% and 23% of Australians skip breakfast regularly.
d. Nine out of 10 kids prefer cereal for breakfast.

THINK	WRITE
a. 43% is an exact value that summarises the sample asked.	a. Sample statistic
b. The information comes from census data. The census questions the entire population.	b. Population parameter
c. 18%–23% is an estimate about the population.	c. Population parameter
d. Nine out of 10 is an exact value. It is unlikely that all kids could have been asked; therefore, it is from a sample.	d. Sample statistic

on Resources

Interactivity Population parameters and sample statistics (int-6442)

13.2.3 Random samples

A good sample should be representative of the population. If we consider our initial interest in the proportion of Year 12 graduates who intend to study Mathematics once they finished school, we could use a Mathematical Methods class as a sample. This would not be a good sample because it does not represent the population — it is a very specific group.

In a **random sample**, every member of the population has the same probability of being selected. The Mathematical Methods class is not a random sample because students who don't study Mathematical Methods have no chance of being selected; furthermore, students who don't attend that particular school have no chance of being selected.

A **systematic sample** is almost as good as a random sample. In a systematic sample, every kth member of the population is sampled. For example, if $k = 20$, a customs official might choose to sample every 20th person who passes through the arrivals gate. The reason that this is almost as good as random sample is that there is an assumption that the group passing the checkpoint during the time the sample is taken is representative of the population. This assumption may not always be true; for example, people flying for business may be more likely to arrive on an early morning flight. Depending on the information you are collecting, this may influence the quality of the data.

In a **stratified random sample**, care is taken so that subgroups within a population are represented in a similar proportion in the sample. For example, if you were collecting information about students in Years 9–12 in your school, the proportions of students in each year group should be the same in the sample and the population. Within each subgroup, each member has the same chance of being selected.

A **self-selected sample**, that is, one where the participants choose to participate in the survey, is almost never representative of the population. This means, for example, that television phone polls, where people phone in to answer yes or no to a question, do not accurately reflect the opinion of the population.

WORKED EXAMPLE 4 Sample sizes

A survey is to be conducted in a middle school that has the distribution detailed in the table below. It is believed that students in different year levels may respond differently, so the sample chosen should reflect the subgroups in the population (that is, it should be a stratified random sample). If a sample of 100 students is required, determine how many from each year group should be selected.

Year level	Number of students
7	174
8	123
9	147

THINK	WRITE
1. Find the total population size.	Total population $= 174 + 123 + 147 = 444$
2. Find the number of Year 7s to be surveyed.	Number of Year 7s $= \frac{174}{444} \times 100$ $= 39.2$ Survey 39 Year 7s.
3. Find the number of Year 8s to be surveyed.	Number of Year 8s $= \frac{123}{444} \times 100$ $= 27.7$ Survey 28 Year 8s.
4. Find the number of Year 9s to be surveyed.	Number of Year 9s $= \frac{147}{444} \times 100$ $= 33.1$ Survey 33 Year 9s.
5. There has been some rounding, so check that the overall sample size is still 100.	Sample size $= 39 + 28 + 33 = 100$ The sample should consist of 39 Year 7s, 28 Year 8s and 33 Year 9s.

on Resources

Interactivity Random samples (int-6443)

13.2 Exercise

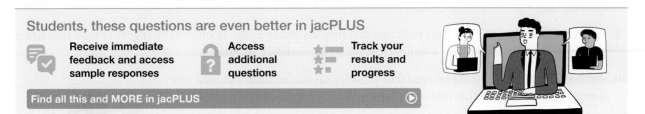

Students, these questions are even better in jacPLUS

Receive immediate feedback and access sample responses

Access additional questions

Track your results and progress

Find all this and MORE in jacPLUS

Technology free

1. WE1 Ms Lane plans to begin her Statistics class each year by telling her students a joke. She tests her joke on this year's class (15 students). She plans to retire in 23 years' time. Identify the population and sample size.

2. Bruce is able to hem 100 shirts per day. Each day he checks 5 to make sure that they are suitable. Identify the population and sample size.

3. WE2 On average, Mr Parker teaches 120 students per day. He asks one class of 30 about the amount of homework they have that night. Identify the population and sample size.

4. Lee-Yin is trying to perfect a recipe for cake pops. She tries 5 different versions before she settles on her favourite. She takes some samples to school and asks 9 friends what they think. Identify the population and sample size.

5. You are trying out a new chocolate pudding recipe. You found 40 volunteers to taste test your new recipe compared to your normal pudding. Half of the volunteers were given a serving the new pudding first, then a serving of the old pudding. The other half were given the old pudding first and then the new pudding. The taste testers did not know the order of the puddings they were trying. The results show that 31 people prefer the new pudding recipe.
   a. Calculate the population size.
   b. Calculate the sample size.

6. You want to test a new flu vaccine on people with a history of chronic asthma. You begin with 500 volunteers and end up with 247 suitable people to test the vaccine.
   a. Calculate the population size.
   b. Calculate the sample size.

7. WE3 Identify the following as either sample statistics or population parameters.
   a. Studies have shown that between 85% and 95% of lung cancers are related to smoking.
   b. About 50% of children aged between 9 and 15 years eat the recommended daily amount of fruit.

8. Identify the following as either sample statistics or population parameters.
   a. According to the 2013 census, the ratio of male births per 100 female births is 106.3.
   b. About 55% of boys and 40% of girls reported drinking at least 2 quantities of 500 mL of soft drink every day.

9. In a recent survey, 1 in 5 students indicated that they ate potato crisps or other salty snacks at least four times per week. State whether this is a sample statistic or a population parameter.

10. Around 25% to 30% of children aged 0 to 15 years eat confectionary at least four times a week. State whether this is a sample statistic or a population parameter.

11. According to the Australian Bureau of Statistics, almost a quarter (24%) of internet users did not make an online purchase or order in 2012–13. The three most commonly reported main reasons for not making an online purchase or order were: 'Has no need' (33%); 'Prefers to shop in person/see the product' (24%); and 'Security concerns/concerned about providing credit card details online' (12%). State whether these are sample statistics or population parameters.

12. According to the 2016 census, there is an average of 2.6 people per household. State whether this is a sample statistic or a population parameter.

Technology active

13. Use CAS technology to produce a list of 10 random numbers between 1 and 100.

14. Use CAS technology to select a random sample from students in your Mathematical Methods class.

15. WE4 A school has 523 juniors and 621 seniors. You are interested in finding out about their attitudes to sport and believe that juniors and seniors may respond differently. If a sample of 75 students is required, determine how many juniors and how many seniors should be selected.

16. In a school, 23% of the students are boarders. For this survey, it is believed that boarders and day students may respond differently. Calculate how many boarders and day students should be selected to select a sample of 90 students.

17. A doctor is undertaking a study about sleeping habits. She decides to ask every 10th patient about their sleeping habits.
    a. Identify what type of sample this is.
    b. Explain whether this is a valid sampling method.

18. A morning television show conducts a viewer phone-in poll and announces that 95% of listeners believe that Australia should become a republic. Comment on the validity of this type of sample.

19. Tony took a survey by walking around the playground at lunch and asking fellow students questions. Explain why this is not the best sampling method.

20. A company has 1500 staff members, of whom 60% are under 50 years of age. 95% of the staff under 50 work full time, and 78% of the staff over 50 work full time. If a sample of 80 staff is to be selected, identify the numbers of full-time staff under 50, part-time staff under 50, full time staff over 50 and part-time staff over 50 who should be included in the sample.

13.2 Exam questions

Question 1 (1 mark) TECH-ACTIVE

MC In a school population census, it is found that 12% of students attending have no access to the internet at home. The '12%' represents the value of a

A. sample **B.** sample statistic **C.** sample parameter
D. population **E.** population parameter

Question 2 (1 mark) TECH-ACTIVE

MC To estimate the ratio of females to males at a tennis club, the coach determines the number of females and males in a particular tennis lesson. The ratio that she then calculates is called a

A. sample **B.** sample statistic **C.** sample parameter
D. population **E.** population parameter

Question 3 (1 mark) TECH-ACTIVE

MC Select the true statement from the following.

A. Sample parameters are used to estimate population statistics.
B. Sample statistics are used to estimate population parameters.
C. Population parameters are used to estimate sample statistics.
D. Population statistics are used to estimate sample parameters.
E. Population proportion is an estimate called a point estimate.

More exam questions are available online.

13.3 The distribution of the sample proportion

LEARNING INTENTION

At the end of this subtopic you should be able to:

- determine the sample proportion, $\hat{p}$
- create a sampling distribution table for $\hat{P}$
- use the binomial distribution to calculate probabilities associated with the sampling distribution
- determine the mean and standard deviation of the sampling distribution.

13.3.1 The sample proportion, $\hat{p}$

Let us say that we are interested in the following collection of balls. As you can see in Figure 1, there are 20 balls and $\frac{1}{4}$ of them are red. This means that the population parameter, p, is $\frac{1}{4}$ and the population size, N, is 20.

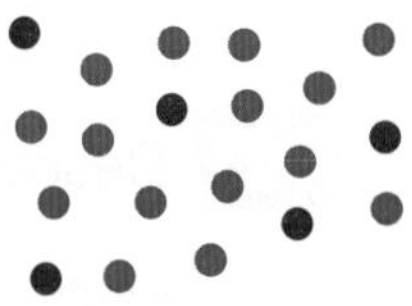

Figure 1

Normally we wouldn't know the population parameter, so we would choose a sample from the population and find the sample statistic. In this case, we are going to use a sample size of 5, that is, $n = 5$.

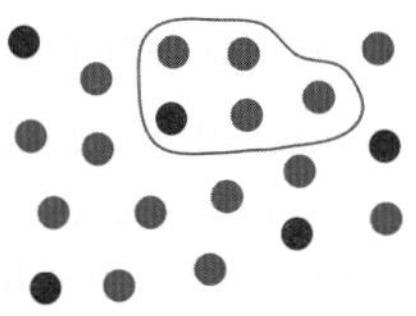

Figure 2

If our sample is the group shown in Figure 2, then as there is 1 red ball, the **sample proportion** would be $\hat{p} = \frac{1}{5}$.

A different sample could have a different sample proportion. In the case shown in Figure 3, $\hat{p} = \frac{2}{5}$.

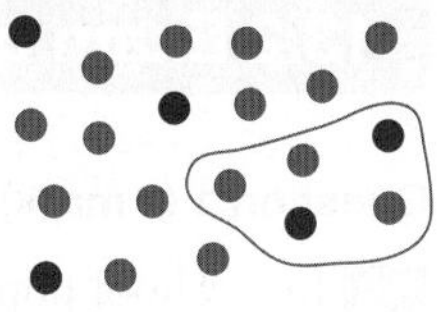

Figure 3

In the case shown in Figure 4, $\hat{p} = 0$.

It would also be possible to have samples for which $\hat{p} = \frac{3}{5}$, $\hat{p} = \frac{4}{5}$ or $\hat{p} = 1$, although these samples are less likely to occur.

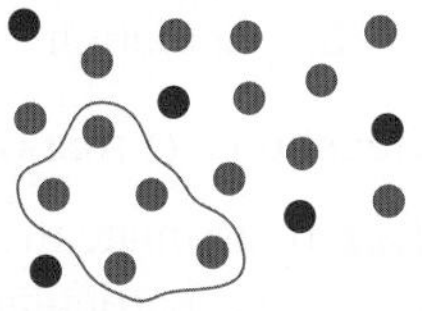

Figure 4

The sample proportion

$$\hat{p} = \frac{\textbf{number of successful outcomes in the sample}}{\textbf{sample size}}$$

It might seem that using a sample does not give a good estimate about the population. However, the larger the sample size, the more likely that the sample proportions will be close to the population proportion.

WORKED EXAMPLE 5 The sample proportion

You are trying out a new chocolate tart recipe. You found 40 volunteers to taste test your new recipe compared to your normal one. The results show that 31 people prefer the new tart recipe. Calculate the sample proportion, $\hat{p}$.

THINK	WRITE
1. There are 40 volunteers. This is the sample size.	$n = 40$
2. 31 people prefer the new recipe.	Number of successes = 31
3. Calculate the sample proportion.	$\hat{p} = \frac{31}{40}$

on Resources

Interactivity Distribution of $\hat{P}$ (int-6444)

Revision of binomial distributions

In a set of binomial data, each member of the population can have one of two possible values. We define one value as a success and the other value as a failure. (A success isn't necessarily a good thing, it is simply the name for the condition we are counting. For example, a success may be having a particular disease and a failure may be not having the disease.)

Population proportion and sample proportion

The proportion of successes in a population is called p and is a constant value.

$$p = \frac{\text{number in the population with the favourable attribute}}{\text{population size}}$$

The proportion of failures in a population is called q, where $q = 1 - p$.

The sample size is called n.

The number of successes in the sample is called x.

The proportion of successes in the sample, $\hat{p}$, will vary from one sample to another.

$$\hat{p} = \frac{\text{number in the sample with the favourable attribute}}{\text{sample size}}$$
$$= \frac{x}{n}$$

13.3.2 Sampling distribution of $\hat{P}$

Normally, you would take one sample from a population and make some inferences about the population from that sample. In this section, we are going to explore what would happen if you took lots of samples of the same size. (Assume you return each sample back to the population before selecting again.)

Consider our population of 20 balls (5 red and 15 blue). There are ${}^{20}C_5 = 15\,504$ possible samples that could be chosen. That is, there are 15 504 possible ways of choosing 5 balls from a population of 20 balls. A breakdown of the different samples is shown in the table, where x is the number of red balls in the sample.

x	$\hat{p}$	Number of samples	Relative frequency
0	0	${}^5C_0\ {}^{15}C_5 = 3003$	0.194
1	$\frac{1}{5}$	${}^5C_1\ {}^{15}C_4 = 6825$	0.440
2	$\frac{2}{5}$	${}^5C_2\ {}^{15}C_3 = 4550$	0.293
3	$\frac{3}{5}$	${}^5C_3\ {}^{15}C_2 = 1050$	0.068
4	$\frac{4}{5}$	${}^5C_4\ {}^{15}C_1 = 75$	0.005
5	1	${}^5C_5\ {}^{15}C_0 = 1$	6.450×10^{-5}
	Total number of samples	**15 504**	

Graphing the distribution of $\hat{p}$ against the relative frequency of $\hat{p}$ results in the following.

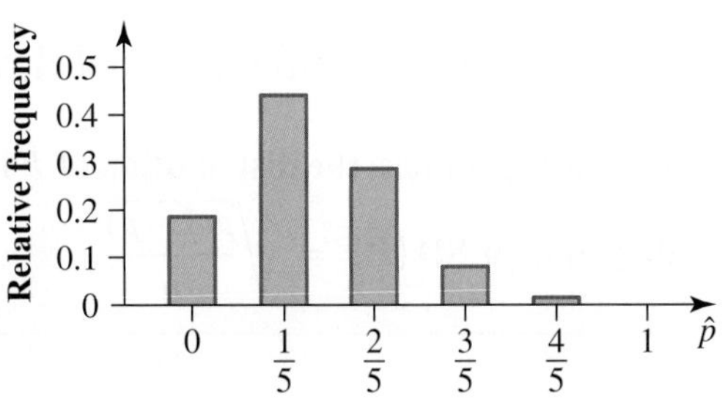

As the value of $\hat{p}$, the sample proportion, varies depending on the sample, these values can be considered as the values of the random variable, $\hat{P}$.

The graph of the distribution of $\hat{P}$ can also be represented in a probability distribution table. This distribution is called a **sampling distribution**.

x	0	1	2	3	4	5
$\hat{p}$	0	$\frac{1}{5}$	$\frac{2}{5}$	$\frac{3}{5}$	$\frac{4}{5}$	1
$\Pr(X=x)=\Pr(\hat{P}=\hat{p})$	0.194	0.440	0.293	0.068	0.005	6.450×10^{-5}

Sampling where the population is large

It was mentioned earlier that larger samples give better estimates of the population.

Expected value

The sample proportion $\hat{p}$ in a large sample conforms to $\hat{P}=\frac{X}{n}$. As the sample is from a large population, X can be assumed to be a binomial variable.

$$\begin{aligned}\therefore \mathrm{E}(\hat{P}) &= \mathrm{E}\left(\frac{X}{n}\right)\\ &= \frac{1}{n}\mathrm{E}(X) \quad \left(\text{because } \frac{1}{n} \text{ is a constant}\right)\\ &= \frac{1}{n}\times np\\ &= p\end{aligned}$$

Variance and standard deviation

The variance and standard deviation can be found as follows.

$$\begin{aligned}\mathrm{Var}(\hat{P}) &= \mathrm{Var}\left(\frac{X}{n}\right)\\ &= \left(\frac{1}{n}\right)^2 \mathrm{Var}(X)\\ &= \frac{1}{n^2}\times npq\\ &= \frac{pq}{n}\\ &= \frac{p(1-p)}{n}\\ \therefore \mathrm{SD}(\hat{P}) &= \sqrt{\frac{p(1-p)}{n}}\end{aligned}$$

Expected value and standard deviation

For large samples, the distribution of $\hat{P}$ has a mean or expected value of $\mathrm{E}\left(\hat{P}\right)=p$ and a standard deviation of $\mathrm{SD}\left(\hat{P}\right) = \sqrt{\frac{p(1-p)}{n}}$.

There are a number of different ways to decide if a sample is large. One generally accepted method that we will adopt for this section is that if $np \geq 10$, $nq \geq 10$ and $10n \leq N$, then the sample can be called large.

Consider the distribution of $\hat{P}$ when $N = 1000$, $n = 50$ and $p = 0.25$.

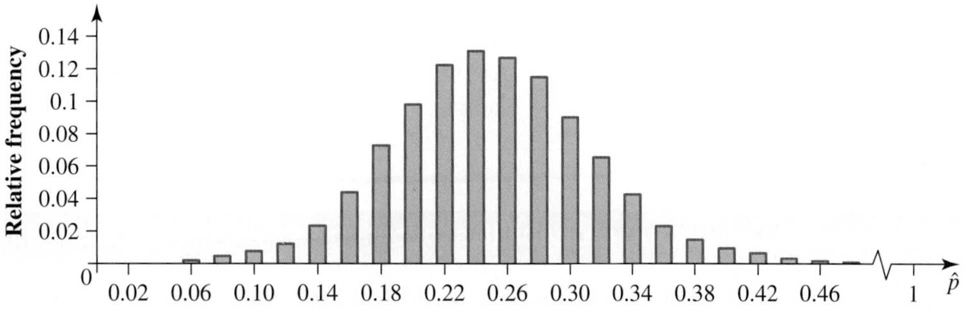

And consider this distribution of $\hat{P}$ when $N = 1000$, $n = 50$ and $p = 0.5$.

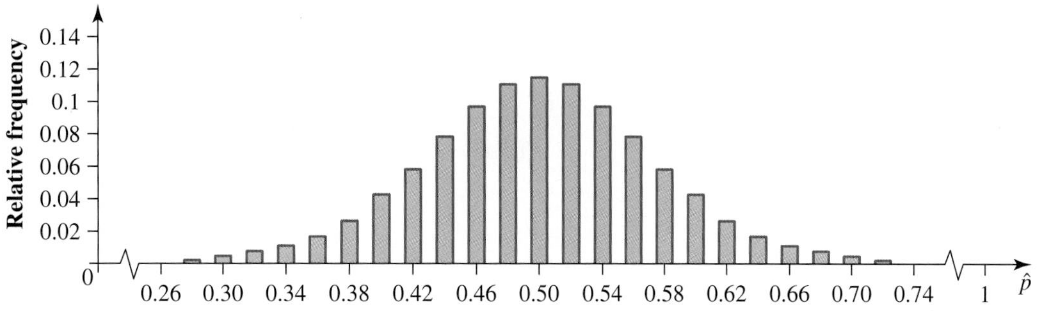

As these graphs show, the value of p doesn't matter. The distribution of $\hat{P}$ is symmetrical about p.

WORKED EXAMPLE 6 Sample size

Consider a population size of 1000 and a sample size of 50. If $p = 0.1$, state whether this would be a large sample. If not, determine how big the sample would need to be.

THINK	WRITE
1. Is $10n \leq N$?	$n = 50$ and $N = 1000$ $10n = 500$ Therefore, $10n \leq N$.
2. Is $np \geq 10$?	$p = 0.1$ $np = 0.1 \times 50$ $= 5$ $5 \ngeq 10$ The sample is not large.
3. Find a value for n to make a large sample by solving $np = 10$.	$np = 10$ $0.1n = 10$ $n = 100$

4. Check the other conditions.

$$\begin{aligned}10n &= 10\times 100\\ &= 1000\\ &= N\\ nq &= 100\times 0.9\\ &= 90\\ nq &\geq 10\end{aligned}$$

A sample size of 100 would be needed for a large sample.

WORKED EXAMPLE 7 Calculating probabilities involving sample proportions

There is a very large bag containing thousands of balls. 15% of the balls are known to be striped. A sample of 20 balls are selected from the bag. If $\hat{P}$ is the random variable of the distribution of sample proportions of striped balls, calculate $\Pr\left(\hat{P}>\frac{3}{20}\right)$, correct to 4 decimal places.

THINK	WRITE
1. A binomial distribution can be used to approximate this sample distribution.	$n=20,\ p=0.15,\ X \sim \text{Bi}(20, 0.15)$
2. Rewrite the probability calculation.	$\Pr\left(\hat{P}>\frac{3}{20}\right)=\Pr(X>3)$ $=\Pr(X\geq 4)$
3. Use the probability functions on your CAS calculator to evaluate.	$=0.3523$

WORKED EXAMPLE 8 Mean and standard deviation

If $N=600$, $n=60$ and $p=0.3$, determine:

a. the mean of the distribution

b. the standard deviation of the distribution, correct to 2 decimal places.

THINK	WRITE
a. The mean is p.	a. $\text{E}(\hat{P})=p$ $=0.3$
b. 1. Write the rule for the standard deviation.	b. $\text{SD}(\hat{P})=\sqrt{\frac{p(1-p)}{n}}$
2. Substitute the appropriate values and simplify.	$=\sqrt{\frac{0.3\times(1-0.3)}{60}}$ $=0.06$

on Resources

Interactivity Sampling distribution of $\hat{P}$ (int-6445)

13.3 Exercise

Students, these questions are even better in jacPLUS

Receive immediate feedback and access sample responses

Access additional questions

Track your results and progress

Find all this and MORE in jacPLUS

Technology free

1. WE5 In a 99-gram bag of lollies, there were 6 green lollies out of the 15 that were counted. Calculate the sample proportion, $\hat{p}$.

2. Hang is interested in seedlings that can grow to more than 5 cm tall in the month of her study period. She begins with 20 seedlings and finds that 6 of them are more than 5 cm tall after the month. Calculate the sample proportion, $\hat{p}$.

3. A car manufacturer has developed a new type of bumper that is supposed to absorb impact and result in less damage than previous bumpers. The cars are tested at 25 km/h. If 30 cars are tested and only 3 are damaged, calculate the proportion of undamaged cars in the sample.
4. A standard warranty lasts for 1 year. It is possible to buy an extended warranty for an additional 2 years. The insurer decides to use the sales figures from Tuesday to estimate the proportion of extended warranties sold. If 537 units were sold and 147 of them included extended warranties, estimate the proportion of sales that will include extended warranties.
5. WE6 Consider a population size of 1000 and a sample size of 50. If $p = 0.9$, state whether this is a large sample. If not, determine how big the sample would need to be.
6. If the population size was 10 000 and $p = 0.05$, determine what a large sample size would be.

Technology active

7. If $N = 1500$, $n = 150$ and $p = 0.15$, use CAS technology to graph the distribution for $\hat{P}$.
8. A Year 12 Mathematical Methods class consists of 12 girls and 9 boys. A group of 4 students is to be selected at random to represent the school at an inter-school Mathematics competition.
   a. Calculate the value of p, the proportion of girls in the class.
   b. Determine the possible values of the sample proportion, $\hat{p}$, of girls.
   c. Use this information to construct a probability distribution table to represent the sampling distribution of the sample proportion of girls in the small group. Give your answer correct to 3 decimal places.
   d. Determine $\Pr(\hat{P} > 0.6)$. That is, find the probability that the proportion of girls in the small group is greater than 0.6. Give your answer correct to 3 decimal places.
   e. Determine $\Pr(\hat{P} > 0.5 \mid \hat{P} > 0.3)$, correct to 3 decimal places.
9. In a particular country town, the proportion of employment in the farming industry is 0.62. Five people aged 15 years and older are selected at random from the town.
   a. Determine the possible values of the sample proportion, $\hat{p}$, of workers in the farming industry.
   b. Use this information to construct a probability distribution table to represent the sampling distribution of the sample proportion of workers in the farming industry. Give your answer correct to 3 decimal places.
   c. Calculate the probability correct to 3 decimal places that the proportion of workers in the farming industry in the sample is greater than 0.5.

10. WE7 There is a very large container containing thousands of pens. Of these, 25% are known to be blue. A sample of 24 pens are selected from the bag. If $\hat{P}$ is the random variable of the distribution of sample proportions of blue pens, calculate $\Pr\left(\hat{P} \geq \frac{5}{24}\right)$, correct to 4 decimal places.

11. There is a very large container containing several thousand coins. Of these, 40% are known to be gold. A sample of 25 coins are selected from the container. If $\hat{P}$ is the random variable of the distribution of sample proportions of gold coins, calculate $\Pr\left(\hat{P} \geq \frac{8}{25} \,\middle|\, \hat{P} \geq \frac{5}{25}\right)$, correct to 4 decimal places.

12. MC Four people are selected for a random sample. $\hat{P}$ is the random variable that represents the proportion who have visited another country.
If $\Pr(X = 4) = \frac{1}{625}$, then $\Pr(\hat{P} < 0.5)$, correct to 4 decimal places, is:

A. 0.2617 B. 0.9492 C. 0.7383 D. 0.0508 E. 0.3125

13. WE8 If $N = 500$, $n = 50$ and $p = 0.5$, determine:
   a. the mean of the distribution
   b. the standard deviation of the distribution, correct to 2 decimal places.

14. If $N = 1000$, $n = 100$ and $p = 0.8$, determine:
   a. the mean of the distribution
   b. the standard deviation of the distribution, correct to 2 decimal places.

15. If $N = 1500$, $n = 150$ and $p = 0.15$, calculate the mean and standard deviation for the distribution of $\hat{P}$. Give your answers correct to 3 decimal places where appropriate.

16. MC If $n = 100$ and $p = 0.75$, calculate the standard deviation for the distribution of $\hat{P}$.

A. 0.043 B. 0.012 C. 0.75 D. 0.25 E. 0.052

17. A distribution for $\hat{P}$ has a mean of 0.12 and a standard deviation of 0.0285. Find the population proportion and the sample size.

18. A distribution for $\hat{P}$ has a mean of 0.81 and a standard deviation of 0.0253. Find the population proportion and the sample size.

19. A distribution for $\hat{P}$ has a standard deviation of 0.015. If the sample size was 510 and $p > 0.5$, calculate the population proportion, correct to 2 decimal places.

20. A distribution for $\hat{P}$ has a standard deviation of 0.0255. If the sample size was 350, calculate the population proportion, correct to 2 decimal places.

13.3 Exam questions

Question 1 (2 marks) TECH-FREE

Source: *VCE 2017, Mathematical Methods Exam 1, Q4; © VCAA.*

In a large population of fish, the proportion of angel fish is $\frac{1}{4}$.

Let $\hat{P}$ be the random variable that represents the sample proportion of angel fish for samples of size n drawn from the population.

Find the smallest integer value of n such that the standard deviation of $\hat{P}$ is less than or equal to $\frac{1}{100}$.

Question 2 (1 mark) TECH-ACTIVE

Source: VCE 2016, Mathematical Methods Exam 2, Section A, Q17; © VCAA.

MC Inside a container there are one million coloured building blocks. It is known that 20% of the blocks are red. A sample of 16 blocks is taken from the container. For samples of 16 blocks, $\hat{P}$ is the random variable of the distribution of sample proportions of red blocks. (Do not use a normal approximation.)

$\Pr\left(\hat{P} \geq \frac{3}{16}\right)$ is closest to

A. 0.6482 **B.** 0.8593 **C.** 0.7543 **D.** 0.6542 **E.** 0.3211

Question 3 (1 mark) TECH-ACTIVE

Source: VCE 2017, Mathematical Methods Exam 2, Section A, Q16; © VCAA.

MC For random samples of five Australians, $\hat{P}$ is the random variable that represents the proportion who live in a capital city. Given that $\Pr\left(\hat{P} = 0\right) = \frac{1}{243}$, then $\Pr\left(\hat{P} > 0.6\right)$, correct to four decimal places, is

A. 0.0453 **B.** 0.3209 **C.** 0.4609 **D.** 0.5390 **E.** 0.7901

More exam questions are available online.

13.4 Confidence intervals

LEARNING INTENTION

At the end of this subtopic you should be able to:

- calculate confidence intervals for various levels of confidence
- determine the margin of error
- calculate the sample size needed for a given type of confidence interval.

We have just learned that different samples can have different values for $\hat{p}$. So what can one sample tell us about a population?

Let us say that you are interested in the proportion of the school who buy their lunch. You decide that your class is a reasonable sample and find out that 25% of the class will buy their lunch today. What can you say about the proportion of the whole school that will buy their lunch today? Assuming that your class is in fact a representative sample, you may say that around 25% of the school will buy their lunch. Is it possible to be more specific? By using **confidence intervals**, it is possible to say how confident you are that a population parameter will lie in a particular interval.

13.4.1 Normal approximation to the distribution of $\hat{P}$

We have learned that when we consider the distributions of $\hat{P}$ with a large sample size, they are normally distributed with $\mu = \mathrm{E}\left(\hat{P}\right) = p$ and $\sigma = \mathrm{SD}\left(\hat{P}\right) = \sqrt{\frac{p(1-p)}{n}}$. As we don't know the exact value for p, the best estimate is $\hat{p}$. This means that the best estimate of the standard deviation is $\sqrt{\frac{\hat{p}(1-\hat{p})}{n}}$.

We know that for normal distributions, $z = \frac{x-\mu}{\sigma}$. This means that, to find the upper and lower values of z, we can use $z = \frac{\hat{p} \pm p}{\sigma_{\hat{p}}}$. Rearranging gives us $p = \hat{p} \pm z\sigma_{\hat{p}}$.

Confidence interval for P

An approximate confidence interval for a population proportion is given by $(\hat{p} - z\sigma_{\hat{p}}, \hat{p} + z\sigma_{\hat{p}})$, where $\sigma_{\hat{p}} = \sqrt{\frac{\hat{p}(1-\hat{p})}{n}}$.

A 95% confidence interval means that 95% of the distribution is in the middle area of the distribution. This means that the tails combined contain 5% of the distribution (2.5% on each end). The z-score for this distribution is 1.96.

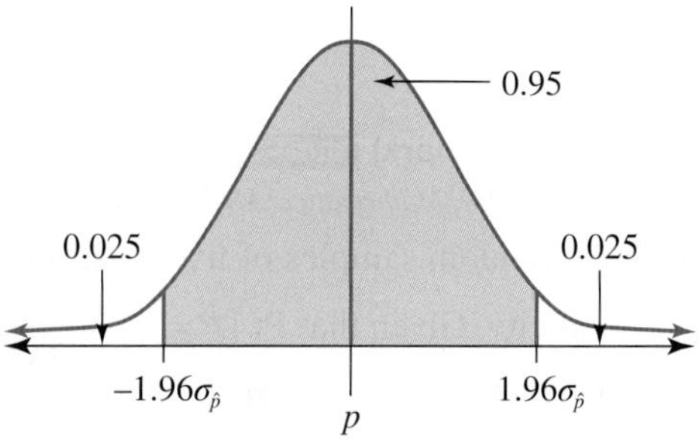

The confidence interval for this distribution can be expressed as

$$\left(\hat{p} - 1.96\sqrt{\frac{\hat{p}(1-\hat{p})}{n}}, \hat{p} + 1.96\sqrt{\frac{\hat{p}(1-\hat{p})}{n}}\right).$$

WORKED EXAMPLE 9 Confidence intervals

There are 20 people in your class and 25% are planning on buying their lunch. Estimate the proportion of the school population who will purchase their lunch today. Calculate a 95% confidence interval for your estimate, given $z = 1.96$. Give your answer correct to 4 decimal places.

THINK	WRITE
1. There are 20 people in the class. This is the sample size. 25% are buying their lunch. This is the sample proportion.	$n = 20$ $\hat{p} = 0.25$
2. For a 95% confidence interval, $z = 1.96$.	$z = 1.96$
3. The confidence interval is $\left(\hat{p} - z\sqrt{\frac{\hat{p}(1-\hat{p})}{n}}, \hat{p} + z\sqrt{\frac{\hat{p}(1-\hat{p})}{n}}\right)$. Find $z\sigma_{\hat{p}}$.	$z\sigma_{\hat{p}} = z\sqrt{\frac{\hat{p}(1-\hat{p})}{n}}$ $= 1.96\sqrt{\frac{0.25 \times 0.75}{20}}$ $= 0.1898$
4. Identify the 95% confidence interval by finding the upper and lower values. CAS technology can also be used.	$\hat{p} - z\sigma_{\hat{p}} = 0.25 - 0.1898$ $= 0.0602$ $\hat{p} + z\sigma_{\hat{p}} = 0.25 + 0.1898$ $= 0.4398$
5. Write the answer.	The confidence interval is (0.0602, 0.4398).

TI | THINK

DISPLAY/WRITE

1. On a Calculator page, press MENU, then select:
   6: Statistics
   6: Confidence intervals
   5: 1-Prop z Interval
   Complete the fields as:
   Successes, x: 5
   n: 20
   C Level: 0.95
   then Select OK.
   Note: The value $x = 5$ represents the sample size, which is 25% of the class of 20 students.

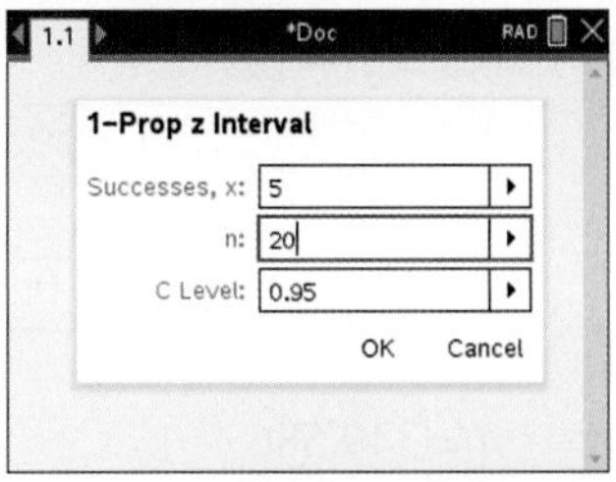

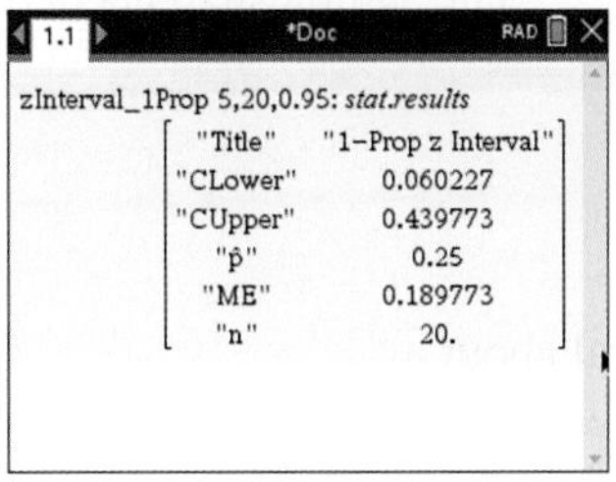

2. The answer appears on the screen.

The confidence interval is (0.0602, 0.4398).

CASIO | THINK

DISPLAY/WRITE

1. On a Statistics screen, select:
   - Calc
   - Interval

   Select the type as Interval, then select One-Prop Z Int from the drop-down menus. Select Next.
   Complete the fields as:
   C-Level: 0.95
   x: 5
   n: 20
   then select Next.
   Note: The value $x = 5$ represents the sample size, which is 25% of the class of 20 students.

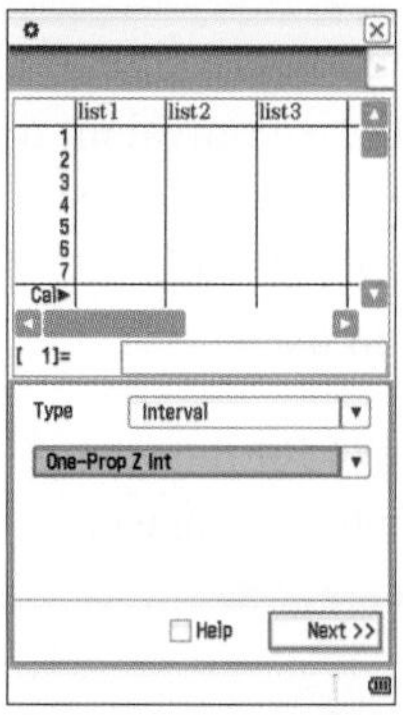

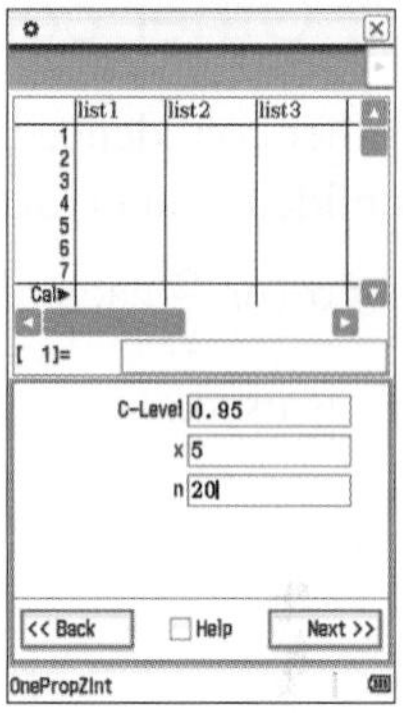

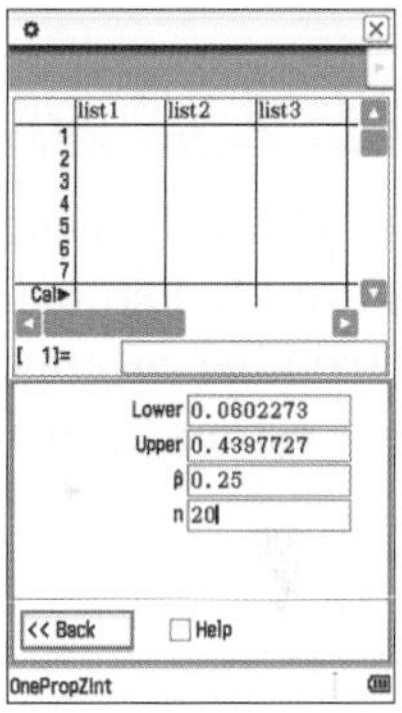

2. The answer appears on the screen.

The confidence interval is (0.0602, 0.4398).

WORKED EXAMPLE 10 Determining sample size

Grow Well are 95% sure that 30% to 40% of shoppers prefer their mulch. Determine the sample size that was needed for this confidence interval.

THINK

WRITE

1. The confidence interval is symmetric about $\hat{p}$: $(\hat{p} - z\sigma_{\hat{p}}, \hat{p} + z\sigma_{\hat{p}})$, so the value of $\hat{p}$ must be halfway between the upper and lower values of the confidence interval.

$$\hat{p} = \frac{30 + 40}{2}$$
$$= 35\%$$
$$= 0.35$$

2. State the z-score related to the 95% confidence interval.

$z = 1.96$

3. The lower value of the confidence interval, 30%, is equivalent to $\hat{p} - z\sigma_{\hat{p}}$. Substitute the appropriate values.
Note: The equation $0.4 = \hat{p} + z\sigma_{\hat{p}}$ could also have been used.

$$0.3 = \hat{p} - z\sigma_{\hat{p}}$$
$$= \hat{p} - z\sqrt{\frac{\hat{p}(1-\hat{p})}{n}}$$
$$= 0.35 - 1.96\sqrt{\frac{0.35(1-0.35)}{n}}$$

4. Solve for n.

$n = 349.586$

5. Write the answer.

The sample size needed was 350 people.

13.4.2 Other confidence intervals

To find other confidence intervals, we can talk in general about a $1 - \alpha$ confidence interval. In this case, the tails combined will have an area of α (or $\frac{\alpha}{2}$ each tail). In this case, the z-score that has a tail area of $\frac{\alpha}{2}$ is used.

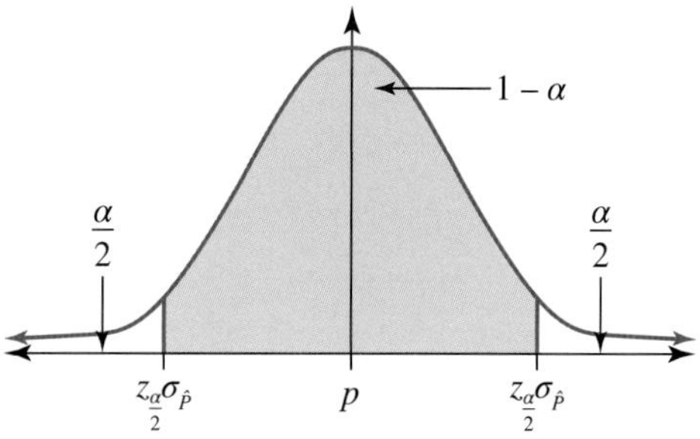

WORKED EXAMPLE 11 Other confidence intervals

Paul samples 102 people and finds that 18 of them like drinking coconut milk. Estimate the proportion of the population that likes drinking coconut milk. Calculate a 99% confidence interval for your estimate, correct to 3 decimal places.

THINK

WRITE

1. There are 102 people in the sample. This is the sample size.
18 people like drinking coconut milk.

$$n = 102$$
$$\hat{p} = \frac{18}{102}$$
$$= \frac{3}{17}$$

2. For a 99% confidence interval, find the z-score using the inverse standard normal distribution.

For the 99% confidence interval, 1% will be in the tails, so 0.5% in each tail. Therefore, the area under the normal distribution curve to the left of z is 0.995.
$z = 2.58$

3. The confidence interval is $(\hat{p} - z\sigma_{\hat{p}}, \hat{p} + z\sigma_{\hat{p}})$. Find $z\sigma_{\hat{p}}$.

$$z\sigma_{\hat{p}} = z\sqrt{\frac{\hat{p}(1-\hat{p})}{n}}$$
$$= 2.58\sqrt{\frac{\frac{3}{17} \times 0.82}{102}}$$
$$= 0.097$$

4. Identify the 99% confidence interval by finding the upper and lower values, correct to 1 decimal place.	$\hat{p} - z\sigma_{\hat{p}} = \frac{3}{17} - 0.097$ $= 0.079$ $\hat{p} + z\sigma_{\hat{p}} = \frac{3}{17} + 0.097$ $= 0.273$
5. Write the answer.	The confidence interval is (0.079, 0.273).

Margin of error

The distance between the endpoints of the confidence interval and the sample estimate is called the **margin of error**, M.

Worked example 10 considered a 95% confidence interval of $(\hat{p} - 1.96\sigma_{\hat{p}}, \hat{p} + 1.96\sigma_{\hat{p}})$. In this case the margin of error would be $M = 1.96\sigma_{\hat{p}}$.

> **Margin of error**
>
> **For a 95% level of confidence,** $M = 1.96\sqrt{\frac{\hat{p}(1-\hat{p})}{n}}$.

Note that the larger the sample size, the smaller the value of M will be. This means that one way to reduce the size of a confidence interval without changing the level of confidence is to increase the sample size.

13.4 Exercise

Technology active

1. WE9 Of 30 people surveyed, 80% said that they like breakfast in bed. Estimate the proportion of the population who like breakfast in bed. Calculate a 95% confidence interval for the estimate. Give your answer correct to 2 decimal places.

2. Of the 60 people at swimming training today, 85% said that their favourite stroke is freestyle. Estimate the proportion of the population whose favourite stroke is freestyle. Calculate a 95% confidence interval for your estimate, correct to 2 decimal places.
3. WE11 Jenny samples 116 people and finds that 86 of them plan to go swimming over the summer holidays. Estimate the proportion of the population who plan to go swimming over the summer holidays. Calculate a 99% confidence interval for your estimate, correct to 2 decimal places.

4. Yuki samples 95 people and finds that 35 of them eat chocolate daily. Estimate the proportion of the population who eat chocolate daily. Determine, correct to 2 decimal places, a 90% confidence interval for your estimate.

The following information relates to question 5 and 6.

Teleco is being criticised for its slow response time when handling complaints. The company claims that it will respond within 1 day. Of the 3760 complaints in a given week, a random sample of 250 was selected. Of these, it was found that 20 of them had not been responded to within 1 day.

5. **MC** The 95% confidence interval for the proportion of claims that take more than 1 day to resolve is:

A. (0.04, 0.11)
B. (0.04, 0.12)
C. (0.06, 0.13)
D. (0.05, 0.11)
E. (0.03, 0.12)

6. Calculate the 99% confidence interval for the proportion of claims that take less than 1 day to resolve. Give your answer correct to 3 decimal places.

7. A sample of 250 blood donors have their blood types recorded. Of this sample, 92 have Type A blood. Calculate the 90% confidence interval for the proportion of Australians who have Type A blood. Give your answer correct to 3 decimal places.

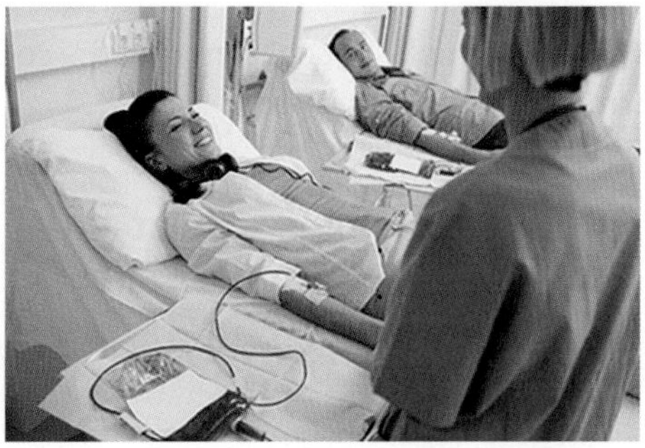

8. It is believed that 65% of people have brown hair. A random selection of 250 people were asked the colour of their hair. Applying the normal approximation, calculate the probability that less than 60% of the people in the sample have brown hair. Give your answer correct to 4 decimal places.

9. Nidya is a top goal shooter. The probability of her getting a goal is 0.8. To keep her skills up, each night she has 200 shots on goal. Applying the normal approximation, calculate the probability that on Monday the proportion of times she scores a goal is between 0.8 and 0.9, given that it is more than 0.65. Give your answer correct to 4 decimal places.

10. **WE10** If Parkers want to be 90% confident that between 75% and 85% of their customers will shop in their store for more than 2 hours, calculate the sample size that will be needed.

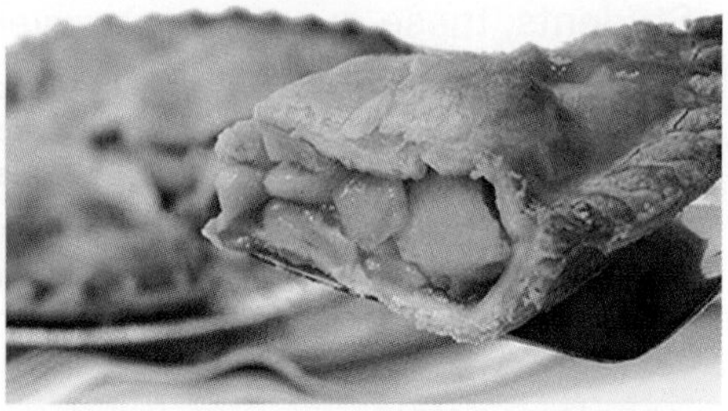

11. **MC** In a country town, the 95% confidence interval for the number of customers that love the homemade apple pie at Edie's Eatery is (0.35, 0.45). Determine the sample size that was needed for this level of confidence.

A. 8 B. 321 C. 369 D. 92 E. 5

12. The 95% confidence interval for the proportion of shoppers that prefer a particular pen at Smooth Writing is (0.25, 0.35). Calculate the sample size that was needed for this level of confidence.

13. The 99% confidence interval for the proportion of students that prefer to use a particular online company is (0.20, 0.30). Calculate the sample size that was needed for this level of confidence.

14. Teleco want to be 95% sure that less than 5% of their complaints take more than 1 day to resolve. Calculate the sample proportion and the size of the sample they need to support this claim.

15. Barton's Dentistry want to be able to claim that 90% to 98% of people floss daily. They would like 99% confidence about their claim. Calculate how many people they need to survey.

16. Tatiana is conducting a survey to estimate the proportion of Year 12 students who will take a gap year after they complete their VCE. Previous surveys have shown the proportion to be approximately 15%. Determine the required size of the sample so that the margin of error for the survey is 3% in a confidence interval of approximately 95% for this proportion.

17. Bentons claim that between 85% and 95% of their customers stay for more than 2 hours when they shop. If they surveyed 100 people, determine how confident they are about that claim.

18. The Hawthorn Football Club claim that between 75% and 80% of their members remain members for at least 10 years. If they surveyed 250 people, determine how confident they are about that claim. Give your answer to the nearest whole number.

13.4 Exam questions

Question 1 (1 mark) TECH-ACTIVE

Source: *VCE 2021, Mathematical Methods Exam 2, Section A, Q3; © VCAA.*

MC A box contains many coloured glass beads.

A random sample of 48 beads is selected and it is found that the proportion of blue-coloured beads in this sample is 0.125.

Based on this sample, a 95% confidence interval for the proportion of blue-coloured glass beads is

A. (0.0314, 0.2186) **B.** (0.0465, 0.2035) **C.** (0.0018, 0.2482)
D. (0.0896, 0.1604) **E.** (0.0264, 0.2136)

Question 2 (1 mark) TECH-ACTIVE

Source: *VCE 2017, Mathematical Methods Exam 2, Section A, Q5; © VCAA.*

MC The 95% confidence interval for the proportion of ferry tickets that are cancelled on the intended departure day is calculated from a large sample to be $(0.039, 0.121)$.

The sample proportion from which this interval was constructed is

A. 0.080 **B.** 0.041 **C.** 0.100 **D.** 0.062 **E.** 0.059

Question 3 (1 mark) TECH-ACTIVE

MC From a survey, 747 out of 1168 Year 12 students said they had obtained their learner's permit. The 99% confidence interval for Year 12 students who have obtained their learner's permit is

A. $(0.598, 0.682)$ **B.** $(0.612, 0.668)$ **C.** $(0.626, 0.654)$ **D.** $(0.630, 0.650)$ **E.** $(0.604, 0.676)$

More exam questions are available online.

13.5 Review

13.5.1 Summary

doc-37043

Hey students! Now that it's time to revise this topic, go online to:

Access the topic summary | Review your results | Watch teacher-led videos | Practise VCAA exam questions

Find all this and MORE in jacPLUS

13.5 Exercise

Technology free: short answer

1. James has heard that 1 in 10 people have been to Alice Springs. He goes to the local supermarket and asks every 10th person if they have been to Alice Springs. He expects that they will all say yes. Describe you would expect.

2. A school has 1100 students, 600 of whom play sport regularly. A random sample of 100 students was chosen, and 70 of those were found to play sport.
   a. Calculate the population and sample sizes.
   b. Calculate the value of the sample proportion, $\hat{p}$.
   c. Calculate the value of the population parameter, p.
3. Natasha believes that she has a biased coin. She tosses the coin 10 000 times and records 5100 Heads.
   a. Calculate the sample size.
   b. Calculate the population size.
   c. Calculate the sample proportion of Heads.
   d. Write an expression for the 95% confidence interval for the likelihood of tossing a Head with this coin.
4. On a particular Friday night, 52 000 people attended the MCG to watch the AFL. Every 25th person entering the stadium was asked who they thought would win. Out of the people asked, 1600 people believed that the Hawks would win.
   a. Calculate the population size.
   b. Calculate the sample size.
   c. Determine $\hat{p}$.

5. Green High has 132 staff members. Every year, the school offers free flu shots to its staff. This year, 120 people decided to have the shot and 12 of them had a sore arm afterwards.

a. Calculate the value of the sample proportion.
b. Write an expression for the 95% confidence interval for the likelihood of getting a sore arm.
c. Write an expression for the margin of error, M, for the 95% confidence interval.
d. If only 60 people had decided to have the flu shot, determine the effect on the margin of error.

6. In a recent voter survey, an approximate 90% confidence interval for the proportion of people who will vote for a republic was $(0.62,\ 0.78)$.

a. Calculate the value of $\hat{p}$ for this confidence interval.
b. Calculate the value of the margin of error.

Technology active: multiple choice

7. **MC** Johansen Enterprises operates for 15 hours per day. It is capable of producing 3000 items per hour. From each hour's output, 10 items are chosen for inspection so that the machinery can be adjusted if necessary. Identify which of the following is correct.

A. $N = 3000$, $n = 10$
B. $N = 3000$, $n = 15$
C. $N = 3000$, $n = 150$
D. $N = 45\,000$, $n = 10$
E. $N = 45\,000$, $n = 150$

8. **MC** Identify which of the following are population parameters.

I. According to the Australian Bureau of Statistics, the unemployment rate is 6.4%.
II. According to the 2011 census, on average there are 1.7 motor vehicles per dwelling.
III. According to a poll in the newspaper *The Age*, 54% of Australians will vote Liberal at the next election.

A. I
B. II
C. I and II
D. II and III
E. I and III

9. **MC** Susan Storm has 350 regular customers. She wants to survey them. She lists the clients in alphabetical order and then assigns each one a customer number. She then uses a random number generator to select 15 customers to survey. This is an example of:

A. a systematic sample
B. a self-selected sample
C. a biased sample
D. a stratified random sample
E. a random sample

10. **MC** State which of the following could be a distribution for $\hat{P}$ for large samples.

A.

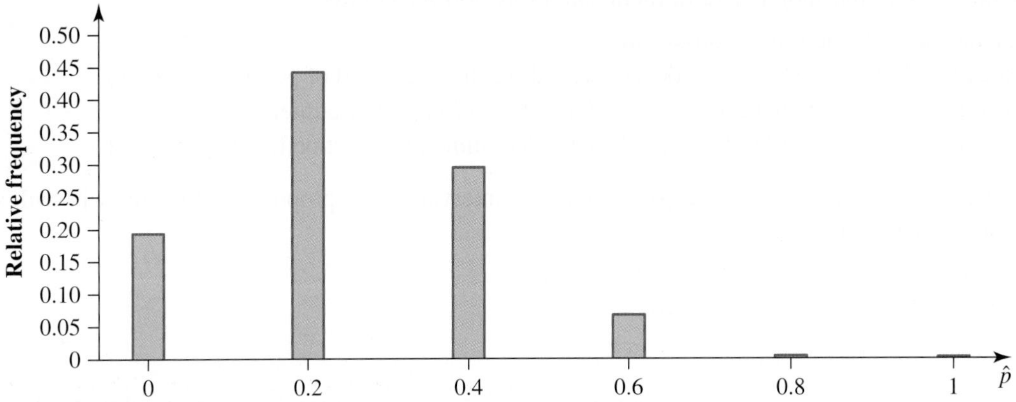

B.

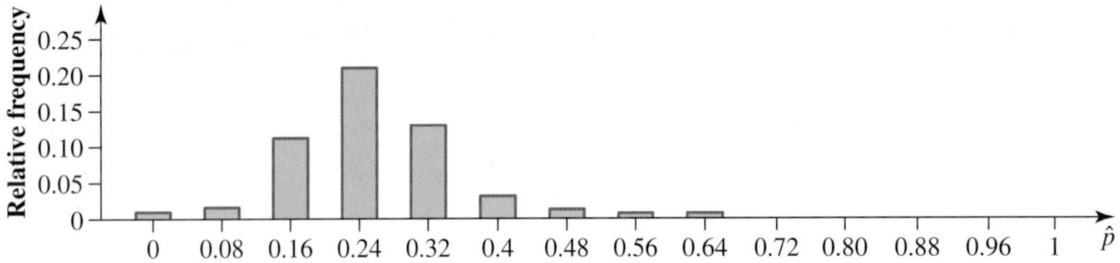

C.

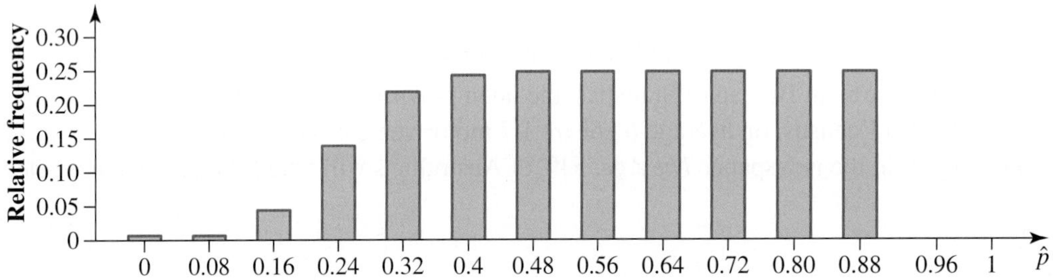

D.

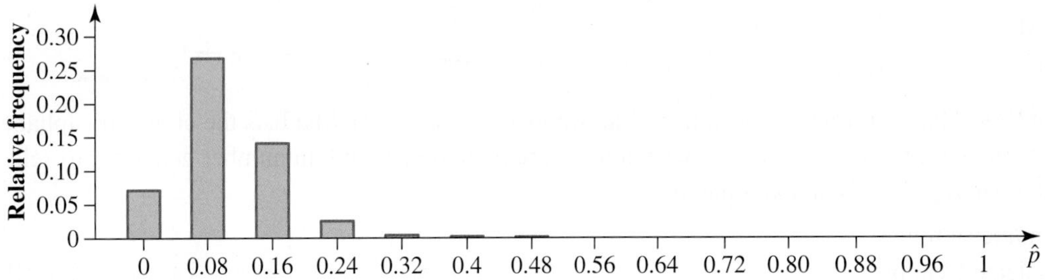

E. None of the above

11. **MC** If the population parameter is believed to be $p = 0.37$ and samples of size 120 are chosen, the standard deviation of $\hat{P}$ is:

A. 5.29 **B.** 0.37 **C.** 0.002 **D.** 0.044 **E.** 0.24

12. **MC** Five people are selected for a random sample. $\hat{P}$ is the random variable that represents the proportion who have earned their driver's licence.

If $\Pr(\hat{P} = 0) = \dfrac{1}{1024}$, then $\Pr(\hat{P} \geq 0.8)$, correct to 4 decimal places, is:

A. 0.3672 **B.** 0.0156 **C.** 0.3955 **D.** 0.6328 **E.** 0.9990

13. **MC** Kei finds a 95% confidence interval. Select the correct statement from the following.

A. There is a 95% chance that the population parameter lies in the interval.
B. 95% of the time, the population parameter is the centre of the interval.
C. In 95% of the samples, the population parameter lies in the interval.
D. 95% of the sample estimates lie within the interval.
E. None of the above

14. **MC** Of 150 people surveyed, 36% said that their favourite colour was blue. A 99% confidence interval for the proportion of the population whose favourite colour is blue is:

A. (0.3, 0.42)
B. (0.28, 0.44)
C. (0.32, 0.40)
D. (0.26, 0.46)
E. (0.20, 0.52)

15. **MC** The Melbourne Vixens claim that between 60% and 70% of their supporters attend at least half of their netball games each year. If 200 people were surveyed, determine how confident can they be about that claim.

A. 1.48%
B. 65%
C. 95%
D. 52%
E. 86%

16. **MC** The 95% confidence interval for the proportion of students who prefer to use a textbook company's resources is (0.65, 0.75). The sample size needed for this level of confidence is:

A. 323
B. 560
C. 81
D. 226
E. 292

Technology active: extended response

17. Every year, thousands of tourists drive the Great Ocean Road. In a recent survey of 50 people, 86% listed seeing the Twelve Apostles as the highlight of their drive. Calculate the proportion of drivers who would rate the Twelve Apostles as the highlight of their drive. Give your answer with a 90% confidence level, correct to 2 decimal places,

18. There is a very large room containing several thousand balls. Of these, 30% are yellow. A sample of 12 balls are selected from the bag. If $\hat{P}$ is the random variable of the distribution of sample proportions of yellow balls, calculate $\Pr\left(\hat{P} \geq \frac{5}{12} \,\middle|\, \hat{P} \geq \frac{1}{4}\right)$, correct to 4 decimal places.

19. The lower limit of a 95% confidence interval is 13%. If 100 people were surveyed, calculate the sample proportion, correct to 2 decimal places.

20. Breanna, Kayley and Teagan spent the day collecting survey results from the same population. They each surveyed 100 people. Breanna found that 23% of people said Yes, Kayley found that 20% of people said Yes, and Teagan found that 19% of people said Yes. They want to obtain an estimate for the population parameter at a 95% confidence interval. Breanna says they should each work out a confidence interval and then average them out to give the population parameter. Kayley says that they should combine their data into one sample and determine the population parameter using that parameter. Teagan says that it doesn't matter, because they will get the same results either way.
Identify which method is better and explain why.

13.5 Exam questions

Question 1 (3 marks) TECH-FREE

Source: *VCE 2019, Mathematical Methods Exam 1, Q6; © VCAA.*

Fred owns a company that produces thousands of pegs each day. He randomly selects 41 pegs that are produced on one day and finds eight faulty pegs.

a. What is the proportion of faulty pegs in this sample? **(1 mark)**

b. Pegs are packed each day in boxes. Each box holds 12 pegs. Let $\hat{P}$ be the random variable that represents the proportion of faulty pegs in a box.

The actual proportion of faulty pegs produced by the company each day is $\frac{1}{6}$.

Find $\Pr\left(\hat{P} < \frac{1}{6}\right)$. Express your answer in the form $a\,(b)^n$, where a and b are positive rational numbers and n is a positive integer. **(2 marks)**

Question 2 (1 mark) TECH-ACTIVE

Source: *VCE 2021, Mathematical Methods Exam 2, Section A, Q12; © VCAA.*

MC For a certain species of bird, the proportion of birds with a crest is known to be $\frac{3}{5}$.

Let $\hat{P}$ be the random variable representing the proportion of birds with a crest in samples of size n for this specific bird.

The smallest sample size for which the standard deviation of $\hat{P}$ is less than 0.08 is

A. 7 **B.** 27 **C.** 37 **D.** 38 **E.** 43

Question 3 (14 marks) TECH-ACTIVE

Source: *Adapted from VCE 2016, Mathematical Methods Exam 2, Section B, Q3; © VCAA.*

A school has a class set of 22 new laptops kept in a recharging trolley. Provided each laptop is correctly plugged into the trolley after use, its battery recharges.

On a particular day, a class of 22 students uses the laptops. All laptop batteries are fully charged at the start of the lesson. Each student uses and returns exactly one laptop. The probability that a student does **not** correctly plug their laptop into the trolley at the end of the lesson is 10%. The correctness of any student's plugging-in is independent of any other student's correctness.

a. Determine the probability that at least one of the laptops is **not** correctly plugged into the trolley at the end of the lesson. Give your answer correct to four decimal places. **(2 marks)**

b. A teacher observes that at least one of the returned laptops is not correctly plugged into the trolley. Given this, find the probability that fewer than five laptops are **not** correctly plugged in. Give your answer correct to four decimal places. **(2 marks)**

c. The time for which a laptop will work without recharging (the battery life) is normally distributed, with a mean of three hours and 10 minutes and standard deviation of six minutes. Suppose that the laptops remain out of the recharging trolley for three hours.
For any one laptop, find the probability that it will stop working by the end of these three hours. Give your answer correct to four decimal places. **(2 marks)**

d. A supplier of laptops decides to take a sample of 100 new laptops from a number of different schools. For samples of size 100 from the population of laptops with a mean battery life of three hours and 10 minutes and standard deviation of six minutes, $\hat{P}$ is the random variable of the distribution of sample proportions of laptops with a battery life of less than three hours.
Find the probability that $\Pr(\hat{P} \geq 0.06 \mid \hat{P} \geq 0.05)$. Give your answer correct to three decimal places. Do not use a normal approximation. **(3 marks)**

e. It is known that when laptops have been used regularly in a school for six months, their battery life is still normally distributed but the mean battery life drops to three hours. It is also known that only 12% of such laptops work for more than three hours and 10 minutes.
Find the standard deviation for the normal distribution that applies to the battery life of laptops that have been used regularly in a school for six months, correct to four decimal places. **(2 marks)**

f. The laptop supplier collects a sample of 100 laptops that have been used for six months from a number of different schools and tests their battery life. The laptop supplier wishes to estimate the proportion of such laptops with a battery life of less than three hours.
Suppose the supplier tests the battery life of the laptops one at a time.
Find the probability that the first laptop found to have a battery life of less than three hours is the third one. **(1 mark)**

g. The laptop supplier finds that, in a particular sample of 100 laptops, six of them have a battery life of less than three hours.

Determine the 95% confidence interval for the supplier's estimate of the proportion of interest. Give values correct to two decimal places. **(1 mark)**

h. The supplier also provides laptops to businesses. The probability density function for battery life, x (in minutes), of a laptop after six months of use in a business is

$$f(x) = \begin{cases} \dfrac{(210 - x)e^{\frac{x-210}{20}}}{400} & 0 \le x \le 210 \\ 0 & \text{elsewhere} \end{cases}$$

Find the **mean** battery life, in minutes, of a laptop with six months of business use, correct to two decimal places. **(1 mark)**

Question 4 (17 marks) TECH-ACTIVE

Source: *VCE 2019, Mathematical Methods Exam 2, Section B, Q4; © VCAA.*

The Lorenz birdwing is the largest butterfly in Town A.

The probability density function that describes its life span, X, in weeks, is given by

$$f(x) = \begin{cases} \dfrac{4}{625}(5x^3 - x^4) & 0 \le x \le 5 \\ 0 & \text{elsewhere} \end{cases}$$

a. Find the mean life span of the Lorenz birdwing butterfly. **(2 marks)**

b. In a sample of 80 Lorenz birdwing butterflies, how many butterflies are expected to live longer than two weeks, correct to the nearest integer? **(2 marks)**

c. What is the probability that a Lorenz birdwing butterfly lives for at least four weeks, given that it lives for at least two weeks, correct to four decimal places? **(2 marks)**

d. The wingspans of Lorenz birdwing butterflies in Town A are normally distributed with a mean of 14.1 cm and a standard deviation of 2.1 cm.
Find the probability that a randomly selected Lorenz birdwing butterfly in Town A has a wingspan between 16 cm and 18 cm, correct to four decimal places. **(1 mark)**

e. A Lorenz birdwing butterfly is considered to be **very small** if its wingspan is in the smallest 5% of all the Lorenz birdwing butterflies in Town A.
Find the greatest possible wingspan, in centimetres, for a **very small** Lorenz birdwing butterfly in Town A, correct to one decimal place. **(1 mark)**

f. Each year, a detailed study is conducted on a random sample of 36 Lorenz birdwing butterflies in Town A. A Lorenz birdwing butterfly is considered to be **very large** if its wingspan is greater than 17.5 cm. The probability that the wingspan of any Lorenz birdwing butterfly in Town A is greater than 17.5 cm is 0.0527, correct to four decimal places.

i. Find the probability that three or more of the butterflies, in a random sample of 36 Lorenz birdwing butterflies from Town A, are **very large**, correct to four decimal places. **(1 mark)**

ii. The probability that n or more butterflies, in a random sample of 36 Lorenz birdwing butterflies from Town A, are **very large** is less than 1%.
Find the smallest value of n, where n is an integer. **(2 marks)**

iii. For random samples of 36 Lorenz birdwing butterflies in Town A, $\hat{P}$ is the random variable that represents the proportion of butterflies that are **very large**.
Find the expected value and the standard deviation of $\hat{P}$, correct to four decimal places. **(2 marks)**

iv. What is the probability that a sample proportion of butterflies that are **very large** lies within one standard deviation of 0.0527, correct to four decimal places? Do not use a normal approximation. **(2 marks)**

g. The Lorenz birdwing butterfly also lives in Town B.
In a particular sample of Lorenz birdwing butterflies from Town B, an approximate 95% confidence interval for the proportion of butterflies that are **very large** was calculated to be (0.0234, 0.0866), correct to four decimal places.
Determine the sample size used in the calculation of this confidence interval. **(2 marks)**

Question 5 (16 marks) TECH-ACTIVE

Source: *VCE 2018, Mathematical Methods Exam 2, Section B, Q4; © VCAA.*

Doctors are studying the resting heart rate of adults in two neighbouring towns: Mathsland and Statsville. Resting heart rate is measured in beats per minute (bpm).

The resting heart rate of adults in Mathsland is known to be normally distributed with a mean of 68 bpm and a standard deviation of 8 bpm.

a. Find the probability that a randomly selected Mathsland adult has a resting heart rate between 60 bpm and 90 bpm. Give your answer correct to three decimal places. **(1 mark)**

b. The doctors consider a person to have a slow heart rate if the person's resting heart rate is less than 60 bpm. The probability that a randomly chosen Mathsland adult has a slow heart rate is 0.1587.
It is known that 29% of Mathsland adults play sport regularly.
It is also known that 9% of Mathsland adults play sport regularly and have a slow heart rate.
Let S be the event that a randomly selected Mathsland adult plays sport regularly and let H be the event that a randomly selected Mathsland adult has a slow heart rate.

i. Find $\Pr(H|S)$, correct to three decimal places. **(1 mark)**

ii. Are the events H and S independent? Justify your answer. **(1 mark)**

c. i. Find the probability that a random sample of 16 Mathsland adults will contain exactly one person with a slow heart rate. Give your answer correct to three decimal places. **(2 marks)**

ii. For random samples of 16 Mathsland adults, $\hat{P}$ is the random variable that represents the proportion of people who have a slow heart rate.
Find the probability that $\hat{P}$ is greater than 10%, correct to three decimal places. **(2 marks)**

iii. For random samples of n Mathsland adults, $\hat{P}_n$ is the random variable that represents the proportion of people who have a slow heart rate.
Find the least value of n for which $\Pr\left(\hat{P}_n > \frac{1}{n}\right) > 0.99$. **(2 marks)**

d. The doctors took a large random sample of adults from the population of Statsville and calculated an approximate 95% confidence interval for the proportion of Statsville adults who have a slow heart rate. The confidence interval they obtained was $(0.102, 0.145)$.

i. Determine the sample proportion used in the calculation of this confidence interval. **(1 mark)**

ii. Explain why this confidence interval suggests that the proportion of adults with a slow heart rate in Statsville could be different from the proportion in Mathsland. **(1 mark)**

e. Every year at Mathsland Secondary College, students hike to the top of a hill that rises behind the school. The time taken by a randomly selected student to reach the top of the hill has the probability density function M with the rule

$$M(t) = \begin{cases} \dfrac{3}{50}\left(\dfrac{t}{50}\right)^2 e^{-\left(\frac{t}{50}\right)^3} & t \geq 0 \\ 0 & t < 0 \end{cases}$$

where t is given in minutes.

Find the expected time, in minutes, for a randomly selected student from Mathsland Secondary College to reach the top of the hill. Give your answer correct to one decimal place. **(2 marks)**

f. Students who take less than 15 minutes to get to the top of the hill are categorised as 'elite'.

Find the probability that a randomly selected student from Mathsland Secondary College is categorised as elite. Give your answer correct to four decimal places. **(1 mark)**

g. The Year 12 students at Mathsland Secondary College make up $\frac{1}{7}$ of the total number of students at the school. Of the Year 12 students at Mathsland Secondary College, 5% are categorised as elite.

Find the probability that a randomly selected non-Year-12 student at Mathsland Secondary College is categorised as elite. Give your answer correct to four decimal places. **(2 marks)**

More exam questions are available online.

Answers

Topic 13 Statistical inference

13.2 Population parameters and sample statistics

13.2 Exercise

1. $n = 15$, population size is unknown.
2. $N = 100, n = 5$
3. $N = 120, n = 30$
4. $n = 9$, population size is unknown.
5. a. The population size is unknown.
   b. $n = 40$
6. a. The population is people who will receive the vaccine in the future. The size is unknown.
   b. $n = 247$
7. a. Population parameter
   b. Sample statistic
8. a. Population parameter
   b. Sample statistic
9. Sample statistic
10. Population parameter
11. Sample statistics
12. Population parameter
13. Use the random number generator on your calculator to produce numbers from 1 to 100. Keep generating numbers until you have 10 different numbers.
14. First, assign every person in your class a number, e.g. 1 to 25 if there are 25 students in your class. Decide how many students will be in your sample, e.g. 5. Then use the random number generator on your calculator to produce numbers from 1 to 25. Keep generating numbers until you have 5 different numbers. The students that were assigned these numbers are the 5 students in your random sample.
15. 34 juniors and 41 seniors
16. 21 boarders, 69 day students
17. a. A systematic sample with $k = 10$
    b. Yes, assuming that the order of patients is random
18. The sample is not random; therefore, the results are not likely to be random.
19. It is probably not random. Tony is likely to ask people who he knows or people who approach him.
20. Full-time staff under 50: 46
    Part-time staff under 50: 2
    Full-time staff over 50: 25
    Part-time staff over 50: 7

13.2 Exam questions

Note: Mark allocations are available with the fully worked solutions online.

1. E
2. B
3. B

13.3 The distribution of the sample proportion

13.3 Exercise

1. $\hat{p} = \frac{2}{5}$
2. $\hat{p} = \frac{3}{10}$
3. $\hat{p} = \frac{9}{10}$
4. $\hat{p} = \frac{147}{537}$
5. This is not a large sample; $n = 100+$ would be a large sample.
6. $n = 200$
7. See the graph at the bottom of the page.*
8. a. $p = \frac{4}{7}$
   b. $0, \frac{1}{4}, \frac{1}{2}, \frac{3}{4}, 1$
   c.

$\hat{p}$	0	$\frac{1}{4}$	$\frac{1}{2}$	$\frac{3}{4}$	1
$\Pr(\hat{P} = \hat{p})$	0.021	0.168	0.397	0.331	0.083

   d. 0.414
   e. 0.510
9. a. $0, \frac{1}{5}, \frac{2}{5}, \frac{3}{5}, \frac{4}{5}, 1$
   b.

$\hat{p}$	0	$\frac{1}{5}$	$\frac{2}{5}$	$\frac{3}{5}$	$\frac{4}{5}$	1
$\Pr(\hat{P} = \hat{p})$	0.008	0.064	0.211	0.344	0.281	0.092

   c. 0.717
10. 0.7534
11. 0.8545
12. C
13. a. $E(\hat{P}) = 0.5$ b. $SD(\hat{P}) = 0.07$
14. a. $E(\hat{P}) = 0.8$ b. $SD(\hat{P}) = 0.04$
15. $E(\hat{P}) = 0.15, SD(\hat{P}) = 0.029$
16. A

*7.

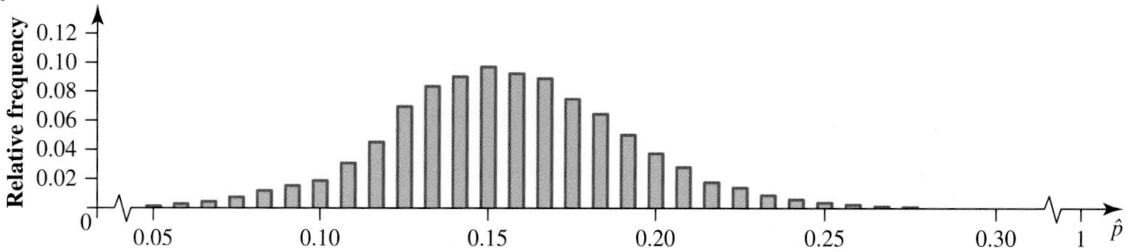

17. $p = 0.12, n = 130$
18. $p = 0.81, n = 240$
19. $p = 0.87$
20. $p = 0.35$ or $p = 0.65$

13.3 Exam questions

Note: Mark allocations are available with the fully worked solutions online.

1. 1875 2. A 3. C

13.4 Confidence intervals

13.4 Exercise

1. (0.66, 0.94)
2. (0.76, 0.94)
3. (0.64, 0.85)
4. (0.29, 0.45)
5. D
6. (0.876, 0.964)
7. (0.318, 0.418)
8. 0.0487
9. 0.4998
10. $n = 173$
11. C
12. $n = 323$
13. $n = 498$
14. $\hat{p} = 2.5\%, n = 150$
15. $n = 235$
16. 544 people
17. 90% sure
18. 66%

13.4 Exam questions

Note: Mark allocations are available with the fully worked solutions online.

1. A 2. A 3. E

13.5 Review

13.5 Exercise

Technology free: short answer

1. Sample responses can be found in the worked solutions in the online resources.
2. a. Population = 1100, sample size = 100
   b. 0.7
   c. $\frac{6}{11}$
3. a. 10 000
   b. The population size (how many times in total the coin will be tossed) is unknown
   c. 0.51
   d. $0.51 \pm 1.96\sqrt{\frac{0.51 \times 0.49}{10\,000}}$
4. a. 52 000 b. 2080 c. $\frac{10}{13}$
5. a. $\frac{1}{10}$
   b. $0.1 \pm 1.96\sqrt{\frac{0.09}{120}}$
   c. $M = 1.96\sqrt{\frac{0.09}{120}}$
   d. M would increase by a factor of $\sqrt{2}$.
6. a. 0.7 b. 0.08

Technology active: multiple choice

7. A 8. B 9. E 10. B 11. D
12. D 13. C 14. D 15. E 16. A

Technology active: extended response

17. (0.78, 0.94)
18. 0.3698
19. 0.21
20. Breanna's method:
    Breanna: (0.1475, 0.3125); Kayley: (0.1216, 0.2784);
    Teagan: (0.1131, 0.2669)
    Average: (0.1274, 0.2859)
    Kayley's method: $X = 62, n = 300$
    Confidence interval: (0.1608, 0.2525)
    Kayley's method is better. Because they actually sampled 300 people, this should be the sample size. Because a larger sample size is more likely to have similar proportions to the population, the confidence interval can be smaller.

13.5 Exam questions

Note: Mark allocations are available with the fully worked solutions online.

1. a. $\frac{8}{41}$ b. $\left(\frac{17}{6}\right)\left(\frac{5}{6}\right)^{11}$
2. D
3. a. 0.9015 b. 0.9311
   c. 0.0478 d. 0.658
   e. 8.5107 f. $\frac{1}{8}$
   g. (0.01, 0.11) h. 170.01
4. a. $3\frac{1}{3}$ weeks b. 73 c. 0.2878
   d. 0.1512 e. 10.6
   f. i. 0.2947 ii. 7
      iii. 0.0372 iv. 0.7380
   g. 200
5. a. 0.838
   b. i. 0.310 ii. Not independent
   c. i. 0.190 ii. 0.747 iii. 39
   d. i. 0.1235
      ii. p(Mathsland) = 0.1587 is not contained within the confidence interval for Statsville, which suggests that the proportions between the two towns differ.
   e. 44.6 minutes
   f. 0.0266
   g. 0.0227

GLOSSARY

absolute maximum a point on a curve where the y-coordinate is greater than that of any other point on the curve

absolute minimum a point on a curve where the y-coordinate is smaller than that of any other point on the curve

acceleration the rate of change of velocity with respect to time

amplitude the distance a sine or cosine graph oscillates up and down from its equilibrium, or mean, position

anti-differentiation the reverse process to differentiation of indefinite integrals

anti-derivative *see* indefinite integral

asymptote a line that a graph approaches but never reaches. A horizontal asymptote shows the long-term behaviour as, for example, $x \to \infty$; a vertical asymptote may occur where a function is undefined such as at $x = 0$ for the hyperbola $y = \frac{1}{x}$.

average value of a function is the height of a rectangle with a width $(b - a)$ that has the same area as the area under the curve $y = f(x)$ for the interval $[a, b]$. Average $= \frac{1}{b-a}\int_a^b f(x)\,dx$

average rate of change of a function f over an interval $[a, b]$ is measured by $\frac{f(b)-f(a)}{b-a}$. This is the gradient of the chord joining the endpoints of the interval on the curve $y = f(x)$.

Bernoulli distribution a discrete probability distribution involving one Bernoulli trial

Bernoulli trial a single event that has only two possible outcomes: success or failure. Each outcome has a fixed probability, the sum of which is 1.

binomial distribution a discrete probability distribution involving n Bernoulli trials. Each trial is an independent trial.

central difference approximation $f'(x) \approx \frac{f(x+h)-f(x-h)}{2h}$ where h is a small change in the x-coordinate

chain rule in differentiation, $\frac{dy}{dx} = \frac{dy}{du} \times \frac{du}{dx}$

co-domain the set of all y-values available for pairing with x-values to form a mapping according to a function rule $y = f(x)$

complementary relationship trigonometric relationships between complementary angles such as $\sin(\theta) = \cos\left(\frac{\pi}{2} - \theta\right)$ and $\cos(\theta) = \sin\left(\frac{\pi}{2} - \theta\right)$

composite function Given the functions f and g, the composite function $f \circ g$ is defined by $(f \circ g)(x) = f(g(x))$.

confidence interval the level of confidence that a population parameter will lie in a particular interval. The 95% confidence interval is given by $\left(\hat{p} - 1.96\sqrt{\frac{\hat{p}(1-\hat{p})}{n}}, \hat{p} + 1.96\sqrt{\frac{\hat{p}(1-\hat{p})}{n}}\right)$.

continuous describes a graph that forms a curve with no breaks

continuous random variable a random variable that assumes quantities that can be measured, such as weights, time, height, in a given range

cubic the function $f: R \to R, f(x) = x^3$. The general rule of a cubic function is $f(x) = ax^3 + bx^2 + cx + d$.

decay The value of the function decreases over time. A general equation to represent exponential decay is given by $A = A_0e^{kt}$, where k is a negative value.

definite integral $\int_a^b f(x)\,dx = F(b) - F(a)$ where $F'(x) = f(x)$. a is the lower terminal and b is the upper terminal

derivative or **gradient function** (of a function f) is the function f' whose rule is defined as $f'(x) = \lim_{h \to 0} \frac{f(x+h) - f(x)}{h}$. For $y = f(x)$, $\frac{dy}{dx} = f'(x)$ where $\frac{dy}{dx} = \lim_{\delta x \to 0} \frac{\delta y}{\delta x}$.

differential calculus the study of instantaneous rates of change of continuous curves

differentiation by first principles requires the derivative to be obtained from its limit definition: either $f'(x) = \lim_{h \to 0} \frac{f(x+h) - f(x)}{h}$ or $\frac{dy}{dx} = \lim_{\delta x \to 0} \frac{\delta y}{\delta x}$

dilations linear transformations that enlarge or reduce the size of a figure by a scale factor k parallel to either axis or both

discrete random variable a random variable that can have only countable numbers, generally integer values

discriminant For the quadratic expression $ax^2 + bx + c$, the discriminant is $\Delta = b^2 - 4ac$.

distance measures the actual length covered over a time period, not the length between start and end points

domain the set of all x-values of the ordered pairs (x, y) that make up a relation

dominant function one of a combination of two other functions that has the greater influence on the shape of the graph for a particular section of the domain

expected value or **mean** a measure of the central tendency of the probability distribution of a random variable. For a discrete random variable, X, $E(X) = \mu = \sum(xp(x))$.

exponent For the number $n = a^p$, the base is a and the power, or index, or exponent, is p.

exponential functions functions of the form $f: R \to R, f(x) = a^x, a \in R^+ \backslash \{1\}$

factor theorem For the polynomial $P(x)$, if $P(a) = 0$, then $(x - a)$ is a factor of $P(x)$.

factorise to convert an expression to a product of factors

fundamental theorem of integral calculus For a continuous function on the interval $[a, b]$ defined as $A(x) = \int_a^x f(t)\,dt$, where $A(x)$ is the area between the curve $y = f(x)$ and the t-axis from $t = a$ to $t = x$, $A'(x) = f(x)$.

gradient measures the steepness of a line as the ratio $m = \frac{\text{rise}}{\text{run}}$. If (x_1, y_1) and (x_2, y_2) are two points on the line, $m = \frac{y_2 - y_1}{x_2 - x_1}$.

gradient function *see* derivative

growth The value of the function increases over time. A general equation to represent exponential growth is given by $A = A_0 e^{kt}$, where k is a positive value.

hyperbola a function with two branches of the form $y = \frac{a}{x-h} + k$

hybrid function a function whose rule takes different forms for different sections of its domain

image a figure after a transformation; for the function $x \to f(x)$, $f(x)$ is the image of x under the mapping f

implied domain the domain over which a relation or function is defined

indefinite integral the **anti-derivative** or primitive of $f(x)$, denoted as $F(x)$; $\int f(x)dx = F(x) + c$ where $F'(x) = f(x)$

index *see* exponent

instantaneous rate of change the derivative of a function evaluated at a given value or instant

integral calculus the study of areas under or between curves

integration the process of calculating the limiting sum of the area of a large number of strips of very small width in order to obtain the total area under a curve

interquartile range the middle 50% of the data. $\text{IQR} = Q_3 - Q_1$ (the value of the upper quartile – the value of the lower quartile)

inverse relation a pair of relations for which the rule of one can be formed from the rule of the other by interchanging x- and y-coordinates. The inverse of a one-to-one function f is given the symbol f^{-1}.

kinematics the study of motion

limit the behaviour of a function as it approaches a point, not its behaviour at that point. If $\lim_{x \to a} f(x) = L$, then the function approaches the value L as x approaches a. The limit must be independent of which direction $x \to a$, so $L^- = L^+ = L$ for the limit to exist.

literal equations equations containing pronumerals rather than numerals as terms or coefficients

logarithm an index or power. If $n = a^x$, then $x = \log_a(n)$ is an equivalent statement.

margin of error the distance between the endpoints of the confidence interval and the sample estimate. For a 95% confidence interval, the margin of error, M, is given by $M = 1.96\sqrt{\dfrac{\hat{p}(1-\hat{p})}{n}}$.

maximal domain *see* implied domain

mean (probability) *see* expected value

mean value (of a function) *see* average value

Newton's method a method of solving an equation of the form $f(x) = 0$ by obtaining an initial approximation x_0 to the solution and improving this value by carrying out the iterative procedure

$x_{n+1} = x_n - \dfrac{f(x_n)}{f'(x_n)}$, $n = 0, 1, 2, \ldots$

normal distribution a continuous probability distribution characterised by a symmetrical bell-shaped curve. It is given by the function $f(x) = \dfrac{1}{\sigma\sqrt{2\pi}} e^{-\frac{1}{2}\left(\frac{x-\mu}{\sigma}\right)^2}$.

Null Factor Law the mathematical law stating that if $ab = 0$ then either $a = 0$ or $b = 0$ or both a and b equal zero. This allows equations for which the product of two or more terms equals zero to be solved.

parameter in algebra, a varying constant in a common equation used to describe a family of polynomials; in statistics, a characteristic of a population

percentile the value below which there is a specified probability that a randomly selected value will fall. The 70th percentile means that 70% of the data lies below this value.

period on a trigonometric graph, the length of the domain interval required to complete one full cycle. For the sine and cosine functions, the period is 2π since $\sin(x + 2\pi) = \sin(x)$ and $\cos(x + 2\pi) = \cos(x)$. The tangent function has a period of π since $\tan(x + \pi) = \tan(x)$.

plane a flat, two-dimensional surface with no thickness, extending infinitely in all directions

polynomial equation an equation in which all powers of the variable are positive whole numbers

population the group that you want information about

population size the number of members of a population, symbol N

position measures both distance and direction from a fixed origin

probability density function a function that represents the probabilities of a given scenario for the interval $x \in [a, b]$. Two conditions must be met for a function to be classified as a probability density function:

$\int_a^b f(x)dx = 1$ and $f(x) \geq 0$ for all $x \in [a, b]$.

probability distribution the possible values of a random variable together with the corresponding probabilities of their occurrence. These may be listed in a table or specified by a function rule. The probability values must satisfy the conditions that $\sum p(x) = 1$ and $0 \leq p(x) \leq 1$.

product function The product of the functions f and g is the function fg defined by $(fg)(x) = f(x)g(x)$, where $x \in d_f \cap d_g$.

product rule in differentiation, if $f(x) = g(x)h(x)$, $f'(x) = g(x)h'(x) + h(x)g'(x)$

Pythagoras' theorem used to calculate lengths for a right-angled triangle: $a^2 + b^2 = c^2$, where c is the hypotenuse

Pythagorean identity For any θ, $\sin^2(\theta) + \cos^2(\theta) = 1$.

quadratic $f: R \to R, f(x) = ax^2 + bx + c$

quantile the value below which there is a specified probability that a randomly selected value will fall. The 0.7 quantile means that 70% of the data lies below this value.

quartic $f: R \to R, f(x) = ax^4 + bx^3 + cx^2 + dx + e$

quartile the 0.25, 0.5 and 0.75 quantiles, splitting the data up into four quarters

quotient rule in differentiation, if $f(x) = \dfrac{u(x)}{v(x)}, f'(x) = \dfrac{v(x)u'(x) - u(x)v'(x)}{[v(x)]^2}$

radians one radian is the measure of the angle subtended at the centre of a circle by an arc of length equal to the radius of the circle

random sample a sample in which every member of a population has the same probability of being selected

random variable in probability theory, a variable whose values are assigned by the outcomes of a random experiment

range (of a function or relation) the set of y-values of the ordered pairs (x, y) that make up a relation; (of a data set) the highest value minus the lowest value

rate of change (instantaneous) the derivative of a function evaluated at a given value or instant

rectilinear motion motion in a straight line

reflections transformations of a figure defined by the line of reflection where the image point is a mirror image of the pre-image point

relation any set of ordered pairs

restricted domain a subset of a function's maximal domain, often due to practical limitations on the independent variable in modelling situations

sample the group within the population that the information is collected from

sample proportion $\hat{p} = \dfrac{\text{number in the sample with the favourable attribute}}{\text{sample size}}$

sample size the number of members in a sample, symbol n

sampling distribution the distribution of $\hat{P}$, the sample proportion

secant a line passing through two points on a curve

self-selected sample one where the participants choose to participate in the sample

simultaneous equations systems of equations for which the solutions for the values of the unknowns must satisfy each equation of the system

speed $\dfrac{\text{distance travelled}}{\text{time taken}}$

standard deviation a measure of spread. $\text{SD}(X) = \sigma = \sqrt{\text{Var}(X)}$. The larger the standard deviation, the more spread out the data. If the standard deviation is small, the data is clumped about the mean.

standard normal distribution the standardised form of the normal distribution where $\mu = 0$ and $\sigma = 1$

stationary points points where a function has a gradient of zero

statistic a characteristic of a sample

stratified random sample a sample where subgroups within a population are represented in a similar proportion

sum and difference functions The sum or difference of the functions f and g are the functions $f \pm g$ defined by $(f \pm g)(x) = f(x) \pm g(x)$, where $x \in d_f \cap d_g$.

systematic sample every kth member of the population is sampled

tangent a line that touches a curve at a single point; for a circle, the tangent is perpendicular to the radius drawn to the point of contact

translations transformations of a figure where each point in the plane is moved a given distance in a horizontal or vertical direction

trapezium rule a rule for approximating the area under a curve by constructing trapeziums of equal width

truncus a function with two branches of the form $y = \dfrac{a}{(x-h)^2} + k$

uniform distribution a probability distribution in which all outcomes are equally likely

variance a measure of spread. $\text{Var}(X) = \sigma^2 = E(X^2) - E[(X)]^2$

velocity the rate of change of position with respect to time

vertical line test a test that determines whether a graph is that of a function; any vertical line that cuts the graph of a function does so exactly once

INDEX